BERGEY'S MANUAL® OF

Systematic Bacteriology

Volume 4

BERGEY'S MANUAL® OF

Systematic Bacteriology

Volume 4

STANLEY T. WILLIAMS
EDITOR, VOLUME 4

M. ELISABETH SHARPE
ASSOCIATE EDITOR, VOLUME 4

JOHN G. HOLT
EDITOR-IN-CHIEF

WITH CONTRIBUTIONS FROM
42 COLLEAGUES

Williams & Wilkins
BALTIMORE • PHILADELPHIA • HONG KONG
LONDON • MUNICH • SYDNEY • TOKYO
A WAVERLY COMPANY

Editor: William R. Hensyl
Associate Editor: Harriet Felscher
Copy Editors: Megan Westerfeld and Bill Cady
Design: Norman W. Och
Illustration Planning: Lorraine Wrzosek
Production: Raymond E. Reter

Printed in the United States of America

Library of Congress Cataloging-in-Publication Data
(Revised for vol. 4)

Bergey's manual of systematic bacteriology.

Based on: Bergey's manual of determinative bacteriology.
Vol. 2: Peter H. A. Sneath, editor; v. 3: James T. Staley, editor; v. 4: Stanley T. Williams, editor.
Includes bibliographies and indexes.
1. Bacteriology—Classification—Collected works. I. Bergey, D. H. (David Hendricks), 1860-1937. II. Krieg, Noel R. III. Holt, John G. IV. Bergey's manual of determinative bacteriology. [DNLM: 1. Bacteriology—Terminology. 2. Bacteria—Classification. QW 4 B832m]
QR81.B46 1984 589.9′0012 82-21760
ISBN 0-683-04108-8 (v. 1)
ISBN 0-683-07893-3 (v. 2)
ISBN 0-683-07908-5 (v. 3)
ISBN 0-683-09061-5 (v. 4)

10 9 8 7 6 5

Contributors

Grace Alderson
School of Biomedical Science, University of Bradford, Bradford, West Yorkshire BD7 1DP, England

N. S. Agre
Institute of Biochemistry and Physiology of Microorganisms, U.S.S.R. Academy of Sciences, Pushchino-on-the-Oka, Moscow Region 142292, U.S.S.R.

Paula M. Colman
SmithKline Diagnostics, Inc., Wayne, Pennsylvania 19087 U.S.A.

Tom Cross
School of Biomedical Science, University of Bradford, Bradford, West Yorkshire BD7 1DP, England

L. I. Evtushenko
Institute of Biochemistry and Physiology of Microorganisms, U.S.S.R. Academy of Sciences, Pushchino-on-the-Oka, Moscow Region 142292, U.S.S.R.

Rhonda M. Ferrin
188 Natchez Trace, Royal Palm Beach, Florida 33411 U.S.A.

Antonio F. Fonseca
Microbiology, School of Medicine, Porto University, 4200 Porto, Portugal

Norman E. Gibbons (Deceased)
Ottawa, Ontario, Canada

Margaret B. Gochnauer
Department of Biology, Carleton University, Ottawa K1S 5B6, Canada

Michael Goodfellow
Department of Microbiology, The Medical School, University of Newcastle-upon-Tyne, Newcastle-upon-Tyne NE1 7RU, England

Morris A. Gordon
State of New York Department of Health, Office of Public Health, Corning Tower, The Governor Nelson A. Rockefeller Empire State Plaza, Albany, New York 12201 U.S.A.

Tōru Hasegawa
Takeda Chemical Industries Ltd., Central Research Division, 17-85, Jusohonmachi 2-chome, Yodogawa-ku, Osaka 532, Japan

Aino Henssen
Fachbereich Biologie der Philipps-Universitität Marburg, Botanik, 3550 Marburg/Lahn, Lahnberge, Federal Republic of Germany

John G. Holt
Department of Microbiology, 205 Sciences I, Iowa State University, Ames, Iowa 50011 U.S.A.

Yuzuru Iwai
The Kitasato Institute, 5-9-1 Shirokane, Minato-ku, Tokyo 108, Japan

John L. Johnson
Department of Anaerobic Microbiology, Virginia Polytechnic Institute and State University, Blacksburg, Virginia 24061 U.S.A.

Kenneth G. Johnson
Division of Biological Sciences, National Research Council of Canada, Ottawa, Ontario, Canada

Dorothy Jones
Department of Microbiology, The Medical School, University of Leicester, Leicester LE1 7RH, England

L. V. Kalakoutskii
Institute of Biochemistry and Physiology of Microorganisms, U.S.S.R. Academy of Sciences, Pushchino-on-the-Oka, Moscow Region 142292, U.S.S.R.

Isao Kawamoto
Kyowa Hakko Kogyo Co. Ltd., Tokyo Research Laboratories, 3-6-6 Asahimachi, Machidashi, Tokyo, Japan

Hans-W. Kothe
Fachbereich Biologie der Philipps-Universitität Marburg, Botanik, 3550 Marburg/Lahn, Lahnberge, Federal Republic of Germany

Noel R. Krieg
Department of Biology, Virginia Polytechnic Institute and State University, Blacksburg, Virginia 24061 U.S.A.

Donn J. Kushner
Department of Biology, Ottawa University, Ottawa, Canada

David P. Labeda
United States Department of Agriculture, Agricultural Research Service, Midwest Area Northern Regional Research Center, 1815 North University Street, Peoria, Illinois 61604 U.S.A.

John Lacey
Plant Pathology Department, Rothamstead Experimental Station, Harpenden, Herts AL5 2JQ, England

Stephen P. Lapage
27 Salisbury Road, Fordingsbridge, Hampshire SP9 1EH, England

Hubert A. Lechevalier
Waksman Institute of Microbiology, State University of New Jersey, Rutgers, P.O. Box 759, Piscataway, New Jersey 08854-0759 U.S.A.

Mary P. Lechevalier
Waksman Institute of Microbiology, State University of New Jersey, Rutgers, P.O. Box 759, Piscataway, New Jersey 08854-0759 U.S.A.

Romano Locci
Istituto Di Difesa Delle Piante, Universita Degli Studi Di Udine, P. le M. Kolbe 4, 33100 Udine, Italy

George M. Luedemann
Sasabe Star Route Box 46E, Tucson, Arizona 85736 U.S.A.

Alan J. McCarthy
Department of Genetics and Microbiology, University of Liverpool, Liverpool L69 3BX, England

Jutta Meyer
Institute of Microbiology and Experimental Therapy, Beutenbergstrasse 11, 6900 Jena, German Democratic Republic

R. G. E. Murray
Department of Bacteriology and Immunology, University of Western Ontario, London, Ontario N6A 5C1, Canada

Hideo Nonomura
Faculty of Engineering, Yamanashi University, Takeda-4, Kofu, Japan

Satoshi Ōmura
The Kitasato Institute, 5-9-1 Shirokane, Minato-ku, Tokyo 108, Japan

Norberto J. Palleroni
47 White Oak Drive, North Caldwell, New Jersey 07006 U.S.A.

Francesco Parenti
Gruppo Lepetit, Via Durando 38, 20158 Milan, Italy

Helmut Prauser
Institute of Microbiology and Experimental Therapy, Beutenbergstrasse 11, 6900 Jena, German Democratic Republic

Richard M. Sayre
Nematology Laboratory, Plant Protection Institute, Beltsville Agricultural Research Center, Agricultural Research Service, United States Department of Agriculture, Beltsville, Maryland 20705 U.S.A.

Geraldine M. Schofield
Unilever Research Colworth Laboratory, Colworth House, Sharnbrook, Bedford MK44 1LQ, England

M. Elisabeth Sharpe
National Institute for Research in Dairying, University of Reading, Shinfield, Reading RG2 9AT, England

Marcia C. Shearer
Senior Scientist, Schering Corporation, 60 Orange Street, Bloomfield, New Jersey 07003 U.S.A.

Peter H. A. Sneath
Department of Microbiology, The Medical School, University of Leicester, Leicester LE1 7RH, England

James T. Staley
Department of Microbiology, University of Washington, Seattle, Washington 98195 U.S.A.

Mortimer P. Starr
Department of Microbiology, University of California, Davis, California 95616 U.S.A. Permanent mailing address: 751 Elmwood Drive, Davis, California 95616 U.S.A.

Yōko Takahashi
The Kitasato Institute, 5-9-1 Shirokane, Minato-ku, Tokyo 108, Japan

Koji Tomita
Bristol-Meyers Research Institute — Tokyo, 2-9-3, Shimomeguro, Meguro-ku, Tokyo, Japan

Gernot Vobis
Fachbereich Biologie der Philipps-Universitität Marburg, Botanik, 3550 Marburg/Lahn, Lahnberge, Federal Republic of Germany

Stanley T. Williams
Botany Department, University of Liverpool, Liverpool L69 3BX, England

Advisory Committee Members

The Board of Trustees is grateful to all who served on the Advisory Committees and assisted materially in the preparation of this edition of the *Manual.* Chairpersons of committees are indicated by an asterisk.

1. *Mycobacteria and Nocardioform Bacteria:* M. Goodfellow, M. P. Lechevalier, H. Prauser, L. G. Wayne*
2. *Actinomycetes with Multilocular Sporangia:* M. A. Gordon, M. P. Lechevalier,* G. M. Luedemann
3. *Sporangiate Actinomycetes:* H. A. Lechevalier,* H. Nonomura, N. J. Palleroni, G. Vobis
4. *Streptomyces and Related Genera:* H. J. Kutzner, R. Locci, T. P. Preobrazhenskaya, A. Seino, S. T. Williams*
5. *Other Genera:* T. Arai, T. Cross,* A. Horan, L. V. Kalakoutskii, H. A. Lechevalier, H. Prauser

Preface to First Edition of *Bergey's Manual® of Systematic Bacteriology*, Volume 4

Many microbiologists advised the Trust that a new edition of the *Manual* was urgently needed. Of great concern to us was the steadily increasing time interval between editions; this interval reached a maximum of 17 years between the seventh and eighth editions. To be useful the *Manual* must reflect relatively recent information; a new edition is soon dated or obsolete in parts because of the nearly exponential rate at which new information accumulates. A new approach to publication was needed, and from this conviction came our plan to publish the *Manual* as a sequence of four subvolumes concerned with systematic bacteriology as it applies to taxonomy. The four subvolumes are divided roughly as follows: (a) the Gram-negatives of general, medical or industrial importance; (b) the Gram-positives other than actinomycetes; (c) the archaeobacteria, cyanobacteria and remaining Gram-negatives; and (d) the actinomycetes. The Trust believed that more attention and care could be given to preparation of the various descriptions within each subvolume, and also that each subvolume could be prepared, published, and revised as the area demanded, more rapidly than could be the case if the *Manual* were to remain as a single, comprehensive volume as in the past. Moreover, microbiologists would have the option of purchasing only that particular subvolume containing the organisms in which they were interested.

The Trust also believed that the scope of the *Manual* needed to be expanded to include more information of importance for systematic bacteriology and bring together information dealing with ecology, enrichment and isolation, descriptions of species and their determinative characters, maintenance and preservation, all focused on the illumination of bacterial taxonomy. To reflect this change in scope, the title of the *Manual* was changed and the primary publication becomes *Bergey's Manual® of Systematic Bacteriology.* This contains not only determinative material, such as diagnostic keys and tables useful for identification, but also all of the detailed descriptive information and taxonomic comments. Upon completion of each subvolume, the purely determinative information will be assembled for eventual incorporation into a much smaller publication that will continue the original name of the *Manual, Bergey's Manual® of Determinative Bacteriology,* which will be a similar but improved version of the present *Shorter Bergey's Manual®*. So, in the end there will be two publications, one systematic and one determinative in character.

An important task of the Trust was to decide which genera should be covered in the first and subsequent subvolumes. We were assisted in this decision by the recommendations of our Advisory Committees, composed of prominent taxonomic authorities to whom we are most grateful. Authors were chosen on the basis of constant surveillance of the literature of bacterial systematics and by recommendations from our Advisory Committees.

The activation of the 1976 Code had introduced some novel problems. We decided to include not only those genera that had been published in the Approved Lists of Bacterial Names in January 1980 or that had been subsequently validly published, but also certain genera whose names had no current standing in nomenclature. We also decided to include descriptions of certain organisms that had no formal taxonomic nomenclature, such as the endosymbionts of insects. Our goal was to omit no important group of cultivated bacteria and also to stimulate taxonomic research on "neglected" groups and on some groups of undoubted bacteria that have not yet been cultivated and subjected to conventional studies.

Some readers will note the consistent use of the stem -var instead of -type in words such as biovar, serovar and pathovar. This is in keeping with the recommendations of the Bacteriological Code and was done against the wishes of some of the authors.

We have deleted much of the synonymy of scientific names that was contained in past editions. The adoption of the new starting date of January 1, 1980 and publication of the Approved Lists of Bacterial Names has made mention of past synonymy obsolete. We have included synonyms of a name only if they have been published since the new starting date, or if they were also on the Approved Lists and, in rare cases, if the mention of an old name would help readers associate the organism with a clinical problem. If the reader is interested in tracing the history of a name we suggest he or she consult past editions of the *Manual* or the *Index Bergeyana* and its *Supplement.* In citations of names we have used the abbreviation *AL* to denote the inclusion of the name on the Approved Lists of Bacterial Names and *VP* to show the name has been validly published.

In the matter of citation of the *Manual* in the scientific literature, we again stress the fact that the *Manual* is a collection of authored chapters and the citation should refer to the author, the chapter title and its inclusive pages, not the Editors.

To all contributors, the sincere thanks of the Trust is due; the Editors are especially grateful for the good grace with which the authors accepted comments, criticisms and editing of their manuscripts. It is only because of the voluntary and dedicated efforts of these authors that the *Manual* can continue to serve the science of bacteriology on an international basis.

A number of institutions and individuals deserve special acknowledgment from the Trust for their help in bringing about the publication of this volume. We are grateful to the University of Liverpool for providing space, facilities and, above all, tolerance for the diverted time taken by the Editor during the preparation of the book. The Department of Microbiology at Iowa State University of Science and Technology continues to provide a welcome home for the main editorial offices and archives of the Trust, and we acknowledge their continued support.

A number of individuals deserve special mention and thanks for their help. Professor Thomas O. MacAdoo of the Department of Foreign Languages and Literatures at the Virginia Polytechnic Institute and State University has given invaluable advice on the etymology and correctness of scientific names. The Editors have been greatly assisted by the advice and experience of Peter Sneath, Dorothy Jones and Nicholas Mair. Those assisting the Editors at the Liverpool office were Carol Langham, Dorothy Lewis, Phyllis Saffer and Katharine Williams, and their help is sincerely appreciated. In the Ames office, we were ably assisted by Cynthia Pease, who had the major responsibility for keying and sorting the list of references and index.

Comments on this edition of the *Manual* will be welcomed and should be addressed to the Bergey's Manual® Trust, c/o Williams & Wilkins, 428 East Preston St., Baltimore, Maryland 21202 U.S.A.

Preface to First Edition of *Bergey's Manual®* *of Determinative Bacteriology*

The elaborate system of classification of the bacteria into families, tribes and genera by a Committee on Characterization and Classification of the Society of American Bacteriologists (1917, 1920) has made it very desirable to be able to place in the hands of students a more detailed key for the identification of species than any that is available at present. The valuable book on "Determinative Bacteriology" by Professor F. D. Chester, published in 1901, is now of very little assistance to the students, and all previous classifications are of still less value, especially as earlier systems of classification were based entirely on morphologic characters.

It is hoped that this manual will serve to stimulate efforts to perfect the classification of bacteria, especially by emphasizing the valuable features as well as the weaker points in the new system which the Committee of the Society of American Bacteriologists has promulgated. The Committee does not regard the classification of species offered here as in any sense final, but merely a progress report leading to more satisfactory classification in the future.

The Committee desires to express its appreciation and thanks to those members of the society who gave valuable aid in the compilation of material and the classification of certain species . . .

The assistance of all bacteriologists is earnestly solicited in the correction of possible errors in the text; in the collection of descriptions of all bacteria that may have been omitted from the text; in supplying more detailed descriptions of such organisms as are described incompletely; and in furnishing complete descriptions of new organisms that may be discovered, or in directing the attention of the Committee to publications of such newly described bacteria.

DAVID H. BERGEY, *Chairman*
FRANCIS C. HARRISON
ROBERT S. BREED
BERNARD W. HAMMER
FRANK M. HUNTOON
Committee on Manual

August, 1923.

Archives of the ASM

DAVID HENDRICKS BERGEY
1860–1937
Bergey set up the Trust on January 2, 1936

History of the *Manual*

The first edition of *Bergey's Manual® of Determinative Bacteriology* was initiated by action of the Society of American Bacteriologists (now called the American Society for Microbiology) by appointment of an Editorial Board consisting of David H. Bergey, Chairman, Francis C. Harrison, Robert S. Breed, Bernard W. Hammer, and Frank M. Huntoon. This Board, under auspices of the Society of American Bacteriologists who, then as now, published the *Journal of Bacteriology* as a service to science, brought the first edition of the *Manual* into print in 1923. The Board, with some changes in membership and Dr. David Bergey as Chairman, published a second edition of the *Manual* in 1925 and a third edition in 1930.

In 1934, during preparation of the fourth edition, Dr. Bergey requested that the Society of American Bacteriologists make available the royalties paid to the Treasurer of the Society from the sale of the earlier editions to defray the expense of preparing the fourth edition for publication. The Society made such provision, but the use of the Society's fiscal machinery proved cumbersome, both to the Society and the Editorial Board. Subsequently, it was agreed by the Society and Dr. Bergey that the Society would transfer to Dr. Bergey all of its rights, title, and interest in the *Manual* and that Dr. Bergey would, in turn, create an educational trust to which all rights would be transferred.

Dr. Bergey was then the nominal owner of the *Manual* and he executed a Trust Indenture on January 2, 1936 designating David H. Bergey, Robert S. Breed, and E. G. D. Murray as the initial trustees, and transferring to the Trustees and their successors the ownership of the *Manual*, its copyrights, and the right to receive the income arising from its publication. The Trust is a nonprofit organization and its income is used solely for the purpose of preparing, editing, and publishing revisions and successive editions of the *Manual* and any supplementary publications, as well as providing for any research that may be necessary or desirable in such activities.

Since the creation of the Trust, the Trustees have published, successively, the fourth, fifth, sixth, seventh, and eighth editions of the *Manual* (dated 1934, 1939, 1948, 1957, and 1974, respectively). In 1977 the Trust published an abbreviated version of the eighth edition, called *The Shorter Bergey's Manual® of Determinative Bacteriology;* this contained the outline classification of the bacteria, the descriptions of all genera and higher taxa, all of the keys and tables for the diagnosis of species, all of the illustrations, and two of the introductory chapters; however, it did not contain the detailed species descriptions, most of the taxonomic comments, the etymology of names, and references to authors.

Other ventures in producing books to assist those engaged in bacteriology and bacterial taxonomy in particular include the *Index Bergeyana* (1966), a *Supplement to Index Bergeyana* (1981), and a planned future volume bringing the lists of published names up to date. The Trust is presently publishing the first edition of *Bergey's Manual® of Systematic Bacteriology,* which has a much broader scope than the previous publications and is intended to act as the amplified source for revision of the determinative *Manual.*

Through the years the *Manual* has become a widely used international reference work for bacterial taxonomy. Similarly, the Bergey's Manual® Trust has become international in its composition, in the location of its meetings and in the breadth of its consultations. In addition to its publication activities, the Trust attempts to foster and support various aspects of taxonomic research. One of the ways in which it does this is by recognizing those individuals who have made outstanding contributions to bacterial taxonomy, through its periodic presentation of the Bergey Award, an effort jointly supported by funds from the Trust and Williams & Wilkins, who have been involved in the production of the *Manual* from its beginning.

The following individuals have served as members of the Editorial Board and Board of Trustees.

On Using the *Manual**

Stanley T. Williams

ARRANGEMENT OF THE *MANUAL*

One important goal of the *Manual* is to assist in the identification of bacteria, but another goal, equally important, is to indicate the relationships that exist between the various kinds of bacteria. The methods of molecular biology have now made it possible to envision the eventual development of a comprehensive classification of bacteria based on their phylogenetic relatedness to one another. Such a general classification scheme would lead, it is hoped, to more unifying concepts of bacterial taxa, to greater stability and predictability, to the development of more reliable identification schemes, and to an understanding of how bacteria have evolved.

Such a general scheme, however, cannot yet be perceived fully. The relatedness within and between some bacterial groups has been intensively studied, but for other groups very little work has been done. Moreover, the relatedness studies that have been done often have involved the use of one or another method without confirmation by other methods. Studies have been done at differing levels of resolution, and the interpretation of the data may not yet be entirely clear. Still another major difficulty is the conflict between "practical" classification vs. strange groupings that may be indicated by molecular biology methods. This is because some of the phenotypic characteristics traditionally used in bacterial classification (e.g. cell shape, flagellar arrangement, fermentative vs. respiratory types of metabolism) do not always correlate well with groups established on the basis of relatedness. This conflict will, one hopes, eventually be relieved by the finding of nontraditional, easily determined, phenotypic characteristics that *do* correlate well with relatedness groups, but much work needs to be done in this regard.

Such considerations have forced the present edition of the *Manual* to adhere largely to traditional characteristics in arranging bacterial taxa. However, the arrangement of the genera included in this volume reflects to some extent the current views on their phylogenetic relationships as indicated by studies of their molecular biology. The present classification must nevertheless be regarded as an interim arrangement pending further data on molecular, chemical, and phenotypic characteristics and analysis of their correlations.

The definition and delimitation of the actinomycetes (order *Actinomycetales*) are similarly transitional, as illustrated by the inclusion of accepted or putative actinomycete genera in both Volumes 2 and 4 of the *Manual.* The genera covered in this volume (including Section 17 of Volume 2) conform mainly to the classical definition of an actinomycete as a Gram-positive bacterium forming transitional or permanent branched filaments and usually producing spores. However, chemical and molecular data support the inclusion in the actinomycetes of some genera in Sections 15 and 16 of Volume 2 of the *Manual,* and the exclusion of *Thermoactinomyces* (Section 32 of this volume).

THE SECTIONS

The *Manual* is presented as various "sections" based on available chemical, molecular, morphological, and other phenetic data (for details see overviews by H. A. Lechevalier and M. Goodfellow). Each section bears an appropriate vernacular name reflecting the criteria used in its definition. Family names have not been used, as their current theoretical and practical relevance is unclear. All accepted genera have been placed in what seems to be the most appropriate section.

Section 26 *(Nocardioform actinomycetes)* consists of Section 17 in Volume 2 of the *Manual,* with the addition of the genera *Actinopolyspora* and *Saccharomonospora.* Genera in this section share many characteristics with some of those in Sections 15 and 16 in Volume 2 of the *Manual.*

Section 33 *(Other Genera)* contains four recently described genera, the relationships of which have yet to be determined.

The expanding interest in the isolation and characterization of actinomycetes has inevitably led to the exclusion of some recently described taxa from this volume. Examples of these are the genera *Amycolata* and *Amycolatopsis,* the *Streptomyces* species recently validated by Russian workers, and the four new genera and numerous species described by Chinese microbiologists (information kindly provided by Dr. Ruan Jisheng, Institute of Microbiology, Beijing, China).

SECTIONS VS. TAXONOMIC NAMES

Each section bears a vernacular name. As indicated previously, no attempt has been made to provide a complete formal hierarchy of higher taxa throughout the *Manual,* and the vernacular names of the sections form the primary basis for the organization of the *Manual;* however, a suggested hierarchy for higher taxa has been proposed in one of the introductory articles (see "The Higher Taxa, or, A Place for Everything. . . ?").

* The material presented in this article is based largely on that prepared by Noel Krieg for Volume 1.

ARTICLES

Each article dealing with a bacterial genus is presented wherever possible in a definite sequence as follows.

(a) *Name of the Genus.* Accepted names are in **boldface,** followed by the authority for the name, the year of the original description, and the page on which the taxon was named and described. The superscript *AL* indicates that the name was included on the Approved Lists of Bacterial Names, published in January 1980. The superscript *VP* indicates that the name, although not on the Approved Lists of Bacterial Names, was subsequently validly published in the *International Journal of Systematic Bacteriology.* Names given within quotation marks have no standing in nomenclature; as of the date of preparation of the *Manual* they had not been validly published in the *International Journal of Systematic Bacteriology,* although they had been "effectively published" elsewhere. Names followed by the term "gen. nov." are newly proposed but will not be validly published until they appear in the *International Journal of Systematic Bacteriology*; their proposal in the *Manual* constitutes only "effective publication," not valid publication.

(b) *Name of Author(s).* The person or persons who prepared the article are indicated. The address of each author can be found in the list of contributors at the beginning of the *Manual.*

(c) *Synonyms.* In some instances a list is given of synonyms that have been used in the past for the same genus. The synonomy may not always be complete, and usually is not given at all, as the Editorial Board believes that the earlier synonyms have been covered adequately in the *Index Bergeyana* or in the *Supplement to the Index Bergeyana.*

(d) *Etymology of the Genus Name.* Etymologies are provided as in previous editions, and many (but undoubtedly not all) errors have been corrected. It is often difficult, however, to determine why a particular name was chosen, or the nuance intended, if the details were not provided in the original publication. Those authors who propose new names are urged to consult a Greek and Latin authority before publishing, in order to ensure grammatical correctness and also to ensure that the name means what it is intended to mean. An excellent authority to communicate with in this regard is Dr. Thomas O. MacAdoo, Department of Foreign Languages, Virginia Polytechnic Institute and State University, Blacksburg, Virginia 24061 U.S.A.

(e) *Capsule Description.* This is a brief resume of the salient features of the genus. The most important characteristics are given in **boldface.** The name of the type species of the genus is also indicated.

(f) *Further Descriptive Information.* This portion elaborates on the various features of the genus, particularly those features having significance for systematic bacteriology. The treatment serves to acquaint the reader with the overall biology of the organisms but is not meant to be a comprehensive review. The information is represented in sequence, as follows:

Morphological characteristics
Colonial morphology and pigmentation
Growth conditions and nutrition
Physiology and metabolism
Genetics, plasmids and bacteriophages
Antigenic structure
Pathogenicity
Ecology

(g) *Enrichment and Isolation Procedures.* A few selected methods are presented, together with the pertinent media formulations.

(h) *Maintenance Procedures.* Methods used for maintenance of stock cultures and preservation of strains are given.

(i) *Procedures for Testing of Special Characters.* This portion provides methodology for testing for unusual characteristics or performing tests of special importance.

(j) *Differentiation of the Genus from Other Genera.* Those characteristics that are especially useful for distinguishing the genus from similar or treated organisms are indicated here, usually in a tabular form.

(k) *Taxonomic Comment.* This summarizes the available information about the taxonomic placement of the genus and indicates the justification for considering genus to be a distinct taxon. Particular emphasis is given to the methods of molecular biology for estimating the relatedness to other taxa, where such information is available. Taxonomic information regarding the arrangement and status of the various species within the genus follows. Where taxonomic controversy exists, the problems are delineated and the various alternative viewpoints are discussed.

(l) *Further Reading.* A list of selected references, usually of a general nature, is given to enable the reader to gain access to additional sources of information about the genus.

(m) *Differentiation of the Species of the Genus.* Those characteristics that are important for distinguishing from one another the various species within the genus are presented, usually with reference to a table summarizing the information.

(n) *List of Species of the Genus.* The citation of each species is given, followed in some instances by a brief list of objective synonyms. The etymology of the specific epithet is indicated. Descriptive information for the species is usually presented in tabular form, but special information may be given in the text. Because of the emphasis on tabular data the species descriptions are usually brief. The type strain of each species is indicated, together with the collection in which it can be found. (Addresses of the various culture collections are given in the chapter "List of Culture Collections.")

(o) *Species Incertae Sedis.* The "List of Species of the Genus" may be followed in some instances by a listing of additional species under the heading, "Species Incertae Sedis." The taxonomic placement or status of such species is questionable, and the reasons for the uncertainty are presented.

(p) *Literature Cited.* All references given in the article are listed alphabetically at the end of the volume rather than at the end of each article.

TABLES

In each article dealing with a genus, there are generally three kinds of tables: (a) those that differentiate the genus from similar or related genera, (b) those that differentiate the species within the genus, and (c) those that provide additional information about the species, with such information not being particularly useful for differentiation. Unless otherwise indicated, the meanings of symbols are as follows:

+, 90% or more of the strains are positive.
d, 11–89% of the strains are positive.
−, 90% or more of the strains are negative.
D, different reactions occur in different taxa (species of a genus or genera of a family).
v, strain instability (NOT equivalent to "d").

Exceptions to use of these symbols, as well as the meaning of additional symbols, are clearly indicated in footnotes to the tables.

USE OF THE *MANUAL* FOR DETERMINATIVE PURPOSES

Entry into the *Manual* is best achieved by studying the titles of the various sections, as listed in the "Contents." These titles provide an elementary, but by no means perfect, key to the various kinds of bacteria. Each section has keys or tables for differentiation of the various taxa contained therein. Suggestions on identification may be found in the article "Identification of Bacteria." For identification of species, it is important to read both the generic and species descriptions because characteristics listed in the generic descriptions are not usually repeated in the species descriptions.

The "Cumulative Index" is useful in locating the names of unfamiliar taxa or in discovering what has been done with a particular taxon. Every bacterial name mentioned in the *Manual* is listed in the "Cumulative Index."

ERRORS, COMMENTS, SUGGESTIONS

As indicated in the "Preface to First Edition of *Bergey's Manual® of Determinative Bacteriology,*" the assistance of all bacteriologists is earnestly solicited in the correction of possible errors in the text. Comments on the presentation will also be welcomed, as well as suggestions for future editions. Correspondence should be addressed to the Bergey's Manual® Trust, c/o Williams & Wilkins, 428 East Preston St., Baltimore, Maryland 21202 U.S.A.

Contents

BACTERIAL CLASSIFICATION I

Classification of Procaryotic Organisms: An Overview

James T. Staley and Noel R. Krieg

CLASSIFICATION, NOMENCLATURE, AND IDENTIFICATION

Classification, nomenclature, and identification are the three separate, but interrelated, areas of taxonomy. **Classification** is the arranging of organisms into taxonomic groups (taxa) on the basis of similarities or relationships. **Nomenclature** is the assignment of names to the taxonomic groups according to international rules. **Identification** is the process of determining that a new isolate belongs to one of the established, named taxa.

There are numerous procaryotic organisms and great diversity in their types. In any endeavor aimed at an understanding of large numbers of entities it is convenient to arrange, or classify, the objects into groups based upon their similarities. Thus, classification has been used to organize the bewildering and seemingly chaotic array of individual bacteria into an orderly framework.

Classification of organisms requires knowledge of their characteristics. For procaryotes, this knowledge is obtained by experimental as well as observational techniques, because biochemical, physiological and genetic characteristics are often necessary, in addition to morphological features, for an adequate description of a taxon.

The process of classification may be applied to existing, named, taxa or to newly described organisms. If the taxa have already been described, named, and classified, either new characteristics about the organisms or a reinterpretation of existing knowledge of characteristics is used to formulate a new classification. However, if the organisms are new, i.e. cannot be identified as existing taxa, they are named according to the rules of nomenclature and placed in an appropriate position in an existing classification.

Taxonomic Ranks

Several levels or ranks are used in bacterial classification. All procaryotic organisms are placed in the kingdom *Procaryotae*. Divisions, classes, orders, families, genera, and species are successively smaller, nonoverlapping subsets of the kingdom, and the names of these subsets are given formal recognition (have "standing in nomenclature"). An example is given in Table I.1.

In addition to these formal, hierarchical taxonomic categories, informal or vernacular groups that are defined by common descriptive names are often used; the names of such groups have no official standing in nomenclature. Examples of such groups are the procaryotes, the spirochetes, dissimilatory sulfate- and sulfur-reducing bacteria, and the methane-oxidizing bacteria.

Species

The basic taxonomic group in bacterial systematics is the species. The concept of a bacterial species is less definitive than for higher organisms. The difference should not seem surprising, because bacteria, being procaryotic organisms, differ markedly from higher organisms. Sexuality, for example, is not used in bacterial species definitions because relatively few bacteria undergo conjugation. Likewise, morphological features alone are usually of little classificatory significance; this is because most procaryotic organisms are too simple morphologically to provide much useful taxonomic information. Consequently, morphological features are relegated to a less important role in bacterial taxonomy in comparison with the taxonomy of higher organisms.

A bacterial species may be regarded as a collection of strains that share many features in common and differ considerably from other strains. (A strain is made up of the descendants of a single isolation in pure culture, and usually is made up of a succession of cultures ultimately derived from an initial single colony.) One strain of a species is designated as the **type strain**; this strain serves as the name-bearer strain of the species and is the permanent example of the species, i.e. the *reference specimen for the name*. (See the chapter on "Bacterial Nomenclature" for more detailed information about nomenclatural types.) The type strain has great importance for classification at the species level, because *a species consists of the type strain and all other strains that are considered to be sufficiently similar to it as to warrant inclusion with it in the species.* This concept of a species obviously involves making subjective judgments, and it is not surprising that some bacterial species have greater phenotypic

Table I.1.
Taxonomic ranks

Formal rank	Example
Kingdom	*Procaryotae*
Division	*Gracilicutes*
Class	*Scotobacteria*
Order	*Spirochaetales*
Family	*Leptospiraceae*
Genus	*Leptospira*
Species	*Leptospira interrogans*

and genetic diversity than others. A more uniform and rigorous species definition would be desirable. For example, the level of DNA homology exhibited among a group of strains might be used as a basis for defining a species, i.e. definition on the basis of a particular degree of genetic relatedness. The advantage of adopting this or a similarly restrictive species definition must be weighed against its potential impact on well-established and accepted bacterial groups. For practical reasons, classifications and nomenclature should remain stable because changes create confusion, particularly at the genus and species levels, and result in costly modifications of identification schemes and texts. However, classifications have *never* remained static and probably never will, because new information bearing on the taxonomy of bacteria is continually being generated by researchers.

Though classification schemes based on genetic relatedness are rather recent, they promise to be quite reliable and stable. This view may have to be reassessed, however, when we more fully understand the impact that transposable elements might have upon the stability of the procaryotic genome. Genetic studies have already resolved many instances of confusion concerning which strains belong to a given species, and DNA homology is increasingly being used for establishing new species and for resolving taxonomic problems at the species level.

Subspecies

A species may be divided into two or more subspecies based on minor but consistent phenotypic variations within the species or on genetically determined clusters of strains within the species. It is the lowest taxonomic rank that has official standing in nomenclature.

Infrasubspecific Ranks

Ranks below subspecies, such as biovars, serovars, and phagovars, are often used to indicate groups of strains that can be distinguished by some special character, such as antigenic makeup, reactions to bacteriophage, or the like. Such ranks have no official standing in nomenclature but often have great practical usefulness. A list of some common infrasubspecific categories is given in Table I.2.

Table I.2.
Infrasubspecific ranks

Preferred name	Synonym	Applied to strains having:
Biovar	Biotype	Special biochemical or physiological properties
Serovar	Serotype	Distinctive antigenic properties
Pathovar	Pathotype	Pathogenic properties for certain hosts
Phagovar	Phagotype	Ability to be lysed by certain bacteriophages
Morphovar	Morphotype	Special morphological features

Genus

All species are assigned to a genus (although not always with a high degree of certainty as to which genus is the best choice). In this regard, bacteriologists conform to the binomial system of nomenclature of Linnaeus in which the organism is designated by its combined genus and species name. The bacterial genus is usually a well-defined group that is clearly separated from other genera, and the thorough descriptions of genera in this edition of *Bergey's Manual*® exemplify the depth to which this taxonomic group is usually known. However, there is so far no general agreement on the *definition* of a genus in bacterial taxonomy, and considerable subjectivity is involved at the genus level. Indeed, what is perceived to be a genus by one person may be perceived as being merely a species by another systematist. The use of genetic relatedness (e.g. ribosomal RNA (rRNA) homology or rRNA oligonucleotide cataloging) offers hope for greater objectivity and has already been useful in several instances.

Higher Taxa

Classificatory relationships at the familial and ordinal levels are even less certain than those at the genus and species levels. Frequently there is little basis for ascription of taxa at these higher levels, except in a few cases (e.g. the family *Enterobacteriaceae*) where there is evidence for genetic relatedness. Thus, rather than formalize families and orders upon uncertain relationships, many systematists frequently adopt a provisional, ad hoc ranking in which purely descriptive and vernacular names for groups are applied (e.g. in this edition of *Bergey's Manual*® see Section 7 on "Dissimilatory Sulfate- or Sulfur-reducing Bacteria"). As more is learned about the similarities among these bacteria, familial and ordinal placements will likely ensue. A recent example that illustrates the effect that increased knowledge has on the taxonomy of groups concerns the methane-producing bacteria. In the eighth edition of the *Manual*, the methanogens were treated as a single family of bacteria, with three genera. Authorities for this group now propose that three *orders* are required for the circumscription of these organisms (Balch et al., 1979).

In this edition of the *Manual* the procaryotes have been classified into four divisions, these being subdivided into classes (see the chapter by Murray on "The Higher Taxa"). There is no general agreement about this or any other arrangement of divisions and classes, however, and even at the kingdom level of classification controversy exists. Recent information based on rRNA oligonucleotide catalogs and biochemical features has led some authorities to propose that not all bacteria are procaryotes, and that some represent a kingdom of life distinct from both procaryotes and eucaryotes (i.e. the so-called *Archaebacteria*) (see Fox et al., 1980, and Woese, 1981, for summaries). That this group possesses a number of unique features is beyond question, and there is strong evidence that it has taken an evolutionary path distinct from that of other bacteria, but so far there is no general agreement as to what level of classification is applicable to the group.

MAJOR DEVELOPMENTS IN BACTERIAL CLASSIFICATION

A century elapsed between Antonie van Leeuwenhoek's discovery of bacteria and Müller's initial acknowledgment of bacteria in a classification scheme (Müller, 1773). Another century passed before techniques and procedures had advanced sufficiently to permit a fairly inclusive and meaningful classification of these organisms. For a comprehensive review of the early development of bacterial classification, readers should consult the introductory sections of the first, second, and third editions of *Bergey's Manual*®. A less detailed treatment of early classifications can be found in the sixth edition of the *Manual* in which post-1923 developments were emphasized.

Two primary difficulties beset early bacterial classification systems. First, they relied heavily upon morphological criteria. For example, cell shape was often considered to be an extremely important feature. Thus, the cocci were often classified together in one group (family or order). In contrast, contemporary schemes rely much more strongly on physiological characteristics. For example, the fermentative cocci are now separated from the photosynthetic cocci, which are separated from the methanogenic cocci, which are in turn separated from the nitrifying cocci, and so forth. Secondly, the pure culture technique which revolutionized microbiology was not developed until the latter half of the 19th century. In addition to dispelling the concept of "polymorphism," this technical development of Robert Koch's laboratory had great impact on the development of modern procedures in bacterial systematics. Pure cultures are analogous to herbarium specimens in botany. However, pure cultures are much more useful because they can be (a) maintained in a viable state, (b) subcultured, (c) subjected indefinitely to experimental tests, and (d) shipped from one laboratory to another. A natural outgrowth of the pure culture technique was the establishment of *type strains* of species which are deposited in repositories referred to as "culture collections" (a more suitable term would be "strain collections"). These type strains can be obtained from culture collections and used as reference strains for direct comparison with new isolates.

Before the development of computer-assisted numerical taxonomy and subsequent taxonomic methods based on molecular biology, the traditional method of classifying bacteria was to characterize them as thoroughly as possible and then to arrange them according to the intuitive judgment of the systematist. Although the subjective aspects of this method resulted in classifications that were often drastically revised by other systematists who were likely to make different intuitive judgments, many of the arrangements have survived to the present day, even under scrutiny by modern methods. One explanation for this is that the systematists usually *knew their organisms thoroughly*, and their intuitive judgments were based on a wealth of information. Their data, while not computer processed, were at least processed by an active mind to give fairly accurate impressions of the relationships existing between organisms. Moreover, some of the characteristics that were given great weight in classification were, in fact, highly correlated with many other characteristics. This principle of *correlation of characteristics* appears to have started with Winslow and Winslow (1908), who noted that parasitic cocci tended to grow poorly on ordinary nutrient media, were strongly Gram-positive, and formed acid from sugars, in contrast to saprophytic cocci, which grew abundantly on ordinary media, were generally only weakly Gram-positive, and formed no acid. This division of the cocci that were studied by the Winslows (equivalent to the present genus *Micrococcus* (the saprophytes) and the genera *Staphylococcus* and *Streptococcus* (the parasites)) has held up reasonably well even to the present day.

Other classifications have not been so fortunate. A classic example of one which was not is that of the genus "*Paracolobactrum*." This genus was proposed in 1944 and is described in the seventh edition of *Bergey's Manual*® in 1957. It was created to contain certain lactose-negative members of the family *Enterobacteriaceae*. Because of the importance of a lactose-negative reaction in *identification* of enteric pathogens (i.e. *Salmonella* and *Shigella*), the reaction was mistakenly given great taxonomic weight in *classification* as well. However, for the organisms placed in "*Paracolobactrum*," the lactose reaction was not highly correlated with other characteristics. In fact, the organisms were merely lactose-negative variants of other lactose-positive species; for example, "*Paracolobactrum coliforme*" resembled *Escherichia coli* in every way except in being lactose-negative. Absurd arrangements such as this eventually led to the development of more objective methods of classification, i.e. numerical taxonomy, in order to avoid giving great weight to any single characteristic.

Phylogenetic Classifications

Classification systems for many higher organisms are based to a large extent upon evolutionary evidence obtained from the fossil record and appropriate sedimentary dating procedures. Such classifications are termed "natural" or "phylogenetic," and are distinguished from "practical" or "artificial" classifications, which are based entirely on phenotypic characteristics. Until about 20 years ago, however, there was no convincing evidence of fossil microorganisms. Now, micropaleontological evidence indicates that microorganisms existed during the Precambrian period. Indeed, many scientists believe that bacteria existed at least 3.5 billion years ago on an earth that is 4.5 billion years old. Of course, the discovery of fossil microorganisms in early sedimentary rocks tells very little about the phylogeny of procaryotic groups. Micropaleontologists are far from reconstructing an evolutionary scheme based upon the presently available fossil record.

Despite the absence of a complete fossil record, proposals have been made since the early part of this century regarding the evolution of bacteria. Until recently, these proposals have been entirely speculative in nature. Orla-Jensen (1909) proposed that autotrophic bacteria were the most primitive group, and he devised an extensive phylogenetic scheme based on this premise. Today, most microbiologists would agree that the premise is probably incorrect, but Orla-Jensen's classification did provide a coherent framework for thinking about the relationships among bacteria. Another notable phylogenetic scheme was that devised by Kluyver and van Niel (1936); in contrast to Orla-Jensen's scheme, which had been based almost entirely on physiological characteristics, Kluyver and van Niel's scheme was based on morphology. The basic premise was that the simplest morphological form, the coccus, was also the most primitive, and from this form developed more complex forms such as spirilla, rods, and branching filaments.

As recently as the seventh edition of *Bergey's Manual*® (i.e. 1957), before convincing evidence of Precambrian microbes had been discovered, the view was expressed that bacteria were a primitive group of organisms, and the classification scheme presented in that edition of the *Manual* claimed to be a natural scheme in which the photosynthetic bacteria were treated first, because they were regarded as the most primitive bacterial group. However, because of the lack of objective evidence for this (or any other) phylogenetic scheme, the eighth edition of the *Manual* abandoned all attempts at a phylogenetic approach to bacterial classification and concentrated instead on providing groupings of organisms under vernacular headings for purposes of recognition and identification; i.e. it was a purely practical and admittedly artificial classification.

Phylogenetic information has increased since the eighth edition, however, largely through the increasing use of methods for measuring genetic relatedness (i.e. DNA/DNA hybridization, DNA/rRNA hybridization, rRNA oligonucleotide cataloging, and protein sequencing). A record of bacterial evolution appears to exist in the amino acid sequences of bacterial proteins and in the nucleotide sequences of bacterial DNA and RNA. Unfortunately, the phylogenetic information is still in a fragmentary form, and it seems probable that the interpretation of the data is still not entirely clear. Not all of the bacterial groups have been surveyed, and it is likely that surprises and strange associations will continue to come from further work. Available phylogenetic information is presented throughout this *Manual* in the "Taxonomic Comments" sections of the various chapters, and some preliminary rearrangements of taxa have already been made based upon phylogenetic information.

Official Classifications

Some microbiologists seem to have the impression that the classification presented in *Bergey's Manual*® is the "official classification" to be used in microbiology. It seems important to correct that impression. **There is no "official" classification of bacteria.** (This is in contrast to bacterial *nomenclature*, where each taxon has one and only one valid name, according to internationally agreed-upon rules, and judicial decisions are rendered in instances of controversy about the validity of a name.) The closest approximation to an "official" classification of bacteria would be one that is widely accepted by the community of microbiologists. A classification that is of little use to microbiologists, no matter how fine a scheme or who devised it, will soon be ignored or significantly modified.

It also seems worthwhile to emphasize something that has often been said before, viz. **bacterial classifications are devised for microbiologists, not for the entities being classified.** Bacteria show little interest in the matter of their classification. For the systematist, this is sometimes a very sobering thought!

Further Reading

Cowan, S.T. 1971. Sense and nonsense in bacterial taxonomy. J. Gen. Microbiol. *67:* 1–8.

An incisive, personal view of bacterial taxonomy, with some "heretical" suggestions.

Cowan, S.T. 1974. Cowan and Steel's Manual for the Identification of Medical Bacteria. Cambridge University Press, Cambridge, England.

Chapters 1 and 9 of this work provide a concise statement of many principles of bacterial taxonomy.

Gerhardt, P., R.G.E. Murray, R.N. Costilow, E.W. Nester, W.A. Wood, N.R. Krieg and G.B. Phillips (Editors). 1981. Manual of Methods for General Bacteriology. American Society for Microbiology, Washington, D.C.

Section V of this book gives a brief introduction to phenotypic characterization, numerical taxonomy, genetic characterization, and classification of bacteria.

Johnson, J.L. 1973. Use of nucleic acid homologies in the taxonomy of anaerobic bacteria. Int. J. Syst. Bacteriol. *23:* 308–315.

This paper proposes a unifying concept of a bacterial species and stresses the importance of correlating nucleic acid homology with phenotypic tests to allow differentiation among species.

Margulis, L. 1968. Evolutionary criteria in Thallophytes: a radical alternative. Science *161:* 1020–1022.

This paper presents the hypothesis that eucaryotic organisms evolved from procaryotic organisms through endosymbioses.

Schopf, J.W. 1978. The evolution of the earliest cells. Sci. Am. *239:* 110–138.

A micropaleontologist's view of microbial evolution.

Schwartz, R.M. and M.O. Dayhoff. 1978. Origins of procaryotes, eucaryotes, mitochondria, and chloroplasts. Science *199:* 395–403.

A discussion of results obtained from the analysis of protein and nucleic acid sequence data as they pertain to the phylogeny of organisms.

Sneath, P.H.A. 1978. Classification of microorganisms. *In* Norris and Richmond (Editors), Essays in Microbiology. John Wiley, Chichester, England, pp. 9/1–9/31.

An excellent general introduction to bacterial classification.

Trüper, H.G. and J. Kramer. 1981. Principles of characterization and identification of prokaryotes. *In* Starr, Stolp, Trüper, Balows and Schlegel (Editors), The Prokaryotes. A Handbook on Habitats, Isolation, and Identification of Bacteria. Springer-Verlag, Berlin, pp. 176–193.

A brief overview of systematic bacteriology, including developments and trends in taxonomy.

Woese, C.R. and G.E. Fox. 1977. Phylogenetic structure of the procaryotic domain: the primary kingdoms. Proc. Natl. Acad. Sci. U.S.A. *74:* 5088–5090.

The authors recognize three distinct groups: the eubacteria, the archaebacteria, and the urcaryotes (cytoplasmic components of eucaryotes).

BACTERIAL CLASSIFICATION II

Numerical Taxonomy

Peter H. A. Sneath

Numerical taxonomy (sometimes called **taxometrics**) developed in the late 1950s as part of multivariate analyses and in parallel with the development of computers. Its aim was to devise a consistent set of methods for classification of organisms. Much of the impetus in bacteriology came from the problem of handling the tables of data that result from examination of their physiological, biochemical, and other properties. Such tables of results are not readily analyzed by eye, in contrast to the elaborate morphological detail that is usually available from examination of higher plants and animals. There was thus a need for an objective method of taxonomic analyses, whose first aim was to sort individual strains of bacteria into homogeneous groups (conventionally species), and which would also assist in the arrangement of species into genera and higher groupings. Such numerical methods also promised to improve the exactitude in measuring taxonomic, phylogenetic, serological, and other forms of relationship, together with other benefits that can accrue from quantitation (such as improved methods for bacterial identification; see the discussion by Sneath of numerical identification on p. 2324 of this *Manual*).

Numerical taxonomy has been broadly successful in most of these aims, particularly in defining homogeneous **clusters** of strains, and in integrating data of different kinds (morphological, physiological, antigenic). There are still problems in constructing satisfactory groups at high taxonomic levels, e.g. families and orders, although this may be due to inadequacies in the available data rather than any fundamental weakness in the numerical methods themselves.

The application of the concepts of numerical taxonomy was made possible only through the use of computers, because of the heavy load of routine calculations. However, the principles can easily be illustrated in hand-worked examples. In addition, two problems had to be solved: the first was to decide how to weight different variables or characters; the second was to analyze similarities so as to reveal the **taxonomic structure** of groups, species, or clusters. A full description of numerical taxonomic methods may be found in Sneath (1972) and Sneath and Sokal (1973). Briefer descriptions and illustrations in bacteriology are given by Skerman (1967), Lockhart and Liston (1970), and Sneath (1978). A thorough review of applications to bacteria is that of Colwell (1973).

It is important to bear in mind certain definitions. Relationships between organisms can be of several kinds. Two broad classes are as follows.

Similarity on Observed Properties. Similarity, or **resemblance**, refers to the attributes that an organism possesses today, without reference to how those attributes arose. It is expressed as proportions of similarities and differences, for example, in existing attributes, and is called **phenetic relationship**. This includes similarities both in phenotype (e.g. motility) and in genotype (e.g. DNA pairing).

Relationship by Ancestry, or Evolutionary Relationship. This refers to the **phylogeny** of organisms, and not necessarily to their present attributes. It is expressed as the time to a common ancestor, or the amount of change that has occurred in an evolutionary lineage. It is not expressed as a proportion of similar attributes, or as the amount of DNA pairing and the like, although evolutionary relationship may sometimes be *deduced* from phenetics *on the assumption* that evolution has indeed proceeded in some orderly and defined way. To give an analogy, individuals from different nations may occasionally look more similar than brothers or sisters of one family: their phenetic resemblance (in the properties observed) may be high though their evolutionary relationship is distant.

Numerical taxonomy is concerned primarily with phenetic relationships. It has in recent years been extended to phylogenetic work, by using rather different techniques: these seek to build upon the assumed regularities of evolution so as to give, from *phenetic data*, the *most probable phylogenetic reconstructions*. Relatively little has been done so far in bacteriology, but a review of the area is given by Sneath (1974).

The basic taxonomic category is the species. It is noted in the chapter on "Bacterial Nomenclature" that it is useful to distinguish a **taxospecies** (a cluster of strains of high mutual phenetic similarity) from a **genospecies** (a group of strains capable of gene exchange), and both of these from a **nomenspecies** (a group bearing a binominal name, whatever its status in other respects). Numerical taxonomy attempts to define taxospecies. Whether these are justified as genospecies or nomenspecies turns on other criteria. It should be emphasized that groups with high genomic similarity are not necessarily genospecies: genomic resemblance is included in phenetic resemblance; genospecies are defined by gene exchange.

Groups can be of two important types. In the first, the possession of certain invariant properties defines the group without permitting any exception. All triangles, for example, have three sides, not four. Such groupings are termed **monothetic**. Taxonomic groups are, however, not of this kind. Exceptions to the most invariant characters are always possible. Instead, taxa are **polythetic**, that is, they consist of assemblages whose members share a high proportion of common attributes, but not necessary any invariable set. Numerical taxonomy produces polythetic groups and thus permits the occasional exception on any character.

LOGICAL STEPS IN CLASSIFICATION

The steps in the process of classification are as follows:

1. Collection of data. The **bacterial strains** that are to be classified have to be chosen, and they must be examined for a number of relevant properties (**taxonomic characters**).
2. The data must be coded and scaled in an appropriate fashion.
3. The **similarity** or **resemblance** between the strains is calculated. This yields a table of similarities (**similarity matrix**) based on the chosen set of characters.
4. The similarities are analyzed for **taxonomic structure**, to yield the groups or clusters that are present, and the strains are arranged into **phenons** (phenetic groups), which are broadly equated with taxonomic groups (**taxa**).
5. The properties of the phenons can be tabulated for publication or further study, and the most appropriate characters (**diagnostic characters**) can be chosen on which to set up **identification systems** that will allow the best identification of additional strains.

It may be noted that those steps must be carried out in the above order. One cannot, for example, find diagnostic characters before finding the groups of which they are diagnostic. Furthermore, it is important to obtain complete data, determined under well-standardized conditions.

Data for numerical taxonomy

The data needed for numerical taxonomy must be adequate in quantity and quality. It is a common experience that data from the literature are inadequate on both counts: most often it is necessary to examine bacterial strains afresh by an appropriate set of tests.

Organisms

Most taxonomic work with bacteria consists of examining individual strains of bacteria. However, the entities that can be classified may be of various forms—strains, species, genera—for which no common term is available. These entities, t in number, are therefore called **operational taxonomic units (OTUs).** In most studies OTUs will be strains. A numerical taxonomic study, therefore, should contain a good selection of strains of the groups under study, together with type strains of the taxa and of related taxa. Where possible, recently isolated strains, and strains from different parts of the world, should be included.

Characters

A **character** is defined as any property that can vary between OTUs. The values it can assume are **character states**. Thus, "length of spore" is a character and "1.5 μm" is one of its states. It is obviously important to compare the same character in different organisms, and the recognition that characters are the same is called the **determination of homology**. This may sometimes pose problems, but in bacteriology these are seldom serious. A single character treated as independent of others is called a **unit character**. Sets of characters that are related in some way are called **character complexes**.

There are many kinds of characters that can be used in taxonomy. The descriptions in the *Manual* give many examples. For numerical taxonomy, the characters should cover a broad range of properties: morphological, physiological, biochemical. It should be noted that certain data are not characters in the above sense. Thus the degree of serological cross-reaction or the percent pairing of DNA are analogous, not to character states, but to similarity measures.

Numbers of Characters

Although it is well to include a number of strains of each known species, numerical taxonomies are not greatly affected by having only a few strains of a species. This is not so, however, for characters. The similarity values should be thought of as estimates of values that would be obtained if one could include a very large number of phenotypic features. The accuracy of such estimates depends critically on having a reasonably large number of characters. The number, n, should be 50 or more. Several hundred are desirable, though the taxonomic gain falls off with very large numbers.

Quality of Data

The quality of the characters is also important. Microbiological data are prone to more experimental error than is commonly realized. The average difference in replicate tests on the same strain is commonly about 5%. Efforts should be made to keep this figure low, particularly by rigorous standardization of test methods. It is very difficult to obtain reasonably reproducible results with some tests, and they should be excluded from the analysis. As a check on the quality of the data, it is useful to reduplicate a few of the strains and carry them through as separate OTUs: the average test error is about half the percentage discrepancy in similarity of such replicates (e.g. 90% similarity implies about 5% experimental variation).

Coding of the Results

The test reactions and character states now need coding for numerical analysis. There are several satisfactory ways of doing this, but for the present purposes of illustration only one common scheme will be described. This is the familiar process of coding the reactions or states into positive and negative form. The resulting table, therefore, contains entries + and − (or 1 and 0, which are more convenient for computation), for t OTUs scored for n characters. Naturally, there should be as few gaps as possible.

The question arises as to what weight should be given to each character relative to the rest. The usual practice in numerical taxonomy is to give each character equal weight. More specifically, it may be argued that unit characters should have unit weight, and if character complexes are broken into a number of unit characters (each carrying one unit of taxonomic information) it is logical to accord unit weight to each unit character. The difficulties of deciding what weight should be given *before* making a classification (and hence in a fashion that does not prejudge the taxonomy) are considerable. This philosophy derives from the opinions of the 18th century botanist Adanson, and therefore numerical taxonomies are sometimes referred to as Adansonian.

Similarity

The $n \times t$ table can then be analyzed to yield similarities between OTUs. The simplest way is to count, for any pair of OTUs, the number of characters in which they are identical (i.e. both are positive or both are negative). These **matches** can be expressed as a percentage or a proportion, symbolized as S_{SM} (for simple matching coefficient). This is the most common coefficient in bacteriology. Other coefficients are sometimes used because of particular advantages. Thus the Gower coefficient S_G accommodates both presence-absence characters and quantitative ones, the Jacquard coefficient S_J discounts matches between two negative results, and the Pattern coefficient S_P corrects for apparent differences that are caused solely by differences between strains in growth rate and hence metabolic vigor. These coefficients emphasize different aspects of the phenotype (as is quite legitimate in taxonomy) so one cannot regard one or another as necessarily the correct coefficient, but fortunately this makes little practical difference in most studies.

The similarity values between all pairs of OTUs yields a checkerboard of entries, a square table of similarities known as a **similarity matrix** or ***S* matrix**. The entries are percentages, with 100% indicating identity and 0% indicating complete dissimilarity between OTUs. Such a table is symmetrical (the similarity of *a* to *b* is the same as that of *b* to *a*), so that usually only one half, the left lower triangle, is filled in.

These similarities can also be expressed in a complementary form, as *dissimilarities*. Dissimilarities can be treated as analogs of distances, when "taxonomic maps" of the OTUs are prepared, and it is a convenient property that the quantity $d = \sqrt{(1 - S_{SM})}$ is equivalent geometrically to a *distance* between points representing the OTUs in a space of many dimensions (a **phenetic hyperspace**).

Taxonomic structure

A table of similarities does not of itself make evident the **taxonomic structure** of the OTUs. The strains will be in an arbitrary order which will not reflect the species or other groups. These similarities therefore require further manipulation. It will be seen that a table of serological cross-reactions, if complete and expressed in quantitative terms, is analogous to a table of percentage similarities, and the same is true of a table of DNA pairing values. Such tables can be analyzed by the methods described below, though in serological and nucleic studies there are some particular difficulties on which further work is needed.

There are two main types of analyses to reveal the taxonomic structure, **cluster analysis** and **ordination**. The result of the former is a treelike diagram or **dendrogram** (more precisely a **phenogram**, because it expresses phenetic relationships), in which the tightest bunches of twigs represent clusters of very similar OTUs. The result of the latter is an **ordination diagram** or **taxonomic map**, in which closely similar OTUs are placed close together. The mathematical methods can be elaborate, so only a nontechnical account is given here.

In cluster analysis, the principle is to search the table of similarities for high values that indicate the most similar pairs of OTUs. These form the nuclei of the clusters and the computer searches for the next highest similarity values and adds the corresponding OTUs onto these cluster nuclei. Ultimately all OTUs fuse into one group, represented by the basal stem of the dendrogram. Lines drawn across the dendrogram at descending similarity levels define, in turn, phenons that correspond to a reasonable approximation to species, genera, etc. The most common cluster methods are the **unweighted pair group method with averages (UPGMA)** and **single linkage**.

In ordination, the similarities (or their mathematical equivalents) are analyzed so that the phenetic hyperspace is summarized in a space of only a few dimensions. In two dimensions this is a scattergram of the positions of OTUs from which one can recognize clusters by eye. Three-dimensional perspective drawings can also be made. The most common ordination methods are **principal components analysis** and **principal coordinates analysis**.

A number of other representations are also used. One example is a similarity matrix in which the OTUs have first been rearranged into the order given by a clustering method and then the cells of the matrix have been shaded, with the highest similarities shown in the darkest tone. In these "shaded *S* matrices," clusters are shown by dark triangles. Another representation is a table of the mean similarities between OTUs of the same cluster and of different clusters (**inter-** and **intragroup similarity table**): if based on S_{SM} with UPGMA clustering, this table expresses the positions and radii of clusters (Sneath, 1979a) and consequently the distance between them and their probable overlap—properties of importance in numerical identification, as discussed later.

For general purposes a dendrogram is the most useful representation, but the others can be very instructive, since each method emphasizes somewhat different aspects of the taxonomy.

The analysis for taxonomic structure should lead logically to the establishment or revision of taxonomic groups. We lack, at present, objective criteria for different taxonomic ranks, that is, one cannot automatically equate a phenon with a taxon. It is, however, commonly found that phenetic groups formed at about 80% *S* are equivalent to bacterial species. Similarly, we lack good tests for the statistical significance of clusters and for determining how much they overlap, though some progress is being made here (Sneath, 1977a, 1979b). The fidelity with which the dendrogram summarizes the *S* matrix can be assessed by the **cophenetic correlation coefficient**, and similar statistics can be used to compare the **congruence** between two taxonomies if they are in quantitative form (e.g. phenetic and serological taxonomies). Good scientific judgment in the light of other knowledge is indispensible for interpreting the results of numerical taxonomy.

Descriptions of the groups can now be made by referring back to the original table of strain data. The better diagnostic characters can be chosen—those whose states are very constant within groups but vary between groups. It is better to give percentages or proportions than to use symbols such as +, (+), *v*, *d*, or − for varying percentages, because significant loss of statistical information can occur with these simplified schemes. It would, however, be superfluous to list percentages based on very few strains. As systematic bacteriology advances, it will be increasingly important to publish the actual data on individual strains or deposit it in archives; such data will show their full value when test methods become very highly standardized.

It is evident that numerical taxonomy (and also numerical identification; see the chapter on "Identification of Bacteria" in this *Manual*) place considerable demands on laboratory expertise. New test methods are continually being devised. New information is continually being accumulated. It is important that progress should be made toward agreed data bases (Krichevsky and Norton, 1974), as well as toward improvements in standardization of test methods in determinative bacteriology, if the full potential of numerical methods is to be achieved.

BACTERIAL CLASSIFICATION III

Nucleic Acids in Bacterial Classification

John L. Johnson

Historically, classification of bacteria has been based on similarities in phenotypic characteristics. Although this method has been quite successful, it has not been precise enough for distinguishing superficially similar organisms or for determining phylogenetic relationships among the bacterial groups. Nucleic acid studies were first applied to such problems in bacterial classification more than 20 years ago and have since become of major importance. There are several advantages to be gained by basing classification on genomic relatedness:

1. A more unifying concept of a bacterial species is possible.
2. Classifications based on genomic relatedness tend not to be subject to frequent or radical changes.
3. Reliable identification schemes can be prepared after organisms have been classified on the basis of genomic relatedness.
4. Information can be obtained that is useful for understanding how the various bacterial groups have evolved and how they can be arranged according to their ancestral relationships.

The purpose of this chapter is to provide an overview of the principles involved in nucleic acid methodology, to give a brief description of the procedures being used, to compare the results obtained by one procedure with those obtained by another, and to indicate how the results are being used in bacterial classification.

PROPERTIES OF NUCLEIC ACIDS

DNA Base Composition

The first unique feature of DNA that was recognized as having taxonomic importance was its mole percent guanine plus cytosine content (mol% G + C). Among the bacteria, the mol% G + C values range from ~25 to 75 and the value is constant for a given organism. Closely related bacteria have similar mol% G + C values. However, it is important to recognize that two organisms that have similar mol% G + C values are not necessarily closely related; this is because the mol% G + C values *do not take into account the linear arrangement of the nucleotides in the DNA.*

Mol% G + C values were initially determined by acid-hydrolyzing the DNA, separating the nucleotide bases by paper chromatography, and then eluting and quantifying the individual bases. Other methods have since become more popular.

Thermal Denaturation Method. During the controlled heating of a preparation of double-stranded DNA in an ultraviolet spectrophotometer, the absorbance increases by ~40%. This is due to the disruption of the hydrogen bonds between the base pairs that link the two DNA strands. *The temperature at the midpoint of the curve obtained by plotting temperature versus absorbance is called the "melting temperature," or T_m.* The T_m is correlated in a linear manner with the mol% G + C content of the DNA (Marmur and Doty, 1962). The higher the T_m, the higher the mol% G + C of the DNA (see Johnson, 1981, for further details).

Buoyant Density Method. When DNA is subjected to centrifugation in a cesium chloride density gradient (isopycnic centrifugation), it will become located in the form of a band at a position where its density exactly matches that of the cesium chloride solution. The higher the density of cesium chloride where the DNA forms a band, the higher is the mol% G + C value of the DNA (Schildkraut et al., 1962; also see Mandel et al., 1968, for further details).

Although these methods are widely used for estimating DNA base composition, technical problems occasionally do arise because of contamination of the DNA preparation by polysaccharides or pigments, or because of excessive fragmentation of the DNA during its purification. Recent developments in high-performance liquid chromatography have resulted in methods that will accurately and rapidly quantify the free bases, nucleosides, or nucleotides of DNA (see, for example, Ko et al., 1977).

DNA Denaturation and Renaturation

A unique physical property of double-stranded (native) DNA is that under certain conditions (high temperature or high pH) the complementary strands will dissociate (denature). When the resulting single-stranded DNA is then subjected to a somewhat lower temperature and a rather high salt concentration, the complementary strands will reassociate (renature) to form double-stranded DNA/DNA structures (duplexes) that are very similar if not identical to the native DNA (Marmur and Doty, 1961). The renaturation rate is inversely proportional to the genome size (see Wetmur and Davidson, 1968, and Wetmur, 1976, for further details).

RNA/DNA Hybrids

Since only one strand of DNA is used by a cell as a template for RNA synthesis, RNA is complementary only to that strand. Since RNA is single-stranded, RNA molecules do not associate with other RNA molecules; however, when mixed with denatured DNA they can pair with a complementary DNA strand (hybridization) (see Galau et al., 1977, for further details).

Heterologous DNA Duplexes or RNA Hybrids

If denatured DNA from one organism is mixed with denatured DNA from a second organism, heterologous duplexes may form (i.e. duplexes consisting of one strand from the first DNA hybridized with one strand from the second DNA). Similarly, heterologous RNA duplexes may be formed when RNA from one organism is mixed with denatured DNA from a second organism. However, in order for heterologous DNA duplexes or RNA hybrids to occur, the two strands must be complementary in their nucleotide base sequence. A perfect match is not required, and estimates of the amount of base pair mismatch that is tolerated range from ~ 8 to 10% (Ullman and McCarthy, 1973). The thermal stability is usually determined by measuring strand separation during stepwise increases in temperature, and the results mimic the optical melting profile previously discussed under "DNA Base Composition." The thermal stability is usually represented by the term "$T_{m(e)}$," which is the midpoint of the thermal stability profile (i.e. analogous to T_m of native DNA). The difference between the $T_{m(e)}$ of a heterologous duplex and that of a homologous duplex is referred to as the $\Delta T_{m(e)}$ and is used as a measure of the degree of base pair mismatching in the heterologous duplex. The $\Delta T_{m(e)}$ values for heterologous duplexes range from 0 (no mismatching) to 18°C (considerable mismatching). In general, as the fractions of the genomes which can form heterologous duplexes decrease, the thermal stabilities of the duplexes that do form also decrease.

DNA AND RNA HOMOLOGY EXPERIMENTS

Such experiments attempt to answer one question: does DNA or RNA from organism A have a base sequence that is sufficiently similar to that from organism B to allow the formation of DNA heteroduplexes or heterologous RNA hybrids?

DNA Homology Values

These are average measurements of similarity in which the *entire genome of one organism is compared with that of another.*

RNA Homology Values

These values are specific for each type of RNA:

Messenger RNA (mRNA) Homology Values. These are similar to those obtained by DNA homology (at least for bacteria) because a large portion of the genome is used for transcribing the mRNA molecules. For this reason, and because mRNA is difficult to label, mRNA homology has not been widely used in bacterial taxonomy.

Ribosomal RNA (rRNA) and Transfer RNA (tRNA) Homology Values. In contrast to mRNA, rRNA and tRNA are coded for by *only a small fraction of the bacterial genome;* therefore, in homology experiments using either of these two types of RNA, only those fractions of the genome are being compared, not the entire genomes. In all groups of bacteria so far studied, the arrangement of nucleotides in the rRNA and tRNA cistrons of the DNA appears to have evolved less rapidly than the bulk of the cistrons in the DNA. This is probably due to their role in determining the structural and functional aspects of the ribosome (Woese et al., 1975).

Therefore, DNA homology experiments are used to detect similarities between *closely related* organisms, whereas RNA homology experiments are used to detect similarities between *more distantly related* organisms.

METHODS FOR HOMOLOGY EXPERIMENTS

Many procedures have been developed for detecting heterologous DNA duplexes or RNA hybrids. A brief description of some of these follows.

Heavy Isotopes

The earliest efforts to quantify the formation of heteroduplexes were made by incorporating a heavy base (5-bromouracil) or a heavy isotope (^{15}N) into one of the DNA preparations. After the labeled and unlabeled DNA preparations were mixed and allowed to reassociate, the mixture was subjected to ultracentrifugation with cesium chloride. This allowed the separation of heteroduplexes (which had an intermediate buoyant density) from the homologous duplexes (which had either a light or heavy density). These experiments were time consuming and worked best only for small genomes such as those of viruses.

Agarose Gels

In 1963, McCarthy and Bolton immobilized high molecular weight denatured DNA in an agarose gel. The gel was then cut into small particles by forcing the agar through a small mesh screen. The agar particles were then incubated with radioactive-labeled RNA or fragmented DNA. The smaller RNA molecules or DNA fragments could diffuse through the agar and form hybrids or duplexes with complementary immobilized DNA. The immobilization of the high molecular weight DNA prevented it from reassociating with other high molecular weight DNA and also provided a means for washing unreacted labeled nucleic acid fragments away from those that had formed hybrids or duplexes with the immobilized DNA. The results from such experiments were quantitative and could be readily applied to broad taxonomic studies (Hoyer et al., 1964).

Binding to Nitrocellulose

In 1963, Nygaard and Hall found that native DNA, denatured DNA, and RNA/DNA hybrids would bind to nitrocellulose whereas RNA would not. This provided another means for immobilizing denatured DNA for use in RNA/DNA hybridization experiments and also for separating RNA/DNA hybrids from free RNA. The parameters for these experiments were worked out in detail by Gillespie and Spiegelman (1965).

In 1966, Denhardt described a procedure for covering the DNA binding sites on nitrocellulose membranes. This made it possible first to immobilize a given amount of denatured DNA on the membrane and then to treat the membrane with a mixture that prevented additional DNA from binding to the membrane (unless it was complementary to the immobilized DNA on the membrane). Thus the membrane procedure became readily applicable to DNA homology experiments and has completely replaced the agarose gel method.

By the use of nitrocellulose membranes, DNA or RNA homology values can be determined by either *direct binding* or *competition* experiments.

Direct Binding Method. In the direct binding method, a given amount of denatured labeled DNA or RNA is incubated under standardized conditions with various single-stranded DNA preparations that have been immobilized on nitrocellulose membranes. After incubation the unbound labeled nucleic acid is washed away and the radioactivity remaining on the membrane (due to duplex or hybrid formation) is measured. The *percent homology* is expressed as the *amount of heterologous binding divided by the amount of homologous binding × 100.* The results are somewhat variable because it is difficult to consistently get the same amounts of DNA on the membranes. This problem is circumvented with the competition method.

Competition Method. In the competition method, unlabeled denatured reference DNA is fixed onto nitrocellulose membranes. A direct binding reaction, used for a reference point, is performed between the homologous denatured labeled DNA in solution and membrane-bound reference DNA. The competitive reactions have the same components as the direct binding reaction but additionally contain high concentrations of unlabeled denatured DNA fragments in solution. If the competitor DNA is homologous to the labeled DNA in solution and to the unlabeled DNA bound to the membrane, the competitor DNA will form duplexes with both the labeled DNA and the immobilized DNA: consequently, the amount of labeled DNA that forms duplexes with the immobilized DNA will be much lower than that occurring in the direct binding reaction. The homologous competition will be ~90% effective. On the other hand, if the competitor DNA is not related, it will not form duplexes with the labeled DNA and immobilized DNA and there will be no competition. The percent homology is the ratio of the heterologous competition to the homologous competition × 100. Such competition experiments give very reproducible results but do require relatively large quantities of DNA (Johnson, 1981).

Free Solution Reassociation

In this method all of the component nucleic acids are in solution rather than being immobilized in some manner. Reassociation of DNA may be monitored optically by ultraviolet spectrophotometry or by means of a labeled probe.

Optical Procedure. In the optical procedure, the rates of reassociation are determined. Since DNA reassociation is a second-order reaction, the rate will be proportional to the square of the concentration. The general procedure for comparing the DNAs from two organisms is to measure the reassociation rates of equivalent concentrations from each of the organisms separately and compare those rates with that of an equal mixture of the two DNA preparations. If the two organisms are identical, the reassociation rates in the three cuvettes will be the same. If the two organisms are unrelated, then each kind of DNA in the mixture will reassociate independently of the other and, since they are each at half the concentration of that used in the cuvettes with a single DNA component, the overall rate will be one half. De Ley et al. (1970) have studied the parameters of the method in detail and have derived equations for calculating the homology values.

Labeled DNA Probe. The most popular procedure for free solution reassociation involves the use of a labeled DNA probe. As discussed above, the rate of DNA reassociation is a function of DNA concentration and, because the labeled probe DNA is used at a very low concentration, very little of it will reassociate. The unlabeled test DNA with which the probe DNA is incubated is at a much higher concentration and most of it will reassociate. Therefore, if the probe DNA is identical with the unlabeled test DNA, it will reassociate with the unlabeled DNA at the rate at which the unlabeled DNA is reassociating. On the other hand, if the two DNAs are unrelated the unlabeled DNA will reassociate, but most of the probe DNA will remain single stranded. To determine the amount of probe DNA that has duplexed with the unlabeled DNA, either *hydroxylapatite* or *S_1 nuclease* is usually used.

Hydroxylapatite is used to separate single-stranded (denatured) DNA from double-stranded DNA. At a phosphate concentration of 0.14 M, only double-stranded DNA will adsorb to hydroxylapatite and single-stranded DNA can be washed away. The double-stranded DNA can then be desorbed by increasing the phosphate concentration. Although originally used as a column chromatography procedure (Miyazawa and Thomas, 1965; Bernardi, 1969a, b), the batch procedure described by Brenner et al. (1969) has been widely used.

Under suitable conditions, S_1 nuclease will have little effect on double-stranded DNA but will hydrolyze single-stranded DNA. Consequently, the extent of duplex formation by the probe DNA can be determined by the amount of S_1 nuclease-resistant (i.e. acid-precipitable) radioactivity (Crosa et al., 1973).

Comparison of the Various Homology Methods

In spite of the diversity of the DNA homology methods, they are all used to measure the same phenomenon and so it is comforting to find that, for the most part, they all give similar results. The major experimental parameters that affect homology results are the sodium ion concentration and the reassociation temperatures. The most commonly used sodium ion concentration is about 0.4 M, although concentrations up to 1 M do not alter the results significantly. The reassociation temperature can have a profound effect on the homology values and therefore a standardized temperature of about 25°C below the T_m (T_m − 25°C) is most commonly used (Marmur and Doty, 1961). The reassociation temperature effect is approximately linear for the membrane competition and the hydroxylapatite procedures: for organisms having less than 50% homology, the homology values will increase by about 20% at 10°C below the T_m − 25°C temperature and decrease by about 20% at 10°C above the T_m − 25°C temperature. Reassociation temperature differences do not have as great an effect on the optical (De Ley et al., 1970) or the S_1 nuclease methods (Grimont et al., 1980).

Under similar conditions of reassociation, the hydroxylapatite, membrane competition, and spectrophotometric methods give very similar results (Kurtzman et al., 1980). The S_1 nuclease procedure results in somewhat lower (15–20%) homology values, particularly between organisms having less than 50% homology.

The rRNA cistrons have been found to be very conserved in all groups of organisms that have been investigated. The nitrocellulose membrane procedures, such as competition, direct binding, and thermal stability of hybrids, have been used for most of the rRNA homology studies. Results from these experiments appear to reflect nucleotide sequence differences that are similar to those found in the DNA homology experiments discussed above.

rRNA OLIGONUCLEOTIDE CATALOGS

Besides the use of RNA/DNA homology experiments for comparison of the rRNA cistrons from various bacteria, rRNA molecules have been compared directly by determining the nucleotide sequences in oligonucleotides. The rRNA preparation is first digested with T_1 ribonuclease, which cleaves between the 3′-guanylic acid and the 5′-hydroxyl group of the adjacent nucleotide. This results in a guanine residue at the 3′ end of each oligonucleotide. The oligonucleotides are then separated by two-dimensional electrophoresis (Sanger et al., 1965; Uchida et al., 1974). The first dimension is on cellulose acetate at pH 3.5. The oligonucleotides are then transferred from the cellulose acetate strip onto DEAE cellulose and electrophoresed in the second dimension in 6.5% formic acid. The oligonucleotide spots form three-to-four series of wedge-shaped patterns (Sanger et al., 1965). Within each pattern the oligonucleotides contain a constant number of uracil residues, and the locations of the spots within a pattern indicate the number of adenine and cytosine residues. Therefore, by inspecting the pattern one can predict the nucleotide sequence of the shorter oligonucleotides and the base compositions for the longer ones. The spots containing the longer nucleotides are then cut out for secondary analysis. After digestion with other ribonucleases they are again electrophoresed on DEAE cellulose. If the nucleotide sequence still is not clear, a tertiary analysis is required. The unique oligonucleotides (usually only one per rRNA molecule) of each organism are entered (cataloged) into computer storage. The oligonucleotide catalog from one organism can then be compared with that of another. The similarity value between two organisms is the number of unique oligonucleotides (in each of their rRNA molecules) that they both share divided by the average total number of unique oligonucleotides. This procedure compares the sequence for a rather large portion of the rRNA molecules.

Most recently, procedures have been developed for rapidly sequencing long segments of DNA and RNA (Maxam and Gilbert, 1977; Sanger et al., 1977; Peattie, 1979). DNAs from several viruses have been sequenced. Sequencing all of the DNA of a bacterium would generate a rather formidable amount of data; however, specific cistrons have been compared by sequence analysis, such as the genes of the tryptophan operon of *Escherichia coli* and *Salmonella typhimurium* (Crawford et al., 1980).

CONTRIBUTIONS OF NUCLEIC ACID STUDIES TO BACTERIAL TAXONOMY

Concept of a Bacterial Species

A major contribution of DNA homology studies has been to provide a more unifying concept of a bacterial species. Although the exact level of DNA homology above which one considers organisms as belonging to the same species is arbitrary, similar homology clusters have been found in all bacterial groups that have been investigated. I have previously suggested what seemed to be reasonable cut-off points for delineating subspecies, species, and closely related species (Johnson, 1973). These are illustrated in Figure III.1. DNA heterogeneity in the species range (*A*) has been found for many bacterial groups that are phenotypically very similar. In some instances the homology values will tend to cluster in the 80–90% homology range (*B*). Examples of this are the clustering of *Propionibacterium acnes* (Johnson and Cummins, 1972) and *Bacteroides uniformis* (Johnson, 1978). In other instances there may also be clustering at the lower end of the species range (*C*). *Bacteroides fragilis*, for example, clusters into two groups where the intergroup homology values are in the range of 60–70% and the intragroup homology values in the 80–90% range (Johnson, 1978). It is important to note that the thermal stabilities of heteroduplexes between organisms in the 80–90% DNA homology range will be very similar to those of homoduplexes ($\Delta T_{m(e)}$ values of 0–3°C), whereas with heteroduplexes between 60 and 70% homology they will be substantially lower ($\Delta T_{m(e)}$ values of 6–9°C). Therefore, it appears that 60–70% homology is a transitional point between genetic events that may be largely cistron rearranging in nature and genetic events where there are also many changes in the base sequences (Johnson, 1973). In other instances, e.g. *Bacteroides ovatus* (Johnson, 1978), multiple groups within the 60–70% homology range make subgrouping at this level rather complicated so that, unless there are other important considerations, such as pathogenicity (Krych et al., 1980), it may not be justified.

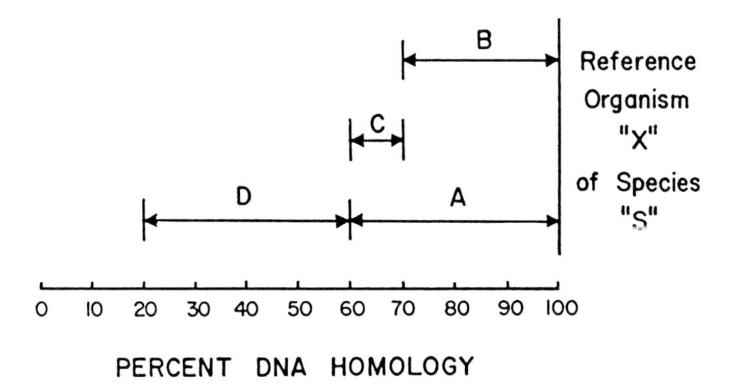

Figure III.1. Proposed taxonomic groupings based upon DNA homology data. *A*, organisms belonging to species "*S*"; *B*, varieties within subspecies to which "*X*" belongs; *C*, other subspecies that belong to "*S*"; *D*, species that are closely related to species "*S*."

The DNA homology groups in the lower homology range (*D* in Fig. III.1) often are quite distinct phenotypically from the species with which they are being compared, although in some instances they may differ only in a few characters (Johnson and Ault, 1978; Johnson, 1981; Mays et al., 1982).

It is important to remember that few bacteria have read Figure III.1; therefore, the exact limits chosen for a given group of organisms will have to remain at the discretion of the individual investigators.

Identification Schemes

A major practical use of DNA homology data is for correlation with individual phenotypic tests. It is common to find variability for a trait among strains within a DNA homology group as well as distinct DNA homology groups that differ from each other by only a few traits (Johnson and Ault, 1978; Johnson, 1980; Holdeman et al., 1982; Mays et al., 1982). Correlating phenotypic test results with DNA homology groups enables investigators to select phenotypic tests that are required for the accurate identification of organisms belonging to these groups.

Concept of a Bacterial Genus and Higher Taxa

Comparisons of rRNA cistrons by rRNA homology experiments and by 16S oligonucleotide catalog similarities are providing data from which a more unifying phylogenetic concept for higher bacterial taxa is possible. De Ley and his associates (De Ley et al., 1978; De Smedt et al., 1980) have proposed the establishment of several genera on the basis of rRNA homology results. On the basis of 16S rRNA oligonucleotide similarity, Woese (in Fox et al., 1980) has proposed the reestablishing of the higher bacterial taxa which were dropped from the eighth edition of *Bergey's Manual*® because it was thought that the higher taxa listed in the seventh edition did not represent phylogenetic relationships. As examples, the 16S rRNA oligonucleotide similarity values have contributed greatly to the present taxonomic scheme of the methanogenic bacteria (Balch et al., 1979) and to the establishment of Division IV *Mendosicutes* in the Kingdom *Procaryotae* (see the chapter on "The Higher Taxa" by Murray in this *Manual*).

BACTERIAL CLASSIFICATION IV

Genetic Methods

Dorothy Jones

The use of genetic characteristics in bacterial classification is comparatively recent. It dates from the mid-1950s when bacterial gene transfer was discovered and Watson and Crick demonstrated the molecular basis of genetic information in the sequence of bases on the DNA molecule. Since that time the development of physicochemical techniques for the analysis of the genetic material, together with the exploitation of bacteria as genetic tools, has resulted in the accumulation of material which has proved significant for bacterial systematics.

In the last two decades it has become clear that the genetic complement of a bacterial cell lies not only in the main chromosome but, in many cases, also in extrachromosomal elements such as plasmids, transposons and lysogenic or temperate phages. All these elements carry genetic material capable of phenotypic expression. What contribution such extrachromosomal entities make to a particular bacterial phenotype, either by direct expression or interaction with the chromosomal DNA of the cell, is only just beginning to be understood (see Broda, 1979; Harwood, 1980; Hardy, 1981).

For the bacterial taxonomist the genetic approach to systematics has great appeal both for its potential to reveal biologically significant, stable groupings (taxa) and for the elucidation of bacterial evolutionary relationships (phylogeny). Consequently several of the newer taxonomic methods have been and are being directed toward the characterization of the genetic complement of bacteria.

Physicochemical methods for the analysis of bacterial genomes have been discussed in the previous chapter. The present chapter is concerned with genetic methods used in bacterial classification, i.e. methods based on the transfer of genes between bacteria.

CHROMOSOMAL GENE EXCHANGE

The three main classes of chromosomal gene exchange are: (a) those in which genes are transferred as soluble DNA molecules, i.e. **transformation**; (b) those involving transfer by bacteriophage, i.e. **transduction**; and (c) those involving cell contact followed by transfer of the whole or part of the bacterial chromosome, i.e. **conjugation**. Of these classes, transformation studies have so far proved the most useful for determining relationships between bacteria.

Transformation

Transformation has been demonstrated usually between different taxospecies and only rarely between taxa presently recognized as different genera. Interspecific transformation has revealed three distinct homology groups among neisseriae and moraxellae. Transformation studies have indicated a close relationship between *Rhizobium leguminosarum* and *Agrobacterium tumefaciens*. Studies with the micrococci have shown a close relationship between *Micrococcus luteus* and *Micrococcus lylae* that, in this case, was confirmed by DNA reassociation studies in vitro. Similar studies have shown a low rate of transformation between *Pasteurella multocida, P. haemolytica, P. ureae*, and *P. pneumotropica*, taxa which are also closely related on phenetic and DNA reassociation criteria. A great deal of transformation work has been done on the genus *Haemophilis. Haemophilus influenzae, H. aegyptius*, and *H. parainfluenzae* all appear to be closely related.

Transformation of chromosomal DNA has been demonstrated also among other taxa and there is no doubt that it is a good indication of the degree of relatedness between different taxospecies and can highlight areas of taxonomic homogeneity and heterogeneity (Jones and Sneath, 1970; Bøvre, 1980).

Transduction

In transduction, host chromosomal material is incorporated into a bacteriophage by several mechanisms and transmitted from one bacteriophage host to another by phage-mediated transduction. Only a small range of bacterial groups are presently known to be susceptible to transduction, e.g. the *Enterobacteriaceae*, the genus *Bacillus*, pseudomonads, and some streptococci. Not much is known about how readily strains of the same species can be transduced, but the host range pattern of the transducing bacteriophages is probably a major limiting factor. It has been suggested also that the greater difficulties associated with transduction are due to the larger sizes of the DNA fragments involved in transduction compared with those involved in transformation, with the larger fragments being less easily integrated into the recipient chromosome. Again this mechanism of genetic transfer appears to have significance for bacterial classification only at the taxospecies level, and its usefulness is further restricted by the host range of bacteriophages (see Jones and Sneath, 1970).

It is appropriate here to mention the other roles of bacteriophages in bacterial classification. As noted earlier, a temperate bacteriophage can lysogenize in a host bacterium and express its genetic information as phenotypic characters different from those typical of the bacterium devoid of phage. The consequences of this for bacterial classification will be dealt with later (see "Extrachromosomal Elements"). Additionally, and this is perhaps their best known feature, virulent bacteriophages infect and lyse bacteria. The process is referred to as **phage lysis.**

The inclusion of phage lysis in a section dealing with genetic methods may cause the reader some surprise. Phage lysis of the bacterial cell (as distinct from the much less specific phage adsorption, or killing of the cell followed by lysis from without) involves phage infection with phage multiplication but without lysogenization. In bacteriophage infection, the genes of the virulent phage are transferred and expressed even though they are not integrated into the host chromosome nor, of course, into the lineage of the recipient. Specific phage receptors are necessary for the adsorption of virulent phage onto the recipient bacterial cell; once in the cell the phage may be repressed if the bacterium is carrying a homologous prophage, or it may be restricted enzymically. The ability of two bacterial strains to support the growth of a given virulent phage may reflect similarity in only one or two host genes. Therefore, the technique has little value for *bacterial classification.* However, the value of phage lysis cross-reactions for *bacterial identification* is high. The reported host range of bacteriophages extends from those specific for very few strains of one taxospecies to those that can lyse bacteria which are currently placed in different bacterial genera, families, and even orders. However, most reports in the literature show that most phages lyse a significant proportion of strains belonging to the same taxospecies as the propagating strain. Phage-typing schemes are playing increasingly important epidemiological and identification roles among a number of bacterial groups, e.g. some pyogenic streptococci, staphylococci, and enterobacteria.

Conjugation

This method of gene exchange refers to the transfer of the whole or a portion of the bacterial chromosome following cell-to-cell contact. The conjugation system is best understood among the coliforms (Curtiss, 1969). Similar systems have been noted among other genera, such as *Pseudomonas, Vibrio, Pasteurella,* and *Rhizobium,* and are known to occur among other groups, but the mechanism is less well understood. In the streptococci there is evidence that in some cases the bacteria make use of sex pheromones to generate cell-to-cell contact (Clewell, 1981). Transfer of bacterial chromosomal material by conjugation has not been reported so frequently as transfer by transformation or transduction. However, evidence suggests that it takes place only between closely related taxa.

Bacterial taxonomy has not, to date, benefited greatly from studies involving genetic exchange of chromosomal material, and the concept of a bacterial genospecies is far from being realized. However, bacteriologists no longer believe that gene transfer is so rare among bacteria that it is of no consequence for natural bacterial populations. In the past two decades it has been recognized that gene transfer, particularly involving phages, plasmids, and transposons, their interaction with each other, and the bacterial chromosome together with the gene transfer mediated by insertion sequences, can be a significant factor in bacterial variation. This variation has obvious consequences for bacterial systematics.

EXTRACHROMOSOMAL ELEMENTS

Plasmids, transposons, and phages are collectively referred to as extrachromosomal elements (Novick, 1969; Broda, 1979; Hardy, 1981). Their transfer between bacteria is essentially by the same mechanisms as those described under "Chromosomal Gene Exchange." Phages play a role in the transduction of all genetic material between bacteria, and it is now recognized that the F' factor is a plasmid. It is therefore probably artificial to make too clear a distinction between the transfer of chromosomal DNA and that of extrachromosomal elements between bacteria. Transformation by chromosomal DNA may or may not be accompanied by plasmid DNA. In transduction phages can carry a portion of the chromosomal or plasmid DNA, and in conjugation plasmid and chromosomal DNA can be transferred at the same time. The situation is far more complex than was previously realized (Novick, 1969; Clewell, 1981; Hardy, 1981).

A range of methods now exists for the isolation of extrachromosomal genetic elements from bacteria and for their analyses by physiochemical methods. A good account is given by Hardy (1981).

The two aspects of extrachromosomal elements which are of prime interest to the bacterial taxonomist are their ability to code for phenotypic traits in a range of bacteria and their significance in evolution.

Phenotypic Traits

Plasmids have been observed in virtually every bacterial genus examined. Many plasmids detected by physical screening methods are not known to code for any phenotypic trait in the host bacterium. They are called **cryptic plasmids**. The fact that their presence has not been correlated with a phenotypic characteristic does not mean that they do not code for such a trait. It may be that their particular phenotypic traits have not been identified.

Phenotypic traits known to be coded for by plasmids include resistance to a variety of antibiotics, heavy metal ions, and ultraviolet light; production of enterotoxin, exfoliate toxin, the surface antigens K88 and K89, hemolysins, proteases, bacteriocins, urease, and H_2S; metabolism of lactose, sucrose, raffinose, and citrate; degradation of a variety of organic compounds such as camphor, octanol, and toluene (at least part of the remarkable diversity shown by pseudomonads in the degradation of organic compounds is due to the presence of degradative plasmids); and nitrogen fixation. Preliminary evidence suggests that the production of gas vacuoles in *Halobacterium* is controlled by a plasmid. There is also evidence that pigment, coagulase, and fibrolysin production in staphylococci are plasmid determined. It also seems highly probable that among the streptococci the production of serum opacity factor, M protein production, nisin production, and the ability to ferment galactose and xylose are plasmid-coded.

Transposons found on the plasmids of Gram-negative bacteria have been shown to code for resistance to a number of antibiotics, lactose fermentation (in *Yersinia enterocolitica*), and heat-stable toxin (in *Escherichia coli*), and doubtless others coding for other phenotypic traits will be found. Full accounts of the phenotypic traits conferred on bacteria by plasmids and transposons are given by Harwood (1980), Clewell (1981), and Hardy (1981).

The classic example of a phage-encoded phenotypic trait is the diphtheria toxin which was shown by Freeman in 1950 to be produced only when *Corynebacterium diphtheriae* is lysogenized by a particular phage. The structural gene for the protein toxin is on the phage chromosome. This phage can lysogenize and synthesize toxin in a number of closely related corynebacteria, viz. *C. diphtheriae, C. ulcerans,* and *C. ovis* (Barksdale, 1970).

Effect of Extrachromosomal Elements on Classification and Identification

Since these elements confer extra phenotypic traits on their hosts they could have a marked effect on bacterial classification if those characters were ones on which the classification was based. Two examples of the presence of plasmids which relate to species nomenclature are the plasmid-coded hemolysin of *Streptococcus faecalis* which resulted in the naming of such plasmid-bearing strains as *Streptococcus faecalis* var. *zymogenes*, and the plasmid-determining citrate utilization in *Streptococcus lactis* which appears to be responsible for the name *Streptococcus lactis* subsp. *diacetylactis*. However, the effect of an extrachromosomal-coded trait on a classification based on a large number of characters would normally be expected to be small, and this has proved to be the case in the few preliminary studies so far conducted.

Such characters can, however, affect the identification of bacteria when the identification is based on a small number of characters and considerable weight is placed on individual features, e.g. lactose fermentation in the identification of enterobacteria. It is best, therefore, if identification schemes are based on stable features chosen as a result of a taxonomic study where a large number of characters have been employed, e.g. computer-assisted classifications (numerical taxonomy). Ideally, computer-based identification matrices derived from such studies should be employed. The risk of a misidentification due to the loss or gain of one or two phenotypic characters is thereby reduced to a minimum.

It has been suggested that strains known to carry extrachromosomal elements should be excluded from taxonomic studies. Such a policy is not practical because present methods do not always detect such strains; further, it is believed that many bacterial populations depend on the presence of these elements for their survival. It has also been suggested that known or suspected extrachromosomal coded characters should be excluded when classifications are constructed. Again, present methods do not allow all such characters to be determined; besides, such characters may have taxonomic relevance.

Bacteriologists should accept that extrachromosomal elements do contribute to bacterial variation. This variation should therefore be recognized, and due allowance made, when bacterial taxa are described and when identification schemes are constructed.

Extrachromosomal Elements and Evolution

At the present time the relative contributions of mutation and recombination to bacterial evolution is difficult to assess. Mutation results in changes in the protein structure of the organism. Recombination leads to the rearrangement of existing genes. Until recently little attention was paid to the possible involvement of gene rearrangement in evolution. The recognition that gene transfer involving extrachromosomal elements can be a significant factor in bacterial variation has led to a view that these elements have played an important role in bacterial evolution. Whether or not the role which these elements play in contemporary bacterial variation and adaptation is one of the major ways in which bacteria have evolved from the earliest times is still not resolved (Reanney, 1976; Cullum and Saedler, 1981; Hardy, 1981; Koch, 1981).

BACTERIAL CLASSIFICATION V

Serology and Chemotaxonomy

Dorothy Jones and Noel R. Krieg

Serology and chemotaxonomy are both methods for investigating the molecular architecture of the bacterial cell, although the methodologies used in the two techniques are quite different.

SEROLOGY

Serological techniques depend on the ability of the chemical constituents of bacterial cells to behave as antigens, i.e. to elicit the production of antibodies in vertebrate animals. The antibodies used in serological studies are the humoral antibodies found in the blood serum and referred to as antiserum. Monoclonal antibodies, highly specific serological agents directed against specific antigenic determinants (epitopes), are now being used increasingly in serological studies (Macario and Conway de Macario, 1985).

Serological techniques used include agglutination, precipitation (including many refinements, e.g. use of gels and electrophoretic techniques), complement fixation, and immunofluorescence. Details of the techniques may be found in a number of immunological or microbiological text books.

Serological studies of value in bacterial taxonomy can be divided into two broad classes: (a) those concerned with detecting differences or similarities between bacteria **on the basis of their cell surface and associated antigenic complement** (e.g. flagella, pili, cell walls, cytoplasmic membranes, capsules, and slime layers) and (b) the use of antisera raised against purified enzymes to assess **structural similarities between homologous proteins from different bacteria.**

Cell Surface and Associated Antigens

On the basis of the antigenic complexity of their surface antigens (cell wall lipopolysaccharide, flagella, and capsule constituents), the genera of the family *Enterobacteriaceae* can be divided into many serovars; e.g. more than 1000 serovars have been detected within the genus *Salmonella* (Kauffmann, 1966). Contrary to the view of Kauffmann (1966) these serovars do not represent separate taxospecies. The information derived from serological studies of a group such as the enterobacteria is now so large, and so many cross-reactions occur that, in the absence of any methods (e.g. computer programs) for analyzing the plethora of data in an objective fashion, it is generally accepted that these techniques are of little value in classification but are valuable in epidemiological studies.

Serological studies of the streptococci based on the use of acid-extracted polysaccharide antigens (Lancefield, 1933, 1934) have resulted in the division of the genus *Streptococcus* into a number (now approaching 30) of serological groups labeled A, B, C, etc. Until fairly recently, very great emphasis was placed on the serological grouping of streptococci for purposes of both classification and identification.

Although some serological groups correspond to distinct taxospecies (e.g. serological group A (*S. pyogenes*) and serological group B (*S. agalactiae*)), other serolological groups comprise more than one taxospecies (e.g. serological groups C, D, and N), while serological groups G, H, and K do not serve to define any good taxa (see Jones, 1978).

Other serological studies of this kind have been based on the use of different classes of antigenic material, e.g. cell walls and spore suspensions. A review of the serochemical specificity and location of antigens in the bacterial cell together with observations on the significance of such serological studies for bacterial classification has been provided by Cummins (1962b). A comprehensive review on the use of monoclonal antibodies as molecular probes for complex structures such as the bacterial cell surface, cell membrane, and spores is that of Macario and Conway de Macario (1985).

Use of Antisera Raised against Purified Proteins

The basis of this approach is that one antiserum raised against a purified enzyme can be used to detect the serological cross-reactions of homologous proteins in crude extracts of other bacteria if the bacteria possess the same enzyme. The use of microcomplement fixation techniques makes this approach a very sensitive one. Comparative studies on purified proteins of known primary structure have indicated that there is a very high correlation between the amino acid sequence of the proteins and the degree of serological similarity (see later section on "Amino Acid Sequences of Various Proteins"). Examples of this approach include studies of the muconate-lactonizing enzymes of the *Pseudomonadaceae* (Stanier et al., 1970), the fructose diphosphate aldolases of the lactic acid bacteria (London and Kline, 1973), and the catalases of staphylococci and micrococci. In the instance of the staphylococci, a very high correlation has been shown to exist between the serological relationships of their catalases and genetic relatedness based on DNA/DNA homology data (see Kandler and Schleifer, 1980).

Similar studies on the transaldolases of several species of bifidobacteria (Sgorbati, 1979; Sgorbati and Scardovi, 1979) indicate that the genus *Bifidobacterium* contains several distinct clusters based on the index of dissimilarity of their respective aldolases. In some instances there was good correlation between the clusters so obtained and clusters formed on the basis of other criteria, but in others the correlation was not so high.

Baumann et al. (1980) and Bang et al. (1981) found that the immunological relationships among glutamine synthetases and superoxide dismutases of *Vibrio* and *Photobacterium* species were in good agreement with relationships based on rRNA/DNA homology experiments. The amino acid sequence of the glutamine synthetases was conserved to a greater extent than was that of the superoxide dismutases, and this supports the idea that the study of proteins having different evolutionary rates can permit the resolution of close, intermediate, and distant relationships among organisms.

It should be noted that serological homology studies of proteins, like many other techniques, have their limitations. There is evidence that the approach is useful only for the study of proteins with relatively high (70% or greater) sequence homologies. Further, serological techniques measure similarities only at the surface of proteins and it is at the protein surface that the greatest number of amino acid changes occur. The results can be influenced also by the number of antigenic sites per protein molecule. Nevertheless, serological techniques of this kind provide a rapid and convenient method for assessing structural similarities between homologous proteins, are useful in the classification of bacteria, and can also cast some light on possible phylogenetic relationships.

The use of monoclonal antibodies to detect particular cell constituents, enzymes and toxins in vitro and in vivo is reviewed by Macario and Conway de Macario (1985).

CHEMOTAXONOMY

During the past 20 years or so, the application of chemical and physical techniques to elucidate the chemical composition of whole bacterial cells or parts of cells has produced information of great value in the classification and identification of bacteria. Indeed, so useful have some of the data generated proved to be that the word "chemotaxonomy," used to describe the classification of bacteria on the basis of their chemical composition, is now firmly entrenched in the literature (see Schleifer and Stackebrandt, 1983; Goodfellow and Minnikin, 1985).

In addition, techniques such as gas chromatography have allowed the more precise analysis of the products of fermentation, and there is a growing awareness of the taxonomic significance of enzyme systems and their regulation as opposed to the detection of individual enzymes.

Cell Wall Composition

The characteristic cell wall polymer of many procaryotes, present in Gram-negative and Gram-positive bacteria and in the cyanobacteria, is peptidoglycan (murein). Peptidoglycan is not found in the mycoplasmas and conventional peptidoglycan (containing muramic acid) is not present in archaeobacteria. The chemical structure of the peptidoglycan of Gram-negative bacteria is, with few exceptions, reasonably uniform. However, the variation in qualitative amino acid and/or sugar composition, especially the variation in the primary structure of the peptidoglycans of various Gram-positive bacteria, has provided information of enormous taxonomic value.

The cell wall composition of Gram-positive bacteria was one of the earliest useful chemotaxonomic characters. On the basis of the analysis of the purified cell walls of Gram-positive bacteria. Cummins and Harris (1956) suggested that the cell wall amino acid composition might prove to be an important taxonomic criterion at the generic level and that the sugar composition might help to distinguish between species. Subsequent studies have indicated this to be the case. The amino acids present in the cell wall are now an accepted important part of the generic description. Information from cell wall analysis of the type done by Cummins and Harris (1956) has proved of special value among the coryneform group of bacteria (see Keddie and Bousfield, 1980, for a comprehensive review).

Information of even greater taxonomic value has resulted from the methods devised by Schleifer and Kandler (1967) and used by them and their associates to determine the peptidoglycan types of a wide range of Gram-positive bacteria (see Schleifer and Kandler, 1972; Kandler and Schleifer, 1980; Schleifer and Seidl, 1985). This approach has revealed differences between bacteria which could not possibly have been detected by qualitative cell wall analysis. The methods for determining differences in peptidoglycan types are, however, quite specialized and cannot be used routinely to screen large numbers of bacteria.

More recently, a number of "rapid" methods have been developed for the routine screening of bacteria to determine those cell wall components which have been shown to be of the greatest discriminatory value in bacterial classification and identification. The review of Keddie and Bousfield (1980), Kandler and Schleifer (1980), Bousfield et al. (1985), and Schleifer and Seidl (1985) contain references to the pertinent literature.

A novel peptidoglycan, the so-called pseudomurein, characterized by the replacement of muramic acid by talosaminouronic acid, has been found to be the typical cell wall constituent of the genus *Methanobacterium,* which taxon is now recognized as a member of the archaeobacteria (see Kandler and Schleifer, 1980). The same cell wall polymer has been detected in the archaeobacterial genus *Methanobrevibacter.* Other of the archaeobacteria possess a thick cell wall composed of heteropolysaccharides or a cell envelope consisting of glycoproteins or proteins. *Thermoplasma* lacks a cell wall but contains a glycoprotein in the cell membrane (see Schleifer and Stackebrandt, 1983).

Lipid Composition

Among the procaryotes there are two quite distinct lipid categories. The archaeobacteria contain isopranyl-branched ether-linked lipids but not aliphatic ester-linked lipids, which are present in other bacteria. The presence of isopranyl-branched ether-linked lipids serves to distinguish the archaeobacteria. Further details of archaeobacterial lipids are given by Kates (1978), Langworthy (1982), and Schleifer and Stackebrandt (1983).

Lipids occur in the cytoplasmic membranes of all bacteria and in the cell wall complex of Gram-negative bacteria and certain Gram-positive bacteria such as the genera *Corynebacterium* and *Mycobacterium.* Bacterial lipids comprise a number of different classes and, in the past decade, it has become increasingly clear that at least some of these lipids have chemotaxonomic potential (see Lechevalier, 1977; Schleifer and Stackebrandt, 1983; Goodfellow and Minnikin, 1985).

The fatty acid composition of the bacterial cell has proved useful in the classification of certain bacteria, and in some cases the fatty acid pattern may be characteristic for a particular taxon (see Lechevalier. 1977). However, it should be noted that the fatty acid patterns obtained may be influenced by a number of factors: composition of growth medium, temperature of incubation, age of culture, and the techniques employed to analyze the sample.

A special category of fatty acids free from the aforementioned limitations are the mycolic acids. These long-chain 3-hydroxy 2-branched acids have been found, so far, only in the taxa *Bacterionema, Corynebacterium, Micropolyspora, Mycobacterium, Nocardia,* and *Rhodococcus.* Differences in the structure of their component mycolic acids have proved to be a valuable criterion in the classification and identification of members of these taxa (see Minnikin and Goodfellow, 1980; Collins et al., 1982a, b; Goodfellow and Minnikin, 1985).

Another class of lipids of recognized chemotaxonomic potential are the polar lipids which occur in all bacteria. The most common polar lipid types are the phospholipids and the glycolipids. Phospholipids occur in many bacteria but certain actinomycetes and coryneform bacteria contain very characteristic phospholipids, the phosphatidylinositol mannosides. Other highly characteristic phospholipids include phosphosphingolipids found in certain Gram-negative taxa, e.g. *Bacteroides* (see Lechevalier, 1977). Glycolipids (glycosyldiacylglycerols) are widely distributed among Gram-positive bacteria and can also be used as chemotaxonomic markers (see Shaw, 1975; Schleifer and Stackebrandt, 1983).

Other lipids with chemotaxonomic potential include hopanoids, hydrocarbons, and carotenoids.

Isoprenoid Quinones

Isoprenoid quinones are a class of terpenoid lipids located in the cytoplasmic membranes of many bacteria. They play important roles in electron transport, oxidative phosphorylation, and, possibly, active transport. Their potential as an aid to the classification of bacteria was recognized by Jeffries et al. (1969), Yamada et al. (1976), and others (see Collins and Jones, 1981). Representatives of one, or more than one, of the three main types, ubiquinones, menaquinones, and demethylmenaquinones, are present in the majority of procaryotes so far examined. The cyanobacteria contain neither ubiquinones nor menaquinones. However, they do contain phylloquinones and plastoquinones, which are indigenous to the plant kingdom but not normally found in bacteria. All the mycoplasmas so far examined contain menaquinones only. Among the archaeobacteria no isoprenoid quinones have been detected in the fastidious anaerobic species *Methanobacterium thermoautotrophicum*, a situation in keeping with that most commonly, but not invariably, found among the strictly anaerobic forms of other bacteria. An unusual terpenoid, caldariellaquinone, has been detected in the extreme acidophile "*Caldariella acidophila.*" The other archaeobacteria examined possess menaquinones.

The majority of the strictly aerobic Gram-negative bacteria produce only ubiquinones, with the exception of cytophagas and myxobacters, which produce only menaquinones. Facultatively anaerobic Gram-negative bacteria contain ubiquinones, menaquinones, or demethylmenaquinones or a combination of the three. Strictly anaerobic Gram-negative bacteria (e.g. the genus *Bacteroides*) produce only menaquinones.

The majority of the aerobic and facultatively anaerobic Gram-positive bacteria produce only menaquinones. Most streptococci do not contain any isoprenoid quinones, but demethylmenaquinones are present in *Streptococcus faecalis*, and menaquinones have been detected in "*S. faecium* subsp. *casseliflavus*" and *S. lactis*. Similarly, members of the genus *Lactobacillus* generally lack isoprenoid quinones, but recently, low levels of an uncharacterized menaquinone have been detected in one strain of *L. brevis*. Uncharacterized menaquinones have also been reported in some strains of the strictly anaerobic genus *Clostridium*, although, in general, this genus lacks quinones.

The current data on isoprenoid quinone structural types in bacteria and their implications for taxonomy are reviewed by Collins and Jones (1981). From the available data on procaryotes in general, it appears that menaquinones have far greater discriminatory value than ubiquinones. Menaquinones possess not only a greater range of isoprenologs but additional modifications such as ring demethylation and partial hydrogenation of the polyprenyl side chain occur. The available data strongly suggest that these compounds will be of considerable value in the classification of micrococci, staphylococci, coryneform bacteria, and certain actinomycetes (see Collins and Jones, 1981, and Collins, 1985).

Cytochrome Composition

Cytochromes are specialized forms of hemoproteins which are involved in a variety of redox processes in the procaryote cell. They can be assigned to four main classes, *a*, *b*, *c*, and *d*, according to the structures of their heme prosthetic groups. Cytochrome *o* is an autoxidizable *b*-type cytochrome.

Two basic methods are available which use cytochromes as an aid to classification and identification of bacteria: the "pattern" and the "structure" approach. The former compares the cytochrome patterns of different bacterial species as compared by conventional difference spectrophotometry (see Meyer and Jones, 1973); the latter compares primary structures and, where possible, the tertiary structures of easily purified cytochrome *c* as determined by amino acid sequence and x-ray diffraction (see Ambler, 1976, and see later section on "Amino Acid Sequences of Various Proteins").

Cytochrome patterns show greater variation among the procaryotes than among eucaryotes and can therefore be a useful aid in bacterial classification. Qualitative analyses of the cytochrome composition of over 200 species of bacteria have now been done. The results indicate that the heterotrophic Gram-positive bacteria comprise a rather homogeneous grouping with cytochromes $bcaa_3o$ forming the predominant pattern. There are some variations, however, and cytochrome *c* is often absent from facultatively anaerobic Gram-positive bacteria. Some lactic acid bacteria, when grown on a heme-containing medium, contain only cytochrome *b*. Propionibacteria exhibit a cytochrome bda_1 pattern. The genus *Clostridium* lacks cytochromes. It is of interest that the cytochrome $bcaa_3o$ pattern of logarithmic growth-phase cells of the aerobe *Arthrobacter globiformis* changes to $bcaa_3od$ when the cells become oxygen limited and lose their ability to retain the crystal violet-iodine complex in the Gram stain. Cytochrome *d* is characteristic of many Gram-negative bacteria.

In contrast, the Gram-negative heterotrophic bacteria form a much less homogeneous group on the basis of cytochrome composition. The majority have the basic pattern $bcdoa_1$, from which *c* may often be absent. Cytochrome c_{co} appears to be characteristic of methylotrophs; however, it is also present in the nonmethylotrophic genus *Chromobacterium*. The phototrophic bacteria contain cytochromes *b* and *c* when grown photosynthetically, but when grown aerobically in the absence of light, there are differences between the taxa. The obligately aerobic chemolithotrophs exhibit the cytochrome pattern $bcaa_3oa$, with some occasional omissions. Neither the phototrophs nor the chemolithotrophs have been shown to produce cytochrome *d*.

There is now sufficient evidence available to indicate that cytochrome patterns, in conjunction with other evidence, are useful guides in bacterial classification. There is, however, little evidence that cytochrome patterns will be useful for purposes of identification, mainly because bacteria contain relatively few types of spectrally distinct cytochromes. A comprehensive review of the use of cytochrome patterns in bacterial classification is that of Jones (1980). More recent information may be found in Schleifer and Stackebrandt (1983) and Carver and Jones (1985).

It should be stressed that when cytochrome patterns are used for taxonomic purposes, the influence of the growth environment should be taken into account. Growth conditions can influence the quantitative and, to a lesser extent, the qualitative cytochrome content of bacteria.

Amino Acid Sequences of Various Proteins

Comparison of the amino acid sequence of specific kinds of proteins, or of properties, such as antigenic reactivity, which reflect the amino acid sequence of these proteins have been used as measures of phylogenetic relationships among organisms. The fundamental concept involved is that most extant proteins are likely to have evolved from a very small number of archetypal proteins by the processes of genetic duplication and modification. In comparing the proteins of any particular group (such as cytochrome *c*, superoxide dismutase, ferredoxin, or other enzymes), the greater the difference in amino acid sequence between the protein of one organism and the corresponding protein of another organism, the greater is believed to be the evolutionary divergence between the two organisms. Conversely, if the amino acid sequence of corresponding proteins from two organisms is very similar, the two organisms are believed to be closely related phylogenetically. Even distant relationships between organisms can be deduced by this approach, and various phylogenetic schemes have been constructed to reflect the perceived evolutionary development of a great variety of organisms, both procaryotic and eucaryotic (see the review by Schwartz and Dayhoff, 1978). Among the proteins that have been used for such studies are ferredoxins, flavodoxins, azurins, plastocyanins, and cytochrome *c*. For example, a remarkable similarity in the structure of cytochrome *c* exists between certain nonsulfur purple photosynthetic bacteria (i.e. *Rhodopseudomonas capsulatus* and *R. sphaeroides*), the nonphotosynthetic respiring bacterium *Paracoccus denitrificans,* and the mitochondria of eucaryotic organisms; this and other kinds of congruent data have led to the view that *P. denitrificans* descended from nonsulfur purple bacteria by loss of photosynthetic properties and that this species is the procaryote that most closely resembles the putative procaryotic ancestor of mitochondria (see the exposition by Dickerson (1980) as well as a critical review of the various theories for the endosymbiont origin of mitochondria and chloroplasts by Gray and Doolittle (1982)). Reservations about many of the conclusions based on cytochrome *c* sequences have been expressed by Ambler et al. (1979 a, b), and an analysis of some of the limitations involved in comparing amino acid sequences of proteins has been given by Doolittle (1981).

Protein Profiles

The basic premise here is that closely related organisms should have similar or identical kinds of cellular proteins. Two-dimensional electrophoretic and isoelectric focusing procedures (O'Farrell, 1975) have made it possible to resolve several hundred proteins from a cell extract. The protein "fingerprint" so obtained for one bacterial strain is a reflection of the genetic background of that strain and can be compared with the "fingerprints" from other strains as a measure of relatedness. For examples of the application of this method, see the comparison of *Rhizobium* strains made by Roberts et al. (1980), and the comparison of *Spiroplasma* strains made by Mouches et al. (1979). The method requires a considerable degree of standardization in order to yield optimum results.

One-dimensional polyacrylamide gel electrophoresis (PAGE) of cellular proteins can yield patterns of up to ~ 30 bands, and although it is not comparable in resolving power to the two-dimensional separation method, it can distinguish related organisms from unrelated organisms. In general, whole cells or cellular membrane fractions are used, and the proteins are solubilized by means of a detergent such as sodium dodecyl sulfate (SDS); however, many studies have employed merely the water-soluble proteins ("soluble" fraction) from disintegrated cells. A few examples of the application of PAGE are: identification of mycoplasmas (Razin and Rottem, 1967); taxonomy of *Haemophilus* strains (Nicolet et al., 1980); comparison of isolates from gingival crevice floras (Moore et al., 1980); and differentiation of isolates of indigenous *Rhizobium* populations (Noel and Brill, 1980). By use of rigorously standardized conditions, extremely reproducible protein patterns can be obtained which are amenable to rapid, computerized, numerical analysis (Kersters and De Ley, 1975, 1980; Jackman, 1985; Kersters, 1985).

In an analysis of the patterns of soluble cellular proteins from strains *Clostridium* species, Cato et al. (1982) found that strains having >80% DNA/DNA homology usually produced identical patterns, strains related by ~ 70% homology showed overall similarity of the total patterns but also showed minor differences, and strains unrelated by DNA homology showed major differences. In many instances, the patterns obtained within 24 h of isolating an organism were sufficiently distinctive so that the identity of the organism could be strongly suspected.

Enzyme Characterization

It is now recognized that the functional and structural patterns displayed by certain bacterial enzymes provide data of use in classification. Good examples are the diverse regulatory and molecular size patterns exhibited by bacterial citrate synthases and succinate thiokinases. Both of these are enzymes of the citric acid (Krebs) cycle, and the near universal occurrence of this cycle in living cells makes it a very suitable pathway for comparative studies between different organisms (see Weitzman, 1980).

In general, the citrate synthases of Gram-negative bacteria are inhibited by reduced NADH, while those of Gram-positive bacteria are not. The citrate synthases of Gram-negative bacteria can be further divided into two classes on the basis of whether their NADH sensitivity is overcome by AMP. Citrate synthases from the majority of strictly aerobic Gram-negative bacteria are reactivated by AMP, while those of the facultatively anaerobic Gram-negative bacteria are not. Citrate synthases of the Gram-negative facultative anaerobes are also inhibited by α-oxoglutarate, but the enzymes from the aerobic Gram-negative bacteria and from Gram-positive bacteria are not. Citrate synthases of the cyanobacteria are not inhibited by NADH, but they are inhibited by α-oxoglutarate and by succinyl-CoA.

Bacterial citrate synthases fall into two groups, "large" and "small," on the basis of molecular size. The majority of Gram-negative bacteria possess large citrate synthases (mol. wt. ~250,000), while the majority of Gram-positive bacteria produce citrate synthases of the small type (mol. wt. ~100,000)

Exceptions to the broad general pattern occur. The citrate synthases of the Gram-negative genus *Acetobacter* do not appear to be inhibited by NADH, although the enzyme is of the large type. On the other hand, the citrate synthases of *Thermus aquaticus* is both insensitive to NADH and of the small type. In both respects it resembles the citrate synthases of the archaeobacterial genus *Halobacterium* and the majority of the citrate synthases of Gram-positive bacteria.

Similar molecular size patterns occur among bacterial succinate thiokinases. All succinate thiokinases from the Gram-positive bacteria so far studied are of the small type (mol. wt. 70,000–75,000), whereas those of Gram-negative bacteria, cyanobacteria, and *Halobacterium* species are of the large type (mol. wt. 140,000–150,000). Bacterial succinate thiokinases can be further subdivided on the basis of their specificity for nucleotide substrates (guanosine diphosphate or inosine diphosphate), and preliminary results point to interesting patterns of enzyme diversity of possible potential in bacterial classification.

Rapid methods are now available for the routine laboratory screening of bacterial citrate synthases, and as further such methods are developed, it is likely that the regulatory and molecular properties of these and other enzymes will prove useful in the classification of bacteria (see Weitzman, 1980).

Fermentation Product Profiles

The use of gas-liquid chromatographic methods to analyze the fatty acids formed as end products of protein or carbohydrate metabolism is particularly useful in the classification and identification of the anaerobic genera *Clostridium, Bacteroides, Eubacterium,* etc. (see Holdeman et al., 1977, and Shah et al., 1985, for references to the literature).

Bacterial Nomenclature

Peter H. A. Sneath

SCOPE OF NOMENCLATURE

Nomenclature has been called the handmaid of taxonomy. The need for a stable set of names for living organisms, and rules to regulate them, has been recognized for over a century. The rules are embodied in international codes of nomenclature. There are separate codes for animals, noncultivated plants, cultivated plants, bacteria, and viruses. But partly because the rules are framed in legalistic language (so as to avoid imprecision), they are often difficult to understand. Useful commentaries are found in Ainsworth and Sneath (1962), Cowan (1978), and Jeffrey (1977).

The nomenclature of the different kinds of living creatures falls into two parts: (a) informal or vernacular names, or very specialized and restricted names, and (b) scientific names of taxonomic groups (taxon, plural taxa).

Examples of the first are vernacular names from a disease, strain numbers, the symbols for antigenic variants, and the symbols for genetic variants. Thus one can have a vernacular name such as the tubercle bacillus, a strain with the designation K12, a serological form with the antigenic formula Ia, and a genetic mutant requiring valine for growth labeled *val*. These names are usually not controlled by the codes of nomenclature, although the codes may recommend good practice for them.

Examples of scientific names are the names of species, genera, and higher ranks. Thus *Mycobacterium tuberculosis* is the scientific name of the tubercle bacillus, a species of bacterium.

These scientific names are regulated by the codes (with few exceptions) and have two things in common: (a) they are all Latinized in form so as to be easily recognized as scientific names, and (b) they possess definite positions in the taxonomic hierarchy. These names are international: thus microbiologists of all nations know what is meant by *Bacillus anthracis*, but few would know it under vernacular names such as Milzbrandbacillus or Bactéridie de charbon.

The scientific names of bacteria are regulated by the International Code of Nomenclature of Bacteria, which is also known as the Revised Code and was most recently published in 1975 (Lapage et al.). This edition authorized a new starting date for names of bacteria on January 1, 1980, and the starting document is the Approved Lists of Bacterial Names (Skerman et al., 1980), which contains all the scientific names of bacteria that retain their nomenclatural validity from the past. The operation of these Lists will be referred to later. The Code and the Lists are under the aegis of the International Committee on Systematic Bacteriology, which is a constituent part of the International Union of Microbiological Societies. The Committee is assisted by a number of Taxonomic Subcommittees on different groups of bacteria, and by the Judicial Commission which considers amendments to the Code and any exceptions that may be needed to specific Rules.

LATINIZATION

Since scientific names are in Latinized form, they obey the grammar of classic or medieval Latin. Fortunately the necessary grammar is not very difficult, and the most common point to watch is that adjectives agree in gender with the substantives they qualify. Some examples are given later. The names of genera and species are normally printed in italics (or underlined in manuscripts to indicate italic font). For higher categories conventions vary: in Britain they are often in ordinary roman type, but in the United States they are usually in italics, which is preferable because this reminds the reader they are Latinized scientific names.

TAXONOMIC HIERARCHY

The taxonomic hierarchy is a conventional arrangement. Each level above the basic level of species is increasingly inclusive. The names belong to successive **categories**, each of which possesses a position in the hierarchy called its **rank**. The lowest category ordinarily employed is that of species, though sometimes these are subdivided into subspecies. The main categories in decreasing rank, with their vernacular and Latin forms, and examples, are shown in Table VI.1.

Additional categories may sometimes be intercalated (e.g. subclass below class, and tribe below family).

Table VI.1.
Ranking of taxonomic categories

Category	Example[a]
Kingdom *(Regnum)*	*Procaryotae*
Phylum *(Phylum)* in zoology or Division (Divisio) in botany and bacteriology	*Gracilicutes*
Class *(Classis)*	*Scotobacteria*
Order *(Ordo)*	*Rickettsiales*
Family *(Familia)*	*Rickettsiaceae*
Genus *(Genus)*	*Coxiella*
Species *(Species)*	*Coxiella burnetii*

[a]Based on the classification given by Murray in the chapter on "The Higher Taxa" in this *Manual.*

FORM OF NAMES

The form of Latinized names differs with the category. The species name consists of two parts. The first is the **genus name**. This is spelled with an initial capital letter, and is a Latinized substantive. The second is the **specific epithet**, and is spelled with a lower case initial letter. The epithet is a Latinized adjective in agreement with the gender of the genus name, or a Latin word in the genitive case, or occasionally a noun in apposition. Thus in *Mycobacterium tuberculosis*, the epithet *tuberculosis* means "of tubercle," so the species name means the mycobacterium of tuberculosis. The species name is called a **binominal name**, or **binomen**, because it has two parts. When subspecies names are used, a trinominal name results, with the addition of an extra **subspecific epithet**. An example is the subspecies of *Lactobacillus casei* that is called *Lactobacillus casei* subsp. *rhamnosus.* In this name, *casei* is the specific epithet and *rhamnosus* is the subspecific epithet. The existence of a subspecies such as *rhamnosus* implies the existence of another subspecies, in which the subspecific and specific epithets are identicial, i.e. *Lactobacillus casei* subsp. *casei.*

One problem that frequently arises is the scientific status of a species. It may be difficult to know whether an entity differs from its neighbors in certain specified ways. A useful terminology was introduced by Ravin (1963). It may be believed, for example, that the entity can undergo genetic exchange with a nearby species, in which event they could be considered to belong to the same **genospecies**. It may be believed that the entity is not phenotypically distinct from its neighbors, in which event they could be considered to belong to the same **taxospecies**. Yet the conditions for genetic exchange may vary greatly with experimental conditions, and the criteria of distinctness may depend on what properties are considered, so that it may not be possible to make clear-cut decisions on these matters. Nevertheless, it may be convenient to give the entity a species name and to treat it in nomenclature as a separate species, a **nomenspecies**. It follows that all species in nomenclature should strictly be regarded as nomenspecies.

Genus names, as mentioned above, are Latinized nouns, and so are subgenus names (now rarely used) which are conventionally written in parentheses after the genus name, e.g. *Bacillus (Aerobacillus)* indicates the subgenus *Aerobacillus* of the genus *Bacillus*. As in the case of subspecies, this implies the existence of a subgenus *Bacillus (Bacillus).*

Above the genus level most names are plural adjectives in the feminine gender, agreeing with the word *Procaryotae*, so that *Brucellaceae* means *Procaryotae Brucellaceae*, for example.

PURPOSES OF THE CODES OF NOMENCLATURE

The codes have three main aims:

1. Names should be stable,
2. Names should be unambiguous,
3. Names should be necessary.

These three aims are sometimes contradictory, and the rules of nomenclature have to make provision for exceptions where they clash. The principles are implemented by three main devices: (a) priority of publication to assist stability, (b) establishment of nomenclatural types to ensure the names are not ambiguous, and (c) publication of descriptions to indicate that different names do refer to different entities. These are supported by subsidiary devices such as the Latinized forms of names, and the avoidance of synonyms for the same taxon.

PRIORITY OF PUBLICATION

In order to achieve stability the first name given to a taxon (provided the other rules are obeyed) is taken as the correct name. This is the **principle of priority**. But to be safeguarded in this way a name obviously has to be made known to the scientific community: one cannot use a name that has been kept secret. Therefore names have to be published in the scientific literature, together with sufficient indication of what they refer to. This is called **valid publication**. If a name is merely published in the scientific literature it is called **effective publication**: to be valid it also has to satisfy additional requirements, which are summarized later.

The earliest names that must be considered are those published after an official starting date. For many groups of organisms this is Linnaeus' *Species Plantarum* of 1753, but the difficulties of knowing to what the early descriptions refer, and of searching a voluminous and growing literature, have made the principle of priority increasingly hard to obey.

The code of nomenclature for bacteria, therefore, has established a new starting date of 1980, with a new starting document, the Approved Lists of Bacterial Names (Skerman et al., 1980). This list contains names of bacterial taxa that are recognizable and in current use. Names not on the lists lost standing in nomenclature on January 1, 1980, although there are provisions for reviving them if the taxa are subsequently rediscovered or need to be reestablished. In order to prevent the need to search the voluminous scientific literature, the new provisions for bacterial nomenclature require that for valid publication new names (including new names in patents) must be published in certain official publications. Alternatively, if the new names were effectively published in other scientific publications they must be announced in the official publications to become validly published. Priority dates from the official publication concerned. At present the only official publication is the *International Journal of Systematic Bacteriology.*

NOMENCLATURAL TYPES

In order to make clear what names refer to, the taxa must be recognizable by other workers. In the past it was thought sufficient to publish a description of a taxon. This has been found over the years to be inadequate. Advances in techniques and in knowledge of the many undescribed species in nature have shown that old descriptions are usually insufficient. Therefore an additional principle is employed, that of **nomenclatural types**. These are actual specimens (or names of subordinate taxa that ultimately relate to actual specimens). These type specimens are deposited in museums and other institutions. For bacteria (like some other microorganisms that are classified according to their properties in artificial culture), instead of type specimens, **type strains** are employed. The type specimens or strains are intended to be typical specimens or strains which can be compared with other material when classification or identification is undertaken, hence the word "type." However, a moment's thought will show that if a type specimen has to be designated when a taxon is *first* described and named, this will be done at a time when little has yet been found out about the new group. Therefore it is impossible to be sure that it is indeed a typical specimen. By the time a completely typical specimen can be chosen the taxon may be so well known that a type specimen is unnecessary: no one would now bother to designate a type specimen of a bird so well known as the common house sparrow.

The word "type" thus does *not* mean it is typical, but simply that it is a **reference specimen for the name**. This use of the word "type" is a very understandable cause for confusion that may well repay attention by the taxonomists of the future.

In recent years other type concepts have been suggested. Numerical taxonomists have proposed the hypothetical median organism (Liston et al., 1963), or the centroid: these are mathematical abstractions, not actual organisms. The most typical strain in a collection is commonly taken to be the **centrotype** (Silvestri et al., 1962), which is broadly equivalent to the strain closest to the center (centroid) of a species cluster. Some workers have suggested that several type strains should be designated. Gordon (1967) refers to this as the "population concept." One strain, however, must be the official nomenclatural type in case the species must later be divided. Gibbons (1974) proposed that the official type strain should be supplemented by reference strains that indicated the range of variation in the species, and that these strains could be termed the "type constellation." It may be noted that some of these concepts are intended to define not merely the center but, in some fashion, the limits of a species. Since these limits may well vary in different ways for different characters, or classes of characters, it will be appreciated that there may be difficulties in extending the type concept in this way. The centrotype, being a very typical strain, has often been chosen as the type strain, but otherwise these new ideas have not had much application to bacterial nomenclature.

Type strains are of the greatest importance for work on both classification and identification. These strains are preserved (by methods to minimize change to their properties) in culture collections from which they are available for study. They are obviously required for new classificatory work, so that the worker can determine if he has new species among his material. They are also needed in diagnostic microbiology, because one of the most important principles in attempting to identify a microorganism that presents difficulties is to compare it with authentic strains of known species. The drawback that the type strain may not be entirely typical is outweighed by the fact that the type strain is by definition authentic.

Not all microorganisms can be cultured, and for some the function of a type can be served by a preserved specimen, a photograph, or some other device. In such instances, these are the nomenclatural types, though it is commonly considered wise to replace them by type strains when this becomes possible.

Sometimes types become lost, and new ones (**neotypes**) have to be set up to replace them: the procedure for this is described in the Code. In the past it was necessary to define certain special classes of types, but most of these are now not needed.

Types of species and subspecies are type specimens or type strains. For categories above the species the function of the type—to serve as a point of reference—is assumed by a *name*, e.g. that of a species or subspecies. The species or subspecies is, of course, tied to its type specimen or type strain.

Types of genera are **type species** (one of the included species) and types of higher names are usually **type genera** (one of the included genera). This principle applies up to and including the category, order. This can be illustrated by the types of an example of a taxonomic hierarchy shown in Table VI.2.

Just as the type specimen, or type strain, must be considered a member of the species whatever other specimens or strains are excluded, so the **type species of a genus must be retained in the genus even if all other species are removed from it.** A type, therefore, is sometimes called a **nominifer** or **name bearer**: it is the reference point for the name in question.

Table VI.2.
Example of taxonomic types

Category	Taxon	Type
Family	*Pseudomonadaceae*	*Pseudomonas*
Genus	*Pseudomonas*	*Pseudomonas aeruginosa*
Species	*Pseudomonas aeruginosa*	American Type Culture Collection strain number 10145

DESCRIPTIONS

The publication of a name, with a designated type, does in a technical sense create a new taxon—insofar as it indicates that the author believes he has observations to support the recognition of a new taxonomic group. But this does not afford evidence that can be readily assessed from the bald facts of a name and designation of a type. From the earliest days of systematic biology it was thought important to describe the new taxon for two reasons: (a) to show the evidence in support of a new taxon, and (b) to permit others to identify their own material with it—indeed this antedated the type concept (which was introduced later to resolve difficulties with descriptions alone).

It is, therefore, a requirement for valid publication that a description of a new taxon is needed. However, just how full the description should be, and what properties must be listed, is difficult to prescribe.

The codes of nomenclature recognize that the most important aspect of a description is to provide a list of properties that distinguish the new taxon from others that are very similar to it, and that consequently fulfill the two purposes of adducing evidence for a new group and allowing another worker to recognize it. Such a brief differential description is called a **diagnosis**, by analogy with the characteristics of diseases that are associated with the same word. Although it is difficult to legislate for adequate diagnoses, it is usually easy to provide an acceptable one: inability to do so is often because insufficient evidence has been obtained to support the establishment of the new taxon.

The Code provides guidance on descriptions, in the form of recommendations. Failure to follow the recommendations does not of itself invalidate a name, though it may well lead later workers to dismiss the taxon as unrecognizable or trivial. The code for bacteria recommends that as soon as minimum standards of description are prepared for various groups, workers should thereafter provide that minimum information; this is intended as a guide to good practice, and should do much to raise the quality of systematic bacteriology. For an example of minimum standards, see the report of the International Committee on Systematic Bacteriology Subcommittee on the Taxonomy of *Mollicutes* (1979).

CLASSIFICATION DETERMINES NOMENCLATURE

The student often asks how an organism can have two different names. The reason lies in the fact that a name implies acceptance of some taxonomy, and on occasion no taxonomy is generally agreed. Scientists are entitled to their own opinions on taxonomies: there are no rules to force the acceptance of a single classification.

Thus opinions may be divided on whether the bacterial genus *Pectobacterium* is sufficiently separate from the genus *Erwinia*. The soft-rot bacterium was originally called *Bacterium carotovorum* in the days when most bacteria were placed in a few large genera such as *Bacillus* and *Bacterium*. As it became clear that these unwieldy genera had to be divided into a number of smaller genera, which were more homogeneous and convenient, this bacterium was placed in the genus *Erwinia* (established for the bacterium of fireblight, *Erwinia amylovora*) as *Erwinia carotovora*. When further knowledge accumulated, it was considered by some workers that the soft-rot bacterium was sufficiently distinct to merit a new genus, *Pectobacterium*. The same organism, therefore, is also known as *Pectobacterium carotovorum*. Both names are correct in their respective positions. If one believes that two separate genera are justified, then the correct name for the soft-rot bacterium is *Pectobacterium carotovorum*. If one considers that *Pectobacterium* is not justified as a separate genus, the correct name is *Erwinia carotovora*.

Classification, therefore, determines nomenclature, not nomenclature classification. Although unprofitable or frivolous changes of name should be avoided, the freezing of classification in the form it had centuries ago is too high a price to pay for stability of names. Progress in classification must reflect progress in knowledge (e.g. no one now wants to classify all rod-shaped bacteria in *Bacillus*, as was popular a century ago). Changes in name must reflect progress in classification: some changes in name are thus inevitable.

CHANGES OF NAME

Most changes in name are due to moving species from one genus to another or dividing up older genera. Another cause, however, is the rejection of a commonly used name because it is incorrect under one or more of the Rules. A much-used name, for example, may not be the earliest, because the earliest name was published in some obscure journal and had been overlooked. Or there may already be another identical name for a different microorganism in the literature. Changes can be very inconvenient if a well-established name is found to be **illegitimate** (contrary to a Rule) because of a technicality. The codes of nomenclature therefore make provision to allow the organizations that are responsible for the codes to make exceptions if this seems necessary. A name thus retained by international agreement is called a **conserved name**, and when a name is conserved the type may be changed to a more suitable one.

When a species is moved from one genus into another, the specific epithet is retained (unless there is by chance an earlier name which forms the same combination, when some other epithet must be chosen), and this is done in the interests of stability. The new name is called a **new combination**. An example has been given above. When the original *Bacterium carotovorum* was moved to *Erwinia*, the species name became *Erwinia carotovora*. The gender of the species epithet becomes the same as that of the genus *Erwinia*, which is feminine, so the feminine ending, *-a*, is substituted for the neuter ending, *-um*.

NAMES SHOULD BE NECESSARY

The codes require that names should be necessary, i.e. **there is only one correct name for a taxon** in a given or implied taxonomy. This is sometimes expressed by the statement that an organism with a given position, rank, and circumscription can have only one correct name.

NAMES ARE LABELS, NOT DESCRIPTIONS

In the early days of biology there was no regular system of names, and organisms were referred to by long Latin phrases which described them briefly, such as *Tulipa minor lutea italica folio latiore*, "the little yellow Italian tulip with broader leaves." The Swedish naturalist Linnaeus tried to reduce these to just two words for species, and in doing so he founded the present *binominal system* for species. This tulip might then become *Tulipa lutea*, just "the yellow tulip." Very soon it would be noted that a white variant sometimes occurred. Should it then still be named "the yellow tulip"? Why not change it to "the Italian tulip"? Then someone would find it in Greece and point out that the record from Italy was a mistake anyway. Twenty years later an orange form would be found in Italy after all. Soon the nomenclature would be confused again.

After a time it was realized that the original name had to be kept, even if it was not descriptive, just as a man keeps his name of Fairchild Goldsmith as he grows older, and even if he becomes a farmer. The scientific names or organisms are today only **labels**, to provide a means of referring to taxa, just like personal names.

A change of name is therefore only rarely justified, even if it sometimes seems inappropriate. Provisions exist for replacement when the name causes great confusion.

CITATION OF NAMES

A scientific name is sometimes amplified by a *citation*, i.e., by adding after it the author who proposed it. Thus the bacterium that causes crown galls is *Agrobacterium tumefaciens* (Smith and Townsend) Conn. This indicates that the name refers to the organism first named by Smith and Townsend (as *Bacterium tumefaciens*, in fact, though this is not evident in the citation) and later moved to the genus *Agrobacterium* by Conn, who therefore created a **new combination**. Sometimes the citation is expanded to include the date (e.g. *Rhizobium* Frank 1889), and more rarely to include also the publication, e.g. *Proteus morganii* Rauss 1936 *Journal of Pathology and Bacteriology* Vol. 42, p. 183.

It will be noted that citation is only necessary to provide a suitable reference to the literature or to distinguish between inadvertent duplication of names by different authors. A citation is *not* a means of giving credit to the author who described a taxon: the main functions of citation would be served by the bibliographic reference without mentioning the author's name. Citation of a name is to provide a **means of referring** to a name, just as a name is a means of referring to a taxon.

SYNONYMS AND HOMONYMS

A homonym is a name identical in spelling with another name but based on a different type, so they refer to different taxa under the same name. They are obviously a source of confusion, and the one that was published later is suppressed. The first published name is known as the **senior homonym**, and later published names are **junior homonyms**. Names of higher animals and plants that are the same as bacterial names are not treated as homonyms of names of bacteria, but to reduce confusion among microorganisms, bacterial names are suppressed if they are junior homonyms of names of fungi, algae, protozoa, or viruses.

A synonym is a name that refers to the same taxon under another scientific name. Synonyms thus come in pairs or even swarms. They are of two kinds:

1. **Objective synonyms** are names with the same nomenclatural type, so that there is no doubt that they refer to the same taxon. These are often called nomenclatural synonyms. An example is *Erwinia carotovora* and *Pectobacterium carotovorum*: they have the same type strain, American Type Culture Collection strain 15713.
2. **Subjective synonyms** are names that are believed to refer to the same taxon but which do not have the same type. They are matters of taxonomic opinion. Thus *Pseudomonas geniculata* is a subjective synonym of *Pseudomonas fluorescens* for a worker who believes that these taxa are sufficiently similar to be included in one species, *P. fluorescens*. They have different types, however (American Type Culture Collection strains 19374 and 13525, respectively) and another worker is entitled to treat them as separate species if he so wishes.

There are senior and junior synonyms, as for homonyms. The synonym that was first published is known as the **senior synonym**, and those published later are **junior synonyms**. Junior synonyms are normally suppressed.

PROPOSAL OF NEW NAMES

The valid publication of a new taxon requires that it be named. The Code insists that an author should make up his mind about the new taxon: if he feels certain enough to propose a new taxon with a new name then he should say he does so propose; if he is not sure enough to make a definite proposal then his name will not be afforded the protection of the Code. He cannot expect to suggest provisional names—or possible names, or names that one day might be justified—and then expect others to treat them as definite proposals at some unspecified future date: how can a reader possibly know when such vague conditions have been fulfilled?

If a taxon is too uncertain to receive a new name it should remain with a vernacular designation (e.g. the marine form, group 12A). If it is already named, but its affinities are too uncertain to move it to another genus or family, it should be left where it is. There is one exception, and that is that a new species should be put into some genus even if it is not very certain which is the most appropriate, or if necessary a new genus should be created for it. Otherwise, it will not be validly published, it will be in limbo, and it will be generally overlooked, because no one else will know how to index it or whether they should consider it seriously. If it is misplaced it can later be moved to a better genus.

The basic needs for publication of a new taxon are four: (a) the publication should contain a new name in proper form that is not a homonym of an earlier name of bacteria, fungi, algae, protozoa, or viruses; (b) the taxon should not be a synonym of an earlier taxon; (c) a description or at least a diagnosis should be given; and (d) the type should be designated. A new species is indicated by adding the Latin abbreviation *sp. nov.*, a new genus by *gen. nov.*, and a new combination by *comb. nov.* The most troublesome part is the search of the literature to cover the first two points. This is now greatly simplified for bacteria, because the new starting date means that one need only search the Approved Lists of Bacterial Names and the issues of the *International Journal of Systematic Bacteriology* from January 1980 onward for all validly published names that have to be considered. However, the new name has to be published in that journal, with its description and designation of type, or—if published elsewhere—the name must be announced in that journal to render it validly published.

Identification of Bacteria

Noel R. Krieg

NATURE OF IDENTIFICATION SCHEMES

Identification schemes are not classification schemes, although there may be a superficial similarity. An identification scheme for a group of organisms can be devised only **after** that group has first been classified (i.e. recognized as being different from other organisms); it is based on one or more characters, or on a pattern of characters, which all the members of the group have and which other groups do not have. The characters used are often not those that were involved in classification of the group; e.g. classification might be based on a DNA/DNA hybridization study, whereas identification might be based on a phenotypic character that is found to correlate well with the genetic information. In general, the characters chosen for an identification scheme should be **easily determinable**, whereas those used for classification may be quite difficult to determine (such as DNA homology values). The characters should also be **few in number**, whereas classification may involve large numbers of characters, such as in a numerical taxonomy study. These ideal features of an identification scheme may not always be possible, particularly with genera or species that are not susceptible to being characterized by traditional biochemical or physiological tests. In such cases, one may need to resort to relatively difficult procedures in order to achieve an accurate identification—procedures such as polyacrylamide gel electrophoresis (PAGE) of cellular proteins, cellular lipid patterns, genetic transformation, or even nucleic acid hybridization.

Serological reactions, which generally have only limited value for classification, often have enormous value for identification. Slide agglutination tests, fluorescent antibody techniques, and other serological methods can be performed simply and rapidly and are usually highly specific; therefore, they offer a means for achieving quick, presumptive identification of bacteria. Their specificity is frequently not absolute, however, and confirmation of the identification by additional physiological or biochemical tests is usually required.

With many genera and species, identification may not be based on only a few tests, but rather on the pattern given by applying a whole battery of tests. The members of the family *Enterobacteriaceae* represent one example of this. To alleviate the need for inoculating large numbers of tubed media, a variety of convenient and rapid multitest systems have been devised and are commercially available for use in identifying various taxa, particularly those of medical importance. A summary of some of these systems has been given by Smibert and Krieg (1981), but new systems are being developed continually. Each manufacturer provides charts, tables, coding systems, and characterization profiles for use with the particular multitest system being offered.

NEED FOR STANDARDIZED TEST METHODS

One difficulty in devising identification schemes is that the results of characterization tests may vary depending on the size of the inoculum, incubation temperature, length of the incubation period, composition of the medium, the surface-to-volume ratio of the medium, and the criteria used to define a "positive" or "negative" reaction. Therefore, the results of characterization tests obtained by one laboratory often do not match exactly those obtained by another laboratory, although the results within each laboratory may be quite consistent. The blind acceptance of an identification scheme without reference to the particular conditions employed by those who devised the scheme can lead to error (and, unfortunately, such conditions are not always specified). Ideally, it would be desirable to standardize the conditions used for testing various characteristics, but this is easier said than done, especially on an international basis. The use of commercial multitest systems offers some hope of increasing the standardization among various laboratories because of the high degree of quality control exercised over the media and reagents, but no one system has yet been agreed on for universal use for any given taxon. **It is therefore always advisable to include strains whose identity has been firmly established** (type or reference strains, available from national culture collections) **for comparative purposes when making use of an identification scheme**, to make sure that the scheme is valid for the conditions employed in one's own laboratory.

NEED FOR DEFINITIONS OF "POSITIVE" AND "NEGATIVE" REACTIONS

Some tests may be found to be based on plasmid- or phage-mediated characteristics; such characteristics may be highly mutable and therefore unreliable for identification purposes. Even with immutable characteristics, certain tests may not be well suited for use in identification schemes because they may not give highly reproducible results (e.g. the catalase test, oxidase test, Voges-Proskauer test, and gelatin liquefaction are notorious in this regard). Ideally, a test should give reproducible results that are clearly either positive or negative, without equivocal reactions. In fact, no such test may exist. The Gram reaction of an organism may be "Gram-variable," the presence of endospores in a strain that makes only a few may be very difficult to determine by staining or by heat resistance tests, acid production from sugars may be difficult to distinguish from no acid production if only small amounts of acid are produced, and a weak growth response may not be clearly distinguishable from "no growth." A precise (although arbitrary) definition of what constitutes a "positive" and "negative" reaction is often important in order for a test to be useful for an identification scheme.

PURE CULTURES

Although a few bacteria are so morphologically remarkable as to make them identifiable without isolation, pure cultures are nearly always a necessity before one can attempt identification of an organism. **It is important to realize that the single selection of a colony from a plate does not assure purity.** This is especially true if selective media are used; live but nongrowing contaminants may often be present in or near a colony and can be subcultured along with the chosen organism. It is for this reason that **nonselective media are preferred for final isolation**, because they allow such contaminants to develop into visible colonies. Even with nonselective media, apparently well-isolated colonies should not be isolated too soon; some contaminants may be slow growing and may appear on the plate only after a longer incubation. Another difficulty occurs with bacteria that form extracellular slime or that grow as a network of chains or filaments; contaminants often become firmly embedded or entrapped and are difficult to penetrate. In the instance of cyanobacteria, contaminants frequently penetrate and live in the gelatinous sheaths that surround the cells, making pure cultures difficult to obtain.

In general, colonies from a pure culture that has been streaked on a solid medium are similar to one another, providing evidence of purity. Although this is generally true, there are exceptions, as in the case of S→R variation, capsular variants, pigmented or nonpigmented variants, etc., which may be selected by certain media, temperatures, or other growth conditions. Another criterion of purity is morphology: organisms from a pure culture generally exhibit a high degree of morphological similarity in stains or wet mounts. Again, there are exceptions, depending on the age of the culture, the medium used, and other growth conditions: coccoid body formation, cyst formation, spore formation, pleomorphism, etc. For example, examination of a broth culture of a marine spirillum after 2 or 3 days may lead one to believe the culture is highly contaminated with cocci unless one is previously aware that such spirilla generally develop into thin-walled coccoid forms following active growth.

APPROACHES TO IDENTIFICATION OF AN ISOLATE

The vernacular headings of the various sections of *Bergey's Manual®* indicate major categories of the procaryotes and are a good starting point for identification. The categories are concerned with such phenotypic characteristics as the Gram-staining reactions, morphology, and general type of metabolism. It is therefore important to establish whether the new isolate is a chemolithotrophic autotroph, a photosynthetic organism, or a chemoheterotrophic organism. Living cells should be examined by phase-contrast microscopy and Gram-stained cells by light microscopy; other stains can be applied if this seems appropriate. If some outstanding morphological property, such as endospore production, sheaths, holdfasts, acid fastness, cysts, stalks, fruiting bodies, budding division, or true branching, is obvious, then further efforts in identification can be confined to those groups having such a property. Whether or not the organisms are motile, and the type of motility (swimming, gliding), may be very helpful in restricting the range of possibilities. Gross growth characteristics, such as pigmentation, mucoid colonies, swarming, or a minute size, may also provide valuable clues to identification. For example, a motile, Gram-negative rod that produces a water-soluble fluorescent pigment is likely to be a *Pseudomonas* species, whereas one that forms bioluminescent colonies is likely to belong to the *Vibrionaceae*.

The source of the isolate can also help to narrow the field of possibilities. For example, a spirillum isolated from coastal sea water is likely to be an *Oceanospirillum*, whereas Gram-positive cocci occurring in grape-like clusters and isolated from the human nasopharynx are likely to belong to the genus *Staphylococcus*.

The relation of the isolate to oxygen (i.e. whether it is aerobic, anaerobic, facultatively anaerobic, or microaerophilic) is often of fundamental importance in identification. For example, a small, microaerophilic vibrio isolated from a case of diarrhea is likely to be a *Campylobacter*, whereas an anaerobic, Gram-negative rod isolated from a wound infection is probably a member of the *Bacteroidaceae*. Similarly, it is important to test the isolate for its ability to dissimilate glucose (or some other simple sugar) to determine if the type of metabolism is oxidative or fermentative, or whether sugars are catabolized at all.

Above all, common sense should be used at each stage where the possibilities are narrowed in deciding what additional tests should be performed. There should be a reason for the selection of each test, in contrast to a "shotgun" type of approach where many tests are used but most provide little pertinent information for the particular isolate under investigation. As the category to which the isolate belongs becomes increasingly delineated, one should follow the specific tests indicated in the particular diagnostic tables or keys that apply to that category.

The following summary is taken from "The Mechanism of Identification" by S. T. Cowan and J. Liston in the eighth edition of the *Manual*, with some modifications:

1. Make sure that you have a pure culture.
2. Work from broad categories down to a smaller, specific category of organism.
3. Use all the information available to you in order to narrow the range of possibilities.
4. Apply common sense at each step.
5. Use the minimum number of tests to make the identification.
6. Compare your isolate to type or reference strains of the pertinent taxon to make sure the identification scheme being used actually is valid for the conditions in your particular laboratory.

If, as may well happen, you cannot identify your isolate from the information contained in the *Manual*, neither despair nor immediately assume that you have isolated a new genus or species; many of the problems of microbial classification are the result of people jumping to this conclusion prematurely. When you fail to identify your isolate, check (a) its **purity**, (b) that you have carried out the **appropriate tests**, (c) that your **methods are reliable**, and (d) that you have used correctly the various keys and tables of the *Manual*. It has been said that the most frequent cause of mistaken identify of bacteria is error in the determination of shape, Gram-staining reaction, and motility. In most cases, you should have little difficulty in placing your isolate into a genus; allocation to a species or subspecies may need the help of a specialized reference laboratory.

On the other hand, it is always possible that you have actually isolated a new genus or species. A comparison of the present edition of the *Manual* with the previous edition indicates that a number of new genera and species have been added. Some prime examples can be found in the family *Legionellaceae*, "Other Genera" of the family *Enterobacteriaceae*, the genus *Azospirillum*, "Dissimilatory Sulfate- and Sulfate-reducing Bacteria," and the genus *Meniscus*. Undoubtedly, there exist in nature a great number of bacteria that have not yet been classified, and therefore cannot yet be identified by existing schemes. Yet, before describing and naming a new taxon, one must **be very sure that it is really a new taxon** and not merely the result of an inadequate identification.

Further Reading

Goodfellow, M. and R.G. Board (Editors). 1980. Microbiological classification and identification. Society for Applied Bacteriology Symposium Series No. 8. Academic Press, London.

Hedén, C. and T. Illéni (Editors). 1975. New Approaches to the Identification of Microorganisms. John Wiley & Sons, New York.

Holding, J.A. and J.G. Colee. 1971. Routine biochemical tests. *In* Norris and Ribbons (Editors), Methods in Microbiology, Vol. 6A. Academic Press, New York, pp. 1-32.

Mitruka, B.J. 1976. Methods of Detection and Identification of Bacteria. CRC Press, Cleveland, Ohio.

Skerman, V.B.D. 1967. A Guide to the Identification of the Genera of Bacteria, 2nd Ed. Williams & Wilkins, Baltimore.

Skerman, V.B.D. 1969. Abstracts of Microbiological Methods. Wiley-Interscience, New York.

Skerman, V.B.D. 1974. A key for the determination of the generic position of organisms listed in the *Manual*. *In* Buchanan and Gibbons (Editors), Bergey's Manual® of Determinative Bacteriology, 8th Ed. Williams & Wilkins, Baltimore, pp. 1098-1146.

Skinner, F.A. and D.W. Lovelock. 1979. Identification methods for microbiologists. Society for Applied Bacteriology Technical Series No. 14. Academic Press, New York.

Smibert, R.M. and N.R. Krieg. 1981. General characterization. *In* Gerhardt, Murray, Costilow, Nester, Wood, Krieg and Phillips (Editors), Manual of Methods for General Bacteriology. American Society for Microbiology, Washington, D.C, pp. 409-443.

NUMERICAL IDENTIFICATION

PETER H. A. SNEATH

The success of numerical taxonomy has in recent years led to the development of a new diagnostic method based upon it, called **numerical identification.** The rapidly growing field is well reviewed by Lapage et al. (1973), and Willcox et al. (1980). The essential principles can be illustrated geometrically (Sneath, 1978) by considering the columns of percent positive test reactions in a new table, a table of q taxa for m diagnostic characters. If an object is scored for two variables, its position can be represented by a point on a scatter diagram. Use of three variables determines a position in a three-dimensional model. Objects that are very similar on the variables will be represented by clusters of points in the diagram or the model, and a circle or sphere can be drawn round each cluster so as to define its position and radius. The same principles can be extended to many variables or tests, which then represent a multidimensional space or "hyperspace." A column representing a species defines, in effect, a region in hyperspace, and it is useful to think of a species as being represented by a hypersphere in that space, whose position and radius are specified by the numerical values of these percentages. The tables form a reference library, or data base, of properties of the taxa.

The operation of numerical identification is to compare an unknown strain with each column of the table in turn, and to calculate a distance (or its analog) to the center of each taxon hypersphere. If the unknown lies well within a hypersphere, this will identify it with that taxon. Further, such systems have important advantages over most other diagnostic systems. The numerical process allows a likelihood to be attached to an identification, so that one can know to some order of magnitude the certainty that the identity is correct. The results are not greatly affected by an occasional aberrant property of the unknown, or an occasional experimental mistake in performing the tests. Furthermore, the system is robust toward missing information, and quite good identifications can be obtained if only a moderate proportion of the tests have been performed.

Numerous applications of numerical identification are now being made. Most commercial testing kits or automatic instruments for microbial identification are based on these concepts, and they require the comparison of results on an unknown strain with a data base using computer software or with printed material prepared by such means. Research sponsored by the Bergey's Manual® Trust (Feltham et al., 1984) shows that these concepts can be extended to a very wide range of genera.

Further Reading

Feltham, R.K.A., P.A. Wood and P.H.A. Sneath. 1984. A general-purpose system for characterizing medically important bacteria to genus level. J. Appl. Bacteriol. *57:* 279-290.

Lapage, S.P., S. Bascomb, W.R. Willcox and M.A. Curtis. 1973. Identification of bacteria by computer. I. General aspects and perspectives. J. Gen. Microbiol. *77:* 273-290.

Sneath, P.H.A. 1978. Identification of microorganisms. *In* Norris and Richmond (Editors), Essays in Microbiology. John Wiley, Chichester, England, pp. 10/1-10/32.

Willcox, W.R., S.P. Lapage and B. Holmes. 1980. A review of numerical methods in bacterial identification. Antonie van Leeuwenhoek J. Microbiol. Serol. *46:* 233-299.

Reference Collections of Bacteria— The Need and Requirements for Type Strains

The Late Norman E. Gibbons
Revised by Peter H.A. Sneath and Stephen P. Lapage

As it became possible to grow bacteria in liquid and solid media, microbiologists began to exchange cultures with their colleagues for information and comparison. Each investigator kept his own isolates, added those received from others and in this way built up his own reference and working collection.

About the turn of the century, Professor František Král of Prague realized the value of a central collection and began to collect cultures which he made available for a fee to other workers. After Král's death in 1911, the collection was acquired by Professor Ernst Pribram and transferred to the University of Vienna in 1915. Pribram brought part of the collection to Loyola University in Chicago some years before the Second World War. He was killed in a car accident in 1940, but the fate of his collection is not known. The cultures left in Vienna were destroyed during World War II.

The next oldest collection—Centraalbureau voor Schimmelcultures—was founded in 1906 by the Association internationale des Botanistes. Although the founding association did not survive the First World War, the collection is still in existence at Baarn under the auspices of The Royal Netherlands Academy of Sciences. This collection provides a holding and distribution center for fungi and an identification service.

Since then many other collections have developed, some general, some specialized, some oriented to service. A full account of the history of culture collections is given by Porter (1976). Some salient developments may be mentioned briefly. About 1946, Professor P. Hauduroy established a centralized information facility at Lausanne, the "Centre de Collections de Types Microbiens" which provided information on which collections held cultures of various bacterial species. In 1947, the Lausanne Centre became associated with the International Association of Microbiological Societies (IAMS, now the International Union of Microbiological Societies, IUMS) and, in cooperation with it, an International Federation of Type Culture Collections was formed. This Federation had ambitious plans which were never realized, and the Federation went out of existence within a few years.

In 1962, therefore, a Conference on Culture Collections (Martin, 1963), held after the VIIth International Congress for Microbiology, asked IAMS to form a Section on Culture Collections. The Section was set up in 1963 and, on the reorganization of IAMS in 1970, became the World Federation of Culture Collections (WFCC). The WFCC is also a multidisciplinary Commission of the International Union of Biological Sciences in the Divisions of Botany and Zoology, linking it with other organizations concerned with problems of biological preservation, such as herbaria, zoological gardens, and museums. It has collected information on several hundred collections throughout the world, and the *World Directory of Collections of Cultures of Microorganisms* (Martin and Skerman, 1972) has been published. This has recently been updated (McGowan and Skerman, 1982). Pridham (1974b) has also compiled a useful list of the acronyms and abbreviations for numerous culture collections. A number of national Federations of Culture Collections have also been formed which are affiliated to the WFCC. The aims of the WFCC include the collection of information on strains held by the collections and more detailed information on the strains themselves.

In the preservation of cultures, satisfactory methods of maintenance, with minimal change in the cultures, are essential, and an accepted system of taxonomy should be used. Particularly stringent standardization is required in the case of cooperative and comparative studies. In pursuit of these and similar aims, the WFCC has held training courses for curators and workers in culture collections at which other important functions of culture collections were also discussed. The WFCC also works in close cooperation with the International Committee on Systematic Bacteriology (ICSB) and its Judicial Commission and other related national or international bodies dealing with all aspects of the preservation of various groups of microorganisms. It has also sponsored a number of international conferences on culture collections. A list of these, together with a summary of other WFCC activities, is given by Lapage (1975). The WFCC supports the development of the World Data Centre at the University of Queensland, Brisbane, Australia, which is collecting cultural, physiological, and other data on strains of microorganisms and is exploring methods of recording such information in a standard format.

NEED FOR CULTURE COLLECTIONS

It is essential for the orderly development of bacteriology that cultures of organisms described or mentioned in publications be available for independent study. Because microbiologists are mortal and their interests vary during their working life, collections are necessary to provide an element of stability and continuity.

While some microbiologists spend a lifetime on one or two groups of organisms and build up large specialized collections, others move from one organism to another, abandoning old favorites. Both approaches generate problems in the preservation of organisms. The specialized collection may become so large and so specialized that it is hard to find a willing successor to the original enthusiastic curator. The worker whose interests are more fickle seldom worries about the systematic aspects, which make the preservation of cultures so desirable to the taxonomist.

Until the 1920s, the main reason for the existence of collections was their value for taxonomic and epidemiological studies. In the 1930s, the burgeoning interest in microbial physiology and biochemistry gave rise to a need for preserving organisms that produced or gave better yields of specific compounds. This greatly increased the value of culture collections.

More recently, studies on bacterial genetics have resulted in the isolation of numerous mutants which have, in turn, necessitated specialized collections. Some of these mutants are concerned with genetic loci useful in studies of nutrients and of biochemical pathways. The 1972 Stockholm Conference on the Environment recognized the importance of genetic pools and of collections of microorganisms.

Current developments in culture collections are diverse. Reviews of these can be found in Lapage (1971), the volume edited by Colwell (1976), and Kirsop (1985). Methods of preservation are undergoing change, with increasing use of storage at very low temperatures to reduce the risk of genetic change. Loss of plasmids is a problem with some methods. Preservation methods are described in Kirsop and Snell (1984). Recent legislation on patents has led to the need for deposition of strains used in industry. A review of requirements for patents is given by Crespi (1982). Cultures are also needed for teaching of microbiology and for quality control in many fields. The growth of numerous new diagnostic aids requires that large sets of strains from numerous species shall be available for establishing the data bases that are needed (Sneath, 1977b). Culture collections in the future may also expand associated activities, such as storage and supply of dried material of microbial origin, standard antisera, and nucleic acid preparations.

TYPE STRAINS

A particularly important function of culture collections is to preserve type strains and make them available to microbiologists who are undertaking taxonomic revisions. The nomenclatural aspects of type strains are discussed in the chapter on "Bacterial Nomenclature," but some related points are briefly summarized here.

Type strains of bacterial species and subspecies are essential for the advance of taxonomy. They are required for comparison with strains that an author may believe belong to a new species or subspecies. Descriptions have never proved to be sufficient, because new techniques in systematics are continually being devised, and there is no substitute for an authentic strain when one wishes to make a critical comparison.

Type strains are of such taxonomic and nomenclatural importance that in this edition as many of them as possible are listed by their designation and catalog number in the main collections; a list of collections mentioned is given in the next chapter.

The new International Code of Nomenclature of Bacteria (or Revised Code) (see Lapage et al., 1975) has made several special provisions for types. It is now a requirement for valid publication of a cultivable new species or subspecies of bacteria that a type strain shall be designated (alternative provisions exist for noncultivable bacteria). The Code urges that a type strain should be deposited in one or more of the permanently established culture collections. The numerous problems caused in the past by taxa for which there were no type strains should thus be largely overcome. Type cultures should, in the future, be available for all cultivable species of bacteria.

In the past it was frequently necessary to distinguish between different classes of type culture, in particular between types and neotypes, but the Revised Code has made most of these distinctions unnecessary. The new starting document for bacterial nomenclature, which came into force on January 1, 1980 (Approved Lists of Bacterial Names; Skerman et al., 1980) lists the type strains for the names of bacterial species that are currently recognized and given in the Lists. In the past, when many species had no type strains, it was necessary to establish **neotypes** for taxa where no type existed or had been lost. A neotype was thus a replacement for a type, and there should rarely be a need in the future for neotypes, except in the case of loss of the types. The procedure for establishing a neotype is given in the Revised Code; of course, the neotype should be deposited in one, or preferably several, of the main culture collections.

While culture collections maintain type strains and neotypes as described above, they also keep typical and atypical strains, reference strains, and strains with particular properties of interest to biochemistry, genetics, serology, bacteriophage studies, and the like; they also carry out many other functions, of which a general account can be found in Lapage (1971). Culture collections are therefore of great value not only to systematists but to all bacteriologists, and are essential to the development of the subject.

List of Culture Collections

There are several hundred culture collections in the world, with the majority being small specialized collections, often collected by one individual. Details of most of these may be found in the *World Directory of Collections of Cultures of Microorganisms* (Martin and Skerman, 1972). This has recently been updated (McGowan and Skerman, 1982). A smaller number of collections are frequently referred to in bacteriological work, and a selection of these is given below with commonly used abbreviations.

AMRC — FAO-WHO International Reference Centre for Animal Mycoplasmas, Institute for Medical Microbiology, University of Aarhus, Aarhus, Denmark.

ATCC — American Type Culture Collection, 12301 Parklawn Drive, Rockville, Maryland 20852, U.S.A.

BKM — See VKM.

BKMW — See VKM.

CBS — Centraalbureau voor Schimmelcultures, Oosterstraat 1, Baarn, The Netherlands.

CCEB — Culture Collection of Entomophagous Bacteria, Institute of Entomology, Czechoslovak Academy of Sciences, Flemingovo N2, Prague 6, Czechoslovakia.

CCM — Czechoslovak Collection of Microorganisms, J. E. Purkyne University, Tr. Obr. Miru 10, Brno, Czechoslovakia.

CDC — Centers for Disease Control, Atlanta, Georgia 30333, U.S.A.

CIP — Collection of the Institut Pasteur, Rue du Dr. Roux, Paris 15, France.

CNC — Czechoslovak National Collection of Type Cultures, Institute of Epidemiology and Microbiology, Srobarova 48, Prague 10, Czechoslovakia.

DSM — Deutsche Sammlung von Mikroorganismen, Grisebachstrasse 8, Gottingen, Federal Republic of Germany.

IAM — Institute of Applied Microbiology, University of Tokyo, Bunkyo-ku, Tokyo, Japan.

ICPB — International Collection of Phytopathogenic Bacteria, University of California, Davis, California 95616, U.S.A.

IFO — Institute for Fermentation, 4-54 Jusonishinocho, Osaka, Japan.

IMET — Institute für Mikrobiologie und Experimentelle Therapie, Deutsche Akademie der Wissenschaften zu Berlin, Beuthenbergstrasse 11, Jena 69, German Democratic Republic.

IMRU — Institute of Microbiology, Rutgers—The State University, New Brunswick, New Jersey 08903, U.S.A.

IMV — Institute of Microbiology and Virology, Academy of Sciences of the Ukrainian S.S.R., Kiev, U.S.S.R.

INA — Institute for New Antibiotics, Bolshaya Pirogovskaya II, Moscow, U.S.S.R.

INMI — Institute for Microbiology, U.S.S.R. Academy of Sciences, Moscow, U.S.S.R.

IPV — Istituto di Patologia Vegetale, Milan, Italy.

KCC — Kaken Chemical Company Ltd., 6-42 Jujodai-1-Chome, Tokyo 114, Japan.

LIA — Museum of Cultures, Leningrad Research Institute of Antibiotics, 23 Ogorodnikov Prospect, Leningrad L-20, U.S.S.R.

LSU — Louisiana State University, Baton Rouge, Louisiana 70803, U.S.A.

LMD — Laboratorium voor Microbiologie, Technische Hogeschool, Julianalaan 67a, 2623 BC Delft, The Netherlands.

NCDO — National Collection of Dairy Organisms, National Institute for Research in Dairying, University of Reading, Shinfield, Reading, England, U.K.

NCIB — National Collection of Industrial Bacteria, Torry Research Station, Aberdeen AB9 8DG, Scotland, U.K.

NCPPB — National Collection of Plant Pathogenic Bacteria, Plant Pathology Laboratory, Hatching Green, Harpenden, England, U.K.

NCTC — National Collection of Type Cultures, Central Public Health Laboratory, Colindale, London NW9 5HT, England, U.K.

NIAID — National Institute of Allergy and Infectious Diseases, Hamilton, Montana 59840, U.S.A.

NIHJ — National Institute of Health, Tokyo, Japan.

NRC — National Research Council, Sussex Drive, Ottawa 2, Canada.

NRL — Neisseria Reference Laboratory, U.S. Public Health Service Hospital, Seattle, Washington 98114, U.S.A.

NRRL — Northern Utilization Research and Development Division, U.S. Department of Agriculture, Peoria, Illinois 61604, U.S.A.

NTHC — North Technical Hogskolles Collection, Department of Biochemistry, Technical University of Norway, Trondheim MTH, Norway.

OEU — Tennoji Branch, Osaka University of Liberal Arts and Education, Minami-Kawabori-Cho, Tennojiku, Osaka, Japan.

PDDCC — Culture Collection of Plant Diseases Division, New Zealand Department of Scientific and Industrial Research, Auckland, New Zealand

TC — Thaxter Collection, Farlow Herbarium, Harvard University, Cambridge, Massachusetts 02138, U.S.A.

TPH Microbiological Culture Collection, Public Health Laboratory, Ontario Department of Health, Toronto 116, Canada.

UMH University of Missouri Herbarium, Columbia, Missouri 65201, U.S.A.

UQM Culture Collection, Department of Microbiology, University of Queensland, Herston, Brisbane 4006, Australia.

VKM Department of Culture Collection, Institute of Biochemistry and Physiology of Microorganisms, U.S.S.R. Academy of Sciences, Pushchino, Moscow region, 142292, U.S.S.R.

VPI Anaerobe Laboratory, Virginia Polytechnic Institute and State University, Blacksburg, Virginia 24061, U.S.A.

WINDSOR Culture Collection, University of Windsor, Windsor, Ontario, Canada

WVU West Virginia University, Department of Microbiology, Medical Center, Morgantown, West Virginia 26506, U.S.A.

The Higher Taxa, or, A Place for Everything . . . ?

R. G. E. Murray

"Quot homines tot sententiae: suo quoque mos." ("So many men, so many opinions: each to his own taste.")

Terence, Phormio

When the eighth edition of *Bergey's Manual*® was in preparation, a major taxonomic concern was the provision of a clear statement of where the bacteria fitted among living things. This was set out in "A Place for Bacteria in the Living World" (Murray, 1974), which summarized the reasons for recognizing the kingdom *Procaryotae*, inclusive of the bacteria and the "blue-green algae" (cyanobacteria). This concept, based on cellular organization, is now a part of the fundamental training of all biologists, and a formal repetition is no longer a necessity. The student who wishes to relive that era should consult the major essays for details and references (Stanier, 1961; Murray, 1962; Stanier and van Niel, 1962; Allsopp, 1969; Stanier, 1970). Taxonomy is not static, however, and new horizons are being explored that give perspective and greater definition to higher taxa as well as the lower categories of genus and species.

The prefatory chapter mentioned above included a tentative proposal of appropriate higher taxa. The arguments and proposals that have arisen since then concern the levels of dissection of the kingdoms of the living world (Woese and Fox, 1977a; Whittaker and Margulis, 1978), the definition and levels of dissection of the major procaryotic groups (Gibbons and Murray, 1978; Whittaker and Margulis, 1978), and the integration of evolutionary information (Stackebrandt and Woese, 1981a). There is a renewal of interest in bacterial taxonomy stimulated by the recognition of novel groups of bacteria that do not fit comfortably into current systematic schemes and by new understanding of the taxonomic utility and phylogenetic significance of molecular and genetic data. The system of "superphyla" and phyla proposed by Whittaker and Margulis (1978) is not sensitive to current interpretations of biochemical relatedness based on wall chemistry or other unique features of well-established groups of procaryotes. There is no advantage, at this stage of our understanding, in debating the relative value of recognizing "super kingdoms" (Whittaker and Margulis, 1978) or "primary kingdoms" (the "Urkingdoms" of Woese and Fox, 1977a) to accommodate views of cellular organization in protists, plants, and animals as well as speculations about the nature of putative progenitors. We should be content for now to deal with the *Procaryotae* and the systematic problems that arise within that circumscription; there must be sufficient time for assimilation and consolidation of the burgeoning data. There are attractive features in the dendrograms generated by C. R. Woese and his colleagues; their time will come when the patterns of associations are less fragmentary. For now it would appear sensible to look to the Gibbons and Murray (1978) proposal as capable of modification as an interim broad classification with a few areas of taxonomic validity.

The conceptual changes deriving from genetic and molecular studies are making inroads into the cherished beliefs of taxonomists and bacteriologists. We have to agree with the moderate statement taken from Stackebrandt and Woese (1981a): " . . . what bacterial classification we have (say up through the eighth edition of *Bergey's Manual*, 1923–1974) is probably not in very good accord with the natural relationships that exist among organisms." This is true enough because it is only in the past decade that sequencing of biopolymers and molecular genetics has provided convincing data on relatedness and because the intent and role of *Bergey's Manual*® has always been to provide a basis for the determination of the identity of a pure culture. It is unfortunate that the expression of nomenclatural decisions, hierarchical arrangements (even if all but abandoned in the eighth edition), and the mask of authority have tended to induce undue confidence in the relationships implied in earlier editions.

Even if our perception of "natural relationships" is flawed by ignorance as well as inadequate information, the practical bacteriologist needs a simple scheme of classification as a framework for recognition. At this stage, for example, the possibility that some micrococci are more closely related to *Arthrobacter* than to other spherical Gram-positive cocci and the implication for a further splitting of the genus *Micrococcus* (Stackebrandt and Woese, 1979, 1981b) would be confusing to the practical bench worker or the physician and is unhelpful. At the higher taxonomic levels some sort of serviceable scheme that can recognize and, to a degree, accommodate the possibilities will be of service, will cushion the shocks to come, and will stimulate appropriate research. This sort of practicality is almost realized in the proposal by Gibbons and Murray (1978). But any scheme will require future modification because it will take time to attain a complete reassessment of the taxonomic significance and validity of those characters that are reasonably easy to determine and apply effectively to each level of identification. The alternative possibility is that we should maintain two entirely independent schemes: a practical taxonomy and an academic (phylogenetic) taxonomy. The dichotomy of these phenotypic and genotypic approaches is with us because people of such persuasions work in semi-isolation and because, as argued by Stackebrandt and Woese (1981a), " . . . the classically defined taxonomic categories, the genera and families, do not correspond to fixed (minimal) S_{AB} values." Furthermore, the genetics of today is beginning to clarify the mechanisms operating in the grand evolutionary experiment, blessed with minimal constraints of time and circumstance, conducted in nature's laboratory. It is clear that point mutations are less important than effective reassociations of determinants with their modifying segments and the mechanisms allowing the exchange, chromosomal integration, and amplification of operative sets of determinants (Campbell, 1981; Cullum and Saedler, 1981). It is conceivable, now, that major complex characters of physiological and taxonomic significance (involving a considerable number of genes) could be transferred between organisms both closely and distantly related in the clonal arborizations of the phylogenetic tree. All that is required is an occasional evolutionarily successful experiment in a time frame measured in thousands, millions, or billions of years (or cell divisions, for that matter).

An overall taxonomic scheme that is capable of incorporating phylogenetic data (as well as providing a primary key) would be helpful in minimizing the dichotomy of interests and understanding among bacteriologists. It is desirable to bridge the growing gap between the practical applied fields and the academic substratum with something more than the perfidy of plasmids and technological legerdemain. The perpetual quandary is how and when to incorporate into systematic bacteriology the generalizations derived by intensive and expensive study of "model" organisms or of a limited set.

The most exciting and evocative of recent explorations of the possibilities of extracting phylogenetic information from highly conserved biopolymers (the "semantides" of Zuckerkandl and Pauling, 1965) are the comparisons of 16S ribosomal RNA (rRNA) catalogs undertaken by Carl Woese and his colleagues (Woese and Fox, 1977b). This approach, together with that involving the functional structural homologies and interchangeability of whole ribosome parts and some protein components, shows promise of providing comparative data with evolutionary significance for the whole living world (Brimacombe et al., 1978; Kandler, 1981). Of most interest to us is the capability of the technique of RNA nucleotide analysis, expensive and slow though it may be, in spanning the widest range of procaryotic clones. The stability of most of this fairly large (1540 residues) molecule is the basis of the application to assessment of taxa at the level of family and higher. The fact that a few variable domains exist in the molecule allows for the partially realized possibility of contributing a degree of resolution at the level of genus. This could add to the data on relations within genus and species generated by utilizing DNA/DNA (Johnson, 1973) and RNA/DNA hybridization (De Smedt and De Ley, 1977). The figures that are generated (either the number of shared oligonucleotides or an association coefficient, the S_{AB} value) are based on a computer comparison of the catalogs of oligonucleotides (liberated from the 16S RNA by ribonuclease T_1) large enough to show individuality (larger than pentamers). This means that a considerable portion of the molecule yielding oligonucleotides smaller than hexamers is not taken into account, and therefore, the sequence data cannot directly be related to all other hybridization data. Nevertheless, the results of comparing some 200 representative bacteria, surprises and all, support the directing thesis that most of the 16S rRNA sequence has drifted only slowly with time. We surmise (q.v. Stackebrandt and Woese, 1981a) that the comparison provides a measure of the "depth" of the separation between the phylogenetic units or branches of the phylogenetic tree. It is an article of faith that the degree of cleavage (a low S_{AB} value) is proportional to time and is reasonable for initial purposes, although the time scale may be different for different major taxa.

Clones giving rise to unique and now recognizable groups of bacteria must have separated at various stages of procaryotic evolution (Fox et al., 1980). Earliest among these departures from the main stem so far detected by this oligonucleotide cataloging are the *Archaeobacteria*,* which comprise the methanogens, halobacteria, and thermoacidophiles: these, it is now known (Kandler, 1981), possess peculiar lipids and either no murein or a pseudomurein in their cell walls. The eight or so major groups of photosynthetic and chemosynthetic bacteria (all designated "eubacteria" in the papers cited) arose somewhat later in this imprecise evolutionary time scale. The data suggest that some genera are truly ancient (e.g. *Clostridium, Spirochaeta*) and older, in fact, than some very complex associations of genera (e.g. the actinomycetes).

Those involved in the comparative studies of 16S rRNA, ribosomes, and ribosomal proteins have come to the enthusiastic conclusion that the procaryotes are made up of two kingdoms, the eubacteria and the *Archaeobacteria.* The molecular and biochemical bases for the separation have been summarized by Woese (1981) and Kandler (1981). There is no doubt whatever that the *Archaeobacteria* are distinguished by a number of specialized characters from the rest of the procaryotes ("eubacteria" has had too many meanings in the past to be a useful term). They include the number of ribosomal proteins, the size and shape of the ribosomal S unit, the proportion of acidic ribosomal proteins, the constitution of transfer RNA initiator, the presence of ether-linked rather than ester-linked lipids, and the absence of muramic acid or the normal form of peptidoglycan from cell walls. These and some other intimate features make for interesting thoughts about eucaryotes, mitochondria, and chloroplasts, as well as the procaryotes. But these distinctions are not suitable to kingdom status. Stackebrandt and Woese (1981a) sum up the situation as follows: "... the general conclusion that seems to be emerging with regard to the differences among archaebacteria, true bacteria and eucaryotes is that all are identical in the basic aspects of their basic processes, yet all differ from one another in the details of these processes." An examination of any of the methanogens, strict halophiles, or thermoacidophiles would place them in the kingdom *Procaryotae* as presently defined. It is not appropriate to separate kingdoms on any basis but a major, reasonably easily determined difference in organization. Therefore, it seems sensible to treat the *Archaeobacteria* as a major taxon within the *Procaryotae* and, if necessary, amend its status at some later date when more evidence is collected and digested. Perhaps we will soon recognize other equally distinctive clones that diverged very early from the stem clones of primitive microbes.

There is a comforting sequel to pondering compilations of articles regarding biochemical evolution (Wilson et al., 1977; Carlile et al., 1981). Although these studies suggest that "strange bedfellows" may be assigned to some of the more complex groups or point to unexpected separations (e.g. among the photosynthetic bacteria, Gibson et al., 1979; among the micrococci, Stackebrandt and Woese, 1979), there are concordant features. Gram-positiveness and Gram-negativeness are still unassailable characters except in what are now known to be phylogenetically and biochemically separate groups, the *Archaeobacteria* (Balch et al., 1979), the radiation-resistant cocci (Brooks et al., 1980) and, of course, the wall-less *Mollicutes.* Among the Gram-positives, it is encouraging to see that the many peptidoglycan types form consistent patterns in the branches of the dendrograms generated by comparison of the oligonucleotide catalogs (Schleifer and Kandler, 1972; Kandler and Schleifer, 1980; Kandler, 1981).

The infinitely diverse groupings of Gram-negative bacteria have yet to be surveyed to an extent that allows of any decisive taxonomic proposals, and many of those included up to now exhibit phylogenetic and phenotypic incoherence. A major surprise arising from the analysis of rRNA oligonucleotides has been that each of the three coherent phylogenetic groupings of anoxygenic photosynthesizers contains a variety of seemingly related, diverse, and well-known nonphotosynthetic genera showing subordinate S_{AB} values (Gibson et al., 1979; Stackebrandt and Woese, 1981a). The implication is that photosynthetic clones may spawn apochlorotic derivatives, a thesis often directed to the cyanobacteria in the past (Pringsheim, 1967) but not yet subjected to this sort of phylogenetic analysis. It would seem wise not to make phototrophism an overriding taxonomic unit until the situation clarifies. This makes for difficulties in more conventional schemes such as that of Gibbons and Murray (1978), which separated in simplistic fashion the photosynthetic (*Photobacteria*) and the nonphotosynthetic (*Scotobacteria*) which are included in the Gram-negative bacteria (*Gracilicutes*). But it may be too early to be either discouraged or encouraged, and there is still room for the exercise of one's prejudices.

Classifications cannot be final, and there are many ways in which bacteria can be classified (Cowan, 1968); this one has no more permanence than those that went before it. But the doubts and criticisms that come to the mind of the reader are the stimuli to further work and a deeper consideration of the taxonomic implications of the new mix of biochemical and phylogenetic data. Changes from the earlier versions (Murray, 1974;

*Equivalent to the term *Archaebacteria.* Because the word is formed by a combination of two Greek words (*archaios*, ancient, and *bakterion*, a small rod), the letter *o* should be used as the combining vowel, hence, *Archaeobacteria.*

Gibbons and Murray, 1978) were inevitable. The hierarchical levels needed to be raised to give greater scope for classifying the range of organisms included at each major level. For example, students of the cyanobacteria, even those most sympathetic to their incorporation into bacterial taxonomy, despair of being able to accommodate their charges within the single order assigned by Gibbons and Murray (1978). As already indicated, it would take more than ordinary taxonomic agility to make a phylogenetically sensitive classification of the phototrophic bacteria and their derivatives in our present state of understanding.

The following is proposed as an arrangement of higher taxa which can serve during this time of taxonomic transition. It involves some amendments of rank and new names.

Kingdom *Procaryotae* Murray 1968, 252.

Division I. *Gracilicutes* Gibbons and Murray 1978, 3.

Class I. *Scotobacteria* Gibbons and Murray 1978, 4.

Class II. *Anoxyphotobacteria* (Gibbons and Murray) classis nov.^{VP†} (Subclassis *Anoxyphotobacteria* Gibbons and Murray 1978, 4.)

Class III. *Oxyphotobacteria* (Gibbons and Murray) classis nov.^{VP} (Subclassis *Oxyphotobacteria* Gibbons and Murray 1978, 3.)

Division II. *Firmicutes* Gibbons and Murray 1978, 5. (*Firmacutes* (sic) Gibbons and Murray 1978, 5.)

Class I. *Firmibacteria* classis nov.^{VP}; L. adj. *firmus* strong; Gr. dim. n. *bakterion* a small rod; M.L. fem. pl. n. *Firmibacteria* strong bacteria, indicative of simple Gram-positive bacteria.

Class II. *Thallobacteria* classis nov.; Gr. n. *thallos* branch; Gr. dim. n. *bakterion* a small rod; M.L. fem. pl. n. *Thallobacteria* branching bacteria.

(These new names are proposed to express the general basis of splitting the division into the simple Gram-positive bacilli and those Gram-positive bacteria showing a branching habit, the actinomycetes and related organisms.)

Division III. *Tenericutes* div. nov.^{VP}; L. adj. *tener* soft, tender; L. fem. n. *cutis* skin; M.L. fem. n. *Tenericutes* procaryotes of pliable, soft nature, indicative of lack of a rigid cell wall.

Class I. *Mollicutes* Edward and Freundt 1967, 267.

(The *Mollicutes* are a distinctive group of wall-less procaryotes of sufficiently diverse phylogenies that separate classes may well be required in the future.)

Division IV. *Mendosicutes* Gibbons and Murray 1978, 2. (*Mendocutes* (sic) Gibbons and Murray 1978, 2.)

Class I. *Archaeobacteria* (Woese and Fox) classis nov.^{VP} (Kingdom *Archaebacteria* (sic) Woese and Fox 1977a, 5089.)

(The *Archaeobacteria* are defined in terms of being procaryotes with unusual walls, membrane lipids, ribosomes, and RNA sequences (Kandler, 1981). The future may bring further classes into the *Mendosicutes* when (and if) truly primitive organisms as envisioned by Woese and Fox (1977b) are isolated and recognized as related to a "universal ancestor or progenote.)

This arrangement of the procaryotes continues to recognize the absence or presence and nature of cell walls as determinative at the highest level. The omission of *Photobacteria* (Gibbons and Murray, 1978) as a class and the elevation of *Oxyphotobacteria* and *Anoxyphotobacteria* to class status is intended to provide more scope than is offered by Gibbons and Murray (1978) for the inevitable arrangement and rearrangement of the lower taxa within these categories as new understanding of lineage and relationships is brought to bear. For instance, the groups of phototrophic bacteria are formed of several major subgroups; separation of these from the level of class would be appropriate to the deep phylogenetic clefts that may be established in both the *Oxyphotobacteria* (the cyanobacteria and the *Prochlorales*) and the *Anoxyphotobacteria*, as pointed out by Stackebrandt and Woese (1981a). Furthermore, it may not be appropriate to place all the nonphotosynthetic, Gram-negative bacteria in the *Scotobacteria* if the molecular evidence points clearly to derivation from photosynthetic ancestors; undoubtedly this would require separation at a high level within the class. The same need for broad scope is apparent in the *Firmicutes* with its two major divisions, the simple Gram-positive bacteria (*Clostridium* and relatives), and the actinomycetes.

The *Tenericutes* would have less support from the molecular phylogenetic evidence as a taxon at the highest level because of their probable origin from Gram-positive bacteria and the possibility that they may not have a single common ancestry (Woese et al., 1980). However, they form a stable and distinctive group; they are not obviously a subset of the *Firmicutes*, and their wall-less state puts them clearly in a division by themselves as long as we base our classification on the presence or absence and character of the cell wall. On the other hand, we must recognize that an organism may lose a component of a very complex wall and still merit consideration as a member of that class: e.g. the members of the genus *Chlamydia* have no muramic acid, but other characters, including some concerned with the relict wall, suggest a relationship to organisms that are definitive members of the *Gracilicutes*.

The *Archaeobacteria*, for their part, are a very diverse group in terms of cell wall attributes (all the way from a complex wall including pseudomurein to wall-less), and there are at least five groupings with S_{AB} values of <0.3. There is sufficient scope within a division for the apparent complexity laid out by Balch et al. (1979) and any extraordinary "primitive" organisms that may be isolated. The sensible approach would seem to be maintenance of the consistency of the higher taxa by including in the class *Mendosicutes* all those procaryotes with a cell wall composition inconsistent with that defined for *Gracilicutes* and *Firmicutes* (e.g. in simplest terms, not possessing muramic acid). This view is supported by Starr and Schmidt (1981). The inclusion of wall-less thermoacidophiles among the *Archaeobacteria* will be necessary even if seemingly inconsistent. There will come a time, without doubt, when we can set up taxa that are precisely defined in terms of molecular genetics, but that time has not yet arrived.

We cannot assume that all possible procaryotic organisms have been observed and isolated for study. The possibility exists that organisms will be found that are even more "primitive" (i.e. separated from the main stem even earlier) than the *Archaeobacteria* and have a constitution revealing more of the nature of the "universal common ancestor" (Woese and Fox, 1977a; Stackebrandt and Woese, 1981a). A class can be formed in the future as a suitable home for any organism whose proteins and genetic translation apparatus do not fit into the line of evolution represented by the procaryotes and eucaryotes studied up to now. They may have characteristics that foreshadow the fundamental eucaryotic cell.

No place is provided in our scheme for the fossil microbes being described from specimens of Precambrian and, possibly, Archean cherts (see Walter, 1977). This is because they can only be described in terms of size, shape, and associations. The oldest are in stratified structures closely resembling the stromatolites and "algal mats" that can be found today, which are complex consortia of cyanobacteria, algae, and bacteria with equally complex layering of metabolic and physiological characteristics. There are several attractive morphological resemblances. Size is about all that distinguishes the interpretation of forms as bacterial or

[†] *VP* denotes that this name, although not on the Approved Lists of Bacterial Names, has been validly published in the official publication, *International Journal of Systematic Bacteriology*.

algal. A further interpretation that a particular form is photosynthetic is entirely circumstantial and assumptive. Names have been assigned in binominal form (usually based on the Botanical Code) as a means of classifying the varied types being observed. This is a legitimate and stimulating activity, but it is not yet helpful in terms of procaryotic classification or evolutionary taxonomy. Because of the uncertainties of alignment, they should be classified for determinative purposes in a separate group of microbiota.

Our view of classification and the taxonomic edifice is based on a century of experience supporting the contention that there is a reasonable degree of fixity in the characters describing a species. Such variation as there is within the clusters (as we now see in computer-assisted studies) can be included in the circumscription offered in describing the species or other categories. Certainly the species is a concept, not an entity (Cowan, 1968); but phenetic studies utilizing large numbers of strains and the widest possible range of characters has, if anything, clarified our concept of taxonomic groups (Sneath, 1978). Despite a growing appreciation of the Adansonian approach to species, the attitude to the definition of higher taxa involves the selection of seemingly single but very complex characters as exemplified by this essay.

The alternative and extreme view that the bacteria are so pleomorphic that the species concept has no reality is misleading and draws the attention away from the important facts of life in clonal populations. Views of this character have been put forward by Sonea (1971) and Sonea and Panisset (1976, 1980). They believe, along with the rest of us, that the procaryotic clones are united by their lineage, but these investigators differ in their interpretations of the stability of taxonomic units. They argue that genetic exchange (and "communications" of all sorts) between diverse clones makes nonsense of the species concept. The extension of their analysis into a concept of unity for the entire world population of bacteria and their interactions with the environment (a sort of global organism as real as a horse or an elephant) is an interesting but philosophical curiosity. The approach stimulates thought, but it seems evident that modern pleomorphists will have to modify and adapt their views to practical necessities dictated by new knowledge as much as our views, expressed in this chapter, will have to bend with the winds of change. A century of bacteriological analysis and studies of cultures can convince one that there is sufficient stability to allow of the recognition of most taxonomic clusters.

No doubt, there are many and variant attitudes to the details of bacterial taxonomy. But the substratum of fact is now beginning to be revealed beneath the veneer of fancy or prejudice and to allow judgment to operate. Perhaps bacteriology is maturing after about 150 years of seeking a stable basis for classification.

> "For now we see through a glass, darkly; but then face to face: now I know in part . . .
>
> First Epistle of Paul
> to the Corinthians (XIII:12)

That we can now perceive, albeit dimly, aspects of classification and phylogeny in the macromolecules of the *Procaryotae* is the legacy of many more scientists than have been cited so far in the prefatory chapters. The views that we espouse today still reflect the prejudices and enthusiams of our teachers, our teachers' teachers, and the influential observers and exponents of each stage in the maturation of bacteriology. The praise and the blame cannot easily be apportioned, but undoubtedly we can recognize the pervasive influence of strong-minded people of the modern era who enjoyed trying to make order out of chaos. These included D. H. Bergey, R. S. Breed, R. E. Buchanan, and N. E. Gibbons, whose efforts have brought *Bergey's Manual*® into print in its various editions. But they, like others before and after them and despite their special interests, not to say prejudices, were collectors of the intellections and arrangements of "authorities." Despite the arguments concerning validity that undoubtedly surfaced at the time, their efforts could not prevent the perpetuation of numerous unstable taxa that we still struggle with today: form genera, color genera, physiological genera, etc., encompassing diverse and probably unrelated species, as genetic and biochemical criteria now force us to realize. But it is more important, perhaps, to realize that the most pervasive influences on all manner of approaches have been the writings of the "Delft school" (M. W. Beijerinck and A. J. Kluyver) and the product of that school, C. B. van Niel. The discussions accompanying "the van Niel course" (attended by an equally remarkable collection of microbiologists) started many thinking about microbes in new ways and set them and their students on productive lines of work. The comparative studies that resulted from these stimuli were of great significance in microbial biochemistry, physiology, and ecology; a high proportion of the studies resulted in significant contributions being made to systematic bacteriology. Not least among those influenced by van Niel was R. Y. Stanier, whose death came just as this volume was being readied for the press. It is obvious that we owe a particular debt to this lineage of bacteriologists. Stanier's arguments in our meetings, when he was a member of the Board of Trustees, were largely responsible for major changes in attitude and format expressed in the eighth and in this edition of the *Manual*; we sharpened our judgments with the help of other well-established heretics such as S. T. Cowan (1970). And now, as strongly expressed and supported in this essay, a new breed of heretic is influencing bacterial systematics, as we had been warned would be the case by another former trustee, A. W. Ravin.

In the end, a reassessment of the diverse characters used in classifications will have as important consequences as the major generalization based on cellular organization used in defining the kingdom. We must identify characters of proven reliability and validity encompassing more of the genome than seems to be the case today. The techniques must allow the comparison of groups whatever their ecological niche or the professional proclivities of those that study them. Happily, we can echo the Rabbi ben Ezra and proclaim: "The best is yet to be."

Editorial note: A publication presenting much of the data and examining the perspectives revealed in research on the *Archaeobacteria* appeared after the completion of this essay. It includes contributions from most if not all of the laboratories engaged in this thrilling task and starts with "an overview" by C. R. Woese. This comprehensive collection of papers is in *Zentralblatt für Bakteriologie, Mikrobiologie und Hygiene*, I Abt. Orig. C.3(1/2):1–345, March/May 1982.

THE ACTINOMYCETES I

Suprageneric Classification of Actinomycetes

Michael Goodfellow

It has long been apparent that reliable suprageneric classifications of bacteria are not possible using traditional approaches based upon a priori weighting of a few morphological and physiological features (Bisset, 1962). Reliance on such properties has resulted in actinomycetes being assigned to suprageneric groups, most of which have turned out to be markedly heterogeneous (Baldacci, 1958; Waksman, 1961; Baldacci and Locci, 1966; Lechevalier and Lechevalier, 1970a; Cross and Goodfellow, 1973; Gottlieb, 1974). Recently, however, approaches to this problem have been revolutionized by molecular methods involving comparative sequencing and reassociation of macromolecules, notably nucleic acids (Schleifer and Stackebrandt, 1983). A most useful method for establishing the relatedness of higher taxa is comparative analysis of the ribonuclease-resistant oligonucleotides of 16S ribosomal RNA (rRNA) (Fox et al., 1977; Stackebrandt and Woese, 1981a). This macromolecule is particularly suitable for generating phenetic data that can be used to determine both close and very distant relationships (Woese and Fox, 1977a; Balch et al., 1979; Stackebrandt and Woese, 1981a) because it is ubiquitous, highly conserved in sequence, and genetically stable.

PARTIAL SEQUENCING OF 16S RIBOSOMAL RNA

Partial sequence analysis of 16S rRNA provides an exacting way of unraveling distant relationships between procaryotes. Purified RNA is digested by T_1 ribonuclease, and the oligonucleotides separated by two-dimensional electrophoresis are sequenced using a combination of endonuclease digestion procedures that give a catalog of sequences characteristic of the test strains. Oligonucleotide sequences are compared and oligomers, of six residues or more, common to any two catalogs are scored using an association coefficient, S_{AB}, that allows for differing lengths among the oligomers. The association coefficient (Fox et al., 1977) is defined as follows:

$$S_{AB} = 2N_{AB}/(N_A + N_B)$$

where N_A and N_B are the total number of residues (>6) in catalogs A and B, respectively, and N_{AB} is the total number of residues (>6) represented by all of the coincident oligomers between the two catalogs. The values of the association coefficient are calculated for every pair of organisms studied. The resulting matrix of S_{AB} values can be sorted using any of a number of standard clustering algorithms and the data produced presented as dendrograms. New catalogs can be compared to all previous catalogs regardless of the laboratory of origin. This expanding data base provides a significant advantage over nucleic acid hybridization methods, which tend to rely on comparisons with a few reference strains. To date, over 400 representatives of conventionally defined species have been characterized by partial sequencing of 16S rRNA (Stackebrandt and Woese, 1981a; Stackebrandt and Schleifer, 1984).

Gram-positive bacteria form one of about 10 major suprageneric groups (Woese et al., 1984). These organisms, however, form two major branches that can be recognized on the basis of DNA base composition (Stackebrandt and Woese, 1981a). The actinomycete branch encompasses organisms with DNA rich in guanine (G) plus cytosine (C) (ca. 55 mol%) and the *Bacillus-Clostridium-Streptococcus* branch encompasses bacteria with a low DNA base composition (<50 mol% G + C). Several genera historically associated with actinomycetes belong to this second group. The genus *Eubacterium* is related to *Clostridium* (Tanner et al., 1981), *Kurthia* to the lactic acid bacteria (Ludwig et al., 1981), and *Thermoactinomyces* to the *Bacillaceae* (Stackebrandt and Woese, 1981a; Stackebrandt et al., 1983c). Conversely the genera *Micrococcus* and *Stomatococcus*, traditionally classified with other Gram-positive bacteria, notably *Staphylococcus*, belong to the actinomycete line (Stackebrandt and Woese, 1979; Stackebrandt et al., 1983d). Similarly, coryneform taxa such as *Arthrobacter, Brevibacterium, Cellulomonas, Corynebacterium*, and *Microbacterium*, which rarely, if ever, form a primary mycelium, are close to certain well-established actinomycete genera (Stackebrandt et al., 1980b). Results such as these challenge the traditional morphological definition of the order *Actinomycetales* Buchanan 1917 and emphasize the dangers of constructing taxonomies on a few morphological properties (Goodfellow and Cross, 1984b; Stackebrandt and Schleifer, 1984).

SUPRAGENERIC CLASSIFICATION OF THE ACTINOMYCETES

It is not yet possible to compile a comprehensive suprageneric classification of actinomycetes from the results of partial sequencing of 16S rRNA (Fox, 1980; Stackebrandt and Woese, 1979, 1981b; Stackebrandt et al., 1980a, 1980b, 1983b, 1983c, 1983d), although the essential features of such a classification are becoming apparent (Fig. XI.1). Good congruence has been found between these sequencing data and the results of rRNA cistron similarity analyses. The latter support the recognition of suprageneric groups corresponding to the actinoplanetes (Stackebrandt et al., 1981), maduromycetes (Stackebrandt et al., 1981), nocardioform actinomycetes (Mordarski et al., 1980b, 1981b), and streptomycetes (Stackebrandt et al., 1981; Gładek et al., 1985), and of discrete subgroups within the actinobacteria (Stackebrandt et al., 1980a, 1980b, Döpfer, 1982). Sequencing studies also show the genus *Actinomadura* to be heterogeneous (Fowler et al., 1985); this finding is strongly supported by the results of chemical (Athalye et al., 1984; Poschner et al., 1985), numerical phenetic (Athalye et al., 1985), and nucleic acid hybridization (Fischer et al., 1983; Poschner et al., 1985) studies. There is similar overwhelming evidence that the family *Dermatophilaceae* encompasses very diverse taxa (Samsonoff et al., 1977; Goodfellow and Pirouz, 1982a; Stackebrandt et al., 1983b) and is in need of taxonomic revision. However, as the taxonomy of actinomycetes unfolds, it becomes increasingly clear that many of the markers previously used to circumscribe suprageneric taxa have little or no predictive value. The genus *Micromonospora* has, for example, been shown to be more closely related to sporangia-forming actinomycetes, such as *Actinoplanes* and *Dactylosporangium*, and sharply distinct from other monosporic genera such as *Thermomonospora* and *Thermoactinomyces*, with which it was recently associated (Küster, 1974a).

The whole panoply of modern taxonomic methods can be used to untangle the fine structure of major suprageneric groups detected in rRNA partial sequencing and DNA/rRNA pairing studies. Data derived from comparative studies on the same group of organisms can then be examined to see which, if any, properties can be weighted as reliable markers. Chemical markers have been shown to be especially good indicators of suprageneric relationships (Stackebrandt and Schleifer, 1984; Goodfellow, 1986; Stackebrandt, 1986). Chemosystematics or chemical taxonomy is a rapidly expanding discipline in which information derived from chemical analyses of whole cells or cell fractions is used for classification and identification. A range of chemical techniques are available for determining DNA base, whole-cell, lipid, wall sugar, and amino acid composition of bacteria (see Goodfellow and Minnikin, 1985; Gottschalk, 1985). Chemical features have been invaluable in establishing suprageneric affinities of actinomycete taxa and in clarifying the taxonomic relationships of the genus *Thermoactinomyces* (Goodfellow and Cross, 1984b; Table XI.1). Nocardioform actinomycetes, for example, have a wall chemotype IV sensu M. P. Lechevalier and Lechevalier (1970b), that is, they have *meso*-diaminopimelic acid (DAP) as the diamino acid of the wall peptidoglycan: arabinose and galactose as major wall sugars; an A1γ peptidoglycan; major amounts of straight-chain saturated and unsaturated fatty acids; diphosphatidylglycerol, phosphatidylinositol, and phosphalidylinositol mannosides as predominant phospholipids; and characteristic 2-alkyl-branched 3-hydroxy acids, the mycolic acids (Goodfellow and Cross, 1984b).

The molecular genetic and chemical variation found within the actinomycetes can be used to assign genera to higher taxonomic groups. Goodfellow and Cross (1984b) recognized eight such groups but considered that it was premature to equate them with higher ranks in the taxonomic hierarchy. The scheme introduced by these workers has been updated in light of recent studies (Table XI.1). Many of the phenotypic properties, notably morphological features such as presence or absence of aerial hyphae, spore vesicles, and spores, motility of spores, and the extent and stability of mycelia, are not always discontinuously distribute along suprageneric lines (Stackebrandt and Schleifer, 1984; Stackebrandt, 1986). It is therefore not surprising that all but one of the eight families, namely the family *Streptomycetaceae*, forming the order *Actinomycetales sensu* Gottlieb (1974) contain diverse taxa.

The suprageneric groupings of actinomycetes recognized in this volume are mainly based on a judicious selection of chemical, morphological, and physiological attributes. For consistency these groupings have been labeled using the terms introduced by Goodfellow and Cross (1984b). However, given the congruence noted earlier between some chemical, numerical phenetic, and molecular genetic data it is perhaps not surprising that the aggregate groupings designated as actinomycetes with multilocular sporangia (Section 27), actinoplanetes (Section 28), streptomycetes (Section 29), maduromycetes (Section 30), and thermomonosporas (Section 31) are generally equivalent to those shown in Figure XI.1. In sharp contrast, the nocardioform actinomycetes (Section 26) is a markedly heterogeneous assemblage that accommodates taxa classified by Goodfellow and Cross (1984b) as actinobacteria (*Oerskovia* and *Promicromonospora*), micropolysporas (*Actinopolyspora, Faenia, Pseudonocardia, Saccharomonospora* and *Saccharopolyspora*), nocardioform actinomycetes (*Nocardia* and *Rhodococcus*) and streptomycetes (*Intrasporangium* and *Nocardioides*).

THE SUPRAGENERIC GROUPS

It is clear that some suprageneric groups (Table XI.1) are more homogeneous than others. The actinoplanetes, maduromycetes, nocardioform actinomycetes, and streptomycetes probably represent distinct suprageneric groups, although additional strains need to be examined using molecular genetic methods to confirm relationships based on chemical and morphological indicators and to clarify apparent anomalies. It is, for instance, necessary to establish the relationships of the genera *Intrasporangium, Kineosporia,* and *Sporichthya* both to one another and to *Streptomyces* and other actinomycetes with an A3γ peptidoglycan. The exact taxonomic position of the genus *Corynebacterium* is open to question given conflicting results presented by Stackebrandt and coworkers. Stackebrandt and Woese (1981a) found *Corynebacterium* to be more closely related to other mycolic acid-containing taxa (*Nocardia* and *Rhodococcus*) than to the mycolateless actinoplanetes (*Actinoplanes, Ampullariella, Dactylosporangium,* and *Micromonospora*), a result in excellent accord with current trends in the taxonomy of these organisms (Minnikin and Goodfellow, 1980; Goodfellow and Minnikin, 1981b; Goodfellow and Cross, 1984b). However, in a subsequent study (Stackebrandt et al., 1983c) representatives of the genera *Mycobacterium, Nocardia,* and *Rhodococcus* were depicted as sharing higher S_{AB} values with actinoplanetes than with *Corynebacterium* (including *C. diphtheriae,* the type species). The confusion caused by these conflicting results (Stackebrandt and Woese, 1981a; Stackebrandt et al., 1983c) needs to be resolved.

A combination of chemical and morphological features can be weighted for the recognition of actinoplanetes (see introduction to Section 28), maduromycetes (see introduction to Section 30), nocardioform actinomycetes (Goodfellow and Cross, 1984b), and streptomycetes (see introduction to Section 29). Nocardioform actinomycetes have many properties in common but show considerable morphological variation.

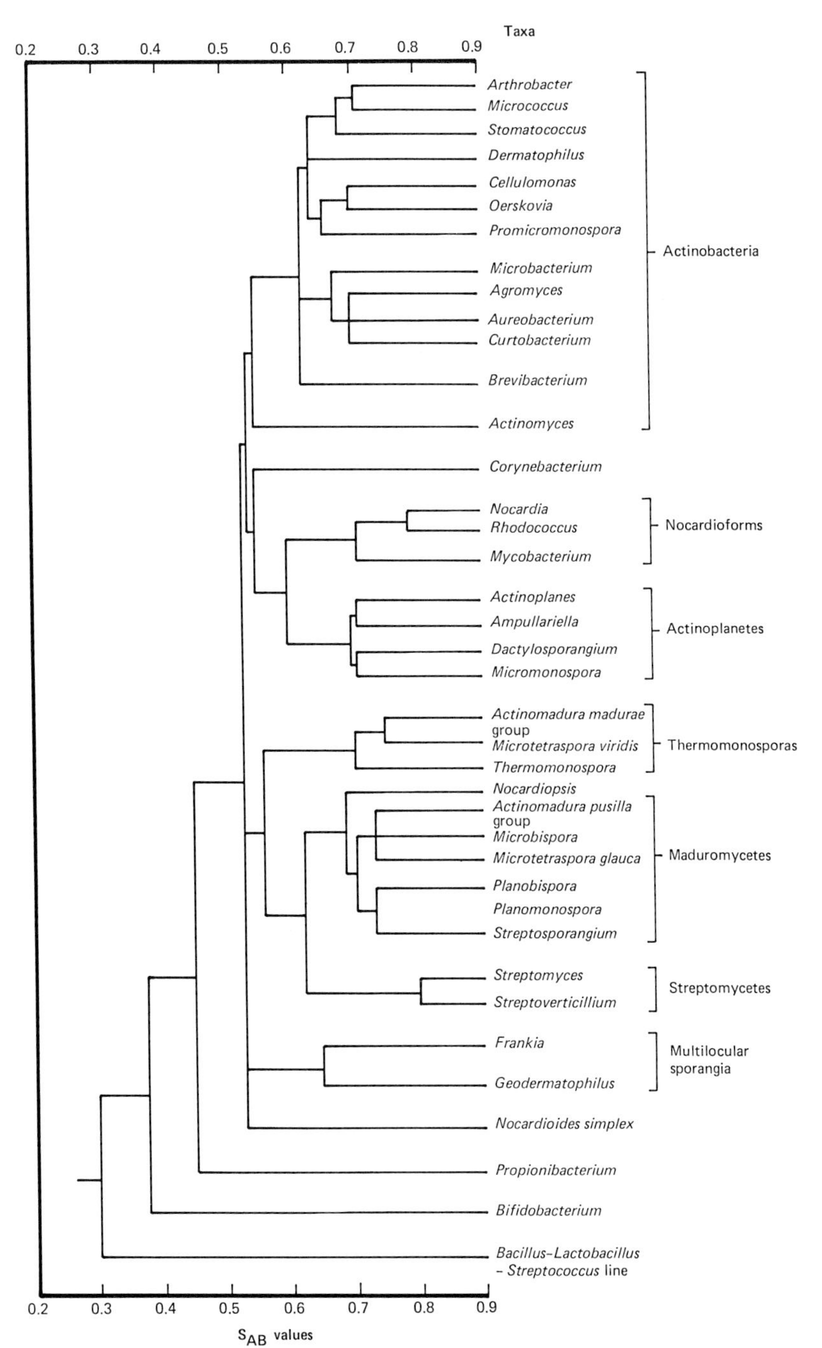

Figure XI.1. Suprageneric relationships of actinomycetes based on partial sequencing of 16S ribosomal ribonucleic acids.

Caseobacters, corynebacteria, many rhodococci, and most mycobacteria are essentially amycelial, but nocardiae, some rhodococci, and a few mycobacteria typically form a branched mycelium that sooner or later fragments into bacillary and coccoid elements that may be markedly pleomorphic. Nocardiae and some mycobacteria also produce aerial hyphae that range from being sparse and invisible to the unaided eye to completely covering the substrate mycelium with a white, powdery down. Nocardioform actinomycetes, however, are most easily recognized and best defined by chemical properties (Table XI.1). It is still difficult to offer all-embracing descriptions of actinobacteria (Goodfellow and Cross, 1984b), actinomycetes with multilocular sporangia (see introduction to Section 27), micropolysporas (Goodfellow and Minnikin, 1985), and thermomonosporas (see introduction to Section 31) because additional comparative studies are needed to establish their suprageneric status. This is also true for the genera that cannot be assigned to any of the suprageneric groups recognized at present (Table XI.1).

Table XI.1.
Suprageneric groups of actinomycetes and some of their chemical properties[a]

Group/genus	Wall chemotype[b]	Whole-cell sugar pattern[c]	Peptido-glycan type[d]	Fatty acid pattern[e]	Major menaquinone (MK)[f]	Phospholipid type[g]	Mol% G + C of DNA
ACTINOBACTERIA							
A. *Agromyces*	VII	—	B2γ	2c	-11, -12, -13	PI	71–77
Aureobacterium	VIII	—	B2β	2c	-11, -12	PI	65–76
Clavibacter	VII	—	B2γ	2c	-9, -10	PI	68–75
Curtobacterium	VIII	—	B2β	2c	-9	PI	68–75
Microbacterium	VI	—	B1′, B1β	2c	-11, -12	PI	69–75
B. *Arthrobacter*				2c	-9(H_2)	PI	59–70
Micrococcus				2c	-7(H_2), -8(H_2), -9(H_2)	PI	64–75
Renibacterium	VI	—	A3α	2c	-9	PI	53–54
Rothia				2c	-7	PI	54–57
Stomatoccus				ND	ND	PI	56–60
C. *Cellulomonas*	VIII	—	A4β	2c	-9(H_4)	PV	71–76
Oerskovia	VI	—	A4α	2c	-9(H_4)	PV	70–75
Promicromonospora	VI	—	A4α	ND	-9(H_4)	PV	70–75
D. *Actinomyces*	V, VI	—	A4α, A4β	1a, 1c	-10(H_2,H_4)	PII	57–69
Arcanobacterium	VI	—	A5α	1a	-9(H_4)	ND	48–52
E. *Arachnia*	I	—	A3γ′	2c	-9(H_4)	PI	63–65
Pimelobacter	I	—	A3γ	2a	-8(H_4)	PI	69–74
F. *Brevibacterium*	III	C	A1γ	2c	-8(H_2)	PI	60–67
G. *Dermatophilus*	III	B	A1γ	1a	-8(H_4)	PI	57–59
ACTINOPLANETES							
Actinoplanes			A1γ′	2d	-9(H_4), -10(H_4)	PII	72–73
Ampullariella			A1γ	2d	-9(H_4), -10(H_4)	PII	72–73
Catellatospora			ND	ND	-9(H_4, -10(H_8)	PII	71–72
Dactylosporangium	II	D	A1γ	3b	-9(H_4,H_6, H_8)	PII	71–73
Micromonospora			A1γ′	3b	-9(H_4), -10 (H_4)	PII	71–73
Pilimelia			A1γ	2d	-9(H_2), -9(H_4)	PII	ND
MADUROMYCETES							
Actinomadura pusilla group		B, C		3c	-9(H_0,H_2,H_4)	PIV	64–69
Microbispora		B, C		3c	-9(H_6,H_2,H_4)	PIV	67–74
Microtetraspora glauca group	III	B, C	A1γ	3c	-9(H_0,H_2,H_4)	PIV	66
Planobispora		B		3c	-9(H_2,H_4)	PIV	70–71
Planomonospora		B		3c	-9(H_2)	PIV	72
Streptosporangium		B		3c	-9(H_2,H_4)	PIV	69–71
MICROPOLYSPORAS							
Actinopolyspora				2c	-9(H_4,H_6)	PIII	64
Amycolata				3e	-8(H_4)	PIII	68–72
Amycolatopsis				3f	-9(H_2,H_4)	PII	66–69
Faenia (Micropolyspora)	IV	A	A1γ	2c	-9(H_4)	PIII	66–71
Kibdelosporangium				3c	ND	PII	66
Pseudonocardia				2b	-8(H_4)	PIII	79
Saccharomonospora				2a	-9(H_4)	PII	69–74
Saccharopolyspora				2c	-9(H_4)	PIII	77
MULTILOCULAR SPORANGIA							
Frankia	III	B, C, E	ND	1	ND	PI	66–71
Geodermatophilus	III	C	A1γ	2b	-9(H_4)	PII	73–76

Table XI.1.—*continued*

Group/genus	Wall chemotype[b]	Whole-cell sugar pattern[c]	Peptido-glycan type[a]	Fatty acid pattern[e]	Major menaquinone (MK)[f]	Phospholipid type[g]	Mol% G + C of DNA
NOCARDIOFORMS							
Caseobacter				1b	-8(H_2), -9(H_2)	ND	65–67
Corynebacterium				1a	-8(H_2), -9(H_2)	PI	51–63
Mycobacterium	IV	A	A1γ	1b	-9(H_2)	PII	62–69
Nocardia				1b	-8(H_4),-9(H_2)	PII	64–72
Rhodococcus				1b	-8(H_2), -9(H_2)	PII	63–72
NOCARDIOIDES							
Nocardioides	I	—	A3γ	3a	-8(H_4)	PI	ND
STREPTOMYCETES							
Intrasporangium				1a	-8	PI	ND
Kineosporia				ND	-9(H_4)	PIII	ND
Sporichthya	I	—	A3γ	3a	-9(H_6,H_8)	ND	ND
Streptomyces				2c	-9(H_6,H_8)	PII	69–78
Streptoverticillium				2c	-9(H_6,H_8)	PII	69–73
THERMOMONOSPORAS							
Actinomadura madurae group		B	A1γ	3a	-9(H_6)	PI	66–69
Actinosynnema		C		3f	-9(H_4), -10(H_4)	PII	71–73
Microtetraspora viridis	III	C	ND	3a	-9(H_4)	PI	67
Nocardiopsis		C		3d	-10(H_2,H_4,H_6)	PIII	64–69
Saccharothrix		C		3f	-9(H_4), -10(H_4)	PII	70–76
Streptoalloteichus		C		ND	-9(H_6), -10(H_6)	ND	ND
Thermomonospora		{ C		{ 3e	{ -10(H_4,H_6)	PII	ND
		{ C		{ 3c	{ -9(H_2,H_4)	PIV	ND
OTHER GENERA							
Glycomyces	II	D	ND	2c	-9(H_4), -10(H_4)	PI	71–73
Kitasatosporia	I, III	C	ND	ND	ND	ND	66–73
Spirillospora	III	B	A1γ	3a	-9(H_4,H_6)	PI, PII	69–71
Thermoactinoyces	III	C	A1γ	2b	-7, -9	ND	53–55

[a] Data from Lechevalier and Lechevalier (1980), Goodfellow and Cross (1984b), Poschner et al. (1985), Asano and Kawamoto (1986), Goodfellow and Williams (1986), Lechevalier et al. (1986), and this volume.

[b] Major constituents in wall chemotypes: *I*, L-DAP and glycine; *II*, *meso*-DAP and glycine; *III*, *meso*-DAP; *IV*, *meso*-DAP, arabinose, and galactose; *V*, lysine and ornithine; *VI* (with variable presence of aspartic acid and galactose); *VII*, diaminobutyric acid and glycine (lysine variable); and *VIII*, ornithine. All wall preparations contain major amounts of alanine, glutamic acid, glucosamine, and muramic acid (Lechevalier and Lechevalier, 1970b, 1980.)

[c] Whole cell sugar patterns of actinomycetes containing *meso*-DAP: *A*, arabinose and galactose; *B*, madurose (3-0-methyl-D-galactose); *C*, no diagnostic sugars; *D*, arabinose and xylose; and –, not applicable (Lechevalier et al., 1971b).

[d] Numbers refer to the variation of cross-linkage. Small Greek letters mark the diversity of amino acids in position 3 of the peptide subunit, and prime (′) indicates the replacement of alanine in position 1 in group A peptidoglycans by glycine (Schleifer and Kandler, 1972; Schleifer and Stackebrandt, 1983; Schleifer and Seidl, 1985).

[e] Fatty acid patterns after Kroppenstedt (1985, unpublished data).

[f] Abbreviations exemplified by MK - 8(H_2), menaquinones having two of the eight isoprene units hydrogenated.

[g] Characteristic phospholipids: *PI*, phosphatidylglycerol (variable); *PII*, only phosphatidylethanolamine; *PIII*, phosphatidylcholine (with phosphatidylethanolamine, phosphatidylmethylethanolamine and phosphatidyglycerol variable, no phospholipids containing glucosamine); *PIV*, phospholipids containing glucosamine (with phosphatidylethanolamine and phosphatidylmethylethanolamine variable); and *PV*, phospholipids containing glucosamine and phosphatidylglycerol; all preparations contain phosphatidylinositol (Lechevalier et al., 1977, 1981).

THE ACTINOBACTERIA

The actinobacteria include several taxa that were considered with the irregular, nonsporing Gram-positive rods in Volume 2 of the *Manual* (Jones and Collins, 1986). The group is clearly heterogeneous because it encompasses a range of chemically, morphologically, and physiologically different organisms, many of which have been classified as coryneform bacteria (Bousfield and Callely, 1978). The group accommodates organisms that form cocci (*Micrococcus*), short rods (*Renibacterium*), irregular rods (*Brevibacterium, Cellulomonas*), and mycelia that fragment (*Actinomyces, Oerskovia,* and *Rothia*). It also contains both motile (*Cellulomonas, Curtobacterium,* and *Oerskovia*) and non-motile (*Actinomyces, Agromyces, Arachnia,* and *Rothia*) organisms. Some taxa are fermentative (*Actinomyces* and *Arachnia*), but most have an oxidative metabolism (*Agromyces, Arthrobacter, Brevibacterium,* and *Curtobacterium*). Lysine is the most common diamino acid in the wall peptidoglycan (*Arthrobacter, Oerskovia, Renibacterium,* and *Stomatococcus*), but ornithine (*Cellulomonas* and *Curtobacterium*), diaminobutyric acid (*Agromyces*), and both the L- (*Arachnia*) and the *meso-* (*Brevibacterium*) forms of diaminopimelic acid occur. Indeed, actinobacteria exhibit eight different peptidoglycan types (Table XI.1). Most strains contain hydrogenated menaquinones, but some have unsaturated components as the major isoprenolog. The first group includes *Actinomyces* (MK-10 [H_2,H_4]), *Arthrobacter* (MK-9[H_2]), *Brevibacterium* (MK-8[H_2]), *Cellulomonas* (MK-9[H_4]), and *Dermatophilus* (MK-8[H_4]), and the second *Agromyces* (MK-12), *Clavibacter* (MK-9, -10), *Microbacterium* (MK-11, -12), and *Rothia* (MK-7).

Molecular systematic information, where available, supports the recognition of subgroups within the actinobacterial line. *Agromyces, Aureobacterium, Clavibacter, Curtobacterium,* and *Microbacterium* together form one such subgroup (Stackebrandt et al., 1980b; Döpfer et al., 1982), and a second grouping comprises the genera *Cellulomonas, Oerskovia,* and *Promicromonospora* (Stackebrandt et al., 1980a). Members of the *Microbacterium* subgroup have the rare group B type peptidoglycan, a type I phospholipid pattern, unsaturated menaquinones, major amounts of *anteiso*-methyl branched-chain fatty acids, and DNA with a high mol% G + C value (Table XI.1). Lipid (Collins et al., 1979; Minnikin et al., 1979), numerical phenetic (Jones, 1975), and nucleic acid pairing (Stackebrandt and Kandler, 1979; Stackebrandt et al., 1980a) data underpin the affinity found between *Cellulomonas, Oerskovia,* and *Promicromonospora*; indeed, Stackebrandt et al. (1982a) proposed the union of the first two of these genera. However, differences in morphology, growth habits, wall composition, fatty acid profiles, and phage host ranges support the integrity of the genera *Cellulomonas* and *Oerskovia* (Prauser, 1986). *Listeria denitrificians* (Seeliger and Jones, 1986) resembles the *Cellulomonas* subgroup in chemical composition (Collins et al., 1983a).

Members of the strictly aerobic genus *Arthrobacter* (*sensu* Keddie et al., 1986) exhibit a marked rod-coccus growth cycle. Chemical (Collins and Kroppenstedt, 1983), nucleic acid pairing (Schleifer and Lang, 1980), and 16S rRNA (Stackebrandt and Woese, 1979, 1981b; Stackebrandt et al., 1980b) studies show that a close relationship exists between true arthrobacters and some micrococci, notably *Micrococcus luteus*, the type species of the genus (Alderson, 1985). Indeed, these micrococci have been considered as degenerate forms of arthrobacters "locked into" a coccoid stage of the arthrobacter life cycle (Stackebrandt and Woese, 1979, 1981a; Stackebrandt et al., 1980b). *Stomatococcus mucilaginosus* (formerly *Micrococcus mucilaginosus*) also belongs to the *Arthrobacter* subgroup (Stackebrandt et al., 1983d). The higher taxonomic affinities of *Renibacterium* and *Rothia* have still to be established, but these taxa display some resemblance to true arthrobacters, notably in chemical features (Table XI.1)

The taxonomic position of bacteria currently designated *Arthrobacter simplex* and *Arthrobacter tumescens* remains unresolved, although it is clear that they do not belong to the genus *Arthrobacter* (Keddie et al., 1986). Indeed, 16S rRNA sequencing studies have shown that *A. simplex* is only distantly related to *A. globiformis*, the type species of the genus *Arthrobacter* (Stackebrandt et al., 1980b). It has been suggested, mainly in the light of lipid data, that *A. tumescens* be reclassified as *Nocardioides simplex* (O'Donnell et al., 1982a), whereas Suzuki and Komagata (1983) proposed that *A. simplex* together with *A. tumescens* be transferred to the new genus *Pimelobacter*. However, the inclusion of *A. tumescens* in the same genus as *A. simplex* is at variance with physiological and chemical data (O'Donnell et al., 1982a; Collins et al., 1983b). *Arthrobacter tumescens* is unusual in containing substantial amounts (~ 20%) of monounsaturated terminally branched long-chain fatty acids (Collins et al., 1983b). A similar fatty acid composition has been detected in *Intrasporangium calvum* (Jones and Collins, 1986). It is clear that further studies are needed to clarify the taxonomic position of these L-DAP-containing actinomycetes.

The suprageneric relationship of the monospecific genus *Arachnia* to other taxa containing an A3γ peptidoglycan has yet to be established. Indeed, *Arachnia* is often assumed to occupy an intermediate position between *Actinomyces* and *Propionibacterium* because it has similarities in morphology and pathogenicity to the former and shares physiological and biochemical features with the latter (Schaal, 1986b). *Arachnia propionica* is found mainly in the oral cavity of man and has been shown to be serologically related to the human propionibacteria. It also resembles propionibacteria in menaquinone and fatty acid composition (O'Donnell et al., 1985). The detection of predominantly methyl branched fatty acids in arachniae is at variance with the report (Amdur et al., 1978) that these organisms contain major amounts of straight-chain and monounsaturated long-chain fatty acids. However, low DNA/DNA homologies (Johnson and Cummins, 1972), an inability to produce catalase (Pine and Georg, 1974), and differences in detailed peptidoglycan structure (Schleifer and Kandler, 1972) do not support an especially close relationship between arachniae and propionibacteria. Numerical phenetic data suggest that *Arachnia* has more in common with *Actinomyces* than with *Propionibacterium* (Schofield and Schaal, 1981).

The genus *Actinomyces* is relatively well circumscribed and currently contains 10 species, all of which are associated with, or cause, disease in animals, including man (Schaal, 1986a). *Arcanobacterium haemolyticium* (formerly *Corynebacterium haemolyticum*), also isolated from animal and human sources, has a similar peptidoglycan to that found in *Actinomyces* (Table XI.1). It also resembles *Actinomyces* in possessing predominantly straight-chain saturated and monounsaturated long-chain fatty acids. The two genera can, however, be separated by menaquinone and DNA base composition. Precise knowledge of the suprageneric relationships of the two genera will require a comparative study of the 16S rRNA sequences of representative strains. The same can be said for the genera *Brevibacterium* and *Dermatophilus*.

MICROPOLYSPORAS, THERMOMONOSPORAS, AND ACTINOMYCETES WITH MULTILOCULAR SPORANGIA

At present, it is not clear whether all of the genera in the micropolyspora aggregate group merit generic status or if they collectively form a distinct branch in the actinomycete line, but they do have many properties in common (Goodfellow and Cross, 1984b; Goodfellow and Minnikin, 1985). They are aerobic, Gram-positive, non-acid-fast sporoactinomycetes that are morphologically somewhat heterogeneous. Thus, single or short chains of spores may be borne either on the aerial mycelium or on both the aerial and substrate mycelium. Furthermore, in addition to having a wall chemotype IV, micropolysporas have major amounts of *iso-* and *anteiso*-methyl branched fatty acids and tetrahydrogenated menaquinones with either eight or nine isoprene units as major isoprenologs. However, they also show variation in polar lipid and fatty acid composition that may be of taxonomic value (Table XI.1). *Actinopolyspora, Amycolata, Faenia, Pseudonocardia thermophila,* and *Saccharopolyspora* contain phosphatidylcholine, whereas *Amycolatopsis, Kibdelosporangium, Pseudonocardia azurae,* and *Saccharomonospora* do not (Embley et al., 1986). Similarly, 10-methyloctadecanoic (tuberculostearic) acid and lower homologs occur in *Amycolata, Faenia, Pseudonocardia,* and *Saccharopolyspora* but not in *Actinopolyspora, Amycolatopsis,* or *Saccharomonospora* (Embley et al., 1987).

The thermomonosporas encompass an apparently varied assortment of aerobic, spore-forming actinomycetes that produce branched vegetative mycelium bearing aerial hyphae. Most strains are mesophilic, but some are thermophilic. Thermomonosporas have a wall chemotype III, that is, they have *meso*-DAP as the diamino acid of the peptidoglycan and whole cell hydrolysates lacking diagnostic sugars. They also have hydrogenated menaquinones with nine or 10 isoprene units. Indeed, *Thermomonospora* and *Nocardiopsis* differ from most other actinomycetes in containing menaquinones with unusually long, partially saturated isoprenyl side chains (Table XI.1; Collins et al., 1982d). Organisms currently assigned to the group show extensive variation both in morphology and in polar lipid type (Table XI.1). Few comparative studies have been carried out to determine the natural relationships of these organisms, but *Actinomadura madurae, Actinomadura verrucosospora* and *Thermomonospora curvata* do share high S_{AB} values (Stackebrandt and Schleifer, 1984; Fowler et al., 1985) and *Actinomadura madurae* and *Microtetraspora viridis* belong to the same rRNA homology group (Stackebrandt et al., 1980b). Additional studies are also needed to determine the suprageneric affinities of *Frankia* and *Geodermatophilus.* These organisms are aerobic, mesophilic actinomycetes that produce multilocular sporangia and have a wall chemotype III. The two taxa can be distinguished by differences in fatty acid and polar lipid composition (Table XI.1). *Frankia* is also noted for its ability to fix nitrogen in vivo and in vitro.

FUTURE TRENDS

Further insights into the suprageneric relationships of actinomycetes will come from comparisons of sequences of 5S rRNA and homologous proteins, in vitro hybridization of nucleic acids, chemometrics, and sequencing of homologous proteins and from parallel developments in data-handling procedures. It is vitally important that the results of such studies be reconciled with existing taxonomies, and this will not always be easy. Even so, generic groups will undoubtedly be formally equated with higher ranks in the taxonomic hierarchy, although care must be taken to follow the International Code of Nomenclature of Bacteria. The delineation of natural groups at and above the genus level will allow new taxonomic data to be interpreted within a context, which will facilitate the discovery of markers for the description of families and genera. Improvements in actinomycete classification will also help to pave the way for the introduction of recommended minimum standards for the recognition of genera and species. There is no doubt that advances in actinomycete systematics will lead to dramatic changes in the classification of the order *Actinomycetales* in future editions of the *Manual.*

THE ACTINOMYCETES II

Growth and Examination of Actinomycetes—Some Guidelines

Tom Cross

Many bacteriologists encounter difficulties when first faced with the task of growing, subculturing, and examining actinomycetes. Familiarity and some years of practice with cultures of *Escherichia coli* or *Bacillus subtilis* do not prepare the laboratory worker for some of the problems that will be encountered, and a mycologist can often find the task slightly easier. The aim here is therefore to suggest some **simple methods and guidelines that will enable the bacteriologist to begin the study of an actinomycete.** It is hoped that experience and confidence will be gained, and perhaps the breathing space to consult papers and articles for the specific methods and media required.

The term "actinomycete" now encompasses a wide range of bacteria. Some workers would include genera such as *Micrococcus, Arthrobacter,* and *Cellulomonas*, which can be handled in the same way as most other oxidative, Gram positive bacteria. However, one must remember that the cells of two of these genera, *Arthrobacter* and *Cellulomonas*, can show **characteristic morphological variation** during exponential and stationary-phase growth in broth culture. *Cellulomonas* and *Oerskovia* strains appear as short motile rods in broth cultures but grow as feathery colonies composed of radiating, branching filaments when cultured on dilute, nutrient-deficient media solidified with agar. One must examine possible actinomycetes growing in broth after incubation for various times, by the phase-contrast microscopy of wet mounts as well as high-magnification observation of stained smears, and examine the appearance of colonies growing on agar media.

Some actinomycete genera are **anaerobic** (e.g. *Actinomyces*) or require very **specialized growth media and incubation conditions** (e.g. *Frankia*). The investigator must consult the relevant sections in this and previous volumes and the specialist papers quoted for specific information on these organisms.

It is the common sporoactinomycetes, bacteria with branching hyphae and specialized spore-bearing structures, and the nocardioform actinomycetes that are now attracting considerable interest from biotechnologists, geneticists, and ecologists. Strains of the genera *Streptomyces* and *Thermomonospora* (sporoactinomycetes; Fig. XII.1), *Nocardia* and *Rhodococcus* (nocardioforms, Fig. XII.2), and *Actinoplanes* and *Dactylosporangium* (actinoplanetes, Fig. XII.3), for example, have been studied intensively by a relatively small group of microbiologists specifically interested in their taxonomy and ability to produce secondary metabolites. There is an extensive but very scattered literature on methods that should be used, but unfortunately it will not be found in the commonly available methods manuals. The above genera will be used as examples of actinomycetes exhibiting differentiation when discussing methods.

Many actinomycetes will grow on the common bacteriological media used in the laboratory, such as nutrient agar, Trypticase soy agar, blood agar, and even brain-heart infusion agar. Indeed, suspected clinical isolates of *Nocardia asteroides* and *Streptomyces somaliensis* should be first cultivated on a rich nutritive medium, but then one should expect to obtain what must be regarded as atypical growth. Tough, leathery colonies will appear, but they usually lack any aerial mycelium so typical of *Streptomyces* strains and it will be impossible to see the spore clusters embedded in the colony that are characteristic of micromonosporae.

Sporoactinomycetes require special media to allow differentiation, and the development of characteristic spores and pigments. Some of these media are not available from commercial suppliers and have to be prepared in the laboratory. Those containing colloidal chitin, soil extract, or decoctions of plant materials require lengthy preparation times but are extremely useful. For example, the pale, shiny, hard colonies of a *Streptomyces* species on nutrient agar can be transformed into bright yellow colonies with a powdery white aerial mycelium and spirals of arthrospores when the organism is subcultured onto a more suitable growth medium such as oatmeal or inorganic salts starch agars. Other species grow thinly but sporulate profusely on tap water agar, where they use the polysaccharide as a carbon source, or on water agar containing trace amounts of yeast extract or peptone.

Outgrowths from a spore or fragments of mycelium (colony-forming units, CFUs), develop into hyphae that penetrate the agar (**substrate mycelium**) and hyphae that branch repeatedly and become cemented together on the surface of the agar to form a **tough, leathery colony**. The density and consistency of the colony will depend on the composition of the medium. Nocardioform actinomycetes exhibit **fragmentation**; the hyphae break up into rods and cocci, thus leading to soft or friable colonies. In strains of certain genera (e.g. *Streptomyces*), the colony becomes covered with **aerial mycelium**: free, erect hyphae surrounded by a hydrophobic sheath that grow into the air away from the colony. These hyphae are initially white but assume a range of colors when spore formation begins. Colonies then appear powdery or velvety and can then be readily distinguished from the more typical bacterial colonies.

Growth can be slow. A branching mycelium growing at the surface of transparent agar can be seen with the aid of a microscope after 24 h, and visible colonies may appear in 3–4 days, but mature aerial mycelium with spores may take 7–14 days to develop, and some very slow growing strains may require up to 1 month's incubation. Lengthy incubation times can result in the evaporation of the medium, so thick agar plates are required. Thermophilic species incubated at 45–55°C require a humid incubator.

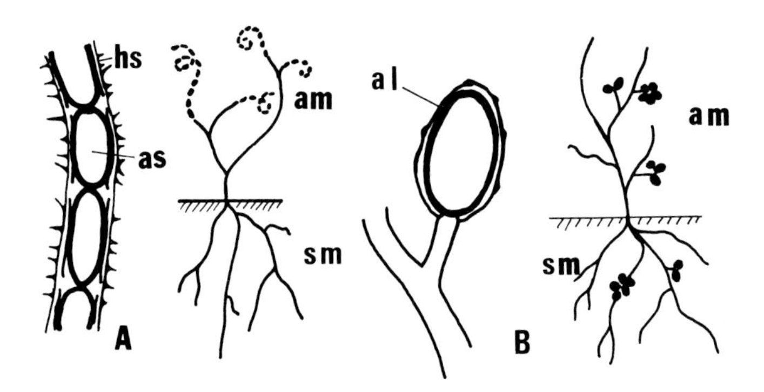

Figure XII.1. Sporoactinomycetes. *A*, *Streptomyces* species have chains of spores on the aerial mycelium (*am*), which are normally absent from the substrate mycelium (*sm*). These spores are arthrospores (*as*), regular segments of hyphae with a thickened spore wall surrounded by a hydrophobic sheath (*hs*) that may bear spines or hairs. *B*, *Thermomonospora* species can have clusters of single spores, aleuriospores (*al*), on both aerial and substrate mycelium.

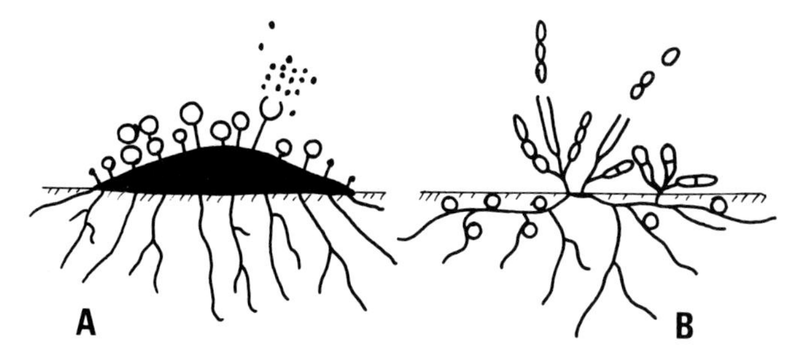

Figure XII.3. Actinoplanetes. *A*, *Actinoplanes* species form globose sporangia on the surface of the colony that liberate motile zoospores. *B*, *Dactylosporangium* species have tubular sporangia containing few potentially motile spores together with aleuriospores (globose bodies) on the substrate mycelium.

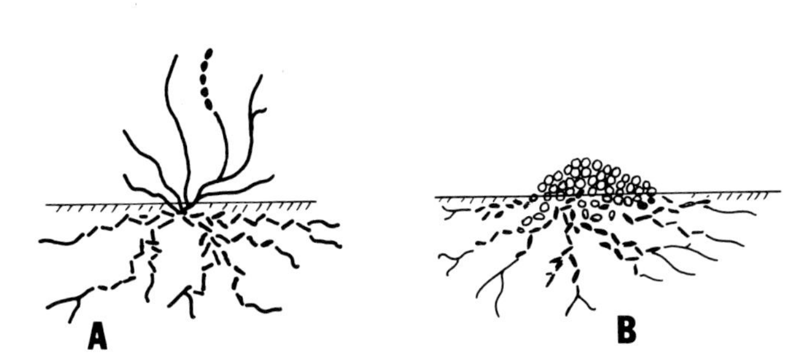

Figure XII.2. Nocardioform actinomycetes. *A*, *Nocardia* species with a fragmenting substrate mycelium and limited aerial mycelium that can bear chains of arthrospores. *B*, *Rhodococcus* species have a rapidly fragmenting substrate mycelium; the segments become rounded and the colony usually consists of a mass of coccoid elements.

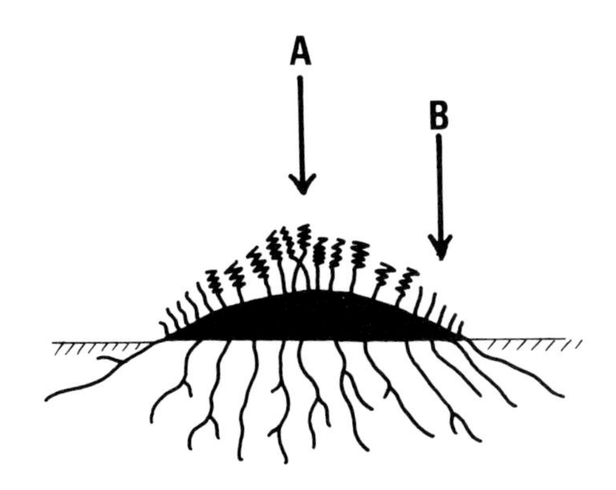

Figure XII.4. The mature, spore-bearing hyphae in the center of the colony (*A*) cannot be seen with a transmitted light microscope because of the density of the colony. Only young, immature aerial hyphae occur at the margins of the colony (*B*) where adequate light levels are possible. Actinomycete hyphae are too small to be seen with a stereomicroscope.

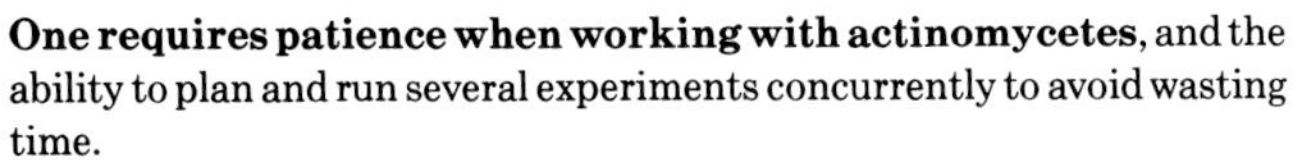

One requires patience when working with actinomycetes, and the ability to plan and run several experiments concurrently to avoid wasting time.

The **morphology** of an actinomycete growing on agar can provide useful and rapid clues to its identity, but viewing isolated colonies can give little worthwhile information (Fig. XII.4). Examine the organism streaked in a cross-hatched pattern on the surface of the agar (Fig. XII.5), first with a stereomicroscope and then with a transmitted light microscope with a × 40 long working distance objective to avoid water condensation on the front lens. The appearance of hyphae within the agar (presence of spores, fragmentation, etc.) and the nature of the spores on the aerial hyphae are important but can only be observed when growth is thin and the medium promotes differentiation.

For detailed light microscope studies on sporulating structures and spore arrangements in some genera such as *Microbispora*, *Microtetraspora*, and *Saccharomonospora* it is advisable to use **slide cultures** (Fig. XII.6) incubated in a moist chamber. The thin squares of nutritive agar or agarose beneath a coverslip can be examined under high magnification and transmitted light not obscured by heavy vegetative growth. Mycelium adhering to coverslips placed at an angle in growing cultures (Fig. XII.7) can be transferred to a slide and also examined at high magnifications. For **scanning electron microscopy**, an agar block carrying a

Figure XII.5. Cross-hatch streak plate. View directly under microscope fitted with a long working distance objective (× 25 and × 40). Mature hyphae with spores should be looked for in the angles of the streaks.

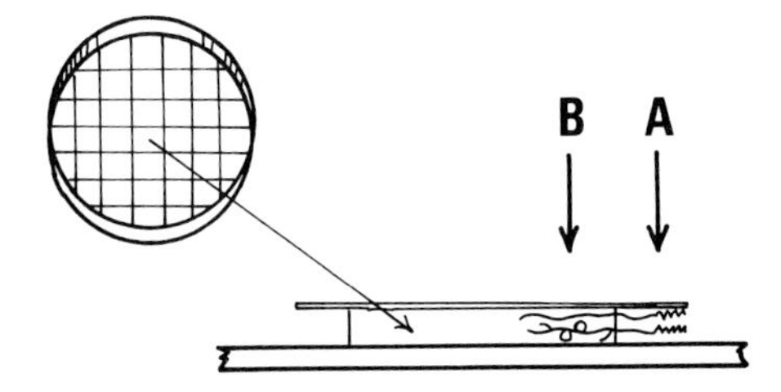

Figure XII.6. Slide culture. Thin agar block, cut from poured plate, is placed on a sterile microscope slide and inoculated, and a sterile coverslip is applied. After incubation in a moist chamber, view slide culture directly on microscope stage when it should be possible to see the aerial mycelium (*A*) and the substrate mycelium (*B*) within the agar.

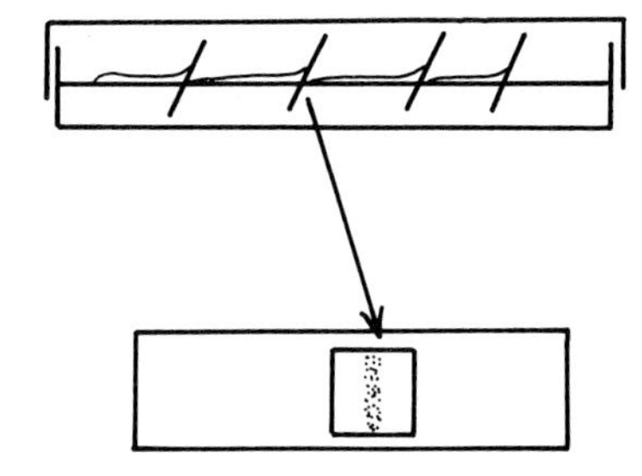

Figure XII.7. Inclined coverslips for observing actinomycete morphology. Inoculate agar plate with coverslips inserted at an angle. After incubation withdraw coverslips and mount, upper surface down, in water containing wetting agent.

sporulating colony should be fixed in osmium tetroxide, gradually dehydrated in alcohol before being subjected to critical point drying, and coating with gold. Sections for **transmission electron microscopy** can also be prepared from agar blocks or from growth scraped off the surface of cellophane strips laid on the surface of agar.

The saprophytic actinomycetes are oxidative and may grow poorly when the air supply is restricted. The screw caps of slope cultures must be loose to allow adequate aeration, and some strains show better growth in tubes with cotton-wool plugs or spring caps that allow better gas exchange. Actinomycetes will grow at a lower a_w than most Gram-negative bacteria, enabling one to use 1.5–2.0% (weight/volume) agar in media and up to 3.0% (w/v) for strains of some genera. This may sound extravagant but can be most useful when one wishes to remove the very hard growth of a strain reluctant to produce aerial mycelium and spores. A rough, stiff loop is then an essential tool for abrading the colony and collecting sufficient mycelial fragments for an efficient transfer.

Actinomycetes will grow in broth but need to be cultivated under specialized conditions. Growth of a streptomycete in a stationary broth tube is usually restricted to a surface pellicle and perhaps a cottony sediment, leaving the broth quite clear. Liquid cultures require considerable **aeration and agitation** to give the even, suspended growth required for most chemotaxonomic studies. Tubes and flasks must be incubated on a shaker at high speeds (e.g. 200–250 rpm) to give the supply of oxygen and mixing necessary for maximum growth. Even, diffuse mycelial growth may require the higher agitation and mixing rates achieved by internal baffles or springs within the flask, but in turn this will only be attained with an **adequate inoculum**. Unlike a typical dividing bacterium that will give a smooth suspension of cells on continued incubation, an actinomycete spore or piece of mycelium will germinate, elongate, intertwine, and grow as a pellet. **High numbers of growth sources** (spores or hyphal fragments) must be introduced into a flask or fermenter in order to achieve heavy, dispersed growth. Hence, there is an initial need for a good sporulation medium or, failing that, a method that will allow homogenization of vegetative mycelium to give many growth sources.

Spore suspensions can be prepared by detaching the spores from aerial hyphae with a loop or scraper and placing them in a suspending medium containing a wetting agent. The arthrospores of streptomycetes are hydrophobic because of the enveloping sheath, and the wetting agent aids their even suspension in the diluent. Free spores may be removed from lawns of aerial mycelium by rolling glass beads or agar cylinders over the surface. Cultures of actinoplanetes bearing sporangia should be flooded with diluent and allowed to stand at room temperature for at least 30 min to permit spore hydration and maturation before the surface is scraped with a loop.

Pure cultures are of vital importance for all taxonomic purposes. **Checks for purity** should be undertaken frequently. The filamentous growth in broths contained in shake flasks or tubes may be a **fungal contaminant** that would be revealed by examining a sample dispersed in lactophenol-cotton blue. Nonactinomycete, **contaminating bacteria** may be hidden within the sporulating mycelial growth on agar only to outgrow the actinomycete in shake flasks. The actinomycete colonies on plate or slope cultures should be examined carefully with a hand lens, and subcultures should be spread on rich agar media or transferred to a broth more suitable for the growth of nonactinomycete contaminants.

Actinomycete contaminants are also a problem. The dry, hydrophobic spores of streptomycetes can be dispersed within a laboratory or sterile room, thus having the potential for contaminating other actinomycetes—an association that is more difficult to detect. One must then streak out to obtain isolated colonies and examine carefully for differences in colony color and size. Finally, in laboratories where soil and litter are being sampled, one must always be on the lookout for **mites** that can wander from plate to plate and among incubators, causing havoc and ruining many experiments.

Morphological characters are still widely used for characterizing genera, for example, the presence or absence of spores on the substrate mycelium, or the formation of zoospores in specialized spore vesicles or sporangia. The spores of *Streptosporangium* species are formed within a sporangium but are nonmotile, whereas those of *Actinoplanes* and *Spirillospora* developing within similar-shaped sporangia are motile. They may not exhibit movement when first released, and it may take 20–60 min suspension in water, or an amino acid solution, or dilute broth plus a wetting agent to induce motility. The ability to produce **motile spores** is more widespread in the actinomycetes than was previously suspected.

Preservation of both sporing and nonsporulating actinomycetes can be achieved by freeze-drying or storage in liquid nitrogen. Freezing suspensions in 20% (v/v) glycerol at −20° to −40°C has proved to be a very useful method in a busy laboratory, and one can even preserve a vegetative inoculum for flasks by freezing a glycerol suspension.

In summary, the following points should be emphasized:

1. Grow actinomycetes on suitable media.
2. Provide sufficient aeration for growth of oxidative species in broth cultures—this means vigorous agitation.
3. All subcultures and particularly broth cultures must receive an adequate inoculum.
4. Morphological examination by the traditional bacteriological stained smear method will invariably give misleading or incomplete information.
5. Examine organisms growing on agar.
6. Be patient!

Most of the methods briefly mentioned in this guide will have to be supplemented with details from specialist papers, but the general books, chapters, and papers listed below should prove useful:

Further Reading

Dietz, A. and D. W. Thayer (Editors). 1980. Actinomycete Taxonomy. Society for Industrial Microbiology, Arlington, Virginia, Special Publication No. 6.
Goodfellow, M. and D. E. Minnikin (Editors). 1984. Chemical Techniques for Bacterial Systematics, Academic Press, London.
Starr, M. P., H. Stolp, H. G. Trüper, A. Balows and H. G. Schlegel (Editors). 1981. Section U, The Actinomycetes. *In* The Prokaryotes, a Handbook on Habitats, Isolation and Identification of Bacteria, Vol II. Springer-Verlag, Berlin, pp. 1915–2123.
Waksman, S. A. 1961. The Actinomycetes, Vol. 2. Classification, Identification and Description of Genera and Species. Williams & Wilkins, Baltimore.
Williams, S. T. and T. Cross. 1971. Actinomycetes. *In* Booth (Editor), Methods in Microbiology, Vol. 4. Academic Press, London, pp. 295–334.

THE ACTINOMYCETES III

A Practical Guide to Generic Identification of Actinomycetes

Hubert A. Lechevalier

This volume of *Bergey's Manual® of Systematic Bacteriology* contains the organisms that are the most filamentous of all bacteria. Some approach, and in some cases may equal, the morphological complexity of the imperfect fungi with which they have been confused in the past. They are aerobic Gram-positive bacteria that form branching filaments or hyphae that may persist as a stable mycelium or may break up into rod-shaped or coccoid elements. Motility, when present, is due to flagellation. The genera in this volume are divided into eight sections. The purpose of this introduction is to guide the user to the most appropriate section(s) and genera.

MORPHOLOGY

Nothing beats morphology for simplicity, but even with considerable experience generic identification based on morphology alone is rarely secure. The tools of the actinomycetic morphologist are a brightfield microscope equipped with a long working distance condenser and objectives. These permit one to observe undisturbed cultures grown in petri dishes. The best media for the observation of the morphology of actinomycetes are lean media such as water agar with traces of nutrients such as casein hydrolysate, yeast extract, or starch. For certain features, such as the determination of the nature of the surfaces of spores, the use of an electron microscope is essential. Morphological features include the following:

1. **Mycelium**, which may be stable or fugacious. If it breaks down, one should observe the shape of the elements and their motility (*Oerskovia* spp. release flagellate elements—Section 26). One should note if both a substrate and an aerial mycelium are formed, or only a substrate mycelium (very common), or only aerial hyphae (extremely rare—*Sporichthya*, Section 29). The mycelium may bear intercalary vesicles that do not contain spores (*Intrasporangium*, Section 26) or contain many spores (*Frankia*, Section 27).
2. **Conidia**, a general term used to refer to any asexual spores that are not intercalary chlamydospores or sporangiospores. Actinomycetes form conidia in a variety of ways.
   a. *Single conidia* are found in several genera. In the genus *Thermoactinomyces* (Section 32) these are bacterial endospores that can be recognized by their thermostability.
      Nonthermostable single conidia are found in the genera *Saccharomonospora* and *Promicromonospora* (Section 26), *Micromonospora* (Section 28), and *Thermomonospora* (Section 31). In addition, members of the genera *Frankia, Dactylosporangium,* and *Intrasporangium* may form terminal vesicles that may be confused with spores. Also many different organisms, such as actinomadurae, will form single vesicles when grown under adverse conditions.
   b. *Pairs of conidia.* Longitudinal pairs of conidia are the key characteristic of the genus *Microbispora* (Section 30) when formed only on the aerial mycelium. Some *Faenia* spp. (Section 26) may form such structures on both aerial and substrate mycelia.
   c. *Short chains of conidia.* Although it is difficult to say how long a short chain of conidia can be, chains of up to 20 spores are usually considered to be "short." Representatives of the following genera may form such chains: *Nocardia, Pseudonocardia, Faenia,* and *Saccharomonospora* (Section 26); *Streptoverticillium* and *Sporichthya* (Section 29); *Actinomadura* and *Microtetraspora* (Section 30); *Streptoalloteichus* (Section 31); and *Glycomyces* (Section 33). This type of morphology may also be encountered in the genera *Amycolata* and *Amycolatopsis* (Lechevalier et al., 1986) and *Catellatospora* (Asano and Kawamoto, 1986), which are not included in the *Manual.* Some of the streptomycetes of the *Microellobosporia* type form extremely short chains of spores, which are surrounded by an envelope that can be seen even by light microscopy.
   d. *Long chains of conidia* are formed by strains belonging to various genera. These may include *Nocardia, Nocardioides, Pseudonocardia, Saccharopolyspora,* and *Actinopolyspora* (Section 26); *Streptomyces* and *Streptoverticillium* (Section 29); *Actinosynnema, Nocardiopsis,* and *Streptoalloteichus* (Section 31); *Kibdelosporangium, Kitasatosporia, Glycomyces,* and *Saccharothrix* (Section 33); and *Amycolatopsis* (Lechevalier et al., 1986).
   a. The conidia-bearing hyphae may be united into synnemata releasing motile spores (*Actinosynnema*, Section 31).
3. **Sporangia** are bags that contain spores. These may be borne (a) on well-developed aerial hyphae or on the surface of colonies with

little or no aerial hyphae: *Actinoplanes, Ampullariella, Pilimelia, Dactylosporangium* (Section 28); *Planobispora, Planomonospora, Spirillospora, Streptosporangium* (Section 30), or (b) mainly within the agar (*Kineosporia*, Section 29).

4. **Other structures.** Some actinomycetes form unusual structures. Already mentioned are the spore-bearing synnemata found in the genus *Actinosynnema*. Organisms grouped in Section 27 form masses of spores that are the result of division in several planes, rather than division perpendicularly to the axis of the hyphae. These spore-bearing structures are called multilocular sporangia.

Many actinomycetes will form spherical structures on their aerial hyphea. These may be nothing more than drops of condensed water that enclose a curled chain of spores, or these structures may contain hyphae embedded in an amorphous matrix (*Kibdelosporangium*, Section 33).

Sclerotia are globose structures formed by some of the streptomycetes of the *Chainia* type (Section 29). These sclerotia contain not spores but cells filled with lipids. They germinate as a whole as do the pseudosporangia of kibdelosporangia.

CELL CHEMISTRY

Lechevalier and Lechevalier (1965) pointed out that actinomycetes can be separated into broad groups on the basis of morphological and chemical criteria. This approach is still the simplest and best for generic identification.

In order to reach the pages of this volume that are most likely to be of interest in the generic identification of a specific actinomycete, two chemical pieces of information are most useful: (a) which diabasic amino acid is present in the cell wall of the unknown, and (b) which diagnostic sugars are present in its whole cell hydrolysate. In both cases, necessary basic information can be obtained using one-way paper chromatography of whole cell hydrolysates (Lechevalier and Lechevalier, 1980). Such a simple method will reveal if the unknown contains no diaminopimelic acid (DAP) or the *meso*- or the L- form of this dibasic acid.

If no DAP is present, two main generic possibilities exist among the organisms covered in this volume: *Oerskovia* and *Promicromonospora* (Section 26). If the unknown does not fit in these two genera and is not an anomolous *Actinoplanes* or *Actinomadura*, an examination of organisms as described in Volume 2 of this *Manual* is recommended.

If the L isomer of DAP (cell wall type I) is present as a major constituent, the number of generic possibilities among the organisms treated in this volume is still limited: *Nocardioides* and *Intrasporangium* (Section 26), all the genera treated in Section 29, and *Kitasatosporia* (Section 33). Also to keep in mind are members of some genera included in Volume 2 (*Arachnia*, Section 15).

Although the number of genera characterized by the presence of cell walls of type I is small, the number of species may be large. Unfortunately, species are still poorly defined in the genus *Streptomyces*, and even with numerical taxonomy identification to species may become a nightmare.

If the organism to be identified contains major amounts of *meso*-DAP, many generic possibilities exist, and a knowledge of its whole cell sugar pattern (WCSP) is essential.

The presence of xylose and arabinose (WCSP of type D) will usually indicate a type II cell wall (*meso*-DAP + glycine) and the organism, if described in this volume, should be in Section 28, or in Section 33 (*Glycomyces*). Also to be considered is *Catellatospora* (Asano and Kawamoto, 1986). The sole exception known to this rule is *Frankia*, where a WCSP of type D corresponds to a cell wall composition of type III.

The presence of madurose (3-*O*-methyl-D-galactose, WCSP of type B) will indicate a cell wall of type III of the *Actinomadura* variety and will lead to Section 30 or, if multilocular sporangia are formed, to some strains of *Frankia* and to the genus *Dermatophilus* (Section 27). The presence of rhamnose without other diagnostic sugars indicates a possible *Saccharothrix*.

Organisms containing fucose in their WCSP may belong to the genus *Frankia*, or to *Actinoplanes*. Some plant pathogenic corynebacteria also contain fucose.

Arabinose and galactose present in the whole cell hydrolysates (WCSP of type A, cell wall type IV) will point to many of the organisms discussed in Section 26, or to the mycobacteria and the corynebacteria covered in Volume 2. In addition, species of *Kibdelosporangium* (Section 33), *Amycolata*, and *Amycolatopsis* (Lechevalier et al., 1986) also have cell walls of type IV.

The large group of organisms with type IV cell walls can be separated into (a) those with mycolic acids such as the corynebacteria, the mycobacteria, and the nocardiae and (b) those without mycolic acids (*Faenia, Pseudonocardia, Saccharomonospora, Saccharopolyspora, Actinopolyspora, Amycolata, Amycolatopsis,* and *Kibdelosporangium*).

On the basis of their molecular weights, the mycolates can be separated into the small corynomycolates, the medium nocardomycolates, and the large mycolates of mycobacteria (Lechevalier, 1982). This volume includes only organisms with nocardomycolic acids (*Nocardia, Rhodococcus*) or those that are mycolateless and belong to the genera listed above.

The absence of any of the above-mentioned sugars in the whole cell hydrolysates is an indication of cell walls of type III without madurose (WCSP of type C). Organisms with this type of chemistry are very diverse and include some with multilocular sporangia (*Geodermatophilus*, Section 27), the organisms covered in Sections 31 and 32, and two of the four genera treated in Section 33 (*Saccharothrix* and *Kitasatosporia*). Strains of *Kitasatosporia* are anomalous in the sense that their mycelium has cell walls of type III but the cell walls of their spores are of type I.

The information discussed above is summarized in Table XIII.1.

Table XIII.1.
Guide to the chemical and morphological properties of genera of actinomycetes

Diagnostic amino acid	Diagnostic sugar	Typical key morphological features and additional chemical properties	Possible generic assignment (section number)
No DAP[a]	NA[b]	Only substrate mycelium formed. Breaks into motile elements.	*Oerskovia* (26)
		Sterile aerial mycelium formed. Substrate mycelium breaking up into nonmotile elements.	*Promicromonospora* (26)
	Xylose	Sporangia with motile spores.	*Actinoplanes* (28)
	Madurose	Short chains of conidia on the aerial mycelium.	*Actinomadura* (30)
L-DAP	NA	Both aerial and substrate mycelia breaking up into fragments.	*Nocardioides* (26)
		Only substrate mycelium formed bearing terminal or subterminal vesicles.	*Intrasporangium* (26)
		Aerial mycelium with long chains of spores.	*Streptomyces* (29) *Kitasatosporia* (33)
		Sclerotia formed (*Chainia* type).	*Streptomyces* (29)
		Very short chains of large conidia formed (*Microellobosporia* type) on aerial and vegetative mycelium.	*Streptomyces* (29)
		Whorls of straight chains of conidia formed.	*Streptoverticillium* (29)
		No aerial mycelium; club-shaped sporangia formed terminally on the vegetative mycelium.	*Kineosporia* (29)
		Aerial mycelium only, motile elements formed.	*Sporichthya* (29)
meso-DAP[c]	Xylose and arabinose	No sporangia. Single conidia formed on substrate mycelia, often in large black mucoid masses.	*Micromonospora* (28)
		No sporangia, short chains of conidia formed protruding from the surface of the colonies.	*Catellatospora*[d]
		Chains of conidia on aerial mycelium.	*Glycomyces* (33)
		Dactyloid oligosporic sporangia protruding from the surface of the colonies. Spores motile.	*Dactylosporangium* (28)
		Sporangia containing spherical motile spores formed on the surface of colonies.	*Actinoplanes* (28)
		Same. Rod-shaped sporangiospores motile by polar flagella.	*Ampullariella* (28)
		Same. Sporangiospores with lateral flagella.	*Pilimelia* (28)
		Multilocular sporangia formed. Spores nonmotile.	*Frankia* (27)
meso-DAP	Madurose	Short chains of conidia on aerial mycelium, often curled into a crozier.	*Actinomadura* (30)
		Chains of conidia with only two spores.	*Microbispora* (30)
		Chains of conidia mainly with four (2-6) spores.	*Microtetraspora* (30)
		Sporangia formed with two motile spores.	*Planobispora* (30)
		Sporangia formed with only one motile spore.	*Planomonospora* (30)
		Spherical sporangia formed on aerial mycelium containing many motile rod-shaped spores.	*Spirillospora* (30)
		Spherical sporangia formed on aerial mycelium containing many aplanospores.	*Streptosporangium* (30)
		Multilocular sporangia formed.	*Dermatophilus* (27) *Frankia* (27)

Table XIII.1.—*continued*

Diagnostic amino acid	Diagnostic sugar	Typical key morphological features and additional chemical properties	Possible generic assignment (section number)
meso-DAP	Fucose	Multilocular sporangia formed.	*Frankia* (27)
		Sporangia with motile spores.	*Actinoplanes* (28)
	Rhamnose and galactose	Both aerial and substrate hyphae fragment into nonmotile elements	*Saccharothrix* (33)
	Rhamnose, galactose, and mannose	*Streptomyces* type of morphology.	*Streptoalloteichus* (31)
	Galactose	*Streptomyces* type of morphology.	*Kitasatosporia* (33)
	Arabinose and galactose	Nocardomycolic acid (NMA) present. Morphology ranging from fugaceous substrate mycelium only to *Streptomyces*-like.	*Nocardia* (26)
		NMA present. Soft, salmon to pink organisms.	*Rhodococcus* (26)
		NMA absent. Short chains of conidia on the substrate and sparse aerial mycelium.	*Faenia* (26)
		NMA absent. Long cylindrical spores on the aerial mycelium. Spores formed by budding.	*Pseudonocardia* (26)
		NMA absent. Single spores formed mainly on the aerial hyphae.	*Saccharomonospora* (26)
		NMA absent. Very long chains of conidia on the aerial mycelium.	*Saccharopolyspora* (26)
		NMA absent. Long chains of conidia on the aerial mycelium. Halophile.	*Actinopolyspora* (26)
		NMA absent. Substrate mycelium tends to break into nonmotile elements. Aerial hyphae may be formed; they may also segment.	*Amycolata*[d] *Amycolatopsis*[d]
		NMA absent. Aerial mycelium bearing curled hyphae embedded in an amorphous matrix.	*Kibdelosporangium* (33)
	No diagnostic sugar	Single conidia formed. These are heat-resistant bacterial endospores.	*Thermoactinomyces* (32)
		Same as above but the spores are not heat-resistant.	*Thermomonospora* (31)
		Long chains of spores formed by the aerial hyphae.	*Nocardiopsis* (31)
		Aerial hyphae, often united into synnemata releasing motile spores.	*Actinosynnema* (31)
		Multilocular sporangia releasing motile spores.	*Geodermatophilus* (27)

[a] *DAP*, diaminopimelic acid.
[b] *NA*, not applicable.
[c] May also contain hydroxy forms of DAP that may even replace *meso*-DAP.
[d] Validly described genera not included in this volume.

SECTION 26

Nocardioform Actinomycetes

Hubert A. Lechevalier

This is an updated, slightly enlarged version of Section 17 of Volume 2 of *Bergey's Manual® of Systematic Bacteriology* (1986).

The term "nocardioform" signifies actinomycetes that form a fugacious mycelium breaking up into rod-shaped or coccoid elements. Like the term "coryneform," it is not one that can be defined satisfactorily, and individual strains of nocardioforms may not exhibit this basic morphological feature. The term "nocardioform" (Prauser, 1967) is intended to bring together, in an informal way, a number of organisms with a similar facies. In no way does it imply that these organisms are closely related, and attempts at defining nocardioforms as a whole in chemical terms have proved futile.

The major characteristics of the 11 genera included in this section are given in Table 26.1. In addition, properties of two recently described genera, not covered in the *Manual*, are included in the table (*Amycolata, Amycolatopsis*; Lechevalier et al., 1986).

All the organisms in this section are Gram-positive and aerobic (except for the oerskoviae, which are facultative anaerobes when grown on certain media). They also have DNAs rich in guanine and cytosine, as do other actinomycetes.

As far as it is known, the major respiratory quinones of all these organisms are menaquinones with either eight or nine isoprene units. Minor variations in the composition of these menaquinones are observed both between and within the genera.

Cell wall composition, whole-cell sugar patterns, and lipid composition furnish excellent markers for the separation of nocardioforms into genera. Most of the nocardioforms have a type IV cell wall composition, with the taxonomically significant constituents *meso*-diaminopimelic acid (*meso*-DAP), arabinose, and galactose. The majority of species (the nocardiae and the rhodococci) contain mycolic acids, as do the corynebacteria and the mycobacteria. However, there are important variations in these mycolates that separate the nocardiae, including the rhodococci, from the mycobacteria and the corynebacteria (Lechevalier, 1982).

Members of the genera *Nocardioides* and *Intrasporangium* are characterized by type I cell walls (L-DAP and glycine). This wall composition is also typical of the genus *Streptomyces* and its satellites (Section 29). The genera *Oerskovia* and *Promicromonospora* do not contain DAP but have lysine in their mureins.

If one is consulting Section 26 to identify an unknown isolate, since morphology is the most rapid source of information, the characteristics listed in Table 26.2 should furnish leads to possible generic assignments. However, it is rare, even with considerable experience, that a firm identification at the genus level can be made on the basis of morphology alone. One's attention should then turn to chemical composition. For a discussion of the methods used, see Lechevalier and Lechevalier (1976, 1980).

Much information can be gained by the study of whole-cell hydrolysates. A one-way paper chromatogram will determine whether an unknown organism contains DAP and whether it is in the L- or the *meso*-form: (the D- form, which is not easily resolved from the *meso*- form, is not known to have taxonomic significance).

If an unknown strain contains L-DAP, there are only two possibilities in Section 26. If the descriptions of *Nocardioides* and *Intrasporangium* do not fit the unknown, one should keep in mind *Streptomyces* and its allies in Section 29 and also some of the organisms in Section 15 (Volume 2) (*Arachnia*, for example).

If the unknown contains *meso*-DAP, a one-way paper chromatogram of its whole-cell sugar hydrolysate will reveal whether arabinose and galactose are present. If they are, a type IV cell wall is confirmed and a study of the mycolates will lead either to the genera *Mycobacterium* (Section 16, Volume 2), *Corynebacterium* and *Caseobacter* (Section 15, Volume 2), or *Nocardia* and *Rhodococcus* in this section. Absence of mycolates suggests *Faenia, Pseudonocardia, Saccharomonospora, Saccharopolyspora, Actinopolyspora, Amycolata*, or *Amycolatopsis*.

The absence of DAP would lead to an examination of the genera *Oerskovia* and *Promicromonospora*, discussed in this section, and to many of the organisms discussed in Section 15. Final identification requires purified cell wall preparations to determine the identity of the dibasic amino acid involved.

For general reviews of the actinomycetes, see Volume II of Starr et al. (1981) and Goodfellow et al. (1984).

Table 26.1.
Differential properties of nocardioform genera[a]

Characteristics	*Nocardia*	*Rhodococcus*	*Faenia (Micropolyspora)*	*Pseudonocardia*	*Saccharomonospora*	*Saccharopolyspora*	*Actinopolyspora*	*Amycolata*	*Amycolatopsis*	*Oerskovia*	*Promicromonospora*	*Nocardioides*	*Intrasporangium*
Marked fragmentation of mycelium in older cultures	d	+	−	+	−	+	−	+	+	+	+	+	d
Aerial mycelium produced	+[b]	+[c]	+[c]	+	+	+[b]	+	d	d	−	+[b]	+	−
Conidia formed	d	−	+	+	+	+[b]	+	d	d	−	−	+	−
Motile elements produced	−	−	−	−	−	−	−	−	−	+[d]	−	−	−
Strictly aerobic	+	+	+	+	+	+	+	+	+	−	+	+	+
Faculative anaerobes	−	−	−	−	−	−	−	−	−	+	−	−	−
Cell wall type[e]	IV	IV	IV	IV	IV	IV	IV	IV	IV	VI	VI	I	I
Mycolic acids present	+[f]	+[f]	−	−	−	−	−	−	−	−	−	−	−
Phospholipid type[g]	PII	PII	PIII	PIII	PII	PIII	PIII	PIII	PII	PV	PV	PI	PIV
Menaquinones	MK-8(H_4), -9(H_2)	MK-8(H_2), -9(H_2)	MK-9(H_4), -9(H_6)	MK-9(H_4)	MK-9(H_4)	MK-9(H_4)	MK-9(H_4)	MK-8(H_2,H_4)	MK-9(H_2,H_4)	MK-9(H_4)	MK-9(H_4)	MK-8(H_4)	MK-8(ND)
Mol% G + C of the DNA	64–72	59–69	66–68	79	69–74	77	64	68–71	66–69	70–75	70–75	66–69	68

[a]Symbols: +, 90% or more of strains positive; −, 10% or less of strains positive; *d*, 11–89% of strains positive; and *ND*, not determined.
[b]Lacking in some strains.
[c]Usually scanty.
[d]Some strains nonmotile (nonmotile oerskoviae (NMO)).
[e]Major constituents in cell walls of types: *I*, L-DAP and glycine; *IV*, *meso*-DAP, arabinose, and galactose; and *VI*, lysine (with variable presence of aspartic acid and galactose).
[f]Nocardiomycolic acids.
[g]Characteristic phospholipids of types, in addition to phosphatidylinositol (which is always present); *PI*, phosphatidylglycerol (variable); *PII*, only phosphatidylethanolamine; *PIII*, phosphatidylcholine (with phosphatidylethanolamine, phosphatidylmethylethanolamine and phosphatidylglycerol variable, no phospholipids containing glucosamine); *PIV*, phospholipids containing glucosamine (with phosphatidylethanolamine and phosphatidylmethylethanolamine variable); and *PV*, phospholipids containing glucosamine and phosphatidylglycerol.

Table 26.2.
Generic assignment of nocardioforms by morphology

Morphological features	Possible generic assignment	Morphological features	Possible generic assignment
No aerial mycelium		Substrate mycelium breaking up into nonmotile elements; aerial hyphae without any special features.	*Nocardia, Rhodococcus, Amycolata, Amycolatopsis*
Substrate mycelium breaks up into motile elements.	*Oerskovia* (Gram-positive); *Mycoplana* (Gram-negative) (see with *Oerskovia*)	Both mycelia breaking up into elements of various shapes.	*Nocardia, Rhodococcus, Promicromonospora, Pseudonocardia, Amycolata, Amycolatopsis*
Substrate mycelium breaks up into nonmotile elements.	*Promicromonospora* (yellow to white); *Rhodococcus* (pinkish, salmon colored); *Mycobacterium* (Section 16, Volume 2); *Amycolata, Amycolatopsis*	Both mycelia bearing single or short chains of conidia.	*Faenia, Nocardia, Saccharomonospora*
Substrate mycelium with ovoid vesicles.	*Intrasporangium* or a culture of many other actinomycetes growing under adverse conditions.	Aerial hyphae bearing single spores that are densely packed along the hyphae.	*Saccharomonospora*
Substrate mycelium without any special features.	*Actinomyces* or relatives (Section 15, Volume 2). *Nocardia, Rhodococcus, Amycolata, Amycolatopsis,* or an atypical form of any of the actinomycetes discussed in Volume 4.	Aerial hyphae bearing short chains of conidia.	*Actinomadura, Microtetraspora, Nocardioides, Nocardia*
		Aerial mycelium with long chains of conidia or arthrospores.	*Saccharopolyspora, Nocardia, Actinopolyspora, Nocardiopsis, Saccharothrix, Glycomyces, Streptomyces, Amycolata, Amycolatopsis.*
Aerial mycelium formed		Both substrate and aerial hyphae divide into segments that may divide again into smaller elements; hyphae often present a zig-zag appearance.	*Nocardia, Saccharopolyspora, Pseudonocardia, Nocardiopsis, Saccharothrix, Amycolata, Amycolatopsis*
Substrate mycelium without any special features, sterile aerial hyphae.	*Nocardia, Rhodococcus,* or atypical forms of *Micropolyspora, Saccharopolyspora, Amycolata, Amycolatopsis, Pseudonocardia, Promicromonospora, Nocardioides,* or any other actinomycetes.	Growth by budding.	*Pseudonocardia*

Genus **Nocardia** *Trevisan 1889, 9*[AL*]

MICHAEL GOODFELLOW AND MARY P. LECHEVALIER

No.car′di.a. M.L. fem. n. *Nocardia*, named after Edmond Nocard, a French veterinarian.

Rudimentary to extensively branched vegetative hyphae, 0.5–1.2 μ in diameter, growing on the surface of, and penetrating, agar media, often fragmenting in situ or on mechanical disruption into **bacteroid, rod-shaped to coccoid, nonmotile** elements. **Aerial hyphae**, at times visible only microscopically, **almost always formed**. Short to long chains of well to poorly formed conidia may occasionally be found on the aerial hyphae and more rarely on both aerial and vegetative hyphae. No endospores, sporangia, sclerotia, or synnemata formed. Nonmotile.

Gram-positive to Gram-variable. Some strains partially acid fast at some stage of growth. **Aerobic.** Mesophilic. **Chemo-organotrophic,** having an oxidative type of metabolism. **Catalase positive.**

Cell wall contains major amounts of ***meso*-diaminopimelic acid** (*meso*-DAP), **arabinose, and galactose**. The organisms contain **diphosphatidylglycerol, phosphatidylethanolamine, phosphatidylinositol and phosphatidylinositol mannosides,** major amounts of **straight-chain, unsaturated,** and **10-methyl (tuberculostearic) fatty acids, mycolic acids with 46–60 carbons and up to three double bonds, and either tetrahydrogenated menaquinone with eight isoprene units (MK-8(H_4))** (most nocardial species) or **dihydrogenated menaquinone with nine isoprene units (MK-9(H_2)) (*N. amarae*) as the predominant isoprenolog. The fatty acid esters released on pyrolysis gas chromatography of mycolic esters contain 12–18 carbon atoms and may be saturated or unsaturated.**

The mol% G + C of the DNA ranges from 64 to 72 (T_m).

Nocardiae are widely distributed and are abundant in soil. Some strains are pathogenic opportunists for man and animals.

Type species: *Nocardia asteroides* (Eppinger) Blanchard 1896, 856, Opinion 58, Judicial Commission 1985, 538.

Further Descriptive Information

Many actinomycetes with a tendency to fragment have been, and continue to be, mistakenly assigned to the genus *Nocardia* on the basis of morphology alone. This has resulted, in the past, in a genus characterized by extreme heterogeneity (Lechevalier, 1976b). Following a certain number of taxonomic reassignments and formation of new genera, such as *Actinomadura, Nocardiopsis, Oerskovia, Rhodococcus, Rothia,* and *Saccharopolyspora* (see section on differentiation from other genera), substantial progress has been made in refining the nocardial concept (Goodfellow and Minnikin, 1981c; Goodfellow and Cross, 1984b).

The genus *Nocardia* is now largely defined on the basis of cell envelope lipid and peptidoglycan composition. At present, only actinomycetes with the following characteristics are considered to be "true" nocardiae: (a) a peptidoglycan composed of *N*-acetylglucosamine, L-alanine, D-alanine, D-glutamic acid with *meso*-DAP as the diamino acid, and muramic acid in the *N*-glycolated form rather than the *N*-acetylated form found in many other actinomycete taxa (Bordet et al., 1972; Uchida and Aida, 1977, 1979; Goodfellow and Cross, 1984b); (b) a polysaccharide fraction of the wall containing arabinose and galactose (i.e. nocardiae have a wall chemotype IV and a whole-cell sugar pattern type A sensu Lechevalier and Lechevalier, 1970c); (c) a phospholipid pattern consisting of diphosphatidylglycerol, phosphatidylethanolamine (taxonomically significant nitrogenous phospholipid), phosphatidylinositol, and phosphatidylinositol mannosides without phosphatidylcholine or

*AL denotes the inclusion of this name on the Approved Lists of Bacterial Names (1980).

phospholipids containing glucosamine (i.e. a phospholipid pattern type PII sensu Lechevalier et al., 1977); (d) a fatty acid profile showing major amounts of straight-chain, unsaturated and tuberculostearic acids (i.e. type IV fatty acid pattern sensu Lechevalier et al., 1977); (e) mycolic acids with 46–60 carbons (Lechevalier and Lechevalier, 1980; Goodfellow and Minnikin, 1981a; Goodfellow and Cross, 1984b; Minnikin et al., 1984a); and (f) a menaquinone fraction containing tetrahydrogenated menaquinone with eight isoprene units or dihydrogenated menaquinone with nine isoprene units as the predominant isoprenolog (Yamada et al., 1976, 1977a; Collins et al., 1977, 1985; Collins and Jones, 1981; Collins, 1984; Kroppenstedt, 1984).

Another group of strains that have been assigned to the genus *Nocardia* over the years is made up of actinomycetes that have a cell wall composition of chemotype IV but that lack mycolic acids (Lechevalier et al., 1971b, 1977; Goodfellow and Minikin, 1981a, 1984). Strains lacking mycolic acids include many of commercial and ecological importance, and their descriptions are included at the end of the genus under "Species Incertae Sedis" for the convenience of the users of the *Manual*. The genera *Amycolata* and *Amycolatopsis* have been proposed to include many of these organisms (Lechevalier et al., 1986).

The only constant morphological feature of nocardiae is a tendency of the aerial or vegetative mycelium to fragment (Locci, 1976; Williams et al., 1976; Nesterenko et al., 1978b). The consistency and composition of the growth medium can affect the growth and stability of both aerial and substrate hyphae (Williams et al., 1976). Unfortunately, fragmentation is not unique to *Nocardia* since members of the genera *Rhodococcus, Oerskovia*, and *Nocardioides*, as well as some *Streptomyces*, show this character. Other morphological features of nocardiae (also shared with other actinomycetes of different genera) include well-developed conidia in *N. brevicatena* (Fig. 26.1) and less well formed spores in some *N. asteroides* strains (Fig. 26.2), in both cases borne on both vegetative and aerial hyphae. Cell pigments may be off-white, gray, yellow, orange, pink, peach, red, tan, brown, or purple. Soluble pigments, when present, are usually undistinguished: brown or yellowish. Aerial mycelium, visible to the naked eye, may be lacking, sparse, or very abundant; thus, colonies may have a superficially smooth appearance or, more commonly, a matte, chalky, or velvety aspect. Aerial hyphae, visible microscopically, are always present. Sections of cells show a typical Gram-positive wall, which in some strains may be quite thick (up to 0.3 μm in *N. amarae*). The thickening of the cell walls of *N. asteroides* strains has been found to be greater in stationary phase cells than in those from the exponential phase (Beaman, 1975). Mesosomes are common. L- Forms of *N. asteroides* and *N. otitidiscaviarum* are known, and there are grounds for believing that they may have a significant role in infectious processes (Beaman et al., 1978; Beaman, 1980, 1984).

Most nocardiae grow readily on a variety of media containing simple nitrogen sources such as ammonium, nitrate, and amino acids as well as on more complex ones containing casein, meat, soy, or yeast peptones and hydrolysates. Glucose, acetate, and propionate are good carbon sources. All nocardiae will grow in a temperature range of 15–37°C; many will grow at higher or lower temperatures. Doubling time, as for most actinomycetes, is appreciably longer than for other bacteria; a generation time of 5.5 h has been reported for certain *N. asteroides* and *N. brasiliensis* strains (Beadles et al., 1980). Some strains will grow to stationary

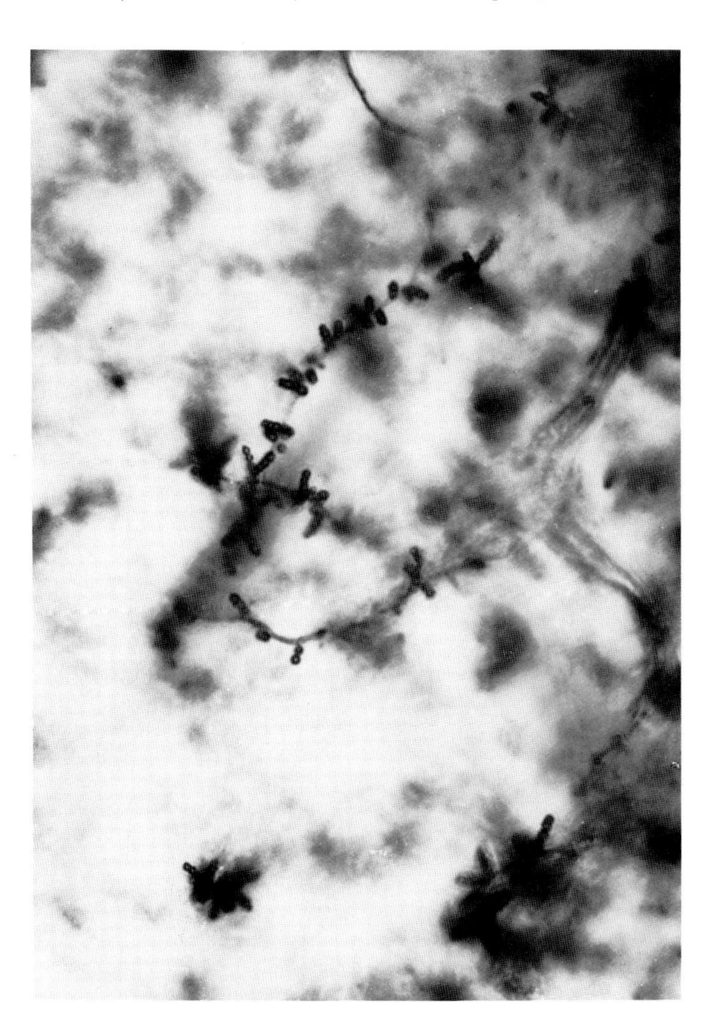

Figure 26.1. Aerial conidial chains of *Nocardia brevicatena* (× 1250).

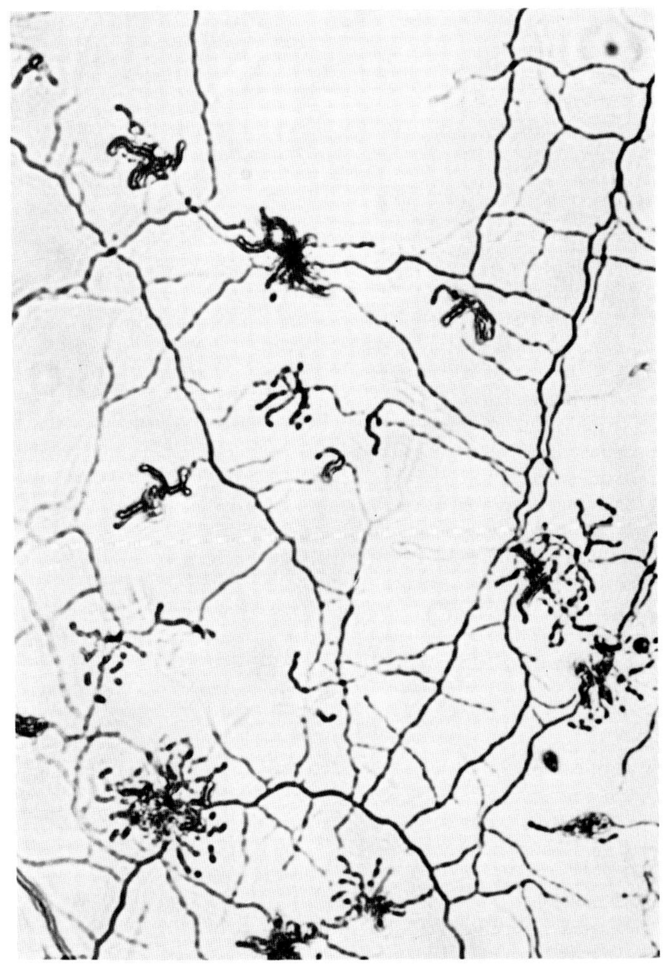

Figure 26.2. Aerial and vegetative conidial chains of *Nocardia asteroides* (K 1250).

phase in 3–7 days; others will grow more slowly. Colonies of nocardiae isolated from clinical sources may take up to several weeks to appear.

Autotrophs are not known among the true nocardiae (Tárnok, 1976). However, *N. autotrophica*, a strain lacking mycolic acids, was reported to grow both hetero- and autotrophically, utilizing a fixed nitrogen source in an atmosphere of CO_2, H_2, and O_2 (Takamiya and Tubaki, 1956); it can also oxidize CO to CO_2 (Bartholomew and Alexander, 1979). A *N. asteroides* strain studied at the same time was negative. *Nocardia saturnea*, another strain lacking mycolic acids, grows on silica gel plates without added carbon or nitrogen, and *N. petroleophila* utilizes CO_2 while oxidizing alkanes (Hirsch, 1960). Because this last strain does not have a wall chemotype IV, its taxonomic position is not clear. *Nocardia opaca* (probably a rhodococcus) had a similar capability (Aggag and Schlegel, 1973). Some nocardiae can grow on nitrogen-free media by scavenging organic substances, such as ammonia, present in the air. *Nocardia hydrocarbonoxydans* (Nolof and Hirsch, 1962), which also lacks mycolic acids, can use such airborne compounds as pyridine as a source of nitrogen.

Little is known of the metabolic pathways and genetics of true nocardiae. Most studies reported on "nocardiae" were actually carried out on strains that are now classed as rhodococci (see Bradley, 1978; Brownell and Denniston, 1984) or on some of the nocardiae lacking mycolic acids (e.g. Schupp et al., 1975). Genetic recombination in *N. asteroides* has been reported (Kasweck and Little, 1982). Plasmids have been found in rhodococci (Brownell, 1978; Reh, 1981) and in *N. asteroides* (Kasweck et al., 1982). Phage (nocardiophage) have been reported for *N. asteroides* (Pulverer et al., 1974; Prauser, 1976a, 1981b; Andrzejewski et al., 1978), *N. brasiliensis* (Pulverer et al., 1974), *N. carnea* (Williams et al., 1980), and *N. otitidiscaviarum* and *N. vaccinii* (Prauser, 1976a). In general, rhodococci and *true* nocardiae are susceptible to nocardiophages, whereas *mycolateless* nocardiae are not (Williams et al., 1980; Prauser, 1981a). Phages active on *N. mediterranei* have been found that were not active on various nocardiae tested (Thiemann et al., 1964; Lechevalier et al., 1986).

Immunodiffusion analysis, using disintegrated cells as antigens, permitted several clusters of nocardiae to be recognized: three groups of *N. asteroides*, one of *N. otitidiscaviarum*, and one of *N. amarae* (Ridell, 1981b). Delayed hypersensitivity reactions were induced by purified wall polysaccharide and ribosomal proteins from *N. asteroides* and *N. brasiliensis* (Ortiz-Ortiz et al., 1976).

Nocardiae are sensitive to antibacterial antibiotics such as aminoglycosides, tetracyclines, and sulfonamides. Cephalosporins, penicillins, and peptide antibiotics are not as active (Bach et al., 1973; Lerner and Baum, 1973; Goodfellow and Orchard, 1974). Therapy with trimethoprim-sulfamethoxazole has been used with success (González-Ochoa, 1976); minocycline (Curry, 1980), fusidic acid, and aminoglycosides such as amikacins, gentamicins, and netilmicin have also been used (Schaal and Beaman, 1984). *N*-Formimidoyl thienamycin and amikacin were the most active antibiotics on *N. asteroides* among 22 β-lactams and 9 aminoglycosides tested (Gutmann et al., 1983). Treatment failures with the trimethoprim-sulfamethoxazole regimen led Dewsnup and Wright (1984) to conclude that of 25 antimicrobial agents tested, doxycycline, minocycline, sulfamethoxazole, or imipenem could be useful alternatives.

Actinomycetes, if pathogenic, are usually opportunists, and nocardiae are no exception to this rule. The basic types of human disease caused are: (a) pulmonary, neural, and/or systemic nocardiosis, usually caused by *N. asteroides* and sometimes by *N. brasiliensis* and *N. otitidiscaviarum*; (b) actinomycotic mycetomas, which are tumor-like growths of the organisms within the tissues and are usually caused by *N. brasiliensis* but sometimes by *N. asteroides* and *N. otitidiscaviarum*; and (c) localized cutaneous or subcutaneous infections, which usually represent primary infections by either *N. asteroides* or *N. brasiliensis* (Gordon, 1974; Gonzáles-Ochoa, 1976; Mariat and Lechevalier, 1977; Pulverer and Schaal, 1978; Schaal and Beaman, 1984). *Nocardia asteroides* strains may be quite infectious but vary in their invasiveness (Beaman, 1976; Schaal and Beaman, 1984). Using inoculation of the footpads of white mice as a test system, Ochoa (1973) concluded that *N. brasiliensis* strains were obligatory pathogens, whereas *N. asteroides* strains were opportunistic pathogens. The occurrence of nocardial infections in the United States (1972–1974) has been reviewed (Beaman et al., 1976).

Lower animals such as birds, cats, dogs, fish, goats, horses, and domestic and wild rodents are also susceptible to nocardial infection (Goodfellow and Minnikin, 1981c), *N. asteroides* being most frequently implicated; however, *N. brasiliensis* may be pathogenic in cats (Kurup et al., 1970; Gonzáles-Ochoa, 1976). *Nocardia vaccinii*, the only well-documented nocardial plant pathogen, causes galls in blueberry (Demaree and Smith, 1952). "*Nocardia rhodnii*" strains are thought to be symbionts of the insect *Rhodnius prolixus* (Cross et al., 1976); the actual identity of the organisms involved remains open.

Nocardiae are widely distributed in nature, being found in soil, water, air, sewage, and, as indicated above, in clinical specimens, insects, and plants (Cross et al., 1976; Goodfellow and Minnikin, 1981c; Goodfellow and Williams, 1983).

Enrichment and Isolation Procedures

Once isolated, nocardiae are not difficult to grow, but their frequency of isolation may be enhanced through the use of selective techniques. *Nocardia asteroides* strains may be isolated from complex mixed communities, such as those found in soil, by plating dilutions on Diagnostic Sensitivity Test Agar (Oxoid) supplemented with the antibiotics demethylchlortetracycline at 5 μg/ml or methacycline at 10 μg/ml and antifungal antibiotics (Orchard et al., 1977). Inoculated plates are incubated for up to 21 days at 25°C. Colonies with a pink to red substrate mycelium, covered to a greater or lesser extent with white aerial hyphae, are characteristic of nocardiae. *Nocardia amarae* strains are most easily isolated from sewage treatment plant foams using Czapek's agar supplemented with 0.4% (w/v) yeast extract (Lechevalier and Lechevalier, 1974). Isolation of nocardiae from clinical specimens may be accomplished by several procedures, including the use of tellurite-containing medium (Schaal, 1972), Sabouraud's dextrose agar, beef heart infusion-blood agar, and Lowenstein-Jensen medium (Schaal, 1984b). Some *N. asteroides* strains survive mycobacterial selection procedures (Gordon, 1974; Goodfellow and Minnikin, 1981c).

Maintenance Procedures

Transfer of cultures every 2–4 months onto a suitable agar medium such as Bennett's or yeast extract-glucose agars (Waksman, 1967; Gordon et al., 1974), with storage at 4°C between transfers, is satisfactory for most nocardial strains. Long-term preservation may be accomplished by lyophilization in skim milk. Shorter term storage may be effected by use of sterile mineral oil overlay of slants kept at 4°C; alternatively, stationary phase cells grown in broth culture may be quick-frozen and stored at −25° to −70°C. Suspensions of mycelial fragments and spores may also be stored in glycerol (20% v/v) at −20°C (Wellington and Williams, 1978).

Procedures for Testing of Special Characters

Procedures for analyses of cell wall composition, whole-cell amino acids and sugars, mycolic acids, and phospholipids are given in detail in Lechevalier and Lechevalier (1980). Other techniques for mycolate analysis will be found in Minnikin et al. (1975, 1980), Hecht and Causey (1976), and Schaal (1984a). Detailed analytical protocols for determining menaquinone composition are also available (Yamada et al., 1976, 1977a; Collins et al., 1977, 1985; Collins, 1985; Kroppenstedt, 1985).

As with all actinomycetes, the most satisfactory method of determining the micromorphology of a nocardial culture is the direct in situ observation, by means of high-dry objective (× 40–60) and a long working distance condenser, of a Petri dish containing undisturbed colonies of the organism grown on a meager medium such as dilute Bennett's or tap water agar. Wet mounts observed under phase or, alternatively, stained preparations are useful for examination of tissue sections or body exudates for the characteristic granules often seen in nocardial infections (Gordon, 1974).

There are two principal systems for testing the physiology of nocardiae, the "Gordon" and "Goodfellow" methods. The former is summarized in Gordon et al. (1974) and Mishra et al. (1980) and the latter in Goodfellow (1971), Lacey and Goodfellow (1975), Goodfellow and Alderson (1977), Goodfellow and Pirouz (1982b), and Goodfellow et al. (1982b). In general, the physiological tests employed by Gordon enable one to recognize a given nocardial species on the basis of about 40 tests. Employing over 150 characters using phenetic techniques, Goodfellow and colleagues found that some of the taxa recognized by Gordon are heterogeneous. This is particularly true for *N. asteroides*, where up to five clusters have been distinguished (Orchard and Goodfellow, 1980). Data on the reactions of various nocardial species in both systems are given below.

Differentiation of the genus **Nocardia** *from other genera*

Nocardia is the type of the family *Nocardiaceae* (Castellani and Chalmers, 1919). The latter currently accommodates aerobic, Gram-positive, actinomycetes that have a wall chemotype IV and form rudimentary to extensive substrate mycelium that usually fragments into bacillary and coccoid elements (Goodfellow and Minnikin, 1981a, 1981c). It is clear that the family is markedly heterogeneous (Goodfellow and Cross, 1984b; Goodfellow and Minnikin, 1984) and can be divided into aggregate groups centered around the genera *Nocardia* and *Faenia (Micropolyspora)*. The first group is characterized by the presence of mycolic acids and major amounts of straight-chain and unsaturated fatty acids, whereas *Faenia* and related taxa lack mycolic acids but contain large amounts of branched-chain *iso-* and *anteiso-* acids (Goodfellow and Cross, 1984b; Goodfellow and Minikin, 1984).

In general, nocardiae and related organisms are either amycelial (*Caseobacter, Corynebacterium, Mycobacterium*) or reproduce by mycelial fragmentation (*Nocardia, Rhodococcus*), whereas the genera in the *Faenia* aggregate show a greater differentiation of sporing structures (Locci, 1976; Locci and Sharples, 1984). However, in practical terms, morphology is of little use in distinguishing most members of the genus *Nocardia* from the taxa in the *Faenia* group since among the nocardiae there is a spectrum of morphological complexity going from an undifferentiated (and often unfragmented) mycelium (e.g. *N. amarae*) to a highly developed system of short conidial chains formed on the aerial and vegetative hyphae (e.g. *N. brevicatena* and some *N. asteroides*) to long chains of well-formed spores on the aerial hyphae (*N. carnea, N. vaccinii*), all morphological types found within the *Faenia* aggregate. Among the latter, only saccharomonosporae are clearly differentiable from nocardiae on the basis of morphology, since the spore pattern of one to two spores on the aerial mycelium only is unique among taxa with a cell wall of chemotype IV.

The nocardioform actinomycetes, i.e. those grouped around *Nocardia*, not only have many properties in common (Goodfellow et al., 1974; Goodfellow and Wayne, 1982; Goodfellow and Cross, 1984b) but also form a recognizable suprageneric group (Mordarski et al., 1980b; Stackebrandt and Woese, 1981b; Stackebrandt et al., 1983c). Nocardiae can, however, be distinguished from the other mycolic acid-containing taxa, and from strains previously assigned to the genus, using a combination of chemical and morphological properties (Tables 26.3 and 26.4). Susceptibility to the antibiotics bleomycin (2.5 μg/ml) and mitomycin (10 μg/ml) has been reported to be useful in distinguishing rhodococci from nocardiae (Tsukamura, 1981b, 1982a).

Much less attention has been paid to the second aggregate group, which currently contains the genera *Actinopolyspora, Faenia, Pseudonocardia, Saccharomonospora,* and *Saccharopolyspora*. It is not yet clear whether all of these taxa merit generic rank or even if they form a distinct suprageneric group, though this seems likely. They do, however, have a number of properties in common (Lechevalier and Lechevalier, 1981b; Goodfellow and Cross, 1984b; Goodfellow and Minnikin, 1984) and can readily be distinguished from *Nocardia* (Table 26.5). The genus *Kibdelosporangium* (Shearer et al., 1986) is chemically related to the "mycolateless nocardiae," but can be distinguished from them on morphological grounds. The standing of the genus *Micropolyspora* has become a matter of dispute as the reclassification of *M. brevicatena* in the genus *Nocardia* as *N. brevicatena* (Goodfellow and Pirouz, 1982a) left the genus without a type species and hence nomenclaturally invalid. McCarthy et al. (1983) requested that the name *Micropolyspora* be conserved with the type species as *M. faeni*, but more recently Kurup and Agre (1983) proposed the generic name *Faenia* for the remaining species of the genus. The Judicial Commission ruled in favor of the name *Faenia* (Wayne, 1986).

Taxonomic Comments

Although the genus *Nocardia* is widely recognized, the name "*Proactinomyces*" is still used in some of the Russian literature to refer to this taxon. The present circumscription of *Nocardia* has been discussed under "Further Descriptive Information" (see above). The redefined genus contains only 9 of the 20 species listed under *Nocardia* in the Approved Lists (AL): *amarae, asteroides, brasiliensis, brevicatena, carnea, farcinica, otitidiscaviarum, transvalensis,* and *vaccinii*. Using more than 50 criteria, both morphological and physiological, Gordon and her colleagues pioneered a broadly based definition of the species *N. asteroides, N. brasiliensis, N. carnea, N. otitidiscaviarum, N. transvalensis,* and *N. vaccinii* (Gordon and Mihm, 1957, 1959, 1962a, 1962b; Gordon et al., 1974, 1978; Mishra et al., 1980). Data summarized from these publications along with results on the two remaining nocardial species, *N. amarae* and *N. brevicatena* (Lechevalier, 1968b, unpublished results; Lechevalier and Lechevalier, 1974; Goodfellow and Pirouz, 1982b), are presented later in Tables 26.6 and 26.8 (Gordon tests). Using numerical taxonomic methods, Goodfellow and colleagues have found that *N. amarae, N. brasiliensis, N. farcinica,* and *N. otitidiscaviarum* are good taxospecies; however, these and other workers have found *N. asteroides*, as defined by Gordon, to be markedly heterogeneous (Goodfellow, 1971; Lechevalier, 1976b; Tsukamura, 1977; Schaal and Reutersberg, 1978; Goodfellow and Minnikin, 1981c; Goodfellow and Cross, 1984b). DNA homology determinations also have shown that many subgroupings are possible in *N. asteroides* (Bradley et al., 1978; Mordarski et al., 1978). Immunological studies also have pointed to heterogeneity in this taxon (Pier and Fichtner, 1971; Magnusson, 1976; Kurup and Schribner, 1981; Ridell, 1981b). However, using more sensitive immunological techniques, Kurup et al. (1983) concluded that the seven immunotypes previously distinguished should be retained within the species *asteroides*, rather than be assigned to separate species as previously proposed by Pier and Fichtner. Additional representatives of the remaining species need to be studied for them to be defined with precision. Differentiating characters culled from Goodfellow's test results (Goodfellow tests) are presented later in Table 26.7.

The principal nomenclatural problem in the redescribed genus *Nocardia* is the status of the former type species, *N. farcinica*. The strain, originally isolated by Nocard (1888) from a case of bovine farcy, was, by the decision of the Judicial Commission in 1954, made the type species of the genus *Nocardia*. It is now known that Nocard's original isolate is represented by two supposedly identical, but actually very different, strains, ATCC 3318 and NCTC 4524. The former contains mycolic acids characteristic of nocardiae, and the latter mycolic acids and mycosides similar to those of mycobacteria (Lanéelle et al., 1971; Lechevalier et al., 1971b). A numerical phenetic study by Orchard and Goodfellow (1980) found that each of these strains fell into a different cluster along with both nocardiae and mycobacteria. Schaal and Reutersberg (1978) found that ATCC 3318 clustered with *N. asteroides* strains, but they did not examine NCTC 4524. Using both serology and physiology, Ridell (1975) noted that the latter strain was grouped with mycobacteria and ATCC 3318 with *N. asteroides*, confirming her previous results using immunodiffusion, which showed that some *N. farcinica* strains were closer to the mycobacteria than to nocardiae (Ridell and Norlin, 1973). Base composition studies also showed that strains ATCC 3318 and NCTC 4524 were different, the mol% G + C values being 68.0 and 71.6, respectively

Table 26.3.
Differential characteristics of the genus **Nocardia** *and related wall chemotype IV taxa containing mycolic acids*[a,b]

Characteristics	*Nocardia*	*Caseobacter*	*Corynebacterium*	*Mycobacterium*	*Rhodococcus*
Morphological characters					
Substrate mycelium	+	−	−	D⁻	D⁺
Aerial mycelium, macroscopic	D	−	−	−	−
Aerial hyphae, microscopic	+	−	−	D⁻	D⁻
Conidia	D⁻	−	−	−	−
Entire colonies	−	+	+	D	D
Lipid characters					
Fatty acids					
Tuberculostearic acid[c]	+	+	−[d]	+[e]	+
Phospholipids					
Phosphatidylethanolamine[f]	+	−[h]	−	+	+
Predominant menaquinone[g]	MK-8(H_4), -9(H_2)	MK-9(H_2)	MK-8(H_2), -9(H_2)	MK-9(H_2)	MK-8(H_2), -9(H_2)
Mycolic acids					
Overall size[h] (number of carbons)	46–60	30–36	22–38	60–90	34–64
Number of double bonds[h]	0–3	0–2	0–2	1–3	0–4
Fatty acid esters released on pyrolysis (number of carbons)[i]	12–18	14–18	8–18	22–26	12–18
Mol% G + C of the DNA	64–72	60–67	51–59	62–70	59–69

[a]Data from Goodfellow and Minnikin (1981a, 1981c, 1984) and Goodfellow and Cross (1984b).
[b]Symbols: +, 90% or more of the strains are positive; −, 90% or more of the strains are negative; *D*, different reactions occur in different taxa (species of a genus or genera of a family); *D*⁻, uncommonly; *D*⁺, more commonly than not.
[c]Determined by gas liquid chromatography (Lechevalier et al., 1977; Collins et al., 1982b; Kroppenstedt, 1985).
[d]*Corynebacterium bovis* contains tuberculostearic acid (Lechevalier et al., 1977; Collins et al., 1982b).
[e]*Mycobacterum gordonae* lacks substantial amounts of tuberculostearic acid (Tisdall et al., 1979; Minnikin et al., 1985).
[f]Determined by thin layer chromatography and chemical analysis (Lechevalier et al., 1977, 1981; Minnikin et al., 1977a; M. P. Lechevalier, unpublished).
[g]Menaquinones detected by chromatographic or physicochemical analysis (Yamada et al., 1976, 1977a; Collins et al., 1977; Collins, 1985). Abbreviations exemplified by MK-8(H_2), menaquinone having two of the eight isoprene units hydrogenated.
[h]Detected by mass spectrometry (Alshamaony et al., 1976; Collins et al., 1982a). In mycobacterial mycolic acids, double bonds may be converted to cyclopropane rings; methyl branches and oxygen functions may be present (Dobson et al., 1984; Minnikin et al., 1984a).
[i]Esters of fatty acids detected by pyrolysis gas chromatography of mycolate esters (Lechevalier et al., 1971b; Goodfellow et al., 1978; Collins et al., 1982a).

(Mordarski et al., 1978), as did phage sensitivity (Prauser, 1981a). In view of the uncertain status of *N. farcinica* (Lechevalier, 1976a) the present type species has been changed to *N. asteroides*, with ATCC 19247 as type (Skerman et al., 1980; Sneath, 1982; Opinion 58, Judicial Commission, 1985) and *N. farcinica* has been retained, with ATCC 3318 as type. An appeal to the Judicial Commission to reject *N. farcinica* as a *nomen dubium* was published (Tsukamura, 1982d); however, the commission has voted to retain the *farcinica* epithet.

Five of the remaining nocardial species listed in the Approved Lists, *N. calcarea, N. coeliaca, N. corynebacteroides, N. globerula*, and *N. restricta*, have been classified with the rhodococci (see article on *Rhodococcus*). One, *N. cellulans*, belongs with the group of oerskoviae referred to as "NMOs" (see article on "*Oerskovia*") and one, *N. petroleophila*, belongs to "*incertae sedis*" because this strain lacks a wall chemotype IV and its taxonomic position is still unclear. The remaining five nocardial species in the Approved Lists, *N. autotrophica, N. hydrocarbonoxydans, N. mediterranei, N. orientalis*, and *N. saturnea*, belong to the "mycolateless" nocardiae. Many chemical data, including lack of mycolic acids, presence of major amounts of *iso-* and *anteiso-* fatty acids, and differences in phage sensitivity (Williams et al., 1980; Goodfellow and Cross, 1984b; Goodfellow and Minnikin, 1984) indicate that these taxa do not belong within the genus *Nocardia*. Other strains lacking mycolic acids that are not listed in the Approved Lists, but that will be considered here because of their importance, include *N. lurida* ATCC 14930 (see *N. orientalis*), *N. rugosa* IMRU 3760, and *N. sulphurea* ATCC 27624. Differentiating characteristics of these mycolateless "nocardiae" are given later in Tables 26.9 and 26.10. Table 26.11 contains their reactions in the carbohydrate utilization tests from the International Streptomyces Project. The Gordon test reactions of the principal nocardioform taxa having a phospholipid pattern of chemotype PIII (phosphatidylcholine as diagnostic phospholipid) are given later in Table 26.12. The genera *Amycolata* and *Amycolatopsis* have recently been proposed for wall chemotype IV actinomycetes lacking mycolic acids (Lechevalier et al., 1986). The genus *Amycolata* includes *A. autotrophica* (type species), *A. hydrocarbonoxydans*, and *A. saturnea*, and the genus *Amycolatopsis* includes four species, *A. orientalis* (type species), *A. mediterranei, A. rugosa*, and *A. sulphurea.*

Two new nocardial species, *N. nova* and *N. paratuberculosis*, have been proposed by Tsukamura (1982c). *Nocardia nova* is the new name for a subgroup of *N. asteroides* defined by numerical taxonomic studies. *Nocardia paratuberculosis* is a substitute name for the controversial *N. farcinica.* Until their status has been clarified, they have been included here as species *incertae sedis*. Another group, also of uncertain status, is the "*aurantiaca* taxon." Originally described by Tsukamura and Mizuno

Table 26.4.
Differential characteristics of the genus **Nocardia** *and taxa previously associated with the genus*[a,b]

Characteristics	*Nocardia*	*Actinomadura*	*Nocardiopsis*	*Oerskovia*	*Rothia*
Morphological characters					
Substrate mycelium					
Fragmentation	+	–	*D*	+	+
Motile spores	–	–	–	D	–
Aerial mycelium	D+	D+	+	–	–
Conidia	D	+	+	–	–
Colony characters					
Entire colonies	–	–	–	D	D
Red-pink colonies	D	D	–	–	–
Metabolism of glucose	O	O	O	O/F	O/F
Wall chemotype[c]	IV	III	III	VI	VI
Whole-cell sugar pattern[c]	A	B	C	Gal	NC
Lipid characters					
Fatty acids[d]					
Unsaturated	+	+	+[e]	–	–
Tuberculostearic	+	+	+	–	–
Iso- and *anteiso-*	–	+[e]	+	+	+
Phospholipid type[f]	II	I/IV	III	V	I
Predominant menaquinone[d]	MK-8(H_4),-9(H_2)	MK-9(H_4,H_6,H_8)	MK-10(H_2,H_4,H_6)	MK-9(H_4)	MK-7
Mycolic acids[d]	+	–	–	–	–
Mol% G + C of the DNA	64–72	65–77	65.7[g]	70–75	65–70

[a]Data taken from Lechevalier et al. (1977, 1981), Collins and Shah (1984), Embley et al. (1984), and Goodfellow and Cross (1984b).
[b]Symbols: see Table 26.3; also *NC,* no characteristic sugars.
[c]See Lechevalier and Lechevalier (1980) for details of methods and explanations of wall chemotypes and sugar patterns.
[d]See footnote *c* in Table 26.3 for references to methods.
*e*Minor components.
[f]Phospholipid types after Lechevalier et al. (1977, 1981).
[g]Data from analysis of one strain (S-C. An, unpublished).

Table 26.5.
Differential characteristics of the genus **Nocardia** *and wall chemotype IV taxa lacking mycolic acid*[a,b]

Characteristics	*Nocardia*	*Actinopolyspora*	*Faenia (Micropolyspora)*	*Pseudonocardia*	*Saccharomonospora*	*Saccharopolyspora*
Morphological characters						
Substrate mycelium						
Fragmentation	+	D	D	D–	–	+
Spores	D	–	+	–	–	–
Aerial mycelium						
Spores	D	+	+	+	+	+
Spores in long chains (>20 spores)	D	+	D	+	–	+
Growth characters						
Extreme halophile	–	+	–	–	–	–
Lipid characters[c]						
Fatty acids						
Unsaturated	+	–	–	D	+	+
Tuberculostearic	+	–	–	D	–	–
Iso- and *anteiso-*	–	+	+	+	+	+
Phospholipid type[c]	II	III	III	III	II	III
Predominant menaquinone	MK-8(H_4),-9(H_2)	MK-9(H_4)	MK-9(H_4,H_6)	MK-8(H_4),-9(H_4)	MK-9(H_4)	MK-9(H_4)
Mycolic acids	+	–	–	–	–	–
Mol% G + C of the DNA	64–72	64	ND	79	74–75	77

[a]Data taken from Lechevalier et al. (1977, 1981), Lechevalier and Lechevalier (1981a); Goodfellow and Cross (1984b), and Goodfellow and Minnikin (1984).
[b]Symbols: see Table 26.3.
[c]See footnote *c* in Tables 26.3 and 26.4 for references to methods.

(1971) as a species of *Gordona*, a genus that is now invalid, *aurantiaca* strains have been shown to have chemical and physiological properties that clearly place them apart from rhodococci, nocardiae, and mycobacteria (see article on "*Rhodococcus*"). They have a cell wall of chemotype IV, mycolates that are intermediate between those of *Nocardia* and *Mycobacterium* (giving rise, on pyrolysis, to C_{20} and C_{22} saturated and unsaturated fatty acid esters), and unusual fully saturated menaquinones having nine isoprene units (MK-9). Numerical taxonomic studies showed the strains of this group to be distinctive and possibly worthy of placement in a new genus (Goodfellow et al., 1978). Certain strains received under the name of *Mycobacterium album* Söhngen have been shown to be related to the *aurantiaca* group in immunodiffusion and lipid analyses (Horan, 1971; Ridell et al., 1985).

Further Reading

Goodfellow, M. and T. Cross, 1984. Classification. *In* Goodfellow, Mordarski and Williams (Editors), The Biology of the Actinomycetes. Academic Press, London pp. 7-164.

Goodfellow, M. and D.E. Minnikin (Editors). 1985. Chemical Methods in Bacterial Systematics. Society for Applied Bacteriology, Technical Series No. 20. Academic Press, London.

Lechevalier, M.P. 1976. The taxonomy of the genus *Nocardia:* some light at the end of the tunnel? *In* Goodfellow, Brownell and Serrano (Editors), The Biology of the Nocardiae. Academic Press, London, pp. 1-38.

Lechevalier, M.P., H. Prauser, D.P. Labeda and J.S. Ruan. 1986. Two new genera of nocardioform actinomycetes: *Amycolata* gen. nov. and *Amycolatopsis* gen. nov. Int. J. Syst. Bacteriol. 36:29-37.

Minnikin, D.E. and M. Goodfellow. 1980. Lipid composition in the classification and identification of acid-fast bacteria. *In* Goodfellow and Board (Editors), Microbiological Classification and Identification. Academic Press, London, pp. 189-256.

Minnikin, D.E. and A.G. O'Donnell. 1984. Actinomycete envelope lipid and peptidoglycan composition. *In* Goodfellow, Mordarski and Williams (Editors), The Biology of the Actinomycetes. Academic Press, London, pp. 337-388.

Williams, S.T., G.P. Sharples, J.A. Serrano, A.A. Serrano and J. Lacey. 1976. The micromorphology and fine structure of nocardioform organisms. *In* Goodfellow, Brownell and Serrano (Editors), The Biology of the Nocardiae. Academic Press, London, pp. 103-140.

Differentiation and characteristics of the species of the genus **Nocardia**

The differential characteristics of the species of the genus *Nocardia* are indicated in Tables 26.6 and 26.7. Other characteristics of the species are listed in Table 26.8.

Table 26.6.
Characteristics differentiating the species of the genus **Nocardia** *(Gordon tests)*[a]

Characteristics	1. *N. asteroides*[b] 2. *N. farcinica*	3. *N. brasiliensis*	4. *N. otitidiscaviarum*	5. *N. amarae*	6. *N. brevicatena*	7. *N. carnea*	8. *N. vaccinii*	9. *N. transvalensis*
Decomposition of								
Adenine	−	−	−	−	−	−	−	d
Casein	−	+	−	−	−	−	−	−
Hypoxanthine	−	+	+	−	−	−	−	+
Tyrosine	−	+	d	−	−	−	−	+
Urea	+	+	+	+	−	−	+	+
Xanthine	−	−	+	−	−	−	−	d
Acid from								
Adonitol	−	−	−	−	−	−	−	+
Arabinose	−	−	d	−	−	−	+	−
Erythritol	−	−	−	−	−	−	−	+
Galactose	d	+	−	−	−	+[c]	+[c]	+[c]
Glucose	+	+	+	+	−	+	+	+
Inositol	−	+	+	+	−	d	d	d
Maltose	−	−	−	+	−	−	d	−
Mannose	d	d	d	+	−	d	−	d
Rhamnose	d	−	−	+	−	−	d	−
Sorbitol	−	−	−	−	−	+	d	d
Decarboxylation of								
Citrate	d	+	d	−	−	d	+	+
Mucate	−	−	−	−	−	−	+	−
Production of								
Nitrate reductase	+	+	+	+	−	+	+	+
Growth at/in								
10°C	d	d	d	−	−	d	−	−
45°C	d	−	d	−	−	d	−	−
Lysozyme broth	+	+	+	−	+	+	+	+
Survival at 50°C/8 h	+	−	+	−	+	+	+	+
Pyrolytic fragments of mycolic acids	S	S	S	U	S	U[c]	U[c]	S

[a]Symbols: see Table 26.3; also *d*, 11-89% of the strains are positive; *S*, saturated, and *U*, unsaturated methyl esters of fatty acids principally released.
[b]*N. asteroides* and *N. farcinica* cannot be distinguished on the basis of Gordon tests.
[c]Based on analysis of the type strain.

Table 26.7.
Characteristics differentiating the species of the genus **Nocardia** *(Goodfellow tests)*[a]

Characteristics	1. *N. asteroides*	2. *N. farcinica*	3. *N. brasiliensis*	4. *N. otitidiscaviarium*	5. *N. amarae*	6. *N. brevicatena*	7. *N. carnea*	8. *N. vaccinii*	9. *N. transvalensis*
Acid fastness	d	d	d	d	−	d	d	d	d
Decomposition of									
Casein	−	−	+	−	−	−	−	−	−
Elastin	−	−	+	−	−	−	−	−	−
Hypoxanthine	−	−	+	+	−	−	−	−	+
Testosterone	+	+	+	−	−	+	+	−	−
Tyrosine	−	−	+	−	−	−	−	−	−
Xanthine	−	−	−	+	−	−	−	−	d
Production of									
Nitrate reductase	+	+	+	+	+	−	+	+	+
Urease	+	+	+	+	+	−	−	+	+
Resistance to lysozyme	+	+	+	+	−	+	+	+	+
Growth on sole carbon source (1% w/v)									
Adonitol	−	−	−	−	−	+	−	−	+
L-Arabinose	−	−	−	−	−	−	+	d	−
Galactose	d	d	+	−	d	+	d	+	−
Inositol	−	−	+	+	+	+	−	+	−
Mannitol	−	−	+	+	+	+	−	+	−
Mannose	+	+	+	d	+	+	−	+	+
Melezitose	−	−	−	−	−	+	−	−	−
Rhamnose	−	+	−	−	+	+	−	+	−
Adipic acid (0.1% w/v)	−	−	−	−	+	−	d	−	+
2,3-Butylene glycol (1% v/v)	−	+	−	−					
Pimelic acid (0.1% w/v)	−	−	−	−	+	−	−	+	−
1,2-Propylene glycol (1% v/v)	−	+	−	−					
Sebacic acid (0.1% w/v)	+	d	−	+	+	−	d	−	−
Predominant menaquinone	MK-8(H_4)	MK-8(H_4)	MK-8(H_4)	MK-8(H_4)	MK-9(H_2)	MK-8(H_4)	MK-8(H_4)	MK-8(H_4)	MK-8(H_4)

[a]Symbols: see Table 26.3; also *d*, 11–89% of the strains are positive.

Table 26.8.
Other characteristics of the species of the genus **Nocardia** *(Gordon tests)*[a]

Characteristics	1. *N. asteroides* 2. *N. farcinica*	3. *N. brasiliensis*	4. *N. otitidiscaviarum*	5. *N. amarae*	6. *N. brevicatena*[b]	7. *N. carnea*	8. *N. vaccinii*	9. *N. transvalensis*
Acid from								
Cellobiose	−	−	−	−	−	−	−	−
Glycerol	+	+	+	+	+	+	+	+
Lactose	−	−	−	−		−	−	−
Mannitol	−	+	d	+	−	+	+	d
Melezitose	−	−	−	−	−	−	−	−
Melibiose	−	−	−	−	−	−	−	−
α-Methyl-D-glucoside	−	−	−	−	−	−	−	−
Raffinose	−	−	−	−	−	−	−	−
Trehalose	d	+	d	+	+	+	d	d
Xylose	−	−	−	−	−	−	d	−
Utilization of								
Benzoate	−	−	−	−	−	−	−	−
Succinate	+	+	+	+	d	+	+	+
Tartrate	−	−	−	−	−	−	−	−
Hydrolysis of								
Esculin	+	+	+	+	+	+	+	+
Potato starch	d	d	d	d	−	d	+	+
Acid fastness	d	d	d	−	Weak	d	+	d

[a]Symbols: see Table 26.3; also *d*, 11–89% of the strains are positive.
[b]Grown at 37°C (all others at 28°C).

List of species of the genus **Nocardia**

1. **Nocardia asteroides** (Eppinger) Blanchard 1896, 856.[AL] (*Cladothrix asteroides* Eppinger 1891, 309.)

as.ter.oi'des. Gr. adj. *asteroides* starlike.

Morphology very diverse in terms of degree of fragmentation, amount of aerial hyphae, colony color, and form. The most typical strains have a salmon-colored to pinkish, or orange-tan matte growth with fringes of white to pinkish, sparse, aerial mycelium and no soluble pigment. More rarely, vegetative growth may be off-white, gray, or brown and a yellowish-brown soluble pigment may be formed. Entire glistening colonies are not found. Although the aerial hyphae are usually sterile or tend to break up into fragments of irregular length, some strains form typical chains of poorly formed spores on the aerial and vegetative mycelium (Fig. 26.1). Usually weakly acid fast.

Chemical characteristics as for true nocardiae. The predominant menaquinone is MK-8(H_4).

Physiological characteristics are given in Tables 26.6–26.8. As previously discussed (see "Taxonomic Comments"), *N. asteroides* cannot be distinguished chemically or physiologically by means of the Gordon tests from *N. farcinica* (as represented by its type strain ATCC 3318).

The mol% G + C of the DNA is 63–70 (T_m).

Some strains are pathogenic for man and animals but most are soil saprophytes.

An extensive review of the ecology of this taxon has been published (Orchard, 1981).

The *N. asteroides* taxon is considered to be heterogeneous by many workers (see text).

Type strain: ATCC 19247.

2. **Nocardia farcinica** Trevisan 1889, 9.[AL] (Nom. cons. Opin. 13 Jud. Comm. 1954, 153; see Int. Code of Nom. 1958, 166 and note in *Index Bergeyana* p. 753.)

far.ci'ni.ca. L. v. *farcio* stuff; L. n. *farciminum* a disease of horses; Fr. n. *farcin* farcy or glanders; M.L. fem. adj. *farcinica* relating to farcy.

See discussion under *N. asteroides*, above.

Physiological characteristics are given in Tables 26.6–26.8.

The mol% G + C of the DNA is 66–71 (T_m).

Some strains are pathogenic for man and animals.

Type strain: ATCC 3318.

3. **Nocardia brasiliensis** (Lindenberg) Pinoy 1913, 936.[AL] (*Discomyces brasiliensis* Lindenberg 1909, 279.)

bra.si.li.en'sis. M.L. adj. *brasiliensis* pertaining to Brazil, South America.

A typical strain will have pinkish or orange-tan to tan or brown vegetative growth with moderate to abundant nonfragmenting aerial hyphae usually off-white to pink-gray in color. Soluble dark pigments occur with very high frequency. Usually weakly acid fast.

Chemical characteristics as for true nocardiae. The predominant menaquinone is MK-8(H_4).

Physiological characteristics are given in Tables 26.6–26.8.

The mol% G + C of the DNA is 67–68 (T_m).

Isolated from various kinds of nocardioses, including mycetoma, and occasionally from soil.

Type strain: ATCC 19296.

4. **Nocardia otitidiscaviarum** Snijders 1924, LXXXVII.[AL]

o.ti'ti.dis.cav.i.ar'um. M.L. n. *otitis* inflammation of the ear; M.L. n. *Cavia* (gen. plur. *caviarum*) generic name of the cavy or guinea pig; M.L. gen. n. *otitidiscaviarum* of ear disease of guinea pigs.

Aerial hyphae usually very sparse and off-white in color, sometimes only visible microscopically, but almost always present. Cream, grayish to peach-tan to purplish vegetative hyphae. Soluble pigments variably present. Hyphae weakly acid fast.

Chemical characteristics of true nocardiae. Predominant menaquinone MK-8(H_4).

Physiological characteristics are given in Tables 26.6–26.8.

The mol% G + C of the DNA is 67–72 (T_m).

Has been isolated from soil but some strains are pathogenic for man and animals.

Type strain: ATCC 14629.

5. **Nocardia amarae** Lechevalier and Lechevalier 1974, 286.[AL]

am.ar'ae. Gr. n. *amara* sewage duct; M.L. gen. n. *amarae* of a sewage duct.

Colonies always tan; aerial hyphae visible only microscopically. No soluble pigments produced. Hyphae usually show banding both in the natural habitat (foam of secondary treatment sewage plants) and in vitro under phase contrast. Characteristic irregular thickenings (up to 0.3 μm) of the cell walls are visible in sections. Not acid fast.

Chemical characteristics of true nocardiae. Like *N. vaccinii* and *N. carnea*, the mycolic acids of *N. amarae* strains have a monounsaturated α-branch. The predominant menaquinone is MK-9(H_2), and in this it differs from other true nocardiae.

Physiological characteristics are given in Tables 26.6–26.8.

The mol% G + C of the DNA is 71 (T_m) (type strain).

Isolated from foam formed on the surface of aeration tanks in activated-sludge sewage-treatment plants.

Type strain: ATCC 27808.

6. **Nocardia brevicatena** (Lechevalier, Solotorovsky and McDurmont) Goodfellow and Pirouz 1982a, 384.[VP†] (Effective publication: Goodfellow and Pirouz 1982b, 523.) (*Micropolyspora brevicatena*, Lechevalier, Solotorovsky and McDurmont 1961, 13.)

bre.vi.cat.e'na. L. adj. *brevis* short; L. n. *catena* chain; M.L. n. *brevicatena* short chain.

Colonies typically orange-tan with off-white, sparse to moderate aerial mycelium. No soluble pigments. Oval to round, **very well-formed spores** in short chains (2–7) borne on the vegetative and aerial mycelium and also at and **just above the surface of the agar.** Weakly acid fast.

Chemical characteristics of true nocardiae. Predominant menaquinone is MK-8(H_4).

Physiological characteristics are given in Tables 26.6–26.8.

The mol% G + C of the DNA is 66–68 (T_m).

Isolated from human sputum. Not pathogenic for mice.

Type strain: ATCC 15333.

7. **Nocardia carnea** (Rossi-Doria) Castellani and Chalmers 1913, 818.[AL] (*Streptothrix carnea* Rossi-Doria 1891, 415.)

car.ne'a. L. fem. adj. *carnea* fleshy.

Cream or peach-colored vegetative growth with sparse to abundant, white to pinkish aerial mycelium. Aerial hyphae are usually sterile, although rudimentary "conidia" have been reported in a few strains. Rarely acid fast. Soluble pigments are uncommon.

The chemical characteristics of the taxon are those of true nocardiae. Like *N. amarae* and *N. vaccinii*, the α-branches of the mycolic acids of *N. carnea* are monounsaturated. The predominant menaquinone type is MK-8(H_4).

Physiological characteristics are given in Tables 26.6–26.8.

The mol% G + C of the DNA is 64–68 (T_m).

Isolated from soil and air.

Type strain: ATCC 6847.

8. **Nocardia vaccinii** Demaree and Smith 1952, 251.[AL]

vac.cin'i.i. M.L. n. *Vaccinium* generic name of the blueberry; M.L. gen. n. *vaccinii* of *Vaccinium*.

Cream to peach-colored colonies with moderate to sparse white aerial

† *VP* denotes that this name has been validly published in the official publication, *International Journal of Systematic Bacteriology.*

mycelium. Conidia are rarely formed on the aerial mycelium. Partially acid fast. Soluble pigments not usually formed.

Chemical characteristics are those of true nocardiae. Like *N. amarae* and *N. carnea*, the α-branches of the mycolic acids of *N. vaccinii* are monounsaturated. The predominant menaquinone is MK-8(H_4).

Physiological characteristics are given in Tables 26.6–26.8.

Strains are pathogenic to blueberry plants, causing galls.

Type strain: ATCC 11092.

9. **Nocardia transvalensis** Pijper and Pullinger 1927, 155.[AL]

trans.val.en'sis. M.L. adj. *transvalensis* pertaining to the Transvaal, South Africa.

Colonies usually pale tannish cream or purplish. Moderate to heavy sterile aerial mycelium. No soluble pigments have been observed. Partially acid fast when grown on glycerol agar.

Chemical characteristics as for true nocardiae. The predominant menaquinone is MK-8(H_4).

Physiological characteristics are given in Tables 26.6–26.8.

The mol% G + C of the DNA is 67.0 (T_m).

Isolated from mycetoma of the foot in South Africa.

Type strain: ATCC 6865.

Species Incertae Sedis

a. **Nocardia orientalis** (Pittenger and Brigham) Pridham 1970, 42.[AL] (*Streptomyces orientalis* Pittenger and Brigham 1956, 642.)

or.i.ent.al'is. L. adj. *orientalis* of the orient.

Cream, yellowish-tan, peach, or brown vegetative growth, aerial hyphae sparse to moderate, white to off-white to cream. Soluble pigments pale yellow-brown, light brown, or greenish yellow. Long chains of smooth, squarish conidia formed by about half the cultures assigned to this taxon by Gordon et al. (1978).

Chemical characteristics of "mycolateless" nocardiae. The predominant menaquinone is MK-9(H_4).

Physiological reactions are given in Tables 26.9–26.11. Except for slightly greater resistance to lysozyme and a weak-to-negative inositol utilization, "*N. lurida*" ATCC 14930 (Grundy et al., 1957), the producer of the antibiotic ristocetin, has the same properties as *N. orientalis*. The type strain of *N. orientalis* produces vancomycin.

The mol% G + C of the DNA of the type strain is 66.0 (T_m).

Isolated from soil, vegetable matter and clinical specimens. Pathogenicity is unknown.

Type strain: ATCC 19795 (ISP 5040).

b. **Nocardia mediterranei** (Margalith and Beretta) Thiemann, Zucco and Pelizza 1969b, 148.[AL] (*Streptomyces mediterranei* Margalith and Beretta 1960b, 321.)

med.i.ter.ra'ne.i. L. neut. n. *mediterraneum* interior of the land; L. gen. *mediterranei*, of the interior of the land, from the Mediterranean area.

Yellowish- to pinkish-tan or orange-brown vegetative mycelium; very sparse, sterile aerial mycelium visible only under the microscope on most media. When formed, the aerial mycelium en masse is white to pinkish. Soluble pigments yellowish pink, orange, or orange-brown. Some strains may form chains of smooth or spiny (short spines) oblong spores.

Cell chemistry as for "mycolateless" nocardiae. Major menaquinone types are MK-9 (H_4, H_6).

Physiological reactions are given in Tables 26.9–26.11.

The mol% G + C of the DNA is 67–69 (T_m).

The DNA can be isolated and restricted in the same way as that of streptomycetes (Kieser et al., 1981).

Most known strains of this taxon produce rifamycins. Isolated from soil.

Type strain: ATCC 13685 (ISP 5501).

Table 26.9.
Differential characteristics of **Nocardia** *species lacking mycolates (Gordon tests)*[a]

Characteristics	a. *N. orientalis*	b. *N. mediterranei*[b]	c. *N. rugosa*[b]	d. *N. sulphurea*[b]	e. *N. saturnea*[b]	f. *N. autotrophica*	g. *N. hydrocarbonoxydans*[b]
Hydrolysis of							
Adenine	–	–	–	–	–	+	–
Casein	+	+	+	+	–	–	–
Starch	+	–	–	–	–	+	+
Urea	+	+	+	–	+	+	–
Xanthine	d	–	+	–	+	d	–
Decarboxylation of							
Benzoate	d	–	+	–	+	d	–
Citrate	+	+	–	+	–	+	–
Acid from							
Adonitol	+	–	+	–	–	+	–
Arabinose	+	+	+	–	+	d	+
Cellobiose	+	+	–		+	d	+
Erythritol	+	–	+	–	–	+	+
Inositol	+	+	–	+	+	d	+
Lactose	+	+	–	–	–	–	+
Maltose	d	+	–	+	+	+	–
Melibiose	d	+	–	–	–	–	–
α-Methyl-D-glucoside	+	+	–	–	+	d	–
Rhamnose	d	+	+	–	–	–	+
Sorbitol	d	–	–	–	–	+	–
Lysozyme resistance	d	d	–	+	–	–	–
Growth at							
10°C	d	+	+	–	+	+	+
45°C	d	–	+	–	–	–	–
Phospholipid pattern	P2	P2	P2	P2	P3	P3	P3

[a]Symbols: see Table 26.3; also *d*, 11–89% of the strains are positive.
[b]Reactions given for the type strain.

Table 26.10.
Other characteristics of **Nocardia** *species lacking mycolates (Gordon tests)*[a]

Characteristics	a. *N. orientalis*	b. *N. mediterranei*[b]	c. *N. rugosa*[b]	d. *N. sulphurea*[b]	e. *N. saturnea*[b]	f. *N. autotrophica*	g. *N. hydrocarbonoxydans*
Hydrolysis of							
Esculin	+	+	+	+	+	+	+
Hypoxanthine	+	+	+	+	+	+[c]	−
Tyrosine	+	+	+	+	+	+[c]	−
Production of							
Nitrate reductase	d	+	−	+	+	d	+
Decarboxylation of							
Mucate	d	−	−	−	−	−	−
Succinate	+	+	+	+	+	+	+
Tartrate	−	−	−	−	−	−	−
Acid from							
Glucose	+	+	+	+	+	+	+
Glycerol	+	+	+	+	+	+	+
Mannitol	+	+	+	+	+	+	−
Mannose	+	+	+	+	+	+	+
Raffinose	d	+	−	−	−	−	−
Trehalose	d	+	+	+	+	+	−
Xylose	+	+	+	−	+	+	+

[a]Symbols: see Table 26.3; also *d*, 11–89% of the strains are positive.
[b]Reactions given for the type strain.
[c]The type strain is negative for both these reactions (M. P. Lechevalier, unpublished).

Table 26.11.
Reactions of **Nocardia** *species lacking mycolates in the ISP carbohydrate utilization tests*[a,b]

Characteristics	a. *N. orientalis*	b. *N. mediterranei*[b]	c. *N. rugosa*	d. *N. sulphurea*	e. *N. saturnea*[b]	f. *N. autotrophica*	g. *N. hydrocarbonoxydans*
Utilization of							
Arabinose	+	+	+	−	−	+	−
Fructose	+	+	+	+	+	+	+
Glucose	+	+	+	+	+	+	+
Inositol	+	+	−	−	+	+	+
Mannitol	+	+	+	+	+	+	+
Raffinose	−	−	−	−	−	+	−
Rhamnose	d	+	+	−	−	+	−
Sucrose	+	+	−	+	+	+	+
Xylose	+	+	+	−	+	+	+

[a]Methods and data from the International Streptomyces Project (ISP) (Shirling and Gottlieb, 1968a, 1972; Pridham and Lyons, 1969) and M. P. Lechevalier (unpublished); reactions given for the type species only.
[b]Symbols: see Table 26.3; also *d*, 11–89% of the strains are positive.

c. **Nocardia rugosa** (ex DiMarco and Spalla 1957) nom. rev.
ru.go'sa. L. adj. *rugosa* full of wrinkles.

Off-white, yellowish to tannish, glistening, pasty vegetative growth. Aerial mycelium totally lacking even microscopically. Vegetative hyphae fragment within 24–48 h on rich media. Brownish soluble pigments sometimes formed. Not, to slightly, acid fast.

Chemical characteristics are those of "mycolateless" nocardiae.

Physiological characteristics are given in Tables 26.9–26.11.

The mol% G + C of the type strain is 68.9 (T_m).

The type strain produces vitamin B_{12}.

Isolated from cattle rumen.

Type strain: IMRU 3760.

d. **Nocardia sulphurea** sp. nov.‡
sul.phu're.a. L. adj. *sulphureus* of sulfur, referring to the yellow color of the vegetative hyphae.

Yellow to brown to blackish-brown vegetative growth. White to yellowish-white sterile aerial mycelium formed copiously on some media. Soluble dark pigment elaborated on most media. Very characteristic cracking and curling occurs in slant culture at base of tube, revealing the vegetative hyphae beneath.

Cell chemistry characteristic of "mycolateless" nocardiae.

Physiological reactions are given in Tables 26.9–26.11.

The mol% G + C of the DNA of the type strain is 66.8 (T_m).

Produces the antibiotic chelocardin.

‡**Editorial note:** The authors of this chapter wish to revive *N. sulphurea* Oliver and Sinclair 1964; however, that name was not validly published in a patent (*Code*, Rule 25(b)5) and, therefore, cannot be revived (*Code*, Rule 28a). Let it be noted that the authors have chosen the same type strain and are, in fact, reviving that name although the citation cannot mention the original authors.

Isolated from soil.
Type strain: ATCC 27624.

e. **Nocardia saturnea** Hirsch 1960, 401.[AL]
sa.turn'e.a. L. n. *saturnus* Saturn, Roman god of seed-sowing; M.L. adj. *saturnea* pertaining to Saturn, referring to the colonies, which have a Saturnian shape.

Yellowish-white to bright butter-yellow vegetative mycelium. White to yellowish abundant aerial hyphae on meager media forming chains of long rectangular fragments. No soluble pigments. Not acid fast.

Cell chemistry typical of "mycolateless" nocardiae.

Physiological reactions are given in Tables 26.9-26.11.

Isolated from air and compost.

Type strain: ATCC 15809.

f. **Nocardia autotrophica** (Takamiya and Tubaki) Hirsch 1960, 405.[AL] (*Streptomyces autotrophicus* Takamiya and Tubaki 1956, 59.)
au.to.tro'phi.ca. G. n. *autos* self; G. part. *trophikos* nursing; M.L. fem. adj. *autotrophica* self-nourishing, referring to the ability to grow at the expense of H_2 and CO_2.

Vegetative hyphae yellow to brownish, tending to give rise to chains of chlamydospore-like cells of unequal size. Aerial mycelium white to cream, with long chains of long oval to cylindrical spores. Pale yellow to yellow-brown soluble pigments sometimes formed late.

Cell chemistry typical of "mycolateless" nocardiae. The predominant menaquinone is MK-8(H_4).

Physiological reactions are given in Tables 26.9-26.11.

Hydrogen utilized in presence of O_2 and CO_2 in a mineral medium. The mol% G + C of the DNA of the type strain is 69.8 (T_m).

Isolated from phosphate buffer solution, aluminum hydroxide gel, vegetable matter, soil, clinical specimens. Pathogenicity is unknown.

Type strain: ATCC 19727.

g. **Nocardia hydrocarbonoxydans** Nolof and Hirsch 1962, 275.[AL]
hy.dro.car.bon.oxy'dans. G. n. *hydro* water, L. n. *carbo* ember, charcoal; Gr. adj. *oxys* sharp, acid; L. part. *dans* giving; M.L. pres. part. *hydrocarbonoxydans* oxidizing hydrocarbons.

Off-white, yellow-white, gold, to brown vegetative growth, sparse to moderate white aerial mycelium. Aerial and vegetative mycelia fragment into long squarish units. No to little soluble pigment formed.

Physiological reactions are given in Tables 26.9-26.11. Utilizes gaseous aliphatic hydrocarbons (C_6-C_{14}) for growth (Nolof, 1962).

Chemistry as for "mycolateless" nocardiae.

The mol% G + C of the type strain is 68.9 (T_m).

Air contaminant, isolated from a silica gel plate.

Type strain: ATCC 15104.

Species Incertae Sedis (Not Further Described)

Nocardia nova Tsukamura 1983, 896.[VP] (Effective publication: Tsukumura 1982b, 1115.)

"*Nocardia paratuberculosis*" Tsukamura 1982b, 1114.

Differentiation and characteristics of "**Nocardia**" *species lacking mycolic acids*

The differential characteristics of these species *incertae sedis* are shown in Table 26.9. Other characteristics of these species are shown in Tables 26.10 and 26.11. Table 26.12 presents a comparison of the differential characteristics of actinomycete taxa having no mycolates and a phospholipid chemotype PIII (phosphatidylcholine as diagnostic phospholipid).

Table 26.12.
Differentiating physiological characteristics of taxa lacking mycolates and containing phosphatidylcholine (phospholipid chemotype PIII) (Gordon tests)[a]

Characteristics	*Nocardia autotrophica*[b]	*Nocardia hydrocarbonoxydans*[c]	*Nocardia saturnea*[d]	*Pseudonocardia thermophila*[e]	*Saccharopolyspora*[f] *hirsuta*	*Faenia*[g] *(Micropolyspora)*
Hydrolysis of						
Adenine	+	−	−	−	+	−
Casein	−	−	−	−	+	+
Hypoxanthine	+	−	+	−	+	−
Tyrosine	+	−	+	+	+	−
Urea	+	−	+	+	+	−
Xanthine	d	−	+	−	+	+
Utilization of						
Benzoate	d	−	+	+	+	+
Citrate	+	−	−	−	+	+
Acid from						
Arabinose	d	+	+	+	−	−
Erythritol	+	+	−	−	+	+
Lactose	−	+	−	+	+	+
Rhamnose	−	+	−	+	+	−
Growth at						
45°C	−	−	−	+	+	+

[a]Symbols: see Table 26.3; also *d*, 11-89% of the strains are positive.
[b]Based on reactions of 31 strains.
[c]Based on reactions of the type strain IMRU 1407 (ATCC 15104).
[d]Based on reactions of type strain IMRU 1181 (ATCC 15809).
[e]Based on reactions of type strain Henssen A 18 (ATCC 19285).
[f]Based on reactions of 24 strains of *S. (Nocardia) hirsuta* (Gordon et al., 1978).
[g]Based on reactions of *Faenia rectivirgula (Micropolyspora faenia)* Lacey A94 (ATCC 33515; type strain).

Genus **Rhodococcus** *Zopf 1891, 28*[AL]

MICHAEL GOODFELLOW

Rho.do.coc′cus. Gr. n. *rhodon* the rose; Gr. n. *coccos* a grain; M.L. neut. n. *Rhodococcus* a red coccus.

Rods to extensively branched substrate mycelium may be formed. In all strains the morphogenetic cycle is initiated with the coccus or short rod stage, with different organisms showing a succession of more or less complex morphological stages by which the completion of the growth cycle is achieved. Thus, **cocci may merely germinate into short rods, form filaments with side projections, show elementary branching** or, in the most differentiated forms, produce **extensively branched hyphae**. The next generation of cocci or short rods are formed by the **fragmentation of the rods, filaments, and hyphae**. Some strains produce feeble microscopically visible aerial hyphae, which may be branched, or aerial synnemata consisting of unbranched filaments that coalesce and project upward. Rhodococci are nonmotile and form neither conidia nor endospores.

Gram-positive. Partially acid-alcohol fast at some stage of growth. **Aerobic. Chemo-organotrophic**, having an oxidative type of metabolism. **Catalase positive**. Most strains grow well on standard laboratory media at 30°C, although some require thiamin. Colonies may be rough, smooth, or mucoid and pigmented buff, cream, yellow, orange, or red, although colorless variants do occur. Rhodococci are arylsulfatase negative, sensitive to lysozyme, and unable to degrade casein, cellulose, chitin, elastin, or xylan. They are able to use a wide range of organic compounds as sole sources of carbon for energy and growth.

The cell wall peptidoglycan contains major amounts of ***meso*-diaminopimelic acid (*meso*-DAP), arabinose, and galactose**. The organisms contain **diphosphatidylglycerol, phosphatidylethanolamine, and phosphatidylinositol mannosides, dihydrogenated menaquinones with either eight or nine isoprene units** as the major isoprenolog, **large amounts of straight-chain, unsaturated, and tuberculostearic acids, and mycolic acids with 32–66 carbons and up to four double bonds. The fatty acid esters released on pyrolysis gas chromatography of mycolic esters contain 12–18 carbons. The mol% G + C** of the DNA ranges from 63 to 72% (T_m).

The organism is widely distributed but is particularly abundant in soil and herbivore dung. Some strains are pathogenic for man and animals.

Type species: *Rhodococcus rhodochrous* (Zopf) Tsukamura 1974a, 43.

Further Descriptive Information

Rhodococci have had a long and confused taxonomic pedigree (Cross and Goodfellow, 1973; Bradley and Bond, 1974; Bousfield and Goodfellow, 1976; Goodfellow and Minnikin, 1981a, 1981c; Goodfellow and Wayne, 1982; Goodfellow and Cross, 1984b). The epithet *rhodochrous* (Zopf, 1889) was reintroduced in 1957 by Gordon and Mihm for actinomycetes that had properties in common with both mycobacteria and nocardiae but that carried a multiplicity of generic and specific names. The taxon was provisionally assigned to the genus *Mycobacterium* but subsequent studies, based on chemical, genetic, and numerical phenetic methods, showed that it merited generic rank and was heterogeneous. The genus *Rhodococcus* was subsequently resurrected and redefined for *rhodochrous* strains (Tsukamura, 1974a; Goodfellow and Alderson, 1977) and currently accommodates 16 species. Many of the latter were circumscribed and described in numerical phenetic surveys (Tacquet et al., 1971; Tsukamura, 1971, 1973, 1974a, 1975b, 1978; Goodfellow, 1971; Goodfellow et al., 1974, 1982a, 1982c; Goodfellow and Alderson, 1977; Rowbotham and Cross, 1977b; Barton and Hughes, 1982; Helmke and Weyland, 1984), and some have been shown to be homogeneous on both chemical and genetic grounds (Bradley and Mordarski, 1976; Mordarski et al., 1976, 1977, 1980a, 1981a; Minnikin and Goodfellow, 1980).

The genus *Rhodococcus* encompasses a wide range of morphological diversity and is defined primarily on the basis of wall envelope composition. Thus, at present, the genus should be restricted to actinomycetes that have: (a) a peptidoglycan consisting of *N*-acetylglucosamine, *N*-glycolylmuramic acid, D- and L-alanine, and D-glutamic acid with *meso*-DAP as the diamino acid; (b) arabinose and galactose as diagnostic wall sugars (i.e. rhodococci have a wall chemotype IV and a whole-cell sugar pattern type A sensu Lechevalier and Lechevalier, 1970b); (c) a phospholipid pattern consisting of diphosphatidylglycerol, phosphatidylethanolamine, phosphatidylinositol, and phosphatidylinositol mannosides; (d) a fatty acid profile containing major amounts of straight-chain, unsaturated, and tuberculostearic acids (i.e. a type IV fatty acid pattern sensu Lechevalier et al., 1977), and mycolic acids with 32–66 carbons (Minnikin and Goodfellow, 1980, 1981); and (e) dihydrogenated menaquinones with either eight or nine isoprene units (Collins et al., 1977, 1985).

Rhodococci have no distinctive morphological features other than the ability of many strains to form hyphae that fragment into rods and cocci, but they do show considerable heterogeneity (Locci, 1976, 1981; Williams et al., 1976; Nesterenko et al., 1982; Helmke and Weyland, 1984; Locci and Sharples, 1984). Thus, *R. bronchialis*, *R. chlorophenolicus*, *R. maris*, *R. rubropertinctus*, *R. sputi*, and *R. terrae* are amycelial, and show a rod-coccus cycle similar to that described for *Arthrobacter* (Clark, 1979), but further morphological development is represented by *R. erythropolis*, *R. globerulus*, *R. rhodnii*, and *R. rhodochrous*, which exhibit elementary branching prior to fragmentation. *Rhodococcus equi* shows traces of elementary branching at early stages of growth and may represent a link between the two groups. *Rhodococcus coprophilus*, *R. fascians*, *R. marinonascens*, and *R. ruber* form a third group that produces well-branched substrate mycelia. The time taken to complete the developmental cycle ranges from 24 h in relatively undifferentiated forms such as *R. equi* to several days for those such as *R. coprophilus* that show the most pronounced morphological differentiation (Locci et al., 1982). However, the timing of the fragmentation process is influenced by environmental factors (Williams et al., 1976), which may act through their effects on growth rates. Rhodococci do not usually form aerial hyphae, but exceptions are *R. coprophilus* and *R. ruber*, which exhibit feeble aerial hyphae, and *R. bronchialis* strains, which produce serial synnemata (Locci and Sharples, 1984). Surfaces of colonies are covered by a few strands to several sheets of slimy extracellular material (Williams et al., 1976). Sections of cells show a typical Gram-positive wall, lipid globules, and polyphosphate granules are characteristic cell inclusions, mesosomes are common, and fibrillar materials are usually visible on the surface of negatively stained or freeze-etched cells (Williams et al., 1976; Beaman et al., 1978).

All members of the genus can be cultivated on standard nutrient media, but they also grow well on very simple substrates, using ammonia, amino acids, and nitrate as nitrogen sources and a host of sugars and organic acids as carbon sources. Glucose, fructose, mannose, raffinose, sucrose, acetate, butyrate, and propionate are used as sole sources of carbon. Some strains require thiamin and others grow well on alkanes (Starr, 1949; Tárnok, 1976; Rowbotham and Cross, 1977b; Nesterenko et al., 1978a). Oligocarbophily has been described for a rhodococcal strain labeled "*Nocardia corallina*" (Tárnok, 1976). Optimum temperatures vary between species, but most strains grow between 15° and 40°C.

Genetic studies have been hampered by the slow growth of rhodococci, their tendency to clump, and their ability to form coenocytic structures. Current developments in the area have been reviewed by Brownell and Denniston (1984). Work has centered on *R. erythropolis*, an organism in which recombination was reported in 1963 by Adams and Bradley. To date, over 60 genetic traits have been used in the development of the *R. erythropolis* linkage map, and temperate phages are available to serve as cloning vectors for the development of a gene cloning system. Genetic recombination has also been demonstrated between rhodococcal strains currently labeled "*Nocardia opaca*" and "*Nocardia restricta*." A transferable plasmid described for a strain of the former carries traits allowing for chemolithoautotrophic growth (aut$^+$), which includes genes for hydrogenase, phosphoribulokinase, and ribulosebisphosphate carboxylase production (Reh, 1981; Reh and Schlegel, 1981). The "*N. opaca*" strain is able to transfer the aut^+ trait to other organisms bearing this name as well as to *R. erythropolis*, and in the presence of CO_2 and H_2 it grows at a generation time of 7 h. Phages isolated by soil enrichment cause true lysis among *Rhodococcus* and *Nocardia* strains but not against representatives of allied taxa (Prauser and Falta, 1968; Prauser, 1976a, 1981b, 1984c).

The catabolic potential of rhodococci includes the ability to assimilate not only proteins and carbohydrates but also unusual compounds such as alicyclic hydrocarbons, nitroaromatic compounds, polycyclic hydrocarbons, pyridine, and steroids (Tárnok, 1976; Cain, 1981). Detergents and pesticides, including warfarin, are also modified (Goodfellow and Williams, 1983). Rhodococci have been implicated in the degradation of lignin-related compounds (Eggeling and Sahm, 1980, 1981; Rast et al., 1980) and humic acid (Cross et al., 1976), and have frequently been isolated from soil polluted with petroleum (Nesterenko et al., 1978a). Members of the genus also produce enzymes that are exploited in the transformation of xenobiotics (Tárnok, 1976; Peczynska-Czoch and Mordarski, 1984). Indeed, many of the transformations traditionally associated with nocardiae are due to rhodococci.

Rhodococci have been examined using serological methods such as agglutination, complement fixation, immunodiffusion, immunoelectrophoresis, and sensitin testing. It has been established that representatives of *Rhodococcus, Corynebacterium, Mycobacterium,* and *Nocardia* have antigens in common (Cummins, 1962a, 1965; Ridell, 1974, 1977; Ridell et al., 1979), and that ribosomes account for many of the cross-reactions (Ridell, 1981b). Analyses of immunodiffusion and immunoelectrophoretic patterns of whole-culture filtrates and cell extracts of rhodococci have proved to be especially useful for defining species (Goodfellow et al., 1974; Ridell, 1974, 1981a, 1984; Lind and Ridell, 1976), as have quantitative comparisons of the skin hypersensitivity reactions of sensitized guinea pigs to crude culture filtrates of homologous and heterologous rhodococci (Hyman and Chaparas, 1977).

Most rhodococci are sensitive to antibacterial antibiotics such as aminoglycosides, cephalosporins, macrolides, penicillins, and tetracyclines, are less sensitive to sulfonamides, but are resistant to most antitubercular compounds (Goodfellow and Orchard, 1974; Rowbotham and Cross, 1977b; Goodfellow et al., 1982a; Helmke and Weyland, 1984). They are also sensitive to lysozyme (Goodfellow, 1971; Mordarska et al., 1978).

Rhodococci are widely distributed in nature and have frequently been isolated from soil, fresh water, and marine habitats, as well as from gut contents of blood-sucking arthropods with which they may form a mutualistic association (Cross et al., 1976; Goodfellow and Williams, 1983). *Rhodococcus coprophilus* grows on herbivore dung (Rowbotham and Cross, 1977a), *R. bronchialis* is associated with sputa of patients with cavitary pulmonary tuberculosis and bronchiectasis (Tsukamura, 1971), *R. fascians* causes leaf gall in many plants and fasciation in sweet peas (Tilford, 1936), and *R. marinonascens* has only been reported from marine sediments (Helmke and Weyland, 1984). *Rhodococcus equi* is an important equine pathogen that can infect other domestic animals, notably cattle and swine, and causes infection in human patients compromised by immunosuppressive drug therapy and/or lymphoma (Barton and Hughes, 1980; Goodfellow et al., 1982a).

Enrichment and Isolation Procedures

Rhodococcus coprophilus has been isolated from both terrestrial and aquatic habitats by plating preheated samples (6 min at 55°C) onto M3 agar and incubating plates at 30°C for 7 days (Rowbotham and Cross, 1977a); *R. chlorophenolicus* from a pentachlorophenol enrichment culture inoculated with lake sediment (Apajalahti et al., 1986); *R. luteus* and *R. maris* from soil and the skin and intestinal contents of carp (*Cyprinus carpio*) on mineral salts agar enriched with *n*-alkanes and incubated at 28°C (Nesterenko et al., 1982); *R. marinonascens* from marine sediments using a number of rich media supplemented with sea water and incubated for 8–12 weeks at 18°C (Weyland, 1969, 1981); *R. erythropolis* and *R. rhodochrous* on mineral salts media supplemented with *m*-cresol or phenol (Gray and Thornton, 1928); and *R. bronchialis, R. rubropertinctus,* and *R. terrae* from sputa and soil using a decontamination method combined with a selective medium (Tsukamura, 1971). Similarly, *R. equi* has been recovered from soil feces, lymph nodes, and the intestinal contents of several animal species using a selective medium supplemented with nalidixic acid, novobiocin, cyclohexamide, and potassium tellurite (Woolcock et al., 1979; Mutimer and Woolcock, 1980); a selective enrichment broth containing nalidixic acid, penicillin, cyclohexamide, and potassium tellurite incubated at 30°C and used in conjunction with Tinsdale medium (Oxoid) and modified M3 medium has also been employed to good effect (Barton and Hughes, 1981). Rhodococci have also been isolated from soil using Czapek's agar (Higgins and Lechevalier, 1969), glycerol agar (Gordon and Smith, 1953), and Winogradsky's nitrite medium (Winogradsky, 1949), and from diseased sweet peas using potato dextrose agar (Tilford, 1936).

Maintenance Procedures

Long-term preservation of *Rhodococcus* strains may be achieved by lyophilization in skim milk. Suspensions of cocci and mycelial fragments can be stored in glycerol (20% v/v) at − 20°C (Wellington and Williams, 1978).

Procedures for Testing of Special Characters

Reliable identification at the generic level is best achieved using a combination of chemical and morphological techniques. Simple and accurate techniques have been introduced for the detection of lipids wall amino acids, and sugars (Goodfellow and Minnikin, 1985). Several methods can be applied to determine whether unknown organisms contain 2,6-DAP, arabinose, and galactose in whole cell hydrolysates, i.e. whether they have a wall chemotype IV characteristic of *Rhodococcus*, and related taxa (Becker et al., 1964; Lechevalier, 1968; Berd, 1973; Staneck and Roberts, 1974; Richter, 1977; Lechevalier and Lechevalier, 1980). The latter also provide detailed protocols for determining wall, mycolic acid, and phospholipid composition.

Fatty acid analysis is the first recommended step in the exploitation of the lipid composition of rhodococci. Indeed, since mycolic acids are restricted to some actinomycetes with a wall chemotype IV their detection (Minnikin et al., 1975, 1980; Hecht and Causey, 1976) dispenses with the need to determine whole-cell amino acid and sugar composition. The simplified procedure of Minnikin et al. (1975) facilitates the rapid analysis of mycolic acid composition. Thus, whole cell methanolysates of rhodococci, corynebacteria, and nocardiae give single spots on thin-layer chromatography in contrast to the multispot pattern produced by mycolates of most mycobacteria (Minnikin et al., 1984a). Mycolic esters can be identified positively on thin layer chromatograms because they are not removed when plates are subsequently washed with methanol-water (5:2 v/v). Mycobacterial mycolates can also be recognized because they are precipitated from ethereal solution by the addition of ethanol (Hecht and Causey, 1976); the mycolic acids from rhodococci, corynebacteria, and nocardiae remain in solution but can be observed by thin-layer chromatography of the supernatant.

Once mycolic acids have been detected, their esters should be isolated and characterized. Partial characterization of mycolic acids can be achieved by pyrolysis gas chromatography of their methyl esters (Lechevalier et al., 1971b; Goodfellow et al., 1978), a procedure that give important information on the length of the fatty acid esters released. In turn, mass spectrometry of mycolic esters allows determinations of overall size, degree of unsaturation, and the size of the chain in the 2 position (Alshamaony et al., 1976; Collins et al., 1982a). Analysis can be taken a stage further by applying combined gas chromatography mass spectrometry of trimethylsilyl ethers of mycolic esters (Yano et al., 1978; Tomiyasu et al., 1981). This technique separates mycolic ester derivatives into their homologous components, each of which can be analyzed by mass spectrometry.

Several techniques based on gas liquid chromatography can be use to detect diagnostic fatty acids such as 10-methyloctadecanoic (tuberculostearic) acid (Lechevalier et al., 1977; Collins et al., 1982b; Bousfield et al., 1983; Kroppenstedt, 1984). In turn, menaquinone composition can be ascertained by chromatographic and physicochemical analysis (Collins et al., 1977; Collins, 1985) and diagnostic polar lipids by applying published procedures (Lechevalier et al., 1977; Minnikin et al., 1977a; O'Donnell et al., 1982a). A simple small-scale procedure for the sequential extraction of isoprenoid quinones and polar lipids (Minnikin et al., 1984b) and an integrated lipid and wall analysis technique may be found useful for the identification of unknown mycolic acid-containing actinomycetes (O'Donnell et al., 1984).

The micromorphology of strains should first be examined unstained on plate, slide, or coverslip culture (Williams and Cross, 1971). Such preparations can also be examined by scanning electron microscopy (Williams and Davies, 1967). The morphology of undisturbed growth can be recorded conveniently on Bennett's agar (Jones, 1949) after 1, 3, and 7 days at 30°C using a long working distance objective (Rowbotham and Cross, 1977b) or in the case of *R. marinonascens* on yeast extract-malt extract-sea water agar at 18°C (Helmke and Weyland, 1984). Strains grown on polycarbonate membranes laid on the top of Bennett's agar can be removed at regular intervals, fixed, and examined using a scanning electron microscope (Locci, 1981).

Identification to the species level is difficult but can be achieved using a combination of morphological and chemical criteria supplemented by biochemical, nutritional, and physiological data. Thus, well-established procedures are available to detect adenine, tyrosine, and urea decomposition (Gordon et al., 1974), acid formation from sugars (Tsukamura, 1966; Goodfellow, 1971), and the ability of rhodococci to grow on sole carbon (Tsukamura, 1966; Goodfellow, 1971), nitrogen, and carbon and nitrogen sources (Tsukamura, 1966), and in the presence of chemical inhibitors (Goodfellow, 1971).

There is evidence that the ability to form amidases (Bönicke, 1962; Tsukamura, 1975a) and esterases (Käppler, 1965; Tsukamura, 1975a), and to grow in the presence of compounds such as ethambutol (Mizuno et al., 1966), sodium nitrite (Tsukamura and Tsukamura, 1968), sodium salicylate (Tsukamura, 1962), picric acid (Tsukamura, 1965), and rifampicin (Tsukamura, 1972) may also yield useful diagnostic data, but tests for properties such as these need to be extended to include representatives of all of the described species of *Rhodococcus*.

Differentiation of the genus **Rhodococcus** *from other closely related taxa*

Rhodococci are closely related to members of the genera *Caseobacter, Corynebacterium, Mycobacterium, Nocardia,* and the *'aurantiaca'* taxon. Features useful for distinguishing *Rhodococcus* from these other mycolic acid-containing taxa are shown in Table 26.13. Most of these taxa have also been distinguished in comparative immunodiffusion studies (Lind and Ridell, 1976; Lind et al., 1980). It is possible to distinguish *Rhodococcus* from *Mycobacterium* and *Nocardia* (Ridell and Norlin, 1973), and from *Corynebacterium* (Ridell, 1977), using precipitinogens α and pα.

Susceptibility to bleomycin (2.5 μg/ml), 5-fluorouracil (20 μg/ml), mitomycin C (10 μg/ml), and β-galactosidase activity have been reported to be useful in distinguishing rhodococci and nocardiae (Tsukamura, 1974b, 1981a, 1981b, 1982a). There is also evidence that rhodococci are more resistant than mycobacteria to prothionamide (Ridell, 1983), and that they can be differentiated from the latter on the basis of the acid-fast stain, arylsulfatase activity, ability to use sucrose as a sole carbon source, and inability to use trimethylenediamine as a simultaneous nitrogen and carbon source (Tsukamura, 1971).

Table 26.13.
Differential characteristics of the genus **Rhodococcus** *and other wall chemotype IV taxa containing mycolic acids*[a,b]

Characteristics	*Rhodococcus*	*Caseobacter*	*Corynebacterium*	*Mycobacterium*	*Nocardia*	*'aurantiaca' taxon*
Morphological characters						
Substrate mycelium	D	–	–	D	+	–
Aerial mycelium	–	–	–	D	D	–
Conidia	–	–	–	–	D	–
Lipid characters						
Fatty acids						
Tuberculostearic acid[c]	+	+	–[d]	+	+	+
Phospholipids						
Phosphatidylethanolamine[e]	+	ND	–	+	+	+
Predominant menaquinone[f]	MK-8(H_2),-9(H_2)	MK-9(H_2)	MK-8(H_2),-9(h_2)	MK-9(H_2)	MK-8(H_4),-9(H_2)	MK-9
Mycolic acids						
Overall size (number of carbons)[g]	34–64	30–36	22–38	60–90	44–60	68–74
Number of double bonds[g]	0–4	0–2	0–2	1–2	0–3	1–5
Ester released on pyrolysis[h]	12–18	14–18	8–18	22–26	12–18	20–22
Mol% G + C of the DNA	63–73	65–67	51–59	62–70	64–72	ND

[a]Data from Goodfellow and Minnikin (1981a, 1981c, 1985), and Goodfellow and Cross (1984b).
[b]Symbols: +, 90% or more of the strains are positive; –, 90% or more of the strains are negative; *D*, different reactions occur in different taxa (species of a genus or genera of a family); *ND*, not determined.
[c]Determined by gas liquid chromatography (Lechevalier et al., 1977; Collins et al., 1982b; Bousfield et al., 1983; Kroppenstedt, 1984).
[d]*Corynebacterium bovis* contains tuberculostearic acid (Lechevalier et al., 1977; Collins et al., 1982b).
[e]Determined by thin-layer chromatography (Lechevalier et al., 1977; Minnikin et al., 1977a; O'Donnell et al., 1982a).
[f]Menaquinones detected by chromatographic or physicochemical analysis (Collins et al., 1977; Collins, 1985). Abbreviations exemplified by MK-8(H_2): Menaquinone having one of the eight isoprene units hydrogenated.
[g]Detected by mass spectrometry (Alshamaony et al., 1976; Collins et al., 1982a).
[h]Esters detected by pyrolysis gas chromatography (Lechevalier et al., 1971b; Goodfellow et al., 1978).

Taxonomic Comments

The application of modern taxonomic methods, notably chemotaxonomy and numerical phenetic taxonomy, led to the reintroduction of the genus *Rhodococcus* for the group of bacteria collectively known as "*Mycobacterium rhodochrous*," the '*rhodochrous*' group or the '*rhodochrous*' complex (Bousfield and Goodfellow, 1976). The genus continues to provide a niche for actinomycetes previously assigned to genera such as *Arthrobacter, Brevibacterium, Corynebacterium, Mycobacterium*, and *Nocardia* (Collins et al., 1982a, 1982b). Rhodococci and the other mycolic acid-containing actinomycetes constitute the nocardioform actinomycetes (Goodfellow and Cross, 1984b), a group of organisms that not only have many properties in common but also form a recognizable suprageneric entity (Mordarski et al., 1980b; Stackebrandt and Woese, 1981b; Stackebrandt et al., 1983c). In addition, rhodococci, mycobacteria, nocardiae, and aurantiaca strains have been recovered in discrete aggregate clusters in numerical phenetic surveys by Goodfellow et al. (1982a, 1982b) and to a lesser extent by Tsukamura et al. (1979).

Further comparative studies are needed to determine the status of the '*aurantiaca*' taxon, which contains strains previously classified as "*Gordona aurantiaca*" (Tsukamura and Mizuno, 1971). Subsequent numerical taxonomic studies underlined the equivocal position of this taxon in the genus *Gordona* (Tsukamura, 1974a, 1975b) and the type strain fell outside the aggregate *Rhodococcus* cluster circumscribed by Goodfellow and Alderson (1977). In a more broadly based investigation (Goodfellow et al., 1978), *aurantiaca* strains formed a numerically well-defined taxon equivalent in rank to phena corresponding to the genera *Corynebacterium, Mycobacterium, Nocardia, and Rhodococcus* and were shown to contain characteristic mycolic acids and unsaturated menaquinones with nine isoprene units. These findings are at variance with the view that the '*aurantiaca*' taxon should be assigned to the genus *Rhodococcus* as *R. aurantiacus* (Tsukamura, 1982c). Four additional species, *R. sputi* (Tsukamura, 1978), *R. aichiense, R. chubuense*, and *R. obuense* (Tsukamura, 1982c), also need to be studied for additional criteria before their status can be accepted. In the mean time these four taxa and *R. aurantiacus* should be considered as *species incertae sedis*.

DNA relatedness studies have shown that *R. bronchialis, R. coprophilus, R. equi, R. erythropolis, R. rhodochrous, R. ruber, R. rubropertinctus*, and *R. terrae* are good species (Mordarski et al., 1976, 1977, 1980a). Rhodococcal species can, however, be divided into two well-circumscribed taxa on the basis of chemical and serological data. Thus, all of the species originally assigned to the genus *Gordona* (Tsukamura, 1971), namely *R. bronchialis, R. rubropertinctus*, and *R. terrae*, have mycolic acids with between 48 and 66 carbon atoms and major amounts of dihydrogenated menaquinones with nine isoprene units, whereas the remaining species have shorter chain mycolic acids and dihydrogenated menaquinones with eight isoprene units as the major isoprenolog (see Table 26.14). The two aggregate groups can also be recognized on the basis of antibiotic sensitivity patterns (Goodfellow and Orchard, 1974), delayed skin reactions on sensitized guinea pigs, and polyacrylamide gel electrophoresis of cell extracts (Hyman and Chaparas, 1977). It is possible that further studies might underline the separation between the two aggregate taxa and thereby raise the question of whether they can be included in the same genus.

The genus *Rhodococcus* includes 5 of the 20 species listed under *Nocardia* in the Approved Lists of Bacterial Names (Skerman et al., 1980). Thus, *N. calcarea* (Metcalf and Brown, 1957) is a synonym of *N. erythropolis* (Goodfellow et al., 1982c); *N. corynebacteroides* (Serrano et al., 1972) and *N. globerula* (Gray 1928) Waksman and Henrici 1948a are synonyms of *R. globerulus* (Goodfellow et al., 1982a); *N. restricta* (Turfitt 1944) McClung 1974 is a synonym of *R. equi* (Goodfellow and Alderson, 1977); and *N. coeliaca* (Gray and Thornton 1928) Waksman and Henrici 1948a has properties (Gordon et al., 1974) consistent with its inclusion in the genus *Rhodococcus*.

There is confusion over the status of *R. rubropertinctus* as defined by Goodfellow and Alderson (1977). In 1973 Tsukamura proposed the name *Gordona rubropertincta* with the type strain ATCC 14343. Since the new taxon included the hypothetical mean organism of *G. rubra* (Tsukamura 1971) this species was considered to be a synonym of *G. rubropertincta* (Hefferan 1904) Tsukamura 1973. The latter was renamed *R. rubropertinctus* by Tsukamura in 1974(a). Goodfellow et al. (1974) recovered strain ATCC 14343 in a subcluster subsequently equated with *R. rhodochrous* (Goodfellow and Alderson, 1977), and on the basis of this reduced *R. rubropertinctus* (Hefferan) Tsukamura to a synonym of the latter species. They also classified strain NCTC 10668 as the type species of a new combination, *R. corallinus* (Goodfellow and Alderson, 1977). Additional strains conforming to Hefferan's original description were included in a further cluster that was named *R. rubropertinctus* (Hefferan) Goodfellow and Alderson with strain ATCC 14352 as the type. The type strains of *R. corallinus* and *R. rubropertinctus* (Hefferan) Goodfellow and Alderson were subsequently found to belong to a single DNA homology group (Mordarski et al., 1980a) and *R. corallinus* became a subjective synonym of *R. rubropertinctus* (Hefferan) Goodfellow and Alderson. This latter name was included in the Approved Lists of Bacterial Names (Skerman et al., 1980) with ATCC 14352 as the type strain. Tsukamura (1973, 1982a) has consistently recovered strain ATCC 14343 in a cluster equated with *R. rubropertinctus* (Hefferan 1904) Tsukamura 1978. He also believes that *R. lentifragmentus* (Kruse 1896) Tsukamura 1978 has priority over *R. ruber* (Kruse, 1896) Goodfellow and Alderson 1979. Such tangled nomenclatural matters reflect the difficulties involved in clarifying the taxonomy of very poorly classified taxa and can only be resolved by action of the Judicial Commission.

Further Reading

Bousfield, I.J and M. Goodfellow. 1976. The "*rhodochrous*" complex and its relationships with allied taxa. *In* Goodfellow, Brownell and Serrano (Editors), The Biology of the Nocardiae. Academic Press, London, pp. 39-65.

Goodfellow, M. and T. Cross. 1984. Classification. *In* Goodfellow, Mordarski and Williams (Editors). The Biology of the Actinomycetes. Academic Press, London, pp. 7-64

Goodfellow, M. and D. E. Minnikin (Editors) 1985. Chemical Methods in Bacterial Systematics, Society for Applied Bacteriology, Technical Series Number 20. Academic Press, London.

Differentiation and characteristics of the species of the genus **Rhodococcus**

The differential characteristics of the species of *Rhodococcus* are indicated in Table 26.14. Other properties of the species are listed in Table 26.15.

List of species of the genus **Rhodococcus**

1. **Rhodococcus rhodochrous** (Zopf) Tsukamura 1974a, 43.[AL] (*Staphylococcus rhodochrous* Zopf 1889, 173; *Rhodococcus rubropertinctus* (Hefferan) Tsukamura 1974a, 43.*)

rho.do.chrous'. Gr. n. *rhodon* the rose; Gr. n. *chrous color*; L. adj. *rhodochrous* rose-colored.

*See "Taxonomic Comments."

Table 26.14.
Characteristics differentiating the species of the genus **Rhodococcus**[a]

Characteristics	1. *R. rhodochrous*	2. *R. bronchialis*	3. *R. chlorophenolicus*	4. *R. coprophilus*	5. *R. equi*	6. *R. erythropolis*	7. *R. fascians*	8. *R. globerulus*	9. *R. luteus*	10. *R. marinonascens*	11. *R. maris*	12. *R. rhodnii*	13. *R. ruber*	14. *R. rubropertinctus*	15. *R. sputi*	16. *R. terrae*
Morphogenetic sequence[b]	EB-R-C	R-C	R-C	H-R-C	R-C	EB-R-C	H-R-C	EB-R-C	EB-R-C	H-R-C	R-C	EB-R-C	H-R-C	R-C	R-C	R-C
Decomposition of																
Adenine	d	−	ND	−	+	+	+	−	ND	−	ND	−	d	−	ND	−
Tyrosine	+	−	ND	−	−	d	+	−	−	d	−	+	+	−	ND	−
Urea	d	+	ND	d	+	+	+	+	+	−	d	+	+	+	ND	+
Growth on sole carbon sources (% w/v)																
Inositol (1.0)	−	+	+	−	−	d	−	−	−	+	−	−	−	−	−	−
Maltose (1.0)	+	+	−	+	−	+	−	+	−	−	−	−	+	+	ND	+
Mannitol (1.0)	+	d	+	−	−	+	+	d	+	−	−	+	+	+	+	+
Rhamnose (1.0)	−	−	+	−	−	−	−	−	−	−	−	−	−	−	−	+
Sorbitol (1.0)	+	+	+	−	−	+	+	d	+	d	−	+	+	+	+	+
p-Cresol (0.1)	+	+	ND	d	−	d	+	−	ND	−	ND	d	+	+	ND	+
m-Hydroxybenzoic acid (0.1)	+	−	ND	+	d	−	+	−	ND	ND	ND	−	+	−	ND	−
Pimelic acid (0.1)	d	+	ND	+	−	d	−	−	ND	−	−	d	+	+	ND	+
Sodium adipate (0.1)	+	+	ND	d	−	+	−	+	−	ND	ND	+	+	+	ND	+
Sodium benzoate (0.1)	+	+	ND	d	−	−	−	d	−	−	−	+	+	+	−	+
Sodium citrate (0.1)	+	d	ND	−	−	+	+	+	+	−	d	d	+	+	+	+
Sodium lactate (0.1)	+	+	ND	d	+	+	+	−	+	−	−	d	d	−	ND	+
Testosterone (0.1)	+	+	ND	+	d	+	−	−	ND	−	ND	−	+	+	ND	+
L-Tyrosine (0.1)	+	−	ND	−	d	−	+	+	ND	ND	ND	+	+	−	ND	+
Growth in the presence of (% w/v)																
Sodium azide (0.02)	−	+	ND	d	d	d	−	d	ND	+	ND	d	+	+	ND	+
Lipid characters																
Mycolic acids (number of carbons)	36–50	54–66	33–43	38–48	30–38	34–38	38–52	ND	ND	ND	ND	38–52	40–50	48–62	ND	52–64
Predominant menaquinone	MK-8(H_2)	MK-9(H_2)	MK-9(H_2)	MK-8(H_2)	MK-8(H_2)	MK-8(H_2)	MK-8(H_2)	MK-8(H_2)	MK-8(H_2)	MK-8(H_2)	MK-8(H_2)	MK-8(H_2)	MK-8(H_2)	MK-9(H_2)	ND	MK-9(H_2)
Mol% G + C of the DNA	67–70	63–65	ND	67–69	70–72	67–71	63–68	63–67	64	65–66	73	66	69–73	67–69	66–68	64–69

[a]Symbols: see Table 26.13; also *d*, 11–89% of strains are positive.
[b]*EB-R-C*, elementary branching-rod-coccus growth cycle; *R-C*, rod-coccus growth cycle; *H-R-C*, hypha-rod-coccus growth cycle.

Table 26.15.
Other characteristics of the species of the genus **Rhodococcus**[a]

Characteristics	1. *R. rhodochrous*	2. *R. bronchialis*	3. *R. chlorophenolicus*	4. *R. coprophilus*	5. *R. equi*	6. *R. erythropolis*	7. *R. fascians*	8. *R. globerulus*	9. *R. luteus*	10. *R. marinonascens*	11. *R. maris*	12. *R. rhodnii*	13. *R. ruber*	14. *R. rubropertinctus*	15. *R. sputi*	16. *R. terrae*
Growth on sole carbon sources (% w/v)																
Ethanol (1.0)	+	+	ND	+	+	+	+	+	ND	−	ND	+	+	+	+	+
Glycerol(1.0)	+	+	−	−	−	+	+	+	+	d	+	ND	+	+		+
Sucrose (1.0)	+	+	+	d	+	+	+	+	+	−	d	+	+	+	+	+
Trehalose (1.0)	+	+	+	d	+	+	+	+	ND	−	ND	+	+	+	+	+
Acetamide (0.1)	d	+	ND	−	d	+	+	d	ND	ND	ND	+	+	−	ND	−
p-Hydroxybenzoic acid (0.1)	+	+	ND	−	+	d	+	d	ND	−	ND	d	+	+	+	+
Sebacic acid (0.1)	+	+	ND	d	−	+	+	+	ND	−	ND	d	+	+	ND	+
Sodium fumarate (0.1)	+	+	ND	d	+	+	+	+	+	ND	ND	d	+	+	+	+
Sodium gluconate (0.1)	d	d	ND	−	−	+	+	−	ND	d	ND	+	d	+	ND	d
Sodium malate (0.1)	+	+	ND	−	+	+	+	+	+	d	+	+	+	+	+	+
Sodium pyruvate (0.1)	+	+	ND	+	+	+	+	+	+	+	+	−	+	−	+	+
Sodium succinate (0.1)	+	+	ND	−	+	+	+	+	+	−	+	+	+	+	+	+
Growth on sole carbon and nitrogen sources (% w/v)																
Acetamide	d	+	ND	−	+	d	−	d	ND	ND	ND	+	d	−	+	−
Serine	−	−	ND	−	−	−	−	−	ND	ND	ND	−	−	−	d	−
Trimethylenediamine	−	−	ND	−	−	−	−	−	ND	ND	ND	−	−	−	−	−
Growth at																
10°C	+	−	−	d	+	+	+	+	ND	+	ND	−	ND	−	ND	−
40°C	+	+	ND	+	+	+	+	+	ND	−	ND	−	+	+	ND	+
45°C	−	−	−	−	−	−	−	−	−	−	−	−	−	−	ND	−
Growth in presence of (% w/v)																
Crystal violet (0.001)	+	+	ND	+	+	d	−	+	ND	+	ND	+	+	+	ND	+
Crystal violet (0.0001)	d	+	ND	+	d	−	+	d	ND	d	ND	d	+	+	ND	+
Phenol (0.1)	+	+	ND	+	−	d	+	d	ND	+	ND	d	+	+	ND	+
Phenyl ethanol (0.3% v/v)	+	+	ND	d	d	d	−	+	ND	+	ND	−	+	+	ND	+
Sodium azide (0.01)	d	+	ND	+	d	d	−	d	ND	+	ND	+	+	+	ND	+
Sodium chloride (5.0)	+	+	ND	+	+	+	−	+	+	+	+	+	+	+	ND	+
Sodium chloride (7.0)	d	+	−	+	d	d	−	+	+	d	+	+	+	+	ND	+

[a]Symbols: see Table 26.13; also *d,* 11–89% of strains are positive.

See the generic description and Tables 26.14 and 26.15 for many of the features of *R. rhodochrous*.

Cocci germinate and give rise to branched filaments that undergo fragmentation into rods and cocci, thereby completing the growth cycle. Rough, orange to red colonies are formed on glucose yeast extract agar, Sauton's agar, and egg media.

Allantoinase, nicotinamidase, and pyrazinamidase negative. β-Esterase weakly positive but negative for α-esterase and β-galactosidase.

Acid is produced from dextrin, ethanol, fructose, glucose, glycerol, maltose, mannitol, mannose, sorbitol, sucrose, and trehalose but not from adonitol, amygdalin, D-arabinose, L-arabinose, cellobiose, dulcitol, galactose, glycogen, inositol, inulin, lactose, melezitose, raffinose, rhamnose, or xylose.

Grows on inulin, *iso*-butanol, 2,3-butylene glycol, DL-norleucine, propyleneglycol, sodium octanoate, and L-tyrosine as sole carbon sources but not on adonitol, amygdalin, D-arabinose, arbutin, cellobiose, dulcitol, galactose, glycogen, lactose, melezitose, raffinose, L-rhamnose, betaine HCl, D-mandelic acid, L-serine, sodium hippurate, sodium malonate, or L-tryptophan.

Resistant to ethambutol (5 μg/ml), rifampicin (25 μg/ml), sodium nitrite (0.2% w/v), sodium salicylate (0.1% w/v), and picric acid (0.2% w/v) but susceptible to 5-fluorouracil (5 μg/ml) and mitomycin C (5 μg/ml).

Isolated from soil.

The mol% G + C of the DNA is 67–70 (T_m).

Type strain: ATCC 13808.

2. **Rhodococcus bronchialis** (Tsukamura) Tsukamura 1974a, 43.[AL] (*Gordona bronchialis* Tsukamura 1971, 22.)

bron.chi′alis. L. adj. *bronchialis* coming from the bronchi.

See the generic description and Tables 26.14 and 26.15 for many of the features of *R. bronchialis*.

A rod-coccus life cycle is shown. Rough brownish colonies formed on glucose yeast extract agar, Sauton's agar, and egg media. Synnemata of vertically arranged coalescing filaments are formed on the surface of colonies after 12–18 h incubation.

Acetamidase, nicotinamidase, pyrazinamidase, and urease positive; allantoinase, benzamidase, isonicotinamidase, malonamidase, salicylamidase, and succinamidase negative; α-esterase and β-galactosidase negative; β-esterase and acid phosphatase positive.

Acid is produced from glucose, inositol, maltose, mannose, and trehalose but not from arabinose, galactose, raffinose, rhamnose, sorbitol, or xylose.

Grows on *iso*-butanol, propanol, and propylene glycol as sole carbon sources. Uses L-glutamate, isonicotinamide, L-methionine, nicotinamide, pyrazinamide, L-serine, succinamide, and urea as sole nitrogen sources, but not benzamide or nitrite. Acetamide and L-glutamate are used as sole sources of carbon and nitrogen, but not benzamide or monoethanolamine.

Resistant to sodium aminosalicylate (0.2% w/v) and picric acid (0.2% w/v) but susceptible to 5-fluorouracil (5 μg/ml) and mitomycin C (5 μg/ml).

Isolated from sputum of patients with pulmonary disease.

The mol% G + C of the DNA is 63–65 (T_m).

Type strain: ATCC 25592.

3. **Rhodococcus chlorophenolicus** Apajalahti, Kärpänoja and Salkinoja-Salonen 1986, 248.[VP]

chlor.o.phen.o′li.cus. N.L. adj. *chloro* containing chlorine; fr. Gr. adj. *chloros* pale green; N.L. n. *pheno* phenol, hydroxybenzene; fr. Gr. *phaino* shine, to be shining, appear; N.L. adj. *chlorophenolicus* relating to chlorophenols.

See generic description and Tables 26.14 and 26.15 for many of the properties of *R. chlorophenolicus*.

A rod-coccus cycle is shown. Orange pigmented and slightly mucoid colonies are formed on yeast extract and rhamnose agars.

Mineralizes pentachlorophenol and degrades 2,3,4,5-tetrachlorophenol, 2,3,4,6-tetrachlorophenol, 2,3,5,6-tetrachlorophenol, 2,3,5-trichlorophenol, and 2,3,6-trichlorophenol.

Grows on *N*-acetylglucosamine (weak), adonitol, L-arabinose, D-arabitol, L-arabitol, arbutin (weak), erythritol, fructose, gluconate, glucose (weak), inositol, mannitol, rhamnose, ribose (weak), sorbitol, sucrose, trehalose, xylitol, and xylose as sole carbon sources but not on amygdalin, D-arabinose, cellobiose, dulcitol, esculin, D-fucose, L-fucose, galactose, gentiobiose, glycogen, glycerol, inulin, 2-ketogluconate, 5-ketogluconate, lactose, D-lyxose, maltose, mannose, melezitose, melibiose, α-methyl-D-glucoside, α-methyl-D-mannoside, α-methyl-D-xyloside, salicin, sorbose, starch, raffinose, D-tagatose, D-turanose, and L-xylose.

Isolated from a pentachlorophenol enrichment culture inoculated from lake sediment.

Type strain: DSM 43826.

4. **Rhodococcus coprophilus** Rowbotham and Cross 1979, 80.[AL] (Effective publication: Rowbotham and Cross 1977b, 136.)

co.pro.ph′lus. Gr. n. *copros* dung; Gr. adj. *philus* loving; M.L. adj. *coprophilus* dung loving.

See the generic description and Tables 26.14 and 26.15 for many of the features of *R. coprophilus*.

Forms a well-developed primary mycelium that fragments into rods and cocci after several days' incubation and aerial hyphae that may be branched. On Bennett's agar, after 2 weeks' incubation at 30°C, rhizoid colonies (2 mm in diameter) are formed with a central orange papilla; growth into the agar also occurs. Pigmentation is enhanced by light. The young microcolonies on Bennett's agar are mycelial, and after 24 h sparse, nonsporulating aerial hyphae are usually present. No macroscopically aerial mycelium, extracellular pigments, or characteristic odors are produced. The central papilli of mature colonies are composed of complex aggregations of Gram-positive, non-acid-fast, nonmotile coccoid elements (1–1.5 μm in diameter). Cystites may occur in the mycelial fringe of the colonies on Bennett's agar. No pellicle is produced on the surface of Bennett's broth, although isolated floating colonies may occur. Growth and pigmentation are reduced on media deficient in thiamin.

Urease positive and acetamidase and nicotinamidase negative. Acid phosphatase and α- and β-esterase negative.

Grows on cetyl alcohol, D-melezitose, D-raffinose, sodium isobutyrate, and sodium valerate as sole carbon sources but not on L-arabinose, galactose, D-glucosamine HCl, lactose, salicin, xylose, acetamide, benzamide, D- and L-alanine, L-asparagine, L-glycine, DL-norleucine, L-phenylalanine, L-proline, L-serine, L-tyrosine, butane-1,3-diol, butane-1,4-diol, sodium γ-aminobutyrate, sodium gluconate, or sodium phenylacetate.

Sensitive to 5-fluorouracil (20 μg/ml) and mitomycin C (5 μg/ml).

Grows on herbivore dung; it has been isolated from the dung of cows, donkeys, goats, horses, and sheep. It is common on grass and in the soil beneath grazed pastures, and is washed into streams and lakes, where it can accumulate in the sediment.

The mol% G + C of the DNA is 67–69 (T_m).

Type strain: ATCC 29080.

5. **Rhodococcus equi** (Magnusson) Goodfellow and Alderson 1977.[AL] (*Corynebacterium equi* Magnusson 1923, 36; *Corynebacterium hoagii* (sic) (Morse) Eberson 1918, 11† *Nocardia restricta* (Turfitt) McClung 1974, 743.)

e′qui. L. n. *equus* horse; L. gen. nov. *equi* of the horse.

See generic description and Tables 26.14 and 26.15 for many of the features of *R. equi*.

A rod-coccus life cycle is shown. Traces of elementary branching may be observed at early stages of growth. Smooth, shiny, orange to red colonies with entire margins are formed on glucose yeast extract agar. Some strains, produce abundant slime that may drop onto the cover of inverted Petri dishes during incubation.

Urease positive, acetamidase and nicotinamidase negative. Neither esculin nor arbutin are degraded. API-ZYM reactions: acid and alkaline phosphatase, esterase lipase (C4), leucine arylamidase, phosphoamidase, and valine arylamidase positive; chymotrypsin, β-glucuronidase, and α-mannosidase negative.

Grows on amyl alcohol, butane-1,3-diol, butan-1-ol, propane-1,2-diol, propan-1-ol, sodium lactate, and sodium octoate as sole carbon sources but not on erythritol, galactose, glycogen, inulin, lactose, raffinose, salicin, xylose, butane-1,4-diol, butane-2,3-diol, propan-2-ol, sodium gluconate, sodium hippurate, sodium malonate, sodium mucate, or tyrosine.

Resistant to ampicillin (20 μg/ml), erythromycin (4 μg/ml), gentamicin (8 μg/ml), lincomycin (64 μg/ml), minocycline (0.125 μg/ml), neomycin (8 μg/ml), novobiocin (4 μg/ml), penicillin (10 μg/ml), picric acid (0.2% w/v), polymixin (256 μg/ml), rifampicin (0.25 μg/ml), streptomycin (5 μg/ml), sulfadiazine (100 μg/ml), tetracycline (50 μg/ml), and tobramycin (8 μg/ml), but susceptible to 5-fluorouracil (5 μg/ml) and mitomycin C (5 μg/ml).

Found in soil, herbivore dung, and the intestinal tract of cows, horses, sheep, and pigs. Causes bronchopneumonia in foals, occasionally infects other domestic animals such as cattle and swine, and can be responsible for infection in human patients compromised by immunosuppressive drug therapy or lymphoma.

The mol% G + C of the DNA is 70–71 (T_m).

Type strain: ATCC 25729 (ATCC 6939).

6. **Rhodococcus erythropolis** (Gray and Thornton) Goodfellow and Alderson 1979, 80.[AL] (Effective publication: Goodfellow and Alderson 1977, 115.) (*Mycobacterium erythropolis* Gray and Thornton 1928, 87; *Nocardia calcarea* Metcalf and Brown 1957, 568.)

e.ry.thro′po.lis. Gr. adj. *erythrus* red; Gr. n. *polis* a city; M.L. n. *erythropolis* red city.

See generic description and Tables 26.14 and 26.15 for many of the features of *R. erythropolis*.

†**Editorial note:** *Corynebacterium hoagii* is considered a synonym; for further information see genus *Corynebacterium* in Volume 2.

Cocci germinate to give filaments that show elementary branching. The growth cycle is completed by the appearance, through fragmentation, of cocci. Rough, orange to red colonies are formed on glucose yeast extract agar and Sauton's agar.

Acid is produced from glucose, glycerol, sorbitol, sucrose, and trehalose but not from adonitol, arabinose, cellobiose, galactose, glycogen, inulin, melezitose, rhamnose, or xylose.

Grows on D-glucosamine, D-salicin, D-alanine, L-asparagine, DL-norleucine, L-phenylalanine, L-proline, L-serine, acetamide, butane-1,3-diol, cetyl alcohol, propane-1,2-diol, sodium γ-aminobutyrate, sodium gluconate, sodium phenylacetate, and stearic acid as sole carbon sources but not on L-arabinose, dextran, galactose, glycogen, lactose, melezitose, benzamide, butane-1,4-diol, glycine, or sodium malonate.

Isolated from soil.

The mol% G + C of the DNA is 67–71 (T_m).

Type strain: ATCC 4277.

7. **Rhodococus fascians** (Tilford) Goodfellow 1984a, 503.[VP] (Effective publication Goodfellow 1984b, 227. (*Phytomonas fascians* Tilford 1936, 394; *Corynebacterium fascians* (Tilford) Dowson 1942, 313.)

fas′ci.ans. L. part. adj. *fascians* binding together, bundling.

See generic description and Tables 26.14 and 26.15 for many of the features of *R. fascians*.

Forms branched hyphae that fragment into rods and cocci. Entire, convex, orange colonies formed on glucose yeast extract agar. Thiamin is required for growth.

Pyrazinamidase positive, esculin and allantoin hydrolyzed.

Acid is produced from dextrin, ethanol, fructose, galactose, glucose, glycerol, mannitol, mannose, ribose, sorbitol, sucrose, and trehalose but not from adonitol, amygdalin, arabinose, arbutin, cellobiose, dulcitol, glycogen, inositol, inulin, lactose, α-methyl-D-glucoside, β-methyl-D-glucoside, raffinose, rhamnose, salicin, or xylose.

Grows on D- and L-alanine, sodium lactate, L-proline, L-serine, and L-tyrosine as sole carbon sources, but not on benzamide, betaine, DL-norleucine, sodium hippurate, sodium malonate, or L-threonine.

Causes leaf gall of many plants and fasciation of sweet peas (*Lathyrus odoratus*).

The mol% G + C of the DNA is 62.9–67.6 (T_m).

Type strain: ATCC 12974.

8. **Rhodococcus globerulus** Goodfellow, Weaver and Minnikin 1982c, 741.[VP] (*Mycobacterium globerulum* Gray 1928, 265; *Nocardia globerula* (sic) (Gray) Waksman and Henrici 1948a, 903; *Nocardia corynebacterioides* (sic) Serrano, Tablante, Serrano, San Blas and Imaeda 1972, 348.)

glo.be′ru.lus. M.L. dim. adj. *globerulus* globular.

See generic description and Tables 26.14 and 26.15 for many of the features of *R. globerulus*.

Cocci give rise to branched filaments that fragment into rods and cocci, thereby completing the growth cycle. Entire, rough, pink to red colonies are formed on glucose yeast extract agar.

Urease positive, allantoinase and benzamidase negative.

Acid is produced from dextrin, ethanol, fructose, glucose, glycerol, maltose, mannitol, mannose, sorbitol, sucrose, and trehalose but not from adonitol, amygdalin, arabinose, cellobiose, dulcitol, galactose, glycogen, inositol, lactose, melezitose, raffinose, rhamnose, or xylose.

Growth on L- and D-alanine, inulin, and L-serine as sole carbon sources but not on benzamide, betaine, inulin, DL-norleucine, L-proline, salicin, sodium hippurate, sodium malonate, sodium octanoate, L-threonine, L-tryptophan, or L-tyrosine.

Sensitive to 5-fluorouracil (20 μg/ml) and mitomycin C (5 μg/ml).

Isolated from soil.

The mol% G + C of the DNA is 63–67 (T_m).

Type strain: ATCC 25714.

9. **Rhodococcus luteus** (Söhngen) (Nesterenko, Nogina, Kasumova, Kvasnikov and Batrakov 1982, 6.[VP] (*Mycobacterium luteum* Söhngen 1913, 599.)

lu′te.us. L. adj. *luteus* golden yellow.

See generic description and Tables 26.14 and 26.15 for many of the features of *R. luteus*.

Straight or slightly curved rods give rise to shorter rods on glycerol (GA) and nutrient (NA) agars and to coccoid elements on wort agar (WA). Abundant, butyrous or mucoid, and intensely yellow growth is produced on GA and WA slants; poor to moderate, butyrous, and pale yellow to pale orange growth on NA slants; and abundant yellow growth on Löwenstein-Jensen medium. A pellicle is formed on the surface of nutrient broth.

Ammonia is formed from peptone, litmus milk is turned alkaline, but *p*-nitrophenoloxidase is not produced. Acetyl methyl carbinol, indole, methyl red, and phosphatase tests are negative.

Acid is produced from arabinose, fructose, galactose, glycerol, glucose, mannitol, mannose, sorbitol, sucrose, and xylose but not from adonitol, cellobiose, dulcitol, inositol, lactose, maltose, α-methyl-D-glucoside, raffinose, rhamnose, or salicin.

Growth occurs in the presence of C_9 to C_{17}, C_{19}, and C_{23} *n*-alkanes but not with C_8 *n*-alkane, ethane, or methane.

Isolated from soil and from skin and intestinal tract of carp (*Cyprinus carpio*).

The mol% G + C of the DNA of the type strain is 64.1 (T_m).

Type strain: IMV 385.

10. **Rhodococcus marinonascens** Helmke and Weyland 1984, 137.[VP]

ma.ri.no.nas′cens. L. adj. *marinus* of the sea; L. part. adj. *nascens* born; M.L. part. adj. *marinonascens* nascent of the sea.

See generic description and Tables 26.14 and 26.15 for many of the properties of *R. marinonascens*.

Forms a well-developed branched primary mycelium that fragments into bacillary and coccoid elements on solid media. Irregularly wrinkled, cream-colored colonies, which are sometimes tinged with pink, are formed on yeast extract–malt extract agar after about 14 days' incubation. Optimal growth is shown in media with a sea water content of 75–100% or with an equivalent salt concentration. Very little, if any, growth occurs on media prepared with distilled water. Optimal growth occurs around 20°C, and fair growth at 5°C, but little, if any, is shown above 30°C.

Oxidase negative but esculin is hydrolyzed.

Acid is produced from fructose, glucose, inositol, inulin, and mannose but not from cellobiose, dulcitol, ethanol, galactose, glycogen, lactose, maltose, mannitol, melezitose, raffinose, rhamnose, salicin, sucrose, trehalose, or xylose.

Grows on α-alanine and L-α-alanine as sole carbon sources but not on glycine, L-proline, or salicylaldehyde.

Quite resistant to lysozyme but highly susceptible to penicillin, rifampicin, and vancomycin.

Isolated from the uppermost layer of marine sediments from the northeast Atlantic Ocean.

The mol% G + C of the DNA is 64.9–66.4 (T_m).

Type strain: 3438W (DSM 43752).

11. **Rhodococcus maris** (Harrison) Nesterenko, Nogina, Kasumova, Kvasnikov and Batrakov 1982, 11.[VP] (*Flavobacterium maris* Harrison 1929, 229.)

mar′is. L. gen. n. *maris* of the sea.

See generic description and Tables 26.14 and 26.15 for many of the features of *R. maris*.

Coccoid cells germinate into short rods that show snapping division and V-forms. Poor to moderate, butyrous, orange growth occurs on glycerol, nutrient, and wort agars. Circular, raised, butyrous, glistening colonies with entire margins are formed on nutrient agar. Growth in nutrient broth is turbid.

Acetyl methyl carbinol, indole, methyl red, *p*-nitrophenoloxidase, and phosphatase tests are negative. Hydrogen sulfide is not produced.

Acid is produced from fructose, glycerol, and glucose but not from adonitol, arabinose, cellobiose, dulcitol, galactose, inositol, lactose, maltose, mannitol, α-methyl-D-glucose, raffinose, rhamnose, salicin, sorbitol, sorbose, sucrose, or xylose.

Growth with C_6 to C_{17}, and C_{23} *n*-alkanes but not with ethane or methane.

Isolated from soil and from skin and intestinal tract of carp (*Cyprinus carpio*).

The mol% G + C of the type strain is 73.2 (T_m).

Type strain: IMV 195.

12. **Rhodococcus rhodnii** Goodfellow and Alderson 1979, 80.[AL] (Effective publication: Goodfellow and Alderson 1977, 117.)

rhod'ni.i. M.L. masc. n. *Rhodnius* generic name of the reduvid bug; M.L. gen. n. *rhodnii* of *Rhodnius*.

See generic description and Tables 26.14 and 26.15 for many of the features of *R. rhodnii*.

Cocci germinate into rods that show limited branching. The growth cycle is completed by the appearance, through fragmentation, of short rods and cocci. Rough, red colonies are formed on glucose yeast extract agar.

Grows on sodium benzoate, sodium octanoate, and L-tyrosine as sole carbon sources, but not on salicin. Grows at 25–37°C.

Isolated from intestine of the reduvid bug, *Rhodnius prolixus*.

The mol% G + C of the DNA is 66.0 (T_m).

Type strain: KCC A-0203.

13. **Rhodococcus ruber** (Kruse) Goodfellow and Alderson 1977, 117.[AL] (*Streptothrix rubra* Kruse 1896, 63.)

rub'er. L. adj. *ruber* red.

See generic description and Tables 26.14 and 26.15 for many of the features of *R. ruber.*

Cocci give rise to multiple branched hyphae that readily fragment into rods and cocci. Single unbranched aerial hyphae are regularly produced. Rough, pink to red colonies are formed on glucose yeast extract agar, Sauton's agar, and egg media. Single aerial hyphae, which may occasionally be branched, are formed on the surface of colonies.

Acetamidase positive but allantoinase and urease negative; α- and β-esterase, galactosidase, and acid phosphatase negative.

Acid is formed from glucose, mannitol, and sorbitol but not from inositol, mannose, rhamnose, or trehalose.

Grows on butan-1-ol, *iso*-butanol, *n*-butanol, inulin, 2-methylpropan-1-ol, propan-1-ol, propane-1,2-diol, *n*-propanol, propylene glycol, sodium malonate, sodium octanoate, and L-tyrosine but not on salicin or xylose.

Resistant to picric acid (0.2% w/v), sodium nitrite (0.1% v/v) and sodium salicylate (0.1% w/v) but not to 5-fluorouracil (5 μg/ml), mitomycin C (5 μg/ml), or rifampicin (25 μg/ml).

Isolated from soil.

The mol% G + C of the DNA is 69–73 (T_m).

Type strain: KCC A-0205.

14. **Rhodococcus rubropertinctus** (Hefferan) Tsukamura 1974a, 43.[AL] (*Bacillus rubropertinctus* Hefferan 1904, 460; *Rhodococcus corallinus* (Bergey et al.) Goodfellow and Alderson 1977, 115.)

rub.ro.per.tinc'tus. L. adj. *ruber* red; L. pref. *per* very; L. part. adj. *tinctus* dyed, colored; M.L. adj. *rubropertinctus* heavily dyed red.

See generic description and Tables 26.14 and 26.15 for many of the features of *R. rubropertinctus.*

Rod-coccus growth cycle. Rough, orange to red colonies formed on glucose yeast extract agar, Sauton's agar, and egg media.

Allantoinase and urease positive; acetamidase, benzamidase, isonicotinamidase, malonamidase, nicotinamidase, pyrazinamidase, salicylamidase, and succinamidase negative. β-Esterase positive; acid phosphatase, α-esterase, and β-galactosidase negative.

Acid is produced from glucose, mannitol, mannose and sorbitol but not from arabinose, galactose, inositol, raffinose, rhamnose, or xylose.

Grows on butan-2,3-diol, 2,3-butylene glycol, butane-2,3-diol, inulin, propan-1-ol, and sodium octanoate as sole carbon sources but not on propane-1,2-diol, propylene glycol, salicin, or sodium malonate. Uses L-glutamate, nitrate, L-serine, and succinamide as sole nitrogen sources, but not nitrite or urea. Glucosamine HCl and monoethanolamine are used as sole sources of carbon and nitrogen, but benzamide is not.

Resistant to ethambutol (5 μg/ml), picric acid (0.2% w/v), rifampicin (25 μg/ml), sodium *p*-aminosalicylate (0.2% w/v), and sodium nitrite (0.1% w/v) but susceptible to 5-fluorouracil (20 μg/ml) and mitomycin C (5 μg/ml).

Isolated from soil.

The mol% G + C of the DNA is 67–69 (T_m).

Type strain: ATCC 14352.

15. **Rhodococcus sputi** (ex Tsukamura 1978) Tsukamura and Yano 1985, 365.[VP]

spu'ti. L. gen. *sputum* discharge from the respiratory tract.

Occurs as short rods and cocci. Rough, pink colonies formed on egg media.

Nicotinamidase, pyrazinamidase, and urease positive, but benzamidase negative. Acid phosphatase positive but negative for β-galactosidase and α-esterase.

Acid is produced from glucose, mannitol, mannose, sorbitol, and trehalose, but not from arabinose, galactose, inositol, rhamnose, or xylose.

Grows on ethanol, mannitol, mannose, sorbitol, sucrose, trehalose, sodium citrate, and sodium fumarate as sole carbon sources, but not on galactose, inositol, rhamnose, propylene glycol, sodium benzoate, or sodium malonate. Uses acetamide, nicotinamide, nitrate, and succinamide as sole nitrogen sources, but not benzamide, isonicotinamide, or nitrite. Acetamide and monoethanolamine used as sole sources of carbon and nitrogen, but not glucosamine or serine.

Resistant to ethambutol (5 μg/ml), *p*-nitrobenzoic acid (0.5 mg), picric acid (0.2% w/v), and rifampicin (25 μg/ml), but not to 5-fluorouracil (20 μg/ml), mitomycin C (5 μg/ml), sodium nitrite (0.1% w/v) or sodium salicylate (0.1% w/v). Grows at 28°C and 37°C but not at 42°C.

Isolated from human sputum.

Type strain: ATCC 29627.

16. **Rhodococcus terrae** (Tsukamura) Tsukamura 1974a, 43.[AL] (*Gordona terrae* Tsukamura 1971, 22.)

ter'res. L. n. *terra* earth; M.L. gen. n. *terrae* of the earth.

See generic descriptions and Tables 26.14 and 26.15 for many of the features of *R. terrae.*

Rod-coccus growth cycle. Produces rough, pink to orange colonies on glucose yeast extract agar, Sauton's agar, and egg media.

Allantoinase, nicotinamidase, pyrazinamidase, and urease positive; acetamidase, benzamidase, isonicotinamidase, malonamidase, and salicylamidase negative. β-Esterase positive but acid phosphatase, α-esterase, and β-galactosidase negative.

Acid is produced from mannitol, rhamnose, sorbitol, and trehalose but not from arabinose, galactose, inositol, or xylose.

Grows on inulin, propanol, and sodium octanoate as sole carbon sources but not on butane-2,3-diol, 1,3-butylene glycol, 1,4-butylene glycol, 2,3-butylene glycol, propane-1,2-diol, or sodium malonate. Uses acetamide, benzamide, isonicotinamide, nicotinamide, nitrate, pyrazinamide, and urea as sole nitrogen sources, but not nitrate; L-glutamate and monoethanolamine are used as sole sources of carbon and nitrogen, but acetamide, benzamide, and serine are not.

Resistant to ethambutol (5 μg/ml), picric acid (0.2% w/v), rifampicin (25 μg/ml), sodium *p*-aminosalicylate (0.2% w/v), and sodium nitrite (0.1%, w/v) but not to 5-fluorouracil (40 μg/ml), mitomycin C (5 μg/ml) or sodium salicylate (0.1% w/v).

Isolated from soil.

The mol% G + C of the DNA is 64–69 (T_m).

Type strain: ATCC 25594.

Species Incertae Sedis

a. **Rhodococcus aichiensis** Tsukamura 1983, 896.[VP] (Effective publication: Tsukamura 1982b, 1116.)

ai'chi.en'sis. M.L. adj. *aichiensis* belonging to Aichi Prefecture, Japan (where the organism was isolated).

Occurs as short rods. Rough, pinkish or orange colonies formed on egg media.

Acetamidase, nicotinamidase, pyrazinamidase, and urease positive, but allantoinase, benzamidase, isonicotinamidase, salicylamidase, and succinamidase negative. Reduces nitrate to nitrite, hydrolyzes Tween 80, and is acid phosphatase positive, but is arylsulfatase, α- and β-esterase, and β-galactosidase negative.

Grows on fructose, glucose, mannose, sucrose, trehalose, *n*-butanol, ethanol, *iso*-butanol, *n*-propanol, sodium acetate, sodium citrate, sodium fumarate, sodium malate, sodium pyruvate, and sodium succinate as sole carbon sources, but not on arabinose, galactose, inositol, mannitol, rhamnose, sorbitol, xylose, 1,3-, 1,4-, and 2,3-butylene glycols, propylene glycol, sodium benzoate, or sodium malonate. Uses acetamide, glutamate, and monoethanolamine as sole sources of carbon and nitrogen.

Resistant to picric acid (0.2% w/v), sodium nitrite (0.1% w/v), and sodium salicylate (0.1% w/v) but not to 5-fluorouracil (20 μg/ml) or mitomycin C (5 μg/ml). Grows at 28°, 37°, and 42°C, but not at 45°C.

Isolated from human sputum.

Type strain: E9028 (ATCC 33611).

b. **Rhodococcus aurantiacus** (ex Tsukamura and Mizuno 1971) Tsukamura and Yano 1985, 365.[VP]

au.ran.ti'a.cus. M.L. n. *aurantium* generic name of orange; *aurantiacus* orange colored.

Shows a rod-coccus life cycle. Rough, cream to orange colonies formed on glucose yeast extract agar and Sauton's agar. Synnemata of vertically arranged coalescing filaments are found on the surface of colonies.

Acetamidase, allantoinase, nicotinamidase, pyrazinamidase, and urease positive, but benzamidase, isonicotinamidase, malonamidase, salicylamidase, and succinamidase negative. β-Galactosidase and unrease positive, but acid phosphatase, arylsulfatase, and α-esterase negative. Nitrate is not reduced. Degrades hypoxanthine, tyrosine, and Tweens 20, 40, 60, and 80, but not adenine, casein, or elastin.

Acid is produced from galactose, glucose, inositol, mannitol, mannose, sorbitol, trehalose, and xylose, but not from rhamnose.

Grows on ethanol, fructose, galactose, glucose, glycerol, inositol, mannitol, mannose, melezitose, sorbitol, sucrose, trehalose, xylose, *n*-butanol, *iso*-butanol, 2,3-butylene glycol, propanol, propylene glycol, sodium acetate, sodium citrate, sodium fumarate, sodium malate, sodium pyruvate, and sodium succinate, but not on adonitol, arabinose, inulin, lactose, raffinose, or rhamnose. Uses acetamide, nicotinamide, nitrate, and urea as sole nitrogen sources, but not benzamide. Acetamide, glutamate, glucosamine HCl, monoethanolamine, and serine are used as sole sources of carbon and nitrogen, but not benzamide or trimethylenediamine.

Resistant to ethambutol (5 μg/ml), 5-fluorouracil (20 μg/ml), mitomycin C (10 μg/ml), picric acid (0.2% w/v), and sodium nitrite (0.2% w/v). Grow at 10°, 28°, and 37°C but not at 40°C.

Isolated from sputum.

Type strain: ATCC 25938.

c. **Rhodococcus chubuensis** Tsukamura 1983, 896.[VP] (Effective publication: Tsukamura 1982b, 1116.)

chu'bu.en'sis. M.L. adj. *chubuensis* belonging to Chubu Hospital, (where the organism was isolated).

Occurs as short rods, rough, pinkish or orange colonies formed on egg media.

Reduces nitrate to nitrite, hydrolyzes Tween 80, and is α- and β-esterase and acid phosphatase positive, but is arylsulfatase and β-galactosidase negative. Usually does not show any amidase activity.

Grows on fructose, glucose, mannitol, mannose, sorbitol, sucrose, trehalose, ethanol, *n*-propanol, sodium acetate, sodium citrate, sodium fumarate, sodium malate, sodium pyruvate, and sodium succinate as sole carbon sources, but not on arabinose, galactose, inositol, rhamnose, xylose, butanols, butylene glycols, propylene glycol, sodium benzoate, or sodium malonate. Uses acetamide and glutamate as sole sources of carbon and nitrogen, but not benzamide, glucosamine, monoethanolamine, or trimethylenediamine.

Resistant to picric acid (0.2% w/v) but sensitive to 5-fluorouracil (20 μg/ml), mitomycin C (5 μg/ml), and sodium nitrite (0.1% w/v). Grows at 28° and 37°C but not at 42°C.

Isolated from human sputum.

Type strain: E6324 (ATCC33609).

d. **Rhodococcus obuensis** Tsukamura 1983, 897.[VP] (Effective publication: Tsukamura 1982b, 1117.)

o.bu'en'sis M.L. adj. *obuensis* belonging to Obu City (where the organism was isolated).

Occurs as rods. Rough, pinkish or orange colonies formed on egg media.

Acetamidase, nicotinamidase, pyrazinamidase and urease positive, but allantoinase, benzamidase, isonicotinamidase, salicylamidase, and succinamidase negative. Reduces nitrate to nitrite, hydrolyzes Tween 80, and is acid phosphatase positive, but is arylsulfatase, α- and β-esterase, and β-galactosidase negative.

Grows on fructose, glucose, mannitol, sucrose, sodium acetate, sodium citrate, sodium malate, sodium pyruvate, and sodium succinate as sole carbon sources, but not on arabinose, galactose, inositol, rhamnose, trehalose, xylose, *iso*-butanol, *n*-butanol, 13-, 14-, and 2,3-butylene glycols, ethanol, *n*-propanol, propylene glycol, sodium benzoate, sodium fumarate, or sodium malonate. Uses glutamate as a sole source of carbon and nitrogen, but not acetamide, benzamide, glucosamine, monoethanolamine, serine, or trimethylenediamine.

Resistant to picric acid (0.2% w/v) and sodium nitrite (0.2% w/v) but sensitive to 5-fluorouracil (20 μg/ml) and mitomycin C (5 μg/ml). Grows at 28° and 37°C but not at 42°C.

Isolated from human sputum.

Type strain: E8179 (ATCC 33610).

Genus **Nocardioides** *Prauser 1976, 61*[AL]

HELMUT PRAUSER

No.car.di.o.i'des. M.L. fem. n. *Nocardia* name of a genus; Gr. suff. *idea* appearance; M.L. mas. n. *Nocardioides* nocardia-like, referring to the similar life cycles of the two genera.

Primary mycelium shows abundantly branching hyphae growing on the surface and penetrating into agar media; they **break up into fragments** that may be irregular or rodlike or coccoid. **Aerial mycelium** consisting of irregular, sparsely and irregularly branching, or unbranched hyphae that **break up into short to elongated rodlike fragments**. The fragments of both the primary and the aerial mycelium give rise to new mycelia. No motile cells. Colonies pasty. **Gram-positive. Non-acid fast. Catalase positive.** Strictly **aerobic**. Chemo-organotrophic. Oxidative catabolism. Grows readily on standard media. **Susceptible to specific phages.** Diagnostic amino acids of the cell wall **L-diaminopimelic acid (L-DAP) and glycine**. Mycolic acids lacking. Diagnostic phospholipids **phosphatidylglycerol** and **acylphosphatidylglycerol**. Phosphatidylethanolamine and other nitrogenous phospholipids lacking. 14-Methylpentadecanoic acid predominating among fatty acids.

Menaquinones primarily of MK-8(H_4) type. Mol% G + C of DNA ranging from 66.1 to 72.7 (T_m). Worldwide in **soil.**

Type species: *Nocardioides albus* Prauser 1976b, 61.

Further Descriptive Information

The information given in the following four paragraphs originates from Prauser (1976b, 1981a) or represents the author's unpublished results.

Fragmentation of the hyphae of the primary mycelium begins in the older parts of the colonies and hyphae (Fig. 26.3). Depending on the size of the fragments they give rise to new mycelia by extruding one, two, or more hyphae. On media rich in organic nitrogen, and in submerged shaken culture, the extent of mycelia and their persistence are reduced. The irregularly shaped and branched hyphae of the aerial mycelium (Fig. 26.4) resemble those of nocardiae. Aerial hyphae break up completely into sporelike fragments (Figs. 26.5 and 26.6). They germinate by producing one or two germ tubes. Motility does not occur at any stage of the life cycle.

Hyphae and fragments show the ultrastructure typical for Gram-positive bacteria (Figs. 26.6 and 26.7). The hyphae of the primary mycelium are irregularly septated. Preceding fragmentation, additional cross-walls are formed. Branching usually takes place near to cross-walls by production of lateral outgrowths (Fig. 26.7). The process of branching can be regarded as restoration of the growth functions in the course of the individualization of the hyphal fragments (Fig. 26.7). The aerial hyphae septate and break up more regularly than those of the primary mycelium, resembling one of the types of spore formation described for streptomycetes (Fig. 26.6). The resulting sporelike fragments display a smooth surface (Fig. 26.5).

Colonies not covered by aerial mycelium have a pasty consistency. The surface of agar cultures is smooth to wrinkled and dull to bright, but usually it is faintly glistening. The aerial mycelium may totally cover the primary mycelium, may be produced only in patches or at the margins of the colonies, or may be lacking, depending on the individual strain, the medium used, and the procedures and the duration of maintenance of strains. The sporelike fragments are rarely recognizable in situ. However, they are seen after gently pressing a coverslip onto the mat of aerial mycelium followed by microscopical observation on a slide covered with a thin film of agar.

The organism grows within 1–2 days on standard media such as nutrient agar or oatmeal agar (Shirling and Gottlieb, 1966); the latter is particularly suited for the production of aerial mycelium. A variety of carbohydrates and alcohols is used as carbon source.

The large number of phages that multiply on *Nocardioides* strains do not affect other nocardioforms, coryneforms, and sporoactinomycetes (Prauser, 1976a, 1984c; Prauser and Falta, 1968; Wellington and Williams, 1981a), except for four phages that also propagate on strains of *Pimelobacter simplex* (*Arthrobacter simplex*) and *Pimelobacter jensenii* (Prauser 1976a, 1984c, and unpublished results). On the other hand, phages that are effective against other nocardioforms, coryneforms, and sporoactinomycetes do not multiply on *Nocardioides* strains (Prauser, 1974a, 1976a, 1984c; Wellington and Williams, 1981a). Some *Streptomyces* phages cause clearing effects (i.e. phage-dependent lysis without phage propagation) on strains of the genus *Nocardioides* (Prauser, 1984c).

Diagnostic amino acids of the cell wall are L-DAP and glycine (cell wall chemotype I *sensu* Becker et al., 1965). L-DAP of one peptide subunit is cross-linked with D-alanine of another via glycine in the interpeptide bridge (peptidoglycan type A3γ sensu Schleifer and Kandler, 1972; Prauser, 1976b). The results on phospholipids are conflicting (Lechevalier et al., 1977, 1981; O'Donnell et al., 1982a; Collins et al., 1983b). According to Lechevalier et al. (1977, 1981), who applied the most conclusive methods, the phospholipid composition of *Nocardioides albus* is unique among actinomycetes. In addition to phosphatidylglycerol, phosphatidylinositol, and traces of phosphatidylinositol mannosides, the organism contains major amounts of acylphosphatidylglycerol. Diphosphatidylglycerol, present in nearly all actinomycetes studied so far, is lacking as well as any nitrogenous phospholipids. The predominating fatty acid is 14-methylpentadecanoic acid. The amounts reported vary between 27 and 51% (O'Donnell et al., 1982a; Collins et al., 1983b; Suzuki and Komagata, 1983). There are conflicting results concerning tuberculostearic acid and its lower homologs (which in contrast to *Pimelobacter simplex* were not found here during the first 24 h of incubation (Schumann, Prauser, et al., unpublished)); saturated straight-chain fatty acids (C_{15}–C_{18}); *iso*- and *anteiso*- branched-chain fatty acids (C_{15}–C_{18}); and the straight-chain unsaturated oleic acid (C_{18}) (O'Donnell et al., 1982a; Collins et al., 1983b; Suzuki and Komagata, 1983). Among menaquinones, tetrahydrogenated ones with eight isoprene units (MK-8(H_4) predominate (O'Donnell et al., 1982a; Collins et al., 1983b; Suzuki and Komagata, 1983). Results concerning minor isoprenologs are conflicting. Mycolic acids are lacking (Prauser, 1976b; O'Donnell et al., 1982a). The mol% G + C of DNA ranges from 66.1 to 72.7 (Prauser, 1966, 1976b; Tille et al., 1978; Suzuki and Komagata, 1983; Schumann, Prauser, et al., unpublished). DNA/DNA homology between *Nocardioides albus* and *Pimelobacter jensenii, P. tumescens, Intrasporangium calvum, Streptomyces griseus*, and other streptomycetes is on the level of experimental noise (Schumann, Prauser, et al., unpublished). Suzuki and Komagata (1983) reported 16% DNA/DNA homology between *N. albus* and *P. jensenii*. DNA/DNA homology between *Nocardioides albus* and *Pimelobacter simplex* was found to range from 9 to 15% (12.5% on average) (Schumann, Prauser, et al., unpublished). Suzuki and Komagata (1983) found 18%.

Pathogenicity for man, animals, and plants has not been reported.

Strains have been isolated from soils of gardens, meadows, arable land, grassland, and savanna and from tephra (Surtsey); the organism predominated in Kaolin prepared for the ceramic industry (Prauser, 1976b).

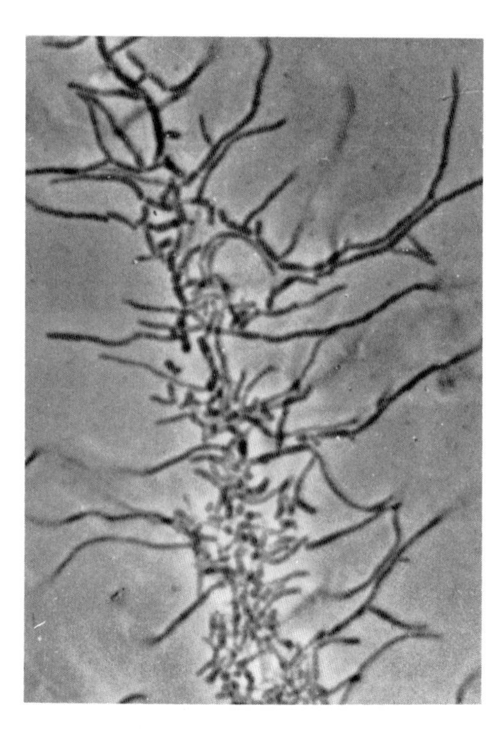

Figure 26.3. *Nocardioides albus.* Fragmentation of hyphae of the primary mycelium in situ on glycerol asparagine agar. Phase-contrast micrograph (× 1600). (Reproduced with permission from H. Prauser, International Journal of Systematic Bacteriology *26:* 58–65, 1976, ©American Society for Microbiology.)

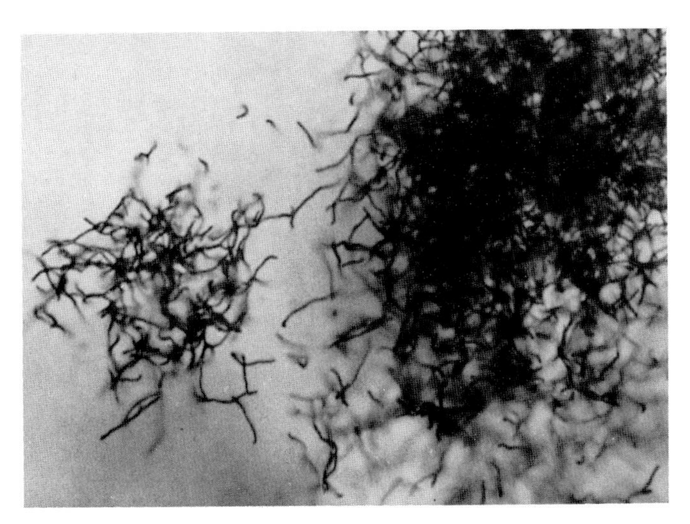

Figure 26.4. *Nocardioides albus.* Aerial mycelium on chitin agar. Phase-contrast micrograph (× 400). (Reproduced with permission from H. Prauser, International Journal of Systematic Bacteriology *26:* 58–65, 1976, ©American Society for Microbiology.)

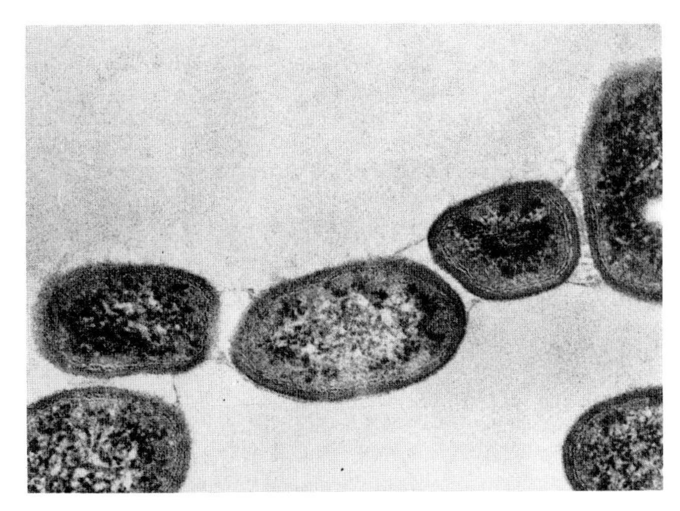

Figure 26.6. *Nocardioides albus.* Developing spore-like elements (arthrospores) still connected by the surface sheath of the aerial hypha. Electron micrograph (× 50,000). (Reproduced with permission from H. Prauser. Actinomycetes. In Schaal and Pulverer (Editors), Proceedings 4th International Symposium on Actinomycetes Biology. Zentralbl. Bakteriol. Hyg. Abt. 1 Orig. B, Supplement *11,* 1981, ©Fischer-Verlag, Stuttgart.)

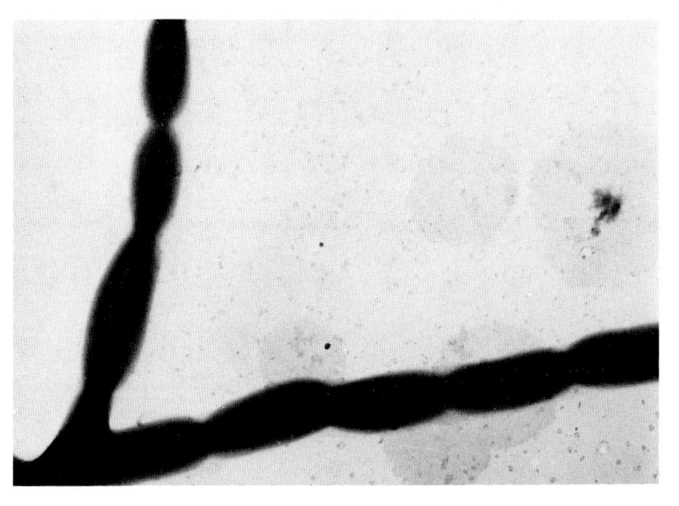

Figure 26.5. *Nocardioides albus.* Aerial hyphae fragmented into sporelike elements. Electron micrograph (× 17,000). (Reproduced with permission from H. Prauser, International Journal of Systematic Bacteriology *26:* 58–65, 1976, ©American Society for Microbiology.)

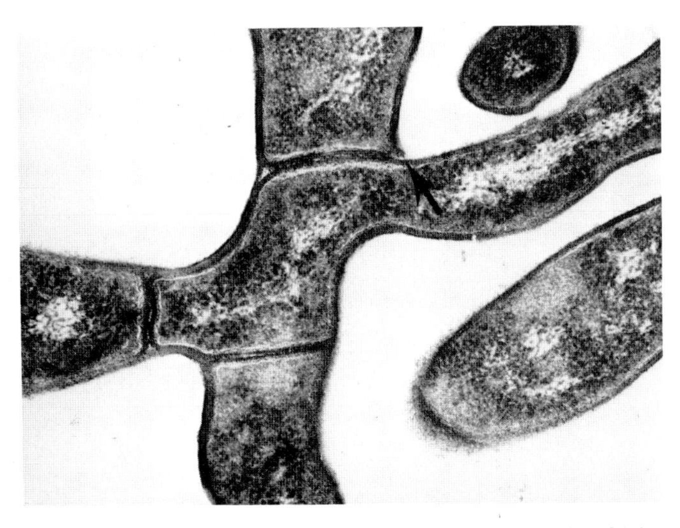

Figure 26.7. *Nocardioides albus.* Part of a hypha (vertical in the figure) of the primary mycelium with branches originating from one segment. Beginning of fragmentation of the hypha at the upper right angle (*arrow*). Electron micrograph (× 40,000). (Reproduced with permission from H. Prauser. Actinomycetes. In Schaal and Pulverer (Editors), Proceedings 4th International Symposium on Actinomycetes Biology. Zentralbl. Bakteriol. Hyg. Abt. 1 Orig. B, Supplement *11,* 1981, ©Fischer-Verlag, Stuttgart.)

Isolation Procedures

Strains can be isolated employing the usual dilution and plating techniques on several agar media appropriate for the isolation of streptomycetes and other actinomycetes. Often a dilute oatmeal agar has been used (Prauser and Bergholz, 1974: 3 g oatmeal, 0.3 g KNO_3, 0.5 g K_2HPO_4, 0.2 g $MgSO_4$, 15 g agar, and 1 liter distilled water; pH = 7).

Maintenance Procedures

Strains may be maintained by serial transfers on oatmeal agar slants (once every 3 months), by lyophilization, and by storage above liquid nitrogen either as suspensions containing 5% dimethylsulfoxide or as minicultures that are likewise suitable for subsequent shipping of the strains (Prauser, 1984b).

Rapid Identification by Phage Typing

All strains of the genus *Nocardioides* hitherto examined are susceptible to at least one of a selected set of five phages: IMET 5013 (X1), IMET 5015 (X3), IMET 5017 (X5), IMET 5057 (X10), and IMET 5056 (X24) (Prauser, 1976b, unpublished results). For all procedures the complex organic medium number 79 (Prauser and Falta, 1968) may be used. Following incubation of an agar slant culture of the strain for 1–2 days the growth is suspended and one drop of the suspension is mixed with 3 ml of molten agar (0.6% agar) and spread over a basal layer (1.5% agar) in a Petri dish. One loopful of each of the five phage suspensions ($<10^9$ plaque-forming units/ml) is placed on the solidified agar at marked positions. Following overnight incubation at 28°C the bacterial lawn is examined for clear or turbid zones. The occurrence of one to five positive reactions identifies the strain under study with the genus *Nocardioides*. Even clearing effects (i.e. phage-dependent lysis without phage propagation) are as a rule taxon specific (Prauser, 1981b, 1984c). Hence, differentiation between true lysis and clearing effects is not essential for correct identification. In addition, the presence of mycelia should be checked microscopically to avoid misidentification as *Pimelobacter* species (see above, and Table 26.14 and genus *Arthrobacter*), although none of the five phages recommended ever displayed any lytic effect against strains of the genus *Pimelobacter.*

Differentiation of the genus **Nocardioides** *from other taxa*

Essential characteristics differentiating the genus *Nocardioides* from other more or less closely related taxa are presented in Table 26.16. The table includes, in addition to the coryneform genus *Pimelobacter* (see genus *Arthrobacter*), all L-DAP containing nocardioforms, most of the remaining nocardioforms, and the family *Streptomycetaceae.*

The data given in Table 26.16 will provide reliable recognition of *Nocardioides* strains, particularly if phage typing is included.

Taxonomic Comments

The first known representative of the genus was strain IMET 7801, isolated from kaolin (Prauser, 1966). The observed type of life cycle was termed "nocardioform," which was proposed for bacteria producing hyphae that break up completely into fragments that give rise to new mycelia (Prauser, 1967, 1978). Additional strains constituted the "IMET 7801 group" (Prauser, 1967), for which the generic designation *Nocardioides* was tentatively introduced (Prauser, 1970). The valid publication was delayed until 1976 (Prauser, 1976b) to give specialists the chance to place this widely distributed organism into an existing taxon.

Nocardioides albus and *Arthrobacter simplex* (*Pimelobacter simplex*) are closely related organisms. The two species share the peptidoglycan type A3 γ (Prauser, 1976b), cross-susceptibility to 4 of 26 specific phages (Prauser, 1970, 1976a, 1984c, unpublished results), similar levels (in the range around 70 mol% G + C) of DNA base composition (Prauser 1966, 1976b; Suzuki and Komagata, 1983), some degree (15–20%) of DNA/DNA homology (Prauser, 1981a, unpublished results; Suzuki and Komagata, 1983), the bending type of cell division (Prauser, 1981a), a similar pattern of fatty acids (O'Donnell et al., 1982a; Collins et al., 1983b) and identical predominating menaquinones (O'Donnell et al., 1982a, Collins et al., 1983b; Suzuki and Komagata, 1983).

On the basis of these data, including the results on phospholipids, O'Donnell et al. (1982) transferred *Arthrobacter simplex* to the genus *Nocardioides* as *Nocardioides simplex* (validated in List 12, Int. J. Syst. Bacteriol., *33*: 896–897, 1983). Consequently, the description of the genus *Nocardioides* had to be emended, i.e. to be extended in order to include these morphologically extremely different organisms. On the other hand, Suzuki and Komagata (1983), including in their comparative biochemical studies the type strains of *Nocardioides albus*, *Arthrobacter globiformis*, and *Arthrobacter simplex*, established the new genus *Pimelobacter* to harbor *Pimelobacter simplex* (formerly *Arthrobacter simplex*) as the type species. This new combination has also been validated (Validation List 11, Int. J. Syst. Bacteriol, *33*: 672–674, 1983). Thus, because they are based on the same type strain, *Arthrobacter sim-*

Table 26.16.
Characteristics differentiating the genus **Nocardioides** *from other taxa*[a,b]

Characteristics	*Nocardioides*	*Pimelobacter*[c]	*Intrasporangium*	*Sporichthya*	*Arachnia*	*Streptomycetaceae*	*Nocardia*	*Rhodococcus*	*Mycobacterium*	*Oerskovia*	*Promicromonospora*
Susceptibile to *Nocardioides* phages[d]	+	–	–	–	–	–	–	–	–	–	–
Primary mycelium on agar media	+[e]	–	+	–	+	+	+	+	–	+	+
Hyphae of primary mycelium break up into fragments[f]	+		+		+	–	+	+	–	+	+
Aerial mycelium[g]	+	–	–	+	–	+	+	–	–	–	+
Acid fast	–	–	–	–	–	–	p	p	+	–	–
Anaerobic growth	–	–	–	–	+	–	–	–	–	+[h]	–
L-Diaminopimelic acid	+	+	+	+	+	+	–	–	–	–	–

[a]For biochemical data differentiating the genera included in this table see "Further Descriptive Information" and the corresponding generic contributions in the present volume and in Volume 2.
[b]Symbols: see Table 26.13; also *p*, partial acid-fastness predominates in *Nocardia* and sometimes occurs in *Rhodococcus.*
[c]See *Arthrobacter* in Volume 2.
[d]The five phages recommended are given in "Rapid Identification by Phage Typing." The phages are available from IMET.
[e]The primary mycelia of *Nocardioides* strains are always extensive and cannot be confused with the short side projections of *Pimelobacter tumescens,* and the very rare ones of other pimelobacters and mycobacteria.
[f]Partial fragmentation (complete fragmentation in *"Nocardia italica"* (sic) under normal growth conditions has been reported for individual strains of a few species of the genus *Streptomyces.* In some nocardiae, or even representatives of other nocardioform genera, fragmentation is not always seen on agar media in situ. Gentle pressure on slide preparations or submerged shaken culture will display fragmentation in nearly all cases.
[g]The data given apply at least to fresh isolates. Aerial hyphae sometimes may be seen only microscopically.
[h]For oerskoviae anaerobic growth has been reported only on trypticase-soy agar (Lechevalier, 1972).

plex,* *Nocardioides simplex*, and *Pimelobacter simplex* are objective synonyms.

In the present contribution the original description of the genus *Nocardioides* Prauser 1976 is retained for several reasons: (a) fundamental morphological differences between *Nocardioides albus* and *Pimelobacter simplex* (*Arthrobacter simplex*) (Table 26.16); (b) differences in the fatty acid and phospholipid patterns (see "Further Descriptive Information"); (c) low DNA/DNA homology between *N. albus* and *P. simplex* (see "Further Descriptive Information"); and (d) the feeling that the inclusion of morphologically highly differing organisms in one genus will render the practical use of classification more difficult.

Prauser (1976b), believing that the introduction of a new family to include *Nocardioides albus* might be appropriate, provisionally placed the genus in the family *Streptomycetaceae*. Lechevalier and Lechevalier (1981a) regarded the genus as "in search of a family." Nesterenko et al. (1985) introduced the family *Nocardioidaceae* (not yet validated) to harbor the genera *Nocardioides* and *Pimelobacter*. This proposal reflects most accurately the relationship of the species of *Nocardioides* and the type species of *Pimelobacter*

Further Reading

O'Donnell et al. 1982. Lipids in the classification of *Nocardioides:* Reclassification of *Arthrobacter simplex* (Jensen) Lochhead in the genus *Nocardioides* (Prauser) emend. O'Donnell et al. as *Nocardioides simplex* comb. nov. Arch. Microbiol. *133:* 323–329.

Prauser, H. 1976. *Nocardioides*, a new genus of the order *Actinomycetales*. Int. J. Syst. Bacteriol. *26:* 58–65.

Prauser, H. 1984. Phage host ranges in the classification and identification of Gram-positive branched and related bacteria. *In* Ortiz-Ortiz et al. (Editors), Biological, Biochemical and Biomedical Aspects of Actinomycetes. Academic Press, London, pp. 617–633.

Suzuki, K-L. and K. Komagata. 1983. *Pimelobacter* gen. nov., a new genus of coryneform bacteria with LL-diaminopimelic acid in the cell wall. J. Gen. Appl. Microbiol. *29:* 59–71.

List of species of the genus Nocardioides

1. **Nocardioides albus** Prauser 1976b, 61.[AL]

al'bus. M.L. adj. *albus* white, referring to the white aerial mycelium.

Hyphae of the primary mycelium 0.5–0.8 μm in diameter, hyphae of the aerial mycelium 0.6–1.0 μm in diameter. Primary mycelium white, whitish, to faintly cream-colored on oatmeal agar, yeast extract–malt extract agar, glucose asparagine agar, and other media. Aerial mycelium thin, dense, and chalky. Surface of the sporelike elements smooth (Fig. 26.5). No soluble pigments, except for a reddish-brown pigment on media containing tyrosine as sole amino acid. Colonies lacking aerial mycelium are pasty, smooth to wrinkled and dull to bright. Growth optimum at about 28°C, good growth at 15° and 37°C; some strains grow at 42°C; no growth at 10° and 50°C.

D-Glucose, L-arabinose, sucrose, D-xylose, D-mannitol, D-fructose, and L-rhamnose utilized as carbon sources; inositol and raffinose not utilized. Acid produced from D-glucose, L-arabinose, sucrose, D-xylose, D-mannitol, D-fructose, and L-rhamnose, and in exceptional cases from adonitol and sorbitol; no acid from inositol and raffinose. Citrate, succinate, and benzoate assimilated; tartrate not assimilated. Hypoxanthine, tyrosine, esculin (most cases), casein, and starch hydrolyzed; adenine and xanthine not hydrolyzed.

Isolated from soil.

The mol% G + C of DNA is 66.4–72.7 (T_m).

Type strain: ATCC 27980 (IMET 7807).

2. **Nocardioides luteus** Prauser 1984a, 647.[VP] ("*Nocardioides flavus*" Ruan and Zhang 1979, 347.)

Displays the morphological, physiological, and biochemical characters of *Nocardioides albus* with a few exceptions (Table 26.17): primary mycelium yellow on oatmeal agar and other media, varying to orange-yellow in aged cultures. Aerial mycelium cream-colored if well developed, otherwise white. Most of the strains do not produce acid from L-rhamnose and sucrose.

Isolated from soil.

The mol% G+C of DNA is 66.1–70.0 (T_m); if the synonym "*Nocardioides flavus*" is included the mol% G + C of the DNA is up to 61.5 (T_m). DNA/DNA homology of IMET 7830 and *Nocardioides albus* IMET 7807 is 49%.

Type strain: IMET 7830.

Comment: The organism was published first as *Nocardioides* sp. IMET 7830 (Tille et al., 1978). The name "*Nocardioides luteus*," accompanied by a brief description, was used first by Prauser in a poster at the International Union of Microbiological Societies Congress, Munich, 1978. The name was used in the literature (Collins et al., 1983b; O'Donnell et al., 1982a). Since the type strain 71-N54 of "*Nocardioides flavus*" obtained from Dr. Ruan was found to be similar to IMET 7830, synonymy was assumed. DNA/DNA homology of the type strain of *N. albus* to *N. luteus* and "*N. flavus*" was found to be 38% and 33%, respectively (Schumann, Prauser, et al., unpublished).

Species Deserving More Study

a. "*Nocardioides fulvus*" Ruan and Zhang 1979, 350.

Life cycle nocardioform. Cell wall containing L-DAP and glycine. Primary mycelium yellow, yellowish brown, to brown; aerial mycelium cream-white or several tinges of yellow, no soluble pigment. Starch, D-mannitol, and D-fructose used as carbon sources; L-arabinose, D-xylose, L-rhamnose, raffinose, mannose, lactose, and inositol not tested. Acid produced from D-glucose; not produced from D-galactose, lactose, L-arabinose, D-xylose, L-rhamnose, raffinose, inositol, and mannitol. H_2S not produced, aminobutyric acid not fermented, D-glucose oxidized, hippuric acid and urea not decomposed. DNA/DNA homology between *N. albus* and "*Nocardioides fulvus*" is 41% (Schumann, Prauser, et al., unpublished). Isolated from soil samples of Beijing.

Type strain: 71-N86.

b. "*Nocardioides thermolilacinus*" Lu and Yan 1983, 224.

Life cycle nocardioform. Cell wall containing L-DAP. Thermophilic, growing between 28° and 55°C. Primary mycelium scarlet, aerial mycelium pink, soluble pigment orange-colored. The two strains (T 505 = IFO 14335, T 511 = IFO 14336) received from Dr. Kusaka, Institute of Fermentation, Osaka, displayed a *Streptomyces*-like life cycle, susceptibility to *Streptomyces* phages, and lack of susceptibility to *Nocardioides* phages (unpublished), and are hence regarded as streptomycetes.

Table 26.17.
Differential characteristics of species of the genus **Nocardioides**[a]

Characteristics	Species	
	1. *N. albus*	2. *N. luteus*
Yellow pigment on oatmeal agar	–	+
Well-developed aerial mycelium	White	Cream
Acid from L-rhamnose	+	–
Acid from sucrose	+	–

[a]Symbols: see Table 26.13.

***Editorial note:** The species is described in Volume 2 under *Arthrobacter*.

Genus **Pseudonocardia** *Henssen 1957, 408*[AL]

AINO HENSSEN

Pseu.do.no.car'di.a Gr. adj. *pseudes* false; M.L. n. *Nocardia* a genus of the actinomycetes; M.L. fem. n. *Pseudonocardia* the false *Nocardia*.

Substrate and aerial mycelium-bearing spores in chains, hyphae segmented, **often zig-zag shaped,** with tendency to form **apical or intercalary swellings.** Hyphal elongation by budding. Segments acting directly as spores or being secondarily divided in sporulation process. Hyphal wall with two layers, cross-walls are interspace septa, **spores without interspore pads. Gram-positive. No motile stages. Aerobic.** Growth on a variety of organic and synthetic media. **Mesophilic or thermophilic.**

Type species: *Pseudonocardia thermophila* Henssen 1957, 408.

Further Descriptive Information

Substrate and aerial hyphae varying in thickness, 0.4–1.0 µm, in swellings to 2.6 µm thick. A characteristic feature is the growth of hyphae by budding (Figs. 26.8–26.10); a constriction is produced behind the tip of the terminal segment, the tip then elongates to form a new segment, another constriction is formed near the tip, and the process repeats (time-lapse photographs, Figs. 1–7, published in Henssen and Schäfer, 1971). The constrictions may be secondarily separated by septa (Fig. 26.9*B*); a septum may also be formed a distance behind a constriction (e.g. in *P. compacta*, Henssen et al., 1983). Side branches usually arise below a septum, more rarely from the center of a segment (Figs. 26.10 and 26.11).

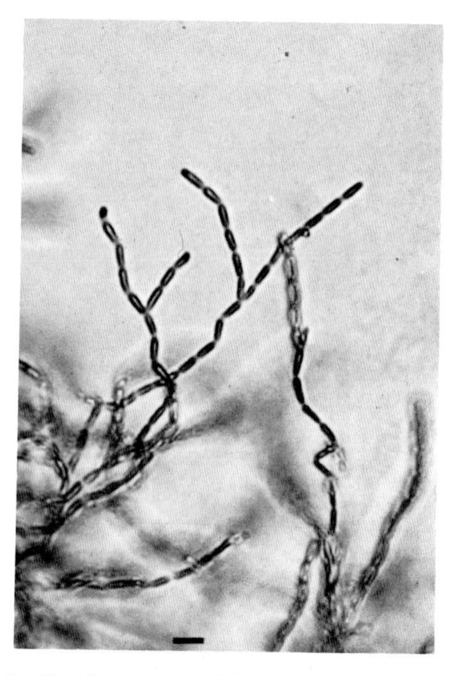

Figure 26.8. Budding, zig-zag-shaped hyphae of *Pseudonocardia spinosa* strain MB SF-1. (Light microscopy. *Bar,* 5 µm.) (Reproduced with permission from A. Henssen and D. Schäfer, International Journal of Systematic Bacteriology *21*: 29–34, 1971, ©American Society for Microbiology.)

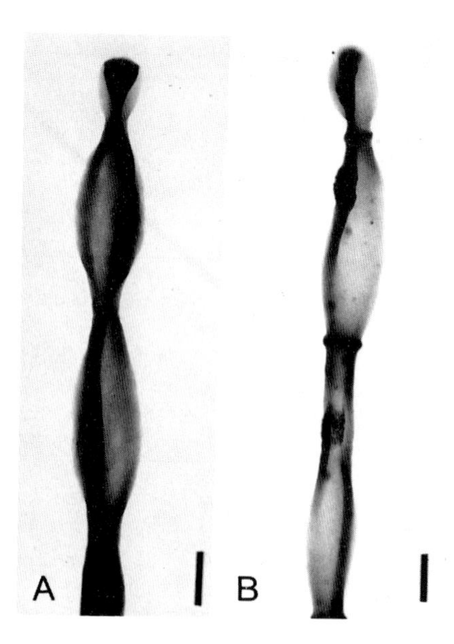

Figure 26.9A and B. Budding aerial hyphae of *Pseudonocardia thermophila* strain MB-A18. *B*, Constrictions separated by septae. (Whole mount silhouettes, transmission electron microscopy (TEM). *Bar,* 0.5 µm.) (Reproduced with permission from A. Henssen and D. Schäfer, International Journal of Systematic Bacteriology *21*: 29–34, 1971, ©American Society for Microbiology.)

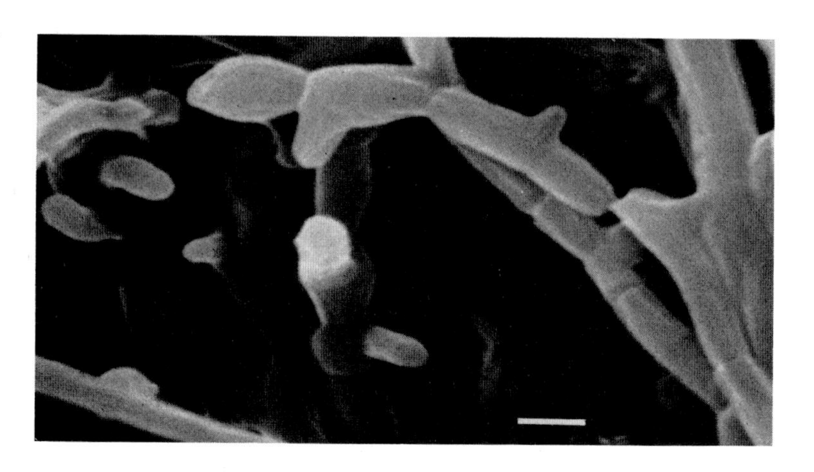

Figure 26.10. Budding substrate and aerial hyphae of *Pseudonocardia compacta* strain MB H-146. (Scanning electron microscopy. *Bar,* 1 µm.) (Reproduced with permission from A. Henssen, C. Happach-Kasan, B. Renner and G. Vobis, International Journal of Systematic Bacteriology *33*: 829–836, 1983, ©American Society for Microbiology.)

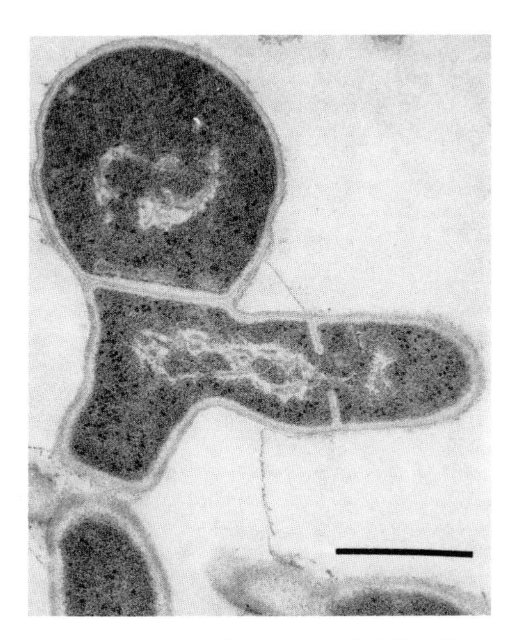

Figure 26.11. *Pseudonocardia thermophila* MB-A18, tip of aerial hyphae with apical swelling and side branch showing septum formation. (TEM, glutaraldehyde/osmium tetroxide fixation. *Bar,* 0.5 μm.)

Spores may be formed on substrate or aerial mycelium in three ways:

1. By successive acropetal formation (the usual way) designated as *Pseudonocardia*-type (Henssen and Schnepf, 1967).
2. By basipetal septation.
3. By irregular spore formation along senescing hyphae.

Spores are smooth walled or spiny, varying greatly in size, usually 0.5–1.0 μm wide by 1.5–3.5 μm long, but varying in length from 1.0 to 4.5 μm.

Cell wall constituents: *meso*-diaminopimelic acid (*meso*-DAP), arabinose, and galactose: type IV (Lechevalier and Lechevalier, 1970b; Henssen and Schäfer, 1971; Henssen et al., 1983), phospholipids type III (Lechevalier et al., 1981; Henssen et al., 1983), and MK-9(H_4) menaquinones (Minnikin and Goodfellow, 1981). No mycolic acid has been found in *P. thermophila* (Goodfellow and Minnikin, 1981a) or *P. compacta* (Henssen et al., 1983).

The hyphal wall in both substrate and aerial mycelium is composed of two layers, an inner electron-transparent, uniformly thick layer, and an outer electron-dense irregular layer (Figs. 26.11–26.13). Cross-walls are interspace septa (Henssen et al., 1981, 1983; cross-wall type 2 in Williams et al., 1973). Hyphal swellings may become subdivided by septa growing inward at different angles. Spore walls are of uniform thickness, and no interspore pads are formed. In spiny spores the ornamentation is formed by folds of the fibrous sheath.

Good growth on casamino-peptone-Czapek agar and yeast starch agar. Substrate mycelium forms a compact mass, colonies are yellow to ochre, pigment with a spectrum characteristic for carotenoids. Aerial mycelium is white, powdery or forming a thick cover.

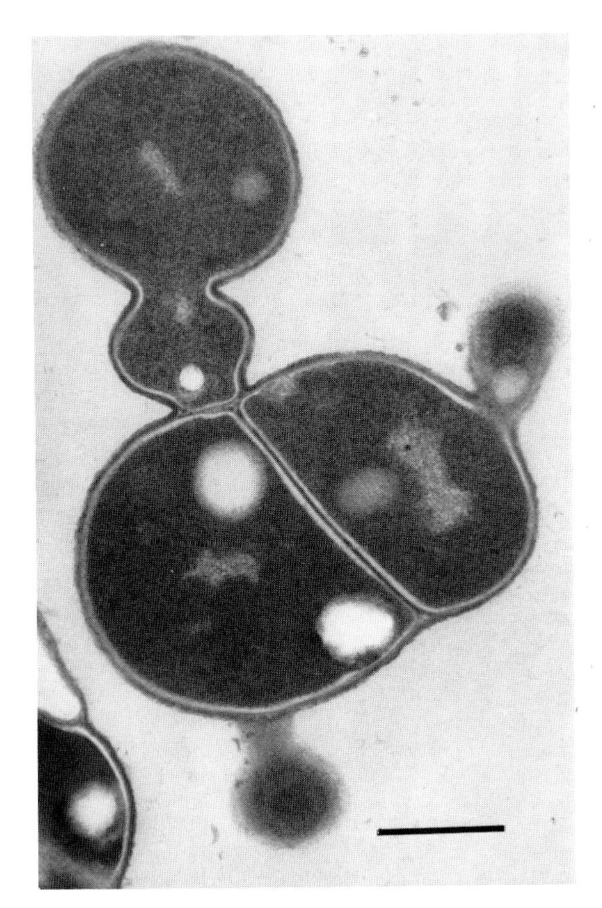

Figure 26.12. Budding cells of *Pseudonocardia spinosa* MB SF-1, the upper bud with constriction (TEM, glutaraldehyde/osmium tetroxide fixation. *Bar,* 0.5 μm.)

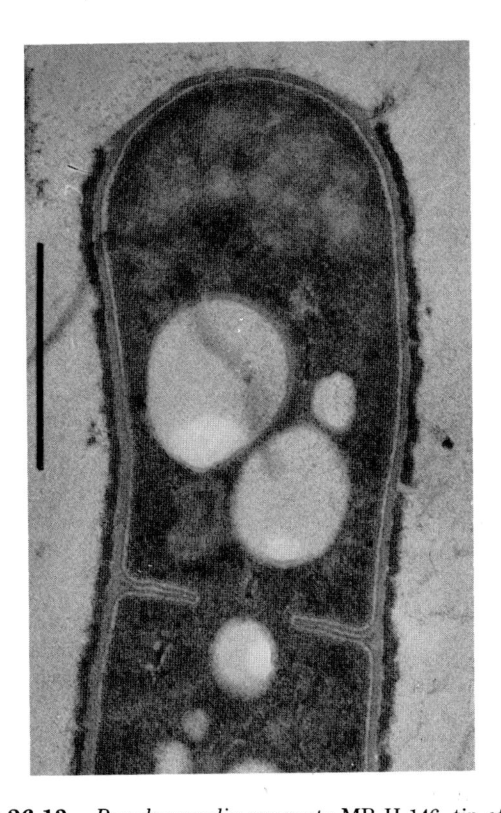

Figure 26.13. *Pseudonocardia compacta* MB H-146, tip of sporulating aerial hypha. Hyphal wall with thick inner and outer layer, cross-wall formed by inward growth of double annulus. (TEM. *Bar,* 0.5 μm). (Reproduced with permission from A. Henssen, C. Happach-Kasan, B. Renner and G. Vobis, International Journal of Systematic Bacteriology *33:* 829–836, 1983, ©American Society for Microbiology.)

Ecological niches for *Pseudonocardia* are rotting plant remains. *P. thermophila* is found in self-heating hay, compost, or manure heaps, and the two mesophilic species occur in cultivated land.

Enrichment and Isolation Procedures

Suitable conditions for the enrichment and isolation of the thermophilic species are cellulose-containing substrates, a temperature of 50°C, and a microaerophilic atmosphere. The mesophilic species are isolated on poorly nutrient media (e.g. an artifical soil agar) under aerobic conditions at 20–30°C (Henssen and Schäfer, 1971).

Maintenance Procedures

Species of *Pseudonocardia* can be lyophilized by common procedures used for aerobes; vigorously sporulating cultures should be used.

Differentiation of the genus Pseudonocardia *from other genera*

The genus *Pseudonocardia* is separated from all other genera that form aerial spores in chains by the budding process of hyphal elongation resulting in segmented hyphae. Species of *Nocardiopsis* with cell walls containing *meso*-DAP and galactose might be misinterpreted as belonging to *Pseudonocardia. Nocardiopsis* species are distinguished with certainty by studies of ultrathin sections. Whereas species of *Pseudonocardia* have a two-layered hyphal wall and interspace septa, species of *Nocardiopsis* have a one-layered hyphal wall and solid septa as cross-walls (Henssen et al., 1981). *Pseudonocardia* differs from the genus *Nocardia* (which also has cell wall type IV and interspace septa) in having a two-layered hyphal wall (one-layered in *Nocardia*, p. 2351), in the lack of mycolic acid, and in having a different type of phospholipid (Lechevalier et al., 1981).

Taxonomic Comments

Precise knowledge of the phylogenetic relationships of the genus *Pseudonocardia* to other genera of *Nocardiaceae*, in which family the genus has been usually included, depend on future studies of its nucleic acids, such as the degree of similarity in ribosomal RNA oligonucleotide sequence.

Further Reading

Henssen, A. and E. Schnepf. 1967. Zur Kenntnis thermophiler Actinomyceten. Arch. Mikrobiol. *57*: 214–231.

Henssen, A. and D. Schäfer. 1971. Emended description of the genus *Pseudonocardia* Henssen and description of a new species *Pseudonocardia spinosa* Schäfer. Int. J. Syst. Bacteriol. *21*: 29–34.

Henssen, A., E. Weise, G. Vobis and B. Renner. 1981. Ultrastructure of sporogenesis in actinomycetes forming spores in chains. *In* Schaal and Pulverer (Editors), Actinomycetes. Proceedings of the Fourth International Symposium on Actinomycete Biology. Fischer-Verlag, Stuttgart, pp. 137–146.

Henssen, A., H. W. Kothe and R. M. Kroppenstedt. 1987. Transfer of *Pseudonocardia azurea* and "*Pseudonocardia fastidiosa*" to the genus *Amycolatopsis*, with emended species description. Int. J. Syst. Bacteriol. *37*: 292–295.

Differentiation of the species of the genus Pseudonocardia

Some differential features of the species. *P. thermophila, P. spinosa*, and *P. compacta* are listed in Table 26.18.

List of species of the genus Pseudonocardia

1. **Pseudonocardia thermophila** Henssen 1957, 408.[AL]

ther.mo'phi.la. Gr. n. *thermus* heat; Gr. adj. *philus* loving; M.L. fem. adj. *thermophila* heat loving.

See Table 26.18 and generic description for some features.

Substrate hyphae septate, often zig-zag shaped, swellings usually present, becoming multidivided. Aerial hyphae often zig-zag shaped, young stages with constrictions, later on septate throughout, swellings rarely present. All three types of spores produced in substrate and aerial mycelium. Inner wall layer in hyphae and spores uniformly thin, not thickened in mature spores.

Good growth on nutrient agar and yeast agar, colonies yellow, thick cover of white aerial mycelium. Good growth on asparagine-glycerol and yeast-glucose agar, colonies yellow, aerial mycelium limited. No growth on oatmeal agar. Optimum temperature 40–50°C, growth slight at 28°C and 60°C.

Isolated from fresh and rotten manure.

Type strain: ATCC 19285 (MB A-18; CBS 277.66).

2. **Pseudonocardia spinosa** Schäfer in Henssen and Schäfer 1971, 31.[AL]

spi.no'sa. L. fem. adj. *spinosa* spiny.

Substrate mycelium compact mass, hyphae irregularly branched, constricted on septate, swellings common, septate. Aerial hyphae constricted or septate. Spores in substrate and aerial mycelium formed by budding or secondary septation of hyphal segments. Inner wall layer of hyphae and spores of varying thickness.

Grows slowly. Moderate growth on asparagine-glycerol agar, colonies yellow, abundant white aerial mycelium. Moderate to good growth on oatmeal agar. Optimum temperature 20–30°C, no growth at 37°C.

Isolated from soil.

Type strain: ATCC 25924 (MB SF-1; CBS 818.70).

3. **Pseudonocardia compacta** Henssen, Happach-Kasan, Renner and Vobis 1983, 834.[VP]

com.pac'ta. L. fem. adj. *compacta* compact.

Substrate mycelium septate, densely branched, swellings common, in part multidivided. Aerial mycelium compact, hyphae constricted or septate, bearing apical and intercalary swellings. Septa formed frequently at a distance behind constrictions. All three types of spores formed in substrate and aerial mycelium. Spores of varying shape and length. Inner layer of hyphal wall of varying thickness. Inner layer of spore wall thick in mature spores.

Growth moderate to good on artificial soil agar, substrate mycelium scanty, white aerial mycelium abundant. Optimum temperature 20–30°C, no growth at 35°C.

Isolated from soil.

Type strain: MB H-146 (CBS 160.82; DSM 43592).

Species to Be Excluded from the Genus

Production of antibiotics has been reported for two actinomycetes included in the genus, "*Pseudonocardia fastidiosa*" (Celmer et al., US Patent 4,031,206, 1977), and *Pseudonocardia azurea* Omura et al. 1983b (effective publication: description in Omura et al., 1979). *P. azurea* has aerial hyphae with swellings and, therefore, resembles *P. compacta*. A re-examination of the type strains of the two species revealed that the two organisms deviate in hyphal growth and fine structure (Henssen et al., 1983).

According to new results (Henssen et al., 1987), both strains lack mycolic acids and have a type IV cell wall composition and phospholipids of type PII, thus corresponding with the newly described genus *Amycolatopsis* Lechevalier et al. (Lechevalier et al., 1986). The new combinations *Amycolatopsis azurea* (Omura et al. 1983b) Henssen et al., and *Amycolatopis fastidiosa* (ex Celmer et al.) Henssen et al. have been made (Henssen et al., 1987).

Table 26.18.
Differential characteristics of the species of the genus **Pseudonocardia**[a]

Characteristics	1. *P. thermophila*	2. *P. spinosa*	3. *P. compacta*
Optimum temperature	40–50°C	20–30°C	20–30°C
Spores	Smooth	Spiny	Smooth
Growth on cellulose agar	+	+	−
Zig-zag hyphae frequent	+	+	−
Hyphal swellings frequent	−	+	+

[a]Symbols: see Table 26.13.

Genus **Oerskovia** *Prauser, Lechevalier and Lechevalier 1970, 534 (Emended Lechevalier 1972, 263[AL])*

HUBERT A. LECHEVALIER AND MARY P. LECHEVALIER

Oer.sko'vi.a. M.L. fem. n. *Oerskovia*, named after Jeppe Ørskov, Danish microbiologist.

Extensively branching vegetative hyphae, about 0.5 µm in diameter, growing on the surface of and penetrating into agar media, **breaking up** into rod-shaped, **motile**, flagellate elements. Growth appears coryneform to bacteroid in smears. **No aerial mycelium** formed. **Mesophilic. Gram-positive** with part of the thallus becoming Gram-negative with age, non-acid fast, catalase positive when grown aerobically. **Facultatively anaerobic** on trypticase-soy medium; catalase negative when grown anaerobically. Glucose metabolized both oxidatively and fermentatively. Rods monotrichous when small and peritrichous when long. Cell wall of type VI (lysine as principal diamino acid) plus major amounts of galactose; aspartic acid may be absent. Phospholipids of type PV with phospholipid fatty acids of type 1. Menaquinones of MK-9(H_4) type. The mol% G + C of the DNA is 70.5–75(T_m) (Sukapure et al., 1970). **Found in soil**, decaying plant materials, brewery sewage (Kaneko et al., 1969), aluminum hydroxide gels, and clinical specimens, including blood samples.

Type species: *Oerskovia turbata* (Erikson) Prauser, Lechevalier and Lechevalier 1970, 534 (emended Lechevalier 1972, 263).

Further Descriptive Information

Nonmotile *Oerskovia*-like strains (NMOs) can be isolated from soil, clinical specimens, and other substrates. The two described species of *Oerskovia* and two types of NMOs so far isolated have the following properties in common. Most strains are yellow; some are buff colored. A few form amorphous reddish crystals in media amended with soil extract. No soluble pigments have been observed. They have extensively branching vegetative hyphae that break up into long or very short rod-shaped elements that are either motile (*Oerskovia*) or nonmotile (NMOs). Colonies may be dull or glistening and develop dense centers with a filamentous fringe (Fig. 26.14). This fringe may be barely visible in age or on rich media. The cells stain Gram-positive to Gram-variable and have a typical Gram-positive peptidoglycan composition with or without galactose (type VI).

Oerskoviae and NMOs produce nitrite from nitrate. They hydrolyze gelatin and starch; they utilize acetate, lactate, and pyruvate but not benzoate, citrate, malate, succinate, and tartrate. Acid is produced from

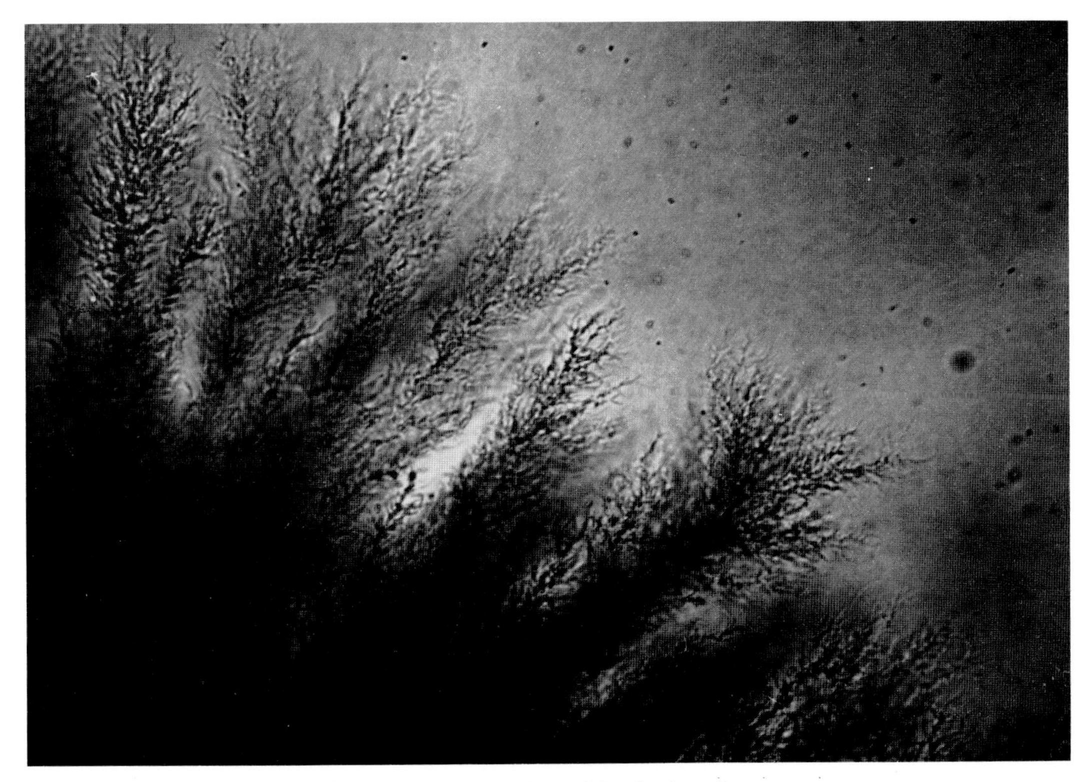

Figure 26.14. Edge of a colony of *Oerskovia turbata* (× 250).

arabinose, cellobiose, dextrin, fructose, galactose, glucose, glycerol, glycogen, α-methyl-D-glucoside, lactose, maltose, mannose, salicin, sucrose, trehalose, xylose, and β-methyl-D-xyloside. No acid is formed from adonitol, erythritol, and sorbose. DNase and β-D-galactosidase are produced but not cytochrome oxidase. No growth takes place in lysozyme (0.05%) broth. These organisms survive exposure for 8 h at 50°C and 4 h at 60°C. Neither sulfate, nitrate, nor fumarate can be used as electron acceptors.

Strains of *Oerskovia* do not contain mycolic acids and have a phospholipid composition of type PV (phosphatidylglycerol, phospholipids of unknown structure containing glucosamine, phosphatidylinositol, and diphosphatidylglycerol). The phospholipid fatty acids are of type 1 (principal components branched chain fatty acids of the *anteiso/iso* series) (Lechevalier et al., 1977), here specifically being *anteiso*-C_{15} with lesser amounts of *anteiso*-C_{17} and *iso*-C_{16}. Menaquinones are MK-9(H_2,H_4) (Minnikin et al., 1978).

In Table 26.19 are listed the properties that differentiate the two described species of *Oerskovia* and the two types of NMOs that have been recognized.

Enzymatic Activity of Oerskoviae and NMOs

Oerskoviae and NMO are prolific producers of a variety of useful enzymes. These include α-mannanases, β-(1,3)-glucanases, β-(1,6)-glucanase, β-glucosidase, proteinases, chitinases (Mann et al., 1978; Scott and Schekman, 1980), dextranase (Hayward and Sly, 1976), and keratinase (Goodfellow, 1971). Strains of these organisms degrade the walls of both live and dead cells of yeast, but there is considerable variation in lytic power from strain to strain. Oerskoviae show chemotactic activity toward yeasts and may be considered true predators of these organisms. The production of yeast-lytic enzymes (principally β-(1,3)-glucanases plus proteinase) is repressed by glucose (Mann et al., 1978; Scott and Schekman, 1980). Crude enzyme preparations from *O. xanthineolytica* are widely used for the preparation of yeast sphaeroplasts.

Oerskoviae and NMOs as Pathogens

Strains of *Oerskovia* have been isolated from various clinical specimens. Sottnek et al. (1977) characterized 35 such isolates and found both species of *Oerskovia* to be represented. Reller et al. (1975) implicated *O. turbata* in a case of endocarditis after homograft replacement of an aortic valve, and Cruickshank and his colleagues (1979) reported on a strain of *Oerskovia* strongly suspected of causing pyonephrosis. Their description, however, indicates that the suspected agent was an NMO. The best regimen for treatment of the infection with *O. turbata* was reported to be a prolonged course of high doses of sulfamethoxazole-trimethoprim combined with amphicillin or amoxicillin (Reller et al., 1975).

Isolation

Oerskoviae are readily isolated on a variety of media ranging from such nutrient-poor media as "tap water agar" (1.5% crude agar in tap water) to very rich media such as trypticase-soy or blood agar. They grow readily at 20–37°C usually in 24–48 h; some will grow at 42°C.

Maintenance Procedures

Good maintenance media include nutrient agar (with or without added soil extract) and trypticase-soy agar. Storage for 3–4 months at 4°C between transfers has proved satisfactory. The strains may be lyophilized in skim milk for preservation.

Table 26.19.
Differential characteristics of the species of the genus **Oerskovia** *and of two types of nonmotile* **Oerskovia**-*like strains (NMOs)*[a]

Characteristics	*O. turbata*	*O. xanthineolytica*	NMOs type A	NMOs type B
Motility	+	+	–	–
Cell wall	Type VI[b] plus galactose	Type VI[b] plus galactose	Type VI[b] plus galactose (variable)	Type VI[b]
Peptidoglycan type	L-Lys-L-Thr-β-D-Asp[c] or L-Lys-L-Thr-γ-D-Glu[d]	L-Lys-D-Ser-β-D-Asp[c]		
Cytochrome a_1	+[c]	–[c]		
Growth at 42°C	–	+[d]	– (66)[e]	–
Dissimilation of				
Casein	+	+	– (66)	–
Xanthine	–	+	+	–
Hypoxanthine	–	+	+	–
Adenine	+[f]	+[g]	+[g]	+[f]
Urea	– (86)	– (88)	– (84)	+
Production of				
Cellulase	–	–	+	+
Phosphatase (24 h)	–	+	+	–
Acid from				
Melibiose	– (86)	+[d]	+ (66)	–
Raffinose	–	+[d]	+ (66)	V
Sorbitol	– (14)	+[d]	– (66)	+

[a]Symbols: see Table 26.13; also *V*, variable.
[b]Lechevalier and Lechevalier (1981b).
[c]Seidl et al. (1980).
[d]Considered variable by Sottnek et al. (1977).
[e]Numbers in parentheses indicate percentages of positive or negative strains in variable tests.
[f]Transformed to hypoxanthine (M. P. Lechevalier, unpublished).
[g]Transformed to 8-hydroxyadenine (Lechevalier et al., 1982a).

Differentiation of the genus **Oerskovia** *from other taxa*

Oerskoviae and NMOs may be differentiated from genera of the nonfilamentous bacteria on the basis of the hyphal fringe of undisturbed colonies of the former growing on nutrient agar. They may be distinguished from other nonsporing actinomycetes and their relatives by the characters given in Table 26.20.

Taxonomic Position of Oerskoviae

Oerskoviae are actinomycetes with many characteristics of rod-shaped bacteria, and their classification has presented many difficulties. They have been placed in various genera such as *Nocardia* (Erikson, 1954), *Corynebacterium* (Sottnek et al., 1977), *Arthrobacter* (Scott and Schekman, 1980), and *Cellulomonas* (Jones and Bradley, 1964; Stackebrandt et al., 1982a). Until recently, they have been considered as forming one of the genera of the *Actinomycetales* "in search of a family" (Lechevalier and Lechevalier, 1981a), and were listed under *Incertae Sedis* in the eighth edition of the *Manual*. On the basis of DNA/DNA reassociation studies supported by comparative analyses of the ribosomal 16S RNA found in oerskoviae and cellulomonads, Stackebrandt et al. (1980a, 1982a) have proposed that species of *Oerskovia* be transferred to the genus *Cellulomonas*. Cellulomonads, oerskoviae, and NMOs are faculative anaerobes and have the same phospholipid and menaquinone patterns (Lechevalier et al., 1981; Yamada et al., 1976), and are undoubtedly related phylogenetically. However, on the basis of differences in morphology (oerskoviae being hyphal and cellulomonads not), chemistry of the cell wall (oerskoviae contain lysine, and cellulomonads ornithine, in their peptidoglycan), and physiology (cellulomonads produce cellulase and oerskoviae do not), the two groups are treated here as belonging to different but closely related genera. NMOs are undoubtedly the bridging group. "*Cellulomonas cartae*" and "*Nocardia cellulans*" are both NMOs of type A (M. P. Lechevalier, unpublished) and are closely related to *O. xanthineolytica* (Seidl et al., 1980), as witnessed by their identical peptidoglycan types. Prauser (1986) has proposed the transfer of "*Cellulomonas cartae*" to *Oerskovia* and has shown that cellulomonads are not susceptible to *Oerskovia* bacteriophage and vice versa, thus providing additional evidence against the case for combining *Cellulomonas* and *Oerskovia*.

List of species of the genus **Oerskovia**

1. **Oerskovia turbata** (Erikson) Prauser, Lechevalier, and Lechevalier 1970, 534.[AL] (Emended Lechevalier 1972, 263.)

tur.ba'ta. L. fem. adj. *turbata*, agitated.

Synonym: "*Nocardia turbata*" Erikson 1954, 206. *Listeria denitrificans* ATCC 14870 is a closely related species (Sottnek et al., 1977). Strains of *O. turbata* dissimilate adenine to hypoxanthine (M. P. Lechevalier, unpublished).

Properties as given above and in Table 26.19.

Type strain: ATCC 25835.

Table 26.20.
Characterisics differentiating oerskoviae and nonmotile **Oerskovia**-*like strain (NMOs) from other actinomycetes and related bacteria*[a]

Taxon	Branching mycelium[b]	Motility	Relation to oxygen[c]	Catalase reaction	Metabolism of glucose[d]	Cell wall type[e]	Phospholipid type[e]
Actinomyces	T	−	A, F, An	−/+	O/F	VI, V	PII
Agromyces	T	−	A, M	−	O	VII	PI
Arachnia	T	−	A, M	−	O/F	I	
Arthrobacter	N	−/+	A	+	O	I, VI	PI/PII
Bacterionema	N, T	−	A, F	+	O/F	IV	PI
Bifidobacterium	N	−	An	−	F	VIII	PI
Cellulomonas	N	−/+	A, F	−/+	O/F	VIII,	PV
Corynebacterium							
Animal spp.	N	−	A, F	+	O/F	IV	PI
Plant spp.	N	−/+	A	+	O	IV, VI, VII	PI, PII
pyogenes	N	−	A, F	−	O/F	VI	PI
Dermatophilus	T	+	A, M	+	O	III	PI
Geodermatophilus	N, T	+	A	+	O	III	PII
Intrasporangium	P	−	A	+	O	I	PIV
Mycobacterium	N, T	−	A	+	O	IV	PII
NMOs	P, T	−	A, F	−/+	O/F	VI	PV
Nocardia	P, T	−	A	+	O	IV	PII
Nocardioides	P	−	A	+	O	I	PI
Oerskovia	P, T	+	A, F	−/+	O/F	VI	PV
Promicromonospora	P	−	A	+	O	VI	PV
Rhodococcus	P, T	−	A	+	O	IV	PII
Rothia	N, T	−	A, M	+	O/F	VI	PI

[a]Symbols: see Table 26.13 and footnotes to individual columns. Where two symbols appear, the first indicates the more common property, i.e. −/+, 11–50% positive; +/−, 51–89% positive.

[b]*T*, transient; *N*, none; *P*, persistent.

[c]*A*, aerobic; *F*, facultative; *An*, anaerobic; *M*, microaerophilic.

[d]*O*, oxidative; *F*, fermentative.

[e]According to Lechevalier and Lechevalier (1981b).

2. **Oerskovia xanthineolytica** Lechevalier 1972, 264.[AL]

xan.thi.ne.o'ly.ti.ca. From xanthine (purine); Gr. adj. *lytos* soluble; M.L. fem. adj. *xanthineolytica* dissolving xanthine.

Synonym: "*Arthrobacter luteus*" Kaneko et al. 1969, 322.

Properties as given above and in Table 26.19.

As indicated in Table 26.19, xanthine and hypoxanthine are lysed by members of this species. Although no novel products of dissimilation of these compounds have been found (xanthine is oxidized to uric acid), *O. xanthineolytica* strains transform adenine to 8-hydroxyadenine, a novel microbial product (Lechevalier et al., 1982a).

Type strain: ATCC 27402.

Comments on **Mycoplana**

The genus *Mycoplana* (Gray and Thornton, 1928) has received little subsequent study and its taxonomic position is uncertain. It was suggested by Sukapure et al. (1970) that it was related to the genus *Oerskovia* and it is included here pending further investigation. Gray and Thornton described two species, *Mycoplana dimorpha* ATCC 4279 (the type species) and *M. bullata* ATCC 4278, both of which are still extant.

The ability of these strains to form branching filaments prior to their fragmentation into motile, irregular rods has led several workers to place the genus in the order *Actinomycetales* (Sukapure et al., 1970; Cross and Goodfellow, 1973; Lechevalier and Lechevalier, 1981a, 1981b). However, *Mycoplana* differs from other actinomycetes in being Gram-negative and having a cell wall of Gram-negative type, containing *meso*-DAP and numerous amino acids (Sukapure et al., 1970; Lechevalier and Lechevalier, 1981b). In the seventh edition of *Bergey's Manual*® it was placed in the *Pseudomonadaceae* (Breed et al., 1957) and it was omitted from the eighth edition of the *Manual* (Buchanan and Gibbons, 1974). Most strains of *Mycoplana* that were tested showed nitrogenase activity (Pearson et al., 1982), and it was concluded from DNA/RNA hybridization studies that *M. dimorpha* and *M. bullata* were remote relatives of the family *Rhizobiaceae* (De Smedt and De Ley, 1977).

A description, based on the observations of Gray and Thornton (1928) and Sukapure et al. (1970), was provided by Lechevalier and Lechevalier (1981a).

Branching filaments breaking into irregular rods (0.5–0.1 µm wide × 1.25–4.5 µm long) that bear subpolar tufts of flagella. Pili may be observed on motile cells. Gram-negative, non-acid fast. Murein contains *meso*-DAP; ribose and glycerol present in whole-cell hydrolysates. DNA G + C content, 64–69 mol%. Minimum growth temperature 24°C, maximum 42°C. No growth anaerobically. Catalase positive. No acid from carbohydrates. No nitrite from nitrate. Gelatin liquefaction variable. Killed by 4 h at 60°C but not by 8 h at 50°C. The main difference between the two species is that *M. dimorpha* hydrolyzes starch whereas *M. bullata* does not.

Genus **Saccharopolyspora** *Lacey and Goodfellow 1975, 77*[AL]

JOHN LACEY

Sac'cha.ro.po'ly.spo.ra. M.L. n. *Saccharum* generic name of sugar cane; Gr. adj. *polus* many; Gr. n. *spora* a seed; M.L. fem. n. *Saccharopolyspora* the many-spored (organism) from sugar cane.

Substrate mycelium well developed, branched, septate 0.4–0.6 µm in diameter, **fragmenting into rod-shaped elements** about 1.0 × 0.5 µm, more often in older parts of the colony and seldom near the growing margins. **Aerial mycelium** 0.5–0.7 µm in diameter, straight or in spirals, characteristically **segmented into beadlike chains of spores,** 0.7–1.3 × 0.5–0.7 µm, **usually separated by lengths of "empty" hypha and retained in a sheath.** Gram-positive, non-acid fast aerobic. **Colonies thin, raised or convex, slightly wrinkled, mucoid or gelatinous** in appearance, with **aerial mycelium sparse, often produced in tufts** and mostly in the older parts (Fig. 26.15). **Able to utilize many organic compounds** as sole sources of carbon and for energy and growth, **to degrade adenine** and other substrates; **resistant to many antibiotics** but **susceptible to lysozyme.** The mol% G + C of the DNA is 77 (T_m).

Type species: *Saccharopolyspora hirsuta* Lacey and Goodfellow 1975, 78.

Further Descriptive Information

Although some hyphae fragment like those of *Nocardia* species, producing long chains of cells in angular opposition (Fig. 26.16), many remain stable (Fig. 26.17). Fragmented hyphae are most abundant in the older parts of colonies but are still usually accompanied by stable hyphae. Aerial hyphae mostly arise in tufts, which soon become differentiated into spore chains. These form loops and loose spirals (Fig. 26.18), but in older parts of the colony may be long and straight between tufts (Fig. 26.19). The spore chains are of indeterminate length and the spores are characteristically separated by short lengths of apparently empty hyphae, giving a beadlike appearance.

The spores are round to oval, 0.7–1.3 × 0.5–0.7 µm, and covered by a sheath. In *S. hirsuta*, this sheath carries tufts of long, straight or curved, brittle hairs (Fig. 26.20). The morphology of the hairs is best seen on lengths of empty sheath (Fig. 26.21) or by scanning electron microscopy (Fig. 26.22). The surface of the sheath between tufts of hairs is smooth.

The cell walls contain major amounts of *meso*-diaminopimelic acid (*meso*-DAP), galactose, and arabinose (type IV; Becker et al., 1965). Whole-cell methanolysates yield no mycolic acids (Minnikin et al., 1975). Phosphatidylcholine is the diagnostic phospholipid found in cells (type III; Lechevalier et al., 1981). Also present are phosphatidylinositol and phosphatidylmethylethanolamine. *S. hirsuta* contains tetrahydrogenated menaquinones with nine isoprene units as a major component (Collins, 1982).

In thin section, hyphae (Fig. 26.23) are bounded by a wall 22–30 nm thick. Within this, a typical unit membrane encloses granular cytoplasm with axial diffuse nuclear material. Electron-transparent vacuoles, up to 0.3 µm in diameter and resembling lipid accumulations in other nocardioform actinomycetes (Williams et al., 1976), were sometimes abundant. Also occasionally present are electron-dense granules, up to 0.1 µm in diameter, resembling polyphosphate or metachromatic granules.

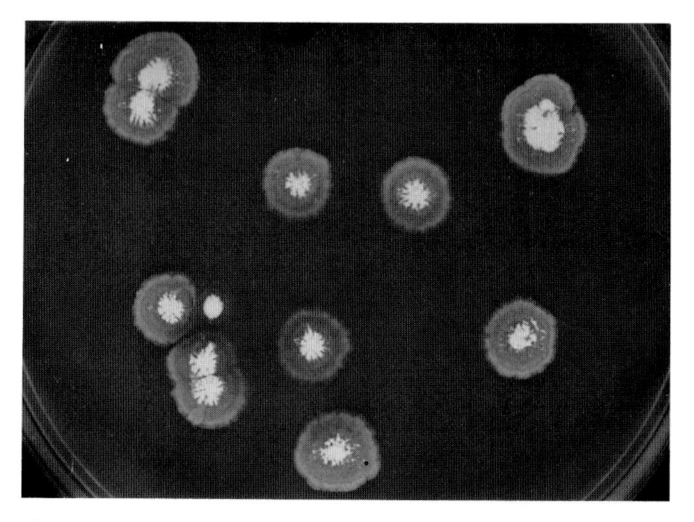

Figure 26.15. Colonies of *Saccharopolyspora hirsuta.* Half-strength nutrient agar, incubation 37°C (× 1). (Reproduced with permission from J. Lacey and M. Goodfellow, Journal of General Microbiology *88:* 75–85, ©Society for General Microbiology, 1975.)

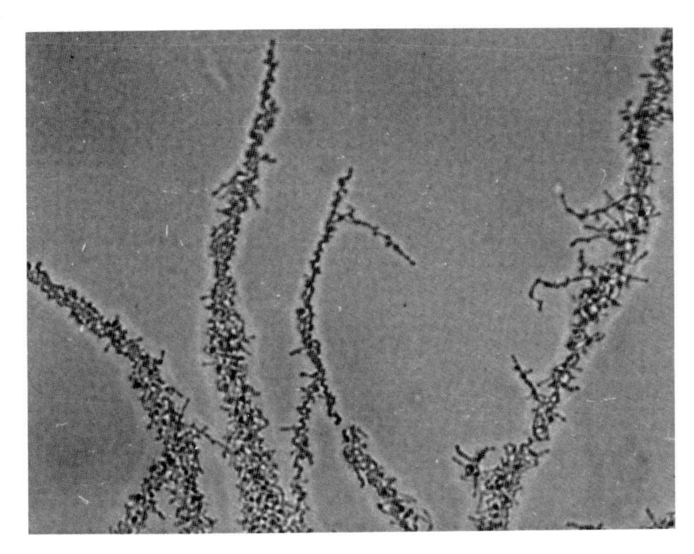

Figure 26.16. Fragmentation of substrate mycelium. Glycerol-asparagine agar, incubation 40°C (× 500). (Reproduced with permission from J. Lacey and M. Goodfellow, Journal of General Microbiology *88:* 75–85, ©Society for General Microbiology, 1975.)

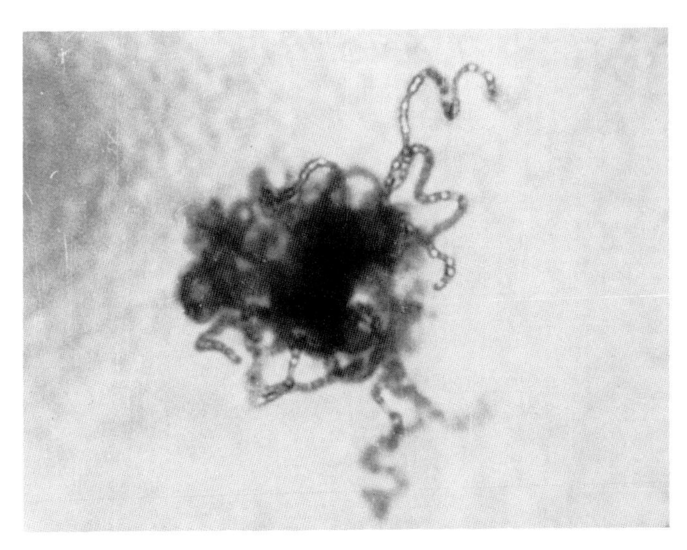

Figure 26.18. Spore chains on aerial mycelium showing tufted appearance and typical curved chains. Half-strength nutrient agar, incubation 40°C (× 800). (Reproduced with permission from J. Lacey and M. Goodfellow, Journal of General Microbiology *88:* 75–85, ©Society for General Microbiology, 1975.)

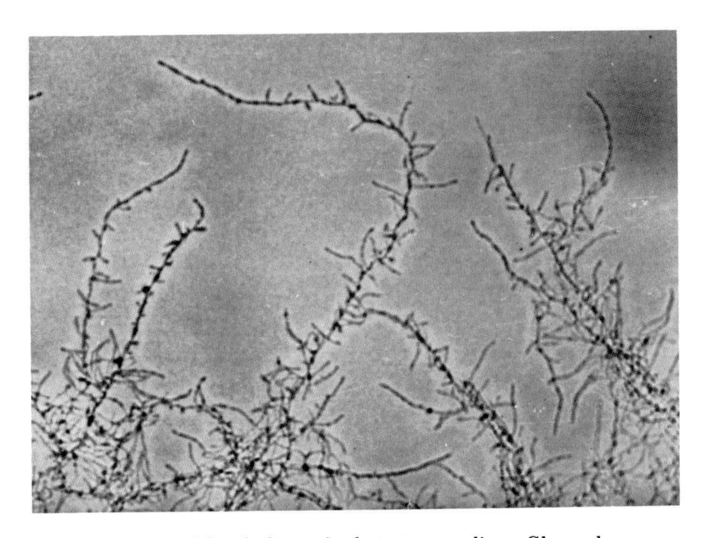

Figure 26.17. Morphology of substrate mycelium. Glycerol-asparagine agar, incubation 40°C (× 550). (Reproduced with permission from J. Lacey and M. Goodfellow, Journal of General Microbiology *88:* 75–85, ©Society for General Microbiology, 1975.)

Figure 26.19. Straight spore chains from older parts of colonies. V-8 juice agar, incubation 37°C (× 650). (Reproduced with permission from J. Lacey and M. Goodfellow, Journal of General Microbiology *88:* 75–85, ©Society for General Microbiology, 1975.)

Septation occurs by double ingrowth of the wall leading to fragmentation (type II; Williams et al., 1973). This may be associated with lamellar mesosomes up to 0.25 μm in diameter.

The sheath surrounding the spores is 18–36 nm thick. It carries tufts of structureless hairs, triangular and 0.2–0.3 μm across at the base, which extend into apical filaments about 20 nm in diameter. Spore walls are thickened uniformly to 50–60 nm, but their internal structure resembles that of hyphae, although they lack many vacuoles (Fig. 26.24).

Colonies of *S. hirsuta* grow to about 1 cm in diameter in 7 days at 40°C, with a central area of white aerial mycelium on an almost colorless substrate mycelium (Fig. 26.15). Good growth is obtained on yeast extract-malt extract agar and V-8 vegetable juice agar, and a yellow soluble pigment is produced on the former.

S. hirsuta can utilize the following as sole sources of carbon for energy and growth: adonitol, cellobiose, erythritol, fructose, galactose, glucose, glycerol, inositol, lactose, maltose, mannitol, mannose, α-methyl-D-glucoside, β-methyl-D-glucoside, raffinose, rhamnose, sorbitol, sucrose, trehalose, xylose, acetate, benzoate, butyrate, citrate, fumarate, H-malate, succinate, sebacic acid, and testosterone; it does not use arabinose, melezitose, salicin, tartrate, or adipic acid. Isolates vary in their ability to utilize propionate and pyruvate. *S. hirsuta* can also degrade adenine, aesculin, casein, elastin, hypoxanthine, keratin, tyrosine, urea, and xanthine, but not xylan. However, it is sensitive to lysozyme. Growth occurs on agar media between 25° and 50°C with an optimum at about 37–40°C. There is no growth at 10°C. Aerial mycelium is produced only close to the optimum temperature.

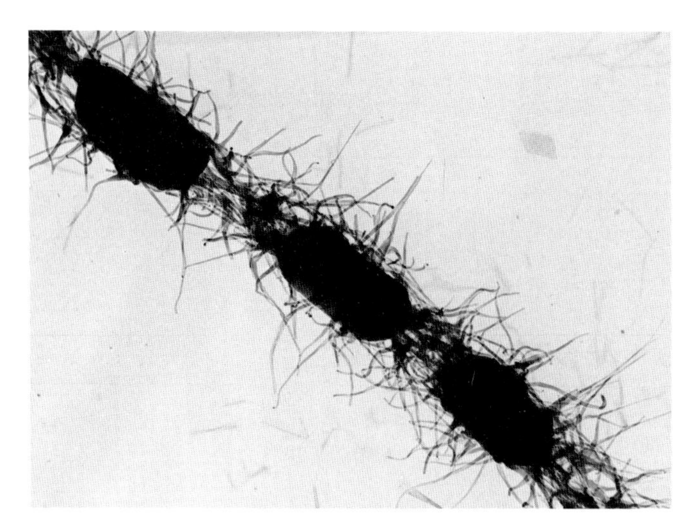

Figure 26.20. Electron micrograph of spore chain (× 18,400). (Reproduced with permission from J. Lacey and M. Goodfellow, Journal of General Microbiology *88:* 75–85, ©Society for General Microbiology, 1975.)

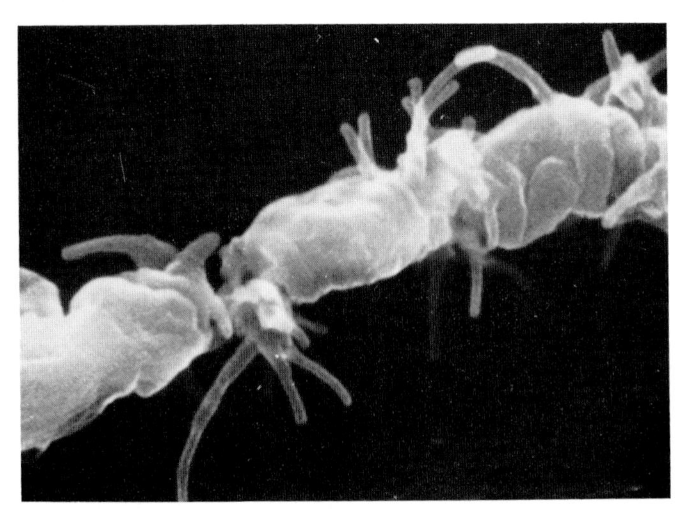

Figure 26.22. Scanning electron micrograph of spores (× 15,000). (Reproduced with permission from J. Lacey and M. Goodfellow, Journal of General Microbiology *88:* 75–85, ©Society for General Microbiology, 1975.)

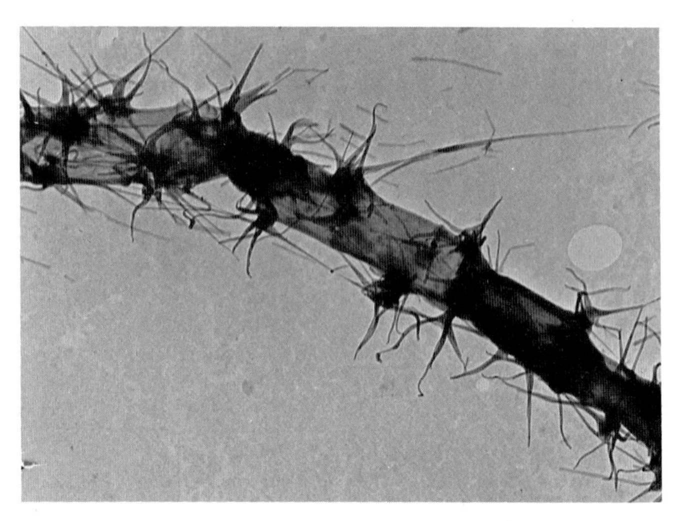

Figure 26.21. Electron micrograph of spore sheath showing tufted production of hairs (× 17,600). (Reproduced with permission from J. Lacey and M. Goodfellow, Journal of General Microbiology *88:* 75–85, ©Society for General Microbiology, 1975.)

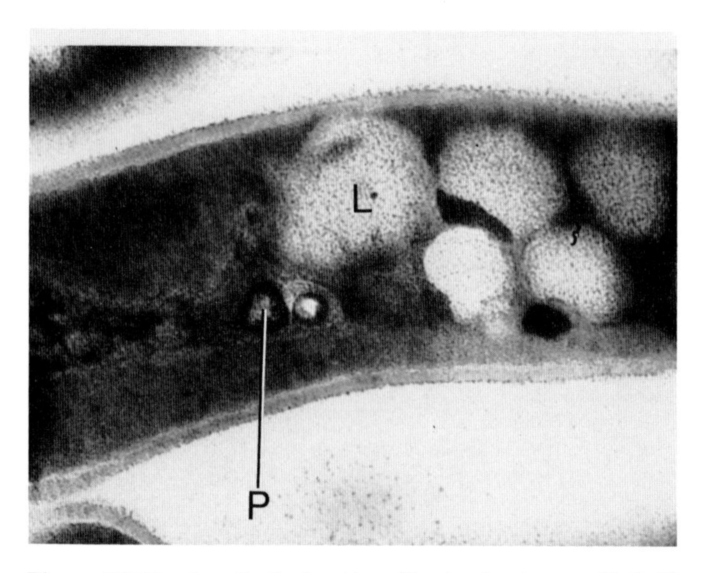

Figure 26.23. Longitudinal section of hypha showing possible lipid accumulation (*L*) and polyphosphate granules (*P*) (× 70,000). (Reproduced with permission from J. Lacey and M. Goodfellow, Journal of General Microbiology *88:* 75–85, ©Society for General Microbiology, 1975.)

Although a single strain of *S. hirsuta* was loosely associated with *Streptomyces* species in a numerical phenetic study (Williams et al., 1981, 1983a), it was resistant to phages isolated from *Streptomyces* species and other wall chemotype I taxa (Wellington and Williams, 1981a).

S. hirsuta is tolerant of a wide range of antibiotics (Table 26.21). However, nearly all strains are susceptible to 500-µg/ml solutions of dapsone and septrin on filter paper disks and most to 50 µg minocycline and 100 µg fusidic acid/ml solution (Lacey and Goodfellow, 1975).

There is no evidence of pathogenicity, but *S. hirsuta* degrades elastin, a property that in *Actinomadura* and *Nocardia* species indicates an ability to cause mycetoma.

S. hirsuta is known mostly from moldy sugar cane bagasse that has heated spontaneously during storage. It was found in 12% of samples originating from Puerto Rico, Trinidad, Jamaica, and India but exceeded 10^5 colony-forming units/g dry weight in only 3% (Lacey, 1974)

Isolation

S. hirsuta has been isolated chiefly from the airborne dust from sugar cane bagasse in a small wind tunnel using an Andersen sampler (Lacey, 1974; Lacey and Goodfellow, 1975). Half-strength nutrient agar (Oxoid CM3) containing 50 µg cycloheximide/ml medium (Gregory and Lacey, 1963) has usually been used for isolation.

Maintenance Procedures

S. hirsuta can be lyophilized by usual procedures.

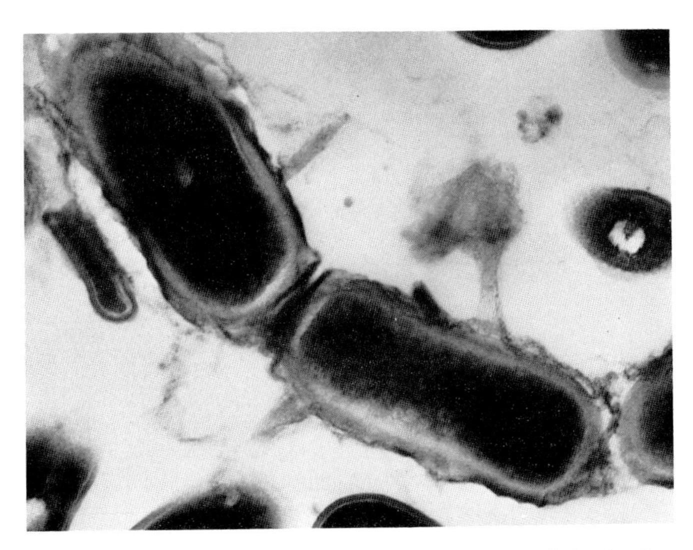

Figure 26.24. Longitudinal section of mature spore chain showing sheath and hair bases (× 34,000). (Reproduced with permission from J. Lacey and M. Goodfellow, Journal of General Microbiology *88:* 75–85, ©Society for General Microbiology, 1975.)

Table 26.21.
Antibiotic sensitivity of **S. hirsuta**[a]

Antibiotic	Concentration[b] (μg/ml)	*S. hirsuta*
Gentamycin	100	−
Kanamycin	10	−
Streptomycin	100	−
Neomycin	50	−
Tobramycin	100	−
Rifampicin	50	−
Erythromycin	50	−
Fusidic acid	100	d
Minocycline	50	d
Vancomycin	50	d
Dapsone	500	+
Septrin	500	+

[a]Symbols: −, tolerant; +, susceptible; *d,* 11–89% of strains are positive.
[b]Concentrations of solution used to soak filter paper disks.

Differentiation of the genus **Saccharopolyspora** *from related genera*

S. hirsuta can be distinguished from related taxa chiefly by the presence of aerial mycelium, the characteristic spores, its wall chemotype IV lacking mycolic acids, carbohydrate utilization pattern, hydrolysis of adenine, casein, and elastin, and resistance to certain antibiotics but not to lysozyme (Table 26.22).

Taxonomic Comments

Only two species of *Saccharopolyspora* have so far been recognized but the genus clearly has affinities with other genera in the *Nocardiaceae* and also with *Nocardiopsis* and *Streptomyces,* although these differ in wall chemotype.

Isolates of *S. hirsuta* have been included in two numerical taxonomic studies. In one study with *Nocardia* species (Goodfellow, Alderson, and Lacey, unpublished data), 31 *S. hirsuta* isolates formed a tight cluster defined at the 92% similarity level, uniting with *Nocardia* at only the 36% similarity level regardless of whether S_{SM} or S_J coefficients were used. In another study (Williams et al., 1981), a single isolate of *S. hirsuta* was loosely associated with *Streptomyces* species while studies of rRNA/DNA pairing showed that *S. hirsuta* was quite closely related to *Streptomyces* and *Nocardia* species phylogenetically (Mordarski et al., 1980b, 1981b).

The morphology and fine structure of spore chains of *S. hirsuta* show similarities with those of *Nocardiopsis dassonvillei* and *Faenia rectivirgula.* All show increased wall thickness in the spores, particularly in the region of cross-walls. However, *F. rectivirgula* has chains up to only 10 spores in length. The hairy sheaths of *S. hirsuta* spores differ in their appearance from *Streptomyces* spores in the separation of spores in the chain, in the appearance of hairs, and in their brittle nature. *S. hirsuta* and *N. dassonvillei* share many nutritional and hydrolytic characters, but *S. hirsuta* shows fewer similarities with *Actinomadura* species.

Further studies of actinomycetes with wall chemotype IV may lead to changes in their taxonomy and modifications in the definition of some genera. *S. hirsuta* occupies an intermediate position between *Nocardia* and *Faenia.* It has never been systematically compared with *Faenia* or with *Saccharothrix aerocolonigenes, Amycolata autotrophica, Amycolatopsis mediterranea,* and *Amycolatopsis orientalis,* which lack mycolic acids. *Saccharopolyspora* could perhaps provide a home for some of these species.

Further Reading

Lacey, J. and M. Goodfellow. 1975. A novel actinomycete from sugar-cane bagasse, *Saccharopolyspora hirsuta* gen. et sp. nov. J. Gen. Microbiol. *88:* 75–85.

Mordarski, M., A. Tkacz, M. Goodfellow, K.P. Schaal, and G. Pulverer. 1981. Ribosomal ribonucleic acid similarities in the classification of actinomycetes. Zentralbl. Bakteriol. Parasitenk. Infektionskr. Hyg., Abt. I, Suppl. *11:* 79–85.

Williams, S.T., G.P. Sharples, J.A. Serrano, A.A. Serrano, and J. Lacey. 1976. The micromorphology and fine structure of nocardioform organisms. *In* Goodfellow, Brownell and Serrano (Editors), The Biology of the Nocardiae. Academic Press, London, pp. 102–140.

List of species of the genus **Saccharopolyspora**

1. **Saccharopolyspora hirsuta** Lacey and Goodfellow 1975, 78.[AL]
hir.sut′a. L. adj. shaggy, bristly, with stiff hairs.
See the generic description and Table 26.21 for features of *S. hirsuta.*
Type strain: ATCC 27875.

The subspecies *Saccharopolyspora hirsuta* subsp. "*kobensis*" Iwasaki and Mori 1979, 185 has been proposed, but not validated, for an isolate that produces the aminoglycoside antibiotic sporaricin (Iwasaki et al., 1979) and the subspecies *Saccharopolyspora hirsuta* subsp. *taberi* Labeda 1987, 21 for an isolate previously known as "*Nocardia taberi.*" Also, the type strain of *Streptomyces erythraeus* (Waksman 1923, 370) Waksman and Henrici 1948a, 938 has been transferred to the genus *Saccharopolyspora* as *S. erythraea* (Waksman 1923, 370) Labeda 1987, 21. Descriptions of these taxa follow.

1a. **Saccharopolyspora hirsuta** subsp. **hirsuta** Lacey and Goodfellow 1975, 78.[AL]
Description as for the species.
Type strain: ATCC 27875.

1b. **Saccharopolyspora hirsuta** subsp. **kobensis** (ex Iwasaki, Itoh and Mori 1979) nom. rev.

Table 26.22.
Characteristics differentiating **Saccharopolyspora** *from related genera*[a,b]

Characteristics	*Saccharo-polyspora*	*Faenia*[c]	*Saccharo-monospora*	*Pseudo-nocardia*[d]	*Nocardia*	*Actino-madura*	*Nocardiopsis*
Morphological characters							
Aerial hyphae	+	+	+	+	+	+	+
Arthrospores	+	+	+	+	d	D	+
Spore surface	Hairy	Smooth	Warty	Smooth	Smooth	Smooth or Warty	Smooth
Acid-fast stain	–	–	–	–	d	–	–
Chemical characters							
Wall type	IV	IV	IV	IV	IV	III	III
Mycolic acid	–	–	–	–	+	–	–
Hydrolysis tests							
Adenine	+	–	–	–	–	–	d
Casein	+	d	+	+	D	D	d
Elastin	+	–	+	+	D	D	+
Sole carbon sources							
Cellobiose	+	d	+	+	–	D	d
Erythritol	+	+	d	+	–	D	–
Lactose	+	+	+	+	–	D	–
α-methyl-D-glucoside	+	+	d	+	–	–	d
Raffinose	+	d	+	+	–	D	–
Sorbitol	+	–	d	+	–	–	–
Benzoate	+	N/D	N/D	N/D	–	–	–
Growth in the presence of							
Lysozyme	–	d	–	+	+	D	–
Gentamycin	+	–	d	+	D	D	–
Rifampicin	+	–	–	–	+	D	d
Streptomycin	+	–	d	–	D	D	d

[a]Symbols: +, 90% or more of strains are positive; –, 90% or more of strains are negative; *d*, 11–89% of strains are positive; *D*, different reactions in different taxa (species of a genus or genera of a family); *ND*, not determined.
[b]Data from: Lacey and Goodfellow (1975), Goodfellow and Alderson (1977), Arden-Jones et al. (1979), Goodfellow et al. (1979), Athalye (1981), and Goodfellow and Pirouz (1982b).
[c]Results given for *F. rectivirgula* only.
[d]*Pseudonocardia thermophila* (type strain) only.

ko.ben′sis. M.L. adj. *kobensis* belonging to Kobe, a city in Japan (where the organism was isolated).

Differs from the species in having yellow to pink substrate mycelium and yellow to red soluble pigment, in reducing nitrate to nitrite, and in not utilizing inositol, rhamnose, sorbitol, or xylose.

Source of the antibiotic sporaricin.

Isolated from soil.

Type strain: ATCC 20501 (= FERM-P 3912 = KC6606)

1c. **Saccharopolyspora hirsuta** subsp. **taberi** Labeda 1987, 21.[VP]

ta′ber.i. M.L. gen. n. *taberi* named after Willard A. Taber (an American microbiologist who first isolated the organism).

No aerial mycelium is produced so that spore chain morphology and spore surface characteristics have not been determined. Differs from species in having colorless to yellow substrate mycelium and orange to red soluble pigment, in reducing nitrate to nitrite, in not decarboxylating benzoate, mucate and tartrate, and in utilizing arabinose and melezitose.

The mol% G + C of the DNA is 77.1% (T_m). DNA hybridization studies show 93% relatedness to *S. hirsuta* subsp. *hirsuta* and 37% relatedness to *S. erythraea*.

Source of the metabolite texazone (Gerber et al., 1983).

Type strain: NRRL B-16173 (= LL-WRAT-210).

2. **Saccharopolyspora erythraea** (Waksman 1923) Labeda 1987, 21.[VP] (*Actinomyces erythraeus* Waksman 1923, 370; *Streptomyces erythraeus* (Waksman 1923, 370) Waksman and Henrici 1948a, 938.[AL])

e.ryth′rae.a. Gr. n. *erythros* red; L. adj. *erythraea* red, referring to colony color.

This isolate is less thermotolerant than *S. hirsuta*, growing only between 20° and 42°C. It is characterized by orange to red colonies with mainly pink to brownish-grey aerial mycelium bearing short spore chains in imperfect spirals or straight to flexuous chains. The spore surface is spiny. Nitrate is reduced and arabinose, erythritol, melibiose, and raffinose are utilized, but not lactose, melezitose or α-methyl-D-glucoside.

Many isolates produce erythromycins A or B.

The mol% G + C is 76.9 (T_m). DNA hydridization studies show 24% relatedness with *S. hirsuta* subsp. *hirsuta* and 37% relatedness with *S. hirsuta* subsp. *taberi*.

Isolated from soil.

Type strain: ATCC 11635 (= NRRL 2338 = ISP5517).

Further Comments

Streptomyces erythraeus (Waksman 1923) Waksman and Henrici 1948a[AL] was listed in the Approved Lists (Skerman et al., 1980) with ATCC 11635 as the type strain. Transfer of this strain to the genus *Saccharopolyspora* constitutes transfer of the species, not the description of a new species. Therefore the correct citation for this species is *Saccharopolyspora erythraea* (Waksman 1923) Labeda 1987, 21.

Genus **Faenia** *Kurup and Agre 1983, 664*[VP] (Micropolyspora *Lechevalier, Solotorovsky and McDurmont 1961, 11*[AL])

JOHN LACEY

Faen.i'a. L. n. *faenum* hay; L. fem. pl. *Faenia* a genus of bacteria associated with hay.

Substrate mycelium well-developed, branched, septate, 0.5–0.8 µm in diameter. **Aerial mycelium** 0.8–1.2 µm in diameter, rising from the substrate mycelium. **Spores in chains** up to 20 spores long, 0.7–1.5 µm long, **on both substrate and aerial hyphae** on short, unbranched lateral or terminal sporophores. **Spore formation basipetal.** Intercalary spores occasionally found. Gram-positive, non-acid fast, aerobic. **Colonies slow-growing, raised, with entire or filamentous margins; aerial mycelium sparse.** Thermoduric, xerotolerant. **Able to utilize a wide range of organic compounds** as sole sources of carbon for energy and growth and to degrade a number of substrates. Resistant to some antibiotics but **susceptible to lysozyme. Wall peptidoglycan containing *meso*-diaminopimelic acid (*meso*-DAP), arabinose, and galactose but not mycolic acid** (wall chemotype IV).

In Section 17 of Volume 2, the genus *Micropolyspora* was described as a genus *incertae sedis.* At the time of that writing a request had been made to the International Committee on Systematic Bacteriology for the conservation of the genus *Micropolyspora,* following the transfer of the type species, *M. brevicatena* Lechevalier, Solotorovsky and McDurmont 1961, 13 [AL], to the genus *Nocardia* (Goodfellow and Pirouz 1982b), with *Micropolyspora faeni* Cross, Maciver and Lacey 1968a, 354[AL] as the type species (McCarthy et al., 1983). Subsequently the Commission ruled that *Faenia* Kurup and Agre 1983, 664 was the legitimate name for this genus, with *Faenia rectivirgula* Kurup and Agre 1983, 664 the type species (Wayne, 1986). Since *F. rectivirgula* and *M. faeni* have already been shown to be synonymous (Arden-Jones et al., 1979; Kurup, 1981), the genus *Micropolyspora* and the species *Micropolyspora faeni* are referred to, respectively, as the genus *Faenia* and the species *Faenia rectivirgula* throughout this section.

The type culture of *F. rectivirgula* is ATCC 33515 (=INMI 683 = VKM-A-810).

Further Descriptive Information

Characteristically, branching of the substrate hyphae is almost at right angles, with chains of spores mostly on short unbranched lateral and terminal sporophores (Fig. 26.25). Aerial hyphae are usually sparse, arising from the substrate hyphae in short tufts with spore chains both lateral and terminal (Fig. 26.26). Although spore chains may be up to 20 spores long (Fig. 26.27), they are usually shorter than 5 spores. The chains are generally straight and spores form basipetally (Dorokhova et al., 1970a). Occasionally intercalary spores may be observed (Fig. 26.28). Spores are round to oval, 0.7–1.5 µm long, with a smooth or irregularly roughened surface in electron micrographs (Fig. 26.29).

The wall peptidoglycan contains *meso*-DAP, arabinose, and galactose as diagnostic constituents (wall chemotype IV) together with glutamic acid, alanine, glucosamine, and muramic acid (Becker et al., 1965). Whole-cell methanolysates yield no mycolic acids (Mordarska et al., 1972), but organisms are rich in *iso*- and *anteiso*- branched-chain fatty acids (Kroppenstedt and Kutzner, 1976, 1978) and have polar lipid contents characterized by large amounts of phosphatidylcholine, diphosphatidylglycerol, phosphatidylglycerol, phosphatidylinositol, and phosphatidylmethylethanolamine (Lechevalier et al., 1977). The major menaquinones are tetrahydrogenated with nine isoprene units, but there are usually smaller amounts of hexa- and octahydrogenated menaquinones also (MK-9(H_4, H_6, H_8); Collins et al., 1977).

Two types of hyphae have been distinguished in thin section, one having walls 19–25 nm thick and the other 11–15 nm thick (Dorokhova et al., 1970a). In general, the cell structure resembles that of other actinomycetes, but in the thicker walled cells the cytoplasm is uniformly fine grained with a large nuclear zone extending the full length of the cell. In thinner walled cells the cytoplasm is less compact and homogeneous and the nuclear zone appears as small areas of low density. Mesosomes are

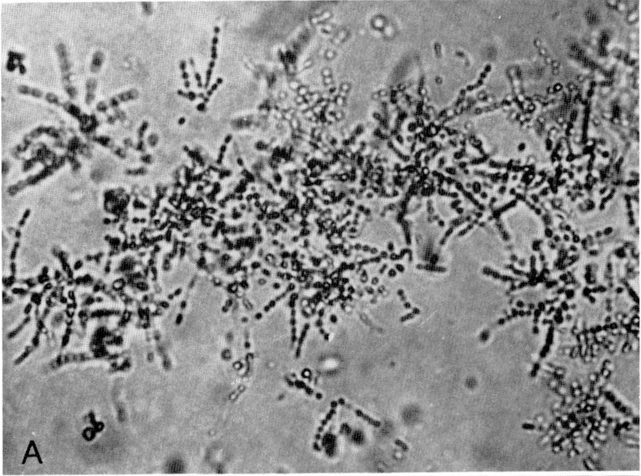

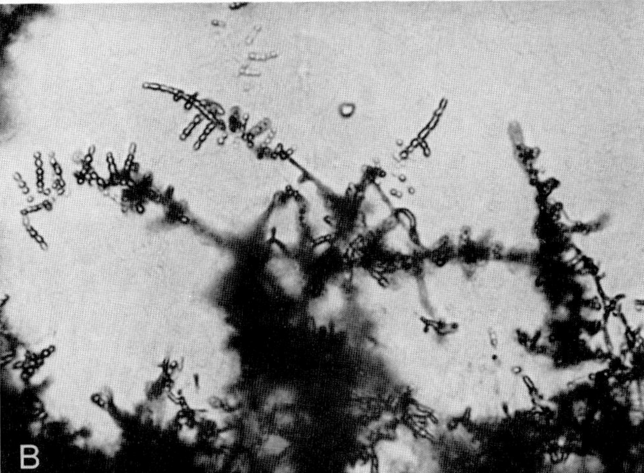

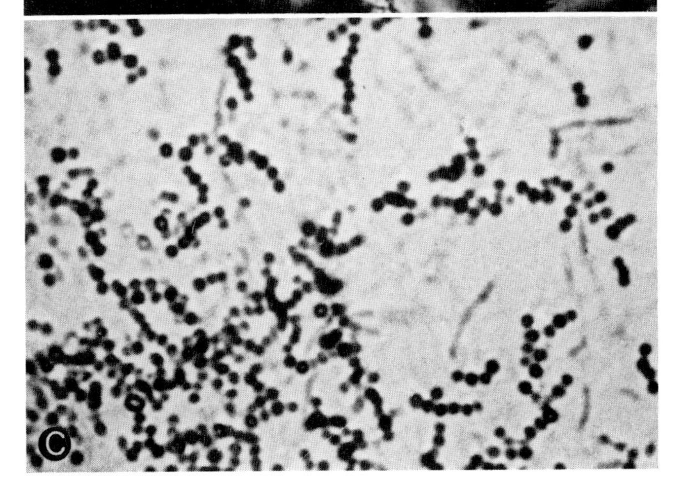

Figure 26.25. Morphology of the substrate mycelium of *Faenia rectivirgula. A,* appearance near growing margin (× 650). *B,* typical right-angle branching (× 650). *C,* spore chains in older part of colony (× 1300). Half-strength nutrient agar, 55°C.

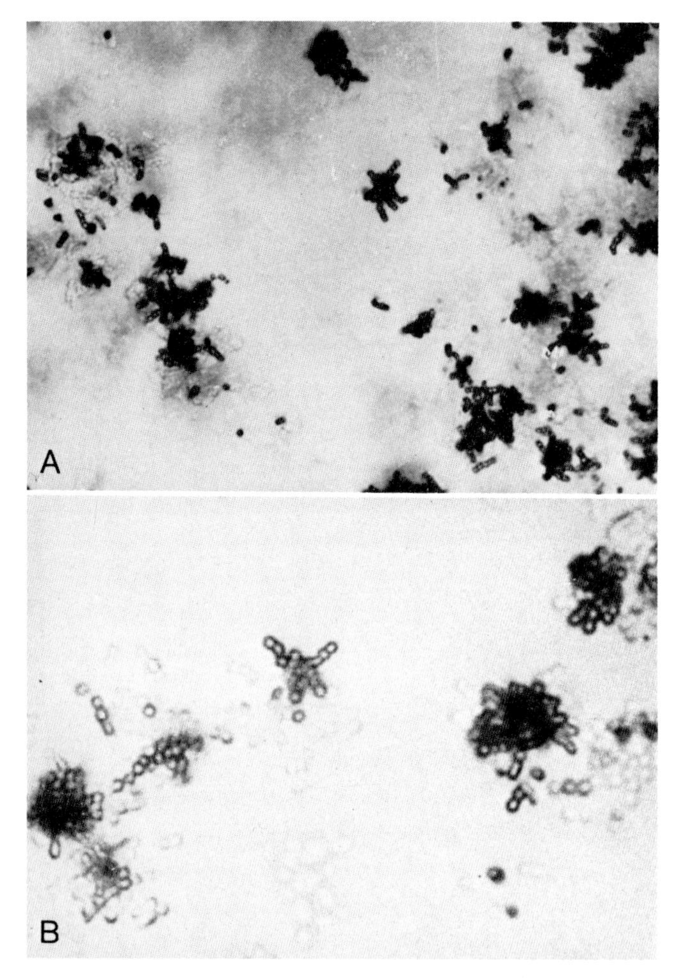

Figure 26.26. Aerial mycelium showing (*A*) sparse, tufted appearance (× 390), and (*B*) formation of spore chains (× 780). Half-strength nutrient agar, 55°C.

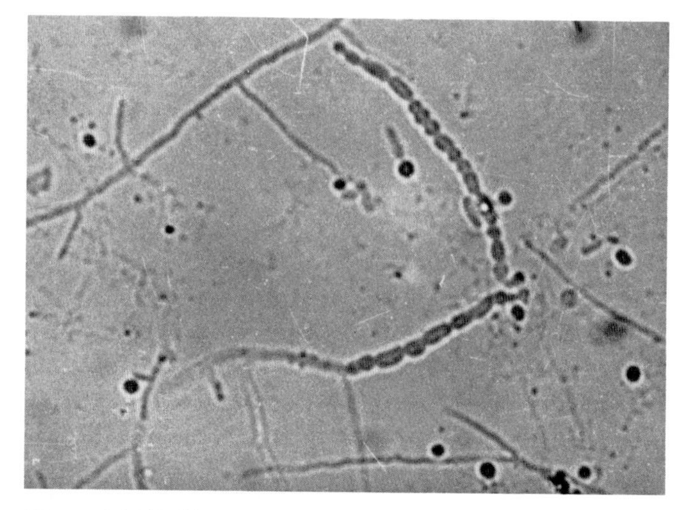

Figure 26.27. Intercalary spore formation or fragmentation of substrate mycelium in slide culture. Half-strength nutrient agar, 55°C (× 1600). (Reproduced with permission from Cross, Maciver and Lacey, Journal of General Microbiology *50:* 351–359, ©Society for General Microbiology, 1968.)

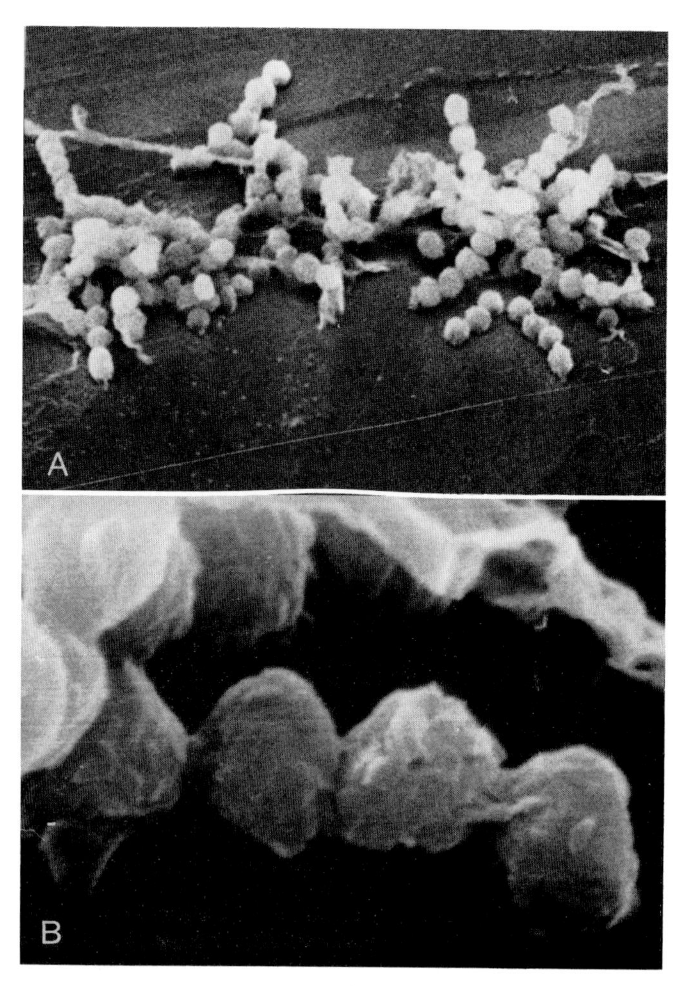

Figure 26.28. Scanning electron micrographs of (*A*) sporulating hyphae (× 3000), and (*B*) spores (× 13,100).

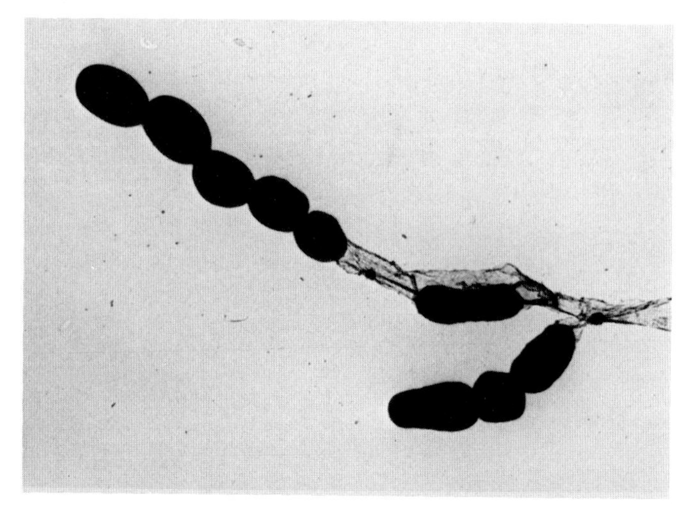

Figure 26.29. Transmission electron micrograph of spore chains (× 9600). (Reproduced with permission from Cross, Maciver and Lacey, Journal of General Microbiology *50:* 351–359, © Society for General Microbiology, 1968.)

also less well developed than in the thicker walled cells. Hyphae tend to autolyze during prolonged incubation at 55°C or at room temperature.

Spore chains are surrounded by a multilayered sheath, although this is less evident on spore chains formed on the substrate mycelium than on those formed on the aerial mycelium (Dorokhova et al., 1969; Williams et al., 1976). The spores are covered by a wall 70–100 nm thick in which two layers may be distinguished (Figs. 26.30 and 26.31), differing in thickness and electron density. Additional thickening of the cross-walls usually occurs, giving characteristic interspore pads (Dorokhova et al., 1969) (Fig. 26.32). These may sometimes be observed by light microscopy of stained preparations as conspicuous nonstaining areas (Cross et al., 1968a), but they may break down as the spores mature (Dorokhova et al., 1969). Plasmodesmata have also been described within these interspore pads. The protoplast is separated from the wall by a membrane and contains small, dark, densely packed ribosomes. Mesosomes are well developed and often adjoin the nuclear material. Although the spores are characteristically round or oval, spores of irregular shape are often seen in sections.

Colonies of *F. rectivirgula*, the only species presently recognized, grow to 5 mm in 7 days at 40–50°C. The substrate mycelium may be colorless, brownish yellow, or orange-yellow. Aerial mycelium is white but often sparse or absent. Good growth is obtained on yeast extract–malt extract agar or casein hydrolysate agar (Cross et al., 1968a), and aerial mycelium production may be enhanced by the addition of 5% (w/v) NaCl to the medium. Spore chain length tends to be greatest on modified Umezawa's medium (Arden-Jones et al., 1979). Usually no soluble pigment is produced, but some brown pigment has been observed on V-8 agar.

F. rectivirgula sometimes gives inconsistent results in carbon utilization tests, but most authors agree that it can utilize the following as sole sources of carbon for energy and growth: amygdalin, D-arabinose, cellobiose, dextrin, erythritol, fructose, galactose, glycerol, lactose, mannitol, mannose, ribose, starch, sucrose, trehalose, and xylose; it does not use L-arabinose, dulcitol, glucosamine, melibiose, melezitose, salicin, and sorbose. Isolates differ in their ability to utilize adonitol, cellobiose, glycogen, inulin, maltose, inositol, raffinose, rhamnose, and sorbitol. Esculin, gelatin, guanine, hypoxanthine, ribonucleic acid, Tween 20, Tween 80, and xanthine are degraded, but not adenine, allantoin, cellulose, chitin, elastin, keratin, starch, tributyrin, or xylan. Degradation of arbutin, casein, deoxyribonucleic acid, hippurate, testosterone, tyrosine, and urea has occasionally been demonstrated. Isolates of *F. rectivirgula* produce catalase, galactosidase, glucosidase, and phosphatase but not oxidase or prodiginins. Nitrates may be reduced to nitrites by some strains. Melanin is not produced from tyrosine nor acid from glucose, but *F. rectivirgula* is sensitive to lysozyme. Growth occurs between about 30° and 63°C, with 50–55°C the optimum. The temperature range for growth may differ with strain, growth conditions, and substrate (Cross et al., 1968a; Arden-Jones et al., 1979; Kurup, 1981; Goodfellow and Pirouz, 1982b).

Phages have been isolated from *Faenia* isolates by Prauser and Momirova (1970) and by Kurup and Heinzen (1978). The latter were examined by electron microscopy and shown to have hexagonal heads 50–60 nm in diameter and long tails 7–8 × 132–145 nm. Their host range was generally restricted to *Faenia* isolates, but Kurup and Heinzen (1978) reported infection of some *Thermoactinomyces* cultures.

The antigenicity of *F. rectivirgula* has received much attention since the organism was implicated as a cause of farmer's lung, the classic form of hypersensitivity pneumonitis (Pepys et al., 1963). Initially three precipitin arcs were recognized in gel diffusion and immunoelectrophoresis tests using extracts of *F. rectivirgula* and sera from farmer's lung patients. Using gel filtration, absorption on columns of DEAE, and elution with crossed immunoelectrophoresis and immunodiffusion, these three have been resolved into up to 75 individual antigenic compo-

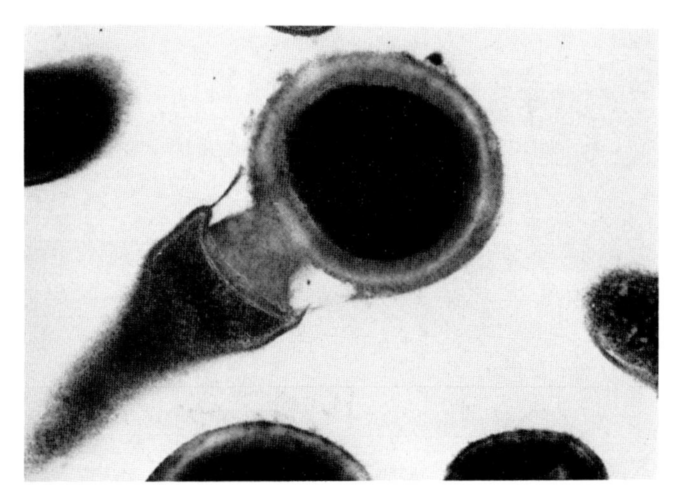

Figure 26.31. Longitudinal section through a developing spore (× 40,000).

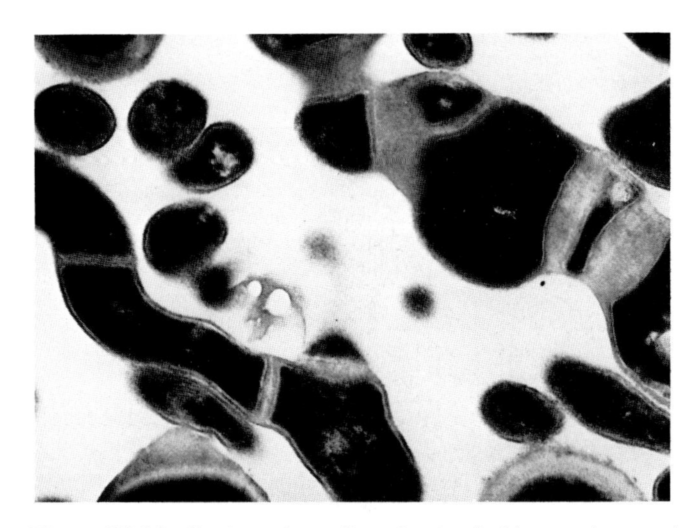

Figure 26.30 Sections of mycelium showing double septa in normal hyphae and irregularly thickened septa in enlarged hyphae or aberrant spore chains (× 25,000).

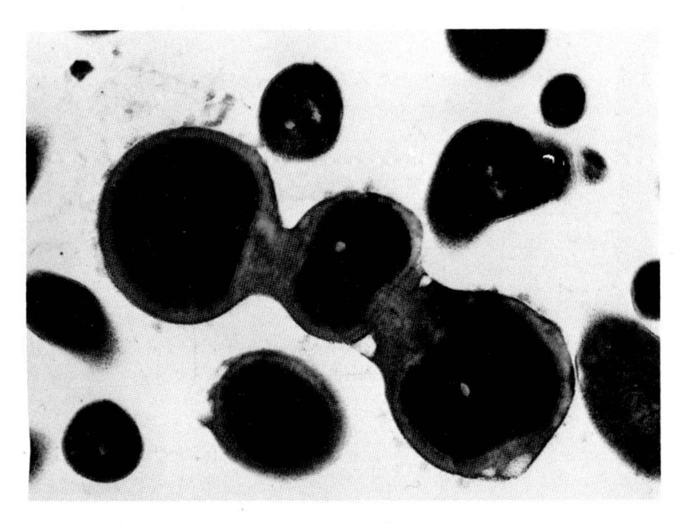

Figure 26.32 Longitudinal section through a spore chain showing interspore pads (× 25,000).

nents (Walbaum et al., 1969, 1973; Fletcher et al., 1970; Edward, 1972b; Hollingdale, 1974; Nicolet and Bannerman, 1975; Bannerman and Nicolet, 1976; Kurup and Fink, 1977; Roberts et al., 1977; Arden-Jones et al., 1979; Kurup et al., 1981b). Many of the components have been characterized and have molecular masses falling within the range 39–265 × 10^3 daltons. They contain protein and carbohydrate, the protein fraction yielding from 4 to 24 amino acids. Sugar contents are from 2 to 98% and their half-lives from 8 h to 50 days. One group of heat-stable antigens consists of glycopeptides with few amino acids and a large sugar content. The remainder, some of which are very heat labile, are protein. Enzymes have been identified in these antigenic fractions, including esterases, lipases, proteases, aminopeptidases, trypsin- and chymotrypsin-like enzymes, catalases, malic dehydrogenases, and peroxidases. The antigen content of "metabolic" or substrate extracts is reported to differ from that of spores and mycelium, and Walbaum et al. (1969) also found that the metabolic extract was more relevant in diagnosis of farmer's lung.

F. rectivirgula tolerates few antibiotics. It is tolerant to lincomycin (at 20 µg/ml medium), neomycin (10 µg/ml) and benzylpenicillin (20 µg/ml) but only to low concentrations of cephaloridine (2 µg/ml), demeclocyclin (2 µg/ml), gentamycin (4 µg/ml), and tobramycin (1 µg/ml). No growth occurred in the presence of rifampicin (2 µg/ml) or streptomycin (4 µg/ml) (Athalye, 1981).

F. rectivirgula is not known to cause infections, but inhalation of its spores can cause farmer's lung (extrinsic allergic alveolitis) in sensitized subjects. This disease occurs widely in Europe, the United States, Canada, and Japan, and there has been a single, unconfirmed case in Sri Lanka. In western Scotland and the Orkneys it may affect up to 8.6% of farm workers (Grant et al., 1971), while in the United States 8.4% of Wisconsin dairy farmers had precipitins to actinomycetes, 86% of these to *F. rectivirgula* (Roberts et al., 1976b). Because precipitins are found in farmer's lung patients to some of the heat-labile antigens that may be destroyed during the spontaneous heating of molding hay, it has been suggested that spores may germinate in the lungs, releasing these antigens, but this is not proven. The spores may also activate complement by the alternative pathway (Edwards, 1972b; Edwards et al., 1974).

F. rectivirgula occurs widely in spontaneously heating vegetable matter. It was first isolated from moldy hay that had been baled wetter than 35% water content and that had heated to 50–65°C (Gregory et al., 1963). Subsequently it has been found in straw, cereal grain, sugar cane bagasse, cotton bales, mushroom compost, soil, and the air over pastures (Lacey, 1978). Inoculum is widespread in grass at the time of cutting, and further contamination may occur in stores from previous crops. Heating of the substrate is initiated by plant cells and mesophilic fungi and bacteria. *F. rectivirgula* commences growth at only 30–35°C and largest numbers are found in hays that heat to about 60°C after baling at about 39% water content. The change in pH from 5.5–6 to 7–8 caused by fungal proteolysis probably favors actinomycete colonization, although *F. rectivirgula* can sometimes grow in hay without pretreatment (Gregory et al., 1963; Festenstein et al., 1965) *F. rectivirgula* may survive in moist grain stored anaerobically in silos, and its spores can survive up to 20 min at 70°C.

Isolation

F. rectivirgula has sometimes been isolated by dilution and direct plating techniques, but enumeration in hay and other vegetable matter is most reliably achieved using an Andersen sampler to isolate airborne spores in a small wind tunnel or sedimentation chamber (Gregory and Lacey, 1963; Lacey and Dutkiewicz, 1976a, 1976b). Half-strength nutrient agar (Oxoid, CM3) or half-strength tryptone-soya agar (Oxoid CM131) + 0.2% casein hydrolysate (Oxoid L41), both containing 50 µg cycloheximide/ml medium, have usually been used for isolation, with incubation at 40–55°C.

Maintenance Procedures

F. rectivirgula can be lyophilized by usual procedures.

Differentiation of the genus **Faenia** *from related genera*

F. rectivirgula can be distinguished from related taxa by the presence of short chains of spores on aerial and substrate mycelium, its wall chemotype IV lacking mycolic acids, carbohydrate utilization pattern, degradation of guanine, hypoxanthine, and xanthine, growth on 10% NaCl, and resistance to certain antibiotics but not to lysozyme (Table 26.23). *Nocardia brevicatena* also bears short chains of spores but the walls contain mycolic acids.

Taxonomic Comments

The nomenclatural history of species of the genera *Faenia* and *Micropolyspora* contains much controversy. The genus *Micropolyspora* was created for actinomycetes bearing short chains of spores on both aerial and substrate mycelia. The type species was designated *M. brevicatena* Lechevalier, Solotorovsky and McDurmont 1961, 13. Later, Lechevalier (1968a) requested the conservation of *Micropolyspora* against "*Micropolispora*" (Shchepkina 1940). The Judicial Commission of the International Committee on Systematic Bacteriology appeared to have made no ruling on this request. However, the inclusion of *Micropolyspora* in the Approved Lists (Skerman et al., 1980) indicated de facto approval.

Other species were included in *Micropolyspora* subsequently. Those in the Approved Lists included *M. faeni* Cross, Maciver and Lacey 1968a, 354 (first described as "*Thermopolyspora polyspora*" Corbaz, Gregory and Lacey 1963); *M. angiospora* (Zhukova, Tsyganov and Morozov 1968, 728); *M. rectivirgula* (Krasil'nikov and Agre 1964b, 106) Prauser and Momirova 1970, 220 (first described as "*Thermopolyspora rectivirgula*," Krasil'nikov and Agre 1964b); and *M. internatus* Agre, Guzeva and Dorokhova 1974, 577. Additional species excluded from the Approved Lists (1980) or never validated include "*M. caesia*" (Kalakoutskii, 1964), "*M. thermovirida*" (Kosmachev, 1964), "*M. viridinigra*" and "*M. rubrobrunea*" (Krasil'nikov et al., 1968), "*M. coerulea*" (Preobrazhenskaya et al., 1973), and "*M. fascifera*" (Prauser, 1974b).

The accepted concept of the genus *Micropolyspora* remained essentially morphological, although the possession of a wall chemotype IV (Lechevalier and Lechevalier, 1970b) became incorporated into the genus definition (Cross and Goodfellow, 1973). Consequently, the genus became heterogeneous (Goodfellow and Pirouz, 1982b). Three species were found to have wall chemotype III. "*M. rubrobrunea*" and "*M. viridinigra*" were transferred to the genus *Excellospora* Agre and Guzeva 1975, 521, and *M. angiospora* may be a species of *Actinomadura* Lechevalier and Lechevalier 1970b, 400 (Lechevalier in Kurup, 1981) or *Excellospora* (Lacey, et al., 1978). Transfer of *M. internatus* and "*M. caesia*" to *Saccharomonospora* Nonomura and Ohara 1971c, 899 has been proposed but not validated (Kurup, 1981) and "*M. coerulea*" and "*M. thermovirida*" also appear to be typical *Saccharomonospora* species. However, there are no isolates of "*M. thermovirida*" extant so that its position cannot be clarified. Mycolic acids have been found in "*M. fascifera*" (Prauser, 1978) but so far no transfer has been proposed. The presence of mycolic acids in *M. brevicatena* together with menaquinones, phospholipids, and fatty acids consistent with *Nocardia* Trevisan 1889, 9 (Collins et al., 1977; Lechevalier et al., 1977; Kroppenstedt and Kutzner, 1978) have led to its reclassification as *Nocardia brevicatena* (Lechevalier, Solorovsky and McDurmont 1961, 13) Goodfellow and Pirouz 1982b, 523. Finally, there was little doubt that the two remaining species, *M. faeni* and *M. rectivirgula*, were synonymous (Dorokhova et al., 1970a; Prauser and Momirova, 1970; Arden-Jones et al., 1979; Kurup, 1981), despite considerable differences between their original descriptions (Krasil'nikov and Agre, 1964b; Cross et al., 1968a).

Thus *Micropolyspora* was a genus without a type species and so nomenclaturally invalid under Rule 37a of the International Code of Nomenclature of Bacteria (Lapage et al., 1975). Only one species remained within the original concept of the genus, for which the epithet *rectivirgula* had priority, although *faeni* was better known, as the cause of

Table 26.23.
Characteristics differentiating **Faenia** *from related genera*[a,b]

Characteristics	*Faenia*	*Saccharo-polyspora*	*Saccharo-monospora*[c]	*Pseudo-nocardia*[d]	*Nocardia*	*Actinomadura*	*Nocardiopsis*
Moprhological characters—spores or fragmentation of:							
Substrate hyphae	+	v	−	−	+	−	+
Aerial hyphae	+	+	+	+	d	D	+
Spore surface	Smooth	Hairy	Warty	Smooth	Smooth	Smooth or warty	Smooth
Chemical characters							
Wall type	IV	IV	IV	IV	IV	III	III
Mycolic acid	−	−	−	−	+	−	−
Hydrolysis tests							
Adenine	−	+	−	−	−	−	d
Guanine	+	+	−	−	D	D	+
Elastin	−	+	+	+	D	D	d
Hypoxanthine	+	+	−	−	D	D	+
Xanthine	+	+	−	−	D	−	+
Sole carbon sources							
Inositol	+	+	+	+	D	D	d
Lactose	+	+	+	+	−	D	−
Maltose	−	+	+	+	D	D	+
Growth in the presence of							
10% NaCl	+	+	−	−	−	D	+
Lincomycin[e]	+	+	d	+	+	D	+
Neomycin[e]	+	+	d	+	+	D	−
Penicillin[e]	+	−	d	−	+	D	d
Lysozyme	−	−	−	+	+	−	−
Growth at 50°C	+	d	d	+	−	D	−
Growth at 60°C	+	−	d	−	−	−	−

[a]Symbols: see Table 26.22; also *v*, strain instability.
[b]Data from: Lacey and Goodfellow (1975), Arden-Jones et al. (1979), Goodfellow et al. (1979), Athalye (1981), Goodfellow and Pirouz (1982b), McCarthy and Cross (1984a).
[c]*S. viridis* only.
[d]*P. thermophila* (type strain only).
[e]Concentrations: lincomycin, 20 µg/ml medium for *Faenia,* filter paper disks soaked in 100 µg/ml solution otherwise; neomycin, 10 µg/ml medium for *Faenia,* filter paper disks soaked in 50 µg/ml otherwise; penicillin, 20 µg/ml medium for *Faenia,* filter paper disks soaked in 10 units/ml otherwise.

farmer's lung and other forms of hypersensitivity pneumonitis. Resolution of this situation gave rise to much controversy, and two proposals were made, one based on a strict interpretation of the International Code of Nomenclature of Bacteria (Lapage et al., 1975) and the other on the Principles on which this Code is based. Kurup and Agre (1983) proposed a new genus, *Faenia*, with *F. rectivirgula* Kurup and Agre 1983, 664 as the type species, while McCarthy et al. (1983) argued for conservation of *Micropolyspora* with *M. faeni* Cross, Maciver and Lacey 1968, 354 as the type species.

The Judicial Commission's recent ruling (Wayne, 1986) establishes *Faenia* as the legitimate name with *F. rectivirgula* (ATCC 33515) as the type species.

Further studies of actinomycetes with wall chemotype IV could lead to changes in taxonomy and modifications in the definition of some genera. The relationships of *Faenia, Saccharopolyspora* Lacey and Goodfellow 1975, 78, *Saccharomonospora*, and wall chemotype IV species lacking mycolic acids (such as *Amycolata autotrophica* (Takamiya and Tubaki 1956, 59) Hirsch 1961 360, *Amycolatopsis mediterranei* (Margalith and Beretta 1960b) Thiemann, Zucco and Pelizza 1969b, 148, *Amycolatopsis orientalis* (Pittenger and Brigham 1956, 642) Pridham 1970, 42, and "*Saccharothrix aerocolonigenes*" (Shinobu and Kawato 1960a, 215) Pridham 1970, which has a wall containing galactose and mannose but not arabinose (Gordon et al., 1978)) need to be established. A possible further species of *Micropolyspora*, "*M. hordei,*" was suggested by Hill and Lacey (1983), but has not yet been described.

Further Reading

Arden-Jones, M.P., A.J. McCarthy and T. Cross, 1979. Taxonomic and serological studies on *Micropolyspora faeni* and *Micropolyspora* strains bearing the specific epithet *rectivirgula* J. Gen. Microbiol. *115:* 343–354.
Cross, T. and M. Goodfellow, 1973. Taxonomy and classification of the actinomycetes. *In* Skyes and Skinner (Editors) Actinomycetales. Characteristics and Practical Importance. Academic Press, London, pp. 11–112.
Cross, T., A. Maciver and J. Lacey, 1968. The thermophilic actinomycetes in mouldy hay: *Micropolyspora faeni* sp. nov. J. Gen. Microbiol. *50:* 351–359.
Kurup, V.P. and N.S. Agre, 1983. Transfer of *Micropolyspora rectivirgula* (Krasil'nikov and Agre 1964) Lechevalier, Lechevalier, and Becker 1966 to *Faenia* gen. nov. Int. J. Syst. Bacteriol. *33:* 663–665.
Lacey, J. 1978. Ecology of actinomycetes in fodders and related substrates. Zentralbl. Bakteriol. Parasitenkd. Infektionskr. Hyg. Abt. 1. Suppl. *6:* 161–170.
Lacey, J. 1981. Airborne actinomycete spores as respiratory allergens. Zentralbl. Bakteriol. Microbiol. Hyg. Abt. 1. Suppl *11:* 243–250.
McCarthy, A.J., T. Cross, J. Lacey and M. Goodfellow, 1983. Conservation of the name *Micropolyspora* Lechevalier, Solotorovsky, and McDurmont and designation of *Micropolyspora faeni* Cross, Maciver, and Lacey as the type species of the genus: request for an opinion. Int. J. Syst. Bacteriol. *33:* 430–433.
Wayne, L.G. 1986. Actions of the Judicial Commission of the International Committee on Systematic Bacteriology on requests for opinions published in 1983 and 1984. Int. J. Syst. Bacteriol. *36:* 357–358.

List of species of the genus **Faenia**

1. **Faenia rectivirgula** (Krasil'nikov and Agre 1964b, 106) Kurup and Agre 1983, 664.[VP] (*Micropolyspora faeni* Cross, Maciver and Lacey 1968, 354[AL]; *Micropolyspora rectivirgula* (Krasil'nikov and Agre 1964b, 106) Prauser and Momirova 1970, 220; *Thermopolyspora rectivirgula* Krasil'nikov and Agre 1964b, 106; *Thermopolyspora polyspora* Corbaz, Gregory and Lacey 1963, 450 non Henssen 1957, 396.)

rect.i.vir'gu.la. L. adj. *rectus* straight; L. dim. n. *virgula* twig; M.L. n. *rectivirgula* straight branch.

For features of *F. rectivirgula* refer to the generic description and Table 26.23.

Type strain: ATCC 33515.

Corbaz et al. (1963) originally identified strains from mouldy hay as "*Thermopolyspora polyspora*" (Henssen, 1957). Subsequently Henssen examined a culture that she considered distinct from "*T. polyspora*," but no strains of the latter were available for comparison. After further study, the strains from hay that had been implicated in farmer's lung (Pepys et al., 1963) were renamed *Micropolyspora faeni* (Cross et al., 1968a). Strains of *M. rectivirgula* were also first identified as a "*Thermopolyspora*" species and described as producing colorless to slightly yellow colonies with abundant yellowish aerial mycelium and straight chains of smooth spores. This contrasted with the sparse white aerial mycelium of *M. faeni*. The status of "*T. rectivirgula*" remained in doubt until Prauser and Momirova (1970) transferred this taxon to *Micropolyspora*.

Earlier Krasil'nikov and his coworkers had questioned the separate status of "*Thermopolyspora*" and its distinction from *Micropolyspora* (Krasil'nikov, 1964; Kalakoutskii et al., 1968). However, they still did not classify *rectivirgula* isolates with *Micropolyspora*, either when they first described it (Krasil'nikov and Agre, 1964b) or when they studied it in more detail (Kalakoutskii et al., 1968). However, Lechevalier et al. (1966a) showed "*T. polyspora*" to have wall chemotype IV and tentatively assigned this taxon to *Micropolyspora*. Then Dorokhova et al. (1969), in using binomial *M. rectivirgula*, finally indicated Russian acceptance that it did belong to this genus but still did not propose formal transfer.

Subsequent studies by Arden-Jones et al. (1979) and Kurup (1981) confirmed the synonymy of *M. faeni* and *M. rectivirgula*, and McCarthy et al. (1983) proposed conservation of *M. faeni* to promote stability of nomeclature and avoid confusion, while Kurup and Agre (1983) described *Faenia rectivirgula* for this taxon.

Genus **Promicromonospora** *Krasil'nikov, Kalakoutskii and Kirillova 1961a, 107*[AL]

L. V. Kalakoutskii, N. S. Agre, Helmut Prauser, L. I. Evtushenko

Pro.mi.cro.mo.no'spo.ra. Gr. pref. *pro* before, primordial; Gr. adj. *micros* small; Gr. adj. *monos* single, solitary; Gr. fem. n. *spora* a seed; M.L. fem. n. *Promicromonospora*; the genus name was coined to reflect the combination of traits then thought to be characteristic of the actinomycete form-genera *Proactinomyces* (the tendency of the mycelium to fragment) and *Micromonospora* (the formation of single spores on the substrate mycelium).

Branching septate **hyphae** (0.5–1.0 µm in diameter) growing on the surface of and penetrating into the agar, which **break up into fragments** of various size and shape. Fragmentation finally results in **nonmotile**, Y- or V-shaped, rodlike, coccoid, chlamydospore-like, and other spore-shaped elements. All of them may give rise to new mycelia. **Growth pasty to leathery.** Aerial hyphae in different strains may vary in abundance (sometimes discernible only microscopically). These are straight to curved, sometimes sparsely branched, usually fragmented into rodlike or elongated coccoid elements. **Gram-positive, non-acid fast, catalase positive. Aerobic.** Chemo-organotrophic. Glucose metabolized oxidatively, rarely also fermentatively. **Mesophilic.** Utilize a wide range of sources and possess a significant spectrum of hydrolytic activities. **Susceptible to taxon-specific phages**; not susceptible to phages of various sets that attack oerskoviae and other nocardioform organisms. **Cell wall chemotype VI** (lysine as principal diagnostic amino acid); peptidoglycan type A3α (L-lysine in position 3). **Mycolic acids lacking.** No wall teichoic acids found. Among fatty acids, branching ones of the *iso-* and *anteiso-* types (*iso-* and *anteiso-*C_{15}:O) predominate. Diagnostic phospholipids represented by phosphatidylglycerol and an unidentified glucosamine-containing phospholipid. **Menaquinones of the MK-9(H_4) type. Mol%** G + C of the DNA in the range of 70–75 (T_m). Mainly **found in soils.**

Type species: *Promicromonospora citrea* Krasil'nikov, Kalakoutskii and Kirillova 1961a, 107.

Further Descriptive Information

Since the genus is monotypic the following more detailed information, as well as the concise genus description given above, originate only from the study of the type species *Promicromonospora citrea*, i.e. from the study of up to 30 strains isolated by various workers from different soils. Colonies develop within 1–2 days, and growth is more abundant on complex peptone-containing media. Some strains respond favorably to the addition of vitamins or require yeast extract for growth on synthetic media (Evtushenko et al., 1981). Colonies usually are concave to wrinkled, but are smooth and pasty with some strains. Colors vary from yellow to white.

The length of hyphae, extent of branching, and persistence of the mycelial state depend on the particular strains, the media employed, and the conditions of cultivation. Branching may be dense (Fig. 26.33*A*) to loose (Fig. 26.33*B*). Fragmentation may begin after 8 h of incubation. Sometimes it is not recognizable at all in situ on and/or below the agar surface. The mycelium frequently has a "barbed wire" appearance resulting from longitudinal growth of fragments after their separation.

The process of fragmentation is more pronounced in submerged shaken cultures on complex media as compared to cultures on solid mineral salt media. Fragmentation results in the formation of coccoid, rodlike, diphtheroid, and chlamydospore-like elements; sometimes enlarged cells (up to 5 µm in diameter) may occur (Fig. 26.33*C*). Spore-shaped elements (Fig. 26.33*D*) are observed mainly in solid surface cultures. By virtue of their regular shape, positioning (often terminal on short side branches), refractility, and response to staining, they were regarded as spores by some authors (Krasil'nikov et al., 1961a; Luedemann, 1974). Aerial hyphae are produced in all strains after 2–4 days; suitable media are oatmeal agar (Shirling and Gottlieb, 1966) and peptone–corn extract agar (Agre, 1964). This capability tends to disappear in the course of maintaining strains by continued serial transfers. In some strains, only extremely few and short aerial hyphae are detectable after thorough microscopical examination. Other strains show a mat of aerial hyphae visible only microscopically or just detectable to the naked eye. Only a few strains produce distinct aerial mycelium. Aerial hyphae are straight to curved and not branched or only sparsely so. On observation in situ, fragments or spore-shaped elements are not visible.

More than 90% of strains were able to utilize L-arabinose, D-galactose, cellobiose, D-fructose, D-maltose, mannose, raffinose, sucrose, trehalose, and D-xylose as a sole carbon source and produce acid from them; and to assimilate glycerol, acetate, fumarate, malate, malonate, and succinate. The strains hydrolyzed esculin, casein, gelatin, and starch, and possessed catalase, DNAase, nitrate reductase, and urease activities. They

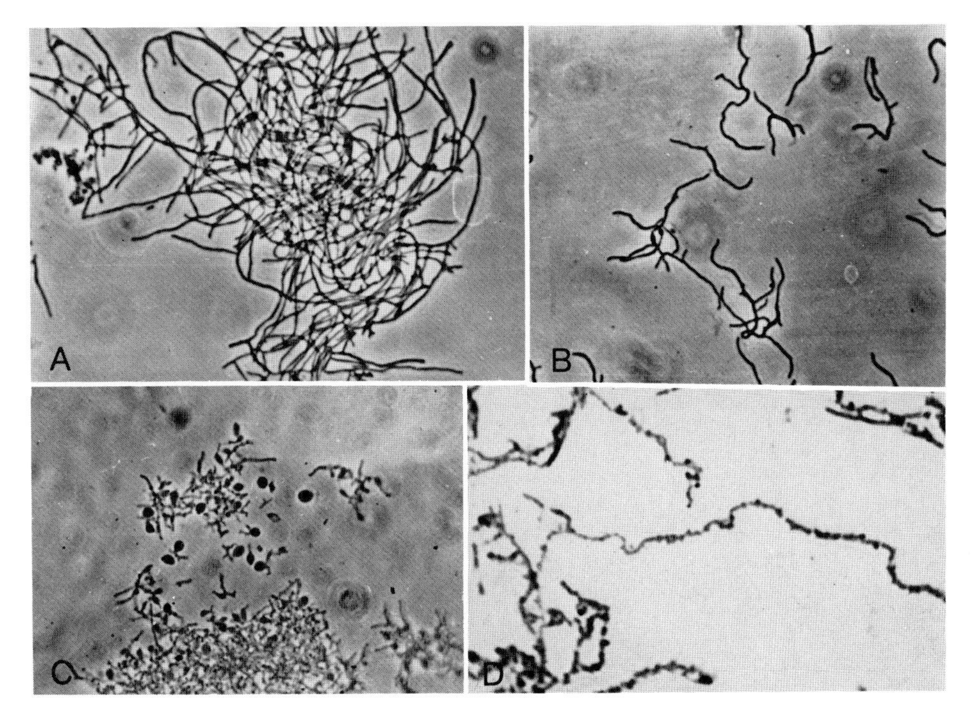

Figure 26.33. *Promicromonospora citrea. A–C*, submerged 2-day (28°C) growth in peptone-yeast extract broth. Phase contrast. *A*, strain VKM Ac 791, mycelium with well-developed hyphae (× 900). *B*, strain LL G 165, mycelium with short hyphae (× 900). *C*, strain VKM Ac 783. Cells of varying shape and size (× 500). *D*, strain VKM Ac 665, surface growth (2 days at 28°C) on potato agar, crystalline violet stain. Spore-shaped cells (× 1200).

were able to grow in the presence of 5% (w/v) NaC1, 7% KI, and 7% KBr, but were sensitive to 7% NaCl, 0.25% phenol, 0.01% thymol, and 0.0001% crystal violet. Not a single strain utilized sorbose, aconitate, benzoate, oxalate, hexadecanoic acid, paraffin, adenine, arbutin, hypoxanthine, or lecithin. There was no growth in lysozyme broth. The following tests were variable (11–89% of strains were positive): utilization of D-arabinose, lactose, melibiose, L-rhamnose, adonitol, dulcitol, inositol, mannitol, sorbitol, citrate, formate, lactate, propionate, cellulose, Tween 20, Tween 40, Tween 60, and xanthine; production of phosphatase; growth at 37°C; and growth at pH 5.0.

No growth under anaerobic conditions was observed on a spectrum of complex and chemically defined media with any of the 30 strains tested. If tryptic-soy agar (Lechevalier, 1972) was employed, a very weak growth was detected with two strains under anaerobic conditions at 28°C. On yeast-peptone agar supplemented with 0.5% (w/v) $CaCO_3$ (Zviagintsev et al., 1980) neither of the strains tested produced clearing zones following aerobic incubation. The key enzymes of the Entner-Doudoroff pathway were not detected in the type strain of *P. citrea* (Kersters and De Ley, 1968).

Cell wall chemotype (Lechevalier and Lechevalier, 1970b) is VI, lysine being the diagnostic amino acid (Yamaguchi, 1965). Many strains contain galactose in the cell wall. The peptidoglycan type (Schleifer and Kandler, 1972) is A3α, displaying L-Lys in position 3 and Ala_2 in the interpeptide bridge (Stackebrandt et al., 1983c). Teichoic acids were not found in wall preparations (Evtushenko et al., 1984a). Branched fatty acids of the *anteiso-* and *iso-* series 12- and 13-methyltetradecanoic (*anteiso-* and *iso-*C_{15}:O) predominate in all preparations (Andreyev et al., 1983). The phospholipid type is PV, represented by phosphatidylglycerol and unknown glucosamine-containing phospholipids (Lechevalier et al., 1977). The phospholipid fatty acids are of type I, characterized by branched-chain fatty acids of the *anteiso-* and *iso-* series (Lechevalier et al., 1977). Tetrahydrogenated menaquinones having nine isoprene units (MK-9(H_4)) are the predominating menaquinones (Collins and Jones, 1981). The mol% G + C of the DNA is in the range of 70–75 (Tsyganov et al., 1966; Yamaguchi, 1967; Evtushenko et al., 1984b).

So far, six taxon-specific phages have been isolated from soils and used for taxonomic purposes. Most of the 40 *Promicromonospora* strains studied are susceptible, at least, to three of these phages (susceptibility to at least one phage being a rare exception). Promicromonosporae are not susceptible to any other phages, including those specific for *Oerskovia* species and related organisms, several nocardioforms, and sporoactinomycetes. None of the *Promicromonospora* phages is effective against strains of other taxa (Prauser and Falta, 1968; Prauser, 1976a, 1984c).

Isolation

Most of the strains were isolated from soils and aluminum hydroxide gel antacid (Lechevalier, 1972). Among the various isolation media employed were several kinds of soil extract agar, an oatmeal agar (Prauser and Bergholz, 1974), and peptone–corn extract agar (Agre, 1964).

Maintenance Procedures

The strains can be maintained by serial transfers (once in 3 months) on agar slants. The following media are recommended: peptone–corn extract agar (Agre, 1964), complex organic agar number 79 of Prauser and Falta (1968), and oatmeal agar (Shirling and Gottlieb, 1966).

The strains survive lyophilization and maintenance above liquid nitrogen, applying routine procedures.

Differentiation of the genus **Promicromonospora** *from related organisms*

Promicromonospora seems to belong to a group of bacteria that contains *Cellulomonas* (including *C. cartae*), *Oerskovia*, the so-called nonmotile organisms (NMOs) of Lechevalier (1972; Lechevalier and Lechevalier, 1981b), and *Nocardia cellulans, Brevibacterium fermentans*, and "*Corynebacterium manihot*." The relationship follows from possession in common of a number of chemotaxonomic traits (mentioned above), comparison of results of 16S ribosomal (r) RNA cataloging (Stackebrandt et al., 1983c; Stackebrandt and Schleifer, 1984), and results of DNA/DNA hybridization at the level of 15–23% (Prauser, 1986).

Promicromonospora may be distinguished from these related bacteria by formation of aerial hyphae, susceptibility to taxon-specific phages, relationship to oxygen on tryptic-soy agar, and peptidoglycan type (Table 26.24).

Taxonomic Comments

Promicromonospora was regarded as an actinomycete genus "in search of a family" (Lechevalier and Lechevalier, 1981a). Stackebrandt and Schleifer (1984), on the basis of comparative cataloging of 16S rRNA, proposed to place *Promicromonospora* with *Oerskovia* (which they united with *Cellulomonas* (Stackebrandt et al., 1982a), *Cellulomonas, Arthrobacter*, and *Micrococcus* in the family *Arthrobacteraceae*. This seems to be a reasonable step toward a natural classification of coryneform and nocardioform bacteria.

Some data seem to indicate the presence of more than one species within the genus. A numerical analysis of 240 phenotypic traits in 30 strains seems to reveal three clusters, perhaps representing different species or subspecies. However, the three clusters thus revealed could not be substantiated by results of DNA/DNA hybridization. DNA/DNA homology among seven strains, including the type strain of *Promicromonospora citrea* as well as representatives of the three clusters, was found to range from 38 to 71%, mainly being in the range 35–57% (Prauser, 1986).

Jager et al. (1983) validly published *Promicromonospora enterophila*. These nocardioform bacteria do not seem to possess features that distinguish the genus *Promicromonospora* (Table 26.24). *P. enterophila* resembles the NMOs that were isolated by Lechevalier (1972). This species belongs to a large and possibly heterogeneous group of millipede intestinal nocardioforms (Szabo et al., 1986). Thus at present *P. enterophila* might be regarded as a species *incertae sedis*.

Further Reading

Andrew, L.V., L.I. Evtushenko, and N.S. Agre. 1983. Fatty acid composition of *Promicromonospora citrea*. Microbiologija *52* 58–63 (in Russian).

Krasil'nikov, N.A. L.V. Kalakoutskii and N.F. Kirillova. 1961. A new genus of ray fungi—*Promicromonospore* gen. nov. Izv. Akad. Nauk SSSR Ser. Biol. *1*: 107–112 (in Russian).

Lechevalier. H.A. and M.P. Lechevalier. 1981. Actinomycete genera "in search of a family." *In* Starr, Stolp, Trüper, Balows and Schlegel (Editors). The Prokaryotes. A Handbook on Habitats, Isolation and Identification of Bacteria. Springer:-Verlag, New York, pp. 2118–2123.

Stackebrandt, E., W. Ludwig, E. Seewalt, and K.-H. Schleifer. 1983. Phylogeny of sporeforming members of the order *Actinomycetales*. Int. J. Syst. Bacteriol. *33*: 173–180.

Differentiation of the species of the genus **Promicromonospora**

1. **Promicromonospora citrea** Krasil'nikov, Kalakoutskii and Kirillova 1961a, 107.[AL]

ci'.tre.a. M.L. adj. *citrea* lemon-yellow.

The species displays the characters of the hitherto monotypic genus. See also "Further Descriptive Information."

Life cycle nocardioform, primary and aerial hyphae fragmenting.

Colors of colonies on oatmeal agar and inorganic salts-starch agar citron-yellow or white to whitish; on peptone-corn extract agar citron-yellow or cream; on the latter media in exceptional cases brown to orange-brown.

In some strains aerial hyphae only microscopically visible, but well developed in others; in the latter case these are thin, white, and chalky.

Optimal temperature 28°C; growth occurs between 6 and 42°C. Physiological features described above.

Table 26.24.
Characteristics distinguishing **Promicromonospora** *from related organisms*[a]

Taxon	Formation of aerial hypae	Susceptible to P phages[b]	Susceptible to O phages[b]	Anaerobic growth on TSA[c]	Peptidoglycan type
Promicromonospora citrea	+	+	−	−	L-Lys-Ala_2[d]
Oerskovia turbata	−	−	+	+	L-Lys-L-Thr-D-Asp or L-Lys-L-Thr-D-Glu[e]
Oerskovia xanthineolytica	−	−	+	+	L-Lys-D-Ser-D-Thr[e]
Promicromonospora enterophila	−	−	+	+	Lys, Ala, Glu, Thr[f]
Cellulomonas cartae	−	−	+	+	L-Lys-D-Ser-D-Asp[g]
Nonmotile organisms[h]	−	−	+	+	
Cellulomonas spp.	−	−	−	−	L-Orn-D-Glu or L-Orn-D-Asp[i]

[a]Symbols: see Table 26.22.
[b]*Promicromonospora*– and *Oerskovia*-specific phages, respectively (Prauser, 1984c, 1986).
[c]Tryptic-soy agar (Lechevalier, 1972).
[d]Stackebrandt et al. (1983c) and Stackebrandt and Schleifer (1984); peptidoglycan of the type L-Lys-Ala-Glu was also reported (Evtushenko et al., 1984b).
[e]Seidl et al. (1980).
[f]Jager et al. (1983); molar ratios in cell wall hydrolysates 1:1:1.08:0.3.
[g]Stackebrandt et al. (1978).
[h]*Sensu* Lechevalier (1972).
[i]Fiedler and Kandler (1973).

Besides characteristics given in Table 26.24, the following features were found to be of value in presumptive differentiation of *P. citrea* from related strains: utilization of malate, malonate, and succinate; production of acid on rhamnose, raffinose, and mannitol; and absence of clearing zones on yeast-peptone agar with $CaCO_3$.

Susceptible to taxon-specific phages. Cell wall chemotype VI. Menaquinone type MK-9(H_4). Mol% G + C above 70 (T_m).

Mainly isolated from soil, also found on aluminum hydroxide gel antacid.

Type strain: ATCC 15908 (INMI 18; RIA 562; VKM Ac 665; KCC A 0051; IMET 7267; DSM 43110).

Species Incertae Sedis

a. **Promicromonospora enterophila** Jager, Marialigeti, Hauck and Barabas 1983, 530.[VP]

en.ter.o'phi.la. Gr. neu. pl. n. *entera* innards, guts; Gr. adj. *philo* loving; M.L. fem. adj. *enterophila* gut loving.

Type strain: HMGB B1078 (DFA 19; see "Taxonomic Comments").

Genus **Intrasporangium** *Kalakoutskii, Kirillova and Krasil'nikov 1967, 79*[AL]

L. V. Kalakoutskii

In.tra.spo.ran'gi.um. L. prep. *intra* within; Gr. n. *spora* a seed; Gr. n. *angeion* a vessel; M.L. neut. n. *Intrasporangium* a name coined to emphasize the possibility of intercalary formation of sporangia in mycelial filaments.

Branching mycelium, about 1.0 µm in diameter, **has a tendency to break into fragments of various size and shape. Aerial mycelium never observed. Oval- and lemon-shaped vesicles** (5–15 µm in diameter) **are formed intercalary and/or at the hyphal apices**. In some of the vesicles (termed sporangia in the original description) in the older cultures one might distinguish up to several round or oval bodies (1.2–1.5 µm in diameter). These are nonmotile, but may undergo a brownian movement within the mature vesicles. When released onto fresh medium the sporelike cells will germinate, giving rise to a branching mycelium. Gram-positive. Non-acid fast. Chemo-organotrophic, having an **oxidative type of catabolism; possess catalase activity**. Aerobic. Grow best at 28–37°C, no growth at 45°C. **Prefer complex media**, especially **containing peptone** and **meat extract**. No growth on the majority of mineral synthetic media routinely employed for actinomycetes. The mol% G + C of the DNA is 68.2 (T_m). **Cell wall chemotype I** (L-diaminopimelic acid (DAP), glycine). **Peptidoglycan of the µ-DAP-Gly type. Phospholipids of the PIV type** (diagnostic of this is an unknown glucosamine-containing phospholipid). **Among cell fatty acids, straight-chain saturated and unsaturated predominate**. Menaquinones of the MK-8 type.

Type species: *Intrasporangium calvum* Kalakoutskii, Kirillova and Krasil'nikov 1967, 79.

Further Descriptive Information

A study of *Intrasporangium* vesicles employing electron microscopy of thin sections failed to reveal spores within them. The vesicles were said to be the result of hyphal swelling and gradual disorganization in response to environmental stress (Lechevalier and Lechevalier, 1969). The cells of *Intrasporangium*, however, were reported (Sukapure et al., 1970) to survive heating at 60°C for 4 h in aqueous suspensions.

The cell wall chemotype is type I (L-DAP, glycine) (Prauser, 1967, personal communication; Sukapure et al., 1970). The peptidoglycan has glycine in the interpeptide bridge (Schleifer and Kandler, 1972). Phospholipids are type PIV (Lechevalier et al., 1977). Among the fatty acids, straight-chain saturated and unsaturated predominate. Mycolic acids, *iso-* and *anteiso-* branched, and 10-methyl-branched fatty acids are not found (Kützner, 1981).

Isolation

The original strain was isolated under nonselective conditions on plates of meat-peptone agar exposed to the atmosphere of a school dining room.

Maintenance Procedures

Survives routine procedure of lyophilization.

Differentiation of the genus **Intrasporangium** *from other genera*

The genus *Intrasporangium* can be separated from several genera of aerobic mycelium-forming actinomycetes having cell wall chemotype I and sometimes assigned to the family *Streptomycetaceae* (Pridham and Tresner, 1974a) on the basis of morphology as well as phospholipid composition (Lechevalier et al., 1977), fatty acid profiles (Kützner, 1981), menaquinone composition (Collins et al., 1984), and susceptibility to the genus-specific phages (Prauser and Falta, 1968; Wellington and Williams, 1981a). Levels of DNA/DNA homology of *I. calvum* with *Arthrobacter simplex, Nocardioides albus*, and *Streptomyces* species are very low (Akimov, unpublished results). Extensive numerical taxonomic studies (Williams et al., 1983a) point to a distinct and separate position of *Intrasporangium* if compared to actinomycetes of the above group.

The formation of an extensive mycelium and intramycelial vesicles, as well as fatty acid spectra (Kützner, 1981) and phospholipid composition (Lechevalier et al., 1977), separate *Intrasporangium* from aerobic nocardioforms and coryneforms (Collins et al., 1983b; O'Donnell et al., 1982a; Suzuki and Komagata, 1983), which are characterized by high mol% G + C in their DNA and L-DAP in the cell walls.

Taxonomic Comments

The taxonomic position of *Intrasporangium* among other actinomycetes has been briefly discussed by Lechevalier and Lechevalier (1981a) as well as by Kützner (1981). The latter author suggested that *Intrasporangium* might even not belong to the whole group of *Actinomycetales*.

Clearly, a comparative study employing more isolates of *Intrasporangium* followed by examination of phylogenetic markers as well as DNA/DNA hybridization in a selected spectrum of strains is in order to understand better the relationship of *Intrasporangium* to other mycelial Gram-positive procaryotes.

Further Reading

Kalakoutskii, L. V., I.P. Kirillova and N.A. Krasil'nikov. 1967. A new genus of the Actinomycetales—*Intrasporangium* gen. nov. J. Gen. Microbiol. *48:* 79–85.

Lechevalier, H.A. and M.P. Lechevalier. 1969. Ultramicroscopic structure of *Intrasporangium calvum* (Actinomycetales). J. Bacteriol. *100:* 522–525.

Lechevalier, H.A. and M.P. Lechevalier. 1981. Actinomycete genera "in search of a family." *In* Starr, Stolp, Trüper, Balows and Schlegel (Editors). The Prokaryotes. A Handbook of Habitats. Isolation and Identification of Bacteria. Springer-Verlag, New York, pp. 2118–2123.

Sukapure, R.S., M.P. Lechevalier, H. Reber, M.L. Higgins, H.A. Lechevalier and H. Prauser. 1970. Motile nocardioid *Actinomycetales*. Appl. Microbiol. *19:* 527–533.

Williams, S.T., M. Goodfellow, G. Alderson, E.M.H. Wellington, P.H.A. Sneath and M.J. Sackin. 1983. Numerical classifications of *Streptomyces* and related genera. J. Gen. Microbiol. *129:* 1743–1813.

Differentiation of the species of the genus Intrasporangium

Intrasporangium calvum Kalakoutskii, Kirillova and Krasil'nikov 1967, 79.[AL]

cal'vum. L. neut. adj. *calvum* bald (referring to the absence of aerial mycelium).

Only this single species is recognized so far. See generic description for most of the features.

The organism grows rather slowly even on peptone-containing media. At 28°C, macroscopically visible colonies appear in 3–5 days of incubation. Colonies on meat extract-peptone agar are round, glistening, and whitish, becoming creamy on aging. The colonial material is viscous, a characteristic that becomes apparent when transfers are being made using wire loop or pipettes. On microscopical examination branching mycelium and intramycelial vesicles can be seen to extend beyond the colonial periphery (Fig. 26-34) onto and into agar.

The vesicles begin to appear following 5–6 days of cultivation on solid media, but are rare under conditions of submerged cultivation. No turbidity develops on cultivation of *I. calvum* in liquid media; growth usually occurs in a sediment form. In liquid cultures, very thin (about 0.1 μm in diameter) mycelium-like threads are occasionally seen, but their fate and origin remain obscure. Vesicles (termed sporangia in the original description) are abundant in older cultures (Figs. 26.35 and 26.36). While some of them have an "empty" appearance in older cultures, in others the progression of differentiation results in formation of round to oval spore-like bodies (Fig. 26.36). These differ from the surrounding sporangial material in refractility and susceptibility to basic dyes, and germinate by one or two germ tubes upon transfer to fresh media.

Temperature range for *I. calvum* grown on Bennett's agar was reported to be 10–42°C; that for another *Intrasporangium* isolate L-12-17 (which might belong to another species) on soil-extract medium was 24–42°C (Sukapure et al., 1970).

Growth of *I. calvum* on complex media is not enhanced upon enrichment of the atmosphere by H_2 and/or CO_2. Nitrates reduced to nitrites. Physiologically not very active.

Intrasporangium calvum was reported (Williams et al., 1983a) to utilize L-arginine, L-cysteine, L-methionine, L-phenylalanine, and L-serine as a sole nitrogen source (0.1% w/v) in a basal medium containing glucose and mineral salts. On ISP medium 9 the following compounds were able to support growth if added (1.0% w/v) as sole carbon compounds: D-glucose, cellobiose, D-fructose, mannitol, L-rhamnose, salicin, trehalose, D-xylose, and sodium pyruvate (0.1% w/v). Degradation of esculin, arbutin, casein, gelatin, elastin, Tween 80, and DNA was noted on modified Bennett's agar and Bacto DNase test agar (Difco). On modified Bennett's agar growth was possible in the presence of any of the following: phenylethanol (0.1% w/v), potassium tellurite (0.01% w/v), thallous acetate (0.001% w/v), cephaloridine (100 μg/ml), gentamicin (100 μg/ml), neomycin (50 μg/ml), or tobramycin (50 μg/ml).

No antibiotic activity was found in standard tests against Gram-positive and Gram-negative bacteria, yeasts, and mycelial fungi.

Type strain: ATCC 23552 (KIP-7; VKM Ac 701; IFO 12982; DSM 43043).

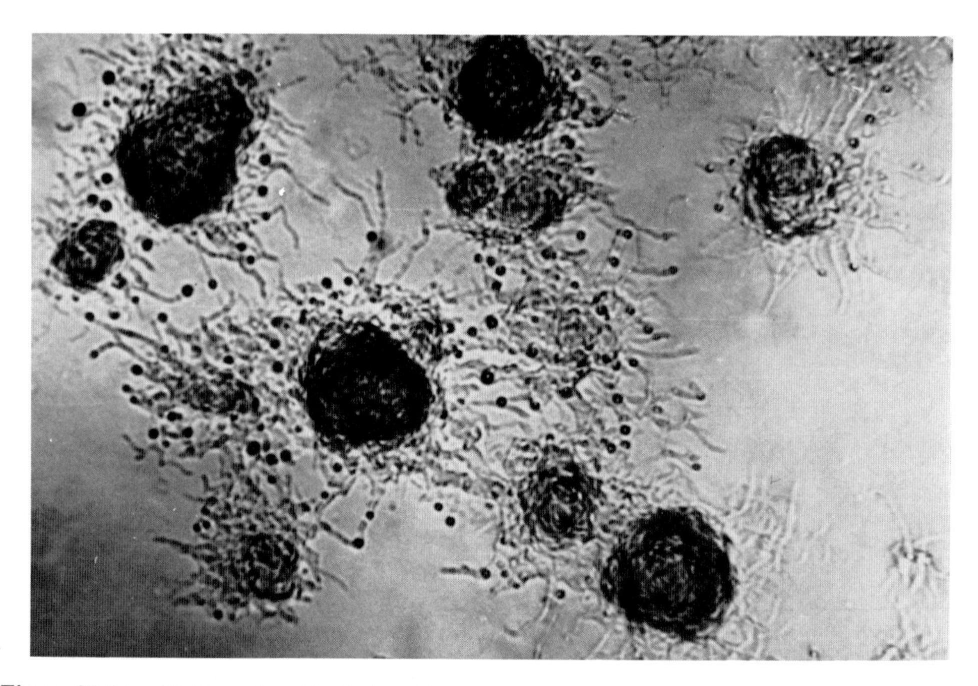

Figure 26.34. Eighteen-day-old colonies of *Intrasporangium calvum* on meat-peptone-glycerol agar (× 300). Many vesicles on mycelial branches ramifying from the colonies' edges are distinguishable. (Reproduced with permission from L. V. Kalakoutskii, I. P. Kirillova and N A. Krasil'nikov, Journal of General Microbiology *48:* 79–85, ©Society for General Microbiology, 1967.)

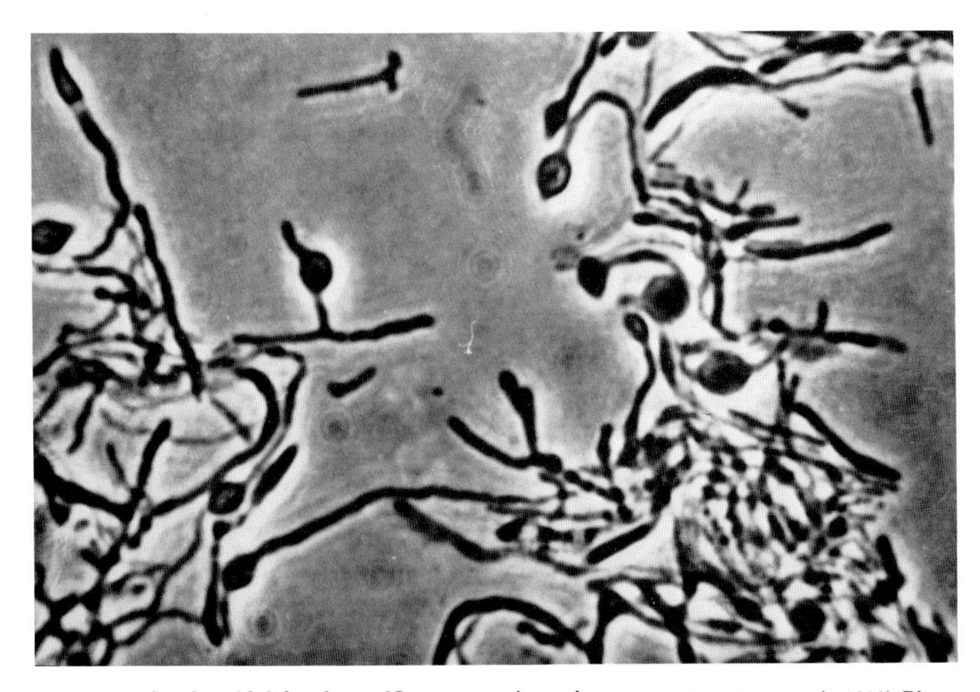

Figure 26.35. Six-day-old slide culture of *Intrasporangium calvum* on meat-peptone agar (× 4000). Phase contrast. Branching mycelium and intramycelial vesicles in both terminal and intercalary position. (Reproduced with permission from L. V. Kalakoutskii, I. P. Kirillova and N. A. Krasil'nikov, Journal of General Microbiology *48:* 79–85, ©Society for General Microbiology, 1967.)

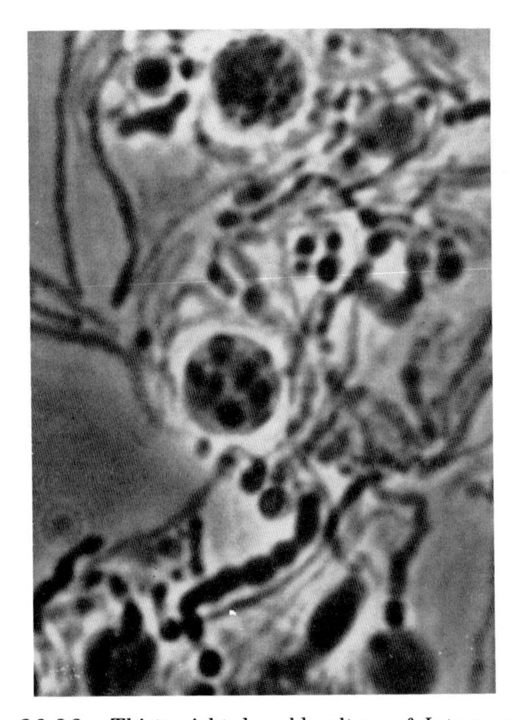

Figure 26.36. Thirty-eight-day-old culture of *Intrasporangium calvum* on meat-peptone agar (× 5000). Phase contrast. Contact preparation. Fragmented mycelium and vesicles with differentiated protoplasm are seen.

Genus Actinopolyspora Gochnauer, Leppard, Komaratat, Kates, Novitsky and Kushner 1975, 1510[AL]

MARGARET B. GOCHNAUER, KENNETH G. JOHNSON, and DONN J. KUSHNER

Ac'ti.no.po.ly.spo'ra. Gr. n. *actis, actinos* ray; Gr. adj. *poly* many; Gr. n. *spora* a seed; M.L. fem. n. *Actinopolyspora* the many-spored ray (fungus).

Branched filaments form extensive substrate mycelium about 1 μm in diameter. Substrate mycelium is mostly unfragmented; fragmentation is occasionally observed near the colony center. **Sporophores containing 20 or more smooth-walled coccobacillary and coccoid spores are produced basipetally on aerial hyphae.** Spores are not observed on substrate mycelium. The organisms are Gram-positive and acid fast. The wall contains *meso*-diaminopimelic acid (*meso*-DAP) with arabinose and galactose. No mycolic acids present. Growth is aerobic. Chemo-organotrophic, utilizing a number of different carbon sources. The one known species is **extremely halophilic**. The mol% G + C of the DNA is 64.2 (spectrophotometric determination).

Type species: *Actinopolyspora halophila* Gochnauer, Leppard, Komaratat, Kates, Novitsky and Kushner 1975, 1510.

Further Descriptive Information

This information deals mainly with the originally described strain of the species, *A. halophila* designated wild type (WT), with some details of a mutant strain ER recently isolated from WT, which is resistant to erythromycin.

Cell Morphology

On agar plates of complex medium (CM)*, branched filaments form extensive substrate mycelium about 1 μm in diameter. Substrate mycelium is mostly unfragmented; fragmentation is occasionally observed near the colony center. Sporophores containing 20 or more smooth-walled coccobacillary and coccoid spores are produced basipetally on aerial hyphae. Spores are not present on substrate mycelium.

Cell Wall Composition

Cell walls contain a peptidoglycan with glutamate, alanine, and *meso*-DAP in a molar ratio of 1:2:1. Equimolar quantities of galactose and arabinose are also present, corresponding to a type IV cell wall (Lechevalier et al., 1971a, 1973b). Almost all the available glycan disaccharide units are peptide substituted in the walls of both the WT and ER strains; peptidoglycan cross-linkage is facilitated by a direct peptide linkage between N^{ϵ}-DAP and COOH-terminal alanine. Peptidoglycans from WT and ER strains are 50 and 67% peptide cross-linked, respectively. The WT strain contains 15.7% non-*N*-substituted muramic acid and 35% non-*N*-substituted glucosamine; corresponding figures for the ER strain are 11 and 48.8%, respectively (Johnson et al., 1986a). The original results (Gochnauer et al., 1975) have been interpreted by Minnikin and Goodfellow (1980) to indicate an absence of mycolic acids. A more recent examination (Kates et al., 1987) also indicated the absence of mycolic acids. The organism contains a number of ester-linked phospholipids, as well as glycolipids and neutral lipids, but no ether-linked phospholipids (Ross et al., 1981). The phospholipids include phosphatidylcholine, *lyso*phosphatidylglycerol, and phosphatidylglyercol. The lipids contain mainly branched-chain fatty acids, especially *iso*-C_{15-17} and *anteiso*-C_{17} acids. Cells possess complex mixtures of isoprenoid quinones, the predominant menaquinone being MK-9(H_4) (Collins et al., 1981; Minnikin and Goodfellow, 1980). No tuberculostearic acid is present.

Fine Structure

(Fine structure was mainly studied by scanning electron microscopy (SEM) after critical point drying; transmission electron microscopy (TEM) studies were performed on whole cells.) The mycelium at the air-agar interface consists of long cylindrical filaments about 1 μm in diameter, terminating in a rounded tip (Fig. 26.37). No cross-walls appear on vegetative filaments, and branching occurs at either right or acute angles to the main filament. Not all branches are the same diameter as the main filament, and the irregular branching pattern includes some opposite branching. Frequently, a filament has several bends or pronounced curvature; some filaments are helical. The sporophores are straight and long (sometimes more than 20 spores) with the spores in single file (Fig. 26.38). Spores vary in size and shape within a single sporophore; some spores are separated by a gap or small plug (Fig. 26.39). Some (possibly immature) sporophores are branched. All spores have a smooth surface; no surface sculpturing or appendages are present (Fig. 26.39). The shape varies from spherical to a short rod with rounded ends. The spore diameter varies from 0.7 to 1.0 μm, and the length varies from 1.0 to 2.0 μm. The spore morphology of cultures grown on agar media containing 15% or 12.5% (w/v) NaCl is similar to those grown with 20% (w/v) NaCl. On 15% NaCl medium, helical filaments are particularly frequent. Spores are absent on 25% (w/v) NaCl medium and infrequent on 10% NaCl medium. Vegetative filaments on 10% (w/v) NaCl frequently form ropelike aggregates (Fig. 26.40), but those on 30% (w/v) NaCl medium have little tendency to aggregate.

Colonial and Cultural Characteristics

The organism grows as a mat on the surface of a stationary liquid medium. Phase-contrast examination of this material reveals a mycelial mode of growth. On solid media growth occurs in the presence of 10–30% NaCl, with optimal growth between 15 and 20% NaCl. No growth occurs on 5% NaCl medium (Fig. 26.41*A–F*). Extensive mycelial formation is seen in microcolonies before production of aerial mycelium. Substrate mycelia do not penetrate the agar surface and do not produce spores. Colonies have a wrinkled appearance and their aerial mycelia are white. The substrate mycelia of colonies growing at the highest NaCl concentrations are buff and become progressively darker as salt concentrations decrease, being black on 10% NaCl medium. At the lower salt concentrations a brown pigment diffuses into the media, but no diffusion of pigment is seen at higher salt concentrations. Examination of cultures on coverslips (after 7 days on CM plus 20% NaCl) demonstrates occasional fragmentation of substrate mycelium into segments 2–3 μm long.

Nutrition and Growth Conditions

Actinopolyspora halophila grows best on CM (Sehgal and Gibbons, 1960; Gochnauer and Kushner, 1969). The organism can utilize glucose, xylose, fructose, sucrose, rhamnose, raffinose, mannitol, glycerol, isopropyl alcohol, sodium succinate, and sodium citrate as carbon sources for growth, and probably other substances as well, with NH_{4+} salts as nitrogen sources in mineral medium or added to Difco yeast nitrogen base medium containing $(NH_4)_2SO_4$. Optimum growth is at 30°C; maximum

*Complex medium (all ingredients except Fe^{2+} given in grams per liter): Difco casamino acids (vitamin free), 7.5; Difco yeast extract, 10.0; $MgSO_4 \cdot 7H_2O$ 10.0; sodium citrate, 3.0; KCl, 1.0; NaCl, 200; Fe^{2+}, prepare 4.98% $FeSO_4 \cdot 7H_2O$ (10,000 ppm Fe^{2+}; 0.001 N in HCl; add 1 ml/liter medium). Adjust to pH 6.8 and sterilize at 15 lbs pressure for 10 min. For tryptone medium substitute 1.5 g Difco tryptone for casamino acids–yeast extract. Also, 27.5 g Difco trypticase-soy broth can be substituted for casamino acids and yeast extract. All support good growth.

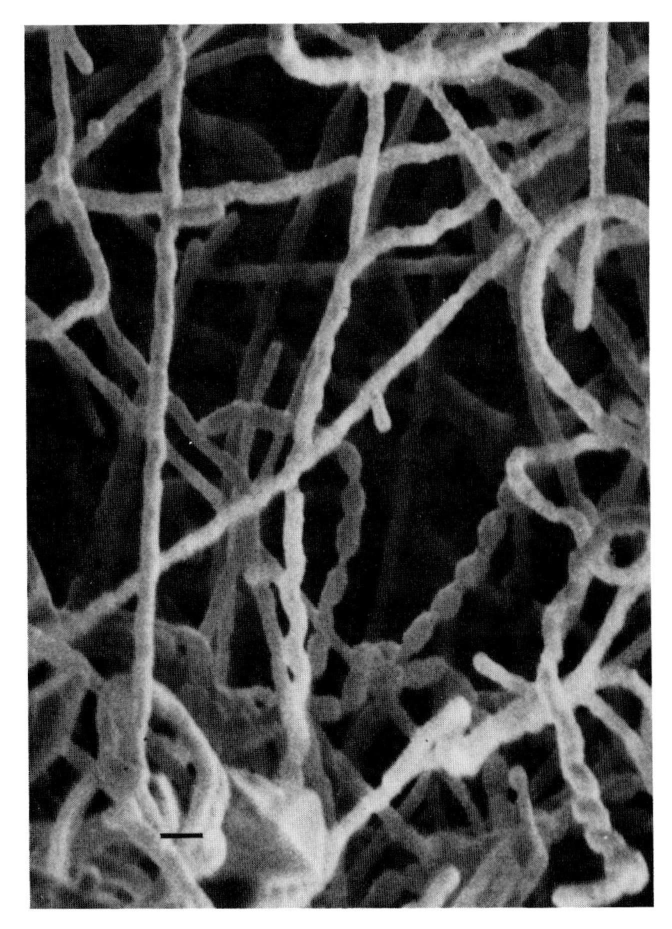

Figure 26.37. Filaments grown on 20% (w/v) NaCl. SEM. *Bar*, 1 μm.

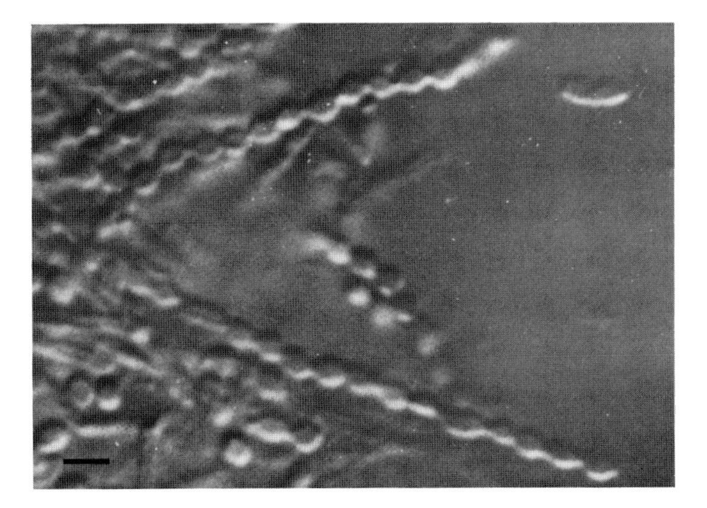

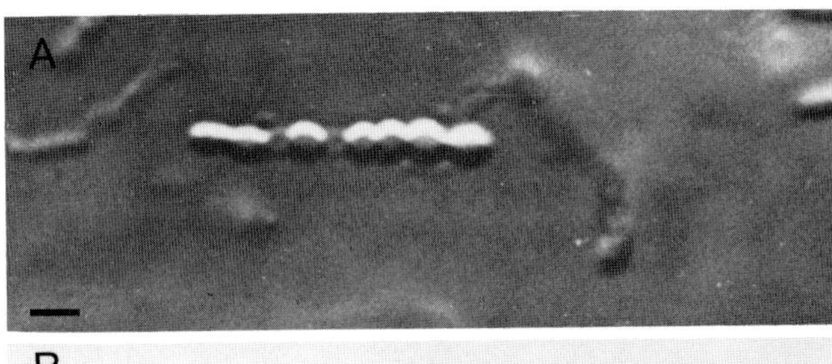

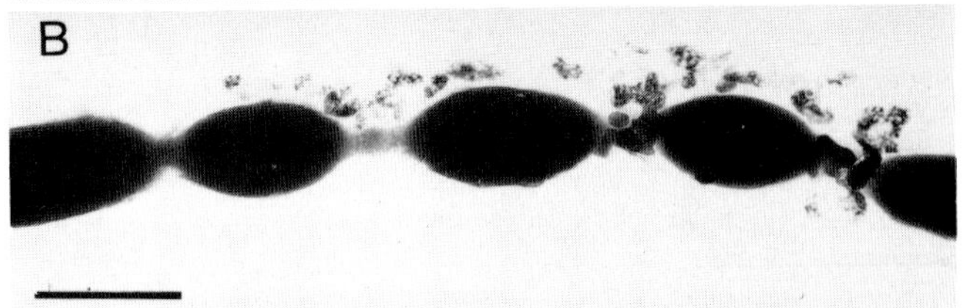

Figure 26.39. *A*, sporophore grown on 20% (w/v) NaCl. Nomarski optics. *Bar*, 2 μm. *B*, sporophore grown on 20% (w/v) NaCl. TEM. *Bar*, 1 μm.

temperature for growth is 42°C, and slow but measureable growth is observed at 10°C.

Strain WT is extremely halophilic, needing at least 12% NaCl in liquid medium, and 10% in agar; it grows optimally in 15–20% NaCl and can grow in 30%. In contrast, the mutant *A. halophila* ER can also grow in 6% (w/v) liquid CM. However, both strains can be washed in distilled water without loss of viability. NaCl cannot be effectively replaced by KCl. On CM agar the WT strain grows and sporulates from pH 6.0 to 8.6. Strain ER can grow from pH 5.0 to 8.6 (poorly at the higher pH); sporulation is slow and occurs only from pH 6.0 to 7.0.

A. halophila grows best in aerobic conditions. If stab-inoculated into soft agar media it grows only at the surface. Growth in liquid media is proportional to the degree of aeration; in stationary culture, growth occurs on the surface as a wrinkled hydrophobic pellicle. It has a very active catalase. It produces extracellular α-amylase, protease, endocellulase, xylanase, phospholipase C, cell-wall lytic activity (tested against purified *Micrococcus lysodeikticus* cell wall), β-galactosidase, β-glucosidase, β-lactamase, phosphatase, and 6-phospho-β-galactosidase activities (Johnson et al., 1986b).

Genetics

Erythromycin resistance in strain ER is genetically stable (Johnson et al., 1986b). Extraction of vegetative cells and subsequent agarose gel electrophoresis (Clewell and Helinski, 1969) did not disclose any plasmids. Plaques that appear most frequently in colonies grown at lower NaCl concentrations suggest the presence of bacteriophages, although such phages have not yet been demonstrated (Gochnauer et al., 1975).

Antibiotic Sensitivity

Patterns of sensitivity are shown in Table 26.25. As noted, both strains are resistant to β-lactam antibiotics and to a number of other antibiotics. The WT strain of *A. halophila* possesses both extracellular and cell-associated β-lactamase activity; enzymes of both kinds have been purified to molecular homogeneity (Johnson and Lanthier, 1986).

Enrichment and Isolation Procedures

The microorganism first appeared as a contaminant in CM containing 25% NaCl from solar salt, and was isolated by plating on this medium. It may well have been a contaminant of solar salt, which suggests that the organism might occur in salt ponds (salterns) where sea salt is harvested. It might also occur in soils and muds juxtaposed to salterns and salt lakes, but so far isolates have not been obtained from such habitats.

A suggested isolation procedure is as follows. Wash cells thoroughly with distilled water to eliminate osmotically sensitive halobacteria and other organisms. Collect cells on membrane filters and incubate them on an agar medium containing 20% NaCl and the basic minerals given for

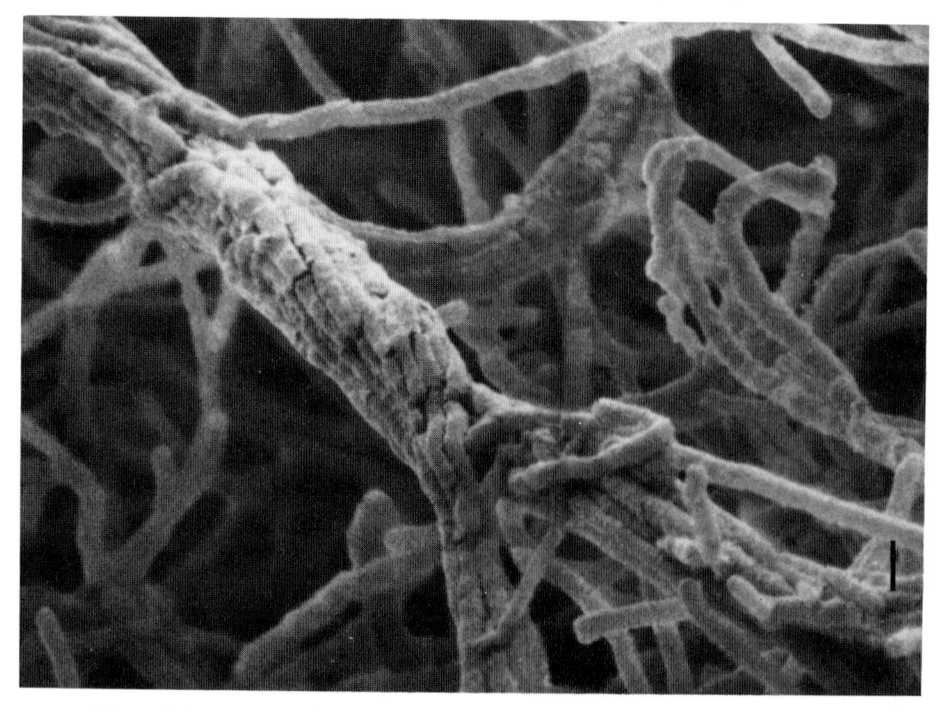

Figure 26.40. Aggregate of filaments grown on 10% (w/v) NaCl. SEM. *Bar*, 1 μm.

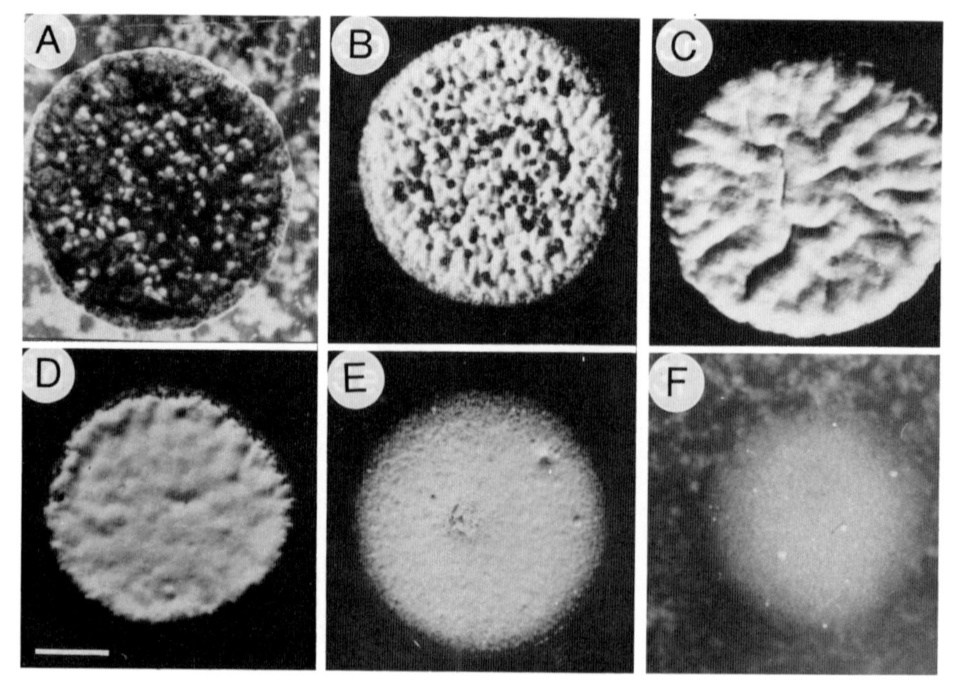

Figure 26.41. Effect of NaCl concentrations on growth of colonies on CM agar. Incubation: 6 days at 37°C. NaCl concentrations (% w/v): *A*, 10; *B*, 12.5; *C*, 15; *D*, 20; *E*, 25; *F*, 30. *Bar*, 1 cm.

Table 26.25.
Antibiotic senstivity of **Actinopolyspora** *strains*[a]

Antibiotic	Concentration (units)	Strains[b]	
		WT	ER
Amoxicillin	30	S	S
Bacitracin	10	S	S
Cefamandole	30	S	R
Cefoxitin	30	S	S
Chloramphenicol	30	S	S
Clindamycin	2	S	R
Erythromycin	15	S	R
Novobiocin	30	S	S
Vancomycin	30	S	S

[a]Paper disks impregnated with the indicated concentrations of antibiotics (General Diagnostics) were placed on lawn cultures growing on CM containing 20% (w/v) NaCl. Incubation at 37°C for 96 h.
[b]*R*, resistant; *S*, sensitive.

CM, plus low concentrations of yeast extract (1–2% w/v) and casamino acids (0.1%). Include cycloheximide to inhibit the growth of salt-tolerant or halophilic eucaryotic cells. Adjust pH to 6.5–7.0, and incubate at 35–40°C. Plates should be enclosed in plastic bags to provide a humid atmosphere. For enrichment, stationary liquid cultures in the same medium are incubated for up to 6 weeks. Typical white mycelial growth on the surface and on the walls at the air-liquid interface should be examined, diluted, and plated on solid medium of the same formula (2% agar), until pure cultures are obtained.

Maintenance Procedures

Cultures can be maintained for at least 1 year in screw-capped tubes on 15–20% NaCl CM agar slopes at 5°C; cultures on CM agar can also be frozen at −70°C and maintained at that temperature. On thawing, cultures should be grown on fresh CM agar for three successive 72–96-h periods before being used for experiments.

For longer term preservation cultures may be frozen in liquid nitrogen and stored at −70°C. They may also be lyophilized. For the latter procedure, cells are grown to the beginning of the stationary phase, centrifuged, and suspended in the following medium: Dextran T-500 (Pharmacia), 5 g; sucrose, 9.5 g; monosodium glutamate, 1.0 g; NaCl, 20 g; and distilled water to a final volume of 100 ml. After the ingredients are dissolved, the medium is dispensed into tubes and sterilized at 121°C for 15 min. After resuspending cells in this medium, 1–2 ml suspension per ampule are frozen in acetone-dry ice. Because of the high NaCl concentration it is especially important to coat the inside of the ampule with a very thin layer of frozen material before applying a vacuum, to ensure complete lyophilization.

Differentiation of the genus **Actinopolyspora** *from other genera*

The genus *Actinopolyspora* seems most closely related to the genus *Saccharopolyspora*, which was originally isolated from sugar cane bagasse. These two genera are grouped together in the micropolysporas by Goodfellow and Cross (1984b), together with the genera *Pseudonocardia* and *Saccharomonospora*, as well as some species previously misclassified in the genera *Nocardia* and *Streptomyces*. All the micropolysporas have type IV cell walls and lack mycolic acids. The lipid composition of *Actinopolyspora* and *Saccharopolyspora* is very similar. They differ in the degree of branching of substrate mycelium and, especially, in that the substrate mycelium of *Saccharopolyspora* fragments into rod-shaped elements, whereas that of *Actinopolyspora* rarely fragments. The most striking difference between the two genera is their mol% G + C content: 64.2 in *Actinopolyspora* (Frédéricq et al., 1961) and 77.0 in *Saccharopolyspora*.

The only species of *Actinopolyspora* is an extreme halophile, requiring 10–12% NaCl for growth. However, a mutant, probably resulting from a single mutation, needs only 6% NaCl for growth.

Acknowledgments

The authors thank Mr. Roger Latta, Curator of the Culture Collection of the National Research Council of Canada, Ottawa, Canada, for his expertise and assistance in preservation of the above cultures.

List of species of the genus **Actinopolyspora**

1. **Actinopolyspora halophila** Gochnauer, Leppard, Komaratat, Kates, Novitsky and Kushner 1975, 1510.[AL]

hal.o.phi′la. Gr. n. *hals* salt; Gr. adj. *philos* loving; M.L. fem. adj. *halophila* salt loving.

The most unusual characteristic of *A. halophila* is its absolute requirement for at least 10–12% NaCl for growth, with an optimal growth concentration of 20% NaCl. When grown in liquid shake culture in CM supplemented with 1% (w/v) starch and 20% NaCl the organism has a pineapple-like odor. On plate culture containing organic nutrients such as casein and yeast extract, 1% (w/v) tryptone, or 1% (w/v) trypticase-soy, the odor is musty.

Nitrate is not used as a terminal electron acceptor for anaerobic growth; an assimilatory nitrate reductase is present. Grows, although poorly, in mineral medium using KNO_3 as the sole nitrogen source and sodium succinate or sodium citrate as carbon source. Nitrate is reduced to NH_3. Produces extracellular gelatinase, caseinase, and lipases that hydrolyze Tweens 20, 40, 60, and 80.

Although growth is optimal in 20% NaCl in CM, the cells can be washed and resuspended in distilled water without loss of viability. If they are allowed to stand in liquid CM plus 20% NaCl they rapidly lyse (Johnson et al., 1986b).

The mol% G + C of the DNA is 64.2.

Type strain: ATCC 27976.

a. *Actinopolyspora halophila ER*

This mutant was found during antibiotic sensitivity screening tests as resistant colonies growing in the sensitive zone for *A. halophila* of 15 μg/ml erythromycin (Johnson et al., 1986b). Its colonial and cell morphology is indistinguishable from the parent (WT). It has the same optimum growth temperature and grows over the same range of NaCl concentrations as WT except that the minimal NaCl concentration required for growth is 6%, whereas that for the type species is 12% (w/v). *A. halophila* ER is resistant to two antibiotics to which WT is sensitive (Table 26.25). Its optimal growth range is pH 6.0–7.0, but unlike WT it grows at pH 5.0.

The ER strain has a few more peptide cross-linkages in its peptidoglycan and is more resistant to lysozyme. Its membrane component lacks two minor protein bands with molecular weights of 49,000 and 13,000 that are present in the WT strain (Johnson et al., 1986a). *A. halophila* ER has been deposited as NRC 2140.

Genus **Saccharomonospora** *Nonomura and Ohara 1971c, 899*[AL]

ALAN J. MCCARTHY

Sac.cha.ro.mon'o.spo.ra. Gr. n. *sacchar* sugar; Gr. adj. *monos* single, solitary; Gr. fem. n. *spora* seed; M.L. fem. n. *spora* a spore; M.L. fem. n. *Saccharomonospora* the sugar (-containing) single-spored (organism).

Produce **predominantly single spores on aerial hyphae**. The spores are heat sensitive, nonmotile aleuriospores, formed at the tips of simple unbranched sporophores of variable length. On agar media, a branched vegetative mycelium forms leathery colonies, usually covered with aerial mycelium in which spores are densely packed along the hyphae. The aerial mycelium is initially white, becoming gray-green to dark green; green pigmentation may also occur on the vegetative mycelium and diffuse into the surrounding medium. Variants that are nonpigmented or lilac colored have also been observed. Spores on the vegetative hyphae and in pairs or short chains on the aerial hyphae are occasionally present. Gram-positive. **The cell wall contains *meso*-diaminopimelic acid (*meso*-DAP) together with the sugars arabinose and galactose. The cell envelope does not contain mycolic acids** but has major amounts of *iso*- and *anteiso*- fatty acids, phosphatidylethanolamine, and menaquinones that are tetrahydrogenated with nine isoprene units (MK-9(H_4)). Aerobic. Chemo-organotrophic. Amino acid and vitamin supplements (e.g. yeast extract) required for good growth. Growth is optimal in the temperature range 35–50°C, and pH range 7.0–10.0. **Growth is not inhibited by NaCl (3% w/v)**. Catalase, deaminase, and phosphatase are produced. Casein, gelatin, starch, xylan, and **tyrosine are degraded. There is no activity against cellulose.** A number of compounds can serve as sources of carbon, among which **the ability to utilize glycerol is characteristic**.

Members of the genus can be isolated from soil, lake sediments, and peat but are more common in manures, composts, and overheated fodders. The mol% G + C of the DNA is 69–74 (T_m).

Type species: *Saccharomonospora viridis* (Schuurmans, Olson and San Clemente 1956) Nonomura and Ohara 1971c, 899.

Further Descriptive Information

The vegetative mycelium is extensively branched and, in most cases, nonfragmenting. The aerial mycelium is usually abundant but can be absent in some strains. The spores are ovoid (0.9–1.1 × 1.2–1.4 μm), with a warty surface, and are rapidly killed at 70°C. Development of a green coloration on the aerial mycelium is an almost universal character but is usually absent on colonies cultured by prolonged incubation at a low temperature (30–35°C). Colonies that lack aerial mycelium and are not pigmented may produce a light brown soluble pigment. There has been considerable disagreement over the occurrence of spore chains and substrate mycelium spores in strains that can now be assigned to the genus *Saccharomonospora*. These morphological features are absent in most *Saccharomonospora* strains and, even when recorded, are not abundant. Misinterpretation of microscopic observations may be a factor (Locci, 1971) but, in any case, the presence of pairs or chains of spores and sporulation on the substrate mycelium should not be used to exclude organisms from this genus.

The cell wall contains *meso*-DAP and an arabinogalactan (wall type IV), which distinguishes *Saccharomonospora* strains from other monosporic actinomycetes. Furthermore, the absence of mycolic acids in the cell envelope places *Saccharomonospora* in a small, poorly defined group of actinomycetes that have a wall chemotype IV but lack mycolic acids (Goodfellow and Minnikin, 1984). Phospholipids detected in *Saccharomonospora* strains include phosphatidylethanolamine, phosphatidylinositol, phosphatidylinositol mannosides, acyl phosphatidylglycerol, and diphosphatidylglycerol, corresponding to a phospholipid type II pattern (Lechevalier et al., 1981; Goodfellow and Cross, 1984b). The fatty acid profile is a mixture of *iso*- and *anteiso*- branched- and straight-chain unsaturated fatty acids (Kroppenstedt and Kutzner, 1978; Goodfellow and Cross, 1984b). Menaquinones with 10 or 11 isoprene units are absent from the cell envelope, which contains major amounts of MK-9(H_4) (Collins et al., 1982d).

Saccharomonospora strains are moderately thermophilic, showing good growth at 50°C but little or no growth at 60°C. In some strains, aerial mycelium production and sporulation are improved at 40–45°C, and most strains can grow down to 30°C. *Saccharomonospora* strains are not particularly fastidious in their pH requirements but generally exhibit a preference for neutral to alkaline pH conditions.

Saccharomonospora strains can utilize D-glucose, D-galactose, D-mannose, D-xylose, maltose, and glycerol as carbon sources. Most strains can utilize D-fructose and trehalose but none can grow on sucrose, lactose, or melezitose (McCarthy and Cross, 1984a). Enzymes produced by *Saccharomonospora* strains include catalase, phosphatase, lipase, xylanase, and amylase. Cellulases, pectinase, chitinase, and oxidase are not produced. Proteolytic activity is evidenced by the ability to degrade casein and gelatin and, in most strains, elastin and keratin. Activity against guanine, hypoxanthine, and xanthine is absent and, although tyrosine is degraded, this does not result in the formation of melanin. Nitrate is not reduced. Tolerance to NaCl often extends up to a concentration of 7% (w/v) in the medium but no resistance is exhibited to a range of antibiotics, including novobiocin (McCarthy and Cross, 1984a). *Saccharomonospora* strains are sensitive to lysozyme, which can be used to prepare stable protoplasts (A.J. McCarthy, unpublished results).

Three antibiotics produced by *Saccharomonospora* strains have been described. The first, thermoviridin, was produced by the original isolate on which the genus is founded (Schuurmans et al., 1956). The remaining two are produced by strains that are here considered as belonging to this genus—an unnamed antitumor agent from "*Micropolyspora coerulea*" (Preobrazhenskaya et al., 1973) and the antibacterial antibiotic primycin, from "*Thermomonospora galeriensis*" (Szabo et al., 1976).

There is no available information on the genetics of *Saccharomonospora* strains, but a number of bacteriophages have been isolated (Prauser, 1984c).

Saccharomonosporas form part of the thermophilic actinomycete population in overheated substrates. They have been isolated from hay, grain, cotton, bagasse, and composts (Lacey, 1973, 1981; McCarthy and Cross, 1984a). They are also present as spores in low-temperature environments and have been isolated from soil, lake muds, peat, and a variety of dusts (Küster and Locci, 1963a; Nonomura and Ohara, 1971c; Johnston and Cross, 1976; Kurup, 1981). *Saccharomonospora* strains are likely to have only a limited role in the primary degradation of plant material by attacking hemicellulose (McCarthy et al., 1985) but not cellulose, lignin, or pectin (McCarthy and Broda, 1984; McCarthy and Cross, 1984a). Their growth in these environments does, however, result in the release of spores that can cause hypersensitivity pneumonitis. On the basis of immunological studies, *Saccharomonospora viridis* is strongly implicated as one of the causative agents of farmer's lung disease (Wenzel et al., 1974; Roberts et al., 1976a; Treuhaft et al., 1980).

Enrichment and Isolation Procedures

Saccharomonospora strains can be isolated by the methods used to recover thermophilic actinomycetes in general. The most efficient isolation methods are those based on the use of a sedimentation chamber and Andersen air sampler (Lacey and Dutkiewicz, 1976a; McCarthy and Broda, 1984). Dried samples are agitated within the chamber to create an aerosol of particles that, after 1–2 h of sedimentation, still contains many actinomycete spores and comparatively few bacteria. Viable actinomycetes are isolated from this spore suspension using an Andersen sampler loaded with half-strength tryptone-soy agar plates with cycloheximide (50 µg/ml) routinely incorporated to prevent the growth of fungi. *Saccharomonospora* strains appear as discrete colonies after 3–5 days at 50°C, and can be recognized by the production of a white aerial mycelium that becomes gray-green to dark green. They may also appear as minute, white colonies on agar plates seeded with dilutions of soil or freshwater sediments and incubated at 30°C for 3 weeks (Johnston and Cross, 1976). Improved recovery of *S. viridis* has been reported when a selective method, incorporating a dry heat treatment and addition of antibiotics, was used (Nonomura and Ohara, 1971c) or, more specifically, when rifampicin (5 µg/ml) was added to the isolation medium (Athalye et al., 1981). Mixtures of moist organic material and soil, incubated in partially sealed polythene bags at 50°C for 7 days, produce samples enriched in thermophilic actinomycetes, including saccharomonosporas (A.J. McCarthy, unpublished results).

Maintenance Procedures

Strains can be maintained as sporulating cultures on Czapek-Dox yeast extract–casamino acids (CYC) agar (see below), stored at 4°C, and subcultured every 4 weeks. For long-term storage, spore suspensions in 10–20% (v/v) glycerol are stable at ≤ -20°C and can also be used as a routine source of inocula. Specialized procedures are not required for the preparation and recovery of lyophilized stains.

Procedures for Testing of Special Characters

Morphology

The morphological features of sporulating colonies are best observed using a microscope fitted with a $\times$ 30 or $\times$ 40 long working distance objective lens. Cultures are grown on CYC agar (Cross and Attwell, 1974) that contains: Czapek-Dox liquid medium (Oxoid), 33.4 g; yeast extract, 2.0 g; casamino acids (Difco), 6.0 g; agar, 18.0 g; and distilled water, 1 liter (pH 7.0–8.0). Strains are incubated at 40–50°C for up to 7 days.

Degradative Tests

Various nutrient media are used (see McCarthy and Cross, 1984a) and insoluble test substrates should be autoclaved separately and added to sterile media as homogeneous suspensions. Prolonged incubation (up to 14 days) may be required to observe zones of clearing, and steps should be taken to prevent desiccation when incubated at 50°C.

NaCl Tolerance

The recommended test medium is CYC agar containing 3, 5, and 7% (w/v) NaCl. Cultures are incubated at 45–50°C for up to 7 days.

Carbon Source Utilization Test

The mineral salts medium contains: sodium nitrate, 2.0 g; potassium chloride, 0.5 g; magnesium glycerophosphate, 0.5 g; ferrous sulfate, 0.01 g; potassium sulfate, 0.35 g; Bacto-agar (Difco), 18.0 g; and distilled water, 1 liter (pH 7.5). The medium is supplemented with 0.5% (w/v) yeast extract (Oxoid). Carbon sources are sterilized by membrane filtration and added to the autoclaved medium to give a final concentration of 1% (w/v). Growth is compared with that on the basal medium alone (negative control) and with that on glucose basal medium (positive control).

Spore Heat Resistance

Sporulating cultures on agar slopes are suspended in 3 ml sterile distilled water and 0.5-ml aliquots are added to tubes containing 4.5 ml 0.1 M phosphate buffer (pH 7.5) preheated to 90°C in a water bath. After 30 min, the tubes are removed and placed in an ice bath for 10 min. Drops of control and heat-treated suspensions are placed on CYC agar plates and incubated for 3–5 days. The heat-treated suspensions do not contain any viable *Saccharomonospora* spores.

Differentiation of the genus **Saccharomonospora** *from other genera*

Table 26.26 lists the major characteristics that distinguish members of the genus *Saccharomonospora* from other mycelial procaryotes forming single, nonmotile spores.

Taxonomic Comments

The genus *Saccharomonospora* (Nonomura and Ohara, 1971c) was erected to accommodate monosporic actinomycetes whose walls contain *meso*-DAP and an arabinogalactan. Originally the single species, *S. viridis*, had been named "*Thermoactinomyces viridis*" (Schuurmans et al., 1956), but it was transferred to *Thermomonospora* and cited as "*Thermomonospora viridis*" (Küster and Locci, 1963b) in the eighth edition of the *Manual*. The reclassification of this species on the basis of its wall composition has since been supported by numerical phenetic data (Goodfellow and Pirouz, 1982b; McCarthy and Cross, 1984a). In the most recent classification of actinomycete genera, *Saccharomonospora* was classified in an aggregate group named Micropolysporas (Goodfellow and Cross, 1984b). These organisms have a similar wall composition and lack mycolic acids, but the relationships between genera have yet to be determined.

The morphological characteristics of *S. viridis* are shared by a number of species previously assigned to the genus *Micropolyspora*. Thermophilic actinomycetes that produced predominantly single spores and a green pigment were often identified as micropolysporas because occasional short chains of spores were observed on aerial and substrate hyphae. Most of the species names assigned to these strains were not included in the Approved Lists of Bacterial Names and need not be considered further. However, one species, *Micropolyspora internatus* (Agre et al., 1974), is nomenclaturally valid, although the application of several

Table 26.26.
Differential characteristics of the genus **Saccharomonospora** *and other monosporic genera*[a]

Characteristics	*Saccharomonospora*	*Thermomonospora*	*Micromonospora*	*Thermoactinomyces*
Aerial mycelium	+	+	−	+
Endospores	−	−	−	+
Glycine in cell wall	−	−	+	−
Sugars in whole-cell hydrolysates:				
Arabinose	+	−	+	−
Galactose	+	−	−	−
Xylose	−	−	+	−

[a]Symbols indicate presence (+) or absence (−) of characteristic in the genus as a whole.

taxonomic criteria has demonstrated that it is clearly a *Saccharomonsopora* strain (Kurup, 1981). In addition to the strains identified as micropolysporas, the type strain of "*Thermomonospora galeriensis*" (Szabo et al., 1976) should also be regarded as a *Saccharomonospora* strain (McCarthy and Cross, 1984a). Numerical phenetic studies (Goodfellow and Pirouz, 1982b; McCarthy and Cross, 1984a) have shown that the internal structure of *Saccharomonospora* is heterogeneous and may contain more than one species. These studies did not include the type strain of *M. internatus* and it is not clear whether this organism can be accommodated within *S. viridis* or represents the nucleus of an additional species (Kurup, 1981).

Further Reading

Cross, T. 1981. The monosporic actinomycetes. *In* Starr, Stolp, Trüper, Balows and Schlegel (Editors), The Prokaryotes, a Handbook on Habitats, Isolation and Identification of Bacteria. Springer-Verlag, Berlin, pp. 2091-2102.

Goodfellow, M. and T. Cross. 1984. Classification. *In* Goodfellow, Mordarski and Williams (Editors), The Biology of the Actinomycetes. Academic Press, London, pp. 7-164.

Kurup, V.P. 1981. Taxonomic study of some members of *Micropolyspora* and *Saccharomonospora*. Microbiologica (Bologna) *4*: 249-259.

McCarthy, A.J. 1986. Developments in the taxonomy and isolation of thermophilic actinomycetes. Front. Appl. Microbiol. *1*: 1-14.

McCarthy, A.J. and T. Cross. 1984. A taxonomic study of *Thermomonospora* and other monosporic actinomycetes. J. Gen. Microbiol. *130*: 5-25.

Lists of species of the genus **Saccharomonospora**

1. **Saccharomonospora viridis** (Schuurmans, Olson and San Clemente 1956) Nonomura and Ohara 1971c, 899.[AL] (*Thermoactinomyces viridis* Schuurmans, Olson and San Clemente 1956, 61.)

vir′i. dis. L. adj. *viridis* green.

The description is that given for the genus.

Type strain: ATCC 15386.

The species includes a number of variants that are nonpigmented, green, or lilac-colored, and, in addition to producing mainly single spores on the aerial hyphae (Fig. 26.42), may form short chains of spores or sporulate on vegetative hyphae. At present, the variation within the genus is accommodated in this single species.

Species Incertae Sedis

a. "**Saccharomonospora internatus**" (*Micropolyspora internatus* Agre, Guzeva and Dorokhova 1974, 679[AL]).

in. ter.nat′us. L. adj. *inter* between; L. part. *natus* having been born; *internatus* born between (intermediate).

The morphological description is that given for the genus. Whole-cell hydrolysates contain *meso*-DAP, arabinose, and galactose (wall type IV) and the cell envelope does not contain mycolic acids.

The optimum temperature for growth is 37-55°C. Physiological and biochemical characteristics are those given for the genus except that "*S. internatus*" does not utilize D-xylose.

Isolated from soil.

Type strain: INMI632.

"*Saccharomonospora internatus*" conforms to the definition of the genus *Saccharomonospora*. Kurup (1981) suggested transfer of *M. internatus* to the genus *Saccharomonospora* on this basis and used immunological cross-reactivity with *S. viridis* to further support his conclusions. The nomenclatural change was not formalized and it has yet to be established that "*S. internatus*" is sufficiently distinct from *S. viridis* to justify its status as a separate species of the genus.

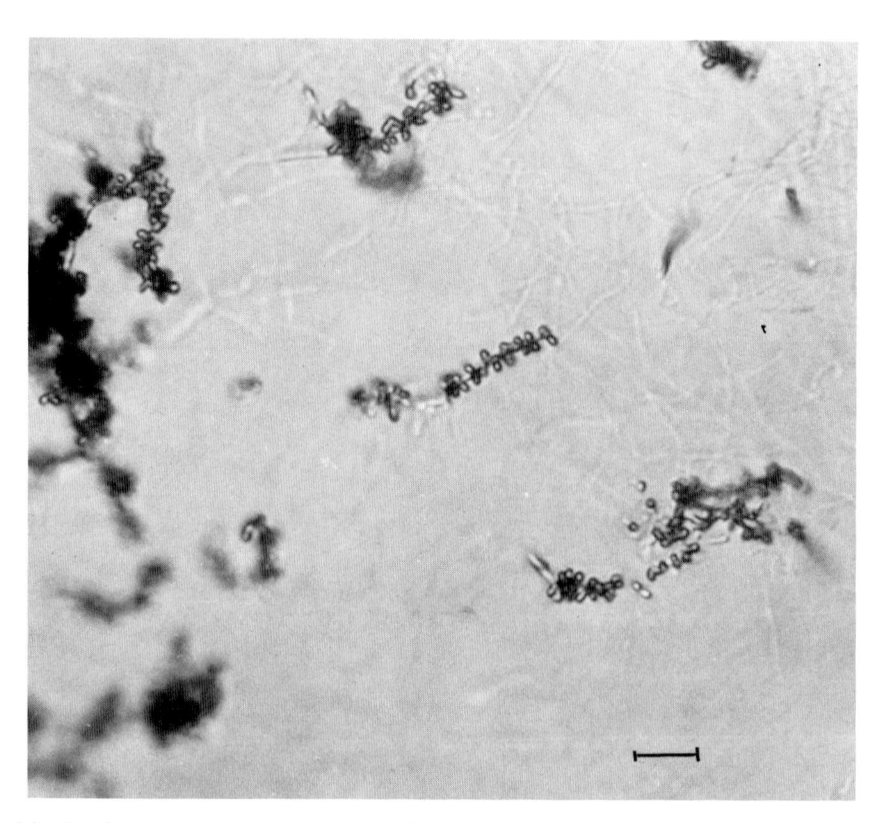

Figure 26.42. *Saccharomonospora viridis*. Aerial hyphae bearing single spores on lateral unbranched sporophores. *Bar*, 10 μm.

SECTION 27

Actinomycetes with Multilocular Sporangia

Mary P. Lechevalier

Despite the many advances in the taxonomic use of cell chemistry, identification of actinomycetes is still very dependent on morphological characters. Thus, organisms having multilocular sporangia, such as those classed in the genera *Dermatophilus, Geodermatophilus*, and *Frankia*, are grouped together here for convenience. Indeed, one may ask: What other characteristics do they have in common? A perusal of Table 27.1 will show that in fact, they do have a number of common features, other than this morphological one.

Morphologically, in the case of *Geodermatophilus* where the filamentous phase is absent or extremely rudimentary, the entire thallus is the sporangium. In *Dermatophilus*, filament formation is moderate to extensive and the filaments ultimately become almost entirely converted to long multilocular sporangia. In *Frankia*, there is extensive filament formation and sporangia are formed either as intercalary swellings, terminally, or on lateral branches. Aerial mycelium is normally absent in all three genera.

Physiologically, all are mesophilic, catalase positive, and aerobic, with dermatophili and frankiae tending toward microaerophily.

Table 27.1.
Differential characteristics of actinomycetes with multilocular sporangia

Characteristics	*Geodermatophilus*	*Dermatophilus*	*Frankia*
Morphology			
Extensive filamentation	−	−	+
Aerial mycelium	−	−	−
Sarcinoid sporangia	+	+	+
Vesicles	−	−	+
Outer spore membrane	−	−	+
Capsule	−	+	−
Physiology			
Spore motile	+	+	−
Temperature range (°C)	10–37	22–37	10–37
Relation to air	Aerobic	Microaerophilic	Microaerophilic
Catalase	+	+	+
Rapidity of growth	Rapid (2–7 days)	Moderate (7–14 days)	Slow (10–60 days)
Fixation of nitrogen	−	−	+
Chemistry			
Cell wall type	III	III	III
Whole cell sugar	C (no characteristic sugars)	B (madurose)	D (xylose), E (fucose), B or other
Phospholipid type[a]	PII	PI	PI
Mycolates	−	−	−
Mol% G + C of the DNA	73–75	57–59	66–71
Habitat	Soil/sea	Mammalian epidermis	Nodules of certain angiosperms/soil

[a] PII, Phosphatidylethanolamine and/or methylethanolamine as characteristic nitrogenous phospholipids; PI, no nitrogenous phospholipids present.

Chemically, they have the same type of cell wall (chemotype III) and lack mycolic acids.

They may be differentiated on the basis of (a) capsule formation (positive in *Dermatophilus*), (b) spore outer membrane and vesicle formation (present in *Frankia*), (c) motility of spores (positive in *Geodermatophilus* and *Dermatophilus*), (d) whole cell sugar and phospholipid patterns, (e) the mol% G + C of DNA, (f) fixation of nitrogen in vivo and in vitro (positive in *Frankia*), and (g) pathogenicity for animals (positive for *Dermatophilus*).

They are also found in different habitats. *Geodermatophilus* species have been found both in soil and in the sea; *Dermatophilus* species have only been found associated with the epidermis of mammals (they have never been isolated from soil) and *Frankia* species are found in nodules of actinorhizal plants. The members of the last genus have also been isolated from soil, but it is not known whether they grow there.

Genus **Geodermatophilus** *Luedemann 1968, 1857[AL*]*

GEORGE M. LUEDEMANN AND ANTONIO F. FONSECA

Ge.o.der.ma.to.phi'lus. Gr. n. *ge* earth; M.L. masc. n. *Dermatophilus* a genus of the *Actinomycetales*; M.L. masc. n. *Geodermatophilus* soil or earth-bound dermatophilus-like organisms.

Produce a **muriform, tuber-shaped, noncapsulated, holocarpic multilocular thallus** containing masses of cuboid cells 0.5–2.0 µm in diameter. The thallus, under favorable environmental conditions, breaks up, releasing cuboid and coccoid nonmotile cells. Some of these cells may develop into **elliptical to lanceolate zoospores** that are propelled by a terminal tuft of long flagella. **Mycelium rudimentary, aerial mycelium not produced.** Gram-positive. Cell wall contains *meso*-diaminopimelic acid (*meso*-DAP), together with glutamic acid, alanine, glucosamine, and muramic acid. Whole-cell hydrolysates do not contain madurose. Phospholipids include phosphatidylethanolamine, phosphatidylinositol mannosides, phosphatidylinositol, phosphatidylglycerol, and diphosphatidylglycerol. Fatty acids of the branched-chain type. Aerobic, Chemo-organotrophic. Mesophilic. Habitat soil. The mol% G + C of the DNA is 72.9–74.6.

Type species: *Geodermatophilus obscurus* Luedemann 1968, 1857.

Further Descriptive Information

The organisms in this genus originally appeared to be related to *Dermatophilus* through an obscure filamentous phase. This filamentous condition may be difficult to discover and sometimes may be induced by using a water culture or dilute broth medium. The filamentous phase does not appear as a true mycelium but rather like the pseudomycelium seen in certain yeasts as loosely united filaments representing elongations of buds, characteristically pinched and constricted at irregular intervals. These filaments or tubes resemble the holocarpic sporangia of fungi producing, within the tubes, spores that, when liberated, may germinate directly or become zoospores. Germinating spores or resting zoospores may divide directly to produce a new thallus or may produce a germ tube and an irregularly constricted and branched filament. The contents of these tubes first divide transversely by septa that do not appear to involve the outer layer of the cell wall, giving rise to a tube of longitudinally compressed cells. Septa are formed later in horizontal and vertical longitudinal planes and give rise to rows of cuboidal cells. Occasionally cells of the thallus may germinate in situ, emerging as lanceolate zoospores, or may produce a germ tube and small filament. The zoospores appear unique in that additional ones may develop as terminal buds, often forming sluggishly motile chains of lanceolate or elliptical cells. The filamentous and zoospore stages may be difficult to observe in some isolates. In other isolates the thallus is reduced to a few cells and may appear reminiscent of miniature fungal dictyospores or sarcinaform packets. The muriform, multilocular thallus, and the large number of morphological forms capable of being produced under different conditions of culture and by various strains, are the most distinctive and constant features of members of this genus. Considering the substantial differences in the mol% G + C, *Geodermatophilus* is no longer considered to be a close relative of *Dermatophilus* (Samsonoff et al., 1977).

The wall composition conforms to type III of Lechevalier et al. (1971a) and the phospholipids to type PII of Lechevalier et al. (1981). Branch-chained fatty acids have been described by Kroppenstedt (1985). The distinction between *Geodermatophilus* and other genera of chemotype III was underlined by the numerical phenetic study of Goodfellow and Pirouz (1982b).

Enrichment and Isolation Procedures

A dilute medium is necessary to prevent the overgrowth of other spreading bacteria and fungi. A medium that has been successful for isolation includes the following ingredients (% w/v): yeast extract, 0.1; glucose, 0.1; soluble starch, 0.1; $CaCO_3$, 0.1; and agar, 1.5.

Soil dilutions are made to give 50 colonies or less per plate when incubated for 2–3 weeks at 28°C. Plates are searched for colonies at weekly intervals under a dissecting microscope and suspected colonies (see colony descriptions of subspecies) are transferred to a growth and maintenance medium composed of (% w/v): yeast extract, 0.5; NZ Amine Type A (Sheffield Chemical Company, Norwich, NY), 0.5; glucose, 1.0; soluble starch, 2.0; $CaCO_3$, 0.1; and agar, 1.5.

It is desirable to attempt to inoculate an area of a centimeter or more in diameter when transferring a colony from the isolation medium to the growth medium. The growth medium is incubated for 1–3 weeks at 28°C and inspected regularly for contamination and colony development. Microscopic examination of developing colonies for the cellular elements typical of members of this genus (see Fig. 27.1) provide a positive indication of isolation of strains of this genus.

Maintenance Procedures

Agar-to-agar transfer is best made with visible amounts of inoculum transferred to fresh medium and distributed over the surface. Colonies usually do not spread but grow in the area of inoculation. Lyophilization in double-strength skim milk has been successful for preservation and is best carried out by using large quantities of cellular material from cultures grown on agar for 3–4 weeks at 28–30°C. Similar suspensions should be used for storage under liquid nitrogen. Some strains may be difficult to maintain (Ishiguro and Fletcher, 1975). Growth in liquid culture is often very light but is preferred for the study of the production of zoospores and pseudomycelium.

Taxonomic Comments

The microscopic study of the cellular elements shows the life cycle of these microbes to be unique when compared to other actinomycetes. The

AL denotes the inclusion of this name on the Approved Lists of Bacterial Names (1980).

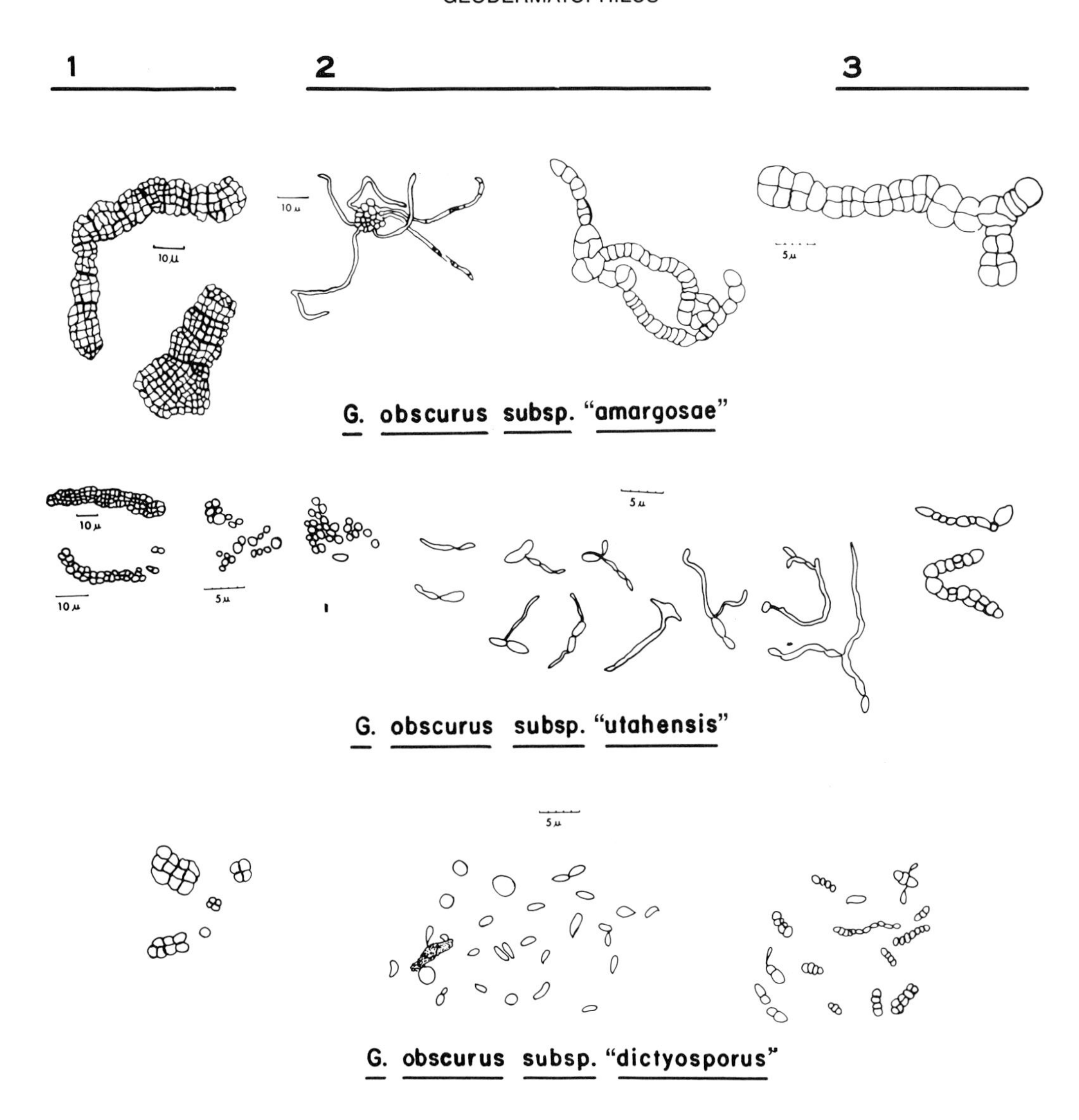

Figure 27.1. Line drawings illustrating the variation in the morphological development of three subspecies of *Geodermatophilus obscurus* and the complexity of the life cycle. *Left to right:* (*1*) variation in size and cell number of multiloculate thalli; (*2*) breakdown of thalli into component cells that may germinate (*top*) directly into filaments, (*center*) to form motile zoospores that then come to rest and form filaments, or (*bottom*) to form motile or nonmotile cells from which small thalli develop directly rather than first developing a filamentous stage; (*3*) early stages in the formation of thalli illustrating muriform septation. (Reprinted from *Bergey's Manual® of Determinative Bacteriology*, ed. 8, 1974, p. 878, with changes added.)

photomicrographs and drawings of cellular elements that appear in the publications of Luedemann (1968), Ahrens and Moll (1970), and Ishiguro and Fletcher (1975) clearly illustrate microbes of a very similar morphological cycle. The latter two publications did not show a mycelial or pseudomycelial stage as illustrated by Luedemann (1968), but Fonseca has recently found this type of mycelium in *Blastococcus aggregatus* (Fonseca, 1983).

The physiological data are based upon the study of more than 20 isolates from soil samples from arid regions of the southwestern United States, which were classified in a single species group, *G. obscurus*. The diagnostic physiological data for *G. obscurus* include: no growth on Brain-Heart Infusion Agar (Difco), no hydrolysis of casein, no hemolysis of red blood cells (10% human blood agar), and growth better at 26°C than at 37°C. Carbohydrate utilization and acid production vary with the strain, as do gelatin hydrolysis and nitrate reduction.

Differentiation and characteristics of the species of the genus **Geodermatophilus**

Details of carbohydrate utilization and acid production patterns of the subspecies are given in Table 27.2.

Table 27.2.
Carbohydrate utilization and acid production in **Geodermatophilus obscurus** *subspecies*[a]

	Subspecies				
Carbohydrate	1a. *obscurus*	1b. *"amargosae"*	1c. *"utahensis"*	1d. *"dictyosporus"*	1e. *"everesti"*
D-Arabinose	0a	0	2a	0a	
L-Arabinose	3A	3A	1	3A	
D-Galactose	3	3A	3	3	
D-Glucose	3a	3A	3a	3a	
Glycerol	3a	3A	1	3A	
Inositol	3	2	2	0	See
β-Lactose	0	1	0	3	text
D-Fructose	3a	3A	2	3A	
D-Mannitol	3	3	3a	3	
Melezitose	0	1	1	3	
L-Rhamnose	1a	2A	1	1A	
D-Ribose	0A	1A	1a	1a	
Sucrose	3	3	3	3	
D-Xylose	3a	3A	3	3A	

[a] *0*, no growth; *1*, poor growth; *2*, fair growth; *3*, good growth. *A*, acid production relatively consistent; *a*, sporadic or transient production of acid. Dulcitol, α-melibiose, and raffinose are not utilized.

List of species of the genus **Geodermatophilus**

1. **Geodermatophilus obscurus** Luedemann 1968, 1857.[AL]

ob.scur'us. L. adj. *obscurus* dark, obscure, indistinct.

Thalli appear greenish-black by transmitted light and vary in size from a few cuboidal cells to many cells arranged in cushion- or tuber-shaped aggregates. Presence and abundance of zoospores, germ tubes, and filaments vary with the strain.

Colonies on agar appear dark brown to black after 30 days' incubation (26–28°C), are flat to plicate, granular in texture, dry; odor is dank. Growth good on yeast extract-starch-sucrose-malt extract agar; poor on Brain-Heart Infusion Agar.

Carbohydrate utilization and acid production varies with the subspecies (Table 27.2). Starch is hydrolyzed.

Casein not hydrolyzed. Gelatin hydrolysis and nitrate reduction vary with the subspecies. Blood not hemolyzed.

Growth optimal at 24–28°C, reduced at 37°C, none at 50°C. In aqueous suspensions, few cells survive 30 min at 60°C.

1a. **Geodermatophilus obscurus** subsp. **obscurus** Luedemann 1968, 1857.[AL]

Thalli varying in size from a few cells to many cells arranged in cushion- or tuber-shaped aggregates. Thalli disintegrate, giving rise to cuboidal and coccoidal cells and elliptical to lanceolate zoospores. Zoospores often bud while motile, giving rise to two- or three-celled motile units. Motile spores may produce germ tubes and filaments, or may undergo enlargement, septation, and thallus formation without an intervening germ tube–filament stage.

Good growth on yeast extract–casein hydrolysate–starch–dextrose agar or yeast extract–glycerol agar; colony plicate, texture characteristically granular-friable, black.

Gelatin not hydrolyzed. Nitrate reduction weak or absent.

Isolated from a soil sample from the Amargosa Desert of Nevada.

Type strain: ATCC 25078.

1b. **Geodermatophilus obscurus** subsp. **"amargosae"** Luedemann 1968, 1857.

a.mar.go'sae. M.L. gen. n. *amargosae* of the Amargosa Desert.

Macromorphology similar to *G. obscurus* subsp. *obscurus*; micromorphology differs in that germinating cells often give rise to long, slender, unbranched filaments that enlarge and undergo septation to produce relatively long, broad, blunt-ended thalli (Fig. 27.1), or the germinating cells may enlarge, become septate, and form new thalli directly. Zoospores have rarely been observed in this isolate.

Gelatin weakly hydrolyzed; nitrate is not reduced.

Isolated from a soil sample from the Amargosa Desert in Nevada.

Type strain: ATCC 25081.

1c. **Geodermatophilus obscurus** subsp. **"utahensis"** Luedemann 1968, 1857.

u.tah.en'sis. M.L. adj. *utahensis* belonging to Utah (one of the United States).

This subspecies is distinct in several respects. The type strain is a prolific producer of zoospores, which bud to produce motile chains of two to three cells; germinating zoospores often produce a well-branched pseudomycelium (Fig. 27.1).

Moderate growth of flat granular colonies on yeast extract–casein hydrolysate–starch–dextrose agar. Growth on agar containing yeast extract and 5–10% (v/v) glycerol is good, but when the glycerol content is reduced to 2%, as in carbohydrate utilization tests, growth is poor.

Carbohydrate utilization and acid production often poor.

Gelatin not hydrolyzed; nitrate is reduced.

Isolated from a soil sample from Zion National Park, Utah.

Type strain: ATCC 25079.

1d. **Geodermatophilus obscurus** subsp. **"dictyosporus"** Luedemann 1968, 1858.

dic.ty.o.spo'rus. Gr. n. *dictyon* a net; Gr. fem. n. *spora* a seed; M.L. n. *dictyosporus* netted spore, referring to longitudinal and vertical septations.

Multilocular thalli vary in size, often small; superficially resemble dictyospores of fungi. Thalli formed directly from resting zoospores are small, a feature characteristic of this isolate (Fig. 27.1). Often a zoospore appears to arise directly from a locule of a small thallus, indicating in situ germination of some spores. Thalli appear olive-green to greenish black in transmitted light.

Good growth on yeast extract-casein hydrolysate-starch-dextrose agar; colony plicate, dark grayish brown. Good growth on yeast extract-glycerol agar; colony granular, plicate, strong brown. Young colonies on some agar media have a salmon pink peripheral border that later turns dark brown or black.

Lactose and melezitose utilized.

Gelatin hydrolyzed; nitrate reduction weak or negative.

Isolated from a soil sample from Westgard Pass, California.

Type strain: ATCC 25080.

1e. **Geodermatophilus obscurus** subsp. **"everesti"** Ishiguro and Fletcher 1975, 106.

ev.er.est'i. M.L. gen. n. *everesti* of Mount Everest.

The characteristic morphogenetic growth cycle involves the following successive stages: (a) highly refractile, nonmotile, irregularly shaped clumps of coccoid cells; (b) cells in the clump forming nonrefractile buds; (c) detachment of buds as actively motile rods; (d) multiplication of the young rods exclusively by budding; (e) late growth stage rods that become refractile and segmented into coccoid units by fission along both transverse and longitudinal planes; and (f) pattern of fission becoming random in all planes, resulting in an irregular aggregate of coccoid cells. Motile rods have single polar tufts of up to five flagella; clumps of coccoid cells have no flagella. Individual coccoid units in clumps measure 0.9–1.2 µm in diameter. Rods variable in size depending on growth phase; young rods about 0.4–0.6 × 1.5–2.5 µm. Aerobic. Optimum growth temperature between 20° and 30°C. Poor growth at 37°C. No growth at 10° and 45°C. Growth rate slow on most media. Colonies punctiform, circular, convex, butyrous, and smooth edged.

Grows in liquid as sediment with slight turbidity; surface growth absent. May produce red, orange, or yellow pigments, or may be nonpigmented.

Gelatin, casein, and urea not hydrolyzed. Starch weakly hydrolyzed by some strains. Nitrate weakly reduced. Acid and gas not produced from D-glucose, D-ribose, D-galactose, D-arabinose, D-xylose, D-rhamnose, sucrose, lactose, inositol, glycerol, D-mannitol, maltose, L-fucose, or cellobiose.

The type strain 22-68 was isolated from a soil sample from the West Ridge of Mt. Everest at an elevation of 27,250 feet (8305.8 m).

Further Comments

The morphologic cycle of the microbes isolated from the Mt. Everest soils is very similar to that of strains of *G. obscurus*. The data of Ishiguro and Fletcher (1975) on the production of acid and gas in Phenol Red Broth Base (Difco) are not comparable to the carbohydrate utilization and acid production reported for strains of *G. obscurus*. The authors reported that they could not extract the DNA from their Mt. Everest isolates in a form that would allow determination of the mol% G + C. Ishiguro and Fletcher mentioned that a budding microbe isolated by Ahrens and Moll (1970) from Baltic Sea water may be identical to their Mt. Everest isolates. The Ahrens and Moll isolate produced pinkish carotenoid-like colony pigments. A similar strain isolated by M.A. Lechevalier (personal communication) from a New Jersey soil produced pinkish-gray colony pigments, and the author obtained an isolate from a Georgia soil that produced orange colony pigments. No dark brown or black pigments comparable to the *G. obscurus* isolates from the western deserts (U.S.A.) were seen in these eastern strains. Isolation and identification of these colony pigments may be useful as species markers. The lighter colored strains may possibly be separable as species from the black-pigmented strains once the range in color variability has been established. Isolates belonging to this group of microbes are worldwide in distribution but remain as a little known but common group of soil microorganisms that await future investigation.

Genus **Dermatophilus** *Van Saceghem 1915, 357, emend. mut. char. Gordon 1964, 521*[AL]

MORRIS A. GORDON

Der.ma.toph'il.us. Gr. n. *derma* skin; Gr. adj. *philos* loving; M.L. masc. n. *Dermatophilus* skin loving.

Aerial mycelium develops in atmospheres containing added CO_2. Substrate mycelium consists of **long tapering filaments, branching laterally at right angles; septa formed in transverse and in horizontal and vertical longitudinal planes, giving rise to up to eight parallel rows of coccoid cells (spores), each of which becomes motile by a tuft of flagella.** Gram-positive. Walls contain *meso*-diaminopimelic acid (*meso*-DAP); madurose is present in whole-cell hydrolysates. Polar lipids include phosphatidylglycerol, diphosphatidylglycerol, and phosphatidylinositol. Chemo-organotropic. **Nonfermentative** but **acid is produced** from certain carbohydrates. **Catalase positive.** Not acid fast. Growth reported only on complex media; minimum nutritional requirements unknown. Aerobic and facultatively anaerobic. Temperature optimum ~ 37°C. Parasitic on mammals, especially the domestic herbivores; pathology usually limited to exudative dermatitis, which may, however, be severe and life-threatening; on rare occasions causes subcutaneous abscesses and lymph node granulomas. Mol% G + C of the DNA is 57–59 (T_m).

Type species: *Dermatophilus congolensis* Van Saceghem 1915, 357, emend. mut. char. Gordon 1964, 521.

Enrichment and Isolation Procedures

Clinical materials containing *D. congolensis*, usually cutaneous crusts or scabs, are heavily contaminated with other bacteria and require a good streaking technique for isolation on beef infusion-horse blood agar plates, which are incubated aerobically at 37°C. Alternative methods are animal passage by cutaneous inoculation and aqueous extraction of ground crusts followed by streaking on antibiotic blood agar (Gordon, 1985). *D. congolensis* may be maintained by lyophilization in skim milk after growth for 3–5 days at 37°C in brain-heart infusion (BHI) broth.

Strains of "*Tonsillophilus suis*" were isolated from porcine tissues in an anaerobic atmosphere with added CO_2. Cultures are maintained on BHI agar or trypticase-soy agar. They are incubated for 5 days at 37°C in an aerobic or anaerobic atmosphere containing 10% CO_2. Such cultures may also be lyophilized after suspension in skim milk.

Table 27.3.
Differentiation of the genus **Dermatophilus** *from other actinomycete genera with multilocular sporangia*

	Dermatophilus	*Geodermatophilus*	*Frankia*
Hyphae	Tapering, branched, septate	Primitive	Slender, sparsely branched
Sporangia	Entire mycelium transforms to spore packets	Muriform, tuber-like	Formed in vesicles on the hyphae
Cell wall type	III	III	III
Sugar pattern	B	C	Variable
Source	Animal lesions	Soil, water	Plant root nodules

List of species of the genus **Dermatophilus**

1. **Dermatophilus congolensis** Van Saceghem 1915, 357, emend. mut. char. Gordon 1964, 521.[AL]

con.go.len′sis. M.L. adj. *congolensis* pertaining to the Congo (named for the Belgian Congo).

Hyphae, 0.5–1.5 µm in diameter, develop from germ tubes. After several transverse and longitudinal divisions, hyphae may be up to 5 µm in diameter, with branches at right angles tapering to nonseptate apices (Fig. 27.2*A*). Become converted entirely into eight-ranked packets of isodiametric segments encased in a gelantinous sheath (Fig. 27.2*B* and *C*). Each segment is released as a motile spore, bearing a tuft of five or more flagella (Fig. 27.2*D*). The spores subsequently lose motility and germinate.

On blood agar, colonies rough, often becoming viscous; adherent through invasion of substrate by hyphae; often white to gray at first, usually becoming orange to yellow; hemolytic on media containing sheep but not horse blood. Good growth on Loeffler's medium, light yellow; medium liquefied by most strains. No growth on Sabouraud glucose agar, Czapek agar, or tomato paste–oatmeal agar. Broth is clear with a flocculent or ropy sediment; sometimes with a surface ring of growth.

Acid from glucose and fructose; transient acid (within 48 h) from galactose; often late production from maltose. Acid not produced from lactose, sucrose, xylose, dulcitol, mannitol, sorbitol, or salicin. Starch hydrolyzed.

Gelatin and casein hydrolyzed. Tyrosine and xanthine not hydrolyzed. Urease and catalase produced. Indole not formed; methyl red and Voges-Proskauer tests negative. Nitrates not reduced.

Aerobic, facultatively anaerobic. In an atmosphere of 10% carbon dioxide at 37°C, growth is accelerated and aerial hyphae are formed, while septation and spore formation are delayed.

Susceptible to a wide range of antibacterial antibiotics, including penicillin, streptomycin, chloramphenicol, erythromycin, and the tetracyclines. Resistant to the antifungal agents griseofulvin, nystatin, and tolnaftate (Roberts, 1981).

Etiological agent of dermatophilosis, an exudative, often severe, dermatitis affecting large numbers of cattle, sheep, goats, and domesticated equines and causing important economic problems in many parts of the world. Occasional infections have been reported in many additional mammalian species, including man, and in certain lizards (Gordon and Salkin, 1977)

Type strain: ATCC 14637.

Genus Incertae Sedis

"*Tonsillophilus suis*" was proposed, but not validly published, by Azuma and Bak (1980) for a group of isolates from swine tonsils that bear a morphological resemblance to the *Dermatophilaceae*. The microorganisms were observed in caseous or gritty granules associated with lesions resembling abscesses of actinomycosis.

In its filamentation, "*T. suis*" appears to be intermediate between *Dermatophilus congolensis* and *Geodermatophilus obscurus*, forming a more primitive mycelium than *D. congolensis*. The relatively short hyphae of varying widths, which undergo transverse and longitudinal septation and ultimately form packets of coccoid cells, show occasional branching, but more often several of these structures radiate from a central mass. A cell wall analysis (M.P. Lechevalier, personal communication) of "*T. suis*" reveals *meso*-DAP and madurose, as does that of *D. congolensis*.

The following description is taken largely from the papers by Azuma and Bak (1980), and Bak and Azuma (1980), with additions and modifications based upon the author's own observations.

"*Tonsillophilus suis*" Azuma and Bak 1980, *348*.

The life cycle begins with a flagellate zoospore that, upon germination and enlargement, undergoes division in transverse and longitudinal planes, ultimately to form packets of coccoid cells at least four cells in diameter. Thalli often consist of two or more short filaments radiating from a common center and occasionally branching. The coccoid elements are freed from these filaments and become zoospores. All elements are Gram-positive.

On BHI agar, growth is better at 37°C than at 25°C and much better anaerobically than aerobically. Aerobic growth is enhanced by 10% CO_2. Colonies are white or off-white and of varying morphologies described as bread-crumb, molar tooth, crateriform, cerebriform, and umbonate. Not hemolytic. No growth on Sabouraud glucose agar. Very good growth in thioglycollate broth. Gelatin is not hydrolyzed. Milk is not coagulated or digested. Nitrate is reduced. Glucose, fructose, galactose, mannose, sucrose, maltose, lactose, trehalose, raffinose, mannitol, and sorbitol are fermented; arabinose, xylose, ribose, rhamnose, adonitol, dulcitol, and glycerine are not. Cellular morphology is shown well by Giemsa and methylene blue stains. In animal tissues, the infecting organisms stain well with periodic acid–Schiff. They appear blue after staining with hematoxylin and eosin.

The taxonomic position of this microorganism is uncertain and still under study.

Type strain: ATCC 35846.

Genus **Frankia** *Brunchorst 1886, 174*[AL]

Mary P. Lechevalier and Hubert A. Lechevalier

Frank′i.a. M.L. fem. noun *Frankia*, named after A.B. Frank (1839–1900), Swiss microbiologist who coined the word "symbiosis."

Vegetative hyphae with limited to extensive branching, 0.5–2.0 µm in diameter and occasionally wider. **No aerial mycelium formed. Round to irregularly shaped multilocular sporangia borne terminally, laterally, or in an intercalary position on the vegetative hyphae.** Lateral sporangia usually borne on sporangiophores; some may be sessile. Sporangia up to 100 µm in length formed by septation in three planes of the cytoplasm of preexisting thin-walled swellings. **Sporangiospores nonmotile,** of irregular (often somewhat polygonal) shape, 1–5 µm in size, colorless to black, showing multilaminar outer membrane-like layers in thin section. Spores not thermally resistant. Sporangiospores do not usually develop and mature simultaneously so that developing sporangia may contain spores of various ages and sizes. **Terminal or laterally borne "vesicles" may be formed.** These are thick-walled swellings that show various irregular septations in thin section and are probably the site of nitrogenase activity in strains fixing atmospheric nitrogen. Vesicles may also be formed under conditions where no nitrogen fixation takes place. All these morphological characteristics may be expressed both in vitro and in planta, although nitrogen-fixing nodules without vesicles are known. Intra- and extracellular pigments common. Gram-positive to Gram-variable. Aerobic to microaerophilic. Will not grow under anaerobic conditions. Catalase positive. Mesophilic. Chemo-organotrophic. **Usually very slow-growing** (doubling time of 1–7 days). **Most strains capable of fixing atmospheric nitrogen in vitro and *in planta*. Cell walls contain *meso*-diaminopimelic acid (*meso*-DAP) glutamic acid, alanine, muramic acid, and glucosamine. No mycolates present. Phospholipids comprised of phosphatidylinositol mannosides, phosphatidylinositol, and diphosphatidylglycerol.** Fatty acids are normal, branched chain, and mono-

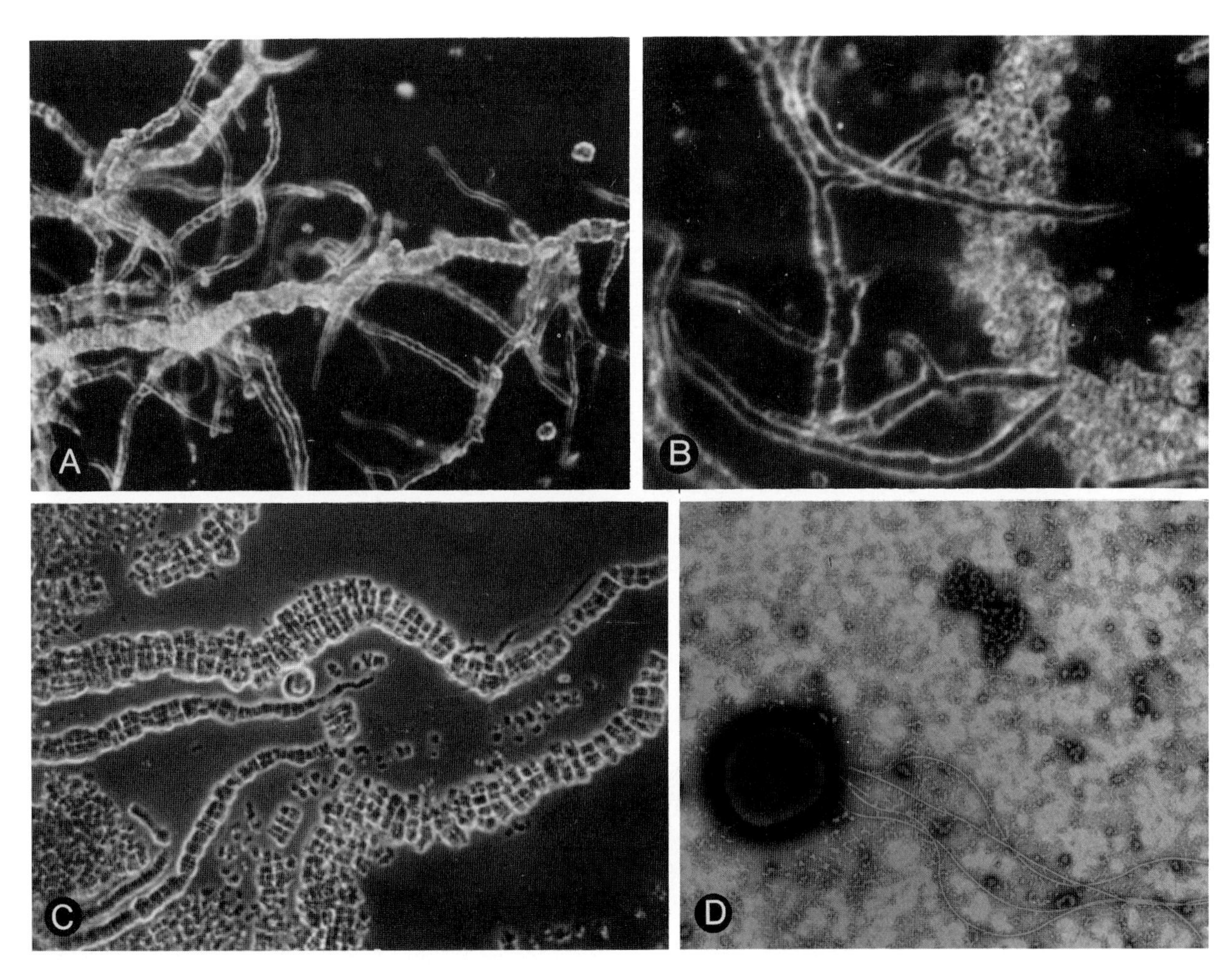

Fig. 27.2. *Dermatophilus congolensis. A*, wet mount of beef infusion-peptone broth culture, showing various stages in development, from fine hyphae to multiseptate coarse filaments transforming into chains of coccal packets. Dark-field microscopy (× 945). *B*, same preparation as in *A*; final transformation of mature hypha into an agglomeration of motile spores, in process of dispersion. Dark-field microscopy (× 2100). (*A* and *B* reprinted with permission from D. J. Dean, M. A. Gordon, C. W. Severinghaus, E. T. Kroll and J. R. Reilly, New York State Journal of Medicine *61*: 1283-1287, 1961, ©The Medical Society of the State of New York.) *C*, fully segmented hyphae forming cubical packets of coccoid spores; wet mount of broth culture. Dark-phase contrast (× 1750). *D*, encapsulated motile spore of *D. congolensis* with tuft of flagella; negative stain (phosphotungstic acid) (× 17,500). (*C* and *D* reprinted by permission from M. A. Gordon, Journal of Bacteriology *88*: 509-522, 1964, ©American Society for Microbiology.) (Reprinted from *Bergey's Manual® of Determinative Bacteriology*, ed. 8, 1974, p. 876.)

unsaturated. **Whole-cell sugar patterns show xylose (without arabinose), madurose, or fucose or cells may contain only glucose or galactose**, sugars not previously found to have taxonomic significance in the *Actinomycetales*. Many strains contain 2-*O*-methyl-D-mannose and most contain rhamnose. **Most strains are symbiotic with certain angiospermous plants, inducing nodules on the roots of suitable hosts.** May be found free in the soil. The mol% G + C of the DNA is 66-71 (T_m).

Type species: *Frankia alni* (Woronin 1866) von Tubeuf 1895, 118.

Further Descriptive Information

Frankia isolates show great diversity in their pigmentation. Cells may be white, yellowish, pink, red, yellow- to greenish-brown, or black. Soluble pigments include yellow, green, red, green-black, and brown-black. Given the recentness of the isolation and maintenance in pure culture of members of this group (the first strain was reported in 1978 by Callaham et al.), strains with other pigments undoubtedly will be found. The red pigments from several strains have been identified as 2-methyl-4,7,9,12-tetrahydroxy-5,6-dihydrobenzo[*a*]napththacene-8,13-dione and its carboxylated derivative (Gerber and Lechevalier, 1984).

Some frankiae have loosely filamentous thalli with relatively little branching and hyphae that are generally 1.0 µm in diameter. Others have finer hyphae (~ 0.5 µm), are more highly branched, and form compact, dense thalli. The hyphal filaments do not fragment, although in one isolate from *Casuarina* the hyphae form elongated, narrow, sporangia-like structures with cross-walls that break down into subunits under pressure (Diem and Dommergues, 1985). Generally the sporangia of the loosely filamentous type of strain are larger than in those forming compact colonies, and their mature spores are also larger (1.5-5 µm). Sporangial shapes are extremely diverse, varying even within the same strain from round to highly irregular (Figs. 27.3 and 27.4). All spores examined to date are smooth surfaced. Mature sporangiospores have thick walls and outer membrane-like layers (Fig. 27.5). Spores germinate at variable rates (usually very low) to give rise to 1-3 germ tubes for most strains. "Vesicles" (Fig. 27.6) are formed on most media by certain strains; others form them only on nitrogen-free media or other special media. Still others have never been observed to form vesicles in vitro, but this may reflect lack of suitable nutrients since sporangia and vesicle formation in vitro has been found to be dependent on nutritional conditions (Tjepkema et al., 1980; Tisa et al., 1983). Sections of hyphae show typical procaryotic,

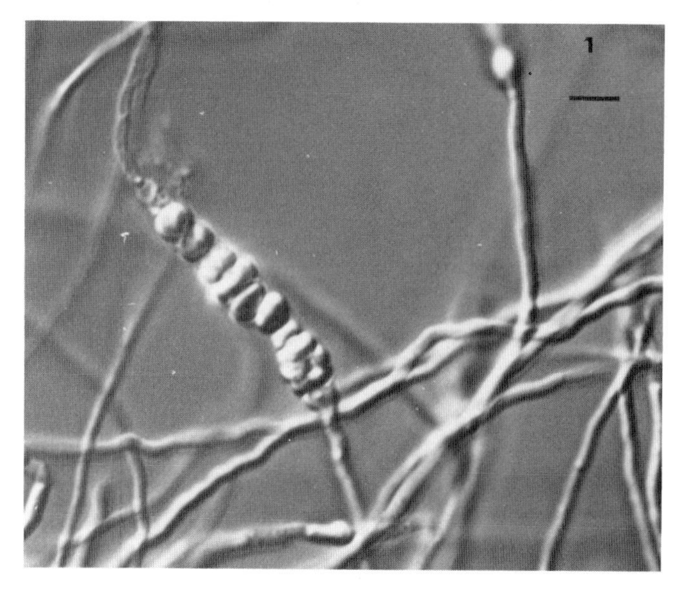

Figure 27.3. Sporangium of *Frankia* sp. WEY 0131395 (isolate from *Alnus rubra*). Interference contrast. *Bar,* 5 μm.

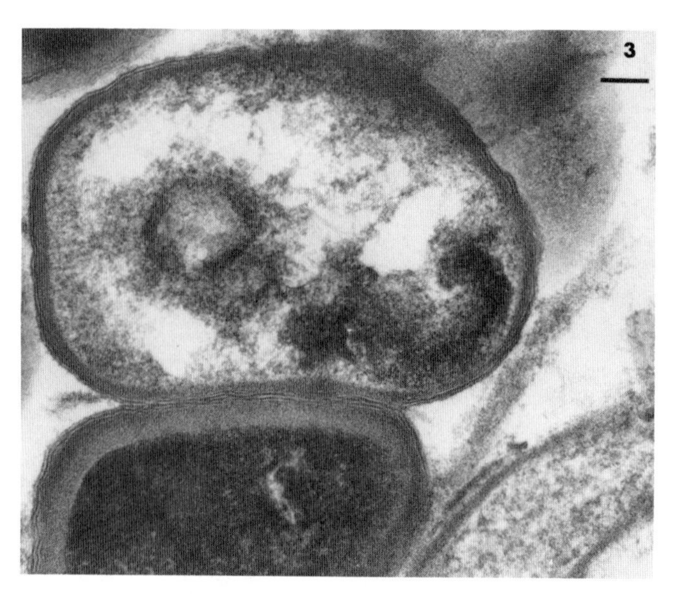

Figure 27.5. Section through spores of *Frankia* sp. DDB 130120 (from *Elaeagnus umbellata*). Note thick walls covered by outer membranes. Transmission electron microscopy. *Bar,* 0.1 μm.

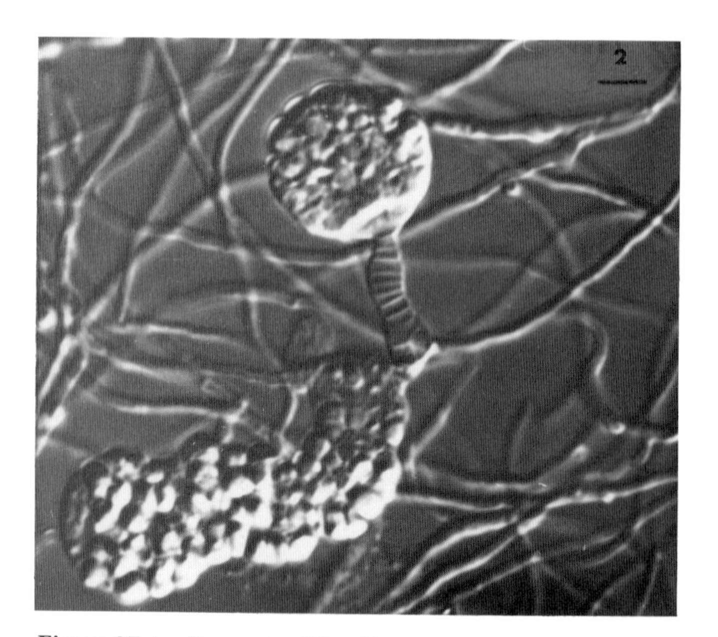

Figure 27.4. Sporangia of *Frankia* sp. WEY 0131395. Interference contrast. *Bar,* 5 μm.

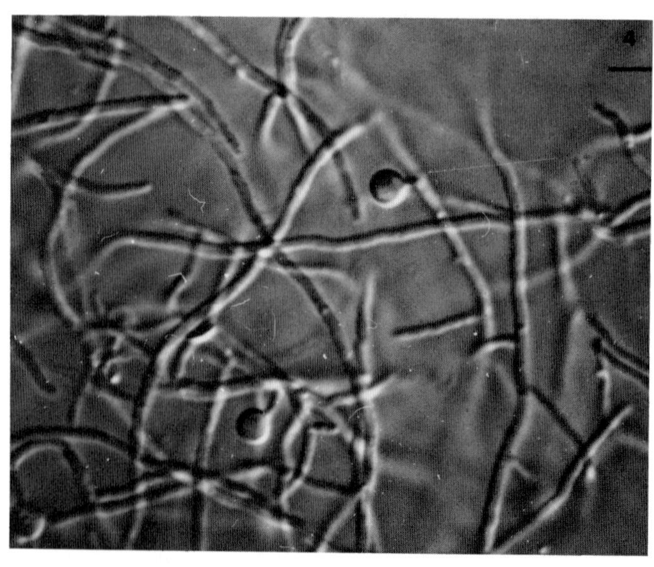

Figure 27.6. Hyphae of *Frankia* sp. LLR 02024 bearing vesicles (isolate from *Casuarina cunninghamiana*). Interference contrast. *Bar,* 5 μm.

Gram-positive features and studies of the in vitro morphogenesis of sporangia (Newcomb et al., 1979; Horrière et al., 1983) have revealed a complex multistaged developmental pattern. Using freeze-fracturing or freeze substitution, vesicles have been demonstrated to have very thick walls composed of many laminae (Torrey and Callaham, 1982; Fontaine et al., 1984; Lancelle et al., 1985). It is hypothesized that these layers play a role as a diffusion barrier in protecting the oxygen-labile nitrogenase from damage. Both vesicles and sporangia as they appear in the plant nodules of *Alnus* are shown in Figures 27.7 and 27.8, respectively. The morphology of the endophytes in various plant hosts has been reviewed by Baker and Seling (1984).

Relative to members of other actinomycete genera having a cell wall of chemotype III, frankiae have unusually diverse whole-cell sugar patterns. The most common sugars encountered in *Frankia* strains are xylose (no to little arabinose) (type D) and fucose (no type assigned). A few strains contain madurose (type B) or sugars having no previous taxonomic value such as galactose or glucose. Many strains, but not all, contain rhamnose and 2-*O*-methyl-D-mannose (Mort et al., 1983). The reason for this diversity is not yet clear, but may be related to plant host specificity (Lechevalier and Lechevalier, 1984). Trehalose was found as a storage product in some strains (Lopez et al., 1983).

Further evidence of the diversity of frankiae has been demonstrated

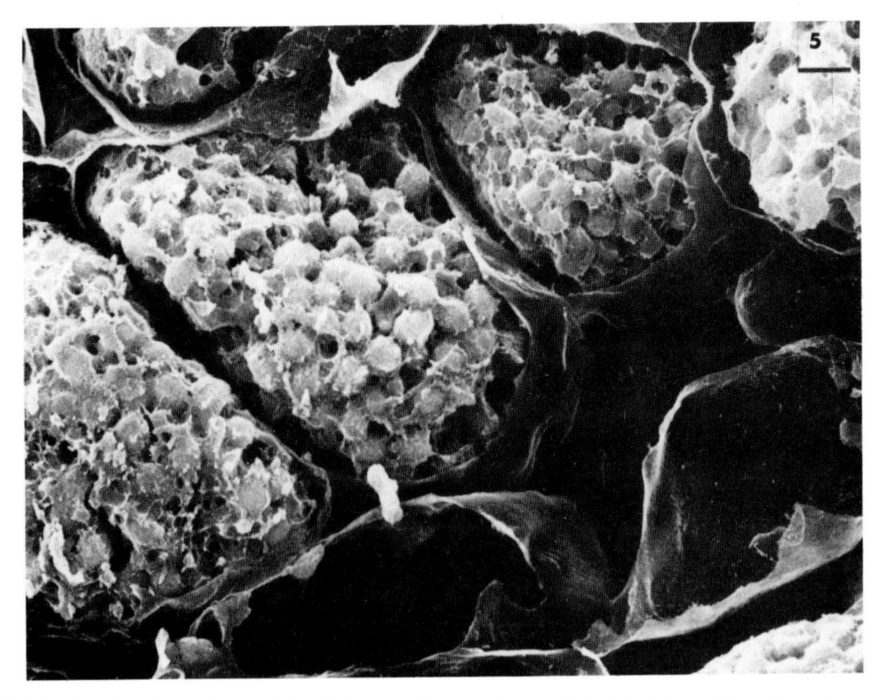

Figure 27.7. Section through a nodule of *Alnus rubra* revealing cells filled with vesicles of *Frankia* sp. LLR 01321 (isolate from *Alnus incana* subsp. *rugosa*). Scanning electron microscopy. *Bar*, 5 μm.

Figure 27.8. Section through a field-collected nodule of *Alnus incana* subsp. *rugosa* containing sporangia. The vesicle cluster has completely disappeared. Scanning electron microscopy. *Bar*, 1 μm.

by other means. Using sodium dodecyl sulfate-polyacrylamide gel electrophoresis, Benson and Hanna (1983) showed differences in whole-cell proteins of isolates obtained from nodules of *Alnus incana* ssp. *rugosa* plants growing in a very restricted area. Further work on protein patterns demonstrated at least three groups (A, B, and D) among isolates from the *Alnus-Myrica-Comptonia* host group. An isolate from *Purshia* was quite different from the others (Benson et al., 1984a). Likewise, Lechevalier and Ruan (1984) discerned morphological, physiological, and chemical differences among strains of frankiae isolated from *Ceanothus americanus* plants growing in a small area. Also, Hafeez et al. (1984) reported quite large differences in isolates from *Alnus nitida* and Zhang et al. (1984b) found at least two types of endophytes in nodules of greenhouse-grown *Casuarina cunninghamiana* inoculated from the same field-collected nodules.

Physiologically, frankiae may be classed into at least two groups (Lechevalier et al., 1983; Lechevalier and Ruan, 1984). The first group (A) is largely a heterogeneous collection of morphologically and chemically diverse strains. In their characteristics they approach normal saprophytic soil actinomycetes. They are relatively aerobic, can be maintained on slants, have pigmented cells, and show rapid (1–3 weeks) growth in the presence of 0.5% carbohydrate in a casein-hydrolysate medium. When grown in Tween-amended media, especially in the presence of carbohydrate, fatty acids are released, often in large amounts, but these are little taken up by the cells and they inhibit the cell growth. Group A strains for the most part are serologically (Baker et al., 1981) and genetically (An et al., 1984) diverse. Most of the strains examined up to the present, which belong to this group, are not infective and effective for their original plant host.

The second group of frankiae (B) is more homogeneous. These have largely colorless cells (although certain of them excrete pigments into the medium). They are microaerophilic and cannot be maintained in slant culture. When transferred to slants, they may grow out once but cannot be recultivated. They take up no carbohydrates at 0.5% (w/v) and show a typical synergistic growth response when Tween 80 is added to media containing carbohydrates (e.g. cell dry weight exceeds that of the sum of the cell weights of strains grown on the carbohydrate and the Tween in separate media). Most of the strains in group B induce effective (nitrogen-fixing) nodules on the roots of their original plant host. Other physiological groups of frankiae exist but further nutritional studies are needed to elucidate their growth characteristics (Lechevalier and Ruan, 1984).

Strains of group A show (variously) protease and amylase activity. Carbohydrates commonly assimilated by this group include arabinose, glucose, maltose, sucrose, trehalose, and xylose (see, for example, Table 27.5, below). Acid may or may not be formed. Generally, glycerol and *O*-methylated sugars are not utilized. Organic acids such as acetate, propionate, pyruvate, and succinate are frequently utilized; fumarate, malate, and tartrate are rarely or not utilized. Besides atmospheric nitrogen, various amino acids, amides including urea, and ammonia can serve as sources of nitrogen.

Group B strains are less physiologically active in many respects than group A types (more symbiotic?). They lack protease activity and do not take up and utilize carbohydrates at a 0.5% (w/v) concentration. However, fatty acids of various types (such as those released from Tweens®), and organic acids such as those listed in the preceding paragraph, can be taken up and utilized. Suitable nitrogen sources are like those for strains in group A. β-glucosidase activity has been reported in both groups (Horrière, 1984). Group B strains are serologically and genetically related to each other and almost all are infective and effective on their original plant host. All strains from both groups examined to date are capable of fixing nitrogen in vitro (Tjepkema et al., 1980; Gauthier et al., 1981b; Torrey et al., 1981). Aside from nitrogenase, other enyzmatic activities involved with *Frankia* nitrogen metabolism include glutamine synthetase and glutamate-oxaloacetate transaminase (Akkermans et al., 1982) and a transport system for methylammonium and ammonia, possibly an NH_4^+ specific permease, which was reported by Mazzucco and Benson (1984). Carbon metabolism enzymes that have been found include some from the Embden-Meyerhof-Parnas and pentose phosphate pathways and also the tricarboxylic acid and glyoxylate cycles. In some cases, the enzyme detection may vary with the strain or the growth conditions used. There is no evidence of the presence of the Entner-Doudoroff pathway (Blom and Harkink, 1981; Akkermans et al., 1982; Lopez and Torrey, 1985a; Stowers et al., 1986); however, propionate may be metabolized via propionyl-CoA and the methylmalonyl pathway (Murry et al., 1984; Stowers et al., 1986). A comparison of the polyacrylamide gel electrophoretic patterns of various enzymes of seven *Frankia* strains isolated from *Casuarina* and *Allocasuarina* showed that the strains were heterogeneous in respect to their phenol oxidases, esterases, and diaphorases; however, their catalase (superoxide dismutase) activity bands all had the same mobility (Puppo et al., 1985). Catalase may thus be a potential taxonomic marker. A trehalase has been isolated and purified from an *Alnus rubra* isolate (Lopez and Torrey, 1985b). The physiology of the endophytes in association with plants has been the subject of several papers (Akkermans et al., 1981; Huss-Danell et al., 1982).

By the use of immunodiffusion, frankiae have been divided into two serological groups (I and II) (Baker et al., 1981; Lechevalier et al., 1983). Serogroup I corresponds to physiological group B; serogroup II is quite heterogeneous and contains strains isolated from such plants as *Alnus, Elaeagnus, Ceanothus, Casuarina* and *Purshia*. All strains in both groups have bands in common, indicating their genetic relationship.

Nothing is known of the genetics of frankiae. The mol% G + C (66–71; average 68.5) is in the normal range for actinomycetes. An examination of the DNA of four isolates revealed no modified bases (An et al., 1983). The mean genome size for two strains representing physiological groups A and B was 4.6×10^9 and 6.2×10^9 daltons, respectively (An et al., 1984). DNA homology studies confirmed the existence of two groups among frankiae (An et al., 1985a). The first group (which corresponds to physiological group B and serological group I) had a high degree of homology with a *Frankia* isolate from *Alnus*. Generally, members of the second group had a low degree of homology to the first and among themselves. The exception was two isolates from *Casuarina* nodules that exhibited extremely high (97%) homology even though isolated from plants thousands of kilometers apart (An et al., 1984). Plasmids ranging in size from 7.1 to 32.2 kilobase pairs have been reported in 4 of 39 isolates examined (Normand et al., 1983). Subsequently, plasmids having 7 to 190 kilobase pairs were detected in 5 of 55 strains of frankiae (Simonet et al., 1984). Evidence for plasmids was also found in *Frankia* vesicles isolated from plants (Dobritsa, 1982). Genes coding for nitrogenase (*nif*) in *Frankia* sp. HFP 070101 (CpI1) were found to hybridize with the *nif* genes from *Klebsiella pneumoniae*, indicating that the DNA sequences coding for nitrogenase are highly conserved (Ruvkun and Ausubel, 1980). Restriction patterns of the genomic DNA of various *Frankia* strains have been published (An et al., 1985b; Dobritsa, 1985) as well as those of certain *Frankia* plasmids (Simonet et al., 1985).

Two groupings within *Frankia* have been proposed on the basis of elevated production of sporangia *in planta* (Van Dijk, 1978; Houwers and Akkermanns, 1981; VandenBosch and Torrey, 1984). The groups are referred to as Sp+ or P (for spore-positive) and Sp− or N (for spore-negative) (Normand and Lalonde, 1982). Despite the term, spore-negative strains can form low numbers of sporangia in the plant and usually form many in vitro. There are varying opinions on whether Sp− endophytes fix greater amounts of nitrogen *in planta* and grow more vigorously in vitro, but there is a consensus that Sp+ types generally produce more nodules by weight (Hall et al., 1979; Normand and Lalonde, 1982). Both Sp+ and Sp− nodules can occur on the same plant. An evaluation of isolates from the two types of nodules showed that certain strains, by some criteria, fell between the two groups (Normand and Lalonde, 1982). The phenomenon is thought to have a genetic rather than a nutritional basis since Sp+ nodules always result from the use of Sp+ nodule inocula (Van Dijk, 1978) and consistent morphological and nutritional differences were seen between Sp+ and Sp− isolates (Normand and Lalonde, 1982). Although it has been proposed as a criterion for defining species of the genus (Normand and Lalonde, 1982), the taxonomic and ecological significance of this phenomenon and the relationships between physiological types A and B and Sp+/Sp− have not been determined.

Frankia sp. CpI1 (HFP 070101) was found not to be pathogenic to rabbits in the skin tests used to assess pathogenicity of *Dermatophilus* (Gordon et al., 1983). The systemic pathogenicity of the frankiae for mammals is probably negligible since use of aerosol inocula for infecting plants has been used for some time in Canada with no known untoward effects on personnel (Lalonde, 1981 and personal communication) and injection of viable *Frankia* cells into mice and rabbits over a 4–5-year period has been without adverse consequences (Baker et al., 1981; D. Baker, personal communication).

Frankiae are sensitive to penicillin and streptomycin (M. P. Lechevalier, unpublished) and to various plant phenolics, including ferulic, *o*-coumaric, and *p*-coumaric acids (Perradin et al., 1983). A strain of *Frankia* has been reported to produce indoleacetic acid, a plant growth factor (Wheeler et al., 1984).

Suitable plant hosts are readily infected by most frankiae, giving rise to nitrogen-fixing (effective) nodules on the roots (Callaham et al., 1978, 1979; Lalonde, 1978; Berry and Torrey, 1983). The association is now referred to as "actinorhizal" (Torrey and Tjepkema, 1983). Some strains can give rise to noneffective nodules and still others are not infective for the host plant from which they were isolated; however, the latter are often infective in plants belonging to the family *Elaeagnaceae* (Gauthier et al., 1981a). It has been demonstrated that the infection process in the *Elaeagnaceae* takes place by intercellular invasion of the root itself (Miller and Baker, 1985); in other plants the infection proceeds through a deformed root hair (Callaham and Torrey, 1977; Lalonde, 1977; Berry and Torrey, 1983). The infection process may be affected by the conditions under which the plant is grown (Blom, 1982; Zhang et al., 1984b). Also, although such infections can occur in axenic culture (Perinet and Lalonde, 1983), there is evidence that eubacterial "helpers" may play an assisting role in the infection process in the environment (Knowlton et al., 1980; Knowlton and Dawson, 1983). The plant host genera from which endophytes have been isolated are listed in Table 27.4 (modified from Baker, 1982). Nodulated species are found worldwide, particularly in temperate and semi-Arctic regions, but occur in the semitropics as well. Such plants are usually "pioneers," growing in nitrogen-poor environments unfavorable for most other plants. A complete listing of actinorhizal plant host species, some 199 belonging to 21 genera, is found in Lechevalier (1983).

Table 27.4.
Plant genera from which ***Frankia*** *strains have been isolated*

Plant		Endophyte(s)	
Family	Genus	isolated[a]	References[b]
Betulaceae	*Alnus*	+	1–7
Casuarinaceae	*Casuarina*	+	1,3,4,7–10
	Allocasuarina	+	11
Coriariaceae	*Coriaria*	−	
Datiscaceae	*Datisca*	+	12
Elaeagnaceae	*Elaeagnus*	+	1,7
	Hippophaë	+	3,7
	Shepherdia	+	13
Myricaceae	*Myrica*	+	1,3,4,7,14
	Comptonia	+	1,15,16
Rhamnaceae	*Ceanothus*	+	16
	Colletia	+	3,17
	Discaria	−	
	Kentrothamnus	−	
	Trevoa	−	
Rosaceae	*Cercocarpus*	+	4
	Chaemabatia	−	
	Cowania	+	4
	Dryas	−	
	Purshia	+	4
	Rubus	−	

[a] −, genera with nodules from which the endophyte has not yet been isolated.
[b] Key: *1,* Baker (1982); *2,* Normand and Lalonde (1982); *3,* Burggraaf and Shipton (1983); *4,* Baker and O'Keefe (1984); *5,* Hafeez et al. (1984); *6,* Horrière (1984); *7,* Huang et al. (1984, 1985); *8,* Gauthier et al., (1981a, 1981b); *9,* Diem et al. (1983); *10,* Zhang et al. (1984b); *11,* Zhang and Torrey (1985); *12,* Hafeez (personal communication); *13,* Perinet et al. (1985); *14,* Lechevalier and Lechevalier (1979); *15,* Callaham et al. (1978); *16,* Lechevalier and Ruan (1984); *17,* Y. Dommergues (personal communication).

Enrichment and Isolation Procedures

Prior to the first reports of isolation and growth in vitro of what we now recognize to be true representatives of the endophytes belonging to the genus *Frankia* (Pommer, 1959; Callaham et al., 1978), Quispel (1955, 1960), using assays of infectivity in plants, observed that ethanol-soluble factors extracted from actinorhizal plant roots might play an important role in the growth of the endophytes. This early prediction turns out to be true for the initial growth in vitro of some Sp− strains (Quispel et al., 1983). Some Sp+ types of frankiae may also require such special nutritional supplements for isolation and continued growth (Quispel and Burggraaf, 1981; Normand and Lalonde, 1982). Some of the active compounds have been identified as "special" fatty acids (Quispel et al., 1983) or as belonging to the porphyrins of the chlorin type (Burggraaf, 1984). However, such compounds are not required for all strains and many isolates have been obtained on a variety of unamended media such as glucose-asparagine (Pommer, 1959), 0.5% yeast extract (Callaham et al., 1978), Q Mod (peptone, yeast extract, phospholipids, and salts; Lalonde and Calvert, 1979), *Frankia* broth (yeast extract, casein hydrolysate, salts and vitamins; Baker and Torrey, 1979), dilute Bennett "S" medium (casein hydrolysate, glucose, salts; Lechevalier et al., 1983), or defined media containing Tween 80, succinate, or pyruvate (Gauthier et al., 1981a; Benson, 1982; Hafeez et al., 1984). In a time study of the outgrowth of frankiae from *Alnus rugosa* nodules, among the compounds tested pyruvate was the best carbon source, followed by succinate and malate; propionate and acetate were less effective (Benson et al., 1984b).

Actinorhizal nodules are best plated out when fresh; however, it has been observed that freezing at −20°C is a satisfactory method of preserving the viability of the endophyte. Other methods of preserving viable nodules include storage in soil or drying over silica. Surface sterilization of the nodules prior to plating out is desirable. Commonly used sterilizing agents include mercuric chloride (Callaham et al., 1978), Chloramine T (Diem et al., 1982), sodium hypochlorite (Baker et al., 1979), osmium tetroxide (Lalonde et al., 1981), and hydrogen peroxide (Zhang et al., 1984b). Isolation procedures include serial dilution of crushed nodule suspensions, microdissection of the nodule with or without pretreatment with such enzymes as cellulase and pectinase, to isolate the vesicle clusters, followed by direct plating (Callaham et al., 1978); filtration through graded membrane filters (Benson, 1982); or sucrose density-gradient centrifugation (Baker and Torrey, 1979). Incubation is carried out at 25–33°C. Colonies on isolation plates may appear as soon as 5–10 days after plating, but more commonly the interval is 4–8 weeks. In a few cases outgrowth may take up to a year (F. Horrière, personal communication). In some strains, colonies may be visible only under the microscope. As previously discussed, the role of media constituents may determine whether outgrowth takes place, and nutritional requirements may differ for different strains, even those isolated from the same plant host. Soil without obvious nodule fragments can be infective for plants, and the definition of the genus (Lechevalier and Lechevalier, 1984) has called for the recognition of free-living members of the genus. A review of the evidence for actual propagation of the endophyte within the soil concluded that it is equivocal (Van Dijk, 1979). *Frankia* strains have been isolated from phenol-treated rhizospheral soil using the sucrose density technique (Baker and O'Keefe, 1984).

Maintenance Procedures

Cultures are best transferred every 4–6 months with storage at room temperature (12–20°C) between transfers. There are no universal maintenance media: each strain must be evaluated for its long-term nutritional requirements. Some isolation media (see previous section) such as Bennett, *Frankia* broth, and Q Mod, are suitable for maintenance; in general, in our experience, complex media are better than defined ones. Fixed sources of nitrogen include ammonia (Blom, 1981), nitrate, amino acids, amides, urea, and proteins (Blom and Harkink, 1981; Lechevalier et al., 1982b; Shipton and Burggraaf, 1982a). Carbon sources include Tweens (Blom et al., 1980), and a Tween 80–bovine serum albumin–salts medium (L/2) has proved to be very useful for the stable maintenance of many strains (Lechevalier et al., 1982b). Other carbon sources are organic acids such as propionate (Blom, 1981); also, fumarate, succinate, malate, acetate, and pyruvate are suitable sources for some strains (Lechevalier et al., 1983; Tisa et al., 1983). As previously discussed, glucose and many other hexoses and pentoses are not suitable carbon sources for many frankiae because of lack of uptake (Blom and Harkink, 1981) except when combined with Tweens (Lechevalier et al., 1983), where a synergistic activity can be seen in type B strains. In contrast, type A strains can take up and utilize a variety of carbohydrates (Table 27.5) (Lechevalier et al., 1982b; Shipton and Burggraaf, 1982a). The vitamins biotin, calcium pantothenate, and riboflavin were stimulatory to growth in some strains, and 0.4 μg/l biotin was reported to enhance the nitrogen-fixing capability of one strain (Shipton and Burggraaf, 1983). Growth of frankiae declines with decreasing water potential (Shipton and Burggraaf, 1982b).

Microaerophilic strains such as those of type B do not grow at the surface of liquid media and cannot be maintained on slants. Strains such as those of type A tend to be more aerobic, form growth rings at the surface of liquid media, and can be maintained on slants. Most frankiae grow best at a neutral or slightly acid pH (6.4–6.8) and, depending on the strain, within a temperature range of 20–36°C. A few grow well at 37°C; some will grow very slowly at 15°C. None is known to grow at 40°C (Burggraaf and Shipton, 1982; Tisa et al., 1983; Moiroud et al., 1984).

Long-term preservation by lyophilization from broth or by storage in sterile soil is satisfactory. Alternately, stocks in growth media may be frozen and kept at −20°C; however, some strains do not survive this treatment. Nodules may be stored at −20°C for up to 5 years and longer without loss of infectivity.

Procedures for Testing of Special Characters

Morphological observations of colonies growing within agar media are easily made directly through the back of a Petri dish using a high-dry objective (20 ×, long working distance metallurgical) and a long working distance condenser. An aliquot of growth from liquid culture may be observed on a slide using high-dry objectives (40–60 ×) or oil immersion, with either light or Nomarski interference contrast optics.

Methods for analyses of cell wall composition and whole-cell amino acids, and sugars, and lipids, including fatty acids and phospholipids, are described in detail in Lechevalier and Lechevalier (1980), Mort et al. (1983), and Lopez et al. (1983). For DNA isolation and analysis consult An et al. (1983).

Physiological testing may be carried out using the methods of Lechevalier et al. (1982b, 1983) and Lechevalier and Ruan (1984). Techniques for demonstration of infectivity and effectivity in plants have been described (Callaham et al., 1978; Lalonde, 1979; Burggraaf and Valstar, 1984; Hafeez et al., 1984; Sellstedt and Huss-Danell, 1984).

Table 27.5.
*Carbohydrate utilization and acid production by **Frankia** species*[a,b]

Host plant	*Frankia* spp.	(Previous designation)	Utilization (U) and acid (A) formation from 0.5%[c]							
			A	G_1	G_2	G_3	M	S	T	X
Alnus rugosa	LLR 01321	(AirI1)	–	–	–	ND	–	–	ND	–
	LLR 01322	(AirI2)	–	U	–	ND	U	U	ND	–
Ceanothus americanus	LLR 03011	(CaI1)	–	U/A	–	–	U^v	U	U/A	U/A
	LLR 03013	(R2)	–	–	–	–	–	–	–	–
	LLR 03014	(R 50)	–	U/A	–	–	–	V	U/A	U/A
	LLR 03015	(R 96)	–	–	–	–	–	–	–	–
	DDB 030210	—	U	U	–	–	U^v	–	U	U
Casuarina equisetifolia	ORS 020604	(G2)	U/A	U	–	ND	U/A	–	U	U/A
	LLR 02021	(R43)[d]	–	–	–	–	–	–	–	–
	HFP 020202	(CcI2)[d]	–	–	–	–	–	–	–	–
	HFP 020203	(CcI3)[d]	–	–	–	–	–	–	–	–
Comptonia peregrina	HFP 070101	(CpI1)	–	–	–	–	–	–	–	–
	LLR 07011	(R 100)	U/A	U/A	–	–	U^v	U	–	U/A
Myrica gale	LLR 161101	(R82)	–	–	–	–	–	–	–	–
	DDB 160210	(2701)	–	U/A	–	–	–	U	–	U/A
	DDB 160410	(51118A)	U/A	U/A	–	–	–	–	–	U/A

[a]Symbols: *U*, Utilized; *U*[v], utilized variably; *A*, acid produced; –, not utilized; *ND*, not determined.
[b]Data from: Lechevalier et al. (1983), Lechevalier and Ruan (1984), and M. P. Lechevalier (unpublished).
[c]*A*, L-arabinose; G_1, D-glucose; G_2, glycerol; G_3, 3-*O*-methyl-D-glucose; *M*, maltose; *S*, sucrose, *T*, trehalose; *X*, D-xylose.
[d]Strains passed through *Casuarina cunninghaminana*.

Differentiation of the genus **Frankia** *from other genera*

A proposed definition of the genus *Frankia* (Lechevalier and Lechevalier, 1984) is as follows:

1. Actinomycetic, nitrogen-fixing, nodule-forming endophytes or endoparasites that have been grown in pure culture in vitro and that:
   a. induce effective or ineffective nodules in a host plant and may be reisolated from within the nodules of that plant, and
   b. produce sporangia containing nonmotile spores in submerged liquid culture, and may also form vesicles.
2. Free-living actinomycetes having no known nodule-forming or nitrogen-fixing capacity, but that show the morphology described in 1(b) above.

In addition to the *distinctive morphology* described above, it is possible to distinguish frankiae from other actinomycete taxa on the following basis:

1. *Distinctive chemistry*: When present, the sugar fucose serves to distinguish members of the genus from other actinomycete taxa. This sugar has been found in a few actinoplanetes (Ruan et al., 1986) and some coryneforms, both of which may be distinguished from frankiae on the basis of morphology. The previous report of fucose from the genus *Chainia* (Maiorova, 1965) was not confirmed for the strains figuring in the original report (M. P. Lechevalier, unpublished). Other chemical markers include 2-*O*-methyl-D-mannose and rhamnose, both of which occur in most strains. An unidentified amino acid that stains blue with ninhydrin ($R_{meso\text{-}DAP} = 0.95$) occurs in about 50% of frankia isolates and may be unique to the genus (Lechevalier and Lechevalier, 1979).
2. *Capacity to infect plants and induce nodule formation.*
3. *Capacity to fix nitrogen in vitro.*

Black-spored frankiae may be distinguished from micromonosporae forming clusters of spores by demonstration of the presence of a sporangial membrane (this is lacking in micromonosporae). Also, cell chemistry is useful: frankiae have a cell wall of chemotype III and a phospholipid composition of type PI; micromonosporae have a cell wall type of type II and phospholipids of type PII.

Taxonomic Comments

No description of the type species, *Frankia alni,* or any other species of the genus *Frankia* is possible at this time. Scientists working with frankiae agree that a description of the genus is relatively straightforward. However, at present, it is not possible to formulate a description of any species (Lechevalier, 1984). When Becking (1970) proposed to resurrect the genus *Frankia*, he based his species on the capability of the endophytes to infect certain plants. Thus, for example, the type species, *F. alni* (Sneath, 1982) was said to infect only the plant genus, *Alnus; F. brunchorstii* infected only *Myrica* (*Gale*) and *Comptonia*; and *F. elaeagni* only *Elaeagnus, Hippophaë*, and *Shepherdia*. Because at that time the first endophyte had not yet been isolated and maintained in pure culture, these host compatibility groupings were based on data from plants inoculated with crushed nodule suspensions. Unfortunately, these compatibility groups have not proven to be clear-cut now that we have pure cultures of the endophytes in hand. Some frankiae from *Alnus*, for example, can induce nodules on both *Elaeagnus* and *Alnus*; others can nodulate *Alnus, Myrica*, and *Comptonia*; thus, the Becking plant host-dependent taxonomic system does not appear to be satisfactory for *Frankia.*

If more conventional actinomycete taxonomic methods such as utilization and/or acid production from carbohydrates are used, it has been found that organisms isolated from the same plant host species differ not only in their host compatibility group but also in tests such as these (Table 27.5). Utilization of organic acids and production of β-glucosidase and urease also differ (Benson et al., 1984b; Horrière, 1984).

Thus, until a more complete understanding of what constitutes a species of *Frankia* is reached, it has been decided that no species names should be used. Rather, strains will be designated by a combination of three letters standing for the laboratory where the strain originated, followed by two numbers representing the plant host genus and up to eight further numbers to be used at the discretion of the person designating the strains. This system will replace the previously used systems of labeling strains such as CpI1, now HFP 070101, (first isolate from *Comptonia peregrina*) or ACN1^{AG} (first Sp− strain from *Alnus crispa* passed through *Alnus glutinosa*). A catalog listing these new designations along with a certain amount of data on each strain will be published periodically. The codes for each plant species and the type of information needed for cataloging are given in Lechevalier (1983).

SECTION 28

Actinoplanetes

Gernot Vobis

The actinoplanetes comprise the members of actinomycetes, which are closely related to the genus *Actinoplanes*. The name was introduced by Nonomura and Takagi (1977) to characterize their adaption to aquatic habitats with a motile stage during the life cycle. The term was used in the classification of Goodfellow and Cross (1984b) to include some genera of the family *Actinoplanaceae* and the genus *Micromonospora*, which have similar chemotaxonomic characters and nucleic acid affinities. They are Gram-positive, non-acidfast organisms growing with nonfragmenting, branched and septate hyphae, 0.2–1.6 (2.6) μm in diameter. Aerial mycelium is rarely developed or only sparse.

Nonmotile spores are produced by the genus *Micromonospora*. They are borne singly, sessile, or on short or long sporophores, occurring often in clusters, and are spherical, ovoid, or ellipsoidal (0.7–1.5 μm), with much-thickened wall layers sometimes bearing blunt spiny ornamentations.

The sporangiate members of the actinoplanetes produce motile spores within sporangia or spore vesicles that are developed at the tip of short or long sporangiophores, usually on the surface of the substrate. The spores are formed within a sporangial envelope by fragmentation of branched or unbranched, straight or coiled sporogenous hyphae. The multisporous sporangia of the genera *Actinoplanes*, *Ampullariella*, and *Pilimelia* have various shapes: cylindrical, bottle shaped, flask shaped, campanulate, lobate, digitate, spherical, subspherical, ovoid, pyriform, or irregular (3–20 × 6–30 μm). Finger-shaped to claviform, oligosporous sporangia (0.6–1.4 × 2.5–6.0 μm) containing a single row of 2–5 spores are produced by *Dactylosporangium*. Additionally, large globose bodies (spores) may be formed on the substrate hyphae. The sporangiospores are spherical to subspherical (0.8–2.0 μm), rod shaped (0.3–1.2 × 1.2–4.0 μm), or oblong, ellipsoidal, or ovoid to pyriform (0.4–1.3 × 0.5–1.8 μm). They are motile by means of flagella, which are generally arranged in polar or lateral tufts (Table 28.1). The spore walls have no ornamentations and consist of only a single layer.

The colonies on agar media are flat to elevated with smooth or wrinkled surfaces. Their color may have all nuances from pale yellow, orange, red, blue, brown, and green to black.

The peptidoglycan of the cell walls contains *meso*- and/or 3-hydroxydiaminopimelic acid and glycine as distinguishing components; xylose and arabinose are present in whole-cell hydrolysates (chemotype II and

Table 28.1.
Diagnostic characteristics of the genera of **Actinoplanetes**[a]

Characteristics	*Actinoplanes*	*Ampullariella*	*Pilimelia*	*Dactylosporangium*	*Micromonospora*
Sporangium	+	+	+	+	–
Oligosporous	–	–	–	+	–
Multisporous	+	+	+	–	–
Sporangiospores (motile)	+	+	+	+	–
Spherical (0.8–2.0 μm); polar flagella	+	–	–	–	–
Oblong, ellipsoidal, ovoid, pyriform (0.4–1.3 × 0.5–1.8 μm); polar flagella	–	–	–	+	–
Rod-shaped (0.5–1.0 × 2.0–4.0 μm); polar flagella	–	+	–	–	–
Rod-shaped (0.3–0.7 × 0.7–1.5 μm); lateral flagella	–	–	+	–	–
Single spores (nonmotile)	–[b]	–[b]	–[b]	+[c]	+
Spherical (0.7–1.5 μm); in clusters	–	–	–	–	+
Spherical (1.7–2.8 μm); not in clusters	–	–	–	+[c]	–
Decompose hairs (keratin)	–	–	+	–	–

[a] Symbols: +, 90% or more of strains are positive; –, 10% or less of strains are positive.
[b] "Conidia" with shape and arrangement similar to those of the sporangiospores or intercalar "chlamydospores" may be produced.
[c] "Globose spores" or "globose bodies" are produced.

sugar pattern D). The phospholipid pattern of the cell membrane corresponds to type PII, with phosphatidylethanolamine present and phosphatidylcholine and the unknown glucosamine-containing phospholipid absent. The pattern of fatty acids is not uniform. *Iso-* and *anteiso-*branched fatty acids are predominant in nearly all genera, with exception of *Pilimelia.* Cyclic acids are not present, unsaturated fatty acids may be present or not, and 10-methyl branched acids are found only in certain strains of *Micromonospora.* The menaquinone composition seems to be heterogeneous. Major amounts of tetra-, hexa-, and/or octa-hydrogenated menaquinones with 9, 10, and/or 12 isoprene units are distributed among the genera and species, and even within subspecies of *Micromonospora.* Their chemotaxonomic relationship is supported by DNA/DNA and DNA/rRNA reassociation studies and oligonucleotide sequencing of 16S rRNA.

Their growth is aerobic and mesophilic, with an optimum between 20° and 30°C and at a pH of about 7. Antibiotic metabolites and pigments may be produced. Intramycelial pigments are in most cases carotenoids, but blue-green and maroon to purple mycelial pigments also may occur. The diffusible pigments range from yellow, amber, pink, red, brown, auburn, purple, and violet to green. The production of melanoid pigments is mainly confined to the genus *Actinoplanes.*

Actinoplanetes have been found in soil, decaying plant material, and freshwater and marine habitats. Strains can be isolated directly from the samples by inoculation onto suitable agar media, with baiting techniques or with chemotactic methods.

The mol% G + C of the DNA is 71–73 (T_m, Bd).

Genus **Actinoplanes** *Couch 1950, 89 (*Amorphosporangium *Couch 1963, 65), emend. mut. char. 1955, 153*[AL*]

NORBERTO J. PALLERONI

Ac.ti.no.pla′nes. Gr. n. *actis, actinos* a ray, beam; Gr. n. *planes* a wanderer; M.L. masc. n. *Actinoplanes* literally, a ray wanderer (intended to signify an actinomycete with swimming spores).

Produce a fine, nonfragmenting, branching mycelium; Gram-positive, although part of the vegetative growth may be Gram-negative. Non-acid-fast. Aerial mycelium scanty or absent. Under certain conditions, many strains have **hyphae arranged in palisade formation. Highly colored.** Diffusible pigments of various colors may be produced. Spores produced within **sporangia** (spore vesicles), which are **spherical or subspherical to very irregular,** 3–20 × 6–30 μm, borne on short sporangiophores or sessile, occasionally within the agar. When palisade formation is present, the sporangia are mainly produced at the tip of the palisade hyphae. **Spores spherical, subspherical, or short rods,** variously arranged within the sporangia, where they are formed by fragmentation of the internal hypha either directly or after one or more ramifications. Upon immersion in water, **motile spores** are released from the sporangia, but in some instances motility begins some time after spore release. **Polar flagella** present in motile spores. Cell wall contains *meso*-diaminopimelic acid (*meso*-DAP) and glycine. D-xylose and L-arabinose are present in whole-cell hydrolysates. Among the phospholipids, phosphatidylethanolamine is present, while phosphatidylmethylethanolamine, phosphatidylcholine, glucosamine-containing phospholipids, and phosphatidylglycerol are absent. Aerobic. The spores have microaerophilic behavior. Chemo-organotrophic, mesophilic or moderately thermophilic. For most, the growth temperature range is 15–35°C. Most strains **do not require organic growth factors.** Mol% G+C of the DNA is 72–73 (T_m).

Type species: *Actinoplanes philippinensis* Couch 1950, 89.

Further Descriptive Information

Morphology

As mentioned by Couch (1963), the vegetative mycelium of *Actinoplanes* formed in water on a variety of plant and animal substrates is usually inconspicuous. On agar media in the laboratory, the aerial mycelium is usually absent, but some species (*A. teichomyceticus, A. rectilineatus*) may produce aerial mycelium that imparts to the colonies a powdery appearance (Lechevalier and Lechevalier, 1975; Parenti et al., 1978). The hyphae are fine, 0.2–1 μm in width, but they may reach 2.6 μm. Usually they do not fragment. In some agar media, straight branches arise from the substrate hyphae and grow toward the surface in parallel arrangements ("palisades"). This property is not universal, but when it occurs, sporangia form at the tip of the palisade hyphae.

Actinoplanes spores form within structures that have been variously named *sporangia* or *spore vesicles.* Although the latter name has been used (Palleroni, 1983a), some practical reasons (e.g. the use of the term "sporangiate actinomycetes") support the continued use of the term *sporangium.* Both *sporangium* and *vesicle* have been used to designate structures different from those characteristic of *Actinoplanes,* but since *sporangium* literally means "a container of spores," it seems to be appropriate.

The sporangium originates as an expansion of the outer sheath of a hyphal tip (Lechevalier and Holbert, 1965; Lechevalier et al., 1966b). The hypha grows inside this sheath, branching one or more times, and finally it fragments into spores. Lechevalier et al. (1966b) observed that the septation that gives origin to the spores is the result of annular ingrowth similar to that observed in other Gram-positive bacteria during division. A first septum forms at the base of the sporogenous hypha, and it is followed by new septa producing large segments. Subsequently, a new septation cycle takes place and these segments fragment into spores (Figs. 28.1 and 28.2).

Growth of the sporangium results in part from the accumulation of an electron-dense material surrounding the hyphal branches. This material becomes less uniformly dense as the sporangium matures, with concomitant clearer demarcation of the spores (Sharples et al., 1974). On dehydration, the thin sporangial envelope collapses over the spore mass, and the spore arrangement appears clearly visible under the scanning electron microscope (Fig. 28.3). The external ornamentation of the sporangium envelope may be taxonomically useful (Seino, 1983).

Spores of *Actinoplanes* are actively motile by means of several polar flagella (Fig. 28.4). Peritrichous flagellation has also been reported occasionally (Willoughby, 1968; Ruan et al., 1976). Some strains produce conidiophore systems as well as sporangia. The two structures may develop in an essentially similar manner (Willoughby, 1966).

Pigmentation

Actinoplanes colonies are usually brightly colored. Orange color is the most common, but some strains have red, yellow, rusty brown, blue, purple, green, or black mycelia. Szaniszlo (1968) has shown that organic solvent extracts from one strain of *A. philippinensis,* four strains of *A. utahensis,* and eight strains of *A. missouriensis* had identical spectral properties, which are those characteristic of carotenoid pigments. Cultures grown in the dark do not synthesize the pigments. Synthesis of the orange pigment of *A. brasiliensis* is not affected by light.

Methanolic extracts of *A. ferrugineus* do not have a characteristic carotenoid spectrum (Palleroni, 1979). A major peak at 408 nm and a shoul-

**AL* denotes the inclusion of this name on the Approved Lists of Bacterial Names (1980).

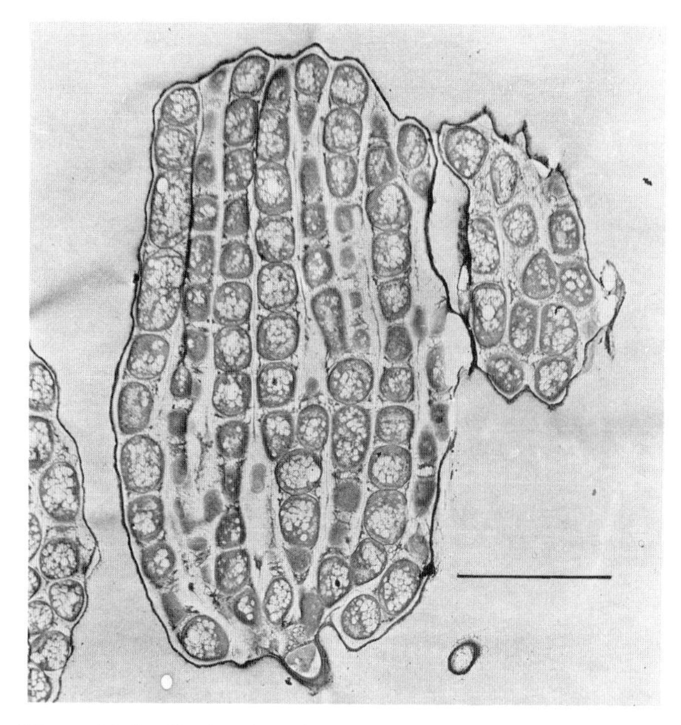

Figure 28.1. Section through sporangia of *Actinoplanes rectilineatus* at a late stage of spore formation. *Bar,* 3 µm. (Courtesy of Dr. H. A. Lechevalier.)

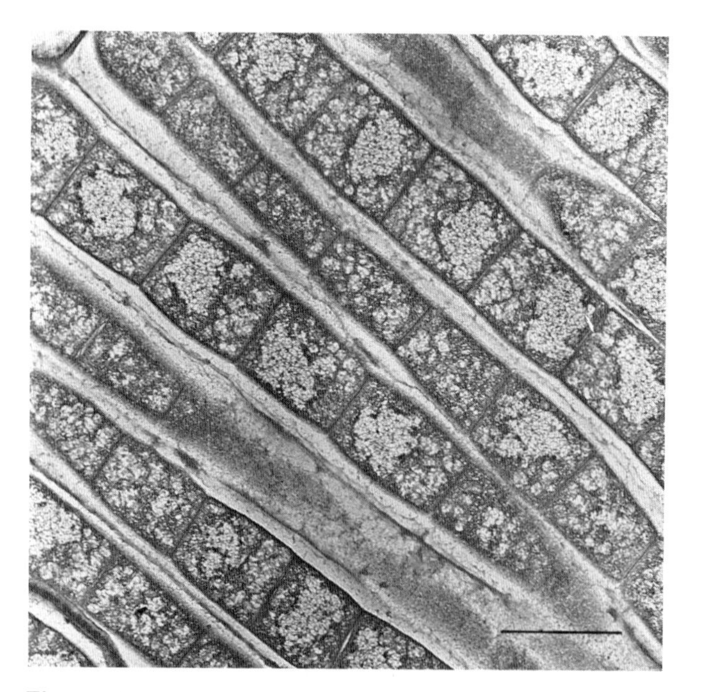

Figure 28.2. Hyphal septation within an *Actinoplanes rectilineatus* sporangium. *Bar,* 1 µm. (Courtesy of Dr. H. A. Lechevalier.)

Figure 28.3. Scanning electron micrograph of an *Actinoplanes* species mature sporangium. *Bar,* 5 µm. (Courtesy of Dr. Akio Seino; reproduced with permission from A. Seino, Hakko to Kogyo (Fermentation and Industry) *41(3):* 3–4, 1983.)

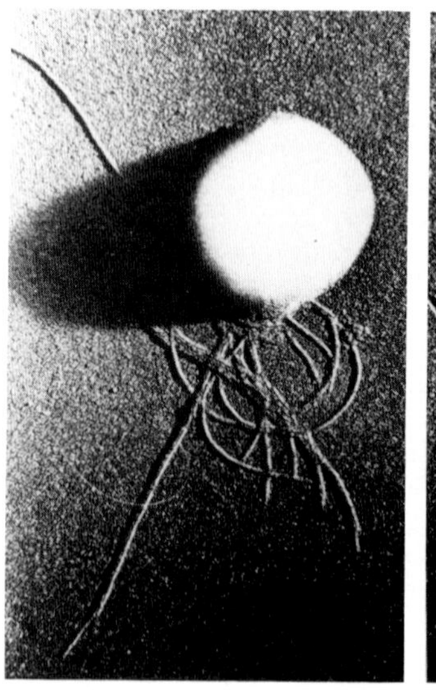

Figure 28.4. Electron micrographs of *Actinoplanes ferrugineus* spores. *Bar,* 1 µm. (Reproduced with permission from N.J. Palleroni, International Journal of Systematic Bacteriology *29:* 51–55, 1979, © American Society for Microbiology, Washington, D.C.)

der at 418 nm are present, but the nature of the pigment has not been further investigated. From two strains of *Actinoplanes* isolated from different soil samples, a red pigment having the structure of an anthraquinone derivative has been isolated and characterized (N. J. Palleroni, unpublished data). A soluble pigment produced in synthetic media by *A. cyaneus* has been shown to belong to the celocomycin-actinorodine group (Terekova et al., 1977).

Cell Wall Composition

The cell wall composition of *Actinoplanes* corresponds to the amino acid pattern of type II (Lechevalier et al., 1971a), characterized by *meso*-DAP and glycine, a sugar composition of type D (xylose and arabinose), and a phospholipid pattern PII (with phosphatidylethanolamine) (Lechevalier and Lechevalier, 1981b). However, the amino acid composition may be considerably more complex, as Szaniszlo and Gooder (1967) have shown. In some instances, DD-DAP can be present, and in others, hydroxy-DAP but none of the other amino acids may appear in the hydrolysates. Some species (e.g. *A. caeruleus*) lack DAP or its derivatives (Horan and Brodsky, 1986). According to Szaniszlo and Gooder (1967), the amino acid composition is highly variable even among strains of the same species, and the patterns may be more useful at the species than at the genus level. Glutamate and alanine are among the cell wall constituents, and in all *Actinoplanaceae* analyzed by these authors, the alanine content was significantly lower than that of glutamate. Among the sugars, xylose and arabinose were found together with galactose, glucose, mannose, and an unidentified deoxyhexose.

The cell wall composition of one strain of each of 19 species of *Micromonospora* were analyzed by Kawamoto et al. (1981), and all except the *M. globosa* strain appeared to contain glycolate in amounts approximately equimolar with DAP. Of 12 other genera examined, 11 strains of different species of *Actinoplanes,* and one strain of each of the species *Amorphosporangium auranticolor, Ampullariella digitata,* and *Dactylosporangium aurantiacum,* also contained appreciable amounts of glycolate. These findings suggest the presence of glycolylmuramic acid instead of acetylmuramic acid in the composition of the peptidoglycan, an idea that is supported by the low sensitivity of these organisms to lysozyme (β-*N*-acetylmuramidase).

Collins et al. (1984) have reported that *Actinoplanes philippinensis* and *A. utahensis,* as well as strains of *Ampullariella, Kineospora,* and *Streptosporangium,* produce menaquinones with nine isoprene units as the major component. Similar compounds were found in *Micromonospora,* although strains of this genus are heterogeneous in menaquinone composition. Kroppenstedt (1985) has reviewed the fatty acid and menaquinone composition of many actinomycetes (including *Actinoplanes*) and related organisms, and has pointed out that the relationships emerging from the results of the analyses agree more closely with the data obtained from other chemotaxonomic approaches, and less with conventional morphological criteria.

DNA Base Composition

The mol% G+C of the DNA of *Actinoplanes* strains is 72–73, in the high range characteristic of actinomycetes (Yamaguchi, 1967; Farina and Bradley, 1970).

Life Cycle

The morphological changes of typical actinoplanetes during their life cycle have been described by Bland and Couch (1981). *Actinoplanes* sporangia form at the tip of short stalks representing part of a rudimentary aerial mycelium, or at the tip of the palisade hyphae. Immersion of the sporangia in water usually results in the release of the spores in approximately 1 h at room temperature. The process can be facilitated by the addition of wetting agents (Tween 80) since the surface of the sporangia is hydrophobic (Higgins, 1967). During immersion in water, the spores swell, becoming more clearly visible within the sporangium. Water also triggers the formation of flagella and, as a general rule, spores show signs of motility before their release. Soon afterward, the motile spores emerge through breaks in the sporangial wall (Lechevalier and Holbert, 1965).

As Higgins (1967) has shown, both swelling and motility of the spores contribute to their release, thus contradicting the suggestion made by Couch (1963) that this may be a consequence of the swelling of the intersporal material. Of the two factors, swelling of the spores is the more important, since in some instances, the sporangium dehiscence gives nonmotile spores.

Motility of the spores suspended in water or in diluted buffers may last for more than 1 day at room temperature. If the liquid contains nutrients, germination may ensue after a few hours, and the emergence of the filamentous stage completes the cycle of morphogenetic changes.

Physiology and Metabolism

Our knowledge of *Actinoplanes* physiology and metabolism is extremely limited. Most investigations have focused on the spores, but aside from these studies, many other basic aspects of the biology of these organisms have been ignored.

Sporangia formation occurs preferentially in certain media, but no single medium can be recommended for all strains; both starch casein (see footnote, page 2422) and the minimal medium (see footnote this page) with certain carbon sources induce copious sporangia formation. Humic and fulvic acids (Willoughby et al., 1968; Willoughby and Baker, 1969) as well as tea infusion (Parenti et al., 1978) have been found to stimulate sporangiogenesis.

Higgins (1967) has observed that spores age rapidly as indicated by the loss of their capacity for acquiring motility soon after hydration, but this property may be restored by the addition of glucose. In other *Actinoplanaceae* Couch and Koch (1962) have observed induction of motility by mixtures of amino acids, and substantial increase of motility by arginine and urea. The effect of arginine is reminiscent of the role of this amino acid as an energy source for motility in other bacteria (Sherris et al., 1959).

New flagella can arise on spores that have been deflagellated, and even though glucose and amino acids are required, reflagellation can occur in the presence of inhibitors of protein and nucleic acid synthesis. Inhibitors such as *p*-chloromercuribenzoate, iodoacetate, iodoacetamide, azide, and 2,4-dinitrophenol inhibit the formation of flagella and also suppress motility (Higgins, 1967). Carbonylcyanide-*m*-chlorophenylhydrazone,which is effective against oxidative phosphorylation, can be added to this list (Palleroni, 1983a).

Sporangia probably survive for long periods in desiccated conditions, such as dry soil and leaf litter (Makkar and Cross, 1982). *Actinoplanes brasiliensis* spores are able to germinate in a minimal medium with glucose as sole carbon source. The proportion of germinating spores is close to 100%, which makes this species a good subject for research. In contrast, the strain used by Higgins (1967) gave a much lower proportion of spores able to germinate under minimal conditions, and normal levels of germination depended on the addition of amino acids. Germination can be inhibited by chloramphenicol and actinomycin D, and even though it is stimulated by 6-azauracil, this compound inhibits further growth (Higgins, 1967).

Spores of *A. brasiliensis,* like those of many other members of the group, can swim in a straight line, and they frequently change the direction of movement in a manner similar to that of other bacteria (Berg and Brown, 1972). No phototactic effect could be observed in the motility of *A. brasiliensis* spores, and many organic compounds appeared ineffective as chemotactic attractants. Quite surprisingly, chloride and bromide ions had a definite positive chemotactic effect. These anions act as attractants at a relatively high concentration (1 mM or higher), and at present it is hard to assess the physiological significance of the phenomenon. As stated in the original paper (Palleroni, 1976b), the effect of the ions may reflect the presence of a low-specificity receptor whose normal attractant is yet to be discovered. Recent investigations on the retinal protein halorhodopsin of *Halobacterium halobium,* which acts as a light-driven chloride pump (Schobert and Lanyi, 1982) may shed some light on the *Actinoplanes* phenomenon. As in other instances of bacterial chemotactic behavior, methionine has a marked stimulatory effect, suggesting the participation of protein methylation (Palleroni, 1983a). Chloride attraction forms the basis of a very useful method for isolation of *Actinoplanes* strains from nature (see below).

Actinoplanes strains can grow in chemically defined media prepared with inorganic salts and a single compound as the sole source of carbon and energy. The medium described by Palleroni and Doudoroff (1972) is adequate.† It contains a small amount of citrate, which prevents the pre-

†Medium of Palleroni and Doudoroff (1972) (grams per liter in 0.33 M Na/K phosphate [pH 6.8]): NH_4Cl, 1.0; $MgSO_4{\cdot}7H_2O$, 0.5; ferric ammonium citrate, 0.05; $CaCl_2$, 0.005. The first two ingredients are added to the buffer and sterilized by autoclaving. The ferric ammonium citrate and $CaCl_2$ are added aseptically from a single stock solution that has been sterilized by filtration.

cipitation of iron, but is not sufficient to support growth. This chemically defined medium is ideal for physiological studies, but in practice minimal media are seldom used routinely for subcultures and maintenance. Most complex media generally recommended for actinomycetes are adequate for *Actinoplanes,* although many of them may be unsuitable for sporulation.

Washed spore suspensions can be easily handled for various physiological studies, including the determination of extensive nutritional spectra similar to those used for the characterization of the pseudomonads (Stanier et al., 1966; Palleroni, 1984).

Genetics

Actinoplanes brasiliensis is the only species of the genus to have been subjected to genetic studies (Palleroni, 1983b). Mutations affecting various characters such as color, sporulation, streptomycin resistance, and amino acid biosynthesis, have been induced by ultraviolet (UV) irradiation. Spores are the material of choice for mutagenesis experiments, probably being uninucleated. A very convenient method for the selection of auxotrophic mutants has been described (Palleroni, 1983b). As with other organisms, glycine inhibits growth, and the use of subinhibitory concentrations of this amino acid in culture media gives rise to cells that are sensitive to lysozyme and can be converted into protoplasts in hypertonic sucrose solutions. Protoplast fusion by polyethylene glycol and regeneration in appropriate media give cell populations that can be subjected to recombination studies and/or strain improvement.

Bacteriophages

Knowledge of *Actinoplanes* phages is extremely limited. In a study on actinophages in fresh water, Willoughby et al. (1972) identified a virus of high specificity for *Actinoplanes.* The authors suggested the possibility of using the identification of phage as an indicator of activity of the host in the corresponding habitat. Host ranges of actinophages have been analyzed by Wellington and Williams (1981a) and by Prauser (1984c). The latter author tested strains of *Actinoplanes, Ampullariella, Dactylosporangium, Micromonospora, Planomonospora,* and *Streptosporangium* for their sensitivity to four *Actinoplanes* and five *Micromonospora* phages, and found that the phages were capable of multiplying on strains of all the above genera, except *Streptosporangium.*

Ecology

The names of many species of *Actinoplanes* (*philippinensis, utahensis, italicus, brasiliensis, nipponensis,* etc.) indicate a worldwide distribution and suggest their presence in many different habitats but, as noted by Parenti and Coronelli (1979), in spite of having isolated some 20,000 strains (probably the largest collection in the world), there was no correlation between their abundance and type of soil.

Soils of neutral pH are good sources (Nonomura and Takagi, 1977). Isolation from desert soils (Couch, 1957), from oven-desiccated soils (Nonomura and Takagi, 1977), and from sand dunes (Palleroni, 1976b) is a good indication of resistance to desiccation. Many strains have been isolated from decaying plant material (Willoughby, 1969a; Makkar and Cross, 1982). Parenti and Coronelli (1979) have suggested the possibility that xylose may be the prime energy source in nature for these organisms, and that they may participate actively in the decomposition of pentosans. The same may be true for chitin, which has been used as a carbon source for the isolation of strains (see below). However, chitin degradation is extremely slow, and the greatest advantage of chitin media is that they discourage the growth of other microorganisms (Willoughby, 1968).

A property of *Actinoplanes* strains that may have interesting ecological implications is a slight but definite capacity for oxidation of CO to CO_2 (Bartholomew and Alexander, 1979). This oxidation is not part of an auxotrophic pathway.

Enrichment and Isolation Procedures

Strains of *Actinoplanes* are found incidentally and in small numbers on agar media streaked with soil suspensions, but the fact that their life cycle includes a stage in which the cells are motile has stimulated the ingenuity of workers interested in isolating these organisms. One of the most ingenious and successful procedures was originally designed for the isolation of water molds and chytrids (Couch, 1939, 1949, 1954). Soil is placed in a Petri dish, charcoal-treated water is added to cover the sample, and bait of various types (seeds, pieces of leaves, bits of filter paper, pollen grains, hair, etc.) are placed on the surface of the water or partially submerged. The plates are left undisturbed for several days at room temperature and inspected from time to time under a dissecting microscope. Occasionally, among the microorganisms colonizing the bait, typical *Actinoplanes* sporangia can be seen attached to the material by short stalks. Pollen grains are the best baiting material, since they are in effect bags of nutrients that always float and do not seriously interfere with microscopic observations. Once pollen grains with *Actinoplanes* growth have been identified, they can be picked up and used for streaking agar plates.

Many workers have used the baiting method with success. A convenient variation consists of making a wet mount with the colonized pollen grains in a drop of water on a slide, and following the spore liberation microscopically. When this occurs, the liquid, now enriched with swimming spores, can be used as an inoculum for isolation. Pollen grains of many different plant species can be used as baits, but Willoughby (1968) and Nonomura and Takagi (1977), among others, considered that *Pinus* pollen gives the best results and it is highly specific for *Actinoplanes.* Willoughby (1969a) used the pollen bait method as a reference, but generally he preferred more orthodox techniques of direct isolation on special media. He emphasized the importance of selecting suitable natural materials, such as plant residues found on the shores of lakes and streams, where these materials are alternatively wetted and dried as the level of the water rises and falls. Before proceeding to the isolation proper, the residues are washed to remove major debris and to reduce the number of contaminants, and are then incubated in humid chambers to encourage sporulation. The isolation media recommended by Willoughby (1968) include starch-casein agar[‡] and chitin agar.[§]

Based on actinoplanetes' ability to withstand desiccation, Makkar and

---

[‡]Starch-casein agar (modified from Waksman, 1961, p. 330) (grams per liter): soluble starch, 10; casein, 1.0, KH_2PO_4, 0.5; $MgSO_4$, 0.5 ($MgSO_4 \cdot 7H_2O$, 1.02); agar, 15; pH 7.0–7.5. Ingredients are dissolved in the following order: phosphate, sulfate, starch, casein. It is helpful to add the casein as a solution in diluted NaOH.

[§]Chitin agar. This medium can be prepared following one of these procedures:

1. Formulation according to Willoughby (1968): 2.5 g purified chitin are placed in a beaker with 70 ml of 50% sulfuric acid. With occasional stirring, the bulk of the chitin dissolves in 90 min or less. The insoluble fraction is filtered off by means of a sintered glass filter, and the chitin solution poured into 1050 ml distilled water. The fine white chitin precipitate is centrifuged and washed repeatedly with water until the pH of the liquid approaches 4.0. The chitin suspension is added to the rest of the components, which are (in grams per liter): $CaCO_3$, 0.02; $FeSO_4 \cdot 7H_2O$, 0.01; KCl, 1.71; $MgSO_4 \cdot 7H_2O$, 0.05; $Na_2HPO_4 \cdot 12H_2O$, 4.11; agar, 18.
2. Formulation according to Makkar and Cross (1982): Crude chitin is washed alternatively with 1 M NaOH and 1 M HCl for 5 periods of 24 h. This is followed by four washes with ethanol (95% v/v). Purified white chitin (15 g) is dissolved in 100 ml concentrated HCl in about 20 min with stirring in an ice bath. After filtration through glass wool, the solution is poured into cold distilled water to precipitate the chitin. The insoluble fraction on the filter is treated again with HCl, and the process repeated until no more precipitate is obtained on dilution in water. The colloidal chitin is allowed to settle overnight and the supernatant decanted. The suspension is neutralized to pH 7.0 with NaOH solution, and the precipitated chitin is centrifuged, washed, and stored as a paste at 4°C. Colloidal chitin agar is prepared by using (in grams per liter of distilled water): colloidal chitin, 2.0 (dry weight); $CaCO_3$, 0.02; $FeSO_4 \cdot 7H_2O$, 0.01; KCl, 1.71; $MgSO_4 \cdot 7H_2O$, 0.005; Na_2HPO_4, 1.63; agar, 18; pH 7.2.

Cross (1982) developed an isolation method in which the source material is dried and subsequently rehydrated under controlled conditions; this achieves a substantial reduction in the proportion of contaminants. As suggested by the authors, the method may be used in combination with the capillary procedure described below. Among the media used for the isolation proper, the authors recommended the colloidal chitin medium (footnote §, page 2422) and soil extract agar.‖

The study of the chemotactic behavior of *A. brasiliensis* spores (Palleroni, 1976b) suggested the basis for another isolation method of *Actinoplanes* from nature. Spores of this species are attracted by chloride, a fact that made the idea particularly appealing as the basis of an isolation method, since these ions have not been reported as attractants for other motile organisms. The chemotactic method (Palleroni, 1980) consists of flooding the soil sample with water, as recommended in the baiting procedure, and after approximately 1 h, immersing a 1-µl capillary (Drummond "Micro-caps," Drummond Scientific Co.) filled with sterile dilute phosphate buffer (5–10 mM, pH 6.8) containing KCl (2 mM) in the liquid phase, keeping the tip of the capillary about 1 mm below the surface of the liquid. The operations are conveniently performed using lucite chambers that are a modification of those recommended for chemotaxis experiments (Palleroni, 1976a). The chambers can be improvised in a number of ways, using, for instance, some commercially available multiwell plates designed for tissue culture work.

One hour after immersion of the capillary, it is removed, the exterior is carefully washed with a few drops of sterile water, and the contents are blown into water or buffer. From the suspension, aliquots may be streaked onto starch-casein agar plates. The method is very effective and demands considerably less time (a few minutes in a total period of about 2 h) than other baiting procedures. At present, however, there is no good explanation for the success of the procedure, since chloride attraction is not universal among *Actinoplanes* species and there is a considerable number of chloride-indifferent strains among those trapped into the capillaries. Other factors that may contribute are the microaerophilic behavior of the spores and negative chemotaxis to repellents that may be present in the soil samples (Palleroni, 1980).

Maintenance Procedures

Actinoplanetes do not differ markedly from other actinomycetes in their maintenance requirements. Good results are obtained by freezing concentrated spore suspensions in dilute phosphate buffer or in liquid media with 5% (v/v) glycerol, at −20°C or below (preferably at liquid nitrogen temperature). Viability seems somewhat better when part of the vegetative mass is also present in the suspension. Both mycelium and sporangia survive lyophilization (Parenti and Coronelli, 1979).

Procedures for Testing of Special Characters

Media and procedures currently used for descriptive studies of *Actinoplanes* species do not differ substantially from those in use for *Streptomyces.* Only a list of the methods followed by most authors will be given here, together with the respective references. Characteristics of growth and sporulation are recorded in the media described by Shirling and Gottlieb (1966). In addition, media described by Waksman (1950, 1961), Cross et al. (1963), and Hickey and Tresner (1952) have been used for descriptive purposes. Color determinations are usually performed with reference to standard color charts (Ridgway, 1912; Maerz and Paul, 1950; *Color Harmony Manual,* ed. 4, Container Corporation of America, Chicago, 1958).

Formation of palisade hyphae, shape and size of the sporangia, and shape and size of the spores are part of standard descriptions. Flagella staining by procedures used for other procaryotes is seldom successful, and by far the best method of flagellar characterization is examination under the electron microscope.

The composition of whole-cell hydrolysates for the determination of cell wall composition is usually done following procedures described by Becker et al. (1964) and Lechevalier and Lechevalier (1970b). When cell wall isolation prior to analysis is preferred, the method described by Yamaguchi (1965) can be followed.

Physiological characters often reported in the literature include growth at various temperatures (usually from 10° to 50°C); hydrolysis of gelatin, casein, and starch; tyrosine degradation (often improperly named tyrosine hydrolysis); production of melanin and of H_2S; nitrate reduction; and NaCl tolerance. Other properties include carbon assimilation tests, usually including several carbohydrates, but which can be extended to many different organic compounds (Palleroni, 1979); and, in special cases, the chemotactic properties of the motile spores (Palleroni, 1976b).

Differentiation of the genus **Actinoplanes** *from other genera*

Spores within sporangia are the main morphological feature defining the actinoplanetes from other actinomycetes. Similar genera include *Actinoplanes, Amorphosporangium, Ampullariella, Pilimelia, Spirillospora, Streptosporangium, Dactylosporangium, Planomonospora,* and *Planobispora* (Bland and Couch, 1981). Phenotypic characters permitting a differentiation among these genera are presented in Table 28.2. The shape of the sporangium, and the shape, motility, flagellation type, and arrangement of the spores within the sporangia, are used for the differentiation of genera.

Taxonomic Comments

The classification obtained by following the above criteria does not reflect the natural relationships among the various genera, and there are major points of discrepancy from taxonomic arrangements based on cell wall composition, nucleic acid homology studies, and 16S rRNA similarities (Farina and Bradley, 1970; Stackebrandt et al., 1981) (Table 28.3). Thus, *Micromonospora,* a nonsporangiate actinomycete, appears to be more closely related to *Actinoplanes* than is *Streptosporangium,* although the latter is a taxon with phenotypic properties very similar to those of *Actinoplanes.* This is one of the many examples of lack of congruence between traditional phenotypic and phylogenetic systems of classification of procaryotic organisms.

The genus *Amorphosporangium* is considered here to be synonymous with *Actinoplanes.* The phenotypic differences between these two genera are mainly the shape of the sporangia and the shape of the spores. *Actinoplanes* has been described as having globose sporangia and globose to subglobose spores, whereas *Amorphosporangium* has very irregular sporangia and the spores are short rods. A clear separation of these two genera is difficult on the basis of these properties. As mentioned by Goodfellow and Cross (1984b), the description of *Amorphosporangium globisporus* (Thiemann, 1967), later to be assigned to *Actinoplanes,* is a good example of the difficulties found in attempting to differentiate between the two genera. Many *Actinoplanes* strains have sporangia approaching a spherical shape together with irregularly shaped ones. In other strains, the irregular shape predominates. The spores of *Actinoplanes* vary from globose to short rods, and a proportion of elongated cells occurs in the spore population of many strains. The presence of a tuft of flagella in a small area of the *Actinoplanes* globose spore indicates a degree of polarity that is compatible with the idea that these cells may be very short rods rather than truly spherical.

‖Soil extract agar (Makkar and Cross, 1982): 150 g garden soil are stirred into 600 ml tap water. The suspension is filtered immediately through Whatman No. 1 filter paper. The extract is made to 1 l with tap water, the pH is adjusted to 7.2, and agar is added (18 g/l). To this and other media used for isolation, actidione (cycloheximide) can be added from a sterile solution after autoclaving, to a final concentration of 50 µg/ml. This compound is tolerated by actinomycetes and other bacteria, and inhibits fungal growth.

Table 28.2.
Main phenotypic characters for the differentiation of the genera of sporangiate actinomycetes

Genus	Shape of sporangium	Shape of spores	Spore number per sporangium	Spore arrangement	Spore motility	Other salient properties
Actinoplanes	Globose to irregular	Spherical, subspherical, rods	Many	Coils, parallel chains	+	
Ampullariella	Bottle, digitate, lobate	Rods,	Many	Parallel chains and coils	+	
Pilimelia	Spherical, cylindrical	Rods	Many	Parallel chains	+	Keratinophilic
Spirillospora	Spherical	Rods, spirals	Many	Coils	+	
Streptosporangium	Spherical, claviform	Spherical, oval, short rods	Many	Coils	−	
Planomonospora	Oval, elongated	Rods	1		+	
Planobispora	Oval, elongated	Straight or curved rods	2	Longitudinal pairs	+	
Dactylosporangium	Finger-like	Oval, pyriform, cylindrical	2–5	Single row	+	

Table 28.3.
Salient properties of sporangiate actinomycetes and related genera

Genus	DNA homology cluster[a]	Peptidoglycan type[b]	Sugar composition	Aerial mycelium[c]	Spore motility[c]	Sporangia[c]
Actinoplanes	I	II	xylose-arabinose	−	+	+
Ampullariella	I	II	xylose-arabinose	−	+	+
Dactylosporangium	I	II	xylose-arabinose	−	+	+
Micromonospora	I	II	xylose-arabinose	−	−	−
Streptosporangium	II	III	3-*O*-methyl-D-galactose	+	−	+
Planomonospora	II	III	3-*O*-methyl-D-galactose	+	+	+
Planobispora	II	III	3-*O*-methyl-D-galactose	+	+	+
Spirillospora		III	3-*O*-methyl-D-galactose	+	+	+
Actinomadura		III	3-*O*-methyl-D-galactose	+	−	−

[a]Stackebrandt et al. (1981).
[b]According to Lechevalier and Lechevalier (1970b).
[c]Symbols: see Table 28.1.

Because modern approaches to the taxonomy of actinomycetes (Stackebrandt et al., 1981; Stackebrandt and Schleifer, 1984) have allowed the creation of an *Actinoplanes* (or *Micromonospora*) cluster, which includes *Amorphosporangium,* the two names should be considered as synonyms.

Acknowledgments

The unconditional help and support of Professor Hubert A. Lechevalier is acknowledged.

Further reading

Goodfellow, M., M. Mordarski, and S.T. Williams (Editors). 1984. The Biology of the Actinomycetes. Academic Press, London.
Palleroni, N.J. 1983. Biology of *Actinoplanes.* Actinomycetes *17:* 46–55.
Parenti, F. and C. Coronelli (1979). Members of the genus *Actinoplanes* and their antibiotics. Annu. Rev. Microbiol. *33:* 389–412.

Differentiation and characteristics of the species of the genus **Actinoplanes**

General phenotypic characteristics of *Actinoplanes* species taken from the literature and, in part, from unpublished observations, are presented in Tables 28.4 through 28.6. Table 28.4 gives morphological and physiological information on several species, and Tables 28.5 and 28.6 summarize the results of carbon assimilation tests. It is interesting to note in Table 28.5 that, aside from glucose, the two pentoses, arabinose and xylose, are almost universal substrates. There is one species (*A. auranticolor*) recorded as giving negative (or extremely slow) growth on xylose (Hanton, 1968).

Nutritional studies are an excellent source of phenotypic data for use in the differentiation of species, and they may also reveal patterns suggestive of more detailed investigations on catabolic pathways. It is clear, however, from Table 28.5 that the carbon assimilation tests, as traditionally performed, can give information of only limited value for species differentiation. A more detailed nutritional analysis, including 135 different carbon compounds, was performed on strains of various species (N.J. Palleroni, unpublished data), and some of the results are summarized in Table 28.6, which only includes the compounds most taxonomically

Table 28.4.
Morphological and physiological characteristics of the species of the genus **Actinoplanes** [a]

Species	Sporangial shape and size (μm)	Spore arrangement shape and size (μm)	Aerial mycelium	Mycelial color	Soluble pigment	Physiological characters[b]
1. *A. philippinensis* (ATCC 12247)	Globose to oval (8–25)	Coils; globose (1–1.2)	Absent	Yellow to orange-brown	Brown	A, b, C, D, E, F, g, h, I, J
2. *A. utahensis* (ATCC 14539)	Irregular (5–18)	Coils; sub globose(1–2)	Absent	Orange to brown-orange	Absent	A, B, C, D, E, F, g, h, I, J
3. *A. missouriensis* (ATCC 14538)	Globose, subglobose irregular (6–14)	Coils; globose (1–1.2)	Absent	Orange	Absent	a, b, C, D, E, F, g, h, I, j
4. *A. brasiliensis* (ATCC 25844	Irregular to umbelliform (3.5–11.5)	Coils subglobose (1.2 × 1.7–2.3)	Absent	Orange	Absent	a, b, c, D, E, F, g, h, I, J
5. *A. italicus* (ATCC 27366)	Globose to oval (6–11)	Coils; globose to oval (1–2)	Absent	Cherry-red	Cherry-red	A, B, C, D, e, F, g, H, I, J
6. *A. rectilineatus* (*ATCC 29234)*	Cylindrical (6–14 × 10–15)	Long rows (1.5–2)	Short sterile	Orange	Absent	B, C, D, f, G, H,I,J
7. *A. deccanensis* (ATCC 21983)	Globose (4–7	Coils; glo-bose (1–1.5)	Absent	Orange	Absent	a, B, d, E, F, g, h, I, J
8. *A. ferrugineus* (ATCC 29868)	Globose to irregular	Coils; globose (0.9–1)	Short sterile	Rusty brown	brown	a, B, C, D, E, F, I, J
9. *A. auranticolor* (ATCC 15330)	Very irregu-lar, lobed (6–25 × 8–15)	Irregular; rods (0.5–0.7 × 1–1.5)	Absent	Orange	Absent yellowish or amber	B, c, D, f, I, J
10. *A. globisporus* (ATCC 23056)	Irregular (3–5 × 4–7)	Coils; globose (0.8–1)	Rudimentary chlamy-dospores	Cream to light orange	Absent	a, b, D, e, f, g, h, I, j
11. "*A. ianthinogenes*" (ATCC 21884)	Globose (8–15)	Coils; subglobose (1.4–1.8)	Absent	Violet	Absent	a, b, C, D, E, F, g, h, I, J
12. "*A. garbadinensis*" (ATCC 31049)	Globose, lobate, irregular (7–12)	Coils; sub-globose (1–1.5)	Rudimentary sterile	Orange	Brown	A, B, C, d, e, F, g, h, I, J
13. "*A. liguriae*" (ATCC 31048)	Globose to oval (15–25)	Coils; globose (1.5–2)	Absent	Orange	Yellow-amber	a, b, c, D, e, f, g, h, I, j
14. "*A. teichomyceticus*" (ATCC 31121)	Globose to oval (15–25)	Coils; globose to oval (1.5–2)	Long sterile	Orange	Absent	A, B, c, D, e, F, g, H, I, J
15. *A. caeruleus* (ATCC 33937)	Globose to irregular (6–16)	Globose to oval (1.3–2)	Absent	Tan to blue	Yellowish brown	b, C, D, F, J

[a]Modified from Parenti and Coronelli (1979).
[b] *A*, H_2S formation; *B*, melanin production; *C*, tyrosine degradation; *D*, casein hydrolysis; *E*, calcium malate degradation; *F*, nitrate reduction; *G*, litmus milk coagulation; *H*, litmus milk peptonization; *I*, starch hydrolysis; *J*, gelatine liquefaction. Positive and negative properties are indicated by capital and small letters, respectively. Letters are missing when the results have not been reported.

Table 28.5.
Carbon assimilation by members of various **Actinoplanes** species[a]

Carbon sources	1. *A. philippinensis* (ATCC 12427)	2. *A. utahensis* (ATCC 14539)	3. *A. missouriensis* (ATCC 14538)	4. *A. brasiliensis* (ATCC 25844)	5. *A. italicus* (ATCC 27366)	6. *A. rectilineatus*[b] (ATCC 29234)	7. *A. deccanensis* ATCC 21983)	8. *A. ferrugineus*[c] (ATCC 29868)	9. *A. auranticolor* (ATCC 15330)	10. *A. globisporus* (ATCC 23056)	11. "*A. ianthinogenes*" (ATCC 21884)	12. "*A. garbadinensis*" (ATCC 31049)	13. "*A. liguriae*" (ATCC 31048)	14. "*A. teichomyceticus*" (ATCC 31121)	15. *A. caeruleus* (ATCC 33937)
D-Xylose	+	+	+	+	+	+	+	+	+	+	+	+	+	+	−
L-Arabinose	+	+	+	+	+	+	+	+	+	+	+	+	+	+	−
D-Glucose	+	+	+	+	+	+	+	+	+	+	+	+	+	+	−
D-Fructose	+	+	+	+	+	+	−	±	+	+	+	+	+	+	+
D-Mannose						+	+	+		+	+	+	+	+	+
L-Rhamnose	+	+	+	+	+	+	+	−	+	+	+	+	+	−	+
m-Inositol	+	−	−	+	+	+	−	−	+	+	−	−	+	−	+
D-Mannitol	+	+	+	+	+	+	−	+	+	+	+	+	−	+	+
Sucrose	+	+	+	+	+	−	+	−	+	+	+	+	−	+	+
Lactose						+	+	+	+	+	−	+	−	±	+
Salicin						+	−	+	+		+	+	−	±	+
Raffinose	+	−	−	−	−	−	−	−	+	−	−	−	−	−	−
Cellulose	+	−	−	+	−	−	−	−	−	−	−	−	−	−	

[a]Symbols: +, positive; −, negative; ±, weak or slow.
[b]Only acid formation is recorded here. Additional results are given by Lechevalier and Lechevalier (1975).
[c]Growth at the expense of many more organic compounds is reported by Palleroni (1979).

useful. Unfortunately, these results, as well as others presented in descriptions available in the literature, refer in most cases to a single strain of each species, and in the absence of a reliable measure of the intraspecific variation, the data can be useful only as a general guide for future studies on a larger collection of strains.

The concept of species in *Actinoplanes* is as ill defined as in many other taxa of procaryotes. It is hoped that a more precise idea may emerge from modern studies at the genetic and molecular levels so that a proliferation of names may be avoided.

List of species of the genus Actinoplanes

The names of the first eight species to be described in this section have been included in the genus *Actinoplanes* in the Approved Lists of Bacterial Names (Skerman et al., 1980). A second group includes two species (9 and 10) which were assigned to the genus *Amorphosporangium* in the Approved Lists. Species within each of the two groups (1–8 and 9–15) have been listed here following a chronological order of publication of the original descriptions.

1. **Actinoplanes philippinensis** Couch 1950, 89.[AL]

phil.ip.pi.nen′sis. M.L. adj. *philippinensis* pertaining to the Philippines.

On Czapek agar growth moderate, flat, or slightly elevated, light buff to tawny, occasionally changing to purplish brown with age. Upper layer of growth with hyphae arranged in palisades. Production of sporangia variable. On peptone Czapek agar, growth is more abundant, and the colonies have a surface ornamented with concentric rings and radial grooves, but there are no palisades. Color is apricot-orange to orange-chrome. Hyphae may fragment in a manner resembling *Nocardia* (Couch and Bland, 1974c).

Sporangia mostly globose, 8–25 μm in diameter. Spores in coils within the sporangia, globose, 1–1.2 μm in diameter, motile with a tuft of flagella, 2–3 μm in length.

Produces macrocyclic lactone antibiotics.

Type strain: ATCC 12427.

2. **Actinoplanes utahensis** Couch 1963, 67.[AL]

u.tah.en′sis. M.L. adj. *utahensis* pertaining to the state of Utah.

Growth on Czapek agar very good, rather flat, with minute bumps. Color apricot-orange to salmon-orange. Few sporangia are formed. On peptone Czapek agar, growth is also very good, rather flat, with minute bumps, and the color is similar, but may become ferrugineus toward the center. No sporangia on these media or on casein and tyrosine agars, but they are produced on starch-casein agar.

Sporangia very irregular in size and shape, 5–18 μm in the largest dimension. Spores subglobose, 1–2 μm in diameter. They are arranged in irregular coils, and they have multitrichous polar flagella.

Produces cyclic peptide antibiotics.

Type strain: ATCC 14539.

3. **Actinoplanes missouriensis** Couch 1963, 69.[AL]

mis.sou.ri.en′sis. M.L. adj. *missouriensis* pertaining to the state of Missouri.

Growth on Czapek agar very good, with elevated center. Palisade formations are rare. Color is mostly ochraceous salmon, with whitish areas where sporangia are abundant. Growth on peptone Czapek agar very good; color zinc-orange to ochraceous orange. Areas with many sporangia on the colonies.

Sporangia globose to subglobose, 6–14 μm; occasionally very irregular. Spores arranged in irregular coils, globose, 1–1.2 μm. Polar multitrichous flagella.

Produces 5-azacytidine.

Type strain: ATCC 14538.

Table 28.6.
Substrates of taxonomic value for the differentiation of some **Actinoplanes** *species*[a,b]

Compounds	1. *A. philippinensis*	2. *A. utahensis*	3. *A. missouriensis* (ATCC 14538)	3. *A. missouriensis* (IMRU 824)	4. *A. brasiliensis*	5. *A. italicus*	8. *A. ferrugineus*
D-Ribose	+	+	−	−	−	+	−
D-Arabinose	−	−	−	−	+	−	+
L-Fucose	−	+	+	−	−	−	+
L-Rhamnose	+	+	+	+	+	+	−
Sucrose	+	+	+	+	+	+	−
Melibiose	−	−	−	−	+	−	−
Melezitose	+	−	−	−	−	+	−
Adipate, pimelate, or suberate	+	+	−	−	−	+	−
D-Malate	−	+	−	−	−	±	−
β-Hydroxybutyrate	+	+	+	+	+	+	−
D,L-Lactate	+	−	+	+	+	+	−
Citrate	−	−	−	−	−	+	−
α-Ketoglutarate	−	−	+	−	−	+	−
Arabitol	−	−	−	+	+	−	+
Xylitol	−	−	−	−	+	−	−
Sorbitol	±	±	+	±	+	±	−
Phenylacetate	−	−	−	+	−	−	±
Quinate	+	+	−	−	−	+	−
β-Alanine	+	+	−	−	−	+	+
L-Asparagine	±	±	−	−	+	−	±
L-Glutamine	−	+	+	+	+	+	−
L-Arginine	+	+	−	−	−	+	−
L-Citrulline	±	+	+	−	−	±	−
L-Ornithine	−	−	−	−	+	±	−
γ-Aminobutyrate	+	−	−	−	+	+	−
L-Histidine	+	−	+	−	−	+	−
L-Proline	+	+	−	+	+	+	+
L-Tyrosine	−	+	−	−	±	+	−
Spermine	−	−	−	−	−	−	+

[a]N. J. Palleroni, unpublished data.
[b]Symbols: +, good growth; ±, poor growth; −, no growth.

4. **Actinoplanes brasiliensis** Thiemann, Beretta, Coronelli and Pagani 1969a, 119.[AL]

bra.si.li.en′sis. M.L. adj. *brasiliensis* pertaining to Brazil.

Very good to moderate growth in ISP media (Shirling and Gottlieb, 1966), with smooth to wrinkled colonies that vary in color from light pink or orange to deep orange. No aerial mycelium is produced.

Sporangia are abundantly produced on soil extract agar, calcium malate agar, and starch-casein agar. They are irregular or umbelliform and occasionally globose, with very wrinkled surface, 3.5–11.5 μm. Spores are subglobose (1.2 μm) to rod shaped (1.2 μm in width × 1.7–2.3 μm in length). They are chemotactically attracted by chloride ions.

Produces the acidic antibiotic A/672.

Type strain: ATCC 25844.

5. **Actinoplanes italicus** Beretta 1973, 37.[AL]

i.ta′li.cus. M.L. adj. *italicus* pertaining to Italy.

Colonies on starch agar have a smooth surface and a dome-shaped center. Cherry-red vegetative mycelium is produced on most media, but on ISP 2 and ISP 7 media (Shirling and Gottlieb, 1966) the color is orange, and on Hickey and Tresner (1952), Bennett (Waksman, 1961), and nutrient agar, the vegetative growth is amber-brown to brown (color according to Maerz and Paul, 1950). Differences from other species include the production of the cherry-red pigment and the production of pink to cherry color pigments diffusing into the medium.

Sporangia abundantly produced on starch and skim milk agar. They are globose to oval or pyriform, 6–11 μm. Spores globose to oval, 1–2 μm.

No antibiotic activity has been detected.

Type strain: ATCC 27366.

6. **Actinoplanes rectilineatus** Lechevalier and Lechevalier 1975, 371.[AL]

rec.ti.li.ne.a′tus. M.L. adj. *rectus* straight; M.L. adj. *lineatus* striped; M.L. adj. *rectilineatus* marked with straight lines.

Hyphae and spores mainly Gram-positive, but sporangia Gram-negative. Color on different agar media may be white, grayish tan, yellow-tan, or yellow-brown, and in some media, aerial mycelium may be formed.

Sporangia cylindrical, 8–15 μm, containing straight, longitudinal rows of spores. Spores globose, 1.5–2 μm.

Type strain: ATCC 29234.

7. **Actinoplanes deccanensis** Parenti, Pagani and Beretta 1975, 248.[AL]

dec.ca.nen′sis. M.L. adj. *deccanensis* pertaining to the Indian locality of Decca.

Growth is abundant on ISP agar media 2, 4, and 7 (Shirling and Gottlieb, 1966), with colors varying from light amber to orange. No growth on calcium malate agar. Growth in various agar media can occur up to a temperature of 42°C, an unusual property among members of the genus.

Sporangia abundantly produced on soil extract agar. They are small and irregular (4–7 µm in diameter). Spores subglobose, 1 × 1.5 µm.

Produces the chloride-containing antibiotic lipiarmycin.

Type strain: ATCC 21983.

8. **Actinoplanes ferrugineus** Palleroni 1979, 52.[AL]

fer.ru.gi'ne.us. M.L. adj. *ferrugineus* of the color of iron rust.

Abundant growth in several media, with colonies of colors varying from amber to deep brown. Aerial hyphae produced on ISP 7 agar (Shirling and Gottlieb, 1966). In this medium, a reddish-brown soluble pigment is also produced. Good production of sporangia on minimal media with various single carbon sources. Sporangia globose to irregular, 4–12 µm. Spores globose, 0.9–1 µm.

Produces the proline analog L-azetidine-2-carboxylic acid.

Type strain: ATCC 29868.

9. **Actinoplanes auranticolor** (Couch 1963, 65) comb. nov. (*Amorphosporangium auranticolor* Couch 1963, 65[AL]; not *Actinoplanes auranticolor* Celmer et al. 1975.)

au.ran'ti.co.lor. M.L. n. *aurantium* a bitter orange; L. n. *color* tint, hue; M.L. adj. *auranticolor* orange colored.

Very good growth on Czapek agar and peptone Czapek agar, frequently elevated and convoluted. Good growth also on casein and tyrosine media. In all cases color of the colonies is in various shades of orange, and a diffusible dark pigment is formed in tyrosine media. In other media, a yellow diffusible pigment may be observed. Conidiospores may be abundant on the tyrosine medium. Good sporangia formation on Czapek agar during the first year of culture.

Sporangia very irregular in shape, multilobed, 6–25 × 8–15 µm, the width usually being greater than the height. Spores rod shaped, 0.5–0.7 × 1–1.5 µm. The original description by Couch (1963) considers the spores to be nonmotile, but later Hanton (1968) reported motility and the presence of polar flagella.

Type strain: 15330.

10. **Actinoplanes globisporus** (Thiemann 1967) comb. nov. (*Amorphosporangium globisporus* Thiemann 1967, 233.[AL])

glo.bi.spo'rus. L. n. *globus* a ball, sphere; Gr. n. *spora* a seed, spore; M.L. *globisporus* round spored.

Colonies are cream to light orange. Chlamydospore-like structures frequently observed, occasionally at the tip of short aerial hyphae. The only species of the genus reported to require organic growth factors.

Highly irregular sporangia, resembling masses of spores not surrounded by a sporangial wall, 4–7 × 3–5 µm. Spores globose, 0.8–1 µm, motile.

No antibiotic activity has been detected.

Type strain: ATCC 23056.

11. "**Actinoplanes ianthinogenes**" Coronelli, Pagani, Bardone and Lancini 1974, 161.

ian.thi.no'ge.nes. Gr. adj. *ianthinus* violet; Gr. v. *gennaio* to produce; M.L. adj. *ianthinogenes* producing violet.

Abundant growth on many agar media with colonies of violet color in some areas and other areas colored light amber, orange, or brown. Aerial mycelium is absent.

Sporangia abundantly produced on oatmeal agar and Czapek glucose agar. They are globose with irregular surface and a diameter of 4–10 µm. Spores are subglobose, 1.4–1.8 µm.

Produces a naphthoquinone antibiotic.

Type strain: ATCC 21884.

12. "**Actinoplanes garbadinensis**" Parenti, Pagani and Beretta 1976, 502.

gar.ba.di.nen'sis. M.L. adj. *garbadinensis* pertaining to the Indian locality of Garbady.

Smooth to very wrinkled colonies formed according to the agar medium, usually from light to deep orange in color. Rudiments of aerial mycelium can be observed in some media.

Sporangia are formed only on calcium malate agar. They are globose and sometimes lobate, 7–12 µm in diameter. Spores subglobose, 1–1.5 µm.

Produces the antibiotic gardimycin.

Type strain: ATCC 31049.

13. "**Actinoplanes liguriae**" Parenti, Pagani and Beretta 1976, 502.

li.gu'ri.ae. M.L. gen. n. *liguriae* pertaining to the Italian region of Liguria.

Growth is poor in Czapek and Czapek glucose agar, but it is very good in ISP 2 and 7 media (Shirling and Gottlieb, 1966). The color of the vegetative mycelium in different media varies from rose-amber to orange, and a yellow-amber diffusible pigment may be produced.

Sporangia formation is good in media ISP 3 and Czapek glucose agar. They are globose to oval, 15–25 µm in diameter. Spores globose, 1.5–2 µm.

Produces the antibiotic gardimycin.

Type strain: ATCC 31048.

14. "**Actinoplanes teichomyceticus**" Parenti, Beretta, Berti and Arioli 1978, 277.

tei.cho.my.ce'ti.cus. Gr. n. *teichos* wall; Gr. n. *myces* fungus; L. adj. suff. *-icus* belonging to; *teichomyceticus* literally belonging to a fungus cell wall (referring to inhibition of cell wall synthesis by teichomycin, produced by the type strain).

Colonies may have a central protuberance or dome, and colors may vary from light to deep orange, and on some media light brown. Usually no soluble pigments are produced. Well-developed aerial mycelium on some media. Sporangia formation can be increased by addition of tea infusion to agar media.

Sporangia abundantly produced on many agar media, mainly on the dome of the colonies. The spores are globose to oval, 1.5–2 µm.

Produces the antibiotics teichomycin A1 and A2. The antibiotic complex has been renamed teicoplanin (Malabarba et al., 1984).

Type strain: ATCC 31121.

15. **Actinoplanes caeruleus** Horan and Brodsky 1986, 189.

cae.ru'le.us. L. adj. *caeruleus* dark blue, azure (referring to the blue vegetative mycelial pigments).

Moderate to good growth in ISP and various other media, with a wide range of colors (yellow, tan, gray, blue). Starch media elicit blue pigmentation of the colonies. Yellow soluble pigments may also be produced. DAP is not present in the cell wall. Nutritional and physiological properties are described by Horan and Brodsky (1986).

Globose to irregular sporangia, 6–16 µm in diameter. Spores globose to ovoid, 1.3–2.0 µm.

Produces heptaene antifungal antibiotic.

Type strain: ATCC 33937.

Genus **Ampullariella** *Couch 1964, 29*[AL]

GERNOT VOBIS AND HANS-W. KOTHE

Am.pul.la.ri.el'la. L. n. *ampulla* flask, bottle; M.L. diminutive suffix *-ella;* M.L. fem. n. *Ampullariella* a small bottle (to indicate bottle-shaped sporangia).

Hyphae of **substrate mycelium** 0.2–1.2 μm in diameter, branched, septate. True aerial mycelium is not developed. **Sporangia** are produced above the surface of the substrate. They are **irregular, cylindrical, lobate, bottle shaped, flask shaped, or digitate,** 5.0–20.0 μm wide and 8.0–30.0 μm long. Numerous **spores** are produced within the sporangium, **arranged in parallel chains.** Spores are **rod shaped** (0.5–1.0 × 2.0–4.0 μm) and motile. Gram-positive, aerobic, chemo-organotrophic and mesophilic; optimum growth temperature 25°C. Colonies on various complex agar media are elevated and convoluted. The **color of substrate mycelium is usually orange, red, brown, or black,** sometimes with white areas. The peptidoglycan of the cell wall contains ***meso*-diaminopimelic acid (*meso*-DAP) and glycine** as distinguishing components, with **xylose and arabinose** as characteristic whole-cell sugars. The mol% G + C of DNA is 72.3 (Bd) to 73.0 (T_m).

Type species: *Ampullariella regularis* (Couch 1963) Couch 1964, 29.

Further Descriptive Information

The sporangia of different species vary in shape and size (Fig. 28.5). Cylindrical or bottle-shaped sporangia are produced by *A. regularis* (Fig. 28.6), which are 5.0–14.0 μm wide and 8.0–30.0 μm long. The size of the sporangia of *A. campanulata* is 5.0–15.0 × 6.0–12.0 μm. They are characteristically bell shaped, frequently lobed, irregular, or pyriform. Because of the varying length of spore chains, the sporangia sometimes appear papillate. *A. lobata* has irregular or lobed sporangia, rarely cylindrical, sometimes papillate. Their size is 4.0–20.0 × 12.0–23.0 μm. The sporangia of *A. digitata* are digitate, subcylindrical, lobed, or bottle shaped, with a size of 3.0–9.0 × 4.0–14.0 μm (Couch, 1963). "Hairy" surfaces of the sporangia of several strains have been reported by Jiang and Ruan (1982) and Seino (1983).

A scheme of sporangial development was proposed by Bland and Couch (1981) presumably following the pattern described for *Actinoplanes rectilineatus* (Lechevalier et al., 1966b).

In all *Ampullariella* strains, the spore chains are arranged in parallel rows within the sporangium. The spores are rod shaped (0.5–1.0 × 2.0–4.0 μm) and motile. The flagella insertion is polar, as shown in *A. regularis* (Kane, 1966; Schäfer, 1973) and in *A. campanulata* (Higgins et al., 1967). Their number ranges from 1 to 12 (Higgins et al., 1967) and they are 3.5–6.0 μm in length (Couch and Bland, 1974b). Peritrichous flagella of sporangiospores of "*A. pekinensis*" is reported by Juan and Zhang (1974). Nonomura et al. (1979) observed polar (lophotrichous and amphitrichous) and peritrichous flagella arrangement in 12 *Ampullariella*-isolates. The flagella are up to 19.0 μm in length, on average 6–12 μm.

The dehiscence of the sporangia is caused by the swelling of an intersporal substance that causes the sporangial wall to rupture. The spores disperse through the resulting slit or hole (Couch, 1963). One hour after sporangia are immersed in water, the spores become motile. Frequently the movement is visible inside the sporangia.

Development of nonmotile spores (conidia) may occur. The conidiophores are brushlike in *A. regularis* and originate at the margin of the colonies; *A. lobata* has oval conidia in a moniliform arrangement that are produced on the substrate mycelium (Couch, 1963).

The peptidoglycan of the cell wall contains *meso*-DAP and glycine as amino acids (Yamaguchi, 1965) and the diagnostic sugars are xylose and arabinose (Szaniszlo and Gooder, 1967). This corresponds to cell wall chemotype II and sugar pattern D in the classification scheme of Lechevalier and Lechevalier (1970b). The phospholipid pattern corresponds to type II of Lechevalier et al. (1981) with phosphatidylethanolamine without phosphatidylcholine or the unknown glucosamine-containing phospholipid (Hasegawa et al., 1979). The fatty acids are *iso-* and *anteiso-* branched and saturated or unsaturated (Kroppenstedt and Kutzner, 1978). The main component of the menaquinones is MK-9 (H_4) (Collins et al., 1984).

Ampullariella strains grow well on various complex media, with colonies up to 28 mm in diameter on Czapek, peptone Czapek, or casein agars after 6 weeks. They are characteristically elevated and convoluted, and frequently have protuberances in the center with ridged or flat marginal areas. The color is usually orange, red, brown, or black (see Table 28.8 below) (Couch, 1963; Nonomura et al., 1979). An orange pigment was identified spectrophotometrically as being associated with carotenoids (Szaniszlo, 1968). Soluble pigments are also produced, ranging from yellowish, greenish, and auburn to dark brown (Couch, 1963); "*A. violaceochromogenes*" produces a purple soluble pigment on various agar media (Nonomura et al., 1979).

D-fructose, D-glucose, sucrose, D-ribose, rhamnose, D-xylose, and glycerol are utilized for growth by all species (see Table 28.8 below) (Nonomura et al., 1979), but diverse strains of *A. regularis* show no uniform pattern of utilization of carbohydrates (Schäfer, 1973). Casein and starch are usually hydrolyzed by all species (Couch, 1963; Schäfer, 1973; Nonomura et al., 1979). Hydrolysis of tyrosine crystals and liquefaction of gelatin can differ. Nitrate is reduced to nitrite and melanoid pigments

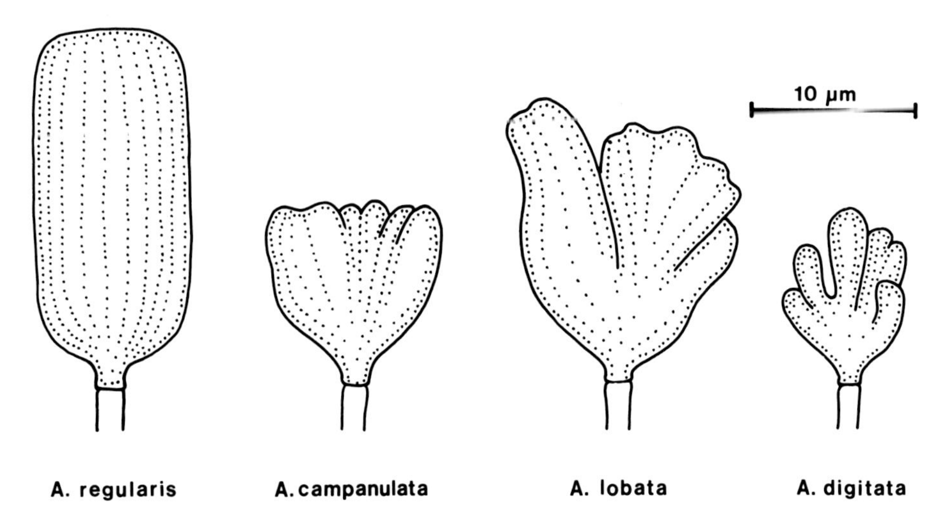

Figure 28.5. Scheme of the different morphological types of sporangia in *Ampullariella* species.

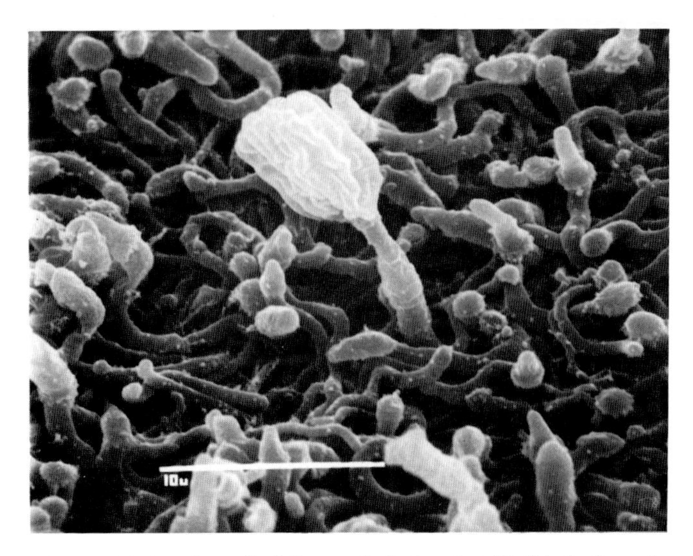

Figure 28.6. *Ampullariella regularis* (strain MB-SE 1). Scanning electron micrograph of substrate mycelium and sporangium developed on yeast extract-starch agar after 14 days of incubation (× 3350).

are produced only by *A. digitata* (see Table 28.8 below) (Nonomura et al., 1979). All strains can grow at temperatures between 18 and 35°C, the optimal temperature being 25°C (Couch and Bland, 1974b).

Strains of *Ampullariella* are widely distributed in soil throughout the world (Couch, 1963). Schäfer (1973) found that about 8% of the isolates of sporangiate actinomycetes were represented by this genus. The pH values of the soil samples varied from 5.1 to 8.5. Willoughby (1969b) isolated *Ampullariella* strains from freshwater habitats.

Enrichment and Isolation Procedures

A method of isolating strains of *Ampullariella* was described by Bland and Couch (1981). A small amount of soil is placed in a sterile Petri plate and flooded with sterile water. Pollen or other natural substances like hair, snake skin, or boiled grass leaves are added as bait, and after 4–7 days' incubation they are examined with a binocular dissecting microscope. Sporangia can be recognized by a white glistening appearance. Pollen baits bearing sporangia are transferred to the surface of a 3% (w/v) agar plate. Individual sporangia are separated from the bait and rolled on the agar surface to free them from contaminating bacteria. The cleaned sporangia are used as an inoculum for pure cultures.

Maintenance Procedures

Subcultures should be made after a period of 12 weeks. For longer preservation, the organisms must be lyophilized by the procedures used for other aerobic actinomycetes, preferably using well-sporulating cultures.

Differentiation of the genus **Ampullariella** *from other genera*

Within the members of the sporangiate actinomycetes having a cell wall chemotype II (Lechevalier and Lechevalier, 1970b), the genus *Ampullariella* may be confused with species of the genus *Pilimelia* (Kane, 1966) or with *Actinoplanes rectilineatus* (Lechevalier and Lechevalier, 1975). All strains of these taxa have parallel rows of spore chains enclosed in sporangia. They can be distinguished by the characteristics given in Table 28.7.

Table 28.7.
Characteristics differentiating the genus **Ampullariella** *from other closely related taxa*[a,b]

Characteristics	*Ampullariella*	*Pilimelia*	*Actinoplanes rectilineatus*
Shape of spores			
Rod	+	+	–
Globose	–	–	+
Size of spores			
0.5–1.0 × 2.0–4.0 µm	+	–	–
0.3–0.7 × 0.7–1.5 µm	–	+	–
1.5–2.0 µm in diameter	–	–	+
Insertion of flagella			
Polar	+	–	+
Lateral	–	+	–
Shape of colonies			
Diffuse	+	–	+
Compact	–	+	–
Decomposing keratinic substances	–	+	–
Pattern of major fatty acids			
Iso	+	+[c]	+
Anteiso	+	–	+
Unsaturated	d	–	+

[a]Data from Couch (1963), Couch and Bland (1974b), Kroppenstedt and Kutzner (1978), Lechevalier and Lechevalier (1975), Lechevalier et al. (1977), Schäfer (1973), and Vobis et al. (1986a).
[b]Symbols: see Table 28.1; *d*, 11–89% of strains are positive.
[c]No C_{16} fatty acids.

Taxonomic Comments

The genus *Ampullariella,* together with the genera *Actinoplanes* (*Amorphosporangium*), *Pilimelia, Dactylosporangium,* and *Micromonospora,* belongs to a group of actinomycetes characterized by the absence of true aerial mycelium and possessing a cell wall chemotype II (Lechevalier and Lechevalier, 1970b). Their close relationship is also demonstrated by DNA/DNA and DNA/rRNA reassociation studies (Stackebrandt et al., 1981), oligonucleotide sequencing of 16S rRNA (Stackebrandt et al., 1983c), and the host ranges of bacteriophage Ap4 (Prauser, 1984c). Because of the high similarity of molecular data Stackebrandt and Kroppenstedt (1987) suggested the union of *Actinoplanes, Amorphosporangium,* and *Ampullariella* into one genus *Actinoplanes.*

The genus *Ampullariella* was originally described as *Ampullaria* (Couch, 1963). Because this name had been used already for a fungal genus, Couch (1964) modified the name to *Ampullariella.*

Further Reading

Couch, J.N. 1963. Some new genera and species of the *Actinoplanaceae.* J. Elisha Mitchell Sci. Soc. *79:* 54–70.
Nonomura, H., S. Iino and M. Hayakawa. 1979. Classification of actinomycetes of genus *Ampullariella* from soils of Japan. J. Ferment. Technol. *57:* 79–85.
Stackebrandt, E., B. Wunner-Füssl, V.J. Fowler and K. -H. Schleifer. 1981. Deoxyribonucleic acid homologies and ribosomal ribonucleic acid similarities among spore-forming members of the order *Actinomycetales.* Int. J. Syst. Bacteriol. *31:* 420–431.

Differentiation and characteristics of the species of the genus **Ampullariella**

For differential characteristics of the species of *Ampullariella* see Table 28.8 and generic descriptions.

Table 28.8.
Characteristics differentiating the species of the genus **Ampullariella**[a]

Characteristics	1. *A. regularis*	2. *A. campanulata*	3. *A. lobata*	4. *A. digitata*
Sporangial shape[b,c]				
Bottle	+	−	−	+
Cylindrical	+	+	+	−
Subcylindrical	−	−	−	+
Bell	−	+	−	−
Lobed	−	+	+	+
Irregular	−	+	+	−
Pyriform	−	+	−	−
Digitate	−	−	−	+
Sporangial size[b,c]				
5–14 × 8–30 μm	+	−	−	−
5–15 × 6–12 μm	−	+	−	−
4–20 × 12–23 μm	−	−	+	−
3–9 × 6–12 μm	−	−	−	+
Color of colonies on				
Czapek agar[b,d]				
Orange	+	−	+	−
Red	+	+	+	−
Brown	−	+	−	+
Black	−	+	−	−
Peptone-Czapek agar[b,d]				
Orange	+	+	+	−
Ochre	+	−	−	−
Red	−	+	+	−
Utilization of[d,e]				
L-Arabinose	+	+	+	−[d],+[e]
D-Fructose	+	+	+	+
D-Glucose	+	+	+	+
Inositol	−	−	−	+
D-Mannitol	−	+	+	−
Raffinose	−	−	+	+
Sucrose	+	+	+	+
D-Ribose	+	+	+	+
Rhamnose	+	+	+	+
D-Xylose	+	+	+	+
Glycerol	+	+	+	+
Hydrolysis of				
Starch[e]	+	+	+	+
Casein[b,e]	+	+	+	+[b],−[e]
Tyrosine[e]	−	−	−	−
Gelatin[e]	+	−	+	v
Reduction of nitrate[e]	+	+	+	+
Skim milk[e]				
Peptonization	−	+	−	+
Coagulation	+	+	+	+
Production of melanoid pigments[e]	−	−	−	+

[a]Symbols: see Table 28.1; *v*, strain instability.
[b]Data from Couch (1963).
[c]Data from Couch and Bland (1974b).
[d]Data from Schäfer (1973)
[e]Data from Nonomura et al. (1979).

List of species of the genus **Ampullariella**

1. **Ampullariella regularis** (Couch 1963) Couch 1964, 29.[AL] (*Ampullaria regularis* Couch 1963, 57.)

reg.u.lar′is. L. adj. *regularis* regular.

Sporangia are mostly cylindrical, measuring 5.0–14.0 × 8.0–30.0 μm (Fig. 28.6). The base of the sporangia is frequently mound shaped and resembles a corked bottle. Formation of sporangia occurs on Czapek agar and casein agar. Spores are arranged in parallel rows inside the sporangia, and are rod shaped, 0.5–1.0 μm wide and 2.0–4.0 μm long, and motile by a polarly inserted tuft of flagella. Brushlike conidiophores may be produced on Czapek agar.

Colonies are up to 10 mm in diameter on Czapek, casein, or tyrosine agars, reaching 20 mm on peptone Czapek agar after 6 weeks. They are flat, frequently convoluted in center with radial ridges on the margin. The color of the substrate mycelium is orange, red, brownish, ochre, or salmon to coral-pink, sometimes with white or gray areas. Soluble yellowish, greenish, or brownish pigments may be produced on various media.

Inositol, D-mannitol, and raffinose are not utilized by the type strain as sole carbon sources. Gelatin is liquified.

The *iso-* and *anteiso-* fatty acids are saturated.

According to their behavior on casein agar and tyrosine agar, three variants are distinguishable (Couch, 1963). The first utilizes casein and tyrosine and darkens both agars. The second variant utilizes casein and tyrosine but darkens only the tyrosine agar. The third variant utilizes casein and tyrosine but darkens only casein agar. The type strain of *A. regularis* is from the first variant.

The mol% G + C of DNA is 72.3 (Bd) (Yamaguchi, 1967).

Type strain: DSM 43151.

2. **Ampullariella campanulata** (Couch 1963) Couch 1964, 29.[AL] (*Ampullaria campanulata* Couch 1963, 59.)

cam.pan.u.la'ta. M.L. dim. n. *campanella* small bell; M.L. adj. *campanulata* bell shaped.

The sporangia are characteristically bell shaped, frequently lobed or irregular, sometimes pyriform or cylindrical. Their size is 5.0–15.0 × 6.0–12.0 μm (Fig. 28.5). They often have a papillate appearance at the proximal end, because the spore chains are of an unequal length. The spores are rod shaped (0.5–1.0 × 2.0–4.0 μm) and motile by a polarly inserted tuft of flagella.

After 6 weeks' incubation colonies measure up to 18 mm in diameter on Czapek, peptone Czapek, or casein agars and up to 10 mm on tyrosine agar. They are elevated, convoluted with ridged areas at the margin. The color of the substrate mycelium is coral-red, coral-pink, orange, brown, or black. Soluble yellowish, greenish, and brownish pigments may be produced on various media. Formation of sporangia occurs on Czapek agar.

D-mannitol is utilized by the type strain; inositol and raffinose are not utilized. Gelatin is not liquified.

The *iso-* and *anteiso-* fatty acids are unsaturated.

Type strain: ATCC 15348.

3. **Ampullariella lobata** (Couch 1963) Couch 1964, 29.[AL] (*Ampullaria lobata* Couch 1963, 59.)

lo.ba'ta. Gr. n. *lobos* lobe; M.L. fem. adj. *lobata* lobed.

Strains of this species have sporangia that vary in size and shape. Characteristically they are bell shaped, frequently lobed or irregular, sometimes pyriform or cylindrical (Fig. 28.5). Their size is 4.0–20.0 × 12.0–23.0 μm. They often have a papillate appearance at the proximal end, because of the varying length of the spore chains. The lobed sporangia sometimes are divided in several parts, giving the appearance of fused sporangia. Formation of sporangia occurs only on Czapek agar. The motile spores are rod shaped (0.5–1.0 × 2.0–4.0 μm). Oval moniliform conidia may also be produced.

Colonies are up to 25 mm in diameter on Czapek agar or peptone Czapek agar and up to 10 mm on casein agar or tyrosine agar after 6 weeks' incubation. They are flattish or convoluted, and frequently ridged. The color of the substrate mycelium is coral-red, coral-pink, or brownish. Soluble yellowish, greenish, and brownish pigments may be produced on various media.

D-mannitol and raffinose are utilized, inositol is not utilized. Gelatin is liquified.

Type strain: ATCC 15350.

4. **Ampullariella digitata** (Couch 1963) Couch 1964, 29.[AL] (*Ampullaria digitata* Couch 1963, 61.)

di.gi.ta'ta. L. fem. adj. *digitata* having fingers.

Sporangia are usually digitate, sometimes subcylindrical, lobed, or rarely bottle shaped. Their size is 3.0–9.0 × 6.0–14.0 μm (Fig. 28.5). Formation of sporangia occurs on Czapek agar. The motile spores are rod shaped (0.5–1.0 × 2.0–4.0 μm).

Colonies reach 5 mm in diameter on Czapek agar, up to 28 mm on peptone Czapek agar, and up to 10 mm on tyrosine agar after 6 weeks' incubation. They are flat, wrinkled, or convoluted, and frequently ridged. The margin is sometimes fimbriate. The color of substrate mycelium is pinkish cinnamon or red-cinnamon to blackish brown, sometimes red with black sectors or very dark with red sectors, flesh-colored, or dirty buff. The marginal areas are usually lighter in color. Soluble yellowish, greenish, and brownish pigments may be produced on various media.

Inositol and raffinose are utilized, but D-mannitol is not utilized. Melanoid pigments are produced.

The *iso-* and *anteiso-* fatty acids are unsaturated.

The mol% G + C of DNA is 73.0 (T_m) (Farina and Bradley, 1970).

Type strain: ATCC 15349.

Species Incertae Sedis

The following species are not included in the Approved Lists of Bacterial Names (Skerman et al., 1980) and not validly published in accordance with rule 27 of the Bacteriological Code (Lapage et al., 1975).

a. "*Ampullariella pekinensis*" Juan and Zhang 1974.

The sporangia are bottle shaped or cylindrical, 4.0–8.0 μm wide and 6.0–13.0 μm long. Hairlike structures have been observed on their surfaces (Seino, 1983). The spores are rod shaped (1.0–2.0 μm) or ellipsoidal (1.0–1.5 μm) and peritrichously flagellated.

The color of substrate mycelium is golden yellow to brownish. A pale yellowish soluble pigment is produced on some media.

Glucose, sucrose, L-rhamnose, D-fructose, D-xylose, and D-mannitol are utilized; raffinose and inositol are not utilized. Starch is hydrolyzed (Jiang and Ruan, 1982).

Type strain: IFO 13662.

b. "*Ampullariella violaceochromogenes*" Nonomura et al. 1979.

The sporangia are bottle shaped or cylindrical, 5.0–11.0 μm wide and 8.0–15.0 μm long. The spores are rod shaped (0.9–1.1 × 2.0–2.6 μm) with peritrichous and polar flagella.

The color of substrate mycelium is brownish purple on Czapek, yellowish orange on peptone Czapek, and pale gray on oatmeal agars. A violet to dark purple soluble pigment is produced on Czapek and oatmeal agars.

D-glucose, sucrose, rhamnose, D-xylose, and L-arabinose are utilized; raffinose, inositol, mannitol, and glycerol are not utilized. Casein and starch are hydrolyzed, nitrate is reduced to nitrite, gelatin is liquified. Melanoid pigments are not produced and tyrosine is not degradated.

Type strain: KCC A-0236.

c. "*Ampullariella regularis*" subsp. "*intermedia*" Nonomura et al., 1979.

This subspecies has spores with peritrichous and polar (lophotrichous and amphitrichous) flagella.

d. "*Ampullariella cylindrica*" Jiang and Ruan 1982.

The sporangia are bell shaped or cylindrical, 6.0–7.0 μm wide and 22.0–32.0 μm long, with short "hair" on the surfaces. The spores are rod shaped (0.4 × 1.1 μm), with a polarly inserted tuft of flagella.

The color of substrate mycelium is white to yellow. No soluble pigments are produced on all media tested.

D-glucose, sucrose, L-rhamnose, D-xylose, raffinose, mannitol, and L-arabinose are utilized; inositol is not utilized. Casein and starch are hydrolyzed.

Type strain: IFO 14264.

e. "*Ampullariella pilifera*" Jiang and Ruan 1982.

The sporangia are bell or bottle shaped, 6.0–7.0 μm wide and 11.0–25.0

µm long, with short "hair" on the surfaces. The spores are rod shaped (0.6 × 2.0 µm), with a polarly inserted tuft of flagella.

The color of substrate mycelium is brown, and a yellowish soluble pigment is produced on Czapek agar.

D-Glucose, L-rhamnose, D-xylose, mannitol, D-fructose, and L-arabinose are utilized; inositol, raffinose, and sucrose are not utilized. Casein and starch are hydrolyzed.

A variant "*A. hainanensis*" with aerial mycelium is described (Jiang and Ruan, 1982)

Type strain: IFO 14265.

Genus **Pilimelia** *Kane 1966, 225*[AL]

GERNOT VOBIS

Pi.li.mel′i.a. L. n. *pilus* a hair; Gr. fem. n. *Melia* a nymph loved by the river god Inachus; M.L. fem. n. *Pilimelia* an aquatic organism growing on hair substrate.

Sporangia are produced **on the surface of the substrate** on sporangiophores. The shape of sporangia is **spherical, ovoid, pyriform, campanulate,** or **cylindrical,** approximately 10–15 µm in size. Sporangia contain **numerous spores in chains** that are arranged in parallel or irregularly swirllike rows. **Spores (zoospores)** are **rod shaped** (0.4 × 1.2 µm) and **motile** by means of a **laterally inserted tuft of flagella.** Nonmotile spores are developed in free chains arranged similarly to the zoospores. The organisms are Gram-positive. Hyphae of substrate mycelium are 0.2–0.8 µm diameter, branched, and septate. True aerial mycelium is not developed. The peptidoglycan of the cell walls contains ***meso*-diaminopimelic acid (*meso*-DAP)** and **glycine,** with **xylose** and **arabinose** as characteristic sugars of whole-cell hydrolysates. Colonies grow only on complex media. They are small, compact, soft pasty, or solid. **Color of substrate mycelia is pale lemon-yellow, golden yellow, orange, or pale,** turning brown to dark with age. Aerobic, chemoorganotrophic, optimal growth at pH 6.5–7.5 and at 20–30°C (minimum 10°C, maximum 38°C). **Strains decompose keratinic substances** (hair of mammals). The mol% G + C of the DNA has not been determined.

Type species: *Pilimelia terevasa* Kane 1966, 225.

Further Descriptive Information

The sporangia or spore vesicles (sensu Cross, 1970) have different shapes and sizes. Each strain is characterized by a special shape of sporangia of varying sizes. Typical cylindrical sporangia are produced by *P. anulata,* reaching lengths of 10–35 µm (Kane, 1966); *P. columellifera* has spherical, oval to pyriform sporangia with diameters of 7–15 µm (Vobis et al., 1986a). Globose sporangia are also developed by *P. terevasa,* their size ranging between 5 and 23 µm (Kane Hanton, 1974). Other strains of *Pilimelia* have campanulate, inverse conical, heart-shaped, or flabelliform sporangia (Karling, 1954; Gaertner, 1955; Rothwell, 1957; Vobis et al., 1986a). Presumably they belong to *P. terevasa.* After repeated subculturing the strains may lose the ability to produce sporangia (Kane Hanton, 1974).

The sporangia are developed at the tip of thickened hyphae that penetrate the surface of the substrate. These hyphae may be called palisade hyphae (Bland, 1968) or sporangiophores (Vobis, 1984). Their diameters are about 1 µm. In *P. terevasa* and *P. anulata,* they are fragmented by double-layered cross-walls (Vobis et al., 1986a). The uppermost fragment can be swollen to a ringlike structure or annulus, the distinctive structure of *P. anulata.* In *P. columellifera,* the sporangiophores are not septate but extended into the lumen of the sporangia to form clearly visible columellae (Fig. 28.7). When sporangia are mature, the cytoplasm of the sporangiophores is generally autolyzed. The ultrastructure of sporangial development was described by Vobis (1984), confirming the scheme of sporangium formation as proposed by Lechevalier and Holbert (1965) for sporangiate actinomycetes. The sporangiospores are either arranged in parallel (*P. terevasa* and *P. anulata*) or swirllike (*P. columellifera*) rows (Fig. 28.7). It is estimated that one sporangium can contain from hundreds to several thousand spores.

The shape of the zoospores is rodlike, occasionally slightly curved or reniform. They are about 0.4 µm in diameter and 1.2 µm in length. The spores bear a laterally inserted tuft of flagella (Vobis et al., 1986a). A single flagellum might reach 5 µm in length with a diameter of about 11 nm. The flagella are frequently bundled and may function as a unit (Schäfer, 1973; Vobis, 1984).

Nonmotile spores (conidia) may be produced, mainly in the aqueous milieu of the substrate (Kane, 1966). The conidia are developed in chains that are arranged in a pattern similar to the spore chains inside the corresponding sporangia (Kane, 1966; Schäfer, 1973). Presumably the formation of conidia is a variation of sporangial development in which the sporangial envelope is not formed.

The peptidoglycan of the cell wall contains *meso*-DAP and glycine with xylose and arabinose as characteristic sugars in whole-cell hydrolysates (sugar pattern D) according to chemotype II of Lechevalier and Lechevalier (1970b) (Szaniszlo and Gooder, 1967; Kroppenstedt, 1979; Vobis et al., 1986a). The phospholipid type of the cell membrane corresponds to type II of Lechevalier et al. (1981), with phosphatidylethanolamine present and phosphatidylcholine and the unknown glucosamine-containing phospholipid absent (Vobis at al., 1986a). The fatty acid pattern of *Pilimelia* strains differs from all other genera of sporangiate actinomycetes. In *Pilimelia,* no *anteiso*- fatty acids are detectable and of the *iso*- fatty acids, *iso*-C_{16} is absent (Kroppenstedt, 1979). *Iso*- branched saturated and unsaturated fatty acids with odd numbers of carbon atoms (i-15:0 and i-17:1) are diagnostically important compounds (R. M. Kroppenstedt, personal communication; B. Renner and E. Stahl, personal communication). Menaquinone composition is predominantly MK-9(H_2) and MK-9(H_4) in the type strain of *P. terevasa* (R. M. Kroppenstedt, personal communication).

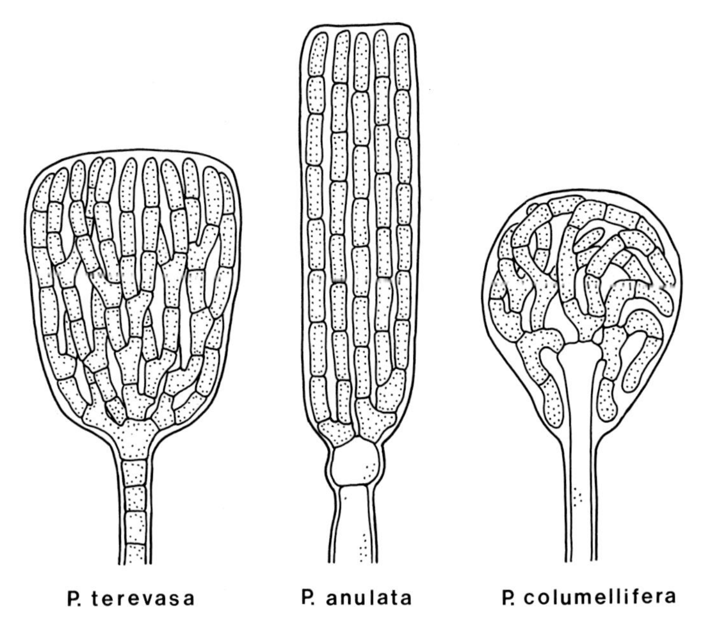

Figure 28.7. Scheme of the morphological types of sporangia in the *Pilimelia* species.

The fine structure of the walls of hyphae and spores of *Pilimelia* is typical for those of Gram-positive cells. A single compact layer surrounds the cytoplasm, which contains irregularly elongated nucleoid regions, ribosomes, and vacuole-like structures. Mesosomes are connected to the nuclear material or involved in the formation of cross-walls (Bland, 1968; Vobis, 1984). The two septation types of Williams et al. (1973) can be recognized. Type 1 occurs in substrate hyphae and type 2 is observable in sporangial and conidial development. Hyphae in contact with the air are additionally covered either by a thin sheath or by a thick layer of perhaps fibrous and mucilagenous material. The sporangial envelopes originate from these outer layers (Vobis, 1984).

Pilimelia strains grow very slowly. After 4–6 weeks of incubation, small colonies of 5 mm in diameter develop. These are compact, with no true aerial mycelia. *Pilimelia terevasa* and *P. anulata* have soft and pasty colonies, colored bright lemon-yellow to yellow-gray. *Pilimelia columellifera* has solid and hard colonies, either golden yellow to orange (subsp. *columellifera*) or pale (subsp. *pallida*) (see Table 28.10 below). Growth occurs only on complex media, such as 50% diluted skim milk agar (Gordon and Smith, 1955), casamino acids–peptone Czapek agar (Henssen and Schäfer, 1971), nutrient-sugar agar (Henssen and Schäfer, 1971), peptone–yeast extract–iron agar (Shirling and Gottlieb, 1966), oatmeal–yeast extract agar (Vobis et al., 1986a), and yeast extract–starch agar (Emerson, 1958). On these nutritive-rich media the colonies do not penetrate deeply into the agar but grow upward. Sporangial development occurs rarely. Nutritive-poor media with addition of keratinic substances promote the production of sporangia. Good examples are artificial soil extract agar (Henssen and Schäfer, 1971) together with hair of white mice (Fig. 28.8) or highly diluted skim milk–mineral agar with the addition of cattle horn meal (Vobis, 1984). On these media the colonies are more flat and the substrate hyphae penetrate deeply into the agar.

Pilimelia strains utilize neither the various carbon sources recommended by Shirling and Gottlieb (1966) (Vobis et al., 1986a) nor individual or combinations of purified amino acids as a sole source of nutrition (Kane Hanton, 1974). The members of *Pilimelia* are mesophilic, although exceptional growth has been observed at 42°C (Schäfer, 1973). Optimal growth occurs at about pH 7.0. For other physiological properties see Table 28.11 (below).

Intramycelial pigments have been reported in *P. terevasa* and *P. anulata*. The properties of absorption spectra are associated with carotenoids, with peaks at 479, 451, and 425 nm and an absorption maximum at 451 nm (Szaniszlo, 1968).

Sporangia with thick envelopes may function as a resistent form and/or as diaspores disseminated by the wind. If sporangia are dipped into water, numerous flagellated spores are released, leaving behind the sporangial envelope. The zoospores also provide adaption for the dissemination in water. The spores seem to swim randomly. On reaching keratinic substances (e.g. hair of mammals), they colonize the new substrate to produce sporangia within 14 days.

Pilimelia strains are not known to be dermatophytes, although they decompose keratinic material. Natural substrates are hair, feathers, or other keratinic substances. *Pilimelia* is distributed worldwide in soil. Schäfer (1973) detected *Pilimelia* strains in 22% of soil samples from various sources.

Enrichment and Isolation Procedures

Members of *Pilimelia* can be baited with natural substrates (Kane, 1966; Schäfer, 1973; Tribe and Abu El-Souod, 1979; Bland and Couch, 1981). Soil samples are placed in Petri dishes and stirred with twice-distilled water. Sterilized hairs are laid upon the surface. After 3–4 weeks of incubation, the hairs are examined microscopically. When they are covered with sporangia, they are transferred onto agar. With a thin needle, slightly hooked at the tip, individual sporangia are removed and rolled in zigzag curves over the surface of agar to remove bacteria, before transferring them to a suitable growth medium, such as 5% diluted skim milk–mineral agar with addition of cattle horn meal (Schäfer, 1973) or

Figure 28.8. Sporangia of *Pilimelia columellifera* developed on mouse hair. Scanning electron micrograph (× 520).

Emerson's yeast extract-starch agar (Kane, 1966). Colonies of 1 mm in diameter can be transferred onto slants after about 3 weeks.

Maintenance Procedures

Colonies growing on agar media must be subcultured after 4–5 weeks, because autolysis of vegetative mycelium may occur after 2 weeks of incubation. For long-term preservation *Pilimelia* strains must be lyophilized.

Procedures for Testing of Special Characters

The ability to decompose keratinic substances can be tested with hair of white mice. Sterilized hairs are incubated together with mycelium and checked microscopically after 2 weeks. Damaged hair segments are stainable with cotton blue; usually the hairs are also deformed and splintered in the longitudinal axis.

Differentiation of the genus **Pilimelia** *from other genera*

Among the sporangiate actinomycetes, characterized by cell wall chemotype II of Lechevalier and Lechevalier (1970b), the genus *Ampullariella* may be confused with *Pilimelia.* Both genera have species with rodlike spores that are arranged in parallel rows inside the sporangia. They are distinguishable from one another by the features listed in Table 28.9.

Taxonomic Comments

The members of the genus *Pilimelia* were discovered by Karling (1954), Gaertner (1955), and Rothwell (1957) studying hair of mammals attacked by unusual, keratinophilic microorganisms. Only prelimary diagnosis could be made, since pure cultures were not obtainable at this time. After successful culturing of single strains, Kane (1966) described the genus *Pilimelia* based on two species, *P. terevasa* and *P. anulata.* According to the original descriptions, the strains appeared to be keratinophilic members of the genus *Ampullariella* (Cross and Goodfellow, 1973). Recent studies on chemotaxonomy and fine structure support the recognition of the genus *Pilimelia.* The pattern of fatty acids is significantly different from other actinoplanetes (Kroppenstedt, 1979). The zoospores differ from those of *Ampullariella* in size (Couch and Bland, 1974a) and in the type of flagellar insertion (Schäfer, 1973; Kane Hanton, 1974; Vobis et al., 1986a) (see Table 28.9).

The proposals to transfer *P. terevasa* and *P. anulata* to *Ampullariella* (Juan and Zhang, 1974), and *P. columellifera* (Schäfer, 1973) to *Spirillospora* (Tribe and Abu El-Souod, 1979) were not considered in the Approved Lists of Bacterial Names (Skerman et al., 1980).

Table 28.9.
Differential characteristics of the genera **Pilimelia** *and* **Ampullariella**[a,b]

Characteristics	*Pilimelia*	*Ampullariella*
Length of zoospores		
0.7–1.5 μm	+	–
2.0–4.0 μm	–	+
Flagellar insertion		
Lateral	+	–
Polar	–	+
Shape of colonies		
Compact	+	–
Diffuse	–	+
Decomposition of keratinic substances	+	–
Pattern of fatty acids		
iso-	+[c]	+
anteiso-	–	+

[a]Data from Couch and Bland (1974a), Kroppenstedt (1979), Schäfer (1973), and Vobis et al. (1986a).
[b]Symbols: see Table 28.1.
[c]No *iso*-C_{16}.

Acknowledgments

I thank Dr. R.M. Kroppenstedt, German Collection of Microorganisms, Darmstadt, F.R.G., and Drs. B. Renner and E. Stahl, University of Hamburg, F.R.G., for making available unpublished biochemical data.

Further Reading

Kane, W.D. 1966. A new genus of *Actinoplanaceae, Pilimelia,* with a description of two species, *Pilimelia terevasa* and *Pilimelia anulata.* J. Elisha Mitchel Sci. Soc. *82:* 220–230.

Kane Hanton, W.D. 1974. Genus *Pilimelia* Kane. *In* Buchanan and Gibbons (Editors), Bergey's Manual of Determinative Bacteriology, 8th Ed. Williams and Wilkins, Baltimore, pp. 718–719.

Vobis, G. 1984. Sporogenesis in the *Pilimelia* species. *In* Ortiz-Ortiz, Bojalil, and Yakoleff (Editors), Biological, Biochemical, and Biomedical Aspects of Actinomycetes. Academic Press Inc., Orlando, FL, pp. 423–439.

Vobis, G., D. Schäfer, H. W. Kothe and B. Renner. 1986a. Descriptions of *Pilimelia columellifera* (*ex* Schäfer 1973) nom. rev. and *Pilimelia columellifera* subsp. *pallida* (*ex* Schäfer 1973) nom. rev. Syst. Appl. Microbiol. *8:* 67–74.

Differentiation and characteristics of the species of the genus **Pilimelia**

The differential characteristics of the species of *Pilimelia* are listed in Table 28.10. Other characteristics of the species and subspecies are indicated in Table 28.11.

List of species of the genus **Pilimelia**

1. **Pilimelia terevasa** Kane 1966, 225.[AL]

ter.e.vas′a. L. adj. *teres* "rounded" (properly: terete, i.e. circular in transverse sections, tapering or narrow cylindric); L. pl. n. *vasa* vessels; M.L. n. *terevasa* indicating "rounded," spherical sporangia.

Sporangia develop on hair of mammals floating on soil-water and on artificial soil extract agar. The shape of sporangia is spherical (applying to the type strain), or more or less campanulate (Fig. 28.7). The sporangia can reach diameters up to 24 μm. The rod-shaped spores (0.3–0.6 × 0.7–1.5 μm) are motile by means of a tuft of laterally inserted flagella. Spores are released after 15–20 min from sporangia when flooded with water. Sporangiophores are septate, approximately 1 μm in diameter.

Colonies on agar media are bright lemon-yellow with rough lobed borders. The surface is tuberculate and warty with curled protrusions. The consistency is soft and pasty.

Type strain: ATCC 25603.

Table 28.10.
Differential characteristics of the species of the genus **Pilimelia**[a]

Characteristics	1. P. terevasa	2. P. anulata	3. P. columellifera
Sporangial shape			
Spherical	+	–	+
Pyriform	–	–	+
Campanulate	+	–	–
Flabelliform	+	–	–
Cylindrical	–	+	–
Sporangiophore			
Septate	+	+	–
Annulate	–	+	–
Extended as columella	–	–	+
Arrangement of spore chains			
Parallel rows	+	+	–
Swirllike	–	–	+
Consistency of colonies			
Solid	–	–	+
Soft	+	+	–
Color of colonies			
Lemon yellow	+	+	–
Yellow gray			
Golden yellow, orange	–	–	+
Pale	–	–	+

[a]Symbols: see Table 28.1.

2. **Pilimelia anulata** Kane 1966, 225.[AL]
an'u.lat.a. L. fem adj. *anulata* having a ring.

Sporangia develop on hair of mammals and on agar media, supported by septate sporangiophores, where the uppermost fragment is swollen to a ringlike structure (Fig. 28.7). The sporangia are cylindrical, 2.8–11.2 µm wide and up to 35 µm long. Inside the sporangium, the spores are arranged in parallel chains. The spores are rod shaped (0.3–0.7 × 0.8–1.3 µm) and equipped with a laterally inserted tuft of flagella.

Colonies on agar media are bright lemon-yellow to yellow-gray and with lobed margins. The surface of the colonies is tuberculate with narrow and often branched, curled protrusions. The consistency is soft and pasty.

Type strain: ATCC 25604.

3. **Pilimelia columellifera** (ex Schäfer 1973) Vobis, Schäfer, Kothe and Renner 1986b, 573.[VP*] (Effective publication: Vobis, Schäfer, Kothe and Renner 1986a, 72.)
co.lu.mel.li'fer.a. L. n. *columella* small column; L. suffix *-fer* carrying; M.L. fem. adj. *columellifera* bearing a small column.

Sporangia are produced on the surface of hair (Fig. 28.8) and on agar media. The shape of sporangia is spherical, ovoid, or pyriform and 7–16 µm in diameter. Zoospores are developed in chains that are arranged like swirls (Fig. 28.7). The nonseptate sporangiophore is extended into the lumen of the sporangium, thus forming a columella. If sporangia are dipped into the water, spores start to swarm after 45 min. Occasionally swarming occurs inside the sporangia until finally the envelope tears and the spores escape. The rodlike zoospores (0.35–0.45 µm wide and 0.8–1.5 µm long) are motile by means of two to four laterally inserted flagella.

Colonies on agar media are small, about 5 mm in diameter after 4

Table 28.11.
Other characteristics of the species of the genus **Pilimelia**[a,b]

Characteristics	1. P. terevasa	2. P. anulata	3. P. columellifera	
			subsp. columell.	subsp. pallida
Hydrolysis of starch	–	–	–	–
Degradation of tyrosine	+	+	–	–
Production of melanoid pigments[c]	v	v	v	+
Liquefaction of gelatin	–	–	+	+
Peptonization of casein	+	+	+	+
Reduction of nitrate	–	–	+	–
pH growth range	6.5–7.6	6.5–7.8	6.5–7.6	5.0–7.5
Temperature growth range (°C)	10–35	15–35	15–35	10–30

[a]Data from studies of type strains (Vobis et al., 1986a).
[b]Symbols: see Table 28.1; *v*, strain instability.
[c]Data from Kane Hanton (1974), Schäfer (1973), and Vobis et al. (1986a).

**VP* denotes that this name has been validly published in the official publication, *International Journal of Systematic Bacteriology.*

weeks of incubation. They are compact, irregular, and pulvinate. The surface is warty and squamous, and the consistency is solid. The color of mycelium is golden yellow to orange or colorless to pale brownish.

Type strain: DSM 43797.

3a. **Pilimelia columellifera** subsp. **columellifera** (ex Schäfer 1973) Vobis, Schäfer, Kothe and Renner 1986b, 573.[VP] (Effective publication: Vobis, Schäfer, Kothe and Renner 1986a, 72.)

The description is as for the species. It differs from the subspecies *pallida* by its golden yellow to orange substrate mycelium and by its ability to reduce nitrate (Table 28.11).

Type strain: DSM 43797.

3b. **Pilimelia columellifera** subsp. **pallida** (ex Schäfer 1973) Vobis, Schäfer, Kothe and Renner 1986b, 573.[VP] (Effective publication: Vobis, Schäfer, Kothe and Renner 1986a, 72.)

pal′li.da. L. fem adj. *pallida* pale.

The subspecies *pallida* is distinguishable from subsp. *columellifera* by a colorless to pale brownish substrate mycelium. The center of the colonies occasionally becomes blackish with increasing age. Colonies are flat or convex to slightly raised and umbonate. Sporangial development occurs occasionally on oatmeal–yeast extract agar or on artificial soil extract agar. Melanoid pigments are produced on peptone–yeast extract iron agar and on tyrosine agar. Nitrate is not reduced to nitrite.

*Type strain:*DSM 43799.

Genus **Dactylosporangium** *Thiemann, Pagani and Beretta 1967a, 43*[AL]

GERNOT VOBIS

Dac.ty.lo.spo.ran′gi.um. Gr. n. *dactylos* finger; Gr. n. *spora* a seed, spore; Gr. n. *argeion* vessel; M.L. neut. n. *Dactylosporangium* an organism with finger-shaped, spore-containing vessels (sporangia).

Finger-shaped to claviform sporangia (0.6–1.4 × 2.5–6.0 µm) are **formed on short sporangiophores on the substrate mycelium.** They are developed singly or in clusters above the surface of the substrate. **Each sporangium contains a single row of normally three to four spores. The spores are oblong, ellipsoidal, ovoid, or slightly pyriform** (0.4–1.3 × 0.5–1.8 µm) and **motile** by means of a polarly inserted tuft of flagella. True aerial mycelium is not formed. Hyphae of the substrate mycelium are 0.5–1.0 µm in diameter, branched, and rarely septate. Large globose spores (1.7–2.8 µm in diameter) are formed on short branches on substrate mycelium. Organisms are Gram-positive and not acid fast. The peptidoglycan of the cell walls contains ***meso*-diaminopimelic acid (*meso*-DAP) and glycine, with xylose and arabinose** as characteristic sugars of whole-cell hydrolysates. Colonies grow on various agar media. They are compact, somewhat tough and leathery, and mostly flat or sometimes elevated with a smooth to slightly wrinkled surface. **The color of the substrate mycelium is pale orange to deep orange, rose or wine-colored to brown.** Aerobic, chemo-organotrophic, with an optimum for growth between 25° and 37°C and a pH optimum of 6.0–7.0. The mol% G + C of the DNA is 71–73 (T_m).

Type species: *Dactylosporangium aurantiacum* Thiemann, Pagani and Beretta 1967, 43.

Further Descriptive Information

The sporangia of the genus *Dactylosporangium* are formed on the surface of the substrate, singly or more frequently in tufts (Fig. 28.9). They are club or finger shaped, 0.6–1.4 µm in diameter and 2.5–6.0 µm in length. The short sporangiophores are 0.5–1.5 µm long and usually branched. Scanning and transmission electron micrographs revealed a collarlike structure at the sporangiophore-sporangium juncture (Ensign, 1978; Shomura et al., 1980; Vobis and Kothe, 1985). Each sporangium normally contains a single straight chain of three to four spores. A minimum of two spores and a maximum of five are produced (Thiemann et al., 1967a; Thiemann, 1974a; Shomura et al., 1980, 1983c, 1985). In addition to "normal" sporangia, long and branched sporangia are also formed by some strains (Thiemann, 1970b). The mode of spore formation has been claimed to be endogenous (Sharples and Williams, 1974; Williams and Wellington, 1980). Further studies on the fine structure and development of sporangia cast doubt on this type of sporogenesis (Ensign, 1978; Vobis and Kothe, 1985) and confirm the scheme of spore formation as proposed by Lechevalier and Holbert (1965) for the sporangiate actinomycetes. The spores are developed by the growth and subsequent septation of a sporogenous hypha inside an expanding envelope (Fig. 28.10). New sporangia may be formed by lateral branches of the sporangiophore (Vobis and Kothe, 1985).

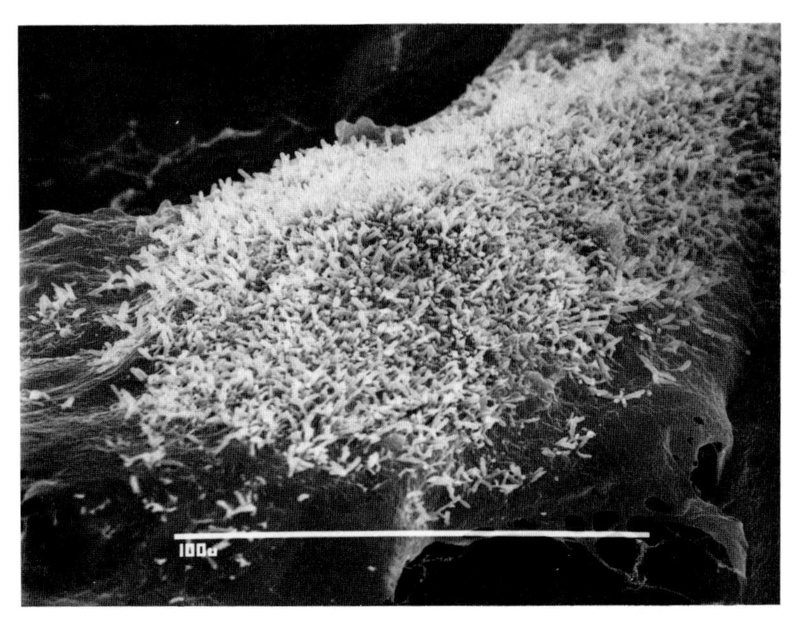

Figure 28.9. *Dactylosporangium* sp. (strain MB-VS 699). The surface of a small, flat colony is completely covered with sporangia formed after 15 days' incubation on mineral salts agar. Scanning electron micrograph (× 700).

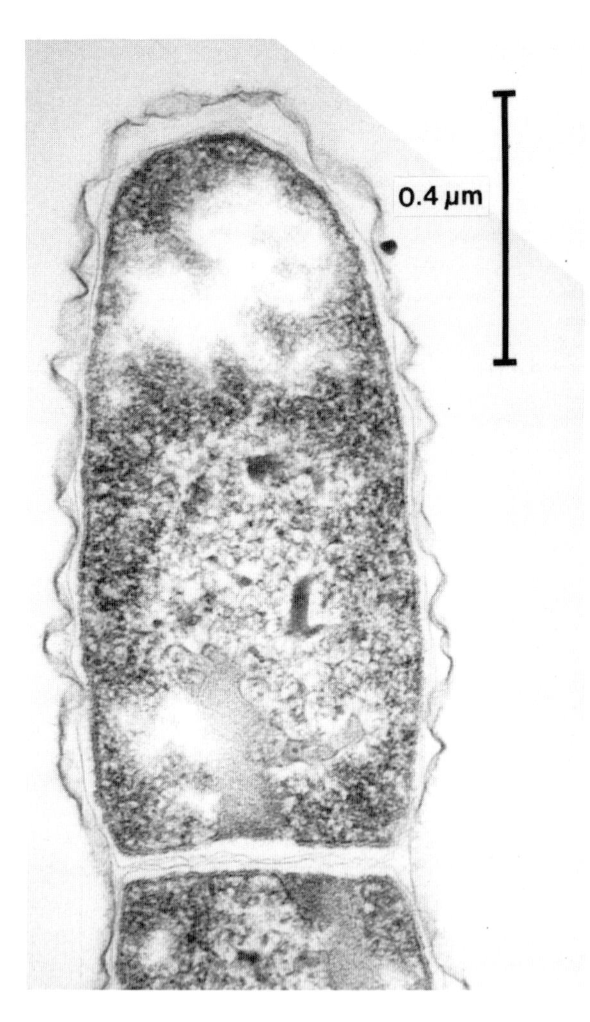

Figure 28.10. *Dactylosporangium* sp. (strain MB-VS 699). Longitudinal section through the tip of a young sporangium with a wavy envelope; the uppermost oblong spore has a very thin cell wall and is separated by a double-layered septum. Transmission electron micrograph (× 90,000).

The spores (zoospores) produced inside the sporangia have a smooth surface and are variable in size and shape. They measure 0.4–1.3 μm in diameter and 0.5–1.8 μm in length. The shape can be oblong, ellipsoidal, ovoid, or slightly pyriform (Thiemann et al., 1967a; Shomura et al., 1980, 1983c, 1985). The zoospores are motile by means of a polar tuft of flagella (Higgins et al., 1967; Lechevalier and Lechevalier, 1970a; Thiemann, 1974a; Shomura et al., 1985).

The zoospores are released after a period of 10–60 min if sporangia are immersed in distilled water, tap water, or soil extract solution. Spore release is probably initiated by the swelling of an intrasporangial substance (Thiemann et al., 1967a). The spore chain is pressed upward and pushed through the apex of the sporangial envelope. High motility starts after a time lag of about 30 min when the spores first separate from one another (Thiemann et al., 1967a; Shomura et al., 1980; Shomura et al., 1983c). Spores are able to swim for up to 30 h. Germination was reported after 24 h on agar medium (Thiemann et al., 1967a).

The substrate hyphae are 0.5–1.0 μm in diameter and irregularly branched. They are rarely septate and do not separate into fragments either in agar or in liquid cultures. A true aerial mycelium is not formed; however, short hyphae in contact with the air are observed occasionally (Thiemann et al., 1967a; Shomura et al., 1983c).

In addition to the zoospores, large, globose-shaped structures (1.7–2.8 μm in diameter) are formed singly on short lateral branches of substrate hyphae. They are also produced in liquid cultures (Thiemann et al., 1967a). These phase-bright spherical bodies embedded in agar or on the surface are considered to be an additional morphological characteristic of the genus (Thiemann, 1970b). Based on cytological studies, Sharples and Williams (1974) considered that the globose bodies are products of abnormal development possibly induced by phage infection or hyphal lysis. More recently, however, Ensign (1978) demonstrated their germination in a 10% (w/v) yeast extract solution and concluded that they were spores.

The peptidoglycan of the cell wall contains *meso*-DAP and glycine, with xylose and arabinose as diagnostic sugars in whole-cell hydrolysates (Lechevalier and Lechevalier, 1970c; Shomura et al., 1980, 1983c, 1985). The chemical composition of cell walls therefore conforms with the chemotype II and sugar pattern D in the classification scheme of Lechevalier and Lechevalier (1970b).

The phospholipid composition is characterized by the presence of phosphatidylinositol mannosides, phosphatidylinositol, phosphatidylethanolamine (PE), and diphosphatidylglycerol and the absence of phosphatidylcholine (PC), phosphatidylmethylethanolamine, acyl phosphatidylglycerol, and the unknown glucosamine-containing phospholipids (GluNu). The presence of phosphatidylglycerol is variable (Lechevalier et al., 1977). Based upon the existence of PE and the absence of PC and GluNu as diagnostic phospholipids, *Dactylosporangium* belongs to the phospholipid type II of Lechevalier et al. (1981).

The analysis of the fatty acids of the cytoplasmatic membranes show that the principal components are branched *iso*- and *anteiso*- fatty acids (Lechevalier et al., 1977; Kroppenstedt and Kutzner, 1978). Cyclopropane and 10-methyl-branched fatty acids are not present (Kroppenstedt, 1979). In some strains small amounts of unknown unsaturated fatty acids are found, which are assumed to have a branched chain (Lechevalier et al., 1977).

The menaquinone composition of *D. aurantiacum* and *D. thailandense* is characterized by possession of major amounts of MK-9(H_6) and MK-9(H_8) (Collins et al., 1984).

The fine structure of *Dactylosporangium* is similar to that of other actinomycetes. The cell walls of the hyphae, the globose bodies, and the zoospores each consist of a single layer. The cross-walls involved in the formation of the globose bodies and in sporangial development belong to type 2 of Williams et al. (1973) (Sharples and Williams, 1974; Vobis and Kothe, 1985). The mesosomes are tubular-vesicular (Williams et al., 1973). The substrate hyphae are not surrounded by a sheath (Sharples et al., 1974). An outer wall layer or sheath occurs only in the form of a sporangial envelope (Fig. 28.10) (Ensign, 1978). The large globose bodies contain nuclear material, large diffuse electron-transparent areas, smaller defined light areas, and possible phage particles. Protein-containing paracrystalline inclusions are conspicuous (Sharples and Williams, 1974). Crystalline phage particles were detected in the cytoplasm of the substrate hyphae of a strain of *D. thailandense* (Higgins and Lechevalier, 1969).

The colonies of *D. aurantiacum* are mostly flat with a smooth surface. The color of substrate mycelium varies from whitish to pale orange to deep orange. The surface of the colonies of *D. thailandense* is usually smooth but can be wrinkled on certain media. The substrate mycelium is light orange, amber, or brownish with a rose tinge (Thiemann et al., 1967a). Colonies of *D. vinaceum* are compact, tough, and somewhat leathery; the mycelia are wine-colored to brown, depending on the medium (Shomura et al., 1983c). The colonial characteristics of *D. matsuzakiense* are similar to *D. vinaceum;* the substrate mycelium is orange (Shomura et al., 1983c). *Dactylosporangium roseum* has a rose-colored substrate mycelium on certain agar media (Shomura et al., 1985). Soluble pigments are produced by *D. vinaceum* (wine-colored to deep red), *D. thailandense* (amber to brown), and *D. matsuzakiense* (light brownish pink) (Shomura et al., 1985). Hydrogen sulfide (H_2S) is produced on peptone-iron agar by *D. aurantiacum* and *D. thailandense* (Thiemann et al., 1967a).

The development of sporangia depends on the agar media used. They can be formed after 2 or 3 days under favorable conditions (Ensign, 1978), although normally they are evident only after 5–15 days of incubation (Fig. 28.9) (Shomura et al., 1983c). Sporangial formation can be promoted by soil agar and calcium malate agar (Thiemann et al., 1967a). Globose bodies appear to be produced mainly on complex agar media that promote the growth of substrate mycelium but not the formation of sporangia (Ensign, 1978).

Good vegetative growth occurs on various agar media, e.g. oatmeal, Bennett, Hickey-Tresner, nutrient, glucose-asparagine, sucrose-nitrate, sucrose–yeast extract, and inorganic salt–starch agars. However, the individual species grow well on only a small spectrum of these media (Thiemann et al., 1967a; Shomura et al., 1980, 1983c, 1985). Different basal media must be used for carbon utilization tests because all strains do not grow on a common basal medium (Thiemann et al., 1967a; Shomura et al., 1983c). *Dactylosporangium* species utilize D-fructose, sucrose, D-glucose, D-xylose, and L-arabinose, but do not utilize glycerol or inositol (see Table 28.13 below) (Thiemann et al., 1967a; Thiemann, 1974a; Shomura et al., 1983c, 1985). Melanoid pigments are not produced. *Dactylosporangium* species are aerobic and mesophilic. The optimum temperature for growth is between 25° and 37°C; no growth occurs at 45°C. For further physiological features see Table 28.13 (below).

Antibiotic metabolites are found in some species. "*Dactylosporangium variesporum*" produces capreomycin, a polypeptide complex (Tomita et al., 1977), and "*D. salmoneum*" a polycyclic ether antibiotic (Celmer et al., 1978). The species *D. matsuzakiense* and *D. vinaceum* produce dactimicin, a pseudodisaccharide antibiotic (Shomura et al., 1980, 1983c). An antibiotic of the orthosomycin group is produced by *D. roseum* (Shomura et al., 1985).

A poorly lytic bacteriophage was discovered in a strain of *D. thailandense* that did not infect other actinomycetes (Higgins and Lechevalier, 1969). Cross-infections with phages of various actinomycetes were not successful (Willoughby et al., 1972; Prauser, 1984c). The mol% G + C of the DNA is 73 for *D. aurantiacum* and 71 for *D. thailandense* (Farina and Bradley, 1970).

The members of the genus *Dactylosporangium* are distributed worldwide in soil. Thiemann (1970b) obtained isolates from about 10% of the various soils sampled. They were present in sandy as well as in loamy soils. No correlation could be established between the type of soil, its pH (4.0–9.0), and the incidence of *Dactylosporangium.* Strains were also found on plant debris (Lechevalier, 1981) and could be isolated from lake sediments (Johnston and Cross, 1976).

Enrichment and Isolation Procedures

The details of the promising isolation technique used by Thiemann et al. (1967a) were never published. Johnston and Cross (1976) isolated a few strains using a spread plate technique on colloidal chitin agar. After dry heat treatment of soil samples, colonies of *Dactylosporangium* were sometimes detectable on an agar medium consisting of humic acid, salts, vitamins, and antibiotics (Nonomura, 1984). We have also obtained strains by the plate method for isolation of *Streptosporangium* (Schäfer, 1969) using an artificial soil agar (Henssen and Schäfer, 1971).

Maintenance Procedures

Slant agar cultures can be stored for several weeks at room temperature or at 6°C. For long-term preservation, the strains must be lyophilized as recommended for the other aerobic actinomycetes.

Differentiation of the genus **Dactylosporangium** *from other genera*

The genus *Dactylosporangium* can be confused with members of the genera *Planomonospora, Planobispora,* and *Micromonospora.* The differentiating morphological and biochemical characteristics are listed in Table 28.12.

Table 28.12.
Characteristics differentiating the genus **Dactylosporangium** *from other genera forming either few-spored sporangia or single spores on the substrate mycelium*[a]

Characteristics	*Dactylosporangium*	*Micromonospora*	*Planomonospora*	*Planobispora*
Aerial mycelium, absent, cell wall chemotype II	+	+	–	–
Aerial mycelium present, cell wall chemotype III	–	–	+	+
Sporangia present	+	–	+	+
Sporangia containing				
1 spore	–	–	+	–
2 spores	+	–	–	+
2–5 spores	+	–	–	–
Spores developed on substrate hyphae	+	+	–	–
0.7–1.5 μm in diam.	–	+	–	–
1.7–2.8 μm in diam.	+	–	–	–

[a]Symbols: see Table 28.1.

Taxonomic Comments

While the morphological characters of *Dactylosporangium* are distinct and useful features to recognize the genus, the species differentiation is difficult (Thiemann, 1970b). Currently species are distinguished on the basis of the color of their substrate mycelium and diffusible pigments. This seems to be unsatisfactory in some cases, since diffusible pigments might have an effect on the mycelium color.

As characterized by cell wall chemotype II and sugar pattern D (Lechevalier and Lechevalier, 1970b), the genus *Dactylosporangium* is related to the genera *Actinoplanes (Amorphosporangium), Ampullariella, Pilimelia,* and *Micromonospora.* The close relationship is also shown by the phospholipid (Lechevalier et al., 1981) and menaquinone (Collins et al., 1984) patterns and, more distinctly, by the molecular genetic data obtained from oligonucleotide sequencing of 16S ribosomal RNA, DNA-ribosomal and RNA cistron similarities, and DNA-DNA hybridization (Stackebrandt et al., 1981, 1983c).

Acknowledgments

I thank Dr. T. Kusaka, Institute for Fermentation, Osaka, Japan, and Dr. R. M. Kroppenstedt, German Collection of Microorganisms, Darmstadt, F.R.G., for providing strains of *Dactylosporangium.*

Further Reading

Ensign, J.C. 1978. Formation, properties, and germination of actinomycete spores. Annu. Rev. Microbiol. *32:* 185–219.

Sharples, G.P. and S.T. Williams. 1974. Fine structure of the globose bodies of *Dactylosporangium thailandense (Actinomycetales).* J. Gen. Microbiol. *84:* 219–222.

Sharples, G.P., S.T. Williams and R.M. Bradshaw. 1974. Spore formation in the *Actinoplanaceae* (*Actinomycetales*). Arch. Microbiol. *101:* 9–20.

Shomura, T., J. Yoshida, S. Miyadoh, T. Ito and T. Niida. 1983. *Dactylosporangium vinaceum* sp. nov. Int. J. Syst. Bacteriol. *33:* 309–313.

Thiemann, J.E., H. Pagani, G. Beretta. 1967. A new genus of the *Actinoplanaceae: Dactylosporangium,*gen. nov. Arch. Mikrobiol. *58:* 42–52.

Vobis, G. and H.-W. Kothe. 1985. Sporogenesis in sporangiate actinomycetes. *In* Mukerji, Pathak and Singh (Editors), Frontiers in Applied Microbiology, Vol. I. Print House (India), Lucknow, pp. 25–47.

Differentiation and characteristics of the species of the genus **Dactylosporangium**

The species of *Dactylosporangium* have essentially the same morphological characteristics. They differ slightly in the pattern of carbon utilization and some physiological properties (Table 28.13). Species differentiation is mainly based on the color of substrate mycelium and on the production of soluble pigments. *D. aurantiacum* has orange substrate mycelium and does not produce pigment. *D. thailandense* has orange to brown mycelium and produces pigments on some media that are amber to brown with a reddish tinge (Thiemann et al., 1967a; Thiemann, 1974a). A wine-colored to deep red diffusible pigment characterizes *D. vinaceum,* and the mycelium of this species is similarly colored. *D. matsuzakiense* has orange colonies and produces a light brownish-pink pigment on tyrosine agar (Shomura et al., 1983c). *D. roseum* is characterized by the rose color of its substrate mycelium (Shomura et al., 1985). The morphology of sporangia, zoospores, substrate hyphae, and globose bodies of the species are as given for the genus.

List of species of the genus **Dactylosporangium**

1. **Dactylosporangium aurantiacum** Thiemann, Pagani and Beretta 1967a, 43.[AL]

au.ran.ti′ac.um. M.L. neut. adj. *aurantiacum* orange colored.

Good sporangial development occurs on soil agar and on calcium malate agar. The zoospores are released after a period of 10–15 min after placing the sporangia in water. They are extremely vigorous swimmers.

Colonies on agar media are mostly flat with a smooth surface. Abundant to good growth occurs on oatmeal agar and nutrient agar. The color of the substrate mycelium is pale orange to orange. On Hickey-Tresner agar the growth is moderate with very pale mycelium. On tyrosine, nutrient, and skim milk agars the colonies are orange to deep orange. Whitish colonies occur on glucose-asparagine agar and on starch agar. On glycerol-asparagine agar and on peptone-iron agar the colonies are hyaline.

D-Mannitol, L-rhamnose, and D-melibiose are utilized for growth; D-ribose is not. NaCl is tolerated up to 3% (w/v). Nitrate is reduced to nitrite. Litmus milk is not coagulated. No soluble pigments are produced on any media.

The mol% G + C of the DNA is 73 (T_m).

Type strain: ATCC 23491.

2. **Dactylosporangium thailandense** Thiemann, Pagani and Beretta 1967a, 49.[AL]

thai.lan.den′se. M.L. neut. adj. *thailandense* pertaining to Thailand (the correct Latin epithet of the species was proposed by Thiemann, 1970a).

Abundant sporangial formation occurs on soil agar, calcium malate agar, and starch agar. Colonies grow on various agar media with a wrinkled or smooth surface. Good growth occurs on oatmeal agar with light orange-brown substrate mycelium, producing a light amber diffusible pigment. On Hickey-Tresner agar the growth is also good; the mycelium is brown with a rose tinge and the pigment is brown to reddish pink. Growth on glycerol-asparagine, glucose-asparagine, and starch agars is moderate, with a light orange to orange substrate mycelium. On nutrient agar growth is good, and the mycelium is pale orange.

D-Ribose, D-mannitol, and L-rhamnose are utilized for growth; D-melibiose is not. The NaCl tolerance ranges up to 1.5% (w/v). Tyrosine is hydrolyzed. Nitrate is not reduced to nitrite. Milk is not coagulated. No growth occurs at 42°C. Production of brown soluble pigments occurs on some media.

The mol% G + C of the DNA is 71 (T_m).

Type strain: ATCC 23490.

3. **Dactylosporangium vinaceum** Shomura, Yoshida, Miyadoh, Ito and Niida 1983c, 312.[VP]

vi.na′ce.um. M.L. neut. adj. *vinaceum* wine colored.

Sporangia are occasionally apparent on Czapek and on oatmeal agar but less apparent on inorganic salts–starch and on calcium malate agar. Colonies grow well on Czapek, glucose-asparagine, inorganic salts–starch, oatmeal, Bennett, and Hickey-Tresner agars. The color of substrate mycelium or reverse color ranges from light to dark wine-colored or occasionally ebony brown. Poor growth occurs on glycerol-asparagine agar with apricot substrate mycelium and no or very light wine-colored pigment production.

Production of a wine red–colored diffusible pigment is conspicuous on various agar media, shaded from rose-wine to dark red or cherry. The pigments of the substrate mycelium and the soluble pigment are stable with changes in pH.

D-Mannitol and L-rhamnose are utilized as carbon source for growth. Gelatin and casein are hydrolyzed; milk is coagulated. Nitrate is not reduced to nitrite. Concentration of NaC1 up to 3% (w/v) is tolerated.

The mol% G + C of the DNA has not been determined.

Type strain: IFO 14181.

4. **Dactylosporangium matsuzakiense** Shomura and Niida *in* Shomura, Kojima, Yoshida, Ito, Amano, Totsugawa, Niwa, Inouye, Ito and Niida 1983b, 672.[VP] (Effective publication: Shomura, Kojima, Yoshida, Ito, Amano, Totsugawa, Niwa, Inouye, Ito and Niida 1980, 928.)

mat.su.za.ki.en′se. M.L. neut. adj. *matsuzakiense* pertaining to Matsuzaki-cho, Ozu Peninsula, Japan.

Abundant production of sporangia occurs on inorganic salts–starch agar, but is rare on Czapek, oatmeal, and tyrosine agars. Formation of globose bodies is not observed.

Colonies grow well on inorganic salts–starch agar, with a russet-orange substrate mycelium. Moderate growth occurs on Czapek agar and on yeast extract–malt extract agar, with an amber to light brown mycelium. Colonies grow moderately on glucose-asparagine agar and on oatmeal agar; the substrate mycelium is russet to orange. Poor growth occurs on glycerol asparagine, nutrient, and calcium malate agars, the color of substrate mycelium varying from light yellow to light orange. Moderate growth occurs on tyrosine agar; the color of colonies is dusty orange to light brown and a light brownish-pink soluble pigment is produced.

D-Mannitol and L-rhamnose are utilized as sole carbon sources; D-melibiose is not. Growth occurs between 15° and 42°C, and the temperature optimum ranges from 25° to 35°C. NaCl tolerance is lower than 1.5% (w/v). Gelatin is not liquified; nitrate is not reduced. Skim milk is neither peptonized nor coagulated.

The mol% G + C of the DNA has not been determined.

Type strain: ATCC 31570.

Table 28.13.
Physiological characteristics of the species of the genus **Dactylosporangium**[a]

Characteristics	1. *D. aurantiacum*	2. *D. thailandense*	3. *D. vinaceum*[b]	4. *D. matsuzakiense*[b,c]	5. *D. roseum*[d]
Utilization of					
D-Fructose	+	+	+	+	+
Sucrose	+	+	+	+	+
D-Glucose	+	+	+	+	+
D-Xylose	+	+	+	+	V
D-Mannitol	+	+	+	+	−
L-Arabinose	+	+	+	+	V
L-Rhamnose	+	+	+	+	−
Raffinose	+[e]; −[b,c]	+[e]; −[b,c]	−	−	−
D-Mannose[e]	+	+	ND	ND	ND
Lactose[e]	+	+	ND	ND	ND
Maltose[e]	+	+	ND	ND	ND
Dextrin[e]	+	+	ND	ND	ND
Inulin[e]	+	+	ND	ND	ND
D-Galactose[e]	+	+	ND	ND	ND
Glycerol	−	−	−	−	ND
D-Sorbitol[e]	−	−	ND	ND	ND
D-Dulcitol[e]	−	−	ND	ND	ND
Inositol	−	−	−	−	−
D-Ribose[e]	−	+	ND	ND	ND
Sorbose[e]	−	ND	ND	ND	ND
D-Melibiose	+	−	ND	−	ND
Hydrolysis of					
Starch	+	+	+	+	−
Cellulose[e]	−	−	ND	ND	ND
Tyrosine	−[e]	+[f]	ND	ND	ND
Gelatin	−[e]; +[b]	+[e]; −[b]	+	−	+
Casein	+[e]; −[b]	+[e]; −[b]	+	−	−
Coagulation of milk	−	−	+	−	−
Digestion of calcium malate[e]	−	−	ND	ND	ND
Reduction of nitrate	+	−	−	−	−
Production of melanoid pigments	−	−	−	−	−
Production of H_2S[e,f]	+	+	ND	ND	ND
Optimum for growth					
pH[e]	6.0–7.0	6.0–7.0	ND	ND	ND
Temperature (°C)	28–37[e]	28–37[e]	25–37	25–37	28–37
NaCl tolerance					
Up to 1.5% (w/v)	+[b]	+[b]	+	−	+
Up to 3.0% (w/v)	+[b]	−[b]	+	−	−

[a]Symbols: see Table 28.1, *v*, strain instability; *ND*, not determined.
[b]Data from Shomura et al. (1983c).
[c]Data from Shomura et al. (1980).
[d]Data from Shomura et al. (1985).
[e]Data from Thiemann et al. (1967a).
[f]Data from Thiemann (1974a).

5. **Dactylosporangium roseum** Shomura, Amano, Tohyama, Yoshida, Ito, Niida 1985, 4.[VP]

ro′se.um. L. neut. adj. *roseum* rose colored, pink.

Sporangia are abundantly formed on chemically defined media such as sucrose-nitrate, glucose-asparagine, glycerol-asparagine, inorganic salts-starch, calcium malate, and tyrosine agars, but rarely on yeast extract-malt extract, nutrient, and Bennett agars. The formation of globose bodies has not been observed.

Colonies grow well on sucrose-yeast extract agar, and moderately on yeast extract-malt extract, inorganic salts-starch, oatmeal, Bennett, and tyrosine agars. The typical rose color of the substrate mycelium occurs on sucrose-yeast extract, inorganic salts-starch, and yeast extract-malt extract agars. On Bennett agar the color of mycelium is pastel orange and on tyrosine agar it is yellowish. Colonies on sucrose-nitrate, glucose-asparagine, glycerol-asparagine, calcium malate, oatmeal, and nutrient agars are colorless. Soluble pigments are not produced.

D-Mannitol and L-rhamnose are not utilized as sole carbon sources. NaCl is tolerated up to 1.5% (w/v). Growth occurs between 20° and 40°C; the temperature optimum ranges from 28° to 37°C. Nitrate is reduced to nitrite and gelatin is liquified. Starch is not hydrolyzed; skim milk is neither peptonized nor coagulated. Melanoid pigments are not produced.

The mol% G + C of the DNA has not been determined.

Type strain: IFO 14352.

Species Incertae Sedis

a. *"Dactylosporangium variesporum"* Tomita, Kobaru, Hanada and Tsukiara 1977, U.S. Patent 4,026,766.

"D. variesporum" was established for a strain producing the antibiotic capreomycin. The species is not included in the Approved Lists of Bacterial Names, and not validated in a List in the *International Journal of Systematic Bacteriology.*

The morphology is typical of the genus. The substrate mycelium is orange to light reddish brown, creamy to light yellowish orange. A yellow, pale brownish- to reddish-orange diffusible pigment is produced on various agar media.

Glycerol, inositol, and D-ribose are utilized as sole carbon sources; L-rhamnose is not utilized. Nitrate is reduced to nitrite.

Considering the color of substrate mycelium and the color of the soluble pigment, the strain is presumably related to *D. thailandense.*

Type strain: ATCC 31203.

b. *"Dactylosporangium salmoneum"* Routien in Celmer, Cullen, Moppett, Routien, Jefferson, Shibakawa and Tone 1978, U.S. Patent 4,081,532.

Polycyclic ether antibiotic-producing strains of *Dactylosporangium* were described as *"D. salmoneum."* The species is not included in the Approved Lists of Bacterial names and not validated in a List in the *International Journal of Systematic Bacteriology.*

The morphological characters are typical of the genus. Sporangia are developed abundantly on calcium malate agar. The color of the substrate mycelium is salmon, pale pink, orange, pink orange, pinkish, or creamish. Soluble pigments are not produced. No growth occurs on starch agar.

This species is very similar to *D. roseum* in its color of substrate mycelium and inability to produce diffusible pigments.

Type strain: ATCC 31222.

Genus **Micromonospora** *Ørskov 1923, 147*[AL]

ISAO KAWAMOTO

Mic.ro.mo.no'spo.ra. Gr. adj. *micros* small; Gr. adj. *monos* single, solitary; Gr. n. *spora* a seed; M. L. fem. n. *Micromonospora* small, single-spored (organism).

Well-developed, branched, septate mycelium averaging 0.5 μm in diameter. Nonmotile spores borne singly, sessile, or on short or long sporophores that often occur in branched clusters. Sporophore development monopodial or in some cases sympodial. **Aerial mycelium absent** or in some cultures appearing irregularly as a restricted white or grayish bloom. Gram-positive. Not acid fast. **Walls contain *meso*-diaminopimelic acid (*meso*-DAP) and/or its 3-hydroxy derivative and glycine. Xylose and arabinose present in cell hydrolysates.** Characteristic phospholipids are phosphatidylethanolamine, phosphatidylinositol, and phosphatidylinositol mannosides. Major menaquinones MK-9(H_4), MK-10(H_4), MK-10(H_6), or MK-12(H_6). Aerobic to microaerobic. Chemo-organotrophic. Sensitive to pH below 6.0. Growth occurs normally between 20° and 40°C but not above 50°C. The mol% G + C of the DNA is 71–73 (Bd).

Type species: *Micromonospora chalcea* (Foulerton 1905) Ørskov 1923, 156.

Further Descriptive Information

Colonies on agar media are initially pale yellow or light orange, becoming orange, red, brown, blue-green, or purple. Mature colonies take on a progressively dark hue along with the production of brown-black, green-black, or black spores, and become mucoid. The yellow-orange-red mycelial pigments appear to offer little diagnostic value in strain or species recognition. There are two characteristic mycelial pigments known to be pH indicators, the maroon-purple pigments of *M. echinospora* subp. *echinospora, M. echinospora* subsp. *ferruginea,* and *M. purpurea,* and the blue-green pigment of *M. coerulea.* Certain other species produce characteristic diffusible pigments.

Mycelial development and sporophore morphology have little diagnostic value for species. For example, the cluster-fragmentation type and the open-web type illustrated by Luedemann (1971c) are seldom seen in pure form; usually a mixed type is observed in the same culture (Sveshnikova et al., 1969). The sympodial sporophore of *M. carbonacea,* a morphological characteristic of the species (Luedemann and Brodsky, 1965), can occasionally be seen to a lesser degree in other *Micromonospora* species (Luedemann, 1971c).

Sporulation appears almost as readily in submerged broth culture as on agar media. The formation of single spores on substrate mycelium is one of the well-defined criteria in the genus *Micromonospora.* Infrequently spores occur in longitudinal pairs and more infrequently as multiple longitudinal spores (Luedemann and Casmer, 1973). Often bizarre-shaped and multiple septate cells that resemble chlamydospores are found among certain enlarged vegetative hyphae. In *M. purpureochromogenes* NRRL B-2671, a sporulation process has been described that closely resembles sporangial formation (Stevens, 1975). This process begins as a hyphal length, is contained within a sporangial wall, becomes septate, and then separates into individual cells or sporangiospores, which do not exhibit the subsequent wall thickening typical of the singly produced spores. Spore surface ornamentation of the strains and the species of *Micromonospora* have been characterized by the terms "smooth," "rough," "warty," or "blunt spiny." These observations were based mainly on examinations by transmission electron microscopy. Recently, it has been confirmed by scanning electron microscopy that the spores of almost all species, including the strains described as having a smooth surface, are blunt spiny-surfaced (unpublished data). The size of the projections varies from strain to strain. Consequently, spore wall ornamentation may not be a useful diagnostic characteristic for differentiation of the species. The protuberances are formed in the outer layer of the spore wall, thus differing in origin from the surface ornamentation found on the sheath of some *Streptomyces* spores (Luedemann and Casmer, 1973). Sporogenesis occurs in two stages: sporulation septum formation and spore maturation (Luedemann and Casmer, 1973; Stevens, 1975; Hardisson and Suárez, 1979). The process is initiated by swelling of the apical end of a hypha before its delimitation by a sporulation septum. Mesosomes are associated with the DNA and with the formation of the sporulation septum. In spore maturation, material is laid down centripetally to form much-thickened wall layers. Spore dehiscence is facilitated by the rupture of the stretched and thinned hyphal wall in the vicinity of the sporulation septum. Therefore, the *Micromonospora* spore has been described as a specialized, broad-based type of aleuriospore (Luedemann and Casmer, 1973) or blastic, holoblastic spore (Stevens, 1975). The spore wall of *Micromonospora* has no inner multilaminar coat as has been observed in *Thermoactinomyces* spores, but has a more stratified deposition than that of *Streptomyces* spores.

The much-thickened walls probably account for their rather high refractility and their relatively high resistance to physical and chemical treatments. Mature spores are unaffected by ultrasonication but mycelia are quickly killed (Johnston and Cross, 1976; Kawamoto et al., 1982). Spore survival of most strains is >50% after 20 min at 60°C in phosphate buffer and <0.5% after 20 min at 80°C (Kawamoto et al., 1982). The spore of *M. chalcea* is absolutely resistant to a temperature below 75°C and its decimal reduction time at 80°C is 12.5 min (Suárez et al., 1980). *Micromonospora* spores are more resistant to ketones than to alcohols and dioxane. More than 10% spore survival is observed after treatment for 30 min at 30°C with a solution containing 60% (v/v) of tert-butyl alcohol, formamide, dimethylformamide, dimethylacetoamide, or acetone (Kawamoto et al., 1982). The spore viability does not change between pH 6.0 and 8.0, but decreases outside this range, particularly at an acidic pH (Kawamoto et al., 1982). Investigation of electrokinetic properties by Douglas et al. (1970) indicated that the spores of *Micromonospora* might have different properties from those of *Streptomyces* and *Thermoactinomyces*. Studies have been made of ultrastructual, physiological, and biological changes during germination of *M. chalcea* spores (Hardisson and Suárez, 1979; Suárez et al., 1980). In general the features are similar to those of *Streptomyces antibioticus* spores, but germination is considerably slower in *M. chalcea*.

Chemotaxonomically, the genus *Micromonospora* is characterized by a cell wall type II (Lechevalier and Lechevalier, 1970b), a whole cell sugar pattern D (Lechevalier and Lechevalier, 1970b), and a phospholipid type PII (Lechevalier et al., 1977). The cell walls of 19 strains have been found to contain glycine, glutamic acid, *meso*-DAP (including its 3-hydroxy derivative), and D-alanine in a molar ratio of 1:1:1:0.6–0.8, and an almost equal molar ratio of glycolic acid to DAP (Kawamoto et al., 1981). Based on this evidence, a primary structure has been proposed that is characterized by the presence of glycine in the first position of the peptide subunit, a direct interbridge between D-alanine and DAP, and the presence of *N*-glycolylmuramic acid. Some *Micromonospora* strains have been reported to be more sensitive to lysozyme (β-*N*-acetylmuramidase) than are *Mycobacterium* strains in which the muramic acid is *N*-glycolated (Mordarska et al., 1978). However, Kawamoto et al. (1981) found that the peptidoglycans of *M. olivasterospora* and "*M. sagamiensis*" were not liquefied by this enzyme. Furthermore, Szvoboda et al. (1980) showed that *Micromonospora* was not effectively protoplasted by lysozyme or lysozyme-EDTA treatment even if very high concentrations (up to 20 mg/ml) of lysozyme were used for 24 h. The pentoses xylose and arabinose are always constituents of the cell wall, although the amounts vary to some extent. Among the hexoses, glucose and galactose are detected more frequently than mannose and rhamnose (Kawamoto et al., 1981). The phospholipids contained in the cells are cardiolipids, phosphatidylethanolamine, phosphatidylinositol, mannosides of phosphatidylinositol, and possibly lysocardiolipids (Lechevalier et al., 1977; Dassain et al., 1983). These conform to the phospholipid type II of Lechevalier et al. (1977). The glycolipids identified by Dassain et al. (1983) are monoglucosyldiglycerides, diglucosyldiglycerides (with a small amount of galactosylglucosyldiglycerides), and esters of fatty acids and trehalose. These authors pointed out that the simultaneous presence of glucosyldiglycerides and trehalose esters might be specific for the genus *Micromonospora*. The predominant cellular fatty acids are *iso*- and *anteiso*-branched fatty acids. Unsaturated or 10-methyl fatty acids may be found in certain strains, but neither mycolic acid nor cyclic fatty acids are present (Kroppenstedt and Kutzner, 1976; Dassain et al., 1983). The menaquinone profile of *Micromonospora* appears complex, having much variation at the subspecies level (Collins et al., 1982d, 1984; unpublished data), although menaquinones MK-10(H_4) or MK-10(H_6) are a major component in most strains. *M. carbonacea* subsp. *carbonacea*, *M. carbonacea* subsp. *aurantiaca*, *M. halophytica* subsp. *halophytica*, and *M. halophytica* subsp. *nigra* contain a large amount of MK-9(H_4), whereas *M. echinospora* subsp. *pallida* contains MK-12(H_4), MK-12(H_6), and MK-12(H_8).

Glucose, maltose, sucrose, and starch are well utilized by all strains, and most strains can utilize D-xylose, cellobiose, D-mannose, and trehalose, whereas there is generally little or no growth with dulcitol, sorbitol, L-sorbose, or melezitose as sole carbon source. The carbohydrate utilization is substantially affected by the basal medium used. With *M. inositola* ATCC 31010, inositol permitted moderate growth on a yeast extract-$CaCO_3$ medium but poor growth on a defined medium (Kawamoto et al., 1983a). Similar discrepancies have been found with some carbohydrates (Hatano et al., 1976; Maehr et al., 1980; Kawamura et al., 1981). When tested with *p*-nitrophenylglycosides, α-glucosidase, β-glucosidase, β-galactosidase, or β-*N*-acetylglucosaminidase activity was almost always present, but there was little α-xylosidase, β-glucuronidase, α-fucosidase, or β-fucosidase activity. The *Micromonospora* strains were found to have a different pattern of α-galactosidase, β-xylosidase, or α-mannosidase activity (Kawamoto et al., 1983a). As nitrogen sources, inorganic ammonium salts appear to be utilized better than nitrate salts, and acidic or basic amino acids utilized more extensively than neutral ones except for L-serine (Kawamoto et al., 1983a). Most strains are strongly proteolytic, cellulolytic, and diastatic. Average NaCl tolerance is 3% (w/v), and the maximum 5% (w/v). No growth occurs at pH 5.0 or 9.5, and certain species cannot grow on a potato plug or slice (pH 5.5–6.2). Most strains do not grow at 45°C.

Genetic recombination has been carried out in *M. chalcea*, *M. echinospora*, *M. purpurea*, and "*M. rosaria*" (Beretta et al., 1971; Kim et al., 1983; Ryu et al., 1983). Two types of phages, monovalent and polyvalent, have been isolated from *M. purpurea* (Kikuchi and Perlman, 1977, 1978).

Organisms of the genus *Micromonospora* occur infrequently in soils, but in relatively high numbers in aquatic habitats such as lake mud and river sediments. Their occurrence in soil was first reported by Jensen (1932) for Australian soils and later by Kriss (1939) for Russian soils. Only 2.5% of 5000 isolates from 16 soil samples were micromonosporae (Lechevalier, 1964), but the genus comprised about 14% of the total actinomycetes isolated from Japanese paddy soils, which are kept waterlogged during summer (Ishizawa et al., 1969). Early work on lakes in Wisconsin showed that micromonosporae comprised 10–50% of the microbial population in the water mass, were the only actinomycetes in mud samples, but were rarely isolated from adjacent soils (Umbreit and McCoy, 1940; Colmer and McCoy, 1943). The occurrence of *Micromonospora* in lake systems has been confirmed by workers in many countries (Potter and Baker, 1956; Corberi and Solaini, 1960; Willoughby, 1969a; Johnston and Cross, 1976). Some workers have also shown that micromonosporae are frequently present in water samples from streams and rivers. For example, Burman (1973) obtained counts of *Micromonospora* ranging from 200 to 1500/ml in stored water from the river Thames, in which counts of *Streptomyces* were 10–500/ml. Micromonosporae have also been isolated from marine environments, such as beach sand (Watson and Williams, 1974), deep marine sediments (Weyland, 1969), and sediments from the White Sea and Black Sea (Solovieva, 1972; Solovieva and Singal, 1972). In addition, a few strains have been reported in the intestinal tract of termites (Hungate, 1946; Sebald and Prevot, 1962) and the rumen of sheep (Maluszynska and Janoto-Bassalik, 1974); these strains are anaerobic and their taxonomy is still uncertain.

Enrichment and Isolation Procedures

A water sample or soil and sediment dilutions, which are often pretreated, are inoculated onto a suitable isolation agar medium. The medium is then incubated at 28–30°C for 2–3 weeks. The following pretreatments have been used: heat treatment at 70°C in a water bath for 10 min (Rowbotham and Cross, 1977a; Sandrak, 1977); and chlorine treatment with 4 ppm ammonia followed by 2 ppm chlorine (as sodium hypochlorite) for 30 min (Burman et al., 1969; Willoughby, 1969a). The most widely used media are colloidal chitin (Hsu and Lockwood, 1975), M3 medium (Rowbotham and Cross, 1977a), starch-casein (Küster and Williams, 1964b; Williams and Davies, 1965), and Kodota's cellulose benzoate medium (Sandrak, 1977). Incorporation of selected antibiotics into the isolation medium has been found to be effective for selective isolation of micromonosporae (novobiocin 25–50 μg/ml, Sveshnikova et al., 1976; gentamicin 1–10 μg/ml, Ivanitskaya et al., 1978; tunicamycin 25–50 μg/ml, Wakisaka et al., 1982).

Maintenance Procedures

Sporulated cultures of most species may be maintained for a few years on a slant at 4°C. Cultures scraped from a slant or submerged cultures can be maintained for longer periods by storage in broth containing 10–15% glycerol at −20°C, by lyophilization, or by liquid drying.

Differentiation of the genus **Micromonospora** *from other genera*

The characteristics that distinguish *Micromonospora* from genera that are similar in morphology and chemical features are given in Table 28.14.

Taxonomic Comments

Results from DNA-DNA pairing and 16S rRNA cataloguing studies have shown that the genus *Micromonospora* is genetically related to *Actinoplanes, Amorphosporangium, Ampullariella,* and *Dactylosporangium* (Stackebrandt et al., 1981, 1983c), thus supporting its chemotaxonomic relationships. The spore morphology of *Micromonospora* distinguishes it from these genera, but it is difficult to identify nonsporing strains having a wall type II.

Further Reading

Cross, T. 1981. The monosporic actinomycetes. In Starr, Stolp, Trüper, Balows, and Schlegel (Editors), The Prokaryotes. A Handbook on Habitats, Isolation, and Identification of Bacteria. Springer-Verlag, New York pp. 2091–2102.

Luedemann, G. M. 1970. *Micromonospora* taxonomy. Adv. Appl. Microbiol. *2:* 101–133.

Luedemann, G. M. 1971c. Species concepts and criteria in the genus *Micromonospora.* Trans. N. Y. Acad. Sci. *33:* 207–218.

Differentiation and characteristics of the species of the genus **Micromonospora**

The differential characteristics of the species of *Micromonospora* are given in Table 28.15.

List of species and subspecies of the genus **Micromonospora**

1. **Micromonospora carbonacea** Luedemann and Brodsky 1965, 47.[AL]

car′bo.na′cea. L. n. *carbo* coal, charcoal; L. suff. *-aceus* quality or nature of; M. L. adj. *carbonacea* charcoal-like (referring to color of spores).

Spores oval to spherical, 0.7–1.0 μm in diameter, brown by transmitted light, and smooth walled when viewed by light microscopy. In broth, sporulating areas are usually sparsely dispersed in definite clumps throughout the mycelia web, which consists mostly of unbranched, long,

Table 28.14.
Differential characteristics of the genus **Micromonospora** *and* related taxa[a]

	Morphological characteristics			Chemotaxonomic characteristics			
	Formation of aerial mycelium	Spores	Spore motility	Cell wall type[b]	Acyl type	Whole cell sugar[b]	Phospholipid type[c]
Micromonospora	−	Single	−	II	Glycolyl	D	PII
Thermoactinomyces	+	Single (endospore)	−	III	ND	C	ND
Thermomonospora	+	Single	−	III	ND	B/C	ND
Saccharomonospora	+	Single	−	IV	Acetyl	A	PII
Kineosporia	−	Single (sporangiospore)	+	I	ND	A/D	PIII
Dactylosporangium	−	Row (3–6) (sporangiospore)	+	II	Glycolyl	D	PII
Actinoplanes	−	Coils (numerous) (sporangiospore)	+	II	Glycolyl	D	PII
Ampullariella	−	Rows (numerous) (sporangiospore)	+	II	Glycolyl	D	PII
Amorphosporangium	−	Coils (numerous) (sporangiospore)	+	II	Glycolyl	D	PII
Pilimelia	−	Rows (numerous) (sporangiospore)	+	II	Acetyl	D	PII

[a]Symbols: see Table 28.1; *ND*, not determined.
[b]Data from Lechevalier and Lechevalier (1970b).
[c]Data from Lechevalier et al. (1977).

Table 28.15.
Differential characteristics of the species of the genus **Micromonospora**[a]

Characteristics	1a. *M. carbonacea* subsp. *carbonacea*	1b. *M. carbonacea* subsp. *aurantiaca*	2a. *M. halophytica* subsp. *halophytica*	2b. *M. halophytica* subsp. *nigra*	3. *M. chalcea*	4. *M. inositola*	5. *M. coerulea*	6. *M. purpureochromogenes*	7. *M. olivasterospora*	8a. *M. echinospora* subsp. *echinospora*	8b. *M. echinospora* subsp. *ferruginea*	8c. *M. echinospora* subsp. *pallida*
Diagnostic mycelial pigment[b]							B-G			Pu	Pu	
Diffusible pigment[b]		(pYF)	RBr		Y			dBr	O-G			(RF)
Growth on												
Czapek-sucrose agar	−	−	+	+	−	−	−	v	+	+	+	+
Potato slice	+	+	−	−	+	+	+	v	−	−	−	−
Carbohydrate utilization												
α-Melibiose	+	+	+	+	+	+	+	+	−	−	−	−
Raffinose	v	v	+	+	+	+	+	+	−	−	−	−
D-Mannitol	−	−	−	−	−	+	+	−	−	−	−	−
L-Rhamnose	−	−	−	−	−	−	−	−	−	+	+	+
Glycerol	−	−	−	−	−	−	−	+	−	−	−	−
Inositol	−	−	−	−	−	v	−	−	−	−	−	−
D-Ribose	−	−	−	−	−	−	−	−	+	−	+	d
Glycosidase activity												
α-Galactosidase	+	+	+	+	+	+	+	+	−	−	−	−
β-Xylosidase	+	+	+	+	+	+	−	−	−	d	d	+
α-Mannosidase	−	−	−	−	−	+	+	−	−	−	−	−
Nitrate reduction	+	−	+	+	v	−	−	−	−	v	−	+
Maximum NaCl tolerance (% w/v)	3	3	4	4	5	1.5	1.5	1.5	3	3	3	3
Cell component												
3-Hydroxydiaminopimelate	+	+	+	+	−	+	−	−	+	+	+	+
Main menaguinone (MK)	9	9	9	9	10	10	10	10	10	10	10	12

[a]Symbols: see Table 28.1; *d,* 11–89% of strains are positive; *υ,* strain instability.
[b]*Pu,* purple; *B,* blue; *G,* green; *R,* red; *F,* fluorescent; *O,* olive; *d,* dark; *Br,* brown; *Y,* yellow; *P,* pale.

loosely woven, fine mycelial strands. Spores remain firmly attached to the sympodial type of sporophore, and are only free in older cultures. The mycelium does not degenerate into polymorphic bodies.

Fair to good growth on Czapek sucrose agar plus 0.1% $CaCO_3$. Only fair growth on autoclaved potato slice but good growth plus $CaCO_3$.

Utilizes L-arabinose, D-galactose, β-lactose, and D-fructose. Poor growth on salicin. Gelatin liquefied. Milk digested. Melanin pigment not produced. Good growth between 27 and 37°C.

Produces the everninomicin antibiotic complex.

Isolated from soil and aquatic environments.

Type strain: ATCC 27114.

1a. **Micromonospora carbonacea** subsp. **carbonacea** Luedemann and Brodsky 1965, 51.[AL]

Colonies raised, folded, initially orange but turning brown to black on sporulation. Spores generally abundant (Fig. 28.11). Characteristic blackish sporulating peripheral sectors similar to alluvial fans are often a macroscopic diagnostic aid. Spore layer moist or dry but not viscid.

Isolated from a soil sample from Olean, New York.

Type strain: ATCC 27114 (=NRRL 2972).

Further comments: In the original description, poor growth occurs on raffinose and salicin. However, *Micromonospora carbonacea* subsp. *carbonacea* NRRL 2972 (=ATCC 27114) utilizes raffinose and salicin in a chemically defined agar medium (Kawamoto et al., 1983a).

1b. **Micromonospora carbonacea** subsp. **aurantiaca** Luedemann and Brodsky 1965, 51.[AL]

au.ran′ti.a′ca. L. fem. adj. *aurantiaca* orange colored.

The orange to orange-red colonies of this subspecies produce few spores, and with occasional small carbon-like specks (spores) on the colony surface.

A pale yellowish diffusible pigment is sometimes produced on mannose and xylose agars.

Type strain: ATCC 27115 (=NRRL 2997).

Further comments: In the original description, poor growth occurs on raffinose and salicin. However, *Micromonospora carbonacea* subsp. *aurantiaca* NRRL 2997 (=ATCC 27115) utilizes raffinose and salicin in a chemically defined medium (Kawamoto et al., 1983a).

2. **Micromonospora halophytica** Weinstein, Luedemann, Oden and Wagman 1968, 436.[AL]

hal′o.phyt′i.ca. M. L. fem. adj. *halophytica* a plant that grows within the influence of salt water.

Spores randomly produced throughout a long, branching mycelium on short or long sporophores or occasionally sessile (Figs. 28.12 and 28.13). Ellipsoidal to spherical, up to 1.2 μm in diameter, appearing dark colored in older cultures. Mycelium does not normally break up into polymorphic elements.

Utilizes L-arabinose, D-galactose, D-fructose, and β-lactose. Poor

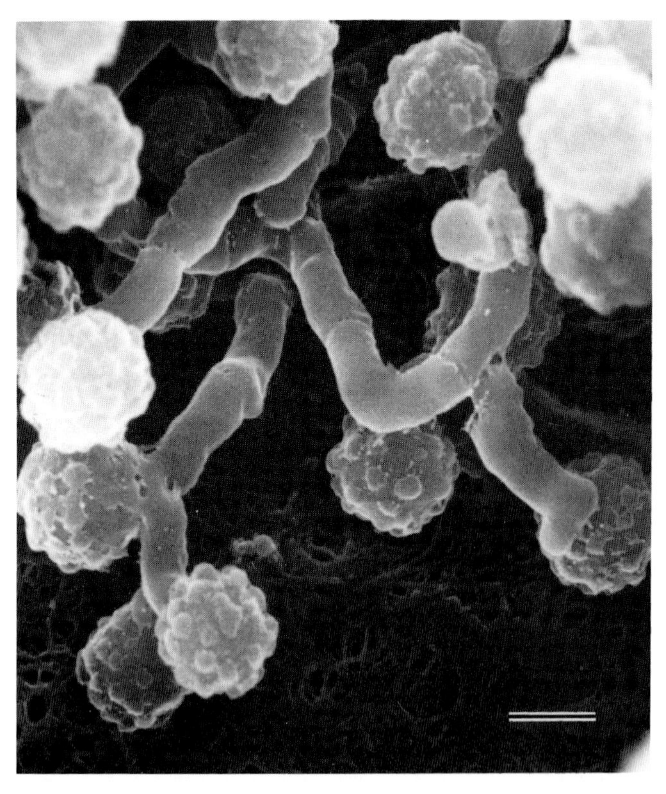

Figure 28.11. Scanning electron micrograph of *M. carbonacea* subsp. *carbonacea* NRRL 2972. *Bar,* 0.5 µm.

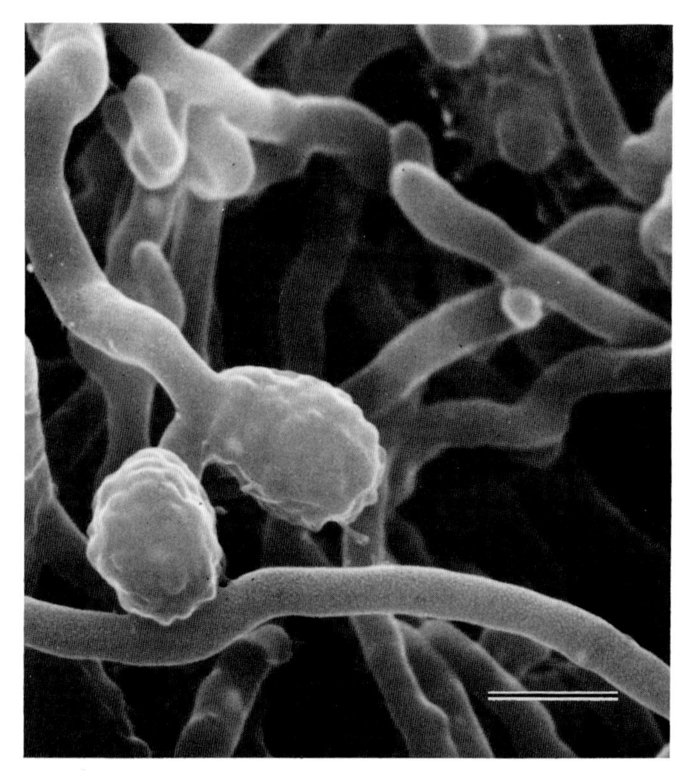

Figure 28.13. Scanning electron micrograph of *M. halophytica* subsp. *nigra* NRRL 3097. *Bar,* 0.5 µm.

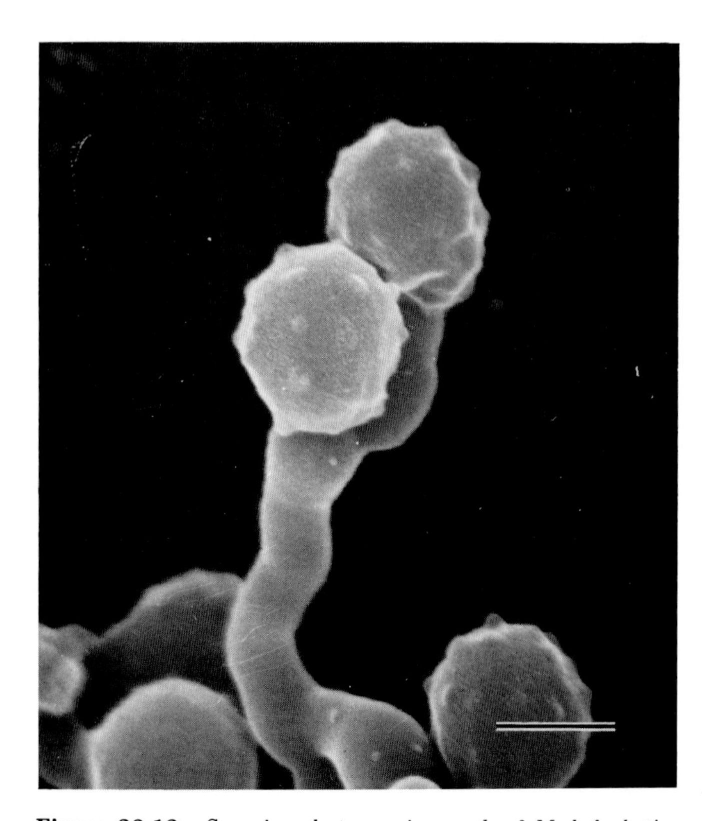

Figure 28.12. Scanning electron micrograph of *M. halophytica* subsp. *halophytica* ATCC 27596. *Bar,* 0.5 µm.

growth on adonitol and cellulose. Starch hydrolyzed. Cellulose decomposed. Nitrate reduced. Melanin pigment not produced. Growth aerobic. Temperature range 18–40°C, none at 50°C.

Produces the halomicin antibiotic complex.

Isolated from soil and aquatic environments.

Type strain: ATCC 27596.

2a. **Micromonospora halophytica** subsp. **halophytica** Weinstein, Luedemann, Oden and Wagman 1968, 437.[AL]

Growth folded, color orange to orange-brown. Sporulation abundant in older, brown-colored colonies. A light reddish brown diffusible pigment often produced on agar media.

Gelatin liquefied. Milk digested.

Isolated from a salt pool in Syracuse, New York.

Type strain: ATCC 27596 (=NRRL 2998).

2b. **Micromonospora halophytica** subsp. **nigra** Weinstein, Luedemann, Oden and Wagman 1968, 437.[AL]

ni′gra. L. fem. adj. *nigra* black.

Reddish-brown diffusible pigment not produced. Sporulation turns the initially orange colony olive-brown and later black.

Gelatin liquefaction and milk digestion variable.

Isolated from a salt pool in Syracuse, New York.

Type strain: ATCC 33088 (=NRRL 3097).

3. **Micromonospora chalcea** (Foulerton 1905) Ørskov 1923, 156.[AL] (*Streptothrix chalcea* Foulerton 1905, 1200.)

chal′ce.a. Gr. fem. adj. *chalcea* copper, bronze.

Young colonies typically raised and folded, reddish orange, turning brown, olive-brown, dark brown, and eventually black on sporulation. Spore layer moist to dry, usually not viscid. Pale yellow fluorescent diffusible pigment often produced on agar media, particularly yeast starch agar.

Spores oval to spherical, 0.7–1.0 µm in diameter, brown to dark brown by transmitted light and smooth walled when viewed by phase-contrast microscopy. Sessile or on short or long sporophores (Fig. 28.14), produced randomly throughout the mycelium. Separate early from their sporophores. Mycelium does not appear to degenerate into polymorphic elements.

Utilize L-arabinose, D-galactose, β-lactose, and D-fructose. Poor growth on salicin. Starch hydrolyzed. Cellulose decomposed. Gelatin liquefied. Milk digested. Melanin pigment not produced. Good growth between 27° and 37°C.

Isolated from air, soil, and aquatic environments.

Neotype strain: ATCC 12452 (Luedemann 1971a, 252).

4. **Micromonospora inositola** Kawamoto, Okachi, Kato, Yamamoto, Takahashi, Takasawa and Nara 1974, 495.[AL]

inos.it′ol.a. M.L. adj. *inositola* of inositol (referring to ability to utilize inositol).

Poor growth on glucose-asparagine agar and inorganic salts-starch agar. Colonies on potato and Bennett agar are folded, and bright orange to orange. No dark spore layer.

Spores oval or spherical, 0.8–1.0 μm in diameter, borne on short sporophores (Fig. 28.15). Generally poor sporulation on most media.

Utilizes D-galactose and D-fructose. No growth on salicin. Gelatin liquefied. Milk slightly coagulated but not peptonized. Cellulose slightly decomposed. Starch hydrolyzed. Melanin pigment not produced. Grows between 25° and 40°C.

Produces the XK-41 antibiotic complex.

Isolated from a soil sample from Hokkaido, Japan.

Type strain: ATCC 21773.

Further comments: Positive utilization of inositol and negative utilization of D-mannitol were indicated in the original description. However, D-mannitol is utilized, whereas inositol is utilized in Luedemann's medium (Luedemann and Brodsky, 1964) but not in a chemically defined medium (Kawamoto et al., 1983a).

5. **Micromonospora coerulea** Jensen 1932, 177.[AL]

coe.ru′le.a. L. adj. *coerulea* dark colored, dark blue, dark green.

Growth on agar media often slow, requiring 3–5 weeks to develop; initially pale yellow-orange, later turning yellow-green and finally dark blue-green or greenish black. The blue-green mycelial pigment is water soluble and pH sensitive.

Spores spherical, 0.8–1.5 μm in diameter, smooth walled by phase-contrast microscopy. Borne often on short or long lateral sporophores (Fig. 28.16). Colonies in shaken broth culture usually composed of long, loosely woven hyphae. Mycelium rarely fragmenting except in old culture.

Utilizes D-galactose, D-fructose, and β-lactose. Poor growth on L-arabinose and D-arabinose. Starch hydrolyzed. Cellulose poorly decomposed. Gelatin and milk poorly hydrolyzed. Grows well between 24° and 37°C.

Isolated from a soil sample obtained from Mt. Haleakala, Island of Maui, Hawaiian Islands.

Type strain: ATCC 27008 (Luedemann 1971b).

6. **Micromonospora purpureochromogenes** (Waksman and Curtis 1916) Luedemann 1971b, 244.[AL] (*Actinomyces purpeo-chromogenes* (sic) Waksman and Curtis 1916, 113; *Micromonospora brunnea* Sveshnikova, Maksimova and Kudrina 1969, 762[AL].)

pur′pur.e.o.chro.mo′ge.nes. L. adj. *purpureus* purple colored; Gr. n. *chroma* color; Gr. v. suff. *-genes* producing; M. L. part. adj. *purpureochromogenes* producing purple color (relating to the color of the diffusible pigment).

Colonies on yeast extract-glucose agar at 21 days are dark brown in color. A dark brown diffusible pigment is produced in most media, which is water soluble, but insoluble in acid.

Spores are 0.8–1.2 μm in diameter, formed singly or in clusters (Fig. 28.17). A predominantly monopodial system of branching occurs in the sporulating hyphae, best observed at the periphery of the colony. In

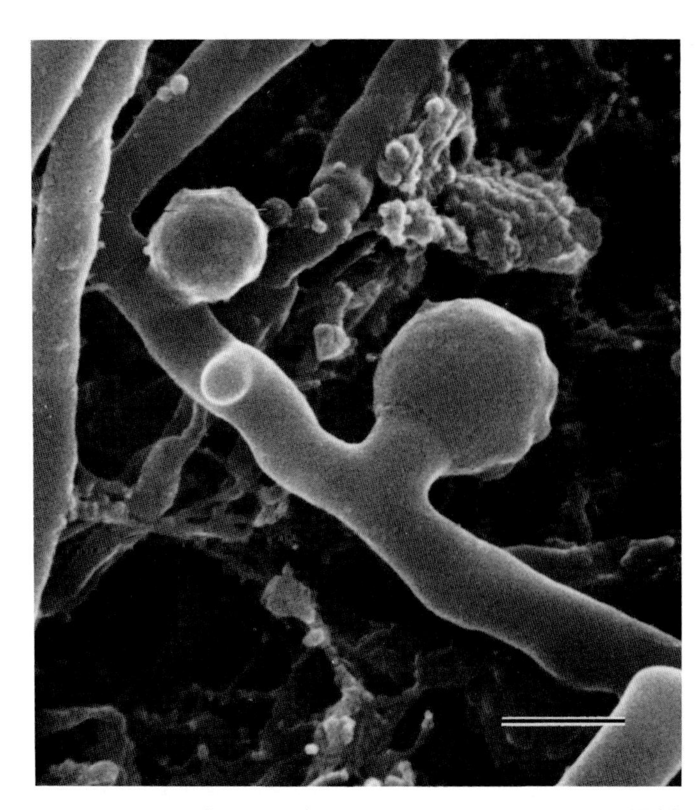

Figure 28.14. Scanning electron micrograph of *M. chalcea* ATCC 12452. *Bar,* 0.5 μm.

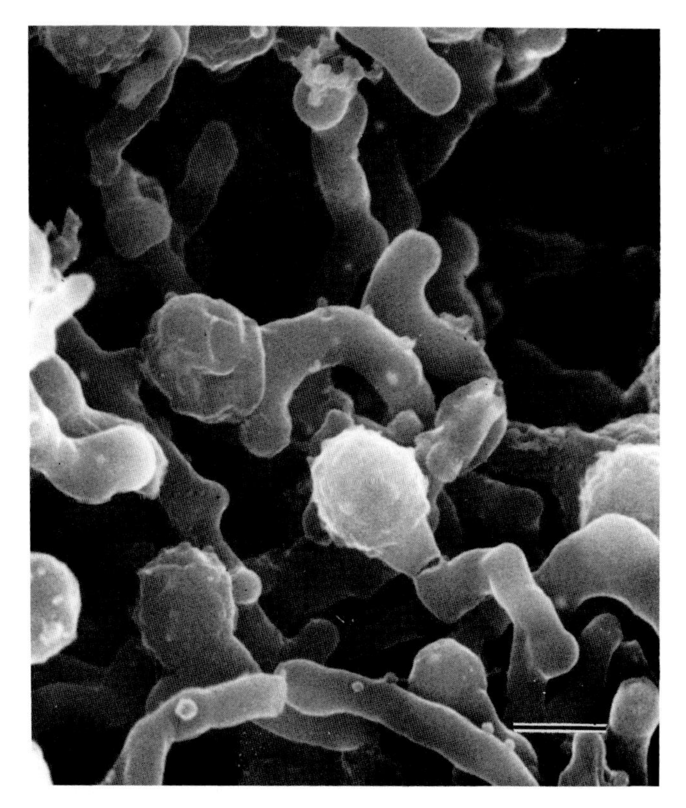

Figure 28.15. Scanning electron micrograph of *M. inositola* ATCC 21773. *Bar,* 0.5 μm.

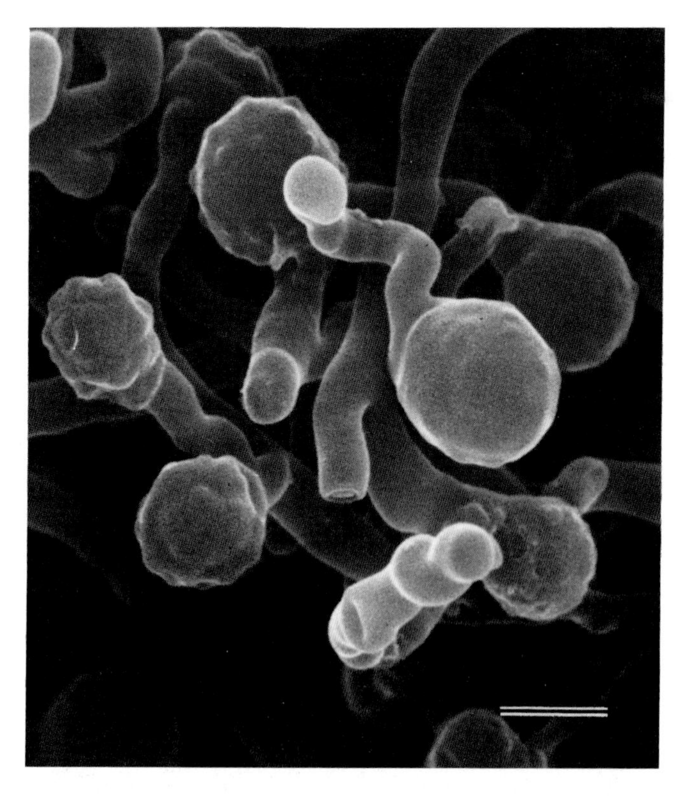

Figure 28.16. Scanning electron micrograph of *M. coerulea* ATCC 27008. *Bar,* 0.5 μm.

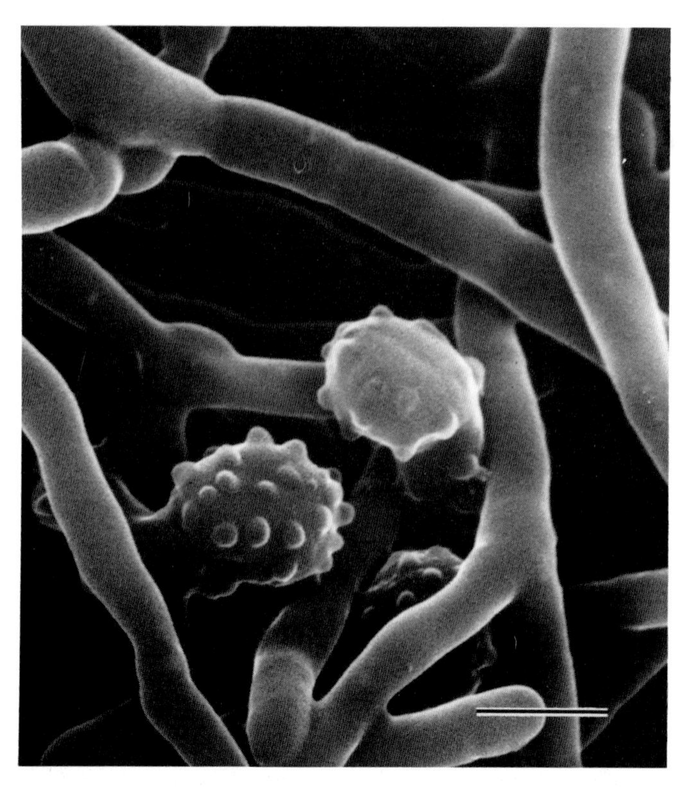

Figure 28.17. Scanning electron micrograph of *M. purpureochromogenes* ATCC 27007. *Bar,* 0.5 μm.

broth, colonies have a tendency to disintegrate (lyse) into mycelial fragments and chlamydospores.

Utilizes D-galactose and D-fructose. Poor growth on D-arabinose and L-arabinose. Milk poorly peptonized. Good growth between 26° and 37°C.

Isolated once from California adobe soil.

Type strain: ATCC 27007.

7. **Micromonospora olivasterospora** Kawamoto, Yamamoto and Nara 1983b, 110.[VP]

o.li.vas.ter.o'spo.ra. L. n. *oliva* an olive; Gr. n. *aster* a star; Gr. n. *spora* a seed; M. L. fem. n. *olivasterospora* olive-colored spore that looks like a star.

Colonies on agar light brown or dark yellow, later covered with an olive or dark green waxy, lustrous layer of spores. Olive-green soluble pigments are produced in ISP media 2 and 3 and are not pH indicators.

Spores spherical to oval, approximately 1.0 μm in diameter, and appearing rough surfaced by phase-contrast microscopy. Blunt spines on spore surface 0.1–0.2 μm long. Borne on short sporophores or sessile, and occur randomly or in clusters (Fig. 28.18). Terminal and intercalary chlamydospore-like swellings are sometimes present.

Utilizes D-fructose and D-galactose. No growth on D-arabinose, lactose, or salicin. Cellobiose is utilized moderately in Luedemann's medium (Luedemann and Brodsky, 1964), but poorly in a chemically defined medium containing NH_4NO_3. Utilizes NH_4NO_3, $(NH_4)_2SO_4$, NH_4Cl, L-serine, L-aspartic acid, L-glutamic acid, L-histidine, or L-arginine as sole nitrogen source. No growth on L-alanine, L-valine, L-homoserine, L-methionine, L-phenylalanine, L-tyrosine, and L-tryptophan. Starch hydrolyzed. Skim milk peptonized but not coagulated. Cellulose weakly decomposed. Melanin pigment not produced. Good growth between 28° and 38°C, and between pH 6.8 and 7.8.

Susceptible to (micrograms per milliliter): rifampicin (MIC, 3.13), chloramphenicol (6.25), chlortetracycline (3.13), mitomycin C (0.39), penicillin G (6.25), ristocetin (1.56), streptomycin A (0.125), neomycins (0.25), spiramycin (25), cycloserine (25), and sagamicin (31.2). Resistant to (micrograms per milliliter): gentamicin C complex (125), fortimicin A (250), kanamycin A (500), seldomycins (500), and spectinomycin (500).

Produces the fortimicin antibiotic complex.

Isolated from a soil sample from Hiroshima, Japan.

Type strain: ATCC 21819.

8. **Micromonospora echinospora** Luedemann and Brodsky 1964, 116.[AL]

e.chi'no.spo.ra. Gr. adj. *echinos* spiny appearance; Gr. n. *spora* seed; M.L. fem. n. *echinospora* spiny spore.

Spores spherical, 1.0–1.5 μm in diameter, dark brown to black, appearing rough walled under phase-contrast microscopy. Blunt spines on spore surface 0.1–0.2 μm long. Sporophores mostly solitary but occasionally in small clusters on the same hypha (Figs. 28.19 and 28.20). Spores adhere firmly to the sporophore until mature.

Utilizes L-arabinose. Poor growth on salicin. Very slow growth on and decomposition of cellulose. Gelatin liquefied. Milk digested. Melanin pigment not produced. Good growth between 27° and 37°C.

Most strains produce the gentamicin antibiotic complex.

Isolated from soil and aquatic environments.

Type strain: ATCC 15837.

8a. **Micromonospora echinospora** subsp. **echinospora** Luedemann and Brodsky 1964, 116.[AL]

Colonies usually folded, orange-brown, maroon, or characteristically dark purple. Maroon to purple mycelial pigments are soluble in acid alcohol and pH-sensitive, being red in the acid range and blue-green and precipitable in the basic range.

Spore layer when present purplish black, waxy to dry, not moist or viscid. Sporulation moderate, often requiring 2 or 3 weeks' incubation. Aerial mycelium absent but occasionally a very short purplish gray bloom devoid of spores occurs on some colonies.

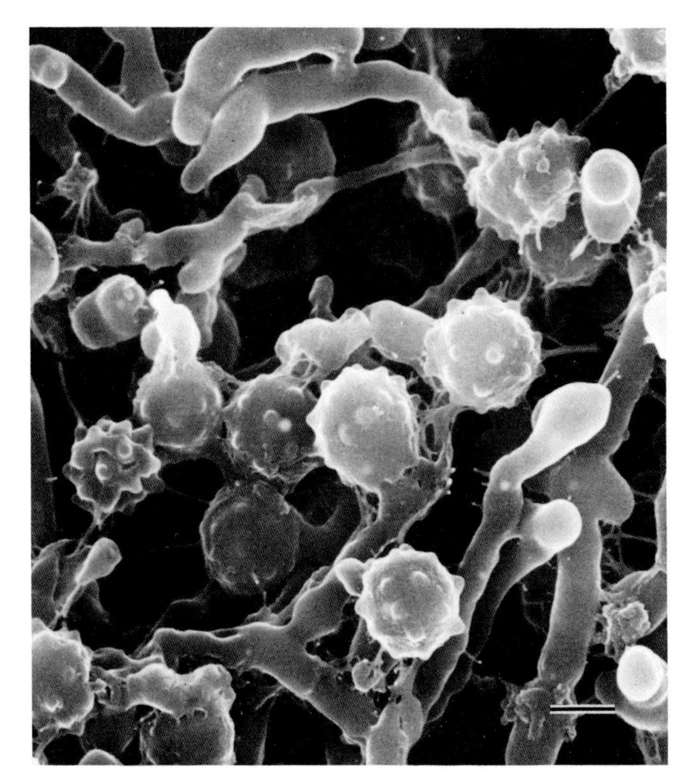

Figure 28.18. Scanning electron micrograph of *M. olivasterospora* ATCC 21819. *Bar,* 0.5 μm.

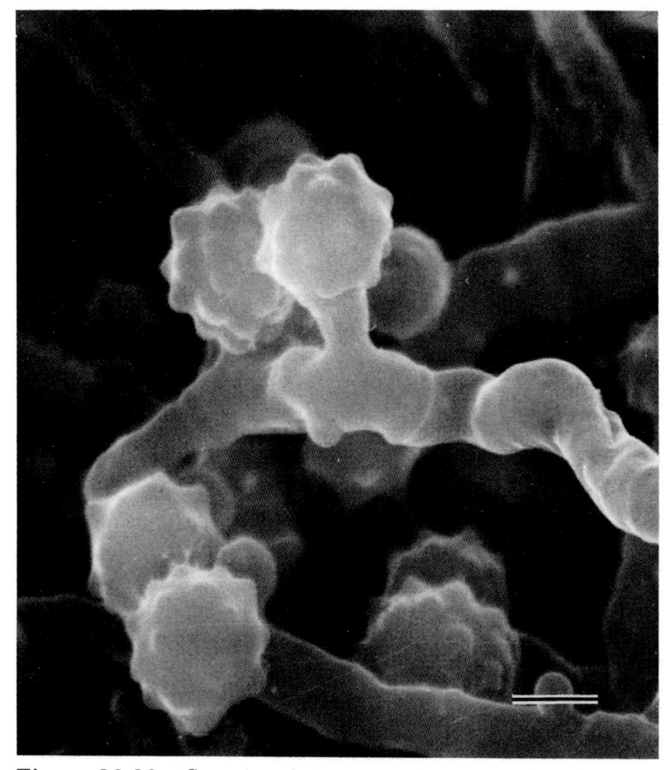

Figure 28.20. Scanning electron micrograph of *M. echinospora* subsp. *pallida* NRRL 2996. *Bar,* 0.5 μm.

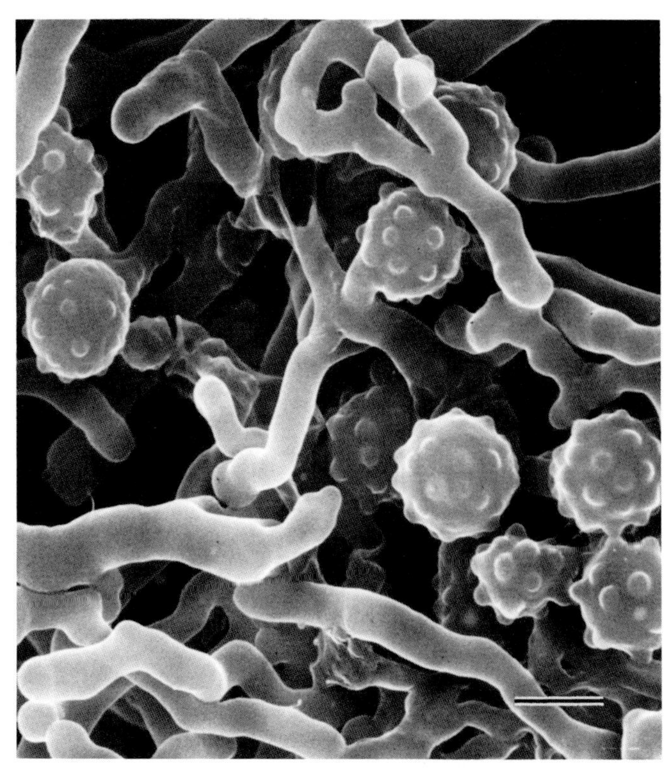

Figure 28.19. Scanning electron micrograph of *M. echinospora* subsp. *echinospora* NRRL 2985. *Bar,* 0.5 μm.

Poor to fair growth on D-galactose, β-lactose, and D-fructose.
Isolated from a soil sample from Jamesville, New York.
Type strain: ATCC 15837 (=NRRL 2985).

8b. **Micromonospora echinospora** subsp. **ferruginea** Luedemann and Brodsky 1964, 116.[AL]

fer′ru.gin′e.a. L. fem. adj. *ferruginea* of the color of iron rust.

Colony color on a number of carbohydrate sources that sustain good growth in shades of orange and maroon; similar in appearance to iron rust. Maroon to purple mycelial pigments produced.

Poor to fair growth on D-galactose, β-lactose, and D-fructose.
Isolated from a soil sample from New York.
Type strain: ATCC 15836 (=NRRL 2995).

8c. **Micromonospora echinospora** subsp. **pallida** Luedemann and Brodsky 1964, 116.[AL]

pal′lid.a. L. fem. adj. *pallida* pale.

Colony color pale, ranging from light ivory to light melon-yellow; dark brown to black on sporulation. Purple mycelial pigments not produced.

Good growth occurs on D-fructose, slight growth on D-ribose but often abundant sporulation.

Isolated from a soil sample from Jamesville, New York.
Type strain: ATCC 15838 (=NRRL 2996).

Further comments: This subspecies may be transferred from *M. echinospora* to a new species in respect to its different menaquinone profile (MK-12) and lack of maroon to purple mycelial pigments.

Species Incertae Sedis

The following species appear on the Approved Lists of Bacterial Names (Skerman et al., 1980) but their taxonomic position is unclear for the reasons given below.

a. **Micromonospora purpurea** Luedemann and Brodsky 1964, 116.[AL]

pur′pur.ea. L. fem. adj. *purpurea* purple colored, dull red with a slight dash of blue.

Spores or sporophores rarely found, sporulation atypical or abortive. Terminal and intercalary chlamydospores often present.

Taxonomic position is uncertain due to lack of descriptive information on single spore formation.

If a sporulating variant of this culture is found and the spore and sporophore appear similar to those of *M. echinospora,* this species would be reduced to a subspecies of the latter species (Luedemann and Brodsky, 1964).

Type strain: ATCC 15835.

b. **Micromonospora rhodorangea** Wagman, Testa, Marquez and Weinstein 1974, 145.[AL]

rho.dor.an′ge.a. M.L. fem. adj. *rhodorangea* red-orange colored.

Macroscopic observations of a 30-day-old culture incubated at 24–26°C in a 3% NZ-amine type A–1% glucose–1.5% agar medium show poor growth with no sporulation. Occasionally, chalmydospores are seen.

Taxonomic position is uncertain due to lack of descriptive information of single spore formation.

Type strain: ATCC 27932.

c. **Micromonospora gallica** (Erikson 1935) Waksman 1961, 296.[AL]
(*Actinomyces gallicus* Erikson 1935, 36.)

gal′li.cus. L. fem. adj. *gallica* of or belonging to the Gauls.

No type strain is available. Strain NCTC 4582 was discarded from the NCTC in about 1952 (information from Dr. L. R. Hill).

Type strain: NCTC 4582.

d. **Micromonospora brunnea** Sveshnikova, Maksimova and Kudrina 1969, 762.[AL]

brun′ne.a. L. fem. adj. *brunnea* brown-colored.

The type strain has 94.2% similarity to *M. purpureochromogenes* ATCC 27007 on the basis of 172 diagnostic features, and has been proposed as a subjective synonym of *M. purpureochromogenes* (Szabo and Fernandez, 1984).

Type strain: ATCC 27334.

e. **Micromonospora aurantiaca** Sveshnikova, Maksimova and Kudrina 1969, 758.[AL]

au.ran′ti.a′ca. L. fem. adj. *aurantiaca* orange colored.

Growth on synthetic media moderate to good, appearing light yellow, orange-yellow, orange to bright orange, and dark. No spore layer. Growth on organic media moderate to good, color pale orange, orange, to bright orange; in some strains grayish on sporulation on some media.

Spores arranged mainly in clusters, round and smooth walled. Sporulation moderate.

Utilizes fructose and arabinose. Cellulose decomposed. Melanin pigment not produced.

The type strain bears globose or subglobose sporangia that contain numerous spores with one polar tuft of flagella. This culture belongs to *Actinoplanes,* but not to *Micromonospora.*

Type strain: ATCC 27029.

SECTION 29

Streptomycetes and Related Genera

Romano Locci

This section includes actinomycetes, which are important both numerically and as producers of useful metabolites. Most actinomycetes isolated from any habitat are likely to be streptomycetes. In addition to *Streptomyces,* three other genera are included in this section. Two of them are monospecific; only five strains of *Sporichthya polymorpha* and one strain of *Kineosporia aurantiaca* have been isolated.

Differential characteristics of the genera are summarized in Table 29.1. Discrimination between them can be achieved without resorting to complex techniques. Even at the macroscopic level useful information can be gained. *Sporichthya* colonies are extremely small so that microscopic examination is suggested for their recognition. Absence of aerial mycelium and the glistening appearance of the colonies of kineosporiae are distinct features. Streptomycetes and streptoverticillia are characterized by the formation of chains of arthrospores on the aerial mycelium; microscopical examination at low magnification reveals the typical verticillate structure of streptoverticillia when sporulation is present. Strains may not spore if they are grown on rich media, and should then be cultivated on poor media favoring sporulation. It is important to ascertain the presence of spore chains before deciding on the attribution of an organism to either *Streptomyces* or *Streptoverticillium.*

Table 29.1.
Diagnostic characteristics of the genera **Streptomyces, Streptoverticillium, Kineosporia,** *and* **Sporichthya**[a]

Characteristics	*Streptomyces*	*Strepto-verticillium*	*Kineosporia*	*Sporichthya*
Colony size	Discrete	Discrete	Small	Microscopic
Substrate mycelium	+	+	+	−
Spores	±	−	−	−
Sporangia	−	−	+	−
Motile spores	−	−	+	−
Aerial mycelium	+	+	−	+
Chains of arthrospores	+	+	−	+
Arthrospores in verticils	−	+	−	−
Spore surface smooth	+	+	−	+
Spore surface hairy, spiny, or warty	+	−	−	−
Motile spores	−	−	−	+
Sugars in cell hydrolysates				
Arabinose, galactose, xylose	−	−	+	−
Lipid characters				
Phospholipid type[b]	PII	PII	PIII	ND
Predominant menaquinones	MK-9(H_6) or MK-9(H_8)	MK-9(H_6) or MK-9(H_8)	MK-9(H_4)	MK-9(H_6) or MK-9(H_8)
Fatty acids				
Saturated straight chain	+	+	ND	+
Iso-/anteiso- branched	+	+	ND	+
Unsaturated	−	−	ND	+
10-Methyl branched	−	−	ND	+
Mol% G + C of DNA	69-78	69-73	ND	ND

[a]Symbols: +, 90% or more of strains are positive; −, 10% or less of strains are positive; *ND,* not determined.
[b]Categories of Lechevalier et al. (1977).

The main feature of *Sporichthya* is its morphology. The absence of a substrate mycelium and the particular arrangement of the aerial hyphae, developing as outgrowths of a holdfast cell, makes this genus unique among actinomycetes. In addition the scarcely branched mycelium splits at maturity into a series of single cells characterized by their motility. Due to the difficulty of growing the organisms in liquid media few chemical data are available. However, sporichthyae do have a type I cell wall (Lechevalier and Lechevalier, 1965) and are extremely rarely isolated. Differentiation from other taxa is therefore not a problem. Possible confusion with representatives of the genus *Pseudonocardia* can be solved by the presence of *meso*-diaminopimelic acid in the cell wall of the latter.

Kineosporia is characterized morphologically by the formation of club-shaped sporangia, borne terminally at the apex of vegetative hyphae, and containing a single spore. The formation of sporangia results in a moist and glossy appearance of the colony, which never shows aerial hyphae. Spores are motile, being characterized by a polar tuft of flagella. Again, the organisms are extremely rare. *Kineosporia* can be distinguished from representatives of *Actinoplanes, Amorphosporangium, Ampullariella, Dactylosporangium, Kitasatoa* (see *Streptomyces*), *Micromonospora, Pilimelia, Planomonospora, Planobispora, Streptoalloteichus, Spirillospora,* and *Streptosporangium* by its colony appearance and localization of the single-spored sporangia on the substrate mycelium. A further distinctive character is its type I (Lechevalier and Lechevalier, 1965) cell wall composition.

Differentiation between *Streptomyces* and *Streptoverticillium* is based essentially on morphological criteria, although this distinction has been supported by phenetic (Locci et al., 1981; Williams et al., 1985a) and genetic (Gladek et al., 1985) data. Both taxa, representing the so-called typical euactinomycetes, are morphologically characterized by the production of a highly branched substrate mycelium (rarely bearing spores) that further develops into a network of aerial hyphae, where sporulation represents the final stage of the developmental cycle. Spores are produced in chains at the tip of sporogenous filaments.

Unless one resorts to time-consuming numerical taxonomy techniques, the alternative for discrimination between the two taxa is achieved by the direct examination of the sporulation process. Due to the particular arrangement of the spore chains, distinction is usually easily made. The cottony appearance of the aerial mycelium of streptoverticillia is a useful macroscopic feature, and the "barbed-wire" structure of aerial hyphae is evident at low magnifications (× 100). Unlike streptomycetes, *Streptoverticillium* species show an aerial sporulating structure consisting of straight filaments that bear, at regular intervals, verticillate branches producing umbels of spore chains at their tips.

Under the electron microscope, streptoverticillia show a peculiar structure of the spore sheath (Attwell et al., 1973), together with a "twisting" of the spore chain in the course of sporogenesis. These phenomena are also detectable in mature spore chains. Unlike streptomycetes, streptoverticillia are mostly resistant to lysozyme and to neomycin.

Genus **Streptomyces** *Waksman and Henrici 1943, 339*[AL*]

Stanley T. Williams, Michael Goodfellow, and Grace Alderson

(*Actinopycnidium* Krasil'nikov 1962, 250[AL]; *Actinosporangium* Krasil'nikov and Yuan 1961, 113[AL]; *Chainia* Thirumalachar 1955, 935[AL]; *Elytrosporangium* Falcão de Morais, Chaves Batista and Massa 1966, 162[AL]; *Kitasatoa* Matsumae and Hata in Matsumae, Ohtani, Takeshima and Hata 1968, 617[AL]; *Microellobosporia* Cross, Lechevalier and Lechevalier 1963, 422[AL].)

Strep.to.my′ces. Gr. adj. *streptos* pliant, bent; Gr. n. *myces* fungus; M.L. masc. n. *Streptomyces* pliant or bent fungus.

Vegetative hyphae (0.5–2.0 μm in diameter) produce an extensively branched mycelium that rarely fragments. **The aerial mycelium at maturity forms chains of three to many spores.** A few species bear short chains of spores on the substrate mycelium. Sclerotia, pycnidial-, sporangia-, and synnemata-like structures may be formed by some species. Spores are nonmotile. **Form discrete and lichenoid, leathery or butyrous colonies.** Initially colonies are relatively smooth surfaced but later they develop a weft of aerial mycelium that may appear floccose, granular, powdery, or velvety. **Produce a wide variety of pigments responsible for the color of the vegetative and aerial mycelia. Colored diffusible pigments may also be formed.** Many strains produce one or more antibiotics. Aerobes. Gram-positive but not acid-alcohol fast. Chemo-organotrophic, having an oxidative type of metabolism. Catalase positive. Generally reduce nitrates to nitrites and degrade adenine, esculin, casein, gelatin, hypoxanthine, starch, and L-tyrosine. **Use a wide range of organic compounds as sole sources of carbon for energy and growth.** Temperature optimum 25–35°C; some species grow at temperatures within the psychrophilic and thermophilic range; optimum pH range for growth 6.5–8.0. The cell wall peptidoglycan contains major amounts of **L-diaminopimelic** acid (**L-DAP**). They lack mycolic acids, contain major amounts of **saturated, *iso*- and *anteiso*-fatty acids,** possess either **hexa- or octahydrogenated menaquinones with nine isoprene units** as the predominant isoprenolog, and have complex polar lipid patterns that typically contain **diphosphatidylglycerol, phosphatidylethanolamine, phosphatidylinositol,** and **phosphatidylinositol mannosides.** They are **widely distributed and abundant in soil,** including composts. A few species are pathogenic for animals and man, others are phytopathogens. The mol% G + C of the DNA is 69–78 (T_m).

Type species: *Streptomyces albus* (Rossi-Doria 1891) Waksman and Henrici 1943, 339.

Further Descriptive Information

Morphology and Fine Structure

Morphology has played a major role in distinguishing *Streptomyces* from other sporing actinomycetes and in the characterization of streptomycete species. The life cycle of a streptomycete offers three features for microscopic characterization: (a) vegetative (substrate) mycelium (on solid and in liquid medium); (b) aerial mycelium bearing chains of arthrospores (sometimes called "sporophores"); and (c) the arthrospores themselves (Kutzner, 1981). It is the last two categories that have produced most diagnostic information for taxonomists.

The genera *Actinopycnidium, Actinosporangium, Chainia, Elytrosporangium, Kitasatoa,* and *Microellobosporia* have been distinguished from *Streptomyces* by morphological criteria (Table 29.2), but they share many phenetic, chemical, molecular, and genetic characters with *Streptomyces* and have therefore been proposed as synonyms of this genus (Goodfellow et al., 1986a, 1986b, 1986c, 1986d, 1986e). The genus *Streptoverticillium* can be distinguished from *Streptomyces* by its verticillate sporophores while sharing many characteristics with streptomycetes, but numerical phenetic (Williams et al., 1983a) and rRNA/DNA similarities (Gladek et al., 1985) studies support the continued recognition of *Streptoverticillium.*

There have been some reports of streptomycetes forming spore chains on their vegetative mycelium in both solid and liquid culture (e.g. Carvajal, 1947; Glauert and Hopwood, 1960; Tresner et al., 1967). However, it remains to be seen if these structures are analogous to the arthro-

**AL* denotes the inclusion of this name on the Approved Lists of Bacterial Names (1980).

Table 29.2.
Morphological features of **Streptomyces** and allied genera[a]

Genus	Substrate mycelium: Chains of spores	Substrate mycelium: Sporangia-like vesicles with few spores	Substrate mycelium: Sclerotia	Aerial mycelium: Long chains of arthro-spores in thin fibrous sheath	Aerial mycelium: Sporangia-like vesicles with few spores	Aerial mycelium: Motility of spores
Streptomyces	±	−	−	+	−	−
Actinopycnidium	−	−	−	+[b]	−	−
Actinosporangium	−	−	−	+[c]	−	−
Chainia	−	−	+	+	−	−
Elytrosporangium	−	+	−	+	−	−
Kitasatoa	−	+	−	+	+	?
Microellobosporia	−	+	−	−	+	−
Streptoverticillium	−	−	−	+	−	−

[a]Symbols: +, positive; −, negative; ±, some species positive; ?, unknown.
[b]Form spore chains in "pycnidia."
[c]Form spore chains in "pseudosporangia."

spores formed on the aerial mycelium rather than deformations of the hyphae produced in staling cultures. More comprehensive studies of the streptomycete life cycle in submerged culture are needed (Kutzner, 1981).

The fine structure and development of the aerial arthrospores have been examined by many workers, and it is clear that they are formed by septation and disarticulation of preexisting hyphal elements within a thin fibrous sheath (Locci and Sharples, 1984). The latter does not appear to play any part in spore formation, the spore wall being formed, at least in part, from wall layers of the parent hypha. This may be termed *holothallic development* (Locci and Sharples, 1984) and it is typical of many other sporing actinomycetes (Williams et al., 1973). The distinction between the streptomycete spore chain and the sporangia-like vesicles of some related genera (Table 29.2) is dubious because available evidence indicates that the latter also have a holothallic development (Williams et al., 1973; Locci and Sharples, 1984). The major differences are in the number and size of the spores produced.

Actinosporangium (Krasil'nikov and Yuan, 1961) and *Actinopycnidium* (Krasil'nivok, 1962) were distinguished by their formation of "pseudosporangia" and "pycnidia," respectively, but subsequent studies have shown that the spore-containing structures are artifacts emanating from dense masses of spore chains and detached spores (e.g. Williams, 1970).

The genus *Chainia* (Thirumalachar, 1955) forms sclerotia embedded in colonies on agar media or in shake culture (Fig. 29.1). The sclerotia may be quite large (up to 100 μm in diameter) and consist of a dense mass of enlarged, branching hyphae containing lipid and cemented together by an amorphous material that contains L-2,3-diaminopropionic acid (Lechevalier et al., 1973a; Sharples and Williams, 1976). After prolonged cultivation on laboratory media the capacity to form sclerotia is gradually lost, while production of aerial mycelium is increased (Thirumalachar and Sukapure, 1964). The latter bears spore chains similar to those of streptomycetes; some species produce angular-shaped spores with pronounced intersporal pads (Sharples and Williams, 1976).

Microellobosporia (Cross et al., 1963) was proposed for species forming club-shaped sporangia that contained a short row of nonmotile spores, borne on both the aerial and substrate mycelium. The sporangial wall may now be equated with the sheath that surrounds streptomycete spore chains (Williams et al., 1973). The spores are larger (1.5–3.5 μm in diameter) than those of streptomycetes and are produced in chains of one to five on both substrate and aerial mycelium (Fig. 29.2).

The genus *Elytrosporangium* (Falcão de Morais et al., 1966) also forms short chains of spores on the substrate mycelium (Fig. 29.3), but produces streptomycete-like chains on the aerial growth (Fig. 29.4). Studies have shown that streptomycetes with a morphology typical of *Elytrosporangium* and *Microellobosporia* can be isolated from soil, together with intermediates with longer chains of irregularly sized spores on the aerial mycelium or even single spores on the substrate mycelium (Cross and Al-Diwany, 1981).

Kitasatoa (Matsumae et al., 1968) was also reported to form club-shaped sporangia on both substrate and aerial hyphae but produced spores that were motile when placed in water. However, type strains of this genus now form chains of nonmotile spores typical of *Streptomyces*.

The configuration of the spore chains (or sporophores) of streptomycetes has played a prominent role in species descriptions for many years. The chains are usually long and often contain over 50 arthrospores. Several genes are involved in their formation (Chater and Merrick, 1979), and cultivation conditions also have an influence. As a result, the range of spore chain morphology is extensive, and some workers have recog-

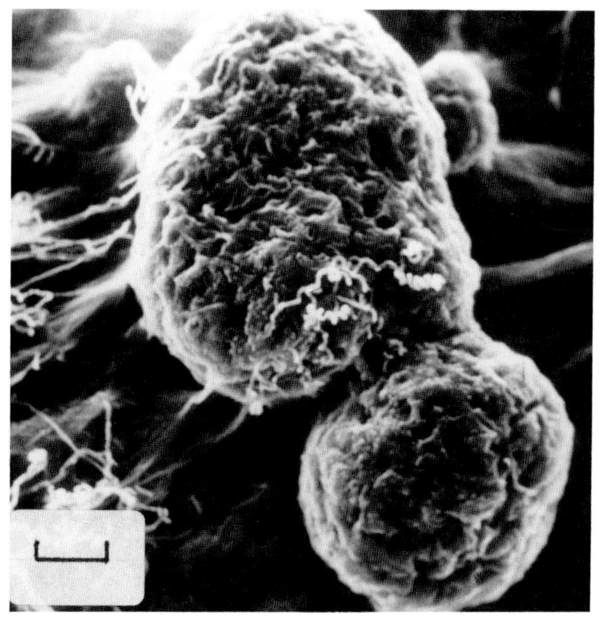

Figure 29.1. Sclerotia of *Streptomyces ochraceiscleroticus (Chainia olivacea)*. Scanning electron microscopy. *Bar*, 5 μm.

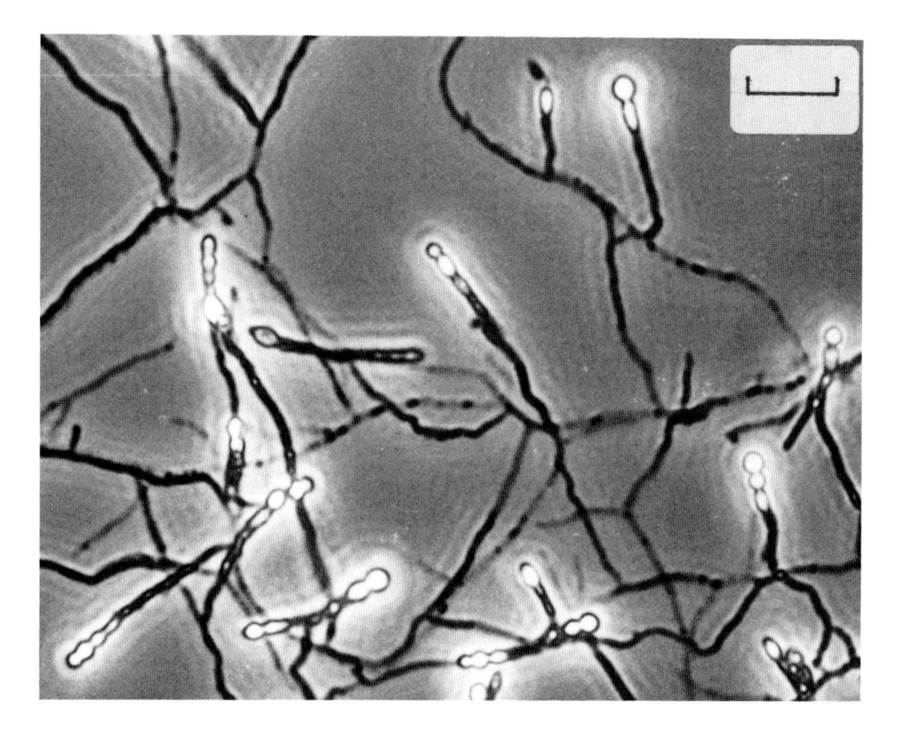

Figure 29.2. Spore chains of *Streptomyces (Microellobosporia)* species on aerial mycelium. Light microscopy. *Bar,* 5 µm. (Courtesy of Prof. T. Cross, University of Bradford, Bradford, U.K.)

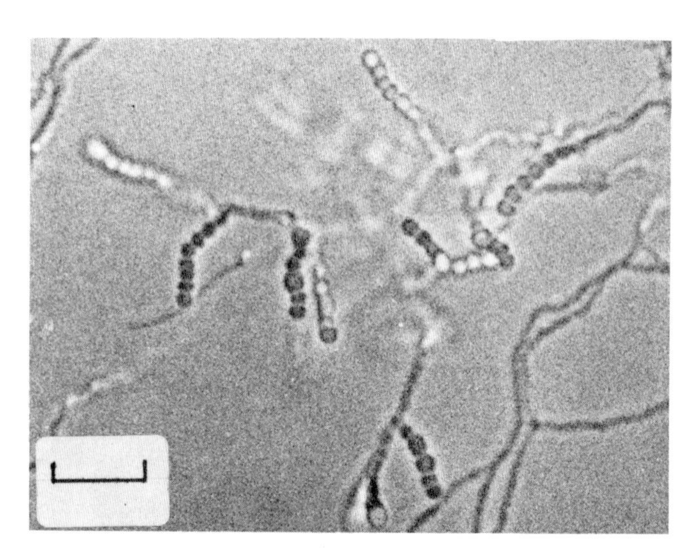

Figure 29.3. Spore chains of *Streptomyces carpinensis (Elytrosporangium carpinense)* on substrate mycelium. Light microscopy. *Bar,* 5 µm. (Courtesy of Prof. T. Cross, University of Bradford, Bradford, U.K.)

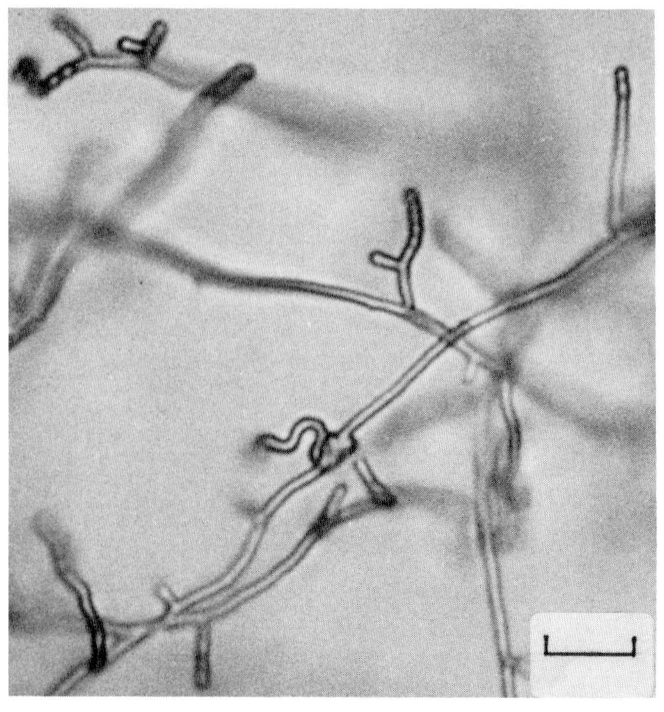

Figure 29.4. Spore chains of *Streptomyces carpinensis (Elytrosporangium carpinense)* on aerial mycelium. Light microscopy. *Bar,* 5 µm. (Courtesy of Prof. T. Cross, University of Bradford, Bradford, U.K.)

nized many categories—for example, Ettlinger et al. (1958a), who grouped strains into 15 morphological types. A much simpler and more practicable scheme was proposed by Pridham et al. (1958), and this was adopted for the International *Streptomyces* Project (ISP; Shirling and Gottlieb, 1966). The three categories recognized were: (a) straight to flexuous (*Rectiflexibles*) (Fig. 29.5); (b) hooks, loops, or spirals with one to two turns (*Retinaculiaperti*) (Fig. 29.6); and (c) spirals (*Spirales*) (Fig.29.7). Even this broad system can present problems because more than one category may be observed in the same culture, and the distinction between *Retinaculiaperti* and *Spirales* is not always clear (Shirling and Gottlieb, 1977; Williams and Wellington, 1980).

The fine structure of streptomycetes at various stages of their life cycle has been widely studied, but it is only spore surface ornamentation that has been adopted as a taxonomic character. The ornaments, which are in fact borne on the spore sheath, were first detected from observations of spore chain silhouettes by transmission electron microscopy (Ettlinger et al., 1958b; Tresner et al., 1961). The categories recognized were

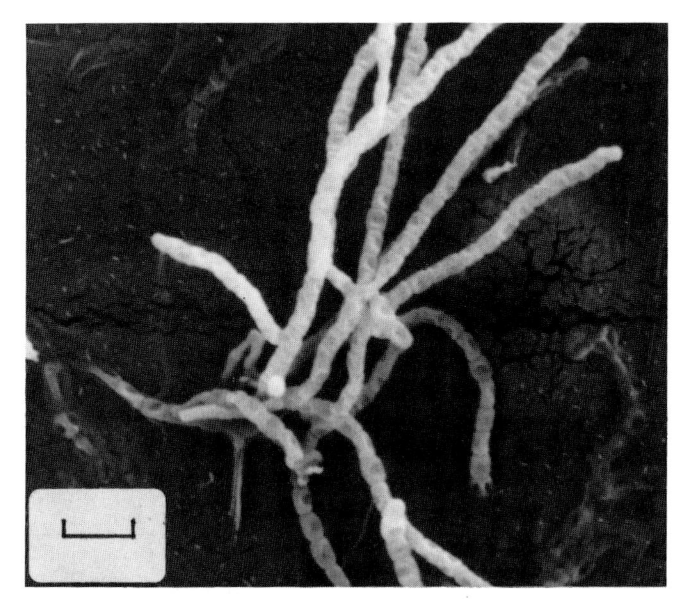

Figure 29.5. Straight to flexuous (*Rectiflexibiles*) spore chains of *Streptomyces griseus*. Scanning electron microscopy. *Bar,* 2 µm.

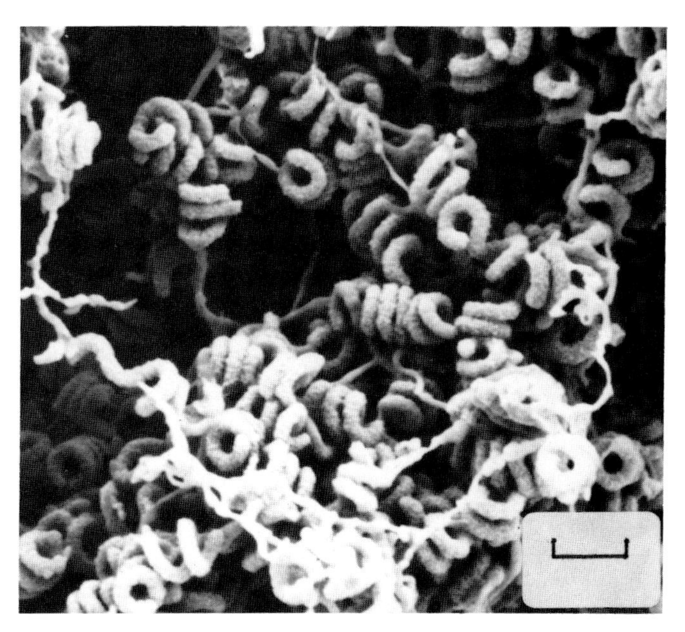

Figure 29.7. Spiral (*Spirales*) spore chains of *Streptomyces hygroscopicus*. Scanning electron microscopy. *Bar,* 5 µm.

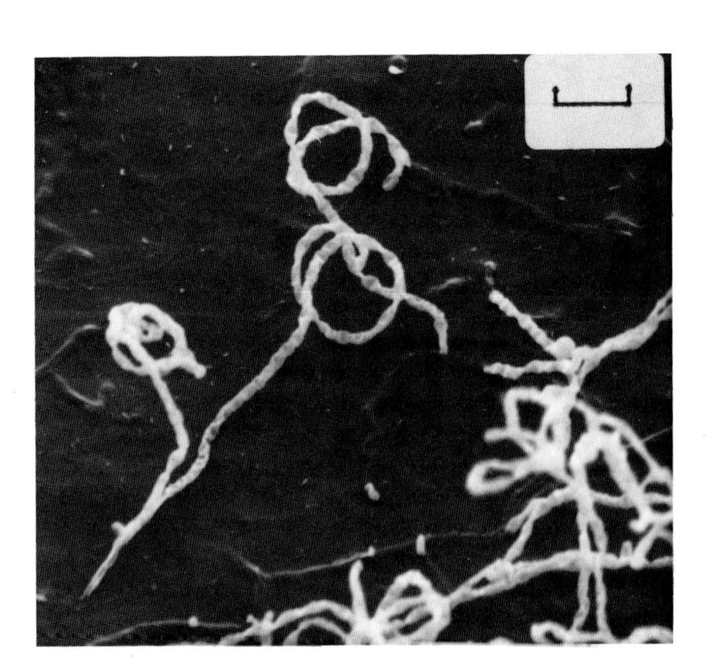

Figure 29.6. Looped (*Retinaculiaperti*) spore chains of *Streptomyces vinaceus*. Scanning electron microscopy. *Bar,* 5 µm.

smooth, spiny, hairy, and warty; a fifth type, rugose, was proposed by Dietz and Mathews (1971) (Figs. 29.8–29.12). Spore surface ornamentation is a stable character, but the distinctions between smooth, warty, and rugose types in silhouette can be difficult. However, these problems are overcome if scanning electron microscopy is used.

Cell Envelope Composition

Traditionally, streptomycete systematics has mainly been based on morphological, pigmentation, and physiological properties, but increasing weight is now given to chemical features, especially for the circumspection of the genus. The genus may now be restricted to actinomycetes that (a) have a peptidoglycan comprised of *N*-acetylglucosamine, *N*-acetylmuramic acid, alanine, D-glutamic acid, and glycine with L-DAP as the diamino acid (Azuma et al., 1970; Lechevalier and Lechevalier, 1970b; Schleifer and Kandler, 1972; Uchida and Aida, 1977, 1979); (b) a fatty acid profile showing major amounts of saturated, *iso*- and *anteiso*-fatty acids (Fig. 29.13) (Kroppenstedt and Kutzner, 1976, 1978; Lechevalier et al., 1977; Kroppenstedt, 1985); and (c) a phospholipid pattern consisting of diphosphatidylglycerol, phosphatidylethanolamine, phosphatidylinositol, and phosphatidylinositol mannosides (i.e. a phospholipid pattern type II sensu Lechevalier et al., 1977), and that contains either hexa- or octahydrogenated menaquinones with nine isoprene units as the predominant isoprenolog (Fig. 29.14) (Collins et al., 1984; Alderson et al., 1985; Kroppenstedt, 1985). The muramic acid is *N*-acetylated (Uchida and Aida, 1977, 1979), but variation has been observed in the amount of the major menaquinones at different stages in the growth cycle (Saddler et al., 1986). Streptomycetes also have an A3y peptidoglycan type (Schleifer and Kandler, 1972) and a wall chemotype I (sensu Lechevalier and Lechevalier, 1970b), contain teichoic acids (Naumova et al., 1978), but lack mycolic acids.

Most chemosystematic studies on streptomycetes have been confined to a visual comparison of qualitative data generated using analytical techniques such as gas chromatography and high-performance liquid chromatography. These investigations have provided valuable chemical markers for separating streptomycetes from other sporoactinomycetes (Goodfellow and Cross, 1984b), but it seems likely that analyses of quantitative data, using appropriate cluster analysis techniques, may yield information for the characterization of streptomycetes at the subgeneric level (O'Donnell, 1986).

Colonial Characteristics

Streptomycetes show a notable array of macroscopic features, such as pigmentation of spores, substrate mycelium, and diffusible exopigments, together with the morphology of colonies, and texture of the aerial mycelium. Pigmentation has been widely used in classification and identification, but colony morphology has generally been regarded as too variable for use as a taxonomic character.

Spore mass color has been widely used in streptomycete taxonomy. Strains have been assigned to "sections," "series," and "species groups" on the basis of their spore color (Burkholder et al., 1954; Flaig and Kutzner,

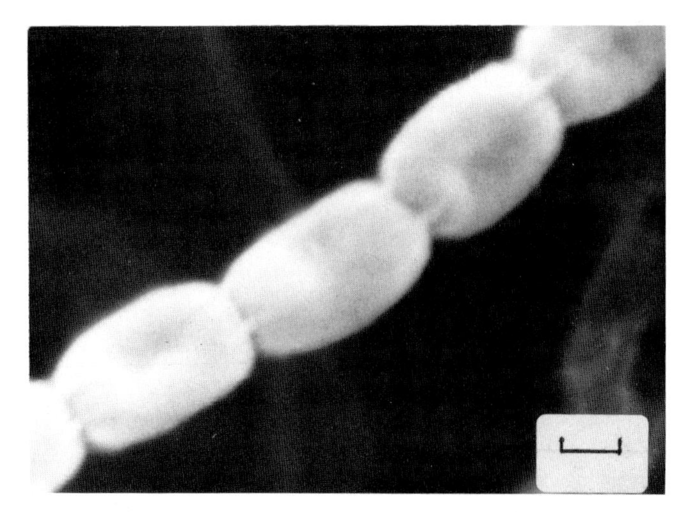

Figure 29.8. Smooth spores of *Streptomyces niveus*. Scanning electron microscopy. *Bar,* 0.25 μm.

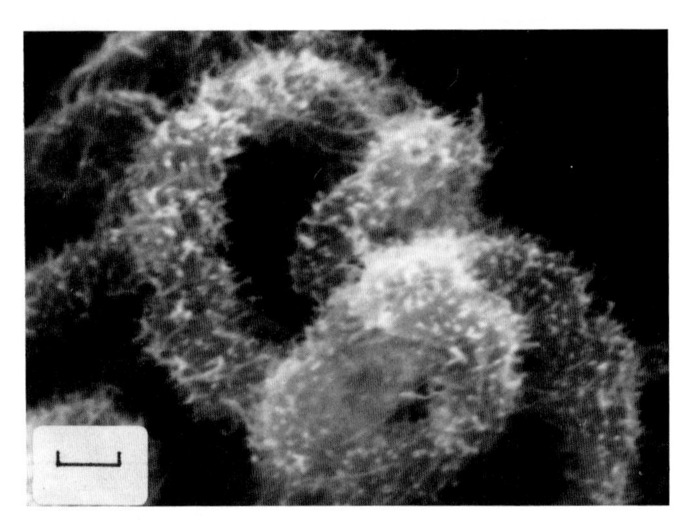

Figure 29.10. Hairy spores of *Streptomyces glaucescens*. Scanning electron microscopy. *Bar,* 0.5 μm.

Figure 29.9. Spiny spores of *Streptomyces viridochromogenes*. Scanning electron microscopy. *Bar,* 0.5 μm.

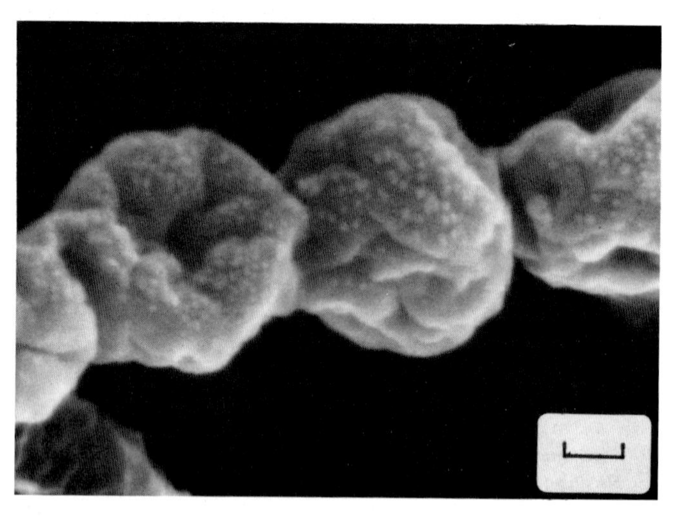

Figure 29.11. Warty spores of "*Streptomyces pulcher.*" Scanning electron microscopy. *Bar,* 0.25 μm.

1954, 1960; Hesseltine et al., 1954; Gauze et al., 1957; Ettlinger et al., 1958a; Pridham et al., 1958; Krasil'nikov, 1960). In the 8th edition of the *Manual* (Pridham and Tresner, 1974b), *Streptomyces* species were assigned to seven color series: blue, gray, green, red, violet, white, and yellow. Subsequently, the series were extended to accommodate additional colors (Kutzner, 1981). The color of the spore mass is still seen to be a useful criterion, but its determination can present some problems (Kutzner, 1981). Thus, the color can be influenced by factors such as the medium, growth regime, and age of the culture. Sometimes the color does not match with any of the established categories. Several useful attempts have been made to standardize the matching of pigments by using selected chips from color manuals (Flaig and Kutzner, 1954, 1960; Tresner and Backus, 1963; Prauser, 1964).

The colors of the substrate mycelium and the soluble pigment are of greatest value when they are striking, e.g. blue, dark green, red, and violet. Substrate mycelium color has been used in the preliminary grouping of streptomycetes (Baldacci et al., 1954; Baldacci, 1958, 1961; Krasil'nikov et al., 1961b), but the expression of the various pigments is often more dependent on the medium, temperature, pH, age of culture, and, in some cases, illumination, than those of the aerial spore mass (Kutzner, 1981). Diffusible pigments and their pH sensitivity have also

Figure 29.12. Rugose spores of *Streptomyces hygroscopicus*. Scanning electron microscopy. *Bar,* 0.5 μm.

been regarded as useful taxonomic characters (Waksman and Curtis, 1916; Jensen, 1930a; Shirling and Gottlieb, 1970), but chemically different pigments may exhibit the same color (Krasil'nikov, 1970a; Kutzner, 1981). Streptomycetes produce anthracyclinglycoside, diazaindophenol, naphthoquinone, phenoxazinone, and prodigiosin pigments (Kutzner, 1981), but more extensive studies are needed before the value of pigment type in streptomycete systematics can be objectively assessed.

Despite these difficulties, color determinations of aerial spore mass, substrate mycelium, and diffusible pigments were used for the descriptions of *Streptomyces* species in the ISP (Shirling and Gottlieb, 1966). The color determinations were based on standardized media and methodology but there was some disagreement on aerial spore color determination, using the Tresner and Backus (1963) color chart. It was also shown that in some cases aerial spore mass color varied on different media (Szabó and Marton, 1976). The diagnostic value of spore color was limited because over half of the test strains were placed in the gray series. Good agreement was found for the determinations of pH sensitivity of the diffusible pigments but not in the interpretation of substrate mycelium color (Shirling and Gottlieb, 1970). It is possible that chemical analyses of spore, mycelial, and diffusible pigments may lead to a better understanding of the nature and taxonomic value of pigmentation.

Nutrition and Growth

Streptomycetes are generally regarded as chemo-organotrophs that are not fastidious, needing only a suitable carbon source, inorganic nitrogen source, and mineral salts for growth (Kutzner, 1981). There does not appear to be any requirement for vitamins or growth factors and hence most strains can be grown on simple, defined media. However, species vary considerably in their carbon and nitrogen source utilization patterns, which are often used as taxonomic characters (e.g. Shirling and Gottlieb, 1966; Pridham and Tresner, 1974b; Williams et al., 1983a). Carbon sources that are widely used include cellobiose, glucose, glycerol, D-mannose, and trehalose; nitrogen sources include ammonium, L-arginine, L-asparagine, and nitrate. Relatively few strains use organic acids, inulin, xylitol, L-methionine, or nitrite. Most can degrade esculin, casein, gelatin, and hypoxanthine.

Although growth will occur on most media, spore production is usually most prolific on those with a high carbon-nitrogen ratio (Kutzner, 1981). Well-known examples are the glucose-yeast extract-malt extract, oatmeal, inorganic salts-starch, and glycerol-asparagine agars used in the ISP (Shirling and Gottlieb, 1966). The environmental requirements and tolerance of streptomycetes have been surveyed in detail by Kutzner (1981). Most grow at temperatures between 10° and 37°C, and are thus regarded as mesophiles. However, several species can grow at temperatures above 37°C, although most are thermotolerant rather than thermophilic in their responses. A variety of type strains studied by Williams et al. (1983a) grew at 10°, 37°, and 45°C, but three species, *"Streptomyces thermoflavus," S. thermonitrificans,* and *S. thermovulgaris,* were thermophilic and did not grow below 37°C. These and several other species were included in a list of those able to grow at 45–55°C (Kutzner, 1981). A few strains grow slowly at 4°C (Williams et al., 1983a).

Streptomycetes generally require a good supply of free water for growth, being unable to develop at high osmotic or matric potentials. Thus growth in soil was severely limited at water tensions above pF 4.0 (Williams et al., 1972b), and in culture there was no growth of isolates below 0.92_{aw} (Fermor and Eggins, 1980). However, some soil isolates were able to grow in media at high osmotic potentials (Wong and Griffin, 1974) and a few strains grow with 13% (w/v) NaCl (Tresner et al., 1968). Survival of streptomycetes in dry conditions is aided by the resistance of their arthrospores to desiccation, their tolerance exceeding that of the vegetative mycelium (Williams et al., 1972b). Streptomycetes are generally regarded as obligate aerobes with a limited capacity for microaerophilic growth in culture (Kutzner, 1981), and dissimilatory reduction of nitrate is common.

Most behave as neutrophiles in culture, growing between pH 5.0 and 9.0 with an optimum close to neutrality. Only a few of the type strains studied by Williams et al. (1983a) were able to grow at pH 4.3, but acidophilic and acidoduric strains have been isolated from acidic soils and other materials (Williams et al., 1971; Khan and Williams, 1975; Hagedorn, 1976). Acidophilic isolates grow in the range from about pH 3.5 to 6.5, with optimum rates at pH 4.5–5.5, but it is clear that a spectrum of pH requirements exists among streptomycetes from acidic environments (Flowers and Williams, 1977b). Acidophiles produce diastases (Williams and Flowers, 1978) and chitinases that have optima at pH levels below the equivalent enzymes from neutrophiles. Study of basophilic streptomycetes has been relatively neglected since the isolation and characterization of *S. caeruleus* by Taber (1960). However, more recently 20 basophilic strains with optimum growth at pH 9.0–9.5 were isolated from soils in Japan by Mikami et al. (1982), who also found that six of the type strains tested were able to grow at pH 11.5. However, nine of the isolates and three of the type strains (including *S. caeruleus*) contained the *meso-* rather than the L-DAP isomer, which is typical of streptomycetes. Alkalophilic strains, which grew between pH 8.0 and 11.5 with optimum at pH 9.0–9.5, were distinguished from alkaline-resistant ones that had an optimum around pH 7.0 with some growth at pH 11.5.

Genetics

This is a well-studied and rapidly developing topic, reflecting the current and potential importance of streptomycetes in biotechnology (Hopwood and Merrick, 1977; Chater, 1979; Chater and Hopwood, 1984). There is, however, comparatively little information on the relevance of genetics to taxonomy, where the degree of congruence between the phenotype and genotype and the influence of genetic exchange mechanisms on phenotypic characters are the prime concerns.

The amount of DNA in the *Streptomyces* genome is unusually large; an estimate of 10.5×10^3 kilobase pairs (kb) was obtained by renaturation analysis (Benigni et al., 1975). *Streptomyces* DNA is also unusual for its high content of guanine and cytosine (G + C), which ranges from 69 to 78 mol% (Goodfellow and Cross, 1984b). The G + C contents of the synonymous or related genera, such as *Actinopycnidium* (72%; Monson et al., 1969), *Actinosporangium* (72%; Yamaguchi, 1967), *Chainia* (72%; Yamaguchi, 1967), *Microellobosporia* (70%; Yamaguchi, 1967), and *Streptoverticillium* (69–73%; Baldacci and Locci, 1974), also fall within this range. Therefore, while G + C contents contribute to the definition of *Streptomyces* and related genera, they are of little use in the delimitation of species.

Relatively few DNA pairing studies have been made on streptomycetes and related genera, and sometimes the interpretation of the results has been impaired by the use of strains that were incorrectly classified or insufficiently characterized. In one of the more successful early studies Monson et al. (1969) showed that strains erroneously designated as *S. coelicolor* in genetic studies were genetically homologous with *S. violaceoruber* rather than the *S. coelicolor* nominifer, thus supporting the earlier phenetic studies of Kutzner and Waksman (1959). They also found a high degree of homology between *Actinopycnidium caeruleum* and *S. violaceoruber.* Farina and Bradley (1970) found that DNA from *S. venezuelae* bound appreciably only to that from *S. albus, Streptoverticillium baldaccii,* and *Microellobosporia flavea,* although the homologies ranged from 30 to 58%. The DNA pairing studies of Kroppenstedt et al. (1981) indicated the relatedness of *Actinoplanes armeniacus* to the *Streptomycetaceae,* thus supporting the transfer of this species to *Streptomyces* (Wellington and Williams, 1981b) and also revealed a high degree of reassociation between the DNA of *Kitasatoa kauaiensis* and the other three species of this genus. Stackebrandt et al. (1981) investigated the relatedness of species from 18 spore-forming actinomycete genera and recognized three DNA homology clusters, one of which included *Chainia antibiotica, Kitasatoa kauaiensis, Streptomyces griseus, Microellobosporia cinerea, Streptoverticillium baldaccii, Elytrosporangium brasiliense,* and *Actinosporangium violaceum.* The first two species also served as sources of labeled DNA, no streptomycetes being included. Homology values ranged from 25 to 41%.

A more detailed study of DNA pairing among phenetically defined *Streptomyces* species was made by Okanishi et al. (1972), who matched DNA from *S. griseus* against that from 25 species that were phenetically

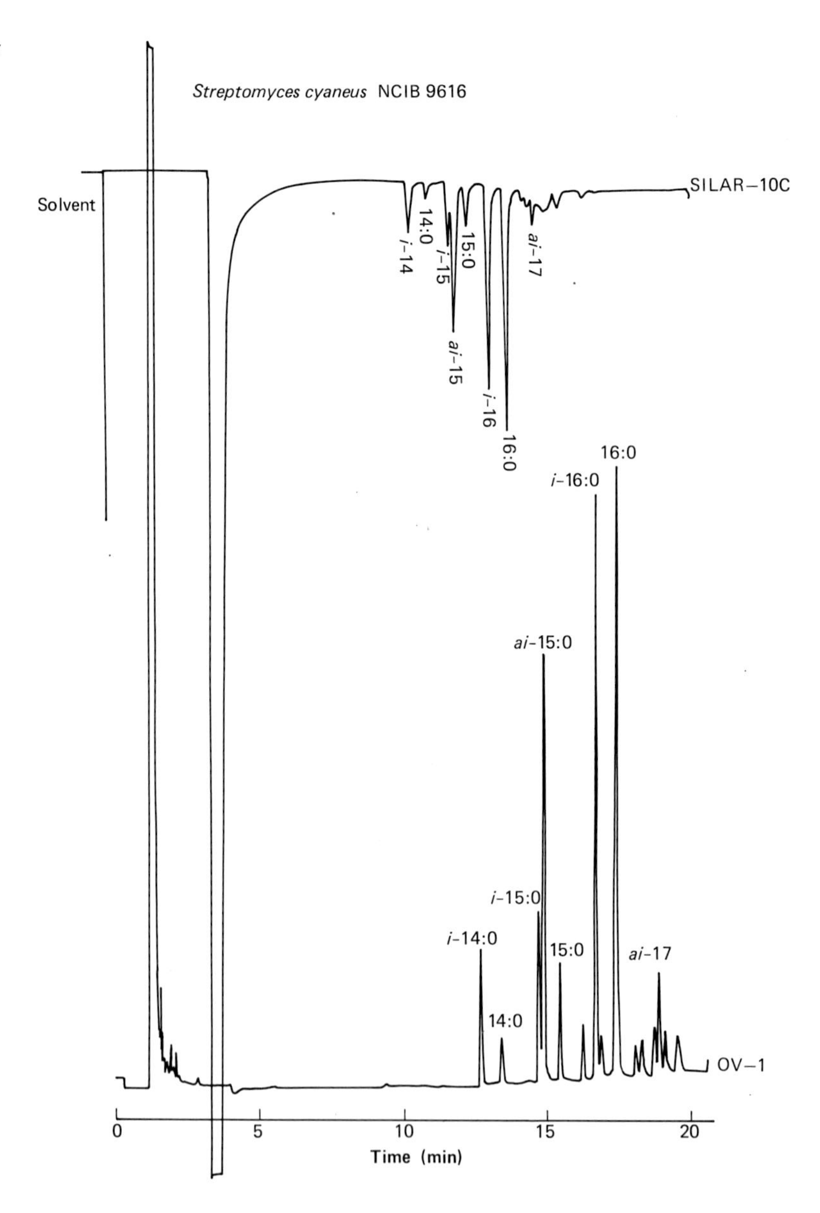

Figure 29.13. Gas chromatographic analysis of the fatty acid methyl esters of *Streptomyces cyaneus* NCIB 9616.

similar on the basis of characters used in the ISP. Although some species showed a high degree of reassociation with *S. griseus,* homologies ranged from 38 to 104% for phenetically similar strains and from 37 to 56% for the phenetically dissimilar strains that were included as controls. A genomic "griseus" group was defined to include all strains with at least 56% homology, but four subgroups were delimited at higher levels of reassociation. The largest of these produced from 71 to 98% reassociation with DNA from its central member, *S. californicus.*

A similar but more extensive study was conducted by Mordarski et al. (1986). The largest cluster defined by the numerical classification of streptomycetes (Williams et al., 1983a) was equivalent to the "griseus" group recognized by the previous workers and could be divided into three subclusters at the 81% S_{SM} similarity level. Reference DNA was prepared from representative strains from each subcluster and paired with that from 35 strains selected from these subclusters together with 12 strains from other clusters. The mean reassociation percentages of all reference strains with members of their own subcluster (37–75%) were higher than with other subclusters or clusters. However, there was a wide range of variation, and no consistent correlation between phenetic similarity and the degree of DNA reassociation was obtained.

The technique of DNA/rRNA association provides a useful means of determining the relationships between actinomycetes that show little similarity in DNA/DNA reassociation assays (Bradley and Mordarski, 1976). Streptomycetes have occasionally been included in such studies. Mordarski et al. (1980b) examined duplexes between labeled rRNA from *Rhodococcus rhodochrous* and DNA from strains of *Streptomyces, Mycobacterium,* and *Nocardia;* they concluded that there was a close phylogenetic relationship between all strains, although each genus occupied fairly distinct areas on the similarity map. Mordarski et al. (1981b) formed duplexes between rRNA from *Streptomyces griseus, Mycobacterium fortuitum,* and *Nocardia asteroides* and DNA from a range of strains that included four streptomycetes. The latter grouped closely on the *S. griseus* similarity map despite their previously determined phenetic differences. Duplexes between rRNA of *Kitasatoa kauaiensis* and DNA from *Streptomyces griseus, Chainia antibiotica, Microellobosporia cinerea, Elytrosporangium brasiliense,* and *Actinosporangium violace-*

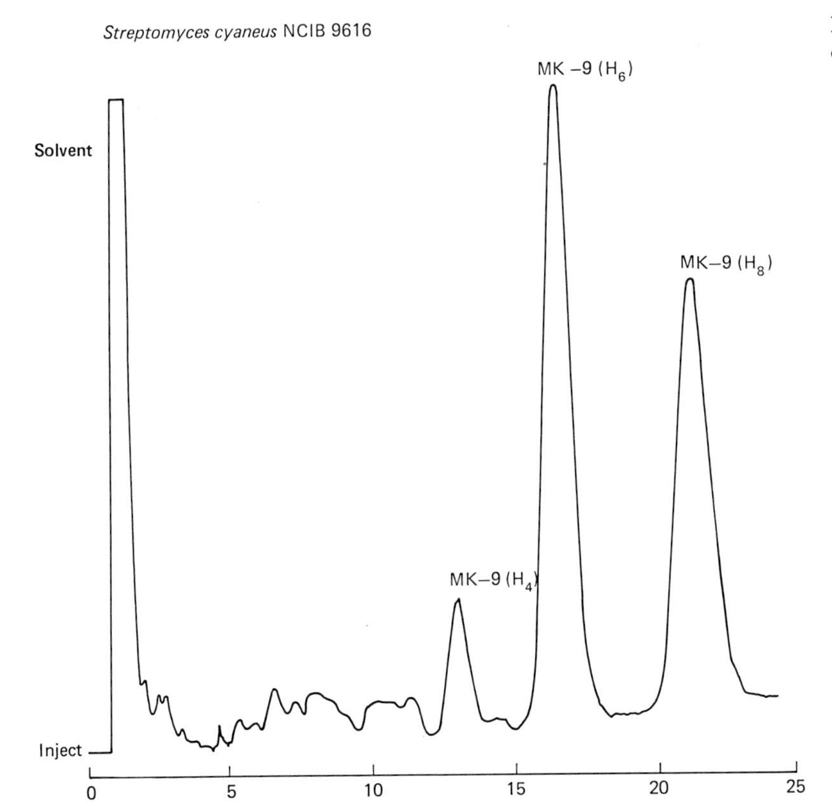

Figure 29.14. High-performance liquid chromatogram of menaquinones from *Streptomyces cyaneus* NCIB 9616.

um formed a distinct group on the similarity map (Stackebrandt et al., 1981). Gładek et al. (1985) examined duplexes between 23S rRNA from the type strains of *Streptomyces albus, S. griseus, S. lavendulae,* and *Streptoverticillium cinnamoneum* and DNA from nine type strains of *Streptomyces* species and two *Streptoverticillium* species, which represented various clusters defined by the numerical phenetic study of Williams et al. (1983a). All of the streptomycetes formed duplexes with high $T_{m(e)}$ values within narrow ranges in each reference system and hence occupied distinct regions in the similarity maps. *Streptoverticillium* strains were considered to be closely related to *Streptomyces,* but could be distinguished from the latter on the similarity map based on the *Streptoverticillium cinnamoneum* reference system, thus supporting the continued separation of these genera.

The method of 16S rRNA oligonucleotide cataloging has been applied to determine suprageneric relationships of actinomycetes and other bacteria (Stackebrandt and Woese, 1981a; Stackebrandt et al., 1981; Stackebrandt and Schleifer, 1984). All results indicate that the family *Streptomycetaceae* forms a distinct suprageneric group.

Phage sensitivity patterns have provided some useful taxonomic information on streptomycetes. Most phage used have been isolated from soil by enrichment techniques and include both virulent and temperate forms; the latter have also been obtained from lysogenic cultures. Attempts to use phage as an aid to species identification have met with little success because most are polyvalent, attacking many species, and others are specific to only some strains of a given species (Prauser and Falta, 1968; Korn et al., 1978; Wellington and Williams, 1981a). Examples of the latter type have been reported for a number of species, including *S. albus, S. griseus, S. griseinus, S. coelicolor, S. glaucescens,* and *S. violaceoruber* (Korn et al., 1978; Wellington and Williams, 1981a). In contrast, phage sensitivity patterns have provided valuable supporting evidence of the relationships between *Streptomyces* and other genera. Thus sets of phage that are family specific for the *Streptomycetaceae (Streptomyces, Streptoverticillium,* and synonomous genera) have been defined (Prauser and Falta, 1968; Wellington and Williams, 1981a; Prauser, 1984c). Cross-infection studies have contributed to the transfer of the genus *Microellobosporia* from the family *Actinoplanaceae* to the *Streptomycetaceae* (Prauser et al., 1967) and to similar transfers of *Elytrosporangium* and *Kitasatoa* (Prauser, 1970). The phage sensitivity pattern of *Actinoplanes armeniacus* stimulated investigation of other characteristics that led to its transfer to *Streptomyces* (Wellington and Williams, 1981b).

Much is known about the various mechanisms of genetic exchange between streptomycetes (see Chater and Hopwood, 1984), but the taxonomic significance of these processes remains to be fully evaluated. Few assessments of the influence of genetic exchange on the phenetic characters used in taxonomy have been made. While such studies do not present any major practical difficulties, interpretation of the results raises some theoretical problems. Evidence of genetic exchange is usually obtained by manipulation of cultures in laboratory conditions designed to enhance the process, as exemplified by the techniques for protoplast fusion and transfer of plasmid or phage DNA. Taxonomists are thus presented with the invidious task of comparing character stability in wild isolates with that in strains subjected to genetic manipulation in the laboratory. This problem is compounded by the fact that negative results from genetic analyses may be due to compatibility, competence, or restriction factors rather than the lack of genetic homology (Jones and Sneath, 1970).

Natural gene exchange between streptomycetes can occur in several ways. Despite the many examples of temperate phage, the only well-documented report of transduction in a streptomycete is that of Stuttard

(1979) on *S. venezuelae* involving the narrow-host-range phage φ SVI (Chater and Hopwood, 1984). Several phenotypically unrelated and genetically distantly linked markers were transduced.

Conjugation between streptomycetes was first reported by Hopwood (1957) and there have subsequently been many similar claims, including conjugation between different species (see Hopwood and Merrick, 1977). Evidence for the role of sex plasmids in conjugation has been obtained for *S. coelicolor* A3(2) and "*S. lividans*" (see Chater and Hopwood, 1984). Frequency of recombination between streptomycete strains and species can be substantially increased by protoplast fusion, which appears to negate the control of the sex plasmids (e.g. Hopwood et al., 1977; Baltz, 1978; Ochi et al., 1979).

Plasmids are probably widespread among streptomycetes. Initial evidence for their presence can be obtained by the phenomenon of lethal zygosis (Bibb et al., 1977). These authors detected plasmids in *S. coelicolor* A3(2) and other species. Evidence for plasmids in a wide range of type strains has been obtained (Ritchie et al., 1986). However, there are, as yet, few conclusive examples of phenotypic characters that are under the control of plasmids. Plasmid-linked genes are involved in the resistance of *S. coelicolor* A3(2) to the antibiotic methylenomycin A, which it produces (Chater and Hopwood, 1984). There is also evidence for the role of plasmids in the synthesis of tyrosinase (Hütter et al., 1981) and agarase (Hodgson and Chater, 1981) by streptomycetes.

Ecology

Streptomycetes are widely distributed in terrestrial and aquatic habitats. Most are strict saprophytes, but some form parasitic associations with plants or animals. Surprisingly little is known about the role of streptomycetes in natural environments, although evidence of their occurrence and numbers in habitats is extensive. Several recent reviews on streptomycete ecology are available (Cross, 1981a; Kutzner, 1981; Goodfellow and Williams, 1983; Williams et al., 1984a; Goodfellow and Simpson, 1987).

Soil, fodder, and composts appear to be the primary reservoirs for streptomycetes. However, the high counts associated with such habitats must be interpreted with care because most colonies on dilution plates originate from spores. Indeed, it appears that streptomycetes exist in soil for long periods as resting arthrospores that germinate given the occasional presence of exogenous nutrients (Mayfield et al., 1972). Thus, when localized organic substrates, such as root fragments and dead fungal hyphae, are available they are rapidly colonized by mycelium with production of spores as nutrients become exhausted. The same investigators estimated the mean doubling time of streptomycetes to be 1.7 days, a figure that probably reflects the short intermittent periods of active growth by streptomycetes rather than supporting the widely held view that the organisms are "slow-growing" microbes. Specific growth rates and doubling times for streptomycetes in laboratory culture are approximately intermediate between those of bacteria and fungi (Flowers and Williams, 1977a).

The survival capacity of streptomycete spores appears to be greater than that of the hyphae (Williams et al., 1972b). The spores usually have thicker walls than the hyphae (Sharples and Williams, 1976) and are more hydrophobic (Ruddick and Williams, 1972) owing to the outer sheath that envelopes the spore wall (Williams et al., 1973). Streptomycete spores have a net negative surface charge except at low pH levels (Douglas et al., 1970) and a relatively low endogenous metabolism (Ensign, 1978), and are generally more resistant to heat than the corresponding hyphae (Goodfellow and Simpson, 1987). Spores are released above soil when particles are disturbed by wind or rain (Lloyd, 1969), whereas dispersal within soil is assisted by movement of water and arthropods (Ruddick and Williams, 1972). Streptomycete spores form up to half of the intestinal microflora of *Bibio marci* L. larvae (Szabó et al., 1967).

While nutrient availability is a major factor governing the distribution and activity of soil streptomycetes, temperature, pH, moisture content, and soil type can also exert an influence, as do season and climate (Williams et al., 1972b; Williams, 1978). Although streptomycetes are usually considered to be strict aerobes, they can grow in sterile soil at low oxygen concentrations, but not when carbon dioxide concentrations exceed 10%. In dry soil, streptomycete counts decrease markedly at moisture tensions above pF 4.0, but their proportion to other bacteria may be higher because their spores are more resistant to desiccation than are the vegetative cells of bacteria. Optimum counts from neutral soil and optimum radial growth of streptomycetes inoculated into sterile soil occur at moisture tensions between pF 1.5 and 2.5. Some streptomycetes from arid soils are able to grow on media at high osmotic potentials (Wong and Griffin, 1974). Halophilic and salt-tolerant streptomycetes have also been reported (Hunter et al., 1981).

It seems unlikely that streptomycetes grow optimally in temperate soils because most strains are mesophilic under laboratory conditions. A variety of mesophilic streptomycetes are involved in the initial stages of decomposition in composts, fodders, and similar substrates, but obligate or facultatively thermophilic strains become active at temperatures above 40°C (Lacey, 1981). One of the most common taxa, *Streptomyces albus,* is mesophilic but can grow at 45°C. Little is known of the role of thermophilic streptomycetes in composts and fodders, although *Streptomyces thermodiastaticus, Streptomyces thermovulgaris,* and *Streptomyces thermoviolaceus* produce cellulases (Crawford and McCoy, 1972).

Because many soils are acidic, pH is clearly an important factor determining the distribution and activity of streptomycetes. Acidophilic and neutrotolerant streptomycetes, the latter growing between pH 3.5 and 7.5 but optimally around pH 5.5, are common in acid soils (Khan and Williams, 1975). The presence of low numbers of neutrophilic streptomycetes in acidic soils can be attributed to their ability to grow in less acidic microsites (Williams and Mayfield, 1971) and to the resistance of their spores to acidity (Flowers and Williams, 1977b). When nitrogen-containing substrates, such as chitin or dead fungal mycelium, are added to poorly buffered acidic soil, a succession of acidophilic to neutrophilic streptomycetes occurs that parallels ammonification and the attendant rise in pH (Williams and Robinson, 1981).

Most soils contain a significant proportion of clay and humic colloids. Such colloidal material can markedly affect microbial activity at the microenvironmental level. Streptomycete spores are readily absorbed to kaolin but not to montmorillonite, except at low pH (Ruddick and Williams, 1972); a similar observation has been made for streptomycete phage (Sykes and Williams, 1978). Addition of calcium montmorillonite or calcium humate to cultures of streptomycetes can accelerate their growth and respiration (Mara and Oragui, 1981). Further, sites of adsorption associated with humic material can lead to microsites of increased pH in acidic soils (Williams and Mayfield, 1971).

The in vitro enzyme-producing and degradative properties of streptomycetes are well known, but their ecological significance has still to be elucidated. Streptomycetes are usually considered to be most active in the more advanced stages of decomposition of plant and other materials, playing an important role in the turnover of relatively complex, recalcitrant polymers (Williams, 1978). It has been demonstrated that some streptomycetes attack both the cellulose and lignin components of lignocelluloses (Crawford and Sutherland, 1979). *Streptomyces flavovirens* decomposed intact cell walls of phloem from Douglas fir (Sutherland et al., 1979) and various strains degraded lignocelluloses of grass, softwood, and hardwood (Antai and Crawford, 1981). Streptomycetes have the potential to degrade other naturally occurring polymers such as chitin, hemicelluloses, keratin, pectin, and fungal cell wall material, and have been implicated in the degradation of herbicides (Percich and Lockwood, 1978), plastics (Sharpell, 1980), polyphenolic tannins (Lewis and Starkey, 1969), and humic acids (Szegi and Gulyas, 1968). In addition, streptomycetes form melanin pigments similar to humic acids (Huntjens, 1972), may contribute to the formation of humus (Kutzner, 1968), and are inhibited by some fungicides (Ou et al., 1978).

The widely held view that rhizosphere streptomycetes can protect plant roots by inhibiting the growth of fungal pathogens is based on their ability to produce antifungal antibiotics in vitro. However, it has been repeatedly stressed that direct evidence for the production of antibiotics in natural soil is lacking (Williams, 1982), although several workers consider antibiotics to be natural products (Hopwood and Merrick, 1977; Martin and Demain, 1980). Nevertheless, there is circumstantial evidence to suggest that streptomycetes contribute to the control of fungal pathogens (Williams, 1982), and some reduction in disease severity has been achieved by inoculating seeds or seedlings with potentially antagonistic streptomycetes (e.g. Sing and Mehrotra, 1980; Rothrock and Gottlieb, 1981). It has also been shown that streptomycetes are implicated in the control of wood-infecting fungi (Blanchette et al., 1981) and of aspergilli in soil of plant pots (Staub et al., 1980). Streptomycete culture filtrates are nematocidal to *Pratylenchus penetrans* under laboratory conditions (Walker et al., 1966).

Streptomycetes are widely distributed in aquatic habitats, but the possibility of their wash-in from surrounding terrestrial habitats must always be considered. The ecology of streptomycetes in aquatic environments has received little attention and many questions remain open (Cross, 1981a; Goodfellow and Haynes, 1984). Most studies have sought to determine the numbers of streptomycetes in aquatic habitats by applying procedures designed for the selective isolation of strains from terrestrial sources. Streptomycete spores are continually washed into freshwater and marine habitats but there is little evidence of their growth in river, lake, or marine sediments. Streptomycetes were found to grow on the chitinous exoskeletons of *Procambarus versutus* immersed in a woodland stream (Aumen, 1980). On balance, streptomycetes are unlikely to form an integral component of the indigenous microbial flora in aquatic ecosystems, but there is no a priori reason to assume that they cannot grow in aquatic habitats given the availability of nutrients required by organotrophic microorganisms and suitable environmental conditions.

The production of earthy tastes and odors in reservoir and water supplies has often been attributed to streptomycetes (Wood et al., 1983). Geosmin (*trans*-1,10-dimethyl-*trans*-9-decalol) and methylisoborneol, common volatile compounds associated with the formation of earthy odors, have been detected in streptomycetes. It is likely that these compounds are not produced by streptomycete spores but arise as secondary metabolites subsequent to hyphal growth (Cross, 1981a). This would explain why attempts to correlate plate counts of streptomycetes in water with periods of odor production have met with limited success. Wood et al. (1983) examined possible sites of odor production by streptomycetes for reservoirs in the northwest of England. The nutrient concentrations in the water mass did not support geosmin production. In contrast, litter on the reservoir banks allowed the growth of streptomycetes and the production of geosmin, so runoff or seepage into the reservoirs was a possible source of contamination.

Very little is known about the distribution of *Streptomyces* species in natural habitats. The poor state of streptomycete systematics and the lack of suitable selective isolation media contribute to this problem. However, a computer-assisted procedure is now available for the identification of unknown streptomycetes (Williams et al., 1983b), and the data base has been used to formulate media designed to isolate selected members of the streptomycete community (Vickers et al., 1984). *Streptomyces albidoflavus, S. diastaticus,* and *S. lydicus* were found to be common in freshwater habitats, the former predominating in marine sediments (Goodfellow and Haynes, 1984; Goodfellow and Simpson, 1987).

The only major plant disease caused by streptomycetes is scab of potato and a variety of other root crops. This disease, which occurs in many potato-growing regions of the world, is generally associated with *Streptomyces scabies.* Common scab causes disruption of potato tubers, and two main types are recognized. Normal scab involves deep or superficial discrete lesions of the tuber, and russet scab a brown roughening of most of the skin of the potato. The incidence of scab is greatest in dry, neutral to alkaline soils and that of russet scab in wet soil. It seems likely that more than one streptomycete species is involved in the production of scab; representatives of implicated species need to be thoroughly characterized and their pathogenicity proven by host reinoculation. The significance and taxonomic status of other possible phytopathogenic streptomycetes is not clear (Young et al., 1978).

Streptomyces somaliensis is pathogenic for man. This organism can readily be distinguished from other causal agents of actinomycetoma (Schaal, 1984b), and the disease can be treated with sulphadiazine or sulphamethoxazole plus trimethoprim (Schaal and Beaman, 1984). Infections caused by *S. somaliensis* have been reported in Mexico as well as in African countries.

The popular view that streptomycetes from clinical material can be dismissed as harmless saprophytes cannot be sustained. Apart from *Streptomyces somaliensis,* strains of *S. albus, S. griseus,* and *S. violaceoruber* have frequently been isolated from clinical material (Kutzner, 1981), and there is evidence that *S. albus* and *S. griseus* can cause infection (Gordon, 1974). Other species of *Streptomyces* isolated from man include "*Streptomyces candidus*" from the purulent exudate of a fractured patella, "*Streptomyces gedaensis*" from sputum and human abscesses, "*Streptomyces horton*" from pus, and *Streptomyces willmorei* from streptothricos of the liver (Kutzner, 1981). Strains identified as *S. griseus* (Gordon, 1974) caused infections in cats (Bakerspigel, 1973) and mycetoma in a captive bottlenose dolphin (Jasmin et al., 1972).

Serology

There have been few serotaxonomic studies of streptomycetes and related genera. The potential taxonomic value of serology was demonstrated by the application of the immunodiffusion technique to streptomycetes by Cross and Spooner (1963), but most of the limited number of subsequent studies have been of little taxonomic value. To some extent this has been due to the lack of a sound classification against which the serological data could be assessed. However, more recently immunodiffusion techniques have been applied to strains of streptomycetes and related genera that had been classified in the numerical phenetic study of Williams et al. (1983a).

Ridell and Williams (1983) found that strains of streptomycetes or streptoverticillia that shared several identifiable precipitinogens also belonged to the same phenetic cluster, while those that were less serologically related usually fell into different clusters. Thus there was evidence for a significant degree of congruence between serological and phenetic similarities, with both indicating the extensive synonymy among described species. Similar conclusions were reached by Ridell et al. (1986), who also obtained serological evidence supporting the close phenetic relationship between streptoverticillia and the *Streptomyces lavendulae* cluster (Williams et al., 1983a).

Enrichment and Isolation Procedures

Detailed reviews of this topic have been provided by Kutzner (1981) and Williams and Wellington (1982a, 1982b); here the general principles and problems are assessed.

As stated previously, streptomycetes as a group do not have specific growth factor requirements and are able to use a wide range of carbon and nitrogen sources. Nevertheless, there is a considerable variation in their nutrient requirements and tolerance of potentially inhibitory media additives, which led Vickers et al. (1984) to conclude that there is no such thing as a "general" isolation medium for streptomycetes. However, the major problem in devising media for the selective isolation of streptomycetes is their general lack of specific nutrient requirements, which excludes the possibility of formulating media on which other bacteria and fungi will not grow. This problem is exacerbated by the fact that many competing bacteria and fungi grow or spread more rapidly than streptomycetes on isolation plates. Therefore, most selective isolation procedures have been designed as much to inhibit growth of competing microbes as to stimulate development of streptomycetes.

Selectivity of isolation procedures may be influenced by: (a) pretreatment of samples, (b) selection of medium nutrient sources, (c) addition of selective inhibitors to the medium, and (d) incubation conditions.

Samples of solid materials such as soil are usually dispersed and diluted in sterile water or saline solution, this being achieved by procedures ranging from mechanical shaking to ultrasonic treatment. A variety of treatments designed to reduce the proportion of other bacteria in soil samples has been applied prior to preparation of suspensions (see

Kutzner, 1981), ranging from enrichment with calcium carbonate (Tsao et al., 1960) to drying at 45°C for 2–16 h (Williams et al., 1972b). Composts and other self-heated materials yield a higher proportion of streptomycetes if their spores are detached by agitation of the dried material and then impacted onto isolation plates (Lacey, 1971a; Lacey and Dutkiewicz, 1976a). Isolation of streptomycetes from water samples usually requires concentration of propagules by membrane filtration. Filters are placed face downward on the surface of a suitable medium and either removed after a few hours (Burman et al., 1969) or left in place to allow streptomycetes to grow through the pores and form colonies on the surface (Trolldenier, 1967). Centrifugation was used to isolate streptomycetes from seawater and marine muds (Okami and Okazaki, 1978).

The range of media that have been used to isolate streptomycetes is extensive (see Kutzner, 1981). Many were designed primarily to discourage growth of other bacteria, hence providing "cleaner" plates, although probably not allowing the development of all streptomycetes in the sample (Vickers et al., 1984). Carbon and nitrogen sources that have been claimed preferentially to encourage growth of streptomycetes include chitin, starch, glycerol, arginine, asparagine, casein, and nitrate. Two of the most widely used media are starch-casein (Küster and Williams, 1964a) and colloidal chitin agar (Hsu and Lockwood, 1975).

Whatever the composition of the isolation medium, it is usually necessary to increase its efficiency by adding selective inhibitors. Competition from fungi is dealt with most effectively by incorporation of antifungal antibiotics in the medium, streptomycetes being generally insensitive to these compounds at concentrations up to 50–100 µg/ml. Cycloheximide (actidione) has been widely used and has the advantage of being heat stable, but others such as pimaricin and mycostatin (nystatin), although heat labile, are more effective (Porter et al., 1960; Williams and Davies, 1965). In contrast, use of antibiotics selectively to isolate streptomycetes in the presence of other bacteria is less effective (Williams and Davies, 1965), because their sensitivity spectra often overlap. However, use of antibacterial antibiotics can facilitate the isolation of particular strains or species if their sensitivity spectra are known (Mayfield et al., 1972; Vickers et al., 1984). Because most streptomycetes are neutrophilic, isolation media are usually adjusted to give a reaction close to neutrality, but acidophiles have been detected using media at pH 4.5 (Williams et al., 1971; Khan and Williams, 1975; Hagedorn, 1976) and alkalophiles at pH 10 to 11 (Mikami et al., 1982).

Isolation plates are normally incubated for about 14 days at 25–30°C; thermotolerant strains are detected at 40°C and thermophiles at 45–55°C for up to 5 days. Distinction between streptomycetes and other bacteria is aided by the former's production of aerial mycelium and pigmentation, as determined by eye, a hand lens, or microscopic examination.

Maintenance Procedures

This topic has received much attention owing to the need to preserve stocks of industrially important strains; useful reviews have been given by Pridham and Hesseltine (1975) and Nisbet (1980). A variety of procedures has been used for both the short- and long-term preservation of streptomycetes (Table 29.3), many of which are also applicable to other sporing actinomycetes.

Table 29.3.
Methods for the preservation of cultures of streptomycetes

Methods for short-term preservation	Slope cultures stored at 4°C (generally used)
	Suspensions in soft agar at 4°C (Kutzner, 1972)
	Suspensions in 10% (v/v) glycerol at −20°C (Wellington and Williams, 1978)
Methods for long-term preservation	Slope cultures stored at −20°C (Tresner et al., 1960)
	Sterile soil culture (Pridham et al., 1973)
	Lyophilization (Hopwood and Ferguson, 1969)
	Storage under liquid nitrogen (Pridham and Hesseltine, 1975)
	Storage in vapor phase of liquid nitrogen (generally used)

Whichever method is used, sporing cultures or a spore inoculum provide the most effective storage material. Therefore cultures should be prepared on a medium conducive to good spore production (see "Nutrition and Growth" under "Further Descriptive Information," above.) If spore suspensions are required, they can be easily prepared by removal of aerial growth with an inoculation loop from heavily sporing cultures on a suitable solid medium. Because streptomycete spores are often hydrophobic, addition of a detergent (e.g. Triton X-100, 0.05%; sodium lauryl sulfonate, 0.01%) to the suspending medium is desirable (Kutzner, 1981). When dealing with poorly or nonsporing cultures, young vegetative mycelium from cultures in liquid medium (e.g. tryptone–yeast extract broth) can be used as inoculum, after being washed and macerated.

For short-term preservation, slope cultures can be preserved in a refrigerator for up to 6 months, but this method has the well-known disadvantages of the need for frequent subculturing, loss of viability of some strains, contamination, and genetic instability. Some of these problems were overcome by storage of suspensions of streptomycetes in soft agar in a refrigerator (Kutzner, 1972). Viability was maintained for at least 3 years, and small volumes of inoculum could be withdrawn as required. Another simple and efficient method for the preservation of inoculum of streptomycetes and other actinomycetes was described by Wellington and Williams (1978). Dense spore or mycelial suspensions in 10% (v/v) glycerol were stored at −20°C. These served as a convenient source of inoculum when thawed and also, in our experience, preserve the viability of most streptomycetes for at least 5 years. If glass beads are incorporated into the glycerol suspensions (Feltham et al., 1978), some of these can be removed directly when inoculum is required, thus eliminating any damage to cells caused by successive freeze-thaw cycles.

Long-term preservation of some streptomycetes has been achieved by freezing slope cultures (Tresner et al., 1960) and by preparation of cultures in sterile soil (Pridham et al., 1973). However, the most widely used methods are lyophilization and storage in liquid nitrogen. The well-established method of lyophilization is initially labor intensive, but, once prepared, the vials require little attention. Most streptomycetes are satisfactorily preserved by this method, and a convenient procedure was described by Hopwood and Ferguson (1969). However, it is acknowledged that lyophilization has the disadvantage of low survival rates, presenting the possibility of inadvertent selection of variants that are more resistant to freezing and drying than the bulk of the parent population. Consequently the most favored method for long-term storage of strains producing interesting and useful metabolites is storage in liquid nitrogen (Nisbet, 1980). Many pharmaceutical companies preserve important strains in sealed vials submerged in liquid nitrogen at −196°C. Alternatively, cultures suspended in a suitable cryoprotectant may be held in the vapor phase over liquid nitrogen, where temperatures range from −196° to −120°C. Selection of an appropriate method clearly depends on the aims of the investigator and the relevant properties of the strain, but it is advisable to preserve important cultures by at least two means.

Procedures for Testing of Special Characters

The need to define and standardize the procedures used to determine the characteristics of streptomycetes was recognized and exemplified by the ISP (Shirling and Gottlieb, 1966). Because most of the species descriptions presented here are based largely on the numerical classification and identification studies of Williams et al. (1983a, 1983b), details of the procedures used by these workers are given (Table 29.4).

Unless stated, all cultures were incubated at 25°C on media at a pH close to neutrality. Incubation times for specific tests are given. Whenever feasible, cultures were grown in multi-inoculated Replidishes (Sneath and Stevens, 1967).

1. Spore Chain Morphology

This was determined by light and scanning electron microscopy of 14-day-old cultures on inorganic salts-starch agar (Shirling and Gottlieb, 1966). However, examination by light microscopy is usually sufficient to determine this character, and other media, such as oatmeal agar and glucose-yeast extract-malt extract agar (Shirling and Gottlieb, 1966), may be used. The three categories of Pridham et al. (1958) were employed with the modified terminology of Shirling and Gottlieb (1968a) (see Figs. 29.5-29.7). If more than one spore chain category was observed in a culture, all were recorded.

2. Spore Surface Ornamentation

This was determined by scanning electron microscopy of the cultures used for the examination of spore chain morphology. Silhouettes of spores examined by transmission electron microscopy (Tresner et al., 1961) provide an alternative, but distinction between smooth, warty, and rugose surfaces is more difficult. Ornaments were assigned to the categories of Tresner et al. (1961) or to the rugose type recognized by Dietz and Mathews (1971) (see Figs. 29.8-29.12).

3. Other Morphological Features

Cultures were also examined by light and scanning electron microscopy for the fragmentation of the substrate mycelium or its production of spores. Formation of sclerotia on the colony surface was also noted.

4. Color of the Spore Mass

This was determined on the cultures used for the morphological examinations. Spore masses were matched against the seven color wheels of Tresner and Backus (1963) as used in the ISP (Shirling and Gottlieb, 1966). Although it is sometimes possible accurately to allocate color categories without using a reference system, spore pigments are very impure (Lyons and Pridham, 1965) and hence the replication between determinations by different workers is not always satisfactory (Pridham, 1965). Variations in lighting conditions obviously contribute to this problem and therefore all color determinations by Williams et al. (1983a, 1983b),

Table 29.4.
Criteria used for the classification and identification of **Streptomyces** *species*[a]

Characters	Character states
1. Spore chain morphology	*Rectiflexibiles, Retinaculiaperti,* or *Spirales*
2. Spore surface ornamentation	Smooth, warty, spiny, hairy, or rugose.
3. Other morphological features	Fragmentation of substrate mycelium, sclerotia formation, sporulation on substrate mycelium.
4. Color of spore mass	Blue, gray, green, red, violet, white, or yellow.
5. Pigmentation of substrate mycelium (colony reverse)	Yellow-brown (no distinctive pigment), blue, green, red-orange, or violet. pH sensitivity of pigments.
6. Diffusible pigments	Yellow-brown, blue, green, red-orange, or violet. pH sensitivity of pigments.
7. Melanin pigment production	On peptone-yeast extract-iron agar and tyrosine agar.
8. Antimicrobial activity	Activity against *Aspergillus niger, Bacillus subtilis, Candida albicans, Escherichia coli, Micrococcus luteus, Pseudomonas fluorescens, Saccharomyces cerevisiae,* and *Streptomyces murinus.*
9. Enzyme activity	Lecithinase, lipolysis, and proteolysis (on egg-yolk medium). Hydrolysis of chitin, hippurate, and pectin. Nitrate reduction. Hydrogen sulfide production. β-Lactamase and β-lactamase inhibitor production.
10. Degradation activity	Adenine, allantoin, arbutin, casein, DNA, elastin, esculin, gelatin, guanine, hypoxanthine, RNA, starch, testosterone, Tween 80, L-tyrosine, urea, xanthine, and xylan.
11. Resistance to antibiotics (μg/ml)	Cephaloridine (100), dimethylchlortetracycline (500) gentamicin (100), lincomycin (100), neomycin (50), oleandomycin (100), penicillin G (10 i.u.), rifampicin (50), streptomycin (100), tobramycin (50) and vancomycin (50).
12. Growth temperatures and pH	4°, 10°, 37°, and 45°C. pH 4.3.
13. Growth in the presence of inhibitory compounds (% w/v)	Crystal violet (0.0001), phenol (0.1), phenylethanol (0.1, 0.3), potassium tellurite (0.001, 0.01), sodium azide (0.01, 0.02), sodium chloride (4, 7, 10, 13), and thallous acetate (0.001, 0.01).
14. Use of nitrogen sources (0.1% w/v)	DL-α-amino-*n*-butyric acid, L-arginine, L-cysteine, L-histidine, L-hydroxyproline, L-methionine, potassium nitrate, L-phenylalanine, L-serine, L-threonine, and L-valine.
15. Use of carbon sources (1.0% w/v)	Adonitol, L-arabinose, cellobiose, dextran, D-fructose, D-galactose, *meso*-inositol, inulin, D-lactose, mannitol, D-mannose, D-melezitose, D-melibiose, raffinose, L-rhamnose, salicin, sucrose, trehalose, xylitol, and D-xylose. Sodium acetate, sodium citrate, sodium malonate, sodium propionate, and sodium pyruvate (these 5 compounds were used at 0.1% w/v)

[a]Data from Williams et al. (1983a, 1983b).

including those on the substrate mycelium and diffusible pigments, were performed under standard lighting of illuminance C, which approximates to average daylight.

5. Pigmentation of Substrate Mycelium (Colony Reverse)

This was determined on the cultures used for characters 1 to 4. Colors were allocated to one of five categories: yellow-brown, which included strains lacking distinctive pigmentation; red-orange; green; blue; or violet.

6. Diffusible Pigments

These were detected on glycerol/asparagine agar (Shirling and Gottlieb, 1966), which encourages pigment production, after 14 days' incubation. Colors were allocated to the five categories used for the substrate mycelium pigmentation.

The pH sensitivity of both the substrate mycelium and diffusible pigments was assessed by noting any color changes induced by the addition of acid or alkali (Shirling and Gottlieb, 1966).

7. Melanin Pigment Production

This was determined after 4 days' incubation on peptone-yeast extract-iron agar and tyrosine agar (Shirling and Gottlieb, 1966). A few strains give a positive reaction on only one of these media.

8. Antimicrobial Activity

The inhibition of the growth of eight test organisms (Table 29.4) was detected using an overlay technique. Spot-inoculated, 5-day-old colonies of streptomycetes on nutrient agar plates were killed by inverting the plates over 1.5 ml chloroform for 40 min. After removal of excess chloroform vapor, the plates were overlaid with 5 ml sloppy agar (0.7% w/v nutrient agar) inoculated with a test organism. Zones of inhibition around the colonies were recorded after 24 h at 30°C.

9. Enzyme Activity

Lecithinase, proteolytic, and lipolytic activities were determined on egg yolk medium (Nitsch and Kutzner, 1969). The slightly modified medium contained (% w/v): bacteriological peptone (Oxoid; 1.0), glucose (0.1), NaCl (1.0), yeast extract (Oxoid; 0.5), agar (1.2), and egg yolk emulsion (Oxoid; 5.0 v/v). Proteolysis was determined after 2 days, and lecithinase and lipolysis at 2, 4, and 6 days.

Pectinolytic activity was determined according to Hankin et al. (1971) using a modified medium made up in three parts containing (% w/v): (a) KH_2PO_4 (0.4) and Na_2HPO_4 (0.6); (b) pectin (from citrus fruit rind, Koch-Light; 0.5); and (c) $(NH_4)_2SO_4$ (0.2), yeast extract (Oxoid; 0.1), $MgSO_4 \cdot 7H_2O$ (0.2), $FeSO_4 \cdot 7H_2O$ (0.0001), $CaCl_2$ (0.0001), and agar (1.0). Hydrolysis zones were detected after 4 and 6 days by flooding plates with an aqueous solution of hexadecyltrimethylammonium bromide (1% w/v).

Chitinolytic activity was detected after 14, 21, and 28 days by the appearance of clear zones around colonies on colloidal chitin agar (Hsu and Lockwood, 1975). Only zones of >4.5 mm were scored as positive because weaker reactions were not reproducible. Hippurate agar (Ziegler and Kutzner, 1973) was used to detect hydrolysis after 6, 12, and 18 days.

Nitrate reduction was determined after 7 and 14 days by addition of 0.2 ml each of Griess-Ilosvay reagents I and II to stab cultures in a sloppy medium composed of nutrient broth (Oxoid) supplemented with KNO_3 (0.2% w/v) and agar (0.6% w/v). Hydrogen sulfide production was detected by inserting lead acetate strips into the necks of the culture tubes used for nitrate reduction determination, any blackening being noted after the same incubation periods.

β-Lactamase production was detected according to O'Callaghan et al. (1972) on 4-day-old cultures grown on a medium containing (% w/v): yeast extract (Oxoid; 0.3), peptone (Oxoid; 0.06), glycerol (1.0), and agar (1.2) and on a medium containing (% w/v): soya peptone (Lab M; 1.0), glucose (2.0), $CaCl_2$ (0.0001), $CaCO_3$ (0.02), Na_2SO_4 (0.1), and agar (1.2). The production of β-lactamase inhibitor was determined on these media by addition of approximately 10 units of *Klebsiella aerogenes* K1 β-lactamase to the agar overlay, the zones of inactivity being noted according to O'Callaghan et al. (1972).

10. Degradation Activity

The degradation of adenine, tyrosine (0.5% w/v), hypoxanthine, xanthine, xylan (0.4% w/v), elastin (0.3% w/v), casein (1% w/v skimmed milk), guanine (0.05% w/v), and testosterone (0.1% w/v) was detected in modified Bennett agar (Jones, 1949), with glucose replaced by glycerol after 7, 14, and 21 days. Clearing of the insoluble compounds around and under growth was scored as positive. Any other basal medium that is clear and supports good growth of streptomycetes may be used for degradation tests.

Gelatin (0.4% w/v) and starch (1.0% w/v) degradation were detected in the same basal medium after 7 days by flooding cultures with acidified $HgCl_2$ and iodine solutions, respectively, to reveal zones of clearance. The degradation of DNA (0.2% w/v) and RNA (0.3% w/v) was observed using DNase test agar (Difco) and tryptone agar (Goodfellow et al., 1979), respectively; cultures incubated for 7 days were flooded with 1 M HCl to visualize the clearance zones. Sierra's (1957) medium supplemented with Tween 80 (1% w/v) was examined for opacity after 3, 7, and 14 days. The degradation of allantoin and urea was recorded after 7, 14, and 21 days using the media and methods of Gordon (1966, 1967).

Esculin (0.1% w/v) and arbutin (0.1% w/v) were studied by the method of Kutzner (1976) using a modified medium containing (% w/v): yeast extract (Oxoid; 0.3), ferric ammonium citrate (0.05), and Lab M agar (0.75). Cultures were examined after 7, 14, and 21 days; blackening of the medium indicated a positive result, any slight production of melanin being indicated in the unsupplemented basal medium that served as a control.

11. Resistance to Antibiotics

Strains were tested for their ability to grow in the presence of 11 antibiotics at concentrations selected to be of diagnostic value (Table 29.4). Filter paper disks, previously soaked in antibiotic solution and then lyophilized, were placed on the surface of modified Bennett agar plates inoculated with 0.1 ml of streptomycete spore suspension. Readings were taken at 1, 2, 3, and 7 days, the first readable result being recorded; resistance was scored as positive.

12. Growth Temperatures and pH

These were tested on modified Bennett agar. Growth at 37°, and 45°C was assessed after 7 and 14 days, that at 4° and 10°C after 2 and 4 weeks. Growth at pH 4.3 was determined after 7 and 14 days.

13. Growth in the Presence of Chemical Inhibitors

A range of potential inhibitors at diagnostic concentrations (Table 29.4) was added to modified Bennett agar. Presence or absence of growth was noted after 7 and 14 days. Other media that support good growth of streptomycetes under optimal conditions may be employed for these and the preceding growth tests.

14. Use of Nitrogen Sources

The ability of strains to use 11 compounds (Table 29.4) was tested, each being incorporated into a basal medium containing (% w/v): D-glucose (1.0), $MgSO_4 \cdot 7H_2O$ (0.05), NaCl (0.05), $FeSO_4 \cdot 7H_2O$ (0.001), K_2HPO_4 (0.1), and agar (1.2). Results were determined after 15 days by comparing the growth on each source with that on the unsupplemented basal medium and on a positive control containing either L-asparagine or L-proline.

15. Use of Carbon Sources

The ability of strains to use 25 compounds (Table 29.4) was tested, each being incorporated into carbon utilization agar (Shirling and Gottlieb, 1966). Results were determined after 7, 14, and 21 days by comparing growth with that on unsupplemented basal medium and on a positive control containing D-glucose.

Differentiation of the genus **Streptomyces** *from other genera*

The genus *Streptomyces* can be separated from other actinomycete genera with a wall chemotype I using a combination of chemical and morphological properties (Table 29.5). It can also readily be distinguished from taxa containing strains that were once assigned to, or were difficult to separate from, the genus (Table 29.6). Although actinomadurae and streptomycetes have essentially similar menaquinone profiles, it may be possible to distinguish them on the basis of the position of the hydrogenated components of the major isoprenologs (Yamada et al., 1982a, 1982b).

Taxonomic Comments

Traditional Approaches to Streptomycete Systematics

Most of the studies between 1916 and 1943 were undertaken by soil microbiologists, who isolated many strains but described relatively few species. The latter were defined mainly on the basis of spore chain morphology, pigmentation, and ecological requirements (Waksman and Curtis, 1916; Waksman, 1919; Jensen, 1930a). The name *Actinomyces* was widely used during this period (Buchanan, 1918; Lieske, 1921; Ørskov, 1923; Krasil'nikov, 1938; Baldacci, 1939). In 1943, Waksman and Henrici introduced the genus *Streptomyces* for aerobic, spore-forming actinomycetes to avoid confusion with the fermentative organisms, which were retained in the genus *Actinomyces.*

The isolation of actinomycin from *S. antibioticus* (Waksman and Woodruff, 1940) stimulated extensive screening for bioactive compounds, with proof of novelty of antibiotics frequently resting on the description of the producer as a new species. This practice, together with the inability of taxonomists to provide reliable features for classification and identification of streptomycetes, led to a proliferation of species, many of which were proposed on trivial differences in morphological and cultural properties. The numbers of streptomycete species rose from about 40 prior to 1940 to over 3000, although many of the latter were merely cited in the patent literature (Trejo, 1970).

In order to cope with this species explosion many attempts were made to delimit infrageneric groups, such as series or species groups. A small number of subjectively chosen characters were used to construct these monothetic groups, the priorities given varying, as did the number of groups recognized (Table 29.7). These artificial classifications enabled unknown isolates to be "identified," but the name obtained was dependent on the scheme used and its information content.

The problems of streptomycete taxonomy were considered by two international cooperative studies carried out between 1958 and 1962. One was conducted by the Subcommittee on Actinomycetes of the Committee on Taxonomy of the American Society of Microbiology (Gottlieb, 1961) and the other by the Subcommittee on the Actinomycetales of the International Committee on Nomenclature of Bacteria of the International Association of Microbiological Societies (Küster, 1961). In each case an attempt was made to evaluate the reliability of characters commonly used in the classification and identification of streptomycetes. The problems of streptomycete systematics were attributed to the use of highly variable and poorly diagnostic tests, inconsistencies in the choice of features for species descriptions, inadequate information on test methods, reliance on tests susceptible to human error, lack of validly published species descriptions, and unavailability of type cultures.

In 1964, the ISP was initiated to furnish reliable descriptions of authentic, extant cultures of *Streptomyces* using characters selected on the basis of results from the earlier cooperative studies. Over 450 species of *Streptomyces* and *Streptoverticillium* were redescribed using a limited number of traditional criteria examined under rigorously standardized conditions. Descriptions were based on spore chain morphology, spore surface ornamentation, color of spores, substrate mycelium and soluble pigments, production of melanin pigment, and utilization of a range of carbon sources (Shirling and Gottlieb, 1968a, 1968b, 1969, 1972), and type or neotype strains were deposited in four internationally recognized culture collections. No attempt was made to ascertain synonyms between species or to group together those with shared features. Similarly, identification schemes were not provided, although ISP data were used by several workers to construct identification schemes (Arai and Mikami, 1969; Küster, 1972; Nonomura, 1974a; Szabó et al. 1975). Nevertheless, the ISP did produce a set of standardly described type cultures that have provided an invaluable reference source in many subsequent taxonomic studies. As such it represented a major development in streptomycete systematics.

The treatment of *Streptomyces* in the 8th edition of the *Manual* (Pridham and Tresner, 1974a, 1974b) relied heavily on ISP data. The 463 *Streptomyces* species described were assigned to 38 groups on the basis of their spore color, spore chain morphology, spore surface ornamentation, and melanin reaction. The distribution of strains within the defined groups was unequal. Thus, the group with gray, smooth spores in spiral chains with a negative melanin reaction (72 species) and the group with yellow, smooth spores in straight to flexuous chains with negative melanin reactions (46 species) accounted for 30% of the species described. In contrast, some groups contained only one species, such as that characterized by red, warty spores in spiral chains with a negative melanin reaction. Delimitation of species within the groups was based on carbon utilization patterns, the diagnostic value of which varied considerably between the groups. The names of 378 validly described streptomycete species were cited on the Approved Lists of Bacterial Names (Skerman et al., 1980).

Numerical Classification of Streptomycetes

Most attempts to classify streptomycetes have been based on a limited number of subjectively chosen features, with heavy emphasis on morphology and pigmentation. Such classifications are artificial, have a narrow data base, and suffer from all of the disadvantages of monothetic groups. In contrast, numerical taxonomy involves examining many strains for a large number of characters prior to assigning the test organisms to clusters based on shared features. The numerically defined taxa are polythetic, so no single property is either indispensable or sufficient to entitle an organism to group membership. Once classification has been achieved, cluster-specific or predictive characters can be selected for identification.

Numerical taxonomy was first applied to *Streptomyces* by Silvestri and his colleagues (Gilardi et al., 1960; Hill et al., 1961; Silvestri et al., 1962), who assigned nearly 200 test strains to 25 centers of variation based on 100 unit characters. Some of the numerically defined groups corresponded closely to those previously recognized using traditional features, but some strains bearing the same name occurred in different groups. Silvestri and his coworkers did not propose any nomenclatural revision, although a probabilistic diagnostic key based on biochemical and physiological features was constructed (Hill and Silvestri, 1962). Surprisingly, the results of these studies had little immediate impact on developments in streptomycete systematics, although several other workers (Woźnicka, 1965, 1967; Gyllenberg et al., 1967; Sundmann and Gyllenberg, 1967; Gyllenberg, 1970; Kurylowicz et al., 1975; Szulga, 1978) applied numerical techniques to relatively narrow data bases. Eventually a large-scale numerical phenetic survey of *Streptomyces* and related taxa was undertaken by Williams et al. (1983a) in order to clarify the infrastructure of the genus and its relationships with other wall chemotype I genera.

The 475 test strains included 394 *Streptomyces* type cultures from the ISP. Each strain was examined for 139 unit characters and the data analyzed using standard resemblance coefficients and an average linkage algorithm. The tests included those used traditionally for streptomycetes as well as newly applied ones (Table 29.4). The resulting classification supported the case for reducing *Actinopycnidium, Actinosporangium, Chainia, Elytrosporangium, Kitasatoa,* and *Microellobosporia* to synonyms of *Streptomyces,* but supported the continued recognition of *Intrasporangium, Nocardioides,* and *Streptoverticillium* as distinct genera in the family *Streptomycetaceae.* The type strains of *Streptomyces* were

Table 29.5.
Differential characteristics of the genus **Streptomyces** *and other wall chemotype I genera*[a,b]

Characteristics	*Streptomyces*	*Strepto-verticillium*	*Arachnia*	*Intra-sporangium*	*Kineosporia*	*Nocardioides*	*Sporichthya*
Morphological characteristics							
Substrate mycelium							
Fragmentation	–	–	+	–	–	+	No substrate mycelium
Chains of spores	±	–	–	–	–	–	
Sporangia containing single zoospores	–	–	–	–	+	–	
Aerial mycelium							
Aerial hyphae with chains of arthrospores	+	+	No aerial mycelium	No aerial mycelium	No aerial mycelium	Aerial mycelium fragments	+
Arthrospores in verticils	–	+					–
Spore surface smooth	+	+					+
Spore surface hairy, spiny, or warty	+	–					–
Spores motile	–	–					+
Sugars in whole-cell hydrolysate							
Arabinose, galactose, xylose[c]	–	–	–	–	+	–	–
Lipid characters							
Phospholipid type[d]	II	II	–[e]	IV	III	I	ND
Predominant menaquinone(s)[f]	MK-9(H_6) or MK-9(H_8)	MK-9(H_6) or MK-9(H_8)	ND	MK-8	MK-9(H_4)	MK-8(H_4)	MK-9(H_6) and MK-9(H_8)
Fatty acids[g]							
Saturated, straight chain	+	+	+	+	ND	+	+
Iso- and *anteiso*- branched	+	+	+	–	ND	+	+
Unsaturated	–	–	–	+	ND	+	+
10-Methyl branched	–	–	–	–	ND	+	+
Mol% G + C of DNA	69–78	69–73	70–72	71	ND	66–72	ND

[a] Symbols: +, positive; –, negative; ±, some species positive; *ND*, not determined.
[b] Data from Alderson et al. (1985), Goodfellow and Cross (1984b), Kutzner (1981), O'Donnell et al. (1984), and Pagani and Parenti (1978).
[c] Determined by thin-layer chromatography (Becker et al., 1965).
[d] Determined by thin-layer chromatography and chemical analysis (Lechevalier et al., 1977, 1981; Minnikin et al., 1977a, 1984b; O'Donnell et al., 1982b, 1984).
[e] Characteristic pattern consisting of diphosphatidylglycerol, phosphatidylglycerol, and two incompletely characterized glycolipids (O'Donnell et al., 1984).
[f] Menaquinones detected by chromatographic or physicochemical analysis (Collins et al., 1977, 1984, 1985; Alderson et al., 1985; Collins, 1985; Kroppenstedt, 1985). Abbreviations exemplified by MK-9(H_6), menaquinone having three of the nine isoprene units hydrogenated.
[g] Determined by gas liquid chromatography (Kroppenstedt and Kutzner 1976, 1978; Lechevalier et al., 1977; O'Donnell et al., 1982b, 1984; Kroppenstedt, 1985).

distributed to 23 major clusters, 20 minor clusters, and 25 single member clusters. The minor and single-member clusters were considered as species and the major clusters were regarded as species groups.

It is instructive to compare the numerical classification with the approach adopted by Pridham and Tresner (1974b) in the 8th edition of the *Manual.* The species groups recognized by these earlier workers were based on four subjectively weighted characters, whereas in the numerical phenetic survey, taxa were defined using 139 unit characters. A comparison between the two studies can be made by examining the distribution of the four "traditional" morphological and pigmentation properties within the major clusters (Table 29.8). It is evident that some clusters such as *S. albus, S. griseoruber,* and *S. violaceusniger* are reasonably consistent in their traditional character states. Others, including *S. chromofuscus, S. cyaneus,* and *S. violaceus,* show considerable variation in these characters. However, it is hardly surprising that polythetic groups defined using 139 unit characters do not always show concordance with the four subjectively chosen characters. A detailed discussion of the relationships of the major clusters to previous classifications was given by Williams et al. (1983a).

Information from the numerical taxonomic data base was used to construct a probability matrix for identification of streptomycetes (Williams et al., 1983b). The matrix, based on 41 traditional and new criteria, was thoroughly tested but did have some imperfections. Thus, it was limited to the major clusters, a few of which were somewhat ill defined. Further, the minimum number of tests needed to distinguish between the clusters was quite large, reflecting the variation within clusters and the necessity of having at least as many tests as taxa in a matrix (Sneath and Chater, 1978). Despite these limitations, the probability matrix has been successfully used for the identification of unknown streptomycetes from a wide range of habitats (Goodfellow and Haynes, 1984; Williams et al., 1984b, 1985b). Identification of strains against the matrix was achieved using three coefficients from the MATIDEN program (Sneath, 1979a), namely the Willcox probability (Willcox et al., 1973), taxonomic distance, and standard error of taxonomic distance. The computer-assisted identification procedure provides a workable system for the identification of streptomycetes, but, unlike previous systems proposed for the recognition of species or species groups (Waksman, 1961; Hütter, 1967; Pridham and Tresner, 1974b; Pridham, 1976a), it is based not on a few, readily determined characters but on a balanced set of features. While the earlier systems have the attractions of speed and simplicity, they can-

Table 29.6.
Differential characteristics of the genus **Streptomyces** *and taxa previously associated with the genus*[a,b]

Characteristics	*Streptomyces*	*Actinomadura*	*Nocardia* Sensu stricto[c]	*Nocardia* Mycolateless species[c]	*Nocardiopsis*
Morphological characters					
Fragmentation of substrate mycelium	−	−	+	±	+
Chains of arthrospores on aerial mycelium	+	+	±	+	+
Wall chemotype[d]	I	III	IV	IV	III
Whole-cell sugar pattern[d]	NC	B	A	A	C
Muramic acid acyl type[e]	Acetyl	Acetyl	Glycolyl	ND	ND
Lipid characteristics					
Fatty acids[f]					
Saturated straight chain	+	+	+	+	+
Iso- and *anteiso*- branched	+	+	−	+	+
Unsaturated	−	+	+	−	+
10-Methyl branched	−	+	+	±	+
Mycolic acids[g]	−	−	+	−	−
Phospholipid type[f]	II	I	II	II or III	III
Predominant menaquinone(s)[f]	MK-9 (H_4,H_6,H_8)	MK-9 (H_4,H_6,H_8)	MK-8(H_4) or MK-9(H_2)	MK-8(H_4) or MK-9(H_4)	MK-10 (H_2,H_4,H_6)
Mol% G + C of DNA	69–78	65–77	64–72	ND	65.7

[a]Symbols: +, present; −, absent; ±, sometimes present; *NC*, no characteristic sugars; *ND*, not determined.
[b]Data from Goodfellow and Cross (1984b), Goodfellow and Minnikin (1984), Kutzner (1981), Lechevalier and Lechevalier (1980), Lechevalier et al. (1981), Minnikin and Goodfellow (1981) and Uchida and Aida (1977, 1979).
[c]See Goodfellow and Lechevalier this volume for explanation. Mycolateless nocardiae include *Nocardia autotrophica, Nocardia mediterranei,* and *Nocardia orientalis.* See also the genera *Amycolata* and *Amycolatopsis* (Lechevalier et al., 1986).
[d]See Lechevalier and Lechevalier (1980) for details of methods and explanations of wall chemotypes and sugar patterns.
[e]See Uchida and Aida (1977, 1979) for method.
[f]References to methods given in footnotes *d* and *f* to Table 29.5.
[g]Data from analysis of one strain (S-C. An, unpublished data).

not accommodate variation and they provide identifications with little information content.

It has been argued here, and in more detailed reviews (Williams et al., 1984b, 1985b), that the application of both well-established and recently devised procedures of numerical taxonomy provides a more objective means of classifying and identifying *Streptomyces* species. Nevertheless these schemes, although adopted in this chapter, merely provide a framework on which further improvements in streptomycete systematics can be built. Their theoretical bases remain to be further tested and improved, and they will not satisfy all of the diverse practical needs of those studying this genus.

Streptomycete taxonomy should be improved by the following developments in classification, nomenclature, and identification:

1. Improved statistical methods for the significance and integrity of groups defined in numerical taxonomy and their application to all validly published species.
2. Assessment of the correlation between numerical phenetic and molecular genetic data using techniques such as DNA/DNA and DNA/rRNA homology.
3. Measurement of genetic exchange potentials between phenetically defined species.
4. Application of more refined techniques of chemotaxonomy in classification and identification; in particular, exploitation of chemical markers at the subgeneric level by development of extraction and analysis procedures and use of improved methods for the analysis of quantitative data.
5. Availability of recommended minimal standards for defining members of the genus *Streptomyces.*
6. Increased observance of and adherence to the rules and procedures of nomenclature by those describing new species.

Table 29.7.
Criteria used for early classifications of **Streptomyces** *species*

Criteria used and their priorities: Spore chains	Spore color	Substrate color	Spore surface	Melanin pigment	No. of groups	Authors
2	1				7	Hesseltine et al. (1954)
	1	2			15	Gauze et al. (1957)
3	2		1	4	34	Ettlinger et al. (1958a)
1	2				42	Pridham et al. (1958)
3	2	1			33	Baldacci (1958)
	1	2			10	Flaig and Kutzner (1960)
	2	3		1	16	Waksman (1961)
3	2		1	4	41	Hütter (1967)

Table 29.8.
Distribution of the characteristics used to delimit species groups in Bergey's Manual® of Determinative Bacteriology (Pridham and Tresner, 1974b) within some major clusters defined by numerical taxonomy (Williams et al., 1983a)

Cluster name	Predominant features			
	Spore chain morphology	Spore surface ornamentation	Spore color	Melanin pigment
Streptomyces albidoflavus	*Rectiflexibiles*	Smooth	Yellow-gray	−
S. albus	*Spirales*	Smooth	White	−
S. chromofuscus	*Spirales/Rectiflexibiles*	Smooth/spiny/hairy	Gray-yellow-white	+/−
S. cyaneus	*Spirales*	Spiny/smooth	Blue-gray-red	+
S. diastaticus	*Spirales/Rectiflexibiles/Retinaculiaperti*	Smooth	Gray-white-red	+/−
S. exfoliatus	*Rectiflexibiles*	Smooth	Red-gray	+/−
S. griseoruber	*Spirales*	Smooth	Gray	−
S. griseoviridis	*Spirales*	Smooth	Red	−
S. lavendulae	*Rectiflexibiles*	Smooth	Red	+
S. violaceusniger	*Spirales*	Rugose/smooth	Gray	−
S. violaceus	*Spirales/Rectiflexibiles/Retinaculiaperti*	Smooth/spiny	Violet-yellow	+

Acknowledgments

The theoretical bases of our numerical studies were provided by Prof. P. H. A. Sneath and valuable practical help was provided by Dr. E. M. H. Wellington and Dr. J. C. Vickers. Our task was considerably assisted by the extensive nomenclatural studies of Dr. T. G. Pridham. Research grants from the Science and Engineering Research Council supported our studies of streptomycete taxonomy.

Further Reading

Chater, K. F. and D. A. Hopwood, 1984. *Streptomyces* genetics. *In* Goodfellow, Mordarski and Williams (Editors), The Biology of the Actinomycetes. Academic Press, London, pp. 229–286.

Goodfellow, M. and T. Cross. 1984. Classification. *In* Goodfellow, Mordarski and Williams (Editors). The Biology of the Actinomycetes. Academic Press, London, pp. 7–164.

Kutzner, H. J. 1981. The family *Streptomycetaceae. In* Starr, Stolp, Trüper, Balows and Schlegel (Editors), The Prokaryotes. A Handbook on Habitats, Isolation and Identification of Bacteria. Springer-Verlag, Berlin, pp. 2029–2090.

Langham, C.P., S.T. Williams, P.H.A. Sneath, and A.M. Mortimer. 1988. New probability matrices for identification of *Streptomyces*. J. Gen. Microbiol. *135:* 121–133.

Williams, S. T., M. Goodfellow, G. Alderson, E.M.H. Wellington, P.H.A. Sneath and M. J. Sackin. 1983a. Numerical classification of *Streptomyces* and related genera. J. Gen. Microbiol. 129: 1743–1813.

Williams, S. T., M. Goodfellow, E.M.H. Wellington, J.C. Vickers, G. Alderson, P.H.A. Sneath, M.J. Sackin and A.M. Mortimer. 1983b. A probability matrix for identification of some streptomycetes. J. Gen. Microbiol. *129:* 1815–1830.

Differentiation and characteristics of the species of the genus **Streptomyces**

As discussed above, many attempts to delimit *Streptomyces* have been made. The approach taken here, while still not definitive, aims to present practical guidelines for the identification of streptomycete species based largely on recent numerical taxonomic studies. In order to simplify nomenclatural problems, emphasis is placed on the species on the Approved Lists of Bacterial Names (Skerman et al., 1980) or those subsequently validly published or validated; subspecies are not included. While such a rigorous selection may be inappropriate for many other genera, it is necessary here because no practical purpose would be served by listing the hundreds of poorly characterized *Streptomyces* "species." Also, it was not possible to include species validated since the completion of this chapter, such as the 45 species included by Russian workers in List No. 22 (Int. J. Syst. Bacteriol. *36:* 573–576, 1986).

Those included have been categorized on the basis of the currently available taxonomic information as follows:

I. Species defined by the major clusters (containing four or more strains) in the numerical classification of Williams et al. (1983a).
II. Species defined by the minor clusters (containing two to three strains) of Williams et al. (1983a).
III. Species defined by the single-member clusters of Williams et al. (1983a).
IV. Species on the Approved Lists of Bacterial Names and those since validly published or validated that were *not* studied by Williams et al. (1983a).

The numerical classification study of Williams et al. (1983a) involved determination of 139 unit characters for each strain; overall similarity was determined by the simple matching coefficient (S_{SM}; Sokal and Michener, 1958) and the Jaccard coefficient (S_J; Sneath, 1957) and clustering was achieved using the unweighted pair group method with arithmetic averages (the UPGMA algorithm of Sneath and Sokal, 1973). Clusters were defined at the 77.5% or 81% S_{SM} and 63% S_J similarity levels, the former coefficient being used to define the clusters. Groupings produced by the two coefficients were generally similar and any significant variations are considered in the species descriptions given below. As a result, 23 major, 20 minor, and 25 single-member clusters were defined, containing 219, 46, and 25 validly described species, respectively. Clusters are named after the earliest validly described species that they contain, following the principle of priority; in most cases the other species within a cluster are listed as *subjective synonyms* of this name. Species that are regarded as not being entirely typical of their cluster, despite their high degree of similarity with other members, are listed as *allied species;* further study of their overall phenetic relationships is required.

The numerical classification study (Williams et al., 1983a) also provided data from which the characters most useful for the identification of the defined species could be selected. The minimum number of characters required to distinguish between the species defined by the major clusters was determined by two computer programs. The CHARSEP program (Sneath, 1979b) includes several separation indices derived from percentage positive character data, and the DIACHAR program (Sneath, 1980b) determines the most diagnostic properties of groups in an identification matrix. Application of these programs determined the minimum 50 characters needed to distinguish between the species defined by the major clusters of Williams et al. (1983a) (Table 29.9). The percentage positive values given are derived not only from species on the Approved Lists of Bacterial Names (Skerman et al., 1980) but also from closely related species excluded from the Lists that were studied by Williams et al. (1983a).

This is done to expand the data base of Table 29.9, which not only provides descriptive information but also constitutes a *probabilistic identification matrix* that can be entered into a computer and accessed by the MATIDEN program (Sneath, 1979a) to identify unknown strains (Williams et al., 1983b). It has been shown that about 80% of isolates from various habitats were identified by this system (Williams et al., 1985b). Despite the large number of character determinations required, this provides a workable, informative means for identifying isolates of this large and diverse genus.

The diagnostic tables for species defined by the minor (Table 29.10) and single-member (Table 29.11) clusters were also derived by the CHARSEP and DIACHAR programs, but the number of strains within the species was insufficient for probabilistic treatment. The DIACHAR program also provided a means for determining which of the 139 unit characters were most typical of each species; details of these form the basis of the test descriptions. Although it is appropriate to present only the most diagnostic and typical characters here, details of all 139 unit characters used by Williams et al. (1983a) for all strains in categories I, II, and III are available on request to the senior author.

Species in category IV, although validly published or validated, have not been subjected to the wide range of tests applied to those in the previous categories. If they were, no doubt some of them would be close to or synonymous with species defined by numerical taxonomy. However, pending further information on these species, they are included with a brief description and review of their commonly determined characters.

Details of antibiotics produced by species included in categories I, II, and III were given in the last edition of the *Manual* and therefore are not repeated here.

The mol% G + C of the DNA of streptomycetes ranges from 69 to 78% (Goodfellow and Cross, 1984b), but relatively few strains have been examined and this property appears to have little diagnostic value within the genus. Therefore details for individual species are not given.

List of species of the genus **Streptomyces**

Category I

Characteristics useful for the differentiation of species are given in Table 29.9, which can also serve as a probabilistic identification matrix (see above). Features in the text descriptions are those most typical of the species, including the morphological and pigmentation characteristics that have formed the basis of most previous classifications.

List of the Species in Category I

1. **Streptomyces albidoflavus** (Rossi-Doria 1891) Waksman and Henrici 1948b, 949.^{AL} (*Streptothrix albido flava* Rossi-Doria 1891, 407.)

al.bi.do.fla′vus. L. adj. *albidus* white; L. adj. *flavus* yellow; M.L. adj. *albidoflavus* whitish yellow.

Spore chains are *Rectiflexibiles* or *Retinaculiaperti;* the spore surface is smooth. The spore mass is yellow, sometimes white or sparse; the reverse is yellow-brown; diffusible pigments are absent or yellow-brown. Melanin pigment is rarely produced.

Most strains show antimicrobial activity against *Candida albicans* and grow in the presence of thallous acetate (0.001% w/v). Raffinose is rarely used.

Type strain: CBS 416.34.

Further comments: This species was clearly defined at the 81% S_{SM} similarity level but is related to *Streptomyces anulatus* and *S. halstedii,* with which it joined at the 77.5% similarity level. It also showed affinity with *S. aburaviensis* at the 63% S_J similarity level.

This species, together with *S. anulatus* and *S. halstedii,* are approximately equivalent to the "*griseus*" groups recognized by Waksman (1959), Hütter (1963), Lyons and Pridham (1966), and other workers. All the species listed below were considered to be synonyms of *S. coelicolor* by Korn et al. (1978), and showed high DNA homologies with this species as a reference strain (Mordarski et al., 1986).

Subjective synonyms:

Streptomyces canescens Waksman 1957a, 768.^{AL} ATCC 19736.

Streptomyces coelicolor (Müller 1908) Waksman and Henrici 1948b, 935.^{AL} CBS 210.27.

Streptomyces felleus Lindenbein 1952, 374.^{AL} ATCC 19752.

Streptomyces gougerotii (Duché 1934) Waksman and Henrici 1948b 947.^{AL} CBS 422.34.

Streptomyces intermedius (Krüger 1904) Waksman in Waksman and Lechevalier 1953, 116.^{AL} CBS 101.21 (ATCC 3329).

Streptomyces limosus Lindenbein 1952, 379.^{AL} ATCC 19778.

Streptomyces odorifer (Rullmann 1895) Waksman in Waksman and Lechevalier 1953, 79.^{AL} ATCC 6246.

Streptomyces rutgersensis (Waksman and Curtis 1916) Waksman and Henrici 1948b, 952.^{AL} ATCC 3350.

Streptomyces sampsonii (Millard and Burr 1926) Waksman in Waksman and Lechevalier 1953, 155.^{AL} ATCC 25495.

2. **Streptomyces anulatus** (Beijerinck 1912) Waksman 1957a, 755.^{AL} (*Actinomyces (Streptothrix) annulatus* (sic) Beijerinck 1912, 7.)

a.nu.la′tus. L. adj. *anulatus* furnished with a ring.

Spore chains are *Rectiflexibles* or occasionally *Spirales;* the spore surface is smooth. The spore mass is yellow, sometimes white; the reverse is yellow-brown; diffusible pigments are absent or yellow-brown. Melanin pigment is generally not produced, but a few strains are positive, particularly on tyrosine agar.

Most strains grow at 4°C and in the presence of thallous acetate (0.001% w/v).

Type strain: CBS 100.18.

Further comments: This species was defined at the 81% S_{SM} similarity level and its relationship to *S. albidoflavus* and *S. halstedii* has been mentioned. Many of the species listed here as synonyms were placed in the *S. griseus* group of Korn et al. (1978) and were regarded as synonyms of *S. griseus* by Hütter (1967). Also many are lysed by phage specific for *S. griseus* (Korn et al., 1978) and form a distinct DNA homology group with an *S. griseus* reference system (Okanishi et al., 1972). High DNA homologies were also obtained using both *S. griseus* and *S. alboviridis* reference DNA against all the species considered here (Mordarski et al., 1986).

Subjective synonyms:

Streptomyces alboniger Porter, Hewitt, Hesseltine, Krupka, Lowery, Wallace, Bohonos and Williams 1952, 409.^{AL} ATCC 12461.

Streptomyces albovinaceus (Kudrina in Gauze et al. 1957) Pridham, Hesseltine and Benedict 1958, 57.^{AL} ATCC 15823.

Streptomyces alboviridis (Duché 1934) Pridham, Hesseltine and Benedict 1958, 74.^{AL} ATCC 25425.

Streptomyces baarnensis Pridham, Hesseltine and Benedict 1958, 74.^{AL} CBS 306.55.

Streptomyces bacillaris (Krasil'nikov 1958) Pridham 1970, 9.^{AL} ATCC 15855.

Streptomyces cavourensis Skarbek and Brady 1978, 52.^{AL} ATCC 14889.

Streptomyces chrysomallus Lindenbein 1952, 369.^{AL} ATCC 11523.

Streptomyces citreofluorescens (Korenyako, Krasil'nikov, Nikitina and Sokolova in Rautenshtein 1960) Pridham 1970, 10.^{AL} ATCC 15858.

Table 29.9.
A percentage positive probability matrix for streptomycete species defined by the major clusters of Williams et al. (1983b)

Characteristics	1. *S. albidoflavus*	2. *S. anulatus*	3. *S. halstedii*	4. *S. exfoliatus*	5. *S. violaceus*	6. *S. fulvissimus*	7. *S. rochei*	8. *S. chromofuscus*	9. *S. albus*	10. *S. griseoviridis*	11. *S. cyaneus*	12. *S. diastaticus*	13. *S. olivaceoviridis*	14. *S. griseoruber*	15. *S. lydicus*	16. *S. violaceusniger*	17. *S. griseoflavus*	18. *S. phaeochromogenes*	19. *S. rimosus*	20. *S. microflavus*	21. *S. antibioticus*	22. *S. lavendulae*	23. *S. purpureus*
Spore chain *Rectiflexibiles*	70	90	77	99	38	67	4	22	1	1	5	42	1	1	1	1	1	1	14	20	20	92	99
Spore chains *Spirales*	1	38	23	1	50	22	64	78	99	99	82	58	86	99	99	99	1	67	57	40	60	17	1
Spore mass red	5	5	1	39	13	66	7	1	1	99	31	16	14	12	9	1	17	1	1	1	1	83	25
Spore mass gray	1	3	99	38	13	11	77	34	1	1	31	48	71	78	91	99	1	1	1	80	99	8	75
Mycelial pigment red-orange	1	1	1	11	1	89	8	1	1	1	21	16	1	67	1	1	1	1	1	20	1	1	1
Diffusible pigment produced	20	11	8	39	1	33	4	11	1	99	33	10	29	99	1	1	1	50	1	20	1	1	99
Diffusible pigment yellow-brown	15	11	1	28	1	1	4	11	1	99	13	5	29	1	1	1	1	33	1	1	1	1	99
Melanin on peptone–yeast–iron agar	5	13	1	61	88	67	4	33	1	1	97	47	14	22	1	1	17	17	1	80	80	99	75
Melanin on tyrosine agar	10	24	1	61	99	1	1	22	1	17	85	47	14	22	1	1	17	33	1	80	60	92	1
Antibiosis against																							
Bacillus subtilis NCIB 3610	5	50	8	56	50	78	35	11	17	99	44	21	1	1	73	67	17	17	99	80	40	92	50
Micrococcus luteus NCIB 196	5	71	8	44	38	89	35	11	17	99	33	21	1	1	99	99	17	17	86	99	20	83	50
Candida albicans CBS 562	85	21	8	1	1	11	19	1	1	33	3	1	1	1	27	17	1	1	57	1	99	67	1
Saccharomyces cerevisiae CBS 1171	80	13	8	1	1	22	15	1	1	33	5	1	29	1	27	50	1	17	86	1	99	67	1
Streptomyces murinus ISP 5091	10	61	15	39	88	99	39	22	17	80	62	5	1	22	99	50	67	1	99	99	80	99	75
Aspergillus niger LIV 131	55	32	1	6	25	22	27	1	17	33	10	11	1	1	99	33	1	1	99	1	80	75	50
Lecithinase activity	15	3	15	50	75	44	4	11	1	50	10	1	1	1	64	1	1	1	86	60	40	99	75
Lipolysis	99	99	99	94	63	99	73	89	99	83	49	26	86	89	18	99	99	83	99	40	1	33	25
Pectin hydrolysis	5	53	77	61	13	1	42	22	1	67	56	68	14	99	1	50	99	50	14	99	40	8	1
Nitrate reduction	20	79	23	83	99	1	27	22	1	33	36	47	14	89	9	83	83	17	86	1	80	50	99
H_2S production	85	90	92	89	63	67	92	89	83	99	90	79	99	99	1	99	99	83	14	99	1	42	75
Hippurate hydrolysis	1	13	77	44	88	44	9	13	1	67	3	28	50	56	30	67	67	83	71	40	40	1	1
Elastin degradation	90	87	85	89	88	67	50	67	50	50	41	37	57	44	36	83	17	99	99	60	1	92	99
Xanthine degradation	95	95	99	94	88	99	96	22	83	83	80	53	29	78	82	1	1	99	86	60	80	83	99
Arbutin degradation	99	99	99	99	99	67	96	99	99	99	54	53	1	99	99	99	50	67	71	99	99	92	75
Resistance to																							
Neomycin (50 μg/ml)	1	1	1	11	25	89	8	1	1	1	1	1	1	1	18	1	1	1	99	1	1	50	75
Rifampicin (50 μg/ml)	40	66	31	11	99	99	89	33	99	83	46	68	86	78	9	83	50	50	99	99	20	33	99
Oleandomycin (100 μg/ml)	70	84	99	44	13	99	46	1	83	17	13	26	1	1	9	50	83	33	71	1	1	8	75
Penicillin G (10 i.u.)	90	74	92	44	99	99	92	44	99	99	64	47	14	78	91	17	83	33	99	20	40	58	25
Growth at 45°C	1	5	23	17	1	22	77	67	99	67	41	16	86	1	1	50	1	67	43	60	80	17	1
Growth with (% w/v)																							
NaCl (7.0)	85	74	92	22	38	22	92	44	99	83	18	32	29	78	55	1	33	83	99	1	20	1	25
Sodium azide (0.01)	75	32	85	23	63	1	62	56	99	99	15	5	57	11	18	50	33	67	71	60	60	1	1
Phenol (0.1)	90	92	85	72	99	44	96	22	17	83	64	95	86	89	9	1	99	99	71	99	80	58	1
Potassium tellurite (0.001)	85	87	99	83	99	56	73	67	50	99	46	74	57	78	55	17	99	99	71	80	60	42	1
Thallous acetate (0.001)	90	87	92	67	63	22	54	54	1	50	13	21	14	33	46	1	50	33	14	1	40	17	1
Utilization of																							
DL-α-Amino-*n*-butyric acid	65	37	54	61	88	89	12	67	1	1	31	32	57	33	9	99	17	67	1	60	20	42	1
L-Cysteine	60	61	69	50	38	44	50	67	1	83	72	79	99	78	46	33	17	17	29	60	80	33	1
L-Valine	35	37	62	50	50	99	15	33	17	17	69	74	86	56	27	33	1	99	57	40	60	17	1
L-Phenylalanine	70	61	77	83	99	89	46	11	17	33	67	16	71	33	99	67	17	83	86	99	40	42	1
L-Histidine	40	74	69	78	25	99	77	78	99	83	85	68	99	99	36	99	17	83	99	40	99	8	1
L-Hydroxyproline	1	37	23	89	88	78	8	1	67	17	28	21	1	22	55	83	17	67	29	20	40	42	1
Sucrose	45	26	23	28	38	34	81	33	33	17	92	74	86	44	73	33	83	83	1	99	80	50	25
meso-Inositol	45	32	23	6	63	99	96	89	33	67	92	84	57	99	91	67	83	99	99	40	80	25	75
Mannitol	90	99	69	1	38	99	99	99	99	99	95	90	99	99	91	99	99	99	99	99	80	8	25
L-Rhamnose	20	82	69	61	38	22	96	67	17	83	92	95	99	99	18	83	83	99	1	80	60	17	1
Raffinose	5	18	31	33	63	89	69	22	33	50	99	84	99	99	82	83	33	99	86	99	60	8	1
D-Melezitose	55	71	46	22	50	44	81	22	67	33	72	26	29	56	82	83	67	99	57	40	80	33	99
Adonitol	50	66	8	1	13	89	35	22	99	67	82	16	14	22	82	67	1	99	99	1	20	8	1
D-Melibiose	25	32	77	28	75	89	96	44	50	33	98	95	99	78	82	83	67	83	99	40	99	17	25
Dextran	20	76	69	6	13	22	89	78	1	17	59	16	14	44	1	17	50	50	14	1	1	1	1
Xylitol	5	21	1	1	1	11	46	1	1	1	21	16	29	1	55	17	1	99	86	40	1	1	25

Table 29.10.
Characteristics useful for the differentiation of minor clusters defined by Williams et al. (1983a)[a]

Characteristics	1. *S. aburaviensis*	2. *S. californicus*	3. *S. cellulosae*	4. *S. aureofaciens*	5. *S. flaveolus*	6. *S. filipinensis*	7. *S. noboritoensis*	8. *S. chattanoogensis*	9. *S. thermovulgaris*	10. *S. longisporoflavus*	11. *S. griseoluteus*	12. *S. pactum*	13. *S. aurantiacus*	14. *S. luridus*	15. *S. xanthochromogenes*	16. *S. misakiensis*	17. *S. psammoticus*	18. *S. fradiae*	19. *S. poonensis*	20. *S. atroolivaceus*
Spore chains *Rectiflexibiles*	+	+	−	+	−	−	+	−	d	−	−	−	−	d	+	+	d	−	−	+
Spore chains *Spirales*	−	−	d	−	+	+	−	+	−	+	−	+	+	d	−	−	d	d	+	−
Spore mass gray	d	d	d	d	d	d	d	−	−	−	+	+	d	d	−	d	−	−	−	d
Diffusible pigment produced	−	+	+	−	d	−	+	d	−	−	−	+	+	−	−	d	d	−	d	−
Antibiosis against																				
Bacillus subtilis NCIB 3610	−	d	d	−	−	d	−	−	−	−	−	d	−	+	d	+	−	+	−	+
Micrococcus luteus NCIB 196	−	d	d	+	−	+	−	−	−	−	d	d	−	+	+	+	d	+	−	+
Streptomyces murinus ISP 5091	−	d	d	d	d	+	−	−	−	−	−	+	−	+	d	d	−	d	−	+
Elastin degradation	d	+	−	+	+	+	d	d	d	+	−	d	−	d	+	+	+	d	d	+
Xanthine degradation	+	+	+	+	+	+	d	−	d	−	d	+	−	+	+	d	d	−	d	+
Resistance to rifampicin (50 μg/ml)	d	+	+	d	d	+	d	−	−	+	d	+	d	d	d	−	−	−	−	+
Resistance to penicillin G (10 i.u.)	+	d	+	d	d	d	d	d	−	d	d	d	d	d	d	−	−	d	+	+
Growth at 45°C	−	−	+	d	+	d	−	−	+	−	d	+	d	d	−	−	−	+	+	−
Growth with (% w/v)																				
NaCl (7.0)	+	+	d	−	−	−	−	d	−	d	d	d	d	−	−	−	−	−	−	d
Sodium azide (0.01)	+	+	+	+	d	+	−	+	+	d	+	d	−	−	d	−	−	−	−	−
Phenol (0.1)	d	+	+	+	d	+	d	+	+	−	+	d	d	d	−	−	−	−	+	d
Potassium tellurite (0.001)	+	+	+	+	−	−	−	+	+	d	+	+	d	−	−	d	+	+	+	d
Utilization of																				
L-Phenylalanine	d	+	−	−	+	−	+	+	−	d	d	d	−	+	d	d	−	d	−	+
L-Histidine	+	+	d	−	d	+	−	+	−	−	+	+	d	d	d	+	d	−	+	d
Sucrose	+	−	d	+	+	+	−	+	d	+	−	−	−	d	d	d	d	−	−	−
meso-Inositol	+	d	+	+	+	d	+	+	d	+	d	d	d	d	−	−	−	−	+	−
Mannitol	+	d	+	+	+	d	+	+	+	+	+	d	−	−	d	−	−	−	+	+
L-Rhamnose	d	d	d	+	+	−	−	−	−	+	−	−	d	−	−	−	−	+	d	d
Raffinose	d	−	−	d	d	+	+	d	+	+	−	−	d	−	d	+	−	d	−	−
D-Melibiose	d	d	+	+	+	d	+	+	+	d	d	−	−	−	d	−	−	d	+	d

[a]Symbols: see Table 29.1; *d*, 11–89% of strains are positive.

Streptomyces cyaneofuscatus (Kudrina in Gauze et al. 1957) Pridham, Hesseltine and Benedict 1958, 58.[AL] ATCC 23619.

Streptomyces fimicarius (Duché 1934) Waksman and Henrici 1948b, 940.[AL] CBS 420.34.

Streptomyces fluorescens (Krasil'nikov 1958) Pridham 1970, 15.[AL] ATCC 15860.

Streptomyces globisporus (Krasil'nikov 1941) Waksman in Waksman and Lechevalier 1953, 39.[AL] ATCC 15864.

Streptomyces griseinus Waksman 1959, 1045.[AL] ATCC 23915.

Streptomyces griseobrunneus Waksman 1961, 220.[AL] ATCC 19762.

Streptomyces griseus (Krainsky 1914) Waksman and Henrici 1948b, 948.[AL] ATCC 23345.

Streptomyces lipmanii (Waksman and Curtis 1916) Waksman and Henrici 1948b, 952.[AL] ATCC 3331.

Streptomyces parvus (Krainsky 1914) Waksman and Henrici 1948b, 939.[AL] NRRL B-1455.

Streptomyces pluricolorescens Okami and Umezawa in Waksman 1961, 259.[AL] ATCC 19798

Streptomyces setonii (Millard and Burr 1926) Waksman in Waksman and Lechevalier 1953, 107.[AL] ATCC 25497.

Streptomyces spheroides Wallick, Harris, Reagan, Ruger and Woodruff 1956, 911.[AL] NRRL 2449.

Streptomyces willmorei (Erikson 1935) Waksman and Henrici 1948b, 966.[AL] ATCC 6867.

Allied species:

Streptomyces cremeus (Kudrina in Gauze et al. 1957) Pridham, Hesseltine and Benedict 1958, 66.[AL] ATCC 19897.

This species also shows affinity with *S. rochei* when matched against probabilistic identification matrix, and with *S. violaceus* at the 63% S_J level.

Streptomyces niveus Smith, Dietz, Sokolski and Savage 1956, 135.[AL] NRRL 2466.

This species also shows affinity with *S. rochei* when matched against the probabilistic identification matrix.

Streptomyces sindenensis Nakazawa and Fujii 1957, 109.[AL] ATCC 23963.

This species also has many properties in common with *S. atroolivaceus.*

Table 29.11.
Characteristics useful for the differentiation of single member clusters defined by Williams et al. (1983a)[a]

Characteristics	1. *S. prunicolor*	2. *S. canus*	3. *S. graminofaciens*	4. *S. viridochromogenes*	5. *S. glaucescens*	6. *S. nogalater*	7. *S. prasinosporus*	8. *S. ochraceiscleroticus*	9. *S. aurantiogriseus*	10. *S. bambergiensis*	11. *S. gelaticus*	12. *S. amakusaensis*	13. *S. varsoviensis*	14. *S. tubercidicus*	15. *S. badius*	16. *S. ramulosus*	17. *S. sulphureus*	18. *S. yerevanensis*	19. *S. massasporeus*	20. *S. alboflavus*	21. *S. bikiniensis*	22. *S. fragilis*	23. *S. lateritius*	24. *S. finlayi*	25. *S. novaecaesareae*
Spore chains *Rectiflexibiles*	+	−	−	−	−	−	−	−	−	−	+	−	−	−	+	−	−	−	−	−	+	−	−	+	−
Spore chains *Spirales*	−	+	+	+	+	−	+	+	−	−	−	+	+	+	−	−	+	+	+	−	−	−	+	−	−
Spore mass gray	−	+	+	−	−	+	−	−	+	−	−	−	−	−	−	+	−	−	−	−	+	−	−	+	−
Diffusible pigment produced	+	−	−	+	+	+	−	−	+	+	−	−	−	−	+	+	−	+	+	−	−	−	+	−	−
Antibiosis against:																									
Bacillus subtilis NCIB 3610	−	−	+	−	+	−	−	−	−	−	−	+	−	−	−	+	−	−	−	+	−	−	+	−	−
Micrococcus luteus NCIB 196	−	−	+	−	+	−	−	−	−	−	−	+	−	−	+	+	−	−	−	+	+	+	+	−	−
Streptomyces murinus 1SP 5091	−	−	+	−	+	+	−	−	−	+	−	+	+	−	+	+	−	−	+	+	−	+	+	−	−
Elastin degradation	−	+	+	+	−	+	−	+	−	+	+	+	−	−	+	+	+	−	−	+	+	+	+	−	−
Xanthine degradation	+	+	+	+	+	+	−	+	−	−	+	−	−	+	+	+	+	−	+	+	−	−	+	−	−
Resistance to rifampicin (50 µg/ml)	+	+	+	+	+	+	−	+	−	−	+	−	+	−	+	+	+	+	+	+	−	−	−	+	+
Resistance to penicillin G (10 i.u.)	−	+	−	−	−	−	−	+	−	−	+	−	+	+	+	+	+	+	−	+	+	−	+	−	−
Growth at 45°C	−	+	−	+	+	+	+	−	+	−	−	−	−	−	−	−	−	−	+	−	+	+	+	−	−
Growth with (% w/v)																									
NaCl (7.0)	−	+	+	−	−	−	−	+	−	−	−	−	+	+	+	+	+	−	−	−	−	−	+	−	−
Sodium azide (0.01)	−	−	−	+	−	−	−	+	−	−	−	−	−	−	+	+	+	−	−	−	−	+	−	−	−
Phenol (0.1)	−	+	+	+	−	−	+	+	−	+	+	−	−	−	+	−	+	+	+	+	−	+	−	+	−
Potassium tellurite (0.001)	−	−	+	+	+	−	−	+	+	+	+	+	+	+	+	−	+	+	−	−	−	+	−	+	+
Utilization of																									
L-Phenylalanine	+	−	+	−	+	+	−	+	−	−	−	+	−	+	−	+	−	−	−	+	+	+	−	+	−
L-Histidine	−	−	−	+	+	+	−	+	+	+	+	−	+	+	+	−	−	+	−	−	+	−	−	+	−
Sucrose	+	+	+	−	+	−	−	+	+	−	+	−	−	−	+	−	−	+	+	−	−	−	−	−	+
meso-Inositol	−	−	+	+	−	+	+	+	+	+	+	+	−	+	−	−	−	+	+	+	−	−	−	−	−
Mannitol	+	+	+	+	+	+	+	+	+	+	−	+	+	−	+	+	+	+	+	+	−	−	−	−	+
L-Rhamnose	+	−	+	+	+	+	+	+	+	+	+	−	−	−	−	+	−	+	+	−	−	−	+	−	−
Raffinose	+	+	+	+	−	+	−	+	+	−	−	+	−	+	−	+	−	+	+	−	−	−	−	−	+
D-Melibiose	+	+	+	+	−	+	−	+	+	+	−	+	−	+	−	+	+	+	+	−	−	−	−	−	+

[a]Symbols: see Table 29.1.

3. **Streptomyces halstedii** (Waksman and Curtis 1916) Waksman and Henrici 1948b, 953.[AL] (*Actinomyces halstedii* Waksman and Curtis 1916, 124.)

hal.ste'di.i. M.L. gen. n. *halstedii* of Halsted (named for Professor Halsted of Rutgers University).

Spore chains are *Rectiflexibiles* or occasionally *Spirales;* the spore surface is smooth. The spore mass is gray; the reverse is yellow-brown, rarely green; diffusible pigments are usually absent. Melanin pigment is not produced.

Most strains grow in the presence of sodium chloride (7% w/v) and thallous acetate (0.001% w/v). They are resistant to oleandomycin (100 µg/ml).

Type strain: ATCC 10897.

Further comments: This species was defined at the 81% S_{SM} similarity level and its relationship to *S. albidoflavus* and *S. anulatus* has been mentioned. Some of the species listed here as synonyms were regarded as being synonymous with *S. olivaceus* by Hütter (1967).

Subjective synonyms:

Streptomyces erythraeus (Waksman 1923) Waksman and Henrici 1948b, 938.[AL] ATCC 11635.

Streptomyces flavogriseus (Duché 1934) Waksman in Waksman and Lechevalier 1953, 55.[AL] CBS 101.34.

Streptomyces flavovirens (Waksman 1923) Waksman and Henrici 1948b, 940.[AL] ATCC 3320.

Streptomyces griseolus (Waksman 1923) Waksman and Henrici 1948b, 938.[AL] ATCC 3325.

Streptomyces nigrifaciens Waksman 1961, 247.[AL] ATCC 19791.

Streptomyces nitrosporeus Okami 1952b, 477.[AL] ATCC 12769.

Streptomyces olivaceus (Waksman 1923) Waksman and Henrici 1948b, 950.[AL] ATCC 3335.

Allied species:

Streptomyces thermodiastaticus (Bergey, Harrison, Breed, Hammer and Huntoon 1925) Waksman in Waksman and Lechevalier 1953, 102.[AL] ATCC 27472.

This species also shows affinity with *S. anulatus* when matched against the probabilistic identification matrix (Table 29.9).

4. **Streptomyces exfoliatus** (Waksman and Curtis 1916) Waksman and Henrici 1948b, 951.[AL] (*Actinomyces exfoliatus* Waksman and Curtis 1916, 116.)

ex.fo.li.a'tus. L. part. adj. *exfoliatus* stripped of leaves.

Spore chains are *Rectiflexibiles;* the spore surface is smooth. The spore mass is gray or red, rarely white or yellow; the reverse is yellow-brown, occasionally red-orange; yellow-brown or red-orange diffusible pigments are produced by a few strains. Melanin pigment is produced by some strains.

Adonitol, *meso*-inositol, and mannitol are not used.

Type strain: ATCC 12627.

Further comments: This species was clearly defined at the 77.5% S_{SM} similarity level. However, it could be divided at the 79% S_{SM} and 63% S_J similarity levels into strains not producing melanin or other diffusible pigments and those that all produced melanin and sometimes other pigments. The former group included the type strain; the latter are listed below as allied species.

Subjective synonyms:

Streptomyces filamentosus Okami and Umezawa in Okami, Okuda, Takeuchi, Nitta and Umezawa 1953, 153.[AL] ATCC 19753.

Streptomyces hydrogenans Lindner, Junk, Nesemann and Schmidt-Thomé 1958, 117.[AL] ATCC 19631.

Streptomyces omiyaensis Umezawa and Okami in Umezawa, Tazaki, Okami and Fukuyama 1950b, 293.[AL] ATCC 27454.

Streptomyces roseolus (Preobrazhenskaya and Sveshnikova in Gauze et al. 1957) Pridham, Hesseltine and Benedict 1948, 61.[AL] ATCC 23210.

Streptomyces roseosporus Falcão de Morais and Dália Maia 1961, 41.[AL] ATCC 23958.

Streptomyces termitum Duché, Heim and Laboureur in Heim 1951, 359.[AL] ATCC 25499.

Allied species:

Streptomyces cinereoruber Corbaz, Ettlinger, Keller-Schierlein and Zähner 1957b, 330.[AL] ATCC 19740.

Streptomyces gardneri (Waksman in Waksman et al. 1942) Waksman 1961, 215.[AL] ATCC 9604.

Streptomyces litmocidini (Ryabova and Preobrazhenskaya in Gauze et al. 1957) Pridham, Hesseltine and Benedict 1958, 65.[AL] ATCC 19914.

This species produces a violet, pH-sensitive, diffusible pigment.

Streptomyces narbonensis Corbaz, Ettlinger, Gäumann, Keller-Schierlein, Kradolfer, Kyburz, Neipp, Prelog, Reusser and Zähner 1955, 935.[AL] ATCC 19790.

Streptomyces nashvillensis McVeigh and Reyes 1961, 312.[AL] ATCC 25476.

Streptomyces roseoviridis (Preobrazhenskaya in Gauze et al. 1957) Pridham, Hesseltine and Benedict 1958, 61.[AL] ATCC 23959.

Streptomyces umbrinus (Sveshnikova in Gauze et al. 1957) Pridham, Hesseltine and Benedict 1958, 61.[AL] ATCC 19929.

Streptomyces violaceorectus (Ryabova and Preobrazhenskaya in Gauze et al. 1957) Pridham, Hesseltine and Benedict 1958, 63.[AL] ATCC 25514.

Streptomyces zaomyceticus Hinuma 1954, 134.[AL] ATCC 27482.

5. **Streptomyces violaceus** (Rossi-Doria 1891) Waksman in Waksman and Lechevalier 1953, 43.[AL] (*Streptothrix violacea* Rossi-Doria 1891, 411.)

vi.o.la′ce.us. L. adj. *violaceus* violet colored.

Spore chains are *Spirales, Rectiflexibiles,* or occasionally *Retinaculiaperti;* the spore surface is smooth or spiny. The spore mass is violet, red, or yellow; the reverse is yellow-brown; diffusible pigments are rarely produced. Melanin pigment is produced by all strains on tyrosine agar.

Growth does not occur at 45°C.

All strains reduce nitrate and use L-phenylalanine.

Type strain: ATCC 15888.

Further comments: This species was defined at the 77.5% S_{SM} similarity level, but three strains, including the type strain, grouped at the 91% level. These all produce *Spirales* chains of spiny, violet spores, unlike the other strains, which are listed below as allied species.

Subjective synonyms:

Streptomyces cellostaticus Hamada 1958, 174.[AL] ATCC 23894.

Streptomyces violascens (Preobrazhenskaya and Sveshnikova in Gauze et al. 1957) Pridham, Hesseltine and Benedict 1958, 68.[AL] ATCC 23968.

Allied species:

Streptomyces michiganensis Corbaz, Ettlinger, Keller-Schierlein and Zähner 1957c, 205.[AL] ATCC 14970.

Streptomyces showdoensis Nishimura, Mayama, Komatsu, Katô, Shimaoka and Tanaka 1964, 150.[AL] ATCC 15105.

Streptomyces venezuelae Ehrlich, Gottlieb, Burkholder, Anderson and Pridham 1948, 467.[AL] NRRL 2277.

The above three species have *Rectiflexibiles* chains of smooth, yellow spores.

Streptomyces spiroverticillatus Shinobu 1958, 93.[AL] ATCC 19811.

Streptomyces vinaceus Jones 1952, 47.[AL] NRRL 2382.

6. **Streptomyces fulvissimus** (Jensen 1930a) Waksman and Henrici 1948b, 946.[AL] (*Actinomyces fulvissimus* Jensen 1930, 66.)

ful.vis′si.mus. L. sup. adj. *fulvissimus* very yellow.

Spore chains are *Rectiflexibiles, Retinaculiaperti,* or *Spirales;* the spore surface is smooth. The spore mass is red or gray; the reverse is yellow-brown or red-orange and its pigment is pH sensitive; diffusible pigments are sometimes produced. Melanin pigment is produced by many strains on peptone-yeast extract-iron agar.

All strains are resistant to oleandomycin (100 µg/ml) and most to neomycin (50 µgml).

All use L-valine.

Type strain: NRRL B-1453.

Further comments: This species was defined above the 80% S_{SM} similarity level and all strains are regarded as synonyms of the type.

Subjective synonyms:

Streptomyces aureoverticillatus (Krasil'nikov and Yuan 1960) Pridham 1970, 9.[AL] ATCC 15854.

Streptomyces longispororuber Waksman in Waksman and Lechevalier 1953, 99.[AL] ATCC 27443.

Streptomyces spectabilis Mason, Dietz and Smith 1961, 118.[AL] NRRL 2494.

7. **Streptomyces rochei** Berger, Jampolsky and Goldberg in Waksman and Lechevalier 1953, 40.[AL]

ro′che.i. M.L. gen. n. *rochei* of Roche.

Spore chains are *Retinaculiaperti* or *Spirales;* the spore surface is smooth, warty, spiny, or hairy. The spore mass is gray, rarely red or yellow; the reverse is yellow-brown or rarely red-orange; diffusible pigments are seldom produced. Melanin pigment is not produced.

Growth occurs in the presence of sodium chloride (7% w/v).

Dextran and L-rhamnose are used.

Type strain: ATCC 10739.

Further comments: This species was clearly defined at the 77.5% S_{SM} and 63% S_J similarity levels. The type strain produces *Spirales* chains of smooth spores, as do about half of the other strains, which are regarded as synonyms. Most of the remaining strains form *Retinaculiaperti* or *Spirales* chains of ornamented spores, and these are listed below as allied species.

Subjective synonyms:

Streptomyces albogriseolus Benedict, Shotwell, Pridham, Lindenfelser and Haynes 1954, 653.[AL] NRRL B-1305.

Streptomyces althioticus Yamaguchi, Nakayama, Takeda, Tawara, Maeda, Takeuchi and Umezawa 1957, 196.[AL] ATCC 19724.

Streptomyces griseoaurantiacus (Krasil'nikov and Yuan 1965) Pridham 1970, 17.[AL] ATCC 19840.

Streptomyces griseofuscus Sakamoto, Kondo, Yumoto and Arishima 1962, 98.[AL] ATCC 23916.

Streptomyces parvulus Waksman and Gregory 1954, 1055.[AL] ATCC 12434.

Streptomyces plicatus Pridham, Hesseltine and Benedict 1958, 65.[AL] NRRL 2428.

Streptomyces pseudogriseolus Okami and Umezawa in Okami, Utahara, Oyagi, Nakamura, Umezawa, Yanagisawa and Tsunematsu 1955, 128.[AL] ATCC 12770.

Streptomyces tendae Ettlinger, Corbaz and Hütter 1958a, 351.[AL] ATCC 19812.

Streptomyces vinaceusdrappus Pridham, Hesseltine and Benedict 1958, 68.[AL] NRRL 2363.

Allied species:

Streptomyces arabicus Shibata, Nakazawa, Miyake, Inoue and Okabori 1957, 32.[AL] ATCC 23881.
Streptomyces calvus Backus, Tresner and Campbell 1957, 533.[AL] ATCC 13382.
Streptomyces griseorubens (Preobrazhenskaya, Blinov and Ryabova in Gauze et al. 1957) Pridham, Hesseltine and Benedict 1958, 65.[AL] ATCC 19909.
Streptomyces matensis Margalith, Beretta and Timbal 1959, 71.[AL] ATCC 23935.
Streptomyces mutabilis (Preobrazhenskaya and Ryabova in Gauze et al. 1957) Pridham, Hesseltine and Benedict 1958, 69.[AL] ATCC 19919.
Streptomyces rubiginosus (Preobrazhenskaya, Blinov and Ryabova in Gauze et al. 1957) Pridham, Hesseltine and Benedict 1958, 70.[AL] ATCC 19927.
Streptomyces variabilis (Preobrazhenskaya, Ryabova and Blinov in Gauze et al. 1957) Pridham, Hesseltine and Benedict 1958, 70.[AL] ATCC 19930.
Streptomyces werraensis Wallhäusser, Huber, Nesemann, Präve and Zepf 1964, 357.[AL] ATCC 14424.
Streptomyces griseomycini (Preobrazhenskaya, Blinov and Ryabova in Gauze et al. 1957) Pridham, Hesseltine and Benedict 1958, 69.[AL] ATCC 23625.

This species also shows affinity with *S. cyaneus* when matched against the probabilistic identification matrix.

Streptomyces indigoferus Shinobu and Kawato 1960b, 49.[AL] ATCC 23924.

This species also shows affinity with *S. halstedii* when matched against the probabilistic identification matrix.

8. **Streptomyces chromofuscus** (Preobrazhenskaya, Blinov and Ryabova in Gauze et al. 1957) Pridham, Hesseltine and Benedict 1958, 68.[AL] (*Actinomyces chromofuscus* Preobrazhenskaya, Blinov and Ryabova in Gauze et al. 1957, 176.)

chro.mo.fus′cus. Gr. n. *chroma* color; L. adj. *fuscus* dark, tawny; M.L. adj. *chromofuscus* dark or tawny colored.

Spore chains are *Spirales,* occasionally *Rectiflexibiles;* the spore surface is usually smooth or sometimes spiny or hairy. Some strains form sclerotia. The spore mass is gray or yellow; the reverse is yellow-brown; diffusible pigments are rarely produced. Melanin pigment is produced by a few strains.

All strains are sensitive to oleandomycin (100 μg/ml).

There is little or no antibiosis against bacteria or fungi.

L-Hydroxyproline is not used and a few strains use L-phenylalanine.

Type strain: ATCC 23896.

Further comments: This rather heterogeneous species was defined at the 77.5% S_{SM} similarity level but split into three subgroups at the 63% S_J level, one of which contained the type strain and *S. galbus.* The other strains are listed below as allied species and include sclerotia-forming strains previously included in the genus *Chainia.*

Subjective synonyms:

Streptomyces galbus Frommer 1959, 195.[AL] ATCC 23910.

Allied species:

Streptomyces argenteolus Tresner, Davies and Backus 1961, 74.[AL] ATCC 11009.
Streptomyces flaviscleroticus (Pridham 1970) Goodfellow, Williams and Alderson 1986a, 574.[VP†](Effective publication: Goodfellow, Williams and Alderson 1986c, 55.) (*Chainia flava* Thirumalachar and Sukapure 1964, 160.) ATCC 199347.
Streptomyces minutiscleroticus (Thirumalachar in Thirumalachar, Rahalkar, Deshmukh and Sukapure 1965) Pridham 1970, 41.[AL] (*Chainia minutisclerotica* Thirumalachar, Rahalkar, Deshmukh and Sukapure 1965, 7.) ATCC 17757.

Both of the latter two species form sclerotia.

Streptomyces viridosporus Pridham, Hesseltine and Benedict 1958, 67.[AL] NRRL 2414.

9. **Streptomyces albus** (Rossi-Doria 1891) Waksman and Henrici 1943, 339.[AL] (*Streptothrix albus* Rossi-Doria 1891, 421.)

al′bus. L. adj. *albus* white.

Spore chains are *Spirales;* the spore surface is smooth. The spore mass is white, rarely yellow; the reverse is yellow-brown and diffusible pigments are not produced. Melanin pigment is not produced.

Growth occurs at 45°C but not at 10°C, and also in the presence of sodium chloride (7.0% w/v) or sodium azide (0.01% w/v).

Adonitol is used.

Type strain: ATCC 3004.

Further comments: This species was clearly defined at the 84% S_{SM} similarity level and all strains are regarded as synonyms of the type.

Subjective synonyms:

Streptomyces almquistii (Duché 1934) Pridham, Hesseltine and Benedict 1958, 74.[AL] ATCC 618.
Streptomyces aminophilus Foster in Oswald et al. in Hütter 1961, 370.[AL] ATCC 14961.
Streptomyces cacaoi (Waksman in Bunting 1932) Waksman and Henrici 1948b, 951.[AL] ATCC 3082.
Streptomyces flocculus (Duché 1934) Waksman and Henrici 1948b, 955.[AL] ATCC 25453.

10. **Streptomyces griseoviridis** Anderson, Ehrlich, Sun and Burkholder 1956, 114.[AL]

gri.se.o.vi′ri.dis. M.L. adj. *griseus* gray; L. adj. *viridis* green; M.L. adj. *griseoviridis* gray-green.

Spore chains are *Spirales;* the spore surface is smooth. The spore mass is red; the reverse is yellow-brown; yellow-brown diffusible pigments are produced. Melanin pigment is rarely produced.

All strains grow in the presence of sodium azide (0.01% w/v) and show antimicrobial activity against *Micrococcus luteus* and *Bacillus subtilis.*

Type strain: NRRL 2427.

Further comments: This species was defined at the 77.5% S_{SM} level. It is clearly homogeneous and all strains are regarded as synonyms of the type.

Subjective synonyms:

Streptomyces chryseus (Krasil'nikov, Korenyako and Nikitina 1965a) Pridham 1970, 10.[AL] ATCC 19829.
Streptomyces daghestonicus (Sveshnikova in Gauze et al. 1957) Pridham, Hesseltine and Benedict 1958, 67.[AL] ATCC 23620.
Streptomyces murinus Frommer 1959, 198.[AL] ATCC 19788.

11. **Streptomyces cyaneus** (Krasil'nikov 1941) Waksman in Waksman and Lechevalier 1953, 42.[AL] (*Actinomyces cyaneus* Krasil'nikov 1941, 14.)

cy.an′e.us. Gr. adj. *cyaneus* dark blue.

Spore chains are *Spirales* or occasionally *Retinaculiaperti;* the spore surface is spiny or smooth. The spore mass is blue, gray, or red; the reverse is yellow-brown, red-orange, or violet and is sometimes pH sensitive; yellow-brown, red-orange, or violet pH-sensitive diffusible pigments are sometimes produced. Melanin pigment is produced by all strains on peptone–yeast extract–iron agar.

D-Melibiose and raffinose are used.

Type strain: ATCC 14923.

Further comments: This species was defined at the 77.5% S_{SM} similarity level and contains 31 strains. Those producing *Spirales* chains of blue, spiny or smooth spores are listed as synonyms of the type strains. General characters of allied species are given below.

Subjective synonyms:

Streptomyces azureus Kelly, Kutscher and Tuoti 1959, 1334.[AL] ATCC 14921.
Streptomyces bellus Margalith and Beretta 1960a, 193.[AL] ATCC 14925.
Streptomyces caelestis DeBoer, Dietz, Wilkins, Lewis and Savage 1955, 831.[AL] NRRL 2418.

† *VP* denotes that this name has been validly published in the official publication, *International Journal of Systematic Bacteriology.*

Streptomyces chartreusis Leach, Calhoun, Johnson, Teeters and Jackson 1953, 4011.^{AL} NRRL 2287.
Streptomyces coeruleofuscus (Preobrazhenskaya in Gauze et al. 1957) Pridham, Hesseltine and Benedict 1958, 67.^{AL} ATCC 23618.
Streptomyces coeruleorubidus (Preobrazhenskaya in Gauze et al. 1957) Pridham, Hesseltine and Benedict 1958, 67.^{AL} ATCC 13740.
Streptomyces coerulescens (Preobrazhenskaya in Gauze et al. 1957) Pridham, Hesseltine and Benedict 1958, 67.^{AL} ATCC 19896.
Streptomyces curacoi Cataldi in Trejo and Bennett 1963, 683.^{AL} ATCC 13385.
Streptomyces lanatus Frommer 1959, 204.^{AL} ATCC 19775.
Allied species:
Streptomyces arenae Pridham, Hesseltine and Benedict 1958, 67.^{AL} NRRL 2377.
Streptomyces echinatus Corbaz, Ettlinger, Gäumann, Keller-Schierlein, Kradolfer, Neipp, Prelog, Reusser and Zähner 1957a, 199.^{AL} ATCC 19748.
Streptomyces griseochromogenes Fukunaga in Fukunaga, Misato, Ishii and Asakana 1955, 181.^{AL} ATCC 14511.
Streptomyces iakyrus de Querioz and Albert 1962, 33.^{AL} ATCC 15375.

The above four species have *Spirales* chains of gray, spiny spores.

Streptomyces afghaniensis Shimo, Shiga, Tomosugi and Kamoi 1959, 1.^{AL} ATCC 23871.

Streptomyces janthinus (Artamonova and Krasil'nikov in Rautenshtein 1960) Pridham 1970, 19.^{AL} ATCC 15870.

Streptomyces longisporus (Krasil'nikov 1941) Waksman in Waksman and Lechevalier 1953, 39.^{AL} ATCC 23931.

Streptomyces purpurascens Lindenbein 1952, 371.^{AL} ATCC 25489.

Streptomyces roseoviolaceus (Sveshnikova in Gauze et al. 1957) Pridham, Hesseltine and Benedict 1958, 68.^{AL} ATCC 25493.

Streptomyces violarus (Artamonova and Krasil'nikov in Rautenshtein 1960) Pridham 1970, 30.^{AL} ATCC 15891.

Streptomyces violatus (Artamonova and Krasil'nikov in Rautenshtein 1960) Pridham 1970, 30.^{AL} ATCC 15892.

The above seven species have *Spirales* or *Retinaculiaperti* chains of red, spiny spores.

Streptomyces cinnabarinus (Ryabova and Preobrazhenskaya in Gauze et al. 1957) Pridham, Hesseltine and Benedict 1958, 62.^{AL} ATCC 23617.

Streptomyces hawaiiensis Cron, Whitehead, Hooper, Heinemann and Lein 1956, 63.^{AL} ATCC 12236.

Streptomyces luteogriseus Schmitz, Deak, Crook and Hooper 1964, 89.^{AL} ATCC 15072.

Streptomyces neyagawaensis Yamamoto, Nakazawa, Horii and Miyake 1960, 268.^{AL} ATCC 27449.

Streptomyces resistomycificus Lindenbein 1952, 376.^{AL} NRRL 2290.

The above five species have *Spirales* or *Retinaculiaperti* chains of gray or yellow, smooth spores.

Streptomyces collinus Lindenbein 1952, 380.^{AL} ATCC 19743.

Streptomyces paradoxus (Krasil'nikov and Yuan 1961) Goodfellow, Williams and Alderson 1986a, 575.^{VP} (Effective publication: Goodfellow, Williams and Alderson 1986b, 62.) (*Actinosporangium violaceum* Krasil'nikov and Yuan 1961, 116.) ATCC 15813.

These two species form *Retinaculiaperti* chains of smooth, gray spores; they joined at the 88% S_{SM} level.

Streptomyces griseorubiginosus (Ryabova and Preobrazhenskaya in Gauze et al. 1957) Pridham, Hesseltine and Benedict 1958, 62.^{AL} ATCC 23627.

Streptomyces pseudovenezuelae (Kuchaeva, Krasil'nikov, Taptykova and Gesheva 1961) Pridham 1970, 24.^{AL} ATCC 23951.

These two species form *Rectiflexibiles* chains of smooth, red or gray spores.

Streptomyces fumanus (Sveshnikova in Gauze et al. 1957) Pridham, Hesseltine and Benedict 1958, 67.^{AL} ATCC 19904.

This species also shows affinity with *S. rochei.*

12. **Streptomyces diastaticus** (Krainsky 1914) Waksman and Henrici 1948b, 939.^{AL} (*Actinomyces diastaticus* Krainsky 1914, 682.)

di.a.sta'ti.cus. M.L. n. *diastasum* the enzyme diastase, hence M.L. adj. *diastaticus* diastatic.

Spore chains are *Spirales* or *Rectiflexibiles,* occasionally *Retinaculiaperti;* the spore surface is smooth. A few strains have short chains of larger spores (1.5-3.5 μm in diameter) on both aerial and substrate mycelium.

The spore mass is usually gray but can be white, red, or sparse; the reverse is yellow-brown or red-orange, diffusible pigments are rarely produced. About half the strains produce melanin pigment.

Most strains grow at 10°C but not in the presence of sodium azide (0.01% w/v).

There is no antimicrobial activity against yeasts or *Streptomyces murinus.*

Lecithinase activity is absent.

L-Rhamnose is used.

Type strain: ATCC 3315.

Further comments: This rather heterogeneous species was defined at the 77.5% S_{SM} level but split into several subgroups at the 63% S_J similarity level. The type strain and those listed as synonyms form *Spirales* chains of spores that are gray or too sparse for colour determination and are melanin negative. General characteristics of allied species, which include two species previously placed in the genus *Microellobosporia,* are given below.

Subjective synonyms:
Streptomyces glomeroaurantiacus (Krasil'nikov and Yuan 1965) Pridham 1970, 17.^{AL} ATCC 15866.
Streptomyces humidus Nakazawa and Shibata in Imamura, Hori, Nakazawa, Shibata, Tatsuoka and Miyake 1956, 648.^{AL} ATCC 12760.
Streptomyces olivochromogenes (Waksman 1923) Waksman and Henrici 1948b, 941.^{AL} ATCC 3336.
Allied species:
Streptomyces flaveus (Cross, Lechevalier and Lechevalier 1963) Goodfellow, Williams and Alderson 1986a, 574.^{VP} (Effective publication: Goodfellow, Williams and Alderson 1986d, 51.) (*Microellobosporia flavea* Cross, Lechevalier and Lechevalier 1963, 421.) ATCC 15332.
Streptomyces vastus Szabó and Marton 1958, 245.^{AL} ATCC 25506.

These two species joined at the 90% S_{SM} level. The former produces short chains of relatively large spores (1.5-3.5 μm in diameter) on both substrate and aerial mycelium. *Streptomyces vastus* forms spore chains that are *Retinaculiaperti* and *Spirales.*

Streptomyces cinereus (Cross, Lechevalier and Lechevalier 1963) Goodfellow, Williams and Alderson 1986a, 574.^{VP} (Effective publication: Goodfellow, Williams and Alderson 1986d, 51.) (*Microellobosporia cinerea* Cross, Lechevalier and Lechevalier 1963, 422.) ATCC 15840.

Streptomyces lincolnensis Mason, Dietz and De Boer 1963a, 555.^{AL} NRRL 2936.

These two species joined at the 86% S_{SM} level. The former produces short chains of relatively large spores (1.5-3.5 μm in diameter) on both substrate and aerial mycelium. *Streptomyces lincolnensis* forms *Rectiflexibiles* spore chains.

Streptomyces bottropensis Waksman 1961, 182.^{AL} ATCC 25435.

Streptomyces diastatochromogenes (Krainsky 1914) Waksman and Henrici 1948b, 941.^{AL} ATCC 12309.

Streptomyces rishiriensis Kawaguchi, Tsukiura, Okanishi, Miyaki, Ohmori, Fujisawa and Koshiyama 1965, 3.^{AL} ATCC 14812.

These three species form *Spirales* chains of gray spores and produce melanin pigment.

Streptomyces achromogenes Okami and Umezawa in Umezawa, Takeuchi, Okami and Tazaki 1953, 268.^{AL} ATCC 12767.

Streptomyces phaeoviridis Shinobu 1957, 63.^{AL} ATCC 23947.

These two species form *Rectiflexibiles* chains of white spores and produce melanin pigment.

Streptomyces galilaeus Ettlinger, Corbaz and Hütter 1958a, 350.^{AL} NRRL 2722.

This species forms *Rectiflexibiles* chains of red spores and produces melanin pigment.

Streptomyces mirabiles Ruschmann 1952, 543.^{AL} ATCC 27447.

This species forms *Retinaculiaperti* spore chains and does not produce melanin pigment.

13. **Streptomyces olivaceoviridis** (Preobrazhenskaya and Ryabova in Gauze et al. 1957) Pridham, Hesseltine and Benedict 1958, 65.[AL] (*Actinomyces olivaceoviridis* Preobrazhenskaya and Ryabova in Gauze et al. 1957, 163.)

o.li.va ce.o.vi′ri.dis. M.L. adj. *olivaceus* olive colored; L. adj. *viridis* green; M.L. adj. *olivaceoviridis* olive-green colored.

Spore chains are *Spirales* or rarely *Reticuliaperti;* some strains also form short spore chains on the substrate mycelium. The spore surface is smooth.

The spore mass is gray or occasionally yellow or red; the reverse is yellow-brown or green; a few strains produce green diffusible pigments, which are otherwise absent. Melanin pigment is rarely produced. Antimicrobial activity is generally absent.

L-Cysteine and raffinose are used but arbutin is not degraded.

Type strain: ATCC 23630.

Further comments: This species was clearly defined at the 81% S_{SM} similarity level. All strains are regarded as synonyms of the type strain, including three species previously included in the genus *Elytrosporangium* that form short spore chains on their substrate mycelium.

Subjective synonyms:

Streptomyces brasiliensis (Falcão de Morais, Chaves Batista and Massa 1966) Goodfellow, Williams and Alderson 1986e, 573.[VP] (Effective publication: Goodfellow, Williams and Alderson 1986d, 50.) (*Elytrosporangium brasiliense* Falcão de Morais, Chaves Batista and Massa 1966, 164.) ATCC 23727.

Streptomyces canarius Vavra and Dietz 1965, 76.[AL] NRRL 2976.

Streptomyces carpinensis (Falcão de Morais, da Silva and Machado 1971) Goodfellow, Williams and Alderson 1986a, 574.[VP] (Effective publication: Goodfellow, Williams and Alderson 1986d, 50.) (*Elytrosporangium carpinense* Falcão de Morais, da Silva and Machado 1971, 204.) ATCC 27116.

Streptomyces corchorusii Ahmad and Bhuiyan 1958, 143.[AL] ATCC 25444.

Streptomyces regensis Gupta, Sobti and Chopra 1963. [AL] ATCC 27461.

Streptomyces spiralis (Falcão de Morais 1970) Goodfellow, Williams and Alderson 1986a, 575.[VP] (Effective publication: Goodfellow, Williams and Alderson 1986d, 50.) (*Elytrosporangium spirale* Falcão de Morais 1970, 79.) ATCC 25664.

14. **Streptomyces griseoruber** Yamaguchi and Saburi 1955, 220.[AL]

gri.se.o.ru′ber. M.L. adj. *griseus* gray; L. adj. *ruber* red; M.L. adj. *griseoruber* grayish red.

Spore chains are *Spirales;* the spore surface is smooth. The spore mass is usually gray; the reverse is red-orange to violet and its pigment is pH sensitive; red-orange to violet diffusible pigments that are also pH sensitive are produced. Melanin pigment is produced by some strains.

Pectin is degraded.

Type strain: ATCC 23919.

Further comments: This homogeneous species was clearly defined at the 77.5% S_{SM} level and all strains are regarded as synonyms of the type.

Subjective synonyms:

Streptomyces coelescens (Krasil'nikov, Sorokina, Alferova and Bezzubenkova 1965b) Pridham 1970, 21.[AL] ATCC 19830.

Streptomyces humiferus (Krasil'nikov 1962) Goodfellow, Williams and Alderson 1986a, 574.[VP] (Effective publication: Goodfellow, Williams and Alderson 1986b, 62.) (*Actinopycnidium caeruleum* Krasil'nikov 1962, 250.) ATCC 15719.

Streptomyces tuirus Albert and Malaquias de Queiroz 1963, 43.[AL] ATCC 19007.

Streptomyces violaceolatus (Krasil'nikov, Sorokina, Alferova and Bezzubenkova 1965b) Pridham 1970, 28.[AL] ATCC 19847.

15. **Streptomyces lydicus** De Boer, Dietz, Silver and Savage 1956, 886.[AL]

ly′di.cus. L. n. *Lydia* an ancient state in Asia Minor; M.L. adj. *lydicus* of Lydia.

Spore chains are *Spirales;* the spore surface is smooth. The spore mass is gray; the reverse is yellow-brown; diffusible pigments are not produced. Melanin pigment is not produced.

All strains show antimicrobial activity against *Micrococcus luteus, Streptomyces murinus,* and *Aspergillus niger.*

Hydrogen sulfide is not produced.

Type strain: NRRL 2433.

Further comments: This homogeneous species was defined at the 77.5% S_{SM} similarity level. With one exception, the strains are regarded as synonyms of the type.

Subjective synonyms:

Streptomyces griseoplanus Backus, Tresner and Campbell 1957, 536.[AL] ATCC 19766.

Streptomyces libani Baldacci and Grein 1966, 196.[AL] CBS 753.72.

Streptomyces nigrescens (Sveshnikova in Gauze et al. 1957) Pridham, Hesseltine and Benedict 1958, 70.[AL] ATCC 23941.

Streptomyces platensis Tresner and Backus 1956, 244.[AL] NRRL 2364.

Streptomyces sioyaensis Nishimura, Okamoto, Mayama, Ohtsuka, Nakajima, Tawara, Shimohira and Shimaoka 1961, 257.[AL] ATCC 13989.

Allied species:

Streptomyces albulus Routien in Pridham and Lyons 1969, 194.[AL] ATCC 12757.

This species produces hairy spores and separated from the other strains of *S. lydicus* at the 63% S_J similarity level.

16. **Streptomyces violaceusniger** (Waksman and Curtis 1916) Pridham, Hesseltine and Benedict 1958, 63.[AL] (*Actinomyces violaceusniger* Waksman and Curtis 1916, 111.)

vi.o.la′ce.us.ni′ger. L. adj. *violaceus* violet; L. adj. *niger* black; M.L. adj. *violaceusniger* violet-black.

Spore chains are *Spirales;* the spore surface is rugose. The spore mass is gray, becoming black and slimy after maturity. The reverse is yellow-brown; diffusible pigments are not produced. Melanin pigment is not produced.

There is no growth at 10°C or in the presence of phenol (0.1% w/v).

Xanthine is not degraded.

Type strain: NRRL B-1476.

Further comments: This homogeneous species was clearly defined at the 77.5% S_{SM} similarity level. All strains are regarded as synonyms of the type.

Subjective synonyms:

Streptomyces endus Anderson and Gottlieb 1952, 302.[AL] NRRL 2339.

Streptomyces hygroscopicus (Jensen 1931) Waksman and Henrici 1948b, 953.[AL] ATCC 27438.

Streptomyces melanosporofaciens Arcamone, Bertazzoli, Ghione and Scotti 1959a, 215.[AL] ATCC 25473.

Streptomyces sparsogenes Owen, Dietz and Camiener 1963, 772.[AL] NRRL 2940.

17. **Streptomyces griseoflavus** (Krainsky 1914) Waksman and Henrici 1948b, 948.[AL] (*Actinomyces griseoflavus* Krainsky 1914, 694.)

gri.se.o.fla′vus. M.L. adj. *griseus* gray; L. adj. *flavus* yellow; M.L. adj. *griseoflavus* grayish yellow.

Spore chains are *Retinaculiaperti;* the spore surface is spiny or hairy. The spore mass is green or occasionally yellow; the reverse is yellow-brown or sometimes green to blue; diffusible pigments are not produced. Melanin pigment is not produced by most strains.

There is no growth in the presence of crystal violet (0.001% w/v).

Xanthine is not degraded.

Type strain: ATCC 25456.

Further comments: This homogeneous species was defined at the 79% S_{SM} similarity level and most strains also grouped at the 89% level. All strains are regarded as synonyms of the type.

Subjective synonyms:

Streptomyces cyanoalbus (Krasil'nikov and Agre in Rautenshtein 1960). Pridham 1970, 13.[AL] ATCC 15859.

Streptomyces hirsutus Ettlinger, Corbaz and Hütter 1958a, 344.[AL] ATCC 19773.

Streptomyces pilosus Ettlinger, Corbaz and Hütter 1958a, 347.[AL] ATCC 19797.

Streptomyces prasinopilosus Ettlinger, Corbaz and Hütter 1958a, 345.[AL] ATCC 19799.

Streptomyces prasinus Ettlinger, Corbaz and Hütter 1958a, 343.[AL] ATCC 19800.

18. **Streptomyces phaeochromogenes** (Conn 1917) Waksman 1957a, 778.[AL] (*Actinomyces phaeochromogenus* (sic) Conn 1917, 16.)

phae.o.chro.mo'ge.nes. Gr. adj. *phaeus* brown; Gr. n. *chroma* color; Gr. v. suff. *-genes* producing; M.L. adj. *phaeochromogenes* producing brown color.

Spore chains are *Spirales* or sometimes *Retinaculiaperti;* the spore surface is smooth. Many strains form sclerotia. The spore mass is white or sparse; the reverse is yellow-brown or sometimes green; diffusible pigments that are yellow-brown or green are sometimes produced. Melanin pigment is produced by a few strains.

There is little or no antimicrobial activity against bacteria or fungi.

L-Valine is used but adonitol, inulin, and xylitol are not.

Type strain: ATCC 3338.

Further comments: This species was defined at the 77.5% S_{SM} similarity level and contains the type plus four strains forming sclerotia, previously placed in the genus *Chainia.* The latter united at the 82% S_{SM} level and were distinguished from the type strain at the 63% S_J level; they are therefore listed as allied species.

Allied species:

Streptomyces niger (Thirumalachar 1955) Goodfellow, Williams and Alderson 1986a, 575.[VP] (Effective publication: Goodfellow, Williams and Alderson 1986c, 57.) (*Chainia nigra* Thirumalachar 1955, 935.) ATCC 17756.

Streptomyces purpurogeneiscleroticus (Thirumalachar and Sukapure 1964) Pridham 1970, 43.[AL] (*Chainia purpurogena* Thirumalachar and Sukapure 1964, 160.) ATCC 19348.

Streptomyces sclerotialus (Thirumalachar 1955) Pridham 1970, 44.[AL] (*Chainia antibiotica* Thirumalachar 1955, 935.) ATCC 15721.

Streptomyces violens (Pridham 1970) Goodfellow, Williams and Alderson 1987, 179.[VP] (Effective publication: Goodfellow, Williams and Alderson 1986c, 59.) (*Chainia violens* Kalakoutskii and Krasil'nikov in Rautenshtein 1960, 55.) ATCC 15898.

19. **Streptomyces rimosus** Sobin, Finley and Kane in Waksman and Lechevalier 1953, 47.[AL]

ri.mo'sus. L. adj. *rimosus* full of fissures.

Spore chains are *Spirales,* sometimes *Rectiflexibiles* or *Retinaculiaperti,* the spore surface is smooth. The spore mass is white or yellow; the reverse is yellow-brown; diffusible pigments are not produced. Melanin pigment is not produced.

Strains are resistant to neomycin (50 µg/ml) and streptomycin (100 µg/ml).

Antimicrobial activity is shown against *Aspergillus niger* and other test organisms.

Adonitol is used but L-rhamnose is not.

Type strain: NRRL 2234.

Further comments: This homogeneous species was defined at the 77.5% S_{SM} level and two strains, *S. anandii* and *S. chrestomyceticus,* joined the type at the 94% S_{SM} level. All strains are regarded as synonyms of the types.

Subjective synonyms:

Streptomyces albofaciens Thirumalachar and Bhatt 1960, 63.[AL] ATCC 25184.

Streptomyces anandii Batra and Bajaj 1965, 242.[AL] ATCC 19388.

Streptomyces chrestomyceticus Canevazzi and Scotti 1959, 248.[AL] ATCC 14947.

Streptomyces kanamyceticus Okami and Umezawa in Umezawa, Ueda, Maeda, Yagashita, Kondo, Okami, Utahara, Osato, Nitta and Takeuchi 1957, 183.[AL] ATCC 12853.

20. **Streptomyces microflavus** (Krainsky 1914) Waksman and Henrici 1948b, 950.[AL] (*Actinomyces microflavus* Krainsky 1914 686.)

mic.ro.fla'vus. Gr. adj. *micros* small; L. adj. *flavus* yellow; M.L. adj. *microflavus* small, yellow.

Spore chains are *Spirales, Retinaculiaperti,* or *Rectiflexibiles;* the spore surface is smooth. The spore mass is usually gray; the reverse is yellow-brown or occasionally red-brown; diffusible pigments are produced by a few strains. Melanin pigment is produced by most strains.

All strains show antimicrobial activity against *Micrococcus luteus.*

Pectin is degraded but allantoin is not.

Type strain: CBS 124.18.

Further comments: This homogeneous species was defined at the 77.5% S_{SM} similarity level. All strains are regarded as synonyms of the type.

Subjective synonyms:

Streptomyces ambofaciens Pinnert-Sindico 1954, 702.[AL] ATCC 23877.

Streptomyces eurythermus Corbaz, Ettlinger, Gäumann, Keller- Schierlein, Kradolfer, Neipp, Prelog, Reusser and Zähner 1957a, 1202.[AL] ATCC 14975.

Streptomyces griseosporeus Niida and Ogasawara 1960, 23.[AL] ATCC 27435.

Streptomyces recifensis (Goncalves de Lima, Machado, Araújo, Falcão de Morais and Biermann 1955) Falcão de Morais, Goncalves de Lima and Dália Maia 1957, 249.[AL] ATCC 19803.

21. **Streptomyces antibioticus** (Waksman and Woodruff 1941) Waksman and Henrici 1948b, 942.[AL] (*Actinomyces antibioticus* Waksman and Woodruff 1941, 246.)

an.ti.bi.o'ti.cus. Gr. pref. *anti-* against; Gr. n. *bios* life; M.L. adj. *antibioticus* against life, antibiotic.

Spore chains are *Spirales,* sometimes *Rectiflexibiles* or *Retinaculiaperti;* the spore surface is usually smooth. The spore mass is gray; the reverse is yellow-brown; diffusible pigments are not produced. Most strains produce melanin pigment.

All strains show antimicrobial activity against *Candida albicans* and *Saccharomyces cerevisiae.*

Hydrogen sulfide is not produced.

There is no lipolytic activity or elastin degradation.

Type strain: ATCC 8663.

Further comments: This rather heterogeneous species was defined at the 77.5% S_{SM} similarity level. At the 63% S_J level, the type and *S. phaeofaciens* grouped separately from the other strains; the latter are listed below as allied species.

Subjective synonyms:

Streptomyces phaeofaciens Maeda, Okami, Taya and Umezawa 1952b, 327.[AL] ATCC 15034.

Allied species:

Streptomyces lucensis Arcamone, Bertazzoli, Canevazzi, di Marco, Ghione and Grein 1957, 119.[AL] ATCC 17804.

Streptomyces misionensis Cercos, Eilberg, Goyena, Souto, Vautier and Widuczynski 1962, 22.[AL] ATCC 14991.

Streptomyces naganishii Yamaguchi and Saburi 1955, 219.[AL] ATCC 23939.

22. **Streptomyces lavendulae** (Waksman and Curtis 1916) Waksman and Henrici 1948b, 944.[AL] (*Actinomyces lavendulae* Waksman and Curtis 1916, 126.)

la.ven'du.lae. M.L. n. *lavendula* lavender; M.L. gen. n. *lavendulae* of lavender color.

Spore chains are *Rectiflexibiles* or occasionally *Spirales;* the spore surface is smooth. The spore mass is red or rarely gray; the reverse is yellow-brown; diffusible pigments are not produced. Melanin pigment is produced.

All strains have lecithinase activity.

Mannitol is not used.

Type strain: ATCC 8664.

Further comments: This homogeneous species was defined at the 77.5% S_{SM} similarity level. All strains are regarded as synonyms of the type.

Subjective synonyms:

Streptomyces colombiensis Pridham, Hesseltine and Benedict 1958, 76.[AL] ATCC 27425.
Streptomyces flavotricini (Preobrazhenskaya and Sveshnikova in Gauze et al. 1957) Pridham, Hesseltine and Benedict 1958, 60.[AL] ATCC 23914.
Streptomyces goshikiensis Niida in Shirling and Gottlieb 1968b, 324.[AL] ATCC 23914.
Streptomyces katrae Gupta and Chopra 1963b, 1.[AL] ATCC 27440.
Streptomyces lavendulocolor (Kuchaeva, Krasil'nikov, Taptykova and Gesheva 1961) Pridham 1970, 20.[AL] ATCC 15871.
Streptomyces polychromogenes Hageman, Penasse and Teillon in Hütter 1964, 615.[AL] ATCC 12595.
Streptomyces racemochromogenes Sugai 1956, 171.[AL] ATCC 23954.
Streptomyces subrutilus Arai, Kuroda, Yamagishi and Katoh 1964, 25.[AL] ATCC 27467.
Streptomyces toxytricini (Preobrazhenskaya and Sveshnikova in Gauze et al. 1957) Pridham, Hesseltine and Benedict 1958, 68.[AL] ATCC 19813.
Streptomyces virginiae Grundy, Whitman, Rdzok, Rdzok, Hanes and Sylvester 1952, 399.[AL] ATCC 19817.
Streptomyces xanthophaeus Lindenbein 1952, 378.[AL] ATCC 19819.

23. **Streptomyces purpureus** (Matsumae and Hata in Matsumae et al. 1968) Goodfellow, Williams and Alderson 1986a, 575.[VP] (Effective publication: Goodfellow, Williams and Alderson 1986e, 65.) (*Kitasatoa purpurea* Matsumae and Hata in Matsumae et al. 1968, 617.)

pur.pur'e.us. M.L. adj. *purpureus* purple colored.

Spore chains are *Rectiflexibiles;* the spore surface is smooth. The spore mass is gray or rarely red; the reverse is yellow-brown; diffusible yellow-brown pigments are produced. Most strains produce melanin pigment on peptone–yeast extract–iron agar but not on tyrosine agar.

L-Histidine is used but not D-lactose or D-xylose.

Type strain: ATCC 27787.

Further comments: This homogeneous species was defined at the 77.5% S_{SM} similarity level. The species *Kitasatoa diplospora* Matsumae, Ohtani and Hata in Matsumae et al. 1968, 621, *K. kauaiensis* Matsumae, Ohtani and Hata in Matsumae et al. 1968, 621, and *K. nagasatiensis* Matsumae and Hata in Matsumae et al. 1968, 622 are regarded as junior synonyms of *Streptomyces purpureus.*

Category II

Characteristics useful for the differentiation of species are given in Table 29.10. Features in the text descriptions are those most typical of the species, including the morphological and pigmentation characteristics that have formed the basis of most previous classifications. All species were defined at the 77.5% S_{SM} similarity level unless stated.

List of the Species in Category II

1. **Streptomyces aburaviensis** Nishimura, Kimura, Tawara, Sasaki, Nakajima, Shimaoka, Okamodo, Shimohira and Isono 1957, 206.[AL]

a.bu.ra.vi.en'sis. M.L. adj. *aburaviensis* of Aburabi, Shiga Prefecture, Japan (the source of soil from which the organism was isolated).

Spore chains are *Rectiflexibiles;* the spore surface is smooth. The spore mass is gray; the reverse is yellow-brown; diffusible pigments are not produced. Melanin pigment is not produced.

Growth occurs at 4°C and in the presence of phenylethanol (0.3% w/v).

Adonitol, dextran, and sodium malonate are used.

Type strain: ATCC 23869.

Further comments: This species shares many characters with *S. albidoflavus, S. anulatus,* and *S. halstedii* and joined the former at the 63% S_J level.

Subjective synonym:

Streptomyces herbaricolor Kawato and Shinobu 1959, 114.[AL] ATCC 23922.

2. **Streptomyces californicus** (Waksman and Curtis 1916) Waksman and Henrici 1948b, 936.[AL] (*Actinomyces californicus* Waksman and Curtis 1916, 122.)

ca.li.for'nic.us. M.L. adj. *californicus* belonging to California (the source of soil from which the organism was isolated).

Spore chains are *Rectiflexibiles;* the spore surface is smooth. The spore mass is red, yellow to gray; the reverse is violet and its pigment is pH sensitive. Violet, pH-sensitive, diffusible pigments are produced. Melanin pigment is not produced.

All strains are resistant to streptomycin (100 μg/ml).

Type strain: ATCC 3312.

Further comments: This species was defined at the 87% S_{SM} similarity level. It shares many characters with *S. fulvissimus,* with which it joined at the 74% S_{SM} level.

Subjective synonyms:

Streptomyces phaeopurpureus Shinobu 1957, 63.[AL] ATCC 23946.
Streptomyces puniceus Patelski in Routien and Hofmann 1951, 387.[AL] ATCC 19801.

3. **Streptomyces cellulosae** (Krainsky 1914) Waksman and Henrici 1948b, 938.[AL] (*Actinomyces cellulosae* Krainsky 1914, 683.)

cel.lu.lo'sae. M.L. n. *cellulosum* cellulose; M.L. gen. n. *cellulosi* of cellulose (presumably relating to the ability to degrade cellulose).

Spore chains are *Retinaculiaperti* or *Spirales;* the spore surface is smooth or spiny. The spore mass is white to gray; the reverse is red-orange and its pigment is pH sensitive. Red-orange, pH-sensitive diffusible pigments are produced. Melanin pigment is not produced.

Adenine is not degraded.

Type strain: CBS 122.18.

Further comments: This species was defined at the 81% S_{SM} similarity level.

Subjective synonym:

Streptomyces griseoincarnatus (Preobrazhenskaya, Ryabova and Blinov in Gauze et al. 1957) Pridham, Hesseltine and Benedict 1958, 69.[AL] ATCC 23623.

4. **Streptomyces aureofaciens** Duggar 1948, 177.[AL]

au.re.o.fa'ci.ens. L. adj. *aureus* golden; L. part. adj. *faciens* producing; M.L. part. adj. *aureofaciens* producing golden (referring to a pigment produced in the vegetative mycelium of the organisms).

Spore chains are *Rectiflexibiles* or *Retinaculiaperti;* the spore surface is smooth. The spore mass is red to gray; the reverse is yellow-brown; diffusible pigments are not produced. Melanin pigment is not produced.

L-Rhamnose is used but L-histidine and L-threonine are not.

Type strain: NRRL 2209.

Further comments: This species was defined at the 79% S_{SM} level.

Subjective synonym:

Streptomyces roseofulvus (Preobrazhenskaya in Gauze et al. 1957) Pridham, Hesseltine and Benedict 1958, 61.[AL]

5. **Streptomyces flaveolus** (Waksman 1923) Waksman and Henrici 1948b, 936.[AL] (*Actinomyces flaveolus* Waksman 1923, 368.)

fla.ve'o.lus. L. adj. *flavus* yellow; L. dim. adj. *flaveolus* somewhat yellow.

Spore chains are *Spirales;* the spore surface is smooth or hairy. The spore mass is gray to green; the reverse is yellow-brown; yellow–brown diffusible pigments are usually produced. Melanin pigment is not produced.

Growth occurs at 45°C but not in the presence of potassium tellurite (0.001% w/v).

All strains have lecithinase activity and use L-rhamnose.

Type strain: ATCC 3319.

Further comments: This is a rather heterogeneous species. The type strain forms hairy, gray spores and a yellow-brown pigment. The other strains show some differences and are listed below as allied species.

Allied species:

Streptomyces chibaenis Suzuki, Nakamura, Okuma and Tomiyama 1958, 81.[AL] ATCC 23895.

This species has smooth, gray spores and does not produce a diffusible pigment.

Streptomyces xantholiticus (Konev and Tsyganov 1962) Pridham 1970, 31.[AL] ATCC 27481.

This species has smooth, gray spores.

6. **Streptomyces filipinensis** Amman, Gottlieb, Brock, Carter and Whitfield 1955, 559.[AL]

fi.li.pi.nen'sis. M.L. adj. *filipinensis* belonging to the Philippines (the source of soil from which the organism was isolated).

Spore chains are *Spirales;* the spore surface is spiny. The spore mass is gray; the reverse is yellow-brown; diffusible pigments are not produced. Melanin pigment is produced.

There is antimicrobial activity against *Candida albicans.*

Adonitol is used.

Type strain: NRRL 2437.

Further comments: This species shares many characters with *S. antibioticus,* which it joined at the 76% S_{SM} similarity level.

Subjective synonyms:

Streptomyces durhamensis Gordon and Lapa 1966, 754.[AL] ATCC 23194.

Streptomyces yokosukanensis Nakamura 1961c, 94.[AL] ATCC 25520.

7. **Streptomyces noboritoensis** Isono, Yamashita, Tomiyama, Suzuki and Sakai 1957, 21.[AL]

no.bo.ri.toen'sis. M.L. adj. *noboritoensis* belonging to noborito (referring to Inada-noborto, Kawasaki City, Kanagawa Prefecture, Japan, the source of soil from which the organism was isolated).

Spore chains are *Rectiflexibiles;* the spore surface is smooth. The spore mass is white to gray; the reverse is yellow-brown; yellow-brown diffusible pigments are produced. Melanin pigment is produced.

There is antimicrobial activity against *Candida albicans.*

DL-α-Amino-*n*-butyric acid is used.

Type strain: ATCC 25477.

Subjective synonym:

Streptomyces melanogenes Sugawara and Onuma 1957, 141.[AL] ATCC 23937.

8. **Streptomyces chattanoogensis** Burns and Holtman 1959, 398.[AL]

chat.ta.noo.gen'sis. M.L. adj. *chattanoogenis* belonging to Chattanooga, Tennessee (the source of soil from which the organism was isolated).

Spore chains are *Spirales* or *Retinaculiaperti;* the spore surface is smooth or spiny. The spore mass is yellow to gray; the reverse is yellow-brown; yellow-brown diffusible pigments are sometimes produced. Melanin pigment is not produced.

Growth occurs in the presence of sodium azide (0.02% w/v).

There is antimicrobial activity against *Candida albicans* and *Saccharomyces cerevisiae.*

Hypoxanthine is not degraded.

Type strain: ATCC 13358.

Further comments: This species was defined at the 81% S_{SM} similarity level and joined *S. thermovulgaris* at the 76% S_{SM} level. The type strain produces spiny, yellow spores and forms a yellow-brown diffusible pigment. The other strain differs in these characteristics and is listed as an allied species.

Allied species:

Streptomyces nodosus Trejo in Waksman 1961, 250.[AL] ATCC 14899.

This species forms smooth, green spores and does not produce a diffusible pigment.

9. **Streptomyces thermovulgaris** Henssen 1957, 391.[AL]

ther.mo.vul.ga'ris. Gr. n. *therme* heat; L. adj. *vulgaris* common; M.L. adj. *thermovulgaris* heat, common.

Spore chains are *Rectiflexibiles;* the spore surface is smooth. The spore mass is white or sparse; the reverse is yellow-brown; diffusible pigments are not produced. Melanin pigment is not produced.

Growth occurs at 37° and 45°C but not at 10°C.

Guanine, hypoxanthine, and urea are not degraded.

L-Serine is not used.

Type strain: ATCC 19284.

Further comments: This species was defined at the 85% S_{SM} similarity level and consists of two thermophilic strains.

Subjective synonym:

Streptomyces thermonitrificans Desai and Dhala 1967b, 137.[AL] ATCC 23385.

10. **Streptomyces longisporoflavus** Waksman in Waksman and Lechevalier 1953, 94.[AL]

lon.gi.spo.ro.fla'vus. L. adj. *longus* long; M.L. n. *spora* a spore; L. adj. *flavus* yellow; M.L. adj. *longisporoflavus* long-spored, yellow.

Spore chains are *Spirales;* the spore surface is smooth or rugose. The spore mass is too sparse for color determination; the reverse is yellow-brown; diffusible pigments are not produced. Melanin pigment is sometimes produced on peptone–yeast extract–iron agar.

Nitrate is reduced.

Pectin is degraded but guanine is not.

L-Rhamnose is used.

Type strain: ATCC 19915.

Further comments: This species shows affinity with *S. griseoflavus,* which it joined at the 75% S_{SM} similarity level. The type strain produces smooth spores and melanin pigment. The other strain differs in these characteristics and separates from the type at the 63% S_J level. It is listed below as an allied species.

Allied species:

Streptomyces bluensis Mason, Dietz and Hanka 1963b, 608.[AL] NRRL 2876.

11. **Streptomyces griseoluteus** Umezawa, Hayano, Maeda, Ogata and Okami 1950a, 112.[AL]

gri'se.o.lu'te.us. M.L. adj. *griseus* gray; M.L. *luteus* yellow; M.L. adj. *griseoluteus* grayish yellow.

Spore chains are *Retinaculiaperti;* the spore surface is smooth. The spore mass is gray; the reverse is yellow-brown or green; diffusible pigments are not produced. Melanin pigment is sometimes produced.

L-Cysteine is used but L-arabinose and D-lactose are not.

Type strain: ATCC 12768.

Subjective synonym:

Streptomyces catenulae Davisson and Finlay in Waksman 1961, 190.[AL] ATCC 12476.

12. **Streptomyces pactum** Bhuyan, Dietz and Smith 1962, 185.[AL]

pac'tum. Gr. adj. *pactos* solid, firm, coagulated; L. adj. *pactus* settled (referring to the compactness of the coiled spore chains).

Spore chains are *Spirales;* the spore surface is smooth or hairy. The spore mass is gray; the reverse is yellow-brown; yellow-brown diffusible pigments are produced. Melanin pigment is not produced.

Strains are resistant to oleandomycin (100 μg/ml).

Sodium citrate and D-xylose are not used.

Type strain: NRRL 2939.

Subjective synonym:

Streptomyces lusitanus Villax 1963, 661.[AL] NCIB 9585.

13. **Streptomyces aurantiacus** (Rossi-Doria 1891) Waksman in Waksman and Lechevalier 1953, 53.[AL] (*Streptothrix aurantiaca* Rossi-Doria 1891, 417.)

au.ran.ti'a.cus. M.L. n. *Aurantium* generic name of the orange; M.L. adj. *aurantiacus* orange colored.

Spore chains are *Spirales;* the spore surface is smooth. The spore mass is gray; the reverse is red-orange and its pigment is pH sensitive. Red-orange, pH-sensitive diffusible pigments are produced. Melanin pigment is sometimes produced.

Type strain: ATCC 19822.

Subjective synonyms:

Streptomyces capoamus Gonçalves de Lima, Albert and Gonçalves de Lima 1964, 317.[AL] ATCC 19006.

Streptomyces thermoviolaceus Henssen 1957, 388.[AL] ATCC 19283.

14. **Streptomyces luridus** (Krasil'nikov, Korenyako, Meksina, Valedinskaya and Veselov 1957) Waksman 1961, 237.[AL] (*Actinomyces luridus* Krasil'nikov et al. 1957, 563.)

lu'ri.dus. L. adj. *luridus* pale yellow, pallid.

Spore chains are *Spirales* or occasionally *Rectiflexibiles;* the spore surface is smooth. The spore mass is red to gray; the reverse is yellow-brown; diffusible pigments are not produced. Melanin pigment is usually produced.

There is no growth in the presence of potassium tellurite (0.001% w/v).

There is antimicrobial activity against *Bacillus subtilis.*

D-Lactose and mannitol are not used.

Type strain: ATCC 19782.

Subjective synonyms:

Streptomyces cirratus Koshiyama, Okanishi, Ohmori, Miyaki, Tsukiura, Matsuzaki and Kawaguchi 1963, 61.[AL] ATCC 14699.

Streptomyces helvaticus (Krasil'nikov, Korenyako and Nikitina 1965a) Pridham 1970, 18.[AL] ATCC 19841.

15. **Streptomyces xanthochromogenes** Arashima, Sakamoto and Sato 1956, 469.[AL]

xan'tho.chro.mo'ge.nes. Gr. adj. *xanthus* yellow; Gr. n. *chroma* color; Gr. suff. *-genes* producing; M.L. adj. *xanthochromogenes* producing yellow color.

Spore chains are *Rectiflexibiles;* the spore surface is smooth. The spore mass is red to gray; the reverse is yellow-brown; diffusible pigments are not produced. Melanin pigment is produced.

There is no growth in the presence of sodium chloride (4% w/v) or potassium tellurite (0.001% w/v).

Type strain: ATCC 19818.

Further comments: This species is closely related to *S. lavendulae.*

Subjective synonym:

Streptomyces albolongus Tsukiura, Okanishi, Koshiyama, Ohmori, Miyaki and Kawaguchi 1964a, 225.[AL] ATCC 27414.

16. **Streptomyces misakiensis** Nakamura 1961a, 86.[AL]

mi.sa.ki.en'sis. M.L. adj. *misakiensis* belonging to misaki (referring to Misakicho, Kanagawo Prefecture, Japan, the source of soil from which the organism was isolated).

Spore chains are *Rectiflexibiles;* the spore surface is smooth. The spore mass is white to gray; the reverse is yellow-brown; yellow-brown diffusible pigments are sometimes produced. Melanin pigment is produced on tyrosine agar but not on peptone–yeast extract–iron agar.

There is antimicrobial activity against *Bacillus subtilis.*

Arbutin and urea are not degraded.

L-Hydroxyproline is used.

Type strain: ATCC 23938.

Subjective synonym:

Streptomyces xanthocidicus Asahi, Nagatsu and Suzuki 1966, 196.[AL] ATCC 27480.

17. **Streptomyces psammoticus** Virgilio and Hengeller 1960, 167.[AL]

psam.mo'ti.cus. Gr. n. *psamma* sand; M.L. adj. *psammoticus* sandy.

Spore chains are *Rectiflexibiles* or *Spirales;* the spore surface is smooth. The spore mass is red or sparse; the reverse is yellow-brown; yellow-brown diffusible pigments are sometimes produced. Melanin pigment is not produced, except for the type strain that forms melanin on peptone–yeast extract–iron agar.

Strains are resistant to dimethylchlortetracycline (500 μg/ml) and vancomycin (50 μg/ml).

Esculin is not degraded.

Potassium nitrate and cellobiose are not used.

Type strain: CBS 175.61.

Further comments: This species was defined at the 82% S_{SM} similarity level.

Subjective synonym:

Streptomyces avellaneus Baldacci and Grein 1966, 195.[AL] ATCC 23730.

18. **Streptomyces fradiae** (Waksman and Curtis 1916) Waksman and Henrici 1948b, 954.[AL] (*Actinomyces fradii* (sic) Waksman and Curtis 1916, 125.).

fra'di.ae. M.L. gen. n. *fradiae* of Fradia (a patronymic).

Spore chains are *Retinaculiaperti* or *Spirales;* the spore surface is smooth. The spore mass is red; the reverse is yellow-brown; diffusible pigments are not produced. Melanin pigment is not produced.

There is antimicrobial activity against *Escherichia coli.*

Hydrogen sulfide is not produced.

L-Tyrosine is not degraded.

Cellobiose and D-fructose are not used.

Type strain: ATCC 10745.

Further comments: This species was defined at the 81% S_{SM} similarity level.

Subjective synonym:

Streptomyces roseolilacinus (Preobrazhenskaya and Sveshnikova in Gauze et al. 1957) Pridham, Hesseltine and Benedict 1958, 68.[AL] ATCC 19922.

19. **Streptomyces poonensis** (Thirumalachar in Rautenshtein 1960) Pridham 1970, 42.[AL] (*Chainia poonensis* Thirumalachar in Rautenshtein 1960, 45.)

poon.en'sis. M.L. adj. *poonensis* belonging to Poona, India (the source of the soil from which the organism was isolated).

The spore chains are *Spirales;* the spore surface is smooth. Sclerotia are formed. The spore mass is white to sparse; the reverse is yellow-brown to red-orange and its pigments are pH sensitive. Diffusible red-orange, pH-sensitive pigments are sometimes produced. Melanin pigment is not produced.

DL-α-Amino-*n*-butyric acid, inulin, and sodium malonate are used.

Type strain: ATCC 15723.

Further comments: This species was defined at the 81% S_{SM} similarity level and consists of strains previously classified in the genus *Chainia.*

Subjective synonym:

Streptomyces roseiscleroticus (Thirumalachar in Thirumalachar et al. 1966) Pridham 1970, 43.[AL] (*Chainia rosea* Thirumalachar in Thirumalachar, Sukapure, Rahalkar and Gopalkrishnan 1966, 10.) ATCC 17755.

20. **Streptomyces atroolivaceus** (Preobrazhenskaya, Blinov and Ryabova in Gauze et al. 1957) Pridham, Hesseltine and Benedict 1958, 68.[AL] (*Actinomyces atroolivaceus* Preobrazhenskaya, Blinov and Ryabova in Gauze et al. 1957, 143.)

at.ro.o.li.va'ce.us. L. adj. *ater* black; M.L. adj. *olivaceus* olive colored; M.L. adj. *atroolivaceus* of a dark olive color.

Spore chains are *Rectiflexibiles;* the spore surface is smooth. The spore mass is gray to white; the reverse is yellow-brown; diffusible pigments are not produced. Melanin pigment is not produced.

Growth occurs at 4°C.

Meso-inositol and D-melezitose are not used.

Type strain: ATCC 19725.

Further comments: The type strain and *S. olivoviridis* joined at the 92% S_{SM} similarity level.

Subjective synonyms:

Streptomyces aureocirculatus (Krasil'nikov and Yuan 1965) Pridham 1970, 8.[AL] ATCC 19823.

Streptomyces olivoviridis (Kuchaeva, Krasil'nikov, Skryabin and Taptykova in Rautenshtein 1960) Pridham 1970, 23.[AL] ATCC 15882.

Category III

Characteristics useful for the differentiation of species are given in Table 29.11. Features in the text descriptions are those most typical of the species, including the morphological and pigmentation characteristics that have formed the basis of most previous classifications. All characteristics are those of the single strains that did not cluster with others at the 77.5% S_{SM} similarity level.

List of Species in Category III

1. **Streptomyces prunicolor** (Ryabova and Preobrazhenskaya in Gauze et al. 1957) Pridham, Hesseltine and Benedict 1958, 63.[AL] (*Actinomyces prunicolor* Ryabova and Preobrazhenskaya in Gauze et al. 1957, 184.)

pru'ni.co.lor. L. n. *prunus* plum; L. n. *color* color; M.L. adj. *prunicolor* plum colored.

Spore chains are *Rectiflexibiles;* the spore surface is smooth. The spore mass is yellow; the reverse is yellow-brown and its pigments are pH sensitive. Red-orange, pH-sensitive, diffusible pigments are produced. Melanin pigment is not produced.

Xanthine is degraded.

Meso-inositol is not used.

Type strain: ATCC 25487.

2. **Streptomyces canus** Heinemann, Kaplan, Muir and Hooper 1953, 1239.[AL]

ca'nus. L. adj. *canus* white, gray.

Spore chains are *Spirales;* the spore surface is spiny. The spore mass is gray; the reverse is yellow-brown; diffusible pigments are not produced. Melanin pigment is produced on tyrosine agar but not on peptone-yeast extract-iron agar.

There is no lipolytic activity.

Adonitol is used but L-arginine is not.

Type strain: ATCC 12237.

3. **Streptomyces graminofaciens** Charney, Fisher, Curran, Machlowitz and Tytell 1953, 1283.[AL]

gra.mi.no.fa'ci.ens. L. n. *gramen* grass; L. part. adj. *faciens* producing; M.L. part. adj. *graminofaciens* grass producing.

Spore chains are *Spirales;* the spore surface is warty. The spore mass is gray; the reverse is yellow-brown; diffusible pigments are not produced. Melanin pigment is not produced.

Growth occurs in the presence of thallous acetate (0.001% w/v).

Adenine and urea are not degraded.

Type strain: ATCC 12705.

4. **Streptomyces viridochromogenes** (Krainsky 1914) Waksman and Henrici 1948b, 942.[AL] *Actinomyces viridochromogenes* Krainsky 1914, 684.)

vi.ri.do.chro.mo'ge.nes. L. adj. *viridis* green; Gr. n. *chroma* color; Gr. v. suff. *-genes* producing; M.L. adj. *viridochromogenes* producing green color.

Spore chains are *Spirales;* the spore surface is spiny. The spore mass is blue; the reverse is green and its pigments are pH sensitive. Green, pH-sensitive, diffusible pigments are produced. Melanin pigment is produced.

Growth occurs in the presence of sodium azide (0.01% w/v).

L-Methionine is used.

Type strain: ATCC 14920.

5. **Streptomyces glaucescens** (Preobrazhenskaya in Gauze et al. 1957) Pridham, Hesseltine and Benedict 1958, 67.[AL] (*Actinomyces glaucescens* Preobrazhenskaya in Gauze et al. 1957, 122.)

glau.ces'cens. L. adj. *glaucus* bluish gray; M.L. adj. *glaucescens* slightly bluish gray.

Spore chains are *Spirales;* the spore surface is hairy. The spore mass is blue; the reverse is red-orange; red-orange diffusible pigments are produced. Melanin pigment is produced.

Proteolytic activity does not occur.

Dextran is used.

Type strain: ATCC 23622.

6. **Streptomyces nogalater** Bhuyan and Dietz 1966, 838.[AL]

no.gal.at'er. Sp. n. *nogal* walnut; L. adj. *ater* black; M.L. adj. *nogalater* black walnut (referring to an odor like that of black walnuts).

Spore chains are *Retinaculiaperti;* the spore surface is smooth. The spore mass is gray; the reverse is yellow-brown; a yellow-brown diffusible pigment is produced. Melanin pigment is not produced.

Antimicrobial activity is shown against *Saccharomyces cerevisiae.*

The strain is resistant to dimethylchlortetracycline (500 μg/ml).

L-Tyrosine is not degraded.

Type strain: NRRL 3035.

7. **Streptomyces prasinosporus** Tresner, Hayes and Backus 1966, 162.[AL]

pra.si.no'spo.rus. L. adj. *prasinus* green; M.L. n. *spora* a spore; M.L. adj. *prasinosporus* green-spored.

Spore chains are *Spirales;* the spore surface is hairy. The spore mass is green; the reverse is yellow-brown; diffusible pigments are not produced. Melanin pigment is produced on peptone-yeast extract-iron agar but not on tyrosine agar.

L-Tyrosine is not degraded.

L-Arginine is not used.

Type strain: ATCC 17918.

8. **Streptomyces ochraceiscleroticus** (Kuznetsov 1962) Pridham 1970, 22.[AL] (*Chainia ochracea* Kuznetsov 1962, 539.)

och.ra'ce.i.scle.ro'ti.cus. Gr. n. *ochre* ochre; M.L. adj. *ochraceus* like ochre, rust colored; L. n. *sclerotium* sclerotium; M.L. adj. *ochraceiscleroticus* sclerotium with rust color.

Spore chains are *Spirales;* the spore surface is smooth. Sclerotia are produced. The spore mass is white; the reverse is yellow-brown; diffusible pigments are not produced. Melanin pigment is not produced.

The strain is resistant to penicillin G (10 i.u.).

Growth occurs in the presence of sodium chloride (7% w/v) and sodium azide (0.01% w/v).

Type strain: ATCC 15814.

9. **Streptomyces aurantiogriseus** (Preobrazhenskaya in Gauze et al. 1957) Pridham, Hesseltine and Benedict 1958, 67.[AL] (*Actinomyces aurantiogriseus* Preobrazhenskaya in Gauze et al. 1957, 74.)

au.ran.ti.o.gri'se.us. L. n. *aurum* gold; M.L. n. *Aurantium* generic name of the orange; M.L. adj. *griseus* gray; M.L. adj. *aurantiogriseus* orange, gray.

Spore chains are *Retinaculiaperti;* the spore surface is smooth. The spore mass is gray; the reverse is yellow-brown; a yellow-brown diffusible pigment is produced. Melanin pigment is produced.

There is no growth at 10°C.

L-Methionine, inulin, and sodium malonate are used.

Type strain: ATCC 19887.

10. **Streptomyces bambergiensis** Wallhäusser, Nesemann, Präve and Steigler 1966, 734.[AL]

bam.ber.gi.en'sis. M.L. adj. *bambergiensis* belonging to Bamberg, Germany (the source of soil from which the organism was isolated).

Spore chains are *Retinaculiaperti;* the spore surface is hairy. The spore mass is green; the reverse is red-orange and its pigments are pH sensitive. Red-orange, pH-sensitive, diffusible pigments are produced. Melanin pigment is not produced.

The strain is sensitive to lincomycin (100 μg/ml).

Type strain: ATCC 13879.

11. **Streptomyces gelaticus** (Waksman 1923) Waksman and Henrici 1948b, 952.[AL] (*Actinomyces gelaticus* Waksman 1923, 356.)

ge.la'ti.cus. L. part. adj. *gelatus* congealed, jellied; M.L. adj. *gelaticus* resembling hardened gelatin.

Spore chains are *Rectiflexibiles;* the spore surface is smooth. The spore mass is too sparse for color determination; the reverse is yellow-brown; diffusible pigments are not produced. Melanin pigment is not produced.

There is no growth at 37°C.

L-Cysteine and L-hydroxyproline are used.

Type strain: ATCC 3323.

12. **Streptomyces amakusaensis** Nagatsu, Anzai, Ohkuma and Suzuki 1963, 209.[AL]

am.a.ku.sa.en'sis. M.L. adj. *amakusaensis* belonging to Amakusa Island, Japan (the source of soil from which the organism was isolated).

Spore chains are *Spirales;* the spore surface is smooth. The spore mass is blue; the reverse is yellow-brown and its pigments are pH sensitive. Diffusible pigments are not produced. Melanin pigment is not produced.

Antimicrobial activity is shown against *Pseudomonas fluorescens, Aspergillus niger,* and *Candida albicans.*

Inulin is used.

Type strain: ATCC 23876.

13. **Streptomyces varsoviensis** Kurylowicz and Woźnicka 1967, 1.[AL]

var.so'vi.en'sis. M.L. *Varsovia* Warsaw; M.L. adj. *varsoviensis* pertaining to Warsaw (named for Warsaw, Poland).

Spore chains are *Spirales;* the spore surface is smooth. The spore mass is white; the reverse is yellow-brown; diffusible pigments are not produced. Melanin pigment is not produced.

Growth occurs in the presence of sodium chloride (10% w/v).

The strain is resistant to gentamicin (100 μg/ml), neomycin (50 μg/ml), and streptomycin (100 μg/ml).

Type strain: ATCC 25505.

14. **Streptomyces tubercidicus** Nakamura 1961b, 90.[AL]

tu.ber.ci'di.cus. L. n. *tuber* nodule; L. comb. form *-cid* from L. v. *caedo* to kill; L. n. stem suff. *-icus* to denote possession; M.L. adj. *tubercidicus* nodule destroying (referring to antitumor activity of an antibiotic produced by the organism).

Spore chains are *Spirales;* the spore surface is smooth. The spore mass is violet to gray; the reverse is yellow-brown and its pigments are pH sensitive. Diffusible pigments are not produced. Melanin pigment is not produced.

Hydrogen sulfide is not produced.

Mannitol is not used.

Type strain: ATCC 25502.

15. **Streptomyces badius** (Kudrina in Gauze et al. 1957) Pridham, Hesseltine and Benedict 1958, 58.[AL] (*Actinomyces badius* Kudrina in Gauze et al. 1957, 87).

ba'di.us. L. adj. *badius* brown.

Spore chains are *Rectiflexibiles;* the spore surface is smooth. The spore mass is yellow; the reverse is yellow-brown; yellow-brown diffusible pigments are produced. Melanin pigment is produced on tyrosine agar but not on peptone-yeast extract-iron agar.

Growth occurs in the presence of sodium chloride (10% w/v).

D-Mannose is not used.

Type strain: ATCC 19888.

16. **Streptomyces ramulosus** Ettlinger, Gäumann, Hütter, Keller-Schierlein, Kradolfer, Neipp, Prelog and Zähner 1958b, 217.[AL]

ram.u.lo'sus. L. adj. *ramulosus* much branched.

Spore chains are *Retinaculiaperti;* the spore surface is smooth. The spore mass is gray; the reverse is yellow-brown to red-orange; red-orange, pH-sensitive, diffusible pigments are produced. Melanin pigment is not produced.

Growth occurs in the presence of sodium chloride (10% w/v).

The strain is resistant to streptomycin (100 μg/ml).

L-Serine is not used.

Type strain: ATCC 19802.

17. **Streptomyces sulphureus** (Gasperini 1894) Waksman in Waksman and Lechevalier 1953, 278.[AL] (*Actinomyces sulphureus* Gasperini 1894, 78.)

sul.phu're.us. L. adj. *sulfureus* of sulfur (referring to the sulfur-yellow color of the spore mass).

Spore chains are *Spirales;* the spore surface is smooth. The spore mass is too sparse for color determination; the reverse is yellow-brown; diffusible pigments are not produced. Melanin pigment is not produced.

Growth occurs in the presence of sodium chloride (13% w/v). No growth at 10°C.

L-Tyrosine is not degraded and L-serine is not used.

Type strain: ATCC 27468.

18. **Streptomyces yerevanensis** (Tsyganov, Zhukova and Timofeeva 1964) Goodfellow, Williams and Alderson 1986a, 575.[VP] (Effective publication: Goodfellow, Williams and Alderson 1986d, 52.) (*Macrospora violaceus* Tsyganov, Zhukova and Timofeeva 1964, 868; *Microellobosporia violacea* (Tsyganov, Zhukova and Timofeeva) Pridham 1974a, 844.)

ye.re.van.en'sis. M.L. adj. pertaining to Yerevan, Armenia, U.S.S.R.

Spore chains are *Rectiflexibiles;* the spore surface is smooth. The spore chains are short and the spores are relatively large (1.5–3.5 μm in diameter); they are produced on both aerial and substrate mycelium. The spore mass is too sparse for color determination; the reverse is violet and its pigments are pH sensitive; violet, pH-sensitive, diffusible pigments are produced. Melanin pigment is not produced.

Type strain: DSM 43167.

19. **Streptomyces massasporeus** Shinobu and Kawato 1959, 283.[AL]

mas.sa.spo're.us. L. n. *massa* mass, lump; M.L. n. *spora* a spore; M.L. adj. *massasporeus* mass, spore (referring to the coalescence of spores into moist masses).

The spore chains are *Spirales;* the spore surface is smooth. The spore mass is violet; the reverse is violet; violet diffusible pigments are produced. Melanin pigment is produced.

The strain is resistant to dimethylchlortetracycline (500 μg/ml).

Gelatin is not degraded.

Type strain: ATCC 19785.

20. **Streptomyces alboflavus** (Waksman and Curtis 1916) Waksman and Henrici 1948b, 954.[AL] (*Actinomyces alboflavus* Waksman and Curtis 1916, 120.)

al.bo.fla'vus. L. adj. *albus* white; L. adj. *flavus* yellow; M.L. adj. *alboflavus* whitish yellow.

Aerial mycelium is sparse or absent. The reverse is yellow-brown; diffusible pigments are not produced. Melanin pigment is not produced.

Growth occurs in the presence of phenylethanol (0.3% w/v).

Antimicrobial activity is shown against *Candida albicans.*

Testosterone is not degraded.

Type strain: ATCC 12626.

21. **Streptomyces bikiniensis** Johnstone and Waksman 1947, 294.[AL]

bi.ki.ni.en'sis. M.L. adj. *bikiniensis* pertaining to Bikini atoll.

Spore chains are *Rectiflexibiles;* the spore surface is smooth. The spore mass is gray; the reverse is yellow-brown; diffusible pigments are not produced. Melanin pigment is produced on peptone-yeast extract-iron agar but not on tyrosine agar.

Growth occurs at 45°C.

Type strain: ATCC 11062.

22. **Streptomyces fragilis** Anderson, Erhlich, Sun and Burkholder 1956, 105.[AL]

fra'gi.lis L. adj. *fragilis* fragile.

Spore chains are *Retinaculiaperti;* the spore surface is smooth. The spore mass is red; the reverse is yellow-brown; diffusible pigments are not produced. Melanin pigment is not produced.

Growth occurs in the presence of phenylethanol (0.3% w/v).

The strain is sensitive to cephaloridine (100 μg/ml).

L-Methionine is used but D-mannose is not.

Type strain: NRRL 2424.

23. **Streptomyces lateritius** (Sveshnikova in Gauze et al. 1957) Pridham, Hesseltine and Benedict 1958, 67.[AL] (*Actinomyces lateritius* Sveshnikova in Gauze et al. 1957, 70.)

la.ter.it'i.us. L. n. *later* brick; M.L. adj. *lateritius* o.v. of *latericius* of bricks, brick red.

Spore chains are *Spirales;* the spore surface is smooth. The spore mass is red; the reverse is red-orange and its pigments are pH sensitive. Red-orange, pH-sensitive, diffusible pigments are produced. Melanin pigment is produced.

The strain is resistant to gentamicin (100 μg/ml), tobramycin (50 μg/ml), and vancomycin (50 μg/ml).

Type strain: ATCC 19913.

24. **Streptomyces finlayi** (Szabó, Marton, Buti and Pártai 1963) Pridham 1970, 35.[AL] (*Actinomyces finlayi* Szabó, Marton, Buti and Pártai 1963, 209.

fin.lay'i. M.L. gen. n. *finlayi* of Finlay (named for A. C. Finlay).

Spore chains are *Rectiflexibiles;* the spore surface is hairy. The spore mass is gray; the reverse is green; diffusible pigments are not produced. Melanin pigment is not produced.

Growth occurs in the presence of potassium tellurite (0.01% w/v).

The strain is sensitive to lincomycin (100 μg/ml).

Lecithinase activity occurs.

Type strain: ATCC 23340.

25. **Streptomyces novaecaesareae** Waksman and Henrici 1948b, 951.[AL]

nov.ae.caes.ar'eae. M.L. adj. *novaecaesareae* pertaining to New Jersey, U.S.A.

The aerial mycelium is sparse or absent. The reverse is red-orange and its pigments are pH sensitive; diffusible pigments are not produced. Melanin pigment is not produced.

Growth does not occur in the presence of sodium chloride (4% w/v).

Type strain: ATCC 27452.

Category IV

The treatment of species in this category is based mainly on criteria used and descriptions provided in the International *Streptomyces* Project (Shirling and Gottlieb, 1966, 1968a, 1968b, 1969, 1972). Morphological and pigmentation characteristics are given in the text descriptions and information on carbon utilization patterns is shown in Tables 29.12 to 29.16. The species are allocated to artificial groups or series based upon the color of the mature sporulated aerial mycelium (spores *en masse*) following the approach of Pridham and Tresner (1974b). Seven color groups are recognized: Gray (gray to brownish); White; Red (tan, pink, and rose shades); Yellow (yellowish to greenish yellow); Blue (bluish to grayish blue shades); Green (greenish to grayish green shades); and Violet.

Gray Series

1. **Streptomyces alanosinicus** Thiemann and Beretta 1966, 158.[AL]

al'an.o.si'ni.cus. Eng. n. *alanosine* name of an antibiotic; L. n. stem. suff. *-icus* belonging to; M.L. adj. *alanosinicus* belonging to alanosine.

Spore chains are *Spirales,* the spore surface is smooth. The spore mass is in the gray series; the reverse is reddish brown to strong brown.

Produces alanosine, an antitumor, antiviral, and antifungal antibiotic; excellent growth on Czapek's solution agar; spores may be spiny.

Type strain: ATCC 15710.

2. **Streptomyces albaduncus** Tsukiura, Okanishi, Ohmori, Koshiyama, Miyaki, Kitazima and Kawaguchi 1964b, 41.[AL]

al.ba.dun'cus. L. adj. *albus* white; L. adj. *uncus* hooked, crooked; M.L. adj. *alba(d)uncus* white, hooked (probably referring to color of aerial mycelium and nature of spore chains of the organism).

Spore chains are *Retinaculiaperti* to *Spirales,* the spore surface is spiny with numerous long spores. The spore mass is in the gray series; the reverse is grayish, yellow to yellow brown. Melanin pigment is not produced.

Produces danomycin, an antibacterial antibiotic; exhibits antifungal activity. Is inhibited by streptomycin.

Type strain: ATCC 14698.

3. **Streptomyces albospinus** Wang, Hamada, Okami and Umezawa 1966, 217.[AL]

al.bo.spin'us. L. adj. *albus* white; L. adj. *spineus* spiny; M.L. adj. *albospinus* white, spiny (referring to the color of the aerial mycelium and nature of spore wall ornamentation).

Spore chains are *Spirales,* the spore surface is spiny. The spore mass is in the gray series. Melanin pigment is not produced.

Produces spinamycin, a nonpolyenic antifungal antibiotic.

Type strain: ATCC 29808.

4. **Streptomyces anthocyanicus** (Krasil'nikov, Sorokina, Alferova and Bezzubenkova 1965b) Pridham 1970, 7.[AL] (*Actinomyces anthocyanicus* Krasil'nikov, Sorokina, Alferova and Bazzubenkova in Krasil'nikov 1965, 118.)

an.tho.cy.an'i.cus. Gr. n. *anthos* a flower; *cyanicus* presumably based on Gr. adj. *cyaneus* dark blue; M.L. adj. *anthocyanicus* presumably referring to dark blue color of anthocyanin.

Spore chains are *Retinaculiaperti,* the spore surface is smooth. The spore mass is in the gray series; the reverse is blue. A diffusible pigment is formed on some media but melanin pigment is not produced.

Shows slight antibacterial activity.

Type strain: ATCC 19821.

5. **Streptomyces antimycoticus** Waksman 1957a, 799.[AL]

an.ti.my.co'ti.cus. Gr. pref. *anti-* against; Gr. n. *myces* fungus; M.L. adj. *antimycoticus* antifungal.

Spore chains are *Spirales,* the spore surface is spiny to warty, i.e. surface ornamentation is intermediate between very short, thick spines and warts. The spore mass is in the gray series; the reverse is pale grayed yellow or grayish-greenish yellow. Melanin pigment is produced.

Produces the endomycin complex (a mixture of polyenic and nonpolyenic antifungal antibiotics designated as endomycins A, B, C, and D); shows antibacterial activity. Is inhibited by streptomycin.

Type strain: NRRL 2421.

6. **Streptomyces atratus** Shibata, Higashide, Yamamoto and Nakazawa 1962, 230.[AL]

a.tra'tus. L. adj. *atratus* clothed in black.

Spore chains are atypical *Retinaculiaperti,* the spore surface is smooth. The spore mass is in the gray series; the reverse is gray to black. Melanin pigment is not produced.

Produces rufomycin A and rufomycin B and shows other antibacterial activity.

Type strain: IFO 3897.

7. **Streptomyces caeruleus** (Baldacci 1944) Pridham, Hesseltine and Benedict 1958, 60.[AL] (*Actinomyces caeruleus* Baldacci 1944, 180.)

cae.ru'le.us. L. adj. *caeruleus* dark blue, azure.

Spore chains are *Rectiflexibiles,* the spore surface is smooth. The spore mass is gray; the reverse is dark blue-violet to black and its pigments are pH sensitive. Bluish-black to greenish-black diffusible pigments are produced on some media. Melanin pigment is not produced. Requires an alkaline medium for optimal growth.

Produces caerulomycin.

Isolated from rice straw.

Type strain: ATCC 27421.

8. **Streptomyces capillispiralis** Mertz and Higgens 1982, 123.[VP]

ca.pil.li.spi.ra'lis. L. n. *capillus* hair; L. adj. *spiralis* spiral or spiraled; M.L. adj. *capillispiralis* hairy-spiraled.

Spore chains are *Spirales,* the spore surface is hairy; coremia may be

Table 29.12.
Utilization of compounds as sole sources of carbon by the type strains of **Streptomyces** *species belonging to the gray series*[a,b]

Species	Carbon-free control	L-Arabinose	D-Fructose	D-Galactose	D-Glucose	*meso*-Inositol	D-Mannitol	Raffinose	L-Rhamnose	Salicin	Sucrose	D-Xylose
1. *S. alanosinicus*		+	+	+	+	+	+	+	−	+	?	+
2. *S. albaduncus*		+	+	+	+	+	+	±	+	+	±	+
3. *S. albospinus*		−	+	+	+	+	+	+	−	+	−	±
4. *S. anthocyanicus*		+	+	+	+	+	−	−	+		−	
5. *S. antimycoticus*	−	+	+		+	+	+	+	+	+	+	+
6. *S. atratus*	−	−	+	+	+	−	−	+	+	+		+
7. *S. caeruleus*	−	−			+				−		?	−
8. *S. capillispiralis*		+	+	+	+	+	+	−	+	?	−	+
9. *S. cinerochromogenes*		+				−	−	−	+	+	+	−
10. *S. clavuligerus*		−	−		−	?	−	−	−		−	−
11. *S. cuspidosporus*	±	+	+	+	+	+	+	+	+	+	+	+
12. *S. djakartensis*[c]												
13. *S. echinoruber*		+	+	+	+	−	−	−	−	−	+	+
14. *S. ederensis*[c]												
15. *S. erumpens*	−	+	+	+	+	+	+	+	+	−	−	−
16. *S. fimbriatus*	−	+	+	+	+	+	+	+	+	−		+
17. *S. gancidicus*	−	+	+	+	+	+	+	−	+	−	−	+
18. *S. geysiriensis*[c]												
19. *S. globosus*	−	+			+				−			+
20. *S. kurssanovii*	−	+	+	+	+	−	−	+	−	−	+	+
21. *S. longwoodensis*		+	+	+	+	+	+	±	−	+	−	+
22. *S. nojiriensis*		−	−	−	+	−	−	−	−	+	−	−
23. *S. noursei*	−	−	+	+	+	+	+	−	−	−	+	−
24. *S. olivaceiscleroticus*	−	+	+		+	+	+	+	+		+	+
25. *S. pulveraceus*	−	−	+	+		−	−	+	+	+	−	+
26. *S. purpeofuscus*	−	+	−	+	+	−	−	−	−		−	+
27. *S. rameus*	−	+	+	+		−	+	+	−	+	+	+
28. *S. roseodiastaticus*	−	+			+				+			+
29. *S. sannanensis*	−	−	−		−	−	−	−	−		−	−
30. *S. tanashiensis*	−	+	−	+	+	−	−	−	−	+	−	+
31. *S. torulosus*		+	+	+	+	+	+	+	+	−		+
32. *S. tricolor*[c]												
33. *S. violaceochromogenes*		+	+		+	+	+	+	+		+	+
34. *S. violaceoruber*	−	+	+	+	+	+	+	−	+	+	−	+
35. *S. viridiviolaceus*		+	+		+	?	+	−	−		?	?
36. *S. viridodiastaticus*		+	+		+	+	+	?	+		?	+

[a]Symbols: +, carbon compound utilized; −, not utilized; ±, very slight utilization; ?, doubtful utilization.
[b]Data were obtained using the basal medium of Pridham and Gottlieb (1948).
[c]No data available.

produced. The spore mass is lightish-brownish gray; the reverse is yellow-brown to brownish black; a brown soluble pigment may be produced. Melanin is not produced.

Casein, hypoxanthine, tyrosine, and xanthine are degraded but esculin, gelatin, and starch are not. Catalase, phosphatase, and urease are formed, but nitrate is not reduced. Resistant to lysozyme.

Growth from 10° to 45°C, optimum 30°C.

Produces a cephalosporin C-4 carboxymethylesterase.

Isolated from soil, Sweden.

Type strain: NRRL 12279.

9. **Streptomyces cinerochromogenes** Miyairi, Takashima, Shimizu and Sakai 1966, 58.[AL]

cin′er.o.chro.mo′ge.nes. L. adj. *cinereus* ashy; Gr. n. *chroma* color; Gr. v. suff. *-genes* producing; M.L. adj. *cinerochromogenes* producing ashy color.

Spore chains are *Spirales*, the spore surface is smooth. The spore mass is in the gray series. Melanin pigment is produced.

Produces the antibacterial antibiotics cineromycin A and cineromycin B.

Type strain: IFO 13922.

Table 29.13.
Utilization of compounds as sole sources of carbon by the type strains of **Streptomyces** *species belonging to the white series*[a]

Species	Carbon-free control	L-Arabinose	D-Fructose	D-Galactose	D-Glucose	*meso*-Inositol	D-Mannitol	Raffinose	L-Rhamnose	Salicin	Sucrose	D-Xylose
1. *S. armeniacus*	−		+	+	±		−	±	+	±	±	±
2. *S. bobili*	−	+	+	+	+	+	−	+	+	−	+	+
3. *S. clavifer*	−	−	+	+	+	−	+	−	+	−		+
4. *S. fumigatiscleroticus*	−	+		+	+	−	+		−	−	−	+
5. *S. gibsonii*	−	+	−		+	−	+	−	−	+		+
6. *S. pseudoechinosporeus*	−	+	+	+	+		+	+	+		+	+
7. *S. rangoon*		?	+		+	−	+	−	−		−	+
8. *S. somaliensis*		−	−	−	+	−	−	−	−	−	−	−
9. *S. vitaminophilus*		−	−		+	−	−	−	+		−	+

[a]For explanation of symbols and basal medium used, see footnotes *a* and *b* to Table 29.12.

Table 29.14.
Utilization of compounds as sole sources of carbon by the type strains of **Streptomyces** *species belonging to the red series*[a]

Species	Carbon-free control	L-Arabinose	D-Fructose	D-Galactose	D-Glucose	*meso*-Inositol	D-Mannitol	Raffinose	L-Rhamnose	Salicin	Sucrose	D-Xylose
1. *S. albosporeus*	−	+	+	+	+	+	+	+	+	−		+
2. *S. cinnamonensis*	−	−	+		+	−	−	−	−	+		−
3. *S. crystallinus*					+						−	
4. *S. erythrogriseus*	−	+	+		+	+	+	−	+	+	−	+
5. *S. gobitricini*	−	+	+	+	+	+	−	−	+	−		+
6. *S. laurentii*		−	−	+	+	−	−	−	−		+	+
7. *S. lavendofoliae*	−	+	?	+	+	+	−	−	−	−	−	+
8. *S. niveoruber*	−	+			+				+			+
9. *S. peucetius*[b]		−	+		+		+	+	−		+	+
10. *S. roseoflavus*	−	+	?	−	+	−	−		−		−	+
11. *S. ruber*		+	+		+		+	−	+			+
12. *S. rubiginosohelvolus*		+	+		+	−	+	−	+		−	+
13. *S. sulfonofaciens*		+	+		+	−	±	−	+		±	+

[a]For explanation of symbols and basal medium used, see footnotes *a* and *b* to Table 29.12.
[b]Carbon utilization basal medium not specified.

10. **Streptomyces clavuligerus** Higgens and Kastner 1971, 327.[AL]
cla.vu.li′ger.us. L. fem. n. *clavula* little club; L. suff. *-igerus* bearing; M.L. adj. *clavuligerus* bearing little clubs.

Spore chains are *Rectiflexibiles;* the spore surface is smooth. The spore mass is dark grayish green on media with abundant sporulation; the reverse is pale yellow to grayish yellow. Melanin pigment is not produced.

Does not reduce nitrate or liquify gelatin.

Isolated from soil.

Type strain: NRRL 3585.

11. **Streptomyces cuspidosporus** Higashide, Hasegawa, Shibata, Mizuno and Akaike 1966, 2.[AL]
cu′spi.do.spo′rus. L. n. *cuspis-idis* point; M.L. n. *spora* a spore; M.L. adj. *cuspidosporus* spore with points or spines.

Spore chains are *Spirales,* the spore surface is spiny. The spore mass is in the gray series. Green to blue to yellowish-green diffusible pigment is formed on some media. Melanin pigment is not produced.

Produces sparsomycin and tubercidin and several other antibiotics.

Type strain: IFO 12378.

Table 29.15.
Utilization of compounds as sole sources of carbon by the type strains of **Streptomyces** *species belonging to the yellow series*[a]

Species	Carbon-free controls	L-Arabinose	D-Fructose	D-Galactose	D-Glucose	*meso*-Inositol	D-Mannitol	Raffinose	L-Rhamnose	Salicin	Sucrose	D-Xylose
1. *S. celluloflavus*	−	−			+				−			−
2. *S. champavatii*	−	+	+	+	+		+	−	−		−	+
3. *S. flavidovirens*	−	+	?		+	?	−	−	?		?	+
4. *S. floridae*		−	+	+	+	−	+	−	−	−		+
5. *S. griseoloalbus*		+	+		+	+	+		+		+	+
6. *S. kunmingensis*		+				−	+	+	+		−	+
7. *S. mediolani*		+	+		+		+		+		+	+
8. *S. praecox*	−	+	+	+	+	−	+	+	+	+	?	+

[a]For explanation of symbols and basal medium used, see footnotes *a* and *b* to Table 29.12.

Table 29.16.
Utilization of compounds as sole sources of carbon by the type strains of **Streptomyces** *species belonging to the blue, green, and violet series*[a]

Species	Carbon-free control	L-Arabinose	D-Fructose	D-Galactose	D-Glucose	*meso*-Inositiol	D-Mannitol	Raffinose	L-Rhamnose	Salicin	Sucrose	D-Xylose
Blue series												
1. *S. inusitatus*		−	−	±	±	−	−	−	−	−	−	−
2. *S. ipomoeae*		+	+		+	+	+	+	+		+	+
3. *S. lomondensis*	−	+	+	+	+	+	+	+	+	−	+	+
Green Series												
1. *S. acrimycini*	−	+	+		+	+	+	−	+		−	+
2. *S. ghanaensis*[c]												
3. *S. griseostramineus*		+	+		+	+	+	?	+		−	+
4. *S. spinoverrucosus*		+	+		+	+	+	+	+		+	+
Violet series												
1. *S. mauvecolor*[b]		+	−	+	+	−	−	+	−	+	−	−

[a]For explanation of symbols and basal medium used, see footnotes *a* and *b* to Table 29.12.
[b]Carbon utilization basal medium, Czapek's solution.
[c]No data available.

12. **Streptomyces djakartensis** Huber, Wallhäusser, Fries, Steigler and Weidenmüller 1962, 1191.[AL]

dja.kart.en′sis. M.L. adj. *djakartensis* belonging to Djakarta, Indonesia (the source of the soil from which the organism was isolated).

Spore chains are *Spirales*. The spore mass is in the gray series. Melanin pigment is produced.

Produces niddamycin (3-desacetylcarbomycin B), an antibacterial macrolide.

Type strain: DSM 40743.

13. **Streptomyces echinoruber** Palleroni, Reichett, Mueller, Epps, Tabenkin, Bull, Schüep and Berger 1981, 382.[VP] (Effective publication: Palleroni, Reichett, Mueller, Epps, Tabenkin, Bull, Schüep and Berger 1978, 1222.)

e.chi′no.ru′ber. Gr. adj. *echinos* spiny appearance; L. adj. *ruber* red; M. L. adj. *echinoruber* spiny red.

Spore chains are *Retinaculiaperti,* the spore surface is spiny. The spore mass is in the gray series; the reverse is rose-brown. A deep cherry-red diffusible pigment is formed on various media. Melanin pigment is not produced.

Adenine, casein, gelatin, and starch are degraded but xanthine is not. Nitrate is reduced.

Isolated from soil collected in Argentina.

Type strain: NRRL 8144.

14. **Streptomyces ederensis** Wallhäusser, Nesemann, Präve and Steigler 1966, 734.[AL]

eder.en′sis. M. L. adj. *ederensis* pertaining to Eder (named for the Eder valley in Germany).

Spore chains are *Rectiflexibiles,* the spore surface is smooth. The spore mass is in the gray series. Melanin pigment is produced.

Produces the moenomycin complex of antibacterial antibiotics comprised of moenomycins A, B_1, B_2, and C.

Type strain: ATCC 15304.

15. **Streptomyces erumpens** Calot and Cercos 1963, 159.[AL]

er.um′pens. L. part. adj. *erumpens* bursting forth.

Spore chains are *Spirales,* the spore surface is smooth. The spore mass is in the gray series. Melanin pigment is not produced.

Shows antibacterial activity; produces the tetraenic antifungal antibiotic 17732 (tetrins A and B and another polyene); is inhibited by streptomycin.

Type strain: ATCC 23266.

16. **Streptomyces fimbriatus** (Millard and Burr 1926) Waksman and Lechevalier 1953, 104.[AL] (*Actinomyces fimbriatus* Millard and Burr 1926, 639.)

fim.bri.a′tus. L. adj. *fimbriatus* fibrous, fringed.

Spore chains are *Spirales,* the spore surface is smooth. The spore mass is in the gray series. Melanin pigment is produced.

Produces septacidin, an antitumor and antifungal purine antibiotic; is not inhibited by streptomycin.

Type strain: ATCC 15051.

Further comments: Millard and Burr's original single isolate, which is no longer extant, was obtained from a case of common potato scab.

17. **Streptomyces gancidicus** Suzuki 1957, 538.[AL]

gan.ci′di.cus. Eng. n. *gancidin* name of an antibiotic; L. n. suff. *-icus* belonging to; M.L. adj. *gancidicus* belonging to gancidin.

Spore chains are *Spirales,* the spore surface is spiny. The spore mass is in the gray series. Melanin pigment is not produced.

Produces the gancidin complex (components A and W) effective against Gram-positive bacteria and tumors; is inhibited by streptomycin.

Type strain: NRRL B-1872.

18. **Streptomyces geysiriensis** Wallhäusser, Nesemann, Präve and Steigler 1966, 734.[AL]

gey′sir.i.en′sis. Icel. *geysir* a geyser; M.L. adj. *geysiriensis* belonging to a geyser (referring to the source of the organism, an Iceland geyser).

Spore chains are *Spirales,* the spore surface is hairy. The spore mass is in the gray series. Melanin pigment is not produced.

Produces the moenomycin complex of antibacterial antibiotics (moenomycins A, B_1, B_2, and C).

Type strain: ATCC 15303.

19. **Streptomyces globosus** (Krasil'nikov 1941) Waksman in Waksman and Lechevalier 1953, 68.[AL] (*Actinomyces globosus* Krasil'nikov 1941, 58.)

glo.bo′sus. L. adj. *globosus* spherical (referring to the shape of the spores when examined with the light microscope).

Spore chains are *Rectiflexibiles;* the spore surface is smooth. The spore mass is in the gray series. Melanin pigment is produced.

Shows antibacterial and antifungal activity; is inhibited by streptomycin.

Type strain: B-2292.

Further comment: None of Krasil'nikov's original notes is extant.

20. **Streptomyces kurssanovii** (Preobrazhenskaya, Kudrina, Ryabova and Blinov in Gauze et al. 1957) Pridham, Hesseltine and Benedict 1958, 69.[AL] (*Actinomyces kurssanovii* Preobrazhenskaya, Kudrina, Ryabova and Blinov in Gauze et al. 1957, 156.)

kurs.san.ov′i.i. M.L. gen. n. *kurssanovii* of Kursanov (possibly named after L. I. Kursanov, a Russian microbiologist).

Spore chains are *Retinaculiaperti,* the spore surface is smooth. The spore mass is in the gray series; the reverse is grayish yellow to yellow-brown. Melanin pigment is produced.

Shows antibacterial activity; is inhibited by streptomycin.

Type strain: ATCC 15824.

Further comments: In the original description the organism is said to form spiral spore chains. Hütter (1967, p. 321) commented that three strains originally assigned to this species are quite different from each other. Characterization, in part, by Shirling and Gottlieb (1968a).

21. **Streptomyces longwoodensis** Prosser and Palleroni 1981, 382.[VP] (Effective publication: Prosser and Palleroni 1976, 322.)

long.wood.en′sis. M.L. gen. n. *longwoodensis* of Longwood (named for Longwood Gardens in Kennett Square, PA, U.S.A., the origin of the soil from which the organism was first isolated).

Spore chains are *Spirales,* the spore surface is smooth. The spore mass is in the gray series; the reverse is brown. Melanin pigment is not produced.

Sensitive to streptomycin (10-µg disk). Produces lysocellin.

Type strain: ATCC 29251.

22. **Streptomyces nojiriensis** Ishida, Kumagai, Niida Hamamoto and Shomura 1967, 64.[AL]

no.jir.i.en′sis. M.L. adj. *nojiriensis* belonging to Nojiri (named for Lake Nojiri at Nagano, Japan, the source of the soil from which the organism was isolated).

Spore chains are *Spirales,* the spore surface is smooth. The spore mass is in the gray series. Melanin pigment is produced.

Produces nojirimycin, an antibacterial antibiotic.

Type strain: ATCC 29781.

Further comments: Two additional strains producing nojirimycin, but not included in this taxon, were reported by Ishida et al. (1967, p. 62).

23. **Streptomyces noursei** Brown, Hazen and Mason 1953, 609.[AL]

nour′se.i. M.L. gen. n. *noursei* of Nourse (referring to the owner of the farm where soil was obtained from which the organism was isolated).

Spore chains are *Spirales,* the spore surface is spiny. The spore mass is in the gray series. Melanin pigment is not produced.

Shows antibacterial activity; produces nystatin and cycloheximide.

Type strain: ATCC 11455.

24. **Streptomyces olivaceiscleroticus** (Thirumalachar and Sukapure 1964) Pridham 1970, 41.[AL] (*Chainia olivacea* Thirumalachar and Sukapure 1964, 160.)

o.li.va′ce.i.scler.o′ti.cus. M.L. adj. *olivaceus* olive colored; L. n. *sclerotium* sclerotium; M.L. adj. *olivaceiscleroticus* sclerotium with olive color.

Spore chains are *Spirales,* the spore surface is smooth. The spore mass is in the gray series; the reverse is grayish yellow to olive-brown. Sclerotia are formed on some media. Melanin pigment is not produced.

Shows antifungal activity; is inhibited by streptomycin.

Type strain: ATCC 15722.

25. **Streptomyces pulveraceus** Shibata, Higashide, Kanzaki, Yamamoto and Nakazawa 1961, 172.[AL]

pul.ver.ac′e.us. L. n. *pulvis,* pulveris powder; M.L. adj. *pulveraceus* powdery.

Spore chains are *Spirales,* the spore surface is smooth. The spore mass is in the gray series. Melanin pigment is produced.

Produces neomycins E and F, paromomycin and paromomycin II, zygomycin B, cycloheximide, and naramycin B.

Type strain: ATCC 13875.

26. **Streptomyces purpeofuscus** Yamaguchi and Saburi 1955, 207.[AL]

pur.pe.o.fus′cus. M.L. var. *purpe-* of L. adj. *purpureus* purple; L. adj. *fuscus* dark, tawny; M.L. adj. *purpeofuscus* dark purple (referring to color of vegatative mycelium).

Spore chains are *Rectiflexibiles,* the spore surface is smooth. The spore mass is in the gray series; the reverse is grayish yellow to yellow-brown but on some media it may be gray-black, violet-black, or violet-brown. Melanin pigment is produced.

Shows antibacterial, antifungal, and antitrichomonal activity; is inhibited by streptomycin.

Produces negamycin.

Type strain: ATCC 23952.

27. **Streptomyces rameus** Shibata 1959, 398.[AL]

ra.me′us. L. n. *ramus* a branch; M.L. adj. *rameus* pertaining to branches.

Spore chains are *Spirales,* the spore surface is smooth. The spore mass is in the gray series. Melanin pigment is produced.

Produces streptomycin.

Type strain: ATCC 21273.

28. **Streptomyces roseodiastaticus** (Duché 1934) Waksman in Waksman and Lechevalier 1953 27.[AL] (*Actinomyces roseodiastaticus* Duché 1934, 329.)

ro.seo.di.a.sta′ti.cus. L. adj. *roseus* rosy; M.L. adj. *diastaticus* diastatic, starch digesting; M.L. adj. *roseodiastaticus* rosy, diastatic.

Spore chains are *Spirales,* the spore surface is smooth. The spore mass is in the gray series. Melanin pigment is not produced.

Inhibited by streptomycin.

Type strain: CBS 102.34.

29. **Streptomyces sannanensis** Iwasaki, Itoh and Mori 1981, 283.[VP]

san.nan.en′sis. M.L. adj. *sannanensis* of Sannan (named for Sannan Town, Hikami District, Hyogo Prefecture, Japan, the source of soil from which the organism was isolated).

Spore chains are *Spirales,* the spore surface is smooth. The spore mass is in the gray series; the reverse is colorless to buff. Melanin pigment is not produced.

Produces sannamycins.

Type strain: ATCC 31530.

30. **Streptomyces tanashiensis** Hata, Ohki and Higuchi 1952, 529.[AL]

ta.na.shi.en′sis. M.L. adj. *tanashiensis* belonging to Tanashi-machi, a town near Tokyo, Japan (the source of the soil from which the organism was isolated).

Spore chains are *Rectiflexibiles,* the spore surface is smooth. The spore mass is in the gray series; the reverse is pale yellow to light brown-olive. Melanin pigment is produced.

Produces luteomycin and shows antifungal activity.

Type strain: ATCC 23967.

31. **Streptomyces torulosus** Lyons and Pridham 1971, 191.[AL]

tor′u.lo′sus. L. n. *torus* any round swelling, a protuberance; L. dim. n. *torulus* a small protuberance; L. adj. suff. *-osus* full of, having; M.L. *torulosus* having small protuberances.

Spore chains are *Spirales,* the spore surface is knobby, that is, intermediate between warty and spiny. The spore mass is in the gray series. Melanin pigment is produced.

Shows antibacterial and antifungal activity; is sensitive to streptomycin.

Type strain: NRRL B-3889.

32. **Streptomyces tricolor** (Wollenweber 1920) Waksman 1961, 158.[AL] (*Actinomyces tricolor* Wollenweber 1920, 13.)

tri′col.or. L. pref. *tri-* three; L. n. *color* color; M.L. *tricolor* of three colors.

The spore chains are *Spirales;* the spore surface is smooth. The spore mass is in the gray series; the reverse is yellow to red to blue. Blue diffusible pigments are produced.

Isolated from flat scab of potato.

Type strain: CBS 103.21.

33. **Streptomyces violaceochromogenes** (Ryabova and Preobrazhenskaya in Gauze et al. 1957) Pridham 1970, 28.[AL] (*Actinomyces violaceochromogenes* Ryabova and Preobrazhenskaya in Gauze et al. 1957, 183.)

vi.o.la.ce.o.chro.mo′ge.nes. L. adj. *violaceus* violet; Gr. n. *chroma* color; Gr. v. suff. *-genes* producing; M.L. adj. *violaceochromogenes* producing violet color.

The spore chains are either *Retinaculiaperti* or *Spirales;* the spore surface is smooth. Short spore chains form incomplete or imperfect spirals, hooks, and flexuous chains. The spore mass is in the gray series; the reverse is grayish yellow to strong brown. When a red reverse pigment is present, it is pH sensitive. Melanin pigment is produced.

May show antimicrobial activity.

Type strain: ATCC 19932.

34. **Streptomyces violaceoruber** (Waksman and Curtis 1916) Pridham 1970, 44.[AL] (*Actinomyces violaceus-ruber* Waksman and Curtis 1916, 127.)

vi.o.la.ce.o.ru′ber. L. adj. *violaceus* violet; L. adj. *ruber* red; M.L. adj. *violaceoruber* violet-red.

The spore chains are *Spirales,* the spore surface is smooth. *Rectiflexibiles* and *Retinaculiaperti* spore chains have also been reported. The spore mass is in the gray series; the reverse is blue or violet and is pH sensitive. pH-sensitive, blue or violet diffusible pigments are produced. Melanin pigment is not produced.

Shows slight antibacterial activity, is inhibited by streptomycin.

Type strain: ATCC 14980.

Further comments: The antibiotic activity may be due to celicomycin. Much genetic work on streptomycetes has been carried out on strains identified as *Streptomyces coelicolor* (Sermonti and Casciano, 1963). These strains and derived mutants are more closely related to *Streptomyces violaceoruber* than to *S. coelicolor* (Kutzner and Waksman, 1959; Pridham et al., 1965).

35. **Streptomyces viridiviolaceus** (Ryabova and Preobrazhenskaya in Gauze et al. 1957) Pridham, Hesseltine and Benedict 1958, 70.[AL] (*Actinomyces viridiviolaceus* Ryabova and Preobrazhenskaya in Gauze et al. 1957, 188.)

vi.ri.di.vi.o.la′ce.us. L. adj. *viridis* green; L. adj. *violaceus* violet; M.L. adj. *viridiviolaceus* green-violet (referring to the greenish color of the aerial mycelium and the violet color of diffusible pigment).

Spore chains are *Spirales,* the spore surface is spiny to hairy. The spore mass is light brownish gray or grayish-yellowish brown; the reverse is reddish or yellowish brown. Orange or red, pH-sensitive, diffusible pigment is formed on some media. Melanin pigment is not produced.

Shows antibacterial and antifungal activity.

Type strain: ATCC 27478.

36. **Streptomyces viridodiastaticus** (Baldacci, Grein and Spalla 1955) Pridham, Hesseltine and Benedict 1958, 67.[AL] (*Actinomyces virido-diastaticus* (sic) Baldacci, Grein and Spalla 1955, 133.)

vi.ri.do.di.a.sta′ti.cus. L. adj. *viridis* green; M.L. n. *diastasum* the enzyme diastase, hence M.L. adj. *diastaticus* diastatic; M.L. adj. *viridodiastaticus* green diastatic.

The spore chains are *Spirales,* but abundant open spirals and flexuous chains also suggest *Retinaculiaperti* morphology. The spore surface is spiny, although spines may not be apparent on some spores. The spore mass is in the gray series; the reverse is yellowish brown to olive-brown. Melanin pigment is not produced.

Type strain: ATCC 25518.

White Series

1. **Streptomyces armeniacus** (Kalakoutskii and Kuznetsov 1964) Wellington and Williams 1981b, 80.[VP] (*Actinoplanes armeniacus* Kalakoutskii and Kuznetsov 1964, 620.)

ar.men.i.a′cus. M.L. adj. *armeniacus* pertaining to Armenia, U.S.S.R.

Spore chains are *Spirales,* the spore surface is smooth. The spore mass is in the white series; the reverse is cream to brown. Melanin pigment is not produced.

The strain is susceptible to chloramphenicol, erythromycin, penicillin, polymixin, and streptomycin but is resistant to kanamycin and neomycin.

Type strain: ATCC 15676.

2. **Streptomyces bobili** (Waksman and Curtis 1916) Waksman and Henrici 1948b, 937.[AL] (*Actinomyces bobili* Waksman and Curtis 1916, 121.)

bo.bi′li. M.L. n. *bobili* named for Bobili (the nickname of an individual).

Spore chains are *Spirales,* the spore surface is smooth. Sparse aerial mycelium, when produced is white; the reverse is grayish yellow to red to reddish gray or pink, reddish brown, or reddish orange, and is pH sensitive. The production of melanin pigment has been reported.

Produces a cinerubin-like antibiotic; inhibited by streptomycin.

Type strain: ATCC 3310.

3. **Streptomyces clavifer** (Millard and Burr 1926) Waksman in Waksman and Lechevalier 1953, 107.[AL] (*Actinomyces clavifer* Millard and Burr 1926, 630.)

cla′vi.fer. L. adj. *clavifer* club-bearing.

Spore chains are *Rectiflexibiles,* the spore surface is smooth. The spore mass is white. Melanin pigment is not produced.

Type strain: CBS 101.27.

4. **Streptomyces fumigatiscleroticus** (ex Pridham 1970) Goodfellow, Williams and Alderson 1986a, 574.[VP] (Effective publication: Goodfellow, Williams and Alderson 1986c, 57.) (*Chainia fumigata* Thirumalachar in Thirumalachar, Sukapure, Rahalkar and Gopalkrishnan 1966, 10.)

fu.mi.ga′ti.scle.rot′i.cus. L. n. *fumus* smoke, steam; L. suff. *-atus* provided with; L. part. adj. *fumigatus* smoked; L. n. *sclerotium* sclerotium; M.L. adj. *fumigatiscleroticus* referring to smoke color and ability to produce sclerotia.

Spore chains are *Spirales.* The spore mass is in the white series. Produces sclerotia. Melanin pigments not formed.

Degrades gelatin, starch, urea, and xanthine. Nitrate is reduced but hydrogen sulfide is not produced.

Growth at 42°C but not at 10°C.

Shows antibacterial activity.

Isolated from soil, Pimpri, India.

Type strain: ATCC 19345.

5. **Streptomyces gibsonii** (Erikson 1935) Waksman and Henrici, 1948b, 963.[AL] (*Actinomyces gibsonii* Erikson 1935, 36.)

gib.so′ni.i. L.gen. n. *gibsonii* of Gibson (named for A. G. Gibson, who first isolated the organism).

Spore chains are *Spirales:* the spore surface is smooth. The spore mass is in the white series. Melanin pigment is not produced.

Degrades esculin, allantoin, hypoxanthine, tyrosine, and urea, but not adenine, chitin, elastin, guanine, starch, testosterone, xanthine, or xylan.

Shows slight antibacterial activity.

Type strain: ATCC 6852.

6. **Streptomyces pseudoechinosporeus** (Konev, Tsyganov, Minbaev and Morogov 1967) Goodfellow, Williams and Alderson 1986a, 575.[VP] (Effective publication: Goodfellow, Williams and Alderson 1986d, 52.) (*Microechinospora grisea* Konev, Tsyganov, Minbaev and Morogov 1967, 309; *Microellobosporia grisea* (Konev, Tsyganov, Minbaev and Morogov) Pridham 1970, 17.)

pseu.do.e.chi′no.spo′reus. Gr. adj. *pseudo* false; Gr. adj. *echinos* spiny appearance; Gr. n. *spora* seed; M.L. adj. *pseudoechinosporeus* false spiny spored.

Single spores or chains of up to three spores form on both the substrate and aerial mycelium. The spores have smooth surfaces and are unequal in size (1.8–3.5 μm in diameter). They were originally thought to be spiny but are now known to be heavily encrusted in needlelike crystals. The spore mass is white to glaucous gray; the reverse is pink to light violet-pink. Melanin pigment is not produced.

Gelatin is degraded but cellulose is not. Does not reduce nitrate or produce hydrogen sulfide.

Growth at 20°, 26°, and 37°C but not at 50°C.

Isolated from sand, Kyzyl-Kum desert, U.S.S.R.

Type strain: ATCC 19618.

7. **Streptomyces rangoon** (Erikson 1935) Pridham, Hesseltine and Benedict 1958, 61.[AL] (*Actinomyces rangoon* Erikson 1935, 37.)

ran.goon′. M.L. n. *rangoon* referring to Rangoon, Burma.

Spore chains are *Spirales,* the spore surface is smooth. The spore mass is in the white series; the reverse is pale yellow to yellowish brown. Melanin pigment is not produced.

Isolated from a fatal case of human pulmonary streptotrichosis.

Type strain: ATCC 6860.

8. **Streptomyces somaliensis** (Brumpt 1906) Waksman and Henrici 1948b, 965.[AL] (*Indiella somaliensis* Brumpt 1906, 555.)

so.mal.ien′sis. M.L. adj. *Somaliensis* pertaining to Somalia.

Neither spore chains nor spores have been accurately determined. Forms white aerial mycelium; the reverse is cream to white.

Resistant to dimethylchlortetracycline, gentamicin, lincomycin, neomycin, penicillin, streptomycin, and vancomycin.

Agent of human actinomycetoma.

Type strain: ATCC 33201.

9. **Streptomyces vitaminophilus** (Shomura, Amano, Yoshida, Ezaki, Ito and Niida 1983a) Goodfellow, Williams and Alderson 1986a, 575.[VP] (Effective publication: Goodfellow, Williams and Alderson 1986b, 63.) (*Actinosporangium vitaminophilum* Shomura, Amano, Yoshida, Ezaki, Ito and Niida 1983, 563.)

vi.ta.mi.no′phi.lus. L. n. *vita* life; M.L. n. *aminum* amine, vitamin; Gr. adj. *philus* loving; M.L. adj. *vitaminophilus* vitamin-loving.

Spores contained in "pseudosporangia"; spore surface is smooth. Aerial mycelium is sparse; the reverse is colorless to pale tan or pale grayish yellow. Growth enhanced by vitamin B_{12}. Melanin pigment is not produced.

Gelatin and starch are degraded, nitrate is reduced.

Growth at 15° and 45°C.

Contains major amounts of hexa- and octahydrogenated menaquinones with nine isoprene units.

Produces antibiotics of the pyrrolomycin complex.

Isolated from soil.

Type strain: ATCC 31673.

Red Series

1. **Streptomyces albosporeus** (Krainsky 1914) Waksman and Henrici 1948b, 954.[AL] (*Actinomyces albosporeus* Krainsky 1914, 649.)

al.bo.spo′re.us. L. adj. *albus* white; M.L. n. *spora* a spore; M.L. adj. *albosporeus* white spored.

Spore chains are *Spirales,* the spore surface is smooth. Very sparse formation of aerial mycelium. The spore mass is in the red series; the reverse is yellow, red, red-brown, violet, or orange colored. Melanin pigment is not produced.

Antimicrobial activity not detected; is inhibited by streptomycin.

Type strain: ATCC 15394.

2. **Streptomyces cinnamonensis** Okami in Maeda et al. 1952a, 572.[AL]

cin.na.mo.nen′sis. Gr. n. *cinnamum* cinnamon; M.L. adj. *cinnamonensis* belonging to cinnamon (referring to the color of the aerial mycelium).

Spore chains are *Retinaculiaperti;* the spore surface is smooth. The spore mass is in the red series. Melanin pigment is produced.

Produces actithiazic acid, a biotin antagonist and antimycobacterial antibiotic; shows antifungal activity.

Type strain: ATCC 12308.

3. **Streptomyces crystallinus** Tresner, Davies and Backus 1961, 74.[AL]

crys.tal.lin′us. L. adj. *crystallinus* of crystal (referring to the crystals formed by the organism in some media).

Spore chains are *Rectiflexibiles;* the spore surface is smooth. The spore mass is in the red series; the reverse is light to dark brown. A brown diffusible pigment is formed on many media. Melanin pigment is produced.

Produces hygromycin A and other antibiotics; shows antifungal activity. Is inhibited by streptomycin.

Type strain: NRRL B-3629.

4. **Streptomyces erythrogriseus** Falcão de Morais and Dália Maia 1959, 64.[AL]

e.ry.thro.gri′se.us. Gr. adj. *erythraeus* red; M.L. adj. *griseus* gray; M.L. adj. *erythrogriseus* red-gray (referring to change in color of aerial mycelium from gray to red).

Spore chains are *Spirales,* but short spore chains may form imperfect or incomplete spirals or crooked hooks or loops. The spore surface has short spines, but smooth spores are also formed. The spore mass is red, gray, or white; the reverse is strong brown, orange-yellow, or reddish orange, and is pH sensitive. A yellow to orange, pH-sensitive, diffusible pigment may be formed. Melanin pigment is not produced.

Produces the antibacterial indicator antibiotic erygrisin; shows antiyeast activity.

Type strain: CBS 485.74.

5. **Streptomyces gobitricini** (Preobrazhenskaya and Sveshnikova in Gauze et al. 1957) Pridham, Hesseltine and Benedict 1958, 67.[AL] (*Actinomyces gobitricini* Preobrazhenskaya and Sveshnikova in Gauze et al. 1957, 34.)

go.bi.tri.ci′ni. Eng. n. *Gobi* the Gobi desert in Mongolia; o.v. *tricini* from Gr. n. *thrix* the hair; M.L. adj. *gobitricini* of Gobi hair (referring to the Gobi desert, the first source of soil from which the organism was isolated, and probably to the formation of a streptothricin-like antibiotic).

Spore chains are *Rectiflexibiles,* although some atypical *Retinaculiaperti* chains may be present. The spore surface is smooth. The spore mass is in the red series. Melanin pigment is produced.

Shows antibacterial and antifungal activity; is inhibited by streptomycin.

Type strain: CBS 123.60

6. **Streptomyces laurentii** Trejo, Dean, Pluscec, Meyers and Brown 1979, 80.[AL] (Effective publication: Trejo, Dean, Pluscec, Meyers and Brown 1977, 639.)

lau.ren′ti.i. Eng. n. *Lawrence;* M.L. n. *laurentii* of Lawrence Township, N.J., U.S.A. (the origin of the soil isolate).

Spore chains are predominantly *Rectiflexibiles,* although some *Retinaculiaperti* chains may be formed. The spore surface is smooth. The spore mass is in the red series; the reverse is yellowish brown. A pink diffusible pigment may be produced. Melanin pigment is not produced.

Mycelium fragments into arthrospores and rods in shaken culture.

Produces thiostrepton.

Type strain: ATCC 31255.

7. **Streptomyces lavendofoliae** (Kuchaeva, Krasil'nikov, Taptykova and Gesheva 1961) Pridham 1970, 19.[AL] (*Actinomyces lavendofoliae* Kuchaeva, Krasil'nikov, Taptykova and Gesheva 1961, 12.)

la.ven′do.fo′li.ae. M.L. n. *lavendula* lavender; L. n. *folium* a leaf; M.L. gen. n. *lavendofoliae* of lavender leaf (referring to the color of the aerial mycelium).

Spore chains are *Retinaculiaperti;* the spore surface is smooth. The spore mass is grayish-yellowish pink or yellowish pink; the reverse is pale yellow to light yellowish brown. Melanin pigment is produced.

Reportedly produces the streptothricin complex; is inhibited by streptomycin.

Type strain: ATCC 15872.

8. **Streptomyces niveoruber** Ettlinger, Corbaz and Hütter 1958a, 350.[AL]

ni′ve.o.ru′ber. L. adj. *niveus* snow white; L. adj. *ruber* red; M.L. adj. *niveoruber* snow white–red (referring to the white color of the aerial mycelium and the red color of the vegetative mycelium).

Spore chains are *Spirales;* the spore surface is smooth. The spore mass is in the red series; the reverse is also red. Melanin pigment is not produced.

Produces cinerubins A and B.

Type strain: ATCC 14971.

9. **Streptomyces peucetius** Grein, Spalla, Di Marco and Canevazzi 1963, 109.[AL]

peu.ce′ti.us. M.L. adj. *peucetius* pertaining to Peucetia (an ancient name for Central Puglia in Italy, the source of the soil from which the organism was isolated).

Spore chains are atypical *Retinaculiaperti;* the spore surface is smooth. Reported to form coremia. The spore mass is in the red series; the reverse is pinkish. Melanin pigment is not produced.

Produces daunomycin, an antibacterial and antitumor antibiotic; produces polyenic antifungal activity.

Type strain: ATCC 29050.

10. **Streptomyces roseoflavus** Arai 1951, 218.[AL]

ro.se.o.fla′vus. L. adj. *roseus* rosy; L. adj. *flavus* yellow; M.L. adj. *roseoflavus* rose-yellow.

Spore chains are *Retinaculiaperti* or *Spirales;* the spore surface is smooth. The spore mass is pale orange-yellow, grayish-yellowish pink, pale yellow, or pale yellowish green; the reverse is moderate to deep orange, grayish yellow to orange-yellow, or yellowish brown. Melanin pigment is not produced.

Produces flavomycin, an antibacterial antibiotic, and mycelin, an antifungal antibiotic.

Type strain: NRRL B-2789.

11. **Streptomyces ruber** (Shirling and Gottlieb 1972) Goodfellow, Williams and Alderson 1986a, 575.[VP‡] (Effective publication: Goodfellow, Williams and Alderson 1986c, 58.) (*Chainia rubra* Shirling and Gottlieb 1972, 347.[AL])

rub′er. L. adj. *ruber* red.

Spore chains are *Spirales;* the spore surface is smooth or with very short spines. The spore mass is grayish-yellowish pink or pale yellow; the

‡**Editorial note:** The validation citation for this name (Goodfellow et al., 1986a) and the Approved Lists of Bacterial Names incorrectly cite Thirumalachar as the author of the basionym. Shirling and Gottlieb should be considered as the authors because they were the first to validly publish the name under the previous Bacteriological Code.

reverse is red. A faint reddish brown or yellow diffusible pigment may be formed. Sclerotia are produced. Melanin pigment is not produced.

Xanthine is degraded but urea is not. Nitrate is reduced.

Contains major amounts of hexa- and octahydrogenated menaquinones with nine isoprene units.

Type strain: ATCC 17754.

12. **Streptomyces rubiginosohelvolus** (sic) (Kudrina in Gauze et al. 1957) Pridham, Hesseltine and Benedict 1958, 59.[AL] (*Actinomyces rubiginosohelvolus* (sic) Kudrina in Gauze et al. 1957, 89.)

ru.bi.gin.o.so.hel'vo.lus. orth.var. of L. adj. *robiginosus* rusty; L. adj. *helvolus* somewhat yellow; M.L. adj. *rubiginosohelvolus* rusty, somewhat yellow.

Spore chains are *Rectiflexibiles;* the spore surface is smooth. The spore mass is pale yellow or pale orange-yellow; the reverse is pale yellow or grayish yellow and is pH sensitive. Melanin pigment is not produced.

Shows antibacterial and antifungal activity.

Type strain: ATCC 19926.

13. **Streptomyces sulfonofaciens** Miyadoh, Shomura, Ito and Niida 1983, 323.[VP]

sul.fon.o.fa'ci.ens. N.L. n. *acidum sulfonicum* sulfonic acid; L. part. adj. *faciens* producing; M.L. part. adj. *sulfonofaciens* producing sulfonic acid.

Spore chains are *Rectiflexibiles;* the spore surface is smooth. The spore mass is grayish-yellowish pink; the reverse is colorless to pale yellowish brown. A light beige diffusible pigment is formed on some media. Melanin pigment is not produced.

Degrades gelatin and starch. Nitrate not reduced.

Growth between 15° and 42°C, optimum is between 27° and 33°C.

Produces a β-lactam antibiotic, which is a carbapenen compound.

Isolated from soil at Nachi-Katsuna. Wakayama Prefecture, Japan.

Type strain: ATCC 31892.

Yellow Series

1. **Streptomyces celluloflavus** Nishimura, Kimura and Kuroya 1953, 64.[AL]

cel.lu.lo.fla'vus. M.L. n. *cellulosum* cellulose; L. adj. *flavus* yellow; M.L. adj. *celluloflavus* cellulose, yellow (intended to refer to the yellow streptomycete that attacks cellulose).

Spore chains are *Rectiflexibiles;* the spore surface is smooth. The spore mass is in the yellow series. Melanin pigment is not produced.

Produces aureothricin; is inhibited by streptomycin.

Type strain: IFO 13780.

2. **Streptomyces champavatii** Uma and Narasimha Rao 1959, 133.[AL]

cham.pa.va'ti.i. M.L. gen. n. *champavatii* of Champavathi (named after the Champavathi River in Andhra Pradesh, India).

Spore chains are *Rectiflexibiles,* the spore surface is smooth. The spore mass is in the yellow series; the reverse is green and a diffusible pigment is produced on some media. Melanin pigment is not produced.

Produces champamycins A and B (heptaenic antifungal antibiotics) and champavatin, a nonpolyenic antifungal antibiotic. Produces vitamin B_{12}.

Type strain: NRRL B-5682.

3. **Streptomyces flavidovirens** (Kudrina in Gauze et al. 1957) Pridham, Hesseltine and Benedict 1958, 66.[AL] (*Actinomyces flavidovirens* Kudrina in Gauze et al. 1957, 90.)

fla.vi.do.vi'rens. L. adj. *flavidus* yellowish; L. part. adj. *virens* being green; M.L. part. adj. *flavidovirens* being yellowish green.

Spore chains are *Retinaculiaperti* or *Rectiflexibiles;* the spore surface is smooth. The preponderance of very flexuous spore chains, some of which appear as imperfect or open spirals together with some straight or slightly flexuous chains, makes this strain difficult to categorize in respect to spore chain morphology. The spore mass is pale yellow; the reverse is pale yellow or light grayish yellow. Melanin pigment is not produced.

Shows antibacterial activity and slight antifungal activity.

Type strain: ATCC 19900.

4. **Streptomyces floridae** Bartz, Ehrlich, Mold, Penner and Smith 1951, 4.[AL]

flo.ri'dae. M.L. gen. n. *floridae* of Florida (the source of the soil from which the organism was isolated).

Spore chains are *Rectiflexibiles;* the spore surface is smooth. The spore mass is in the yellow series; the reverse is dull violet to red-brown and a diffusible pigment is produced on some media. Melanin pigment is not produced.

Produces the viomycin complex; is inhibited by streptomycin.

Type strain: NRRL 2423.

5. **Streptomyces griseoloalbus** (Kudrina in Gauze et al. 1957) Pridham, Hesseltine and Benedict 1958, 58.[AL] (*Actinomyces griseoloalbus* Kudrina in Gauze et al. 1957, 112.)

gri.se'o.lo.al'bus. M.L. adj. *griseolus* somewhat gray; L. adj. *albus* white; M.L. adj. *griseoalbus* somewhat grayish white.

Spore chains are *Rectiflexibiles;* the spore surface is smooth. The spore mass is pale yellow or white; the reverse is light or grayish yellow to orange-yellow. A trace of yellow to orange-yellow diffusible pigment may be formed on some media. Melanin pigment is not produced.

Produces the grisein (albomycin) complex, shows antifungal activity.

Type strain: ATCC 23624.

6. **Streptomyces kunmingensis** (Ruan, Lechevalier, Jiang and Lechevalier 1985) Goodfellow, Williams and Alderson 1986a, 574.[VP] (Effective publication: Goodfellow, Williams and Alderson 1986c, 57.) (*Chainia kunmingensis* Ruan, Lechevalier, Jiang and Lechevalier 1985, 168.)

kun.ming.en'sis. M.L. adj. *kunmingensis* pertaining to Kunming (a province of South China, the source of the soil from which the organism was isolated).

Spore chains are *Spirales.* The spore mass is colorless to light yellowish tan; the reverse is light grayish yellow to dark yellowish brown. Melanin pigment is not produced.

Adenine, esculin, casein, hypoxanthine, starch, tyrosine, and xanthine are degraded but gelatin and urea are not. Nitrate is reduced and phosphatase produced.

Growth at 10° and 37°C but not at 42°C.

Type strain: ATCC 35682.

7. **Streptomyces mediolani** Bianchi, Grein, Julita, Marnati and Spalla 1970, 243.[AL§]

med.i.ol'ani. L. n. *Mediolanum* Milan; M.L. gen. adj. *mediolani* of Milan, Italy.

Spore chains are *Rectiflexibiles;* the spore surface is smooth. The spore mass is yellow-vanilla, yellow-rose, or yellow-beige; the reverse is gold-yellow to orange-yellow. Melanin pigment is not produced.

Tyrosine is degraded and hydrogen sulfide is produced.

Produces peptide antibiotics and carotenoid pigments.

Isolated from soil.

Type strain: NCIB 10969.

---

§**Editorial note:** Contrary to the citation in the Approved Lists, Arcamone et al. (1969) did not validly publish this name.

8. **Streptomyces praecox** (Millard and Burr 1926) Waksman in Waksman and Lechevalier 1953 107.[AL] (*Actinomyces praecox* Millard and Burr 1926, 633.)

prae'cox. L. adj. *praecox* premature, precocious.

Spore chains are predominantly *Rectiflexibiles,* but open spirals have been reported. The spore surface is smooth. The spore mass is pale yellow or pale yellow-green; the reverse is pale or grayish yellow to light olive-brown. A trace of yellow diffusible pigment is found in some media. Melanin pigment is not produced.

Shows slight antimicrobial activity; is inhibited by streptomycin.

Isolated from unruptured knoblike scab of potatoes.

Type strain: ATCC 3374.

Blue Series

1. **Streptomyces inusitatus** Hasegawa, Yamano and Yoneda 1978b, 409.[AL]

in.usi.ta'tus. L. adj. *inusitatus* unusual.

Spore chains are *Spirales;* the spore surface is smooth. The spore mass is blue-gray to gray-blue; the reverse is colorless or pale brown. Melanin pigment is not produced.

Gelatin and starch are degraded.

Growth optimal between 28° and 30°C.

Produces oxamicetin.

Isolated from soil.

Type strain: IFO 13601.

2. **Streptomyces ipomoeae** (Person and Martin 1940) Waksman and Henrici 1948b, 948.[AL] (*Actinomyces ipomoea* (sic) Person and Martin 1940, 923.)

i.po.moe'ae. M.L. n. *Ipomoea* generic name of sweet potato; M.L. gen. n. *ipomoeae* of *Ipomoea* (referring to the source of the organism).

Spore chains are *Spirales;* the spore surface is smooth. The spore mass is in the blue series; the reverse is pale yellow or grayish yellow. Traces of yellow or green diffusible pigment may be present on some media.

Causes soft rot of sweet potatoes (*Ipomoea* sp.).

Type strain: ATCC 25462.

3. **Streptomyces lomondensis** Johnson and Dietz 1969, 755.[AL]

lo.mon.den'sis. M.L. adj. possibly pertaining to Loch Lomond, Scotland.

Spore chains are *Rectiflexibiles, Retinaculiaperti,* or *Spirales.* The spore surface is warty to spiny. The spore mass is blue; the reverse is brick red, rust-brown, or straw colored. A brown or pink diffusible pigment is produced on some media. Melanin pigment is produced.

Starch, tyrosine, and xanthine are degraded but casein is not. Nitrate is reduced.

Growth optimal at 37°C.

Produces lomofungin, which inhibits bacteria and fungi.

Isolated from soil.

Type strain: NRRL 3252.

Green Series

1. **Streptomyces acrimycini** (Preobrazhenskaya, Blinov and Ryabova in Gauze et al. 1957) Pridham, Hesseltine and Benedict 1958, 65.[AL] (*Actinomyces acrimycini* Preobrazhenskaya, Blinov and Ryabova in Gauze et al. 1957, 190.)

a.cri.my.ci'ni. L. adj. *acer* sharp, keen, pungent; M.L. adj. suff. *-mycin* for antibiotic names; M.L. gen. adj. *acrimycini* of the sharp antibiotic.

Spore chains are *Spirales,* with many loose spirals, but *Retinaculiaperti* and flexuous chains have also been reported. The spore surface is hairy. The spore mass is in the green color series. Melanin pigment is not produced.

Shows antibacterial and antifungal activity; is inhibited by streptomycin.

Type strain: ATCC 19885.

2. **Streptomyces ghanaensis** Wallhäusser, Nesemann, Präve and Steigler 1966, 734.[AL]

ghan.a.en'sis. M.L. adj. *ghanaensis* belonging to Ghana (the source of the soil from which the organism was isolated).

Spore chains are *Spirales;* the spore surface is spiny. The spore mass is in the green series. Melanin pigment is not produced.

Produces the moenomycin complex of antibacterial antibiotics (components A, B, B_1, and C).

Type strain: ATCC 14672.

3. **Streptomyces griseostramineus** (Preobrazhenskaya, Kudrina, Blinov and Ryabova in Gauze et al. 1957) Pridham, Hesseltine and Benedict 1958, 65.[AL] (*Actinomyces griseostramineus* Preobrazhenskaya, Kudrina, Blinov and Ryabova in Gauze et al. 1957, 155.)

gri.se.o.stra.mi'ne.us. M.L. adj. *griseus* gray; L. adj. *stramineus* straw colored; M.L. adj. *griseostramineus* gray, straw colored (referring to the gray aerial mycelium and straw-yellow vegetative mycelium on chemically defined medium).

Spore chains are *Spirales,* but hooks and loops suggestive of *Retinaculiaperti* spore chains may also be found. The spore surface is hairy to spiny. The spore mass is in the green series; the reverse is grayish yellow or grayish yellow modified by green. Melanin pigment is produced.

Shows antibacterial and antifungal activity.

Type strain: ATCC 23628.

4. **Streptomyces spinoverrucosus** Dias and Al-Gounaim 1982, 331.[VP]

spi.no.ver.ru.co'sus. L. adj. *spinosus* thorny; L. adj. *verrucosus* warty; M.L. adj. *spinoverrucosus* spiny and warty (referring to the spiny and warty surface).

Spore chains are *Spirales;* the spore surface is spiny or spiny and warty. The spore mass can be placed in the green, yellow, red, or gray series, depending on the medium; the reverse is faint yellow or brownish red, and the latter is pH sensitive. Brown, red, or reddish-brown diffusible pigments are formed. Melanin pigment is produced.

Isolated from air sample, Kuwait City.

Type strain: NCIB 11666.

Violet Series

1. **Streptomyces mauvecolor** Okami and Umezawa in Murase, Hikiji, Nitta, Okami, Takeuchi and Umezawa 1961, 114.[AL]

mau've.co.lor. L. n. *malva* mallow, a plant with violet-colored petals, whence Fr. *mauve;* L. n. *color;* M.L. adj. *mauvecolor* mauve colored.

Spore chains are *Spirales;* the spore surface is spiny. The spore mass is in the violet series. Melanin pigment is produced.

Produces peptimycin, a peptidic antitumor antibiotic.

Type strain: ATCC 29835.

Genus **Streptoverticillium** *Baldacci 1958, 15, emend. mut. char. Baldacci, Farina and Locci 1966, 168*[AL]

ROMANO LOCCI AND GERALDINE M. SCHOFIELD

Strep.to.ver.ti.cil'li.um. Gr. adj. *streptos* pliant, twisted; L. n. *verticillus* whorl, whirl of a spindle; M.L. neut. n. *Streptoverticillium* a whorled actinomycete.

Substrate mycelium, 0.8–1.2 μm in diameter, branching. **The aerial mycelium consists of long, straight filaments bearing at more or less regular intervals branches (3–6) arranged in whorls (verticils). The appearance at ~100 × magnification is of "barbed**

wire." Each branch of the verticil produces at its apex an umbel that consists of two to many chains of spherical to ellipsoidal spores. Spore chains may be straight or flexuous or terminate in hooks. **Spiral spore chains have not been observed.** Reproduction occurs either from fragments of substrate and/or aerial mycelium or from germination of spores. Spores are smooth surfaced to slightly rough. **Neither spiny nor hairy spores have been observed.** On primary isolation colonies are small and discrete, developing a weft of aerial mycelium that typically appears cottony. Chemo-organotrophic. **Media low in available carbohydrates are particularly suitable for spore production.** Most species produce soluble pigments and colored substrate and aerial mycelium. They are resistant to lysozyme and to neomycin; they produce compounds that exhibit antifungal, antibacterial, antiprotozoal, and antitumour activity and are sensitive to antibacterial agents and actinophages. Aerobic, substrate growth may develop under reduced oxygen tensions or increased carbon dioxide concentrations. Optimum growth between 26° and 32°C at pH 6.5–8.0. Mesophilic. Mostly saprophytes in soil. Gram-positive. Cell walls contain **L-diaminopimelic acid** (L-**DAP**). Cells contain major amounts of saturated, *iso*- and *anteiso*- fatty acids, MK-9(H_6) and MK-9(H_8) menaquinones, and the phospholipids (diphosphatidylglycerol, phosphatidylethanolamine, phosphatidylinositol, and phosphatidylinositol mannosides). The mol% G + C of the DNA is 69–73 (T_m).

Type species: *Streptoverticillium baldaccii* Farina and Locci 1966, 48.

Further Descriptive Information

Terminology is particularly important at this level. So-called biverticillate forms have been described. This term should be restricted to structures that are of rare occurrence when the verticil, instead of giving rise directly to umbels of spore chains, branches again and produces terminal umbels of spores (Locci et al., 1969; Locci and Petrolini Baldan, 1970).

Morphology and Development (Figs. 29.15–29.22)

The sporophore structure of *Streptoverticillium* consists of a main axis having at regular intervals sets of three or more side branches. These branches in turn bear terminal umbels of spore chains; such formations should be interpreted as umbellate monoverticillate. Although rare, biverticillate structures do exist (Locci et al., 1969), and this term must be restricted to forms where a second series of verticils appears on the side branches themselves. To summarize, streptoverticillia are characterized by having verticils of sporophore filaments bearing terminal umbels of spore chains. It is essential, therefore, that spores should be present before a correct definition of the sporophore structure is attempted (Locci and Petrolini Baldan, 1970).

Spore germination takes place by means of one to three germ tubes. The process can occur even when spores are still joined together in chains. The primary mycelium does not bear any spores. Secondary mycelium, which can arise from any point of the primary growth, consists of a network of interwoven filaments and is characterized by the development of long, straight hyphae. It is these formations that partly explain the height of some *Streptoverticillium* colonies and their cottony consistency.

On the main axes the initiation of side-branches is not simultaneous in each verticil. In extreme cases one branch may be well developed and even have started umbel formation while the others are still in the initial phase. The reasons for this are not clear, and these structures should be regarded as developmental malformations caused by local unfavorable conditions. In some species an enlargement of the main hypha at the point of insertion of the verticil may be observed. Similarly the formation of umbels and spore chains is not simultaneous in all branches of the verticil. Mature spore chains and nonsporulated hyphae may be found side by side.

Prior to sporulation, filaments destined to become spore chains show a twisting along their main axis (Locci and Petrolini Baldan, 1970). Folds are evident on the filament surface and they can persist even in mature spores. The phenomenon appears to be unique for streptoverticillia. The development is not always simultaneous; in the same umbel some filaments are still at the twisting stage while others are completely sporulated. This process of twisting occurs only in filaments that will give rise to spore chains; there is no twisting of the verticil branches.

Ultrastructure

Previous investigations have dealt mainly with the structure of the spore sheath. According to Cross et al. (1973), the chain of spores is surrounded by a thin sheath that contains a conspicuous and continuous helical thickening. Segments of the sheath still envelop individual spores when the latter separate. When segments are viewed in transparency, after the shrinkage of the enclosed spore, the numerous parallel sheath fibers, in the form of a multiple helix, give the illusion of basketwork. Individual fibers are about 10 nm in diameter and at least 10 such fibers are involved in the helical pattern.

It has been postulated that the amorphous material of the sheath, or the fibrous thickening, contracts during spore maturation or during

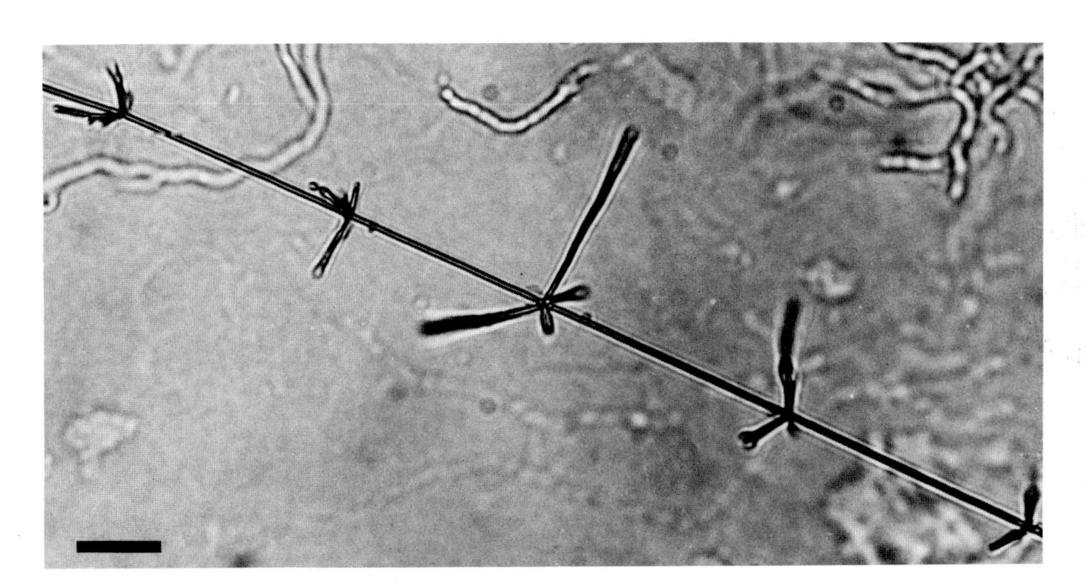

Figure 29.15. Aerial filament and initial verticils of *Streptoverticillium cinnamoneum.* Light microscopy. *Bar,* 20 μm.

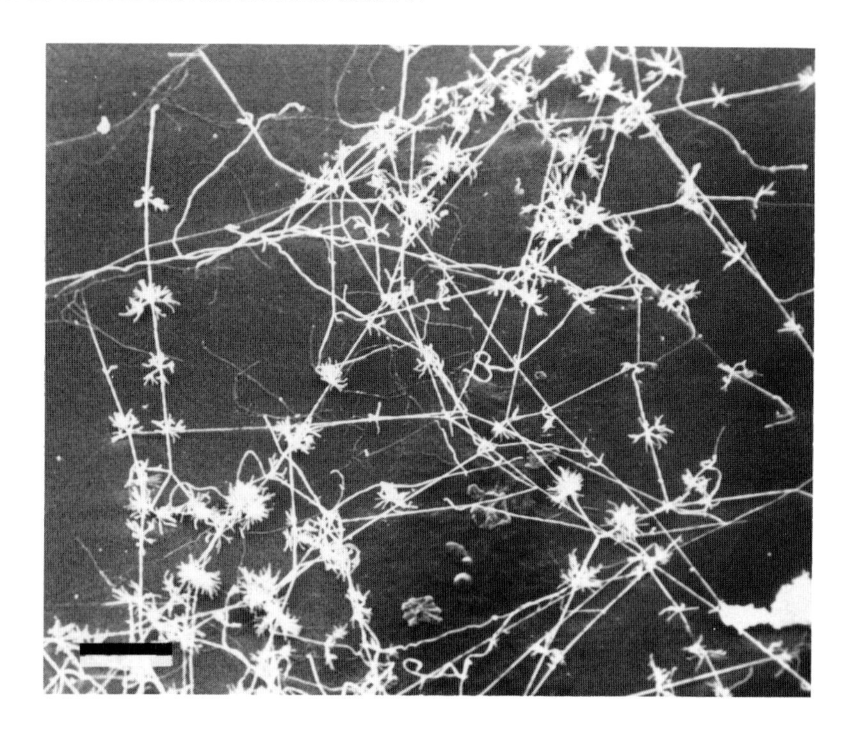

Figure 29.16. Aerial mycelium of *Streptoverticillium baldaccii.* Scanning electron microscopy. *Bar,* 50 μm.

Figure 26.17 Sporulated aerial filament of *Streptoverticillium roseoverticillatum* subsp. *albosporum.* Light microscopy. *Bar,* 20 μm.

preparation for scanning electron microsocopy, thus giving the twisted form to the sporulating hyphae. The characteristic appearance of the *Streptoverticillium* sheath is quite unlike that described for the *Streptomyces* species. The absence of short rodlets from the sheath may also explain the absence of conspicuous surface appendages. It has been suggested that these rod subunits appear to aggregate to form the spines and hairlike projections seen in many *Streptomyces* species (Wildermuth, 1972; Williams et al., 1972a).

Colonial Characteristics

A detailed study of colonial characteristics has not been carried out, but the characteristics of streptoverticillia do not differ significantly from those of streptomycetes. However, a distinctive feature in most species is the cottony consistency of the aerial growth, and this can be of use as a preliminary screening of isolation plates for the presence of streptoverticillia.

Nutrition and Growth Conditions

Care must be taken to ensure adequate inocula for such tests; false-negative results may be caused by lack of growth. General carbon, nitrogen, and mineral requirements do not differ greatly from those of streptomycetes, but most *Streptoverticillium* species can utilize only a limited number of carbon sources (Locci et al., 1969; Kutzner, 1981).

Vegetative growth develops under low oxygen tension (R. Locci and G. M. Schofield, unpublished results).

A verticillate thermophilic species (*Actinomyces* III) was referred to by Waksman et al. (1939) and was supposedly similar to strain 89 isolated by Lieske (1921) from animal excreta. No further records of thermophilic streptoverticillia can be found in the literature. However, until recently preferential isolation procedures were virtually unknown, making ecological studies difficult. The preferred pH is close to neutrality, but improved growth of *S. ehimense* and *S. thioluteum* has been recorded at pH 8.0–9.0 (Nomi, 1960).

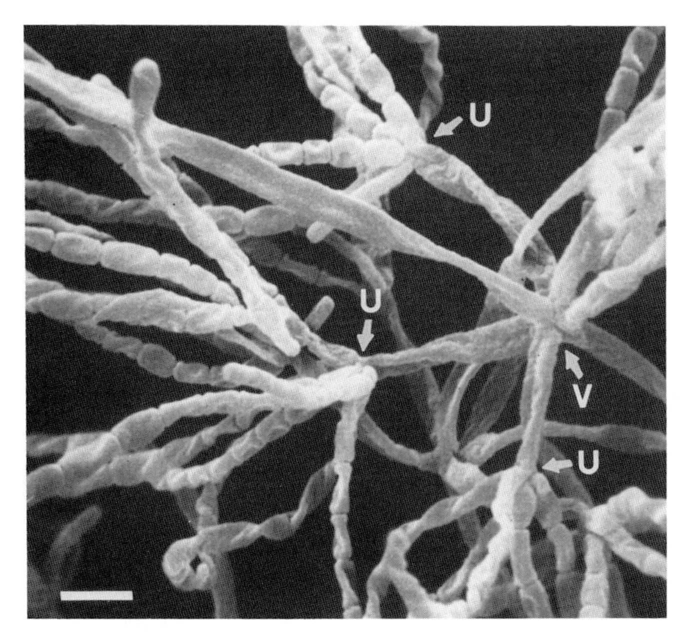

Figure 29.18. Verticil (*V*) and umbels (*U*) of spore chains of *Streptoverticillium kentuckense.* Scanning electron microscopy. *Bar,* 2 µm. (Reproduced by permission from R. Locci and B. Petrolini Baldan, Rivista di Patologia Vegetale 7 *(Suppl.):* 3–19, 1971.)

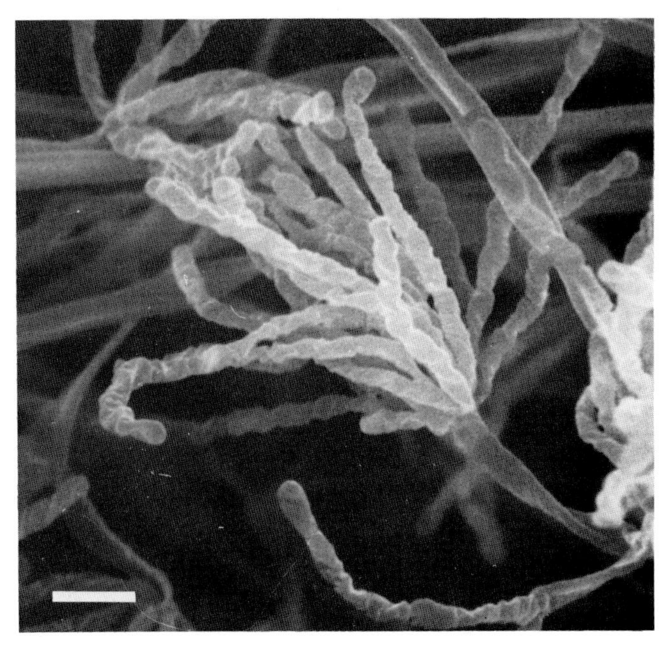

Figure 29.20. Mature umbel of spore chains of *Streptoverticillium baldaccii.* Scanning electron microscopy. *Bar,* 2 µm.

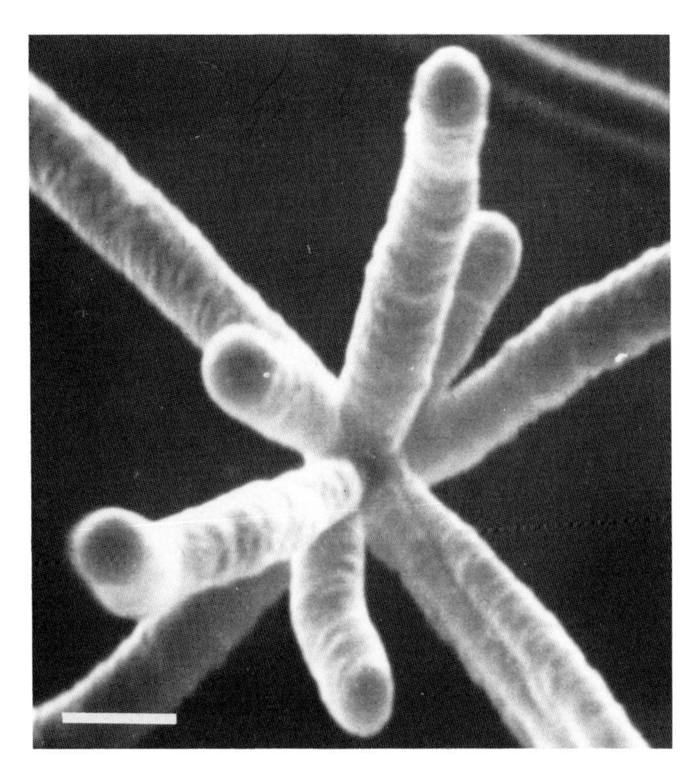

Figure 29.19. Verticil formation in *Streptoverticillium cinnamoneum.* Scanning electron microscopy. *Bar,* 1 µm. (Reproduced by permission from R. Locci and B. Petrolini Baldan, Rivista di Patologia Vegetale 7 *(Suppl.):* 3–19, 1971.)

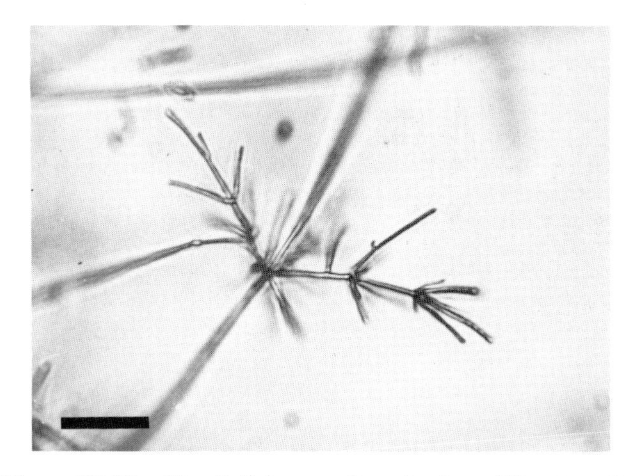

Figure 29.21. Biverticillate sporophore structure of *Streptoverticillium ehimense.* Light microscopy. *Bar,* 20 µm.

Chemical Characters

Streptoverticillium species share with streptomycetes a cell wall chemotype I (Lechevalier and Lechevalier, 1970b), G + C-rich DNA, high intraspecies DNA homology values (Kroppenstedt et al., 1981), and similar lipids (Collins et al., 1977; Lechevalier et al., 1977, 1981) and are lysed by the same phages (Prauser, 1976a, 1984c; Wellington and Williams, 1981a). The streptoverticillia have a type II polar lipid pattern (Lechevalier et al., 1977, 1981), major amounts of *iso-* and *anteiso-* fatty acids (Kroppenstedt and Kutzner, 1976; Kroppenstedt, 1985), and hexa- and octahydrogenated menaquinones with nine isoprene units (Collins et al., 1985; Kroppenstedt, 1985). Immunodiffusion techniques (Ridell et al., 1986) stress the affinity between *Streptomyces* and *Streptoverticillium,* in particular the close relationship between streptoverticillia and *Streptomyces lavendulae.*

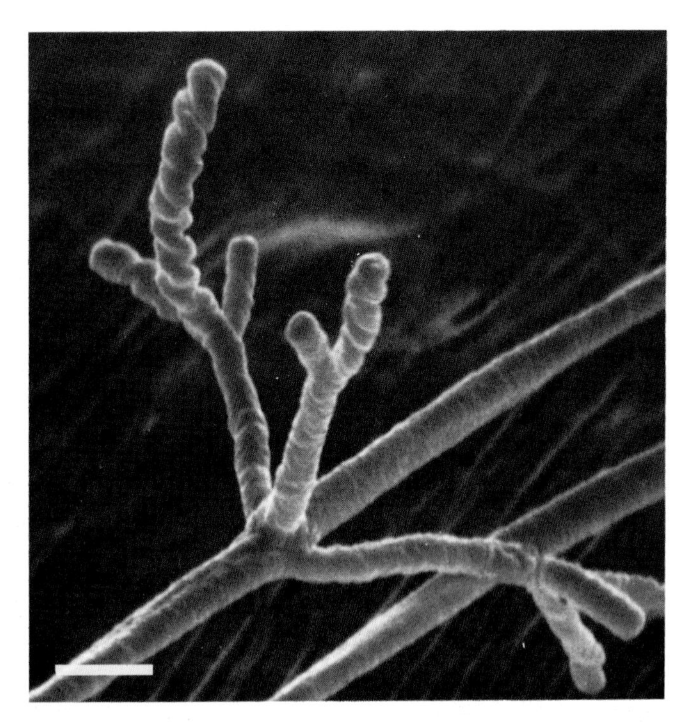

Figure 29.22. Twisting of developing spore chains of *Streptoverticillium salmonis.* Scanning electron microscopy. *Bar,* 2 µm. (Reproduced by permission from R. Locci and B. Petrolini Baldan, Rivista di Patologia Vegetale *7 (Suppl.):* 3-19, 1971.)

A low degree of DNA homology was observed when comparing the DNA of 26 species of *Streptoverticillium* with three reference DNAs, confirming the heterogeneity of the species (Toyama et al., 1974). However in a numerical study strains with the highest relative reassociation with *S. kentuckense,* i.e. *S. flavopersicum, S. netropsis,* and *S. hiroshimense,* were all recovered in the same cluster (Locci et al., 1981).

Support for distinguishing between *Streptomyces* and *Streptoverticillium* comes from DNA/RNA pairing studies (Gladek et al., 1985). *Streptoverticillium* species are also much more resistant than *Streptomyces* to lysozyme (Kutzner et al., 1978) and to neomycin (Goodfellow and Cross, 1984b).

Enrichment and Isolation Procedures

Interest in antibacterial and antitumor agents produced by streptoverticillia has stimulated attempts to isolate these organisms from soils. Most isolation media used have favored the streptomycetes, perhaps at the expense of the streptoverticillia. Heat treatment of soils to eliminate most bacteria is a useful first step (Rowbotham and Cross, 1977b). Elimination of most nonfilamentous organisms can be achieved by membrane filtration. This should be followed by plating the soil filtrate on a semidefined agar containing antifungal agents suitable for the cultivation of streptoverticillia. Recent work has shown that *Streptoverticillium* species are able to grow in the presence of lysozyme, whereas most *Streptomyces* are sensitive (Attwell et al., 1985). In addition, the incorporation of oxytetracycline, at a final concentration of 25 µg/ml, allows the growth of streptoverticillia but suppresses the growth of several actinomycete genera commonly found in soil, including the abundant streptomycetes (Hanka et al., 1985).

Maintenance Procedures

Long-term preservation of *Streptoverticillium* is achieved by lyophilization in skim milk. Spore suspensions can also be stored in glycerol (20% v/v) at −20°C (Wellington and Williams, 1978). The spores of *Streptoverticillium* are characterized by very low levels of germination so that the amount of material required for preservation of the stock must be large. It is not clear whether the spores are nonviable or whether some particular stimulus is required for germination.

A common occurrence is the loss of cultures that are maintained only on agar slopes in the laboratory. Subculturing requires a heavy transfer inoculum.

Procedures for Testing of Special Characters

Species description and identification are derived from numerical classification and identification studies (Locci et al., 1981; Williams et al., 1985a). Details are given of the methods used to determine the 41 characters listed in Table 29.17. In all cases, the inoculum consisted of dense spore and/or mycelial suspensions in sterile water, incubation was at 27°C unless stated otherwise, and all media were adjusted to pH 7.0. Negative results were retested to reduce the possibility of false readings due to poor viability of the inoculum.

Colony Appearance and Pigmentation (Characters 1–4)

These were determined from 14-day-old cultures on inorganic salts-starch medium (ISP medium 4, Difco; Shirling and Gottlieb, 1966). The appearance of the aerial growth was designated as cottony or powdery and the spore mass color was determined against the color wheels of

Table 29.17.
Criteria used for the classification and identification of **Streptoverticillium** *species*[a]

Category	Character	
	Numbers	States
Colony appearance and pigmentation	1–4	
Colony appearance		Cottony, powdery.
Color of spore mass		Yellow, white.
Melanin pigment production		On peptone iron agar.
Utilization of carbon and nitrogen sources	5–13	Mannitol, D-melibiose, raffinose, sorbitol (1.0% w/v). Coumarin (0.025%, w/v), L-methionine, L-proline, shikimic acid (0.1% w/v), DL-α-aminobutyric acid (1% w/v).
Acid production from sugars	14–18	D-Galactose, *meso*-inositol, D-fructose, D-ribose, D-trehalose (1% w/v).
Degradation activity	19–26	Esculin, citrate, DNA, hypoxanthine, L-tyrosine, Tween-20. NO_3^- reduction, H_2S production.
Growth temperatures	27	12°C.
Growth inhibition	28–38	
Growth in the presence of inhibitory compounds		NaCl (5% w/v), 1-phenylethanol (0.3% w/v), potassium tellurite (0.01% w/v), crystal violet (0.01% w/v), malachite green (0.01% w/v).
Resistance to antibiotics (µg/ml)		Azlocillin (30), carbenicillin (100), cephaloridine (30), cephalotin (30), cephamandole (30), colistin (30).
Antibiosis	39–41	*Aspergillus niger, Bacillus subtilis, Candida albicans.*

[a]Data from Williams et al. (1985).

Tresner and Backus (1963). Production of melanin pigment was determined after 4 days on peptone iron agar (ISP medium 6, Difco; Shirling and Gottlieb, 1966).

Utilization of Carbon and Nitrogen Sources (Characters 5–13)

Tests were carried out in Replidishes (Sterilin, Teddington, Middlesex, U.K.). Carbon and nitrogen sources were filter sterilized and added to the basal media. Sugars (5-8) were incorporated into carbon utilization medium (ISP medium 9, Difco; Shirling and Gottlieb, 1966) at a concentration of 1% (w/v). Other substrates (9-12) were added to the basal medium of Goodfellow (1971), consisting of Bacto yeast nitrogen base (Difco) supplemented with 10 mg/l casamino acids (Difco), to give concentrations of 0.1% (w/v), with the exception of coumarin, which was used at 0.025% (w/v). DL-α-Aminobutyric acid was incorporated into a basal medium containing (per liter): 10 g glucose, 1 g K_2HPO_4, 0.5 g $MgSO_4·7H_2O$, 0.5 g NaCl, and 15 g agar to give a concentration of 1% (w/v) (Williams et al., 1983a). Cultures were incubated for 14 days, the ability of a strain to utilize a source being determined by comparison of its growth with that on the unsupplemented medium.

Acid Production from Sugars (Characters 14–18)

The basal medium contained (per liter): 1 g $(NH_4)_2HPO_4$, 0.02 g KCl, 0.2 g $MgSO_4·7H_2O$, 15 g agar, and 15 ml of a 0.04% (w/v) solution of bromcresol purple (Gordon, 1968). Filter-sterilized carbohydrates were added separately to give a concentration of 1% (w/v). Acid production, as shown by the pH indicator, was determined after 14 days.

Degradation (Characters 19–26)

Esculin degradation (0.1% w/v) was determined after Kutzner (1976) with a basal medium containing (per liter): 3 g yeast extract (Oxoid), 0.5 g ferric ammonium citrate, and 7.5 g Lab M agar. Cultures were examined after 7 days for blackening of the medium. The degradation of DNA (0.2% w/w) was observed on Bacto DNase test agar (Difco), plates being flooded with 1 M HCl to reveal clearance zones after 7 days. Hypoxanthine (0.4% w/v) and L-tyrosine (0.5% w/v) degradation were detected using modified Bennett agar (Williams et al., 1983a) that contained (per liter): 1 g yeast extract (Oxoid), 0.8 g Lab Lemco (Oxoid), 10 g glycerol, 2 g NX amine A, and 15 g agar. Cultures were examined for clearing of the medium after 14 days. Sierra's (1957) medium supplemented with Tween 20 was examined for opacity around colonies after 7 days. Citrate degradation was determined after Gordon (1968) with a medium containing (per liter): 2 g sodium citrate, 1 g NaCl, 0.2 g $(NH_4)_2HPO_4$, 0.1 g KH_2PO_4, 15 g agar, and 20 ml of a 0.04% (w/v) solution of phenol red. A positive reaction was established by the alkaline color of the indicator after 7 days.

Nitrate reduction was determined in broth consisting of (per liter): 5 g peptone (Oxoid), 3 g beef extract (Oxoid), and 1 g KNO_3. After 7 days nitrate was detected using Griess-Ilosvey reagents I and II. Hydrogen sulfide production was detected by insertion of lead acetate paper strips into the mouths of slope cultures on peptone iron agar (Difco) that had been incubated for 14 days.

Inhibition of Growth (Characters 27–38)

Presence or absence of growth at 12°C was determined after incubation for 14 days. Modified Bennett agar (Williams et al., 1983a) was used as the basal medium to test for growth in the presence of 5% (w/v) NaCl, 0.3% (w/v) phenylethanol, 0.01% (w/v) potassium tellurite, 0.01% (w/v) crystal violet, and 0.01% (w/v) malachite green. Plates were examined for growth after 14 days.

Resistance to antibiotics was determined by placing four different sensitivity disks (Oxoid) of each antibiotic onto a preseeded medium that contained (per liter): 30 g glycerol, 2 g K_2HPO_4, 1 g $MgSO_4·7H_2O$, 0.5 KCl, 0.01 g $FeSO_4·7H_2O$, and 12 g agar (Williams, 1967). The disks contained either carbenicillin (100 μg), azlocillin, cephaloridine, cephalotin, cephamandole, or colistin (30 μg). The zones of inhibition were noted after 4 days.

Antibiosis (Characters 39–41)

Plates of nutrient agar (Difco) in glass Petri dishes were spot inoculated with streptoverticillia and incubated for 7 days. They were then inverted, and 1.5 ml chloroform were added to the lids and left for 1 h to kill the colonies. After removal of excess vapor in a laminar flow cabinet, plates were overlaid with 10 ml sloppy agar (0.7% w/v nutrient agar) seeded with the test organism. The test organisms used were *Aspergillus niger* strain CBS 131.52, *Bacillus subtilis* NCIB 3610, and *Candida albicans* HIK 183. Overlaid cultures were examined for zones of inhibition after 3 days.

Differentiation of the genus **Streptoverticillium** *from other genera*

The streptoverticillia are closely related to streptomycetes but can be differentiated using several criteria. The attribution of an actinomycete isolate to the genus *Streptoverticillium* may be made using the following characters:

1. Macroscopically most streptoverticillia are characterized by the cottony appearance of their aerial growth.
2. Under low-power magnification (100 ×) the "barbed wire" structure of the aerial mycelium is immediately evident. This consists of long, straight filaments showing, at regular intervals, side branches.
3. Complete recognition of the genus requires examination, at greater magnifications, of the fully mature sporulation structures. These consist of verticillate spore-bearing filaments, producing at their tips umbels of spore chains.
4. Unlike streptomycetes, most streptoverticillia are resistant to lysozyme and to neomycin.

When ultrastructural observation techniques are available some additional discriminatory parameters can be of use:

5. Transmission electron microscopy examination reveals the peculiar structure of the spore sheath.
6. The twisting of the developing or mature spore chains, unique among actinomycetes, is easily dectable by scanning electron microscopy.

Taxonomic Comments

Comparisons between the previous and current taxonomic approaches to the genus *Streptoverticillium* are difficult. In addition, the number of strains available for each described species is limited, only one strain of each species usually being available. The scarcity of information on preferential isolation techniques has not only impeded ecological studies but has had a strong influence on the systematics of the genus.

Early morphological investigations of some aerobic, aerial spore-forming actinomycetes revealed the presence of verticillate sporophores. Waksman and Curtis (1916) noted that *Actinomyces reticuli* had an aerial mycelium "consisting of filaments having no branches or very short ones" and in the *Key to the Identification of Actinomyces* an alternative description was given: "aerial mycelium thin, rare, net-like." Three years later (Waksman, 1919) the following diagnosis was proposed for the species: "whirl formation characteristic of this as well as the following species (*A. reticulus-ruber.*) This species represents a *distinct group* of organisms, *widely separated* from the other actinomycetes" (emphasis added). Following these descriptions the number of species characterized by the formation of verticils increased very slowly, and in a key of Waksman and Lechevalier (1953) in Section A II.8 ("Aerial Mycelium Forms Whorls") five species were listed: *Streptomyces reticuli, Streptomyces netropsis, Streptomyces verticillatus, Streptomyces circulatus,* and *Streptomyces rubrireticuli.*

However, interest in the potential taxonomic significance of the verticillate structure of these actinomycetes increased with a concomitant rise in the number of proposed classifications. The formal introduction of the genus *Streptoverticillium* by Baldacci (1958) in order to accommodate *"actinomycetes cum sporophora opposita et verticillata"* (actinomycetes with opposite and verticillate sporophores) or, as specified in the key, characterized by "verticillate spore-bearing hyphae" was not universally favored. Reluctance to accept the status of *Streptoverticillium* was due mainly to a misunderstanding of the brief description of the morphological characteristics of the new genus. Most confusion arose from the initial, improper, attribution to the new genus of streptomycetes with spore chains arranged in so-called clusters, tufts, whorls, whirls, etc., without due attention to the typical characteristics of the genus. In addition, in Baldacci's paper typical streptoverticillia such as *S. abikoense* and *S. hachijoense* were listed in the *Albidoflavus* series of the genus *Streptomyces,* while the first series of the newly proposed genus was *Circulatus,* which included typical streptomycetes. The effect of different media on growth and sporulation convinced Waksman (1967) that "the justification of this genus is . . . debatable." Further confusion occurred because in the original single-line Latin diagnosis a clear distinction was not made between sporophore and spore chain arrangements; also no type species of the genus was designated.

New streptoverticillate forms were recognized but, according to the different authors, were assigned either to places in the new genus or among the streptomycetes. However, emphasis was always placed on their unusual morphology. The primary interest is usually in the potential for antibiotic production, and very few organisms devoid of this potential have been described. As the number of available species became large enough to allow a structured study of their taxonomic status, the need for an improved description of the original species definition was realized. Baldacci et al. (1966) proposed an emendation of the characteristics of the genus that stressed the distinction, in morphological terms, from the genus *Streptomyces.* A tentative evaluation of 14 species was produced and a type species designated (Farina and Locci, 1966). This was followed by a taxonomic study of 50 named species potentially belonging to the genus (Locci et al., 1969). This resulted in the introduction of the Series, a series of infrageneric taxa, in an attempt to overcome the problem of the large number of species that had been described for patent reasons. At that time, although it was recognized that this proposed system was artificial in construction, it did indicate that in reality the genus contained far fewer species than had been described. A drastic reduction in species number had previously been proposed by Hütter (1962), reducing the species to *cinnamoneus, griseocarneus, netropsis,* and *reticuli,* which would have been retained within the genus *Streptomyces.* Pridham (1976b) proposed only the two species *Streptoverticillium recticulum* and *S. netropsis,* while Konev (1981) put forward the idea of two subgenera of *Streptoverticillium,* namely *Reticulicompactum* and *Reticuliremotum.* These subgenera were based on the relative compactness of the verticils, with subspecies described on the basis of cultural, biochemical and antibiotic characteristics.

New taxonomic techniques appeared to confirm both the separation of the genera *Streptomyces* and *Streptoverticillium* and the necessity of a reduction in the number of species. In a numerical taxonomic study of streptomycetes (Silvestri et al., 1962) the 10 streptoverticillia included were recovered in a single cluster (Group A, Sphere II). Further support for the separation of the two genera was provided by the work of Kurylowicz et al. (1975). Most streptoverticillia included in a phenetic study by Williams et al. (1983a) were distributed among six clusters. The clearest separation of the two groups was achieved in the investigation by Locci et al. (1981), where 185 unit characters were used in a study of 111 strains of streptoverticillia and streptomycetes that produced pseudoverticillate sporophores. The streptoverticillia were distributed among the 10 multimembered and 14 single-membered clusters. These results confirmed that the number of species currently defined could not be supported by the taxonomic evidence.

Acknowledgments

The authors thank the British Council for provision of travel grants and Drs. R. Attwell and T. Cross for information on the newly developed isolation techniques.

Further Reading

Baldacci, E., G. Farina and R. Locci. 1966. Emendation of the genus *Streptoverticillium* Baldacci (1958) and revision of some species. G. Microbiol. *14:* 153–171.

Locci, R. 1985. New combinations and validation of some taxa of the genus *Streptoverticillium.* Ann. Microbiol. Enzimol. *35:* 231–234.

Williams, S.T., R. Locci, J. Vickers, G.M. Schofield, P.H.A. Sneath and A.M. Mortimer. 1985. Probabilistic identification of *Streptoverticillium* species. J. Gen. Microbiol. *131:* 1681–1689.

Differentiation and characteristics of the species of the genus **Streptoverticillium**

The species list is based on a numerical taxonomic approach to classification and identification. These methods may be able to clarify some of the problems within the genus, as outlined in "Taxonomic Comments." A posteriori approaches are easy to criticize, but the present classification is also artificial, owing to the selected nature of the strains studied. The taxonomy involved in handling a limited set of strains is basically devoid of any rationale because of lack of possibility of determining natural variation, ignorance about the frequency in natural habitats, and the impossibility of recovery of the organisms from different environments.

The numerical classification was based on 185 unit characters for each strain, with overall similarity being determined by the simple matching coefficient (S_{SM}; Sokal and Michener, 1958) using the unweighted pair group method with the arithmetic averages (UPGMA) algorithm (Sneath and Sokal, 1973). At the 76% similarity level a group of strains were clearly linked and separated from the rest of those included in the study. This group of strains consisted of species classified in the genus *Streptoverticillium* by Baldacci et al. (1966) and those not previously examined but that showed the typical morphology of the genus (Locci et al., 1969). Also included in this group were a few strains lacking in aerial growth and/or sporulation and that could not be attributed to either *Streptomyces* or *Streptoverticillium.* The second group of organisms consisted of well-described *Streptomyces* species, including the genus type species, *Streptomyces albus,* but also included were some organisms previously attributed to *Streptoverticillium* (examples being *Streptomyces aureocirculatus* and *S. caespitosus*) but that did not comply with the description of the genus (Baldacci et al., 1966). No culture with the typical conformation of the streptoverticillia appeared in this second group.

Within the true *Streptoverticillium* group 24 clusters were defined at the 84% similarity level. Ten clusters were multimembered, and the other 14 were each composed of a single species. These groupings have been used as a basis for the infrageneric structure of the streptoverticillia. The clusters are named after the earliest described species that they contained, the rest being listed as subjective synonyms.

The classification matrix was used to provide data useful for identification. The diagnostic value of the 185 unit characters was determined and ranked using the CHARSEP program of Sneath (1979b), which produces separation indices for each test. These selected diagnostic tests were progressively reduced to the minimum number that still provided good identification scores for the Hypothetical Median Organism of each cluster, as determined by the MOSTTYP program (Sneath, 1980a). Finally the character selection was checked by applying the DIACHAR program (Sneath, 1980b), which ranks the diagnostic scores of each character for each group in an identification matrix and also gives the sum of scores of all characters for each taxon. These data have been used to propose a set of characters (Table 29.17) for use in the identification of *Streptoverticillium* species. These are newly developed methods that have the advantages of possibly eradicating single characters for evaluation and discrimination.

Among the 10 multimembered clusters there are a few that can be described as nonfocused groups that showed some overlap (clusters 1 and 5, 2 and 4, and 3 and 4). Close, tight groups with little overlap are those that are correctly focused.

Identification of an unknown with a cluster was assessed by using the MATIDEN program (Sneath, 1979a), which provided the best scores for a known or unknown strain against the matrix. It also lists scores for the two next-best alternatives, properties of the unknown that are atypical of the best-fit taxon and characters that distinguish the unknown from the two nearest taxa.

A set of 41 characters were selected to enable species discrimination (Williams et al., 1985a). Included in this is the possibility of evaluating new taxa by providing descriptions of the single-species clusters. The taxonomic value of the single-species taxa will have to be determined a posteriori. The number of single-membered clusters probably does not represent a defect in the methodology of numerical taxonomy, but serves to stress the bias to which these organisms have been subjected. The high number of single-membered clusters is probably not an irregularity but a sign of incompleteness of the material available for investigation. These clusters should not be ignored. With the development of new isolation techniques specific for streptoverticillia and with tests enabling differentiation from the streptomycetes these strains may be "condensation centers" for future isolates.

List of species of the genus **Streptoverticillium**

I. Species Defined by the Major Clusters of Locci et al. (1981)

Characters useful for the differentiation of the species are given in Table 29.18.

1. **Streptoverticillium baldaccii** Farina and Locci 1966, 48.[AL]

bal.dac'ci.i. M.L. n. *baldaccii* of Baldacci (named for Prof. E. Baldacci, who introduced the genus *Streptoverticillium.*)

Off-white to reddish spore mass. Melanin pigment is commonly produced. Most strains do not degrade esculin, are sensitive to azlocillin, cephamandole, and cephalotin, grow at 12°C, and produce acid from trehalose.

Type strain exhibits antibacterial and antifungal activity.

Type species of the genus.

Type strain: ATCC 23654.

Subjective synonyms:

S. biverticillatum (Preobrazhenskaya in Gauze et al. 1957) Farina and Locci 1966, 49.[AL] CBS 211.62.
This type strain shows antifungal activity.

S. distallicum Locci, Baldacci and Petrolini Baldan 1969, 42.[AL] NRRL 2886.
Type strain produces distamycins A, B and C.

S. fervens (De Boer, Dietz, Evans and Michaels 1960) Locci, Baldacci and Petrolini Baldan 1969, 23.[AL] NRRL 2755.
Type strain produces fervenulin and exhibits antifungal activity.

S. flavopersicum (Oliver, Goldstein, Bower, Holper and Otto 1961) Locci, Baldacci and Petrolini Baldan 1969, 41.[AL] ATCC 19756.
The type strain produces actinospectacin and exhibits antifungal activity.

S. griseoverticillatum (Shinobu and Shimada 1962) Locci, Baldacci and Petrolini Baldan 1969, 39.[AL] OEU 722.
The type strain produces takacidin (monazomycin).

S. hiroshimense (Shinobu 1955) Farina and Locci 1966, 49.[AL] ATCC 19772.
The type strain exhibits antibacterial and antifungal activity.

S. kentuckense (Barr and Carman 1956) Baldacci, Farina and Locci 1966, 160.[AL] ATCC 12691.
The type strain produces raisnomycin.

S. netropsis (Finlay, Hochstein, Sobin and Murphy 1951) Baldacci, Farina and Locci 1966, 161.[AL] ATCC 23940.
The type strain produces netropsin and exhibits antifungal activity.

S. roseoverticillatum (Shinobu 1956) Farina and Locci 1966, 49.[AL] ATCC 19807.
The type strain exhibits antibacterial and antifungal activity.

"*S. roseoverticillatum* subsp. *albosporum*" (ex Locci, Baldacci and Petrolini Baldan 1969) Locci 1985, 231. ATCC 25189.
The type strain produces streptorubin A and B.

"*S. rubrochlorinum*" (ex Locci, Baldacci and Petrolini Baldan 1969), Locci 1985, 231. LIA 0084.
The type strain produces rubrochlorin, tetraene 51-10, pentaene 51-10, and heptaene 51-10, and exhibits antibacterial activity.

"*S. rubroverticillatum*" (ex Locci, Baldacci and Petrolini Baldan 1969) Locci 1985, 231. INA 3517.
The type strain produces pentaene antibiotics.

2. **Streptoverticillium cinnamoneum** (Benedict, Dvonch, Shotwell, Pridham and Lindenfelser 1952) Baldacci, Farina and Locci 1966, 158.[AL] (*Streptomyces cinnamoneus* Benedict, Dvonch, Shotwell, Pridham and Lindenfelser 1952, 591.)

cin.na.mo'ne.um. L. n. *cinnamum* cinnamon; M.L. adj. *cinnamoneum* cinnamon-colored (after the color of the aerial mycelium).

Whitish-yellow to gray spore mass. Melanin pigment is rarely produced. Most strains produce acid from inositol and ribose, utilize sorbitol, degrade citrate but not hypoxanthine, and are sensitive to azlocillin.

The type strain produces cinnamycin, fungichromin, and leucinamycin.

Type strain: ATCC 11874.

Subjective synonyms:

S. blastmyceticum (Watanabe, Tanaka, Fukuhara, Miyairi, Yonehara and Umezawa 1957) Locci, Baldacci and Petrolini Baldan 1969, 43.[AL] ATCC 1973.
The type strain produces blastomycin (antimycin A2) and exhibits antibacterial activity.

S. cinnamoneum subsp. *albosporum* Thirumalachar in Rahalkar and Thirumalachar 1968, 96.[AL] ATCC 25186.
The type strain produces pentaene HA-145.

S. cinnamoneum subsp. *lanosum* Thirumalachar in Rahalkar and Thirumalachar 1968, 96.[AL] ATCC 25187.
The type strain produces pentaene HA-176.

S. cinnamoneum subsp. *sparsum* Thirumalachar in Rahalkar and Thirumalachar 1968, 96.[AL] ATCC 25185.
This type strain produces pentaene HA-106.

S. eurocidium (Okami, Utahara, Nakamura and Umezawa 1954) Locci, Baldacci and Petrolini Baldan 1969, 36.[AL] NRRL B-1676.
The type strain produces azomycin, eurocidin, and tertiomycin A and B.

S. lavenduligriseum Locci, Baldacci and Petrolini Baldan 1969, 43.[AL] ATCC 13306.
This type strain produces narangomycin.

"*S. mediocidicum* subsp. *multivertillatum*" (ex Kusakabe, Yamaguchi, Nagatsu, Abe, Akasaki and Shirato 1969) Locci 1985, 232. KCC S-0285.
The type strain produces citromycin.

S. olivoreticuli (Arai, Nakada and Suzuki 1957) Baldacci, Farina and Locci 1966, 162.[AL] ATCC 23943.
The type strain produces heptaene 100, olivomycins, and the viomycin complex.

S. parvisporogenes Locci, Baldacci and Petrolini Baldan 1969, 37.[AL] ATCC 12568.
The type strain produces antibiotic PA-150 (compound 616) and exhibits antibacterial activity.

"*S. paucisporogenes*" (ex Hagemann, Nomine and Penasse 1958) Locci, 1985, 232. ATCC 12596.
The type strain produces antifongine (heptaene), antifungin, and antifungone.

S. septatum Prokop 1964, 434.[AL] NRRL 2974.
The type strain produces antibiotic M-741.

"*S. sporiferum*" (ex Thirumalachar 1968) Locci 1985, 232. ATCC 25188.
This type strain shows antifungal and antibacterial activity.

Table 29.18.
A percentage positive probability matrix for **Streptoverticillium** *clusters*

Cluster[a] / Character:	1. Aerial mycelium cottony	2. Spores yellow	3. Spores white	4. Melanin produced	Utilization of: 5. Mannitol	6. D-Melibiose	7. Raffinose	8. Sorbitol	9. Coumarin	10. L-Methionine	11. L-Proline	12. Shikimic acid	13. DL-α-Amino-butyric acid	Acid production from: 14. D-Galactose	15. *meso*-Inositol	16. D-Fructose	17. D-Ribose	18. D-Trehalose
1. *S. baldaccii* (17)	99	35	15	55	10	10	20	1	30	10	99	15	35	25	80	25	85	99
2. *S. cinnamoneum* (14)	94	67	22	28	6	6	17	1	39	44	94	11	50	22	99	11	94	72
3. *S. griseocarneum* (13)	93	13	67	40	13	1	40	27	1	7	87	13	47	13	93	20	80	87
4. *S. hachijoense* (2)	99	1	99	1	1	50	99	50	50	1	50	1	1	1	50	1	1	99
5. *S. salmonis* (3)	99	1	1	99	1	1	1	33	1	1	99	1	67	1	99	1	99	99
6. *S. ladakanum* (2)	50	1	50	1	1	1	1	1	50	99	50	1	50	1	1	99	99	1
7. *S. mobaraense* (3)	75	1	1	1	1	1	50	1	1	25	99	25	50	25	50	1	99	25
8. "*S. morookaense*" (2)	99	1	1	1	99	50	50	1	50	99	99	99	99	99	99	99	99	99
9. *S. abikoense* (5)	99	80	20	80	80	1	1	60	60	80	99	60	80	40	99	80	99	1
10. "*S. olivoreticulum* subsp. *cellulophilum*" (2)	99	1	99	50	99	99	99	99	1	99	99	1	1	1	99	1	50	50
11. *S. albireticuli* (1)	99	99	1	99	99	1	1	1	99	1	99	99	1	1	99	99	99	99
12. "*S. alboverticillatum*" (1)	99	1	99	1	1	1	1	1	99	99	99	1	99	99	99	1	1	99
13. *S. album* (1)	99	99	1	1	1	1	1	1	1	99	99	99	1	1	1	1	99	99
14. *S. kashmirense* (1)	99	1	1	99	1	99	99	99	1	1	1	1	1	1	99	1	1	99
15. *S. kishiwadense* (1)	99	1	99	1	1	1	1	1	99	1	99	1	1	1	99	99	99	1
16. *S. lilacinum* (1)	99	1	1	99	1	1	1	1	99	1	1	1	1	1	1	99	1	99
17. *S. orinoci* (1)	1	99	1	1	1	1	99	1	1	1	1	99	99	99	1	1	1	99
18. *S. rectiverticillatum* (1)	99	1	1	99	99	99	1	1	1	1	99	1	1	99	99	99	99	99
19. "*S. reticulum* subsp. *protomycicum*" (1)	1	1	1	1	1	1	1	1	1	1	1	1	99	1	99	1	99	99
20. "*S. sapporonense*" (1)	99	1	1	1	1	1	1	1	1	99	1	99	99	99	99	1	1	1
21. *S. thioluteum* (1)	1	99	1	1	1	99	1	99	99	99	99	1	1	1	1	1	99	99
22. "*S. verticillium* subsp. *quintum*" (1)	1	1	1	1	1	99	99	99	99	99	99	1	99	99	1	1	1	1
23. "*S. verticillum* subsp. *tsukushiense*" (1)	99	1	1	1	99	99	99	99	99	1	99	1	99	99	99	1	99	1
24. "*S. viridoflavum*" (1)	1	1	99	1	1	1	1	1	99	99	99	99	1	1	99	1	99	99

[a]The number of strains in each cluster is given in parentheses.

Table 29.18.—*continued*

	Degradation of:									Growth with:					Resistance to:						Antibiosis to:		
Cluster[a] / Character:	19. Esculin	20. Citrate	21. DNA	22. Hypoxanthine	23. L-Tyrosine	24. Tween 20	25. NO_3^- reduction	26. H_2S production	27. Growth at 12°C	28. NaCl (5.0% w/v)	29. 1-Phenylethanol (0.3% w/v)	30. Potassium tellurite (0.01% w/v)	31. Crystal violet (0.01% w/v)	32. Malachite green (0.01% w/v)	33. Azlocillin (30 µg/ml)	34. Carbenicillin (100 µg/ml)	35. Cephaloridine (30 µg/ml)	36. Cephalotin (30 µg/ml)	37. Cephamandole (30 µg/ml)	38. Colistin (30 µg/ml)	39. *Aspergillus niger*	40. *Bacillus subtilis*	41. *Candida albicans*
1. *S. baldacci* (17)	1	99	60	40	70	65	25	75	95	85	45	80	20	10	1	20	10	1	1	5	20	15	25
2. *S. cinnamoneum* (14)	28	99	83	1	61	67	17	83	61	39	33	6	28	22	1	39	17	17	17	6	89	44	67
3. *S. griseocarneum* (13)	1	80	99	33	53	1	13	93	67	80	73	67	53	20	33	60	73	73	73	40	73	27	73
4. *S. hachijoense* (2)	1	99	99	1	99	99	1	99	99	1	1	50	1	1	50	50	1	50	50	1	99	99	99
5. *S. salmonis* (3)	1	99	99	99	99	67	99	33	99	99	33	99	1	1	33	67	33	1	1	67	99	67	33
6. *S. ladakanum* (2)	99	1	99	1	50	50	50	99	99	1	1	50	1	50	99	99	99	99	99	1	1	50	1
7. *S. mobaraense* (3)	99	25	75	1	25	75	99	75	50	99	99	1	25	25	50	99	50	1	50	25	99	25	25
8. "*S. morookaense*" (2)	99	99	1	1	50	99	99	99	1	99	99	50	99	1	50	50	1	1	1	50	99	1	1
9. *S. abikoense* (5)	60	99	99	60	99	40	1	60	20	99	20	40	99	99	1	99	80	99	99	20	80	1	80
10. "*S. olivoreticulum* subsp. *cellulophilum*" (2)	1	99	99	1	99	99	1	50	99	50	1	1	99	50	50	99	99	99	99	1	99	50	99
11. *S. albireticuli* (1)	1	99	1	1	1	1	1	99	99	99	1	99	1	1	1	1	1	1	1	1	99	1	99
12. "*S. alboverticillatum*" (1)	1	99	99	1	99	99	1	99	99	99	1	1	1	1	99	99	1	1	99	99	1	99	1
13. *S. album* (1)	99	1	99	1	99	99	1	1	1	1	99	1	1	1	1	99	1	1	1	1	1	1	1
14. *S. kashmirense* (1)	1	99	99	1	1	99	1	1	1	1	99	99	99	99	1	99	99	1	99	99	1	1	1
15. *S. kishiwadense* (1)	1	99	99	99	1	1	1	99	1	1	99	1	99	99	1	1	1	1	1	1	99	99	99
16. *S. lilacinum* (1)	1	99	1	1	1	1	1	1	1	99	99	99	99	1	99	99	1	1	1	1	1	1	1
17. *S. orinoci* (1)	99	1	1	1	1	1	1	99	1	99	99	1	1	1	99	99	99	99	1	99	99	99	1
18. *S. rectiverticillatum* (1)	1	99	99	99	99	99	99	99	99	99	1	1	1	1	1	99	1	1	1	1	99	99	1
19. "*S. reticulum* subsp. *protomycicum*" (1)	1	1	1	1	1	99	99	99	99	1	1	1	99	99	1	1	1	99	1	1	1	1	1
20. "*S. sapporonense*" (1)	1	99	1	1	1	99	1	99	99	99	1	1	1	1	1	99	99	99	1	99	99	99	1
21. *S. thioluteum* (1)	99	1	1	1	99	99	1	1	99	1	99	1	1	1	1	1	99	1	1	1	99	1	99
22. "*S. verticillium* subsp. *quintum*" (1)	99	99	1	1	99	1	1	99	1	99	1	1	99	1	99	99	1	1	99	99	1	99	1
23. "*S. verticillum* subsp. *tsukushiense*" (1)	99	99	1	99	1	1	1	1	1	99	1	1	1	99	1	1	99	1	1	1	99	1	1
24. "*S. viridoflavum*" (1)	1	99	99	99	1	99	1	99	99	1	1	1	1	1	1	1	1	99	99	99	99	1	99

[a]The number of strains in each cluster is given in parentheses.

3. **Streptoverticillium griseocarneum** (Benedict, Stodola, Shotwell, Borud and Lindenfelser 1950) Baldacci, Farina and Locci 1966, 158.[AL] (*Streptomyces griseocarneus* Benedict, Stodola, Shotwell, Borud and Lindenfelster 1950, 77.)

gri.se.o.car'ne.um. L. adj. *griseus* gray; L. adj. *carneum* pertaining to flesh; M.L. adj. *griseocarneum* gray flesh-colored.

White to reddish spore mass. Melanin pigment is rarely produced. Most strains do not degrade esculin and Tween 20 but are DNA positive, and are unable to utilize coumarin, methionine, or melibiose.

The type strain produces hydroxystreptomycin and rotaventin.

Type strain: ATCC 12628.

Subjective synonyms:

S. ardum (De Boer, Dietz, Lummis and Savage 1961) Locci, Baldacci and Petrolini Baldan 1969, 34.[AL] NRRL 2817.

The type strain produces porfiromycin and exhibits antifungal activity.

"*S. cinnamoneum* subsp. *azacolutum*" (ex Locci, Baldacci and Petrolini Baldan 1969) Locci 1985, 231. S-205; ATCC 12686.

The type strain produces azacolutin A and B and duramycin.

S. luteoverticillatum (Shinobu 1956) Locci, Baldacci and Petrolini Baldan 1969, 28.[AL] ATCC 23933.

The type strain produces neutramycin and exhibits antifungal activity.

S. mashuense (Sawazaki, Suzuki, Nakamura, Kawasaki, Yamashita, Isono, Anzai, Serizawa and Sekiyama 1955) Locci, Baldacci and Petrolini Baldan 1969, 36.[AL] ATCC 23934.

The type strain produces monozomycin and streptomycin.

S. olivoverticillatum (Shinobu 1956) Baldacci, Farina and Locci 1966, 163.[AL] ATCC 25480.

The type strain produces trichomycin and exhibits antibacterial and antifungal activity.

"*S. pentaticum* subsp. *jenense*" (ex Locci, Baldacci and Petrolini Baldan 1969) Locci 1985, 231. IMET JA4495.

The type strain produces fervenulin (planomycin) fungichromin and prodigiosin.

"*S. triculaminicum*" (ex Suzuki, Asahi, Nagatsu, Kawashima and Susuki 1967) Locci 1985, 232. KCC S-0242.

The type strain produces triculamin.

"*S. tropicalense*" (ex Gupta 1965) Locci 1985, 232. ATCC 17963.

The type strain produces pentaenes and heptaenes.

4. **Streptoverticillium hachijoense** (Hosoya, Komatsu, Soeda and Sonoda 1952) Locci, Baldacci and Petrolini Baldan 1969, 34.[AL] (*Streptomyces hachijoensis* Hosoya, Komatsu, Soeda and Sonoda 1952, 505.)

ha.chi.jo.en'se. M.L. adj. *hachijoense* from Hachijo (named for the place of origin, Hachijo Jima, a small island in the Pacific Ocean).

White spore mass. Melanin pigment is not produced. Most strains grow in the presence of 5% (w/v) NaCl, do not utilize methionine or produce acid from ribose but utilize raffinose, and have an inhibitory effect on *Bacillus subtilis* and *Candida albicans.*

The type strain produces soedomycin and trichomycin A, B, and C.

Type strain: ATCC 19769.

Subjective synonym:

S. hachijoense, strain IPV 2057.

5. **Streptoverticillium salmonis** (Baldacci, Farina and Locci) Locci, Baldacci and Petrolini Baldan 1969, 27.[AL] (*Streptoverticillium salmonicida* (Rucker) Baldacci, Farina and Locci 1966, 164.)

sal.mon'is. L. n. *salmonis* of the salmon.

Gray to reddish spore mass. Melanin pigment is produced. Methionine and coumarin are not utilized, whereas nitrate is reduced and hypoxanthine and tyrosine are degraded. Most strains grow in the presence of 0.01% (w/v) potassium tellurite.

The type strain exhibits antibacterial and antifungal activity.

Type strain: NRRL B-1472.

Subjective synonym:

S. aureoversales Locci, Baldacci and Petrolini Baldan 1969, 24.[AL] ATCC 15853. (The name *Actinomyces aureoversales* was used in a thesis (Yuan, 1962, p. 188), which was abstracted but has not been validly published.) In the "Corrigenda to the Approved List" (Hill et al., 1984) the correction into *aureoversile* (sic) of the specific epithet is most unfortunate. First, *aureoversale* is not the neuter form, and in the process an additional orthographic error was introduced.

The reference strain produces tetraene 380 and pentaene 380 and exhibits antibacterial activity.

6. **Streptoverticillium ladakanum** Hanka, Evans, Mason and Dietz 1967, 620.[AL]

la.da.ka'num. M.L. adj. *ladakanum* from Ladislav Hanka, who isolated the strain.

Off-white spore mass. Melanin pigment is not produced. Most strains do not produce acid from inositol or trehalose but are fructose positive; citrate is not utilized. Most strains are resistant to cephalotin, and have no antifungal effect against *Aspergillus niger.*

The type strain produces 5-azacytidine.

Type strain: ATCC 27441.

Subjective synonym:

"*S. verticillum*" (ex Takita 1959) Locci 1985, 232. ATCC 15003.

The type strain produces bleomycin A and B and phleomycins.

7. **Streptoverticillium mobaraense** (Kubo, Suzuki and Tamura 1964) Locci, Baldacci and Petrolini Baldan 1969, 42.[AL] (*Streptomyces mobaraenisis* Kubo, Suzuki and Tamura 1964, 47).

mo.bar.a.en'se. M.L. adj. *mobaraense* from Mobara (named after the place of origin, Mobara City, Chiba Prefecture, Japan).

Off-white to gray spore mass. Melanin pigment is not produced. Most strains reduce nitrate, degrade esculin, and will grow in the presence of 0.3% (w/v) phenylethanol and 5% (w/v) NaCl, but are sensitive to cephalotin and will not utilize coumarin.

The type strain produces detoxin, piericidin A, and piericidin B, and exhibits antibacterial activity.

Type strain: IPV 2058.

Subjective synonym:

"*S. luteoreticuli*" (*ex* Kato and Arai, 1957) Locci 1985, 232. ATCC 27446.

The type strain produces aureothricin, mycometoxin A and B, and thiolutin.

8. **"Streptoverticillium morookaense"** (ex Niida, Hamamoto, Tsuruoka and Hara 1963) Locci 1985, 232.

mo.rook.a.en'se. M.L. adj *morookaense* from Moro-oka (possibly the isolation place).

Whitish to gray spore mass. Melanin pigment is not produced. Most strains produce acid from fructose and galactose, utilize mannitol and shikimic acid, degrade esculin, and reduce nitrate.

This type strain shows antifungal activity.

Type strain: ATCC 19166.

Subjective synonym:

"*S. aspergilloides*" (ex Rao, Marsh and Renn 1967) Locci 1985, 232. ATCC 14804.

The type strain produces antibiotic BA-181314.

9. **Streptoverticillium abikoense** (Umezawa, Tazaki and Fukuyama 1951) Locci, Baldacci and Petrolini Baldan 1969, 35.[AL] (*Streptomyces abikoensis* Umezawa, Tazaki and Fukuyama 1951, 333.)

a.bi.ko.en'se. M.L. adj. *abikoense* from Abiko (named after Abiko in Japan).

White to yellow spore mass. Most cultures produce melanin pigment. Acid is not produced from trehalose and tyrosine is not degraded. Most strains grow in the presence of malachite green (0.01% w/v), crystal violet (0.01% w/v), cephamandole, and cephalotin.

The type strain produces abikoviromycin, actinoleukin, trichomycin, and viomycin.

Type strain: ATCC 12766.

Subjective synonyms:

S. ehimense (Shibata, Honso, Tokui and Nakazawa 1954) Locci, Baldacci and Petrolini Baldan 1969, 40.[AL] ATCC 23903.

The type strain produces candimycin and exhibits antibacterial activity.
"*S. rimofaciens*" (ex Niida 1965) Locci 1985, 232. ATCC 22166.
The type strain produces destomycin A and B.
"*S. takataense*" (ex Anonymous 1962) Locci 1985, 232. ATCC 27469.
The type strain produces takamycin.
"*S. waksmanii*" (ex Baldacci, Farina and Locci 1966) Locci 1985, 231. ATCC 12629.
The type strain produces antibiotic F-20.

10. **"Streptoverticillium olivoreticulum** subsp. **cellulophilum"** (ex Anonymous 1973) Locci 1985, 232.

o.liv.o.re.ti'cul.um. L. n. *oliva* olive; L. n. *reticulum* small net; M.L. adj. *cellulophilum* lover of cellulose.

Off-white spore mass. Some strains produce melanin pigment. Most species utilize mannitol, melibiose, sorbitol, and raffinose. Resistant to cephalotin and cephamandole.

The type strain produces destomycins.

Type strain: ATCC 21632.

Subjective synonym:

"*S. hachijoense* subsp. *takahagiense*" (ex Arai 1976) Locci 1985, 232. T. Niida strain.
The type strain produces leucopeptin.

II. Species Defined by Single-Member Clusters of Locci et al. (1981)

Characters useful for the differentiation of the species are given in Table 29.18.

11. **Streptoverticillium albireticuli** (Nakazawa 1955) Locci, Baldacci and Petrolini Baldan 1969, 37.[AL] (*Streptomyces albireticuli* Nakazawa 1955, 649.)

al.bi.ret.i'cu.li. L. adj. *albus* white; L. n. *reticulum* a small net; M.L. n. *albireticuli* of the small white net.

Yellow spore mass. Melanin pigment is produced. Mannitol and shikimic acid are utilized but DNA is not degraded. Acid is produced from fructose, 0.01% (w/v) potassium tellurite is tolerated, and *Candida albicans* is inhibited.

The type strain produces carbomycin, enteromycin, eurocidin, and tertiomycin A and B.

Type strain: ATCC 19721.

12. **"Streptoverticillium alboverticillatum"** (ex Arai 1976) Locci 1985, 232.

al.bo.ver.ti.cil'la.tum. L. adj. *albus* white; M.L. adj. *verticillatum* forming whorls; M.L. adj. *alboverticillatum* white and forming whorls.

Off-white spore mass. Melanin pigment is not produced. Acid is not produced from ribose. Resistant to cephaloxidine, colistin, and cephamandole. *Aspergillus niger* is not inhibited, whereas *Bacillus subtilis* is inhibited.

The type strain produces alboverticillin.

Type strain: IPV 2254.

13. **Streptoverticillium album** Locci, Baldacci and Petrolini Baldan 1969, 40.[AL]

al'bum. From L. adj. *album* white (with reference to the aerial mycelium color).

Yellow spore mass. Melanin pigment is not produced. Citrate is not degraded, and H_2S is not produced. Shikimic acid is utilized and esculin degraded. *Aspergillus niger* is not inhibited.

The type strain produces acetopyrrothine (thiolutin).

Type strain: NRRL 2401.

14. **Streptoverticillium kashmirense** (Gupta and Chopra 1963a) Locci, Baldacci and Petrolini Baldan 1969, 45.[AL] (*Streptomyces kashmirensis* Gupta and Chopra 1963a, 112.)

kash.mir.en'se. M.L. adj. *kashmirense* from Kashmir (named after the place of origin, Jammu in Kashmir).

Reddish spore mass. Melanin pigment is produced. Melibiose, raffinose, and sorbitol are utilized, acid is not produced from ribose. H_2S is not produced and growth is not inhibited in the presence of 0.01% (w/v) malachite green. Proline is not utilized.

The type strain exhibits antibacterial and antifungal activity.

Type strain: NRRL B-3103.

15. **Streptoverticillium kishiwadense** (Shinobu and Kayamura 1964) Locci, Baldacci and Petrolini Baldan 1969, 35.[AL] (*Streptomyces kishiwadensis* Shinobu and Kayamura 1964, 176.)

ki.shi.wad.en'se. M.L. adj. *kishiwadense* from Kishiwada (named after the place of origin, Kishiwada City, Japan).

Whitish spore mass. Melanin pigment is not produced. Acid is produced from fructose but not from trehalose; hypoxanthine is degraded and resistance is shown to malachite green and crystal violet. Antibiotic effect against *Candida albicans*.

The type strain shows antibacterial and antifungal activity.

Type strain: ATCC 25464.

16. **Streptoverticillium lilacinum** (Nakazawa, Tanabe, Shibata, Miyake and Takewaka 1956) Locci, Baldacci and Petrolini Baldan 1969, 44.[AL] (*Streptomyces lilacinus* Nakazawa, Tanabe, Shibata, Miyake and Takewaka 1956, 81.)

li.la.cin'um. L. adj. *lilacinum* lilac-colored.

Grayish to violet spore mass. Melanin pigment is produced. Acid is produced from fructose but not from inositol or ribose. Proline is not utilized, and H_2S is not produced. Grows in the presence of 0.01% (w/v) potassium tellurite.

The type strain produces cladomycin and exhibits antifungal activity.

Type strain: ATCC 23930.

17. **Streptoverticillium orinoci** Cassinelli, Grein, Orezzi, Pennella and Sanfilippo 1967, 358.[AL]

o.ri.no'ci. M.L. n. *orinoci* of Orinoco (named after the Orinoco River (S. America), from whose banks it was isolated).

Whitish-yellow spore mass. Melanin pigment is not produced. Acid is not produced from inositol or ribose, proline is not utilized, and citrate is not degraded. Raffinose and shikimic acid are utilized.

The type strain produces neoantimycin, neoaureothin, and ochramycin.

Type strain: ATCC 23202.

18. **Streptoverticillium rectiverticillatum** (Krasil'nikov and Yuan 1965) Locci, Baldacci and Petrolini Baldan 1969, 41.[AL] (*Actinomyces rectiverticillatus* Krasil'nikov and Yuan 1965, 49.)

rect.i.ver.ti.cil.la'tum. L. adj. *rectus* straight; L. adj. *verticillatum* forming whorls; M.L. adj. *rectiverticillatum* verticillate with straight spore chains.

Reddish spore mass. Melanin pigment is produced. Mannitol and melibiose are utilized, hypoxanthine is degraded, nitrate is reduced, and acid produced from fructose and galactose.

The type strain exhibits antibacterial and antifungal activity.

Type strain. ATCC 19845.

19. **"Streptoverticillium reticulum** subsp. **protomycicum"** (ex Arai 1976) Locci 1985, 232.

re.ti'cu.lum. L. n. *reticulum* small net (and with reference to production of protomycin).

Poor aerial mycelium. Melanin pigment is not produced. Proline is not utilized and citrate is not degraded. Nitrate is reduced. Growth takes place in the presence of malachite green and cephalotin. *Aspergillus niger* is not inhibited.

The type strain produces protomycin.

Type strain: KCC S-0180.

20. **"Streptoverticillium sapporonense"** (ex Arai 1976) Locci 1985, 232.

sap.po.ron.en′se. M.L. adj. *sapporonense* from Sapporo (named after the place of origin of the strain, Sapporo, Japan).

Reddish spore mass. Melanin pigment is not produced. Acid produced from galactose but not from trehalose or ribose. Shikimic acid is utilized but not proline. Resistant to cephalotin.

The type strain produces bicyclomicin.

Type strain: ATCC 21532.

21. **Streptoverticillium thioluteum** (Okami 1952a) Baldacci, Farina and Locci 1966, 165.[AL] (*Streptomyces thioluteus* Okami 1952, 30.)

thi.o.lu′te.um. G. n. *thion* sulfur; L. adj. *luteum* yellow; L. adj. *thioluteum* sulfur-yellow in color.

Poor sporulation on most media. Spore mass yellow. Melanin pigment is not produced. Melibiose and sorbitol are utilized, and esculin is degraded whereas citrate is not; H_2S is not produced. Acid is not produced from inositol.

The type strain produces aureothin and propriopyrrothin (aureothricin).

Type strain: ATCC 12310.

22. "**Streptoverticillium verticillium** subsp. **quintum**" (ex Arai 1976) Locci 1985, 232.

quin′tum. L. adj. *quintum* the fifth. Etymology uncertain.

Poor aerial mycelium. Melanin pigment is not produced. Melibiose, raffinose, and sorbitol are utilized. Acid is not produced from inositol, trehalose, or ribose.

The strain shows antibacterial activity.

Type strain: Y. Okami strain.

23. "**Streptoverticillium verticillum** subsp. **tsukushiense**" (ex Arai 1976) Locci 1985, 232.

Name of subspecies possibly derived from the place of isolation.

Poor aerial mycelium. Melanin pigment is not produced. Mannitol, melibiose, and sorbitol are utilized. Hypoxanthine is degraded. Acid is not produced from trehalose. Growth in the presence of 0.01% (w/v) malachite green.

The strain produces XK-19-2 antibiotic.

Type strain: ATCC 21633.

24. "**Streptoverticillium viridoflavum**" (ex Waksman and Taber 1953) Locci 1985, 232.

vir.i.do.flav′um. L. adj. *viridis* green; L. adj. *flavum* yellow; L. adj. *viridoflavum* green-yellow.

Poor off-white aerial vegetation. White spore mass. Melanin pigment is not produced. Shikimic acid is utilized and hypoxanthine is degraded. Resistance shown to cephalotin, colistin, and cephamandole. *Candida albicans* is inhibited.

The strain shows antifungal activity.

Type strain: Y. E. Konev strain.

Species Incertae Sedis

Streptomyces avidinii Stapley, Mata, Miller, Demny and Woodruff 1964, 20.[AL]

This species clustered with streptoverticillia in the numerical classification of Williams et al. (1983a) and produces verticillate sporophores. It also showed close immunological affinity with streptoverticillia (Ridell et al., 1986). Its generic status therefore requires further study. However, in the International *Streptomyces* Project *S. avidinii* is described as a typical streptomycete with RF sporophores (Shirling and Gottlieb, 1972).

Type strain: NRRL 3077.

Genus **Kineosporia** *Pagani and Parenti 1978, 401[AL]*

FRANCESCO PARENTI

Ki.ne.o.spo′ri.a. Gr. n. *kinesis* motion; Gr. n. *sporos* a seed; M.L. n. *spora* a spore; M.L. fem. n. *Kineosporia* motile spore.

Small colonies of 0.5–2 mm in diameter, colorless for the first few days of growth, turning cream to orange thereafter. Surface of the colony becomes moist and of glossy appearance with age. Colonial morphology variable even on the same agar plate, ranging from conical-crateriform to cerebriform. The vegetative mycelium consists of fine (1 µm in diameter), slightly branched hyphae. **Aerial mycelium has never been observed** on any medium tested. **Sporangia appear as elongated, club-shaped vesicles of 1–2** µm in diameter, borne at the terminal end of the vegetative hyphae. Each sporangium contains a single spore. The spores are pleomorphic, their shape ranging from nearly spherical to oval or pyriform. The long axes of the spores vary between 1 and 2 µm. Upon sporangium dehiscence, the spores are **motile** for several hours, with a **polar tuft of flagella.** Gram-positive and not acidfast. **The peptidoglycan of the cell wall contains L-diaminopimelic acid** (L-DAP) **a very small amount of *meso*-DAP, and glycine.** Traces of lysine are also found. Aerobic. Chemo-organotrophic, using a range of simple sugars. Optimum temperature for growth and sporulation is 20–30°C, with no growth at or above 37°C. Optimum pH 7.0. Nonpathogenic to mice, occurs free-living in soil. The mol% G + C of the DNA is unknown.

Type species: *Kineosporia aurantiaca* Pagani and Parenti 1978, 401.

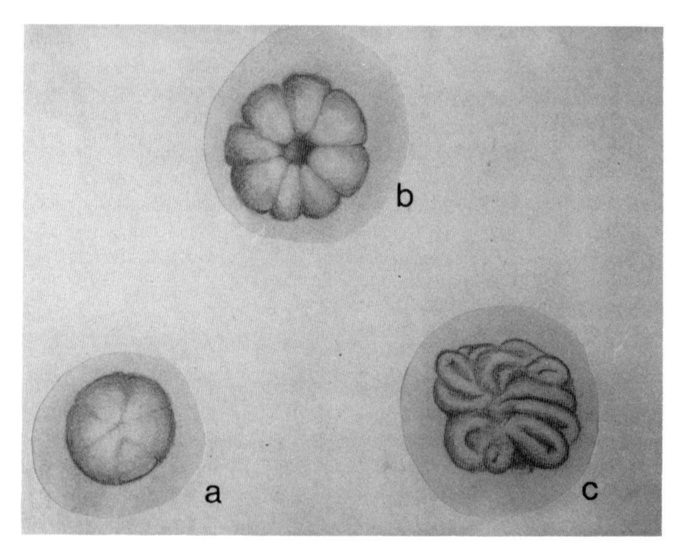

Figure 29.23. Illustrations of colonial morphology of *Kineosporia aurantiaca*. *A*, conical. *B*, crateriform. *C*, cerebriform. Growth on oatmeal agar (× 9).

Further Descriptive Information

Colonies are pleomorphic on all media, particularly on yeast extract-malt agar. The wet, glossy appearance that is evident after 7 days is due to abundant sporangia formation. Growth occurs by spreading of vegetative mycelium into the agar and by thickening of the layer formed by the vegetative mycelium above the agar. This mode of growth generates complex colony morphologies ranging from conical to crateriform to cerebriform (Fig. 29.23).

In a liquid medium with a readily utilizable carbon source, incubated for 4 days with shaking at 30°C, abundant growth occurs, consisting of mycelium highly fragmented into small rods.

Sporangia are found at the ends of hyphae (Fig. 29.24) and, by electron microscopy (EM), appear inside the terminal cell (Fig. 29.25). The sporangium is therefore the tip of the hypha, which becomes enlarged with a spherical to ovoid shape, with a diameter of less than 2 μm and containing a single spore. Dehiscence of the sporangium occurs soon after the sporangium-bearing mycelium is placed in water. The sporangium always ruptures at the terminal, or upper, end. When dehiscence is completed the cell wall of the sporangium is no longer visible by light microscopy.

Electron microscopy suggests the possibility of sequential formation of sporangia, i.e. a second sporangium may develop just below the one at the tip of the hypha. However, the absence of a "sausagelike" configuration of the terminal part of hypha indicates that the subterminal sporangium develops only upon dehiscence of the terminal one. The single planospore is round to pyriform and possesses a tuft of flagella originating from a single area of the cell (Fig. 29.26). The free spore is enveloped by a substance having affinity for the tungsten stain used in EM, with a weblike pattern apparently resulting from the breakdown of the sporangium cell wall sheath (Bernard and Parenti 1983).

The cultural characteristics on selected media are given in Table 29.19. Cell wall composition is type I (Lechevalier and Lechevalier, 1970b). Glucose, fructose, rhamnose, xylose, and arabinose are readily utilized. Growth is poor on sucrose and abundant on starch. Lactose, salicin, raffinose, cellulose, proteins (litmus milk, casein, and gelatin), and tyrosine are not hydrolyzed. Calcium malate is not solubilized, citrate is not reduced, and H_2S is not produced.

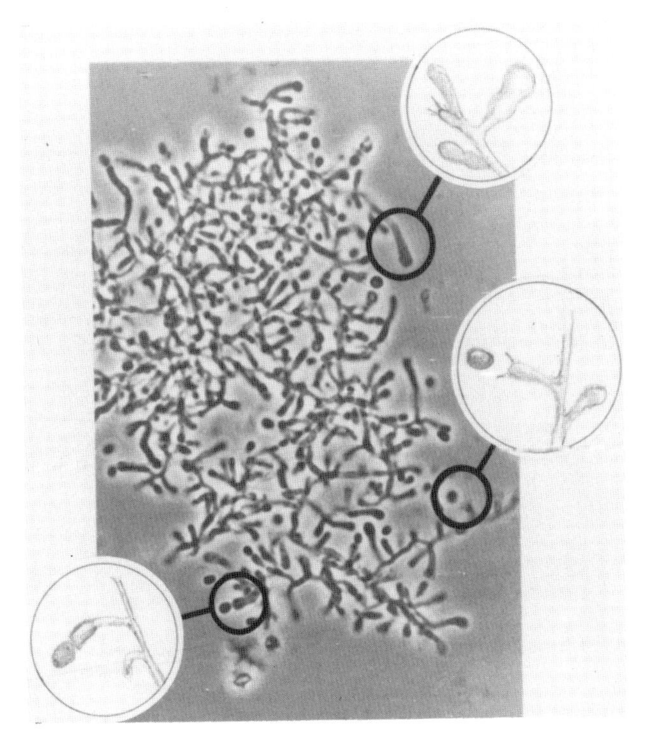

Figure 29.24. Vegetative mycelium of *Kineosporia aurantiaca* on oatmeal agar (× 1200). In circles are drawings of sporangia at different developmental stages.

Kineosporia is highly susceptible to low concentrations (0.1–2 μg/ml) of the antibacterial antibiotics streptomycin, tetracycline, rifampicin, bacitracin, and chloramphenicol; it is moderately susceptible to penicillin (20 μg/ml) and resistant to the antifungal antibiotics cycloheximide and amphotericin B (Pagani and Parenti, 1978). All information given is based on the single strain of *K. aurantiaca* (Pagani and Parenti, 1978).

Enrichment and Isolation Procedures

The type strain was isolated from a garden soil sample from India that was subjected to an enrichment and isolation procedure used for the selective isolation of members of the *Actinoplanaceae* family (Parenti and Coronelli, 1979).

Although an extensive isolation program produced over 20,000 isolates of strains belonging to the genus *Actinoplanes,* since the first discovery of *K. aurantiaca* no new strains of this species have been isolated (Parenti and Coronelli, 1979).

Maintenance Procedures

Kineosporia can be maintained by growing it on oatmeal agar slants at 28°C for 10 days, followed by storage at 5°C for 3–4 weeks with little loss of viability. For lyophilization the surface of a 10-day-old slant is scraped with a Pasteur pipette and the cells are suspended in 6 ml of a 20% (w/v) suspension of skim milk (Difco). For recovery, the lyophilized cells are suspended in distilled water and transferred onto oatmeal agar slants.

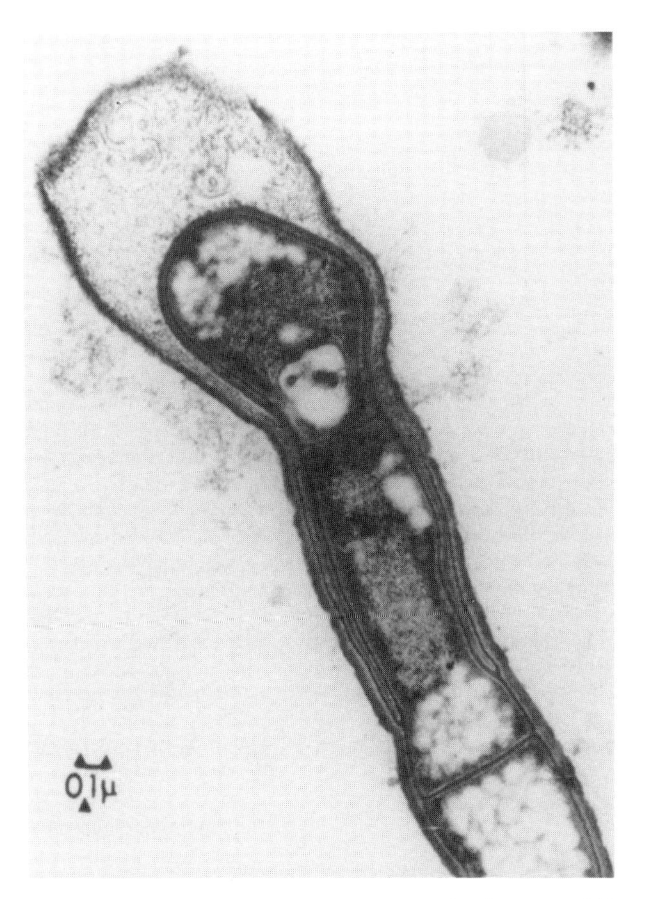

Figure 29.25. Electron micrograph of spore development within a terminal hyphal cell. (Reproduced by permission from S. D. Bernard and F. Parenti, Current Microbiology *8:* 173–176, 1983, © Springer-Verlag, New York.)

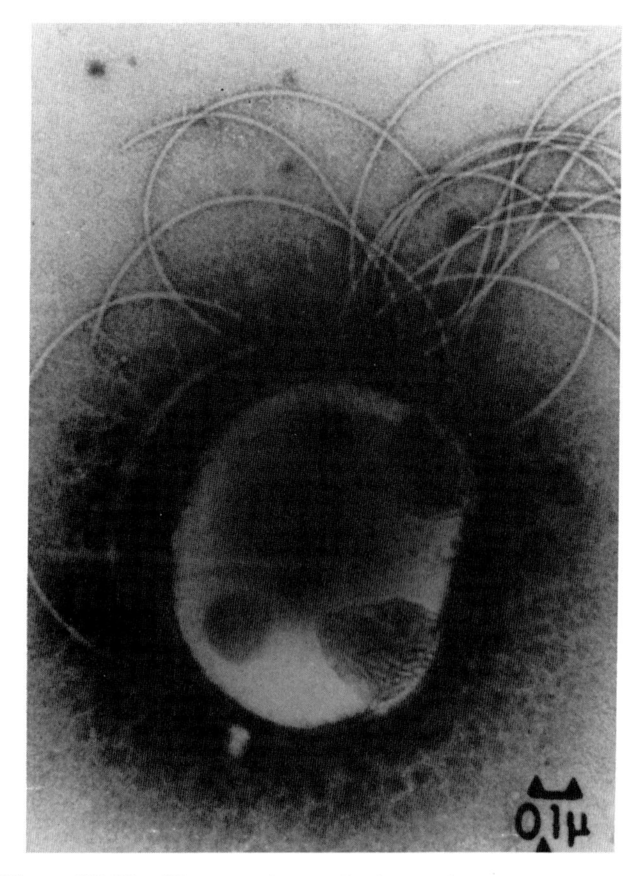

Figure 29.26. Electron micrograph of a motile spore. (Reproduced by permission from S. D. Bernard and F. Parenti, Current Microbiology *81:* 173–176, 1983, Springer-Verlag, New York.)

Table 29.19.
Cultural characteristics of **Kineosporia aurantiaca**

Medium	Cultural characteristics at 30°C
ISP 2[a]	Abundant growth, wrinkled surface, opaque, hard, amber-orange
ISP 3[a]	Moderate growth, smooth surface, moist, bright, light orange
Oatmeal agar	Abundant growth, smooth surface, moist, bright, soft, deep orange, slightly yellowish pigment
Nutrient agar	Moderate growth, smooth surface, opaque, hyaline to cream
Skim milk agar	Abundant growth, wrinkled surface, opaque, deep orange

[a]Shirling and Gottlieb (1966).

Differentiation of the genus **Kineosporia** *from other genera*

The absence of an aerial mycelium confers on colonies of *K. aurantiaca* the same gross morphological appearance as those of *Actinoplanes* and *Micromonospora.* However, the colonies of *Kineosporia* are more complex and have a distinct glossy appearance due to the large number of highly refractile sporangia.

The localization of the sporangium on the substrate mycelium helps to differentiate *Kineosporia* from *Planobispora, Planomonospora, Spirillospora, Kitasatoa,* and *Streptosporangium,* in which sporangia are borne on a well-developed aerial mycelium.

The shape, size, connection to substrate mycelium, and number of spores of the sporangium help to distinguish *Kineosporia* from the genera with sporangia borne directly on the substrate mycelium (*Actinoplanes, Ampullariella, Dactylosporangium, Pilimelia, Streptoallotheicus,* and *Amorphosporangium*).

In addition, *Kineosporia* cell wall composition, like that of *Streptomyces,* is type I (Lechevalier and Lechevalier, 1970b) (prevalence of L-DAP and glycine), whereas *Actinoplanes, Ampullariella, Dactylosporangium,* and *Micromonospora* have a cell wall type II (*meso*-DAP and glycine) and *Planobispora, Planomonospora, Spirillospora,* and *Streptosporangium*) have cell wall type III (*meso*-DAP, no glycine).

Acknowledgments

We thank L. Acquati for drawing Figures 29.23 and 29.24.

Further Reading

Bernard, D.S. and F. Parenti. 1983. Ultrastructural morphology of the genus *Kineosporia.* Curr. Microbiol. *8:* 173–176.

Pagani, H. and F. Parenti. 1978. *Kineosporia,* a new genus of the Order *Actinomycetales.* Int. J. Syst. Bacteriol. *28:* 401–406.

List of species of the genus **Kineosporia**

1. **Kineosporia aurantiaca** Pagani and Parenti 1978, 401.[AL]
au.ran.ti'a.ca. M.L. adj. *aurantiaca* orange color.

Features are as given for the genus. Cultural characteristics are presented in Table 29.19. *K. aurantiaca* is the only species isolated and described.

Type strain: ATCC 28727.

Genus **Sporichthya** *Lechevalier, Lechevalier and Holbert 1968, 279*[AL]

MARY P. LECHEVALIER AND HUBERT A. LECHEVALIER

Spor.ich'thy.a. Gr. n. *sporos* seed; Gr. n. *ichthys* fish; M.L. fem. n. *Sporichthya* an organism with fishlike spores.

Very short, sparse aerial mycelium composed of hyphae 0.5–1.0 µm in diameter that grow on the surface of solid media. **The aerial hyphae are maintained upright at the surface of the medium by holdfasts, which are outgrowths of the wall of the basal cell. Primary (substrate) mycelium not formed. The sparingly branched aerial mycelium divides into rod-shaped to coccoid spores, which,** in the presence of water, **may become polarly flagellate and motile.** On nitrogen-rich media the spores swell to give rise to swollen elements of various shapes, including fish-shaped structures, which may be motile. **Gram-variable.** Young cells tend to be Gram-negative; older cells are mainly Gram-positive. Cell sections examined with an electron microscope reveal a **Gram-positive type of cell wall. Facultatively anaerobic.** Mesophilic. Chemo-organotrophic. Cell wall preparations contain major amounts of L-diaminopimelic acid (L-DAP), glycine, alanine, glutamic acid, glucosamine, and muramic acid. Because of the difficulties of obtaining enough cells for analysis, there are no data on phospholipid, menaquinone, or DNA composition. Sporichthyae are rare; all those isolated to date were from cultivated soil.

Type species: *Sporichthya polymorpha,* Lechevalier, Lechevalier and Holbert, 1968, 279 (the only species described so far).

Further Descriptive Information

All strains of *Sporichthya* isolated to date have similar physiological characteristics: complex carbohydrate substrates are utilized (both amylases and β-glucosidases are produced) but no utilization of or acid production from monosaccharides, disaccharides, or trisaccharides is observed. The report of acid production from fructose in the original paper (Lechevalier et al., 1968) was in error. Among organic acids, only pyruvate and acetate are utilized by all strains. Few nitrogen-containing complex substrates are utilized and/or hydrolyzed: adenine, hypoxanthine, tyrosine, urea, and xanthine are not attacked and casein is hydrolyzed by only a few strains; the latter is usually inhibitory to growth of others. Nitrate reductase is produced by a few isolates and the reaction is dependent on the medium used in the test.

Although they are readily recognized by their distinctive morphology and are easy to maintain on solid media, little growth of sporichthyae takes place in liquid culture; consequently, obtaining cell masses for chemical analysis has proved to be difficult.

Enrichment and Isolation Procedures

Sporichthyae are rare: only five strains have been isolated in our laboratory during 20 years. To our knowledge, no other isolates have been reported in the literature. All were isolated from cultivated soil by spreading soil dilutions on the surface of plates of tap water agar or dilute (1/10th strength) Czapek agar. Colonies of sporichthyae are hard to see with the naked eye and the use of a 10 × to 40 × microscope lens with a long working distance condenser in examining plates is a useful stratagem.

Maintenance Procedures

Transfer of slant-maintained cultures every 3–4 months and storage at 4°C between transfers is satisfactory. Bennett or Czapek agars are used for maintenance. Long-term preservation by lyophilization from water suspensions of cells is best for most strains. Use of skim milk is also possible, but may inhibit outgrowth of some sensitive strains.

Procedures for Testing of Special Characters

Cell wall and whole cell analyses are carried out by the techniques described in Lechevalier and Lechevalier (1980). Physiological tests are those described in Lechevalier et al. (1968) and Lechevalier (1972).

Differentiation of the genus **Sporichthya** *from other genera*

Because members of the genus *Pseudonocardia* form short chains of irregularly sized aerial spores and produce scanty vegetative mycelium, strains of this genus may be mistaken for *Sporichthya* species. Whole-cell analysis of the DAP isomer present in the cell walls is sufficient to distinguish the two genera: sporichthyae contain the L isomer and pseudonocardiae, the *meso* form.

Taxonomic Comments

The genus *Sporichthya* was originally placed in the family *Streptomycetaceae* because its cell wall composition was like that of other members of the family. However, because of its unusual morphology and the difficulties of determining the chemistry of its phospholipids, menaquinones, and DNA, it has more recently been referred to as "a genus in search of a family" (Lechevalier and Lechevalier, 1981a).

List of species of the genus **Sporichthya**

1. **Sporichthya polymorpha** Lechevalier, Lechevalier and Holbert 1968, 279.[AL]

po.ly.mor'pha. Gr. adj. *polys* numerous; Gr. n. *morphe* form; M.L. fem. adj. *polymorpha* having numerous shapes.

Aerial hyphae chalky white, 0.5–1.2 µm in diameter, 10–25 µm long, sparingly branched, growing on the surface of solid media to which they adhere by crampons, which are outgrowths of the cell wall of the basal cells of the hyphae. Division takes place within the hyphae by annular ingrowth of the cell wall, resulting in the formation of chains of rod-shaped and coccoid segments on nitrogen-poor media (Czapek). After division, the cell walls of the segments thicken at the end that is directed toward the substrate, forming a ring that looks like the bud scar of a yeast. Hyphae are hydrophobic and seem to be encrusted with a hard substance of unknown composition, since they crackle when pressed between two glass surfaces. The surface of the hyphae and its segments is smooth; segments do not become flagellated if not pretreated with water or a liquid nutritive solution. In cultures 2–4 days old, such treatment results in the rapid development of one to three flagella located at the end of the cell having the thickened ring. In older cultures motility may appear only several hours after the addition of liquids. When grown on nitrogen-rich media (such as Bennett) the segments become distended and pisciform (up to 6 µm long) without losing their flagella and motility. Under these conditions the colony surface is hydrophilic and its appearance is mucoid, hyphae being formed only at the edge of the colonies.

Gram-variable: 95% of the cells in old (more than 5 days) cultures are Gram-positive. Younger cultures, especially those formed on nutritionally poor media, contain a high proportion of Gram-negative cells that are presumed to represent the immature forms. Young, Gram-negative cells are more hydrophobic than old, Gram-positive cells. Examination of cross-sections reveals cell walls with a Gram-positive architecture. Cells of *S. polymorpha* are readily colored by methylene blue, carbol fuchsin, and safranin. Sudan black reveals no lipid deposits, and nigrosin no capsular materials.

Production of pigments has not been observed; when grown on poor media, such as Czapek agar, the appearance of the colonies is dry chalky white; on rich media such as Bennett agar, the appearance is humid dirty white to beige. Viewed with a brightfield microscope, the growth may appear black, blue, red, brown, or colorless as a consequence of various light diffraction and interference phenomena. In general, highly hydrophobic growth appears black and old hydrophilic cells are colorless.

Growth is rapid (visible in 24 h) on Czapek, Bennett, and yeast extract–glucose agars. There is no growth in shaken liquid culture; however, under static conditions, a thin white pellicle is formed on the surface of the medium. This growth usually adheres to and climbs up the glass walls of the flask. As indicated above, the cells of *S. polymorpha* grown on poor media, such as Czapek agar, seem to be encrusted with a hard substance that may be responsible for their resistance to desiccation and vacuum deformation (Williams, 1970).

Compared with many other actinomycetes, *S. polymorpha* is physiologically quite inactive. Starch and esculin are hydrolyzed and casein is variably so, but gelatin, adenine, hypoxanthine and tyrosine are not. No acid is produced from adonitol, arabinose, cellobiose, dextrin, fructose, galactose, glucose, glycerol, inositol, lactose, maltose, mannitol, mannose, melibiose, α-methyl-D-mannoside, raffinose, rhamnose, trehalose, xylose, and β-methyl-D-xyloside. Acetate and pyruvate are utilized; lactate and succinate are utilized by some strains and benzoate, citrate, malate, oxalate, propionate, and tartrate are not utilized. Nitrates are reduced on some media. Phosphatase production varies with the strain, and urease is negative. Growth on 5% (w/v) NaCl also varies with the strain; no growth occurs in salicylate broth. *S. polymorpha* strains grow at 10–37°C with an optimum range of 28–37°C. Some strains grow at 42°C but no growth takes place at 45°C. Light and gravity do not affect growth and morphology.

S. polymorpha grows best in the presence of oxygen, but reduced growth can take place on agar media incubated in an atmosphere of hydrogen and carbon dioxide (Gaspak). The anaerobically grown cells cannot be recultivated anaerobically, but will grow out aerobically.

The principal steps in the life cycle of *S. polymorpha* are illustrated in Figure 29.27. A flagellate element (*a* or *b*) settles on the surface of the agar medium and the wall of its ringlike structure becomes transformed into crampons or hooks that penetrate the medium slightly and anchor the organism. Growth takes place by the upward elongation of such cells into filaments (Figs. 29.28 and 29.29), which eventually branch. Septa are set down at various levels within the hypha, and develop into a chain of coccoid to rodlike spores. Passage through a lophotrichous flagellated stage is not necessary for germination of the spores since this life cycle had been observed on dry solid media. On all culture media, but especially on rich media such as Bennett agar, distended cells are formed (Fig. 29.27 *f, g*) that are also capable of reproducing the species as indicated above; however, the largest cells are the least likely to germinate.

Type strain: ATCC 23823.

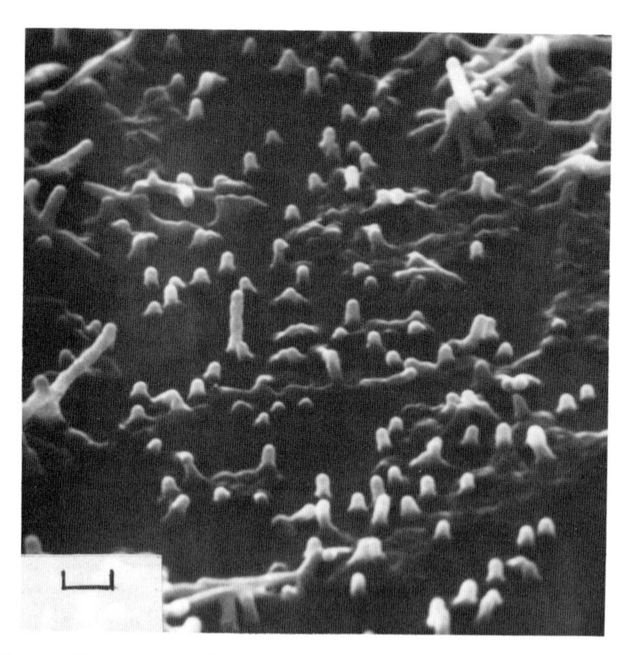

Figure 29.28. *S. polymorpha,* general view of developing upright cells. Scanning electron microscopy. *Bar,* 1 μm. (Reproduced with permission from S. T. Williams, Journal of General Microbiology *62:* 67–73 1970.)

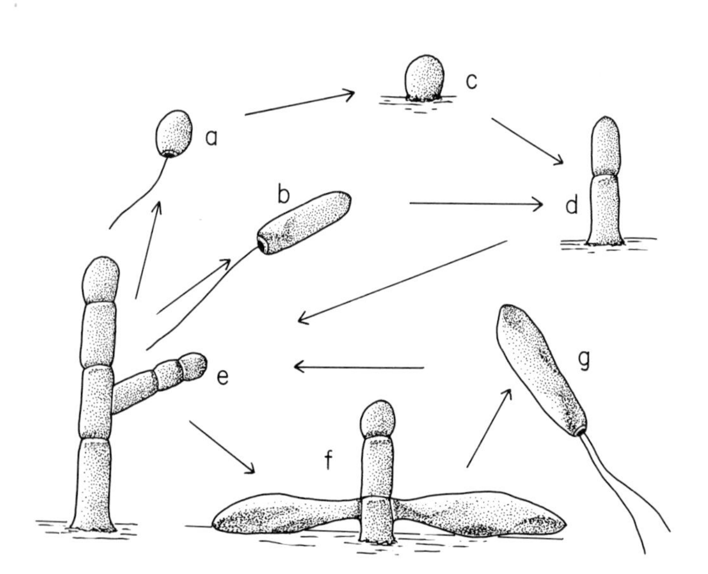

Figure 29.27. Diagrammatic representation of the life cycle of *S. polymorpha. a* and *b,* flagellated motile elements. *c, d,* and *e,* growing filaments attached to the surface of the agar. *f* and *g,* production of large, distended cells on rich media.

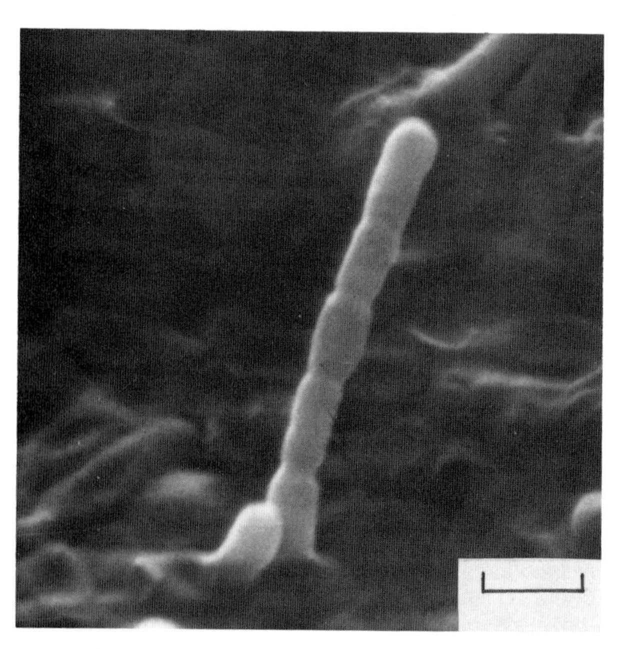

Figure 29.29. *S. polymorpha,* chain of cells with enlarged basal one attached to the substrate. *Bar,* 1 μm. (Reproduced with permission from S. T. Williams, Journal of General Microbiology *62:* 67–73, 1970.)

Section 30

Maduromycetes

Michael Goodfellow

Proposed by Goodfellow and Cross (1984b), the term "maduromycetes" was introduced for a number of poorly circumscribed sporoactinomycetes in need of taxonomic revision. Although maduromycetes were not conceived as a natural group, they do have a number of chemical and morphological features in common. They are aerobic, Gram-positive actinomycetes that form a branched substrate mycelium that does not carry spores but bears aerial hyphae that can differentiate either into short chains of arthrospores or into sporangia containing one to many spores. Further, they have *meso*-diaminopimelic acid (*meso*-DAP) in a wall peptidoglycan that lacks characteristic sugars (wall chemotype III; Lechevalier and Lechevalier, 1970b), yield whole-cell hydrolysates that contain madurose (3-*O*-methyl-D-galactose; Lechevalier and Gerber, 1970), produce major amounts of straight- and branched-chain fatty acids, and have partially hydrogenated menaquinones with nine isoprene units as major isoprenologs. All of them have DNA rich in G + C, as do most other actinomycetes.

The maduromycetes currently include sporangiate actinomycetes (*Planobispora, Planomonospora, Spirillospora,* and *Streptosporangium*) and those forming paired or short chains of spores on aerial hyphae (*Actinomadura, Microbispora,* and *Microtetraspora*). The major characteristics of the seven genera included in this section are given in Table 30.1.

The removal of the genera *Planobispora, Planomonospora, Spirillospora,* and *Streptosporangium* from the family *Actinoplanaceae* (Couch and Bland, 1974a) is indisputable because all four taxa can be distinguished from the remaining sporangiate actinoplanetes (*Actinoplanes, Ampullariella, Dactylosporangium,* and *Pilimelia*) on the basis of morphology, wall composition, DNA/DNA homology, and rRNA cistron similarity (Stackebrandt et al., 1981). Representatives of the sporangia-producing wall chemotype III taxa were found to be related by Farina and Bradley (1970) but in more extensive nucleic acid pairing experiments (Stackebrandt *et al.* 1981) *Spirillospora albida* was found to be distantly related to actinoplanetes and other maduromycetes studied. 16S rRNA sequencing data support the sharp separation of *Streptosporangium* from sporangiate-forming wall chemotype II actinomycetes (Stackebrandt et al., 1983c; Stackebrandt and Schleifer, 1984). The close association between the genera *Planobispora, Planomonospora,* and *Streptosporangium* is also supported by chemical data (Table 30.1).

The genera *Actinomadura, Microbispora,* and *Microtetraspora* are heterogeneous given data from chemotaxonomic (Fischer et al., 1983; Athalye et al., 1984), molecular genetic (Fischer et al., 1983; Poschner et al., 1985), and numerical phenetic (Goodfellow and Pirouz, 1982b; Athalye et al., 1985) surveys. Clearly, these genera cannot be adequately circumscribed by a few a priori weighted chemical and morphological properties, although the distribution of major fatty acids and menaquinones is in good agreement with the classification derived from the nucleic acid pairing and sequencing studies. Data from rRNA sequencing analyses (Fowler et al., 1985) not only underline the heterogeneity of the nonsporangiate maduromycete taxa but show that *Actinomadura madurae,* the type species of *Actinomadura,* and related species are more closely related to *Thermomonospora curvata* than to a second group containing *Actinomadura pusilla* and related taxa. The *A. pusilla* group, *Microbispora, Microtetraspora, Planobispora, Planomonospora,* and *Streptosporangium* form a suprageneric group equivalent to those formed by other sporoactinomycetes, including actinoplanetes (Section 28) and streptomycetes (Section 29).

The genus *Actinomadura* should be restricted to *Actinomadura madurae* and related species (Fischer et al., 1983; Athalye et al., 1985; Poschner et al., 1985). The latter include *Actinomadura citrea, A. coerulea, A. cremea, A. malachitica, A. pelletieri, A. verrucosospora, Microbispora echinospora,* and *Microtetraspora viridis.* The *Actinomadura pusilla* group, which is also worthy of generic status, has still to be described. This taxon currently contains *Actinomadura fastidiosa, A. ferruginea, A. libanotica, A. roseola, A. roseoviolacea, A. salmonea, Micropolyspora angiospora,* and *Nocardiopsis africana.* Similarly, the term "maduromycetes" should be used only for the natural group encompassing *Microbispora* sensu stricto, *Microtetraspora* sensu stricto, *Planobispora, Planomonospora,* and *Streptosporangium,* and the *Actinomadura pusilla* group.

Unknown isolates can usually be identified as maduromycetes using a judicious selection of chemical and morphological features. Morphology is the most rapid source of information, but accurate identification at the genus level can rarely be made on the basis of morphology alone. Indeed, it can be very difficult to distinguish maduromycetes from other oligosporic and sporangiate actinomycetes solely on morphological criteria. For an account of the chemical methods that can be used to this end, see Lechevalier and Lechevalier (1980), Gottschalk (1985) and Goodfellow and Minnikin (1985).

It is important not to destroy the arrangement of hyphae, spores, and sporangia when examining the morphology of maduromycetes. The in situ examination of cultures growing on agar plates is facilitated by the use of long working distance condensers and objectives. Particular attention needs to be paid to the formation and arrangement of conidia, the presence of sporangia (called "spore vesicles" by Cross, 1970 and Sharples et al., 1974) and the release of motile elements. Care must also be taken to distinguish between sporangia and "pseudosporangia"; the latter, formed by some strains of *Actinomadura* (Nonomura and Ohara, 1971d), are sporangia-like in appearance but are covered with slimy material, not sporangial membranes as in streptosporangia and related actinomycetes.

Much information can be gained by the study of whole-cell hydrolysates. Unidimensional chromatography will determine whether an or-

Table 30.1.
Differential properties of **Maduromycetes** *and other wall chemotype III taxa containing madurose*[a,b]

Characteristics	Maduromycetes								Other madurose-containing taxa	
	I. *Actinomadura madurae* group	I. *Actinomadura pusilla* group	II. *Microbispora*	III. *Microtetraspora*	IV. *Planobispora*	V. *Planomonospora*	VI. *Spirillospora*	VII. *Streptosporangium*	*Dermatophilus*	*Frankia*
Morphological characters										
Substrate hyphae dividing in more than one plane	−	−	−	−	−	−	−	−	+	+
Spores-aerial mycelium										
Absent or short chains	+	+	−	−	−	−	−	−	−	−
Paired	−	−	+	−	−	−	−	−	−	−
Mostly in chains of four	−	−	−	+	−	−	−	−	−	−
Sporangiospores	−	−	−	−	−	+	+	+	+[c]	+[c]
Spores per sporangium	ND	ND	ND	ND	Two	One	Many	Many	Many[c]	Many[c]
Spore motility	−	−	−	−	+	+	+	−	+	−
Symbionts in plant nodules	−	−	−	−	−	−	−	−	−	+
Chemical characters										
Fatty acid type[d]	3a	3c	3c	3c	3c	3c	3a	3c	1a	ND
Phospholipid pattern[e]	PI	PIV	PIV	PIV	PIV	PIV	PI,II	PIV	PI	PI
Predominant menaquinone (MK-)	$9(H_6)$	$9(H_0)$	$9(H_0)$	$9(H_0)$	$9(H_2)$	$9(H_0)$	$9(H_4)$	$9(H_2)$	$8(H_4)$	ND
		$9(H_2)$	$9(H_2)$	$9(H_2)$	$9(H_4)$	$9(H_2)$	$9(H_6)$		$9(H_4)$	
		$9(H_4)$	$9(H_4)$	$9(H_4)$		$9(H_4)$				
Mol% G + C of DNA	66–69	64–69	67–74	66	70–71	72	71–73	69–71	57–59	66–71

[a]Data from Fischer et al. (1983), Collins et al. (1984), Goodfellow and Cross (1984), Athalye et al. (1985), Poschner et al. (1985), and Goodfellow and Williams (1986).
[b]Symbols: +, present; −, absent; *ND,* not determined.
[c]Aerial mycelium not formed, multilocular sporangia borne on the substrate mycelium.
[d]Fatty acid types: *1a,* saturated and unsaturated acids; *3a,* saturated, unsaturated, *iso-* (variable) and methyl-branched acids; *3c,* saturated, unsaturated, *iso-*, *anteiso-* (variable), and methyl-branched acids (Kroppenstedt, 1985).
[e]Characteristic phospholipids: *PI,* phosphatidylglycerol (variable); *PII,* only phosphatidylethanolamine; *PIII,* phosphatidylcholine (with phosphatidylethanolamine, phosphatidylmethylethanolamine, and phosphatidylglycerol variable, no phospholipids containing glucosamine); *PIV,* phospholipids containing glucosamine (with phosphatidylethanolamine and phosphatidylmethylethanolamine variable); *PV,* phospholipids containing glucosamine and phosphatidylglycerol. All preparations contain phosphatidylinositol (Lechevalier et al., 1977, 1981).

ganism contains DAP and whether it is in the L- or *meso-* form. The presence of L-DAP will lead to *Streptomyces* and related genera (Section 29). The detection of *meso-*DAP and madurose with the absence of characteristic sugars serves to distinguish maduromycetes from nocardioform actinomycetes (Section 26), actinoplanetes (Section 28), and *Thermomonospora* and related genera (Section 31) but not from the genera *Dermatophilus* and *Frankia.* The latter are readily distinguished from maduromycetes on morphological grounds (Table 30.1). To date, the presence of madurose is associated with some wall chemotype III actinomycetes, although there is an unsubstantiated report of this compound from a wall chemotype I actinomycete with a streptomycete morphology (H. Weyland, 1982, Proceedings of the V International Symposium on Actinomycete Biology, Mexico). The discovery of 3-*O*-methylgalactosyl (madurosyl) units in the structure of teichoic acids of an *Actinomadura carminata* strain (Naumova et al., 1986) is also interesting because madurose is not considered to be a cell wall constituent (Lechevalier and Lechevalier, 1981b).

Wall chemotype III sporangiate actinomycetes can be identified to the genus level using morphological criteria (Table 30.1), although care needs to be taken to distinguish between the genera *Spirillospora* and *Streptosporangium.* In contrast, additional chemical analyses are needed to separate the oligosporic maduromycetes. The *A. madurae* and *A. pusilla* groups each contain characteristic menaquinones and polar lipids. A small-scale procedure is available for the sequential extraction of those two lipid types (Minnikin et al., 1984b).

Genus Actinomadura *Lechevalier and Lechevalier 1970a, 400*[AL*]

JUTTA MEYER

Ac.ti.no.ma.du'ra. Gr. n. *actis, actinos* a ray; *Madura* name of a province in India; L. fem. n. *Actinomadura* referring to a microorganism first described as the causative agent of "Madura foot" disease.

Extensively branching vegetative hyphae, forming a dense, **non-fragmenting substrate mycelium; aerial mycelium moderately developed** or absent. When aerial mycelium is lacking colonies have a leathery or cartilaginous appearance. The **aerial mycelium** at maturity forms **short or occasionally long chains of arthrospores. Spore chains straight, hooked** (open loops), or **irregular spirals** (1–4 turns). **Spore surface smooth or warty.** Color of mature sporulated aerial mycelium white, gray to brownish, yellow, red (pink and rose), blue, greenish, or violet. Aerobes, chemo-organotrophic. Temperature range 20–45°C. Some species grow at temperatures within the thermophilic range (up to 55°C). Gram-positive. **Cell wall contains *meso*-2, 6-diaminopimelic acid** (*meso*-DAP) as principal diamino acid and major amounts of galactose in most strains. Whole-cell hydrolysates contain the sugar madurose. Mycolic acids absent. **Menaquinones** are predominantly of the **MK-9(H_4)** and **MK-9(H_6)** type. Additional menaquinones of the MK-9(H_2) and MK-9(H_8) types occur. Isolated from soils of different regions. Has also been found in clinical specimens. The mol% G + C of the DNA is 65–69.

Type species: *Actinomadura madurae* (Vincent 1894) Lechevalier and Lechevalier 1970a, 400.

Further Descriptive Information

Colonial Characteristics and Morphology

In comparison with streptomycetes the actinomadurae are slow-growing organisms. Strains from clinical sources mostly lack aerial mycelium. Colonies exhibit a cartilaginous or leathery appearance. The majority of species, however, form a spore-bearing, powdery aerial mycelium on suitable media (e.g. oatmeal agar, yeast extract–malt extract agar, starch–mineral salts agar) after incubation for 10–14 days. For further information about media and growth conditions see Shirling and Gottlieb (1966).

Aerial mycelium at maturity (e.g. with well-developed spore chains) is of white, cream, pale yellow, pink, blue, and gray or greenish color, thus differing little from streptomycetes. Superficial similarity to streptomycetes is also reinforced by the morphology of the sporophores. *Actinomadura roseola* and *A. spiralis* may easily be mistaken for streptomycetes because of straight spore chains or more or less well-developed spirals, respectively. Several species that form so-called pseudosporangia, i.e. tightly spiraled spore chains embedded in a dry to slimy mass, thus simulate the colonial appearance of a *Streptosporangium* when examined microscopically at low magnification.

Confusion with streptomycetes can be avoided only by direct microscopic comparison of the culture under question with a *Streptomyces* culture. Mostly *Actinomadura* strains are conspicuous for the size of their spores. The diameter of their spores noticeably exceeds the diameter of the hyphae, whereas in streptomycetes spores and hyphae are of similar diameter.

Cell Wall Composition

The genus *Actinomadura* was defined chemically (Lechevalier and Lechevalier, 1970a) to accommodate former *Nocardia* species with a cell wall type III, i. e. with *meso*-DAP in the peptidoglycan, which, in contrast to *Nocardia* species, is not associated with an arabinogalactan polymer. Some species have also been found to contain small amounts of L-DAP.

Whole-cell hydrolysates contain madurose (3-*O*-methyl-D-galactose). This sugar also occurs in other wall type III taxa, including *Dermatophilus, Microbispora, Spirillospora, Streptosporangium, Planobispora,* and *Planomonospora.* Differentiation from these genera is therefore based on morphological criteria.

Phages

Although *Actinomadura* species have been isolated mainly from soil, there are no reports of isolation of *Actinomadura*-specific phages.

Pathogenicity

Actinomadura madurae and *A. pelletieri* are known to cause human actinomycetoma in tropical and subtropical areas, particularly on the African and American continents. The natural habitat of these pathogenic actinomadurae is thought to be the environment, mainly the surface layers of the soil, from where they invade the human body via contaminated soil or dust particles that penetrate into wounds in the lower extremities. This is the most likely explanation of the etiology of "Madura foot" because the main localization of this disease is the lower extremities, although other sites of the human body may also be affected.

Information on pathogenicity is contradictory. Most cases of actinomycetomas caused by *A. madurae* or *A. pelletieri* occur in warm, humid areas such as Africa, Mexico, and India; this localization of infection may be explained by the tendency to go barefoot in these regions. The route of infection is percutaneous through skin lesions, hence in a special selective situation these relatively rare forms of actinomadurae become opportunistic pathogens. According to Pulverer and Schaal (1978) it has not been possible to demonstrate the pathogenicity of these actinomadurae for laboratory animals. Rippon (1968), however, reported that virulent strains of *A. madurae* produce a collagenase that has a significant role in the pathogenicity of the organism.

Pigments

Prodigiosin-like pigments were obtained from *A. madurae* and *A. pelletieri* (Gerber, 1971, 1973; Lechevalier et al., 1971a) that are similar to those of *Serratia marcescens.* Of interest, perhaps from an ecological point of view, is that all the actinomadurae (both *A. madurae* and *A. pelletieri*) strains that were found to produce prodiginines were strains isolated from patients. The pigments are characterized by a tripyrrole skeleton and have been identified as nonylprodiginine, undecylprodiginine, and cyclononylprodiginine.

Ecology

Most of the *Actinomadura* species recently described originated from soil (Nonomura and Ohara, 1971d; Lavrova et al., 1972: Preobrazhenskaya et al., 1975b; Meyer, 1979; Galatenko and Preobrazhenskaya, 1981). *Actinomadura pelletieri* has been found only in clinical specimens, whereas *A. madurae* seems to be widespread in soil. Isolates of the latter from soil lack the red endopigment of clinical isolates and sporulate more readily. Information on the occurrence and frequency of other *Actinomadura* species in different soils has been provided mainly by Preobrazhenskaya and her coworkers (Preobrazhenskaya et al., 1978; Chormonova and Preobrazhenskaya, 1981; Galatenko and Preobrazhenskaya, 1981). A comparison of three different soil types (chernozem, dark chestnut soil, and sierozem) in Kazakhstan and Turkmenistan revealed that the total number of *Actinomadura* was higher in cultivated than in uncultivated soils. The highest number of actinomadurae was isolated from the chernozem. *Actinomadura citrea* was the most frequent species in cultivated as well as in uncultivated soils of these types, followed by *A. cremea* and *A. verrucosospora.* Dark chestnut soil (cultivated) contained only half the number of actinomadurae as did chernozem, but the variety of species in both soil types was the same. The lowest number was isolated from sierozem. The frequency of actinomadurae appears to depend on the humus content of the soil.

Occurrence of "*Actinomadura*" strains in self-heated stored barley

AL denotes the inclusion of this name on the Approved Lists of Bacterial Names (1980).

grain and slightly heated and molded hay was reported by Lacey (1978). However, these strains proved to have a cell wall chemotype IV, which excludes them from *Actinomadura* (Goodfellow et al., 1979).

Enrichment and Isolation Procedures

Many different media have been employed for isolation of *Actinomadura* strains, especially from soil samples. The media that have proved to be most suitable are; oatmeal agar (ISP No. 3, Shirling and Gottlieb, 1966); yeast extract–malt extract agar (ISP No. 2); starch–mineral salt agar (ISP No. 4); Bennett sucrose agar; and glycerol-asparagine agar (ISP No. 5). Strains may be isolated from agar plates by dilution techniques, after incubation for 14–21 days. Enrichment of actinomadurae in isolation procedures from soil can be achieved by relatively simple methods. Nonomura and Ohara (1971d) reduced numbers of unwanted microbes by air drying the soil sample and then applying dry heat at 100°C for 1 h before plating on several media. Lavrova et al. (1972) and Preobrazhenskaya et al. (1975b) were able to increase considerably the numbers of actinomadurae isolated by addition of antibiotics to the isolation media to inhibit growth of bacteria and the more frequently occurring streptomycetes, thus providing more favorable conditions for the slow-growing and rare actinomadurae. Streptomycin (0.5, 1.0, or 2.0 μg/ml), rubomycin (5.0, 10.0, or 20.0 μg/ml), and bruneomycin (0.5, 1.0, or 2.0 μg/ml) were the most successful antibiotics used.

Maintenance Procedures

Serial transfer from agar slants of appropriate media (see above) every 2 months is the most convenient method for short-term storage. The tubes should be tightly closed by cotton-wool plugs dipped in melted paraffin wax.

For long-term preservation lyophilization or storage in liquid nitrogen can be used. For lyophilization the spore suspension or vegetative mycelium is suspended in a suitable fluid, e.g. 7.5% (w/v) glucose serum, or skim milk + 7.5% (w/v) glucose. For storage in liquid nitrogen the microorganisms are inoculated into small test tubes containing an appropriate medium and incubated until satisfactory growth is visible. The tubes are then tightly closed with cotton-wool plugs dipped in melted paraffin wax and placed in a liquid-nitrogen container.

Differentiation of the genus **Actinomadura** *from the other genera*

Table 30.2 shows the primary characteristics that can be used to differentiate the genus *Actinomadura* (as defined in this chapter) from morphologically and biochemically similar taxa. At present it is impossible to delimit the genus *Actinomadura* properly from phenotypically similar strains of *Microtetraspora*, e.g. strains of the latter with two to six spores per chain, since both genera are biochemically characterized by the presence of the whole-cell sugar madurose, a peptidoglycan of the *meso*-DAP type, and menaquinones of MK-9(H_4,H_6,H_8). Usually members of *Microtetraspora* can be distinguished from members of *Actinomadura* by an aerial mycelium in which spores are arranged in tetrads. The discrepancies in the phospholipid type within both genera as shown by Lechevalier et al. (1981) are partly due to misidentified strains, e.g. certain strains of *Microtetraspora viridis* var. *intermedia* require reidentification.

Taxonomic Comments

The historical starting point of the actinomadurae was in 1894 when Vincent described the causative agent of the "Madura foot disease" as *Streptothrix madurae.* Since the combination proved to be illegitimate, Blanchard (1896) transferred the organism into the genus *Nocardia* Trevisan; thus *Nocardia madurae* (Vincent) Blanchard is the oldest legitimate name of this microorganism. In 1906 Laveran described another pathogenic, mycetoma-causing microorganism, *Micrococcus pelletieri,* which he thought to be a micrococcus because the colonies all frag-

Table 30.2.
Differential characteristics of the genus **Actinomadura** *and morphologically or biochemically similar taxa*[a]

Characteristics	*Actinomadura*	*Nocardiopsis*	*Saccharothrix*	*Streptomyces*	*Nocardia*	*Microtetraspora*
Fragmentation of substrate mycelium	–	+	+	–	+	–
Sporulation of aerial mycelium	Short spore chains	Total sporulation	Total sporulation	Long various spore chains	Absent or total sporulation	Chains of 4 spores
Cell wall chemotype	III	III	III	I	IV	III
Characteristic sugars	Madurose	–	Galactose, mannose, rhamnose	–	Arabinose, galactose	Madurose
Mycolic acids	–	–	–	–	+	–
Predominant menaquinone	MK-9(H_4,H_6,H_8)	MK-10(H_4,H_6,H_8)	MK-9(H_4) MK-10(H_4)	MK-9(H_4,H_6,H_8)	MK-8(H_4)	MK-9(H_4,H_6,H_8)
Phospholipid type[b]	I (IV)	III, IV[c]	II	II	II	IV
Predominating phospholipid[d]	PIM, PI, DPG (or IV: PE and Glu NU)	PC, PG, APG	PE	PE	PE	Glu NU, PI, DPG

[a]Symbols: see Table 30.1.
[b]According to Lechevalier et al. (1981).
[c]According to Shearer et al. (1983a).
[d]Abbreviations of phospholipids: *APG,* acylphosphatidylglycerol; *DPG,* diphosphatidylglycerol; *PC,* phosphatidylcholine; *PE,* phosphatidylethanolamine; *PG,* phosphatidylglycerol, *PI,* phosphatidylinositol; *PIM,* phosphatidylinositol mannosides; *Glu NU,* phospholipids of unknown structure containing glucosamine.

mented into cocci. Later Pinoy (1912) included it in the genus *Nocardia*. However, the genus *Nocardia* at this time was based mainly on fragmentation of the substrate mycelium, a characteristic that was lacking in both *N. madurae* and *N. pelletieri.* Thus both species were atypical of *Nocardia* and were sometimes regarded as "degenerative streptomycetes." Waksman and Henrici (1948b) transferred *N. pelletieri* and *N. madurae* into the genus *Streptomyces.* Their inclusion in both *Nocardia* and *Streptomyces* was controversial until it was shown by Becker et al. (1965) that whole-cell hydrolysates contained *meso*-DAP and a hitherto unknown sugar madurose, later identified as 3-*O*-methyl-D-galactose by Lechevalier and Gerber (1970). In contrast, *Nocardia* species contained *meso*-DAP, galactose, and arabinose and *Streptomyces* species contained L-DAP. The consequence was the genus *Actinomadura,* described by Lechevalier and Lechevalier in 1970(a) to include "aerobic actinomycetes forming a branching primary (substrate) mycelium. Aerial (secondary) mycelium may or may not be formed. When it is formed it may bear chains of conidia (arthrospores). Cell wall preparations are of type III according to Becker et al. (1965)." The type species was *Actinomadura madurae* (Vincent) Lechevalier and Lechevalier. Furthermore, the former *Nocardia pelletieri* and a third species, *Nocardia dassonvillei,* were included within *Actinomadura.* While *A. madurae* and *A. pelletieri* were regarded as closely related species by the authors, they also pointed out the morphological and chemical differences of *A. dassonvillei,* thus anticipating the later exclusion of this taxon and the description of a separate genus *Nocardiopsis* by Meyer (1976).

Although eight additional species of *Actinomadura* were subsequently described by Nonomura and Ohara (1971d) and Lavrova et al. (1972), McClung (1974), in the eighth edition of the *Manual* considered the genus among the "Genera Incertae Sedis," thus neglecting the accumulated knowledge of this group. Numerical taxonomic studies by Goodfellow (1971), Alderson and Goodfellow (1979), Goodfellow et al. (1979), and Goodfellow and Pirouz (1982b) strongly supported the validity of the genus. The menaquinone analysis (see "Further Comments," below) also confirmed the integrity of the genus *Actinomadura.*

Since the establishment of the genus, 37 additional species (excluding those names mentioned only in patent literature) have been added to it primarily on the basis of morphology and wall chemotype, most of them being isolated during the search for new antibiotics.

The present circumscription of the genus *Actinomadura* is based on cell wall chemotype, morphology of aerial mycelium, and menaquinone pattern. It comprises on the whole 26 species, 25 of which are listed under *Actinomadura* in the Approved Lists of Bacterial Names (Skerman et al., 1980) or validated in subsequent validation lists. Four of the remaining *Actinomadura* species (*A. africana, A. coeruleofusca, A. flava,* and *A. longispora*) listed in the Approved Lists have been reclassified with *Nocardiopsis* (Preobrazhenskaya et al., 1982; Preobrazhenskaya and Sveshnikova, 1985).

The issue of the position of *Actinomadura* in a family is still unresolved. Lechevalier and Lechevalier (1970a) placed the genus within the family *Thermoactinomycetaceae* (Baldacci and Locci) Lechevalier and Lechevalier. Since this family was later limited to actinomycetes forming endospores, a new family *Thermomonosporaceae* (Cross and Goodfellow, 1973) was created to harbor *Actinomadura* together with *Microbispora, Microtetraspora,* and other genera. This placement again proved to be questionable. To escape the difficulties in the classification of taxa of higher rank (i.e. actinomycete genera) Goodfellow and Cross (1984a) proposed "aggregate groups," the composition of which is supported by criteria derived from modern taxonomic methods, notably chemotaxonomy and numerical taxonomy. One of these nine "aggregate groups" is the "Maduromycetes," comprising the genera *Actinomadura, "Excellospora," Microbispora, Microtetraspora, Planobispora, Planomonospora, Spirillospora,* and *Streptosporangium.* It is already clear that this aggregate group is artificial.

Another attempt to elucidate phylogenetic relatedness was that of Stackebrandt et al. (1981) with DNA/rRNA reassociation experiments and by Stackebrandt and Schleifer (1984) by comparative cataloging of 16S rRNA. Four species were included in the latter studies: *A. madurae, A. verrucosospora, A. pusilla,* and *A. roseoviolacea.* Whereas in the former experiments only *A. madurae* was studied, *Actinomadura* was found to be unrelated to most of the actinomycete genera examined, the results of the comparative cataloging of 16S rRNA revealing the genus to be genetically heterogeneous. *Actinomadura madurae* and *A. verrucosospora* were remotely related to *Thermomonospora curvata.* On the other hand the similarity coefficients (S_{AB} values) of *A. pusilla* and *A. roseoviolacea* placed these two species in the family *Streptosporangiaceae* (Stackebrandt and Schleifer, 1984) together with *Streptosporangium* and *Nocardiopsis.* Although only a few species were studied there is obviously some evidence for the heterogeneity of the genus *Actinomadura,* confirmed by the results of other authors (Fischer et al., 1983; Poschner et al., 1985).

Further Comments

Chemotaxonomic Studies

The genus *Actinomadura* has only recently been included in extensive chemotaxonomic investigations. In the comparative studies reported (polar lipids: Lechevalier et al., 1977, 1981; Minnikin et al., 1977b; Mordarska et al., 1983; isoprenoid quinones: Komura et al., 1975; Collins et al., 1977; Yamada et al., 1977b; Collins and Jones, 1981; fatty acids: Kroppenstedt and Kutzner, 1978; Minnikin and Goodfellow, 1981; Minnikin and O'Donnell, 1984) usually only *A. madurae, A. pelletieri,* and *Nocardiopsis dassonvillei* (the former *Actinomadura dassonvillei*) were included. The results obtained underlined the differences between the genera *Actinomadura* and *Nocardiopsis,* but little information about the intrageneric grouping or homogeneity of the genus *Actinomadura* was obtained.

Komura et al. (1975) and Minnikin et al. (1977b) considered phospholipid composition to be a useful criterion for differentiation and classification of *Actinomadura* species. However, only a few strains of the two species were studied. With *Actinomadura madurae,* which was studied by both groups, the results are inconsistent. Lechevalier et al. (1977, 1981) defined five phospholipid types in actinomycetes. *Actinomadura madurae* and *A. pelletieri* were assigned to type PI, characterized by the prevalence of phosphatidylinositol mannosides, phosphatidylinositol, and diphosphatidylglycerol, whereas two other *Actinomadura* species were assigned to type P IV, containing in addition phosphatidylethanolamine, phosphatidylmethylethanolamine, and unknown glucosamine-containing phospholipids. From these results Lechevalier et al. (1977) concluded that the genus *Actinomadura* might be heterogeneous. Since the genus *Microtetraspora* also appears heterogeneous with respect to phospholipid pattern (Type IV and I), it is obvious that misclassified strains have prevented the clear delineation of both genera.

Actinomadurae contain a complex mixture of hydrogenated menaquinones with nine isoprene units (MK-9(H_2,H_4,H_6,H_8), whereas strains of *Nocardiopsis* show menaquinones with MK-10(H_2,H_4,H_6) or MK-9(H_4,H_6). Differentiation of other cell wall chemotype III genera such as *Microbispora* or *Microtetraspora* from *Actinomadura* by means of menaquinone composition is impossible because they have identical profiles (Athalye et al., 1984; Poschner et al., 1985).

Fischer et al. (1983) studied 24 strains of 13 species by means of high-performance liquid chromatography and Athalye et al. (1984) studied 30 strains of 19 species by means of high-performance thin-layer chromatography and mass spectroscopy. Where identical strains were included in both studies the results are in good agreement. *Actinomadura madurae, A. citrea, A. malachitica, A. verrucosospora,* and *A. luteofluorescens* a. o. (corresponding to the "*A. madurae*-cluster" of Fischer et al., 1983) exhibit hexahydrogenated menaquinone main components with nine isoprene units. In contrast, *A. pusilla, A. ferruginea, A. roseola, A. roseoviolacea,* and *A. spiralis* (corresponding to the "*A. pusilla*-cluster" of Fischer et al., 1983) exhibit tetrahydrogenated menaquinone main components with nine isoprene units. Athalye et al. (1984) reported major amounts of MK-9(H_8) and substantial amounts of MK-9(H_6) in *A. pelletieri,* as did Collins et al. (1977). However, Fischer et al. (1983) and Poschner et al. (1985) found MK-9(H_6) predominating in *A. pelletieri.*

Different results have also been obtained for *A. libanotica.* MK-9(H_4) was found by Fischer et al. (1983) and by Poschner et al. (1985), whereas Athalye et al. (1984) detected MK-9(H_6) units. These differences may be due to the use of different separation methods. Therefore it is difficult to generalize about species-specific menaquinone patterns expressed as degree of hydrogenation.

The fatty acids of various *Actinomadura* species have been studied by Guzeva et al. (1973), Agre et al. (1975), and Fischer et al. (1983), but the results are not comparable, because different methods and different species and strains were used. Qualitative and quantitative differences in fatty acid composition between species are clear. Poschner et al. (1985) observed that in the *A. pusilla* subgroup 10-methyl branched fatty acids with 17 carbon atoms exceeded those with 18 carbon atoms, while in the *A. madurae* subgroup 10-methyl branched fatty acids with 18 carbon atoms predominated.

DNA and RNA Homologies

Fischer et al. (1983) studied the phylogenetic coherence of the genus *Actinomadura* and its relationship to *Nocardiopsis, Streptomyces,* and *Streptosporangium* by determining the melting points of DNA/rRNA duplexes ($T_{m(e)}$ values) obtained with ^{3}H-labeled 23S rRNA of *A. madurae* and DNA from the genera and species in question. In terms of the thermal stability of DNA/rRNA duplexes, strains of *Actinomadura* formed two clusters, thus revealing genetic heterogeneity of the genus. The first cluster, comprising *A. madurae, A. pelletieri, A. verrucosospora, A. malachitica, A. citrea,* and *A. kijaniata,* showed high $T_{m(e)}$ values (79.0–84.0°C). The thermal stability of the duplexes in the second cluster, containing *A. ferruginea, A. roseola, A. pusilla, A. roseoviolacea, A. libanotica, A. spiralis,* and *A. spadix,* was markedly lower (73.1–75.8°C). These $T_{m(e)}$ values were in the same range as those found for duplexes formed by *A. madurae* with strains of *Nocardiopsis dassonvillei, Streptosporangium roseum,* and *Streptomyces griseus.*

Fischer et al. (1983) subjected strains of these two *Actinomadura* rRNA clusters to DNA/DNA hybridization. Homology values ranged from 18 to 100%, the highest values being between *A. madurae* strains (96%), between various strains of *A. pelletieri* (85–100%), and between two strains of *A. verrucosospora.* Homology values found between strains of *A. madurae, A. pelletieri, A. verrucosospora,* and *A. citrea* ranged from 25 to 44%. Representatives of the second *Actinomadura* rRNA cluster and strains of *Nocardiopsis* and *Streptosporangium* revealed lower homology values (7–12%) to *A. madurae* or *A. pelletieri,* respectively. DNA homologies among *A. pusilla, A. roseoviolacea, A. libanotica, A. spiralis, A. spadix, A. ferruginea,* and *A. roseola* ranged from 8 to 33%, thus demonstrating moderate to slight relatedness between these species.

These results have been expanded by Poschner et al. (1985) by inclusion of further *Actinomadura* species in DNA reassociation and chemotaxonomic studies. Summarizing all results so far examined, it can be concluded that the genus *Actinomadura* can at present be subdivided into two groups on the basis of DNA/rRNA and DNA/DNA hybridization and menaquinone pattern. The first group (the so-called *A. madurae* group) comprises *A. madurae, A. pelletieri, A. verrucocospora, A. citrea, A. coerulea, A. malachitica,* and *A. cremea.* The second group (the so-called *A. pusilla* group) contains *A. pusilla, A. roseoviolacea, A. ferruginea, A. roseola, A. salmonea, A. fastidiosa, A. libanotica,* and *A. spiralis.* DNA homologies between species of different groups do not exceed 15%, thus indicating no relationship, whereas values between species within one group range between 15 and 48%, implying a moderate degree of relationship. These results have been supported by ribosomal RNA analysis (Fowler et al., 1985), indicating that these two subgroups of *Actinomadura* may be considered as separate genera (Poschner et al., 1985).

Numerical Taxonomy

Actinomadura strains have been included in many numerical analyses (Goodfellow, 1971; Alderson and Goodfellow, 1979; Goodfellow et al., 1979; Goodfellow and Pirouz, 1982b). Goodfellow (1971) confirmed the sharp separation between *Actinomadura* and *Nocardia.* Further numerical taxonomic studies of *Actinomadura* and related actinomycetes containing *meso*-DAP in the cell wall (Goodfellow et al., 1979; Goodfellow and Pirouz, 1982b) supported the transfer of the *A. dassonvillei* to the genus *Nocardiopsis* (Meyer, 1976). Moreover, *A. madurae* and *A. pelletieri,* although represented by many strains, turned out to be heterogeneous taxa, being recovered in more than one phenon.

In a later study on *Actinomadura* and *Nocardiopsis* (Athalye et al., 1985), representatives of 20 *Actinomadura* species were compared, using the simple matching (S_{SM}) and the pattern coefficients (D_p). Clustering was achieved by using the unweighted pair group method with averages (UPGMA). The dendrograms derived from D_p-UPGMA as well as S_{SM}-UPGMA analysis showed that most actinomadurae were assigned to one of two aggregate clusters. The composition of the aggregate groups coincided fairly well in both analyses. In the D_p-UPGMA analysis, aggregate cluster 1 contained strains of *A. luteofluorescens, A. pusilla, A. roseola, A. roseoviolacea, A. salmonea, A. spadix,* and *A. spiralis.* The second aggregate cluster contained *A. citrea, A. malachitica, A. madurae, A. pelletieri, A. coerulea, A. cremea,* and *A. verrucosospora.* These two aggregate clusters correspond quite well with the genetically and chemically defined groups recognized by Fischer et al. (1983) and Poschner et al. (1985). Major differences, however, are discernable between the positions of *A. ferruginea, A. libanotica,* and *A. spiralis* in the dendrograms compared with the genetically defined groups. Also the different placement of *A. rubra, A. helvata,* and *A. spadix* in the dendrograms is questionable because it may be due to the slow and poor growth of these species and the fact that only single strains were included in the numerical phenetic surveys. Moreover, significant differences in the cluster composition pertain to *A. madurae* and *A. pelletieri.* The heterogeneity of these species has already been pointed out by genetic and chemical studies (Fischer et al., 1983).

Further Reading

Athalye, M., M. Goodfellow, J. Lacey and R. P. White. 1985. Numerical classification of *Actinomadura* and *Nocardiopsis.* Int. J. Syst. Bacteriol. *35:* 86–98.

Athalye, M., M. Goodfellow and D. E. Minnikin. 1984. Menaquinone composition in the classification of *Actinomadura* and related taxa. J. Gen. Microbiol. *130:* 817–823.

Fischer, A., R. M. Kroppenstedt and E. Stackebrandt. 1983. Molecular-genetic and chemotaxonomic studies on *Actinomadura* and *Nocardiopsis.* J. Gen. Microbiol. *129:* 3433–3446.

Poschner, J., R. M. Kroppenstedt, A. Fischer and E. Stackebrandt. 1985. DNA-DNA reassociation and chemotaxonomic studies on *Actinomadura, Microbispora, Microtetraspora, Micropolyspora* and *Nocardiopsis.* Syst. Appl. Microbiol. *6:* 264–270.

Williams, S. T. and E. M. H. Wellington. 1981. The genera *Actinomadura, Actinopolyspora, Excellospora, Microbispora, Micropolyspora, Microtetraspora, Nocardiopsis, Saccharopolyspora* and *Pseudonocardia. In:* Starr, Stolp, Trüper, Balows and Schlegel (Editors), The Prokaryotes. A Handbook on Habitats, Isolation and Identification of Bacteria. Springer-Verlag, Berlin, pp. 2103–2117.

Differentiation and characteristics of the species of the genus **Actinomadura**

The species of *Actinomadura* may be distinguished by means of spore chain morphology, spore wall ornamentation, color of mature sporulated aerial mycelium (AM), and color of substrate mycelium (SM) (Table 30.3). Only strains studied in detail are described here, other species being considered "incertae sedis." The latter pertains to *A. carminata,* although this species is on the Approved Lists. Additional physiological characters (see Table 30.4) are of minor value in differentiation, because in most cases only one (the type) strain or a few strains have been studied. Even when many strains had been studied (e.g. *A. pelletieri, A. madurae*), the results proved to be variable or were inconsistent when results from the literature were compared. Recent knowledge obtained from molecular genetic and chemotaxonomic studies about intrageneric

Table 30.3.
Characteristics differentiating the species of the genus **Actinomadura**[a]

	1. *A. madurae*	2. *A. aurantiaca*	3. *A. citrea*	4. *A. coerulea*	5. *A. cremea*	6. *A. fastidiosa*	7. *A. ferruginea*	8. *A. flexuosa*	9. *A. helvata*	10. *A. kijaniata*	11. *A. libanotica*	12. *A. livida*	13. *A. luteofluorescens*
Spore chains[b]	h,s	h,s	h	h,s	h,s	sp	h,s	h,s	psp	sp	h	h,s	h
Spore surface[c]	w	w	u	w	w	u	f	w	s	s	f	u	w
Interspore pads	–	+	–	–	–	–	–	–	–	–	+	–	–
Medium ISP 3[d]													
AM[e]	tr	c-p	w-bl	p-bl	w-y	w-p	w-p	w-y	tr	tr	w-p	tr	y-bl
SM[e]	d	y	y	d	d	d	p	b	y-b	g	y-b	gy-b	y-g
SP[e]	–	–	y	–	–	–	–	–	–	–	–	v	y-g
Medium ISP 2[d]													
AM[e]	–	w	–	–	w	tr	o-p	–	w-y	tr	p	–	w-y
SM[e]	p-b	y-b	y-b	b	b	b	o	–	y-b	g	y-b	b	y
SP[e]	–	–	–	–	–	–	–	–	–	–	–	–	y
Medium ISP 4[d]													
AM[e]	–	w	y-bl	bl	c-p	d-p	–	–	w-c	–	–	tr	tr
SM[e]	w-gy	c-o	y-b	d	d	d	d-b	–	y-b	g	gy	y-b	o-b
SP[e]	–	–	y	–	–	–	–	–	–	g	–	v	y

	14. *A. macra*	15. *A. malachitica*	16. *A. pelletieri*	17. *A. pusilla*	18. *A. roseola*	19. *A. roseoviolacea*	20. *A. rubra*	21. *A. salmonea*	22. *A. spadix*	23. *A. spiralis*	24. *A. verrucosospora*	25. *A. vinacea*	26. *A. yumaensis*
Spore chains[b]	h,s	str	h,s	psp	str	psp	h,s	h,s	psp	sp	h,s	str	h
Spore surface[c]	s	s	w	s	f	s	u	w	s	f	w	u	s
Interspore pads	–	–	–	–	–	–	–	–	–	–	–	–	
Medium ISP 3[d]													
AM[e]	c-p	g	–	w-c	p	p-v	tr	p	y-b	w-y	p-bl	–	gy-y
SM[e]	c-p	y-w	p-b	gy-b	b-r	v	o-r	r	y-r-b	y-b	o-p	p-r	gy-y
SP[e]	–	–	–		–	v	r	–	–	–	–	–	–
Medium ISP 2[d]													
AM[e]	w-gy	c	tr	tr	p	w-p	tr	c-p	–	w-y	w	–	y-gy
SM[e]	a	d	p-b	b-r	r-b	pu-r	r-b	b	b	y-b	y	b-r	y-b
SP[e]	b	–	–	–	–	–	–	–	b	–	–	r	o
Medium ISP 4[d]													
AM[e]	–	c	–	tr	–	w	–	tr	–	–	w	w-gy	w
SM[e]	d-gy	y-b	–	d-b	–	w-p	–	b	–	y-b	y	d	d
SP[e]	–	–	–	–	–	–	–	–	–	–	–	p	–

[a]Symbols: see Table 30.1.
[b]Abbreviations: *h*, hooks, curled; *psp*, pseudosporangia; *s*, spirals of 1–2 turns; *sp*, spirals of 2–4 turns; *str*, straight.
[c]Abbreviations: *f*, folded; *s*, smooth; *u*, irregular, uneven; *w*, warty.
[d]Abbreviations: *AM*, aerial mycelium; *SM*, substrate mycelium; *SP*, soluble pigment.
[e]Abbreviations: *a*, black; *b*, brown; *bl*, blue; *c*, cream; *d*, colorless; *g*, green; *gy*, gray; *o*, orange; *p*, pink; *pu*, purple; *r*, red; *tr*, traces of aerial mycelium, only microsopically visible; *v*, violet; *w*, white; *y*, yellow.

Table 30.4.
Differential characteristics of the species of the genus **Actinomadura**[a]

Characteristics	1. *A. madurae*	2. *A. aurantiaca*	3. *A. citrea*	4. *A. coerulea*	5. *A. cremea*	6. *A. fastidiosa*	7. *A. ferruginea*	8. *A. flexuosa*	9. *A. helvata*	10. *A. kijaniata*	11. *A. libanotica*	12. *A. livida*	13. *A. luteofluorescens*	14. *A. macra*	15. *A. malachitica*	16. *A. pelletieri*	17. *A. pusilla*	18. *A. roseola*	19. *A. roseoviolacea*	20. *A. rubra*	21. *A. salmonea*	22. *A. spadix*	23. *A. spiralis*	24. *A. verrucosospora*	25. *A. vinacea*	26. *A. yumaensis*
Reduction of nitrate	+	+	+	+	+	+	+	+	+	−	+	+	+	+	−	+	+	+	+	+	+	+	+	−	−	+
Hydrolysis of																										
Casein	d	−	+	+	+	+	+	+	−	+	−	+	+	−	+	−	−	−	−	−	+	+	−	+	+	+
DNA	d	v	−	−	v	+	+	+	−	+	−	+	+	ND	−	−	+	−	+	−	+	ND	+	+	−	ND
Esculin	+	−	+	−	+	+	−	+	+	+	+	+	−	ND	−	d	+	+	+	−	+	+	+	+	+	+
Gelatin	+	+	+	+	+	+	−	+	−	+	−	+	+	+	+	+	−	+	+	+	+	−	−	+	+	+
Hypoxanthine	+	−	+	+	−	+	+	−	−	+	−	−	+	ND	+	d	+	+	+	+	+	−	−	+	−	d
Starch	+	−	+	−	−	−	+	+	−	+	+	−	+	−	−	d	−	−	−	+	−	+	−	+	−	+
Testosterone	+	+	+	+	+	−	+	+	+	−	+	+	+	ND	+	−	+	+	+	+	+	ND	+	+	−	ND
Tyrosine	+	−	+	−	−	−	+	+	−	+	−	+	−	+	+	+	+	+	−	+	+	ND	+	+	−	+
Xanthine	−	−	−	−	−	−	−	−	−	+	−	−	−	−	−	−	−	−	−	−	−	−	−	−	−	+

[a]Symbols: +, 90% or more of strains are positive; −, 10% or less of strains are positive; *d*, 11–89% of strains are positive; *v*, strain instability; *ND*, not determined.

grouping in *Actinomadura* has not been taken into consideration in the present classification since it is of no help at present in the description of species. Even morphologically similar species have turned out to be only moderately related.

List of species of the genus **Actinomadura**

1. **Actinomadura madurae** (Vincent 1894) Lechevalier and Lechevalier 1970a, 400.[AL] (*Streptothrix madurae* Vincent 1894, 132.)

ma.du'rae. M.L. gen. of n. *Madura* name of a district in India.

Spore chains short, hooked or curled, in clusters directly emerging from the agar surface or borne on long aerial hyphae. Three to 12 spores per chain. Spores elliptical to round. Spore surface warty (Fig. 30.1).

Oatmeal agar†: good growth, surface leathery, aerial mycelium absent or sparse white specks, substrate mycelium colorless center, edge often red, no diffusible pigment.

Yeast extract–malt extract agar: moderate growth, surface cartilaginous, aerial mycelium absent, substrate mycelium dark pink to brownish violet, no diffusible pigment.

Inorganic salts–starch agar (ISP 4): poor growth, surface granular, aerial mycelium absent, substrate mycelium grayish white, no diffusible pigment.

Peptone-glucose medium: poor growth, surface cartilaginous, aerial mycelium absent, substrate mycelium dark pink to red, no diffusible pigment.

Growth from 10° to 45°C, optimum between 28° and 37°C.

Isolated from clinical specimens (mycetoma) and from soil.

Mol% G + C of the DNA is 66.0–68.2 (T_m) (type strain 68.2).

Type strain: NCTC 5654.

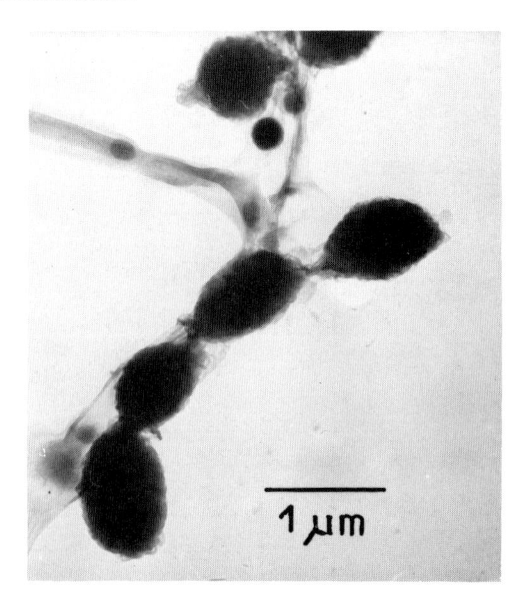

Figure 30.1. *Actinomadura madurae,* strain RG 1091. Spore chain, electron micrograph (× 16,500).

2. **Actinomadura aurantiaca** Lavrova and Preobrazhenskaya 1975, 485.[AL]

au.ran.ti'a.ca. M.L. fem. adj. *aurantiaca* referring to the gold-colored substrate mycelium.

Spore chains short, hooked or in spirals (one turn), in small clusters emerging directly from the agar. Four to eight spores per chain. Spore surface warty. Spores connected by interspore pads.

†Media used are the following: yeast extract–malt extract agar (ISP 2), oatmeal agar (ISP 3), inorganic salts–starch agar (ISP 4), glycerol-asparagine agar (ISP 5), and tyrosine agar (ISP 7) (see Shirling and Gottlieb, 1966); oatmeal-nitrate agar (see Prauser and Bergholz, 1974); peptone-glucose medium (see Prauser and Falta, 1968); inorganic salts–starch agar (Gauze 1) (see Gauze et al., 1957); and glucose-asparagine agar (see Lindenbein, 1952).

Oatmeal agar: moderate growth, surface farinaceous, aerial mycelium cream colored to pink, substrate mycelium yellowish white, no diffusible pigment.

Yeast extract–malt extract agar: poor growth, sparsely scattered dots of sporulating hyphae, aerial mycelium whitish, substrate mycelium yellowish brown, no diffusible pigment.

Inorganic salts–starch agar (ISP 4): moderate growth, sparsely scattered dots of sporulating hyphae, aerial mycelium whitish, substrate mycelium cream colored to orange, no diffusible pigment.

Peptone-glucose medium: moderate growth, surface farinaceous, aerial mycelium cream-colored, substrate mycelium yellow to orange, no diffusible pigment.

Optimum temperature: 28–30°C.

Found in soil.

Type strain: INA 1933.

3. **Actinomadura citrea** Lavrova, Preobrazhenskaya and Svesnnikova 1972, 967.[AL]

ci'tre.a. M.L. fem. adj. *citrea* referring to the lemon-yellow color of the substrate mycelium.

Spore chains short, hooked or curved, in sparse clusters on the moderately branched aerial mycelium. Three to nine spores per chain. Spore surface warty or irregular.

Oatmeal agar: growth abundant, surface leathery, aerial mycelium only traces and yellowish white turning blue, substrate mycelium lemon yellow, diffusible pigment yellow.

Yeast extract–malt extract agar: moderate growth, surface leathery, aerial mycelium filmy cover of sterile hyphae, substrate mycelium yellowish brown, no diffusible pigment.

Inorganic salts–starch agar (ISP 4): moderate growth, surface leathery, aerial mycelium only traces and yellow turning blue with age, substrate mycelium yellowish brown, diffusible pigment yellow.

Peptone-glucose medium: good growth, surface leathery, aerial mycelium filmy cover of sterile and coremia-like hyphae, substrate mycelium yellow, no diffusible pigment.

Optimum temperature: 28–30°C.

Mol% G + C of the DNA is 67.6 (T_m).

Type strain: ATCC 27887.

4. **Actinomadura coerulea** Preobrazhenskaya, Lavrova, Ukholina and Nechaeva 1975b, 404.[AL]

coe.ru'le.a. L. fem. adj. *coerulea* referring to the blue aerial mycelium.

Spore chains curved or hooked or in spirals of one turn, arranged in tufts on long aerial hyphae. Spore surface warty.

Oatmeal agar: good growth, surface slightly farinaceous, aerial mycelium pale pink turning blue at maturity, substrate mycelium colorless or pale pink, no diffusible pigment.

Yeast extract–malt extract agar: moderate growth, surface granular with a filmy cover of aerial mycelium consisting of sterile hyphae and coremia, substrate mycelium pale brown, no diffusible pigment.

Inorganic salts–starch agar (Gauze 1): poor growth, surface slightly farinaceous, aerial mycelium pale blue, substrate mycelium colorless, no diffusible pigment.

Oatmeal-nitrate agar: good growth, surface slightly farinaceous, aerial mycelium white, substrate mycelium white, no diffusible pigment.

Mol% G + C of the DNA is 67.0 (T_m).

Type strain: INA 765.

5. **Actinomadura cremea** Preobrazhenskaya, Lavrova, Ukholina and Nechaeva 1975b, 404.[AL]

cre'me.a. M. L. fem. adj. *cremea* cream-colored (referring to the color of the aerial mycelium).

Spore chains short, in hooks or spirals of one turn, 3–8 spores per chain, arranged in clusters. Spore surface warty.

Oatmeal agar: moderate growth, surface farinaceous, aerial mycelium white to yellowish white, substrate mycelium colorless, no diffusible pigment.

Yeast extract–malt extract agar: moderate growth, surface farinaceous, aerial mycelium white, substrate mycelium light brown, no diffusible pigment.

Inorganic salts–starch agar (ISP 4): poor growth, surface granular, aerial mycelium cream colored to pale pink, substrate mycelium colorless, no diffusible pigment.

Peptone-glucose medium: good growth, surface granular, aerial mycelium white to cream colored, substrate mycelium brown, no diffusible pigment.

Mol% G + C of the DNA is 68.0 (T_m).

Type strain: INA 292.

Further comments: Gauze et al. (1975) isolated a rifamycin-producing strain of this species (INA 1349). Since production of this antibiotic had also been found in the genera *Streptomyces* and *Nocardia,* they described *A. cremea* subsp. *rifamycini,* which has since been validly published (Gauze et al., 1987). However, there are no characteristics, except production of rifamycin, that distinguish the variety from *A. cremea* itself.

6. **Actinomadura fastidiosa** Soina, Sokolov and Agre 1975, 883.[AL]

fas.tid.i.o'sa. L. fem. adj. *fastidiosa* fastidious (referring to the difficulties in growing the organisms).

Spore chains irregular spirals of one to two, more rarely three to four, turns arranged monopodially on long aerial hyphae. Spore surface smooth or slightly irregular (Fig. 30.2).

Oatmeal agar: good growth, surface farinaceous, aerial mycelium white to pale pink, substrate mycelium colorless or pale brownish, no diffusible pigment.

Yeast extract–malt extract agar: moderate growth, surface granular with spots of whitish aerial mycelium at the edges of colonies, substrate mycelium pale brown, no diffusible pigment.

Inorganic salts–starch agar (Gauze 1): moderate growth, aerial mycelium sparse and cream colored to pinkish, substrate mycelium colorless, no diffusible pigment.

Temperature range: 23–55°C. Optimum temperature 30–45°C.

Mol% G + C of the DNA is 67.0 (T_m).

Type strain: INMI 104.

7. **Actinomadura ferruginea** Meyer 1981, 215.[VP‡] (Effective publication: Meyer 1979, 41.)

fer.ru.gi'ne.a. L. fem. adj. *ferruginea* rusty brown (referring to the orange-brown-colored substrate mycelium).

Spore chains short, hooked or irregular spirals of one to two turns, arranged monopodially on long aerial hyphae. Four to nine spores per chain. Spore surface smooth or irregularly folded (Fig. 30.3).

Oatmeal agar: growth abundant, surface farinaceous, aerial mycelium white to pale pink, substrate mycelium pink, no diffusible pigment.

Yeast extract–malt extract agar: growth abundant, surface farinaceous, aerial mycelium orange-pink, substrate mycelium bright orange-brown, no diffusible pigment.

Inorganic salts–starch agar (ISP 4): poor growth, surface leathery, aerial mycelium absent, substrate mycelium colorless to brownish, no diffusible pigment.

Oatmeal-nitrate agar: growth abundant, surface dusty, aerial mycelium white to pale pink, substrate mycelium orange-pink, no diffusible pigment.

Peptone-glucose medium: growth moderate, surface cartilaginous, aerial mycelium absent, substrate mycelium light brown, no diffusible pigment.

Optimum temperature: 28–30°C.

Mol% G + C of the DNA is 68.1 (T_m).

Type strain: IMET 9567.

‡*VP* denotes that this name has been validly published in the official publication, *International Journal of Systematic Bacteriology.*

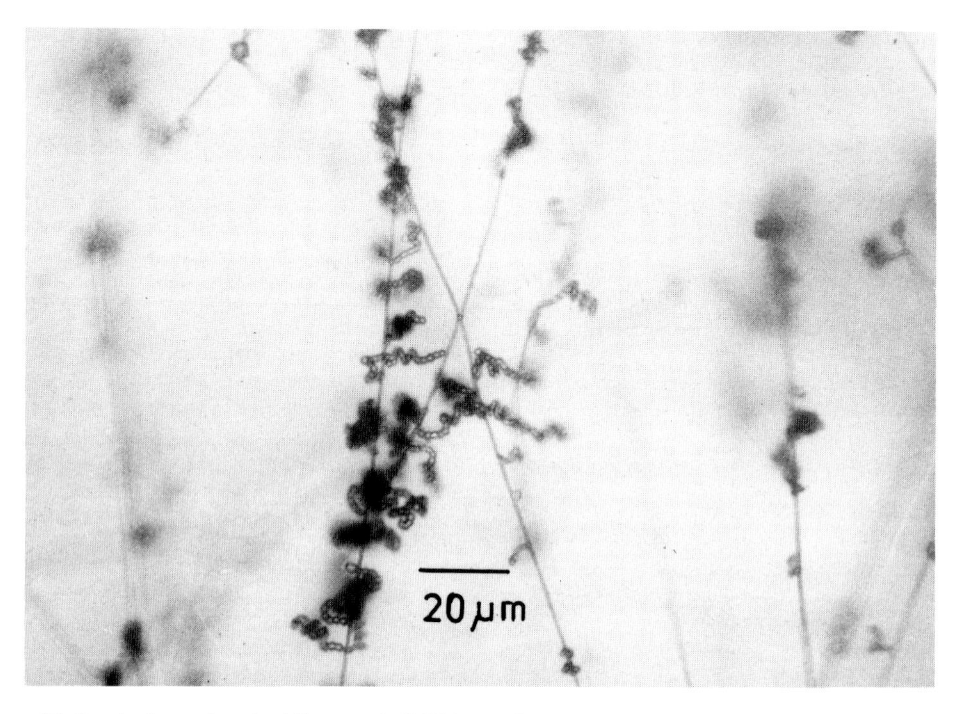

Figure 30.2. *Actinomadura fastidiosa,* strain IMET 9614. Sporulating aerial mycelium. Oatmeal agar, 28°C, 21 days (× 600).

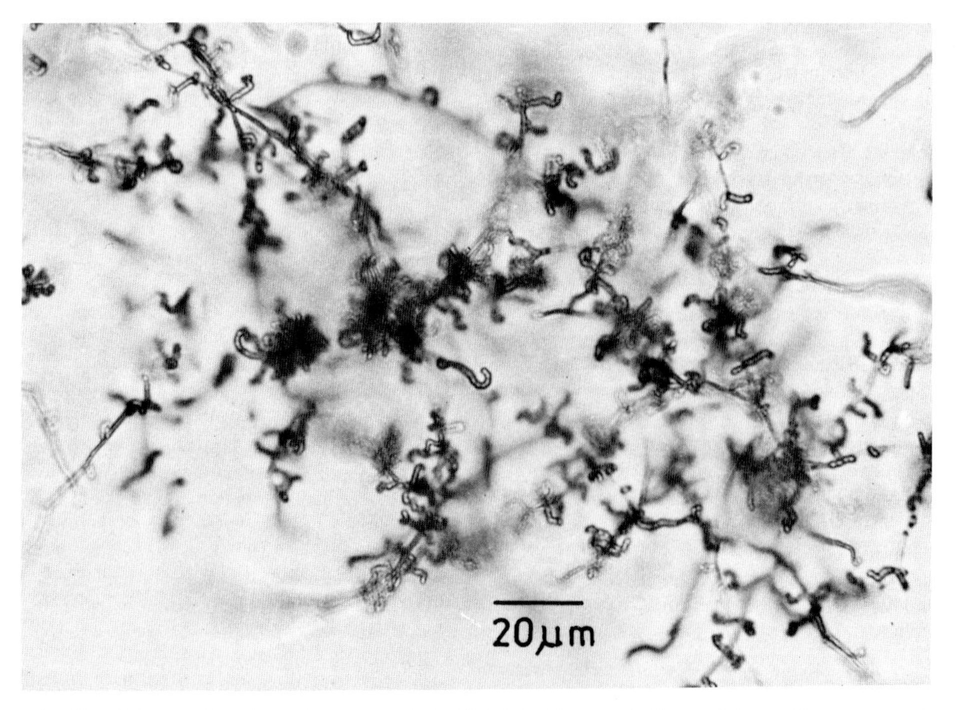

Figure 30.3. *Actinomadura ferruginea,* type strain IMET 9567. Sporulating aerial mycelium. Oatmeal agar, 28°C, 18 days (× 600).

8. **Actinomadura flexuosa** (ex Cross and Goodfellow 1973) nom. rev. (*Actinomadura flexuosa* (Krasil'nikov and Agre 1964b) Cross and Goodfellow 1973, 82; *Thermopolyspora flexuosa* Krasil'nikov and Agre 1964b, 105.)

flex.u'o.sa. M.L. fem. adj. *flexuosa* referring to the morphology of the spore chains.

Spore chains short, hooked or irregular spirals, arranged in clusters on long, moderately branched aerial hyphae. Four to 10 spores per chain. Spore surface warty.

Oatmeal agar: good growth, surface with a filmy cover of farinaceous aerial mycelium that is white to yellowish white, sometimes sectors with intensified development of aerial mycelium, substrate mycelium light brown, no diffusible pigment.

Glycerol-asparagine agar and Bennett sucrose agar: poor growth, aerial mycelium lacking, substrate mycelium brownish, no diffusible pigment.

Peptone-glucose medium: moderate growth, aerial mycelium lacking, substrate mycelium brown, no diffusible pigment.

Temperature range from 40° to 60°C. Optimum temperature 45-55°C.

Type strain: K 1132 (IMET 9552, received from H. A. Lechevalier).

Further comments: On the basis of cell wall composition Becker et al. (1965) transferred *Thermopolyspora flexuosa* Krasil'nikov and Agre 1964b to *Nocardia flexuosa,* since the type of sporulation was the same as that of *Nocardia madurae.* Later Kalakoutskii et al. (1968) proposed to include the strain K 1132 in question in the genus *Micropolyspora* on the grounds that thermophily alone should not be used as a criterion for differentiation of a genus. They insisted, however, that they had observed spore formation on aerial *and* substrate mycelium. Other authors were unable to confirm a clear sporulation of the substrate mycelium.

There are other thermophilic taxa (see "*Genus Incertae Sedis*" "*Excellospora*" Agre and Guzeva 1975, below) that display a considerable similarity to *A. flexuosa* in spore chain morphology and degradation tests, while differing in color of the aerial and substrate mycelium.

9. **Actinomadura helvata** Nonomura and Ohara 1971d, 904.[AL]

hel.va'ta. M.L. fem. adj. *helvata* honey-yellow (referring to the color of the substrate mycelium).

Spore chains in hooks and pseudosporangia on long aerial hyphae. About 10 spores per chain. Spore surface smooth.

Oatmeal agar: moderate growth, surface leathery, aerial mycelium a filmy cover of sterile hyphae, substrate mycelium yellowish brown, no diffusible pigment.

Yeast extract-malt extract agar: good growth, surface farinaceous, aerial mycelium yellowish white, substrate mycelium yellowish brown, no diffusible pigment.

Inorganic salts-starch agar (ISP 4): moderate growth, surface farinaceous, aerial mycelium white to cream colored, substrate mycelium yellowish brown, no diffusible pigment.

Oatmeal-nitrate agar: good growth, surface farinaceous, aerial mycelium white, substrate mycelium cream colored, no diffusible pigment.

According to Nonomura and Ohara (1971d) B vitamins are essential for growth.

Type strain: ATCC 27295.

10. **Actinomadura kijaniata** Horan and Brodsky 1982, 195.[VP]

ki.ja'nia.ta. M.L. fem. adj. *kijaniata* derived from "kijani" (the Swahili word for green).

Spore chains in long, open spirals with 10 or more spores per chain. Spores elliptical, 1.0-1.5 μm in diameter by 1.5-2.0 μm long. Spore surface smooth.

Oatmeal agar: moderate growth, surface flat to granular, aerial mycelium only white specks, substrate mycelium dark pine-green, no diffusible pigment.

Yeast extract-malt extract agar: good growth, surface raised and folded, aerial mycelium only white specks, substrate mycelium center dark jade-green and periphery biscuit colored, no diffusible pigment.

Inorganic salts-starch agar (ISP 4): moderate growth, surface flat and granular, aerial mycelium absent, substrate mycelium center slate-green and periphery light tawny, faint green diffusible pigment.

Nutrient agar: good growth, surface raised and folded, aerial mycelium absent, substrate mycelium evergreen, faint dull-green diffusible pigment.

Tyrosine agar: good growth, surface raised and folded, aerial mycelium abundant white, substrate mycelium lead gray, no diffusible pigment.

Mol% G + C of the DNA is 69.7 (T_m).

Type strain: ATCC 31588.

Further comments: Actinomadura kijaniata differs from other species of *Actinomadura* in that it forms deep green substrate mycelium and spore chains of considerable length on the aerial mycelium. Whole-cell hydrolysates contain *meso*-DAP together with a trace of the L- isomer (Horan and Brodsky, 1982). Furthermore, this species produces a complex of novel acid enol antibiotics, the major component of which was designated kijanimicin. This antibiotic possesses an unusual in vitro spectrum of activity against Gram-positive and anaerobic microorganisms. In vivo it has shown activity against *Plasmodium berghei* and *P. chabaudi* in mice (Waitz et al., 1981).

11. **Actinomadura libanotica** Meyer 1981, 215.[VP] (Effective publication: Meyer 1979, 41.)

li.ba.no'ti.ca. M.L. fem. adj. *libanotica* referring to Lebanon (the country in which the soil sample was taken).

Spore chains in hooks or curled, arranged in clusters on short hyphae of the aerial mycelium, 5-12 spores per chain, spores subspherical, spore surface folded or warty. The spores are connected by interspore pads (Figs. 30.4 and 30.5).

Oatmeal agar: abundant growth, surface farinaceous, aerial mycelium white to pale pink, substrate mycelium yellowish brown, no diffusible pigment.

Yeast extract-malt extract agar: abundant growth, surface farinaceous, aerial mycelium pale pink, substrate mycelium yellowish brown, no diffusible pigment.

Inorganic salts-starch agar (ISP 4): moderate growth, only a filmy cover of aerial mycelium, substrate mycelium grayish, no diffusible pigment.

Oatmeal-nitrate agar: abundant growth, surface farinaceous, aerial mycelium white to pale pink, substrate mycelium yellowish brown, no diffusible pigment.

Peptone-glucose medium: moderate growth, surface granular, aerial mycelium white, substrate mycelium yellowish brown, no diffusible pigment.

Optimum temperature: 28-30°C.

Mol% G + C of the DNA is 66.2 (T_m).

Type strain: IMET 9616.

12. **Actinomadura livida** Lavrova and Preobrazhenskaya 1975, 483.[AL]

li'vi.da. L. fem. adj. *livida* livid (referring to the grayish-violet color of the substrate mycelium).

Short spore chains in hooks or spirals of one turn. Spore surface irregular.

Oatmeal agar: good growth, surface cartilaginous dull, aerial mycelium only microscopically visible hyphae, substrate mycelium pale grayish pink, pale violet diffusible pigment.

Yeast extract-malt extract agar: moderate growth, surface cartilaginous, aerial mycelium absent, substrate mycelium pale brownish, no diffusible pigment.

Inorganic salts-starch agar (ISP 4): good growth, surface cartilaginous glistening, aerial mycelium only microscopically visible hyphae, substrate mycelium yellowish brownish, pale violet diffusible pigment.

Peptone-glucose medium: good growth, surface cartilaginous, aerial mycelium absent, substrate mycelium yellowish brown, no diffusible pigment.

Type strain: INA 1678.

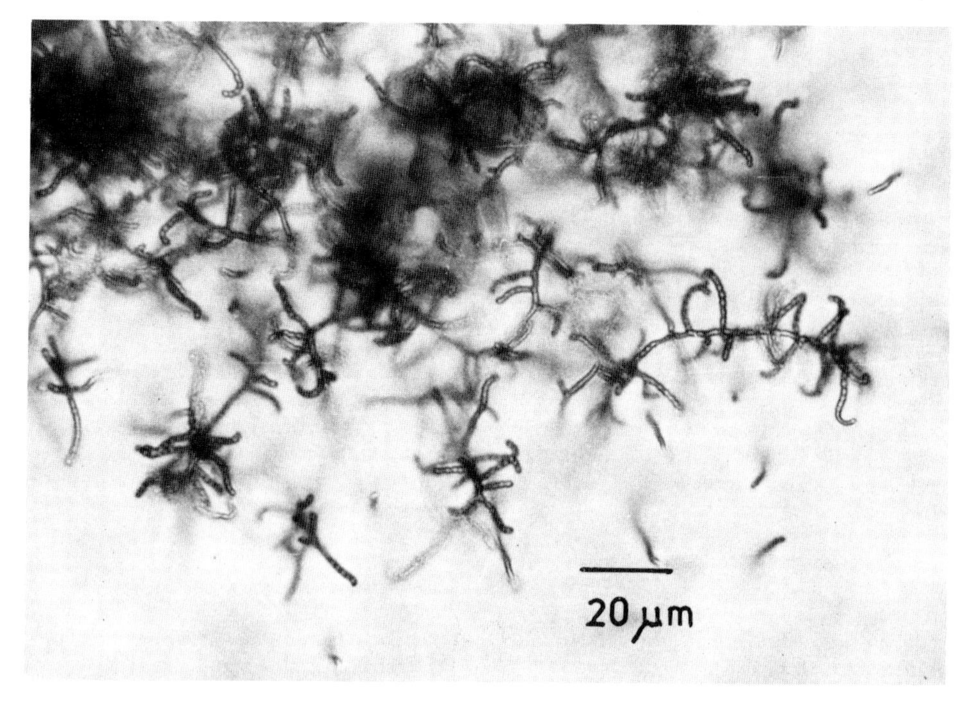

Figure 30.4. *Actinomadura libanotica,* strain IMET 9618. Sporulating aerial mycelium. Oatmeal-nitrate agar, 28°C, 12 days (× 600).

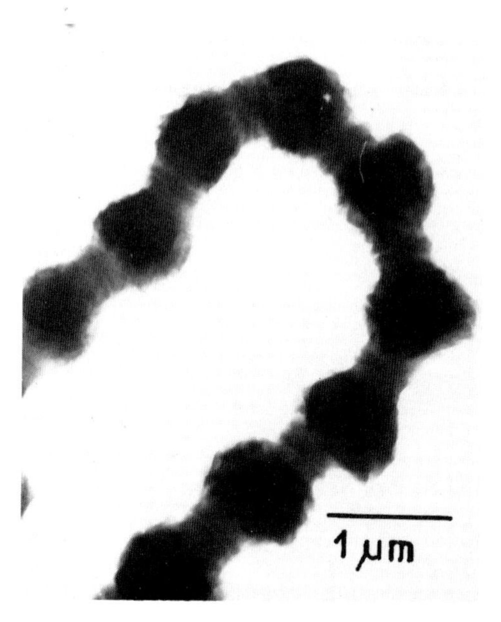

Figure 30.5. *Actinomadura libanotica,* strain IMET 9618. Spore chain with "interspore pads." Electron micrograph (× 16,500).

13. **Actinomadura luteofluorescens** (Shinobu 1962) Preobrazhenskaya and Lavrova in Preobrazhenskaya, Lavrova and Blinov 1975a, 526.[AL] (*Streptomyces luteofluorescens* Shinobu 1962, 115.)

lu'te.o.fluo.res.cens. M.L. comb. fem. adj. *luteus* yellow and *fluorescens* referring to the greenish tinge of the diffusible pigment produced by the organism.

Short spore chains in hooks or curled, arranged in clusters. Spore surface warty.

Oatmeal agar: good growth, surface farinaceous, aerial mycelium yellowish pink turning blue with age, substrate mycelium greenish yellow, greenish-yellow diffusible pigment.

Yeast extract-malt extract agar: moderate growth, surface farinaceous, aerial mycelium yellowish white, substrate mycelium dark yellow, pale yellow diffusible pigment.

Inorganic salts-starch agar (ISP 4): poor growth, surface leathery, aerial mycelium microscopically visible hyphae, substrate mycelium orange to brown, yellow diffusible pigment.

Peptone-glucose medium: good growth, surface wrinkled, leathery, aerial mycelium absent, substrate mycelium orange to brown, yellow diffusible pigment.

Type strain: IFO 13057.

14. **Actinomadura macra** Huang 1980, 565.[VP]

ma'cra. L. fem. adj. *macra* lean (referring to the poor, thin growth of this organism).

Sporulation is extremely rare and delayed. Only on Jensen (1930b) isolation agar and Czapek sucrose agar a few short spore chains have been observed after 5 weeks of incubation. They contained 4–15 spores and were straight, flexuous, hooked or coiled (once or twice). The spores were oval to elliptical (0.8–1.0 × 1.2–2.0 μm). Spore surface smooth.

Oatmeal agar: moderate growth, surface smooth, aerial mycelium scant and cream to faint pink, substrate mycelium cream to faint pink, no diffusible pigment.

Yeast extract-malt extract agar: good growth, surface raised and wrinkled, aerial mycelium white to grayish, substrate mycelium black, brown diffusible pigment.

Inorganic salts-starch agar (ISP 4): growth very scant, surface smooth, aerial mycelium absent, substrate mycelium colorless or pale grayish, no diffusible pigment.

Nutrient agar: poor to moderate growth, surface smooth, aerial mycelium absent, substrate mycelium cream colored, no diffusible pigment.

Czapek sucrose-nitrate agar: poor growth, surface smooth, aerial mycelium scant and pale cream colored, substrate mycelium colorless, no diffusible pigment.

Jensen (1930b) isolation agar: poor to moderate growth, surface smooth, aerial mycelium sparse and colorless to cream, substrate mycelium colorless to cream, no diffusible pigment.

Type strain: ATCC 31286.

15. **Actinomadura malachitica** Lavrova, Preobrazhenskaya and Sveshnikova 1972, 969.[AL]

ma.la.chi.ti'ca. M.L. fem. adj. *malachitica* from malachite, a green-colored mineral (referring to the bright green color of the aerial mycelium).

Spore chains short, straight, arranged in dense whorls on the main hyphae. Two to eight spores per chain. Spore surface smooth.

Oatmeal agar: good growth, surface farinaceous to granular, aerial mycelium pale greenish, substrate mycelium yellowish white, no diffusible pigment.

Yeast extract-malt extract agar: moderate to poor growth, surface farinaceous, aerial mycelium cream colored, substrate mycelium colorless, no diffusible pigment.

Inorganic salts-starch agar (ISP 4): moderate growth, surface farinaceous to granular, aerial mycelium cream colored, substrate mycelium yellowish brown, no diffusible pigment.

Inorganic salts-starch agar (Gauze 1): poor growth, aerial mycelium light green, substrate mycelium colorless, no diffusible pigment.

Peptone-glucose medium: poor growth, aerial mycelium absent, substrate mycelium yellowish brown, no diffusible pigment.

Mol% G + C of the DNA is 68.1 (T_m).

Type strain: ATCC 27888 (INA 1920).

16. **Actinomadura pelletieri** (Laveran 1906) Lechevalier and Lechevalier 1970a, 400.[AL] (*Micrococcus pelletieri* Laveran 1906, 341.)

pel.le.tier'i. M.L. gen. n. *pelletieri* named after T. Pelletier, who first isolated this species.

Spore chains short, hooked or in spirals of two to three turns. Two to six spores per chain. Spores subspherical, spore surface warty.

Oatmeal agar: moderate growth, surface cartilaginous, aerial mycelium absent, substrate mycelium pink to brownish red, no diffusible pigment.

Yeast extract-malt extract agar: moderate growth, surface cartilaginous, aerial mycelium traces of sterile hyphae, substrate mycelium pink to brownish red, no diffusible pigment.

Oatmeal-nitrate agar: moderate growth, surface cartilaginous, aerial mycelium only traces of sporulating hyphae, substrate mycelium colorless, no diffusible pigment.

Peptone-glucose medium: Moderate growth, surface cartilaginous, aerial mycelium absent, substrate mycelium brownish red, no diffusible pigment.

Mol% G + C of the DNA is 65.5-67.3 (T_m).

Type strain: NCTC 4162.

Further comments: Cummins (1962a), who was the first to isolate *meso*-DAP from this species, also found a small amount of L-DAP and glycine in some strains.

17. **Actinomadura pusilla** Nonomura and Ohara 1971d, 909.[AL]

pu.sil'la. L. fem. adj. *pusilla,* dwarfish (referring to the aerial mycelium of the organism).

Spore chains in tightly closed spirals, forming so-called pseudosporangia, i.e. spores are embedded in a slimy mass. Pseudosporangia 3-6 µm in diameter. Spore surface smooth (Fig. 30.6).

Oatmeal agar: abundant growth, surface farinaceous, aerial mycelium white to cream colored, substrate mycelium pale grayish brown, no diffusible pigment.

Yeast extract-malt extract agar: good growth, surface cartilaginous, aerial mycelium only microscopically visible, substrate mycelium dark brown with a red tinge, no diffusible pigment.

Inorganic salts-starch agar (ISP 4): scant growth, aerial mycelium microscopically visible, substrate mycelium colorless to brownish, no diffusible pigment.

Oatmeal-nitrate agar: good growth, surface granular to farinaceous,

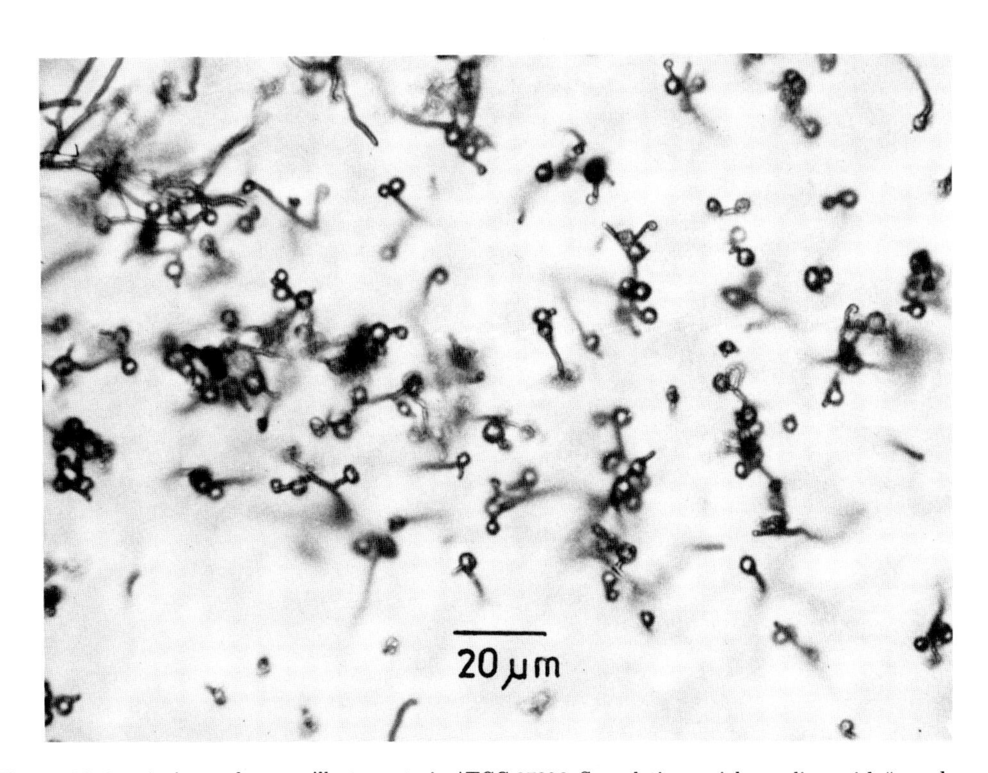

Figure 30.6. *Actinomadura pusilla,* type strain ATCC 27296. Sporulating aerial mycelium with "pseudosporangia." Yeast extract-malt extract agar, 28°C, 9 days (× 600).

aerial mycelium white, substrate mycelium colorless or white, no diffusible pigment.

Glycerol-asparagine agar: good growth, aerial mycelium white, substrate mycelium grayish pink, no diffusible pigment.

Mol% G + C of the DNA is 68.3 (T_m).

Type strain: ATCC 27296 (Nonomura A-118).

Further comments: Actinomadura pusilla can be easily mistaken for a *Streptosporangium* by superficial microscopic examination, and can only be differentiated by exact proof of a sporangial wall, which indicates a *Streptosporangium.*

Tamura et al. (1973b) isolated a sulfur-containing peptide antibiotic from *A. pusilla,* named actinotiocin.

18. **Actinomadura roseola** Lavrova and Preobrazhenskaya 1975, 483.[AL]

ro.se.o'la. L. fem. adj. dim. of *roseus* rosy (referring to the rose-colored aerial mycelium).

Spore chains straight or slightly bent, short or moderately long with up to 30 spores per chain. Spore chains may be stalked or in clusters emerging directly from the agar. Spores elliptical, spore surface folded (Fig. 30.7).

Oatmeal agar: abundant growth, surface farinaceous, aerial mycelium pink, substrate mycelium brownish red, no diffusible pigment.

Yeast extract-malt extract agar: abundant growth, surface farinaceous, aerial mycelium pink, substrate mycelium rusty brown, no diffusible pigment.

Glycerol-asparagine agar: good growth, surface farinaceous, aerial mycelium white to pale pink, substrate mycelium pink to orange colored, no diffusible pigment.

Czapek sucrose agar: good growth, surface farinaceous, aerial mycelium white, substrate mycelium colorless, no diffusible pigment.

Peptone-glucose medium: good growth, surface leathery, aerial mycelium absent, substrate mycelium yellow to orange colored, no diffusible pigment.

Mol% G + C of the DNA is 66.2 (T_m).

Type strain: INA 1671.

19. **Actinomadura roseoviolacea** Nonomura and Ohara 1971d, 909.[AL]

ro.se.o.vi.o.la'ce.a. L. fem. adj. *roseus* rose-colored; L. fem. adj. *violaceus* violet; M.L. fem. adj. *roseoviolacea* referring to the color of the substrate mycelium.

Spore chains in tightly closed spirals, forming pseudosporangia, i.e. spores are embedded in a slimy mass. Pseudosporangia about 4 μm in diameter, arranged on long aerial hyphae. Spore surface smooth.

Oatmeal agar: good growth, surface farinaceous, aerial mycelium pale pink to violet, substrate mycelium pale violet, pale violet diffusible pigment.

Yeast extract-malt extract agar: good growth, surface farinaceous to granular, aerial mycelium white to pale pink, substrate mycelium dark purple, no diffusible pigment.

Inorganic salts-starch agar (Gauze 1): good growth, surface farinaceous, aerial mycelium white, substrate mycelium white to pale pink, no diffusible pigment.

Glycerol-asparagine agar: good growth, surface leathery, aerial mycelium absent, substrate mycelium brick red, no diffusible pigment.

Glucose-asparagine agar: moderate growth, surface slightly farinaceous on the edge of the colonies, substrate mycelium brownish red, pale violet diffusible pigment.

Mol% G + C of the DNA is 68.5 (T_m).

Type strain: ATCC 27297.

Further comments: In their original description Nonomura and Ohara (1971d) reported on antibacterial activity of the type strain against *Staphylococcus aureus.* Further investigation of this strain (J. Meyer, unpublished) later revealed an active agent similar to daunomycin.

Tamura et al. (1973a) described "*A. roseoviolacea* var. *rubescens,*" which differs from the type strain of *A. roseoviolacea* by the following characters:

A. roseoviolacea (A-5) reverse side of colony changes to violet with 0.05 N NaOH. Hydrolysis of DNA and hypoxanthine positive.

"*A. roseoviolacea* var. *rubescens*" changes to red with 0.05 N NaOH. Hydrolysis of DNA and hypoxanthine negative.

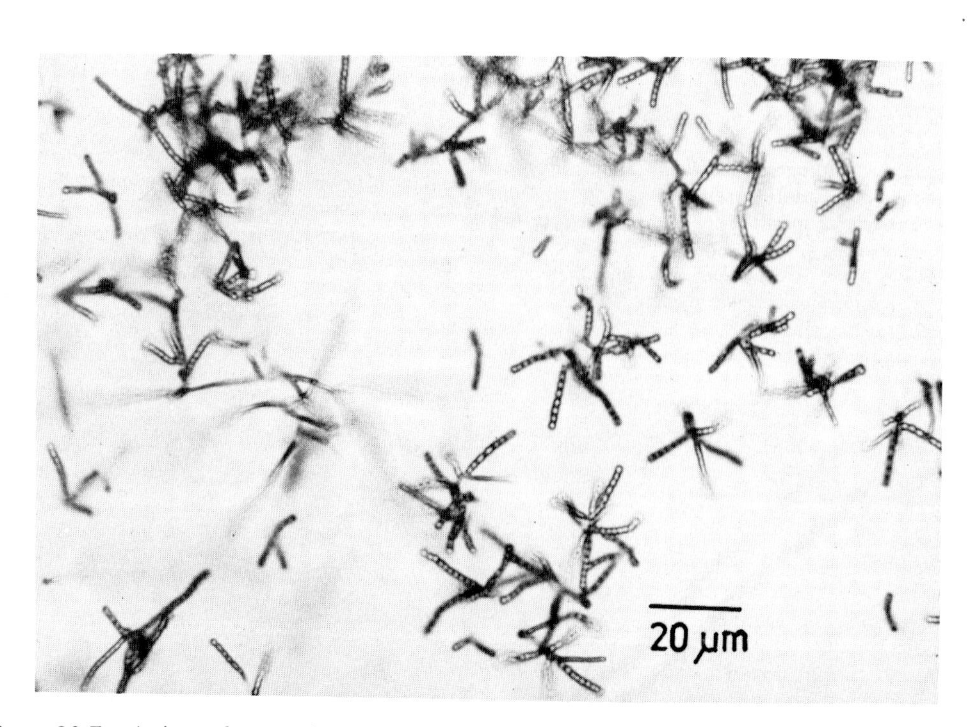

Figure 30.7. *Actinomadura roseola,* type strain INA 1671. Sporulating aerial mycelium. Oatmeal-nitrate agar, 28°C, 12 days (× 600).

"*A. roseoviolacea* var. *rubescens*" produces an antibiotic AB-64 that seems to be a heteroaromatic compound as shown by its ultraviolet, infrared, and nuclear magnetic resonance spectra.

Nakagawa et al. (1983) isolated four active components related to the carminomycins from a broth culture of another strain of *A. roseoviolacea.* The pigments A, B, and C were confirmed to be identical with carminomycin II and carminomycin I, whereas pigment D yielded 13-dihydrocarminomycin.

20. **Actinomadura rubra** (Sveshnikova, Maksimova and Kudrina 1969) Meyer and Sveshnikova 1974, 167.[AL] (*Micromonospora rubra* Sveshnikova, Maksimova and Kudrina 1969, 883.)

ru'bra. L. fem. adj. *rubra* red-colored (referring to the color of substrate mycelium).

Short, hooked spore chains or spirals of one to three turns, occasionally up to five turns and arranged on sporophores. Three to 12 spores per chain. Spores elliptical, surface smooth or slightly irregular.

Sporulating aerial mycelium mostly only in microscopically visible traces after 4–6 weeks of incubation on oatmeal agar, oatmeal-nitrate agar, and Bennett sucrose agar.

Oatmeal agar: good growth, surface wrinkled, aerial mycelium very poorly developed and mostly sterile or coremia-like, substrate mycelium bright orange-red, pale red diffusible pigment.

Yeast extract–malt extract agar: moderate growth, surface cartilaginous glistening, aerial mycelium only traces of sterile hyphae, substrate mycelium dark red to brown, no diffusible pigment.

Bennett sucrose agar: good growth, surface cartilaginous or wrinkled, aerial mycelium only traces and pinkish gray, substrate mycelium red to orange, red diffusible pigment.

Oatmeal-nitrate agar: moderate growth, surface cartilaginous glistening, aerial mycelium only traces and pinkish-gray, substrate mycelium red to orange, no diffusible pigment.

Type strain: ATCC 27031.

Further comments: A. rubra produces the antibiotic maduramycin, a red pigment with indicator properties (pH < 7.0 yellow, pH > 7.0 red) which possesses a strong antimicrobial activity against Gram-positive bacteria (Fleck et al., 1978).

The cell wall composition of *A. rubra* shows a slight difference to type III/B of Lechevalier and Lechevalier (1970a). In addition to *meso*-DAP, minor amounts of L-DAP and glycine have been found.

21. **Actinomadura salmonea** Preobrazhenskaya, Lavrova, Ukholina and Nechaeva 1975b, 408.[AL]

sal.mo'ne.a. M.L. fem. adj. *salmonea* from L. n. *salmo* salmon, salmon colored (referring to the color of substrate mycelium).

Spore chains in hooks or irregular spirals of one or occasionally up to four turns, arranged laterally on long hyphae. Spore surface warty.

Oatmeal agar: good growth, surface farinaceous, aerial mycelium pale pink, substrate mycelium dark red, no diffusible pigment.

Yeast extract–malt extract agar: moderate growth, surface cartilaginous, aerial mycelium cream colored to pinkish, substrate mycelium light brown, no diffusible pigment.

Inorganic salts–starch agar (ISP 4): moderate growth, surface wrinkled, aerial mycelium only traces of cream colored mycelium, substrate mycelium pale brown, no diffusible pigment.

Mol% G + C of the DNA is 66.0 (T_m).

Type strain: INA 2488.

22. **Actinomadura spadix** Nonomura and Ohara 1971d, 911.[AL]

spa'dix. M.L. adj. *spadix* derived from Gr. n. *spadix* chestnut colored (referring to the color of the substrate mycelium).

Short spore chains, 5–10 spores, in small, round spore masses (pseudosporangia) on long aerial hyphae. Sporulation was observed on soil extract agar and partly on oatmeal-yeast-glucose agar. Spore surface smooth.

Oatmeal agar: aerial mycelium light grayish-yellowish brown, substrate mycelium grayish brown, reddish-gray or grayish brown diffusible pigment.

Yeast extract–malt extract agar: aerial mycelium absent, substrate mycelium dark brown, brown diffusible pigment.

Glycerol-asparagine agar: aerial mycelium absent, substrate mycelium grayish brown, reddish-gray or grayish-brown diffusible pigment.

According to Nonomura and Ohara (1971d) B vitamins are essential for growth.

Mol% G + C of the DNA is 66.4 (T_m).

Type strain: ATCC 27298.

23. **Actinomadura spiralis** Meyer 1981, 215.[VP] (Effective publication: Meyer 1979, 39.)

spi.ra'lis. L. fem. adj. from L. n. *spira* coil (referring to the morphology of the spore chains).

Spore chains in spirals of two to five turns, closely packed or more or less loose spirals in pseudoverticillate arrangement along short straight aerial hyphae. Ten to 15 spores per chain. Spores spherical to subspherical. Spore surface folded (Figs. 30.8 and 30.9).

Oatmeal agar: good growth, surface woolly to farinaceous, aerial mycelium white to yellowish white, substrate mycelium yellow to yellowish brown, no diffusible pigment.

Yeast extract–malt extract agar: moderate growth, surface farinaceous, aerial mycelium white to yellowish, substrate mycelium yellow to yellowish brown, no diffusible pigment.

Inorganic salts–starch agar (ISP 4): poor growth, surface leathery, aerial mycelium absent, substrate mycelium yellow to yellowish brown, no diffusible pigment.

Oatmeal-nitrate agar: good growth, surface farinaceous, aerial mycelium white, substrate mycelium whitish, no diffusible pigment.

Peptone-glucose medium: poor growth, surface leathery, aerial mycelium absent, substrate mycelium yellow to yellowish brown, no diffusible pigment.

Mol% G + C of the DNA is 68.1 (T_m).

Type strain: IMET 9621.

24. **Actinomadura verrucosospora** Nonomura and Ohara 1971d, 908.[AL]

ver.ru.co.so.spo'ra. L. fem. adj. *verrucosa* warty; Gr. n. *spora* seed; M.L. adj. *verrucosospora* referring to the warty surface of the spores.

Spore chains hooked or curved or spirals of one turn, arranged in tufts on long aerial hyphae. Five to 12 spores per chain. Spores elliptical, spore surface warty (Fig. 30.10).

Oatmeal agar: good growth, surface farinaceous, aerial mycelium pink turning blue with maturity, substrate mycelium orange to pink, no diffusible pigment.

Yeast extract–malt extract agar: moderate to good growth, surface farinaceous, aerial mycelium white, substrate mycelium yellowish, no diffusible pigment.

Inorganic salts–starch agar (ISP 4): poor growth, surface farinaceous to granular, aerial mycelium white, substrate mycelium yellowish white, no diffusible pigment.

Oatmeal-nitrate agar: good growth, surface slightly farinaceous, aerial mycelium white to grayish, substrate mycelium white to pink, no diffusible pigment.

Glycerol-asparagine agar: moderate growth, aerial mycelium absent, substrate mycelium bright red, no diffusible pigment.

Mol% G + C of the DNA is 69.0 (T_m).

Type strain: ATCC 27299 (Nonomura A-184).

25. **Actinomadura vinacea** Lavrova and Preobrazhenskaya 1975, 486.[AL]

vi.na'ce.a. L. adj. derived from L. neut. n. *vinum* wine; M.L. adj. *vinacea* referring to the brownish-red color of the substrate mycelium.

Short, straight spore chains, arranged in small clusters on moderately branched hyphae of the aerial mycelium. Spore surface irregular.

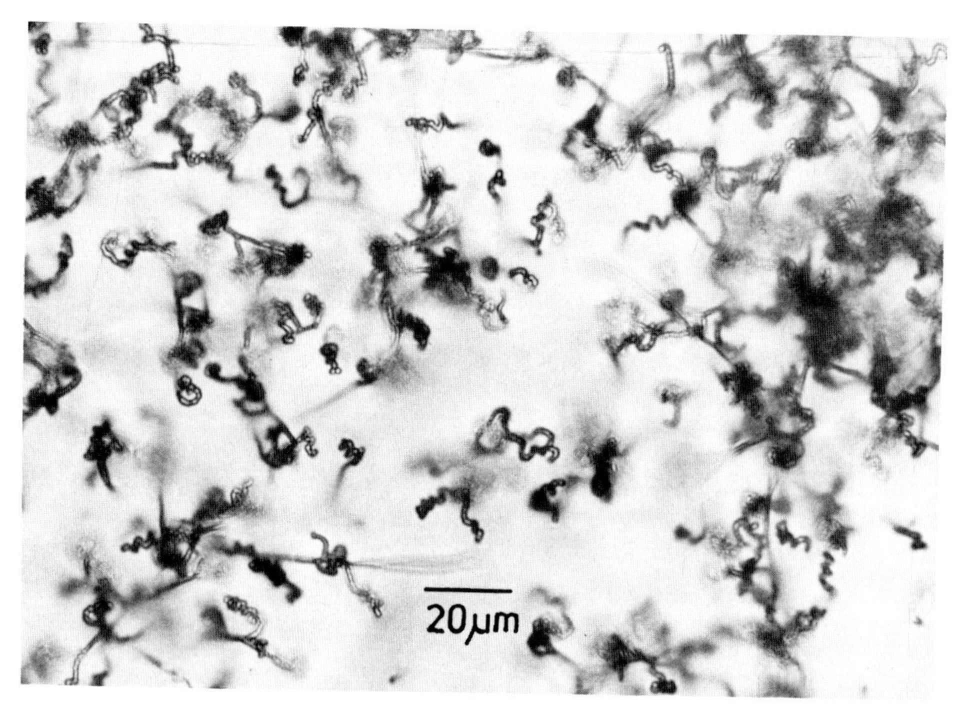

Figure 30.8. *Actinomadura spiralis,* type strain IMET 9621. Sporulating aerial mycelium. Oatmeal-nitrate agar, 28°C, 18 days (× 600).

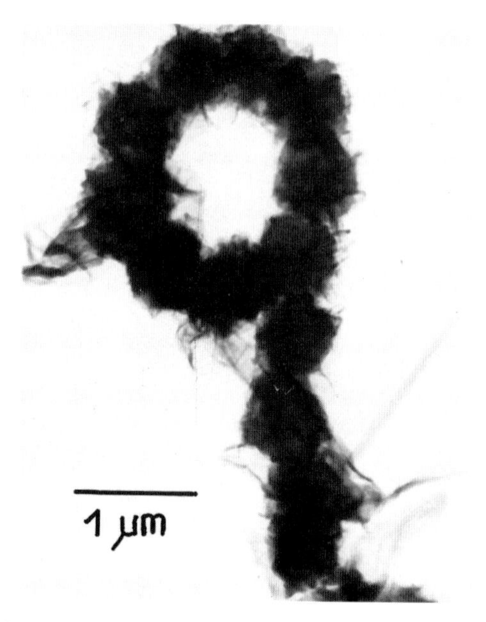

Figure 30.9. *Actinomadura spiralis,* type strain IMET 9621. Spore chain, electron micrograph (× 16,500).

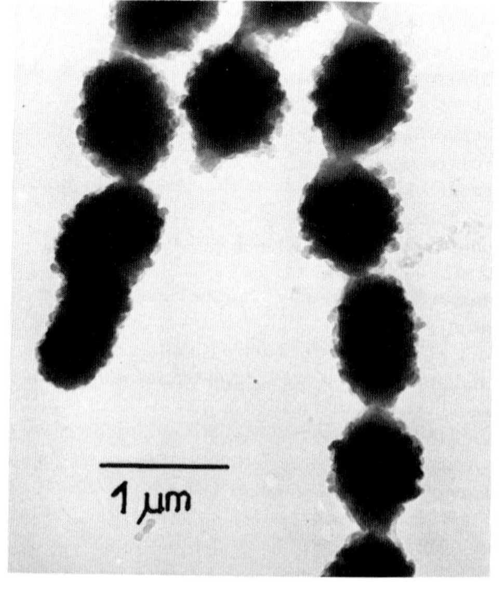

Figure 30.10. *Actinomadura verrucosospora,* type strain ATCC 27299. Spore chain, electron micrograph (× 16,500).

Oatmeal agar: poor growth, surface dull leathery, aerial mycelium absent, substrate mycelium pink to red, no diffusible pigment.

Yeast extract–malt extract agar: poor growth, surface leathery, aerial mycelium absent, substrate mycelium dark brown-red, dark red diffusible pigment.

Inorganic salts–starch agar (Gauze 1): poor growth, surface slightly velvety, aerial mycelium white to gray, substrate mycelium colorless, pale pink diffusible pigment.

Czapek sucrose agar: poor growth, aerial mycelium pink or absent, substrate mycelium brownish red, pale lilac to red diffusible pigment.

Peptone-glucose medium: moderate to poor growth, surface cartilaginous, aerial mycelium absent, substrate mycelium brownish red, dark red diffusible pigment.

Type strain: INA 1682.

26. **Actinomadura yumaensis** Labeda, Testa, Lechevalier and Lechevalier 1985a, 333.[AL]

yu'ma.en.sis. M.L. adj. *yumaensis* derivative of Yuma County, Arizona, U.S.A. (the source of the soil sample from which the type strain was isolated).

Spore chains short, loosely coiled, usually borne on branched, almost verticillate aerial sporophores. Spores are ovoid and measure 0.6–0.8 × 1.0–1.4 μm. Spore surface smooth.

Oatmeal agar: moderate growth, surface flat waxy, grayish-yellow; moderate white aerial mycelium, substrate mycelium grayish yellow, no soluble pigment.

Yeast extract–malt extract agar: good growth, surface raised waxy, convoluted colonies yellowish gray to grayish-yellowish brown, no aerial mycelium, substrate mycelium dark yellowish brown, orange soluble pigment.

Inorganic salts–starch agar (ISP 4): poor growth, white aerial mycelium, substrate mycelium colorless, no soluble pigment.

Bennett sucrose agar: good growth, no aerial mycelium, convoluted vegetative growth yellowish gray to grayish-yellowish brown, substrate mycelium dark grayish-yellowish brown, orange soluble pigment.

Optimum temperature: 37°C.

Isolated from soil.

Type strain: NRRL 12515 (= LL-C23024).

Species Incertae Sedis

1. *Actinomadura carminata* Gauze, Sveshnikova, Ukholina, Gavrilina, Filicheva and Gladkikh 1973, 675.[AL]

Although included in the Approved Lists (Skerman et al., 1980) the type strain was not available. The original description, however, is too meagre to establish the exact range of this species. From the data available from literature (color of aerial and substrate mycelium, spore chains in pseudosporangia and smooth spores) *A. carminata* is probably synonymous with *A. roseoviolacea.*

Produces antitumor antibiotic carminomycin.

Type strain: INA 4281.

2. "*Actinomadura azurea*" Nakamura and Isono 1983, 1468.

Produces cationomycin.

Type strain: ICM 2033.

3. *Actinomadura coeruleoviolacea* Preobrazhenskaya, Terekhova, Laiko, Selezneva, Zenkova and Blinov 1987, 179.[VP] (Effective publication: Preobrazhenskaya et al. 1976, 779.)

Type strain: INA 3564.

4. *Actinomadura oligospora* Mertz and Yao 1986, 179.[AL]

Produces a polyether antibiotic.

Type strain: NRRL 15878.

5. *Actinomadura fulvescens* Terekhova, Galatenko and Preobrazhenskaya 1987, 179.[VP] (Effective publication: Terekhova et al. 1982, 87.)

Type strain: INA 3321.

6. "*Actinomadura luzonensis*" Tomita, Hoshino, Sasahira and Kawaguchi 1980b, 1098.

Produces an antitumor antibiotic complex BBM-928.

Type strain: ATCC 31491 (G455-101).

7. *Actinomadura polychroma* Galatenko, Terekhova, and Preobrazhenskaya 1987, 179.[VP] (Effective publication: Galatenko et al. 1981, 803.)

Type strain: INA 2755.

8. "*Actinomadura pulveracea*" Iwami, Kiyoto, Nishikawa, Terano, Kohsaka, Aoki and Imanaka 1985, 835.

Produces antitumor antibiotics FR-900405 and FR-900406.

Type strain: No. 6049.

9. *Actinomadura recticatena* Gauze, Terekhova, Galatenko, Preobrazhenskaya, Borisova and Fedorova 1987, 179.[VP] (Effective publication: Gauze et al. 1984, 3.)

Type strain: INA 308.

10. *Actinomadura turkmeniaca* Terekhova, Galatenko and Preobrazhenskaya 1987, 179.[VP] (Effective publication: Terekhova et al. 1982, 87.)

Type strain: INA 3344.

11. *Actinomadura umbrina* Galatenko, Terekhova and Preobrazhenskaya 1987, 179.[VP] (Effective publication: Galatenko et al. 1981, 803.)

Type strain: INA 2309.

12. "*Actinomadura albolutea*" Tohyama, Miyadoh, Ito, Shomura, Ito and Ishikawa 1984, 1144.

The placement of this species in the genus *Actinomadura* is doubtful, because aerial mycelium exhibits total sporulation, and spores have a smooth surface. Thus *A. albolutea* morphologically resembles *Nocardiopsis* rather than *Actinomadura.* But presence of madurose in whole-cell hydrolysates, phospholipids of type PIV, and MK-9(H_4) as major menaquinones indicate it is clearly differentiated from *Nocardiopsis.*

Produces an indole-*N*-glycoside antibiotic SF-2140.

Type strain: FERM-BP 386 (= SF-2140).

Genus Incertae Sedis

Excellospora Agre and Guzeva 1975, 521.[AL]

Ex.cel'lo.spo.ra. M.L. fem. adj. from *excellens,* L. pr. part of *excello* prominent; Gr. n. *spora* seed (referring to the special structure of the spores).

Substrate and aerial mycelium well developed, nonfragmenting. Spore chains in hooks or spirals, single spores or spores in pairs, not only on the aerial mycelium but also on the substrate mycelium.

Cell wall chemotype III (*meso*-DAP, madurose) with high content of branched fatty acids with C_{16}, C_{17}, and C_{18}.

Type species: *Excellospora viridilutea* Agre and Guzeva 1975, 522.

The genus was created to accommodate mainly thermophilic actinomycetes, a newly isolated strain named *E. viridilutea* and two species that previously had been placed in the genus *Micropolyspora* ("*M. viridinigra*" and "*M. rubrobrunea,*" Krasil'nikov et al., 1968). According to Agre and Guzeva (1975) the species forms short spore chains or single or paired spores on both substrate and aerial mycelium (thus differing from the genus *Actinomadura*). The authors put emphasis on the predominance of C_{16}, C_{17}, and C_{18} branched fatty acids in *Excellospora* to separate the genus from *Actinomadura.* This character, however, seems unreliable, because Agre et al. (1975) reported on the heterogeneity of the genus *Actinomadura* with respect to fatty acid types (see also "Further Comments," above). Another variable character is the occurrence of spore chains on the substrate mycelium. Some actinomadurae, when grown at higher temperatures, develop spore chains directly on the agar surface, which may be mistaken for sporulation of the substrate mycelium.

The genus *Excellospora* has been included in the Approved Lists with its type species *E. viridilutea.* The definition of *Excellospora* in comparison with *Actinomadura,* however, is not yet clear-cut and convincing. At present the *Excellospora* species may be regarded as thermophilic actinomadurae until DNA hybridization and menaquinone analysis have proved their separate generic position.

Species Incertae Sedis

a. *Excellospora viridilutea* Agre and Guzeva 1975, 522.[AL]

Displays orange-yellow-colored colonies.

Type strain: INMI 187.

b. "*Excellospora rubrobrunea*" Agre and Guzeva 1975, 522.

Displays reddish-brown colonies.

No type strain designated.

c. "*Excellospora viridinigra*" Agre and Guzeva 1975, 522.
Displays greenish-brown colonies.
No type strain designated.

All three species are morphologically uniform. Their temperature range of growth is from 40° to 60°C. There is a tendency for sporulating hyphae to autolyze. They have similar physiological characteristics and are differentiated only by the color of aerial and substrate mycelium.

Genus **Microbispora** *Nonomura and Ohara 1957, 307*[AL]

HIDEO NONOMURA

Mi.cro.bi'spo.ra. Gr. adj. *micros* small; Gr. adj. *bis* two; Gr. n. *spora* a seed; M.L. fem. n. *Microbispora* the small two-spored (organism).

Stable branched mycelium producing **spores in characteristic longitudinal pairs on aerial mycelium; spores not usually formed on the substrate mycelium.** Spores sessile or on short sporophores, spherical to oval (usually 1.2–1.6 μm in diameter), and nonmotile. **Cell walls contain N-acetylated muramic acid** and **major amounts of *meso*-diaminopimelic acid** (*meso*-**DAP) but no characteristic sugars. Whole-cell hydrolysates contain madurose.** Menaquinones partially saturated, with MK-9(H_4) as major isoprenolog. Major phospholipids include phosphatidylcholine and unknown glucosamine-containing compounds, but no phosphatidylglycerol. Gram-positive. Aerobic. Mesophilic and thermophilic. Chemo-organotrophic. Most species require B vitamins, particularly thiamin, for growth. Natural habitat soil. The mol% G + C of the DNA is 71.3–73.

Type species: *Microbispora rosea* Nonomura and Ohara 1957, 307.

Further Descriptive Information

Microbisporae produce characteristic paired spores on the aerial hyphae. Initially buds are formed on the aerial hyphae; later the bud or the tip of a side branch swells and is separated by a cross-wall, giving rise to two spherical or oval spores, 1.0–2.0 μm (usually 1.2–1.6 μm) in diameter. The spores are produced either directly on the aerial hyphae or on short sporophores. The latter may be so short that the spores appear to be sessile. The sporophore of *M. rosea* partially encloses the basal spore, giving the appearance of a ball-and-socket joint (Williams, 1970). Mature spores are easily detached from the sporophores and from each other when placed in water. As a rule spores are not formed on substrate mycelium, but occasional production has been reported for one unidentified strain (Lechevalier and Lechevalier, 1967).

Spores of most species of *Microbispora* are produced on the entire aerial mycelium, and the spore surface is smooth (Figs. 30.11 and 30.12). However, in the two species, *M. echinospora* and *M. viridis,* the spores are produced only on side branches from axial hyphae that do not bear spores (Figs. 30.13 and 30.14), and the spore surface is spiny in the former and rugose in the latter. A strain similar to these species but with warty spores has been isolated from soil (H. Nonomura et al., unpublished data).

Three species of *Microbispora, M. aerata, M. amethystogenes,* and *M. parva,* produce bronze-violet irridescent crystals of iodinin (1,6-phenazine-5,10-dioxide) on oatmeal agar supplemented with yeast extract, or release the compound as a red soluble pigment in submerged culture. *M. aerata* produces four other related pigments (Gerber and Lechevalier, 1964, 1965).

The genus *Microbispora* includes mesophilic and thermophilic species. Most of mesophilic strains produce a pale to distinct pink aerial spore mass; the reverse side of colonies is yellowish brown or orange. Only *M. viridis* produces a grayish-green spore mass. *M. echinospora* and *M. parva* grow best at around 35°C but poorly at 28°C. Thermophilic strains produce a white or pale yellowish-brown to pale pinkish-brown aerial mass; the reverse side of colonies is either pale yellowish brown or yellow brown.

Most microbisporae require B vitamins for growth on synthetic media. The following nutritional properties have been reported for *M. rosea* (Nonomura and Ohara, 1957): Inorganic nitrogen compounds are not utilized as sole source of nitrogen; phosphate concentrations over 0.5% (w/v) inhibit growth; Fe^{2+} are favorable for growth whereas Cu^{2+} are injurious even at 2 ppm; and biotin controls the pigmentation of the substrate mycelium. In addition, the optimal pH for growth is below neutral and chlamydospores may be produced abundantly on rich media such as Emerson's agar (Waksman, 1957b).

Microbisporae have a wall chemotype III, and whole-cell hydrolysates

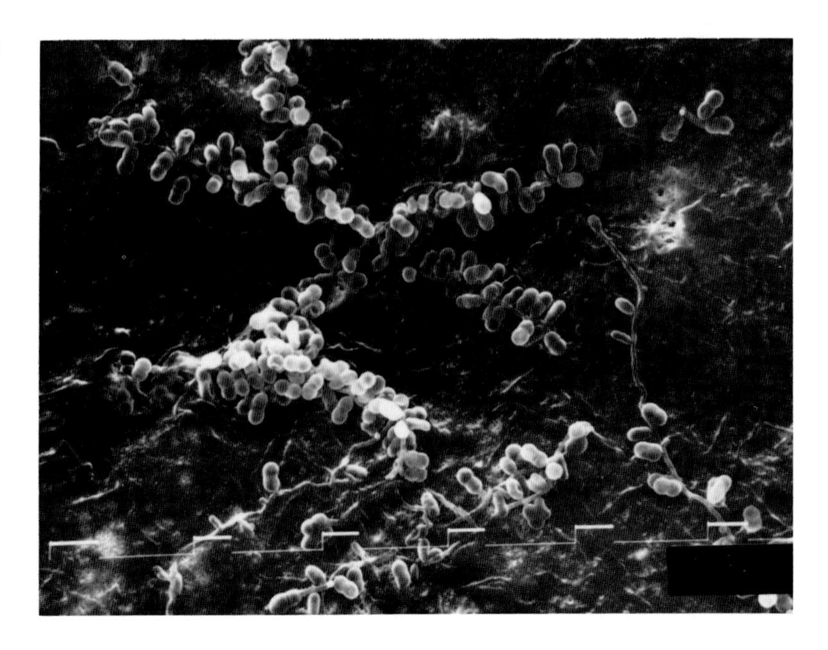

Figure 30.11. Morphology of *Microbispora rosea* ATCC 12950 on oatmeal agar: spores on entire mycelium. Scanning electron micrograph (SEM); gold spattered. *Bar interval,* 10 μm.

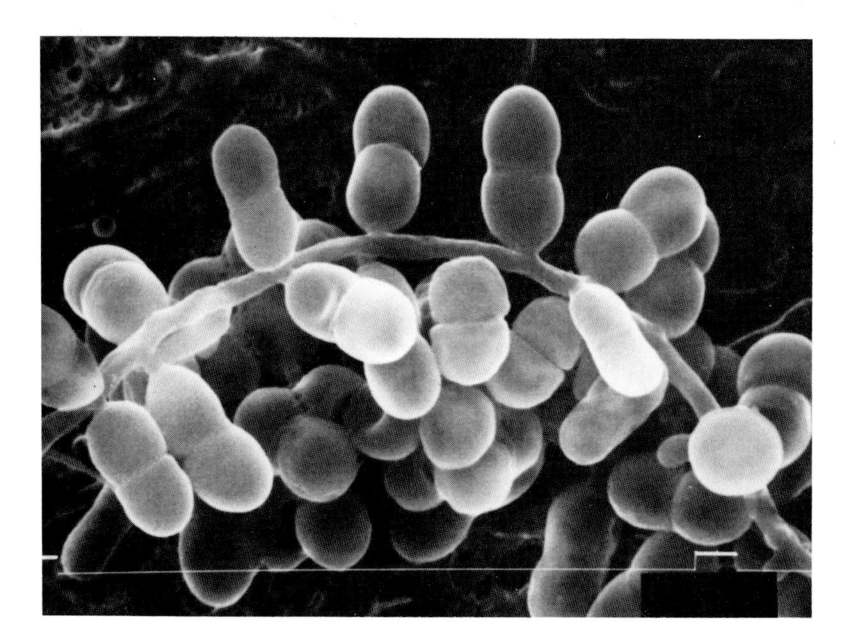

Figure 30.12. *Microbispora rosea:* paired spores on hyphae. *Bar interval,* 10 μm.

Figure 30.13. Morphology of *Microbispora echinospora* ATCC 27300 on inorganic salts–starch agar-V (SEM; gold spattered): spores on side branches from axial hyphae, which do not bear spores. *Bar interval,* 10 μm. Oval granules are an artifact.

contain madurose (Lechevalier and Lechevalier, 1970b; Nonomura and Ohara, 1971b; Miyadoh et al., 1985). The peptidoglycan is acetylated (Kawamoto et al., 1981). *M. rosea* contains tetrahydrogenated menaquinones with nine isoprene units, MK-9 (H_4), as the predominant isoprenolog (Athalye et al., 1984). All *Microbispora* species examined contain tuberculostearic acid and its analogs, but lack mycolic acid (Lechevalier, 1977). *M. amethystogenes* and *M. rosea* contain the phospholipids phosphatidylethanolamine, phosphatidylinositol, diphosphatidylglycerol, and phosphatidylinositol mannosides (Lechevalier et al., 1977). The mol% G + C of the DNA of *M. rosea* is 71.3 (Yamaguchi, 1967) and 73.7 (Lechevalier et al., 1971a).

Microbispora is common in soils. Using selective isolation procedures counts of between 10^4 and 10^6 colony-forming units/g dry weight of soil have been reported from various Japanese soils. Larger populations have been found in slightly acid (pH 5–6), humus-rich garden soils (Nonomura and Ohara, 1969a; Nonomura, 1984). One species, *M. rosea,* has been implicated in a case of pericarditis and pleuritis in man (Louria and Gordon, 1960).

Enrichment and Isolation Procedures

The use of selective synthetic media, together with the practice of heating air-dried soil, makes it possible to enumerate and isolate some rare actinomycetes in soil. The following procedures have been developed for the selective isolation of the genera *Microbispora* and *Streptosporangium* (Nonomura and Ohara, 1969a). These procedures can be modified for the selective isolation of strains of other genera, including *Actinomadura, Dactylosporangium, Elytrosporangium, Microtetraspora, Thermoactinomyces, Thermomonospora,* and *Saccharomonospora* (Nonomura and Ohara, 1971a, 1971b, 1971c, 1971d; Hayakawa and Nonomura, 1984; Nonomura, 1984).

Figure 30.14. *Microbispora echinospora:* spores are spiny. Sporulation in clusters (*arrow*). *Bar interval,* 10 µm.

Pretreatment

Soil samples are dried slowly at room temperature for a week, passed through a 2-mm sieve, ground slightly in a mortar, spread on filter paper, and heated in a hot air oven at 120°C for 1 h. This procedure reduces drastically the number of bacteria, and significantly the number of streptomycetes, but leads to an increased isolation frequency of *Microbispora* and *Streptosporangium.*

If necessary, *Microbispora* strains alone can be more preferentially isolated on agar plates by treating the dry-heated samples with 1.5% phenol for 20 min (H. Nonomura et al., unpublished data).

*Selective Media**

Initially AV agar was designed for the selective isolation of microbisporae, but two additional media, chitin-V and HV agars, are currently being developed for this purpose. Generally, HV agar gives the highest counts and clearest plates, but MGA-SE agar is usually useful for the isolation or enumeration of *M. echinospora.*

Plate Culture

Molten selective agar media are added to Petri dishes containing the pretreated and diluted soil suspension. Alternatively, the latter is spread over solidified selective agar in Petri dishes using a sterile glass rod. The inoculated plates are incubated for 4–6 weeks at 30°C, cooled at room temperature, and examined using a light microscope with a long working distance lens. Plates are incubated for 4–6 weeks at 38–40°C for the isolation of *M. echinospora,* and for 2–3 weeks at 50°C for the isolation of thermophilic strains.

Soil Culture

A small amount (0.5 g) of the dry-heated soil is placed, in a star shape, on an AV agar plate, and incubated for 3–5 weeks at 30°C. Microbisporae can usually be observed using a light microscope. Sometimes they develop around and over the soil particles. When such soil particles are placed on a MGA-SE agar plate and incubated at 40°C, growth of *M. echinospora* may be observed.

Maintenance Procedures

For preservation, agar slants carrying fully sporulating microbisporae may be kept at 5°C. The following media are recommended[†]: oatmeal agar-YG for mesophilic strains but inorganic salts-starch agar-V for *M. echinospora,* and glycerol agar (the C-1 medium[‡] supplemented with 0.5% (w/v) glycerol) for thermophilic strains. For long-term preservation, lyophilization is preferable.

Procedures for Testing of Special Characters

Iodinin production is observed on either oatmeal agar plus 0.1% (w/v) yeast extract or oatmeal agar-YG after 1 month's incubation at 30°C in

---

*Media composition as follows:

AV agar (Nonomura and Ohara, 1969a): arginine, 0.3 g; glucose, 1 g; glycerol, 1 g; K_2HPO_4, 0.3 g; $MgSO_4{\cdot}7H_2O$, 0.2 g; NaCl, 0.3 g; agar, 15 g; distilled water, 1000 ml; B vitamins (thiamin HCl, riboflavin, niacin, pyridoxine HCl, inositol, calcium panthothenate, *p*-aminobenzoic acid, each 0.5 mg; and biotin, 0.25 mg); trace salts ($Fe_2(SO_4)_3$, 10 mg; $CuSO_4{\cdot}5H_2O$, 1 mg; $ZnSO_4{\cdot}7H_2O$, 1 mg; $MnSO_4{\cdot}7H_2O$, 1 mg); antibiotics (Acti-Dione, 50 mg; nystatin, 50 mg; polymixin B, none or 4 mg; penicillin G, none or 0.8 mg); pH 6.4.

HV agar (Hayakawa and Nonomura, 1984; Nonomura, 1984): humic acid, 1 g (used as an alkaline solution; artificial humic acid prepared from glucose and urea may be used, as may natural humic acid from soil humus, but the pale brown humic acid designated as Rp type gives the best result); Na_2HPO_4, 0.5 g; $MgSO_4{\cdot}7H_2O$, 0.05 g; KCl, 1.7 g; $FeSO_4{\cdot}7H_2O$, 0.01 g; $CaCO_3$, 0.02 g; agar, 18 g; distilled water, 1000 ml; B vitamins, as for AV agar; Acti-Dione, 50 mg; pH 7.2.

Chitin-V agar (Hayakawa and Nonomura, 1984): colloidal chitin, 2 g (dry wt.); K_2HPO_4, 0.35 g; KH_2PO_4, 0.15 g; $MgSO_4{\cdot}7H_2O$, 0.2g; NaCl, 0.3 g; $CaCO_3$, 0.02 g; $FeSO_4{\cdot}7H_2O$, 10 mg; $ZnSO_4{\cdot}7H_2O$, 1 mg; $MnCl_2$, 1 mg; agar, 18 g; distilled water, 1000 ml; B vitamins, as for AV agar; Acti-Dione, 50 mg; pH 7.2.

MGA-SE agar (Nonomura and Ohara, 1971b): glucose, 2 g; asparagine, 1 g; K_2HPO_4, 0.5 g; soil extract, 200 ml; agar, 20 g; distilled water, 800 ml; antibiotics (Acti-Dione, 50 mg; nystatin, 50 mg; polymixin B, 4 mg; penicillin G, 0.8 mg); pH 8.0.

[†]*-YG,* with yeast extract, 1 g; glucose, 2 g; glycerol, 2 g/l. *-V,* with B vitamins as given for the formula of C-1 medium.

[‡]*C-1 medium* (Nonomura and Ohara, 1969b): casamino acids, 2 g; K_2HPO_4, 0.3 g; $MgSO_4{\cdot}7H_2O$, 0.5 g; NaCl, 0.3 g; agar, 20 g; distilled water, 1000 ml; trace salts ($FeSO_4{\cdot}7H_2O$, 10 mg; $MnSO_4{\cdot}7H_2O$, 1 mg; $ZnSO_4{\cdot}7H_2O$, 1 mg; $CuSO_4{\cdot}5H_2O$, 1 mg); B vitamins (thiamin HCl, riboflavin, niacin, pyridoxine HCl, inositol, calcium pantothenate, *p*-aminobenzoic acid, each 0.5 mg; and biotin, 0.25 mg); pH 7.2.

the case of mesophilic strains, and after 2 weeks' incubation at 50°C for thermophiles. Utilization of carbon sources is determined by comparing growth on a given compound in C-1 or C-2 medium[§] with two controls, one consisting of the medium and the other of the basal medium supplemented with 0.5% (w/v) glucose. Most strains of *Microbispora* do not grow on ISP Carbon Utilization Medium (Shirling and Gottlieb, 1966). Starch hydrolysis is detected on starch agar supplemented with 0.05% (w/v) yeast extract; nitrate reduction to nitrite on Bacto-nitrate broth plus 0.2% (w/v) yeast extract; gelatin liquefaction in gelatin medium (gelatin, 200 g; peptone, 5 g; yeast extract 2 g; glucose, 2 g; distilled water, 1000 ml); and milk peptonization in litmus milk poured on well-grown agar slant cultures.

Differentiation of the genus **Microbispora** *from other genera*

Three genera of maduromycetes, *Actinomadura, Microbispora,* and *Microtetraspora,* form paired or short chains of spores on aerial hyphae, and they all have a cell wall chemotype III, and also contain the sugar madurose (3-*O*-methyl-D-galactose) and menaquinone main components with MK-9. These genera can be morphologically differentiated from each other. *Microbispora* forms characteristic longitudinal pairs of spores (Table 30.5).

Table 30.5.
Differential characteristics of the genus **Microbispora** *and other closely related genera*[a]

Spores on aerial hyphae	*Microbispora*	*Microtetraspora*	*Actinomadura*
Paired	+	−	−
Most in chains of 4	−	+	−
Absent or short chains	−	−	+

[a]Symbols: see Table 30.4.

Taxonomic Comments

The genus *Microbispora* was described in 1957 by Nonomura and Ohara, and in the same year the genus *Waksmania* was proposed by Lechevalier and Lechevalier. It was later found that these genera were identical and priority was given to *Microbispora* (Lechevalier, 1965). *Thermopolyspora* was also considered to be a synonym of *Microbispora* (Lechevalier, 1965), and a species described as *Thermopolyspora bispora* Henssen 1957 was transferred to *Microbispora* as *M. bispora* (Lechevalier 1965).

Microbispora strains were originally distinguished by their ability to produce paired spores on aerial hyphae and the presence of *meso*-DAP and lack of characteristic sugars in the wall peptidoglycan (wall chemotype III sensu Lechevalier and Lechevalier, 1970b). However, in a preliminary phenetic survey the genus was found to be heterogeneous, with *M. echinospora* (morphologically atypical species), *M. rosea* (mesophilic), and *M. thermodiastatica* (thermophilic) having little affinity for one another (Goodfellow and Pirouz, 1982b). The results of this study add weight to the view that it is questionable to separate *Microbispora* species on the basis of a few physiological properties (Williams and Wellington, 1981).

It should not be difficult to obtain additional microbisporae for comparative taxonomic studies given the availability of suitable selective isolation techniques.

Further Reading

Lechevalier, H. A. and M. P. Lechevalier. 1981. Introduction to the *Actinomycetales*. *In* Starr, Stolp, Trüper, Balows and Schlegel (Editors), The Prokaryotes; A Handbook on Habitats, Isolation and Identification of Bacteria, Volume II. Springer-Verlag, Berlin, pp. 1915–1922.

Williams, S. T. and E. M. H. Wellington. 1981. The genera *Actinomadura, Actinopolyspora, Excellospora, Microbispora, Micropolyspora, Microtetraspora, Nocardiopsis, Saccharopolyspora* and *Pseudonocardia*. *In* Starr, Stolp, Trüper, Balows and Schlegel (Editors), The Prokaryotes; A Handbook on Habitats, Isolation and Identification of Bacteria, Volume II. Springer-Verlag, Berlin, pp. 2103–2117.

Differentiation and characteristics of the species of the genus **Microbispora**

The differential characteristics of the species of *Microbispora* are given in Table 30.6. Other characteristics of the species are listed in Table 30.7.

List of species of the genus **Microbispora**

1. **Microbispora rosea** Nonomura and Ohara 1957, 307.[AL]

ro'se.a. L. fem. adj. *rosea* rose colored.

Aerial mycelium is white at first, becoming pale pink with the formation of spores on oatmeal–yeast extract agar. Substrate mycelium is distinctly orange on oatmeal agar with 0.1% (w/v) peptone.

Type strain: ATCC 12950.

2. **Microbispora chromogenes** Nonomura and Ohara 1960a, 404.[AL]

chro.mo'ge.nes. Gr. n. *chromo* color; Gr. v. suff. *-genes* producing; M.L. adj. *chromogenes* producing color.

Aerial mycelium is distinct pink, soluble pigment is dark purple-gray on oatmeal–yeast extract agar. On yeast-starch agar the soluble pigment is yellowish green to olive-green. Substrate mycelium is orange on oatmeal-peptone agar.

Type strain: CBS 304.61; DSM 43165.

3. **Microbispora diastatica** Nonomura and Ohara 1960a, 404.[AL]

di.a.sta'ti.ca. M.L. fem. adj. *diastatica* starch hydrolyzing.

Aerial mycelium is distinct pink on oatmeal–yeast extract agar. Substrate mycelium is pale yellowish brown on oatmeal-peptone agar, yellow orange on yeast-starch agar.

Type strain: CBS 305.61; KCC A-0023.

4. **Microbispora amethystogenes** Nonomura and Ohara 1960a, 403.[AL]

am.e.thys.to'ge.nes. L. adj. *amethystinus* amethyst colored; Gr. v. suff. *-genes* producing; M.L. adj. *amethystogenes* producing violet-colored (crystals).

Aerial mycelium is pale pink to pink and abundant crystals of iodinin are present on oatmeal–yeast extract agar. Substrate mycelium is light brown to brownish gray on oatmeal-peptone agar.

Type strain: CBS 303.61; DSM 43164.

[§]*C-2 medium* (Nonomura and Ohara, 1971b): casamino acids, 0.5 g, and asparagine, 0.5 g, instead of casamino acids, 2 g, in the composition of C-1 medium. The other components are the same as for the C-1 medium.

Table 30.6.
Characteristics differentiating the species of the genus **Microbispora**[a]

Species	1. *M. rosea*	2. *M. chromogenes*	3. *M. diastatica*	4. *M. amethystogenes*	5. *M. parva*	6. *M. bispora*	7. *M. thermorosea*	8. *M. thermodiastatica*	9. *M. aerata*	10. *M. echinospora*	11. *M. viridis*
Nonsporogenous axial aerial hyphae	−	−	−	−	−	−	−	−	−	+	+
Spore surface											
Smooth	+	+	+	+	+	+	+	+	+	−	−
Spiny	−	−	−	−	−	−	−	−	−	+	−
Rugose	−	−	−	−	−	−	−	−	−	−	+
Spore mass color											
Pink	+	+	+	+	+	−	+	+	+	+	−
White	−	−	−	−	−	+	−	−	−	−	−
Green	−	−	−	−	−	−	−	−	−	−	+
Growth at											
25°C	+	+	+	+	+	−	−	−	−	+w	+
50°C	−	−	−	−	+	+	+	+	+	−	−
55°C	−	−	−	−	−	+	+	+	+	−	−
Optimal growth at 35–40°C	−	−	−	−	+	−	−	−	−	+	−
Iodinin production	−	−	−	+	+w	−	−	−	+	−	−
Starch hydrolysis	−	+	+	−	d	−	−	+	+	−	+
Nitrite from nitrate	d	+	−	d	−	−	−	−	+	−	−
Soluble pigments[b]	−	+	−	−	−	−	−	−	−	−	−
Utilization of											
Arabinose	+	+	+	+	+	−	+	+	+	+	+
Glycerol	+	+	+	d	+	−	+	+w	+	+	+
Inositol	−	+	−	+w	−	+	−	−	−	−	−
Rhamnose	d	−	+	−	+w	+w	−	−	−	−	+

[a]Symbols: see Table 30.4; +*w*, weak reaction.
[b]Other than pale yellow-brown.

5. **Microbispora parva** Nonomura and Ohara 1960a, 403.[AL]
par′va. L. fem. adj. *parva* poor (growth).

Growth is poor on oatmeal–yeast extract agar at 30°C, but good on oatmeal agar-YG at 35°C with white to pale pink aerial mycelium. A few crystals of iodinin may be present after 30 days at 30°C. Substrate mycelium is pale yellowish brown to light brown on oatmeal-peptone agar. Optimal temperatures for growth are 35–40°C.

Type strain: CBS 306.61; KCC A-0024.

6. **Microbispora bispora** (Henssen 1957) Lechevalier 1965, 141.[AL] (*Thermopolyspora bispora* Henssen 1957, 395.)
bi.spo′ra. Gr. adj. *bis* two; Gr. n. *spora* a seed; M.L. n. *bispora* two spores.

Aerial mycelium is chalk white and the substrate mycelium is pale yellowish brown on oatmeal–yeast extract agar. Thermophilic.

Type strain: ATCC 19993.

7. **Microbispora thermorosea** Nonomura and Ohara 1969b, 707.[AL]
ther.mo.ro′se.a. Gr. adj. *thermus* hot; L. adj. *roseus* rose colored; M.L. fem. adj. *thermorosea* heat (loving) rose colored.

Poor growth without aerial mycelium on oatmeal–yeast extract agar and yeast-starch agar. On glycerol agar, pale pinkish-white aerial mycelium is formed. Substrate mycelium is pale yellowish brown. Thermophilic.

Type strain: ATCC 27099.

8. **Microbispora thermodiastatica** Nonomura and Ohara 1969b, 706.[AL]
ther.mo.di.a.sta′ti.ca. Gr. adj. *thermus* hot; M.L. adj. *diastaticus* starch hydrolyzing; M.L. fem. adj. *thermodiastatica* starch hydrolyzing, heat (loving).

Poor or moderate growth on oatmeal agar, but good growth on yeast-starch agar. The aerial mycelium is white to pale yellowish brown. On C-1 medium supplemented with glucose (0.5% w/v), pale pink aerial mycelium is formed. Substrate mycelium is pale yellowish brown. Thermophilic.

Type strain: ATCC 27098.

9. **Microbispora aerata** (Gerber and Lechevalier 1964) Cross 1974, 859.[AL] (*Waksmania aerata* Gerber and Lechevalier 1964. 598.)
aer.a′ta. L. fem. adj. *aerata* covered with bronze.

Aerial mycelium is off-white to pale pink, abundant crystals of iodinin present on oatmeal agar. Substrate mycelium yellowish brown. Thermophilic.

Type strain: ATCC 15448.

10. **Microbispora echinospora** Nonomura and Ohara 1971b, 891.[AL]
e.chi.no.spo′ra. L. n. *echinus* sea urchin; Gr. n. *spora* a seed; M.L. fem. n. *echinospora* spiny spore.

Aerial mycelium differentiates into axial hyphae and spore-bearing side branches. Spiny spores are formed in clusters. The aerial mycelium

Table 30.7.
Other characteristics of the species of the genus **Microbispora**[a]

Species	1. *M. rosea*	2. *M. chromogenes*	3. *M. diastatica*	4. *M. amethystogenes*	5. *M. parva*	6. *M. bispora*	7. *M. thermorosea*	8. *M. thermodiastatica*	9. *M. aerata*	10. *M. echinospora*	11. *M. viridis*
Good sporulation[b] on											
Oatmeal–yeast extract agar	+	+	+	+	–	–	–	–	–	d	+
Oatmeal agar-YG	+	+	+	+	+	+	–	–	–	d	+
Glycerol agar	–	–	–	–	–	+	+	+	+	+	
Spore size 1.2–1.6 μm	+	+	+	+	+	+	+	+	+	+	–
Color of substrate mycelium											
Orange	+	+	+	–	–	–	–	–	+	+	–
Yellow-brown	+	+	+	+	+	+	+	+	+	+	+
Yellow	–	–	–	–	–	–	–	–	–	–	+
Melanoid pigment	–	±[c]	–	–	–	–	–	–	±	–	–
Soluble pigments[d]	–	+w	–	–	–	–	–	–	–	+	–
B vitamins required	+	+	+	+	+	+	+	+	+	+	–
Optimal growth at											
pH 6.0–7.0	+	+	+	+	+					–	
pH 7.0–8.0	–	–	–	–	–					+	
Gelatin liquefaction	d	d	d	d	d	–	+	+	+	+	+
Milk peptonization	–	d	+	+	d	–	+	+	+	+	+
Utilization of											
Fructose	+	+	+	+	+					+	+
Sucrose	+	+	+	+	+					+	+
Xylose	+	+	+	+	+					+	–

[a]Symbols: see Table 30.4.
[b]The best sporulation is obtained on inorganic salts–starch agar–V at 35–40°C.
[c]Not distinct. May be negative sensu "melanoid pigment" designated by ISP.
[d]Other than pale yellow-brown.

is yellowish pink and the substrate mycelium yellowish brown on inorganic salts–starch agar-V. On oatmeal agar, the substrate mycelium is yellowish orange to dark yellow. Yellow or yellowish-brown soluble pigment is produced. Optimal temperatures for growth are 35–40°C; very poor growth at 25°C. Optimal pH for growth is 7.0–8.0.

Type strain: ATCC 27300.

11. **Microbispora viridis** Miyadoh, Tohyama, Amano, Shomura and Niida 1985, 281.[VP]

vi'ri.dis. L. adj. *viridis* green.

Aerial mycelium differentiates into axial hyphae and spore-bearing side branches. Spores are oval (1.0–1.2 × 1.4–2.0 μm) and rugose (vertically ridged). The aerial mycelium is grayish green and the substrate mycelium pastel yellow on oatmeal agar. On inorganic salts–starch agar, the substrate mycelium is pale yellowish brown. Mesophilic. Produces the antibiotic SF-2240.

Type strain: IFO 14382.

Genus **Microtetraspora** *Thiemann, Pagani and Beretta 1968b, 296*[AL]

HIDEO NONOMURA

Mi.cro.tet'ra.spo.ra. Gr. adj. *micros* small; Gr. adj. *tetra* four; Gr. n. *spora* a seed; M.L. fem. n. *Microtetraspora* the four-spored (organism).

Form a stable, branched substrate and aerial mycelium. **Spore chains, typically containing four spores, are borne on the short aerial mycelium only.** Spores smooth, spherical (1.2–1.5 μm in diameter) or oval to short cylindrical (1.0–1.4 × 1.2–1.7 μm), nonmotile. Gram-positive. **Cell wall peptidoglycan is the acetyl type containing** ***meso*****-diaminopimelic acid** (*meso*-DAP) with no characteristic sugars. Whole-cell hydrolysates contain madurose. Menaquinones partially saturated, with MK-9(H_4) as major isoprenolog. Major phospholipids usually include phosphatidylcholine and unknown glucosamine-containing compounds, but no phosphatidylglycerol. Aerobic, mesophilic, chemo-organotrophic. Some species require B vitamins for growth. Natural habitat soil.

Type species: *Microtetraspora glauca* Thiemann, Pagani and Beretta 1968b, 296.

Further Descriptive Information

Although chains containing four spores are typical for this genus (Figs. 30.15–30.19), occasional chains of two to three spores and more rarely five spores have been observed. Peculiar branches of the spore chains have been revealed in *M. niveoalba* (Fig. 30.19). The spores of *M. fusca*

Figure 30.15. Morphology of *Microtetraspora glauca* on oatmeal agar. Scanning electron micrograph (SEM); gold spattered. *Bar interval,* 10 μm.

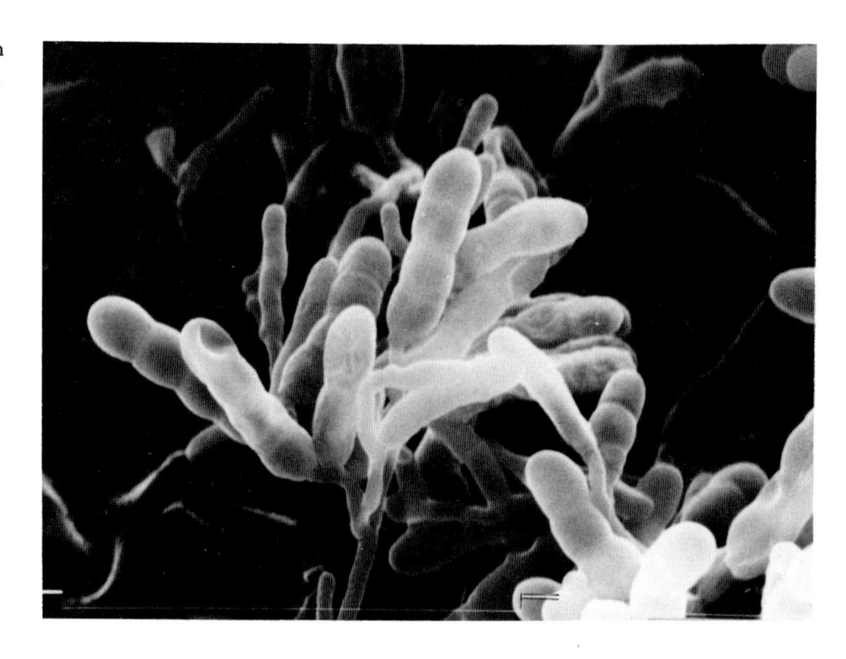

Figure 30.16. *Microtetraspora glauca* ATCC 27645 on oatmeal agar (SEM, gold spattered). *Bar interval,* 10 μm.

and *M. viridis* tend to fuse into a spore mass as the culture ages. The cell wall peptidoglycan, basically of wall chemotype III, can contain traces of LL-DAP, lysine, and glycine (Thiemann et al., 1968b; Nonomura and Ohara, 1971a). An acetyl type of glycan was reported for *M. niveoalba* and *M. viridis* (Kawamoto et al., 1981). The menaquinone composition of two species, *M. glauca* and *M. niveoalba,* has been reported (Athalye et al., 1984): both contained major amounts of tetrahydrogenated menaquinones with nine isoprene units.

Strains of *M. viridis* do not grow on many usual media such as oatmeal agar and glucose-asparagine agar. They sporulate moderately only on inorganic salts-starch, yeast-malt, and Hickey-Tresner* (Waksman, 1961) agars, three out of the 18 media recommended for actinomycetes. Special media—GAC, MC, and MGA agars—have been designed for this species (Nonomura and Ohara, 1971a).†

The identification of species within the genus *Microtetraspora* is still largely based on the color of the aerial and substrate mycelia (Nonomura, 1974b).

Microtetrasporae are widely distributed in soils (Thiemann et al., 1968b; Nonomura and Ohara, 1971a, 1971b).

**Hickey-Tresner agar* (Waksman, 1961): dextrin, 10.0 g; yeast extract, 1.0 g; beef extract, 1.0 g; N-Z-amine A, 2.0 g; $CaCl_2 \cdot 7H_2O$, 0.02 g; agar, 20.0 g; distilled water, 1000 ml; pH 7.3.

†*GAC agar* for the double-layered plate (Nonomura and Ohara, 1971a). Upper layer agar contains casamino acids (Difco), 0.5 g; agar, 2.0 g; nystatin, 200 mg; cycloheximide, 200 mg; polymixin B, 16 mg; benzyl penicillin, 3.2 mg; distilled water 1000 ml. The basal agar contains glucose, 1.0 g; asparagine, 1.0

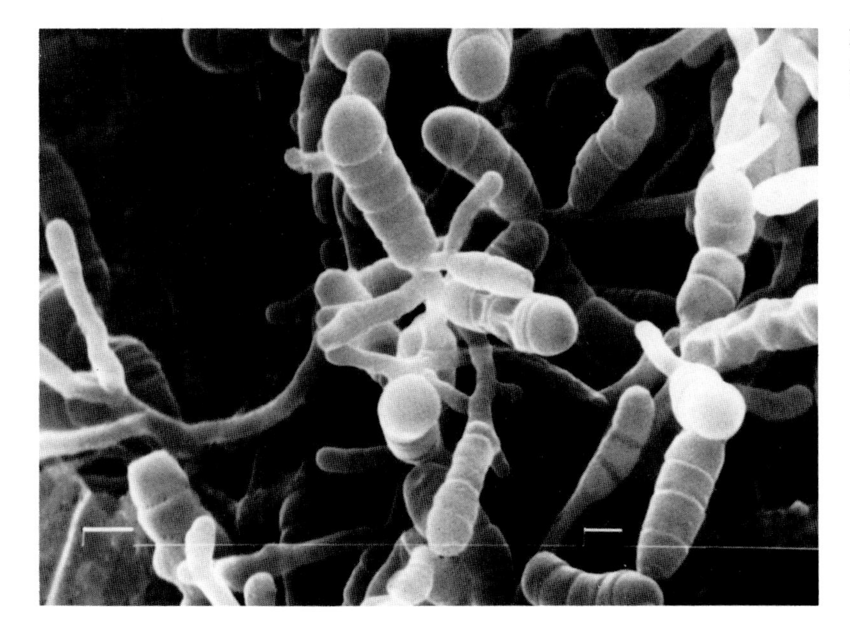

Figure 30.17. Morphology of *Microtetraspora viridis* ATCC 27103 on inorganic salts-starch agar: spores spherical (SEM, gold spattered). *Bar interval,* 10 μm.

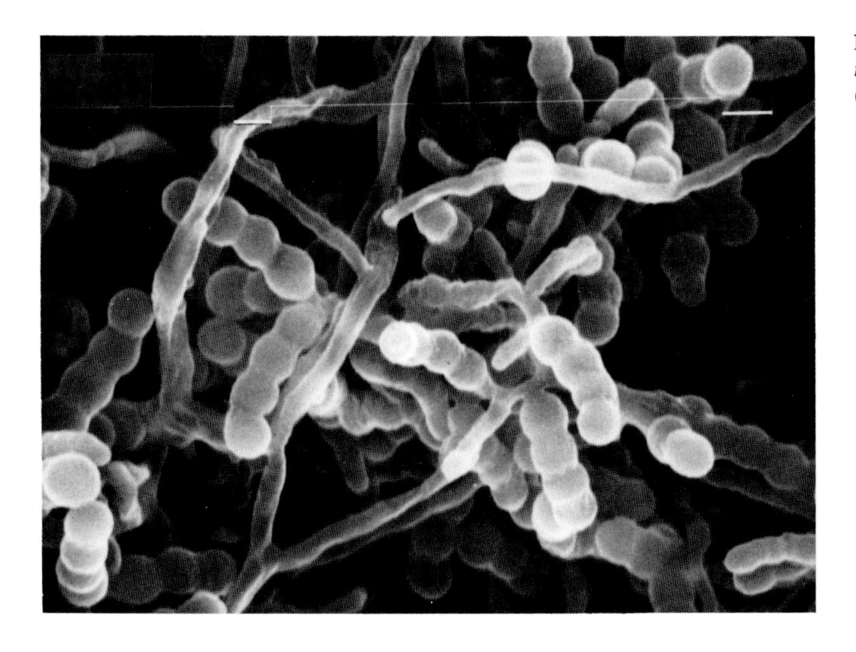

Figure 30.18. Morphology of *Microtetraspora niveoalba* ATCC 27301 on inorganic salts-starch agar (SEM, gold spattered). Bar, 10 μm.

Enrichment and Isolation Procedures

Details of the method originally used for isolating strains of *M. glauca* and *M. fusca* (Thiemann et al., 1968b) have not been disclosed. However, a pretreatment of dry heating air-dried soil samples, developed for the isolation of *Microbispora* and *Streptosporangium* (Nonomura and Ohara, 1969a) has proved useful for isolating most of the species in this genus and is recommended. Dry spores appear to be particularly resistant to dry heat of 100–120°C and such treatment significantly reduces the associated actinomycetes and bacteria in soil and allows the slow-growing *Microtetraspora* to develop into recognizable colonies on isolation plates.

To isolate *M. viridis* from soil a sample is first air dried, ground in a

g; K_2HPO_4, 0.3 g; $MgSO_4 \cdot 7H_2O$, 0.3 g; NaCl, 0.3 g; $FeSO_4 \cdot 7H_2O$, 10.0 mg; $MnSO_4 \cdot 7H_2O$, 1.0 mg; $CuSO_4 \cdot 5H_2O$, 1.0 mg; $ZnSO_4 \cdot 7H_2O$, 1.0 mg; nystatin, 100 mg; cycloheximide, 100 mg; polymixin B, 8.0 mg; benzyl penicillin, 1.6 mg; agar, 20.0 g; distilled water, 1000 ml, pH 7.4. The upper layer (4 ml) is poured onto the solidified basal layer.

MC agar (Nonomura and Ohara, 1971a): glucose, 2.0 g; $NaNO_3$, 0.5 g; K_2HPO_4, 0.3 g; $MgSO_4 \cdot 7H_2O$, 0.3 g; KCl, 0.3 g; $FeSO_4 \cdot 7H_2O$, 10 mg; $MnSO_4 \cdot 7H_2O$, 1.0 mg; $CuSO_4 \cdot 5H_2O$, 1.0 mg; $ZnSO_4 \cdot 7H_2O$, 1.0 mg; nystatin, 50 mg; cycloheximide, 50 mg; polymixin B, 4.0 mg; benzyl penicillin, 0.8 mg; agar, 20.0 g; distilled water, 1000 ml; pH 7.4.

MGA agar (Nonomura and Ohara, 1971a): glucose, 2.0 g; asparagine, 1.0 g; K_2HPO_4, 0.5 g; $MgSO_4 \cdot 7H_2O$, 0.5 g; trace salts (same as for C-2 medium); agar, 20.0 g; distilled water, 1000 ml; pH 7.2.

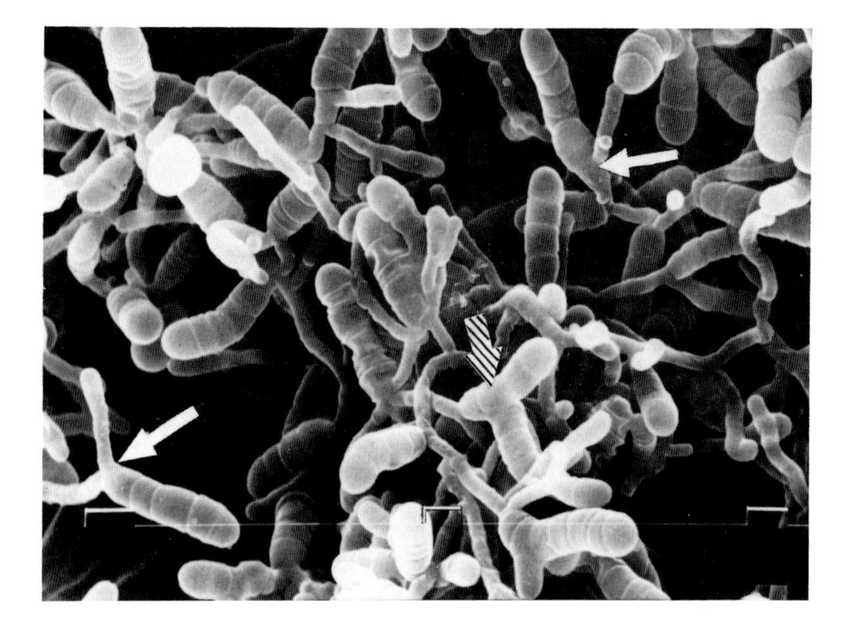

Figure 30.19. Morphology of *Microtetraspora niveoalba. Striped arrow,* branched spore chains; *white arrows,* branches at broad base of spore chains. (SEM, gold spattered). *Bar interval,* 10 µm.

mortar, and then heated in a hot-air oven at 100°C for 1 h. A particle of the soil is then placed at the center of a plate containing double-layered GAC agar and incubated at 32°C for 3–4 weeks. The green colonies of *M. viridis* develop preferentially at the margin of the deposited soil and can be detected with the aid of a microscope. Usually two plates per soil sample are sufficient to reveal colonies of this species. To obtain isolated colonies and to enumerate this species, dilutions (1:1000) of dry-heated soils are spread on MC agar at 30°C for 3–4 weeks. The plates must then be examined microscopically for the characteristic green colonies. One to three colonies of this species on five replicate plates were isolated from eight out of 10 Japanese soil samples (Nonomura and Ohara, 1971a).

M. niveoalba was isolated from soil that had been dry-heated at 120°C for 1 h. Particles of soil were sprinkled sparingly on plates containing MGA-SE agar[‡] and incubated at 38–39°C for 1 month. To isolate pure cultures, fragments of a colony seen on the primary isolation plates were streaked onto MGA-SE agar. This species was found to be widely distributed in soils but numbers were less than 10^3/g of dry soil (Nonomura and Ohara, 1971b).

Strains of *M. glauca* were isolated infrequently on MGA-SE plates at 30°C, but *M. fusca* was not obtained, possibly because of the high pH (8.0) of the medium.

Maintenance Procedures

Suitable media for sporulation of each species are given in the list of the species of *Microtetraspora,* and the optimal temperatures for growth are given in Table 30.10 (below). Sporulated slant cultures can be stored at 5°C. For long-term preservation, lyophilization is preferable.

Procedures for Testing of Special Characters

Utilization of carbon sources can be determined by comparing growth on a given carbon source in a basal medium with control media. MC agar lacking glucose is used as the basal medium for the strains of *M. viridis,* and C-2 medium[§] is used for *M. niveoalba, M. glauca,* and *M. fusca.* The morphology of *M. viridis* can be observed on MGA agar or inorganic salts–starch agar (ISP 4).

Differentiation of the genus **Microtetraspora** *from other genera*

Three genera of maduromycetes, *Microbispora, Microtetraspora,* and *Actinomadura,* form paired or short chains of spores on aerial hyphae and they all have a cell wall type III, and contain madurose and menaquinone main components with MK-9 (Athalye et al., 1984). In addition, numerical phenetic studies have shown representatives of these taxa to be closely related (Athalye et al., 1985).

The genus *Microtetraspora* has been morphologically differentiated from the others by its characteristic chains of four spores on distinct sporophores (Table 30.8).

Taxonomic Comments

The strains given variety status and named "*M. viridis* var. *intermedia*" (Nonomura and Ohara, 1971a) have well-developed aerial hyphae and streptomycete-like spore chains in addition to short chains of four spores (often six spores). The phospholipid composition of the type strain Mt-2 would place it in the PI group in contrast to other species of *Microtetraspora,* which have a PIV phospholipid pattern (Lechevalier et al., 1981b). "*M. viridis* var. *intermedia*" may prove to be a member of the genus *Actinomadura.*

"*Microtetraspora caesia*" (Tomita et al., 1980a) formed spores either singly or in short chains of one to eight spores on the aerial hyphae and vesicles containing a single or short chain of spores on the substrate mycelium. Some spores from the vesicles were motile in water. The inclusion of this species would widen the original circumscription of the taxon (Goodfellow and Cross, 1984b). Pending further studies on actinomycetes with a wall chemotype III, "*M. caesia*" is considered to be a *species incertae sedis.*

[‡]*MGA-SE agar* (Nonomura and Ohara, 1971b): glucose, 2.0 g; asparagine, 1.0 g; K_2HPO_4, 0.5 g; soil extract, 200 ml; agar, 20.0 g; nystatin, 50 mg; cycloheximide, 50 mg; polymixin B, 4.0 mg; benzyl penicillin, 0.8 mg; distilled water, 800 ml; pH 8.0. The soil extract was prepared by autoclaving 1000 g of soil in 1000 ml water for 30 min before decanting and filtering.

[§]*C-2 medium* (Nonomura and Ohara, 1971b): casamino acids (Difco), 0.5 g; asparagine, 0.5 g; K_2HPO_4, 0.3 g; $MgSO_4 \cdot 7H_2O$, 0.3 g; agar, 20.0 g; distilled water, 1000 ml; trace salts ($FeSO_4 \cdot 7H_2O$, 10 mg; $MnSO_4 \cdot 7H_2O$, 1 mg; $ZnSO_4 \cdot 7H_2O$, 1 mg; $CuSO_4 \cdot 5H_2O$, 1 mg); B vitamins (thiamin HCl, riboflavin, niacin, pyridoxine HCl, inositol, calcium pantothenate, and *p*-aminobenzoic acid, each 0.5 mg; biotin, 0.25 mg); pH 7.2.

Table 30.8.
Differential characteristics of the genus **Microtetraspora** *and closely related genera*[a]

Spores on aerial hyphae	*Micro-bispora*	*Micro-tetraspora*	*Actino-madura*
Paired	+	–	–
Most in chains of 4	–	+	–
Absent or short chains	–	–	+

[a]Symbols: see Table 30.4.

Further Reading

Lechevalier, H. A. and M.P. Lechevalier. 1981. Introduction to the order *Actinomycetales*. *In* Starr, Stolp, Trüper, Balows and Schlegel (Editors), The Prokaryotes; A Handbook on Habitats, Isolation and Identification of Bacteria, Vol. II. Springer-Verlag, Berlin, pp. 1915–1922.
Williams, S. T. and E. M.H. Wellington. 1981. The genera *Actinomadura, Actinopolyspora, Exellospora, Microbispora, Micropolyspora, Microtetraspora, Nocardiopsis, Saccharopolyspora* and *Pseudonocardia*. *In* Starr, Stolp, Trüper, Balows and Schlegel (Editors), The Prokaryotes; A Handbook on Habitats, Isolation, and Identification of Bacteria, Vol. II. Springer-Verlag, Berlin, pp. 2103–2117.

Differentiation and characteristics of the species of the genus Microtetraspora

The differential characteristics of the species of *Microtetraspora* are indicated in Table 30.9. Other characteristics are listed in Table 30.10.

List of species of the genus Microtetraspora

1. **Microtetraspora glauca** Thiemann, Pagani and Beretta 1968b, 296.[AL]

glau'ca. L. fem. adj. *glauca* blueish gray.

Good sporulation on Hickey-Tresner, oatmeal, and soil agars.[‖] The aerial spore mass color is blue-gray. Substrate mycelium blue-green to yellowish green on Hickey-Tresner agar.

Type strain: ATCC 23057.

2. **Microtetraspora fusca** Thiemann, Pagani and Beretta 1968b, 296.[AL]

fus'ca. L. fem. adj. *fusca* dark, tawny.

Good sporulation on Hickey-Tresner, glucose-asparagine,[#] glycerol-asparagine (ISP 5), and soil agars but none on oatmeal agar.

The aerial spore mass color is gray. Substrate mycelium brown-violet on Hickey-Tresner agar.

Type strain: ATCC 23058.

3. **Microtetraspora viridis** Nonomura and Ohara 1971a, 5.[AL]

vi'ri.dis. L. adj. *viridis* green.

Good sporulation on inorganic salts-starch (ISP 6) and MGA agars but none on oatmeal, soil, or Czapek agars.

The aerial spore mass color is green. Substrate mycelium yellowish brown on Hickey-Tresner agar, yellowish brown to green on inorganic salts-starch (ISP 4) and MGA agars.

Type strain: ATCC 27103.

4. **Microtetraspora niveoalba** Nonomura and Ohara 1971b, 872.[AL]

ni.ve.o.al'ba. L. adj. *niveus* snowy; L. adj. *albus* white; M.L. fem. adj. *niveoalba* snow white.

Good sporulation on oatmeal agar, yeast-starch agar, and glycerol-asparagine agar supplemented with B vitamins.

The aerial spore mass color is white. Substrate mycelium pale yellowish brown on Hickey-Tresner agar. Branched spore chains are present. Optimal temperature for growth is 35–40°C.

Type strain: ATCC 27301.

Table 30.9.
Characteristics differentiating the species of the genus **Microtetraspora**[a]

Characteristics	1. *M. glauca*	2. *M. fusca*	3. *M. viridis*	4. *M. niveoalba*
Branched spore chains	–	–	–	+
Spore mass color				
White	–	–	–	+
Green	–	–	+	–
Blue-gray	+	–	–	–
Gray	+	+	+	–
Color of substrate mycelium				
Green	–	–	+	–
Green-blue	+	–	–	–
Purplish	–	+	–	–
Nitrite from nitrate	+	–	–	+
Starch hydrolysis	+w	–	+	+w
Utilization of				
Rhamnose	+w	–	+	+
Inositol	+w	–	–	+

[a]Symbols: see Table 30.4; +*w*, weak reaction.

[‖]*Soil agar* (Thiemann et al., 1968b): air-dried garden soil, 30.0 g; agar, 20.0 g; tap water to 1000 ml; pH 7.0.

[#]*Glucose-asparagine agar* (Waksman, 1961): glucose, 10.0 g; asparagine, 0.5 g; K_2HPO_4, 0.5 g; agar, 15.0 g; distilled water, 1000 ml; pH 7.4.

Table 30.10.
Other characteristics of the species of the genus **Microtetraspora**[a]

Characteristics	1. *M. glauca*	2. *M. fusca*	3. *M. viridis*	4. *M. niveoalba*
Optimal temperature for growth				
25–30°C	+	+	–	–
30–35°C	+	+	+	–
35–40°C	–	–	–	+
Optimal pH for growth				
6.0–7.0	+	+	–	–
7.0–8.0	+	–	+	+
Soluble pigments[b]	–	–	–	–
Melanoid pigment	–	–	–	–
Utilization of				
Raffinose	–	–	–	–
Sucrose	–	–	+	+w
Fructose	+	–	+	+
Mannitol	+w	–	+	+
Glycerol	–	–	+w	+
Arabinose	+	+	+	+
Galactose	+	+	+	+
Trace of L-DAP	+	+	+	–
Growth on				
Czapek agar	–	–	+	–
oatmeal agar	+	–	–	+
glycerol asparagine agar	+w	+	–	–
Hickey-Tresner agar	+	+	+	+
B vitamins requirement	+	–	–	+

[a]Symbols: see Table 30.4; +*w*, weak reaction.
[b]Other than yellowish brown.

Genus **Planobispora** *Thiemann and Beretta 1968, 157*[AL]

GERNOT VOBIS

Pla.no.bi′spo.ra. poetic late Gr. n. *planos* wanderer, tramper; L. adv. *bis* twice (double); Gr. n. *spora* a seed, spore; M.L. fem. n. *Planobispora* a motile, double-spored organism.

Substrate and aerial mycelia are developed on agar media. Substrate hyphae (0.5–1.0 µm in diameter) are irregularly branched, occasionally septate, and nonfragmenting. Aerial hyphae (1.0 µm in diameter) are sparsely branched and rarely septate. Gram-positive and not acid fast. **Cylindrical to clavate sporangia** (1.0–1.2 µm wide × 6.0–8.0 µm long), each containing a **longitudinal pair of spores,** are formed singly or in bundles on short ramifications of the **aerial hyphae.** The **spores** (zoospores) are **straight or slightly curved** with rounded ends and are **motile** by means of peritrichous flagella. Colonies grown on agar media are flat or occasionally elevated. The **substrate mycelium either is without distinctive color or is rose colored.** The **aerial mycelium,** which is developed only on certain agar media, is **white or with a light rose tinge.** The peptidoglycan of the cell wall contains ***meso*-diaminopimelic acid** (*meso*-DAP) and **madurose** is the characteristic sugar of whole-cell hydrolysates. Aerobic, chemo-organotrophic, and mesophilic, growing well between 28° and 40°C and from pH 6.0 to 9.0. The mol% G + C of the DNA is 70–71 (T_m).

Type species: *Planobispora longispora* Thiemann and Beretta 1968, 157.

Further Descriptive Information

The hyphae of the substrate mycelium are 0.5–1.0 µm in diameter, irregularly branched, and occasionally septate. Fragmentation of hyphae is not apparent either on solid agar media or in liquid cultures (Thiemann and Beretta, 1968). The hyphae of the aerial mycelium are 1.0 µm in diameter, long, slender, and wavy, with few lateral branches, and grow more or less parallel to the surface of the substrate. The aerial mycelium is extremely hydrophobic (Thiemann, 1970b).

Cylindrical to clavate sporangia are formed on short side branches of the hyphae of the aerial mycelium. They can be arranged singly or in bundles (Fig. 30.20). Sporangiophores (1.0–3.0 µm long) are very fragile and collapse easily (Thiemann and Beretta, 1968). Each sporangium contains a longitudinal pair of spores (Fig. 30.21). The average size of a sporangium is 6.0–8.0 µm in length and 1.0–1.2 µm in width (Thiemann, 1974b). Since ultrathin sections of the sporangial development have not been studied in detail, the question as to whether the spores originate endogenously, as claimed by Williams and Wellington (1980), or by simple transformation of sporogenous hyphae (Bland and Couch, 1981) cannot be answered at present. The fine structure has been examined by whole-mount preparations in transmission electron microscopy. The sporangial envelope is smooth (Thiemann and Beretta, 1968) and contains fibrillar elements (Vobis and Kothe, 1985) that resemble those present in *Planomonospora* (Sharples et al., 1974). A transverse septum connected to the sporangial envelope divides the two spores (Thiemann, 1970b). Because this septum is not part of the actual cell wall it has recently been termed a "diaphragm" (Vobis and Kothe, 1985).

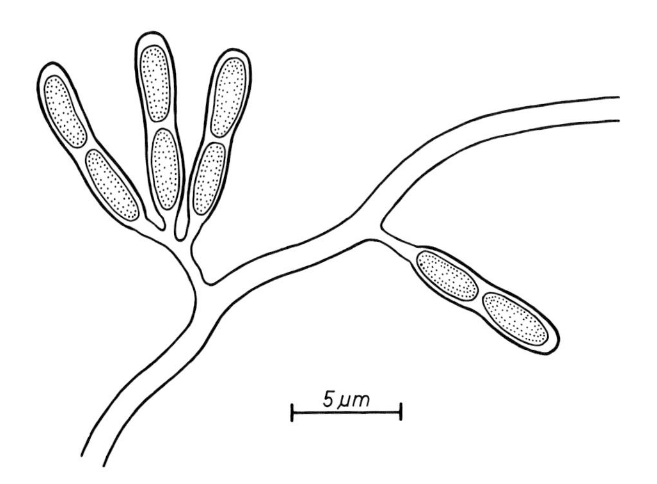

Figure 30.20. Aerial hypha of *Planobispora* species bearing single and bundled sporangia.

The spores almost fill the sporangia (Fig. 30.21). They measure 1.0–1.2 μm in diameter and 2.6–4.0 μm in length, and they are oblong with rounded ends, occasionally slightly curved. The spores (zoospores) are motile by means of peritrichous flagella (Thiemann, 1974b). If sporangia are dipped into water, the upper spore is pushed out at first through the tip of the sporangium. Frequently the basal spore is not released (Thiemann, 1970b; Locci and Petrolini Baldan, 1971). The spores become motile only some time after being dispersed. In spite of good culture conditions, only 2–5% of the spores become motile. The spores germinate with one or two polar germ tubes; occasionally lateral germination is observed (Thiemann and Beretta, 1968).

The peptidoglycan of the cell wall contains *meso*-DAP with madurose (3-*O*-methyl-D-galactose) as the characteristic sugar of whole-cell hydrolysates (Kroppenstedt and Kutzner, 1976, 1978). This chemical cell wall composition is in accordance with cell wall type III and sugar pattern B of the classification scheme of Lechevalier and Lechevalier (1970b). The phospholipids of the cell membranes consist of unknown glucosamine-containing phospholipids, phosphatidylinositol, phosphatidylethanolamine, and diphosphatidylglycerol (Hasegawa et al., 1979), in accordance with phospholipid type IV of Lechevalier et al. (1981). The fatty acids consist of straight-chain acids, *iso*- and 10-methyl branched acids, and unsaturated fatty acids. *Anteiso*- branched fatty acids are not present (Kroppenstedt and Kutzner, 1978).

Substrate and aerial mycelia develop on a variety of agar media. The colonies are mostly flat or occasionally elevated and have a smooth surface. On yeast extract–malt extract agar and on Bennett agar the colonies of *P. longispora* are crusty (Thiemann and Beretta, 1968); *P. rosea* has slightly wrinkled surfaces on Bennett agar and Hickey-Tresner agar (Thiemann, 1970b). No distinctive color of the mycelium is formed by *P. longispora* (Thiemann and Beretta, 1968); *P. rosea* produces rose-colored substrate mycelium and white aerial mycelium with a rose tinge (Thiemann, 1970b). For *P. longispora* the development of the aerial mycelium and formation of sporangia is promoted by oatmeal, calcium malate, and soil agars (Thiemann and Beretta, 1968). Hickey-Tresner agar and glycerol-asparagine agar have the same effect on *P. rosea* (Thiemann, 1970b).

Members of *Planobispora* are aerobic, growing well on the various standard culture media recommended by Waksman (1961) and Shirling and Gottlieb (1966). The type strains of *P. longispora* and *P. rosea* can utilize L-arabinose, cellobiose, fructose, glucose, glycogen, inositol, maltose, mannitol, and starch as sole carbon sources. Adonitol, D-arabinose, erythritol, ethanol, glycerol, inulin, lactose, mannose, α-methyl-D-glucoside, raffinose, sorbitol, and sucrose are not utilized (Thiemann, 1974b; Goodfellow and Pirouz, 1982b) (see Table 30.11, below). Acetamide and serine are not utilized as sole carbon and nitrogen sources (Goodfellow and Pirouz, 1982b).

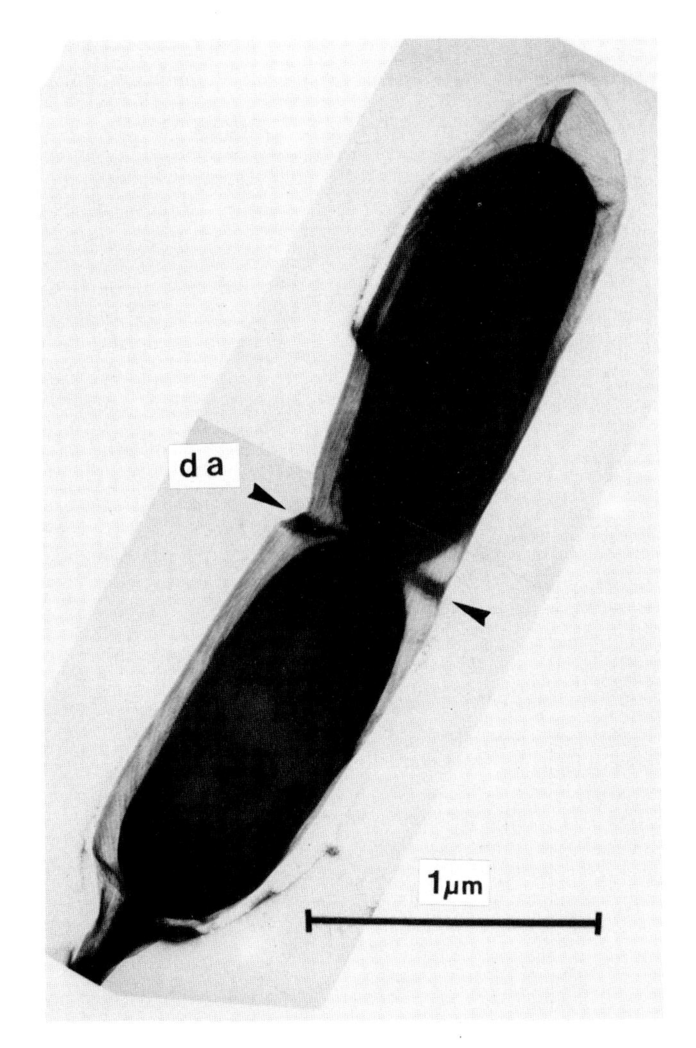

Figure 30.21. Sporangium of *Planobispora rosea* (strain MB-SE 893) with a diaphragm (*da*) between the two spores and longitudinal fibrillar elements in the sporangial envelope. Negative stained; transmission electron micrograph (× 33,000).

Strains of *Planobispora* are mesophilic. Good growth occurs between 28° and 40°C with no growth at 20° or 45°C (Thiemann, 1974b; Goodfellow and Pirouz, 1982b). Cultures grow well between pH 6.0 and 9.0. As an exception, *P. longispora* grows at pH 5.0 on oatmeal agar (Thiemann and Beretta, 1968). Strains grow in the presence of brilliant green (up to 0.01% w/v), crystal violet (up to 0.001% w/v), and pyronine (0.01% w/v) but they do not tolerate lysozyme (0.005% w/v) or NaCl (3.0% w/v) (Goodfellow and Pirouz, 1982b). Nitrate is reduced to nitrite and phosphatase is produced. The type strains of *P. longispora* and *P. rosea* are able to degrade arbutin, casein, elastin, DNA, gelatin, keratin, RNA, starch, tyrosine, and Tween 20, 40, 60, and 80. The following compounds are not degraded: adenine, cellulose, guanine, hippurate, testosterone, xanthine, and xylan (Goodfellow and Pirouz, 1982b) (see Table 30.11, below).

P. longispora and *P. rosea* are sensitive to the antibiotics kanamycin, neomycin, novobiocin, tobramycin, vancomycin, and penicillin. They are not sensitive to cephalosporin, lincomycin, rifampicin, and streptomycin. *P. longispora* is sensitive to dimethylchlortetracycline, and *P. rosea* is sensitive to gentamicin (Goodfellow and Pirouz, 1982b).

The genus *Planobispora* is a very rare microorganism, known from only two localities in the world. Several strains, including the type strains, were isolated from two soil samples collected from a river bank in Venezuela. The pH values of the two samples were 5.3 and 7.6 (Thiemann, 1970b). Further strains were isolated from a soil sample collected near Windhoek, Namibia (D. Schäfer, personal communication).

Enrichment and Isolation Procedure

Unfortunately, Thiemann and coworkers never published the isolation procedure they used. Schäfer (personal communication) isolated strains by the baiting technique with pollen and hair as described by Bland and Couch (1981). He observed that the two-spored sporangia of *Planobispora* developed on long aerial hyphae growing between the baits. Single sporangia were picked up with a thin needle and placed on a suitable agar medium in small Petri plates. Minute colonies could be transferred after 2 weeks to slant cultures.

Maintenance Procedures

For some weeks or even months cultures can be stored at room temperature in slant culture tubes on agar media that support good growth of substrate and aerial mycelium. For long-term preservation, the strains must be lyophilized by the procedures described for aerobic actinomycetes.

Differentiation of the genus **Planobispora** *from other genera*

The genus *Planobispora* may be confused with members of genera that can produce pairs of spores on aerial and/or substrate mycelia. The possession of a cell wall chemotype III distinguishes *Planobispora* from *Dactylosporangium* (chemotype II) and from *Elytrosporangium, Kitasatoa,* and *Microellobosporia* (chemotype I) (Cross and Goodfellow, 1973; Kutzner, 1981). The genus *Microbispora* also has a chemotype III and sporulating aerial mycelium, but the paired spores of *Microbispora* are neither enclosed in a sporangial envelope nor elongated and motile as are those of *Planobispora.*

Taxonomic Comments

Based on the chemical composition of the cell wall (presence of *meso*-DAP and madurose; chemotype III), the genus *Planobispora* is affiliated with the sporangiate genera *Planomonospora, Spirillospora,* and *Streptosporangium* and with the nonsporangiate genera *Actinomadura, Microbispora,* and *Microtetraspora* (Goodfellow and Williams, 1983). All members of this "aggregated group" are also characterized by 10-methyl branched and *iso*- branched fatty acids (Kroppenstedt, 1979). The close relationship between the genera *Planobispora, Planomonospora,* and *Streptosporangium* is indicated by DNA homologies and ribosomal RNA similarities (Farina and Bradley, 1970; Stackebrandt et al., 1981). Studies of numerical classification show that strains of *Planobispora* form a distinct and homogeneous cluster (Goodfellow and Pirouz, 1982b).

Acknowledgment

I thank Dr. D. Schäfer, who made his isolation technique of sporangiate actinomycetes available to me.

Further Reading

Thiemann, J.E. 1970. Study of some new genera and species of the *Actinoplanaceae. In* Prauser (Editor), The Actinomycetales. VEB Gustav Fischer Verlag, Jena, pp. 245–257.

Thiemann, J.E. 1974. Genus *Planobispora* Thiemann and Beretta. *In* Buchanan and Gibbons (Editors), Bergey's Manual of Determinative Bacteriology 8th ed. Williams & Wilkins, Baltimore, pp. 720–721.

Thiemann, J.E. and G. Beretta. 1968. A new genus of the *Actinoplanaceae: Planobispora,* gen. nov. Arch. Mikrobiol. *62:* 157–166.

Differentiation and characteristics of the species of the genus **Planobispora**

The two species presently known do not differ decisively in their morphology. Tests for utilization of 28 carbon sources and degradation of 21 compounds all show only 10% deviation between the two species (Table 30.11). Besides other distinguishing cultural characteristics, the substrate mycelium of *P. longispora* is hyaline to creamish, and the aerial mycelium is white, whereas *P. rosea* has a rose-colored substrate mycelium and an aerial mycelium with a light rose tinge.

List of species of the genus **Planobispora**

1. **Planobispora longispora** Thiemann and Beretta 1968, 157.[AL]

lon.gi.spo′ra. L. adj. *longus* long; Gr. n. *spora* a seed, spore; M.L. adj. *longispora* long spored.

Sporangial development is supported by soil, calcium malate, and oatmeal agars. Spores are straight to slightly curved with rounded ends, measuring 1.0–1.2 × 2.6–4.0 μm. They are motile by peritrichous flagella.

No specific color occurs either in the aerial or in the substrate mycelium. The aerial mycelium is white and the substrate mycelium hyaline to creamy colored. No soluble pigments are produced.

Good growth with abundant aerial mycelium occurs on oatmeal agar. On yeast extract-malt extract, Hickey-Tresner, Bennett, and peptone-beef extract agars, colonies also grow well but aerial mycelium is not developed.

Amygdalin is used for growth; galactose, melezitose, and salicin are not. Hypoxanthine is degraded; esculin and chitin are not degraded. Melanoid pigments are not produced.

Litmus milk is coagulated and peptonized.

The mol% G + C of the DNA is 71 (T_m).

Type strain: ATCC 23867.

2. **Planobispora rosea** Thiemann 1970b, 251.[AL]

ro′se.a. L. fem. adj. *rosea* rose colored.

Sporangial development is promoted by all media on which aerial mycelium is formed, e.g. on soil and Hickey-Tresner agars. The spores are elongated and fusiform, with rounded ends, 1.0–1.2 μm wide and 3.0–3.5 μm long. They are motile by peritrichous flagella.

Substrate mycelium in most media is rose colored; if aerial mycelium is developed, it always has a light rose tinge.

On Bennett, peptone–beef extract, and potato plug agars the colonies grow well, are slightly wrinkled or flat and rose colored; no aerial mycelium is developed. Good growth occurs also on Hickey-Tresner agar and the colonies are slightly wrinkled and yellow-amber colored with abundant aerial mycelium. Colonies grow well on oatmeal agar; they are smooth and rose colored and aerial mycelium is moderately developed showing a rose tinge. On glycerol-asparagine agar, growth is moderate and the colonies are smooth, flat, and hyaline; abundant aerial mycelium, white with a rose tinge, is formed.

Galactose, melezitose, and salicin are used for growth; amygdalin is not. Esculin and chitin are degraded, hypoxanthine is not hydrolyzed. Litmus milk is neither coagulated nor peptonized.

The mol% G + C of the DNA is 70 (T_m).

Type strain: ATCC 23866.

Table 30.11.
Characteristics of the species of the genus **Planobispora**[a]

Characteristics	1. *P. longispora*	2. *P. rosea*
Utilization of carbohydrates		
Adonitol	−	−
Amygdalin	+	−
L-Arabinose	+	+
D-Arabinose	−	−
Cellobiose	+	+
Erythritol	−	−
Ethanol	−	−
Fructose	+	+
Galactose	−	+
Glucose	+	+
Glycerol	−	−
Glycogen	+	+
Inositol	+	+
Inulin	−	−
Lactose	−	−
Maltose	+	+
Mannitol	+	−[b], +[c]
Mannose	−	−
Melezitose	−	+
α-Methyl-D-glucoside	−	−
Raffinose	−	−
Rhamnose	+	−
Salicin	−	+
Sorbitol	−	−
Starch	+	+
Sucrose	−	−
Trehalose	−	−
Xylose	+	+

Characteristics	1. *P. longispora*	2. *P. rosea*
Degradation of compounds		
Adenine	−	−
Arbutin	+	+
Calcium malate	−	−
Casein	+	+
Cellulose	−	−
Chitin	−	+
Elastin	+	+
Esculin	−	+
DNA	+	+
Gelatin	+	+
Guanine	−	−
Hippurate	−	−
Hypoxanthine	+	−
Keratin	+	+
RNA	+	+
Starch	+	+
Testosterone	−	−
Tween 20, 40, 60, 80	+	+
Tyrosine	−[d], +[c]	+
Xanthine	−	−
Xylan	−	−
Production of		
Diffusible pigments	−	−
Melanoid pigments	−	ND
H_2S	−	ND
Phosphatase	+	+
Reduction of nitrate	+	+
Coagulation of litmus milk	+	−
Peptonization of litmus milk	+	−

[a]Symbols: see Table 30.1.
[b]Data from Thiemann (1970b).
[c]Data from Goodfellow and Pirouz (1982b).
[d]Data from Thiemann and Beretta (1968).

Genus **Planomonospora** *Thiemann, Pagani and Beretta 1967b, 29*[AL]

GERNOT VOBIS

Pla.no.mo.no′spo.ra. poetic late Gr. n. *planos* wanderer, vagabond; Gr. adj. *monos* alone, single; Gr. n. *spora* a seed, spore; M.L. fem. n. *Planomonospora* a motile, single-spored organism.

Substrate and aerial mycelia develop on various agar media. Substrate hyphae (0.6–1.0 μm in diameter) are irregularly branched, occasionally septate, and nonfragmenting. Aerial hyphae (0.5–1.0 μm in diameter) are sparsely branched and rarely septate. Organisms are Gram-positive and not acid fast. **Cylindrical to clavate sporangia** (1.0–1.5 μm wide × 3.5–5.5 μm long), each **containing a single spore,** are formed only **on the aerial mycelium.** The **spores** (zoospores) are fusiform and **motile by peritrichous flagella.** Colonies grown on complex agar media are raised or flat with rugose or smooth surface. **The color of substrate mycelium is either rose to light orange or brown-violet to light brown. The aerial mycelium is white with a rose tinge or grayish white.** The peptidoglycan of the cell walls contains ***meso*-diaminopimelic acid (*meso*-DAP),** and **madurose** is the characteristic sugar of whole-cell hydrolysates. Aerobic, chemoorganotrophic, and mesophilic, growing well between 28° and 37°C and at a pH from 7.0 to 8.0. The mol% G + C of the DNA is 72 (T_m).

Type species: *Planomonospora parontospora* Thiemann, Pagani and Beretta 1967b, 29.

Further Descriptive Information

Members of the genus *Planomonospora* develop substrate and aerial mycelia on solid media. The hyphae of the substrate mycelium are 0.6–1.0 μm in diameter. They are irregularly branched, occasionally septate, and do not fragment on agar media or in liquid-submerged cultures (Thiemann, 1970b). Twisting and swelling of the hyphae can be observed in *P. parontospora* (Thiemann et al., 1967b, 1968a). The substrate hyphae grow profusely within the agar medium and form a compact layer on its surface (Thiemann, 1974b). The diameter of the aerial hyphae is 0.5–0.6 μm in *P. venezuelensis* and about 1.0 μm in *P. parontospora* (Thiemann, 1974b). The aerial hyphae are sparsely branched and in *P. venezuelensis* usually long, wavy, and slender (Thiemann, 1970b).

Sporangia are formed only on aerial mycelium, each containing a single spore. Two morphological groups of sporangial arrangements can be recognized. In *P. parontospora,* the sporangia are sessile and occur in double parallel rows on a typically curved sporangiophore (Fig. 30.22). A single sporangiophore can bear up to 60 sporangia. Mature sporangia are

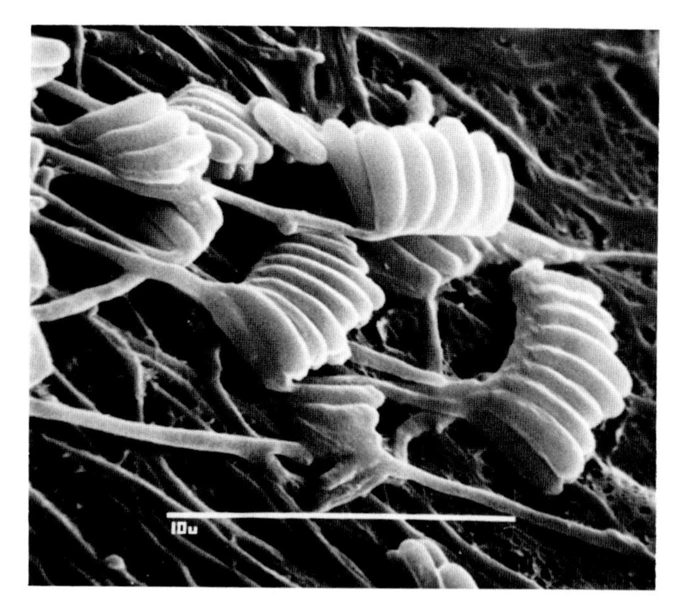

Figure 30.22. *Planomonospora parontospora* (strain ATCC 23863); numerous monosporous sporangia in double parallel rows, sitting directly on bent aerial hyphae. Colonies were cultivated for 10 days on soil agar. Scanning electron micrograph (× 3800).

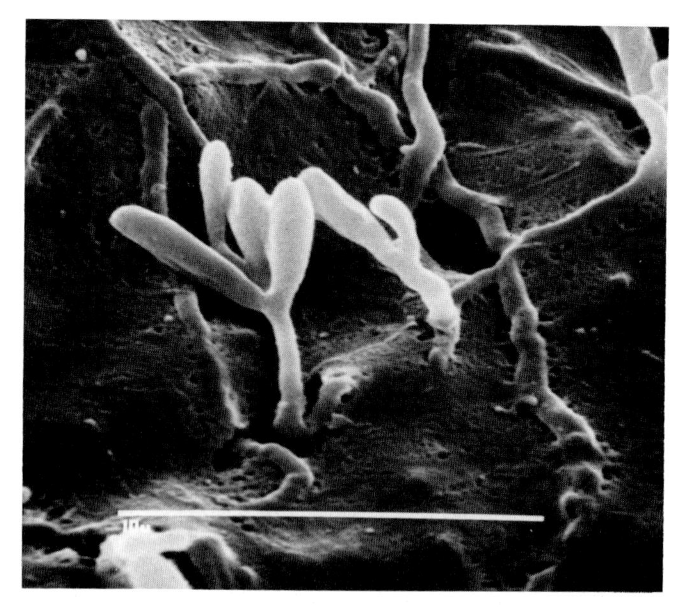

Figure 30.23. *Planomonospora venezuelensis* (strain ATCC 23865); monosporous sporangia on aerial hyphae in young stages of formation of the palm leaf pattern. Cultivation as in Figure 30.22. Scanning electron micrograph (× 6400).

cylindrical and usually 1.5 μm wide and 3.5–4.5 μm long (Thiemann et al., 1967b). In *P. venezuelensis,* the sporangia are developed singly or in groups on short lateral branches, approximately 0.5 μm long, that originate from a main trailing hypha (Williams, 1970). They form a characteristic palm leaf pattern (Fig. 30.23). The sporangia are claviform, elongated, 1.0 μm in diameter, and 4.5–5.5 μm in length.

The spores of *P. parontospora* are fusiform and slightly curved. They are 1.0–1.5 μm in diameter and 3.5–4.5 μm in length. Flagellation is described as lophotrichous (Lechevalier and Lechevalier, 1970a) or peritrichous (Thiemann, 1974b). Large clavate spores, 10–15 μm long, moving slowly, may be observed occasionally (Thiemann et al., 1967b). *Planomonospora venezuelensis* also has fusiform spores, slightly thickened at the terminal end. They are 1.0 μm wide and 3.0–3.5 μm long, filling the sporangium almost completely, and are motile by peritrichous flagella (Thiemann, 1970b, 1974b).

The fine structure of spore formation was investigated in *P. venezuelensis* by Sharples et al. (1974). The authors observed that the young sporogenous hyphae had a double-layered wall. The spore wall was formed by the inner layer and the sporangial wall originated from the outer layer. The sporangium was delimited by a cross wall at the base. A sheath composed of fibrillar elements could be shown on the surface of the sporangium. Based on spore formation within a parental hypha, this type of sporogenesis has been called "endogenous" (Williams et al., 1973; Sharples et al., 1974). Vobis and Kothe (1985) observed that in *P. parontospora* the sporangial development begins with the growth of a sporogenous hypha inside a thin expanding sheath. After reaching the full length, the initial growth is separated from the sporangiophore by a double-layered cross wall. Through thickening, the sheath becomes a massive sporangial envelope. A "diaphragm" between sporangiophore and the single spore is additionally formed (Fig. 30.24). Scanning electron microscopical studies demonstrated that the tips of the sporangia may be rostellate. In older sporangia, pore-shaped openings have been observed (Williams, 1970; Locci and Petrolini Baldan, 1971).

The process of spore release begins immediately after the sporangia are placed in water. The sporangia change their optical characteristics and become highly opaque (Thiemann, 1970b). Probably due to swelling of material located at the base of the sporangium, the spore is pushed upward through the tip. The spores, by means of peritrichous flagella, be-

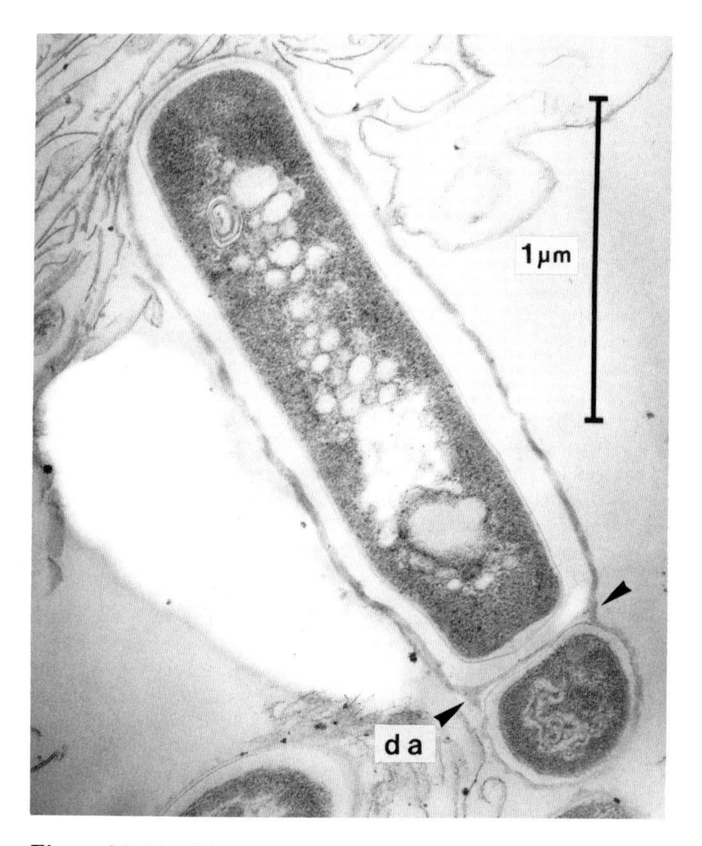

Figure 30.24. *Planomonospora parontospora* (strain ATCC 23863); ultrathin section through a young sporangium; the massive sporangial envelope encloses the single spore and the sporangiophore, divided by a diaphragm (*da*). Cultures were incubated for 8 days on soil agar. Transmission electron micrograph (× 42,000).

come motile 30–40 min after they have been expelled (Thiemann et al., 1967b). They remain motile for 5–24 h, during which time spore germination may begin (Thiemann, 1970b).

The peptidoglycan of the cell walls contains *meso*-DAP (Lechevalier and Lechevalier, 1970a). The characteristic sugar of whole-cell hydrolysates is madurose (3-*O*-methyl-D-galactose) (Kroppenstedt and Kutzner, 1978). *Planomonospora* has cell wall chemotype III and sugar pattern B (Lechevalier and Lechevalier, 1970b).

The two species of *Planomonospora* have different menaquinone profiles. Di- and tetrahydrogenated menaquinones with nine isoprene units (MK-9(H_2) and MK-9(H_4)) are the major compounds in *P. parontospora,* whereas tetrahydrogenated menoquinones with eight isoprene units (MK-8(H_4)) predominate in *P. venezuelensis* (Collins et al., 1984).

The phopholipids of *P. parontospora* consist of phosphatidylinositol, phosphatidylethanolamine, unknown glucosamine-containing phospholipids, and diphosphatidylglycerol (DPG) plus lyso-DPG (Hasegawa et al., 1979). This composition corresponds to phospholipid type IV of Lechevalier et al. (1981).

In the fatty acid spectrum of the type strains of *P. parontospora* and *P. venezuelensis,* straight and *iso-* branched acids are present but there are no *anteiso-* branched fatty acids. Unsaturated fatty acids and 10-methyl branched fatty acids are also present (Kroppenstedt and Kutzner, 1978; Kroppenstedt, 1979).

Colonies on agar media are flat or elevated with smooth surfaces, occasionally wrinkled or slightly crustose. Abundant aerial mycelium is observed on only a few media, such as oatmeal agar for *P. parontospora* (Thiemann et al., 1967b). Generally the aerial mycelium is developed only moderately or in traces. The color of the aerial mycelium is white with a rose tinge in *P. parontospora* and white to grayish white in *P. venezuelensis* (Thiemann, 1974b). The substrate mycelium of *P. parontospora* is light rose to rose color on oatmeal, starch, tyrosine, and skim milk agars. On Bennett, potato plug, and nutrient agars the color of the substrate mycelium is yellowish or creamish to light orange; on Hickey-Tresner and Czapek glucose agars it is hyaline (Thiemann et al., 1967b, 1968a). The substrate mycelium of *P. venezuelensis* has a violet color on oatmeal and skim milk agars, and a brown-violet color on Bennett, Hickey-Tresner, glucose-asparagine, starch, and yeast extract–malt extract agars. On nutrient agar, the color of substrate mycelium is light brown, on potato plug agar it is gray, and it is hyaline on tyrosine, glycerol-asparagine, and calcium malate agars (Thiemann, 1970b).

Species of *Planomonospora* are aerobic, mesophilic organisms, growing well on various complex agar media (Thiemann, 1974b). Their type strains can utilize a number of compounds as sole carbon sources (See Table 30.12, below). Melezitose is used by *P. parontospora,* but not by *P. venezuelensis* (Goodfellow and Pirouz, 1982b). As sole carbon and nitrogen sources, acetamide is not utilized and serine is utilized by only some strains (Goodfellow and Pirouz, 1982b).

Planomonospora strains can grow between 20° and 40°C, with temperature optima from 28° to 37°C. No growth takes place at 10° or 45°C. Growth of *P. parontospora* is consistently good between 22° and 37°C (Thiemann et al., 1967b; Thiemann, 1974; Goodfellow and Pirouz, 1982b). pH values from 6.0 to 9.0 are tolerated by all species, with no growth at pH 5.0. *Planomonospora parontospora* grows only sparsely at pH 6.0. The optimum is between pH 7.0 and 8.0 (Thiemann et al., 1967b; Thiemann, 1974b; Goodfellow and Pirouz, 1982b). Strains of *Planomonospora* can grow in the presence of NaCl (3% w/v); no growth occurs at 5% (w/v). They tolerate also the presence of pyronine (0.01% w/v). The strains show different sensitivities to brilliant green and crystal violet. The presence of lysozyme (0.005% w/v) is tolerated only partially (Goodfellow and Pirouz, 1982b). For further physiological features and degradative abilities see Table 30.12 (below).

The production of pigments is restricted in *P. venezuelensis* to traces of a brown-violet soluble pigment on oatmeal agar and amber to brown on Hickey-Tresner and glucose-asparagine agars (Thiemann, 1970b). A very faint yellow pigment is formed by *P. parontospora* on oatmeal agar (Thiemann et al., 1967b). On tyrosine agar, *P. parontospora* subsp. *antibiotica* produces a light brown diffusible pigment (Thiemann et al., 1968a). Under submerged conditions an antibacterial agent called sporangiomycin is produced by *P. parontospora* subsp. *antibiotica* (Thiemann et al., 1968a). A protease inhibitor has been isolated from a strain of *P. parontospora* (Wingender et al., 1975).

The type strains of *P. parontospora* and *P. venezuelensis* are sensitive to the antibiotics kanamycin, neomycin, and tobramycin. They are not sensitive to cephaloridine, gentamicin, lincomycin, novobiocin, rifampicin, streptomycin, vancomycin, or penicillin. *Planomonospora venezuelensis* is sensitive to demethylchlortetracycline, but *P. parontospora* is not (Goodfellow and Pirouz, 1982b).

Members of *Planomonospora* have a worldwide distribution in the soil of temperate, arid, and tropical regions. Thiemann (1970b) isolated 42 strains that occurred in 10 of the 454 soil samples tested. These samples originated from Argentina, Chile, India, Italy, Peru, and Venezuela (Thiemann, 1974b) and their pH values ranged from 5.3 to 7.8 (Thiemann, 1970b). A further 35 strains were isolated by D. Schäfer (personal communication) from soil samples collected in Ceylon, Egypt, France, Greece, Italy, Mexico, Namibia, Tunisia, Turkey, and the United States (Arizona, Florida and Texas).

Enrichment and Isolation Procedures

Using the baiting technique described by Bland and Couch (1981), it is possible to enrich for *Planomonospora.* Sporulating aerial hyphae develop on pollen grains as natural substrates (D. Schäfer, personal communication). Single sporangia or bundles of sporangia can be picked up with a thin needle and placed onto the surface of agar medium in small Petri plates. After 2–4 weeks, the young colonies can be transferred into slant culture tubes.

Maintenance Procedures

Cultures on agar slants can be stored at room temperature for several weeks. For long-term preservation, the strains must be lyophilized by the procedures recommended for aerobic actinomycetes.

Differentiation of the genus **Planomonospora** *from other genera and taxonomic comments*

The genus *Planomonospora* might be confused morphologically with *Kineosporia,* a genus that also has monosporous sporangia and motile spores. Since *Kineosporia* has wall chemotype I (Pagani and Parenti, 1978), it belongs to the streptomycetes (Goodfellow and Cross, 1984b). Based upon its wall chemotype III, *Planomonospora* is related to the sporangiate genera *Planobispora, Spirillospora,* and *Streptosporangium* and to the nonsporangiate genera *Actinomadura, Microbispora,* and *Microtetraspora.* All these genera are complied into the aggregated group maduromycetes (Goodfellow and Williams, 1983). Studies on DNA reassociation showed the close relationship of *Planomonospora, Planobispora, Streptosporangium,* and *Spirillospora* (Farina and Bradley, 1970). They are additionally related by production of 10-methyl branched fatty acids (Kroppenstedt and Kutzner, 1978). Based upon DNA/DNA homologies and ribosomal RNA cistron similarities, *P. parontospora* and *Planobispora longispora* have been shown to be closely related to *Streptosporangium roseum* (Stackebrandt et al., 1981). It was suggested that *Planomonospora, Planobispora,* and *Streptosporangium* form a single taxon of genus status (Stackebrandt and Schleifer, 1984).

Acknowledgments

I thank Dr. D. Schäfer, who made his isolation technique and information pertaining to the origin of his *Planomonospora* strains available to me. I am also indebted to Miss G. Traxler, for reviewing the English texts of our descriptions of the genera *Ampullariella, Dactylosporangium, Pilimelia, Planobispora, Planomonospora,* and *Spirillospora* in this volume.

Further Reading

Sharples, G.P., S.T. Williams and R.M. Bradshaw. 1974. Spore formation in the *Actinoplanaceae (Actinomycetales)*. Arch. Microbiol. *101:* 9-20.

Thiemann, J.E. 1970. Study of some new genera and species of the *Actinoplanaceae*. *In* Prauser (Editor), The Actinomycetales. VEB Gustav Fischer Verlag, Jena, pp. 245-257.

Thiemann, J.E., H. Pagani and G. Beretta. 1967. A new genus of the *Actinoplanaceae: Planomonospora,* gen. nov. G. Microbiol. *15:* 27-38.

Vobis, G. and H.W. Kothe. 1985. Sporogenesis in sporangiate actinomycetes. *In* Mukerji, Pathak and Singh (Editors), Frontiers in Applied Microbiology, Vol. I. Print House (India), Lucknow, pp. 25-47.

Differentiation and characteristics of the species of the genus **Planomonospora**

The two species, *P. parontospora* and *P. venezuelensis,* form a tight cluster in the numerical phenetic study of Goodfellow and Pirouz (1982b). They can be distinguished by the morphological arrangement of the sporangia (Figs. 30.22 and 30.23), different menaquinone compositions (Collins et al., 1984), and the characteristic color of the mycelium (Thiemann, 1974b). For further differences see the genus description and Table 30.12.

List of species of the genus **Planomonospora**

1. **Planomonospora parontospora** Thiemann, Pagani and Beretta 1967b, 29.[AL]

pa.ron.to'spo.ra. Gr. v. *pareimi* to be side by side; Gr. n. *spora* a seed, spore; M.L. *parontospora* spores side by side.

The hyphae of the substrate mycelium are 0.6–0.8 μm in diameter, occasionally septate, branched, twisted, frequently with swellings. They grow profusely into the medium. The color of substrate mycelium is rose to light orange.

Table 30.12.
Characteristics of the species of the genus **Planomonospora**[a]

Characteristics	1. *P. parontospora* subsp. *parontospora*	1. *P. parontospora* subsp. *antibiotica*	2. *P. venezuelensis*
Utilization of			
Adonitol	+	ND	+
Amygdalin	−	ND	−
L-Arabinose	+	+	+
D-Arabinose	−	ND	−
Cellobiose	+	ND	+
Dextrin	+	ND	ND
Dulcitol	−	ND	ND
Erythritol	−	ND	−
Ethanol	−	ND	−
Fructose	+	+	+[b], D[c]
Galactose	+	ND	+
Glucose	+	+	+
Glycerol	−	ND	−
Glycogen	+	ND	+
Inositol	−	−	−[b], D[c]
Inulin	+[d], −[b]	ND	−
Lactose	+	ND	+
Maltose	−[d], +[b]	ND	+
Mannitol	+	+	+
Mannose	+[d], −[b]	ND	−
Melezitose	+	ND	−
α-Methyl-D-glucoside	−	ND	−
Raffinose	−	−	−[b], D[c]
Rhamnose	+	+	+
Ribose	−	ND	ND
Salicin	−	ND	−
Sorbitol	−	ND	−
Sorbose	−	ND	ND
Starch	+	+	+
Sucrose	−[d], +[b]	+	+
Trehalose	+	ND	+
Xylose	+	+	+

Characteristics	1. *P. parontospora* subsp. *parontospora*	1. *P. parontospora* subsp. *antibiotica*	2. *P. venezuelensis*
Degradation tests			
Adenine	−	ND	−
Arbutin	+	ND	+
Calcium malate	−	−	−
Casein	+	+	−[c], +[b]
Cellulose	−	v[e]	−
Chitin	+	ND	+
Elastin	+	ND	+
Esculin	+	ND	+
DNA	+	ND	+
Gelatin	−[d], +[b]	+	+
Guanine	−	ND	−
Hippurate	−	ND	−
Hypoxanthine	+	ND	+
Keratin	+	ND	+
RNA	+	ND	+
Starch	+	+	+
Testosterone	−	ND	−
Tween (20-80)	+	ND	+
Tyrosine	−[d], +[b]	+	+
Xanthine	−	ND	−
Xylan	−	ND	−
Physiological properties			
Nitrate reduced	+[d], −[b]	+	+[c], −[b]
Phosphatase produced	+	ND	+
H_2S produced	ND	+	+
Melanin produced	−	+	ND
Litmus milk			
Coagulated	−	−	−
Peptonized	+	−	−
Gelatin liquified	−	+	+

[a]Symbols: see Table 30.1; *v*, strain instability; *D*, 33-66% of the strains are positive.
[b]Data from Goodfellow and Pirouz (1982b).
[c]Data from Thiemann (1970b).
[d]Data from Thiemann et al. (1967b).
[e]Data from Thiemann et al. (1968a).

The hyphae of the aerial mycelium are about 1.0 μm in diameter, sparsely branched, growing away from the agar surface. The aerial mycelium is white, always with a light rose tinge. Sporangia are formed on aerial mycelium only, which is developed abundantly on oatmeal, Hickey-Tresner, and soil agars. The monosporous sporangia are arranged in double parallel rows, attached directly to a characteristically bent sporangiophore. Mature sporangia are 1.5 μm wide and 3.5–4.5 μm long. Sporangiospores are motile, fusiform, and slightly curved, measuring 1.0–1.5 × 3.5–4.5 μm.

On oatmeal, Bennett, Hickey-Tresner, and glucose-asparagine agars growth is very good. The surface of the colonies is smooth, with abundant to moderate development of aerial mycelium. Moderate growth occurs on Czapek-glucose, nutrient, and soil agars, with poor growth on glycerol-asparagine, starch, skim milk, and tyrosine agars; no growth occurs on cellulose, peptone-iron and calcium malate agars.

Melanoid pigments are not produced; gelatin is not liquified; litmus milk is peptonized. Growth occurs between 22° and 37°C; pH optimum 7.0–8.0.

The mol% G + C of the DNA is 72 (T_m).

Type strain: ATCC 23863.

1a. **Planomonospora parontospora** subsp. **parontospora** Thiemann, Pagani and Beretta 1967b, 29.[AL]

Description is as for the species.

Type strain: ATCC 23863.

1b. **Planomonospora parontospora** subsp. **antibiotica** Thiemann, Coronelli, Pagani, Beretta, Tamoni and Arioli 1968a, 528.[AL]

an.ti.bi.o′ti.ca. Gr. pref. *anti-* against; Gr. n. *bios* life; M.L. adj. *antibiotica* producing antibiotic.

Description as for the species. Good growth on inorganic salts–starch and nutrient agars. Gelatin is liquified, litmus milk is not peptonized, and tyrosine is degraded with production of a light brown diffusible melanoid pigment. H_2S is produced.

Produces the antibiotic sporangiomycin.

Type strain: ATCC 23864.

2. **Planomonospora venezuelensis** Thiemann 1970b, 247.[AL]

ve.ne.zuel.en′sis. M.L. *venezuelensis* pertaining to Venezuela.

Hyphae of the substrate mycelium are 1.0 μm in diameter, irregular, and frequently branched, and septate. The substrate mycelium is violet-brown on most agar media.

Aerial hyphae are sparsely branched, long, wavy, and slender, 0.5–0.6 μm in diameter. The color of aerial mycelium is white to grayish white. Monosporous sporangia are developed along aerial hyphae on short lateral ramifications, singly or in bundles, producing a characteristic palm leaf pattern. Sporangia are cylindrical to clavate, 1.0 μm wide and 4.5–5.5 μm long. They are produced abundantly on Hickey-Tresner agar. Spores are motile and fusiform, measuring 1.0 μm in diameter and 3.0–3.5 μm in length.

Good growth occurs on several complex agar media. On Bennett agar, the colonies are raised with a wrinkled surface. Rugose surfaces occur also on yeast extract–malt extract and Hickey-Tresner agars. Flat colonies with a smooth surface are developed on nutrient and potato plug agars. Growth of colonies is moderate on oatmeal, tyrosine, skim milk, and glycerol-asparagine agars. Slight growth occurs on glucose-asparagine, starch, calcium malate, and peptone–yeast extract–iron agars; no growth occurs on Czapek-glucose agar. Aerial mycelium is developed on most of the above-mentioned media, but only moderately or in traces.

Traces of brown-violet soluble pigment are produced on oatmeal agar, and amber to amber-brown on Bennett and Hickey-Tresner agars. Gelatin is liquified and tyrosine is degraded. Casein is not hydrolyzed. Litmus milk is neither coagulated nor peptonized. The temperature optimum is between 28° and 37°C.

Type strain: ATCC 23865.

Genus Spirillospora *Couch 1963, 61*[AL]

GERNOT VOBIS AND HANS-W. KOTHE

Spi.ril.lo.spo′ra. Gr. n. *speira* coil, Gr. n. *spora* a seed, spore; M.L. fem. n. *Spirillospora* an organism with spores in spirals.

Spherical to vermiform sporangia (5.0–24.0 μm in diameter) are produced **on the aerial mycelium.** The sporangial envelope encloses numerous spores that are arranged in **coiled and branched spore chains.** The **spores are rod shaped or curved** (0.5–0.7 × 2.0–6.0 μm) and **motile** by means of one to seven subpolarly inserted flagella. The hyphae of the substrate and aerial mycelium are 0.2–1.0 μm thick, branched, and septate. The **color of substrate mycelium is white to pale yellow or pale buffy pink to red; the aerial mycelium is usually white.** Gram-positive, aerobic, chemo-organotrophic, and mesophilic; the original temperature of growth is 25°C, the range is 18–35°C. The peptidoglycan of the cell wall contains ***meso*-diaminopimelic acid** (*meso*-DAP), and **madurose** is the characteristic sugar of whole-cell hydrolysates. The mol% G + C of DNA is 71.0–73.0 (T_m, Bd).

Type species: *Spirillospora albida* Couch 1963, 61.

Further Descriptive Information

The members of the genus *Spirillospora* develop substrate and aerial mycelium. The hyphae are 0.2–1.0 μm in diameter, branched, and septate (Couch and Bland, 1974d).

The sporangia are produced on the aerial mycelium. They are usually spherical with a diameter of 5.0–24.0 μm; the average diameter is 10 μm (Fig. 30.25). Subspherical to elongated or club-shaped and vermiform sporangia are also formed (Couch, 1963). On initiation of the sporangial development, the end of an aerial hypha winds into a coil that is enclosed in a common sheath (Lechevalier et al., 1966b; Locci and Petrolini Baldan, 1971; Bland and Couch, 1981). In some cases it seems that the first coils are temporarily free (Vobis and Kothe, 1985). A cross wall, dividing the sporangiophore from the young sporangium, is not visible. The coiled sporogenous hyphae are branched and fragment into spore-size segments (Lechevalier et al., 1966b).

The spores are short to long rods, frequently curved (0.5–0.7 × 2.0–6.0 μm) (Couch and Bland, 1974d). A subpolarly inserted tuft of one to seven flagella gives them a slight motility that can be more vigorous when an energy source is provided (Higgins et al., 1967).

In addition to the spores enclosed in sporangial envelopes, free, exposed spores in regular or irregular coils may be found among the aerial hyphae. When flooded with water, the coils break up into rod-shaped to curved spores that subsequently become motile. Besides these coiled spore chains, conidia-like structures in moniliform arrangement may be produced by the substrate mycelium (Couch, 1963).

The fine structure of *Spirillospora* corresponds to that described for other Gram-positive bacteria. The cell walls consist of a single compact layer. The septa involved in spore formation are double layered (cross wall type 2 of Williams et al., 1973). Aerial hyphae are additionally covered with a thin sheath, from which the sporangial envelope originates (Lechevalier et al., 1966b; Vobis and Kothe, 1985).

The peptidoglycan of the cell walls contains *meso*-DAP, and madurose is the diagnostic sugar of whole-cell hydrolysates (Yamaguchi, 1965; Lechevalier and Lechevalier, 1970c). This corresponds to cell wall chemotype III and sugar pattern B in the classification scheme of Lechevalier and Lechevalier (1970b). The phospholipid pattern of the type strain of *Spirillospora albida* is type I of Lechevalier et al. (1981).

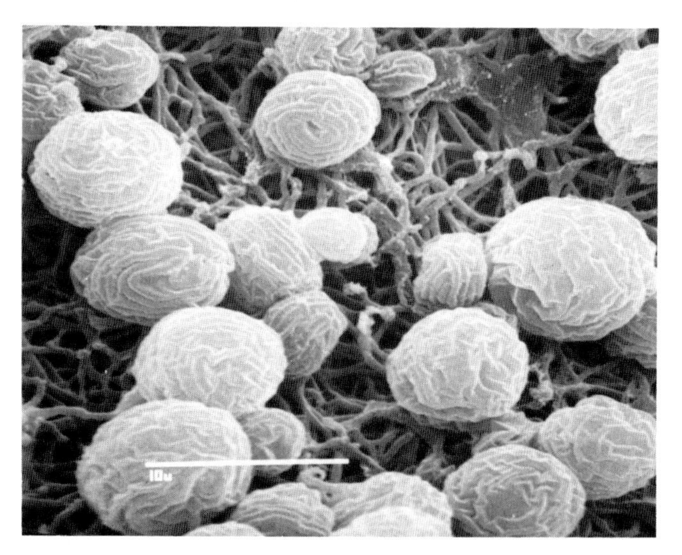

Figure 30.25. *Spirillospora albida,* strain ATCC 15331. Scanning electron micrograph of spherical sporangia formed on the aerial mycelium after 15 days' incubation (× 3200).

Phosphatidylinositol mannosides, phosphatidylinositol, and diphosphatidylglycerol are present and phosphatidylcholine, phosphatidylethanolamine (PE), and the unknown glucosamine-containing phospholipid are absent. The phospolipids of *Spirillospora albida* ATCC 14541 were found to be of the PII type (Hasegawa et al., 1979). This means that PE may be variably present in organisms of this genus, as is phosphatidylglycerol. The fatty acids of the cell membrane are *iso-* and *anteiso-*branched; 10-methyl fatty acids are also present (Kroppenstedt and Kutzner, 1978). The main components of menaquinones are MK-9 (H_4) and MK-9(H_6) (Collins et al., 1984).

Strains of *Spirillospora* can be cultured on various complex media. The colonies grow well on Czapek, peptone-Czapek, and oatmeal agars and moderately on casein or tyrosine agar. They are compact and elevated, sometimes with protuberances on peptone-Czapek, casein, and tyrosine agar. On Czapek and oatmeal agars, the colonies are flattish or confluent. The substrate mycelium is usually white, pale yellow, pale buffy pink, or red to reddish brown; the aerial mycelium is white (Couch, 1963; Schäfer, 1973).

One period of the life cycle is adapted to the aqueous milieu. The zoospores are motile and are able to colonize pollen grains that float on the surface of the water and constitute their natural substrate. Mycelia are developed, forming sporangia when in contact with the air. The sporangia are considered to be resistant stages against desiccation. When flooded with water, the zoospores are released from the sporangia through a rupture in the envelope or through a large, irregular pore (Couch, 1963).

Some strains produce a blue soluble pigment called spirillomycin, which exhibits antibiotic activity against some Gram-positive bacteria (Domnas, 1968; McInnis and Domnas, 1970). The pigment-producing strains utilize glucose, galactose, sucrose, maltose, and amylopectin as carbon sources for growth (Domnas, 1970). *Spirillospora* grows well between 18° and 35°C (Couch and Bland, 1974d).

Members of the genus *Spirillospora* occur in soil but only infrequently. Using the baiting technique, Schäfer (1973) found that only 0.5% of the isolates of sporangiate actinomycetes was represented by this genus.

Enrichment and Isolation Procedures

A method to isolate strains of the genus *Spirillospora* was described by Bland and Couch (1981). A small amount of soil is placed in a sterile Petri plate and flooded with sterile water. Pollen or other natural substances like hair, snakeskin, or boiled grass leaves are added as baits and examined with a binocular dissecting microscope after 1–4 weeks of incubation. Sporangia and aerial mycelium can be recognized by their white, glistening appearance. Pollen baits bearing sporangia are transferred to the surface of a 3% (w/v) agar plate. Individual sporangia are separated from the bait and rolled on the agar surface to free them from contaminating bacteria. The cleaned sporangia are used as an inoculum for pure cultures. Schäfer (1973) also isolated strains of *Spirillospora* with keratinic substances as bait, but these substances were presumably not necessary for growth.

Maintenance Procedures

Subcultures should be made after a period of 12 weeks. For longer preservation, the organisms must be lyophilized by the procedures used for other aerobic actinomycetes, preferably using well-sporulating cultures.

Differentiation of the genus Spirillospora *from other genera*

Within the members of the sporangiate actinomycetes with a cell wall chemotype III of Lechevalier and Lechevalier (1970b), the genus *Spirillospora* may be confused with species of the genus *Streptosporangium.* All strains belonging to these taxa have multispored, usually spherical, sporangia produced by true aerial mycelium. They can be distinguished by various morphological and biochemical features (Table 30.13).

Table 30.13.
Characteristics differentiating the genus **Spirillospora** *from* **Streptosporangium**[a,b]

Characteristics	*Spirillospora*	*Streptosporangium*
Shape of spores		
Rod	+	+
Globose	−	+
Motility of spores	+	−
Phospholipid pattern		
Type II (or type I)	+	−
Type IV	−	+
Menaquinone pattern		
MK-9(H_2)	−	+
MK-9(H_4)	+	+
MK-9(H_6)	+	−

[a]Data from Couch and Bland (1974a), Hasegawa et al. (1979), and Collins et al. (1984).
[b]Symbols: see Table 30.1.

Taxonomic Comments

Goodfellow and Williams (1983) grouped all actinomycetes with cell wall chemotype III and madurose as the characteristic sugar into an "aggregate group" called maduromycetes. This "group" comprises both sporangiate actinomycetes (*Planobispora, Planomonospora, Streptosporangium,* and *Spirillospora*) and nonsporangiate actinomycetes (*Actinomadura, Microbispora,* and *Microtetraspora*). However, studies of DNA/DNA and DNA/rRNA reassociation show that *Spirillospora* has an extremely isolated position within this "group" (Stackebrandt et al., 1981). This is also confirmed by numerical classification methods (Goodfellow and Pirouz, 1982b).

Further Reading

Bland, C.E. and J.N. Couch. 1981. The family *Actinoplanaceae. In* Starr, Stolp, Trüper, Balows and Schlegel (Editors). The Prokaryotes; A Handbook on Habi-

tats, Isolation and Identification of Bacteria, Vol. 2. Springer Verlag, Berlin, pp. 2004–2010.
Couch, J.N. 1963. Some new genera and species of the *Actinoplanaceae.* J. Elisha Mitchell Sci. Soc. *79:* 54–70.
Lechevalier, H.A., M.P. Lechevalier and P.E. Holbert. 1966. Electron microscopical observation of the sporangial structure of strains of *Actinoplanaceae.* J. Bacteriol. *92:* 1228–1235.

Differentiation and characteristics of the species of the genus **Spirillospora**

The two species presently known do not differ decisively in their morphology. They can be distinguished by the characteristic color of substrate mycelium. The colonies of *S. albida* are white to pale yellow; the colonies of *S. rubra* are red to reddish brown.

List of species of the genus **Spirillospora**

1. **Spirillospora albida** Couch 1963, 65.[AL]

al'bi.da. L. fem. adj. *albida* whitish.

Spherical sporangia are produced at the tip of aerial hyphae. They are 5.0–24.0 µm in diameter, on average 10µm. Subspherical to elongated or club-shaped and vermiform sporangia are also formed.

The spores are rod shaped, frequently curved in shape (0.5–0.7 × 2.0–6.0 µm) and weakly motile.

The color of substrate mycelium is white to pale yellow or buffy pink; the aerial mycelium is white.

After 8 weeks, the colonies reach diameters up to 13 mm on Czapek, peptone-Czapek, and oatmeal agars and diameters from 5 to 10 mm on casein and tyrosine agars.

Formation of sporangia occurs on Czapek, peptone-Czapek, and oatmeal agars.

Casein and tyrosine are hydrolyzed. Good growth occurs between 18° and 35°C, the optimal temperature is 25°C.

A pale yellowish soluble pigment is produced on casein agar and a clay-colored pigment on tyrosine agar. Some strains produce a blue soluble pigment on peptone-Czapek agar.

The mol% G + C of the DNA is 71.0–72.0 (T_m) for the type strain (Farina and Bradley, 1970) and 72.9 (Bd) for strain UNCC 761 (Yamaguchi, 1967).

Type strain: ATCC 15331.

2. **Spirillospora rubra** (ex Schäfer 1973) nom. rev. (*Spirillospora rubra* Schäfer 1973, 199.)

ru'bra. L. fem. adj. *rubra* red.

Hyphae of substrate mycelium are branched and 0.4–0.9 µm in diameter. Aerial hyphae are also branched and 0.6–1.2 µm in width. The sporangia are produced at the tip of aerial hyphae. They are spherical with a diameter of 10.0–25.0 µm. The spores are rod shaped, frequently slightly curved (0.8 µm thick and 1.8–2.8 µm long), and weakly motile.

The color of substrate mycelium is red to reddish brown; the aerial mycelium is white.

Good growth occurs on yeast extract-starch and casein agars, and moderate growth on starch, oatmeal-yeast extract, peptone, and tyrosine agars.

Sporangia are formed on artificial soil agar (Henssen and Schäfer, 1971) and on cornmeal-soil agar.*

Tyrosine crystals are not hydrolyzed and melanoid pigments are not produced on tyrosine agar. Casein is hydrolyzed; nitrate is not reduced to nitrite.

Good growth occurs between 20° and 37°C.

Type strain: CBS 571.75

Genus **Streptosporangium** *Couch 1955, 148*[AL]

HIDEO NONOMURA

Strep.to.spo.ran'gi.um. Gr. adj. *streptos* twisted; Gr. n. *spora* a seed; Gr. n. *angeion* a vessel; M.L. neut. n. *Streptosporangium* spore coiled within a sporangium.

Stable branched mycelium, producing **globose sporangia** (usually 10 µm in diameter) **on aerial mycelium. Sporangiospores are formed by septation of a coiled, unbranched hypha within the sporangium;** they are spherical, oval, or rod shaped, 0.2–1.3 × 0.2–3.5 µm (usually 1.2 × 1.5 µm), and nonmotile. **Cell walls contain *N*-acetylated muramic acid, *meso*-diaminopimelic acid (*meso*-DAP) but no characteristic sugars. Whole-cell hydrolysates contain madurose.** Major phospholipids include phosphatidylcholine and unknown glucosamine-containing compounds, but no phosphatidylglycerol. Gram-positive. Aerobic. Mesophilic, a few species thermotolerant. Chemo-organotrophic. Some species require B vitamins for growth. Natural habitat soil. The mol% G + C of the DNA is 69.5–71 (T_m).

Type species: *Streptosporangium roseum* Couch 1955, 151.

Further Descriptive Information

The following morphological groups can be distinguished within the genus *Streptosporangium:*

Group 1. Aerial mycelium is well developed and cottony. Sporangia large, 10–48 µm in diameter. Long sporangiophores (exceeding 50 µm) present. The sporangial walls are thick and strong. The sporangia are not easily disrupted in wet mounts even by pressing the cover glass onto the slide. When intact sporangia of *S. viridogriseum* subsp. *kofuense* are transferred to nutrient broth and incubated for a few days, numerous germ tubes arise through a split in the wall, or from the surface of the spore mass, which retains a constant exterior by means of a spore sheath. The wall frequently disrupts into a few petal-shaped fragments (see Fig. 30.31) (Nonomura and Ohara, 1969b). *S. viridogriseum, S. viridogriseum* subsp. *kofuense,* and *S. albidum* are included in this morphological group (Figs. 30.26–30.32).

Group 2a. Aerial mycelium is rather short in length and much branched. Sporangia of moderate size (6–15 µm in diameter). Sporangiophores short (up to 10 µm in length). The sporangial membranes are thin and readily disrupted in water. *Streptosporangium album, S. amethystogenes, S. longisporum, S. nondiastaticum, S. pseudovulgare, S. roseum, S. violaceochromogenes, S. viridialbum,* and *S. vulgare* belong to this morphological group (Figs. 30.33 and 30.34).

Group 2b. Sporangia of moderate size (6–12 µm in diameter), but the sporangial membrane is very fragile and its presence cannot be detected by light microscopy (Shearer et al., 1983b). Relatively long sporangiophores may be observed. *S. fragile* is in this group.

Group 3. Sporangia are small (1–5 µm in diameter). Club-shaped spore vesicles are also formed, which contain short straight chains of spores. Ridged spores are present. *S. corrugatum* is in this group.

*Difco cornmeal agar, 8.5 g; agar, 7.5 g; dried, sterile garden soil, 50 g; distilled water, 1000 ml.

Figure 30.26. Morphological group 1. *Streptosporangium viridogriseum* subsp. *kofuense* ATCC 27102 on oatmeal agar-YG. Scanning electron micrograph (SEM), gold spattered. *Bar interval,* 10 µm.

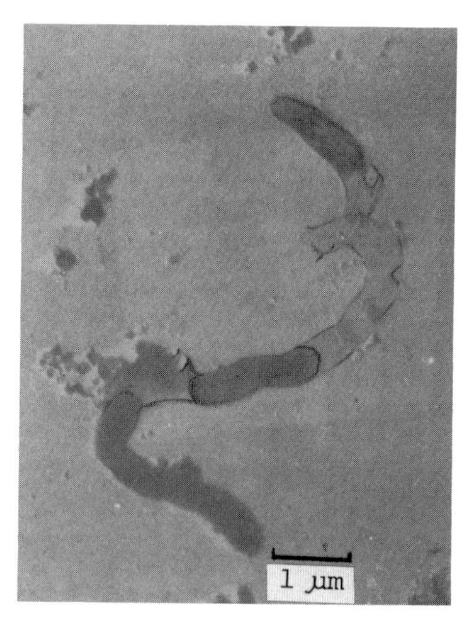

Figure 30.29. *Streptosporangium viridogriseum* subsp. *kofuense:* the sporangium contains a chain of sheathed spores. Electron micrograph. (Reproduced by permission from H. Nonomura and Y. Ohara, Journal of Fermentation Technology *47:* 468, 1969.)

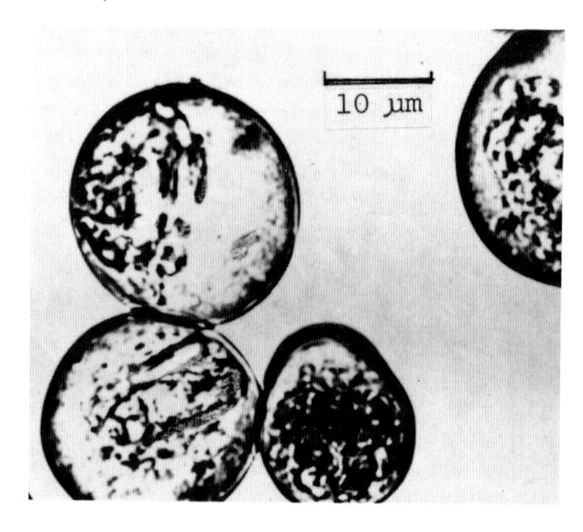

Figure 30.27. *Streptosporangium viridogriseum* subsp. *kofuense:* sporangial walls are thick. Light microscope (LM), stained with methylene blue. (Reproduced by permission from H. Nonomura and Y. Ohara, Journal of Fermentation Technology *47:* 648, 1969.)

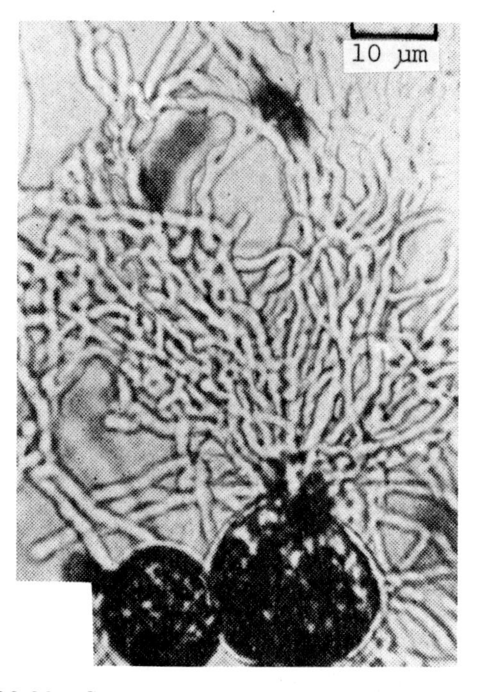

Figure 30.30. *Streptosporangium viridogriseum* subsp. *kofuense:* germination tubes from sporangia on agar medium. LM. (Reproduced by permission from H. Nonomura and Y. Ohara, Journal of Fermentation Technology *47:* 468, 1969.)

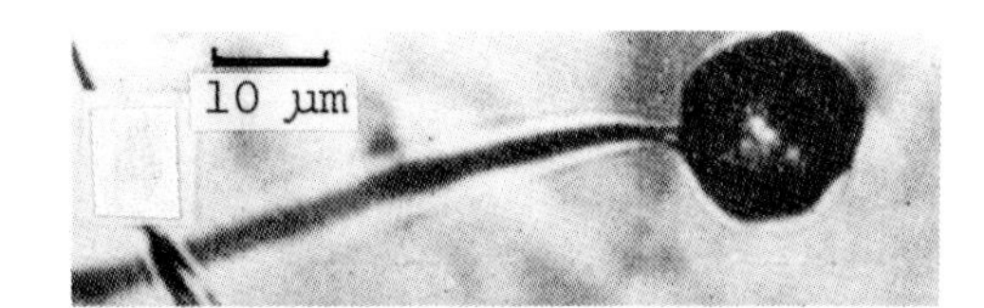

Figure 30.28. *Streptosporangium viridogriseum* subsp. *kofuense:* sporangiospores are long. LM. (Reproduced by permission from H. Nonomura and Y. Ohara, Journal of Fermentation Technology *47:* 468, 1969.)

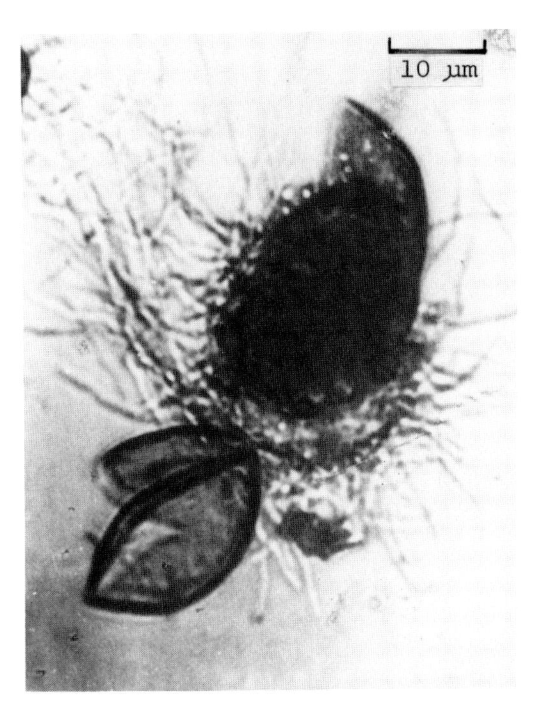

Figure 30.31. *Streptosporangium viridogriseum* subsp. *kofuense:* a ruptured sporangial wall in liquid medium. LM. The spores are not easily released from the sporangium. (Reproduced by permission from H. Nonomura and Y. Ohara, Journal of Fermentation Technology *47:* 468, 1969.)

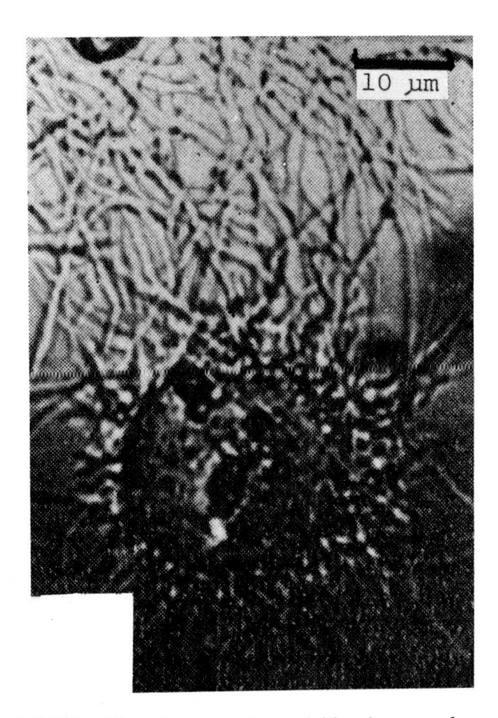

Figure 30.32. *Streptosporangium viridogriseum* subsp. *kofuense:* germination tubes from sporangia in liquid medium. LM. (Reproduced by permission from H. Nonomura and Y. Ohara, Journal of Fermentation Technology *47:* 468, 1969.)

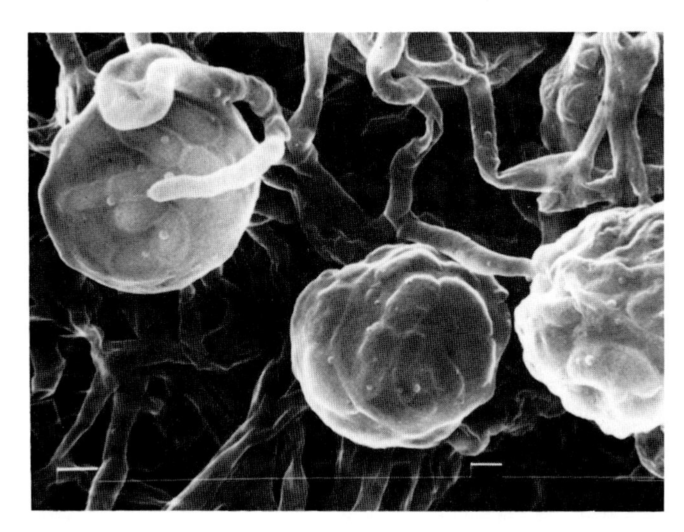

Figure 30.33. Morphological group 2. *Streptosporangium album* CBS 426.61 on oatmeal agar. SEM, gold spattered. *Bar interval,* 10 µm. Sporangial walls (membranes) are thin.

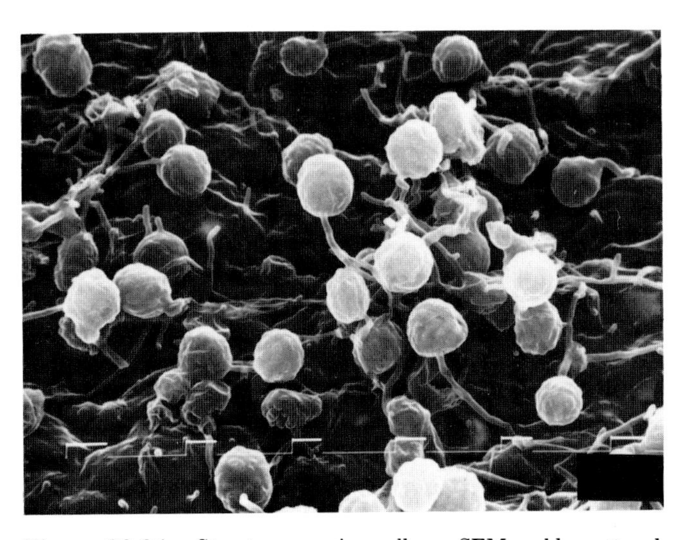

Figure 30.34. *Streptosporangium album.* SEM, gold spattered. Sporangiophores are short. *Bar interval,* 10 µm.

The aerial mass color of streptosporangia is pink, white, pale greenish gray, or dusty green. Most species grow best at 28–30°C, but some are thermotolerant. Some species require B vitamins for growth. *S. amethystogenes* produces violet crystals of iodinin on oatmeal agar-YG* and certain other media.

The mol% G + C of the DNA is 69.5–71 (T_m) in four strains (Jones and Bradley, 1964; Tsyganov et al., 1966; Yamaguchi, 1967; Farina and Bradley, 1970). *Streptosporangia* have a cell wall chemo-type III, a whole-cell sugar pattern B (Lechevalier and Lechevalier, 1970b), and a phospholipid type PIV (Lechevalier et al., 1981). The cell wall peptidoglycan of streptosporangia is an acetyl type (Kawamoto et al., 1981).

Strains of *Streptosporangium* were isolated only infrequently from soil and dung. (Couch, 1955) or leaf litter (van Brummelen and Bent, 1957; Potekchina, 1965) before the introduction of a specific isolation technique (Nonomura and Ohara, 1969a) showed that these organisms were a significant component of the actinomycete population in soils. The number of streptosporangia in various soils in Japan has been estimated at 10^4–10^6 colony-forming units/g dry weight of soil (Nonomura and Ohara, 1969a; Nonomura, 1984). Slightly acid, humus-rich garden soils are a favorable habitat. Species have also been isolated from lake sediments (Johnston and Cross, 1976) and beach sand (Williams and Sharples, 1976). "*Streptosporangium bovinum*" was isolated from infected bovine hooves (Chaves Batista et al., 1963).

Enrichment and Isolation Procedures

Dry-heat treatment of air-dried soil samples and dilution plate culture with selective synthetic media are useful for the preferential isolation and enumeration of streptosporangia in soils.

Pretreatment

Soil samples are dried slowly at room temperature for a week, passed through a 2-mm sieve, ground slightly in a mortar, spread on filter paper, and heated in a hot-air oven at 120°C for 1 h. This procedure reduces drastically the number of bacteria, and significantly the number of streptomycetes, but leads to an increased isolation frequency of *Streptosporangium* and *Microbispora* strains.

Selective Media†

Initially, AV agar was designed for the selective isolation of streptosporangia and microbisporae, but two additional media, HV and chitin-V agars, are currently being developed for this purpose. Generally HV agar gives the highest counts and clearest plates.

Plate Culture

Molten selective agar media are added to Petri dishes containing the pretreated and diluted soil suspension. Alternatively, the latter is spread over solidified selective agar in a Petri dish using a sterile, bent glass rod. The inoculated plates are incubated for 4–6 weeks at 30°C, cooled at room temperature, and examined using a light microscope with a long working distance lens.

Soil Culture

Growth of streptosporangia around soil particles can be observed when a small amount (0.5 g) of pretreated soil is placed sparsely on an agar plate and incubated for 1 month at 30°C.

Maintenance Procedures

For preservation, organisms are inoculated on oatmeal agar-YG slants and incubated until the cultures are observed to produce sporangia; the tubes are then stored at 5°C. For long-term preservation, lyophilization is preferable.

Procedures for Testing of Special Characters

Iodinin production is determined after culturing on oatmeal agar plus 0.1% (w/v) yeast extract for 1 month at 28–30°C. Utilization of carbon sources is determined by comparing growth on a given carbon source in a basal C-1 or C-2 medium‡ with two controls, growth on the basal medium alone and growth on the basal medium plus 0.5% (w/v) glucose. Most species of *Streptosporangium* do not grow on ISP Carbon Utilization Medium (Shirling and Gottlieb, 1966). Other useful physiological tests are hydrolysis of starch on starch agar plus yeast extract (0.5 g/l), nitrate reduction to nitrite in Bacto-nitrate broth (Difco) plus yeast extract (2 g/l), and gelatin liquefaction in gelatin medium (gelatin, 200 g; peptone, 5 g; yeast extract, 2 g; glucose, 2 g; distilled water, 1000 ml).

Differentiation of the genus **Streptosporangium** *from other genera*

Three genera of maduromycetes, *Streptosporangium, Planomonospora,* and *Planobispora,* produce sporangia on aerial hyphae and they all have a cell wall chemotype III and contain the sugar madurose (3-*O*-methyl-D-galactose). Among these genera, *Streptosporangium* is readily distinguished from the others by sporangia containing numerous and nonmotile spores (Table 30.14).

Spirillospora is closely related to *Streptosporangium* but the sporangia of the former are produced on a substrate mycelium; in addition, the spores are formed from branched hyphae within the sporangium. Some strains of *Actinomadura* form "pseudosporangia" (Nonomura and Ohara, 1971d), similar to sporangia in appearance but different in being covered with a slimy substance instead of sporangial membranes.

Actinoplanetes such as *Actinoplanes, Ampullariella,* and *Pilimelia* can be differentiated from maduromycetes on chemotaxonomic grounds.

Taxonomic Comments

It has been suggested that the term "spore vesicle" should replace "sporangium," because true endogenous sporulation does not occur in strains

---

*-*YG*, with yeast extract, 1 g; glucose, 2 g; glycerol, 2 g/l.

†*AV agar* (Nonomura and Ohara, 1969a): arginine, 0.3 g; glucose, 1 g; glycerol, 1 g; K_2HPO_4, 0.3 g; $MgSO_4{\cdot}7H_2O$, 0.2 g; NaCl, 0.3 g; agar, 15 g; distilled water, 1000 ml; B vitamins (thiamin HCl, riboflavin, niacin, pyridoxine HCl, inositol, calcium pantothenate, *p*-aminobenzoic acid, each 0.5 mg; biotin, 0.25 mg); trace salts ($Fe_2(SO_4)_3$, 10 mg; $CuSO_4{\cdot}5H_2O$, 1 mg; $ZnSO_4{\cdot}7H_2O$, 1 mg; $MnSO_4{\cdot}7H_2O$, 1 mg); antibiotics (Acti-Dione, 50 mg; nystatin, 50 mg; polymixin B, none or 4 mg; penicillin G, none or 0.8 mg); pH 6.4.

HV agar (Hayakawa and Nonomura, 1984; Nonomura, 1984): humic acid, 1g (used as an alkaline solution; artificial humic acid prepared from glucose and urea may be used, as may natural humic acid from soil humus, but the pale brown humic acid designated as Rp type gives the best result); Na_2HPO_4, 0.5 g; $MgSO_4{\cdot}7H_2O$, 0.05 g; KCl, 1.7 g; $FeSO_4{\cdot}7H_2O$, 0.01 g; $CaCO_3$, 0.02 g; agar, 18 g; distilled water, 1000 ml; B vitamins, as for AV agar; Acti-Dione, 50 mg; pH 7.2.

Chitin-V agar (Hayakawa and Nonomura, 1984): colloidal chitin, 2 g (dry wt.); K_2HPO_4, 0.35 g; KH_2PO_4, 0.15 g; $MgSO_4{\cdot}7H_2O$, 0.2 g; NaCl, 0.3 g; $CaCO_3$, 0.02 g; $FeSO_4{\cdot}7H_2O$, 10 mg; $ZnSO_4{\cdot}7H_2O$, 1 mg; $MnCl_2$, 1 mg; agar, 18 g; distilled water, 1000 ml; B vitamins, as for AV agar; Acti-Dione, 50 mg; pH 7.2.

‡*C-1 medium* (Nonomura and Ohara, 1969b): casamino acids, 2 g; K_2HPO_4, 0.3 g; $MgSO_4{\cdot}7H_2O$, 0.5 g; NaCl, 0.3 g; agar, 20 g; distilled water, 1000 ml; trace salts ($FeSO_4{\cdot}7H_2O$, 10 mg; $MnSO_4{\cdot}7H_2O$, 1 mg; $ZnSO_4{\cdot}7H_2O$, 1 mg; $CuSO_4{\cdot}5H_2O$, 1 mg); B vitamins (thiamine HCl, riboflavin, niacin, pyridoxine HCl, inositol, calcium pantothenate, *p*-aminobenzoic acid, each 0.5 mg; biotin, 0.25 mg); pH 7.2.

C-2 medium (Nonomura and Ohara, 1971b): casamino acids, 0.5 g, and asparagine, 0.5 g, instead of casamino acids, 2 g, in the composition of C-1 medium. The other components are the same as for the C-1 medium.

Table 30.14.
Differential characteristics of the genus **Streptosporangium** *and other related genera*[a]

Characteristics	*Plano-monospora*	*Plano-bispora*	*Strepto-sporangium*	*Spirillo-spora*
Sporangia				
On aerial mycelium	+	+	+	−
On substrate mycelium	−	−	−	+
Spores per sporangium				
One	+	−	−	−
Two	−	+	−	−
20–numerous	−	−	+	+
Spore motility	+	+	−	+

[a]Symbols: see Table 30.1.

of *Streptosporangium* (Cross, 1970; Sharples et al., 1974). Studies on the process of spore formation in this genus have shown that spores in both sporangia and chains are formed in essentially the same way. In both cases, spores are produced by fragmentation of a hypha within its sheath, which either expands to form the sporangial envelope or remains around the spore chain (Lechevalier et al., 1966b; Sharples et al., 1974; Williams and Sharples, 1976). Indeed, the sporangial membrane of *S. fragile* is so thin and fragile that it cannot be detected by light microscopy (Shearer et al., 1983b). This may lead to difficulty in distinguishing between *Streptosporangium* and *Actinomadura,* because some species of the latter produce "pseudosporangia" covered with a slimy substance (Nonomura and Ohara, 1971d). However, the structure of the sporangia of *S. viridogriseum* and the related species (morphological group 1) seems very different from those mentioned above. The sporangium of *S. viridogriseum* has a well-developed, thick sporangial wall, and a "sheathed" chain of spores within it (Nononura and Ohara, 1969b). This may be the most elaborate sporangial type in this genus. The recently described genus *Kibdelosporangium* (Shearer et al., 1986) bears a close morphological resemblance to *Streptosporangium* species in group 1, but has a wall chemotype IV. The wall chemotype of these streptosporangia needs to be determined to clarify their relationship to *Kibdelosporangium.*

Streptosporangium indianensis Gupta 1965 is excluded from the genus *Streptosporangium,* because, according to Schäfer (1969), it does not form true sporangia, and is probably a *Streptomyces* species with spore aggregates resulting from the autolysis of sporulating aerial hyphae (Goodfellow and Cross, 1984b). "*Streptosporangium album* subsp. *thermophilum*" (Manachini et al., 1965) is a thermophile, but it has been shown to be in the genus *Thermoactinomyces* (Goodfellow and Cross, 1984b).

Further Reading

Lechevalier, H.A. and M.P. Lechevalier. 1981. Introduction to the order *Actinomycetales. In* Starr, Stolp, Trüper, Balows and Schlegel (Editors), The Prokaryotes; A Handbook on Habitats, Isolation and Identification of Bacteria, Vol. II. Springer-Verlag, Berlin, pp. 1915–1922.

Differentiation and characteristics of the species of the genus **Streptosporangium**

The differential characteristics of the species of *Streptosporangium* are given in Table 30.15.

List of species of the genus **Streptosporangium**

1. **Streptosporangium roseum** Couch 1955, 151.[AL]

ro′se.um. L. neut. adj. *roseum* rose colored.

Aerial mycelium is pink and the substrate mycelium yellowish brown to orange on oatmeal–yeast extract agar. Soluble pigment reddish to purplish brown. Sporangia are usually 8–10 μm in diameter but larger ones, up to 20 μm, are found in some strains. Spores spherical, short or bent rod shaped.

Type strain: ATCC 12428.

2. **Streptosporangium album** Nonomura and Ohara 1960b, 407.[AL]

al′bum. L. neut. adj. *album* white.

Aerial mycelium is white and the substrate mycelium pale yellow on oatmeal–yeast extract agar. Starch not hydrolyzed.

Type strain: CBS 426.61; DSM 43023.

3. **Streptosporangium viridialbum** Nonomura and Ohara 1960b, 407.[AL]

vi.ri.di.al′bum. L. adj. *viridis* green; L. adj. *album* white; M.L. neut. adj. *viridialbum* greenish white.

Aerial mycelium is greenish white to pale yellowish gray and the substrate mycelium pale yellow on oatmeal–yeast extract agar.

Type strain: CBS 432.61; KCC A-0027.

4. **Streptosporangium amethystogenes** Nonomura and Ohara 1960b, 407.[AL]

am.e.thys.to′ge.nes. L. adj. *amethystinus* amethyst colored; Gr. v. suff. *-genes* producing; M.L. adj. *amethystogenes* producing violet-colored (crystals).

Aerial mycelium is pink, and violet crystals of iodinin are produced after 1 month at 30°C on oatmeal–yeast extract agar. The substrate mycelium is pale brownish gray.

Type strain: CBS 430.61; KCC A-0026.

5. **Streptosporangium vulgare** Nonomura and Ohara 1960b, 407.[AL]

vul.ga′re. L. neut. adj. *vulgare* common.

Aerial mycelium is pink and the substrate mycelium yellowish brown to orange on oatmeal–yeast extract agar. Neither reddish soluble pigment nor crystals of iodinin are produced. No growth at 42°C.

Type strain: CBS 431.61; KCC A-0028.

6. **Streptosporangium pseudovulgare** Nonomura and Ohara 1969b, 708.[AL]

pseu.do.vul.ga′re. Gr. adj. *pseudes* false; *vulgare* specific epithet *vulgare;* M.L. neut. adj. *pseudovulgare* similar in appearance to strain of *S. vulgare.*

Table 30.15.
Characteristics differentiating the species of the genus **Streptosporangium**[a]

Characteristics	1. *S. roseum*	2. *S. album*	3. *S. viridialbum*	4. *S. amethystogenes*	5. *S. vulgare*	6. *S. pseudovulgare*	7. *S. nondiastaticum*	8. *S. longisporum*	9. *S. violaceochromogenes*	10. *S. fragile*	11. *S. corrugatum*	12. *S. albidum*	13a. *S. viridogriseum*	13b. *S. viridogriseum* subsp. *kofuense*
Sporangium														
1–5 µm	−	−	−	−	−	−	−	−	−	−	+	−	−	−
6–10 µm	+	+	+	+	+	+	−	+	+	+	−	−	−	−
11–15 µm	(+)	−	−	−	−	−	+	+	−	+	−	(+)	−	+
16–20 µm	(+)	−	−	−	−	−	−	(+)	−	−	−	(+)	−	+
21–30 µm	−	−	−	−	−	−	−	−	−	−	−	+	+	−
31–50 µm	−	−	−	−	−	−	−	−	−	−	−	−	+	−
Sporangiophores														
Short (10 µm)	+	+	+	+	+	+	+	+	+	+	+	−	−	+
Long (50 µm)	−	−	−	−	−	−	−	−	−		−	+	+	+
Spores														
Spherical-oval[b]	+	+	+	+	+	+	+	−	+	+	+	+	+	−
Rods	−	−	−	−	−	−	−	+	−	−	−	−	−	+
Color of spore mass														
White	−	+	−	−	−	−	−	−	−	−	+	+	−	−
Greenish gray	−	−	+	−	−	−	−	−	−	−	−	−	+	+
Pink	+	−	−	+	+	+	+	+	+	+	−	−	−	−
Color of substrate mycelium														
Red, orange	+	−	−	−	+	+	+	+	−	−	−	−	−	−
Yellowish brown–brown	+	+	+	+	+	+	+	−	+	−	+	+	+	+
Brown-black	−	−	−	−	−	−	−	−	−	+	−	−	−	−
Soluble pigments[c]	+	−	−	−	−	−	−	−	+	+	−	−	−	−
Iodinin production	−	−	−	+	−	−	−	−	−	−		−	−	−
Starch hydrolysis	+	−	+	+	+	+	−	+	+	+		−	+	+
Nitrite from nitrate	+	−	d	+	−	+	+	+w	+	+		+	+	−
Gelatin liquefaction	+	+	d	−	d	+	+		+w	−		−	+	+
Utilization of														
Rhamnose	+	−	d	+	d	−	−		+w	+			+	+w
Inositol	+	−	+	+	d	−	−		+w				+	+
B vitamins required	+	+	+	+	+	+	+	−	−	−		−	−	−
Growth at														
42°C	−	−	−	−	−	+	+	−	−	+		−	+	+
50°C	−	−	−	−	−	−	−	−	−	−		−	+	+w

[a]Symbols: see Table 30.4; (+), rarely present; +*w*, weak.
[b]Or short rods.
[c]Other than pale yellow-brown.

Aerial mycelium is pink and the substrate mycelium yellowish brown to orange on oatmeal-yeast extract agar. Good growth at 42°C. Sporangia, 7–10 µm in diameter. Starch hydrolyzed.

Type strain: ATCC 27100.

7. **Streptosporangium nondiastaticum** Nonomura and Ohara 1969b, 708.[AL]

non.di.as.ta′ti.cum. M.L. pref. *non-* not; M.L. *diastaticus* diastatic; M.L. neut. adj. *nondiastaticum* not starch digesting.

Aerial mycelium is pink and the substrate mycelium yellowish brown to orange on oatmeal-yeast extract agar. Good growth at 42°C. Sporangia, 10–15 µm in diameter. Starch is not hydrolyzed.

Type strain: ATCC 27101.

8. **Streptosporangium longisporum** Schäfer 1969, 368.[AL]

lon.gi.spo′rum. L. adj. *longus* long; M.L. n. *spora* a seed; M.L. neut. adj. *longisporum* long spored.

Aerial mycelium is sparse and the substrate mycelium is red on oatmeal agar. Pink aerial mycelium is formed on starch agar. Spores rod shaped, 0.6–0.9 × 1.5–3.5 µm.

Type strain: ATCC 25212.

9. **Streptosporangium violaceochromogenes** Kawamoto, Takasawa, Okachi, Kohakura, Takahashi and Nara 1975, 358.[AL]

vi.o.la.ce.o.chro.mo′ge.nes. L. adj. *violaceus* violet; Gr. n. *chroma* color; Gr. v. suff. *-genes* producing; M.L. adj. *violaceochromogenes* producing violet color.

Aerial mycelium is pink and the substrate mycelium pale yellow on oatmeal agar. Violet soluble pigment is produced on nutrient, Bennett (Waksman, 1961) and Emerson (Waksman, 1957b) agars. Produces the antibiotic victomycin.

Type strain: ATCC 21807.

10. **Streptosporangium fragile** Shearer, Colman and Nash 1983b, 364.[VP]

fra′gi.le. L. adj. *fragile* fragile, easily broken (referring to sporangial membrane).

Aerial mycelium is pink and the substrate mycelium black on oatmeal agar. Soluble pigment distinct light brown. Sporangial membrane is so fragile that it is not detected by light microscopy. Grows at 42°C.

Type strain: ATCC 31519.

11. **Streptosporangium corrugatum** Williams and Sharples 1976, 45.[VP]

cor.ru.ga′tum. L. neut. adj. *corrugatum* ridged (spores).

Aerial mycelium is white and the substrate mycelium pale buff on oatmeal agar. Sporangia small (1-5 μm in diameter). Club-shaped spore vesicles (3.5-8 μm long) present. Thick-walled, ridged spores are observed.

Type strain: ATCC 29331.

12. **Streptosporangium albidum** Furumai, Ogawa and Okuda 1968, 174.[AL]

al′bi.dum. L. neut. adj. *albidum* white.

Aerial mycelium is white, cottony on glucose-Czapek and starch agars. Sporangia large (10-30 μm in diameter). Sporangiophores long (30-120 μm). The sporangial walls are thick and strong. Spores oval. Grows to 38°C. Produces sporoviridin-like antibiotics.

Type strain: ATCC 25243.

13. **Streptosporangium viridogriseum** Okuda, Furumai, Watanabe, Okugawa and Kimura 1966, 126.[AL]

vi.ri.do.gri′se.um. L. adj. *viridis* green; L. adj. *griseus* gray; M.L. neut. adj. *viridogriseum* greenish gray.

Aerial mycelium is olive-gray, cottony on glucose-Czapek or starch agar. Partly segmented aerial hyphae may be observed on old cultures. Long sporangiophores (25-65 μm) present. The sporangial walls are thick and very strong.

13a. **S. viridogriseum** subsp. **viridogriseum** Okuda, Furunai, Watanabe, Okuyawa and Kimura 1966, 126.[AL]

Sporangia are very large (20-48 μm in diameter). Spores small, spherical (0.2-0.6 μm in diameter). Grows to 55°C. Produces the antibiotic sporoviridin.

Type strain: ATCC 25242.

13b. **S. viridogriseum** subsp. **kofuense** Nonomura and Ohara 1969b, 708.[AL]

ko.fu.en′se. M.L. adj. *kofuense* belonging to Kofu (district).

Sporangia are smaller (10-20 μm in diameter). Spores small, rod shaped (0.4 × 1.0 μm). Good growth on oatmeal-yeast extract-glucose agar and yeast-malt agars. Grows to 50°C. Segmented aerial hyphae may be observed. Produces chloramphenicol (Tamura et al., 1971).

Type strain: ATCC 27102.

Species Incertae Sedis

Streptosporangium indianensis Gupta 1965 appears on the Approved Lists of Bacterial Names (Skerman et al., 1980) but is excluded from the genus *Streptosporangium,* because, according to Schäfer (1969), it does not form true sporangia.

SECTION 31

Thermomonospora and Related Genera

Alan J. McCarthy

These are aerobic spore-forming actinomycetes that produce a branched vegetative mycelium bearing aerial hyphae. The cell wall contains *meso*-diaminopimelic acid, but whole-cell hydrolysates do not contain diagnostic sugars (wall chemotype III). Mycolic acids are absent. Menaquinones typically contain nine or 10 isoprene units (MK-9, MK-10). Spore arrangement and morphology are distinctive for each genus (Table 31.1).

This is an artificial grouping of four actinomycete genera that have a common wall chemotype but are morphologically extremely diverse. Some degree of uniformity in menaquinone profiles provides further evidence for chemotaxonomic homology between these genera. However, nucleic acid hybridization, and possibly 16S rRNA sequence data, are required to determine whether the four genera are naturally related. Values for the mol% G + C of the DNA of *Thermomonospora* and *Streptoalloteichus* strains could also be helpful. To date, comparative studies have been restricted to the genera *Thermomonospora* and *Nocardiopsis,* but they do provide a case for classifying these two genera in the same group. Strains representing *Nocardiopsis* and *Thermomonospora* form a single distinct cluster in the dendrogram of actinomycete 16S rRNA sequences (Goodfellow and Cross, 1984b), and strains of both genera contain menaquinones with unusually long partially saturated isoprenyl side chains (MK-10(H_4–H_8)) (Yamada et al., 1977a; Collins et al., 1982d; Fischer et al., 1983). Strains that exhibit a marked preference for alkaline (>pH 9.0) growth conditions can also be found in both genera (Miyashita et al., 1984; McCarthy and Cross, 1984a).

The genus *Thermomonospora* is largely composed of thermophilic actinomycetes that produce single spores on aerial, and in some cases, substrate hyphae. In morphology, wall chemotype, and natural habitat (composts and overheated fodders) they are similar, but not related, to thermoactinomycetes. Members of the genus *Thermomonospora* have an important role in the primary degradation of organic material, and are one of the most active groups of lignocellulose-degrading procaryotes. The genus *Nocardiopsis* has its origins in *Actinomadura* and was originally created to accommodate strains of the latter that lacked the characteristic whole-cell sugar madurose (Meyer, 1976). Its distinction from the genus *Actinomadura* has since been supported by a range of taxonomic criteria, although *Nocardiopsis* itself is heterogeneous. Unlike *Thermomonospora, Nocardiopsis* is characterized by the production of chains of arthrospores on the aerial mycelium, and a substrate mycelium that exhibits a tendency to fragment.

The morphology of *Streptoalloteichus* and *Actinosynnema* is much more complex, and differs fundamentally from *Thermomonospora* and *Nocardiopsis* in that motile spores are produced. In *Streptoalloteichus* these originate from vesicles or sporangia on the substrate hyphae, whereas in *Actinosynnema* peritrichously flagellated zoospores originate from aerial spore chains borne on synnemata. *Streptoalloteichus* also produces chains of *Streptomyces*-like arthrospores on the aerial mycelium. Both *Streptoalloteichus* (Tomita et al., 1978) and *Actinosynnema* (Hasegawa et al., 1978a) were proposed as new genera to describe morphologically unusual isolates. *Streptoalloteichus* remains a monospecific genus, but *Actinosynnema* contains species and subspecies that, together with morphologically similar but unclassified isolates (Willoughby, 1969b; Makkar and Cross, 1982), may be the nucleus of a heterogeneous taxon.

Table 31.1.
Morphological characteristics of the genus **Thermomonospora** *and related genera*[a]

Characteristics	Genus I *Thermomonospora*	Genus II *Actinosynnema*	Genus III *Nocardiopsis*	Genus IV *Streptoalloteichus*
Single spores	+	–	–	–
Chains of arthrospores	–	+	+	+
Sporangia-like structures	–	–	–	+
Synnemata	–	+	–	–
Motile spores	–	+	–	+

[a]Symbols: +, positive; –, negative.

Genus **Thermomonospora** *Henssen 1957, 398*[AL*]

ALAN J. MCCARTHY

Ther.mo.mon'o.spo.ra. Gr. n. *thermos* heat; Gr. adj. *monos* single, solitary; Gr. fem. n. *spora* seed; M.L. fem. n. *spora* a spore; M.L. fem. n. *Thermomonospora* the heat (-loving) single-spored (organism).

Produce **single, heat sensitive, nonmotile aleuriospores on aerial hyphae.** On agar media, a branched, nonfragmenting vegetative mycelium forms leathery colonies usually covered with aerial mycelium. Spores may be sessile but are more often formed at the tips of simple unbranched or branched sporophores; in many strains repeated sporophore branching leads to the formation of spore clusters. Spores may also be produced on the substrate hyphae. Gram-positive. **The cell wall contains *meso*-diaminopimelic acid (*meso*-DAP) but no other diagnostic amino acids or sugars.** Aerobic. Chemo-organotrophic. Amino acid and vitamin supplements, e.g. yeast extract, required for good growth. **All strains can grow in the temperature range 40–48°C** and pH range 7.0–9.0. Aerial mycelium production and sporulation often optimal at pH >8.0. Catalase, deaminase, β-glucosidase, and β-galactosidase are produced. Esculin, xylan, casein, gelatin, and **carboxymethylcellulose are degraded. Spores are killed by treatment at 90°C for 30 min in aqueous suspension** and all strains are **sensitive to novobiocin (50 μg/ml).** Can be isolated from soil but are more **common in manures, composts, and overheated fodders.**

Type species: *Thermomonospora curvata* Henssen, 1957, 401.

Further Descriptive Information

The substrate mycelium is composed of extensively branched nonfragmenting hyphae. The aerial mycelium may be simple or branched and of variable abundance. Spores, single and oval to round (0.5–2.0 μm in diameter), are borne on simple or branched sporophores on the aerial hyphae. *Thermomonospora mesophila* spores are sessile or on very short unbranched sporophores, whereas *T. fusca* produces dense clusters of spores as a result of repeated sporophore branching. In *T. chromogena*, spore clusters are formed by sequential sporulation on incurving hyphae. Spores on the substrate mycelium or in submerged liquid culture can be observed in some strains of *T. fusca* and *T. alba.* Spore arrangement on the aerial mycelium can be influenced by medium composition and incubation temperature (Cross and Lacey, 1970). The spores are heat-sensitive, thick-walled aleuriospores but may have a fine structure resembling that of bacterial endospores (Mach and Agre, 1970). The spore surface is smooth (*T. fusca, T. alba,* and *T. mesophila*) or spiny (*T. curvata* and *T. chromogena*).

The cell wall peptidoglycan contains *meso*-DAP (wall chemotype III) (Lechevalier and Lechevalier, 1970b); a trace of L-DAP may be detected in whole-cell hydrolysates of *T. fusca.* The cell envelope does not contain mycolic acids and there is no diagnostic information on the phospholipid content. The long-chain fatty acid composition has been reported to be a mixture of straight-chain and branched-chain *iso-* and *anteiso-* forms (Goodfellow and Cross, 1984b). The predominant menaquinones in white *Thermomonospora* strains (*T. curvata, T. alba,* and *T. fusca*) have side chains with 10 isoprene units that are hexa- or octahydrogenated (MK-10(H_6, H_8)); substantial amounts of the uncommon menaquinones MK-11(H_6–H_{10}) are also present. This pattern is not found in *T. chromogena,* which possesses major amounts of MK-9(H_4) (Collins et al., 1982d). The mol% G + C of the DNA has not been determined for any member of the genus except *T. formosensis,* which is included as a *species incertae sedis.*

With the exception of *T. mesophila, Thermomonospora* strains are moderately thermophilic, showing good growth at 50°C. Some strains may grow up to 60°C, but aerial mycelium production and sporulation are often poor at high temperatures. *Thermomonospora mesophila* and most *T. alba* strains grow optimally at 40–45°C, whereas, *T. chromogena* strains show little or no growth below 40°C. Optimum growth and sporulation of all *Thermomonospora* strains occurs on slightly alkaline nutrient media incubated under aerobic conditions. For many white *Thermomonospora* strains, the pH of the medium may be increased to 11.0 without inhibiting growth or sporulation.

All *Thermomonospora* strains can utilize D-glucose and D-mannose but not L-sorbose, D-raffinose, sorbitol, dulcitol, or inulin as carbon sources. Most strains of all species can utilize trehalose and D-fructose. Glycerol utilization is unique to a single strain of *T. fusca* (McCarthy and Cross, 1984a). *Thermomonospora mesophila* can utilize methoxylated aromatic ring compounds and shows activity against grass lignin (McCarthy and Broda, 1984).

A number of enzymes are produced by all *Thermomonospora* strains, including catalase, β-glucosidase, β-galactosidase, deaminase, xylanase, and carboxymethylcellulase. Proteolytic activity is evidenced by the ability of all strains to attack casein, gelatin, and usually keratin. Esculin and testosterone are also degraded by all strains. Chitinase activity is universally absent. The ability to degrade cellulosic substrates is an important property of many strains, and *Thermomonospora* cellulases (Crawford and McCoy, 1972; Stutzenberger, 1972; Hägerdal et al., 1978) and xylanases (McCarthy et al., 1985) have been partially characterized.

Most *Thermomonospora* strains are sensitive to a range of antibiotics; however, *T. chromogena* strains are usually resistant to the aminoglycoside antibiotics gentamicin, kanamycin, and tobramycin (McCarthy and Cross, 1981, 1984a). Resistance to lysozyme is found in some white *Thermomonospora* strains.

There is no available information on the genetics of *Thermomonospora* species; however, many cultures are infected with bacteriophage and plasmids have been isolated from *T. chromogena* strains (A. J. McCarthy, unpublished results) and *Thermomonospora* isolates (Pidcock et al., 1985). *Thermomonospora fusca* mutants with enhanced production of cellulase have also been described (Fennington et al., 1982, 1984).

Thermophilic *Thermomonospora* strains are common in overheated substrates such as manures, composts, and fodders. They are particularly abundant in mushroom compost (Fergus, 1964; Lacey, 1973; McCarthy and Cross, 1981) and have an important role in the primary degradation of native cellulose (Stutzenberger, 1971; Crawford, 1974; McCarthy and Broda, 1984). Both mesophilic and thermophilic strains can also be isolated from soil (Krasil'nikov and Agre, 1965; Locci et al., 1967; Nonomura and Ohara, 1971c, 1974).

The growth of a number of thermophilic actinomycete species in high-temperature environments results in the release of spores that can cause hypersensitivity pneumonitis. As yet, there is no evidence that implicates the spores of *Thermomonospora* species in such respiratory disorders (Lacey, 1981).

Enrichment and Isolation Procedures

Thermophilic *Thermomonospora* strains can be isolated by dilution plating on nonselective agar media, but recovery is poor owing to the rapid competing growth of *Bacillus* and *Thermoactinomyces* isolates. The most efficient isolation methods are those based on the use of a sedimentation chamber and Andersen air sampler (Lacey and Dutkiewicz, 1976a; McCarthy and Cross, 1981; McCarthy and Broda, 1984). Dried samples are agitated within the chamber to create an aerosol of particles that, after 1–2 h of sedimentation, still contains many actinomycete spores and comparatively few bacteria. Viable actinomycetes are isolated from this spore suspension using an Andersen sampler loaded with half-strength tryptone soy agar plates with cycloheximide (50 μg/ml) routinely incorporated to prevent the growth of fungi; *Thermomonospora* colonies can usually be identified after 3–5 days' incubation at 50°C. Re-

*AL denotes the inclusion of this name on the Approved Lists of Bacterial Names (1980).

covery of white *Thermomonospora* strains may be further improved by adjusting the isolation medium to pH 11.0 (Cross, 1982) and cellulolytic isolates can be identified by incorporating cellulose powder or ball-milled straw in the agar (Stutzenberger et al., 1970; McCarthy and Broda, 1984).

Thermomonospora chromogena is readily isolated on highly selective media containing kanamycin (25 µg/ml) (McCarthy and Cross, 1981) or rifampicin (5 µg/ml) (Athalye et al., 1981). Mesophilic *Thermomonospora* strains have been isolated from soil samples using a selective method that includes dry heat at 100°C as a pretreatment (Nonomura and Ohara, 1971c, 1974). However, excluding *T. mesophila,* strains isolated at 30°C are usually able to grow at 50°C. Mixtures of moist organic material and soil, incubated in partially sealed polythene bags at 50°C for 7 days, produce samples enriched in thermophilic actinomycetes, including thermomonosporas (A. J. McCarthy, unpublished results).

Maintenance Procedures

Strains of all species can be maintained as sporulating cultures on Czapek-Dox–yeast extract–casamino acids (CYC) agar at pH 8.0 (see below), stored at 4°C, and subcultured every 4 weeks. For long-term storage, spore suspensions in 10–20% (v/v) glycerol are stable at ≤ -20°C and can also be used as a routine source of inocula. Specialized procedures are not required for the preparation and recovery of lyophilized strains.

Procedures for Testing of Special Characters

Morphology

The morphological features of sporulating colonies are best observed using a microscope fitted with a × 32 or × 40 long working distance objective lens. Cultures are grown on CYC agar (Cross and Attwell, 1974) that contains: Czapek-Dox liquid medium (Oxoid), 33.4 g; yeast extract, 2.0 g; casamino acids (Difco), 6.0 g; agar, 18.0 g; distilled water, 1 l; pH 8.0. Strains are incubated at 40–50°C for up to 7 days.

Growth Temperature Range

Determined on CYC agar slopes incubated for 7 days.

pH Tolerance

The pH of CYC agar is adjusted with NaOH or HCl and cultures are incubated at 50°C (40°C for *T. mesophila*) for up to 7 days.

Degradative Tests

Various nutrient media are used (see McCarthy and Cross, 1984a) and insoluble test substrates should be autoclaved separately and added to sterile media as homogeneous suspensions. Prolonged incubation (up to 14 days) may be required to observe zones of clearing, and steps should be taken to prevent desiccation when incubated at 50°C. The agar medium for testing pectin degradation must be made up in 0.1 M phosphate buffer (pH 7.6).

Nitrate Reduction

Production of nitrite from nitrate is determined by adding 1-ml amounts of reagent (Follett and Ratcliffe, 1963) to 1-ml samples of 4-day-old stationary CYC broth cultures. Presence of nitrate is indicated by the appearance of a pale orange to red color.

Carbon Source Utilization Test

The mineral salts medium contains: sodium nitrate, 2.0 g; potassium chloride, 0.5 g; magnesium glycerophosphate, 0.5 g; ferrous sulfate, 0.01 g; potassium sulfate, 0.35 g; Bacto-agar (Difco), 18.0 g; distilled water, 1 l; pH 7.5. This medium is supplemented with 0.1% (w/v) vitamin-free casamino acids (Difco) plus 0.1 µg biotin/ml and 0.1 µg thiamin/ml. Carbon sources are sterilized by membrane filtration and added to the autoclaved medium to give a final concentration of 1% (w/v). Growth is compared with that on the basal medium alone (negative control) and with that on glucose basal medium (positive control).

Spore Heat Resistance

Sporulating cultures on agar slopes are suspended in 3 ml sterile distilled water and 0.5-ml aliquots are added to tubes containing 4.5 ml 0.1 M phosphate buffer (pH 7.5) preheated to 90°C in a water bath. After 30 min, the tubes are removed and placed in an ice bath for 10 min. Drops of control and heat-treated suspensions are placed on CYC agar plates and incubated for 3–5 days. The heat-treated suspensions will not contain any viable *Thermomonospora* spores.

Differentiation of the genus **Thermomonospora** *from other genera*

Table 31.2 lists the major characteristics that distinguish members of the genus *Thermomonospora* from other mycelial procaryotes forming single, nonmotile spores.

Taxonomic Comments

The circumscription of the genus has been improved considerably since the eighth edition of the *Manual,* in which wall composition was cited as a variable character and only two valid species were recognized. One of these species, "*T. viridis,*" formed the basis for recognition of a new genus, *Saccharomonospora* (Nonomura and Ohara, 1971c), erected to accommodate monosporic actinomycetes whose walls contained an arabinogalactan. This has since been supported by numerical phenetic data (Goodfellow and Pirouz, 1982b; McCarthy and Cross, 1984a). Invalidation of the genus "*Actinobifida*" (Cross and Goodfellow, 1973; Skerman et al., 1980) led to the transfer of two species to *Thermomonospora,* in which sporulation on substrate mycelium is now accepted as a variable characteristic.

A detailed numerical taxonomic study of *Thermomonospora* and related organisms (McCarthy and Cross, 1984a) identified five species. Resulting nomenclatural changes included the reduction of *T. mesouviformis* (Nonomura and Ohara 1974) to a synonym of *T. alba* (Locci et al. 1967 emend. Cross and Goodfellow 1973), the revival of *T. fusca* (ex Henssen 1957), and the validation of *T. chromogena,* originally described as an "*Actinobifida*" species (Krasil'nikov and Agre, 1965). Organisms previously identified as white *Thermomonospora* strains comprised three species, *T. fusca, T. curvata,* and *T. alba,* of which the last may prove to be a collection of low-temperature variants of the other two species. There is a long history of misidentification between white strains of *Thermomonospora* and *Thermoactinomyces,* which, although superficially similar in morphology, are taxonomically unrelated (Stackebrandt and Woese, 1981b; McCarthy and Cross, 1984a). Growth in the presence of novobiocin (50 µg/ml) and survival of spores after heat treatment at 90°C for 30 min (see above) are routinely applicable tests for which thermoactinomycetes are positive and thermomonosporas negative.

Thermomonospora chromogena and *T. mesophila* are distinct species, biochemically unrelated to the other three species but nevertheless conforming to the genus definition. Menaquinone analysis revealed a uniform pattern in the white *Thermomonospora* group, which *T. chromogena* did not share (Collins et al., 1982d) and together with numerical phenetic data (Goodfellow and Pirouz, 1982b; McCarthy and Cross, 1984a) may suggest future reclassification of *T. chromogena* outside the genus. Furthermore, menaquinones with unusually long partially saturated isoprenyl side chains (MK-10), a feature of the white *Thermomonospora* species, have also been found in the genus *Nocardiopsis* (Yamada et al., 1977b; Fischer et al., 1983), which comprises polysporic actinomycetes with a wall structure similar to that of *Thermomono-*

Table 31.2.
Differential characteristics of the genus **Thermomonospora** *and other monosporic genera*[a]

Characteristics	*Thermomonospora*	*Saccharomonospora*	*Micromonospora*	*Thermoactinomyces*
Aerial mycelium	+	+	−	+
Endospores	−	−	−	+
Glycine in cell wall	−	−	+	−
Sugars in whole-cell hydrolysates:				
Arabinose	−	+	+	−
Galactose	−	+	−	−
Xylose	−	−	+	−

[a]Symbols: see Table 31.1.

spora. A relationship between these two genera is also suggested by 16S rRNA sequence data (Goodfellow and Cross, 1984b) and may be another example of true taxonomic relatedness masked by morphological dissimilarity. In comparative 16S rRNA sequencing studies where the genus is represented by the type species *T. curvata,* and *Nocardiopsis* strains have not been included, *Thermomonospora* appears to be phylogenetically distinct from other actinomycete groups (Stackebrandt et al., 1983b, 1983c).

The genus has been classified in a number of families in the *Actinomycetales,* and in the eighth edition of the *Manual, Thermomonospora* was one of six genera in the family *Micromonosporaceae.* However, it was recognized that these genera had few common features, and family assignments have since been avoided. In a recent classification of the *Actinomycetales* (Goodfellow and Cross, 1984b), four genera including *Thermomonospora* and *Nocardiopsis* form the aggregate group named thermomonosporas. This is an artifical grouping of actinomycetes whose walls contain *meso*-DAP and no other characteristic sugars or amino acids. The extent of the relationship between *Thermomonospora* and the other genera in this group has yet to be revealed.

Further Reading

Goodfellow, M. and T. Cross. 1984. Classification. *In* Goodfellow, Mordarski and Williams (Editors), The Biology of the Actinomycetes. Academic Press, London, pp. 7–164.

McCarthy, A.J. and T. Cross. 1984. A taxonomic study of *Thermomonospora* and other monosporic actinomycetes. J. Gen. Microbiol. *130:* 5–25.

McCarthy, A.J. and T. Cross, 1984. Taxonomy of *Thermomonospora* and related oligosporic actinomycetes. *In* Ortiz-Ortiz, Bojalil and Yakoleff (Editors), Biological, Biochemical and Biomedical Aspects of Actinomycetes. Academic Press, San Diego, pp. 521–536.

Stackebrandt, E., R.M. Kroppenstedt and V.J. Fowler. 1983. A phylogenetic analysis of the family *Dermatophilaceae.* J. Gen. Microbiol. *129:* 1831–1838.

Differentiation and characteristics of the species of the genus Thermomonospora

Species of *Thermomonospora* can be differentiated using the characteristics described in Table 31.3. Other characteristics of the species are listed in Table 31.4.

List of species of the genus Thermomonospora

1. **Thermomonospora curvata** Henssen 1957, 401.[AL]

cur.va'ta. L. v. *curvo* to curve; L. part. fem. adj. *curvata* curved.

Colonies on agar media have a yellow to orange reverse color and bear an abundant white aerial mycelium. Sporophores do not branch repeatedly to form spore clusters (Fig. 31.1) and spores are not formed on the substrate mycelium.

In Henssen's original description of the genus (Henssen, 1957), *T. curvata* was the only species isolated in pure culture and is the only current valid species also included in the eighth edition of the *Manual.* The species cannot be identified by morphology alone, and the assignment of strains to *T. curvata* on this basis (Fergus, 1964; Balla, 1968; Stutzenberger, 1971) is therefore not valid.

Type strain: ATCC 19995.

2. **Thermomonospora fusca** (ex Henssen 1957) McCarthy and Cross 1984b, 356.[VP†] (Effective publication: McCarthy and Cross 1984a, 22.)

fus'ca. L. fem. adj. *fusca* dark, tawny.

Colonies on agar media have an abundant white aerial mycelium usually bearing large numbers of spores in dense clusters (Fig. 31.2).

Thermomonospora fusca was not originally isolated in pure culture (Henssen, 1957) and the unavailability of strains led to the citation of this species as a *nomen dubium* in the eighth edition of the *Manual.* Furthermore, a wholly inaccurate description of *T. fusca* Henssen, given by Waksman (1961), led to the use of this name for strains that probably belonged to *T. chromogena* (Fergus, 1964; Locci et al., 1967; Nonomura and Ohara, 1969b; Cross and Lacey, 1970). This confusion was removed by the subsequent description of what is now the type strain of *T. fusca* (Crawford, 1975), although this did not result in validation of the name. Proposals that *T. fusca* Crawford 1975 be regarded as a synonym of *T. alba* (Cross, 1981b) or vice versa (Kurup, 1979) are not in agreement with numerical phenetic data (McCarthy and Cross, 1984a), which recovered the respective type strains in separate clusters.

The majority of *Thermomonospora* strains isolated from overheated fodders and composts are members of this species.

Type strain: ATCC 27730.

3. **Thermomonospora alba** (Locci, Baldacci and Petrolini 1967) Cross and Goodfellow 1973, 87.[AL] (*Actinobifida alba* Locci, Baldacci and Petrolini 1967, 88; *Thermomonospora mesouviformis* Nonomura and Ohara 1974, 11.[AL])

al'ba. L. fem. adj. *alba* white.

The morphology is as given for *T. fusca,* although repeated sporophore branching to give clusters of spores may be rare or absent. Sporulation

† *VP* denotes that this name has been validly published in the official publication, *International Journal of Systematic Bacteriology.*

Table 31.3.
Characteristics differentiating the species of the genus **Thermomonospora**[a,b]

Characteristics	1. *T. curvata*	2. *T. fusca*	3. *T. alba*	4. *T. chromogena*	5. *T. mesophila*
Colony reverse color	Yellow/orange	Pale yellow	Pale yellow	Brown	Brown
Growth at					
30°C	d	−	d	−	+
53°C	d	+	d	+	+
pH 11.0	+	+	+	−	−
Growth in					
Crystal violet (0.00002% w/v)	+	+	−	+	+
Tetrazolium chloride (0.002% w/v)	d	+	−	d	−
Novobiocin (10 μg/ml)	−	−	−	−	+
Kanamycin (25 μg/ml)	−	−	−	+	−
Degradation of					
Tyrosine, xanthine, hypoxanthine	−	−	−	+	+
Starch	+	+	+	−	+
Pectin	−	+	+	d	+
Elastin	−	+	d	+	+
Growth on[a]					
D-Galactose	−	+	d	+	+
D-Ribose	+	−	−	d	−
Sucrose	+	+	+	−	−
Lactose	−	+	d	−	−
Nitrate reduction	+	−	d	+	+
Oxidase	−	−	−	+	+
Tween (20 and 80) hydrolysis	+	+	+	+	−

[a]Based on data from McCarthy and Cross (1984a).
[b]Symbols: +, 90% or more of strains are positive; −, 10% or less of strains are positive; *d*, 11–89% of strains are positive.
[c]Utilization as carbon source at 1.0% (w/v).

and, often, growth are improved at lower incubation temperatures (40–45°C).

A distinction between *T. alba* and *T. mesouviformis* has been maintained only because the latter was regarded as a mesophilic species incapable of growth at 50–55°C (Nonomura and Ohara, 1974; Kurup, 1979; Cross, 1981b). In fact, the type strain of *T. mesouviformis* can show poor growth at 50°C, and the two species should be considered synonymous. However, *T. alba* is not well circumscribed (see Tables 31.3 and 31.4), and whether it is truly a center of variation or a collection of low-temperature *T. fusca* and *T. curvata* strains has yet to be determined (see McCarthy and Cross, 1984a).

Type strain: IPV 1900 (=NCIB 10169).

4. **Thermomonospora chromogena** (ex Krasil'nikov and Agre 1965) McCarthy and Cross 1984b, 356.[VP] (Effective publication: McCarthy and Cross 1984a, 22.)

chro.mo.ge'na. Gr. n. *chroma* color; Gr. v. *gennaio* produce; M.L. adj. *chromogenes* color producing.

Colonies on agar media are small, entire, raised, and dark reddish brown to light brown in color. Aerial mycelium white to light brownish, turning blue-gray on prolonged incubation. A dark brown soluble pigment is often produced. Spore clusters on the aerial mycelium are formed by sequential sporulation on incurving hyphae (Fig. 31.3). Spores are not formed on the substrate mycelium.

The reclassification of "*Actinobifida chromogena*" Krasil'nikov and Agre, 1965 in the genus *Thermomonospora* was delayed by the equivocal demonstration of an endospore fine structure (Mach and Agre, 1970) suggesting a more appropriate classification in the genus *Thermoactinomyces*. It was subsequently shown, but not published, that *T. chromogena* spores are heat sensitive and not formed endogenously (Atwell, 1973). Although *T. chromogena* conforms to the genus definition of *Thermomonospora*, it has little in common with the other species (Collins et al., 1982d; McCarthy and Cross, 1984a).

Type strain: Agre no. 577 (=NCIB 10212).

5. **Thermomonospora mesophila** Nonomura and Ohara 1971c, 899.[AL]

me.so.phi'la. Gr. n. *mesos* middle; Gr. adj. *philus* loving; M.L. fem. adj. *mesophila* middle (temperature) loving.

Colonies on agar media are small, entire, and raised with a brown reverse and abundant white aerial mycelium. Spores sessile or on very short sporophores (Fig. 31.4). Optimum temperature for growth and sporulation is 35–40°C.

Commonly isolated from Japanese soils using a selective method (Nonomura and Ohara, 1971c), but only the type strain has been available for study.

Type strain: ATCC 27303.

Table 31.4.
Other characteristics of the species of the genus **Thermomonospora**[a,b]

Characteristics	1. *T. curvata*	2. *T. fusca*	3. *T. alba*	4. *T. chromogena*	5. *T. mesophila*
Aerial mycelium spore arrangement					
Clusters	–	+	d	+	–
Branched and unbranched sporophores	+	–	d	–	–
Short sporophores/sessile	–	–	–	–	+
Spores on substrate mycelium	–	d	d	–	–
Pellicle formed in liquid culture	+	+	+	–	–
Growth at					
25°C	–	–	–	–	+
35°C	+	d	+	–	+
51°C	+	+	d	+	–
60°C	–	d	–	–	–
pH 6.0	–	d	d	d	+
pH 10.0	+	+	+	d	–
pH 12.0	–	d	–	–	–
Growth in					
NaCl (3% w/v)	–	d	d	d	–
NaCl (5% w/v)	–	–	–	–	–
Bile salts (0.05% w/v)	d	+	d	–	–
Brilliant green (0.001% w/v)	d	d	–	–	+
Sodium azide (0.02% w/v)	d	d	–	+	+
Thallous acetate (0.001% w/v)	d	+	–	d	+
Lysozyme (625 units/ml)	d	d	d	–	–
Degradation of					
Agar	+	+	+	–	+
Cellulose powder (MN300)	+	+	+	–	–
Arbutin	+	+	+	+	+
Guanine	–	–	–	d	–
Acid produced from D-glucose	d	+	+	d	–
Deoxyribonuclease	–	–	–	–	+
Phosphatase	+	+	+	d	–
Growth on[c]					
D-Fructose	+	d	d	d	+
L-Rhamnose	–	–	–	d	–
L-Arabinose	–	–	–	–	+
Maltose	+	+	+	d	+
D-Xylose	d	–	d	d	+
Melezitose	d	+	d	–	–
Mannitol	–	–	–	d	–
meso-Inositol	–	–	–	d	–

[a]Based on data from McCarthy and Cross (1984a).
[b]Symbols: see Table 31.3.
[c]Utilization as carbon source at 1.0% (w/v).

Species Incertae Sedis

a. **Thermomonospora formosensis** Hasegawa, Tanida and Ono 1986, 22.[AL]

for.mo.sen′sis. L. adj. pertaining to Formosa.

Colonies on agar media are usually pink to light orange with sparse white aerial mycelium. Single, warty spores (~1 μm in diameter) are formed on unbranched sporophores on both aerial and substrate hyphae. The spores are heat sensitive and do not contain dipicolinic acid. Whole-cell hydrolysates contain *meso*-DAP with galactose and madurose as diagnostic sugars. The growth temperature range is 23–41°C.

Cultures hydrolyze gelatin and milk, but not starch. Nitrate is not reduced. The following compounds are utilized as carbon sources: D-glucose, D-mannitol, D-galactose, and sucrose. Little or no growth is supported by D-sorbitol, L-arabinose, L-rhamnose, raffinose, D-mannose, maltose, trehalose, lactose, *i*-inositol, D-xylose, D-fructose, or glycerol.

Growth is absent on media containing >2% (w/v) NaCl and is inhibited by the antibiotics penicillin, tetracycline, chloramphenicol, erythromycin, and streptomycin. The type strain produces rifamycins O and S and is resistant to rifampicin.

The type strain was isolated from soil and is the only strain described.

The mol% G + C of the DNA is 72.0 (T_m).

Type strain: IFO 14204.

Figure 31.1. *Thermomonospora curvata.* Single spores borne laterally along aerial hyphae on branched and unbranched sporophores. *Bar,* 10 μm.

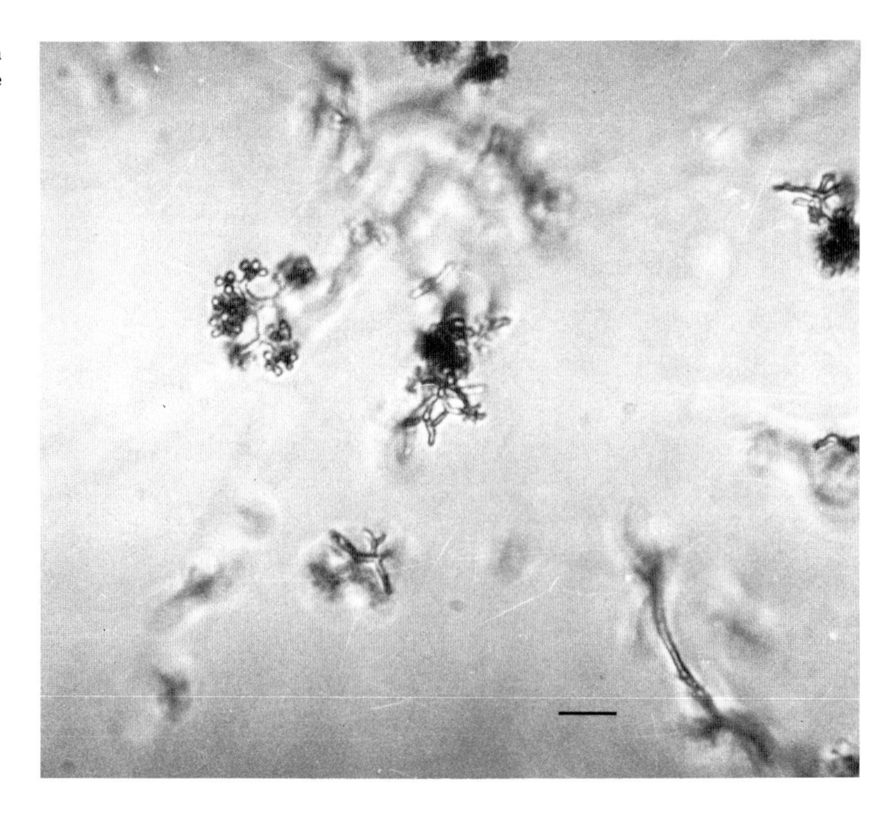

Figure 31.2. *Thermomonospora fusca.* Spores on repeatedly branched sporophores resulting in spore clusters on aerial hyphae. *Bar,* 10 μm.

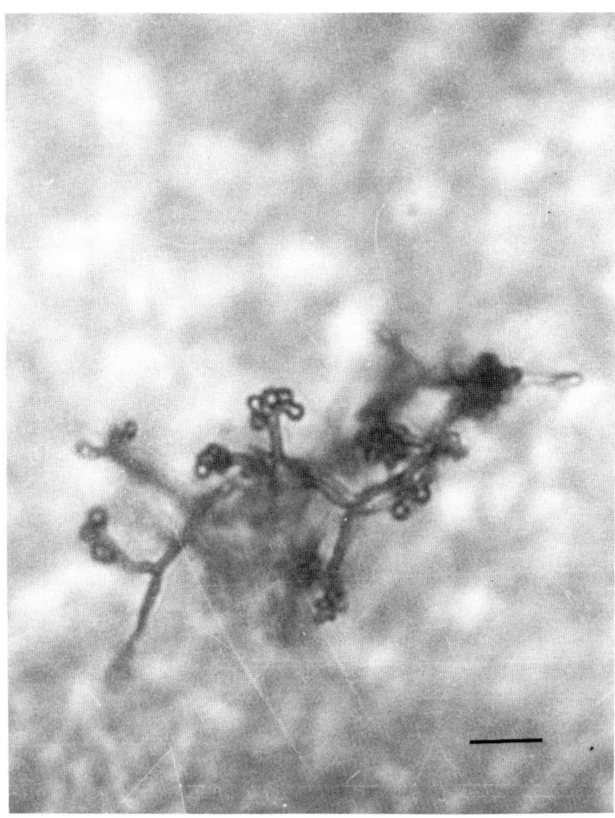

Figure 31.3. *Thermomonospora chromogena.* Single spores produced sequentially on incurving aerial hyphae. *Bar,* 10 μm.

Further comments: This species deviates from the genus definition only in the presence of madurose and galactose in whole-cell hydrolysates. Madurose is characteristic of a number of actinomycete genera, tentatively grouped as the maduromycetes (Goodfellow and Cross, 1984b), but *T. formosensis* is morphologically distinct from these organisms. However, it may be significant that the unusual combination of madurose and galactose in whole-cell hydrolysates of *T. formosensis* has now been detected in many *Actinomadura* species (Terekhova et al., 1986).

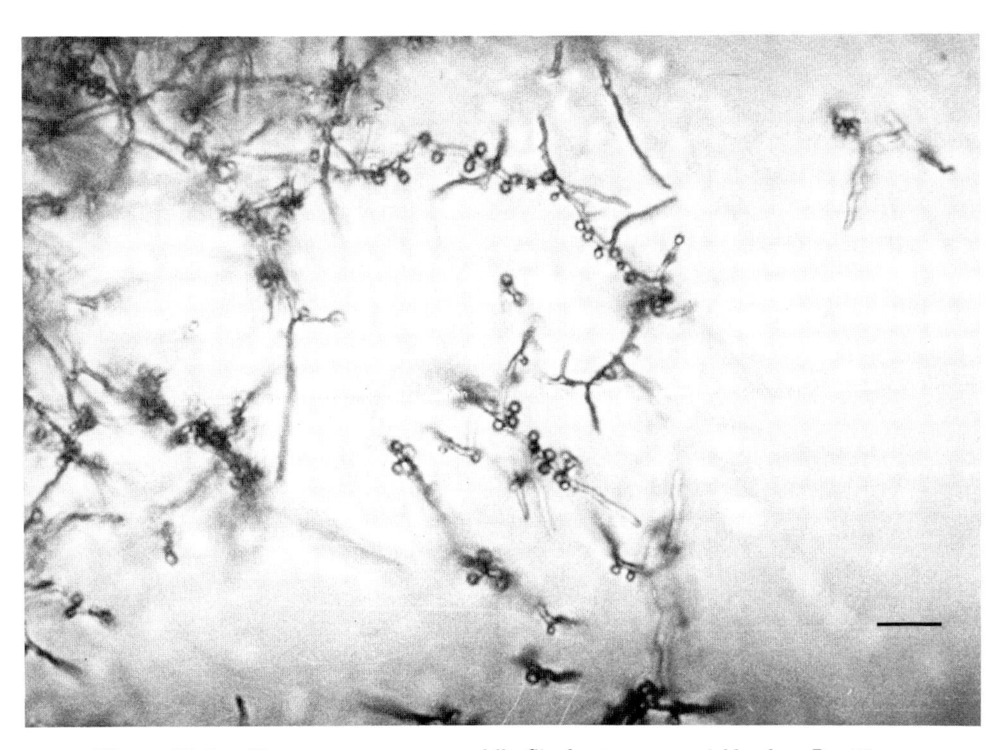

Figure 31.4. *Thermomonospora mesophila.* Single spores on aerial hyphae. *Bar,* 10 μm.

Genus **Actinosynnema** *Hasegawa, Lechevalier and Lechevalier, 1978a, 304*[AL]

TŌRU HASEGAWA, MARY P. LECHEVALIER, AND HUBERT A. LECHEVALIER

Ac.ti.no.syn'ne.ma Gr. n. *actis, actinos* ray; Gr. adv. *syn* together; Gr. n. *nema, nematos* thread; *Actinosynnema* indicates a synnema-forming actinomycete.

Fine hyphae (about 0.5 μm in diameter) are differentiated into (a) **substrate hyphae,** long branching filaments that penetrate the agar and also grow into and **form synnemata, dome-like bodies, or flat colonies** on the surface of agar; and (2) **aerial hyphae** (0.5–1.0 μm in diameter) that **arise from synnemata, dome-like bodies, or flat colonies. The hyphae bear chains of spores capable of forming flagella** in an aqueous environment. **The cell wall contains** major amounts of ***meso*-diaminopimelic acid (*meso*-DAP),** glutamic acid, alanine, glucosamine, and muramic acid. **The whole-cell hydrolysate has no characteristic sugars.** Phospholipids are phosphatidylinositol mannosides, phosphatidylinositol, phosphatidylethanolamine, and diphosphatidylglycerol. Fatty acids are of the normal and branched-chain types. Menaquinones are tetra- and hexahydrogenated with nine isoprene units (MK-9(H_4, H_6)). Gram-positive. Not acid fast. Catalase positive. Aerobic. Mesophilic. Chemo-organotrophic. Most strains were isolated directly from plant tissue such as grass blades found at riversides. The G + C mol% of the DNA of the type species is 73.0 ± 1.0 (T_m).

Type species: *Actinosynnema mirum* Hasegawa, Lechevalier and Lechevalier, 1978b, 304.

Further Descriptive Information

Synnemata (Fig. 31.5) or dome-like bodies are at first whitish on most media and then become yellowish or yellowish orange. In young cultures, the tips of the aerial hyphae often curl, and the long, branching hyphae are white to pale yellow and sparse on most media. Regular septation in the mature hyphae results in chains of spores. The septa are not apparent under the light microscope; in electron microscope preparations they appear as "swellings," which give the hyphae a bamboolike appearance. If synnemata with aerial hyphae are mounted beneath a coverslip in broth such as brain-heart infusion, zoospores can be observed after 30 min to 1 h at room temperature. The rod-shaped spores (0.4–1.1 × 1.5–3.0 μm) bear 5–15 peritrichous flagella. Motile elements may also be observed in static culture in other media such as Actinomycete Broth (Difco) or thin potato-carrot broth. During germination, one to three germ tubes are formed. Most strains produce a pale yellow-brown or purple-brown soluble pigment on tyrosine agar (ISP 7) and a pale greenish one on oatmeal agar (ISP 3) (Shirling and Gottlieb, 1966). They do not produce any diffusible pigments on other media.

Some hyphae (0.5–1.2 μm) of the substrate mycelium of *A. pretiosum* grown in TYG broth* divide into rod-shaped, irregular or branching fragments after 42–66 h incubation at 30°C, and a few of these elements become motile. No satisfactory electron micrograph of these elements has been obtained. A mutant derived from the parent strain has a high fragmentation rate and lacks aerial hyphae. When the mutant is incubated for 30 h in TYG broth at 30°C, most of the substrate hyphae fragment into rod-shaped elements (0.6–0.8 × 4.0–4.5 μm) with peritrichous flagella 9–11 μm long. The motile elements of the mutant strain are morphologically similar to those of the parent strain (Hasegawa et al., 1983).

The cell wall contains *meso*-DAP, alanine, glutamic acid, muramic acid, and glucosamine (type III of Lechevalier and Lechevalier, 1965). In whole-cell hydrolysates, galactose and mannose are present, but madurose, xylose, and arabinose are absent (whole-cell sugar pattern of Type C; Lechevalier, 1968b). The lipids include straight- and branched-chain fatty acids. Major fatty acid components are a C_{17} acid (margaric acid) and 10-methyloctadecanoic acid (tuberculostearic acid). Mycolic acids are not present (Hasegawa et al., 1979, 1983). The phospholipid composition is of type PII (Lechevalier et al., 1977), the most common phospholipid type in actinomycetes, characterized by having major amounts of phosphatidylethanolamine (diagnostic constituent) with phosphatidylinositol mannosides, phosphatidylinositol, and diphosphatidylglycerol (Hasegawa et al., 1979, 1983). The major menaquinones of *A. mirum* were found to be MK-9(H_4) and MK-9(H_6) (Goodfellow and Cross, 1984b).

No growth occurs on any medium under anaerobic conditions, but growth is moderate in a 10% (v/v) CO_2 atmosphere. The temperature range for growth is 10–38°C. The best growth occurs at pH 7–7.5. Growth occurs in 0.1% lysozyme broth. Most strains give positive reactions in the following tests: peptonization of milk; hydrolysis of casein, starch, esculin, and tyrosine; utilization of xylose, glucose, fructose, galactose, mannose, sucrose, and trehalose as sole carbon sources; tolerance to 2% (w/v) NaCl; and resistance to 100 μg of amphotericin B/ml and 100 μg of sulbenicillin/ml. They give negative reactions in the following tests: coagulation of milk; hydrolysis of urea, xanthine, and hypoxanthine; utilization of sorbitol or inositol as a sole carbon source; production of H_2S; and resistance to 20 μg of streptomycin/ml, 20 μg of tetracycline/ml, or 20 μg of

Figure 31.5. Scanning electron micrograph of a synnema growing on a blade of *Carex* species *Bar,* 100 μm.

**TYG medium* (grams per liter): tryptone (Difco), 10; yeast extract, 6; glucose, 10; pH 7.0.

chloramphenicol/ml. None of them produces melanoid pigments on ISP media, but they produce melanoid pigments when dioxyphenylalanine is used as the substrate. This indicates a lack of activity of tyrosinase, an enzyme that oxidizes tyrosine to dioxyphenylalanine and is indispensable for the formation of melanoid pigments.

Enrichment and Isolation Procedures

A grass blade is placed on yeast extract agar† and incubated for 2–3 weeks at 28°C. At the end of this period, the surface of the agar will be covered with various types of microbial growth, but on the grass itself small synnemata (Fig. 31.5) may be observed with a stereoscopic microscope. With a flamed loop, some of the synnemata are transferred to a sterile plate of the same medium.

Most strains of *Actinosynnema* rapidly lose their typical morphology when maintained on common laboratory media but may often be induced to grow typically when subcultivated on grass. Typical morphology is often enhanced by a nutritionally poor growth medium, which more closely resembles the real nutritional situation in nature. Thin potato-carrot agar and tyrosine agar are the best media found so far for producing sporulating synnemata.

Maintenance Procedures

For short preservation (several weeks), plate and tube cultures will usually survive with subcultivation. Survival is often better at room temperature than at 4°C. For longer preservation (several months), cells grown on agar media are suspended in 40% (v/v) glycerol and stored at −20°C. The most satisfactory means of long-term preservation (over 5 years) is by lyophilization.

Differentiation of the genus **Actinosynnema** *from other genera*

Characteristics useful for differentiating *Actinosynnema* from other morphologically and chemically similar genera are listed in Table 31.5.

Taxonomic Comments

Synnemata, also called coremia, are compacted groups of erect hyphae that are often fused and that bear spores at their apex only, or at both their apex and on their sides. Sterile columnar hyphal structures are sometimes also called synnemata or coremia. In 1976, Kalakoutskii and Agre noted that sterile synnemata were known to be produced by actinomycetes but that no signs of special transformations of hyphae in these structures had been reported up to that time.

In the genus *Sporichthya* aerial chains of motile spores are formed, but the two genera may be differentiated on the basis of the formation of substrate mycelium and synnemata in the genus *Actinosynnema* and their absence in the genus *Sporichthya*. In addition, members of these two genera are very different in their cell wall compositions (type I for *Sporichthya*, type III for *Actinosynnema*). Some actinomycetes with a type III cell wall are known to produce motile elements. These include the sporangia-forming, madurose-containing members of the genera *Planomonospora, Planobispora,* and *Spirillospora.* Actinomycetes of this cell wall type that form multilocular sporangia with motile elements include the madurose-containing members of the genus *Dermatophilus* and their relatives in the genus *Geodermatophilus* that lack madurose. However, there is no morphological similarity between these last genera, which do not form aerial hyphae but which produce masses of motile cocci by the division of hyphae in three planes, and *A. mirum,* with its aerial chains of motile, rod-shaped spores. These differences are summarized in Table 31.5.

The nocardicin A producing actinomycete named "*Nocardia uniformis* subsp. *tsuyamanensis*" ATCC 21806, WS 1571 (Aoki et al., 1976) has characteristics in common with members of the genus *Actinosynnema* and might have to be relocated in that genus.

Further Reading

Goodfellow, M. and T. Cross. 1984. Classification. *In* Goodfellow, Mordarski and Williams (Editors), The Biology of the Actinomycetes. Academic Press, London, pp. 116–117.

Lechevalier, H. A. and M. P. Lechevalier. 1981. Actinomycete genera "in search of a family." *In* Starr, Stolp, Trüper, Balows and Schlegel (Editors), The Prokaryotes: A Handbook on Habitats, Isolation and Identification of Bacteria. Springer-Verlag, Berlin, pp. 2118–2123.

Differentiation and characteristics of the species of the genus **Actinosynnema**

Characteristics useful for differentiating *Actinosynnema* species are listed in Table 31.6. In general, macromorphological characteristics are not useful.

Table 31.5.
Characteristics that differentiate **Actinosynnema** *from other genera of actinomycetes that are morphologically and/or chemically similar*[a]

Genus	Cell wall type[b]	Whole-cell sugar pattern[c]	Aerial mycelium	Aerial mycelium with motile spores	Sporangia with motile spores
Actinosynnema	III	C	+	+	−
Sporichthya	I	C	+	+	−
Planomonospora	III	B	+	−	+
Planobispora	III	B	+	−	+
Spirillospora	III	B	+	−	+
Dermatophilus	III	B	−	−	+
Geodermatophilus	III	C	−	−	+

[a]Symbols: see Table 31.1.
[b]Data from Lechevalier and Lechevalier (1965).
[c]Data from Lechevalier (1968b).

†*Yeast extract agar* (grams per liter): yeast extract, 0.2; agar, 15.0.

List of species of the genus **Actinosynnema**

1. **Actinosynnema mirum** Hasegawa, Lechevalier and Lechevalier 1978a, 304.[AL]

mi'rum. L. neut. adj. *mirum* marvelous.

The characteristics are as described for the genus.

The mol% G+C of the DNA is 73 ± 1 (T_m).

Isolated from a sedge blade.

Type strain: ATCC 29888 (IMRU 3971, strain Hasegawa 101).

Reference strain: ATCC 31896 (NR 0364), β-lactam antibiotic producer isolated from a fallen leaf (Watanabe et al., 1983).

2. **Actinosynnema pretiosum** Hasegawa, Tanida, Hatano, Higashide and Yoneda 1983, 314.[VP]

pre.ti.o'sum. L. neut. adj. *pretiosum* precious.

The mol% G+C of the DNA is 71 ± 1 (T_m).

Isolated from a sedge blade.

Type strain: ATCC 31281 (IFO 13726, strain C-15003[N-1]).

Reference strain: ATCC 31280 (IFO 13723, strain C-14919[N-2001]).

2a. **Actinosynnema pretiosum** subsp. **pretiosum** Hasegawa, Tanida, Hatano, Higashide and Yoneda 1983, 314.[VP]

Colonies pale yellow to pale orange yellow. Branched hyphae in liquid culture break up into motile elements with peritrichous flagella. The aerial mycelium consists of long, straight, helical or, rarely, branching hyphae that appear white to pale yellow. These hyphae form in tufts at the tips of synnemata or on the surface of domelike or irregular structures on the surface of the colonies. They have a bamboolike appearance and become transformed into chains of peritrichously flagellate spores.

Produces the antibiotics ansamitocins and tomaymycin.

Type strain: ATCC 31281 (IFO 13726, strain C-15003[N-1]).

Table 31.6.
Characteristics that distinguish **Actinosynnema mirum** *from* **A. pretiosum**[a]

Characteristics	1. *A. mirum*	2. *A. pretiosum*
Growth at		
10°C	+	–
38°C	–	+
Utilization of		
Melibiose	–	+
Raffinose	–	+
Fragmentation of substrate hyphae in liquid media	–	+
Antibiotics produced	Nocardicins	Ansamitocins Dnacins Macbecins Nocardicins Tomaymycin

[a]Symbols: see Table 31.1.

2b. **Actinosynnema pretiosum** subsp. **auranticum** Hasegawa, Tanida, Hatano, Higashide and Yoneda 1983, 314.[VP]

au.ran'ti.cum. M.L. neut. adj. *auranticum* orange.

The characteristics of this subspecies are similar to those of the subspecies *pretiosum.* Colonies tend to be darker yellow to orange colored and differ from those of subsp. *pretiosum* in appearance and texture.

Produces the antibiotics dnacins and ansamitocins.

Type strain: ATCC 31309 (IFO 13725, strain C-14482[N-1001]).

Genus **Nocardiopsis** *Meyer 1976, 487*[AL]

JUTTA MEYER

No.car.di.op'sis. M.L. fem. n. *Nocardia* a genus of the actinomycetes, G. n. *opsis* appearance; M.L. fem. n. *Nocardiopsis* an organism that has the appearance of *Nocardia.*

Substrate mycelium well developed, **hyphae long and densely branched,** 0.5–0.8 μm in diameter; **fragmentation** into coccoid and bacillary elements **may occur. Aerial mycelium** usually well developed and abundant, **hyphae long** and **moderately branched, straight, flexuous,** or **irregularly zig-zagged, completely fragmenting into spores** of various lengths. **Spores oval to elongated, spore surface smooth.** When aerial mycelium is lacking, surface of colonies coarsely wrinkled or folded; otherwise covered by a woolly or farinaceous aerial mycelium. Gram-positive, aerobic, chemo-organotrophic, not acid fast. Temperature range from 10° to 45°C. The **cell wall** contains ***meso*-2, 6-diaminopimelic acid (*meso*-DAP)** with no diagnostic sugars. **Mycolic acids absent.** Menaquinones are predominantly of the MK-10(H_2,H_4,H_6) or of the MK-9(H_4,H_6) type. Found in soil, mildewed grain, and clinical material of animal and human origin. The mol% G + C of the DNA is 64–69 (T_m).

Type species: *Nocardiopsis dassonvillei* (Brocq-Rousseau) Meyer 1976, 487.

Further Descriptive Information

Colonial Characteristics and Morphology

Cultures grown on oatmeal, yeast extract–malt extract, and yeast extract–glucose agar media* produce abundant growth. The production of aerial mycelium varies from sparse to abundant. The aerial hyphae formed by some strains are not visible to the unaided eye, but at least on peptone-glucose and oatmeal agars the cultures show microscopically a sparse coating or a few whitish patches. Other cultures are thickly covered with a powdery to velvety aerial mycelium, thus simulating, because of the yellowish-gray color, cultures of *Streptomyces griseus* (cf. Gordon and Horan, 1968).

Aerial hyphae are long, moderately branched, straight, or flexuous. They fragment completely into spores. Initiation of sporulation is often characterized by twisted hyphae, which, by examination at higher magnification, reveal a zig-zag arrangement of the developing spores. The process of sporulation finally results in subdivision into spores of varying lengths. Carbon replication of spores of *N. dassonvillei* by Williams et al. (1974) showed that they were covered by a sheath with a distinct pattern that resembled the patterns observed in *Streptomyces griseus.*

The process of spore formation in *N. dassonvillei* was studied by Williams et al. (1974). They observed that the process is initiated by a single ingrowth of the hyphae wall to produce a cross-wall. The order of formation of cross-walls in the hyphae is irregular and the cross-walls increase considerably in thickness. The first elements delimited are often long and are sometimes subdivided by further cross-wall formation. Spores of various lengths are finally delimited by cleavage of the thickened cross-walls. The process is completed by the disruption of the

*For descriptions of media, see: yeast extract–malt extract agar (ISP 2), oatmeal agar (ISP 3), inorganic salts–starch agar (ISP 4), and glycerol-asparagine agar (ISP 5), Shirling and Gottlieb (1966); peptone-glucose medium, Prauser and Falta (1968); inorganic salts–starch agar (Gauze 1), Gauze et al. (1957); Czapek sucrose agar and Bennett sucrose agar, Waksman (1961).

sheath between the spores. From these observations it seems likely that the zig-zag arrangment of developing spore chains is caused by lateral displacement of spores within the sheath.

In contrast to *Streptomyces*, where spores are delimited almost simultaneously, in *N. dassonvillei* the cross-walls are formed in a relatively uncoordinated manner, resulting in spores of various lengths. In this respect the process in *N. dassonvillei* resembles the fragmentation of *Nocardia* substrate hyphae into irregularly sized elements. However, the macroscopic resemblance to *Streptomyces griseus* and also the similarity with the fragmentation of *Nocardia* species are only superficial likenesses.

The substrate mycelium of *Nocardiopsis* shows some tendency to fragmentation into rods or coccoid elements according to species or depending on the medium and age of the cultures. In most cases the organisms have been examined on solid media where fragmentation may be delayed. Fragmentation in submerged culture was reported for *N. mutabilis* by Shearer et al. (1983). *Nocardiopsis dassonvillei*, on the other hand, shows a relatively stable substrate mycelium with a barbed wire-like appearance in submerged culture.

Cell Wall Chemotype

Lechevalier and Lechevalier created the genus *Actinomadura* in 1970(a) and included the former *Nocardia dassonvillei* in the new genus. It became obvious that these organisms were unsatisfactorily placed because they lacked the characteristic whole-cell sugar madurose. The introduction of whole-cell sugar patterns type A, B, C, and D by Lechevalier et al. (1971a) led to the separation of cell wall chemotype III/B (characterized by *meso*-DAP and madurose) and III/C (i.e. *meso*-DAP without diagnostic sugars). The creation of the genus *Nocardiopsis* (Meyer, 1976) was based not only on morphology but also on this difference in whole-cell sugar composition. The importance of the characteristic whole-cell sugar pattern, i.e. type C without diagnostic sugars, may be debatable, since Shearer et al. (1983) described a second species of *Nocardiopsis, N. mutabilis,* that contained *meso*-DAP but no diagnostic sugars in purified cell wall preparations, although whole-cell hydrolysates contained galactose, glucose, mannose, ribose, and rhamnose. Moreover, Labeda et al. (1984) established the genus *Saccharothrix,* related to *Nocardiopsis,* which is, if the question of significance of a different phospholipid type and menaquinones is not taken into consideration, based on the whole-cell sugar pattern. The delimitation between these genera seems at present to be unclear.

Phage

Prauser (1981b) reported the isolation of four *Nocardiopsis* phages: however, to date information on their taxon specificity is lacking. Clearing effects caused by soil-isolated polyvalent *Streptomyces* phages on *N. dassonvillei* gave rise to some speculation about the relationship of these two taxa.

Pathogenicity

Although many strains of *N. dassonvillei* have been isolated from human or animal clinical material, *Nocardiopsis* species, like most other actinomycete pathogens, are opportunistic rather than invasive pathogens. According to Gordon and Horan (1968) and Schaal (1977) they have been implicated in human infection, especially in cutaneous and subcutaneous forms and also in respiratory disease.

Ecology

Nocardiopsis dassonvillei was originally isolated from mildewed grain (Brocq-Rousseu, 1904). According to Gordon and Horan (1968), about one third of the *N. dassonvillei* strains distributed in laboratories throughout the world originated from soil. Lacey (1978) reported on the frequent occurrence of *N. dassonvillei* in cotton waste, and its occasional isolation from hay. While two recently described strains were isolated from soil in a temperate climate, Abyzov et al. (1983b) described another species isolated from the ice sheet of the central Antarctic glacier.

Enrichment and Isolation Procedures

No specific enrichment or isolation procedures have been recommended. The organisms grow readily on a variety of media, especially those recommended by the International Streptomyces Project (ISP) (Shirling and Gottlieb, 1966), at 28–30°C. Miyashita et al. (1984) described a group of *Nocardiopsis* strains that required alkalinity for growth. These organisms show optimum growth at pH values of 9.0–10.0 and no or scant growth at pH 7.0.

Maintenance Procedures

Serial transfer from agar slants of suitable media every 2 months is the most convenient method for short-term storage. The tubes should be tightly sealed by cotton-wool plugs dipped in melted paraffin wax.

There are two useful methods for long-term preservation. First, spore or mycelial suspensions suspended in a suitable fluid may be lyophilized. Second, mature slope cultures in small tubes are sealed with cotton-wool plugs dipped in melted paraffin wax and stored under liquid nitrogen.

Differentiation of the genus **Nocardiopsis** *from other genera*

Table 31.7 shows the primary characteristics that distinguish the genus *Nocardiopsis* from morphologically and biochemically similar taxa.

Taxonomic Comments

The organism that gave its name to the type species of the genus *Nocardiopsis* was originally isolated from mildewed grain and named *Streptothrix dassonvillei* by Brocq-Rousseu (1904). This strain has been lost. Some years later, Liegard and Landrieu (1911) isolated a strain from a case of ocular conjunctivitis that they found to be similar to Brocq-Rousseu's *S. dassonvillei,* but they proposed to include it within the genus *Nocardia. Nocardiopsis* species may have been overlooked or mistaken for streptomycetes for the reasons given by Gordon and Horan (1968). Their separation from *Nocardia* and inclusion in *Actinomadura* (Lechevalier and Lechevalier, 1970a) was only a transitory solution for their uncertain affiliation.

The separation of the genus *Nocardiopsis* from *Actinomadura* was based primarily on morphological and biochemical criteria (Meyer, 1976). Subsequent analysis of polar lipids (Lechevalier et al., 1977; Minnikin et al., 1977b) as well as menaquinone patterns (Collins et al., 1977; Yamada et al., 1977b; Fischer et al., 1983; Athalye et al., 1984) and numerical taxonomic analysis (Goodfellow et al., 1979; Goodfellow and Pirouz, 1982b; Athalye et al., 1985) strongly supported the creation of the genus *Nocardiopsis,* as represented by *N. dassonvillei.*

Despite a superficial resemblance in the peptidoglycan type and the mol% G + C of the DNA between the genera *Actinomadura* and *Nocardiopsis,* genetic studies by Fischer et al. (1983) revealed very low homology values among *N. dassonvillei* strains and *Actinomadura* species (7–12%).

The genus *Nocardiopsis* in its present form, however, is characterized by a considerable heterogeneity. Using DNA/DNA hybridization, Poschner et al. (1985) found that *N. coeruleofusca* and *N. longispora* were moderately related to each other (45% homology). *Nocardiopsis africana* and *N. longispora,* however, were only remotely related (11–13% homology), while *N. africana* showed a substantial degree of relationship to certain *Actinomadura* species (20–64% homology), especially *A. roseola* (64%). Furthermore, according to Poschner et al., neither *N. africana, N. longispora,* or *N. coeruleofusca* showed any relationship to the type species *N. dassonvillei* (8–12% homology). In addition, *N. dassonvillei* was heterogeneous, as shown by Fischer et al. (1983), who studied four strains of this species by DNA/DNA hybridization. The homology

Table 31.7.
Differentiation of the genus **Nocardiopsis** *from other genera of actinomycetes*

Characteristics	*Nocardiopsis*	*Saccharothrix*	*Actinomadura*	*Nocardia*	*Streptomyces*
Morphology of sporulation of aerial mycelium	Long chains	Long chains	± short spore chains, straight, hooks, or spirals	Mostly lacking or short and long chains	± long chains, straight or spirals
Fragmentation of substrate mycelium	+ / ±	+	−	+	−
Cell wall chemotype	III	III	III	IV	I
Characteristic whole-cell sugar pattern	− (mannose)	Galactose, rhamnose, mannose	Madurose	Arabinose, galactose	−
Mycolic acids	−	−	−	+	−
Predominating menaquinones	MK-10 (H_4,H_6,H_8) or MK-9(H_4)	MK-9(H_4) and MK-10(H_4)	MK-9(H_4,H_6, H_8)	MK-8(H_4) or MK-9(H_2)	MK-9 (H_4,H_6,H_8)
Phospholipid type	PIII,[b] PIV[c]	PII	PI, PIV	PII	PII
Predominating phospholipid[d]	PC, PG, APG (PIII)	PE	PIM, PI, DPG (PI) or PE and Glu NU (PIV)	PE	PE

[a]Symbols: see Table 31.1; ±, weak.
[b]Phospholipid type according to Lechevalier et al. (1977), referring to *N. dassonvillei.*
[c]Phospholipid type according to Shearer et al. (1983), referring to *N. mutabilis.*
[d]Abbreviations of phospholipids: *APG,* acylphosphatidylglycerol; *DPG,* diphosphatidylglycerol; *PC,* phosphatidyline; *PE,* phosphatidylethanolamine; *PG,* phosphatidylglycerol; *PI,* phosphatidylinositol; *PIM,* phosphatidylinositol mannosides; *Glu NU,* phospholipids of unknown structure containing glucosamine.

values obtained were 98–100% or 36–38%, respectively. The heterogeneity of this species was also reflected by differences in the mol% G + C of their DNA and their menaquinone patterns.

This situation demonstrates the necessity of further comparative biochemical and genetic studies of all members of the genus, including *N. mutabilis* and the recombined *N. flava,* as well as the recently described genus *Saccharothrix* (Labeda et al., 1984), all of which share many morphological and biochemical properties.

In view of the above biochemical and genetic results, any discussion about the phylogenetic placement of the genus *Nocardiopsis* seems premature.

Further Reading

Athalye, M., M. Goodfellow, J. Lacey and R. P. White. 1985. Numerical classification of *Actinomadura* and *Nocardiopsis.* Int. J. Syst. Bacteriol. *35:* 86–98.
Fischer, A., R. M. Kroppenstedt and E. Stackebrandt. 1983. Molecular-genetic and chemotaxonomic studies on *Actinomadura* and *Nocardiopsis.* J. Gen. Microbiol. *129:* 3433–3446.
Poschner, J., R. M. Kroppenstedt, A. Fischer and E. Stackebrandt. 1985. DNA-DNA reassociation and chemotaxonomic studies on *Actinomadura, Microbispora, Microtetraspora, Micropolyspora,* and *Nocardiopsis.* Syst. Appl. Microbiol. *6:* 264–270.

Differentiation and characteristics of the genus Nocardiopsis

The species of *Nocardiopsis* may be distinguished by means of the color of their mature aerial mycelium and the color of the substrate mycelium. Additional physiological characteristics, listed in Table 31.8, may be helpful. Since in most cases only one (the type) strain has been studied, their value in species differentiation is restricted. The same applies to the ability to utilize different carbon sources. Since the species specificity of carbon source utilization has not been proved to be reliable and available data are incomplete for all taxa concerned, these data have been omitted from Table 31.8.

List of species of the genus Nocardiopsis

1. **Nocardiopsis dassonvillei** (Brocq-Rousseu 1904) Meyer 1976, 487.[AL] (*Streptothrix dassonvillei* Brocq-Rousseu 1904, 228.)

das.son.vil'lei. M.L. gen. n. *dassonvillei,* named after Charles Dassonville (a French microbiologist and veterinarian at the Pasteur Institute).

Aerial hyphae (see Figs. 31.6 and 31.7) are long, moderately branched, and, at the beginning of sporulation, more or less zig-zag shaped. They then divide into long segments that subsequently subdivide into smaller spores of irregular size. Spores are elongated and smooth.

Good or abundant growth is found on peptone-glucose medium, yeast extract–malt extract agar, oatmeal agar inorganic salts–starch agar (ISP 4), and Bennett sucrose agar (Waksman, 1961). Aerial mycelium, when present, is farinaceous, white or yellowish to grayish. Colonies have dense, filamentous margins. Color of substrate mycelium on the media mentioned is either yellowish-brown or olive-colored to dark brown. Gordon and Horan (1968) reported on production of yellowish, greenish-yellow, or brown diffusible pigment by some strains. Colonies of a few strains display purple-colored crystals identified by Gerber (1966) as crystals of iodinin.

Nitrite may or may not be produced from nitrate.

Adenine, esculin, casein, DNA, gelatin, guanine, hypoxanthine, starch, tyrosine, and xanthine are degraded but testosterone is not.

Table 31.8.
Differentiating characteristics of the species of the genus **Nocardiopsis**[a]

Species	Color of aerial mycelium[b]	Color of substrate mycelium[b]	Nitrate reduced to nitrite	Degradation of: Esculin	Casein	DNA	Gelatin	Guanine	Hypoxanthine	Starch	Testosterone	Tyrosine	Xanthine
1. *N. dassonvillei*	W, Cr, Y, Gr	Y, O, Br	d	d	+	+	+	+	+	+	−	+	+
2. *N. africana*[c]	Bl	O, Br	ND	ND	ND	ND	ND	ND	ND	ND	ND	ND	ND
3. *N. coeruleofusca*[c]	Bl	Br	−	−	+	+	+	−	−	+	−	+	−
4. *N. flava*[c]	W	Y, O	−	+	−	ND	ND	ND	ND	ND	ND	ND	−
5. *N. longispora*[b]	Bl	Y, Br, O	+	+	+	+	+	−	+	+	+	+	−
6. *N. mutabilis*[c]	W, Y, O	Y, Br	+	+	+	ND	+	ND	+	+	ND	+	−
7. *N. syringae*[c]	V	Br	−	+	+	+	+	−	+	+	ND	+	−

[a]Symbols: see Table 31.3; *ND*, no data.
[b]Abbreviations: *W*, white; *Y*, yellow; *Cr*, cream; *O*, orange; *Gr*, gray; *Bl*, blue; *Br*, brown; *V*, violet.
[c]Only the type strain has been studied.

Optimum growth temperature 28–37°C.

The mol% G + C of the DNA is 64.5–69.0 (T_m) (type strain 69.0%).

Type strain: ATCC 23218.

1a. **Nocardiopsis dassonvillei** subsp. **dassonvillei** (Brocq-Rousseu) Meyer 1976, 487.[AL]

The description is as for the species. It differs from subsp. *prasina* by its ability to utilize maltose and rhamnose as sole carbon source; by its white, yellowish, or grayish aerial mycelium; and by growth at pH 6.0–8.0 (optimum growth at pH 8.0).

Type strain: ATCC 23218.

1b. **Nocardiopsis dassonvillei** subsp. **prasina** Miyashita, Mikami and Arai 1984, 405.[VP]

pra.si.na. Gr. adj. *prasina* leek green (referring to the color of the mature aerial mycelium).

This subspecies possesses the same morphological characteristics as the species. It differs by its green aerial mycelium on yeast extract–malt extract agar and especially on oatmeal agar, by its higher optimum pH (9.0–10.0) for growth, and by its lack of ability to utilize maltose and rhamnose as sole carbon sources.

Type strain: JCM 3336.

Further Comments

Besides the above division into two subspecies based on morphological and physiological criteria, there is evidence from menaquinone and DNA base composition that *N. dassonvillei* is heterogeneous (Fischer et al., 1983; Athalye et al., 1984). There was a difference in the DNA G + C content of 5% among the four strains studied. Further, DNA/DNA homology values for the same four strains of *N. dassonvillei* divided this taxon into two pairs that were either highly related (98–100%) or moderately related (36–38%) to each other (Fischer et al., 1983). Also, although these same strains synthesized menaquinones with a long isoprenylic side chain (MK-10(H_2,H_4,H_6)), two strains contained in addition remarkably large amounts of the MK-9(H_4,H_6) types.

Comprehensive studies of a greater number of strains are required to elucidate the heterogeneity of this taxon.

2. **Nocardiopsis africana** (Preobrazhenskaya and Sveshnikova 1974) Preobrazhenskaya and Sveshnikova 1985. 224.[VP] (Effective publication: Preobrazhenskaya, Sveshnikova and Gauze 1982, 111.) (*Actinomadura africana* Preobrazhenskaya and Sveshnikova 1974, 865.)

a.fri.ca'na. M.L. fem. adj. *africana* referring to Africa (where the soil sample was taken).

Aerial hyphae moderately long, straight or slightly flexuous, moderately branched, totally fragmenting into spores. Spore surface smooth.

On oatmeal agar aerial mycelium is dark grayish blue, substrate mycelium brownish orange colored, with no diffusible pigment. On inorganic salts–starch agar (Gauze 1) aerial mycelium grayish blue, substrate mycelium brownish red to brownish violet, with grayish-violet diffusible pigment; on peptone-glucose agar aerial mycelium is absent or scant and cream colored, substrate mycelium yellowish brown, with no diffusible pigment.

The mol% G + C of the DNA is 64.0 (T_m) (Poschner et al., 1985).

Type strain: INA 1839.

3. **Nocardiopsis coeruleofusca** (Preobrazhenskaya and Sveshnikova 1974) Preobrazhenskaya and Sveshnikova 1985, 224.[VP] (Effective publication: Preobrazhenskaya, Sveshnikova and Gauze 1982, 111.) (*Actinomadura coeruleofusca* Preobrazhenskaya and Sveshnikova 1974, 864.)

coe.ru.le.o.fus'ca. L. adj. *coeruleus* blue; M.L. adj. *fuscus* brown; M.L. adj. *coeruleofuscus* blue-brown (referring to the color of aerial and substrate mycelium).

Aerial hyphae (see Fig. 31.8) long, straight or slightly bent, moderately branched, totally fragmenting into spores. Spore surface smooth.

On oatmeal agar aerial mycelium is dark grayish blue, substrate mycelium yellow or yellowish brown, with no diffusible pigment. On inorganic salts–starch agar (Gauze 1), aerial mycelium dark grayish blue, substrate mycelium dark yellow to orange-brown, with no diffusible pigment. On peptone-glucose medium, aerial mycelium absent or whitish to cream colored, substrate mycelium brownish yellow, with no diffusible pigment.

Nitrite not produced from nitrate.

Casein, gelatin, starch, and tyrosine are degraded but esculin, DNA, guanine, hypoxanthine, testosterone, and xanthine are not.

The mol% G + C of the DNA is 67.0 (T_m) (Poschner et al., 1985).

Type strain: INA 1335.

4. **Nocardiopsis flava** (Gauze, Maksimova, Olkhovatova, Sveshnikova, Kochetkova and Ilchenko 1974) Gauze and Sveshnikova 1985, 224.[VP] (Ef-

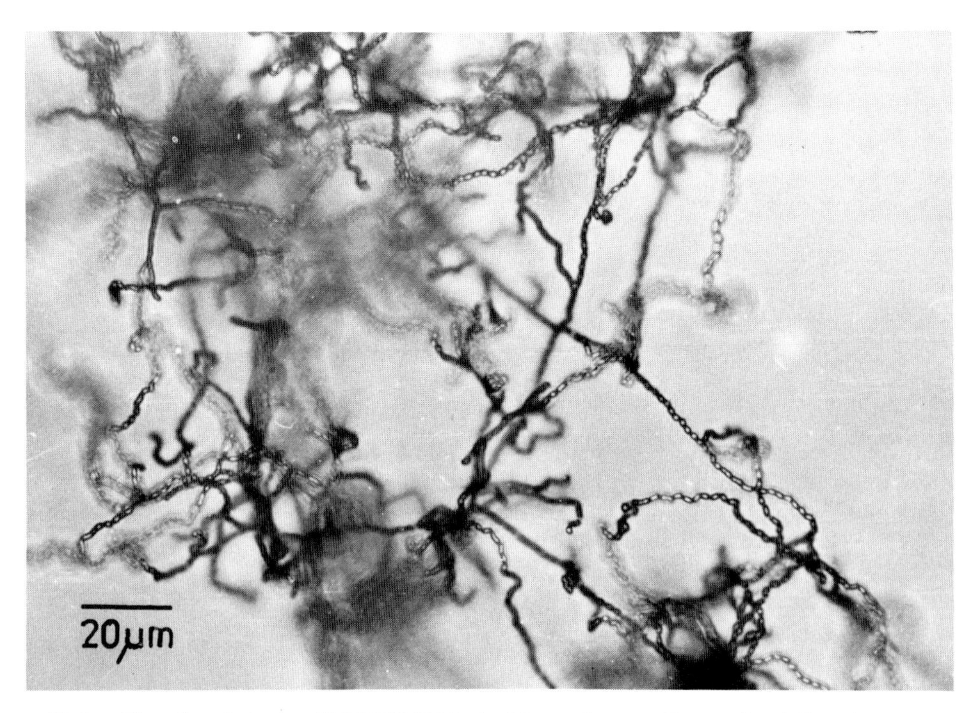

Figure 31.6. *Nocardiopsis dassonvillei* ATCC 23219. Aerial mycelium totally sporulated. Czapek sucrose agar, 7 days.

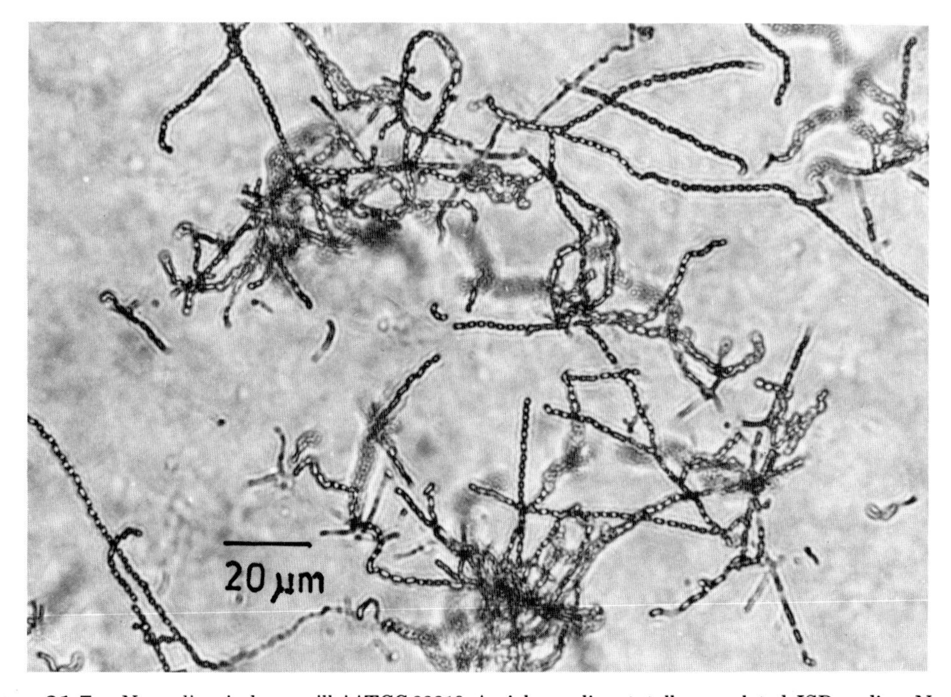

Figure 31.7. *Nocardiopsis dassonvillei* ATCC 23219. Aerial mycelium totally sporulated. ISP medium No. 4, 4 weeks.

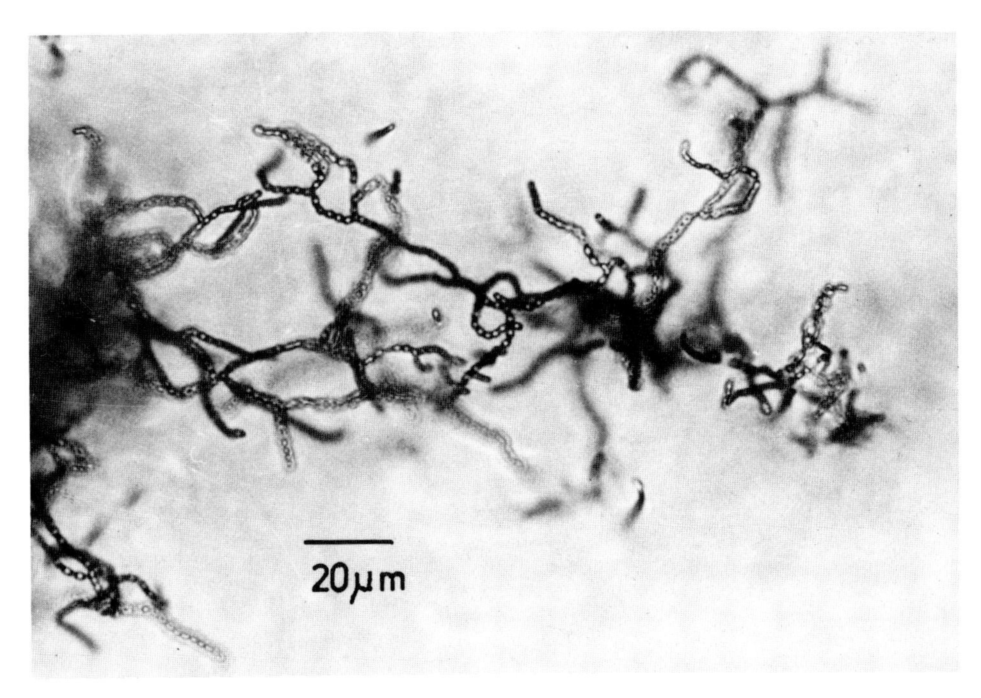

Figure 31.8. *Nocardiopsis coeruleofusca* type strain INA 1335. Aerial mycelium totally sporulated. Oatmeal agar, 6 weeks.

fective publication: Preobrazhenskaya, Sveshnikova and Gauze 1982, 111.) (*Actinomadura flava* Gauze, Maksimova, Olkhovatova, Sveshnikova, Kochetkova and Ilchenko 1974, 771.)

fla′va. M.L. fem. adj. *flava* yellow (referring to the color of the substrate mycelium).

Aerial hyphae long, straight. Spore surface smooth.

On inorganic salts–starch agar (Gauze 1), aerial mycelium is white, substrate mycelium pale yellow to orange-yellow, with no soluble pigment. On oatmeal agar aerial mycelium is white, substrate mycelium yellow to orange-yellow, with no soluble pigment. On glucose-asparagine agar, aerial mycelium scant and whitish, substrate mycelium yellow to orange-yellow, with no soluble pigment. On Czapek-sucrose agar, aerial mycelium white, substrate mycelium pale yellow to orange-yellow, with no soluble pigment.

Nitrite not produced from nitrate.

Esculin is degraded but xanthine is not; other reactions not tested.

Produces madumycin (Gauze et al., 1974).

Type strain: INA 2171 (= ATCC 29533).

5. **Nocardiopsis longispora** (Preobrazhenskaya and Sveshnikova 1974) Preobrazhenskaya and Sveshnikova 1985, 224.[VP] (Effective publication: Preobrazhenskaya, Sveshnikova and Gauze 1982, 111.) (*Actinomadura longispora* Preobrazhenskaya and Sveshnikova 1974, 866.)

lon.gi.spo′ra. M.L. fem. adj. *longispora* referring to the oblong shape of the spores.

Aerial hyphae (see Fig. 31.9) long, straight or slightly flexuous, moderately branched, totally fragmenting into oblong spores. Spore surface smooth.

On oatmeal agar aerial mycelium scant and white turning blue, substrate mycelium colorless or yellow to orange colored, with no soluble pigment. On inorganic salts–starch agar (Gauze 1), development of aerial mycelium delayed, with pale blue color; substrate mycelium yellow to orange colored, with no diffusible pigment. On peptone-glucose medium, aerial mycelium absent or cream colored, substrate mycelium brownish orange, with no soluble pigment.

Nitrite produced from nitrate.

Esculin, casein, DNA, gelatin, hypoxanthine, starch, testosterone, and tyrosine are degraded but guanine and xanthine are not.

The mol% G + C of the DNA is 68.0 (T_m) (Poschner et al., 1985).

Type strain: INA 10222.

6. **Nocardiopsis mutabilis** Shearer, Colman and Nash 1983.[VP]

mu.ta′bi.lis. M.L. adj. *mutabilis* changeable, variable, inconstant (referring to the variety of colonial morphology observed).

Aerial hyphae either unbranched or moderately and irregularly branched, straight or irregularly curved. Prior to sporulation hyphae show nocardioid zig-zag formation, and completely fragment into spores of irregular length. Spore surface smooth.

On oatmeal agar growth fair to good; aerial mycelium abundant but thin, and white, substrate mycelium yellow to yellowish brown, with no soluble pigment. On yeast extract–malt extract agar growth excellent; raised, wrinkled, aerial mycelium abundant and white, substrate mycelium yellow to yellowish brown, with no soluble pigment. On inorganic salts–starch agar (ISP 4) good growth; aerial mycelium abundant but thin and white, substrate mycelium yellow to yellowish brown, light brown soluble pigment variably present. On glycerol-asparagine agar good growth; aerial mycelium abundant but thin and white, substrate mycelium bright yellow, light brown soluble pigment variably present.

Nitrite produced from nitrate.

Esculin, casein, gelatin, hypoxanthine, starch, and tyrosine are degraded but adenine and xanthine are not.

Temperature range of growth, 15–45°C.

Produces the antibiotic polynitroxin (Shearer et al., 1983).

Type strain: ATCC 31520.

Further comments: The description of the species was based on a single strain that displayed some morphological peculiarities. Particularly on rich media, the strain showed a variety of colony textures; there were two basic colony types with a continuous spectrum of intermediates. One basic type was light yellow butyrous, and convex with a smooth surface; this type never produced any visible aerial mycelium. The second colony

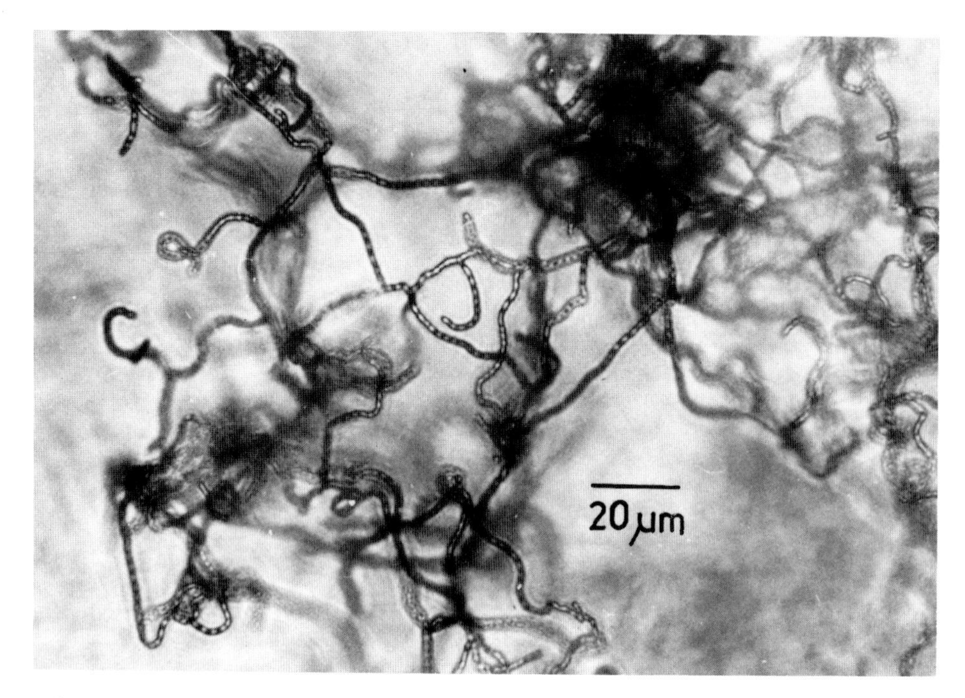

Figure 31.9. *Nocardiopsis longispora* type strain INA 10222. Aerial mycelium totally sporulated. Oatmeal agar, 6 weeks.

type was yellowish brown, leathery, and conical-crateriform; this type became completely covered with a white aerial mycelium (Shearer et al., 1983). The substrate mycelium was well developed, moderately to densely branched on solid media. The degree of fragmentation depended on medium and age of culture. The yellow butyrous colonies usually began to fragment after 4 days, whereas substrate mycelium of the yellowish-brown leathery colonies remained relatively unfragmented after 21 days or fragmented after 4–6 days depending on the medium. Growth in submerged cultures consisted largely of very short to relatively long rods. Branched hyphae were rarely observed.

As already stressed by Shearer et al., (1983), at present the generic placement of *N. mutabilis* presents problems, since this species possesses a type PIV phospholipid pattern that differs from that of *N. dassonvillei* as well as the recently described genus *Saccharothrix* (see "Taxonomic Comments").

7. **Nocardiopsis syringae** Gauze and Sveshnikova 1985, 224.[VP] (Effective publication: Gauze, Sveshnikova, Ukholina, Komarova and Bazhanov 1977, 483.)

sy.rin′gae. M.L. fem. n. gen. *syringa* lilac (referring to the color of aerial mycelium).

Aerial hyphae completely fragmenting into spores. Spore surface smooth.

On oatmeal agar aerial mycelium lilac colored, substrate mycelium brown with a violet tinge, with brownish diffusible pigment. On inorganic salts–starch agar (Gauze 1), aerial mycelium pale to dark lilac colored, sometimes with a brownish tinge; substrate mycelium dark brown. On Czapek and glucose-asparagine agars, aerial mycelium lilac colored to dark brownish violet, substrate mycelium dark brown with a violet tinge, with brown diffusible pigment.

Nitrite not produced from nitrate.

Esculin, casein, DNA, gelatin, hypoxanthine, starch, and tyrosine are degraded but guanine and xanthine are not.

Produces nocamycin (Gauze et al., 1977).

Type strain: INA 2240.

Species Incertae Sedis

1. *Nocardiopsis antarcticus* Abyzov, Phillippova, and Kuznetsov 1983a, 91.[VP] (Effective publication: Abyzov, Phillippova and Kuznetsov 1983b, 559.)

Type strain: VKM A-836.

In its cultural characteristics this species is rather similar to *N. dassonvillei.* The description of this new species is mainly based on the chromogenicity and the capability of growing within a range of 4–42°C (optimum temperature 37°C). Since the strain has not yet been distributed for further study and the differences seem to be insignificant to create a new species, the taxon is considered "*incertae sedis.*"

2. "*Nocardiopsis streptosporus*" Liu, Ruan and Yan 1984, 26.

Type strain not designated.

Description: Liu, Z., J. Ruan and X. Yan. 1984. Acta Microbiol. Sin. *24:* 26–29 (in Chinese).

3. "*Nocardiopsis trehalosei.*" Nomen nudum.

Only cited by Dolak et al. (1980) for antibiotic production. Strain NRRL 12026.

Genus Streptoalloteichus *Tomita, Nakakita, Hoshino, Numata and Kawaguchi 1987, 211*[VP]

KOJI TOMITA

Strep.to.al.lo.tei'chus. Gr. adj. *streptos* bent, turned; Gr. adj. *allos* different; Gr. n. *teichos* wall; M.L. masc. n. *Streptoalloteichus* streptomycete with different wall.

Slender, well-branched hyphae. **The aerial mycelium bears chains of 5–50 spores,** 0.5–1.2 μm in diameter. The spore chain and spore morphology is indistinguishable from that of the genus *Streptomyces.* Fragmentation of the substrate mycelium is absent. The vegetative hyphae bear oval or spherical **sporangium-like vessels,** which envelope one spore or a single row of two to four spores. The spores are **motile** with a single polar flagellum. Aerobic. Chemo-organotrophic. Mesophilic or thermotolerant. Gram-positive **Cell walls contain *meso*-diaminopimelic acid (*meso*-DAP). Whole-cell hydrolysates contain galactose, mannose, and rhamnose but no madurose. Phospholipids include phosphatidylethanolamine but not phosphatidylglucosamine or phosphatidylcholine. Major menaquinones MK-9(H_6) and MK-10(H_6).** Habitat soil.

Type species: Streptoalloteichus hindustanus Tomita, Nakakita, Hoshino, Numata and Kawaguchi 1987, 211.

Further Descriptive Information

Both substrate and aerial mycelia are well developed and branched, averaging 0.5 μm in diameter. Chains of spores are formed only at the tip of the aerial mycelium. These may be either long (10–50 spores in a chain) or short hooked, branching spore chains (Figs. 31.10 and 31.11). The arthrospores in the long chains are oval to cylindrical, 0.5–2.0 μm in diameter, and the conidiospores in the short chains are barrel shaped. Both types of spores have a smooth surface. The substrate mycelium does not fragment.

Single, oval or spherical sporangium-like vessels enveloping one spore or a single row of two to four spores are randomly formed among the vegetative hyphae (Fig. 31.12). The sporangia measure 1.5–4.5 × 2.7–7.0 μm, and the sporangiospores are 0.9–1.5 × 1.2–4.0 μm in size, oval or rod shaped, with a single polar flagellum. Globose dense bodies consisting of coalesced vegetative mycelium, and sclerotia, may be formed on the aerial mycelium.

Purified cell wall preparations contain *meso*-DAP (Becker et al., 1965). The *L*- isomer and the hydroxy analogs of DAP are not found. Whole-cell hydrolysates contain galactose, mannose, and rhamnose, but no diagnostic sugars such as madurose, arabinose, or xylose (Lechevalier, 1968b). The phospholipids include phosphatidylethanolamine but not phosphatidylglucosamine or phosphatidylcholine, and hence belong to type PII (Lechevalier et al., 1977). Mass spectra show that the major menaquinones are MK-9(H_6) and MK-10(H_6) (Collins et al., 1977).

The substrate mycelium penetrates the agar and is thin, especially in chemically defined media; it is covered with a thick mass of white aerial mycelium that turns to pale yellow after sporulation.

Gelatin and starch are hydrolyzed; skim milk is coagulated and slightly peptonized. Melanoid pigments are not formed in ISP media 1, 6, and 7 (Shirling and Gottlieb, 1966). Tyrosinase reaction is negative. Abundant growth at 30–50°C, but none at 56°C. Growth occurs with 5% (w/v) NaCl or less, but not with 7% (w/v). Resistant to kanamycin, gentamicin, ampicillin, cephalothin, erythromycin, and tetracycline at 100 μg/ml, and to novobiocin at 25 μg/ml. Less resistant to chloramphenicol and rifampicin.

Enrichment and Isolation Procedures

Streptoalloteichus hindustanus was isolated from dry soil samples collected in Gujarat or adjacent states where the natural vegetation is dry tropical forest and scrub. A pulverized dry soil sample was transferred with a nylon sponge to yeast extract–malt extract agar (ISP 2) or Bennett agar, supplemented with kanamycin or gentamicin at 10 μg/ml and nystatin at 100 μg/ml. The agar plates were incubated 1–2 weeks at 43°C. The colonies were distinguishable from those of most species of the

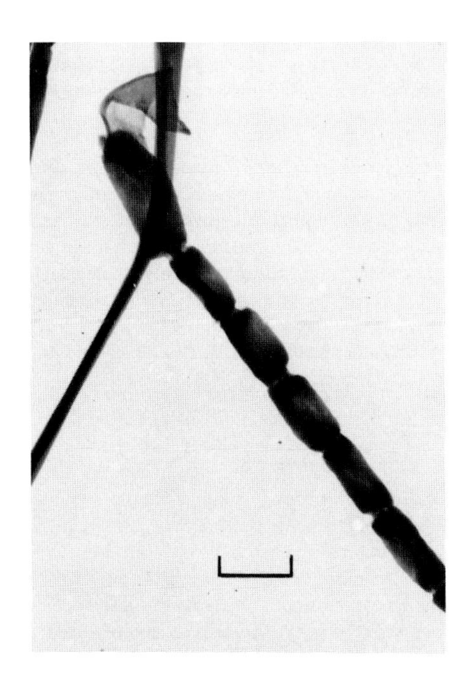

Figure 31.10. *Streptoalloteichus hindustanus:* A fragment of a long spiral chain of arthrospores with a smooth surface. Transmission electron micrograph. *Bar,* 1 μm.

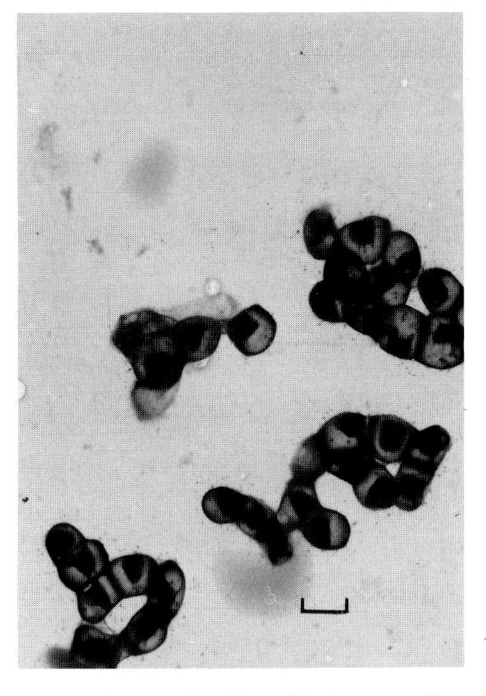

Figure 31.11. *Streptoalloteichus hindustanus:* Fragments of branched short chains of barrel-shaped conidiospores. Transmission electron micrograph. *Bar,* 1 μm.

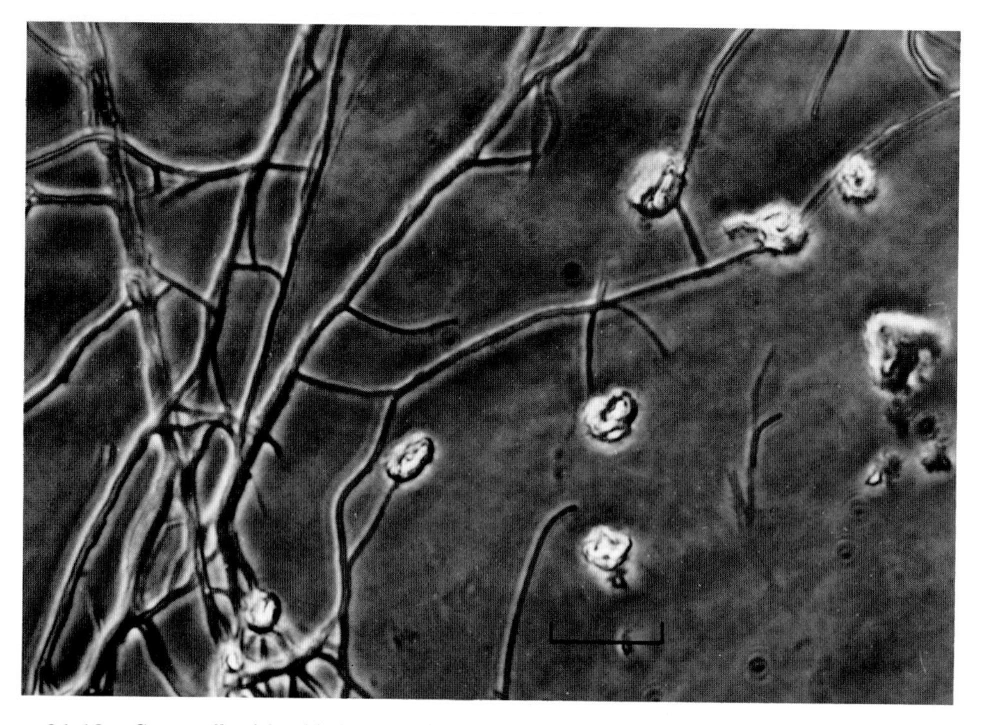

Figure 31.12. *Streptoalloteichus hindustanus:* Small oval sporangia-like vessels enveloping a single spore or a row of two to four spores, which are formed on the sporangiophore of the substrate mycelium. *Bar,* 10 μm.

genus *Streptomyces* by pale pinkish-yellow aerial mycelium, and the lack of a distinct reverse-side color.

Maintenance Procedures

The sporulated portion of aerial mycelium, obtained from growth on yeast extract-malt extract agar slants, is suspended in autoclaved 10% (w/v) skim milk. Screw-capped tubes containing the spore suspension can be stored at −80°C. Lyophilization of the spore suspension is carried out by standard procedures for actinomycetes, and the lyophilized culture is preserved under a vacuum of between 0.05 and 0.005 mm Hg.

Procedures for Testing of Special Characters

Formation and Observation of Sporangia

The organism is grown at 28°C for 3–4 weeks on yeast extract-malt extract agar or glycerol-asparagine agar. The sporangia and other structures in the vegetative mycelium are investigated by microscopic observation of a thin-layer culture prepared by the coverslip technique (Kawato and Shinobu, 1959).

The following procedures are also applicable to observe sporangia. Glycerol-asparagine broth (ISP 5 without agar) or soil extract broth* are used. Three milliliters of the medium in metal-capped tubes (18×180 mm in size) are autoclaved, inoculated heavily with a spore suspension, and incubated as slants for 4–6 weeks at 28°C. The sporangia are observed at hyphal tips of mycelial masses.

Differentiation of the genus **Streptoalloteichus** *from other genera*

Based on the data of follow-up studies after the descriptions of Tomita et al. (1978), the taxonomic position of the genus *Streptoalloteichus* has been reconsidered (Tomita et al. 1987). Although strains of the genus *Streptoalloteichus* form sporangium-like vessels enveloping one to four motile spores in the vegetative mycelium, aerial spore chains similar to those of *Streptomyces* species are predominantly formed. The cellular chemical compositions comprising the amino acids and sugars of the cell wall, the phospholipids, and menaquinones, as well as the formation of arthrospore chains, indicate that the genus *Streptoalloteichus* has some similarities to the genera *Microtetraspora* and *Nocardiopsis.* However, the spore chain morphology of the genera *Microtetraspora* and *Nocardiopsis* is definitely different from that of the genus *Streptoalloteichus.* The genus *Microtetraspora* bears chains of four spores by basipetal budding, and the genus *Nocardiopsis* sporulates in total parts of the aerial mycelium. The genus *Streptoalloteichus* is similar to the genus *Actinomadura* but differs from the latter in its spore-chain morphology and the cellular chemical compositions; the latter contains madurose in cell hydrolysates, no phosphatidylethanolamine in phospholipids, and no MK-10 analog in its menaquinones. Among the species of the genus *Actinomadura,* two distinct types of spore chain are reported. Most species

**Soil extract broth:* glucose, 1.0 g; K_2HPO_4, 0.5 g; soil extract, 1000 ml; pH 6.8–7.0. Preparation: Soil extract is prepared by heating 1 kg of garden soil with 1000 ml of tap water in an autoclave for 30 min. About 0.5 g of calcium carbonate is added and the soil suspension is filtered through double paper filters until clear. The extract may be bottled and sterilized in 1000-ml quantities. One-third strength of this medium is preferred.

form only hooked spore chains, whereas *Actinomadura kijaniata* Horan and Brodsky 1982 and "*A. albolutea*" Tohyama et al. 1984 bear the *Nocardiopsis*-type spore chains.

The morphology and cellular chemical compositions support the distinction of the genus *Streptoalloteichus* from the other members of spore chain-forming actinomycetes (Table 31.9).

Taxonomic Comments

The spore chain cluster, which is one of two spore chain types of *Streptoalloteichus hindustanus,* consists of curved or L-shaped conidiophore chains with many branches and often develops into a thick mass. The spore chains contain 10–20 spores per chain. This type of spore chain is also formed by some species of the genus *Streptomyces,* such as *S. massasporeus, S. ramulosus, S. catenulae,* and *S. antimycoticus.* However, these four species have type I cell wall composition, containing *L*-DAP and glycine, and no diagnostic sugar component (Tomita et al., 1978).

The species most similar to *S. hindustanus* is "*Streptomyces tenebrarius*" ATCC 17920 (Higgens and Kastner, 1967). They have several important characteristics in common, such as the cell wall type, short spore chain cluster, sclerotium formation, pale pinkish-yellow aerial mycelium, carbohydrate utilization profile, and thermoduric property. On the other hand, *S. hindustanus* may be differentiated from "*S. tenebrarius*" by the following characteristics. The sporangia-like vessels enveloping flagellate spores are not described in "*S. tenebrarius*" (Higgens and Kastner, 1967). Light sensitivity for formation of aerial mycelium was not observed in *S. hindustanus.* The NaCl tolerance was 5% (w/v) in *S. hindustanus* and 8% (w/v) in "*S. tenebrarius.*" *Streptoalloteichus hindustanus* produces tallysomycins A, B, and C as well as nebramycin factors II, IV′, and V′, whereas "*S. tenebrarius*" forms nebramycin factors I to XIII.

Further Comments

Streptoalloteichus hindustanus has the ability to bear both nonmotile conidiospores and motile sporangiospores. This morphology may be derived from the adaptation to the original environments of this organism, which have distinct dry and rainy seasons in a year.

Further Reading

Tomita, K., Y. Uenoyama, K. Numata, T. Sasahira, Y. Hoshino, K. Fujisawa, H. Tsukiura and H. Kawaguchi. 1978. *Streptoalloteichus*, a new genus of the family *Actinoplanaceae*. J. Antibiot. *31:* 497–510.

List of species of the genus **Streptoalloteichus**

1. **Streptoalloteichus hindustanus** Tomita, Nakakita, Hoshino, Numata and Kawaguchi 1987, 211.[VP]

hin.du.stan′us. M.L. adj. *hindustanus* referring to source of isolate in Western India.

The morphology and cell chemistry of this species are as given for the genus.

The aerial mycelium is abundant on most diagnostic media. After sporulation, the aerial mycelium is a pale color that combines yellow, pink, and gray. The vegetative mycelium is colorless to light yellowish brown. Tyrosinase reaction is negative. Among the diagnostic carbon sources, hexoses are utilized, but pentoses and sugar alcohols are not.

The temperature range for growth is 20–54°C, and no growth is observed at 56°C. The optimal temperature for growth is 45°C.

The cultural and physiological characteristics of *S. hindustanus* are shown in Tables 31.10 and 31.11, respectively.

Type strain: ATCC 31217.

Table 31.9.
Differential characteristics of the genus **Streptoalloteichus** *and four related genera*[a]

Characteristics	*Streptoalloteichus*	*Thermomonospora*	*Actinosynnema*	*Nocardiopsis*	*Saccharothrix*[b]
Morphology					
Monospores	–	+	–	–	–
Chains of arthrospores	+	–	+	+	+
Sporulation in whole parts of aerial mycelium	–	+	–	+	+
Sporangia-like vessels	+	–	–	–	–
Synnemata	–	–	+	–	–
Motile spores	+	–	+	–	–
Physiology					
Species growing at 50°C or more	+	+	–	–	–
Cell chemistry					
Whole-cell sugars	Galactose, mannose, rhamnose	None or glucose, ribose[c]	Galactose, mannose	None or galactose, mannose, rhamnose	Galactose, mannose, ribose[d]
Phospholipids	PII	PII[d]	PII[e]	PIII or PIV[f]	PII
Menaquinones	MK-9(H_6), MK-10(H_6)	MK-10(H_4)[g]	MK-9(H_4,H_6)[h]	MK-10(H_{0-6})[i] or MK-9(H_4)[j]	MK-9(H_4), MK-10(H_4)

[a]Symbols: see Table 31.1.
[b]Data from Labeda et al. (1984).
[c]Data from McCarthy and Cross (1984a).
[d]K. Tomita and Y. Nakakita, unpublished data.
[e]Data from Hasegawa et al. (1979).
[f]Data from Shearer et al. (1983).
[g]Data from Collins et al. (1982d).
[h]Data from Hasegawa (1985).
[i]Data from Kuraishi (1985).
[j]Data from Takahashi et al. (1986).

Table 31.10.
Cultural characteristics of **Streptoalloteichus hindustanus**

Medium	Aerial mycelium	Vegetative mycelium	Diffusible pigment
Yeast extract-malt extract (ISP 2)	Thick, pale yellowish pink (31)[a]	Light yellowish brown (76)	None
Inorganic salts-starch (ISP 4)	Pale pinkish yellow	Thin, colorless	None
Glycerol-asparagine (ISP 5)	Patches, white later yellowish gray (93)	Thin, colorless to grayish yellow (90)	None
Peptone-yeast extract-iron (ISP 6)	Scant, white	Moderate brown (58)	Light yellowish brown (76)

[a]ISCC-NBS centroid color chart code (Kelly and Judd, 1976).

Table 31.11.
Physiological characteristics of **Streptoalloteichus hindustanus**[a]

Hydrolysis of	
Gelatin	+
Starch	+
Casein	+
Production of	
Tyrosinase	–
Catalase	+
Nitrate reductase	+
Tolerance to	
5% (w/v) NaCl	+
7% (w/v) NaCl	–
Utilization of[b]	
L-Arabinose	–
D-Fructose	+
D-Galactose	+
D-Glucose	+
Inositol	–
Lactose	+(w)
D-Mannitol	–
D-Mannose	+
D-Raffinose	–
L-Rhamnose	–
Sucrose	+(w)
D-Sorbitol	–
D-Xylose	–
Growth at	
52°C	+
56°C	–
Antibiotics produced	Tallysomycins A, B and C and nebramycin factors II, IV′, and V′

[a]Symbols: see Table 31.1; +(*w*), weak.
[b]Basal medium: Pridham-Gottlieb medium (ISP 9), supplemented with 0.01% (w/v) yeast extract.

SECTION 32

Thermoactinomycetes

John Lacey

The thermoactinomycetes comprise only one genus, consisting mostly of thermophilic organisms. The genus *Thermoactinomyces* is among the earliest known bacteria to have been classified with the actinomycetes. They were discovered almost simultaneously with other thermophilic bacteria, and *Thermoactinomyces vulgaris* was described by Tsiklinsky in 1899. However, recent studies have suggested that they should no longer be classified in the *Actinomycetales* but placed in the family *Bacillaceae* (Stackebrandt and Woese, 1981b). Their spores are true endospores, with the same stages of development, heat resistance, and other properties as bacterial endospores (Cross et al., 1971; Lacey and Vince, 1971); their G + C contents are lower than those of other actinomycetes and their 16S rRNA oligonucleotide sequences also suggest a close relationship with *Bacillus* species (Stackebrandt and Woese, 1981b). However, like true actinomycetes, they produce a well-developed mycelium and they can resemble *Thermomonospora,* at least superficially, in their morphology. It is therefore convenient for the present to consider *Thermoactinomyces* together with other actinomycetes.

Aerial mycelium is produced by all species. This is yellow in *T. dichotomicus* and white in all other species. In some species, notably *T. sacchari* and *T. putidus,* the aerial mycelium may be transient, rapidly autolyzing and depositing the spores on the surface of the substrate. The endospores are formed singly on both aerial and substrate mycelia, either sessile (*T. thalpophilus, T. intermedius,* and *T. peptonophilus*) or on unbranched (*T. vulgaris, T. sacchari, T. putidus,* and *T. intermedius*) or dichotomously branched (*T. dichotomicus*) sporophores. All species are aerobic and saprophytic and most are thermophilic, growing at 30–60°C. Only one species, *T. peptonophilus,* is mesophilic, growing only below 45°C (Nonomura and Ohara, 1971c). Cell walls of *Thermoactinomyces* species contain *meso*-diaminopimelic acid (*meso* DAP) but no characteristic sugars or other amino acids (wall chemotype III; Becker et al., 1965). Menaquinones are unsaturated with, predominantly, seven or nine isoprene units (MK-7 or MK-9; Collins et al., 1982c). The spores contain dipicolinic acid (Cross et al., 1971; Lacey, 1971b; Attwell, 1978). Species have a DNA base composition averaging 53 mol% G + C (Craveri and Manachini, 1966; Craveri et al., 1966b; Fritzsche, 1967). The thermophilic species, but not *T. peptonophilus,* tolerate up to 200 μg novobiocin/ml (Cross, 1968). Those species that have been tested are resistant to lysozyme (Kurup et al., 1980; Goodfellow and Pirouz, 1982b).

Thermophilic species of *Thermoactinomyces* are found widely in nature. They are most numerous in molding hay and cereal grains and in decaying vegetable materials that have heated spontaneously to temperatures of 50°C or more. They can perhaps also grow in plant litter heated by insolation and can be found in soil, river, lacustrine, and marine sediments, in dairy products, in water, and even in ice on Mount Kilimanjaro. They may be isolated from the lungs and sputum of man and animals that have been exposed to moldy fodders (Wenzel et al., 1967), and will pass through the digestive system of cattle to be excreted and contaminate milk and soil (Kosmachev, 1963; Cross, 1968; Falkowski, 1978). They may also be found in the humidifiers of air-conditioning systems and have been implicated in various forms of hypersensitivity pneumonitis (extrinsic allergic alveolitis), especially farmer's lung and bagassosis (Pepys et al., 1963; Lacey, 1971b; Kurup et al., 1975, 1980). Spores of some *Thermoactinomyces* species have been shown to survive for extremely long periods, being found in soil cores from Roman sites (Cross and Attwell, 1974). By contrast, *T. peptonophilus* has rarely been isolated and only from soil, probably because it has much more specialized nutritional requirements than other species (Nonomura and Ohara, 1971c).

The long history of the genus and changing concepts in the definition of species have led to taxonomic confusion. The thermophilic species, apart from *T. dichotomicus,* which was previously classified in the genus *Actinobifida,* are clearly closely related, sharing up to 85% S_{SM} similarity in numerical taxonomic studies (Goodfellow and Pirouz, 1982b). As a consequence, they were, for a time, considered to be a single, variable species (Kuster and Locci, 1964), and *T. vulgaris* was allocated characters that were at variance with the original description. The epithet *vulgaris* has now been used in different senses by different authors (Kurup et al., 1975; Cross and Unsworth, 1981). In the following chapter, the approach of Cross and Unsworth (1981) has been followed, with *T. vulgaris* having the characteristics of Tsiklisky's original description. *Thermoactinomyces peptonophilus* has been little studied; although it is reported to produce endospores (Attwell, 1978), it differs from other *Thermoactinomyces* species in a number of respects, not least in its temperature and nutritional requirements and intolerance of novobiocin. A more detailed comparison is required to confirm its classification in this genus.

Genus **Thermoactinomyces** *Tsiklinsky 1899, 501[AL*]*

JOHN LACEY AND TOM CROSS

Ther.mo.ac.ti.no.my'ces. Gr. adj. *thermos* hot; Gr. n. *actis, actinos* a ray; Gr. n. *myces* fungus; M.L. masc. n. *Thermoactinomyces* heat (loving) ray fungus.

Substrate mycelium well-developed, branched, septate, 0.4–0.8 µm in diameter. **Aerial mycelium** 0.5–1.0 µm diameter, variable in amount, sometimes transient lysing to leave layer of spores. **Spores formed singly on aerial and substrate hyphae, sessile or on simple or branched sporophores,** globose, often ridged, 0.5–1.5 µm in diameter, **with the structure and properties of bacterial endospores.** Gram-positive, not acid fast, chemo-organotrophic, aerobic. **Most species thermophilic, fast growing,** giving flat or ridged colonies with entire or filamentous margins. Resistant to lysozyme. **Wall peptidoglycan containing *meso*-diaminopimelic acid (*meso*-DAP) but no characteristic sugars.** Major menaquinones unsaturated with seven or nine isoprene units. Habitats are soil, molding and decaying plant materials, and composts, often with spontaneous heating. Mol% G + C of the DNA is 52.0–54.8 (T_m).

Type species: *Thermoactinomyces vulgaris* Tsiklinsky, 1899, 501.

Further Descriptive Information

Cell Morphology

The substrate mycelium consists of stable, branched, septate hyphae, from which aerial hyphae arise, first vertically but, in most species then forming a loose network of almost straight hyphae over the substrate (Locci, 1972). Intercalary growth of the primary substrate mycelium has been reported for *T. thalpophilus* (Kretschmer, 1984b). Aerial hyphae of *T. vulgaris* are hydrophobic and may be found within 6.5 h of spore germination (Kretschmer, 1984a). The aerial hyphae may autolyze, within 2–4 days in *T. vulgaris* or more rapidly in *T. sacchari,* depositing spores on the agar surface (Küster and Locci, 1963a; Lacey, 1971b). Spores are formed singly on both substrate (Fig. 32.1) and aerial mycelium and may be either sessile or on sporophores (Figs. 32.2–32.6). In *T. dichotomicus,* both sporophores and mycelium may be dichotomously branched (Fig. 32.6). The spores are spheroidal, 0.5–1.5 µm in diameter, with a ridged surface that gives an angular appearance by transmission electron microscopy. By scanning electron microscopy, the ridges can be seen to form pentagonal and hexagonal areas on the spore surface. Most species resemble that studied by McVittie et al. (1972) in having ~12 pentagonal and 12 hexagonal faces (Fig. 32.7), although *T. dichotomicus* has only ~12 faces in total (Fig. 32.8). The spores are refractile and phase-bright by light microscopy, staining only with endospore stains.

Cell Wall Composition

The wall peptidoglycan contains *meso*-DAP but no diagnostic sugars (wall chemotype III; Becker et al., 1965). Menaquinones are unsaturated, with seven (*T. vulgaris, T. dichotomicus,* and *T. sacchari*) or nine (*T. putidus* and *T. thalpophilus*) isoprene units in major amounts (Collins et al., 1982c). Organisms contain straight-chain and *iso-* and *anteiso-*branched-chain fatty acids, but phospholipids of diagnostic value have not been studied (Goodfellow and Cross, 1984b). As with bacterial endospores, spores of *Thermoactinomyces* species contain dipicolinic acid: 6.5–7% (w/w) in *T. vulgaris* and *T. sacchari,* 3.6% in *T. dichotomicus,* but only 0.6% in *T. peptonophilus* (Cross et al., 1968b; Lacey, 1971b; Attwell, 1978). They also contain large amounts of calcium and magnesium ions (Kalakoutskii et al., 1969; Kalakoutskii and Agre, 1973, 1976). The DNA base composition of *Thermoactinomyces* species is lower in guanine plus cytosine (G + C) than in other actinomycete genera and similar in that respect to the DNA of *Bacillus* species. Values of mol% G + C (T_m) of 52.0 (Fritzsche, 1967), 53.4–54.4 (Craveri and Manachini, 1966), and 54.1 and 54.8 (Craveri et al., 1966b) have been reported.

Fine Structure

In thin section, hyphae of *Thermoactinomyces* species are 0.3–0.6 µm diameter and bounded by a wall about 22 nm thick, similar to that of other Gram-positive bacteria. The cytoplasm is bounded by a plasma membrane, a typical unit membrane about 10 nm thick, that bears tubular vesicular mesosomes 0.1–0.3 µm in diameter. Septa are formed by a single ingrowth of plasma membrane and hyphal wall, often bearing mesosomes (type 1; Williams et al., 1973). Nuclear material occurs as an axial filament, and ribosomes, about 12 nm in diameter, are usually present.

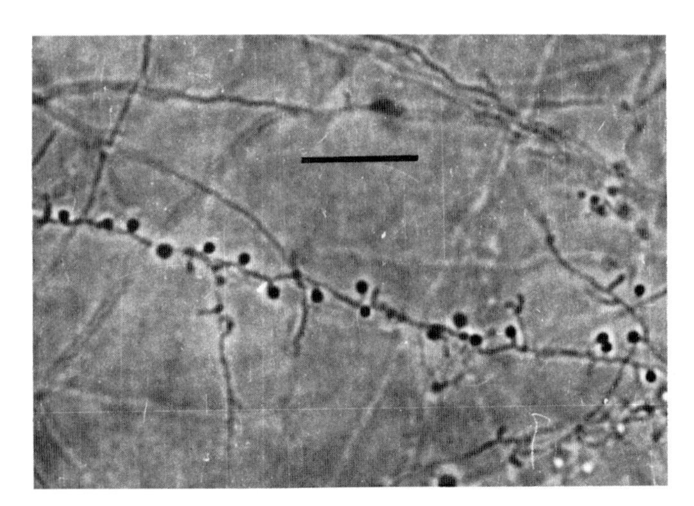

Figure 32.1. Substrate mycelium spores of *Thermoactinomyces sacchari.* Half-strength nutrient agar, incubation 55°C. *Bar,* 10 µm.

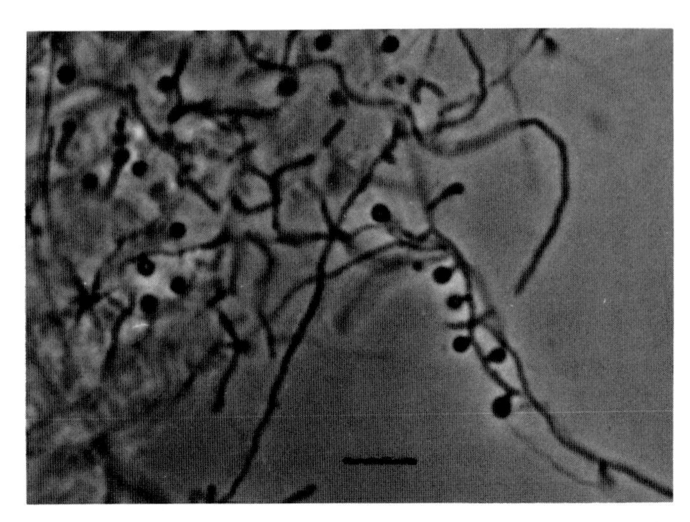

Figure 32.2. Inclined coverslip preparation of growth of *T. vulgaris.* CYC agar, incubation 50°C. *Bar,* 5 µm.

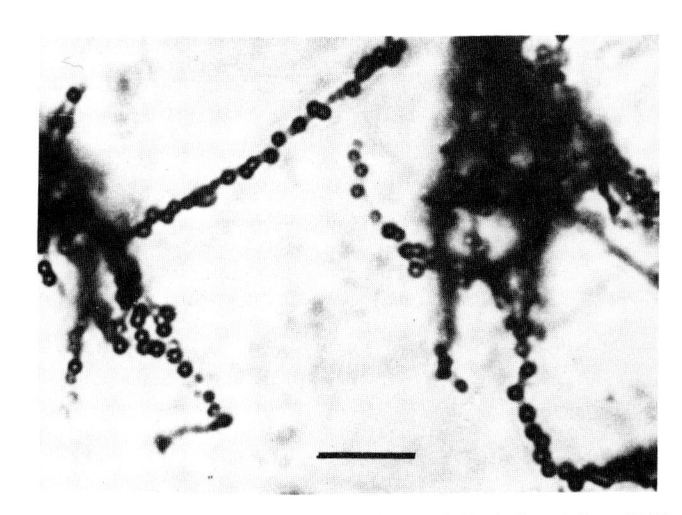

Figure 32.3. Aerial mycelium spores of *T. thalpophilus*. Half-strength nutrient agar, incubation 55°C. *Bar,* 10 μm.

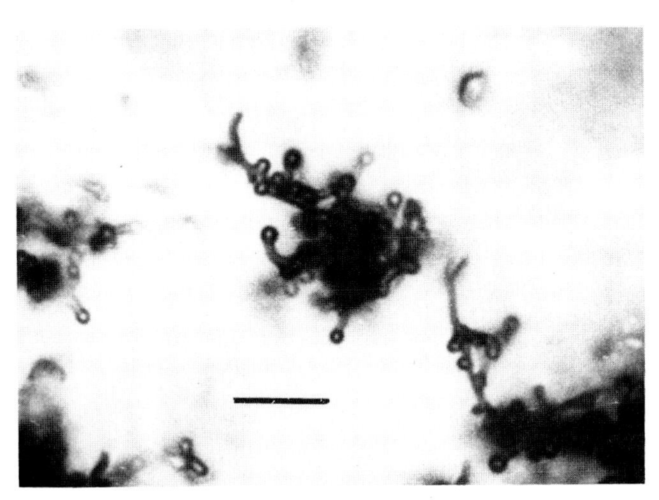

Figure 32.4. Aerial mycelium spores of *T. sacchari*. Yeast-malt agar, incubation 55°C. *Bar,* 5 μm.

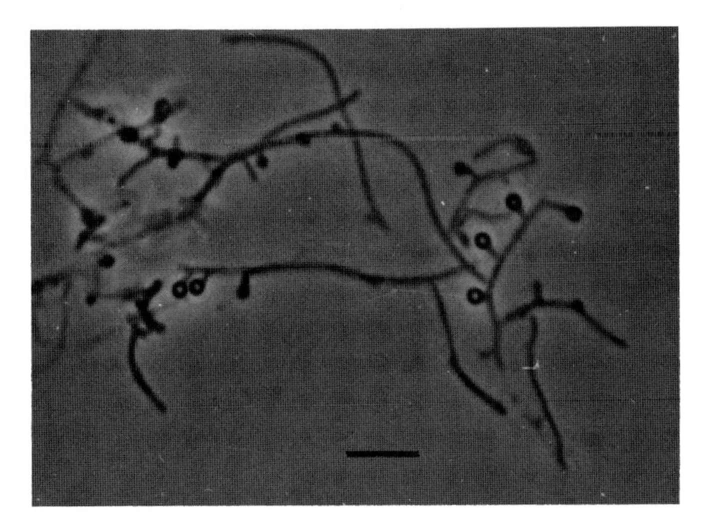

Figure 32.5. Inclined coverslip preparation of growth of *T. putidus*. CYC agar, incubation 50°C. *Bar,* 10 μm.

Figure 32.6. Aerial mycelium spores of *T. dichotomicus*. Half-strength nutrient agar, incubation 55°C. *Bar,* 20 μm.

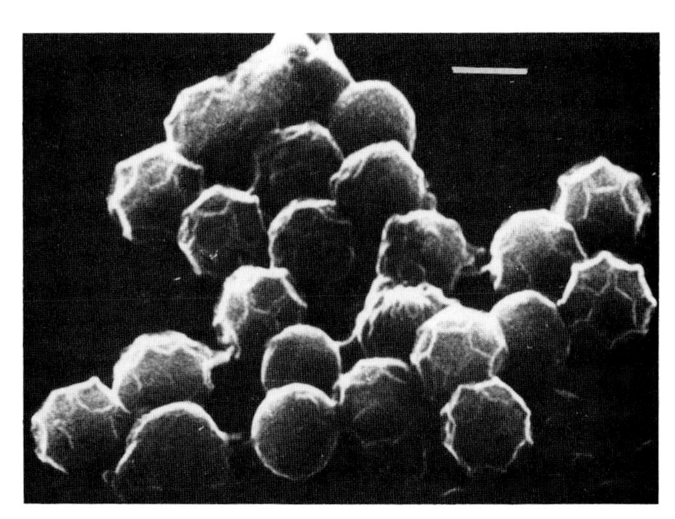

Figure 32.7. Scanning electron micrograph of spores of *T. sacchari*. *Bar,* 1 μm.

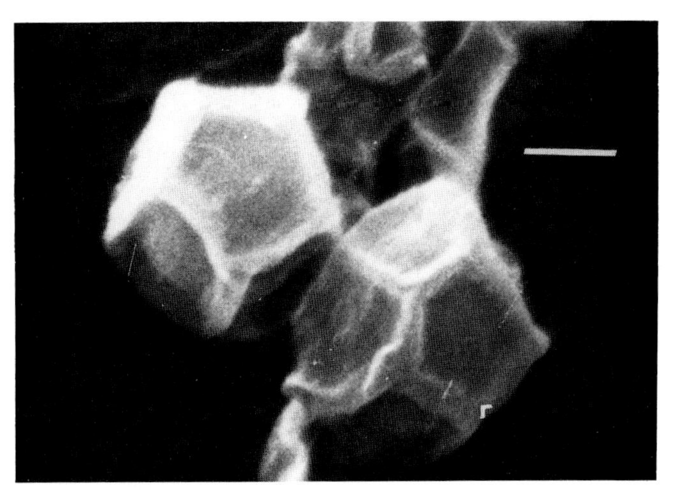

Figure 32.8. Scanning electron micrograph of spores of *T. dichotomicus*. *Bar,* 0.5 μm.

Such structures have been observed in *T. thalpophilus, T. sacchari,* and *T. peptonophilus* (Cross et al., 1968b; Dorokhova et al., 1968, 1970b, 1970c; Lacey and Vince, 1971; Attwell, 1978), but in *T. sacchari,* although the cytoplasm is uniformly dense in young cells, it soon becomes coarsely granular and less dense and is then released by lysis of the cells (Lacey and Vince, 1971).

Spore formation follows the same stages of development as with *Bacillus* endospores. The first stage is the formation of a double membrane across the cell, forming a spore septum, about 18 h after inoculation. The septum lengthens as the area of attachment to the plasma membrane progressively decreases, engulfing cytoplasm and nuclear material from the mother cell. Mesosomes are closely associated with this process. Eventually this membrane closes and breaks away from the plasma membrane to form a forespore, one to a cell, which moves to the hyphal wall and out into a lateral sporangium, sessile in *T. thalpophilus* and terminal on a short sporophore in *T. sacchari.* A cortex is formed up to 0.17 µm thick, with a dense inner layer, a thick, pale middle layer, and a dark, granular outer layer. The inner spore coat is multilayered, with 6–10 alternate light and dark layers about 27 nm thick, forming first in discrete zones. The two-layered outer spore coat consists of an inner, ridged layer 25–70 nm thick surrounded by a dense layer 8 nm thick (Fig. 32.9). After maturity, often no sporangium is recognizable, probably due to lysis, but a thin, dark membrane may surround the spore, perhaps the remains of the sporangium or of an exosporium (Fig. 32.10). The outer coat of the mature spore consists of parallel rows of fibrils, each measuring about 5 nm. Nuclear material is often poorly differentiated from the cytoplasm but may appear irregular, U-shaped, or as separate areas at opposite poles of the spore. Ribosome granules, up to 15 nm in diameter, are frequent in the cytoplasm, and the middle cortex may show radial striations (Cross et al., 1971; Lacey and Vince, 1971; McVittie et al., 1972). Spore

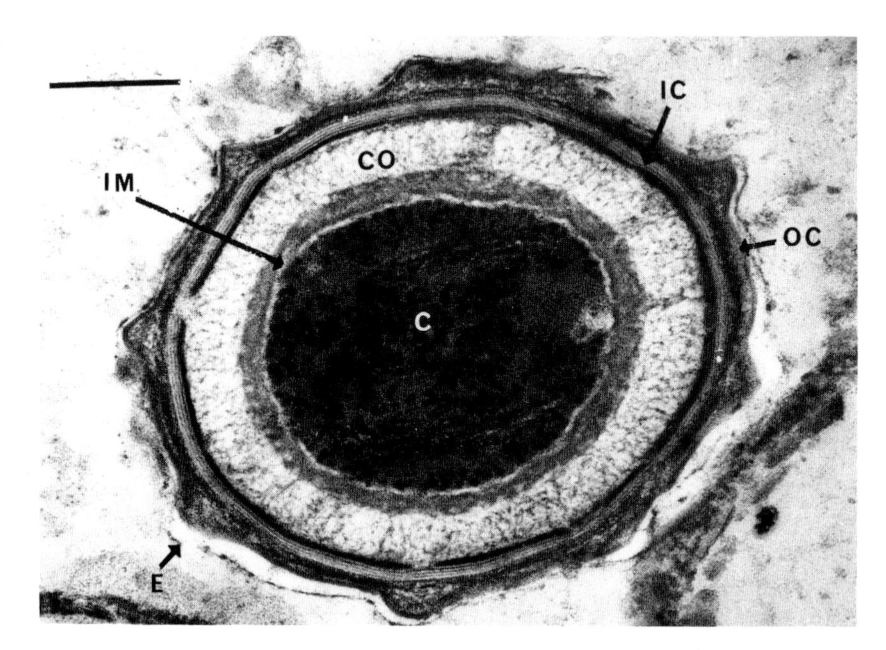

Figure 32.9. Thin section of *T. sacchari* endospore. *Bar,* 0.1 µm.

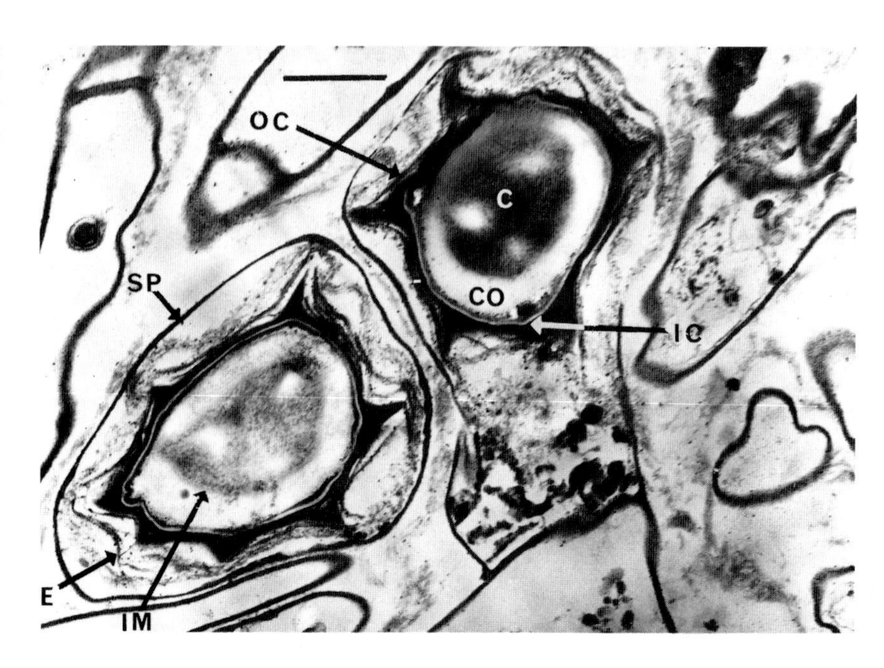

Figure 32.10. Thin section of *T. dichotomicus* endospore. *Bar,* 0.4 µm. *C,* core; *CO,* cortex; *IC,* inner spore coat; *IM,* inner forespore membrane; *OC,* outer spore coat; *E,* possible exosporium; *SP,* sporangial wall.

development in *T. dichotomicus* and *T. peptonophilus* is essentially the same, differing chiefly in the degree of sporophore development and in the possible presence of an exosporium in *T. dichotomicus* (Cross et al., 1968b; Dorokhova et al., 1970b; Attwell, 1978).

Spore germination may occur within 3 h of incubation but frequently takes longer. Enhanced CO_2 levels (Kretschmer and Jacob, 1983) and a pH greater than 7.0 are reported to be essential to germination. Optimum pH is 8.0–10.0 (Foerster, 1975). Activation and germination may be stimulated by calcium dipicolinate, L-alanine, L-leucine, L-α-aminobutyric acid, inosine, and adenosine; by heating briefly to 100°C; or by cooling to 20°C. However, heating may also deactivate spores (Agre et al., 1972a, 1972b; Attwell et al., 1972; Kirillova et al., 1974; Foerster, 1975; Kalakoutskii and Agre, 1976). Agre et al. (1972a, 1972b) observed variation in the ability of spores to germinate, identifying in aqueous spore suspensions three types consisting of those germinating without previous activation, those requiring mild activation treatment, and those requiring severe activation treatment. Activated spores either had germinated, were activated, or could be deactivated by heating at 80°C. Ability to germinate was decreased by drying. Kokina and Agre (1977a, 1977b) found that cultures degenerated with repeated subculture, resulting in less sporulation and poor spore germination.

On germination, the cortex disappears, the core swells to fill the space, and U-shaped fibrillar areas of nuclear material arise. The core swells to about 2 μm diameter with destruction of the spore coats, although the outermost layer may remain more or less intact. Finally, a germ tube is formed with an axial nuclear area of coarsely granular material and the spore contents then senesce (Dorokhova et al., 1968; Lacey and Vince, 1971).

Cultural Characteristics

Colonies are usually fast growing at 55°C on nutrient or Czapek-yeast-casein (CYC)[†] agars, flat or lightly ridged, and usually with white aerial mycelium. However, in *T. sacchari*, aerial mycelium is often transient and colonies may appear "bacterial," whereas in *T. dichotomicus* the aerial mycelium is yellow. In *T. putidus* and *T. sacchari* lysis may deposit spores on the agar surface (Lacey, 1971b; Unsworth, 1978); *T. peptonophilus* is exceptional in being mesophilic and having a requirement for peptone and B vitamins for growth while aerial mycelium production is favored by 0.02% (w/v) glycerol or glucose as supplements in glycerol-asparagine, oatmeal, yeast-starch and peptone-yeast extract (PY)[‡] agars (Nonomura and Ohara, 1971). Soluble pigments are produced by some species and *T. thalpophilus*, *T. putidus*, and *T. intermedius* produce brown, water-soluble melanin pigments with 0.5% L-tyrosine in CYC agar (Lacey, 1971b; Unsworth, 1978; Kurup et al., 1980).

Life Cycles

Alternative life cycles for a *Thermoactinomyces* species, *T. thalpophilus*, have been suggested (Kretschmer, 1984a). The primary mycelium formed on germination, with septa about every 10 μm, continues to grow, without producing endospores, until growth conditions become limiting, at which time two types of secondary mycelium, substrate or aerial, may form alternatively depending on the composition of the substrate, while older hyphae may lyse. Hydrophobic aerial mycelium may develop as soon as nutrients become limiting, while secondary substrate mycelium develops after a lag phase. Coincidentally, the interval between septa decreases to 0.75–0.8 μm and endospores form. Whether secondary substrate or aerial mycelia are formed depends on the nature of the limitation and differs between different agar media. The low phosphate in corn steep liquor (CSL) agar[§] favors secondary aerial mycelium production, whereas the low nitrogen in Czapek-Dox (CD) agar[‖] favors the production of secondary substrate mycelium. Changes in gene expression occur during the lag period before secondary substrate mycelium is formed, resulting in the production of extracellular proteases. Since spore formation starts as soon as aerial mycelium is produced, mature spores are produced by this route ~5 h earlier than by the secondary substrate mycelium route. Intercalary extension of cells may sometimes be observed, giving a zig-zag appearance, but separation of cells as in *Nocardia* never occurs (Kretschmer, 1984c).

Nutrition and Growth Conditions

Nutritional requirements of *Thermoactinomyces* species are incompletely known. Table 32.2 (below) shows characters distinguishing the different species, and Table 32.3 (below) shows other characteristics of the four most studied species. All the thermophilic species tested degrade casein and, except for some strains of *T. sacchari*, gelatin, but none degrades hypoxanthine or xanthine. *Thermoactinomyces peptonophilus* degrades neither gelatin nor casein. *Thermoactinomyces vulgaris* isolates differ greatly in their ability to utilize individual carbon sources. Few isolates of any species reduce nitrate to nitrite. Water-soluble melanin pigments are produced by *T. thalpophilus*, *T. putidus*, and *T. intermedius* on media containing tyrosine provided that these do not contain inhibitory peptones (Unsworth, 1978). Biotin and methionine are reported to be essential for growth of at least some *Thermoactinomyces* isolates, and no aerial growth is obtained in the absence of methionine (Webley, 1958). Most *T. putidus* isolates produce acid from sucrose, and up to 30% of *T. thalpophilus* strains produce acid from fructose, glycerol, mannitol, mannose, ribose, sucrose, and trehalose. A few isolates of *T. vulgaris* produce acid from these sugars, whereas only 20% of *T. sacchari* produce acid from glycerol but no other sugar. Acid production by other species has not been tested (Unsworth, 1978). *Thermoactinomyces vulgaris*, *T. intermedius*, *T. thalpophilus*, *T. dichotomicus*, and *T. sacchari* are all resistant to lysozyme but the resistance of other species is not known (Kurup et al., 1980; Goodfellow and Pirouz, 1982b).

Thermophilic *Thermoactinomyces* species all grow between 35° and 58°C but *T. putidus* and *T. thalpophilus* grow down to 29°C, *T. sacchari* grows up to 60°C, and *T. vulgaris* and *T. dichotomicus* up to 62°C. *Thermoactinomyces peptonophilus* grows poorly at 25°C and optimally at 35°C, and fails to grow at 45°C (Lacey, 1971b; Nonomura and Ohara, 1971c; Unsworth, 1978).

Production of aerial mycelium may be inhibited when air is replaced by pure oxygen, perhaps through the inactivation of essential -SH groups in thiol enzymes (Webley, 1954). Respiration is greatest in 1–2-day-old cultures bearing aerial mycelium. Vegetative mycelium more than 1 day old and spores respire little (Erikson and Webley, 1953), perhaps a consequence of low cytochrome *a* content (Taptykova et al., 1969).

Enzyme Production

Because of their activity at high temperatures, enzymes of *Thermoactinomyces* species, especially their proteases, lipases, and amylases, have attracted much interest. Thermitase is an extracellular endopeptidase that is reportedly well suited for use in the food industry. It is the principal component of proteases produced in culture medium; temperature of optimum activity against peptide esters is 60°C, against peptide *p*-nitroanilides 65–75°C, and against casein 90°C, although it is rapidly inactivated at this temperature (Behnke et al., 1982). Optimum production

[†]*CYC agar:* Czapek-Dox agar powder (Oxoid CM 97), 33.4 g; yeast extract, 2.0 g; vitamin-free casamino acids, 6.0 g; pH 7.2. Novobiocin (Albamycin, Upjohn; 25 μg/ml) and cycloheximide (Acti-Dione, Koch-Light; 50 μg/ml) added after autoclaving.

[‡]*PY agar:* peptone, 20 g; yeast extract, 20 g; glycerol, 2 g; $MgSO_4 \cdot 7H_2O$, 0.3 g; agar, 20 g; water 1 l; pH 7.6.

[§]*CSL agar:* corn steep liquor (50% dry matter), 5 g; vitamin-free casamino acids (Difco), 3 g; sucrose, 15 g; corn starch, 5 g; $NaNO_3$, 1 g; NaCl, 2.5 g; KCl, 0.025 g; $CaCl_2$, 0.25 g; $FeSO_4 \cdot 7H_2O$, 0.005 g; magnesium glycerophosphate, 0.25 g; agar, 20 g; distilled water, 1 l; pH 7.2.

[‖]*CD agar:* sucrose, 10 g; vitamin-free casamino acids (Difco), 1.5 g; casein peptone (Serval), 1.5 g; NaCl, 1 g; KH_2PO_4, 1.2 g; $Na_2HPO_4 \cdot 2H_2O$, 2.8 g; $CaCO_3$, 0.005 g; $FeSO_4 \cdot 7H_2O$, 0.01 g; $MgSO_4$, 0.5 g; agar, 20 g; water, 1 l; pH 6.3.

of thermitase is at 50–52°C with a pH of 6.6–7.3 (Leuchtenberger et al., 1979). Protease production begins at the transition from exponential to linear growth phases after about 5 h of incubation. During linear growth, up to 45% of the hyphae are lysed. After 22 h of fermentation the enzyme comprises three components, forming 33%, 64%, and 3% of the total, differing in their retention in Sephadex gel and also in the protein bands on polyacrylamide gel electrophoresis. Enzyme production ceases after 10 h of growth when easily utilizable carbohydrates are exhausted, leading to decreased respiration (Taufel et al., 1979; Kretschmer et al., 1982). Enzyme synthesis is promoted by adding rape-oil to the medium as an antifoam agent because lysis of the mycelium is decreased (Leuchtenberger and Ruttloff, 1983). Autolysis is closely associated with heat inactivation of the enzyme thermitase.

Serine proteases reported by Roberts et al. (1977) from both *T. vulgaris* and *T. thalpophilus* were similar, but *T. sacchari* shows no proteolytic activity. However, zymograms of the first two species differ, with *T. vulgaris* showing two intense proteolytic bands at rA 0.3 and 1.4 and *T. thalpophilus* three to four bands between rA 0.1 and 0.4. Proteolytic enzymes, with molecular weights of 27500 and 23800 and produced optimally by some *Thermoactinomyces* isolates at 60–70°C and pH 9.0, cause lysis of heat-inactivated cells of fungi, mycobacteria, and Gram-positive and, especially, Gram-negative bacteria (Desai and Dhala, 1966, 1967a, 1969; Golovina et al., 1973). Strong lipolytic activity may be found in *T. thalpophilus* (Elwan et al. 1978a, 1978b). Lipase is produced optimally at 55°C and pH 6.8 in shake culture in a Czapek medium containing corn oil and 0.2% (w/w) yeast extract, incubated for 24–36 h. Activity of the isolated enzyme is greatest at 55°C and pH 8.0. Inactivation occurs in 45 min at 80°C and in 5 min at 90° and 100°C. Kuo and Hartman (1966) first reported α-amylase activity by a *Thermoactinomyces* species, probably *T. thalpophilus*, and subsequently this was confirmed by Allam et al. (1975), Shimizu et al. (1978), and Obi and Odibo (1984). Although the organism used by Allam et al. (1975) was described as *Thermomonospora vulgaris*, illustrations of it in Hussain et al. (1975) are consistent with *T. thalpophilus*. However, more than one enzyme may be involved. Kuo and Hartman (1966, 1967) found a neutral α-amylase with optimum activity at 60°C and pH 5.9–7.0, whereas other α-amylases with different temperature and pH optima of 70°C and pH 5.0 and 80°C and pH 7.0, respectively, were found in other studies (Shimizu et al., 1978; Obi and Odibo, 1984). The amylase of Shimizu et al. (1978) also had pullulan-hydrolyzing ability, giving panose as the main product, unlike other pullulanases. Cellulolytic activity has been reported in a *Thermoactinomyces* isolate by Hagerdal et al. (1978) but not by Fergus (1969), nor has cellulose utilization ever been found in taxonomic tests. It is possible that the isolate used by Hagerdal et al. (1978) was in fact a species of *Thermomonospora*, a genus in which cellulolytic activity is well known (McCarthy and Cross, 1984a).

Enzymatic profiles of double-dialysis antigens (Edwards, 1972a) on API-ZYM strips (A.P.I. Systems, S.A., Montalieu Vercieu, France) show similarities between *T. vulgaris* and *T. sacchari* in that both possess alkaline phosphatase, C_4 esterase, and C_8 esterase-lipase, all of which were absent from *T. thalpophilus* antigens. P. Boiron (personal communication) has additionally found C_{14} lipase, leucine aryl-amidase, acid phosphatase, and naphthol AS-BI phosphohydrolase in *T. sacchari* preparations. However, *T. thalpophilus* is the only species to possess α-glucosidase. Weak acid phosphatase activity is present in *T. vulgaris* and phosphoamidase activity in *T. thalpophilus*. C_4 esterase, leucine aminopeptidase, and chymotrypsin activity are present in *T. thalpophilus* when whole cells are used (Hollick, 1982).

Antimicrobial Activity

Antimicrobial activity has been reported in cultures of *T. thalpophilus* and *T. dichotomicus* (Krasil'nikov and Agre, 1964a; Craveri et al., 1964; Cross and Unsworth, 1976). Both these species can inhibit growth of *T. vulgaris*, which accounts for the so-called autoinhibition phenomenon (Locci, 1963), observed between isolates when *T. vulgaris* was considered to be a single variable species. *Thermomonospora chromogena* (Krasil'nikov and Agre 1965) McCarthy and Cross 1984b is also inhibited by *Thermoactinomyces dichotomicus*, whereas thermorubin (Moppet et al., 1972) isolated from cultures of *T. thalpophilus* (then named *T. antibioticus*) is more inhibitory to Gram-positive than to Gram-negative bacteria but it is also highly toxic to mammals (Kosmachev, 1962; Terao et al., 1965).

Genetics

Recombination, with substitution of small fragments of homologous segments of genetic material, has been observed with mutants obtained from a *T. thalpophilus* isolate (Hopwood and Ferguson, 1970). It was also shown that *T. thalpophilus*, like other prokaryotes, has partially diploid zygotes. Most recombinants differed from one or another parent strain by only a single marker, irrespective of the coupling of the markers. The process was characterized as transformation that occurred when agar containing a constant amount of DNA was tested with $>10^5$ spores, leading to confluent mycelial growth (Hopwood and Wright, 1972). The transformation frequency has been estimated at 1×10^{-3} to 1×10^{-4} (Kretschmer and Sarfert, 1980). This high transformation frequency in *T. thalpophilus* implies that there is little extracellular DNase activity. Transformation was inhibited if DNase was added during the first 6–8 h of incubation but not if it was added after 7–9 h of growth. Competence develops at the time when aerial mycelium first appears (Hopwood and Wright, 1972).

Phages

Phages to *Thermoactinomyces* species have been reported frequently (Agre, 1961; Patel, 1969; Prauser and Momirova, 1970; Treuhaft, 1977; Kurup and Heinzen, 1978; Sarfert et al., 1979; Kretschmer and Sarfert, 1980; Kretschmer, 1982), mostly from *T. thalpophilus* but also from *T. sacchari* and *T. vulgaris*. The size and structure of phages from different sources have differed, having heads 60–72 × 62–84 nm and tails 5–20 × 90–120 nm (Agre, 1961; Patel, 1969; Kurup and Heinzen, 1978). In some the tails had a helical structure with about 29 turns (Patel, 1969); in two others, the DNA content was 28.8×10^6 and 37×10^6 daltons per phage (Kretschmer, 1982). Infectivity was lost after 10 min at 70°C and pH 3.6 or when treated with H_2O_2 (Kurup and Heinzen, 1978).

Phages differed in their species specificity. Those from *T. sacchari* were species specific but some from *T. thalpophilus* and *T. vulgaris* could infect the other species (Treuhaft, 1977; Kurup and Heinzen, 1978). None infected other genera. Plaque morphology, however, differed among hosts. When *T. thalpophilus* phage infected *T. thalpophilus* confluent lysis was characteristic, but when infecting *T. vulgaris* only small plaques were formed (Treuhaft, 1977). Seven host range/plaque type groups were distinguished using two isolates of *T. vulgaris* and one of *T. thalpophilus* (Treuhaft, 1977), and three types of interaction between phage and host were found in *T. thalpophilus* (Kretschmer, 1982). Multiplication of phages occurred only in the primary mycelium and they decreased in number in secondary mycelium and during sporulation. The phage genome was incorporated into spores early in their formation in a heat-stable state, and only multiplied on germination. Growing secondary substrate mycelium was competent to take up exogenous DNA, but transfection did not occur (Kretschmer, 1980).

Antigenicity

Species of *Thermoactinomyces* differ antigenically, although there are common components that give some cross-reactivity. *Thermoactinomyces vulgaris*, *T. sacchari*, *T. dichotomicus*, and *T. putidus* are serologically homogeneous, whereas *T. thalpophilus* isolates are heterogenous (Arden Jones and Cross, 1980). *Thermoactinomyces sacchari* and *T. intermedius* both cross-react with *T. vulgaris* and *T. thalpophilus* but not with one another (Lacey, 1971b; Kurup et al., 1976a, 1980). Distinctive protein bands are found on polyacrylamide gel electrophoresis of double-dialysis antigens (Edwards, 1972a). *Thermoactinomyces vulgaris* gives

10–16 bands, 5 of which are glycoprotein. There are major bands at rA 0.42, 0.66, and 1.32 that differ from a more diffuse band at rA 0.54 and a prominent solitary band ar rA 0.97 in *T. thalpophilus. Thermoactinomyces sacchari* combined the features of both preceding species but two bands are distinctive (Roberts et al., 1977; Hollick et al., 1979). Pyridine extracts revealed only 8–11 protein bands, 3 of which were glycoprotein. Crossed immunoelectrophoresis of *T. vulgaris* antigen revealed 15 immunogens when tested against homologous antiserum and 19 bands by flat bed isoelectric focusing with isoelectric points between 3.5 and 5.7. Most were heat labile and unaffected by pronase (Hollick and Larsh, 1979). Most components of *T. sacchari* antigen are heat labile and partially sensitive to pronase (Lehrer and Salvaggio, 1978), perhaps because this species lacks serine protease (Roberts et al., 1977).

Antibiotic Sensitivity

There is little information on antibiotic sensitivity of *Thermoactinomyces* species. However, all the thermophilic species, but not the mesophilic *T. peptonophilus,* are tolerant to nalidixic acid and up to 200 μg novobiocin/ml medium (Cross, 1968; Cross and Attwell, 1975). Most isolates are sensitive to ampicillin, benzylpenicillin, cephaloridine, chloramphenicol, colistin sulfate, demethylchlortetracycline, erythromycin, gentamycin, kanamycin, neomycin, nitrofurantoin, oleandomycin, penicillin, streptomycin, and tetracycline; *T. dichotomicus* isolates are also sensitive to lincomycin and vancomycin and *T. vulgaris* to chloramphenicol. Isolates of some species differ in their sensitivity (Goodfellow and Pirouz, 1982b).

Pathogenicity

Thermoactinomyces species have often been implicated as causes of extrinsic allergic alveolitis (hypersensitivity pneumonitis). *Thermoactinomyces vulgaris, T. thalpophilus,* and *T. dichotomicus* have all been implicated in farmer's lung disease, although *Faenia rectivirgula* (Cross et al., 1968a) is the major source of the antigen. *Thermoactinomyces putidus* has been identified from a lung biopsy of a patient (Pepys et al., 1963; Wenzel et al., 1967, 1974; Molina, 1974; Cross and Unsworth, 1976; Pether and Greatorex, 1976; Unsworth, 1978; Terho and Lacey, 1979). *Thermoactinomyces sacchari* is a principal source of bagassosis antigen (Lacey, 1971b) and *T. vulgaris* is reported to cause humidifier fever (Banaszak et al., 1970; Sweet et al. 1971). Often the role of individual species has not been clear because *T. vulgaris* and *T. thalpophilus* have not been differentiated. More farmers have been found with precipitins to *T. vulgaris* than to *T. thalpophilus* (Greatorex and Pether, 1975; Terho and Lacey, 1979), but most screening has been done with antigens prepared from *T. thalpophilus.* Although *T. vulgaris* is much more abundant than *T. thalpophilus* in hay (Terho and Lacey, 1979), isolates of *T. thalpophilus* were chosen for antigen production because they were regarded, at the time, as more vigorously growing variants of *T. vulgaris,* following the species concept of Küster and Locci (1964). Commercial antigens, labeled "*T. vulgaris,*" have represented both species and it is necessary that these be identified and the role of *Thermoactinomyces* species in farmer's lung be reevaluated (Lacey, 1981). *Thermoactinomyces vulgaris* and *T. thalpophilus,* at least, should be present in panels of antigens used for screening (Terho and Lacey, 1979).

Ecology

Thermoactinomyces species are most abundant in moldy fodders and other vegetable matter, including cotton, straw, cereal grains, hay, composts, and manures (Forsyth and Webley, 1948; Henssen, 1957; Gregory and Lacey, 1963; Gregory et al., 1963; Fergus, 1964; Craveri et al., 1966a; Desai and Dhala, 1966; Lacey, 1973, 1978; Lacey and Lacey, 1987). They are favored by spontaneous heating to temperatures up to 70°C, often resulting in production of more than 10^7 spores/g dry weight. The spores easily become airborne when the substrate is disturbed. However, growth may be limited where aeration is restricted. Growth in agar cultures was halved by decreasing the oxygen concentrations in air to 1% (v/v) and although some growth occurred with 0.1% oxygen, little or no sporulation occurred with less than 1% (Deploey and Fergus, 1975). *Thermoactinomyces vulgaris* is usually the most abundant species, but *T. thalpophilus* is also common (Terho and Lacey, 1979) and *T. dichotomicus* has been isolated from mushroom composts. *Thermoactinomyces sacchari* is most abundant in heated sugar cane bagasse, where it occupies a niche similar to that of *T. vulgaris* and *T. thalpophilus* in moldy hay. All three species have also been isolated from soil and peat, although usually in small numbers seldom exceeding 10^4/g dry weight of soil (Küster and Locci, 1963a; Goodfellow and Cross, 1974). Many originate from manure, sewage, or dung added to the soil (Cross, 1968; Diab, 1978), but deposition of airborne spores from moldy hay on farms is also possible, while some growth may occur on vegetation heated by the sun. Even in temperate regions solar heating may raise the temperature of soil and litter to more than 30°C (Eggins et al., 1972). Erosion of soil may result in the accumulation of spores in lake muds and marine sediments, giving counts of 10^4–10^6 spores/g dry weight (Cross and Johnston, 1971). The occurrence of *Thermoactinomyces* species in deep mud cores and in archaeological excavations suggests that spores may remain viable for over 2000 years (Cross and Attwell, 1974; Seaward et al. 1976; Unsworth et al., 1977). *Thermoactinomyces putidus* and *T. peptonophilus* have also been isolated from soil (Nonomura and Ohara, 1971c), but *T. intermedius* has been found only in air conditioners, humidifiers, house dust, and grass compost (Kurup et al., 1980), where it occurs with other *Thermoactinomyces* species (Kurup et al., 1976b).

Heat Resistance

Spores of *Thermoactinomyces* species are characteristically heat resistant, surviving up to 4 h at 100°C in sucrose solution or 15 h of dry heat at this temperature (Fergus, 1967). Survival curves and $D_{100°C}$ values have been calculated for *T. thalpophilus* (11 min), *T. dichotomicus* (77 min), and *T. sacchari* (59 min) (Cross et al., 1968b; Lacey, 1971b). *Thermoactinomyces peptonophilus* spores are less heat resistant ($D_{90°C}$ = 45 min) than *T. thalpophilus* ($D_{90°C}$ = 128 min) (Atwell, 1978). A report that heat resistance may be lost in 24 h at 4°C (Kirillova et al., 1973) was not confirmed by Foerster (1978). Heat shock at 100°C or low-temperature storage may also sometimes decrease germination or kill spores (Attwell and Cross, 1973; Ensign, 1978; Foerster, 1978; Kirillova et al., 1973).

Enrichment and Isolation Procedures

Isolation of most thermophilic *Thermoactinomyces* species may be achieved on agar media containing 25 μg novobiocin/ml and 50 μg cycloheximide/ml incubated at 50–55°C. Suitable media include CYC agar and half-strength nutrient or tryptone soya-casein agars (Lacey and Dutkiewicz, 1976b). However, the last two contain NaCl and are unsuitable for *T. putidus.* Isolation of *T. sacchari* is best achieved on yeast malt agar (Shirling and Gottlieb, 1966) containing 25 μg novobiocin and 50 μg cycloheximide/ml incubated at 55°C (Lacey, 1971b). Samples may be suspended in an aqueous diluent containing gelatin (0.5 g/l) or agar (0.2 g/l) and suitable dilutions spread on agar in prepoured plates, or spores may be suspended in the air of a small wind tunnel or sedimentation chamber and plated using an Andersen sampler (Gregory and Lacey, 1963; Lacey and Dutkiewicz, 1976a). *Thermoactinomyces peptonophilus* has been isolated from dry heat-treated (100°C) soil samples using either MGA-SE agar[#] or salts starch agar + B vitamins + antibiotics at pH 8.0 and incubated at 40°C. Growth on MGA-SE agar occurred only on colonies of other actinomycetes because of the restricted nutritional requirement of this species (Nonomura and Ohara, 1971c).

[#] *MGA-SE agar:* glucose, 2.0 g; L-asparagine, 1.0 g; K_2HPO_4, 0.5 g; $MgSO_4 \cdot 7H_2O$, 0.5 g; soil extract, 200 ml; penicillin 1 mg; polymyxin B, 5 mg; cycloheximide, 50 mg; nystatin, 50 mg; agar, 20.0 g; water, 800 ml; pH 8.0. (*Soil extract:* 1000 g soil autoclaved with 1 l water for 30 min, decanted, and filtered.)

Maintenance Procedures

Thermophilic species may be maintained on the same media as used for isolation, incubating for 2–3 days at 50–55°C. Agre (1964) recommended a maize-starch medium** for *T. dichotomicus.* Incubation may be continued for up to 1 week if plates are enclosed in polyethylene bags or placed in sealed containers with some water. Transfer of *T. sacchari* is aided by transfer of agar bearing the culture and by sealing the Petri dish with a broad rubber band (Lacey, 1971b). *Thermoactinomyces peptonophilus* must be grown on glycerol-asparagine agar (Shirling and Gottlieb, 1966) containing 10 g yeast extract/l or on PY agar with incubation at 35°C (Nonomura and Ohara, 1971c).

Cultures can be maintained on agar slopes in screw-capped bottles at room temperature or 4°C, but for long-term preservation lyophilization is preferred, with spores suspended in double-strength skimmed milk or other media.

Differentiation of the genus **Thermoactinomyces** *from other genera*

Some *Thermomonospora* species often appear similar to *Thermoactinomyces* species on isolation plates, growing well at 55°C, producing white aerial mycelium, and having a wall chemotype III. However, the sporophores of the latter show differing degrees of dichotomous branching, causing the spores to appear clustered, although they are formed singly, giving the colony a granular appearance. Also, *Thermomonospora* spores are usually ovoid, with a smooth or spiny surface, they are not endospores, and they are killed at 70°C (Cross and Lacey, 1970; McCarthy and Cross, 1984a). *Thermomonospora* species will not grow in the presence of 25 μg novobiocin/ml. *Saccharomonospora viridis* (Schuurmans, Olson and San Clemente 1956) Nonomura and Ohara 1971c also produces single oval spores that are not endospores and are heat sensitive. Colonies of this taxon are characteristically blue-green but may remain white when grown at suboptimal temperatures. This genus has a wall chemotype IV.

Differences between monosporic actinomycetes genera are summarized in Table 32.1.

Taxonomic Comments

Until recently, the genus *Thermoactinomyces* and its type species, *T. vulgaris,* were considered to be among the oldest actinomycete taxa, having been first described by Tsiklinsky in 1899. However, recent studies have cast doubts on their status as true actinomycetes. Not only do 16S rRNA oligonucleotide sequences suggest a close relationship with the genus *Bacillus,* but also *Thermoactinomyces* species, like *Bacillus* species, produce endospores and have lower G + C contents than other actinomycetes. Consequently classification of the genus *Thermoactinomyces* in the family *Bacillaceae* has been proposed (Stackebrandt and Woese, 1981b). However, because of their morphology, it is still convenient to consider *Thermoactinomyces* with other actinomycetes.

The eighth edition of the *Manual* (Küster, 1974b) listed only two species of *Thermoactinomyces (T. vulgaris* and *T. sacchari)* but there have been many changes since its publication. Thus, the Approved Lists of Bacterial Names (Skerman et al., 1980) included five species and this chapter recognizes seven. The characteristics of phenon 11 of Goodfellow and Pirouz (1982b) suggest that other *Thermoactinomyces* species still await description. Changes since the eighth edition include the transfer of *T. dichotomicus* from the genus *Actinobifida* on the basis of endospore formation, the description of *T. candidus, T. intermedius, T. peptonophilus,* and *T. putidus* as new species, and the revival of *T. thalpophilus.* Numerical studies have shown the genus *Thermoactinomyces* to be defined at the 70% similarity (S_{SM} coefficient) level, although *T. intermedius* and *T. peptonophilus* were not included (Unsworth, 1978). Individual species were defined at the 79% similarity (S_{SM}) level or greater, while in another study *T. vulgaris, T. thalpophilus,* and *T. sacchari* formed an aggregate cluster at the 85% S_{SM} level of similarity (Goodfellow and Pirouz, 1982b).

The status of *T. vulgaris, T. candidus,* and *T. thalpophilus* is confused. There can be little doubt that these species are closely related. Indeed, the epithet *vulgaris* has been used in the literature in three senses: synonymous with *T. candidus,* synonymous with *T. thalpophilus,* and for an aggregate species comprising all the thermophilic taxa except *T. sacchari* and *T. dichotomicus.*

The confusion arises because of changing concepts of *Thermoactinomyces* species. Prior to 1964, six species had been described. Of these, three, "*T. glaucus*" Henssen 1957, "*T. thermophilus*" (Berestnev) Waksman 1961, and "*T. monosporus*" (Lehmann and Schutze) Waksman 1953 (in Waksman and Corke, 1953), are *nomina dubia;* one, "*T. viridis*" Schuurmans, Olson and San Clemente 1956, is now placed in the genus *Saccharomonospora* Nonomura and Ohara 1971c; and the remaining two, *T. vulgaris* and *T. thalpophilus,* were placed in synonymy by Küster and Locci (1964). *Thermoactinomyces vulgaris* was considered to be a variable species, a conclusion supported by Flockton and Cross (1975), but it was concluded that the variation was insufficient to justify creation of additional taxa. As a consequence, *T. vulgaris* acquired characters that were not present in the original concept of Tsiklinsky (1899). A prime example is the ability to utilize starch. Tsiklinsky (1899) stated clearly "il ne donne pas d'amylase," but later Kuo and Hartman (1966) described isolates that produced amylase, and this character is found in the description of *T. vulgaris* in the eighth edition of the *Manual.* Kurup et al. (1975) placed isolates producing amylase or not into two species, supported also by differences in their ability to utilise tyrosine, hypoxanthine, esculin, and arbutin. They named isolates lacking amylase as *T. candidus* and those producing amylase as *T. vulgaris,* and also noted that isolates of *T. candidus* produced spores on short sporophores while those of *T. vulgaris* were mostly sessile. Sporophores are also described by Tsiklinsky (1899) in her description of *T. vulgaris.*

It is therefore appropriate that such isolates should remain the type species of *Thermoactinomyces* rather than those considered by Kurup et al. (1975) to be *T. vulgaris.* Type cultures of *T. vulgaris* are not extant and the neotype proposed for the genus *Thermoactinomyces* and for *T. vulgaris* is the oldest strain. This was isolated by Erikson (1953) as "*Micromonospora vulgaris*" strain D, and is listed in the Approved Lists of Bacterial Names as KCC A-0162. This corresponds to Tsiklinsky's original concept of *T. vulgaris,* as does also *T. candidus.* The two species should therefore be regarded as synonymous and, in accordance with the Code of Bacteriological Nomenclature, the oldest legitimate epithet retained. Thus *T. vulgaris* remains the legitimate name for this taxon. Isolates that Kurup et al. (1975) named as *T. vulgaris* are thus left without a name, but correspond to the aggregate cluster for which Unsworth and Cross (1980) proposed reviving the name *T. thalpophilus* Waksman and Corke 1953. The original strains of Waksman and Corke are not extant, but a strain isolated by Henssen (1957), which she considered identical with strains from Waksman, has been designated as the neotype. However, this occurs at the margin of the *T. thalpophilus* phenon of Unsworth and Cross (1980) and additional reference strains have been specified. Isolations from hay, cotton, and other substrates suggest that isolates unable to produce amylase and with the characters of *T. vulgaris sensu* Unsworth and Cross (1980) are more abundant than those producing amylase and therefore most likely to be the type isolated by Tsiklinsky (1899) (Terho and Lacey, 1979; Cross and Unsworth, 1981; Lacey and Lacey, 1987). "*Thermoactinomyces antibioticus*" Craveri, Coronelli,

*******Maize-starch agar:* split maize, 50 g, boiled in 1 l water for 30 min, then filtered before adding starch, 10 g; NaCl, 5 g; $CaCl_2$, 0.5 g; peptone (Oxoid), 5.0 g; agar, 20.0 g; pH 7.2.

Table 32.1.
Differentiation of the genus **Thermoactinomyces** *from other monosporic actinomycete genera*[a,b]

Characteristics	*Thermo-actinomyces*	*Thermo-monospora*	*Saccharo-monospora*	*Micro-monospora*
Wall chemotype	III	III	IV	II
Aerial mycelium				
White	D[c]	D[e]	–	–
Yellow	D[d]	–	–	–
Brown	–	D[f]	–	–
Blue-green	–	–	+	–
Substrate mycelium				
White	D[c]	D[e]	–	–
Yellow	D[d]	D[e]	–	D
Orange	D[d]	–	–	D
Brown	D[g]	D[h]	–	D
Blue-green	–	–	+	D
Spores on				
Aerial mycelium	+	+	+	–
Substrate mycelium	+	D[i]	–	+
Sporophores				
Absent	D[j]	–	+	–
Simple	D[k]	D	–	–
Branched	D[d]	D	–	+
Spores endospores	+	–	–	–
Spore surface				
Ridged	+	–	–	–
Smooth, warty, or spiny	–	+	+	+
Growth at 50°C	D[l]	D[m]	+	–
Tolerant of novobiocin (25 μg/ml medium)	D[l]	–	–	–
Melanin pigments produced	D[n]	–	–	–
Degradation of				
Starch	D[o]	D[e]	+	NT
Arbutin	D[p]	+	+	NT
Esculin	D[p]	+	+	NT

[a]Data from Cross and Lacey (1970), Cross and Goodfellow (1973), Unsworth (1978), Goodfellow and Pirouz (1982b), and McCarthy and Cross (1984a).
[b]Symbols: +, 90% or more of strains are positive; –, 10% or less of strains are negative; *D*, different reactions in different species; *NT*, not tested.
[c]Only *T. dichotomicus* negative.
[d]Only *T. dichotomicus* positive.
[e]Only *T. chromogena* negative.
[f]Only *T. chromogena* positive.
[g]*Thermoactinomyces vulgaris* and *T. dichotomicus* negative; other species may be positive.
[h]*Thermomonospora chromogena* and *T. mesophila* positive; other species negative.
[i]*Thermomonospora fusca* only positive.
[j]*Thermoactinomyces thalpophilus, T. peptonophilus,* and some *T. intermedius* positive; other species negative.
[k]*Thermoactinomyces vulgaris, T. putidus, T. sacchari,* and some *T. intermedius* positive; other species negative.
[l]Only *T. peptonophilus* negative.
[m]Only *T. mesophila* negative.
[n]*Thermoactinomyces thalpophilus, T. putidus,* and *T. intermedius* positive; other species negative.
[o]*Thermoactinomyces vulgaris, T. intermedius,* and some *T. sacchari* and *T. putidus* positive.
[p]*Thermoactinomyces vulgaris, T. sacchari, T. intermedius,* and some *T. putidus* positive.

Pagani and Sensi 1964 is a synonym of *T. thalpophilus* and "*T. albus*" Orlowska 1969 of *T. vulgaris*.

Thermoactinomyces peptonophilus differs from other *Thermoactinomyces* in that it is mesophilic, is sensitive to novobiocin, has less heat-resistant spores, and has special nutrient requirements. However, it produces endospores and the differences are insufficient to create a new genus.

Acknowledgment

We are grateful to Dr. Bridget A. Unsworth for the use of unpublished data.

Further Reading

Cross, T. and B.A. Unsworth. 1981. The taxonomy of the endospore-forming actinomycetes. *In* Berkely and Goodfellow (Editors), the Aerobic Endospore-

Forming Bacteria: Classification and Identification. London, Academic Press, pp. 17–32.
Kalakoutskii, L.V. and N.S. Agre. 1976. Comparative aspects of development and differentiation in actinomycetes. Bacteriol. Rev. *40:* 469–524.
Lacey, J. 1978. Ecology of actinomycetes in fodders and related substrates. Zentralbl. Bakteriol. Mikrobiol. Hyg. 1 Abt. Suppl. *6;* 161–170.
Lacey, J. 1981. Airborne actinomycete spores as respiratory allergens. Zentralbl. Bakteriol. Mikrobiol. Hyg. 1 Abt. Suppl. *11;* 243–250.
Lacey, J. and D.A. Vince. 1971. Endospore formation and germination in a new *Thermoactinomyces* species. *In* Barker, Gould and Wolf (Editors), Spore Research 1971. Academic Press, London, pp. 181–187.
Stackebrandt, E. and C.R. Woese. 1981. Towards a phylogeny of actinomycetes and related organisms. Curr. Microbiol. *5;* 131–136.

Differentiation and characteristics of the species of the genus **Thermoactinomyces**

The differential and other characteristics of *Thermoactinomyces* species are listed in Tables 32.2 and 32.3.

List of species of the genus **Thermoactinomyces**

1. **Thermoactinomyces vulgaris** Tsiklinsky 1899, 501.[AL] (*Thermoactinomyces albus* Orlowska 1969, 25; *Thermoactinomyces candidus* Kurup, Barboriak, Fink and Lechevalier 1975, 152.[AL])

vul.ga′ris. L. adj. *vulgaris* common.

Colonies fast growing, flat on nutrient and CYC agars at 55°C, with a moderate covering of white mycelium and, often, a feathery margin on CYC agar. Endospores are produced on short, unbranched sporophores (Fig. 32.2). The colony reverse is white or cream, never pink or brown. No soluble pigments are produced. This is the only *Thermoactinomyces* species unable to produce amylase. Grows on CYC agar + 5% (w/w) NaCl.

Frequently isolated from soils and muds, vegetable composts, moldy hay, straw, cereal grains, sugar cane bagasse, cotton, mushroom compost, humidifiers, and air conditioning units and from air.

A probable cause of extrinsic allergic alveolitis (hypersensitivity pneumonitis), but its importance has probably been underestimated because isolates used in testing patients antisera have often been *T. thalpophilus.*

Type strain: KCC A-0162.

2. **Thermoactinomyces thalpophilus** (ex Waksman and Corke 1953, 378) nom. rev.

thal.po′phi.lus. Gr. n. *thalpos* heat; Gr. adv. *philos* loving; M.L. masc. adj. *thalpophilus* heat loving.

Colonies fast growing, flat or lightly ridged and furrowed on nutrient or CYC agars at 55°C, with moderate to dense covering of white aerial mycelium that may appear cream or pinkish from the underlying substrate mycelium, often with an entire margin on CYC agar. Endospores are sessile on the aerial and substrate hyphae (Fig. 32.3). Substrate mycelium may be brown and colony reverse cream to pinkish brown to dark brown. Pinkish-brown soluble pigment may be produced on CYC agar or rose-red on nutrient agar + 1% (w/v) glucose. On CYC agar + 0.5% (w/v) L-tyrosine, brown water-soluble melanin pigments are formed. Starch and tyrosine are degraded. Will not grow on 5% (w/v) NaCl but will grow on 1%. An antibiotic, thermorubin, is produced that inhibits Gram-positive bacteria and also *T. vulgaris.* Heterogeneity occurs within the *T. thalpophilus* phenon, with seven subclusters differing by a number of characters (Table 32.4) that could be the nuclei of new species.

Occurs commonly in soil, hay, cereal grain, sugar cane bagasse, cotton, grass compost, mushroom compost, peat, air conditioners, and manure. Also isolated from sputum and lung biopsies of farmer's lung patients.

Type strain: CBS 319.66.

Reference strains: CBS 422.63, CBS 109.62, ATCC 14570.

3. **Thermoactinomyces sacchari** Lacey 1971b, 327.[AL]

sac′cha.ri. M.L. n. *Saccharum* generic name of sugar cane; M.L. gen. n. *sacchari* of sugar cane.

Differs from other *Thermoactinomyces* species in having lightly ridged colonies, often "bacterial" in appearance and olive-buff in color. A sparse, transient, tufted aerial mycelium (Fig. 32.4) is produced that rapidly autolyzes, depositing the endospores in a thick layer on the surface of yeast-malt or nutrient + 1% (w/v) glucose agars. Growth on nutrient agar poor, restricted, thin with no aerial mycelium and few spores. Endospores are produced on sporophores up to 3 μm long. Yellow-brown soluble pigment may be formed.

Isolated from sugar cane, self-heated sugar cane bagasse, filter press muds, sugar mills, and soil. Implicated in bagassosis, a form of extrinsic allergic alveolitis.

Type strain: ATCC 27375.

4. **Thermoactinomyces dichotomicus** (Krasil'nikov and Agre 1964a) Cross and Goodfellow 1973, 77.[AL] (*Actinobifida dichotomica* Krasil'nikov and Agre 1964a, 939.)

di.chot′o.mi.cus. Gr. adj. *dichotomos* cut in two parts, forked; *-ica* M.L. adj. ending of intensification; M.L. masc. adj. *dichotomicus* dichotomous (branches).

Distinctive fast-growing yellow to orange colonies with dichotomously branched mycelium and sporophores on nutrient or CYC agars at 55°C (Fig. 32.6), margins entire on CYC agar. The presence of an exosporium surrounding the spore has been suggested (Fig. 32.10).

Isolated from soil and mushroom compost and may be implicated in extrinsic allergic alveolitis (hypersensitivity pneumonitis).

The specific epithet *thermolutea* predates *dichotomicus,* but although the description lists many typical properties of *T. dichotomicus,* others regarded as essential for *Thermoactinomyces* species are omitted and cultures are unavailable for comparison. *Theremoactinomyces dichotomicus* was included in the Approved Lists of Bacterial Names and it seems desirable that this name should be retained.

Type strain: INMI 114.

Syntype: KCC A-0055.

5. **Thermoactinomyces putidus** n. sp.

pu′ti.dus. L. masc. adj. *putidus* stinking, fetid.

Colonies often very wrinkled and puckered with endospores formed on short, unbranched sporophores (Fig. 32.5). Aerial mycelium white but may appear cream, pale yellow, or yellowish brown due to the yellowish-brown substrate mycelium. A greyish-yellow soluble pigment may be produced, and on CYC agar + 0.5% (w/v) L-tyrosine a brown, water-soluble melanin pigment is formed. The sporing hyphae lyse quickly, leaving spores on the surface of the agar. Some strains are sensitive to 0.5% (w/v) NaCl and all to 1.0%. Grows between 36° and 58°C, optimum 48°C. Cultures characteristically produce a distinctive, unpleasant smell. Isolated rarely from a narrow range of substrates, including soil and deep mud cores (Unsworth et al., 1977) and once from a lung biopsy of a farmer's lung patient. Two subgroups were obtained within the *T. putidus* phenon in a numerical phenotypic analysis (Unsworth, 1978), but the differences were insufficient to consider these as separate species. Most isolates in the smaller group degrade esculin and arbutin and fail to produce melanin on CYC + tyrosine to which trypticase (>5 g/l) had been added, contrasting with the larger group.

Type strain: NCIB 12324.

Table 32.2.
Characteristics differentiating the species of the genus **Thermoactinomyces** [a,b]

Characteristics	1. *T. vulgaris*	2. *T. thalpophilus*	3. *T. sacchari*	4. *T. dichotomicus*	5. *T. putidus*	6. *T. intermedius*	7. *T. peptonophilus*	8. *T.* species[c]
Aerial mycelium								
Abundant	+	+	−	+	+	+	*d*	+
Transient	−	−	+	−	+	−	−	−
White	+	+	+	−	+	+	+	+
Yellow	−	−	−	+	−	−	−	−
Soluble pigment								
Pink-red	−	+	−	−	−	−	−	−
Yellowish grey	−	−	−	−	+	−	−	−
Growth at								
30°C	−	d	−	−	d	−	+	+
55°C	+	+	+	+	d	+	−	NT
Spores sessile	−	+	−	−	−	d	+	NT
On unbranched sporophores	+	−	+	−	+	d	−	NT
On dichotomously branched sporophores	−	−	−	+	−	−	−	NT
Growth on								
25 µm/ml novobiocin	+	+	+	+	+	+	−	+
1.0% (w/v) NaCl	+	+	−	−	d	NT	NT	NT
Melanin production on CYC agar with 0.5% (w/v) L-tyrosine	−	+	−	−	+	+	NT	NT
Degradation of								
Arbutin	+	−	d	−	d	+	NT	−
Casein	+	+	+	+	+	+	−	+
Chitin	−	+	−	NT	d	−	NT	−
Esculin	+	−	+	−	d	+	NT	−
Gelatin	+	+	d	+	+	+	−	+
Guanine	−	−	−	+	NT	NT	NT	−
Hypoxanthine	−	−	−	+	NT	−	NT	−
Starch	−	+	+	−	+	−	NT	+
Testosterone	−	−	−	+	NT	NT	NT	d
Tyrosine	−	+	−	−	+	+	NT	−
Xanthine	−	−	−	+	NT	−	NT	−
Utilization as carbon source								
Fructose	d	+	+	NT	−	NT	NT	−
Mannitol	d	+	+	+	−	NT	NT	−
Sucrose	d	d	−	+	+	NT	NT	−
Trehalose	d	+	d	NT	−	NT	NT	−

[a]Data from Krasil'nikov and Agre (1964a), Lacey (1971b), Unsworth (1978), and Goodfellow and Pirouz (1982b).
[b]Symbols: see Table 32.1; *d*, 11–89% of strains are positive.
[c]Phenon 11 of Goodfellow and Pirouz (1982b).

6. **Thermoactinomyces intermedius** Kurup, Hollick and Pagan, 1981a, 216[VP††] (Effective publication: Kurup, Hollick and Pagan 1980, 107).

in.ter.me′di.us. L. masc. adj. *intermedius* intercalated, intermediate.

Colonies have white aerial mycelium and yellowish to yellowish-brown substrate mycelium. Brown, water-soluble melanin pigments produced on CYC agar + 0.5% (w/v) L-tyrosine. Endospores are sessile or produced on short sporophores. Growth is good at 50–55°C but poor at 37°C. The maximum temperature for growth has not been determined. Few physiological and nutritional characters have been determined. Isolated from air conditioner filters.

Type strain: ATCC 33205.

7. **Thermoactinomyces peptonophilus** Nonomura and Ohara 1971c, 902.[AL]

pep.to.no′phi.lus. Gr. adj. *peptos* cooked; Gr. adv. *philos* loving; M.L. adj. *peptonophilus* peptone loving.

The only mesophilic species in the genus, with endospores less heat re-

††*VP* denotes that this name has been validly published in the official publication, *International Journal of Systematic Bacteriology.*

Table 32.3.
Other characteristics of **Thermoactinomyces** *species*[a,b]

Characteristics	1. *T. vulgaris*	2. *T. thalpophilus*	3. *T. sacchari*	4. *T. dichotomicus*	5. *T. putidus*
Degradation of					
Adenine	–	–	–	–	NT
Cellulose	–	–	–	–	NT
Elastin	+	+	+	+	NT
DNA	+	+	+	+	–
Hippurate	NT	NT	NT	–	NT
Keratin	–	–	–	–	NT
RNA	+	+	+	+	NT
Tween 20	+	+	+	+	NT
Tween 40	+	+	+	+	NT
Tween 60	+	+	+	+	NT
Tween 80	+	+	+	+	NT
Utilization as carbon source					
L-Arabinose	NT	–	+	+	NT
Cellulose	NT	–	–	NT	NT
D-Galactose	NT	NT	NT	+	NT
D-Glucose	+	+	+	+	+
Glycerol	d	d	d	+	–
meso-Inositol	NT	–	–	+	NT
Lactose	NT	NT	NT	+	NT
Maltose	NT	NT	NT	+	NT
D-Mannose	d	d	–	NT	–
D-Raffinose	NT	–	–	+	NT
L-Rhamnose	NT	–	–	+	NT
D-Ribose	d	d	d	NT	–
D-Sorbitol	NT	NT	NT	+	NT
Starch	–	NT	NT	+	NT
D-Xylose	NT	–	–	+	NT
Enzyme production[c]					
Alkaline phosphatase	d	–	+	+	+
C_4 esterase	+	+	+	+	+
C_8 lipase	+	–	+	+	+
Acid phosphatase	d	d	–	+	+
Phosphoamidase	+	d	–	–	d
α-Glucosidase	–	(+)	–	–	d
β-Glucosidase	–	+	–	–	–
β-Glucuronidase	–	+	–	–	–
Leucine aminopeptidase	d	(+)	–	–	+
Chymotrypsin	–	(+)	–	–	+
Sensitivity to					
Ampicillin (25)[d]	+	+	+	+	+
Cephaloridine (100)	+	+	+	+	NT
Chloramphenicol (50)	+	+	+	+	+
Coliston sulfate 10)	+	+	+	+	+
Demethylchlortetracycline (500)	+	+	+	+	NT
Erythromycin (10)	+	+	+	+	+
Gentamycin (50)	+	+	+	+	NT
Kanamycin (100)	+	+	+	+	NT
Lincomycin (100)	NT	NT	NT	+	NT
Nalidixic acid (30)	–	–	–	–	–
Neomycin (50)	+	+	+	+	NT
Nitrofurantoin (200)	+	+	+	+	+
Oleandomycin (5)	+	+	+	+	+
Penicillin (10 I.U.)	+	+	+	+	+
Rifampicin (50)	NT	NT	NT	d	NT
Streptomycin (100)	+	+	+	+	+
Sulfafurazole (500)	+	+	+	+	+
Sulfafurazole (100)	–	–	–	–	–
Tobramycin (100)	NT	NT	NT	+	NT
Vancomycin (50)	+	+	+	+	NT

Table 32.3.—*continued*

Characteristics	1. *T. vulgaris*	2. *T. thalpophilus*	3. *T. sacchari*	4. *T. dichotomicus*	5. *T. putidus*
Growth in the presence of					
Lysozyme	+	+	*NT*	+	NT
NaCl 5.0% (w/v)	d	−	−	−	−
NaCl 1.0% (w/v)	+	+	−	−	d
NaCl 0.5% (w/v)	+	+	d	+	d
Pyronine 0.01% (w/v)	−	−	−	d	NT

[a]Data from Krasil'nikov and Agre (1964a), Lacey (1971b), Goodfellow and Pirouz (1982b), P. Boiron (unpublished), and T. Cross (unpublished).
[b]Symbols: see Table 32.1; *d*, 11–89% of strains are positive.
[c]Enzyme production assessed on API-ZYM strips (API Systems, S.A., Montalieu Vercieu, France); + indicates detection in whole-cell preparations but not in double-dialysis antigens.
[d]Figure in parentheses indicates concentration of antimicrobial agent (μg/ml) used to soak filter paper disks.

Table 32.4.
Characteristics distinguishing subgroups of **T. thalpophilus**[a,b]

Characteristics	(a)	(b)	(c)	(di)	(d)	(e)	(f)
Thick brown substrate mycelium	d(−)[c]	−	−	+	d	d	d
Colony reverse light pinkish-dark brown	d(+)	d	d	d(+)	d	+	+
Good melanin production in CYC + trypticase	d(+)	+	−	−	−	−	−
Good melanin production in CYC + protease	d	+	d	+	d	−	d(−)
Weak melanin production in CYC + protease	d	d	d	−	d	+	d(+)
DNase produced	d	d	d	−	d	d	d(−)
Enhanced growth with							
Trehalose	d(−)	+	+	+	+	d(+)	+
Glycerol	d(−)	+	+	+	+	d(+)	+
Mannitol	−	+	+	+	+	d(+)	+
Acid from							
Trehalose	d(−)	d	−	d(+)	d	d(+)	d
Glycerol	d(−)	−	−	d(+)	d	d(+)	d(+)
Mannitol	−	d	d	d(+)	d	d	+

[a]Data from Unsworth (1978).
[b]Symbols: see Table 32.1; *d*, 11–89% of strains are positive.
[c]In groups smaller than 10 isolates with only 1 deviant culture, the reaction of the majority is given in parentheses.

sistant ($D_{90°C}$ = 45 min) than those of other species. Endospores sessile on flexuous branches of the aerial mycelium and on the substrate mycelium. Aerial mycelium white and substrate mycelium white to yellowish brown. Unlike other species, it is not resistant to novobiocin. B vitamins and high concentrations of peptone (3% w/v) in the substrate are essential for good growth. Aerial mycelium production is favored by glycerol or glucose (0.2% w/v) and is best on supplemented glycerol-asparagine, oatmeal, yeast-starch, and PY agars at 35°C. Growth occurs between 25° and 45°C with an optimum at 35°C and pH 7.8. No growth occurs below pH 5.0. Isolated from heat-treated soil and grows only on other actinomycete colonies on the isolation medium used.

Type culture: ATCC 27302.

SECTION 33

Other Genera

Tom Cross

In the past few years several new actinomycete genera have been described and four have been placed together in this final section because as yet there has not been time to discover their relationships, if any, to other actinomycete groups. *Glycomyces, Kibdelosporangium, Kitasatosporia,* and *Saccharothrix* have been proposed as new genera on the basis of extensive and detailed morphological and chemotaxonomic data.

Colonies belonging to species of these genera would be very difficult to recognize on isolation plates inoculated with dilutions of soil and carrying many colonies of *Streptomyces* species and other sporulating nocardioform actinomycetes. One suspects that strains were initially studied in great detail because of their ability to produce a potentially useful secondary metabolite. The trained and experienced microbiologist might then be able to recognize similar colonies because of some characteristic morphological trait or be alerted when other strains were suspected of an ability to produce a similar antibiotic. It is interesting to note that colonies of two genera, *Glycomyces* and *Kibdelosporangium,* were isolated on agar media supplemented with antibiotics to reduce the numbers of competing bacteria, including the common soil actinomycetes. It is almost certain that similar selective pressures applied to new soil samples or the use of alternative selective procedures will reveal actinomycetes with different permutations of properties leading to new taxonomic proposals in time for the next edition of the *Manual.*

The studies on the genus *Kitasatosporia* have raised some interesting questions about the very heavy reliance now being placed on wall composition for classifications. Most workers have used the mycelium grown in shake flasks for determinations of peptidoglycan composition and this has proved to be satisfactory and reproducible. However, organisms with extensive spore production on the substrate mycelium may give different combinations of key amino acids. It is becoming apparent that there is a need to check the assumed agreement between spore and vegetative wall composition in other actinomycete genera.

The key properties of the four genera described in this section have been compared with those of other actinomycete genera also able to form chains of spores on their aerial mycelium and lacking mycolic acids (Table 33.1). It is realized that this comparison might be misleading because of the initial weighting given to a morphological character, but the table might help an investigator when making first attempts to identify an unfamiliar isolate.

Finally, it must be noted that a further generic name, "*Streptomycoides,*" has been proposed for actinomycetes morphologically similar to strains of the genus *Streptomyces* but having walls containing *meso*-diaminopimelic acid and glycine as the diagnostic amino acids, and with galactose as the major whole-cell sugar (Zhang et al. 1984a). More information is required on lipid and menaquinone composition before preliminary comparisons can be made. Cultures must then be deposited in recognized culture collections so that organisms can be studied side by side and alternative chemotaxonomic methods applied.

A fifth genus, *Pasteuria,* is also included in this section. This genus has some morphological similarities to actinomycetes, forming a mycelium that produces sporangia and endospores. These bacteria are endoparasites of invertebrates and have not yet been cultivated axenically. Therefore it is not possible to determine their true taxonomic relationships.

Genus **Glycomyces** *Labeda, Testa, Lechevalier and Lechevalier, 1985b, 419*[VP*]

David P. Labeda

Gly'co.my.ces. Gr. adj. *glykys* sweet; Gr. n. *mykes* fungus; M.L. n. *Glycomyces* sweet (glycolipid-containing) fungus.

Branching vegetative hyphae, approximately 0.40 μm in diameter, with aerial mycelium produced on some media. **The vegetative mycelium does not fragment, but short chains of square-ended conidia develop on the aerial hyphae.** Gram-positive. Lysozyme sensitive. Catalase positive and aerobic. Chemo-organotrophic. **Cell wall contains major amounts of *meso*-diaminopimelic acid (*meso*-DAP) and glycine. Whole-cell sugar pattern consists of xylose and arabinose.** No nitrogenous phospholipids, but significant amounts of phosphatidylinositol mannosides are present. No mycolic acids are present. Principal menaquinones are MK-10(H_2) and MK-10(H_6). The mol% G + C of the DNA is from 71 to 73 (T_m).

Type species: *Glycomyces harbinensis* Labeda, Testa, Lechevalier and Lechevalier 1985b, 420.

Further Descriptive Information

The genus *Glycomyces* can be defined as containing strains producing chains of spores on aerial sporophores and having a peptidoglycan composed of D-glycine, L-alanine, D-alanine, D-glutamic acid, *meso*-DAP, glu-

* *VP* denotes that this name has been validly published in the official publication, *International Journal of Systematic Bacteriology.*

Table 33.1.
Comparison of **other genera** *with genera that form aerial spore chains and lack mycolic acids*

Genus	Cell wall type[a]	Whole-cell sugars[b]	Phospholipid group[c]	Principal menaquinone(s)	Mol% G + C of DNA	Fragmentation of substrate mycelium[d]
Actinomadura	III	mad	PI	MK-9(H_4,H_6)	66–70	−
Actinopolyspora	IV	ara, gal	PIII	MK-9(H_4)	64	−
Amycolata	IV	ara, gal	PIII	MK-8(H_2,H_4)	68–71	+
Amycolatopsis	IV	ara, gal	PII	MK-9(H_2,H_4)	66–69	+
Faenia	IV	ara, gal	PIII	MK-9(H_4)	66–68	±
Glycomyces	II	xyl, ara	PI	MK-10(H_2,H_6)	71–73	−
Kibdelosporangium[e]	IV	ara, gal	PII	ND[f]	66	+
Kitasatosporia	III[g]	gal	PII	ND	66–73	−
Microtetraspora	III	mad	PI, PIV	MK-9(H_4)	ND	−
Nocardioides	I	none	PI	MK-8(H_4)	66–67	+
Nocardiopsis	III	none	PIII	MK-10(H_4)	64–69	+
Pseudonocardia	IV	ara, gal	PIII	MK-9(H_4)	79	+
Saccharopolyspora	IV	ara, gal	PIII	MK-9(H_4)	77	+
Saccharothrix	III	rha, gal	PII	MK-9(H_4)	70–76	+
Streptomyces	I	none	PII	MK-9(H_6,H_8)	69–78	−

[a]Cell wall composition classified according to Lechevalier and Lechevalier (1970b).
[b]*mad,* madurose; *ara,* arabinose; *gal,* galactose; *xyl,* xylose; *rha,* rhamnose.
[c]Phospholipid grouping according to Lechevalier et al. (1977).
[d]Symbols: +, positive; −, negative; ±; weak.
[e]Sporangium-like structures on aerial hyphae.
[f]*ND,* no data available.
[g]Spores formed on both aerial and substrate mycelium contain L-diaminopimelic acid and glycine (type I wall).

cosamine, and muramic acid. The whole-cell sugar pattern consists of xylose and arabinose (sometimes in extremely small quantities) as diagnostic sugars. These observations correspond to a type II cell wall and whole-cell sugar pattern D *sensu* Lechevalier and Lechevalier (1970b). The phospholipid pattern contains no nitrogenous phospholipids, but diphosphatidylglycerol, phosphatidylinositol, and phosphatidylinositol mannosides are generally present (i.e., a type PI phospholipid pattern *sensu* Lechevalier et al., 1977). The menaquinones present consist predominantly of those with 10 isoprenoid units, which are tetra- to hexahydrogenated (Labeda et al., 1985b).

Enrichment and Isolation Procedures

Strains of *Glycomyces* have been successfully isolated from soil samples by conventional spread plating methods on media such as Gauze No. 1 agar (Gauze, 1957) or Czapek agar (Pridham et al., 1957) amended with antifungal antibiotics and novobiocin or a combination of novobiocin and streptomycin (D. P. Labeda, unpublished data). After 2–3 weeks of incubation, small ridged white to yellowish-white colonies can be observed macroscopically and subcultured to a rich medium such as ATCC Medium No. 172 agar (American Type Culture Collection (ATCC) 1982). White aerial mycelium is sometimes produced by these colonies on isolation media. The taxonomic identity of isolates must be confirmed by the analysis for the presence of *meso*-DAP and the sugars xylose and arabinose in whole-cell hydrolysates. Determination of the lack of nitrogenous phospholipids in cell extracts (phospholipid type PI) further verifies the identification.

Maintenance Procedures

Glycomyces strains may be maintained by bimonthly transfer on a suitable agar medium such as ATCC Medium No. 172 agar (ATCC, 1982) or yeast extract–malt extract agar (Shirling and Gottlieb, 1966) with storage at 4°C between transfers. Long-term preservation of strains is best achieved by lyophilization in beef serum, in sucrose and gelatin, or in skim milk, and subsequent storage of the resulting ampules at −20° to 4°C. Strains have also been successfully stored for shorter periods as quick-frozen stationary-phase broth cultures or mycelial suspensions in 20% (v/v) glycerol, both held at −20° to −72°C.

Procedures for Testing of Special Characters

A detailed description of procedures for the analyses of cell wall composition, whole-cell amino acids and sugars, and phospholipids is given in Lechevalier and Lechevalier (1980). Methods for the determination of menaquinone composition are also readily available (Collins et al., 1977; Kroppenstedt, 1985; Yamada et al., 1976, 1977a).

The physiology of these strains has been characterized using the methods of Gordon et al. (1974) and Mishra et al. (1980), as well as tests suggested by Goodfellow (1971) and Kurup and Schmitt (1973).

Differentiation of the genus **Glycomyces** *from other genera*

Strains belonging to the genus *Glycomyces* can be clearly differentiated chemotaxonomically from other actinomycete genera that form aerial spore chains. This can be seen in Table 33.2, where *Glycomyces* appears as the only genus exhibiting this type of sporulation morphology that has a type II cell wall *sensu* Lechevalier and Lechevalier (1970b) and contains xylose and arabinose as the characteristic whole-cell sugars. The type PI phospholipid pattern (Lechevalier et al., 1977), in which no nitrogen-containing phospholipids are found in whole-cell extracts, is typical of *Glycomyces* strains because it is observed in only two other genera producing spores in chains, *Actinomadura* and *Nocardiopsis,* and the cell wall composition of these genera is different from that of *Glycomyces.*

Table 33.2.
A comparison of **Glycomyces** *with actinomycete genera that form aerial spore chains*

Genus	Cell wall type	Whole-cell sugars	Phospholipid group	Principal menaquinones	Mol% G + C of DNA
Glycomyces	II	Xylose, arabinose	PI	MK-10(H_2,H_6)	71–73
Actinomadura	III	Madurose	PI	MK-9(H_4,H_6)	66–70
Actinopolyspora	IV	Arabinose, galactose	PIII	MK-9(H_4)	64
Microtetraspora	III	Madurose	PI, PIV	MK-9(H_4)	NA[a]
Nocardia	IV	Arabinose, galactose	PII	MK-8(H_4), MK-9(H_2)	64–72
Nocardioides	I	None	PI	MK-8(H_4)	66–67
Nocardiopsis	III	None	PIII	MK-10(H_4,H_6)	64–69
Pseudonocardia	IV	Arabinose, galactose	PIII	MK-9(H_4)	79
Saccharopolyspora	IV	Arabinose, galactose	PIII	MK-9(H_4)	77
Saccharothrix	III	Rhamnose, galactose	PII	MK-9(H_4)	70–76
Streptomyces	I	None	PII	MK-9(H_6,H_8)	69–78

[a]*NA,* data not available.

Taxonomic Comments

The type species of *Glycomyces harbinensis* was discovered during the screening for new antibiotics and displayed morphology similar to that of *Actinomadura,* i.e. it produced short chains of conidia borne on the aerial hyphae. Analysis of the cell wall chemistry of this strain yielded unexpected results in that the walls contain *meso*-DAP and glycine, as well as glutamic acid, alanine, muramic acid, and glucosamine, i.e. a type II cell wall *sensu* Lechevalier and Lechevalier (1970b). Moreover, the whole-cell sugar pattern for this strain consists of arabinose and xylose, which is a type D sugar pattern *sensu* Lechevalier and Lechevalier (1970b). This type of cell wall chemistry is typical of members of the *Actinoplanaceae* and *Micromonosporaceae,* but had not been reported for actinomycetes producing conidia in chains. The unique combination of spore chain morphology and cell wall chemistry in addition to the PI phospholipid pattern observed for this strain precludes its inclusion in any described genus of the *Actinomycetales.* Thus, a new genus, *Glycomyces,* was described to accommodate strains having this combination of chemical composition and morphology. Additional strains exhibiting these same morphological and cell chemistry characteristics have been subsequently isolated, and a second species, *G. rutgersensis,* has been described.

Further Reading

Labeda, D. P., R. T. Testa, M. P. Lechevalier and H. A. Lechevalier. 1985. *Glycomyces,* gen. nov.: a new genus of the *Actinomycetales.* Int. J. Syst. Bacteriol. *35:* 417–421.

Differentiation and characteristics of the species of the genus Glycomyces

Reassociation studies have shown that the percentage of homologous DNA between the type strains of *G. harbinensis* and *G. rutgersensis* ranges from 21.5 to 30.2% based on four separate determinations (Labeda et al., 1985b). The physiological characteristics of *Glycomyces* species are given in Table 33.3. The major physiological differences are utilization of citrate, lactate, succinate, and tartrate, and acid production from adonitol and melibiose.

List of species of the genus Glycomyces

1. **Glycomyces harbinensis** Labeda, Testa, Lechevalier and Lechevalier 1985b, 420.[VP]

har.bin.en′sis. M.L. adj. *harbinensis* referring to Harbin, Peoples Republic of China (the source of the soil sample from which the organism was first isolated).

Pale yellow to yellowish-white vegetative mycelium with very sparse production of white aerial mycelia on some media (i.e. Czapek sucrose agar). Ability to produce aerial mycelia appears to be lost on repeated transfer. Short chains of square-ended conidia produced on the aerial hyphae. Soluble pigments are rarely produced, but are of yellowish shades when present. A fetid odor is produced during growth on some rich growth media.

Temperature range for growth 20–42°C.

The mol% G + C of the DNA is 71 (T_m).

Produces azaserine and antibiotic LL-DO5139-beta.

Isolated from soil.

Type strain: NRRL 15337.

2. **Glycomyces rutgersensis** Labeda, Testa, Lechevalier and Lechevalier 1985b, 420.[VP]

Table 33.3.
Physiological characteristics of species of the genus **Glycomyces**[a]

Characteristics	1. *G. harbinensis*	2. *G. rutgersensis*
Decomposition of		
Adenine	+	+
Casein	+	+
Cellulose	−	−
Esculin	+	+
Gelatin	−	w
Hypoxanthine	+	+
Potato starch	+	+
Tween 80	+	+
Tyrosine	−	−
Urea	−	−
Xanthine	−	−
Acid from		
Adonitol	−	+
Arabinose	+	+
Cellobiose	+	+
Dextrin	+	+
Dulcitol	−	−
Erythritol	−	−
Fructose	+	+
Galactose	+	+
Glucose	+	+
Glycerol	+	+
Inositol	−	−
Lactose	+	w
Maltose	+	+
Mannose	+	+
Melibiose	−	−
Melezitose	−	w
α-Methyl-D-glucoside	+	+
β-Methylxyloside	+	+
Raffinose	+	+
Rhamnose	+	+
Salicin	+	+
Sorbitol	+	+
Sucrose	−	−
Trehalose	+	+
Xylose	+	+
Utilization of		
Acetate	+	+
Benzoate	−	−
Citrate	+	−
Lactate	−	+
Malate	+	+
Mucate	−	−
Oxalate	−	−
Propionate	+	+
Succinate	+	−
Tartrate	−	−
Production of		
Nitrate reductase	w	+
Phosphatase	+	+
Growth in presence of		
5.0% NaCl	−	w
Lysozyme	−	−
Growth at		
10°C	+	+
37°C	+	+
42°C	−	+
45°C	−	−

[a]Symbols: see Table 33.1; *w*, weak positive.

rut.gers.en′sis. M.L. adj. *rutgersensis* referring to Rutgers University (where the organism was first isolated).

Yellowish-white to tan vegetative mycelium, dependent upon the growth medium. Moderate to abundant white aerial mycelium produced on many media. Yellowish to brown soluble pigments produced. A fetid odor is produced during growth on several rich growth media.

Temperature range for growth 20–37°C.

Principal menaquinones present MK-10(H_2) and MK-10(H_6).

The mol% G + C of the DNA is 73 (T_m).

Isolated from soil.

Type strain: NRRL B-16106.

Genus **Kibdelosporangium** *Shearer, Colman, Ferrin, Nisbet and Nash 1986, 48*[VP]

MARCIA C. SHEARER, PAULA M. COLMAN, AND RHONDA M. FERRIN

Kib.del'o.spo.ran.gi.um. Gr. adj. *kibdelos* false, ambiguous; Gr. n. *spora* seed; Gr. n. *angeion* a vessel; M.L. neut. n. *Kibdelosporangium* false or ambiguous sporangium.

Filamentous, with well-developed, moderately branched vegetative hyphae (0.4–1.0 μm in diameter) that penetrate the agar and form a compact layer on top of the agar. **The vegetative hyphae may exhibit varying degrees of fragmentation and frequently bear specialized structures that appear to be dichotomously branched, septate hyphae radiating from a common stalk. The aerial hyphae bear both long chains of conidia and sporangium-like structures. These structures are surrounded by a well-defined wall but contain hyphae embedded in an amorphous matrix rather than spores. They germinate directly.** No motile elements, sclerotia, or synnemata are present. Gram-positive. Not acid fast. Aerobic. Mesophilic. Chemo-organotrophic, having an oxidative metabolism. **Cell walls contain *meso*-diaminopimelic acid** (*meso*-DAP), **glutamic acid, alanine, muramic acid, glucosamine, galactose, and a minor amount of arabinose. Whole-cell hydrolysates contain galactose and arabinose; traces of madurose are also usually present.** Contain diphosphatidylglycerol, phosphatidylethanolamine, phosphatidylinositol, and phosphatidylinositol mannosides; phosphatidylmethylethanolamine may also be present. **No mycolic acids are present.** Mol% G + C of the DNA is 66 (T_m).

Widely distributed, but with low frequency, in soil.

Type species: *Kibdelosporangium aridum* Shearer, Colman, Ferrin, Nisbet and Nash 1986, 48.

Further Descriptive Information

When viewed from above, the surface of a *Kibdelosporangium* colony is frequently covered with straight to flexuous chains of rod-shaped spores (Fig. 33.1). The sporangium-like structures tend to be borne nearer to the surface of the agar, beneath the long chains of spores. Sporangium-like structures and chains of spores may be produced on the same aerial hypha and are borne apically on branched or unbranched hyphae as well as terminally on short lateral hyphal branches.

The sporangium-like structures originate as small round swellings at the tips of the hyphae. These continue to enlarge and, at maturity, are usually round (9–35 μm in diameter). Sporangium-like structures that are flattened in one axis or very irregularly shaped may also be present. They are surrounded by a well-defined wall and contain septate, branched hyphae embedded in an amorphous matrix. When placed on agar, these structures germinate, usually within 24–48 h, by the production of one or more germ tubes.

The specialized structures present on the vegetative mycelium consist of dichotomously branched, septate hyphae radiating from a common stalk. They appear to be "naked" sporangium-like structures analogous to the conidial structures that Couch (1963) described on the substrate hyphae of the *Actinoplanaceae.*

Although only one *Kibdelosporangium* strain has been characterized to the species level, 14 others have been isolated from soil. All 14 strains produce characteristic crystals in the agar, but the size, shape, and color of the crystals varies from strain to strain.

Kibdelosporangium strains are not difficult to grow. However, they tend to lose their capacity to produce sporangium-like structures on serial passage. Suitable nitrogen sources are ammonium, amino acids, casein, and peptones. Nitrates are not usually reduced. Glucose, glycerol, and sucrose are good carbon sources. Oatmeal agar (ISP 3; Shirling and Gottlieb, 1966), thin potato-carrot agar (Higgins et al., 1967), soil extract agar (Shearer et al., 1983a), and tap water agar are good media for typical morphological development. Temperature growth range is approximately 15–42°C, but above 35°C the sporangium-like structures tend to lose their characteristic size and shape.

Many of the *Kibdelosporangium* cultures produce glycopeptide antibiotics. Most of the strains that have been tested are resistant to a wide variety of antibiotics, including glycopeptides, penicillins, cephalosporins, and aminoglycosides. They are usually sensitive to rifampin and tetracyclines (Shearer et al., 1986).

Enrichment and Isolation Procedures

Kibdelosporangium species may be isolated from soil by plating dilutions on starch-casein-nitrate agar (Küster and Williams, 1964b) or arginine-glycerol-salt agar (El-Nakeeb and Lechevalier, 1963) containing either vancomycin (1–25 μg/ml) or gentamicin (2.5–5.0 μg/ml) and antifungal antibiotics. Heat-treating the air-dried soil (120°C for 60 min) before preparing the dilutions may also prove helpful. Inoculated plates are incubated at 28°C for 4 weeks.

Since *Kibdelosporangium* colonies closely resemble colonies of the mycolateless *Nocardia* and some *Streptomyces,* they are not easily identified on isolation plates. Checking sporulating colonies under the microscope at 400 × for typical structures in the aerial mycelium is the only way to identify *Kibdelosporangium* colonies on isolation plates.

Maintenance Procedures

Because *Kibdelosporangium* species tend to lose their ability to produce the sporangium-like structures on serial transfer, this method of maintenance is best avoided. A suitable method for short-term preserva-

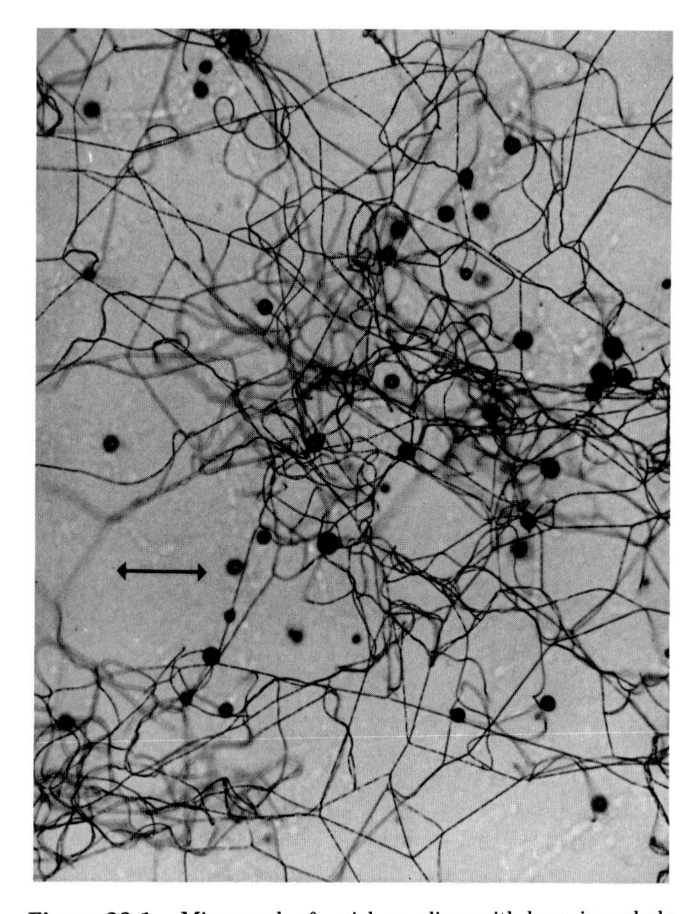

Figure 33.1. Micrograph of aerial mycelium with long, irregularly curved chains of spores and, nearer the agar surface, sporangium-like structures (*K. aridum;* 26-day-old culture on water agar). *Bar,* 55 μm.

tion is storage of log-phase broth cultures at −70°C (suspending medium: glucose-yeast extract broth (Waksman, 1961) with 10% v/v glycerol). Long-term preservation may be accomplished by either lyophilization in skim milk or quick freezing and storage in the vapor phase of liquid nitrogen (suspending medium: glucose-yeast extract broth with 10% v/v glycerol).

Procedures for Testing of Special Characters

Procedures for analyses of cell wall composition, whole-cell amino acids and sugars, mycolic acids, and phospholipids were those of Lechevalier and Lechevalier (1980). The DNA base ratios were determined by T_m (Marmur and Doty, 1962).

The micromorphology of *Kibdelosporangium* strains is best determined by direct in situ observation of undisturbed Petri dish colonies with a long working distance high-dry objective (40 ×). A nutrient-poor medium such as potato-carrot agar, soil extract agar, or tap water agar permits optimum development and observation of the characteristic morphological structures. Stained coverslip cultures prepared by the method of Shearer et al. (1986) may also be useful.

Methods used in testing the physiological characteristics of *Kibdelosporangium* are those of Gordon (1966, 1967) and Gordon and Mihm (1962a). Melanoid pigments and phosphatase activity are determined by the methods of Shirling and Gottlieb (1966) and Kurup and Schmitt (1973), respectively.

Differentiation of the genus **Kibdelosporangium** *from other genera*

Morphologically, kibdelosporangia most closely resemble those sporangium-producing genera that produce a well-developed aerial mycelium. However, they can be differentiated from all such genera on the basis of cell wall type (Table 33.4). In addition, mature sporangia of these genera contain aplanospores or zoospores that are released by dissolution or rupture of the sporangial wall, whereas the mature sporangium-like structures of *Kibdelosporangium* contain no spores but septate, branched hyphae embedded in an amorphous matrix, and they germinate directly.

Chemotaxonomically, kibdelosporangia appear to be most closely related to the genera that have a type IV cell wall (Lechevalier et al., 1971a), no mycolic acids, and phospholipids of either type PII or PIII (Cross and Goodfellow, 1973; Lechevalier and Lechevalier, 1981b). However, none of these chemotaxonomically related genera produce sporangia or sporangium-like structures and they are therefore easily differentiated from *Kibdelosporangium* on the basis of morphology.

Table 33.4.
Cell wall types and whole-cell sugar patterns of **Kibdelosporangium** *and genera producing sporangia*[a]

Genus	Cell wall type	Whole-cell sugar pattern
Actinoplanes[b]	II	D
Amorphosporangium[b]	II	D
Ampullariella[b]	II	D
Dactylosporangium[b]	II	D
Pilimelia[c]	II	D
Planobispora[d]	III	B
Planomonospora[b]	III	B
Spirillospora[b]	III	B
Streptoalloteichus[e]	III	C
Streptosporangium[b]	III	B
Kibdelosporangium	IV	A

[a]Reproduced with permission from M. C. Shearer, P. M. Colman, R. M. Ferrin, L. J. Nisbet and C. H. Nash, International Journal of Systematic Bacteriology *36:* 47–54, 1986.
[b]Data from Lechevalier and Lechevalier (1970b).
[c]Data from Szaniszlo and Gooder (1967).
[d]Data from Lechevalier and Lechevalier (1981b).
[e]Data from Tomita et al. (1978).

Taxonomic Comments

This recently described genus presently contains one described species, *K. aridum,* and its relationships to other actinomycete genera remain to be determined. Studies of DNA/DNA homology and DNA/rRNA hybridization are needed to determine the relationship of *Kibdelosporangium* to other actinomycete genera.

List of species of the genus **Kibdelosporangium**

1. **Kibdelosporangium aridum** Shearer, Colman, Ferrin, Nisbet and Nash 1986, 48.[VP]

ar′id.um. L. adj. *aridum* dry, arid.

Physiological characteristics are given in Table 33.5.

The substrate mycelium is off-white to grayish yellow-brown. It is well developed with moderately branching, septate hyphae that are 0.4–1.0 μm in diameter. The substrate hyphae do not fragment into rods and cocci during smear preparation, but fragmentation without hyphal displacement frequently occurs in plate cultures. Specialized structures are present on the substrate hyphae, both deep in the agar and just below its surface (Figs. 33.2 and 33.3). Characteristic crystals are produced in the agar on many media. No pigments other than melanin or yellow-brown soluble pigments are produced.

The aerial mycelium is white, sparse to moderate. It bears both chains of spores and sporangium-like structures. The spore chains are straight to flexuous and frequently contain more than 50 spores per chain, although chains of 10 spores or less may also be present. The rod-shaped spores are smooth walled (Fig. 33.4) and somewhat irregular in length (0.4 μm wide × 0.8–2.8 μm long). The sporangium-like structures (Figs. 33.5–33.8) are usually round (9–22 μm in diameter), but may be irregularly shaped.

Cell wall is type IV with a type A whole-cell sugar pattern and a type PII phospholipid pattern. No mycolic acids are present. The DNA base composition is 66 mol% G + C (T_m).

Produces the glycopeptide antibiotics aridicins A, B, and C (Shearer et al., 1985; Sitrin et al., 1985).

Type strain: ATCC 39323.

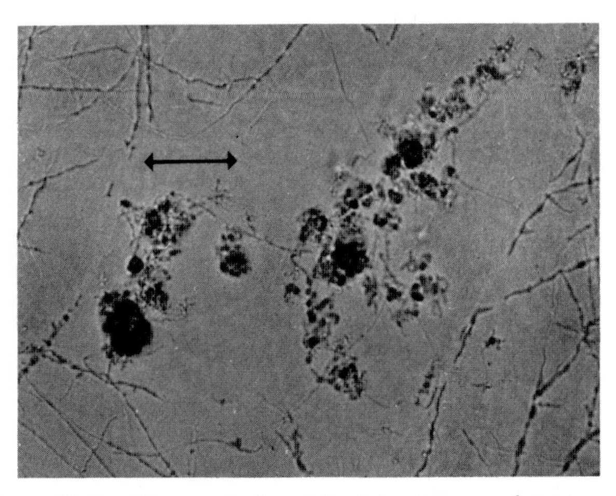

Figure 33.2. Micrograph of specialized structures on substrate mycelium (24-day-old culture on soil extract agar). *Bar,* 28 μm.

Table 33.5.
Physiological characteristics of **Kibdelosporangium aridum**[a]

Acid from		Utilization of	
Adonitol	−	Acetate	+
L-Arabinose	+	Benzoate	−
D-Cellobiose	+	Citrate	+
Dextrin	+	Formate	+
Dulcitol	−	Lactate	+
i-Erythritol	−	Malate	+
D-Fructose	+	Oxalate	+
D-Galactose	+	Propionate	+
Glucose	+	Pyruvate	+
Glycerol	+	Succinate	+
Glycogen	+	Tartrate	−
i-Inositol	+		
Inulin	−	Decomposition of	
Lactose	+	Adenine	−
Maltose	+	Casein	+
D-Mannitol	+	L-Tyrosine	+
D-Mannose	+	Hypoxanthine	+
D-Melezitose	+	Guanine	+
Melibiose	+	Elastin	+
α-Methyl-D-glucoside	+	Testosterone	+
α-Methyl-D-mannoside	+	Xanthine	−
Raffinose	+	Cellulose (Avicel)	−
Rhamnose	+	Urea	+
D-Ribose	+	Esculin	+
Salicin	v	Hippurate	+
D-Sorbitol	−	Allantoin	+
L-Sorbose	−		
Sucrose	+	Growth at/in	
Trehalose	+	10°C	v
D-Xylose	+	15°C	+
		42°C	+
Production of		45°C	Trace
Nitrate reductase	−	Survival at 50°C/8 h	+
Catalase	+	Lysozyme broth	−
Phosphatase	+	4% NaCl	+
Hydrogen sulfide	+	5–7% NaCl	v
Melanin	+	8% NaCl	−
Hydrolysis of			
Potato starch	−		
Gelatin	+		

[a]Symbols: see Table 33.1; *v*, variable.

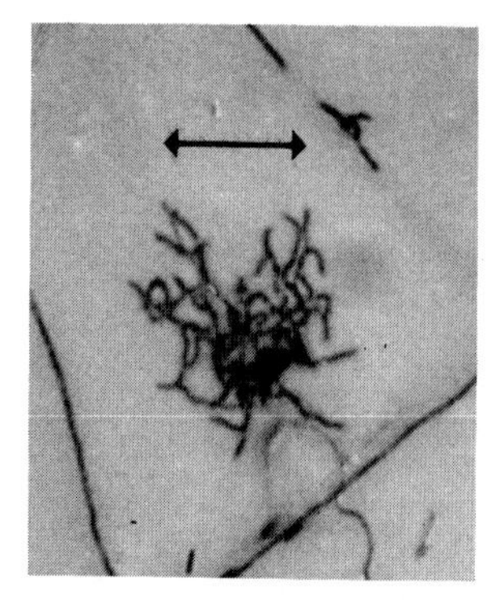

Figure 33.3. Micrograph of "naked" sporangium-like structure (24-day-old coverslip culture on soil extract agar). *Bar,* 7 μm. (Reproduced with permission from M. C. Shearer, P. M. Colman, R. M. Ferrin, L. J. Nisbet and C. H. Nash, International Journal of Systematic Bacteriology *36:* 47–54, 1986.)

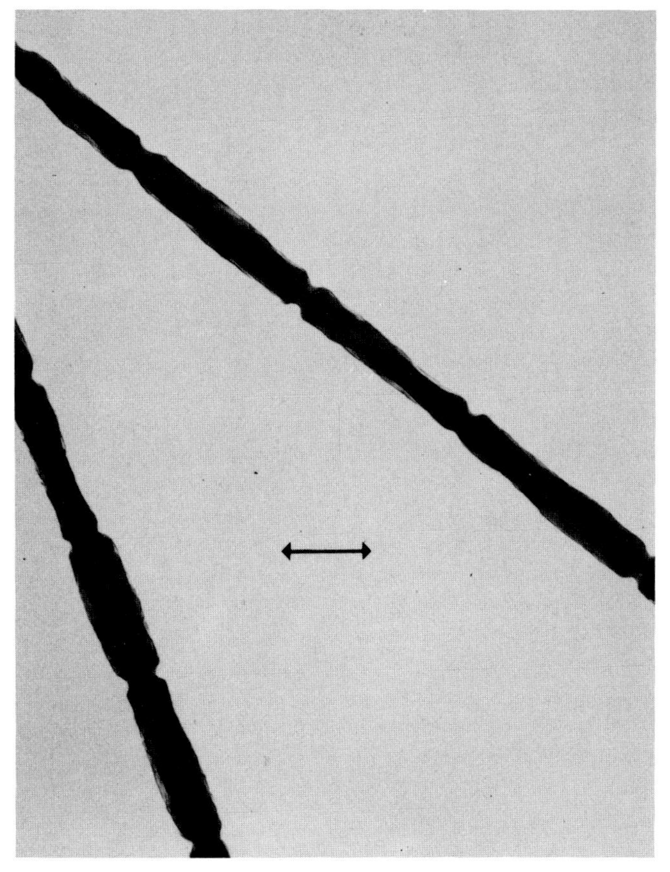

Figure 33.4. Transmission electron micrograph of spore chains (19-day-old culture on Bennett agar). *Bar,* 1 μm. (Reproduced with permission from M. C. Shearer, P. M. Colman, R. M. Ferrin, L. J. Nisbet and C. H. Nash, International Journal of Systematic Bacteriology *36:* 47–54, 1986.)

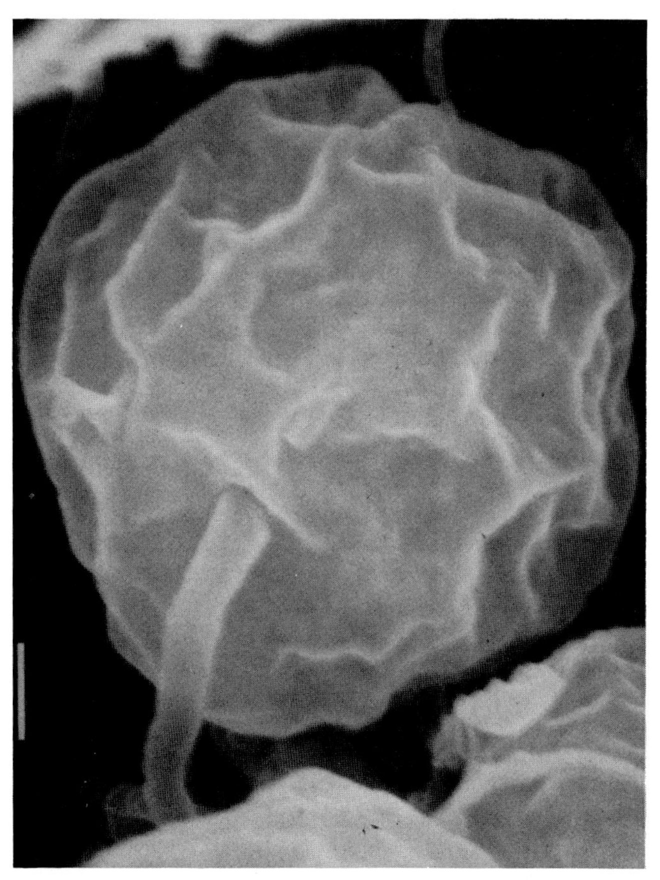

Figure 33.6. Scanning electron micrograph of sporangium-like structure (2½-week-old culture on thin Pablum agar). *Bar,* 1 μm. (Reproduced with permission from M. C. Shearer, P. M. Colman, R. M. Ferrin, L. J. Nisbet and C. H. Nash, International Journal of Systematic Bacteriology *36:* 47–54, 1986.)

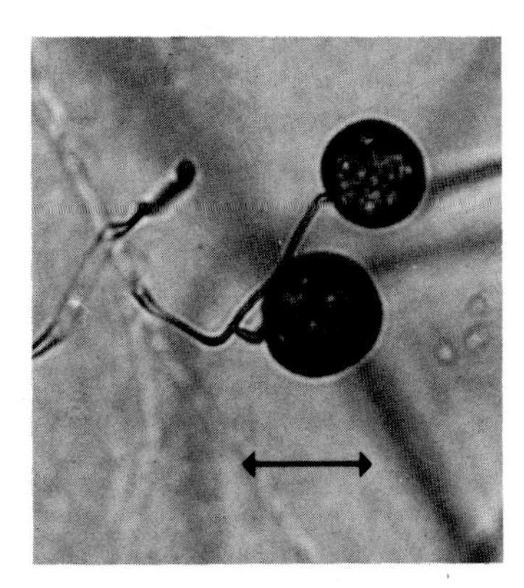

Figure 33.5. Micrograph of sporangium-like structures (8-week-old culture on water agar). *Bar,* 14 μm. (Reproduced with permission from M. C. Shearer, P. M. Colman, R. M. Ferrin, L. J. Nisbet and C. H. Nash, International Journal of Systematic Bacteriology *36:* 47–54, 1986.)

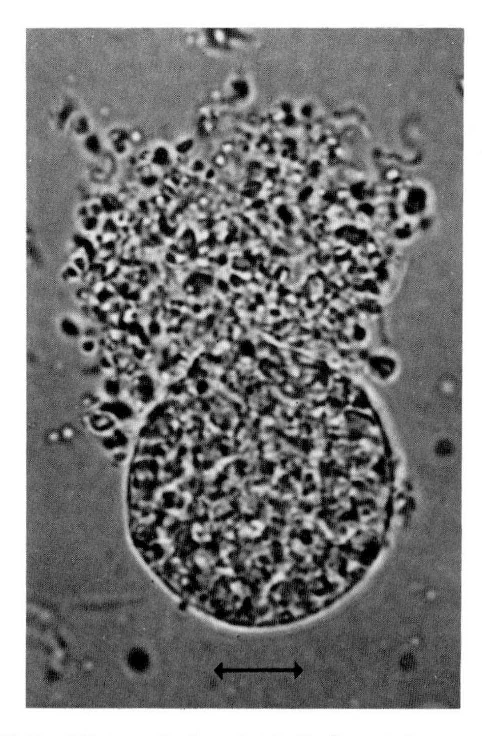

Figure 33.7. Micrograph of mechanically disrupted sporangium-like structure (50-day-old culture on oatmeal agar). *Bar,* 7 μm. (Reproduced with permission from M. C. Shearer, P. M. Colman, R. M. Ferrin, L. J. Nisbet and C. H. Nash, International Journal of Systematic Bacteriology *36:* 47–54, 1986.)

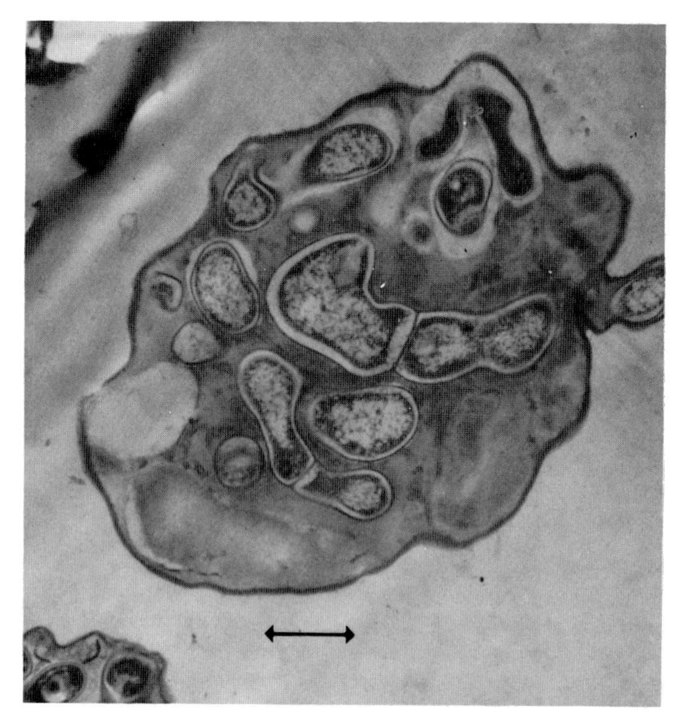

Figure 33.8. Transmission electron micrograph; thin section of sporangium-like structure (2½-week-old culture on thin potato-carrot agar). *Bar,* 1.2 μm. (Reproduced with permission from M. C. Shearer, P. M. Colman, R. M. Ferrin, L. J. Nisbet and C. H. Nash, International Journal of Systematic Bacteriology *36:* 47–54, 1986.)

Genus Kitasatosporia *Ōmura, Takahashi, Iwai and Tanaka 1983a, 672*[VP]

SATOSHI ŌMURA, YŌKO TAKAHASHI, AND YUZURU IWAI

Ki.ta.sa.to.spo′ria. *Kitasato* named for S. Kitasato, a Japanese bacteriologist (1852–1931); M.L. fem. n. *sporia* spore; M.L. fem. n. *Kitasatosporia* Kitasato spore.

Colony leathery. **Aerial hyphae bear long spore chains of more than 20 spores.** No fragmentation of vegetative mycelia. No sporangia, synnemata, sclerotia, or zoospores formed. **Major constituents of cell wall are glycine, galactose, and L- or *meso*-diaminopimelic acid (DAP) depending on the type of cell analyzed. DAP type of cells grown on agar media: aerial spores, L-DAP; vegetative mycelia, *meso*-DAP. DAP type of cells grown on liquid media: submerged spores, L-DAP; filamentous mycelia, *meso*-DAP.** Whole-cell hydrolysates contain galactose, but lack arabinose, xylose and madurose. Major phospholipids are diphosphatidylglycerol, phosphatidylethanolamine, phosphatidylinositol, and phosphatidylinositol mannosides. Glycolate test, negative. Gram-positive. Aerobic. Chemo-organotrophic. Range for growth, 15–42°C and pH 5.5–9.0. The mol% G + C of the DNA is 66–73 (hydrolyzed with formic acid and analyzed by high-performance liquid chromatography (HPLC)).

Type species: *Kitasatosporia setae* Ōmura, Takahashi, Iwai and Tanaka 1985, 221.

Further Descriptive Information

On solid media, *Kitasatosporia* strains produce aerial hyphae bearing chains of spores (Fig. 33.9*A*) and a nonfragmenting vegetative mycelium (Fig. 33.9*B*). They also produce both submerged spores and mycelium in liquid culture (Fig. 33.10*D*). The submerged spores contain L-DAP, while the mycelium contains mainly *meso*-DAP with a trace of the L- isomer (Takahashi et al., 1983, 1984a). Figure 33.10 shows morphological changes during submerged culture in which only submerged spores (Fig. 33.10*A*) were used as the inoculum: the spores begin to elongate in 1 h (Fig. 33.10*B*); mycelium develops in 9 h (Fig. 33.10*C*); spores appear again in 15 h (Fig. 33.10*D*); and thereafter both spores and mycelium are observed. Only L-DAP is detected at 0 h, but *meso*-DAP increases rapidly during 5–11 h of incubation. Similar amounts of L- and *meso*-DAP are detected after 50 h of incubation (Takahashi et al., 1983).

The cell wall of the submerged spore is about 80 nm thick, whereas that of the mycelium is about 10 nm thick (Fig. 33.11).

The aerial spores contain L-DAP, whereas the vegetative mycelium contains *meso*-DAP (Ōmura et al., 1981; Takahashi et al., 1984a). Thus both spore types contain L-DAP, whereas their mycelia contain mainly *meso*-DAP. The cell morphology of *Kitasatosporia* is therefore clearly associated with the DAP isomeric type.

The walls of all *K. setae* cells contain DAP, alanine, glutamic acid, and glycine as main components. The respective molar ratios of these amino acids in submerged and aerial spores are similar: 1.0:1.5:0.7:0.8 and 1.0:1.6:0.5:1.0, respectively. Also the molar ratios in mycelia in liquid and solid culture are similar: 1.0:1.6:0.7:0.4 and 1.0:1.6:0.7:0.2, respectively (Takahashi et al., 1984b).

Both spore types are resistant to sonication, but are sensitive to lysozyme digestion and moist heat (Takahashi et al., 1984b).

Isolation Procedures

Plates containing inorganic salts–starch agar (Shirling and Gottlieb, 1966) mixed with a soil suspension are incubated for 1 week in aerobic conditions at 27°C. Colonies are small and leathery, and similar to those of *Streptomyces* strains.

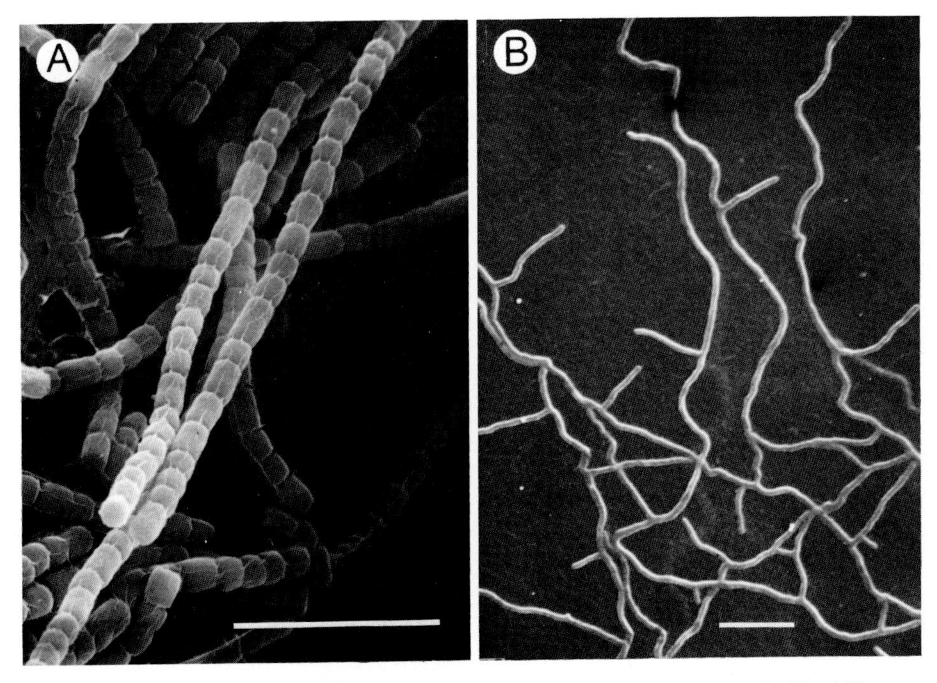

Figure 33.9. Scanning electron micrographs of aerial spores (*A*) and vegetative mycelia (*B*) of *Kitasatosporia setae* grown on inorganic salts-starch agar. *Bars,* 5 μm. (Reproduced with permission from Y. Takahashi, Y. Iwai and S. Ōmura, Journal of General and Applied Microbiology 29: 464, 1983.)

Maintenance Procedures

Kitasatosporia strains well grown on inorganic salts-starch agar at 27°C for 2 weeks can be stored at room temperature.

For long-term preservation, strains can be lyophilized; 10% (w/v) skim milk is satisfactory as a dispersing agent.

Procedures for Testing of Special Characters

Preparation of Submerged Spores and Filamentous Mycelia

Mycelia and spores of a strain are transferred into a medium consisting of 1% (w/v) yeast extract (Difco) and 1% (w/v) glucose, and incubated with reciprocal shaking at 27°C. The 3-day culture (stationary phase) is filtered through a Toyo Roshi No. 1 filter. The filtrate is centrifuged (2940 × *g*, 20 min) to obtain submerged spores. The residue is washed several times with water to separate mycelium. The purity of each preparation is confirmed by microscopic observation.

Preparation of Aerial Spores and Vegetative Mycelia

The aerial mycelia, well grown on inorganic salts-starch agar at 27°C for 2 weeks, are scraped off with sterile glass beads to obtain aerial spores. After the agar surface has been washed thoroughly with water, the agar is melted in a boiling water bath and filtered through cotton gauze. The residue is washed with hot water to give vegetative mycelium.

Differentiation of the genus **Kitasatosporia** from other genera

Table 33.6 provides the primary characteristics that can be used to differentiate the genus *Kitasatosporia* from morphologically similar genera.

Taxonomic Comments

"*Kitasatosporia melanogena*" Shimazu et al. 1984 has been reported. This is similar to other *Kitasatosporia* strains but differs in the chemical composition of the cells. The aerial spores and vegetative mycelium of this strain on agar media contain L- and *meso*-DAP, respectively, but the cells in submerged culture contain only *meso*-DAP. Further studies are required for the adequate classification of this strain.

Differentiation and characteristics of the species of the genus **Kitasatosporia**

The differential characteristics of the species of *Kitasatosporia* are indicated in Table 33.7. Other characteristics of the spores are listed in Table 33.8.

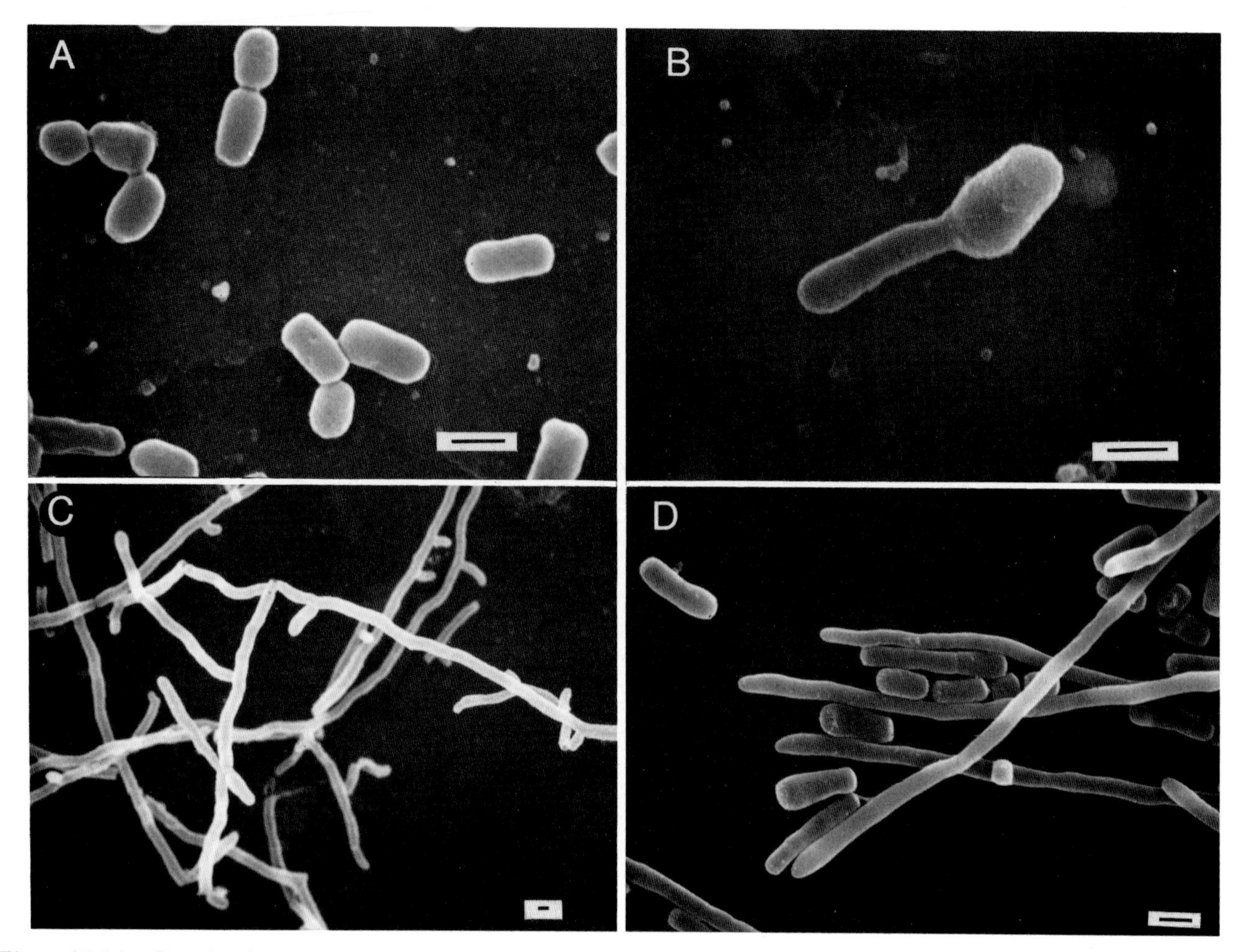

Figure 33.10. Scanning electron micrographs of *Kitasatosporia setae* grown in a submerged culture at 0 (*A*), 1 (*B*), 9 (*C*), and 15 h (*D*). *Bars,* 1 μm.

Table 33.6.
Differential characteristics of the genus **Kitasatosporia** *and morphologically similar genera*[a]

Characteristics	*Kitasatosporia*	*Streptomyces*	*Nocardioides*	*Actinomadura*	*Nocardiopsis*	*Pseudonocardia*	*Nocardia*	*Saccharothrix*
Cell wall								
Type	X[b]	I	I	II	III	IV	IV	III
meso-DAP	+	−	−	+	+	+	+	+
L-DAP	+	+	+	−	−	−	−	−
Glycine	+	+	+	−	−	−	−	NR
Arabinose	−	−	−	−	−	+	+	−
Galactose	+	−	−	−	−	+	+	+
Sporulation of aerial mycelium	+	+	+	+	+	+	−[c]	+
Fragmentation of vegetative mycelium	−	−	+	−	+	+[d]	+	+

[a]Symbols: +, present; −, absent; *NR*, not reported.
[b]Proposed type number.
[c]Some species were reported to produce aerial mycelium.
[d]No fragmentation of vegetative mycelium was reported in some species.

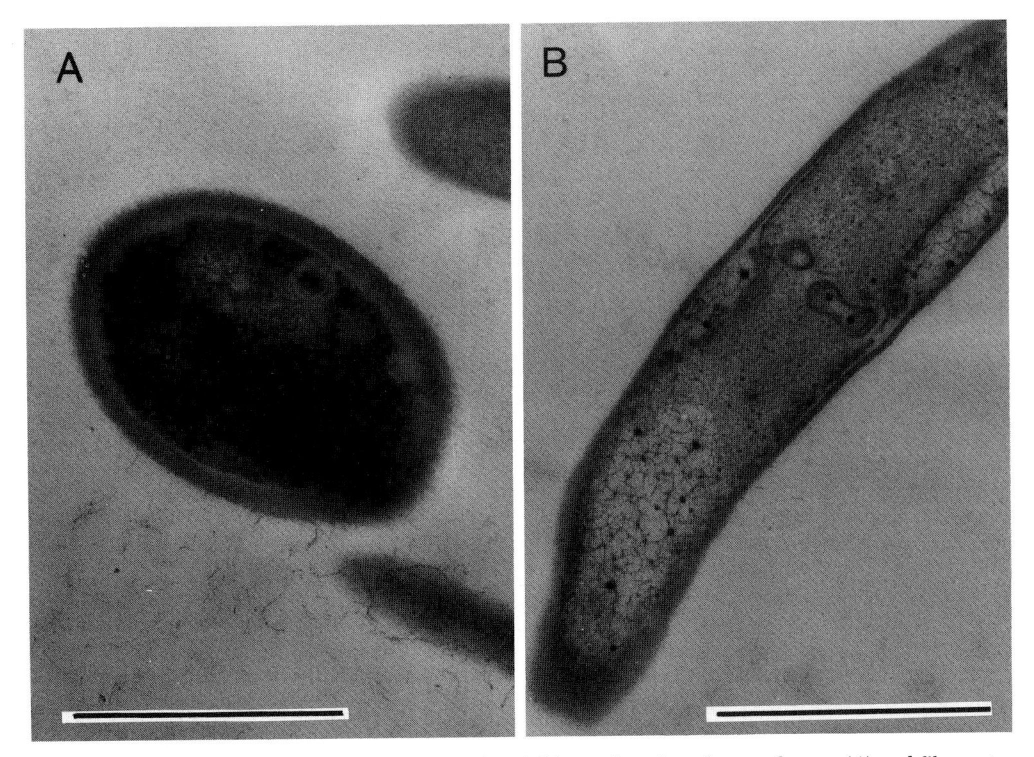

Figure 33.11. Transmission electron micrographs of thin section of a submerged spore (*A*) and filamentous mycelium (*B*) of *Kitasatosporia setae*. *Bars*, 1 μm.

Table 33.7.
Characteristics differentiating the species of the genus **Kitasatosporia**[a]

Characteristics	1. *K. setae*	2. *K. griseola*	3. *K. phosalacinea*
Spore shape	Cylindrical[b]	Cylindrical	Cylindrical
Spore surface	Smooth[c]	Smooth	Smooth
Aerial mycelium	White or light gray	Gray	White or light gray
Vegetative mycelium	Pale yellow or light ivory	Golden olive or parchment	Pale yellowish brown
Diffusible pigment	Yellow maple	Pink	Light tan
Nitrate reduction	−	−	+
Coagulation of milk	+	+	−
Carbon source utilization			
Raffinose	−	+	+
D-Fructose	−	−	+
L-Rhamnose	−	−	+
Sucrose	−	−	+
Mol% G + C of DNA	73.1	66.0	66.6

[a]Symbols: +, positive or utilized; −, negative or not utilized.
[b]Septum is not clear.
[c]Spore surface has a wrinkled form.

Table 33.8.
Other characteristics of the species of the genus **Kitasatosporia**[a]

Characteristics	1. *K. setae*	2. *K. griseola*	3. *K. phosalacinea*
Melanin formation	–	–	–
Tyrosinase reaction	–	–	–
Hydrolysis of starch	+	+	+
Liquefaction of gelatin	–	–	–
Peptonization of milk	+	+	+
H_2S production	–	–	–
Carbon source utilization			
D-Glucose	+	+	+
L-Arabinose	+	+	+
D-Xylose	+	+	+
Melibiose	–	–	–
D-Mannitol	–	–	–
i-Inositol	–	–	–
Cellulose	–	–	–
Temperature range for growth (°C)	15–37	15–37	15–42
pH range for growth	5.5–9.0	5.5–9.0	5.5–9.0
NaCl tolerance (% w/v)	<1.5	<2.0	<2.0

[a]+, positive or utilized; –, negative or not utilized.

List of species of the genus **Kitasatosporia**

1. **Kitasatosporia setae** Ōmura, Takahashi, Iwai and Tanaka 1985, 221.[VP] (Effective publication: *Kitasatosporia setalba* (sic) Ōmura, Takahashi, Iwai and Tanaka 1982, 1013.)

se.tae′. M.L. gen. n. *setae* of Seta.

Spore chain morphology *Rectiflexibiles* type. Color of vegetative mycelia pale yellow to light ivory on yeast extract–malt extract and glycerol-asparagine agars (Shirling and Gottlieb, 1966). Aerial mass color white or light gray on yeast extract–malt extract and inorganic salts–starch agars. Forms yellow diffusible pigment on inorganic salts–starch agar.

Produces setamycin, an antibiotic active against trichomonads and bacteria.

The mol% G + C of the DNA is 73.1 (hydrolyzed with formic acid, and analyzed by HPLC).

Type strain: KM-6054 (ATCC 33774, IFO 14216, JCM 3304).

2. **Kitasatosporia griseola** Takahashi, Iwai and Ōmura 1985, 535.[VP] (Effective publication: Takahashi, Iwai and Ōmura 1984a, 377.)

gri.se′o.la. M.L. fem. adj. *griseola* gray.

Spore chain morphology *Rectiflexibiles* type. The aerial spores are poorly septate and the spore surface is somewhat wrinkled. Color of vegetative mycelia golden-olive to parchment on yeast extract–malt extract and glycerol-asparagine agars. Aerial mass color gray or silvery-gray on most media. Forms a pinkish diffusible pigment on oatmeal (Shirling and Gottlieb, 1966) and glucose-asparagine agars.

Produces setamycin, an antibiotic active against trichomonads and bacteria.

The mol% G + C of the DNA is 66.0 (hydrolyzed with formic acid and analyzed by HPLC).

Type strain: AM-9660 (IFO 14371, JCM 3339).

3. **Kitasatosporia phosalacinea** Takahashi, Iwai and Ōmura 1985, 535.[VP] (Effective publication: Takahashi, Iwai and Ōmura 1984a, 377.)

phos.a.la.cin′e.a. M.L. fem. adj. *phosalacinea* pertaining to the name of phosalacine (an antibiotic produced by the organism).

Spore chain morphology *Rectiflexibiles* type. Color of vegetative mycelia pale yellowish-brown on most media. Aerial mass color white or light gray on oatmeal and inorganic salts–starch agars. Forms yellowish-brown diffusible pigment on some media.

Produces phosalacine, a herbicidal antibiotic; activity in vitro against *Bacillus subtilis* on a synthetic medium, which is reversed by glutamine.

The mol% G + C of the DNA is 66.6 (hydrolyzed with formic acid and analyzed by HPLC).

Type strain: KA-338 (IFO 14372, JCM 3340).

Genus Saccharothrix *Labeda, Testa, Lechevalier and Lechevalier 1984, 429*[VP]

DAVID P. LABEDA

Sac'cha.ro.thrix. Gr. neut. n. *sacchar* sugar; Gr. fem. n. *thrix* hair; M.L. n. *Saccharothrix* sugar-containing hair.

Branching vegetative mycelium, approximately 0.5 μm in diameter. Aerial mycelium produced on some growth media. **Both vegetative and aerial hyphae fragment into ovoid to bacillary nonmotile elements.** Gram-positive. Lysozyme resistant. Catalase positive and aerobic. Chemo-organotrophic. Optimum growth temperature 28–32°C. **Cell wall contains *meso*-diaminopimelic acid (*meso*-DAP) but not glycine. Whole-cell sugar pattern has galactose and rhamnose in major amounts as the diagnostic components.** Phosphatidylethanolamine is the major phospholipid. **No nocardomycolic acids are present in cells.** The mol% G + C of the DNA is 70–76 (T_m).

Type species: *Saccharothrix australiensis* Labeda, Testa, Lechevalier, and Lechevalier 1984, 429.

Further Descriptive Information

The genus *Saccharothrix* can be defined by the lack of mycolic acids, the composition of the peptidoglycan, whole-cell sugar content, and phospholipid content. The peptidoglycan is composed of L-alanine, D-alanine, D-glutamic acid, *meso*-DAP, glucosamine, and muramic acid. The whole-cell sugar pattern consists of galactose (in the absence of arabinose) and rhamnose as the diagnostic sugars. The phospholipid pattern consists of phosphatidylethanolamine (as the diagnostic nitrogenous phospholipid), diphosphatidylglycerol, phosphatidylinositol, and phosphatidylinositol mannosides (i.e. a type PII phospholipid pattern *sensu* Lechevalier et al., 1977). The menaquinones present predominantly consist of those with nine isoprenoid units, although menaquinones with 10 isoprenoid units are also found, and the majority appear to be tetrahydrogenated (Labeda et al., 1984).

Enrichment and Isolation Procedures

At present, there appears to be no selective procedure for the isolation of *Saccharothrix* species. The strains that have been isolated from soil thus far have come from platings of soil dilutions on media normally used for the general isolation of *Streptomyces* strains and other soil actinomycetes. Media such as inorganic salts–starch agar (Shirling and Gottlieb, 1966) or Gauze No. 1 agar (Gauze, 1957) amended with antifungal antibiotics will permit the isolation of *Saccharothrix* strains, if present, but careful microscopic observation is necessary to select appropriate colonies. These exhibit the "zig-zag" morphology of aerial hyphae during sporulation that is similar to that exhibited by members of the genus *Nocardiopis,* along with fragmentation of the substrate mycelium. Chemotaxonomic analysis of isolates is essential to determine if they are actually *Saccharothrix* strains.

Maintenance Procedures

Saccharothrix strains may be maintained by transfer every 2 months on a suitable agar medium, such as ATCC Medium No. 172 agar (American Type Culture Collection, 1982) or yeast extract–malt extract agar (Shirling and Gottlieb, 1966), with storage at 4°C between transfers. Long-term preservation of strains is best achieved by lyophilization in beef serum, in sucrose and gelatin, or in skim milk, and subsequent storage of the resulting ampules at −20° to 4°C. Strains have also been successfully stored for shorter periods as quick-frozen stationary-phase broth cultures or mycelial suspensions in 20% (v/v) glycerol, both held at −20° to −72°C.

Procedures for Testing of Special Characters

A detailed description of procedures for the analyses of cell-wall composition, whole-cell amino acids and sugars, and phospholipids is given by Lechevalier and Lechevalier (1980). Methods for the determination of menaquinone composition are also readily available (Collins et al., 1977; Kroppenstedt, 1985; Yamada et al., 1976, 1977a).

The micromorphology of *Saccharothrix* strains can be easily determined by observation of colonial growth on a relatively dilute medium, such as tap water or Czapek agar, using a high-dry (40–60 ×) objective. Stained smears or wet mounts observed by phase-contrast microscopy can be used to detect fragmentation.

The physiology of these strains has been characterized by the methods of Gordon et al. (1974) and Mishra et al. (1980), together with those of Goodfellow (1971) and Kurup and Schmitt (1973).

Differentiation of the genus Saccharothrix *from other genera*

Strains of the genus *Saccharothrix* closely resemble some other nocardioform genera in morphological appearance. They can, however, be clearly distinguished from members of these other genera on the basis of chemotaxonomic criteria (Table 33.9). The lack of nocardomycolic acids clearly differentiates the genus *Saccharothrix* from *Nocardia* and *Rhodococcus,* and the absence of madurose as a characteristic whole-cell sugar differentiates it from *Actinomadura. Nocardioides* and *Streptomyces* strains have a type I cell wall (L-DAP rather than *meso*-DAP), which differs from the type III cell wall of *Saccharothrix. Nocardiopsis* species have a PIII phospholipid pattern (phosphatidylcholine present as nitrogenous phospholipid), whereas *Saccharothrix* strains have a PII phospholipid pattern (phosphatidylethanolamine as nitrogenous phospholipid). Furthermore, the whole-cell sugar pattern of *Nocardiopsis* does not contain the taxonomically significant rhamnose marker observed in *Saccharothrix* hydrolysates.

Taxonomic Comments

Both species of the genus *Saccharothrix* were at one time described as *Nocardia* species, based on their morphological appearance. Improved chemotaxonomic methods have resulted in the creation of several new genera, such as *Actinomadura, Nocardiopsis,* and *Rhodococcus,* from strains that were originally grouped together in the genus *Nocardia.* Members of the genus *Saccharothrix* appear most closely related morphologically to members of the genus *Nocardiopsis,* but can be distinguished from this taxon by chemotaxonomic criteria (see Table 33.9).

Further Reading

Labeda, D. P., R. T. Testa, M. P. Lechevalier and H. A. Lechevalier. 1984. *Saccharothrix:* a new genus of the *Actinomycetales* related to *Nocardiopsis.* Int. J. Syst. Bacteriol. *34:* 426–431.

Labeda, D. P. 1986. Transfer of "*Nocardia*" *aerocolonigenes* (Shinobu and Kawato 1960) Pridham 1970 into the genus *Saccharothrix* Labeda, Testa, Lechevalier, and Lechevalier (1984) as *Saccharothrix aerocolonigenes* sp. nov. Int. J. Syst. Bacteriol. *36:* 109–110.

Differentiation and characteristics of the species of the genus Saccharothrix

Physiological characteristics of *Saccharothrix* species are given in Table 33.10.

Table 33.9.
A comparison of **Saccharothrix** *with other nocardioform actinomycetes that lack mycolic acids*

Genus	Cell wall type	Whole-cell sugars	Phospholipid group	Principal menaquinones	Mol% G + C of DNA
Saccharothrix	III	Rhamnose, galactose	PII	MK-9(H_4)	70–76
Actinomadura	III	Madurose	PI	MK-9(H_4, H_6)	66–70
Nocardioides	I	None	PI	MK-8(H_4)	66
Nocardiopsis	III	None	PIII	MK-10(H_4, H_6)	64–69

Table 33.10.
Physiological characteristics of species of the genus **Saccharothrix**[a]

Characteristics	1. *S. australiensis*	2. *S. aerocolonigenes*
Decomposition of		
Adenine	−	−
Allantoin	−	−
Casein	+	+
Cellulose	−	−
Esculin	+	+
Gelatin	+	+
Hippurate	−	−
Hypoxanthine	−	+
Potato starch	−	+
Tyrosine	+	+
Urea	−	v
Xanthine	−	−
Acid from		
Adonitol	−	−
Arabinose	−	+
Cellobiose	+	+
Dextrin	+	+
Dulcitol	−	−
Erythritol	+	−
Fructose	+	+
Galactose	+	+
Glucose	+	+
Glycerol	+	+
Inositol	−	+
Lactose	−	+
Maltose	+	+
Mannose	+	+
Melibiose	−	+
α-Methyl-D-glucoside	−	v
β-Methylxyloside	−	v
Raffinose	−	+
Rhamnose	−	+
Salicin	−	+
Sorbitol	+	−
Sucrose	−	+
Trehalose	+	+
Xylose	−	+
Utilization of		
Acetate	+	+
Benzoate	−	−
Citrate	−	+
Lactate	v	+
Malate	+	+
Mucate	−	−
Oxalate	−	+
Propionate	+	+
Succinate	+	+
Tartrate	−	+
Production of		
Nitrate reductase	+	+
Phosphatase	−	+
Growth in presence of NaCl		
1.5%	+	+
3.0%	+	+
4.0%	+	+
5.0%	−	w
Growth at		
10°C	+	+
37°C	+	+
42°C	+	v
45°C	+	−
55°C	−	−

[a]Symbols: see Table 33.1; *v*, variable; *w*, weak.

List of species of the genus **Saccharothrix**

1. **Saccharothrix australiensis** Labeda, Testa, Lechevalier and Lechevalier 1984, 430.[VP]

aus.tral.i.en′sis. M.L. adj. *australiensis* referring to Australia (the location of the soil sample from which the organism was first isolated).

Branching substrate mycelium, diameter approximately 0.5 μm; on some media aerial mycelium is produced. Vegetative and aerial hyphae fragment into coccoid elements.

The typical strain has brownish to grayish-yellow substrate mycelium with sparse to abundant white to yellowish-gray aerial mycelium produced on many media. Brownish soluble pigments are produced on several media.

Temperature range for growth 10–45°C.

The mol% G + C of the DNA is 73 (Bd) to 76 (T_m).

Isolated from soil.

Type strain: NRRL 11239 (=ATCC 31947).

2. **Saccharothrix aerocolonigenes** (Shinobu and Kawato 1960a) Labeda 1986, 109.[VP] (*Streptomyces aerocolonigenes* Shinobu and Kawato 1960a, 215.)

ae.ro.co.lo'ni.ge.nes. Gr. masc. gen. n. *aeros* of air; L. fem. n. *colonia* settlement, colony; Gr. v. *gennao* to produce; M.G. adj. *aerocolonigenes* producing aerial colonies.

Branching substrate mycelium produced, diameter approximately 0.5 μm; on some media extremely sparse aerial hyphae produced. Some strains may appear to form clumps of interwoven hyphae or "colonies" in the aerial mycelium. Substrate and aerial mycelia fragment. Strains usually lose capacity to form aerial mycelium during subcultivation.

Yellowish to brownish substrate mycelium with extremely sparse white aerial hyphae. Yellowish to brownish soluble pigment produced on several media.

Temperature range for growth 10–45°C.

The mol% G + C of the DNA is 70–71 (T_m).

Isolated from soil.

Type strain: NRRL B-3298 (=ATCC 23870, = IFO 3837).

Genus **Pasteuria** *Metchnikoff 1888, 166,*[AL*] *emend. Sayre and Starr 1985, 149, Starr and Sayre 1988a, 27 (Nom. Cons. Opin. 61 Jud. Comm. 1986, 119. Not* **Pasteuria** *in the sense of Henrici and Johnson (1935), Hirsch (1972), or Staley (1973); see Starr et al. (1983) and Judicial Commission (1986))*

RICHARD M. SAYRE AND MORTIMER P. STARR

Pas.teu'ri.a. M.L. gen. n. *Pasteuria* of Pasteur (named for Louis Pasteur, French savant and scientist).

Gram-positive, dichotomously branching, septate mycelium, the terminal hyphae of which enlarge to **form sporangia and eventually endospores. Vegetative colonies shaped like cauliflower florets or elongated grapes in clusters or small elongate clusters;** daughter colonies are formed by **fragmentation.** The sporogenous cells at the periphery of the colonies are usually attached by **narrow "sacrificial" intercalary hyphae that lyse,** causing arrangement of the developing **sporangia in quartets, then in doublets, and finally as single teardrop-shaped or cup-shaped or rhomboidal mature sporangia.** The rounded end of the sporangium encloses a **single refractile endospore,** 1.0–3.0 μm in major dimension, an oblate spheroid, **ellipsoidal or almost spherical in shape, resistant to desiccation and elevated temperatures** (one species has somewhat limited heat tolerance). **Nonmotile.** Sporangia and microcolonies are **endoparasitic in the bodies of freshwater, plant, and soil invertebrates. Has not been cultivated axenically,** but can be **grown in the laboratory with the invertebrate host.**

Type species: *Pasteuria ramosa* Metchnikoff 1888, 166.

Further Descriptive Information

A complex of errors has confused our understanding of the genus *Pasteuria* Metchnikoff 1888. Stated briefly, Metchnikoff (1888) described an endospore-forming bacterial parasite of cladocerans; he named this bacterium *Pasteuria ramosa.* Metchnikoff presented drawings and photomicrographs of the life stages of this parasite as they occurred in the hemolymph of the water fleas *Daphnia pulex* Leydig and *D. magna* Strauss; he was, however, unable to culture the organism in vitro. Subsequent workers (Henrici and Johnson, 1935; Hirsch, 1972; Staley, 1973), who were looking in cladocerans for Metchnikoff's unique bacteria, reported on a different bacterium with only superficial resemblance to certain life stages of *P. ramosa.* Their investigations led to the axenic cultivation of a budding bacterium that is occasionally found on the exterior surfaces of *Daphnia* species. Unlike Metchnikoff's organism, this bacterium divides by budding, it forms a major nonprosthecate appendage (a fascicle), it does not form endospores, it is not mycelial or branching, its staining reaction is Gram-negative, and it is not an endoparasite of cladocerans.

After searching for but not finding in water fleas the bacterial endoparasite as described by Metchnikoff, the erroneous conclusion was reached that this budding bacterium that occurs on the surfaces of *Daphnia* species was the organism Metchnikoff had described in 1888. As a result, a budding bacterium (strain ATCC 27377) was mistakenly designated (Staley, 1973) as the type culture for *Pasteuria ramosa* Metchnikoff 1888, the type (and, then, sole) species of the genus *Pasteuria.* This confusion between Metchnikoff's cladoceran parasite and the quite different budding bacterium has only recently been resolved (Starr et al., 1983). The budding bacterium (with strain ATCC 27377 as its type culture) has been named *Planctomyces staleyi* Starr, Sayre and Schmidt 1983; conservation of the original descriptions of the genus *Pasteuria* and *P. ramosa,* as updated, was recommended (Starr et al., 1983) and approved (Judicial Commission, 1986). Unfortunately, vestiges of this nomenclatural disorder are still with us; for example, in a recent essay certain evolutionary and taxonomic inferences (Woese, 1987) regarding the genus *Pasteuria* were based upon data concerning bacteria belonging to the *Blastocaulis-Planctomyces* group of budding and nonprosthecately appendaged bacteria rather than the mycelial and endospore-forming invertebrate parasites that actually comprise the genus *Pasteuria.*

Confusion of a different kind has occurred in the nomenclature of those bacterial parasites of nematodes now known to belong to the genus *Pasteuria.* The first report of this kind of microorganism was by Cobb (1906), who found numerous highly refractile bodies infecting specimens of the nematode *Dorylaimus bulbiferous.* He erroneously viewed these bodies as "perhaps monads" of a parasitic sporozoan. The incorrect placement in the protozoa of an organism now known to be a bacterial parasite of nematodes has persisted for nearly 70 years. A more precise but still incorrect placement was suggested by Micoletzky (1925), who found a nematode parasite similar in size and shape to those reported by Cobb; Micoletzky suggested that this organism belonged to the genus *Duboscqia* Perez 1908, another sporozoan group (Perez, 1908). Later, Thorne (1940) described in detail a new parasite from the nematode *Pratylenchus pratensis* (later, the nematode's identity was corrected to *Pratylenchus brachyurus* by Thorne; see Sayre et al., (1988). On the assumption that this organism was similar to the nematode parasite described by Micoletzky, it also was assigned by Thorne (1940) to the protozoan genus *Duboscqia,* as *D. penetrans.*

Thorne's description and nomenclature persisted until 1975, even though other investigators (Canning, 1973; Williams, 1960), who had examined this nematode parasite in some detail, questioned this placement. It was not until the nematode parasite was reexamined using electron microscopy that its true affinities to the bacteria rather than to the protozoa were recognized and the name "*Bacillus penetrans*" (Thorne 1940) Mankau 1975 was applied to it (Mankau, 1975; Mankau and Imbriani, 1975; Imbriani and Mankau, 1977). Because "*Bacillus penetrans*" was not included in the Approved Lists (Skerman et al., 1980), it has no nomenclatural standing. Although this microorganism forms endospores of the sort typical of the genus *Bacillus* Cohn 1872, its other traits (e.g., mycelial habit, endoparasitic associations with plant-parasitic nematodes) suggest it does not belong in the genus *Bacillus.* Rather, it is closely related to *P. ramosa* (see Table 33.11 below) and it has

*AL denotes the inclusion of this name on the Approved Lists of Bacterial Names (1980).

more properly been assigned (Sayre and Starr, 1985) to the genus *Pasteuria* Metchnikoff 1888 as *Pasteuria penetrans.*

The group of nematode parasites to which *P. penetrans* belongs is widespread and diverse. Members of this group have been reported from about 70 nematode genera and 175 nematode species, and from a dozen states of the United States as well as roughly 40 other countries on five continents and on various islands in the Atlantic, Pacific, and Indian Oceans (Sayre and Starr, 1988). This bacterial group is by no means a uniform entity; rather, it is an assemblage of numerous pathotypes and morphotypes that may also comprise a multiplicity of taxa with boundaries and categorial levels that are by no means clear. To alleviate some of the confusion in referring to these organisms, we recommend that the unmodified name *Pasteuria penetrans* (or its earlier synonyms, "*Duboscqia penetrans*" or "*Bacillus penetrans*") in prior literature be treated as a vernacular name meaning "member(s) of the *Pasteuria penetrans* group" or a similar locution.

To clarify the characteristics of the genus *Pasteuria,* the meanings of the terms "endospore" and "sporangium" must be modified slightly from their usual senses so as to be reasonably applicable to this genus. Metchnikoff observed the several stages of endosporogenesis that occurred in *Pasteuria ramosa.* In his discussion, he noted within each sporangium a single refractile body that stained with difficulty; he called this structure, as we do nowadays, an endospore.

However, the *Pasteuria* endospore is not entirely typical of those found in *Bacillus* or *Clostridium.* For one thing, the *Pasteuria* endospore has a mass of fibrous outgrowths emanating from the outermost cortical wall. These microfibrillar strands (usually called parasporal fibers), which surround the central body of the endospore, are organelles involved in attachment of the endospore to its invertebrate host (Fig. 33.12). Albeit an integral part of the endospore, the parasporal fibers are arrayed differently in the three *Pasteuria* species discussed here. It is difficult to include these somewhat amorphous fibrous masses in any precise measurements of endospores. For this reason, the measurements reported here include only the major and minor axes of the central body of the endospore, the endospore proper, and explicitly exclude the parasporal fibers. Measurements and descriptions of parasporal fibers are presented separately.

As stated, the attachments of the endospores to their invertebrate hosts is mediated by the parasporal fibers. The attached endospore is often overlayed by seemingly nonfunctional remnants of the old sporangium (Fig. 33.13). The presence or absence of these sporangial remnants may be due in part to the length of time the sporangium had been subjected to degradative processes in the soil or the amount of abrasion received as the migratory host moves through the environment (Fig. 33.14). When such sporangial material was significant, it was included in the measurements of the endospores reported here. *Pasteuria* endospores, at least in the case of the nematode parasites, also differ from those of *Bacillus* in that the former germinate by a fine hyphal strand that penetrates the host and initiates bacterial colonization. Even though sporangial material is sometimes seen, we have adopted the convention of calling the infectious unit on the nematode's surface an endospore.

The type-descriptive material (Sayre and Starr, 1985) of *Pasteuria penetrans* (ex Thorne 1940) Sayre and Starr 1985 refers to the bacteria occurring on the root-knot nematode *Meloidogyne incognita.* Hence, the name *Pasteuria penetrans sensu stricto* must refer to that organism (Starr and Sayre, 1988a); members of the *Pasteuria penetrans* group are demonstrably different from it and must be assigned to other taxa. The first such assignment has been for the bacterium from the root-lesion nematode *Pratylenchus brachyurus,* to which the name *Pasteuria thornei* Starr and Sayre 1988a was affixed (Starr and Sayre, 1988a).

Enrichment and Isolation Procedures

Attempts to devise methods for the isolation and cultivation of any species of *Pasteuria* apart from its host have never been reported to be successful. Neither the factors governing germination of endospores nor those influencing vegetative growth are known. However, the distinctive

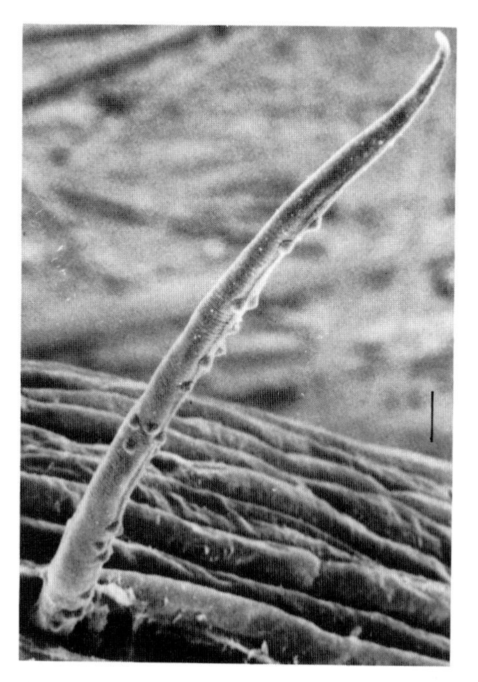

Figure 33.12. Scanning electron micrographs of endospores of *P. penetrans sensu stricto* attached to a larva of a root-knot nematode that has partially penetrated a tomato root. The endospores carried on the larva will germinate inside the plant and penetrate the developing nematode, completing their life cycle in synchrony with their host. On decay of the plant root in the soil, the endospores developed within the parasitized nematode are released. *Bar,* 10 μm.

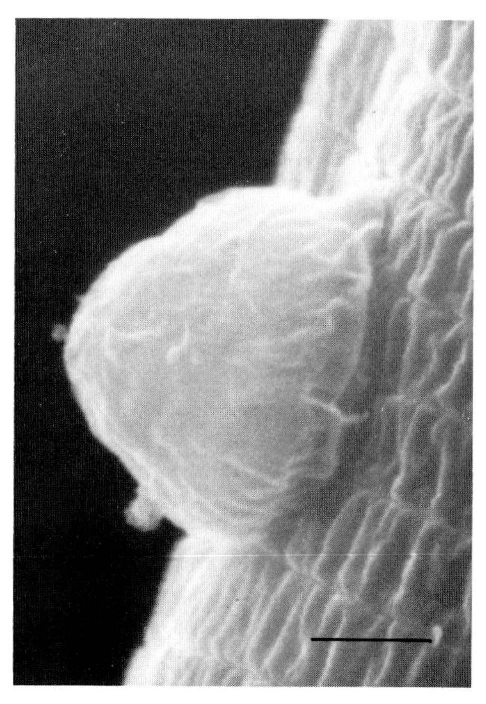

Figure 33.13. Scanning electron micrograph of an endospore of *P. penetrans sensu stricto* on cuticle of a second-stage larva of *Meloidogyne incognita.* This endospore has retained its exosporium, resulting in the appearance of a crinkled surface. *Bar,* 1.0 μm.

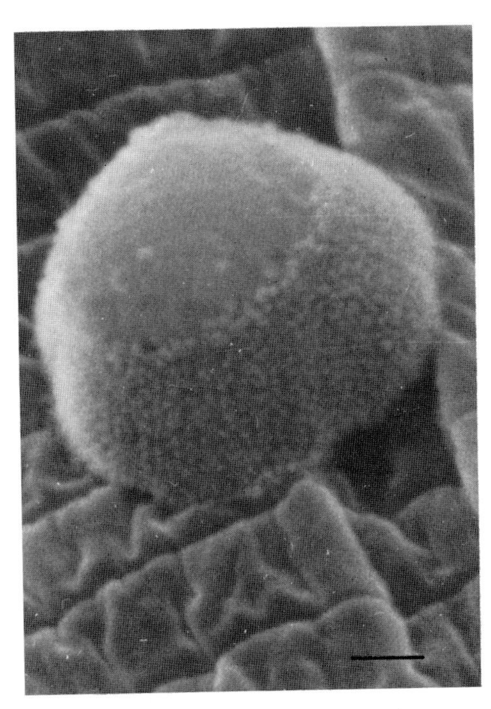

Figure 33.14. Endospore of *P. penetrans sensu stricto* attached to the cuticle along the lateral field of the larva of a root-knot nematode. The exosporium has been sloughed, exposing the central dome of the endospore; the peripheral fibers can be distinguished. *Bar,* 0.5 μm.

morphology and the unique relationship to invertebrate hosts shared by the three named *Pasteuria* species (see Table 33.11 below) suggest that this commonality may be the harbinger of similarities in their physiological requirements for endospore germination, vegetative growth, and eventually axenic cultivation. Based on microscopic observations, the physiological and physical requirements for their growth in vivo would appear to be similar. Summarized briefly, vegetative growth seems linked through the environment provided in the coelom of the different host invertebrates. The hemolymph of the cladoceran or the pseudocoelomic fluid of the nematode allows for the exchange of nutrients and waste products, and the coeloms provide the space for development of the mycelial colonies, which fragment after the colonies reach a critical size. Finally, it is reasoned that various factors, possibly the accumulation of bacterial biomass and metabolites, as well as the onset of senility in the invertebrate hosts, trigger sporulation of the bacteria. The similar physical conditions found in the separate host species suggests that the bacteria might have common nutritional requirements; hence, when the requirements for axenic cultivation of one *Pasteuria* species become known, they may, with slight modification, apply to the other bacterial species.

Since isolation and cultivation of *P. ramosa* apart from its cladoceran host has not yet been achieved, this bacterium is usually studied in field samples of the infected invertebrate host. The following procedures may increase the chances of finding *P. ramosa* in its natural habitat.

1. Cladocerans should be collected when their populations are highest, usually during the late summer; this is close to the time (i.e. in early August) when the parasites also are expected to be most prevalent.
2. Because the frequency of occurrence of the parasite in a cladoceran population generally tends to be low (usually less than 10%), a large number of living specimens needs to be examined to increase the odds for detecting the parasite.
3. The internal parasites are most easily identified by using an inverted microscope at magnifications of 100–250 ×.

Similarly, axenic cultivation of the *P. penetrans* group has not been reported, and most investigations have been limited to studies on naturally or artificially infected nematode hosts (Mankau, 1975; Mankau and Imbriani, 1975; Imbriani and Mankau, 1977; Sayre and Wergin, 1977; Sayre et al., 1983, 1988; Starr and Sayre, 1988a) and exploration of the bacterium's potential as a biological control agent against plant-parasitic nematode populations (Prasad, 1971; Mankau, 1973; Sayre, 1980; Stirling and Wachtel, 1980; Stirling, 1984; Brown et al., 1985). Consequently, the studies have depended on finding, maintaining, and manipulating nematode populations infected with these bacteria. Because of this direct dependence on host-nematode populations, the procedures and methods used in maintaining members of the *P. penetrans* group are by-and-large those used in maintaining the nematodes (Zuckerman, Mai, and Harrison, 1985; Southey, 1986).

A few generalizations may be useful.

1. Members of the *P. penetrans* group will most likely be found in soils where nematode populations have been consistently high and are causing crop damage. Consequently, planting of susceptible crops is necessary for maintenance of the nematode populations. Generally, members of the *P. penetrans* group are more likely to be associated with pest nematodes in greenhouse situations than in the field, perhaps because of more intensive cultivation practices in greenhouses and longer growing seasons of nematode-susceptible crops in greenhouses.
2. To find the bacteria, nematodes are extracted from the suspected soil; e.g. by a sugar-flotation procedure (Caveness and Jensen, 1955). Other separation methods, relying on the nematode's motility (e.g. Baermann, 1917), may not yield those nematodes heavily encumbered with endospores, since such nematodes are partially immobilized. Addition of healthy nematodes to soils, together with subsequent extraction and examination, is the only reliable bioassay currently available for determining the presence of members of the *P. penetrans* group in soil.
3. Occurrence of endospores on the surfaces of nematodes is most easily observed by means of an inverted microscope at 250–400 × magnification.
4. Direct confirmation of the presence of members of the *P. penetrans* group inside the nematode can be made by microscopic examination of the esophageal region of nematodes, also at 250–400 × magnification; both larvae and adults should be examined for the characteristic mycelial colonies and endospores. In the uninfected nematode, this anterior region is normally clear, with internal structures visible; it will be cloudy and filled with endospores in nematodes parasitized by the bacterium.
5. Detection of the bacterium in the sedentary endoparasitic nematodes depends on manual (Thorne, 1961) and/or enzymatic (Hussey, 1971) removal of root tissues from around the adult female nematodes. The freed adult females are placed on glass slides and crushed, and their body contents are examined microscopically for vegetative stages and endospores of the bacterium.

Maintenance Procedures

Since none of the *Pasteuria* species has been cultivated axenically, these bacterial parasites are maintained by cocultivation with the respective invertebrate host.

Pasteuria ramosa can be grown in a two-membered system with laboratory-reared cladocerans (Sayre et al., 1979). Cells of an alga, *Scenedesmus* sp., grown in greenhouse aquaria, provided a suitable food source for the cladocerans. At feeding time, these algal cells were supplemented with a water-suspension of Trout Chow®, a finely ground alfalfa meal, and Tetra-Min®, a tropical fish food. Culture dishes (60 × 100 mm), each containing 150 ml of pond water, were used as aquaria for populations of 200–500 water fleas. Algal cells (approximately 3×10^6 cells/dish) were added daily to the aquaria, along with the supplements.

Sporangia and spores used to inoculate healthy cladocerans were obtained from two sources: (a) the crushed bodies of living, parasitized cladocerans; and (b) sediments from the bottoms of aquaria in which dead and parasitized cladocerans had accumulated. The inoculum was added to 10 ml of pond water containing a few hundred water fleas and allowed to remain for 3 h, thus affording ample opportunity for the cladocerans to acquire the parasite before they were returned to the rearing aquaria containing 140 ml of pond water. Each aquarium was then placed under a bank of cool-white fluorescent tubes; the water temperature in the aquarium was regulated by placing it on a thermostatically controlled slide-warming tray. At 25° and 31°C, the parasite was observed in some cladocerans at 6 and 3 days, respectively.

In general, members of the *P. penetrans* group can be maintained in a system consisting of the immediate nematode host and the host plant of this nematode. A good example is the system consisting of *P. penetrans sensu stricto, Meloidogyne incognita,* and tomato plants (Sayre and Wergin, 1977). To initiate and increase a bacterial population, dried bacterial endospore preparations (e.g., Stirling and Wachtel, 1980) are mixed into soils heavily infested with larvae of *M. incognita.* The larvae migrating in the soil become encumbered with the bacterial endospores as a result of random contact (see Fig. 33.12). The endospores preferentially adhering to the cuticular surfaces of the nematode are carried by the larvae into the tomato roots, where they then germinate. This germination occurs shortly after the nematode establishes its plant-parasitic relationship. Endospores germinating on the nematodes' surface penetrate cuticles by means of infection tubes (see Fig. 33.22 below), which extend into the pseudocoelomic cavities. The bacteria then enter into their vegetative endoparasitic developmental stages (see Figs. 33.23 and 33.24 below). Finally, the bacteria form, in the mature and moribund host nematodes, sporangia that contain endospores (see Figs. 33.25–33.27 below).

Endospores of *P. penetrans sensu stricto* growing on *M. incognita* can be harvested by two methods. The simplest procedure is to allow the nematode-infested plant roots to decay in situ in the soil; during such decay, approximately 2×10^6 endospores are released from each adult female nematode. The soil containing the endospores is air dried, mixed, and stored. Such preparations have yielded bacterial endospores that can attach to the larvae of their respective host nematode even after a few years in storage (Mankau, 1973). Stirling and Wachtel (1980) suggested a second method for obtaining a more concentrated preparation of the bacterial endospores. Larvae were heavily encumbered by endospores by placing them in aqueous suspensions containing infected adult nematodes and endospores; these encumbered larvae were then allowed to penetrate roots of tomato seedlings. After the life cycle of the nematode was completed in soil, the roots of the tomato seedlings were harvested, washed, air dried, ground into a fine powder, and stored. Such preparations provided adequate sources of endospores for use in bioassays and other procedures.

Pasteuria thornei can be maintained in similar systems, except for the involvement of a different host nematode, *Pratylenchus brachyurus* rather than *M. incognita.* This host nematode is collected from the roots of infected peanut plants using a procedure suggested by Chapman (1957). Another useful source of *P. thornei* endospores is soil that has been repeatedly cropped by nematode-infested peanuts. Millions of bacterial sporangia and endospores are liberated upon decay of the plant roots and the remnants of the infested nematodes. Healthy larvae of *Pratylenchus brachyurus* migrating through such soils would become encumbered with sporangia of *P. thornei* and repeat the infective cycle, with not only maintenance of the bacterium but also a net increase in the endospore content of the soil.

Although methodological details have not been published, a number of investigators (including both authors of this essay) do cultivate members of the *P. penetrans* group in three-membered systems consisting of plant tissue cultures, gnotobiotically reared nematodes, and the desired bacteria free, it is hoped, of contaminating microbes. Once perfected, these cultural methods are useful for maintenance of these bacteria, at least until methods for their axenic cultivation become generally available.

Differentiation of the genus **Pasteuria** *from other genera*

Table 33.11 summarizes the characteristic features of the species of the genus *Pasteuria.*

Taxonomic Comments

De Toni and Trevisan (1889) provided the first generic diagnosis of *Pasteuria;* it followed quickly and closely Metchnikoff's (1888) original description of *P. ramosa.* However, some other early investigators, particularly those interested in taxonomic coherence, not having observed this enigmatic organism and relying solely on descriptions, rejected both the generic and specific concepts (Lehmann and Neumann, 1896; Migula, 1904; Smith, 1905). Laurent (1890) suggested that a bacteroid-forming species from nodules on leguminous plants, together with *P. ramosa,* comprised the new family he erected, *Pasteuriaceae.* Similarly, Vuillemin (1913) believed a generic relationship existed between *Nocardia* and *Pasteuria.* De Toni and Trevisan (1889) speculated that *P. ramosa,* because of its ability to form endospores, should be placed in the subtribe *Pasteurieae* of the tribe *Bacilleae.* The unusual morphology of *P. ramosa* became the basis for numerous suppositions about its affinities to other bacterial groups (Buchanan, 1925), a speculative process that continues up to the present (Sayre et al., 1983, 1988; Sayre and Starr, 1985; Starr and Sayre, 1988a). The taxonomic status of this genus will remain in flux until axenic cultures of these organisms become generally available and are thoroughly characterized. The genus is included in this volume on the basis of its production of mycelia, sporangia, and endospores, but the true extent of its affinity with the actinomycetes remains to be determined.

A significant change made since the eighth edition of the *Manual* is reduction in the confusion between Metchnikoff's Gram-positive, endospore-forming, and mycelial cladoceran parasite with certain Gram-negative, budding, and nonprosthecately appendaged aquatic bacteria. Conservation of *Pasteuria ramosa sensu* Metchnikoff 1888 on the basis of type-descriptive material, as well as rejection of ATCC 27377 as the type of *Pasteuria ramosa* Metchnikoff 1888 because it actually is a quite different organism (*Planctomyces staleyi* Starr, Sayre and Schmidt 1983), have been recommended (Starr et al., 1983) and approved (Judicial Commission, 1986). The detailed drawings, photomicrographs, and lengthy description offered by Metchnikoff (1888) provided a sound basis for comparing his species with the current cladoceran parasites. These recently rediscovered bacteria are identical to *P. ramosa* in essentially all features noted in the original description (Sayre et al., 1983).

One reason investigators (Henrici and Johnson, 1935; Hirsch, 1972; Staley, 1973) have been interested in reexamining *P. ramosa* stems from the questions raised by Metchnikoff's assertion that the bacterium divides longitudinally. Metchnikoff suggested that cells of *P. ramosa* undergo a longitudinal fission, giving rise to a branched structure in which the two daughter cells remain attached at their tips. Based on a hypothesis about the evolution of division patterns in bacteria, he concluded that *P. ramosa,* a fairly primitive bacterium in his view, divided longitudinally.

Observations on the rediscovered *P. ramosa* (Sayre et al., 1979) indicate that cleavage indeed occurs in three planes, as evidenced by the spherical mycelial colonies. However, no common plane of division was found, as illustrated in drawing 2 of Metchnikoff's (1888) paper, starting at the surface of the cauliflower-like growth and ending at its interior. The prominent bifurcations in the mycelium (see Fig. 33.16, page 2609), which Metchnikoff also observed and which may have prompted his longitudinal fission theory, are not products of atypical fission but rather are probably the branched distal or terminal cells of the mycelium undergoing rapid enlargement during formation of endogenous spores. These terminal cells develop into sporangia.

Table 33.11.
Characteristics held in common by **Pasteuria ramosa** *Metchnikoff 1888,* **Pasteuria penetrans sensu stricto** *emend. Starr and Sayre 1988a, and* **Pasteuria thornei** *Starr and Sayre 1988a*

Category	Subcategory	Characteristics
Morphological similarities as observed by light microscopy		
	Vegetative cells	Microcolonies consist of dichotomously branched mycelium.
		Diameter of mycelial filaments similar.
		Mycelial filaments are seen in host tissues only during early stages of infection.
		Daughter microcolonies seem to be formed by lysis of "sacrificial" intercalary cells.
		Nearly all vegetative mycelium eventually lyses, leaving only sporangia and endospores.
	Endospores	Terminal hyphae or peripheral cells of the colony elongate and swell, giving rise to sporangia.
		A single endospore is produced within each sporangium.
		Endospores are in the same general size range.
		Refractility of endospores, as observed in the light microscope, increases with maturity.
	Staining reaction	Gram-positive.
Ultrastructural similarities		
	Vegetative cells	Mycelial cell walls are typical of Gram-positive bacteria.
		Mycelial filaments divided by septa.
		Double-layered cell walls.
		Where they occur, mesosomes are similar in appearance and seem to be associated with division and septum formation.
	Endospores	Typical endogenous spore formation.
		Identical sequences in endospore formation: (a) septa form within sporangia; (b) sporangium cytoplast condenses to form forespore; (c) endospore walls form; (d) final endospore matures; and (e) "light" areas adjacent to endospore give rise to extrasporal fibers.
Similar sequences of life stages		Microcolonies.
		Fragmentation of microcolonies.
		Quartets of sporangia.
		Doublets of sporangia.
		Single sporangia.
		Free endospores.
Host-bacterium relationships		All parasitize invertebrates.
		Colonies first observed in the host are sedentary and located in the host's musculature.
		Growth in muscle tissue eventually leads to fragmentation and entry of microcolonies into the coelom or pseudocoelom of the respective host.
		Microcolonies carried passively by body fluids.
		Colonization of hemolymph or pseudocoelomic fluid by the parasite is extensive.
		Host ranges are very narrow: *Pasteuria ramosa* occurs only in cladoceran water fleas, *P. penetrans sensu stricto* only in the root-knot nematode *Meloidogyne incognita,* and *P. thornei* only in the root-lesion nematode *Pratylenchus brachyurus.*
		Host is completely utilized by the bacteria; in the end, the host becomes little more than a bag of bacterial endospores.
Survival mechanisms		Survive in field soils and at bottoms of ponds.
		Resist desiccation.
		Moderately to strongly resistant to heat.

Of some historical interest are a few scattered reports in the literature about organisms similar in appearance to *P. ramosa.* In these reports, each author came to a different taxonomic decision about the organism, as follows: *Torula* or other yeast species (Rühberg, 1933); two different genera of microsporidians (Jirovec, 1939; Weiser, 1943); and a possible intermediate stage in the life cycle of a *Dermatocystidium* species (Sterba and Naumann, 1970). These results probably stem from the inability of these investigators to cultivate the particular organism they observed, taken together with their dependence solely on morphology and staining reactions as the basis for their descriptions and classifications. Surprisingly, Metchnikoff (1888), the first person to report on this organism, recognized it correctly as a bacterium, while the later workers did not and, moreover, were apparently unaware of Metchnikoff's work.

Until rather recently, taxonomic studies of the *P. penetrans* group were carried out mainly by nematologists. Cobb (1906), a distinguished nematologist, erroneously designated such microbes as protozoan parasites of nematodes. At first glance, it would appear from the literature that subsequent workers (Steiner, 1938; Thorne, 1940) had independently come to the same conclusion. However, their conclusions were probably by consensus. Members of the *P. penetrans* group were then, and largely still are, essentially known only to the community of plant nematologists. When Thorne described *Duboscqia penetrans* as a protozoan, he could not have realized its bacterial nature because electron microscopes were not then available. Later, Williams (1960) studied a similar organism in a population of root-knot nematodes from sugarcane, presented an interpretation of its life stages, and indicated some reservations about Thorne's identification. Canning (1973) also doubted the identification as a protozoan and stressed the organism's fungal characteristics. Finally, Mankau (1975) and Imbriani and Mankau (1977) established the bacterial nature of the organism and brought its attendant taxonomic problems to the attention of the bacteriologists. Some results of the ensuing interdisciplinary enterprise are summarized elsewhere (Sayre and Starr, 1988; Sayre et al., 1988; Starr and Sayre, 1988a).

Acknowledgments

We thank R. E. Davis for calling our attention to the genus *Pasteuria* and for his help in initiating its study. Thanks go, also, to W. P. Wergin, J. R. Adams, C. Pooley, R. Reise, and S. Ochs for sustained technical advice and assistance, particularly in the preparation of the figures involving transmission and scanning electron microscopy. Thanks are due R. L. Gherna for supplying strain ATCC 27377. We are grateful to J. M. Schmidt for her collaboration during the period when the taxonomic confusion between *Pasteuria* and *Planctomyces* was being corrected. Skillful bibliographic and redactional assistance were provided by P. B. Starr.

Further Reading

Mankau, R. 1975. *Bacillus penetrans* n. comb. causing a virulent disease of plant-parasitic nematodes. J. Invertebr. Pathol. *26:* 333–339.

Metchnikoff, E. 1888. *Pasteuria ramosa* un reprèsentant des bactèries à division longitudinale. Ann. Inst. Pasteur, Paris *2:* 165–170.

Sayre, R. M., R. L. Gherna and W. P. Wergin. 1983. Morphological and taxonomic reevaluation of *Pasteuria ramosa* Metchnikoff 1888 and "*Bacillus penetrans*" Mankau 1975. Int. J. Syst. Bacteriol. *33:* 636–649.

Sayre, R. M. and M. P. Starr. 1985. *Pasteuria penetrans* (ex Thorne 1940) nom. rev., comb. nov., sp. nov., a mycelial and endospore-forming bacterium parasitic in plant-parasitic nematodes. Proc. Helminthol. Soc. Wash. *52:* 149–165.

Starr, M. P., and R. M. Sayre. 1988. *Pasteuria thornei* sp. nov. and *Pasteuria penetrans sensu stricto* emend, mycelial and endospore-forming bacteria parasitic, respectively, on plant-parasitic nematodes of the genera *Pratylenchus* and *Meloidogyne*. Ann. Inst. Pasteur/Microbiol. *139:* 11–31.

Starr, M. P., R. M. Sayre and J. M. Schmidt. 1983. Assignment of ATCC 27377 to *Planctomyces staleyi* sp. nov., and conservation of *Pasteuria ramosa* Metchnikoff 1888 on the basis of type descriptive material. Request for an Opinion. Int. J. Syst. Bacteriol. *33:* 666–671.

Thorne, G. 1940. *Duboscqia penetrans* n. sp. (Sporozoa, Microsporidia, Nosematidae), a parasite of the nematode *Pratylenchus pratensis* (de Man) Filipjev. Proc. Helminthol. Soc. Wash. *7:* 51–53.

Differentiation and characteristics of the species of the genus Pasteuria

The differential characteristics of *Pasteuria* species are given in Table 33.12.

List of species of the genus Pasteuria

1. **Pasteuria ramosa** Metchnikoff 1888, 166.[AL] (Nom. Cons. Opin. 61 Jud. Comm. 1986, 119. Not *P. ramosa* in the sense of Henrici and Johnson (1935), Hirsch (1972), and Staley (1973); see Starr et al. (1983) and Judicial Commission (1986).)

ra.mo′sa. L. adj. *ramosus* much-branched.

Gram-positive. Sporangia and microcolonies are parasitic in the hemocoele of cladocerans, water fleas of the genera *Daphnia* and *Moina*. Usually occur attached to one another at pointed ends of the teardrop-shaped sporangia, forming quartet and doublet configurations. The rounded end of the sporangium encloses a single refractile endospore, having axes of 1.37–1.61 × 1.20–1.46 μm, narrowly elliptic in cross-section. Nonmotile. Resistant to dessication but with only limited heat tolerance. Vegetative stages are cauliflowerlike, septate, mycelial growths that branch dichotomously and fragment to form microcolonies. Has not been cultivated axenically, but can be grown in the laboratory with the invertebrate host. The type-descriptive material consists of descriptions and illustrations in Metchnikoff's original publication (Metchnikoff, 1888) and elsewhere (Sayre et al., 1979, 1983; Starr et al., 1983; Starr and Sayre, 1988a).

2. **Pasteuria penetrans** (ex Thorne 1940) Sayre and Starr 1986, 355.[VP] (Effective publication: Sayre and Starr 1985, 163.) (Emend. Sayre, Starr, Golden, Wergin and Endo 1988, 28; *Duboscqia penetrans* Thorne 1940, 51.)

pen′e.trans. L. v. *penetro* pres. part. *penetrans* to enter.

Gram-positive vegetative cells. Mycelium is septate; hyphal strands, 0.2–0.5 μm in diameter, branch dichotomously. The sporangia, formed by expansion of hyphal tips, are cup shaped, ~2.26–2.60 μm in height with a diameter of 3.0–4.0 μm. Each sporangium is divided into two unequal sections. The smaller proximal body is not as refractile as the larger, rounded, cup-shaped portion, which encloses an ellipsoidal endospore broadly elliptic in section having axes of 0.99–1.21 × 1.30–1.54 μm. Endospores seem to be of the kind typical of the genus *Bacillus;* they are resistant to both heat and desiccation. Nonmotile. Sporangia and vegetative cells are found as parasites in the pseudocoelomic cavities of plant-parasitic nematodes. The epithet is now restricted to members of the *P. penetrans* group with cup-shaped sporangia and ellipsoidal endospores broadly elliptic in section occurring primarily as parasites of root-knot nematodes belonging to the genus *Meloidogyne,* particularly *M. incognita;* they also may parasitize other kinds of plant-parasitic nematodes. Has not been cultivated axenically; the type-descriptive material consists of the text and photographs in Sayre and Starr (1985) and Starr and Sayre (1988a). *Pasteuria penetrans sensu stricto* differs from *P. thornei* sp. nov. and other members of the *P. penetrans* group in host specificity, in size and shape of sporangia and endospores, and in other morphological and developmental characteristics.

3. **Pasteuria thornei** Starr and Sayre 1988b, 328.[VP] (Effective publication: Starr and Sayre 1988a, 28.)

thor′ne.i. M.L. gen. n. *thornei* of Thorne; named for Gerald Thorne, an American nematologist, who described and named as a protozoan this parasite of nematodes belonging to the genus *Pratylenchus*.

Gram-positive vegetative cells. Mycelium is septate; hyphal strands, 0.2–0.5 μm in diameter, branch dichotomously. Sporangia, formed by expansion of hyphal tips, are rhomboidal in shape, ~2.22–2.70 μm in diameter and 1.96–2.34 μm in height. Each sporangium is divided into two almost equal units. The smaller unit, proximal to the mycelium, is not refractile, and contains a granular matrix interspersed with many fibrillar strands. The refractile apical unit is cone shaped; it encloses an ellipsoidal endospore, sometimes almost spherical, having axes of 0.96–1.20 × 1.15–1.43 μm, with cortical walls about 0.13 μm in thickness except for an additional inner sublateral wall that gives the endospore a somewhat triangular appearance in cross-section. The tapering outer cortical wall at the base of the endospore forms an opening approximately 0.13 μm in diameter. Sporangia and endospores are found as parasites of root-lesion nematodes belonging to the genus *Pratylenchus;* they also may parasitize other kinds of plant-parasitic nematodes. Has not been cultivated axenically; the type-descriptive material consists of the text and photographs in Starr and Sayre (1988a) and Sayre et al. (1988). *Pasteuria thornei* differs from *P. penetrans sensu stricto* and other members of the *P. penetrans* group in host specificity, size and shape of sporangia and endospores, and other morphological and developmental traits.

Further Descriptive Information

Pasteuria ramosa

When using light microscopy, the earliest visible growth stages of *P. ramosa* in the particular cladoceran host we use, the water flea *Moina rectirostris* (Leydig) 1860, are the cauliflowerlike microcolonies usually

Table 33.12.
Comparison of **Pasteuria ramosa, Pasteuria penetrans sensu stricto** *emend., and* **Pasteuria thornei**

Characteristics	1. *P. ramosa*	2. *P. penetrans*	3. *P. thornei*
Colony shape	Like cauliflower floret	Spherical to cluster of elongated grapes	Small, elongate clusters
Sporangia			
Shape	Teardrop	Cup	Rhomboidal
Diameter (μm)[a]	2.12–2.77	3.0–3.9	2.22–2.70
Height (μm)[a]	3.40–4.35	2.26–2.60	1.96–2.34
Fate of sporangial wall at maturity of endospore	Remains rigidly in place; external markings divide sporangium in three parts	Basal portion collapses inward on the developed endospore; no clear external markings	Remains essentially rigid, sometimes collapsing at bases; no clear external markings
Exosporium	Not observed	Present	Present
Stem cell	Remains attached to most sporangia	Rarely seen; attachment of a second sporangium sometimes observed	Neither stem cell nor second sporangium seen
Endospore			
Shape	Oblate spheroid, an ellipsoid, narrowly elliptic in section	Oblate spheroid, an ellipsoid, broadly elliptic in section	Oblate spheroid, an ellipsoid sometimes almost spherical, narrowly elliptic in section
Orientation of major axis to sporangium base	Vertical	Horizontal	Horizontal
Dimensions (μm)[a]	1.37–1.61 × 1.20–1.46	0.99–1.21 × 1.30–1.54	0.96–1.20 × 1.15–1.43
Wall thickness (μm)[a] 0.17–0.23	0.28–0.34	0.22–0.26	
Protoplast	Contains pronounced stranded inclusions	Stranded inclusions sometimes seen	Stranded inclusions observed
Partial middle spore wall	Not observed	Surrounds endospore laterally, not in basal or polar areas	Surrounds endospore somewhat sublaterally
Pore			
Occurrence	Absent	Present	Present
Characteristics	—	Basal annular opening formed from thickened outer wall	Basal cortical wall thins to expose inner endospore
Diameter (μm)[a]	—	0.28 ± 0.11	0.13 ± 0.01
Parasporal structures			
Fibers, origin and orientation	Long primary fibers arise laterally from cortical wall, bending sharply downward to yield numerous secondary fibers arrayed internally toward the granular matrix	Fibers arise directly from cortical wall, gradually arching downward to form an attachment layer of numerous shorter fibers	Long fibers arise directly from cortical wall, bending sharply downward to form an attachment layer of numerous shorter fibers
Matrix, at maturity	Persists as fine granular material	Becomes coarsely granular; lysis occurs; sporangial wall collapses; base is vacuolate	Persists, but more granular; some strands are formed and partial collapse may occur
Host	Cladocerans: *Daphnia; Moina*	Nematodes: *Meloidogyne incognita*	Nematodes: *Pratylenchus brachyurus*
Completes life cycle in nematode larvae	—	No, only in adult	Yes, in all larval stages and adult
Location in host	Hemocoele and musculature; sometimes found attached to coelom walls	Pseudocoelom and musculature; no attachment to coelom walls seen	Pseudocoelom and musculature; no attachment to coelom walls seen
Attachment of spores on host	Spores not observed to attach or accumulate on surface of cladoceran	Spores accumulate in large numbers on cuticular surface	Spores accumulate in large numbers on cuticular surface
Mode of penetration of host	Not known; suspected to occur through gut wall	Direct penetration of nematode cuticle by hyphal strand	Direct penetration suspected but not seen
Source of host	Pond mud, freshwater	Soil, plants	Soil, plants

[a]Measurements are based on preparations examined by transmission electron microscopy. Somewhat different apparent sizes are obtained by phase-contrast light microscopy and scanning electron microscopy.

found on the inner wall of the invertebrate's carapace (Sayre et al., 1979; Fig. 33.15). The next detectable stage consists of quartets of sporangia carried in the hemolymph throughout the body of the cladoceran. Later, as the parasite develops, the hemolymph is noticeably clouded by the myriad immature sporangia of *P. ramosa,* mainly singles or doublets.

Electron micrographs of the "cauliflower" stage reveal circular patterns of septate hyphal strands of an actinomycete-like organism (Fig. 33.16). The hyphae measure ~0.67 μm in width. The hyphal wall, which is fairly homogeneous in density, is 14.5–15 nm thick. The periplasmic region between the wall and the cell membrane is about 8.7–9.0 nm in width. The cell membrane measures 5.8 nm in thickness. Mesosomes (circular membrane complexes) are often found associated with the septa (Sayre et al., 1979; Fig. 33.17).

The distal or terminal mycelial cells of the microcolonies enlarge to form teardrop-shaped sporangia that, when viewed by light microscopy, measured approximately 4.8–5.7 × 3.3–4.1 μm in diameter. Similar materials prepared for transmission electron microscopy and sectioned gave measurements of 3.40–4.35 μm in height and 2.12–2.77 μm in diameter. Early in this process, two septa form; these septa divide the sporangium into anterior, middle, and stem sections (Metchnikoff, 1888; Fig. 33.18). The partitioning is reflected in the outer wall of the sporangia (Fig. 33.19).

The anterior section, the upper two thirds of the sporangium, gives rise to the forespore. Within the anterior section, granular material condenses to form an electron-dense endospore, slightly ellipsoidal to almost spherical in shape, having axes measuring 1.37–1.61 × 1.20–1.46 μm. It is comprised of a multilayered central cytoplasm, containing numerous doubled fibrillar strands (Fig. 33.20). Structural changes also occur within the median section, where electron-transparent areas appear to expand and attach laterally to the multilayered endospore wall to form fibrous appendages. The mode of penetration of spores into host cladocerans was not determined.

Parasitized water fleas (*M. rectirostris*) taken from a pond in College Park, Maryland, were found to yield about 2×10^5 *P. ramosa* sporangia per host individual. Generally, the parasite was found in mature females with no young in their egg pouches. *Daphnia* species reported previously (Metchnikoff, 1888) as hosts of this organism were not parasitized by this bacterial strain (Sayre et al., 1979). Neither sporangia from crushed water fleas nor sediments from the rearing aquaria of parasitized *M. rectirostris* (Sayre et al., 1977) resulted in their becoming infected when added to healthy populations of *Daphnia magna* Strauss 1820 or *D. pulex* Leydig 1860. These two *Daphnia* species were listed by Metchnikoff (1888) as hosts for *P. ramosa.* However, because *M. rectirostris* is in the same family (*Daphnidae*) as these two *Daphnia* species, this result may suggest only a very marked host specificity in the particular strain of *P. ramosa* available to us, perhaps at the level of a *forma specialis,* the situation in which one form of a parasite reproduces only in one host species and not in others that are closely related taxonomically. The descriptive type material of this taxon (Metchnikoff, 1888) is attached to the form on *Daphnia* species. If later work should show differences warranting separation at the specific or subspecific level, the taxon on *Moina* would of course have to be given a name different from *P. ramosa.*

The influence of water temperature on the occurrence of *P. ramosa* in *M. rectirostris* in nature was observed over a 3-year period. The parasite was not found until the surface water temperature in the pond reached 26°C or higher, usually about mid-July in the College Park, Maryland, area. The apparent temperature requirement was confirmed in laboratory tests in which water in the aquaria was held at constant temperatures; the parasite was found in 6 and 3 days at 26° and 31°C, respectively, but not at all at 21°C.

Although endospores of *P. ramosa* appear to withstand desiccation, they have only limited resistance to heat. Air-dried sporangia, which were stored for 6 months, were capable of infecting healthy populations of cladocerans. However, when air-dried aquarium sediments were heated to 40°, 60°, or 80°C on a block heater for 10 min and then added to cultures of healthy cladocerans, the parasite developed in cladoceran cultures to which sediments heated for 10 min at 40° and 60°C were added, but not when the additions were sediments heated at 80°C. Similarly, when sporangia taken directly from crushed, parasitized cladocerans were heated in glass capillaries to 40°, 60°, or 80°C on a block heater for 10 min and then added to cultures of healthy cladocerans, parasite development occurred in the unheated sporangia and when heated to 40°C but not when heated at 60° or 80°C (Sayre et al., 1979).

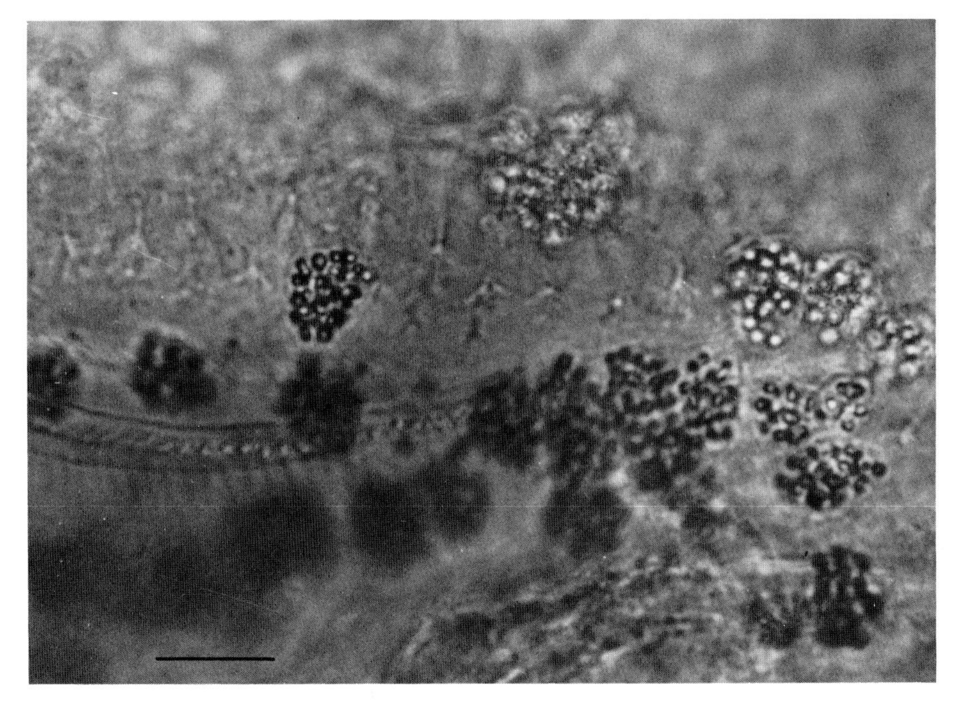

Figure 33.15. Cauliflowerlike, branching, mycelial colony of *P. ramosa* attached to the inner walls of the carapace of the cladoceran *Moina rectirostris. Bar,* 10 μm.

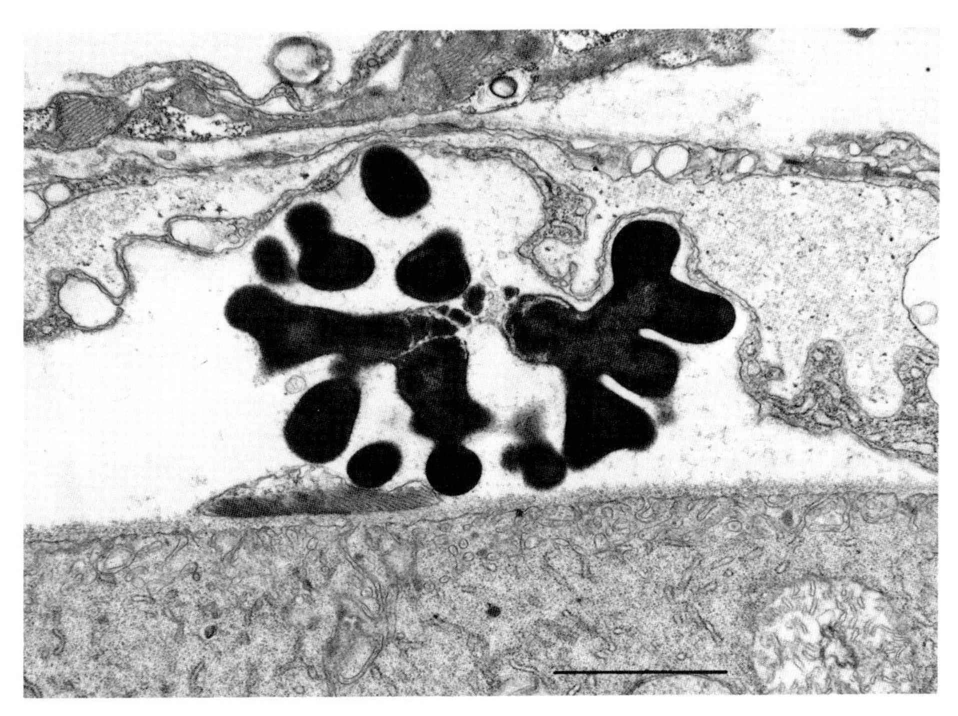

Figure 33.16. Cross-section of a fragmenting mycelium of *P. ramosa* in the body cavity of a cladoceran. *Bar,* 2.0 µm.

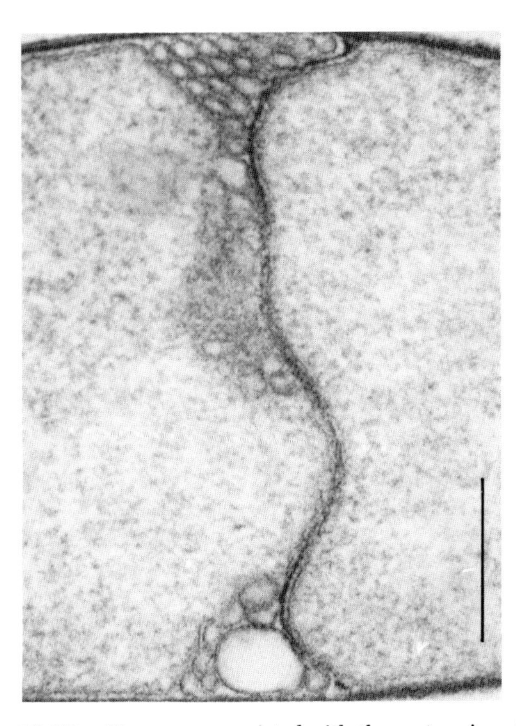

Figure 33.17. Mesosome associated with the septum in a dividing cell of *P. ramosa. Bar,* 0.5 µm.

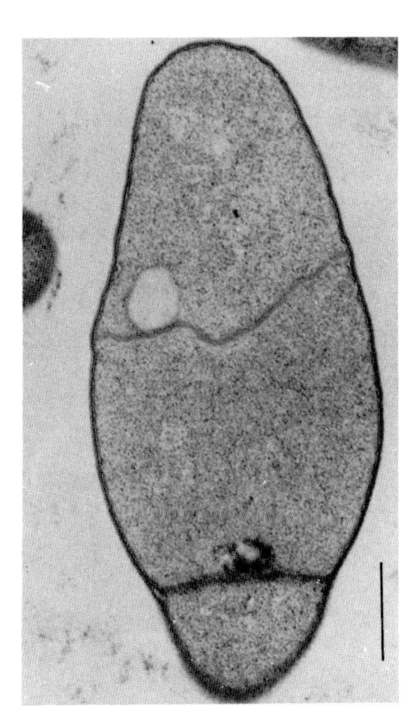

Figure 33.18. Septa divide the immature sporangium of *P. ramosa* into three parts. *Bar,* 0.5 µm.

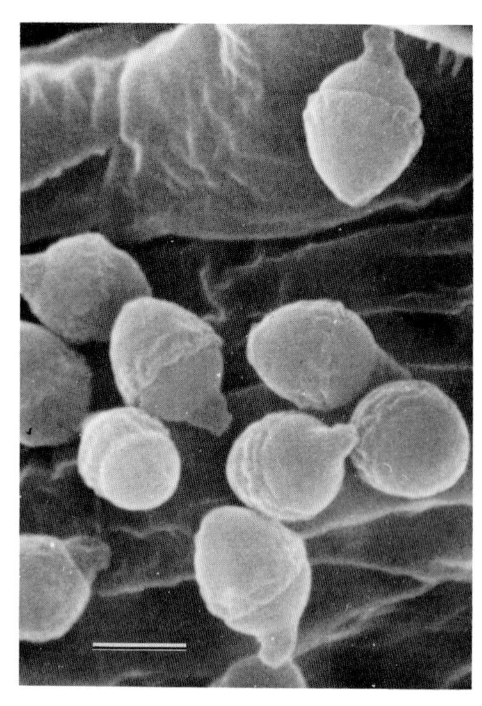

Figure 33.19. Scanning electron micrograph of sporangia of *P. ramosa.* External ridges mark boundaries of the endospore, middle section, and stem of a mature sporangium. *Bar,* 2.5 μm.

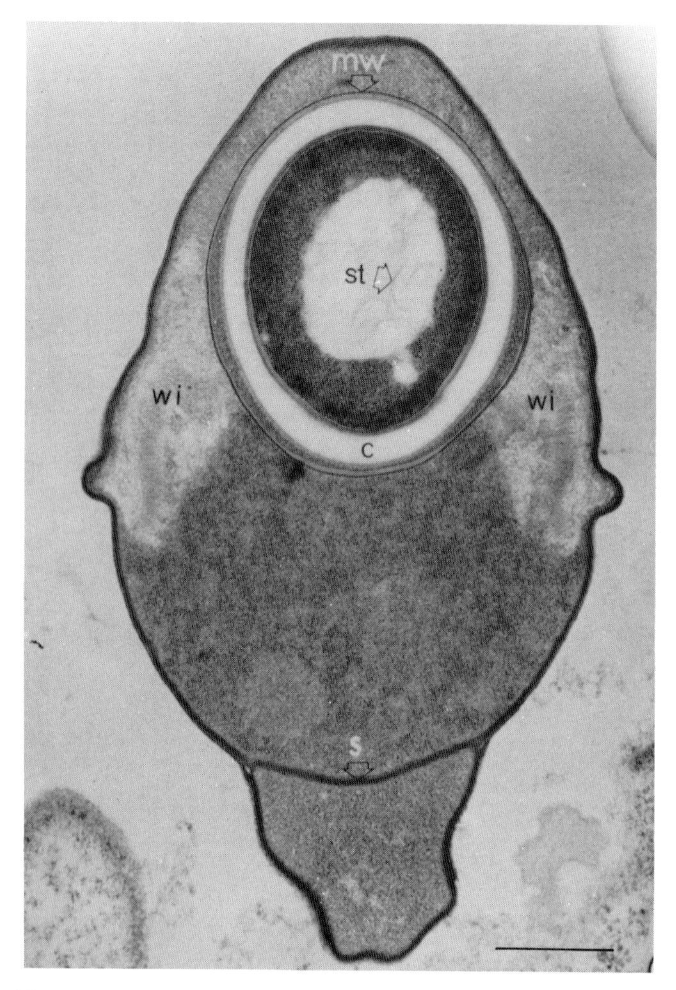

Figure 33.20. Mature sporangium of *P. ramosa* containing an endospore, made up of multilayered spore walls (*mw*), cortex (*c*), and cytoplast with stranded inclusions (*st*). Septum (*s*) separates the stem from the middle section. The function of the fibrous apendages (*wi;* wing-shaped light areas) is not known. *Bar,* 0.5 μm.

Pasteuria penetrans

Members of the *P. penetrans* group and *P. ramosa* share several morphological, ultrastructural, and ecological features (Starr et al., 1983; Sayre and Starr, 1985; Starr and Sayre, 1988a; Table 33.11; Fig. 33.21): all are Gram-positive; all form endospores; all form mycelial, septate, dichotomously branched, vegetative cells; and all parasitize invertebrates (Table 33.11). The members of the *P. penetrans* group differ in many respects from *P. ramosa:* colony shape, shape and size of sporangia and endospores, and host relations (Table 33.12). Upon recognition (Sayre and Starr, 1985) of its relationship to the genus *Pasteuria,* the first of these nematode parasites to receive such taxonomic attention was renamed *Pasteuria penetrans* (ex Thorne) Sayre and Starr 1985. Subsequently, the name *Pasteuria penetrans sensus stricto* (i.e. in the strict sense) was limited in scope to the bacterium parasitic on root-knot nematodes of the genus *Meloidogyne* and particularly *M. incognita* (Starr and Sayre, 1988a). A second species, *P. thornei,* was erected for parasites of the root-lesion nematodes of the genus *Pratylenchus* and particularly *P. brachyurus* (Starr and Sayre, 1988a).

Much of the following information stems from studies of *P. penetrans sensu stricto. Pasteuria thornei,* which has received much less study than its relative, is similar in most respects examined; where substantial differences obtain, they are noted below. Cross-sections viewed by transmission electron microscopy reveal (Imbriani and Mankau, 1977; Sayre and Wergin, 1977; Sayre and Starr, 1985; Starr and Sayre, 1988a) that the endospore of *P. penetrans sensu stricto* consists of a central, highly electron-opaque core surrounded by an inner and an outer wall composed of several distinct layers (see Fig. 33.27 below). When observed with the transmission electron microscope, the peripheral matrix of the spore is fibrillar. Fine microfibrillar strands, about 1.5 nm thick, extend outward and downward from the sides of the endospore to the cuticle of the nematode, where they become more electron dense.

A mature endospore of *P. penetrans sensu stricto* attaches to the surface of a nematode so that a basal ring of wall material lies flatly against the cuticle. A median section through the endospore and perpendicular to the surface of the nematode bisects this basal ring. As a result, the ring appears as two protruding pegs, which are continuous with the outer layer of the spore wall and rest on the cuticular surface of the nematode (Fig. 33.22).

The peripheral fibers of the endospore also are closely associated with the nematode's cuticle. These fibers, which encircle the endospore, lie along the surface of the nematode and follow the irregularities of the cuticular annuli. They seem not to penetrate the cuticle. The germ tube of the endospore of *P. penetrans sensu stricto* emerges through the central opening of the basal ring, penetrates the cuticle of the nematode, and enters the hypodermal tissue (Fig. 33.22). Hyphae were initially encountered beneath the cuticle of the nematode near the site of germ-tube penetration. From this site, they apparently penetrate the hypodermal and muscle tissues and enter the pseudocoelom.

Mycelial colonies of *P. penetrans sensu stricto* up to 10 μm in diameter are formed in the pseudocoelom, where they are observed after the diseased larvae penetrate plant roots (Fig. 33.23). The hyphae comprising the colony are septate. A hyphal cell, which is 0.40–0.50 μm in cross-section, is bounded by a compound wall, 0.12 μm thick, composed of an outer and an inner membrane. The inner membrane of the wall forms the septation and delineates individual cells. In addition, this membrane is continuous with a membrane complex or mesosome that is frequently as-

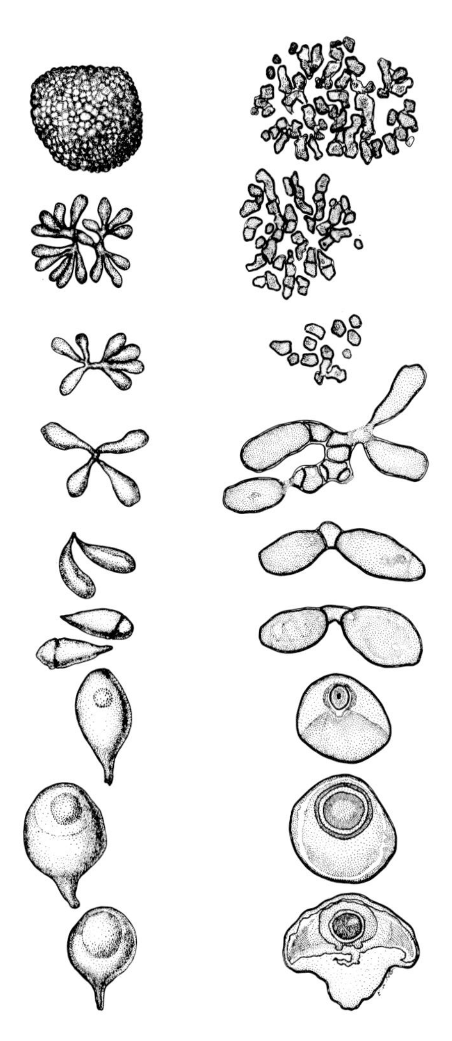

Figure 33.21. Drawings of *P. penetrans sensu stricto (left)* based on electron micrographs are compared with those of *P. ramosa (right)* as drawn by Metchnikoff (1888). Starting at the top of the left column is a vegetative colony of *P. penetrans* followed by daughter colonies, quartets of sporangia, doublets of sporangia, single sporangia, and finally at the bottom the mature endospore within the old sporangial wall. The drawings of *P. ramosa* on the right are placed in order of their occurrence in the life cycle of the parasite as reported by Metchnikoff (1888). (Reproduced by permission from R. M. Sayre, W. P. Wergin and R. E. Davis, Canadian Journal of Microbiology *23:* 1573–1579, 1977.)

sociated with the septum (Fig. 33.24). Because of the sinuous and branching growth habit, cell length cannot be determined from thin-section electron micrographs.

Sporulation of *P. penetrans sensu stricto* is a synchronously initiated process in the adult female nematode host feeding on its plant host. As the process begins, the terminal hyphal cells of the mycelium bifurcate and enlarge from typical hyphal cells to ovate cells measuring 2.0 × 4.0 µm. Structure and content of the cytoplast change from a granular matrix, which contains numerous ribosomes as found in the hyphal cells, to one lacking particulate organelles. During these changes, the developing sporangia separate from their parental hyphae, which cease to grow and eventually degenerate.

After these early structural alterations, a membrane forms within the sporangium and separates the upper third of the cell or forespore from its lower or parasporal portion (Fig. 33.25). The granular matrix confined within the membrane then condenses into an electron-opaque body, 0.6 µm in diameter, which eventually becomes encircled by a multilayered wall. The resulting discrete structure is an endospore.

Coincident with the formation of an endospore in *P. penetrans sensu stricto* is the emergence of the parasporal fibers. These fine fibers, which form around the base of the spore, differentiate from an electron-transluscent, granular substance. They appear to connect with and radiate from the external layer of the wall of the endospore (Fig. 33.26). During development of the parasporal fibers, the formation of another membrane, the exosporium, isolates the newly formed endospore within the sporangium. At this later stage of spore development, the granular content of the paraspore becomes less dense, degenerates, and disappears. As a result, the mature sporangium contains a fully developed endospore enclosed within the exosporium (Fig. 33.27).

The cell wall of the sporangium of *P. penetrans sensu stricto* remains intact until the remnants of the infested nematode are disrupted, after which event the endospores are released. The exosporium apparently remains associated with the endospore until contact is made with a new nematode and the infection cycle restarts. The vermiform larval stages of the nematodes are encumbered by the parasite as they migrate through soils infested by the endospores of *P. penetrans sensu stricto,* and the infectious cycle is repeated.

Much morphological, ultrastructural, developmental, and host diversity is evident in the *P. penetrans* group (Sayre and Starr, 1985; Sayre et al., 1988). We believe this diversity speaks for the existence of more than one taxon within the group. *Pasteuria thornei* represents the first case in which a new species was erected on substantial grounds (Table 33.12) for another member of the group (Sayre et al., 1988; Starr and Sayre, 1988a). One striking difference between these two members of the *P. penetrans* group is the simultaneous occurrence in individual *Pratylenchus brachyurus* larvae of all developmental stages of *P. thornei,* from mycelial microcolonies to mature endospores (Fig. 33.28). Unlike the development of *P. penetrans sensu stricto* in *Meloidogyne incognita,* where the parasite and host develop synchronously, beginning with the bacterial vegetative stage in the early molts of the host nematode and ending with the sporangial stages in the adult *Meloidogyne* female (Sayre and Starr, 1985), there appeared to be no synchrony among the various stages of *P. thornei,* nor could this bacterium's development be tied to the development of its nematode host.

Endospore formation in both species begins with the swelling of terminal cells in the microcolonies. During the subsequent formation of the single endospore within each sporangium, there was little difference in the development and sequence of events to differentiate between the two species (e.g. compare Figs. 33.25 and 33.29). However, the final mature sporangia released from the two host nematodes are distinctly different from each other in several respects. The sporangia of *P. thornei* are smaller in diameter than those of *P. penetrans sensu stricto* and slightly less in height (Table 33.12). These dimensions reflect the respective shapes; sporangia of *P. thornei* are rhomboidal (Fig. 33.30) and sporangia of *P. penetrans sensu stricto* are cup shaped (Fig. 33.27). The endospores of the two also are different in size and shape (Table 33.12). The central endospore of *P. thornei* is nearly spherical; that of *P. penetrans sensu stricto* is ellipsoidal, broadly elliptic in section. Compared with *P. penetrans sensu stricto,* the *P. thornei* endospore is surrounded by thinner cortical walls; and it has a smaller basal germinal pore not clearly set off by the cortical walls as in *P. penetrans sensu stricto.* The parasporal fibers surrounding the endospore provide further morphological features to separate the two species. The number of fibers radiating out from the cortical walls appeared to be fewer in number in *P. thornei;* in transverse sections, their angle of attachment gave the endospores of *P. thornei* the appearance of an arrowhead. This arrangement is markedly different from the saucer-shaped transverse sections of the endospores of *P. penetrans sensu stricto.* The drawings (Fig. 33.31) of mature sporangia of all three *Pasteuria* species illustrate some ultrastructural features useful in identification.

The relationships of both species with particular nematodes are quite specific. Host specificity of members of the *P. penetrans* group is usually scored by a procedure involving quantifying attachments of bacterial

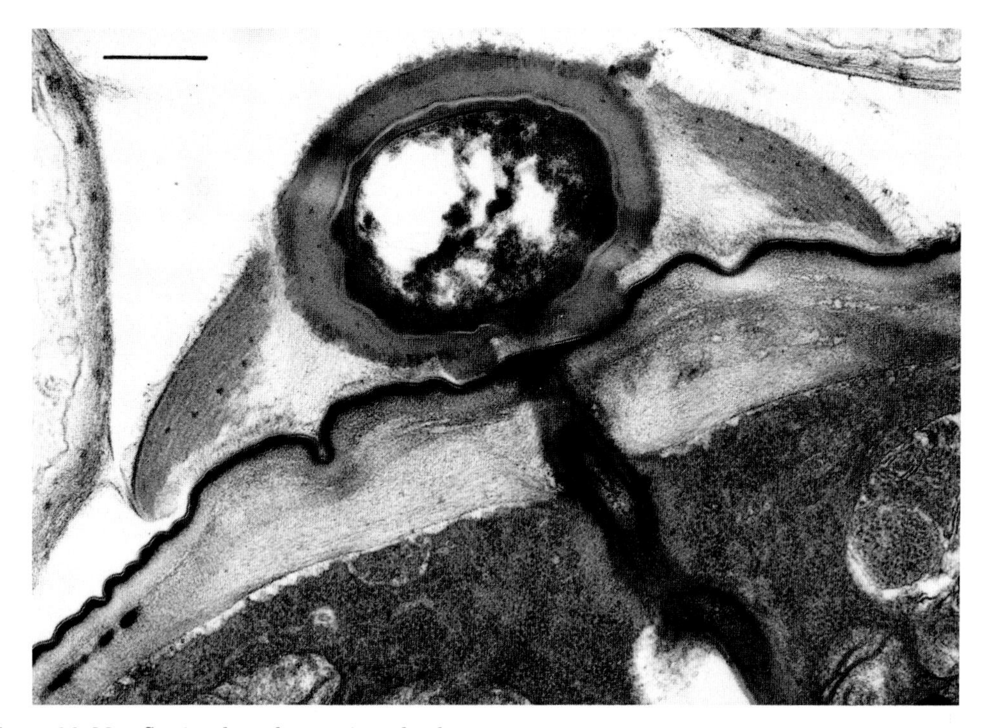

Figure 33.22. Section through a germinated endospore of *P. penetrans sensu stricto;* the penetrating germ tube follows a sinuous path as it traverses the cuticle and hypodermis of the nematode. *Bar,* 0.5 μm.

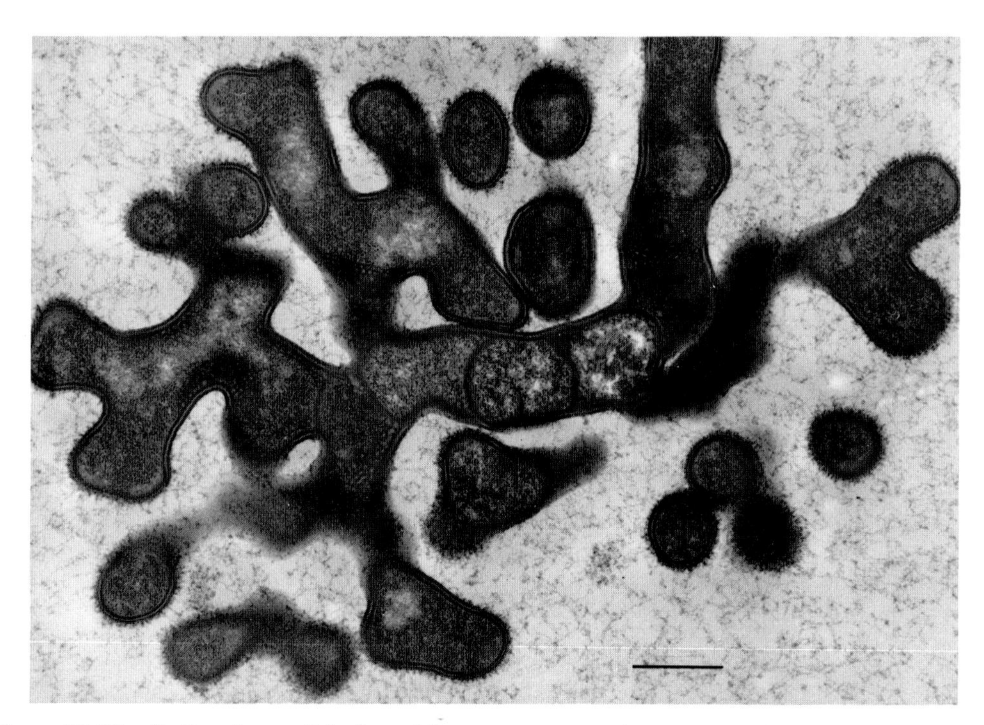

Figure 33.23. Section of a mycelial colony of *P. penetrans sensu stricto* in the pseudocoelom of the nematode. The septate hyphae appear to bifurcate at the margins of the colony. *Bar,* 0.5 μm.

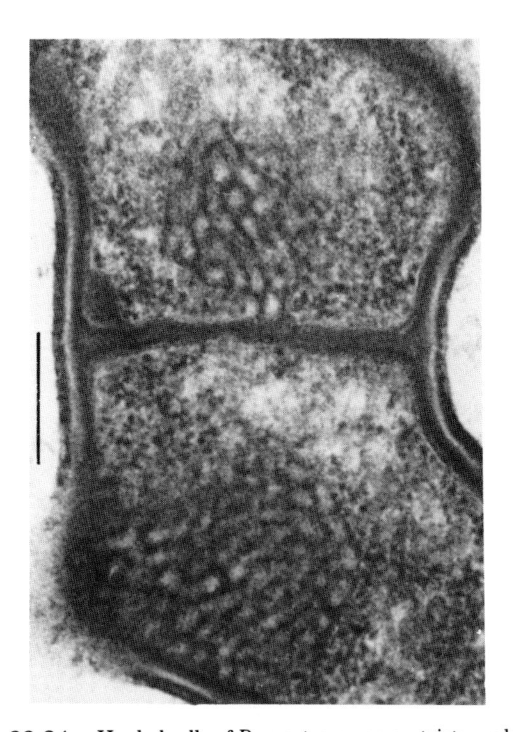

Figure 33.24. Hyphal cells of *P. penetrans sensu stricto* are bounded by a compound wall consisting of a double membrane. A mesosome is associated with the septum. *Bar,* 0.5 μm.

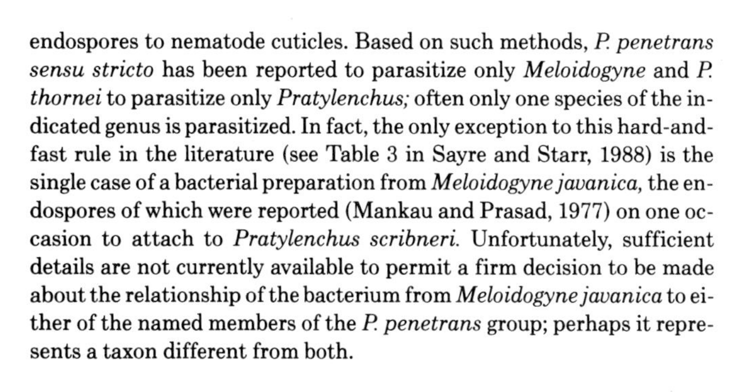

endospores to nematode cuticles. Based on such methods, *P. penetrans sensu stricto* has been reported to parasitize only *Meloidogyne* and *P. thornei* to parasitize only *Pratylenchus;* often only one species of the indicated genus is parasitized. In fact, the only exception to this hard-and-fast rule in the literature (see Table 3 in Sayre and Starr, 1988) is the single case of a bacterial preparation from *Meloidogyne javanica,* the endospores of which were reported (Mankau and Prasad, 1977) on one occasion to attach to *Pratylenchus scribneri.* Unfortunately, sufficient details are not currently available to permit a firm decision to be made about the relationship of the bacterium from *Meloidogyne javanica* to either of the named members of the *P. penetrans* group; perhaps it represents a taxon different from both.

Figure 33.25. An early stage of endospore development in *P. penetrans sensu stricto* is shown in this median section through a sporangium. An electron-opaque body has formed with the forespore; the body is surrounded by membranes that will condense and contribute to the multilayered wall of the mature endospore. *Bar,* 0.5 μm.

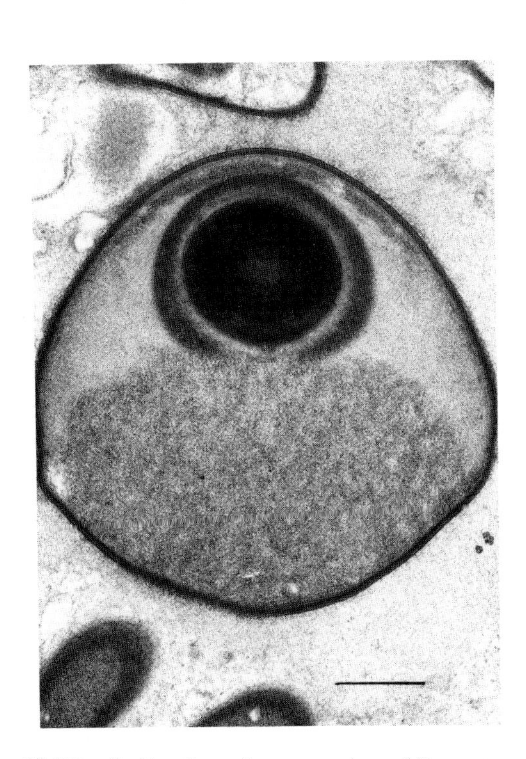

Figure 33.26. Section through a sporangium of *P. penetrans sensu stricto* with an almost mature endospore. The lateral regions (light areas) will differentiate into parasporal fibers. *Bar,* 0.5 μm.

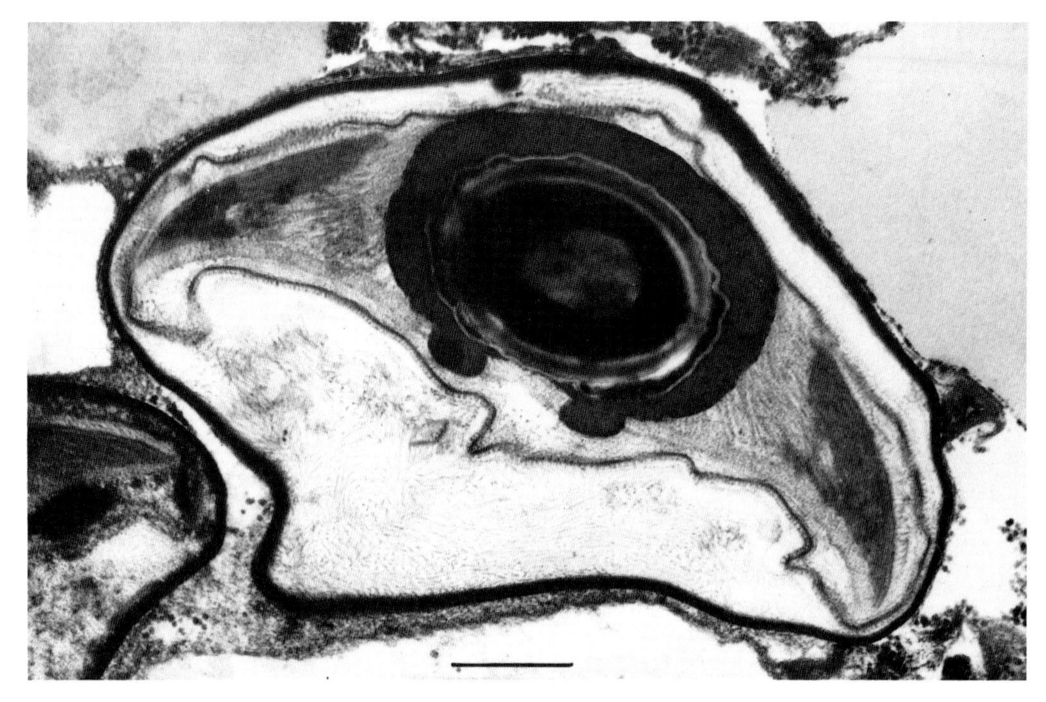

Figure 33.27. Median section through a sporangium of *P. penetrans sensu stricto* containing a fully mature endospore. Final stages of endospore differentiation include formation of an encircling membrane or exosporium and emergence of parasporal fibers within the granular material that lies laterally around the spore. *Bar,* 0.5 μm.

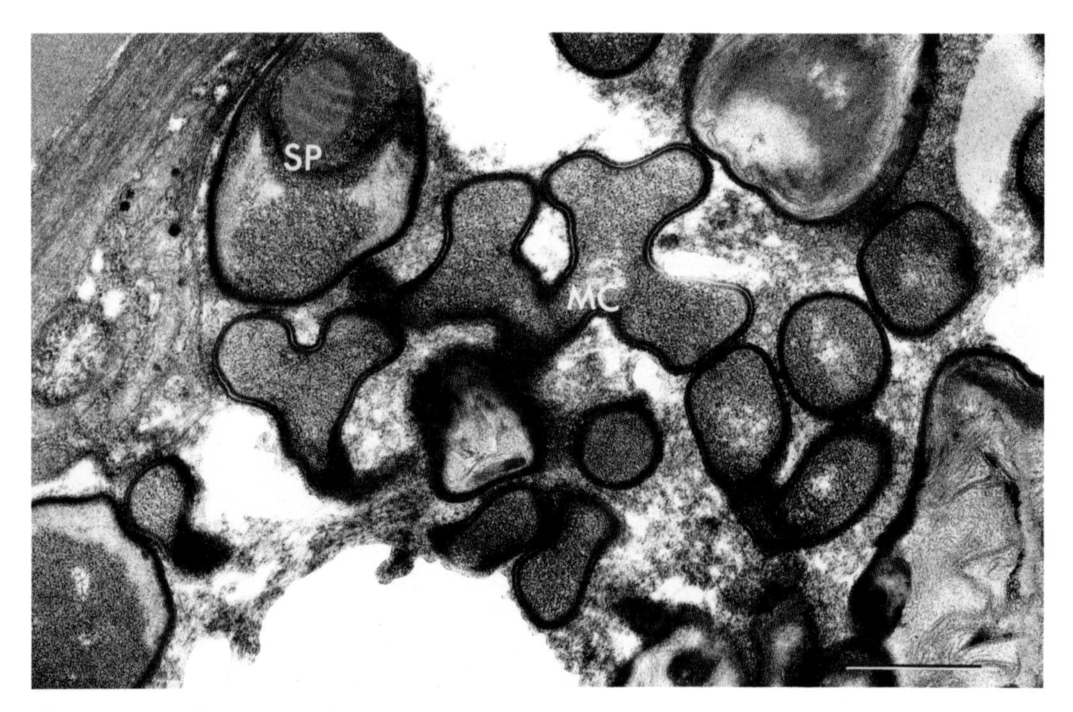

Figure 33.28. Transmission electron micrograph of *P. thornei* parasitizing a larva of the root-lesion nematode *Pratylenchus brachyurus,* showing simultaneous occurrence of vegetative microcolonies (*MC*) and sporangia (*SP*) of the bacterium in the nematode's pseudocoelomic cavity. *Bar,* 1.0 μm.

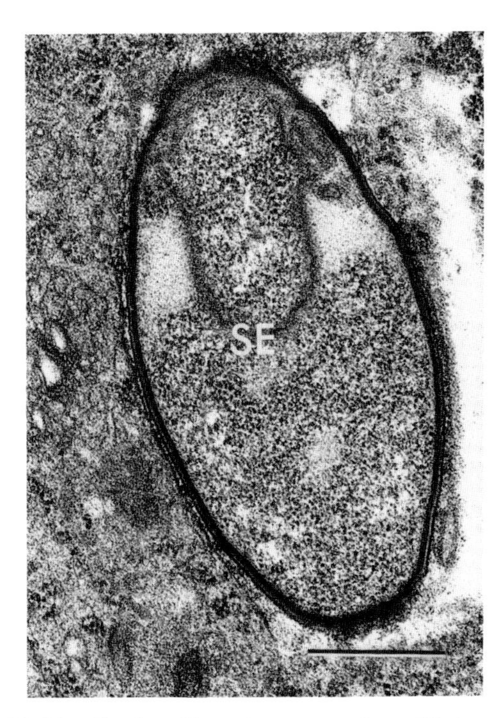

Figure 33.29. Section of immature sporangium of *P. thornei*, showing the developing septum (*SE*) separating the polar area that condenses into an endospore from the parasporal matrix; the light areas sublateral to the polar area will develop into parasporal fibers. *Bar*, 0.5 µm.

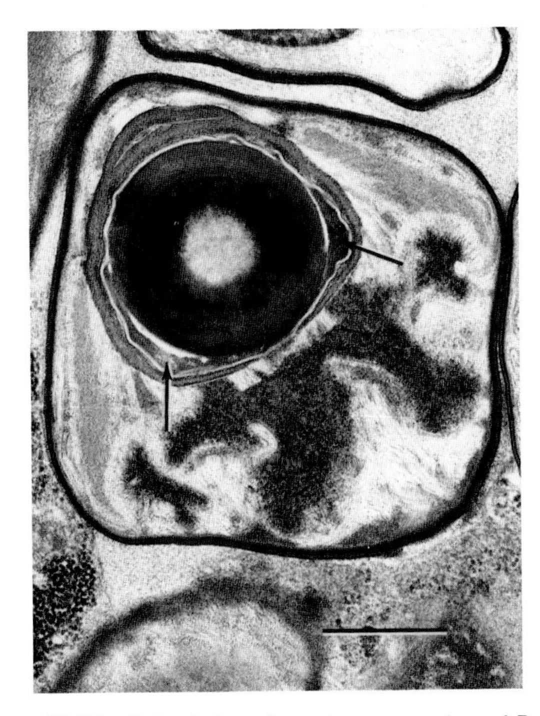

Figure 33.30. Lateral view of a mature sporangium of *P. thornei*, showing inner sublateral cortical wall (*arrows*) that gives the endospore an angular appearance. The basal cortical wall of the endospore thins to provide a germinal pore. The basal portion of the sporangium contains an irregular granular matrix intermingled with fibrillar strands. *Bar*, 0.5 µm.

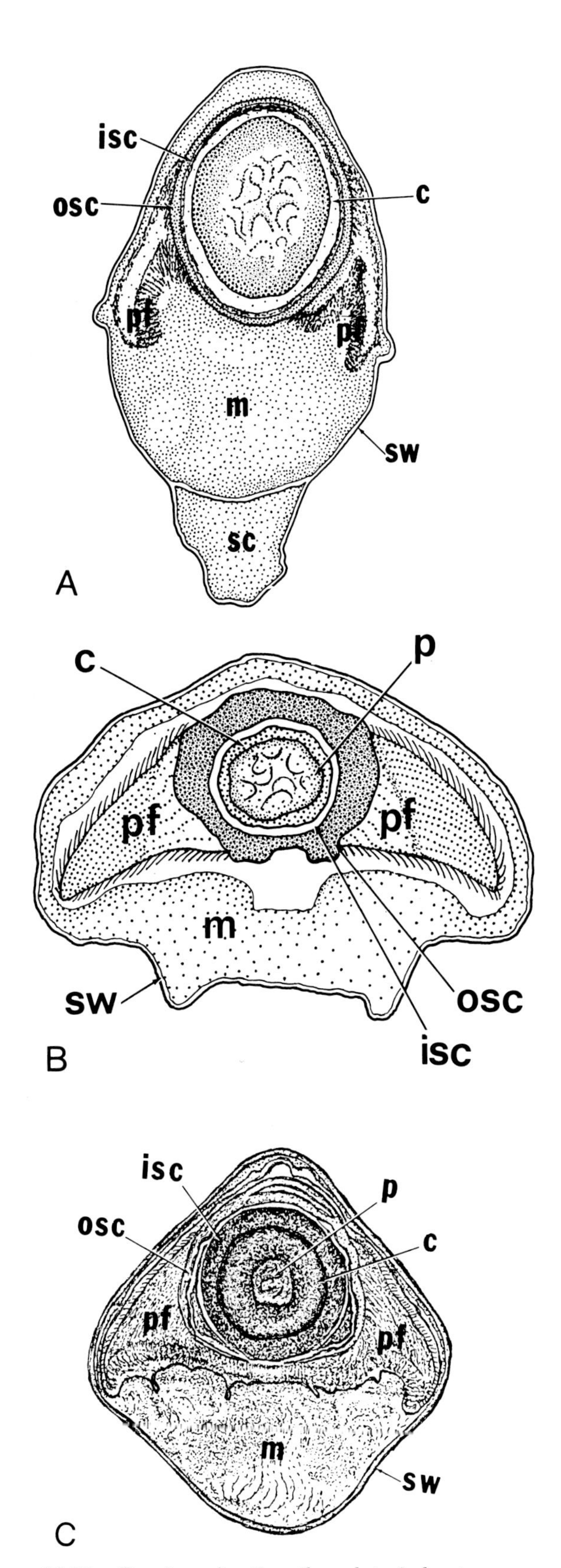

Figure 33.31. Drawings of sections through typical mature sporangia of the three *Pasteuria* species, emphasizing ultrastructural features useful in identification. *A*, *P. ramosa*, parasite of cladoceran water fleas. *B*, *P. penetrans sensu stricto*, parasite of *Meloidogyne incognita*. *C*, *P. thornei*, parasite of *Pratylenchus brachyurus*. Labels: protoplast (*p*) containing stranded inclusions; multilayered structures including a cortex (*c*) and inner (*isc*) and outer (*osc*) spore coats; parasporal fibers (*pf*); granular matrix (*m*); sporangial wall (*sw*); in *P. ramosa*, a stem cell (*sc*) remains attached to the sporangium.

ADDENDUM

Some Novel Actinomycete Taxa That Have Been Validly Published or Validated since Volume 4 of *Bergey's Manual®* of *Systematic Bacteriology* Went to Press

Actinomadura atramentaria sp. nov. Miyadoh et al. 1987. Int. J. Syst. Bacteriol. *37:* 342–346.

Actinoplanes cyaneus sp. nov. Terekhova, Sadikova and Preobrazhenskaya 1977. Antibiotiki *22:* 1059–1063. (Validation List No. 23, Int. J. Syst. Bacteriol. *37:* 179–180, 1987.)

Actinoplanes minutisporangius sp. nov. Ruan, Lechevalier, Jiang and Lechevalier 1986. The Actinomycetes *19:* 163–175. (Validation List No. 22, Int. J. Syst. Bacteriol. *36:* 573–576, 1986.)

Amycolata gen. nov. Lechevalier, Prauser, Labeda and Ruan 1986. Int. J. Syst. Bacteriol. *36:* 29–37.

Amycolata autotrophica comb. nov. Lechevalier, Prauser, Labeda and Ruan 1986. Int. J. Syst. Bacteriol. *36:* 29–37.

Amycolata saturnea comb. nov. Lechevalier, Prauser, Labeda and Ruan 1986. Int. J. Syst. Bacteriol. *36:* 29–37.

Amycolata hydrocarbonoxydans comb. nov. Lechevalier, Prauser, Labeda and Ruan 1986. Int. J. Syst. Bacteriol. *36:* 29–37.

Amycolatopsis gen. nov. Lechevalier, Prauser, Labeda and Ruan 1986. Int. J. Syst. Bacteriol. *36:* 29–37.

Amycolatopsis orientalis comb. nov. Lechevalier, Prauser, Labeda and Ruan 1986. Int. J. Syst. Bacteriol. *36:* 29–37.

Amycolatopsis mediterranei comb. nov. Lechevalier, Prauser, Labeda and Ruan 1986. Int. J. Syst. Bacteriol. *36:* 29–37.

Amycolatopsis orientalis subsp. *lurida* nom. rev., comb. nov., subsp. nov. Lechevalier, Prauser, Labeda and Ruan 1986. Int. J. Syst. Bacteriol. *36:* 29–37.

Amycolatopsis rugosa sp. nov. Lechevalier, Prauser, Labeda and Ruan 1986. Int. J. Syst. Bacteriol. *36:* 29–37.

Amycolatopsis sulphurea sp. nov. Lechevalier, Prauser, Labeda and Ruan 1986. Int. J. Syst. Bacteriol. *36:* 29–37.

Catellatospora gen. nov. Asano and Kawamoto 1986. Int. J. Syst. Bacteriol. *36:* 512–517.

Catellatospora citrea sp. nov. Asano and Kawamoto 1986. Int. J. Syst. Bacteriol. *36:* 512–517.

Catellatospora citrea subsp. *methionotrophica* subsp. nov. Asano and Kawamoto 1988. Int. J. Syst. Bacteriol. *38:* 326–327.

Catellatospora ferruginea sp. nov. Asano and Kawamoto 1986. Int. J. Syst. Bacteriol. *36:* 512–517.

Dactylosporangium fulvum sp. nov. Shomura et al. 1986. Int. J. Syst. Bacteriol. *36:* 166–169.

Kibdelosporangium aridum subsp. *largum* Shearer et al. 1986. J. Antibiot. *39:* 1386–1394. (Validation List No. 24, Int. J. Syst. Bacteriol. *38:* 136–137, 1988.)

Kibdelosporangium philippinese sp. nov. Mertz and Yao 1988. Int. J. Syst. Bacteriol. *38:* 282–286.

Kitasatosporia mediocidica sp. nov. Labeda 1988. Int. J. Syst. Bacteriol. *38:* 287–290.

Microbispora indica sp. nov. Rao et al. 1987. Int. J. Syst. Bacteriol. *37:* 181–185.

Microbispora karnatakensis sp. nov. Rao et al. 1987. Int. J. Syst. Bacteriol. *37:* 181–185.

Micromonospora rosaria sp. nov. nom. rev. Horan and Brodsky 1986. Int. J. Syst. Bacteriol. *36:* 478–480.

Nocardia seriolae sp. nov. Kudo, Hatai and Seino 1988. Int. J. Syst. Bacteriol. *38:* 173–178.

Saccharomonospora azurea sp. nov. Runmao 1987. Int. J. Syst. Bacteriol. *37:* 60–61.

Streptomyces aculeolatus sp. nov. Shomura et al. 1987. J. Antibiot. *40:* 732–739. (Validation List No. 24. Int. J. Syst. Bacteriol. *38:* 136–137, 1988.)

Streptomyces albidochromogenes sp. nov. Preobrazhenskaya 1983. *In* Gauze et al. A guide for determination of actinomycetes. Nauka, Moscow. (Validation List No. 22, Int. J. Syst. Bacteriol. *36:* 573–576, 1986.)

Streptomyces carpaticus sp. nov. Maximova and Terekhova 1983. *In* Gauze et al. A guide for determination of actinomycetes. Nauka, Moscow. (Validation List No. 22, Int. J. Syst. Bacteriol. *36:* 573–576, 1986.)

Streptomyces cinereorectus sp. nov. Terekhova and Preobrazhenskaya 1983. *In* Gauze et al. A guide for determination of actinomycetes. Nauka, Moscow. (Validation List No. 22, Int. J. Syst. Bacteriol. *36:* 573–576, 1986.)

Streptomyces cinereospinus sp. nov. Terekhova, Preobrazhenskaya and Gauze 1983. *In* Gauze et al. A guide for determination of actinomycetes. Nauka, Moscow. (Validation List No. 22, Int. J. Syst. Bacteriol. *36:* 573–576, 1986.)

Streptomyces ciscaucasicus sp. nov. Sveshnikova 1983. *In* Gauze et al. A guide for determination of actinomycetes. Nauka, Moscow. (Validation List No. 22, Int. J. Syst. Bacteriol. *36:* 573–576, 1986.)

Streptomyces coeruleoflavus sp. nov. Preobrazhenskaya and Maximova 1983. *In* Gauze et al. A guide for determination of actinomycetes. Nauka, Moscow. (Validation List No. 22, Int. J. Syst. Bacteriol. *36:* 573–576, 1986.)

Streptomyces coeruleoprunus sp. nov. Preobrazhenskaya 1983. *In* Gauze et al. A guide for determination of actinomycetes. Nauka, Moscow.

(Validation List No. 22, Int. J. Syst. Bacteriol. *36:* 573–576, 1986.)

Streptomyces flavidofuscus sp. nov. Preobrazhenskaya 1983. *In* Gauze et al. A guide for determination of actinomycetes. Nauka, Moscow. (Validation List No. 22, Int. J. Syst. Bacteriol. *36:* 573–576, 1986.)

Streptomyces fulvorobeus sp. nov. Vinogradova and Preobrazhenskaya 1983. *In* Gauze et al. A guide for determination of actinomycetes. Nauka, Moscow. (Validation List No. 22, Int. J. Syst. Bacteriol. *36:* 573–576, 1986.)

Streptomyces graminearus sp. nov. Preobrazhenskaya 1983. *In* Gauze et al. A guide for determination of actinomycetes. Nauka, Moscow. (Validation List No. 22, Int. J. Syst. Bacteriol. *36:* 573–576, 1986.)

Streptomyces levis sp. nov. Sveshnikova 1983. *In* Gauze et al. A guide for determination of actinomycetes. Nauka, Moscow. (Validation List No. 22, Int. J. Syst. Bacteriol. *36:* 573–576, 1986.)

Streptomyces lienomycini sp. nov. Gauze and Maximova 1983. *In* Gauze et al. A guide for determination of actinomycetes. Nauka, Moscow. (Validation List No. 22, Int. J. Syst. Bacteriol. *36:* 573–576, 1986.)

Streptomyces macrosporus nom. rev. Goodfellow, Lacey and Todd 1987. J. Gen. Microbiol. *133:* 3135–3149. (Validation List No. 26, Int. J. Syst. Bacteriol. *38:* 328–329, 1988.)

Streptomyces monomycini sp. nov. Gauze and Terekhova 1983. *In* Gauze et al. A guide for determination of actinomycetes. Nauka, Moscow. (Validation List No. 22, Int. J. Syst. Bacteriol. *36:* 573–576, 1986.)

Streptomyces mutomycini sp. nov. Gauze and Maximova 1983. *In* Gauze et al. A guide for determination of actinomycetes. Nauka, Moscow. (Validation List No. 22, Int. J. Syst. Bacteriol. *36:* 573–576, 1986.)

Streptomyces speleomycini sp. nov. Preobrazhenskaya and Szabo 1983. *In* Gauze et al. A guide for determination of actinomycetes. Nauka, Moscow. (Validation List No. 22, Int. J. Syst. Bacteriol. *36:* 573–576, 1986.)

Streptomyces thermolineatus sp. nov. Goodfellow, Lacey and Todd 1987. J. Gen. Microbiol. *133:* 3135–3149. (Validation List No. 26, Int. J. Syst. Bacteriol. *38:* 328–329, 1988.)

Streptomyces variegatus sp. nov. Sveshnikova and Timuk 1983. *In* Gauze et al. A guide for determination of actinomycetes. Nauka, Moscow. (Validation List No. 22, Int. J. Syst. Bacteriol. *36:* 573–576, 1986.)

Streptomyces violaceorubidus sp. nov. Terekhova 1983. *In* Gauze et al. A guide for determination of actinomycetes. Nauka, Moscow. (Validation List No. 22, Int. J. Syst. Bacteriol. *36:* 573–576, 1986.)

Streptomyces virens sp. nov. Gauze and Sveshnikova 1983. *In* Gauze et al. A guide for determination of actinomycetes. Nauka, Moscow. (Validation List No. 22, Int. J. Syst. Bacteriol. *36:* 573–576, 1986.)

Streptoverticillium olivomycini sp. nov. Gauze and Sveshnikova 1983. *In* Gauze et al. A guide for determination of actinomycetes. Nauka, Moscow. (Validation List No. 22, Int. J. Syst. Bacteriol. *36:* 573–576, 1986.)

Bibliography

Abyzov, S.S., S.N. Phillippova and V.D. Kuznetsov. 1983a. *In* Validation of the publication of new names and new combinations previously effectively published outside the IJSB. List No. 13. Int. J. Syst. Bacteriol. *34:* 91–92.

Abyzov, S.S., S.N. Phillippova and V.D. Kuznetsov. 1983b. *Nocardiopsis antarcticus*, a new species of *Actinomyces*, isolated from the ice sheet of the central antarctic glacier. Izv. Akad. Nauk SSR, Ser. Biol. *4:* 559–568.

Adams, J.N. and S.G. Bradley. 1963. Recombination events in the bacterial genus *Nocardia*. Science *140:* 1392–1394.

Aggag, M. and H.G. Schlegel. 1973. Studies on a Gram-positive hydrogen bacterium, *Nocardia opaca* strain 1 b. Description and physiological characterization. Arch. Mikrobiol. *88:* 299–318.

Agre, N.S. 1961. The phage of the thermophilic *Micromonospora vulgaris*. Mikrobiologiya *30:* 414–417.

Agre, N.S. 1964. A contribution to the technique of isolation and cultivation of thermophilic actinomycetes (in Russian). Mikrobiologiya *33:* 913–917.

Agre, N.S., T.N. Efimova and L.N. Guzeva. 1975. Heterogeneity of the genus *Actinomadura* Lechevalier and Lechevalier. Mikrobiologiya *44:* 253–257.

Agre, N.S. and L.N. Guzeva. 1975. New genus of the actinomycetes: *Excellospora* gen. nov. Mikrobiologiya *44:* 518–522.

Agre, N.S., L.N. Guzeva and L.A. Dorokhova. 1974. A new species of the genus *Micropolyspora—Micropolyspora internatus*. Mikrobiologiya *43:* 679–685.

Agre, N.S., I.P. Kirillova and L.V. Kalakoutskii. 1972a. Spore germination in thermophilic actinomycetes. I. Preliminary observations with *Thermoactinomyces vulgaris* and *Actinobifida dichotomica*. Zentralbl. Bakteriol. Parasitenkd. Infektionskr. Hyg. Abt. II *127:* 525–538.

Agre, N.S., I.P. Kirillova and L.V. Kalakoutskii. 1972b. Spore germination in thermophilic actinomycetes. II. Germinal changes in water suspensions of *Thermoactinomyces vulgaris* spores. Zentralbl. Bakteriol. Parasitenkd. Infektionskr. Hyg. Abt. II *127:* 539–544.

Ahmad, K. and J.A.M. Bhuiyan. 1958. A new antifungal *Streptomyces* species. *Streptomyces corchorusii*. Pak. J. Biol. Agric. Sci. *1:* 137–143.

Ahrens, R. and G. Moll. 1970. Ein neues knospendes Bakterium aus der Ostee. Arch. Mikrobiol. *70:* 243–265.

Ainsworth, G.C. and P.H.A. Sneath (Editors). 1962. Microbial classification: Appendix I. Symp. Soc. Gen. Microbiol. *12:* 456–463.

Akkermans, A.D.L., J. Blom, K. Huss-Danell and W. Roelofsen. 1982. The carbon and nitrogen metabolism of *Frankia* in pure culture and in root nodules. *In*: Second National Symposium on the Biology of Nitrogen-Fixation, Helsinki, 1982, pp. 169–179.

Akkermans, A.D.L., K. Huss-Danell and W. Roelofsen. 1981. Enzymes of the tricarboxylic cycle and the malate-aspartate shuttle in the N_2-fixing endophyte of *Alnus glutinosa*. Physiol. Plant. *53:* 289–294.

Albert, C.A. and V.M. Malaquias de Queiroz. 1963. *Streptomyces tuirus* nov. sp., productor do antibiotico tuoromicina. Rev. Inst. Antibiot. Univ. Recife *5:* 43–51.

Alderson, G. 1985. The current taxonomic status of the genus *Micrococcus*. *In* Jeljaszewicz (Editor), The Staphylococci, Gustav Fischer Verlag, Stuttgart, pp. 117–124.

Alderson, G. and M. Goodfellow. 1979. Classification and identification of actinomycetes causing mycetoma. Postepy. Hig. Med. Dosw. *33:* 109–124.

Alderson, G., M. Goodfellow and D.E. Minnikin. 1985. Menaquinone composition in the classification of *Streptomyces* and other sporoactinomycetes. J. Gen. Microbiol. *131:* 1671–1679.

Allam, A.M., A.M. Hussein and A.M. Ragab. 1975. Amylase of a thermophilic actinomycete *Thermomonospora vulgaris*. Z. Allg. Mikrobiol. *15:* 393–398.

Allsopp, A. 1969. Phylogenetic relationships of the Procaryota and the origin of the eucaryotic cell. New Phytol. *68:* 591–612.

Alshamaony, L., M. Goodfellow and D.E. Minnikin. 1976. Free mycolic acids as criteria in the classification of *Nocardia* and the 'rhodochrous' complex. J. Gen. Microbiol. *92:* 188–199.

Ambler, R.P. 1976. Amino acid sequences of prokaryotic cytochromes and proteins, Vol. 3. *In* Fasman (Editor), Handbook of Biochemistry and Molecular Biology, 3rd ed., CRC Press, Cleveland, Ohio, pp. 292–307.

Ambler, R.P., M. Daniel, J. Hermoso, T.E. Meyer, T.G. Bartsch and M.D. Kamen. 1979a. Cytochrome c_2 sequence variation among the recognized species of purple nonsulphur photosynthetic bacteria. Nature (London) *278:* 659–660.

Ambler, R.P., T.E. Meyer and M.D. Kamen. 1979b. Anomalies in amino acid sequences of small cytochromes *c* and cytochromes c_1 from two species of purple photosynthetic bacteria. Nature (London) *278:* 661–662.

Amdur, B.H., E.I. Szabo and S.S. Socransky. 1978. Fatty acids of Gram-positive bacterial rods from human dental plaque. Arch. Oral Biol. *23:* 23–29.

American Type Culture Collection. 1982. Catalog of Strains I, 15th ed. American Type Culture Collection, Rockville, Maryland.

Amman, A., D. Gottlieb, T.D. Brock, H.E. Carter and G.B. Whitfield. 1955. Filipin, an antibiotic effective against fungi. Phytopathology *45:* 559–563.

An, C.S., W.S. Riggsby and B.C. Mullin. 1984. DNA relatedness of *Frankia* isolates. *In* Veeger and Newton (Editors), Advances in Nitrogen Fixation Research, Martinus Nijhoff/Dr. W. Junk Pubs., The Hague, p. 369.

An, C.S., W.S. Riggsby and B.C. Mullin. 1985a. Relationships of *Frankia* isolates based on deoxyribonucleic and homology studies. Int. J. Syst. Bacteriol. *35:* 140–146.

An, C.S., W.S. Riggsby and B.C. Mullin. 1985b. Restriction pattern analysis of genomic DNA of *Frankia* isolates. Plant Soil *87:* 43–48.

An, C.S., J.W. Wills, W.S. Riggsby and B.C. Mullin. 1983. Deoxyribonucleic acid base composition of 12 *Frankia* isolates. Can. J. Bot. *61:* 2859–2862.

Anderson, H.W. and D. Gottlieb. 1952. Plant disease control with antibiotics. Econ. Bot. *6:* 294–308.

Anderson, L.E., J. Ehrlich, S.H. Sun and P.R. Burkholder. 1956. Strains of *Streptomyces*, the sources of azaserine, elaiomycin, griseoviridin and viridogrisein. Antibiot. Chemother. *6:* 100–115.

Andreyev, L.V., L.I. Evtushenko and N.S. Agre. 1983. Fatty acid composition of *Promicromonospora citrea* (in Russian). Mikrobiologiya *52:* 58–62.

Andrzejewski, J., G. Müller, E. Röhrscheidt and D. Pietkiewicz. 1978. Isolation, characterization and classification of a *Nocardia asteroides* bacteriophage. Zentralbl. Bakteriol. Parasitenkd. Infektionskr. Hyg. I Abt., Suppl. *6:* 319–326.

Anonymous. 1962. *In* French Patent 1,425,264 May 1962, Cited in Arai, T. 1976. Actinomycetes: The Boundary Microorganisms. Toppan Co. Ltd., Tokyo, pp. 1–651.

Anonymous. 1973. *In* Japanese patent 30,400. October 1973.

Antai, S.P. and D.L. Crawford. 1981. Degradation of softwood, hardwood and grass lignocelluloses by two *Streptomyces* strains. Appl. Environ. Microbiol. *42:* 378–380.

Aoki, H., H. Sakai, M. Kohsaka, T. Konomi, J. Hosoda, Y. Kubochi, E. Iguchi and H. Imanaka. 1976. Nocardicin A, a new monocyclic β-lactam antibiotic. I. Discovery, isolation and characterization. J. Antibiot. *29:* 492–500.

Apajalahti, J.H.A., P. Kärpänoja and M.S. Salkinoja-Salonen. 1986. *Rhodococcus chlorophenolicus* sp. nov., a chlorophenol-mineralising actinomycete. Int. J. Syst. Bacteriol. *36:* 246–251.

Approved Lists of Bacterial Names. 1980. American Society for Microbiology, Washington, D.C. (Reprinted from Int. J. Syst. Bacteriol. *30:* 225–420.)

Arai, T. 1951. Studies of flavomycin: Taxonomic investigations on the strain, production of the antibiotic and application of cup method to the assay. J. Antibiot. (Tokyo) *4:* 215–221.

Arai, T. 1976. Actinomycetes: The Boundary Microorganisms. Toppan Co. Ltd., Tokyo, pp. 1–651.

Arai, T., S. Kuroda, S. Yamagishi and Y. Katoh. 1964. A new hydroxystreptomycin source, *Streptomyces subrutilus*. J. Antibiot. (Tokyo) Ser. A *17:* 23–28.

Arai, T. and Y. Mikami. 1969. Identification keys for antibiotic producing *Streptomyces*. I. Antifungal antibiotic producers. Annu. Rep. Inst. Food Microbiol. Chiba Univ. *22:* 59–79.

Arai, T., T. Nakada and M. Suzuki. 1957. Production of viomycin-like substance by a *Streptomyces*. Antibiot. Chemother. *7:* 435–442.

Arashima, M., J.M. Sakamoto and T. Sato. 1956. Studies on an antibiotic *Streptomyces* No. 689 strain. Part I. Taxonomic studies. J. Agric. Chem. Soc. Jpn. *30:* 469–471.

Arcamone, F.M., C. Bertazzoli, G. Canevazzi, A. di Marco, M. Ghione and A. Grein. 1957. La etruscomicina, nuovo antibiotico antifungino prodotto dallo *Streptomyces lucensis* n. sp. G. Microbiol. *4:* 119–128.

Arcamone, F.M., C. Bertazzoli, M. Ghione and T. Scotti. 1959a. Melanosporin and elaiophylin, new antibiotics from *Streptomyces melanosporus* (sive *melanosporofaciens*) n. sp. G. Microbiol. *7:* 207–216.

Arcamone, F., F. Bizioli, G. Canevazzi and A. Grein. 1959b. Methods for production and preparation of antibiotics. German Patent 1039198.

Arcamone, F., B. Camerino, E. Cotta, G. Franceschi, A. Grein, S. Penco and C.

Spalla. 1969. New carotenoids from *Streptomyces mediolani* n. sp. Experientia *25:* 241-242.
Arden-Jones, M. and T. Cross. 1980. Antigenic variation within and between species of *Thermoactinomyces*. Abstr. Annu. Meet. Am. Soc. Microbiol. *1980:* 89.
Arden-Jones, M.P., A.J. McCarthy and T. Cross. 1979. Taxonomic and serological studies on *Micropolyspora faeni* and *Micropolyspora* strains from soil bearing the specific epithet *rectivirgula*. J. Gen. Microbiol. *115:* 343-354.
Asahi, K., J. Nagatsu and S. Suzuki. 1966. Xanthocidin, a new antibiotic. J. Antibiot. (Tokyo) Ser. A *19:* 195-199.
Asano, K. and I. Kawamoto. 1986. *Catellatospora*, a new genus of *Actinomycetales*. Int. J. Syst. Bacteriol. *36:* 512-517.
Athalye, M. 1981. Classification and isolation of actinomadurae. Ph.D. Thesis, University of Newcastle upon Tyne, England.
Athalye, M., M. Goodfellow, J. Lacey and R.P. White. 1985. Numerical classification of *Actinomadura* and *Nocardiopsis*. Int. J. Syst. Bacteriol. *35:* 86-98.
Athalye, M., M. Goodfellow and D.E. Minnikin. 1984. Menaquinine composition in the classification of *Actinomadura* and related taxa. J. Gen. Microbiol. *130:* 817-823.
Athalye, M., J. Lacey and M. Goodfellow. 1981. Selective isolation and enumeration of actinomycetes using rifampicin. J. Appl. Bacteriol. *51:* 289-297.
Attwell, R.W. 1973. Actinomycete spores. Ph.D. thesis, University of Bradford, England.
Attwell, R.W. 1978. The spores of *Thermoactinomyces peptonophilus*. Actinomycetes *13:* 30.
Attwell, R.W. and T. Cross. 1973. Germination of actinomycete spores. *In* Sykes and Skinner (Editors), Actinomycetales: Characteristics and Practical Importance. Soc. Appl. Bacteriol. Symp. Ser. No. 2, Society for Applied Bacteriology, London, pp. 197-207.
Attwell, R.W., T. Cross and G.W. Gould. 1972. Germination of *Thermoactinomyces vulgaris* endospores: microscopic and optical density studies showing the influences of germinants, heat treatment, strain differences and antibiotics. J. Gen. Microbiol. *73:* 471-481.
Attwell, R.W., T. Cross and R. Locci. 1973. Fine structure of the spore sheath in *Streptoverticillium* species. J. Gen. Microbiol. *71:* 421-424.
Attwell, R.W., A. Surrey and T. Cross. 1985. Lysozyme sensitivity in *Streptoverticillium* and *Streptomyces* species. Zentralbl. Bakteriol. Mikrobiol. Hyg. C *3:* 239-242.
Aumen, N.G. 1980. Microbial succession on a chitinous substrate in a woodland stream. Microb. Ecol. *6:* 317-327.
Azuma, I., D.W. Thomas, A. Adam, J.M. Ghuysen, R. Bonaly, J.F. Petit and E. Lederer. 1970. Occurrence of *N*-glycolyl-muramic acid in bacterial cell walls. A preliminary survey. Biochim. Biophys. Acta *208:* 444-451.
Azuma, R. and U.B. Bak. 1980. Isolation of *Dermatophilus*-like microorganisms from swine tonsils. *In* Proceedings of the International Pig Veterinary Society Congress, Copenhagen, Denmark, p. 348.
Bach, M.C., L.D. Sabath and M. Finland. 1973. Susceptibility of *Nocardia asteroides* to 45 antimicrobial agents in vitro. Antimicrob. Agents Chemother. *3:* 1-8.
Backus, E.J., H.D. Tresner and T.H. Campbell. 1957. The nucleocidin and alazopeptin producing organisms: Two new species of *Streptomyces*. Antibiot. Chemother. *7:* 532-541.
Baermann, G. 1917. Eine einfache Methode zur Affindung von Ankylostomun (Nematoden) Larven in Erdproben. Geneesk. Tijdschr. Nederl. Indiüe *57:* 131-137.
Bak, U.B. and R. Azuma. 1980. *Dermatophilus* infection of the porcine tonsils. *In* Proceedings of the International Pig Veterinary Society Congress, Copenhagen, Denmark, p. 349.
Baker, D. 1982. A cumulative listing of isolated frankiae, the symbiotic nitrogen-fixing actinomycetes. Actinomycetes *17(1):* 35-42.
Baker, D. and D. O'Keefe. 1984. A modified sucrose fractionation procedure for the isolation of frankiae from actinorhizal root nodules and soil samples. Plant Soil *78:* 23-28.
Baker, D., W. Pengelly and J.G. Torrey. 1981. Immunochemical analysis of relationships among isolated frankiae (Actinomycetales). Int. J. Syst. Bacteriol. *31:* 148-151.
Baker, D. and E. Seling. 1984. *Frankia*: new light on an actinomycete symbiont. *In* Ortiz-Ortiz, Bojalil and Yakoleff (Editors), Biological, Biochemical and Biomedical Aspects of Actinomycetes, Academic Press, New York, pp. 562-574.
Baker, D. and J.G. Torrey. 1979. The isolation and cultivation of actinomycetous root nodule endophytes. *In* Gordon, Wheeler and Perry (Editors), Symbiotic Nitrogen Fixation in the Management of Temperate Forests, Oregon State University, Corvallis, pp. 38-56.
Baker, D., J.G. Torrey and G.H. Kidd. 1979. Isolation by sucrose-density fractionation and cultivation *in vitro* of actinomycetes from nitrogen-fixing nodules. Nature (London) *281:* 76-78.
Bakerspigel, A. 1973. An unusual strain of *Nocardia* isolated from an infected cat. Can. J. Microbiol. *19:* 1361-1365.
Balch, W.E., G.E. Fox, L.J. Magrum, C.R. Woese and R.S. Wolfe. 1979. Methanogens: Reevaluation of a unique biological group. Microbiol. Rev. *43:* 260-296.
Baldacci, E. 1939. Introduzione allo studio degli Attinomiceti. Mycopathologia *2:* 84-106.
Baldacci, E. 1944. Contributo alla sistematica degli attinomiceti: X-XVI. *Actinomyces madurae; Proactinomyces ruber; Proactinomyces pseudomadurae; Proactinomyces polychromogenus; Actinomyces violaceus; Actinomyces caeruleus;* con un elenco alfabetico delle specie e delle varietàga finora studiate. Atti Ist. Bot. Univ. Lab. Crittogram Pavia, Ser. 5, *3:* 139-193.
Baldacci, E. 1958. Development in the classification of actinomycetes. G. Microbiol. *6:* 10-27.
Baldacci, E. 1961. The classification of actinomycetes in relation to their antibiotic activity. Adv. Appl. Microbiol. *3:* 257-278.
Baldacci, E., G. Farina and R. Locci. 1966. Emendation of the genus *Streptoverticillium* Baldacci (1958) and revision of some species. G. Microbiol. *14:* 153-171.
Baldacci, E. and A. Grein. 1966. *Streptomyces avellaneus* and *Streptomyces libani:* two new species characterized by a hazelnut brown (*avellaneus*) aerial mycelium. G. Microbiol. *14:* 185-198.
Baldacci, E., A. Grein and C. Spalla. 1955. Studio di una 'Serie' di specie di attinomiceti: *A. diastaticus*. G. Microbiol. *1:* 127-143.
Baldacci, E. and R. Locci. 1966. A tentative arrangement of the genera in *Actinomycetales*. G. Microbiol. *14:* 131-139.
Baldacci, E. and R. Locci. 1974. Genus II. *Streptoverticillium* Baldacci. *In* Buchanan and Gibbons (Editors), Bergey's Manual of Determinative Bacteriology, 8th Ed., The Williams and Wilkins Co., Baltimore, pp. 829-842.
Baldacci, E., C. Spalla and A. Grein. 1954. The classification of the *Actinomyces* species (*Streptomyces*). Arch. Mikrobiol. *20:* 347-357.
Balla, P. 1968. Contributions to the knowledge of thermophilic actinomycetes occurring in champignon compost. Ann. Univ. Sci. Budap. Rolando Eotvos. Nominatae Sect. Biol. *9:* 27-35.
Baltz, R.H. 1978. Genetic recombination in *Streptomyces fradiae* by protoplast fusion and cell regeneration. J. Gen. Microbiol. *107:* 93-102.
Banaszak, E.F., W.H. Thiede and J.N. Fink. 1970. Hypersensitivity pneumonitis due to contamination of an air conditioner. N. Engl. J. Med. *283:* 271-276.
Bang, S.S., L. Baumann, M.J. Woolkalis and P. Baumann. 1981. Evolutionary relationships in *Vibrio* and *Photobacterium* as determined by immunological studies of superoxide dismutase. Arch. Microbiol. *130:* 111-120.
Bannerman, E.M. and J. Nicolet. 1976. Isolation and characterization of an enzyme with esterase activity from *Micropolyspora faeni*. Appl. Environ. Microbiol. *32:* 138-144.
Barksdale, L. 1970. *Corynebacterium diphtheriae* and its relatives. Bacteriol. Rev. *34:* 378-422.
Barnes, E.M. and H.S. Goldberg. 1968. The relationship of bacteria within the family *Bacteroidaceae* as shown by numerical taxonomy. J. Gen. Microbiol. *51:* 313-324.
Barr, F.S. and P.E. Carman. 1956. *Streptomyces kentuckensis*, a new species, the producer of raisnomycin. Antibiot. Chemother. *6:* 286-289.
Bartholomew, G.W. and M. Alexander. 1979. Microbial metabolism of carbon monoxide in culture and in soil. Appl. Environ. Microbiol. *37:* 932-937.
Barton, M.D. and K.L. Hughes. 1980. *Corynebacterium equi*: a review. Vet. Bull. *50:* 65-80.
Barton, M.D. and K.L. Hughes. 1981. Comparison of three techniques for isolation of *Rhodococcus* (*Corynebacterium*) *equi* from contaminated sources. J. Clin. Microbiol. *13:* 219-221.
Barton, M.D. and K.L. Hughes. 1982. Is *Rhodococcus equi* a soil organism? J. Reprod. Fertil., Suppl. *32:* 481-489.
Bartz, Q., J. Ehrlich, J.D. Mold, M.A. Penner and R.M. Smith. 1951. Viomycin, a new tuberculostatic antibiotic. Am. Rev. Tuberc. Pul. Dis. *63:* 4-6.
Batra, S.K. and B.S. Bajaj. 1965. *Streptomyces anandii*—a new species of *Streptomyces* isolated from soil. Ind. J. Exp. Biol. *3:* 240-242.
Baumann, L., S.S. Bang and P. Baumann. 1980. Study of relationship among species of *Vibrio, Photobacterium* and terrestrial enterobacteria by an immunological comparison of glutamine synthetase and superoxide dismutase. Curr. Microbiol. *4:* 133-138.
Beadles, T.A., G.A. Land and D.J. Knezek. 1980. An ultrastructural comparison of the cell envelopes of selected strains of *Nocardia asteroides* and *Nocardia brasiliensis*. Mycopathologia *70:* 25-32.
Beaman, B.L. 1975. Structural and biochemical alterations of *Nocardia asteroides* cell walls during its growth cycle. J. Bacteriol. *123:* 1235-1253.
Beaman, B.L. 1976. Possible mechanisms of nocardial pathogenesis. *In* Goodfellow, Brownell and Serrano (Editors), The Biology of the Nocardiae, Academic Press, London, pp. 386-417.
Beaman, B.L. 1980. Induction of L-phase variants of *Nocardia caviae* within intact murine lungs. Infect. Immun. *29:* 244-251.
Beaman, B.L. 1984. Actinomycete pathogenesis. *In* Goodfellow, Mordarski and Williams (Editors), The Biology of the Actinomycetes, Academic Press, London, pp. 457-479.
Beaman, B.L., J. Burnside, B. Edwards and W. Causey. 1976. Nocardial infections in the United States, 1972-1974. J. Infect. Dis. *134:* 286-289.
Beaman, B.L., J.A. Serrano and A.A. Serrano. 1978. Comparative ultrastructure within the nocardiae. Zentralbl. Bakteriol. Parasitenkd. Infektionskr. Hyg. I Abt., Suppl. *6:* 201-220.
Becker, B., M.P. Lechevalier, R.E. Gordon and H.A. Lechevalier. 1964. Rapid differentiation between *Nocardia* and *Streptomyces* by paper chromatography of whole-cell hydrolysates. Appl. Microbiol. *12:* 421-423.
Becker, B., M.P. Lechevalier and H.A. Lechevalier. 1965. Chemical composition of cell wall preparations from strains of various form genera of aerobic actinomycetes. Appl. Microbiol. *13:* 236-243.
Becking, J.H. 1970. *Frankiaceae* fam. nov. (Actinomycetales) with one new combination and six new species of the genus *Frankia* Brunchorst 1886, 174. Int J.

Syst. Bacteriol. *20:* 201-220.

Behnke, U., H. Ruttloff and R. Kleine. 1982. Preparation and characterization of proteases from *Thermoactinomyces vulgaris.* V. Investigations on autolysis and thermostability of the purified protease. Z. Allg. Mikrobiol. *22:* 511-519.

Beijerinck, M.W. 1912. Mutation bei Mikroben. Folia Mikrobiol. (Delft) *1:* 4-100.

Benedict, R.G., W. Dvonch, O.L. Shotwell, T.G. Pridham and L.A. Lindenfelser. 1952. Cinnamycin, an antibiotic from *Streptomyces cinnamoneus* nov. sp. Antibiot. Chemother. *2:* 591-594.

Benedict, R.G., O.L. Shotwell, T.G. Pridham, L.A. Lindenfelser and W.C. Haynes. 1954. The production of the neomycin complex by *Streptomyces albogriseolus,* nov. sp. Antibiot. Chemother. *4:* 653-656.

Benedict, R.G., F.H. Stodola, O.L. Shotwell, A.M. Borud and L.A. Lindenfelser. 1950. A new streptomycin. Science (Washington) *112:* 77-78.

Benigni, R., P.P. Antonov and A. Carere. 1975. Estimate of the genome size by renaturation studies in *Streptomyces.* Appl. Microbiol. *30:* 324-326.

Benson, D.R. 1982. Isolation of *Frankia* strains from alder actinorhizal root nodules. Appl. Environ. Microbiol. *44:* 461-465.

Benson, D.R., S.E. Buchholz and D.G. Hanna. 1984a. Identification of *Frankia* strains by two-dimensional polyacrylamide gel electrophoresis. Appl. Environ. Microbiol. *47:* 489-494.

Benson, D.R. and D. Hanna. 1983. *Frankia* diversity in an alder stand as estimated by sodium dodecyl sulfate-polyacrylamide gel electrophoresis of whole cell proteins. Can. J. Bot. *61:* 2919-2923.

Benson, D.R., C.E. Mazzucco and T.J. Browning. 1984b. Physiological aspects of *Frankia. In* Ludden and Burris (Editors), Nitrogen Fixation and Carbon Dioxide Metabolism, American Elsevier Publ. Co., New York, pp. 175-182.

Berd, D. 1973. Laboratory identification of clinically important aerobic actinomycetes. Appl. Microbiol. *25:* 665-681.

Beretta, G. 1973. *Actinoplanes italicus,* a new red-pigmented species. Int. J. Syst. Bacteriol. *23:* 37-42.

Beretta, M., M. Betti and M. Polsinelli. 1971. Genetic recombination in *Micromonospora.* J. Bacteriol. *107:* 415-419.

Berg, H.C. and D.A. Brown. 1972. Chemotaxis in *Escherichia coli* analyzed by three-dimensional tracking. Nature *239:* 500-504.

Bergey, D.H., F.C. Harrison, R.S. Breed, B.W. Hammer and F.M. Huntoon. 1923. Bergey's Manual of Determinative Bacteriology, 1st Ed., The Williams and Wilkins Co., Baltimore, pp. 1-442.

Bergey, D.H., F.C. Harrison, R.S. Breed, B.W. Hammer and F.M. Huntoon. 1925. Bergey's Manual of Determinative Bacteriology, 2nd Ed., The Williams and Wilkins Co., Baltimore, pp. 1-462.

Bergey, D.H., F.C. Harrison, R.S. Breed, B.W. Hammer and F.M. Huntoon. 1930. Bergey's Manual of Determinative Bacteriology, 3rd Ed., The Williams and Wilkins Co., Baltimore, pp. 1-589.

Bernard, D.S. and F. Parenti. 1983. Ultrastructural morphology of the genus *Kineosporia.* Curr. Microbiol. *8:* 173-176.

Bernardi, G. 1969a. Chromatography of nucleic acids on hydroxyapatite. I. Chromatography of native DNA. Biochim. Biophys. Acta *174:* 423-434.

Bernardi, G. 1969b. Chromatography of nucleic acids on hydroxyapatite. II. Chromatography of denatured DNA. Biochim. Biophys. Acta *174:* 435-448.

Bernstein, A. and H.E. Morton. 1934. A new thermophilic actinomyces. J. Bacteriol. *27:* 625-628.

Berry, A. and J.G. Torrey. 1983. Root hair deformation in the infection process of *Alnus rubra.* Can. J. Bot. *61:* 2863-2876.

Bhuyan, B.K. and A. Dietz. 1966. Fermentation, taxonomic and biological studies on nogalamycin. Antimicrob. Agents Chemother. *1965:* 836-844.

Bhuyan, B.K., A. Dietz and C.G. Smith. 1962. Pactamycin, a new antitumor antibiotic. I. Discovery and biological properties. Antimicrob. Agents Chemother. *1961:* 184-190.

Bianchi, M.L., A. Grien, P. Julita, M.P. Marnati and C. Spalla. 1970. *Streptomyces mediolani* (Arcamone et al.) *emend.* Bianchi et al. and its production of carotenoids. Z. Allg. Mikrobiol. *10:* 237-244.

Bibb, M.J., R.F. Freeman and D.A. Hopwood. 1977. Physical and genetic characterization of a second sex factor, SCP2, for *Streptomyces coelicolor* A3(2). Mol. Gen. Genet. *154:* 155-166.

Bisset, K.A. 1962. The phylogenetic concept in bacterial taxonomy. *In* Ainsworth and Sneath (Editors), Microbial Classification. Symp. Soc. Gen. Microbiol. *12:* 361-373.

Bisset, K.A. and F.W. Moore. 1950. *Jensenia,* a new genus of the *Actinomycetales.* J. Gen. Microbiol. *4:* 280.

Blanchard, R. 1896. Parasites végetaux à l'exclusion des bactéries. *In* Bouchard (Editor), Traité de Pathologie Générale, Vol II, G. Masson, Paris, pp. 811-932.

Blanchette, R.A., J.B. Sutherland and D.L. Crawford. 1981. Actinomycetes in discoloured wood of living silver maple. Can. J. Bot. *59:* 1-7.

Bland, C.E. 1968. Ultrastructure of *Pilimelia anulata (Actinoplanaceae).* J. Elisha Mitchell Sci. Soc. *84:* 8-15.

Bland, C.E. and J.N. Couch. 1981. The family *Actinoplanaceae. In* Starr, Stolp, Trüper, Balows and Schlegel (Editors), The Prokaryotes, A Handbook on Habitats, Isolation and Identification of Bacteria, Springer-Verlag, New York, pp. 2004-2010.

Blom, J. 1981. Utilization of fatty acids and NH_4^+ by *Frankia* AvcI1. FEMS Microbiol. Lett. *10:* 143-145.

Blom, J. 1982. Carbon and nitrogen source requirements of *Frankia* strains. FEMS Microbiol. Lett. *13:* 51-55.

Blom, J. and R. Harkink. 1981. Metabolic pathways for gluconeogenesis and energy generation in *Frankia* AvcI1. FEMS Microbiol. Lett. *11:* 221-224.

Blom, J., W. Roelofsen and A.D.L. Akkermans. 1980. Growth of *Frankia* AvcI1 on media containing Tween 80 as C source. FEMS Microbiol. Lett. *9:* 131-135.

Bönicke, R. 1962. L'identification des mycobactéries à l'aide de methodes biochimiques. Bull. Int. Union Tuberc. *32:* 13-76.

Bordet, C., M. Karahjoli, O. Gateau and G. Michel. 1972. Cell walls of nocardiae and related actinomycetes: Identification of the genus *Nocardia* by cell wall analysis. Int. J. Syst. Bacteriol. *22:* 251-259.

Bousfield, I.J. and A.G. Callely (Editors). 1978. Coryneform Bacteria, Academic Press, London, pp. 315.

Bousfield, I.J. and M. Goodfellow. 1976. The "rhodochrous" complex and its relationships with allied taxa. *In* Goodfellow, Brownell and Serrano (Editors), The Biology of the Nocardiae, Academic Press, London, pp. 39-65.

Bousfield, I.J., R.M. Keddie, T.R. Dando and S. Shaw. 1985. Simple rapid methods of cell wall analysis as an aid in the identification of aerobic coryneform bacteria. *In* Goodfellow and Minnikin (Editors), Chemical Methods in Bacterial Systematics, Academic Press, London, pp. 221-236.

Bousfield, I.J., G.L. Smith, T.R. Dando and G. Hobbs. 1983. Numerical analysis of total fatty acid profiles in the identification of coryneform, nocardioform and some other bacteria. J. Gen. Microbiol. *129:* 375-394.

Bøvre, K. 1980. Progress in classification and identification of *Neisseriaceae* based on genetic affinity. *In* Goodfellow and Board (Editors), Microbial Classification and Identification, Academic Press, London, pp. 55-72.

Bradley, S.G. 1978. Physiological genetics of nocardiae. Zentralbl. Bakteriol. Parasitenkd. Infektionskr. Hyg. I Abt., Suppl. *6:* 287-302.

Bradley, S.G. and J.S. Bond. 1974. Taxonomic criteria for mycobacteria and nocardiae. Adv. Appl. Microbiol. *18:* 131-190.

Bradley, S.G., L.W. Enquist and H.E. Scribner, III. 1978. Heterogeneity among deoxyribonucleotide sequences of actinomycetales. *In* Freerksen, Tárnok and Thumim (Editors), Genetics of the Actinomycetales, Gustav Fischer Verlag, Stuttgart, pp. 207-224.

Bradley, S.G. and M. Mordarski. 1976. Association of polydeoxyribonucleotides of deoxyribonucleic acids from nocardioform bacteria. *In* Goodfellow, Brownell and Serrano (Editors), The Biology of the Nocardiae, Academic Press, London, pp. 310-336.

Breed, R.S. 1953. The families developed from *Bacteriaceae* Cohn with a description on the family *Brevibacteriaceae* Breed, 1953. VI Congr. Intern. Microbiol. Rome *1:* 10-15.

Breed, R.S., E.G.D. Murray and N.R. Smith. 1957. Bergey's Manual of Determinative Bacteriology, 7th Ed., The Williams and Wilkins Co., Baltimore.

Brenner, D.J., G.R. Fanning, A.V. Rake and K.E. Johnson. 1969. Batch procedure for thermal elution of DNA from hydroxyapatite. Anal. Biochem. *28:* 447-459.

Brimacombe, R., G. Staffer and H.G. Wittmann. 1978. Ribosome structure. Annu. Rev. Biochem. *47:* 217-249.

Brocq-Rousseu, D. 1904. Sur un *Streptothrix* cause de l'alteration des avoines moisies. Rev. Bot. *16:* 219-230.

Broda, P. 1979. Plasmids. W.H. Freeman Co., London and San Francisco.

Brooks, B.W., R.G.E. Murray, J.L. Johnson, E. Stackebrandt, C.R. Woese and G.E. Fox. 1980. Red-pigmented micrococci: A basis for taxonomy. Int. J. Syst. Bacteriol. *30:* 627-646.

Brown, D.W. and W.E.C. Moore. 1960. Distribution of *Butyrivibrio fibrisolvens* in nature. J. Dairy Sci. *43:* 1570-1574.

Brown, R., E.L. Hazen and A. Mason. 1953. Effect of fungicidin (nystatin) in mice injected with lethal mixtures of aureomycin and *Candida albicans.* Science (Washington) *117:* 609-610.

Brown, S.M., J.L. Kepner and G.C. Smart, Jr. 1985. Increased crop yields following application of *Bacillus penetrans* to field plots infested with *Meloidogyne incognita.* Soil Biol. Biochem. *17:* 483-486.

Brownell, G.H. 1978. Plasmid transfer between *Nocardia erythropolis* and other nocardioform organisms. Zentralbl. Bakteriol. Parasitenkd. Infektionskr. Hyg. I Abt., Suppl. *6:* 313-317.

Brownell, G.H. and K. Denniston. 1984. Genetics of the nocardioform bacteria. *In* Goodfellow, Mordarski and Williams (Editors), The Biology of the Actinomycetes, Academic Press, London, pp. 201-228.

Brumpt, E. 1906. Les Mycétomes. Arch. Parasitol. *10:* 489-527.

Brunchorst, J. 1886. Ueber einige wurzelanschwellungen besonders die jenigen von *Alnus* und den *Elaeagnaceen.* Botanisches Institut zur Tubingen *2:* 151-177.

Bryant, M.P. 1972. Commentary on the Hungate technique for culture of anaerobic bacteria. Am. J. Clin. Nutr. *25:* 1324-1328.

Bryant, M.P. 1973. Nutritional requirements of the predominant rumen cellulolytic bacteria. Fed. Proc. *32:* 1809-1813.

Bryant, M.P., B.F. Barrentine, J.F. Sykes, I.M. Robinson, C.B. Shawver and L.W. Williams. 1960. Predominant bacteria in the rumen of cattle on bloat provoking ladino clover pasture. J. Dairy Sci. *43:* 1435-1444.

Bryant, M.P. and I.M. Robinson. 1962. Some nutritional characteristics of predominant culturable ruminal bacteria. J. Bacteriol. *84:* 605-614.

Bryant, M.P., I.M. Robinson and I.L. Lindahl. 1961. A note on the flora and fauna in the rumen of steers fed a feedlot bloat-provoking ration and the effect of penicillin. Appl. Microbiol. *9:* 511-515.

Bryant, M.P. and N. Small. 1956a. Characteristics of two new genera of anaerobic curved rods isolated from the rumen of cattle. J. Bacteriol. *72:* 22-26.

Bryant, M.P. and N. Small. 1956b. The anaerobic monotrichous butyric acid-producing curved rod-shaped bacteria of the rumen. J. Bacteriol. *72:* 16-21.

Buchanan, R.E. 1917. Studies in the nomenclature and classification of the bacteria. II. The primary subdivisions of the Schizomycetes. J. Bacteriol. *2:* 155–164.

Buchanan, R.E. 1918. Studies in the classification and nomenclature of the bacteria VIII. The subgroups and genera of the *Actinomycetales*. J. Bacteriol. *3:* 403–406.

Buchanan, R.E. 1925. General Systematic Bacteriology. The Williams and Wilkins Co., Baltimore.

Buchanan, R.E. and N.E. Gibbons (Editors). 1974. Bergey's Manual of Determinative Bacteriology, 8th Ed., The Williams and Wilkins Co., Baltimore.

Bunting, R.H. 1932. Actinomyces in cacao-beans. Ann. Appl. Biol. *19:* 515–517.

Burggraaf, A.J.P. 1984. Isolation, cultivation and characterization of *Frankia* strains from actinorhizal root nodules. Doctoral Thesis, Univ. of Leiden, Netherlands, p. 44.

Burggraaf, A.J.P. and W.A. Shipton. 1982. Estimates of *Frankia* growth under various pH and temperature regimes. Plant Soil *69:* 135–147.

Burggraaf, A.J.P. and W.A. Shipton. 1983. Studies on the growth of *Frankia* isolates in relation to infectivity and nitrogen fixation. Can. J. Bot. *61:* 2774–2782.

Burggraaf, A.J.P. and J. Valstar. 1984. Heterogeneity within *Frankia* sp. LD Agpl studied among clones and reisolates. Plant Soil *78:* 29–43.

Burkholder, P.R., S.H. Sun, J. Ehrlich and L. Anderson. 1954. Criteria of speciation in the genus *Streptomyces*. Ann. N.Y. Acad. Sci. *60:* 102–123.

Burman, N.P. 1973. The occurrence and significance of actinomycetes in water supply. *In* Sykes and Skinner (Editors), *Actinomycetales:* Characteristics and Practical Importance, Academic Press, London, pp. 219–230.

Burman, N.P., C.W. Oliver and J.K. Stevens. 1969. Membrane filtration techniques for the isolation from water, of coli-aerogenes, *Escherichia coli*, faecal streptococci, *Clostridium perfringens*, actinomycetes and microfungi. *In* Shapton and Gould (Editors), Isolation Methods for Microbiologists, Academic Press, London, pp. 127–134.

Burns, J. and D.F. Holtman. 1959. Tennecetin: A new antifungal antibiotic. Antibiot. Chemother. *9:* 398–405.

Cain, R.B. 1981. Regulation of aromatic and hydroaromatic catabolic pathways in nocardioform actinomycetes. Zentralbl. Bakteriol. Mikrobiol. Hyg. I Abt. Orig., Suppl. *11:* 335–354.

Callaham, D., P. del Tredici and J.G. Torrey. 1978. Isolation and cultivation *in vitro* of the actinomycete causing root nodulation in *Comptonia*. Science *199:* 899–902.

Callaham, D., W. Newcomb, J.G. Torrey and R.L. Peterson. 1979. Root hair infection in actinomycete-induced root nodule initiation in *Casuarina*, *Myrica* and *Comptonia* (*Myricaceae*). Bot. Gaz. *140(Suppl):* S1–S9.

Callaham, D. and J.G. Torrey. 1977. Prenodule formation and primary nodule development in roots of *Comptonia* (*Myricaceae*). Can. J. Bot. *55:* 2306–2318.

Calot, L. and A.P. Cercos. 1963. *Streptomyces ornatus*, nov. sp. et *Streptomyces erumpens*, nov. sp. producteurs d'ornamicine et antibiotique 17732. Ann. Inst. Pasteur (Paris) *105:* 159–161.

Campbell, A. 1981. Evolutionary significance of accessory DNA elements in bacteria. Annu. Rev. Microbiol. *35:* 55–83.

Canevazzi, G. and T. Scotti. 1959. Descrizione di uno streptomicete (*Streptomyces chrestomyceticus*). sp. nova, produttore del nuovo antibiotico amminosidina. G. Microbiol. *7:* 242–250.

Canning, E.U. 1973. Protozoal parasites as agents for biological control of plant-parasitic nematodes. Nematologica *19:* 342–348.

Carlile, M.J., J.F. Collins and B.E.B. Moseley (Editors). 1981. Molecular and Cellular Aspects of Microbial Evolution, Cambridge University Press, Cambridge, England.

Carvajal, F. 1947. The production of spores in submerged cultures by some *Streptomyces*. Mycologia *39:* 426–440.

Carver, M.A. and C.W. Jones. 1985. The detection of cytochrome patterns in bacteria. *In* Goodfellow and Minnikin (Editors), Chemical Methods in Bacterial Systematics. Academic Press, London, pp. 383–399.

Cassinelli, G., A. Grein, P. Orezzi, P. Pennella and A. Sanfilippo. 1967. New antibiotics produced by *Streptoverticillium orinoci*, n. sp. Arch. Mikrobiol. *55:* 358–368.

Castellani, A. and A.J. Chalmers. 1913. Manual of Tropical Medicine, 2nd Ed., Balliere, Tindall and Cox, London.

Castellani, A. and A.J. Chalmers. 1919. Manual of Tropical Medicine. 3rd Ed., Williams, Wood and Co., New York.

Cato, E.P., D.E. Hash, L.V. Holdeman and W.E.C. Moore. 1982. Electrophoretic study of *Clostridium* species. J. Clin. Microbiol. *15:* 688–702.

Caveness, F.E. and H.J. Jensen. 1955. Modification of the centrifugal-flotation technique for the isolation and concentration of nematodes and their eggs from soil and plant tissue. Proc. Helminthol. Soc. Wash. *22:* 87–89.

Celmer, W.D., W.P. Cullen, C.E. Moppett, J.B. Routien, R. Shibakawa and J. Tone. 1975. Mélange d'antibiotiques produit par une espèce due genre *Actinoplanes*. Patent BE 827.935 (Belgium) International Classification C12D/A61K.

Celmer, W.D., W.P. Cullen, C.E. Moppett, J.R. Oscarson, J.B. Routien, R. Shibakawa and J. Tone. 1977. Mixture of antibiotics produced by a new species of *Micromonospora*. U.S. Patent 4,032,631, June 28, 1977.

Celmer, W.D., W.P. Cullen, C.E. Moppett, J.B. Routien, M.T. Jefferson, R. Shibakawa and J. Tone, Pfizer Inc. 1978. Polycyclic ether antibiotic produced by new species of *Dactylosporangium*. U.S. Patent 4,081,532. March 28.

Cercos, A.P., B.L. Eilberg, J.G. Goyena, J. Souto, E.E. Vautier and I. Widuczynski. 1962. Misionina: antibiotico polienico producido por *Streptomyces misionensis* n. sp. Rev. Invest. Agric. *16:* 5–27.

Chalmers, A.J. and J.B. Christopherson. 1916. A Sudanese actinomycosis. Ann. Trop. Med. Parasitol. *10:* 223–282.

Chapman, R.A. 1957. The effects of aeration and temperature on the emergence of species of *Pratylenchus* from roots. Plant Dis. Rep. *41:* 836–841.

Charney, J., W.P. Fisher, C. Curran, R.A. Machlowitz and A.A. Tytell. 1953. Streptogramin, a new antibiotic. Antibiot. Chemother. *3:* 1283–1286.

Chater, K.F. 1979. Some recent developments in *Streptomyces* genetics. *In* Sebek and Laskin (Editors), Genetics of Industrial Microorganisms, American Society for Microbiology, Washington, pp. 123–133.

Chater, K.F. and D.A. Hopwood. 1984. *Streptomyces* genetics. *In* Goodfellow, Mordarski and Williams (Editors), The Biology of the Actinomycetes, Academic Press, London, pp. 229–286.

Chater, K.F. and M.J. Merrick. 1979. Streptomycetes. *In* Parish (Editor), Development Biology of Prokaryotes, Blackwell Scientific Publications, Oxford, pp. 93–114.

Chaves Batista, A., S.K. Shome and J. Americo de Lima. 1963. *Streptosporangium bovinum* sp. nov. from cattle hooves. Derm. Trop. *2:* 49–54.

Chen, M. and M.J. Wolin. 1979. Effect of Monensin and Lasalocid-sodium on the growth of methanogogenic and rumen saccharolytic bacteria. Appl. Environ. Microbiol. *38:* 72–77.

Cheng, K.-J. and J.W. Costerton. 1977. Ultrastructure of *Butyrivibrio fibrisolvens*: A gram-positive bacterium? J. Bacteriol. *129:* 1506–1512.

Cheng, K.-J., D. Dinsdale and C.S. Stewart. 1979. Maceration of clover and grass leaves by *Lachnospira multiparus*. Appl. Environ. Microbiol. *38:* 723–729.

Cheng, K.-J., G.A. Jones, F.J. Simpson and M.P. Bryant. 1959. Isolation and identification of rumen bacteria capable of anaerobic rutin degredation. Can. J. Microbiol. *15:* 1365–1371.

Chormonova, N.T. and T.P. Preobrazhenskaya. 1981. Occurrence of *Actinomadura* in Kazakhstan soils. Antibiotiki *26:* 341–345.

Clark, J.B. 1979. Sphere-rod transitions in *Arthrobacter*. *In* Parish (Editor), Developmental Biology of Prokaryotes, Blackwell Scientific Publications, Oxford, pp. 73–92.

Clewell, D.B. 1981. Plasmids, drug resistance and gene transfer in the genus *Streptococcus*. Microbiol. Rev. *45:* 409–436.

Clewell, D.B. and D.R. Helinski. 1969. Supercoiled circular DNA protein complexes in *Escherichia coli:* Purification and induced conversion to an open-circular DNA form. Proc. Nat. Acad. Sci. (U.S.A.) *62:* 1159–1166.

Cobb, N.A. 1906. Fungus maladies of the sugar cane, with notes on associated insects and nematodes, 2nd Ed. Hawaiian Sugar Planters Assoc., Expt. Sta. Div. Path. Physiol. Bull. *5:* 163–195.

Collins, M.D. 1982. A note on the separation of natural mixtures of bacterial menaquinones using reverse-phase high-performance liquid chromatography. J. Appl. Bacteriol. *52:* 457–460.

Collins, M.D. 1985. Isoprenoid quinone analyses in bacterial classification and identification. *In* Goodfellow and Minnikin (Editors), Chemical Methods in Bacterial Systematics, Society for Applied Bacteriology, Technical Series No. 20, Academic Press, London, pp. 267–287.

Collins, M.D. 1985. Isoprenoid quinone analyses in bacterial classification and identification. *In* Gottschalk (Editor), Methods in Microbiology, Vol. 19, Academic Press, London, pp. 267–287.

Collins, M.D., M. Faulkner and R.M. Keddie. 1984. Menaquinone composition of some sporeforming actinomycetes. Syst. Appl. Microbiol. *5:* 20–29.

Collins, M.D., S. Feresu and D. Jones. 1983a. Cell wall, DNA base composition and lipid studies on *Listeria denitrificans*. (Prévot). FEMS Microbiol. Lett. *18:* 131–134.

Collins, M.D., M. Goodfellow and D.E. Minnikin. 1979. Isoprenoid quinones in the classification of coryneform and related bacteria. J. Gen. Microbiol. *110:* 127–136.

Collins, M.D., M. Goodfellow and D.E. Minnikin. 1982a. A survey of the structure of mycolic acids in *Corynebacterium* and related taxa. J. Gen. Microbiol. *128:* 129–149.

Collins, M.D., M. Goodfellow and D.E. Minnikin. 1982b. Fatty acid composition of some mycolic acid-containing coryneform bacteria. J. Gen. Microbiol. *128:* 2503–2509.

Collins, M.D., M. Goodfellow, D.E. Minnikin and G. Alderson. 1985. Menaquinone composition of mycolic acid-containing actinomycetes and some sporoactinomycetes. J. Appl. Bacteriol. *58:* 77–86.

Collins, M.D. and D. Jones. 1981. Distribution of isoprenoid quinone structural types in bacteria and their taxonomic implications. Microbiol. Rev. *45:* 316–354.

Collins, M.D., R.M. Keddie and R.M. Kroppenstedt. 1983b. Lipid composition of *Arthrobacter simplex*, *Arthrobacter tumescens* and possibly related taxa. Syst. Appl. Microbiol. *4:* 18–26.

Collins, M.D. and R.M. Kroppenstedt. 1983. Lipid composition as a guide to the classification of some coryneform bacteria containing an A4α type peptidoglycan (Schleifer and Kandler). Syst. Appl. Microbiol. *4:* 95–104.

Collins, M.D., G.C. Mackillop and T. Cross. 1982c. Menaquinone composition of members of the genus *Thermoactinomyces*. FEMS Microbiol. Lett. *13:* 151–153.

Collins, M.D., A.J. McCarthy and T. Cross. 1982d. New highly saturated members of the vitamin K_2 series from *Thermomonospora*. Zentralbl. Bakteriol. Mikrobiol. Hyg. I Abt. Orig. C *3:* 358–363.

Collins, M.D., T. Pirouz, M. Goodfellow and D.E. Minnikin. 1977. Distribution of menaquinones in actinomycetes and corynebacteria. J. Gen. Microbiol. *100:* 221-230.

Collins, M.D., H.N.M. Ross, B.J. Tindall and W.D. Grant. 1981. Distribution of isoprenoid quinones in halophilic bacteria. J. Appl. Bacteriol. *50:* 559-565.

Collins, M.D. and H.N. Shah. 1984. Fatty acid menaquinone and polar lipid composition of *Rothia dentocariosa*. Arch. Microbiol. *137:* 247-249.

Collins, M.D., H.N. Shah and D.E. Minnikin. 1980. A note on the separation of natural mixtures of bacterial menaquinones using reverse-phase partition thin-layer chromatography. J. Appl. Bacteriol. *48:* 277-282.

Colmer, A.R. and E. McCoy. 1943. *Micromonospora* in relation to some Wisconsin lakes and lake population. Trans. Wis. Acad. Sci. Arts Lett. *35:* 187-200.

Colwell, R.R. 1973. Genetic and phenetic classification of bacteria. Adv. Appl. Microbiol. *16:* 137-175.

Colwell, R.R. (Editor). 1976. The Role of Culture Collections in the Era of Molecular Biology, American Society for Microbiology, Washington, D.C.

Conn, H.J. 1917. Soil flora studies V. Actinomycetes in soil. Bull. N.Y. State Agric. Expt. Sta. *60:* 3-25.

Corbaz, R., L. Ettlinger, E. Gäumann, W. Keller-Schierlein, F. Kradolfer, E. Kyburz, L. Neipp, V. Prelog, R. Reusser and H. Zähner. 1955. Stoffwechselprodukte von Actinomyceten. I. Narbomycin. Helv. Chim. Acta *38:* 935-942.

Corbaz, R., L. Ettlinger, E. Gäumann, W. Keller-Schierlein, F. Kradolfer, L. Neipp, V. Prelog, P. Reusser and H. Zähner. 1957a. Stoffwechselprodukte von Actinomyceten. 7. Echinomycin. Helv. Chim. Acta *40:* 199-204.

Corbaz, R., L. Ettlinger, W. Keller-Schierlein and H. Zähner. 1957b. Zur Systematik der Actinomyceten 1. Über Streptomyceten mit rhodomycinastigen Pigmenten. Arch. Mikrobiol. *25:* 325-332.

Corbaz, R., L. Ettlinger, W. Keller-Schierlein and H. Zähner. 1957c. Zur Systematik der Actinomyceten. 2. Über Actinomycin bildende Streptomyceten. Arch. Mikrobiol. *26:* 192-208.

Corbaz, R., P.H. Gregory and M.E. Lacey. 1963. Thermophilic and mesophilic actinomycetes in mouldy hay. J. Gen. Microbiol. *32:* 449-455.

Corberi, E. and A. Solaini. 1960. Osservazioni sulla microflora della acque del Lago Maggiore. Ann. Microbiol. *10:* 36-56.

Coronelli, C., H. Pagani, M.R. Bardone and G.C. Lancini. 1974. Purpuromycin, a new antibiotic isolated from *Actinoplanes ianthinogenes* n. sp. J. Antibiotics *27:* 161-168.

Couch, J.N. 1939. Technique for collection, isolation and culture of chytrids. J. Elisha Mitchell Sci. Soc. *55:* 208-214.

Couch, J.N. 1949. A new group of organisms related to *Actinomyces*. J. Elisha Mitchell Sci. Soc. *65:* 315-318.

Couch, J.N. 1950. *Actinoplanes* a new genus of the *Actinomycetales*. J. Elisha Mitchell Sci. Soc. *66:* 87-92.

Couch, J.N. 1954. The genus *Actinoplanes* and its relatives. Trans. N.Y. Acad. Sci. *16:* 315-318.

Couch, J.N. 1955. A new genus and family of the *Actinomycetales*, with a revision of the genus *Actinoplanes*. J. Elisha Mitchell Sci. Soc. *71:* 148-155.

Couch, J.N. 1957. A new horizon in soil microbiology. Proc. Natl. Acad. Sci. India *27:* 69-73.

Couch, J.N. 1963. Some new genera and species of the *Actinoplanaceae*. J. Elisha Mitchell Sci. Soc. *79:* 53-70.

Couch, J.N. 1964. A proposal to replace the name *Ampullariella*. J. Elisha Mitchell Sci. Soc. *80:* 29.

Couch, J.N. and C.E. Bland. 1974a. Family *Actinoplanaceae* Couch. *In* Buchanan and Gibbons (Editors), Bergey's Manual of Determinative Bacteriology, 8th Ed., The Williams and Wilkins Co., Baltimore, pp. 706-708.

Couch, J.N. and C.E. Bland. 1974b. Genus *Ampullariella* Couch. *In* Buchanan and Gibbons (Editors), Bergey's Manual of Determinative Bacteriology, 8th Ed., The Williams and Wilkins Co., Baltimore, pp. 717-718.

Couch, J.N. and C.E. Bland. 1974c. Genus I. *Actinoplanes. In* Buchanan and Gibbons (Editors), Bergey's Manual of Determinative Bacteriology, 8th Ed., The Williams and Wilkins Co., Baltimore, pp. 708-710.

Couch, J.N. and C.E. Bland. 1974d. Genus *Spirillospora* Couch. *In* Buchanan and Gibbons (Editors), Bergey's Manual of Determinative Bacteriology, 8th Ed., The Williams and Wilkins Co., Baltimore, p. 711.

Couch, J.N. and W.J. Koch. 1962. Induction of motility in the spores of some *Actinoplanaceae (Actinomycetales)*. Science *138:* 987.

Cowan, S.T. 1968. A Dictionary of Microbial Taxonomic Usage, Oliver and Boyd, Edinburgh.

Cowan, S.T. 1970. Heretical taxonomy for bacteriologists. J. Gen. Microbiol. *61:* 145-154.

Cowan, S.T. 1978. A Dictionary of Microbial Taxonomy (Edited by L.R. Hill). Cambridge University Press, Cambridge, United Kingdom.

Craveri, R., C. Coronelli, H. Pagani and P. Sensi. 1964. Thermorubin, a new antibiotic from a thermoactinomycete. Clin. Med. *71:* 511-521.

Craveri, R., A. Guicciardi and N. Pacini. 1966a. Distribution of thermophilic actinomycetes in compost for mushroom production. Ann. Microbiol. Enzimol. *16:* 111-113.

Craveri, R. and P.L. Manachini. 1966. Base composition of DNA in *Streptomyces argenteolus* and *Thermoactinomyces vulgaris* cultivated at different temperatures (in Italian). Ann. Microbiol. Enzimol. *16:* 1-3.

Craveri, R., P.L. Manachini and N. Pacini. 1966b. Deoxyribonucleic acid base composition of actinomycetes with different temperature requirements for growth. Ann. Microbiol. Enzimol. *16:* 115-117.

Crawford, D.L. 1974. Growth of *Thermomonospora fusca* on lignocellulosic pulps of varying lignin content. Can. J. Microbiol. *20:* 1069-1072.

Crawford, D.L. 1975. Cultural, morphological and physiological characteristics of *Thermomonospora fusca* (strain 190Th). Can. J. Microbiol. *21:* 1842-1848.

Crawford, D.L. and E. McCoy. 1972. Cellulases of *Thermomonospora fusca* and *Streptomyces thermodiastaticus*. Appl. Microbiol. *24:* 150-152.

Crawford, D.L. and J.B. Sutherland. 1979. The role of actinomycetes in the decomposition of lignocellulose. Dev. Ind. Microbiol. *20:* 143-151.

Crawford, I.P., B.P. Nichols and C. Yanofsky. 1980. Nucleotide sequence of the trpB gene in *Escherichia coli* and *Salmonella typhimurium*. J. Mol. Biol. *142:* 489-502.

Crespi, R.S. 1982. Patenting in the Biological Sciences. John Wiley & Sons, New York.

Cron, M.J., D.F. Whitehead, I.R. Hooper, B. Heinemann and J. Lein. 1956. Bryamycin, a new antibiotic. Antibiot. Chemother. *6:* 63-67.

Crosa, J.H., D.J. Brenner and S. Falkow. 1973. Use of a single-strand specific nuclease for analysis of bacterial and plasmid deoxyribonucleic acid homo- and heteroduplexes. J. Bacteriol. *115:* 904-911.

Cross, T. 1968. Thermophilic actinomycetes. J. Appl. Bacteriol. *31:* 36-53.

Cross, T. 1970. The diversity of bacterial spores. J. Appl. Bacteriol. *33:* 95-102.

Cross, T. 1974. Genus V. *Microbispora* Nonomura and Ohara 1957. *In* Buchanan and Gibbons (Editors), Bergey's Manual of Determinative Bacteriology, 8th Ed., The Williams and Wilkins Co., Baltimore, pp. 856-860.

Cross, T. 1981a. Aquatic actinomycetes: A critical survey of the occurrence, growth and role of actinomycetes in aquatic habitats. J. Appl. Bacteriol. *50:* 397-423.

Cross, T. 1981b. The monosporic actinomycetes. *In* Starr, Stolp, Trüper, Balows and Schlegel (Editors), The Prokaryotes, A Handbook on Habitats, Isolation and Identification of Bacteria, Springer-Verlag, Berlin, pp. 2091-2102.

Cross, T. 1982. Actinomycetes: A continuing source of new metabolites. Dev. Ind. Microbiol. *23:* 1-18.

Cross, T. and L.J. Al-Diwany. 1981. Streptomycetes with substrate mycelium spores: The genus *Elytrosporangium*. Zentralbl. Bakteriol. Mikrobiol. Hyg. I Abt. Suppl. *11:* 59-65.

Cross, T. and R.W. Attwell. 1974. Recovery of viable thermoactinomycete endospores from deep mud cores. *In* Barker, Gould and Wolf (Editors), Spore Research 1973. Academic Press, London, pp. 11-20.

Cross, T. and R.W. Attwell. 1975. Actinomycete spores. *In* Gerhardt, Costilow and Sadoff (Editors), Spores. VI. American Society for Microbiology, Washington, D.C., pp. 3-13.

Cross, T., R.W. Attwell and R. Locci. 1973. Fine structure of the spore sheath in *Streptoverticillium* species. J. Gen. Microbiol. *75:* 421-424.

Cross, T., F.L. Davies and P.D. Walker. 1971. *Thermoactinomyces vulgaris*. I. Fine structure of the developing endospores. *In* Barker, Gould and Wolf (Editors), Spore Research 1971. Academic Press, London, pp. 175-187.

Cross, T. and M. Goodfellow. 1973. Taxonomy and classification of the actinomycetes. *In* Sykes and Skinner (Editors), Actinomycetales: Characteristics and Practical Importance, Academic Press, London, New York, pp. 11-112.

Cross, T. and D.W. Johnston. 1971. *Thermoactinomyces vulgaris*. II. Distribution in natural habitats. *In* Barker, Gould and Wolf (Editors), Spore Research 1971, Academic Press, London, pp. 315-330.

Cross, T. and J. Lacey. 1970. Studies on the genus *Thermomonospora*. *In* Prauser (Editor), The Actinomycetales, VEB Gustav Fischer Verlag, Jena, pp. 211-219.

Cross, T., M.P. Lechevalier and H.A. Lechevalier. 1963. A new genus of the *Actinomycetales: Microellobosporia*. J. Gen. Microbiol. *31:* 421-429.

Cross, T., A. Maciver and J. Lacey. 1968a. The thermophilic actinomycetes in mouldy hay: *Micropolyspora faeni* sp. nov. J. Gen. Microbiol. *50:* 351-359.

Cross, T., T.J. Rowbotham, E.N. Mishustin, E.Z. Tepper, F. Antoine-Portaels, K.P. Schaal and H. Bickenbach. 1976. The ecology of nocardioform actinomycetes. *In* Goodfellow, Brownell and Serrano (Editors), The Biology of the Nocardiae. Academic Press, London, pp. 337-371.

Cross, T. and D.F. Spooner. 1963. The serological identification of streptomycetes by agar gel diffusion techniques. J. Gen. Microbiol. *33:* 275-282.

Cross, T. and B.A. Unsworth. 1976. Farmer's lung: A neglected antigen. Lancet *1:* 958-959.

Cross, T. and B.A. Unsworth. 1981. Taxonomy of the endospore-forming actinomycetes. *In* Berkeley and Goodfellow (Editors), The Aerobic Endospore Forming Bacteria: Classification and Identification, Academic Press, London, pp. 17-32.

Cross, T., P.D. Walker and G.W. Gould. 1968b. Thermophilic actinomycetes producing resistant endospores. Nature (Lond.) *220:* 352-354.

Crowle, A.J. 1962. *Corynebacterium rubrum* nov. spec.: A Gram-positive non acid-fast bacterium of unusually high lipid content. Antonie Leeuwenhoek J. Microbiol. Serol. *28:* 182-192.

Cruickshank, J.G., A.H. Gawler and C. Shaldon. 1979. *Oerskovia* species: Rare opportunistic pathogens. J. Med. Microbiol. *12:* 513-515.

Cullum, J. and H. Saedler. 1981. DNA rearrangements and evolution. *In* Carlile, Collins and Moseley (Editors), Molecular and Cellular Aspects of Microbial Evolution, Symposium No. 32 of the Society for General Microbiology, Cambridge University Press, London, pp. 131-150.

Cummins, C.S. 1962a. Chemical composition and antigenic structure of cell walls of *Corynebacterium, Mycobacterium, Nocardia, Actinomyces* and *Arthrobacter*. J. Gen. Microbiol. *28:* 35-50.

Cummins, C.S. 1962b. Immunochemical specificity and the location of antigens in the bacterial cell. *In* Ainsworth and Sneath (Editors), Microbial Classification, 12th Symposium of the Society for General Microbiology, Cambridge University Press, Cambridge, United Kingdom.

Cummins, C.S. 1965. Chemical and antigenic studies on cell walls of mycobacteria, corynebacteria and nocardias. Am. Rev. Resp. Dis. *92:* 63-72.

Cummins, C.S. and H. Harris. 1956. The chemical composition of the cell wall in some Gram-positive bacteria and its possible value as a taxonomic character. J. Gen. Microbiol. *14:* 583-600.

Curry, W.A. 1980. Human nocardiosis. A clinical review with selected case reports. Arch. Intern. Med. *140:* 818-826.

Curtiss, R. 1969. Bacterial conjugation. Annu. Rev. Microbiol. *23:* 69-136.

Dassain, M., G. Tiraby, M.-A. Laneelle and J. Asselineau. 1983. Étude comparative de la composition en lipides de sept espéces de *Micromonospora*. Ann. Microbiol. *134A:* 9-17.

De Boer, C., A. Dietz, J.S. Evans and R.M. Michaels. 1960. Fervenulin, a new crystalline antibiotic. I. Discovery and biological activities. Antibiot. Annu. 1959-1960: 220-226.

De Boer, C., A. Dietz, N.E. Lummis and G.M. Savage. 1961. Porfiromycin, a new antibiotic. I. Discovery and biological activities. Antimicrob. Agents Annu. 1960-1961: 17-22.

De Boer, C., A. Dietz, W.S. Silver and G.M. Savage. 1956. Streptolydigin, a new antimicrobial antibiotic. 1. Biologic studies of streptolydigin. Antibiot. Annu. *1955-1956:* 886-892.

De Boer, C., A. Dietz, J.R. Wilkins, O.N. Lewis and G.M. Savage. 1955. Celesticetin—a new crystalline antibiotic. I. Biologic studies of celesticetin. Antibiot. Annu. *1954-1955:* 831-836.

Dehority, B.A. 1969. Pectin-fermenting bacteria isolated from the bovine rumen. J. Bacteriol. *99:* 189-196.

De la Fuente, R., G. Suarez and K.H. Schleifer. 1985. *Staphylococcus aureus* subsp. *anaerobius* subsp. nov., the causal agent of abscess disease of sheep. Int. J. Syst. Bacteriol. *35:* 99-102.

Delaporte, B. 1969. Une nouvelle espèce de *Fusosporus*: *Fusosporus minor* n. sp. C.R. Acad. Sci. Paris Ser. D Sci. Nat. *268:* 1454-1455.

De Ley, J., H. Cattoir and A. Reynaerts. 1970. The quantitative measurement of DNA hybridization from renaturation rates. Eur. J. Biochem. *12:* 133-142.

De Ley, J., P. Segers and M. Gillis. 1978. Intra- and intergeneric similarities of *Chromobacterium* and *Janthinobacterium* ribosomal ribonucleic acid cistrons. Int. J. Syst. Bacteriol. *28:* 154-168.

Demaree, J.B. and N.R. Smith. 1952. *Nocardia vaccinii* n. sp. causing galls on blueberry plants. Phytopatholoty *42:* 249-252.

den Dooren de Jong, L.E. 1927. Über protaminophage Bakterien. Zentralbl. Bakteriol. Parasitenkd. Infektionskr. Hyg. Abt. II *71:* 193-232.

Denhardt, D.T. 1966. A membrane-filter technique for the detection of complementary DNA. Biochem. Biophys. Res. Comm. *23:* 641-646.

Dennis, S.M., T.G. Nagaraja and E.E. Bartley. 1981. Effects of Lasolocid or Monensin on lactate-producing or -using rumen bacteria. Anim. Sci. *52:* 418-426.

Deploey, J.J. and C.L. Fergus. 1975. Growth and sporulation of thermophilic fungi and actinomycetes in O_2-N_2 atmospheres. Mycologia *67:* 780-797.

de Querioz, V.M. and C.A. Albert. 1962. *Streptomyces iakyrus* nov. sp., produtor dos antibioticos Iaquirina I, IIe, III. Rev. Inst. Antibiot. Univ. Recife *4:* 33-46.

Desai, A.J. and S.A. Dhala. 1966. Isolation and study of thermophilic actinomycetes from soil, manure and compost from Bombay. Indian J. Microbiol. *6:* 54-58.

Desai, A.J. and S.A. Dhala. 1967a. Bacteriolysis by thermophilic actinomycetes. Antonie van Leeuwenhoek J. Microbiol. Serol. *33:* 56-62.

Desai, A.J. and S.A. Dhala. 1967b. *Streptomyces thermonitrificans* sp. n., a thermophilic streptomycete. Antonie Leeuwenhoek J. Microbiol. Serol. *33:* 137-144.

Desai, A.J. and S.A. Dhala. 1969. Purification and properties of proteolytic enzymes from thermophilic actinomycetes. J. Bacteriol. *100:* 149-155.

De Smedt, J., M. Bauwens, R. Tytgat and J. De Ley. 1980. Intra- and intergeneric similarities of ribosomal ribonucleic acid cistrons of free-living, nitrogen-fixing bacteria. Int. J. Syst. Bacteriol. *30:* 106-122.

De Smedt, J. and J. De Ley. 1977. Intra- and intergeneric similarities of *Agrobacterium* ribosomal ribonucleic acid cistrons. Int. J. Syst. Bacteriol. *27:* 222-240.

De Toni, J.B. and V. Trevisan. 1889. Schizomycetaceae Naeg. *In* P.A. Saccardo (Editor), Sylloge Fungorum *8:* 923-1087.

Dewsnup, D.H. and D.N. Wright. 1984. *In vitro* susceptibility of *Nocardia asteroides* to 25 antimicrobial agents. Antimicrob. Agents Chemother. *25:* 165-167.

Diab, A. 1978. Studies on thermophilic microorganisms in certain soils in Kuwait. Zentralbl. Bakteriol. Parasitenkd. Infektionskr. Hyg. Abt. II *133:* 579-587.

Dias, A. and M.Y. Al-Gounaim. 1982. *Streptomyces spinoverrucosus*, a new species from the air of Kuwait. Int. J. Syst. Bacteriol. *32:* 327-331.

Dickerson, R.E. 1980. Cytochrome *c* and the evolution of energy metabolism. Sci. Amer. *242:* 137-153.

Diem, H.G. and Y. Dommergues. 1985. *In vitro* production of specialized torulose hyphae by *Frankia* strain ORS 021001 isolated from *Casuarina junghuhniana* root nodules. Plant Soil *87:* 17-29.

Diem, H.G., D. Gauthier and Y.R. Dommergues. 1982. Isolation of *Frankia* from nodules of *Casuarina equisetifolia*. Can. J. Microbiol. *28:* 526-530.

Diem, H.G., D. Gauthier and Y. Dommergues. 1983. An effective strain of *Frankia* from *Casuarina* sp. Can. J. Bot. *61:* 2815-2821.

Dietz, A. and J. Mathews. 1971. Classification of *Streptomyces* spore surfaces into five groups. Appl. Microbiol. *21:* 527-533.

DiMarco, A. and C. Spalla. 1957. La produzione di cobalamine da fermentazione con una nuova specie di *Nocardia*: *Nocardia rugosa*. G. Microbiol. *4:* 24-30.

Dobritsa, S.V. 1982. Extrachromosomal circular DNA's in endosymbiont vesicles from *Alnus glutinosa* root nodules. FEMS Microbiol. Lett. *15:* 87-91.

Dobritsa, S.V. 1985. Restriction analysis of the *Frankia* spp. genome. FEMS Microbiol. Lett. *29:* 123-128.

Dobson, G., D.E. Minnikin, S.M. Minnikin, J.H. Parlett, M. Goodfellow, M. Ridell and M. Magnusson. 1984. Systematic analysis of complex mycobacterial lipids. *In* Goodfellow and Minnikin (Editors), Chemical Methods in Bacterial Systematics, Society for Applied Bacteriology, No. 20, Academic Press, London, pp. 237-265.

Dolak, L.A., T.M. Castle and A.L. Laborde. 1980. 3-Trehalosamine, a new disaccharide antibiotic. J. Antibiot. *33:* 690-694.

Domnas, A. 1968. Pigments of the *Actinoplanaceae*. I. Pigment production by *Spirillospora* 1655. J. Elisha Mitchell Sci. Soc. *84:* 16-23.

Domnas, A. 1970. Pigment production in the *Actinoplanaceae* as affected by cultural conditions. *In* Prauser (Editor), The Actinomycetales, VEB Gustav Fischer Verlag, Jena, pp. 259-263.

Doolittle, R.F. 1981. Similar amino acid sequences: Chance or common ancestry? Science (Washington) *214:* 149-159.

Döpfer, H., E. Stackebrandt and F. Fiedler. 1982. Nucleic acid hybridization studies on *Microbacterium, Curtobacterium, Agromyces* and related taxa. J. Gen. Microbiol. *128:* 1697-1708.

Dorokhova, L.A., N.S. Agre, L.V. Kalakoutskii and N.A. Krasil'nikov. 1968. Fine structure of spores in a thermophilic actinomycete, *Micromonospora vulgaris*. J. Gen. Appl. Microbiol. *14:* 295-303.

Dorokhova, L.A., N.S. Agre, L.V. Kalakoutskii and N.A. Krasil'nikov. 1969. Fine structure of sporulating hyphae and spores in a thermophilic actinomycete, *Micropolyspora rectivirgula*. J. Microsc. Biol. Cell. *8:* 845-854.

Dorokhova, L.A., N.S. Agre, L.V. Kalakoutskii and N.A. Krasil'nikov. 1970a. A study on the morphology of two cultures belonging to the genus *Micropolyspora*. Mikrobiologiya *39:* 95-100.

Dorokhova, L.A., N.S. Agre, L.V. Kalakoutskii and N.A. Krasil'nikov. 1970b. Comparative study on spores of some actinomycetes with special reference to their thermoresistance. *In* Prauser (Editor), The Actinomycetales, Gustav Fischer Verlag, Jena, pp. 227-232.

Dorokhova, L.A., N.S. Agre, L.V. Kalakoutskii and N.A. Krasil'nikov. 1970c. Electron microscopic study of spore formation in *Micromonospora vulgaris*. Mikrobiologiya *39:* 680-684.

Douglas, H.W., S.M. Ruddick and S.T. Williams. 1970. A study of the electrokinetic properties of some actinomycete spores. J. Gen. Microbiol. *63:* 289-295.

Dowson, W.J. 1942. On the generic name of the Gram-positive bacterial plant pathogens. Trans. Br. Mycol. Soc. *25:* 311-314.

Duché, J. 1934. Les Actinomyces du groupe Albus. Encycl. Mycol. *VI:* 1-375.

Duggar, B.M. 1948. Aureomycin: A product of the continuing search for new antibiotics. Ann. N.Y. Acad. Sci. *51:* 177-181.

Eberson, F. 1918. A bacteriologic study of the diphtheroid organisms with special reference to Hodgkin's disease. J. Infect. Dis. *23:* 1-42.

Edward, D.G., and E.A. Freundt. 1967. Proposal for *Mollicutes* as name of the class established for the order *Mycoplasmatales*. Int. J. Syst. Bacteriol. *17:* 267-268.

Edwards, J.H. 1972a. The double dialysis method of producing farmer's lung antigens. J. Lab. Clin. Med. *79:* 683-688.

Edwards, J. H. 1972b. The isolation of antigens associated with farmer's lung. Clin. Exp. Immunol. *11:* 341-355.

Edwards, J.H., J.T. Baker and B.H. Davies. 1974. Precipitin test negative farmer's lung—activation of the alternative pathway of complement by mouldy hay dusts. Clin. Allergy *4:* 379-388.

Eggeling, L. and H. Sahm. 1980. Degradation of coniferyl alcohol and other lignin-related aromatic compounds by *Nocardia* sp. DSM 1069. Arch. Mikrobiol. *126:* 141-148.

Eggeling, L. and H. Sahm. 1981. Degradation of lignin-related aromatic compounds by *Nocardia* spec. DSM 1069 and specificity of demethylation. Zentralbl. Bakteriol. Mikrobiol. Hyg. I Abt. Orig., Suppl. *11:* 361-366.

Eggins, H.O.W., A. von Szilvinyi and D. Allsopp. 1972. The isolation of actively growing thermophilic fungi from insulated soils. Int. Biodeterior. Bull. *8:* 53-58.

Ehrlich, J., D. Gottlieb, P.R. Burkholder, L.E. Anderson and T.G. Pridham. 1948. *Streptomyces venezuelae*, n. sp., the source of chloromycetin. J. Bacteriol. *56:* 467-477.

El-Nakeeb, M.A. and H.A. Lechevalier. 1963. Selective isolation of aerobic actinomycetes. Appl. Microbiol. *11:* 75-77.

Elwan, S.H., S.A. Mostafa, A.A. Khodair and O. Ali. 1978a. Identity and lipase activity of an isolate of *Thermoactinomyces vulgaris*. Zentralbl. Bakteriol. Parasitenkd. Infektionskr. Hyg. Abt. II *133:* 713-722.

Elwan, S.H., S.A. Mostafa, A.A. Khodair and O. Ali. 1978b. Lipase productivity of a lipolytic strain of *Thermoactinomyces vulgaris*. Zentralbl. Bakteriol. Parasitenkd. Infektionskr. Hyg. Abt. II *133:* 706-712.

Embley, T.M., M. Goodfellow, D.E. Minnikin and A.G. O'Donnell. 1984. Lipid and wall amino acid composition in the classification of *Rothia dentocariosa*. Zentralbl. Bakteriol. Mikrobiol. Hyg. I. Abt. A *257:* 285-295.

Embley, T.M., M. Goodfellow, A.G. O'Donnell, D. Rose and D.E. Minnikin. 1986. Chemical criteria in the classification of some mycolateless wall chemotype IV actinomycetes. *In* Szabó, Biró and Goodfellow (Editors), Biological, Biochemical and Biomedical Aspects of Actinomycetes, Academiai Kiado, Budapest,

pp. 553-556.
Embley, T.M., R. Wait, G. Dobson and M. Goodfellow. 1987. Fatty acid composition in the classification of *Saccharopolyspora hirsuta*. FEMS Microbiol. Lett. *41:* 131-135.
Emerson, R. 1958. Mycological organization. Mycologia *50:* 589-621.
Emery, R.S., C.K. Smith and L.F. To. 1957. Utilization of inorganic sulfate by rumen microorganisms. II. The ability of single strains of rumen bacteria to utilize inorganic sulfate. Appl. Microbiol. *5:* 363-366.
Ensign, J.C. 1978. Formation, properties, and germination of actinomycete spores. Annu. Rev. Microbiol. *32:* 185-219.
Eppinger, H. 1891. Über eine neue pathogene *Cladothrix* und eine durch sie hervorgerufene Pseudotuberculosis (Cladothrichia). Beitr. Pathol. Anat. Allg. Pathol. *9:* 287-328.
Erikson, D. 1935. The pathogenic aerobic organisms of the actinomyces group. Med. Res. Counc. (Great Br.) Spec. Rep. Ser. *203:* 5-61.
Erikson, D. 1953. The reproductive pattern of *Micromonospora vulgaris*. J. Gen. Microbiol. *8:* 449-454.
Erikson, D. 1954. Factors promoting cell division in a "soft" mycelial type of Nocardia: *Nocardia turbata* n. sp. J. Gen. Microbiol. *11:* 198-208.
Erikson, D. and D.M. Webley. 1953. The respiration of a thermophilic actinomycete, *Micromonospora vulgaris*. J. Gen. Microbiol. *8:* 455-463.
Ettlinger, L., R. Corbaz and R. Hütter. 1958a. Zur Systematik der Actinomyceten. 4. Eine arteinteilung der gattung *Streptomyces* Waksman et Henrici. Arch. Mikrobiol. *31:* 326-358.
Ettlinger, L., E. Gäumann, R. Hütter, W. Keller-Schierlein, F. Kradolfer, L. Neipp, V. Prelog and H. Zähner. 1958b. Stoffwechselprodukte von Actinomyceten. 12. Mitteilung über die Isolierung und Characteristierung von Acetomycin. Helv. Chim. Acta *41:* 216-219.
Evtushenko, L.I., N.A. Janushkene, G.M. Streshinskaya, I.B. Naumova and N.S. Agre. 1984a. Occurence of teichoic acids in representatives of Actinomycetales. Dokl. Akad. Nauk SSSR *278:* 237-239.
Evtushenko, L.I., G.F. Levanova and N.S. Agre. 1984b. Nucleotide composition of DNA and amino acid composition of A 4a peptidoglycan in *Promicromonospora citrea*. Mikrobiologiya *53:* 519-520.
Evtushenko, L.I., D.T. Pataraya, H. Prauser and N.S. Agre. 1981. Physiological and biochemical features of *Promicromonospora citrea*. Dep. VINITI, No. 4282-81.
Falcão de Morais, J.O. 1970. *Elytrosporangium spirale:* nova especie de *Actinoplanaceae* do genero *Elytrosporangium*. Rev. Microbiol. *1:* 79-84.
Falcão de Morais, J.O, A. Chaves Batista and D.M.G. Massa. 1966. *Elytrosporangium*: a new genus of the *Actinomycetales*. Mycopathol. Mycol. Appl. *30:* 161-171.
Falcão de Morais, J.O. and M.H. Dália Maia. 1959. *S. erythrogriseus*: novo *Streptomyces* produtor de antibiotico. Rev. Inst. Antibiot. Univ. Recife *2:* 63-67.
Falcão de Morais, J.O. and M.H. Dália Maia. 1961. Uma contribuicãa ao estudo taxonômico do gênero *Streptomyces*—Una tentativa de simplicacão. Rev. Inst. Antibiot. Univ. Recife *3:* 33-60.
Falcão de Morais, J.O., J. da Silva and C. Machado. 1971. Uma terceira especie de *Actinomycetales* do genero *Elytrosporangium, E. carpinense* sp. nov., isolada de solo em. Pernambuco. Rev. Microbiol. Brazil *2:* 203-206.
Falcão de Morais, J.O., O. Goncalves de Lima and M.H. Dália Maia. 1957. Novo estudo sôbre *Nocardia recifei* Lima et al., e sua designacão como *Streptomyces recifensis*. An. Soc. Biol. Pernambuco *15:* 239-253.
Falkowski, J. 1978. Occurrence and thermoresistance of thermophilic actinomycetes in milk and dairy products. Zentralbl. Bakteriol. Parasitenkd. Infektionskr. Hyg. Abt. I Orig. Reihe B *167:* 171-176.
Farabaugh, P.J., V. Schmeissner, M. Hofer and J.H. Miller. 1978. Genetic studies of the lac repressor. VII. On the molecular nature of spontaneous hotspots in the lacI gene of *Escherichia coli*. J. Mol. Biol. *126:* 847-863.
Farina, G. and S.G. Bradley. 1970. Reassociation of deoxyribonucleic acids from *Actinoplanes* and other actinomycetes. J. Bacteriol. *102:* 30-35.
Farina, G. and R. Locci. 1966. Contributo allo studio di *Streptoverticillium:* descrizione di una nuova specie (*Streptoverticillium baldaccii* sp. nov.) ed esame di alcune specie precedentemente dilineate. G. Microbiol. *14:* 33-52.
Feltham, R.K.A., A.K. Power, P.A. Pell and P.H.A. Sneath. 1978. A simple method for storage of bacteria at −76°C. J. Appl. Bacteriol. *44:* 313-316.
Feltham, R.K.A., P.A. Wood and P.H.A. Sneath. 1984. A general-purpose system for characterizing medically important bacteria to genus level. J. Appl. Bacteriol. *57:* 279-290.
Fennington, G., D. Lupo and F.J. Stutzenberger. 1982. Enhanced cellulase production in mutants of *Thermomonospora curvata*. Biotech. Bioeng. *24:* 2487-2497.
Fennington, G., D. Neubauer and F.J. Stutzenberger. 1984. Cellulase biosynthesis in a catabolite repression-resistant mutant of *Thermomonospora curvata*. Appl. Environ. Microbiol. *47:* 201-204.
Fergus, C.L. 1964. Thermophilic and thermotolerant molds and actinomycetes of mushroom compost during peak heating. Mycologia *56:* 267-284.
Fergus, C.L. 1967. Resistance of spores of some thermophilic actinomycetes to high temperature. Mycopath. Mycol. Appl. *32:* 205-208.
Fergus, C.L. 1969. The cellulolytic activity of thermophilic fungi and actinomycetes. Mycologia *61:* 120-129.
Fermor, T.R. and H.O.W. Eggins. 1980. The effect of water activity on growth of *Streptomyces* species. Int. Biodeterior. Bull. *16:* 95-101.
Fernandez, F., M.D. Collins and M.J. Hill. 1984. Production of vitamin K by human gut bacteria. Biochem. Trans. *13:* 223-224.
Festenstein, G.N., J. Lacey, F.A. Skinner, P.A. Jenkins and J. Pepys. 1965. Self-heating of hay and grain in Dewar flasks and the development of farmer's lung hay antigens. J. Gen. Microbiol. *41:* 389-407.
Fiedler, F. and O. Kandler. 1973. Die Mureintypen in der Gattung *Cellulomonas* Bergey et al. Arch. Microbiol. *89:* 41-50.
Finlay, A.C., F.A. Hochstein, B.A. Sobin and F.X. Murphy. 1951. Netropsin, a new antibiotic produced by a *Streptomyces*. J. Am. Chem. Soc. *73:* 341-343.
Fischer, A., R.M. Kroppenstedt and E. Stackebrandt. 1983. Molecular-genetic and chemotaxonomic studies on *Actinomadura* and *Nocardiopsis*. J. Gen. Microbiol. *129:* 3433-3446.
Flaig, W. and H.J. Kutzner. 1954. Zur Systematik der Gattung *Streptomyces*. Naturwissenschaften *41:* 287.
Flaig, W. and H.J. Kutzner. 1960. Beitrag zur Systematik der Gattung *Streptomyces* Waksman et Henrici. Arch. Mikrobiol. *35:* 105-138.
Fleck, W.F., D.G. Strauss, J. Meyer and G. Porstendorfer. 1978. Fermentation, isolation and biological activity of maduramycin: A new antibiotic from *Actinomadura rubra*. Z. Allg. Mikrobiol. *18:* 389-398.
Fletcher, S.M., C.J.M. Rondle and I.G. Murray. 1970. The extracellular antigens of *Micropolyspora faeni*: Their significance in farmer's lung disease. J. Hyg. *68:* 401-409.
Flockton, H.I. and T. Cross. 1975. Variability in *Thermoactinomyces vulgaris*. J. Appl. Bacteriol. *38:* 309-313.
Flowers, T.H. and S.T. Williams. 1977a. Measurements of growth rates of streptomycetes: Comparison of turbidimetric and gravimetric techniques. J. Gen. Microbiol. *98:* 285-289.
Flowers, T.H. and S.T. Williams. 1977b. The influence of pH on the growth rate and viability of neutrophilic and acidophilic streptomycetes. Microbios *18:* 223-228.
Foerster, H.F. 1975. Germination characteristics of some of the thermophilic actinomycete spores. *In* Gerhardt, Costilow and Sadoff (Editors), Spores, Vol. VI, American Society for Microbiology, Washington, D.C., pp. 36-43.
Foerster, H.F. 1978. Effects of temperature on the spores of thermophilic actinomycetes. Arch. Microbiol. *118:* 257-264.
Follett, M.J. and P.W. Ratcliffe. 1963. Determination of nitrate and nitrite in meat products. J. Sci. Food Agric. *14:* 138-144.
Fonseca, A.F. 1983. Ultrastructural and biochemical data that support the inclusion of *Blastococcus aggregatus* in Genus *Geodermatophilus*. *In* Proceedings of the XVIII Meeting of the Sociedade Portuguesa de Microscopia Electronica, Porto, Abstract No. 75.
Fontaine, M.S., S.A. Lancelle and J.G. Torrey. 1984. Initiation and ontogeny of vesicles in cultured *Frankia* sp. strain HFP ArI 3. J. Bacteriol. *160:* 921-927.
Ford, W.W. 1927. Textbook of Bacteriology, W.B. Saunders, Philadelphia.
Forsyth, W.G.C. and D.M. Webley. 1948. The microbiology of composting. II. A study of the aerobic thermophilic bacterial flora developing in grass composts. Proc. Soc. Appl. Bacteriol. *3:* 34-39.
Foulerton, A.G.R. 1905. New species of *Streptothrix* isolated from the air. Lancet *1:* 1199-1200.
Fowler, V.J., W. Ludwig and E. Stackebrandt. 1985. Ribosomal ribonucleic acid cataloguing in bacterial systematics: The phylogeny of *Actinomadura*. *In* Goodfellow and Minnikin (Editors), Chemical Methods in Bacterial Systematics, Academic Press, London, pp. 17-40.
Fox, G.E., K.R. Pechman and C.R. Woese. 1977. Comparative cataloguing of 16S ribosomal ribonucleic acid: Molecular approach to prokaryotic systematics. Int. J. Syst. Bacteriol. *27:* 44-57.
Fox, G.E., E. Stackebrandt, R.B. Hespell, J. Gibson, J. Maniloff, I. Dyer, R.S. Wolfe, W. Balch, R. Tanner, L. Magrum, L.B. Zablen, R. Blakemore, R. Gupta, K.R. Luehrsen, L. Bonen, B.J. Lewis, K.N. Chen and C.R. Woese. 1980. The phylogeny of prokaryotes. Science *209:* 457-463.
Frédéricq, E., A. Oth and F. Fontaine. 1961. The ultraviolet spectra of deoxyribonucleic acids and their constituents. J. Mol. Biol. *3:* 11-17.
Fritzsche, H. 1967. Infra-red studies of deoxyribonucleic acids, their constituents and analogues. II. Deoxyribonucleic acids with different base compositions. Biopolymers *5.* 863-870.
Frommer, W. 1959. Zur Systematik der Actinomycin bildenden Streptomyceten. Arch. Mikrobiol. *32:* 187-206.
Fukunaga, K., T. Misato, I. Ishii and M. Asakawa. 1955. Blasticidin, a new antiphytopathogenic fungal substance. Part I. Bull. Agric. Chem. Soc. Jpn. *19:* 181-188.
Furumai, T., H. Ogawa and T. Okuda. 1968. Taxonomic study on *Streptosporangium albidum* sp. nov. J. Antibiotics (Tokyo) *21:* 179-181.
Gadek, A., M. Mordarski, M. Goodfellow and S.T. Williams. 1985. Ribosomal ribonucleic acid similarities in the classification of *Streptomyces*. FEMS Microbiol. Lett. *26:* 175-180.
Gaertner, A. 1955. Über zwei ungewöhnliche keratinophile Organismen aus Ackerböden. Arch. Mikrobiol. *23:* 28-37.
Galatenko, O.A. and T.P. Preobrazhenskaya. 1981. *Actinomadura* isolated from sierozem soil samples of Turkmenistan and their antagonistic properties. Antibiotiki *26:* 723-727.
Galatenko, O.A., L.P. Terekhova and T.P. Preobrazhenskaya. 1981. New *Actinomadura* species isolated from Turkmen soil samples and their antagonistic properties. Antibiotiki *26:* 803-807.
Galatenko, O.A., L.P. Terekhova and T.P. Preobrazhenskaya. 1987. *In* Validation of the publication of new names and new combinations previously effectively

published outside the IJSB. List No. 23. Int. J. Syst. Bacteriol. *37:* 179-180.
Galau, G.A., R.J. Britten and E.H. Davidson. 1977. Studies on nucleic acid reassociation kinetics: rate of hybridization on excess RNA with DNA, compared to the rate of DNA renaturation. Proc. Nat. Acad. Sci. USA *74:* 1020-1023.
Gasperini, G. 1892. Ricerche morfologiche e biologiche sul genere *Actinomyces*—Harz come contributo allo studio delle relative micosi. Ann. 1st Igiene sper. Univ. Roma (N.S.) *2:* 167-231.
Gasperini, G. 1894. Ulteriori ricerche sul genere *Actinomyces*. Processi verbali della Societa Toscana di Scienze naturali di Pisa. *9:* 64-89.
Gauthier, D., H.G. Diem and Y. Dommergues. 1981a. Infectivité et effectivité de souches de *Frankia* isolées de nodules de *Casuarina equisetifolia* et d'*Hippophaë* rhamnoides. C. R. Acad. Sci. (Paris) *293:* 489-491.
Gauthier, D., H.G. Diem and Y. Dommergues. 1981b. *In vitro* nitrogen fixation by two actinomycete strains isolated from *Casuarina* nodules. Appl. Environ. Microbiol. *41:* 306-308.
Gauze, G.F. 1957. *In* Problems in the Classification of Antagonists of Actinomycetes, Medgiz, Moscow, p. 22.
Gauze, G.F., T.S. Maksimova, O.L. Olkhovatova, M.A. Sveshnikova, G.V. Kochetkova and G.B. Ilchenko. 1974. Production of madumycin, an antibacterial antibiotic by *Actinomadura flava* sp. n. Antibiotiki *19:* 771-774.
Gauze, G.F., T.P. Preobrazhenskaya, E.S. Kudrina, N.O. Blinov, I.D. Ryabova and M.A. Sveshnikova. 1957. Problems in the Classification of Antagonistic Actinomycetes, State Publishing House for Medical Literature (in Russian), Medzig, Moscow.
Gauze, G.F., T.P. Preobrazhenskaya, N.V. Lavrova, R.S. Ukholina, G.V. Kochetkova, N.P. Nechaeva, N.V. Konstantinova and I.V. Tolstykh. 1975. *Actinomadura cremea* var. *rifamycini*, a rifamycin-producing organism. Antibiotiki *20:* 963-966.
Gauze, G.F. and M.A. Sveshnikova. 1985. *In* Validation of the publication of new names and new combinations previously effectively published outside the IJSB. List No. 17. Int. J. Syst. Bacteriol. *35:* 223-225.
Gauze, G.F., M.A. Sveshnikova, R.S. Ukholina, G.N. Gavrilina, V.A. Filicheva and E.G. Gladkikh. 1973. Production of antitumor antibiotic carminomycin by *Actinomadura carminata* sp. nov. Antibiotiki *18:* 675-678.
Gauze, G.F., M.A. Sveshnikova, R.S. Ukholina, G.N. Komarova and W.S. Bazhanov. 1977. Production of a new antibiotic, nocamycin, by *Nocardiopsis syringae* sp. nov. Antibiotiki *22:* 483-486.
Gauze, G.F., L.P. Terekhova, O.A. Galatenko, T.P. Preobrazhenskaya, V.N. Borisova and G.B. Fedorova. 1984. *Actinomadura recticatena* sp. nov., a new species and its antibiotic properties. Antibiotiki *29:* 3-7.
Gauze, G.F., L.P. Terekhova, O.A. Galatenko, T.P. Preobrazhenskaya, V.N. Borisova and G.B. Federova. 1987. *In* Validation of the publication of new names and new combinations previously effectively published outside the IJSB. List No. 23. Int. J. Syst. Bacteriol. *37:* 179-180.
Gee, J.M., B.M. Lund, G. Metcalf and J.L. Peel. 1980. Properties of a new group of alkalophilic bacteria. J. Gen. Microbiol. *117:* 9-17.
Gerber, N.N. 1966. Phenazines and phenoxazines from some novel *Nocardiaceae*. Biochemistry *5:* 3824-3829.
Gerber, N.N. 1971. Prodigiosin-like pigments from *Actinomadura (Nocardia) pelletieri*. J. Antibiot. *24:* 636.
Gerber, N.N. 1973. Minor prodiginine pigments from *Actinomadura madurae* and *Actinomadura pelletieri*. J. Heterocycl. Chem. *10:* 925.
Gerber, N.N. and M.P. Lechevalier. 1964. Phenazines and phenoxazinones from *Waksmania aerata* sp. nov. and *Pseudomonas iodina*. Biochemistry *3:* 598-602.
Gerber, N.N. and M.P. Lechevalier. 1965. 1,6-Phenazinediol-5-oxide from microorganisms. Biochemistry *4:* 176-180.
Gerber, N.N. and M.P. Lechevalier. 1984. Novel benzo(a)naphacene quinone, from an actinomycete; *Frankia* G2 (ORS 020604). Can. J. Chem. *62:* 2818-2821.
Gerber, N.N., H.L. Yale, W.A. Taber, I. Kurobane and L.C. Vining. 1983. Structure and syntheses of texazone, 2-(*N*-methylamine)-^{3}H-phenoxazin-3-one-8-carboxylic acid, an actinomycete metabolite. J. Antibiot. *36:* 688-694.
Gibbons, N.E. 1974. Reference collections of bacteria—the need and requirements for type and neotype strains. *In* Buchanan and Gibbons (Editors), Bergey's Manual of Determinative Bacteriology, 8th ed., The Williams and Wilkins Co., Baltimore, pp. 14-17.
Gibbons, N.E. and R.G.E. Murray. 1978. Proposals concerning the higher taxa of bacteria. Int. J. Syst. Bacteriol. *28:* 1-6.
Gibson, J., E. Stackebrandt, L.B. Zablen, R. Gupta and C.R. Woese. 1979. A genealogical analysis of the purple photosynthetic bacteria. Curr. Microbiol. *3:* 59-66.
Gilardi, E., L.R. Hill, M. Turri and L.G. Silvestri. 1960. Quantitative methods in the systematics of Actinomycetales. G. Microbiol. *8:* 203-218.
Gill, J.W. and K.W. King. 1958. Nutritional characteristics of a butyrivibrio. J. Bacteriol. *75:* 666-673.
Gillespie, D. and S. Spiegelman. 1965. A quantitative assay for DNA-RNA hybrids with DNA immobilized on a membrane filter. J. Mol. Biol. *12:* 829-842.
Głądek, A., M. Mordarski, M. Goodfellow and S.T. Williams. 1985. Ribosomal ribonucleic acid similarities in the classification of *Streptomyces*. FEMS Microbiol. Lett. *26:* 175-180.
Glauert, A.M. and D.A. Hopwood. 1960. The fine structure of *Streptomyces coelicolor*. I. The cytoplasmic membrane system. J. Biophys. Biochem. Cytol. *7:* 479-488.
Gochnauer, M.B. and D.J. Kushner. 1969. Growth and nutrition of extremely halophilic bacteria. Can. J. Microbiol. *15:* 1157-1165.
Gochnauer, M.B., G.G. Leppard, P. Komaratat, M. Kates, T. Novitsky and D.J. Kushner. 1975. Isolation and characterization of *Actinopolyspora halophila*, gen. et sp. nov., an extremely halophilic actinomycete. Can. J. Microbiol. *21:* 1500-1511.
Golovina, I.G., E.P. Guzhova, T.I. Bogdanova and L.G. Loginova. 1973. Lytic enzymes produced by thermophilic actinomycetes *Micromonospora vulgaris* PA 11-4. Mikrobiologiya *62:* 620-626.
Gonçalves de Lima, O., M.P. Machado, L.A. de Araújo, J.O. Falcão de Morais and H. Biermann. 1955. Novo espécie do gênero *Nocardia*: *N. recefei* sua ativadade antagonista. Antibiótico produzido. An. Soc. Biol. Pernambuco *13:* 21-36.
Gonçalves de Lima, V.O., C.A. Albert and O. Gon;alcalves de Lima. 1964. *Streptomyces capoamus* nov. sp. produtor da ciclamicina e das ciclacidinas A & B. An. Acad. Brazil Ciênc. *36:* 317-322.
González-Ochoa, A. 1976. Nocardiae and chemotherapy. *In* Goodfellow, Brownell and Serrano (Editors), The Biology of the Nocardiae. Academic Press, New York, pp. 429-450.
González-Ochoa, A. and M.A. Sandoval. 1955. Caracteristicas de los actinomicetes patogenos mas comunes. Rev. Inst. Salubr. Enferm. Trop. *16:* 149-161.
Goodfellow, M. 1971. Numerical taxonomy of some nocardioform bacteria. J. Gen. Microbiol. *69:* 33-80.
Goodfellow, M. 1984a. *In* Validation of the publication of new names and combinations previously effectively published outside the IJSB. List No. 10. Int. J. Syst. Bacteriol. *34:* 503-504.
Goodfellow, M. 1984b. Reclassification of *Corynebacterium fascians* (Tilford) Dowson in the genus *Rhodococcus*, as *Rhodococcus fascians* comb. nov. Syst. Appl. Microbiol. *5:* 225-229.
Goodfellow, M. 1986. Actinomycete systematics: Present state and future prospects. *In* Szabó, Biró and Goodfellow (Editors), Biological, Biochemical and Biomedical Aspects of Actinomycetes, Akademiai Kiado, Budapest, pp. 487-496.
Goodfellow, M. and G. Alderson. 1977. The actinomycete genus *Rhodococcus*: A home for the 'rhodochrous' complex. J. Gen. Microbiol. *100:* 99-122.
Goodfellow, M. and G. Alderson. 1979. *In* Validation of the publication of new names and combinations previously effectively published outside the IJSB. List No. 2. Int. J. Syst. Bacteriol. *29:* 79-80.
Goodfellow, M., G. Alderson and J. Lacey. 1979. Numerical taxonomy of *Actinomadura* and related actinomycetes. J. Gen. Microbiol. *112:* 95-111.
Goodfellow, M., A.R. Beckham and M.D. Barton. 1982a. Numerical classification of *Rhodococcus equi* and related actinomycetes. J. Appl. Bacteriol. *53:* 199-207.
Goodfellow, M. and T. Cross. 1974. Actinomycetes. *In* Dickinson and Pugh (Editors), Biology of Plant Litter Decomposition, Academic Press, London, pp. 269-302.
Goodfellow, M. and T. Cross. 1984a. *Actinomadura* (H.A. Lechevalier and M.P. Lechevalier 1970). *In* Goodfellow, Mordarski and Williams (Editors). The Biology of the Actinomycetes, Academic Press, London, pp. 105-107.
Goodfellow, M. and T. Cross. 1984b. Classification. *In* Goodfellow, Mordarsky and Williams (Editors), The Biology of the Actinomycetes, Academic Press, London, pp. 7-164.
Goodfellow, M. and J.A. Haynes. 1984. Actinomycetes in marine sediments. *In* Ortiz-Ortiz, Bojalil and Yakoleff (Editors), Biological, Biochemical and Biomedical Aspects of Actinomycetes, Academic Press, Orlando, pp. 453-472.
Goodfellow, M., J. Lacey and C. Todd. 1987. Numerical classification of thermophilic streptomycetes. J. Gen. Microbiol. *133:* 3135-3149.
Goodfellow, M., A. Lind, H. Mordarska, S. Pattyn and M. Tsukamura. 1974. A cooperative numerical analysis of cultures considered to belong to the '*rhodochrous*' taxon. J. Gen. Microbiol. *85:* 291-302.
Goodfellow, M. and D.E. Minnikin. 1981a. Classification of nocardioform bacteria. Zentralbl. Bakteriol. Mikrobiol. Hyg. Suppl. *11:* 7-16.
Goodfellow, M. and D.E. Minnikin. 1981c. The genera *Nocardia* and *Rhodococcus*. *In* Starr, Stolp, Trüper, Balows and Schlegel (Editors), The Prokaryotes: A Handbook on Habitiats, Isolation and Identification of Bacteria, Springer-Verlag, Berlin, pp. 2016-2017.
Goodfellow, M. and D.E. Minnikin. 1981b. Introduction to the coryneform bacteria. *In* Starr, Stolp, Trüper, Balows and Schlegel (Editors), The Prokaryotes: A Handbook on Habitats, Isolation and Identification of Bacteria, Springer-Verlag, Berlin, pp. 1811-1826.
Goodfellow, M. and D.E. Minnikin. 1984. A critical evaluation of *Nocardia* and related taxa. *In* Ortiz-Ortiz, Bojalil and Yakoleff (Editors), Biological, Biochemical and Biomedical Aspects of Actinomycetes, Academic Press, Orlando, Florida, pp. 583-596.
Goodfellow, M. and D.E. Minnikin (Editors). 1985. Chemical Methods in Bacterial Systematics. Society for Applied Bacteriology, Technical Series No. 20, Academic Press, London.
Goodfellow, M., D.E. Minnikin, C. Todd, G. Alderson, S.M. Minnikin and M.D. Collins. 1982b. Numerical and chemical classification of *Nocardia amarae*. J. Gen. Microbiol. *128:* 1283-1297.
Goodfellow, M., M. Mordarski and S.T. Williams (Editors). 1984. The Biology of the Actinomycetes, Academic Press, London.
Goodfellow, M. and V.A. Orchard. 1974. Antibiotic sensitivity of some nocardioform bacteria and its value as a criterion for taxonomy. J. Gen. Microbiol. *83:* 375-387.
Goodfellow, M., P.A.B. Orlean, M.D. Collins, L. Alshamaony and D.E. Minnikin.

1978. Chemical and numerical taxonomy of some strains received as *Gordona aurantiaca*. J. Gen. Microbiol. *109:* 57–68.

Goodfellow, M. and T. Pirouz. 1982a. *In* Validation of the publication of new names and new combinations previously effectively published outside the IJSB. List No. 9. Int. J. Syst. Bacteriol. *32:* 384–385.

Goodfellow, M. and T. Pirouz. 1982b. Numerical classification of sporoactinomycetes containing *meso*-diaminopimelic acid in the cell wall. J. Gen. Microbiol. *128:* 503–527.

Goodfellow, M. and K.E. Simpson. 1987. Ecology of streptomycetes. Front. Appl. Microbiol. *2:* 97–125.

Goodfellow, M. and L.G. Wayne. 1982. Taxonomy and nomenclature. *In* Ratledge and Stanford (Editors), The Biology of the Mycobacteria, Vol. 1, Academic Press, London, pp. 471–521.

Goodfellow, M., C.R. Weaver and D.E. Minnikin. 1982c. Numerical classification of some rhodococci, corynebacteria and related organisms. J. Gen. Microbiol. *128:* 731–745.

Goodfellow, M. and E. Williams. 1986. New strategies for the selective isolation of industrially important bacteria. Biotechnol. Genet. Eng. Rev. *4:* 213–262.

Goodfellow, M. and S.T. Williams. 1983. Ecology of actinomycetes. Annu. Rev. Microbiol. *37:* 189–216.

Goodfellow, M., S.T. Williams and G. Alderson. 1986a. *In* Validation of the publication of new names and new combinations previously effectively published outside the IJSB. List No. 22. Int. J. Syst. Bacteriol. *36:* 573–576.

Goodfellow, M., S.T. Williams and G. Alderson. 1986b. Transfer of *Actinosporangium violaceum* Krasil'nikov and Yuan, *Actinosporangium vitaminophilum* Shomura et al. and *Actinopycnidium caeruleum* Krasil'nikov to the genus *Streptomyces*, with emended descriptions of the species. Syst. Appl. Microbiol. *8:* 61–64.

Goodfellow, M., S.T. Williams and G. Alderson. 1986c. Transfer of *Chainia* species to the genus *Streptomyces* with emended description of species. Syst. Appl. Microbiol. *8:* 55–60.

Goodfellow, M., S.T. Williams and G. Alderson. 1986d. Transfer of *Elytrosporangium brasiliense* Falcão de Morais et al., *Elytrosporangium carpinense* Falcão de Morais et al., *Elytrosporangium spirale* Falcão de Morais et al., *Microellobospora cinerea* Cross et al., *Microellobosporia flavea* Cross et al., *Microellobosporia grisea* (Konev et al.) Pridham and *Microellobosporia violacea* (Tsyganov et al.) Pridham to the genus *Streptomyces* with emended descriptions of the species. Syst. Appl. Microbiol. *8:* 48–54.

Goodfellow, M., S.T. Williams and G. Alderson. 1986e. Transfer of *Kitasatoa purpurea* Matsumae and Hata to the genus *Streptomyces* as *Streptomyces purpureus* comb. nov. Syst. Appl. Microbiol. *8:* 65–66.

Goodfellow, M., S.T. Williams and G. Alderson. 1987. *In* Validation of the publication of new names and new combinations previously effectively published outside the IJSB. List No. 23. Int. J. Syst. Bacteriol. *37:* 179–180.

Gordon, M.A. 1964. The genus *Dermatophilus*. J. Bacteriol. *88:* 509–522.

Gordon, M.A. 1974. Aerobic pathogenic *Actinomycetaceae*. *In* Lennette, Spaulding and Truant (Editors), Manual of Clinical Microbiology, American Society for Microbiology, Washington, D.C., pp. 175–188.

Gordon, M.A. 1985. Aerobic pathogenic *Actinomycetaceae*. *In* Lennette, Balows, Hausler and Shadomy (Editors), Manual of Clinical Microbiology, 4th Ed. American Society for Microbiology, Washington, D.C., pp. 249–262.

Gordon, M.A. and E.W. Lapa. 1966. Durhamycin, a pentaene antifungal antibiotic from *Streptomyces durhamensis* sp. n. Appl. Microbiol. *14:* 754–760.

Gordon, M.A., M.P. Lechevalier and E.W. Lapa. 1983. Nonpathogenicity of *Frankia* sp. CpI1 in the *Dermatophilus* pathogenicity test. Actinomycetes *18:* 50–53.

Gordon, M.A. and I.F. Salkin. 1977. *Dermatophilus* dermatitis enzootic in deer in New York state and vicinity. J. Wildl. Dis. *13:* 184–190.

Gordon, R.E. 1966. Some criteria for the recognition of *Nocardia madurae* (Vincent) Blanchard. J. Gen. Microbiol. *45:* 355–364.

Gordon, R.E. 1968. The taxonomy of soil bacteria. *In* Gray and Parkinson (Editors), The Ecology of Soil Bacteria, Liverpool University Press, Liverpool, England, pp. 293–321.

Gordon, R.E., D.A. Barnett, J.E. Handerhan and C.H.-N. Pang. 1974. *Nocardia coeliaca, Nocardia autotrophica* and the nocardin strain. Int. J. Syst. Bacteriol. *24:* 54–63.

Gordon, R.E. and A.C. Horan. 1968. *Nocardia dassonvillei*, a macroscopic replica of *Streptomyces griseus*. J. Gen. Microbiol. *50:* 235–240.

Gordon, R.E. and J.M. Mihm. 1957. A comparative study of some strains received as nocardiae. J. Bacteriol. *73:* 15–27.

Gordon, R.E. and J.M. Mihm. 1959. A comparison of *Nocardia asteroides* and *Nocardia brasiliensis*. J. Gen. Microbiol. *20:* 129–135.

Gordon, R.E. and J.M. Mihm. 1962a. Identification of *Nocardia caviae* (Erikson) *nov. comb.* Ann. N.Y. Acad. Sci. *98:* 628–636.

Gordon, R.E. and J.M. Mihm. 1962b. The type species of the genus *Nocardia*. J. Gen. Microbiol. *27:* 1–10.

Gordon, R.E., S.K. Mishra and D.A. Barnett. 1978. Some bits and pieces of the genus *Nocardia*: *N. carnea, N. vaccinii, N. transvalensis, N. orientalis* and *N. aerocolonigenes*. J. Gen. Microbiol. *109:* 69–78.

Gordon, R.E. and M.M. Smith. 1953. Rapidly growing acid-fast bacteria. I. Species description of *Mycobacterium phlei* Lehmann and Neumann and *Mycobacterium smegmatis* (Trevisan) Lehmann and Neumann. J. Bacteriol. *66:* 41–48.

Gordon, R.E. and M.M. Smith. 1955. Proposed group of characters for the separation of *Streptomyces* and *Nocardia*. J. Bacteriol. *69:* 147–150.

Gottlieb, D. 1961. An evaluation of criteria and procedures used in the description and characterization of streptomycetes. A co-operative study. Appl. Microbiol. *9:* 55–65.

Gottlieb, D. 1974. Order *Actinomycetales* Buchanan 1917. *In* Buchanan and Gibbons (Editors), Bergey's Manual of Determinative Bacteriology, 8th Ed., The Williams and Wilkins Co., Baltimore, pp. 657–659.

Gottschalk, G. (Editor). 1985. Methods in Microbiology, Vol. 18, Academic Press, London, p. 383.

Grant, J.B.W., W. Blyth, V.E. Wardrop, R.M. Gordon, J.C.G. Pearson and A. Mair. 1972. Prevalence of farmer's lung in Scotland: A pilot survey. Br. Med. J. *1:* 530–534.

Gray, M.W. and W.F. Doolittle. 1982. Has the endosymbiont hypothesis been proven? Microbiol. Rev. *46:* 1–42.

Gray, P.H.H. 1928. The formation of indigotin from indol by soil bacteria. Proc. Roy. Soc. Ser. B *102:* 263–280.

Gray, P.H.H. and H.G. Thornton. 1928. Soil bacteria that decompose certain aromatic compounds. Zentralbl. Bakteriol. Parasitenkd. Infektionskr. Hyg. Abt. II *73:* 74–96.

Greatorex, F.B. and J.V.S. Pether. 1975. Use of a serologically distinct strain of *Thermoactinomyces vulgaris* in the diagnosis of farmers' lung disease. J. Clin. Path. *28:* 1000–1002.

Gregory, P.H. and M.E. Lacey. 1963. Mycological examination of dust from mouldy hay associated with farmers' lung disease. J. Gen. Microbiol. *30:* 75–88.

Gregory, P.H., M.E. Lacey, G.N. Festenstein and F.A. Skinner. 1963. Microbial and biochemical changes during the moulding of hay. J. Gen. Microbiol. *33:* 147–174.

Grein, A., C. Spalla, A. Di Marco and G. Canevazzi. 1963. Descrizione e classificazione di un attinomicete (*Streptomyces peucetius* sp. nova) produttore di una sostanza attivita antitumorale: La daunomicina. G. Microbiol. *11:* 109–118.

Grimont, P.A.D., M.Y. Popoff, F. Grimont, C. Coynault and M. Lemelin. 1980. Reproducibility and correlation study of three deoxyribonucleic acid hybridization procedures. Curr. Microbiol. *4:* 325–330.

Grundy, W.E., A.C. Sinclair, R.J. Theriault, A.W. Goldstein, C.J. Rickher, H.B. Warren, Jr., T.J. Oliver and J.C. Sylvester. 1957. Ristocetin, microbiologic properties. Antibiot. Annu. *1956–1957:* 687–792.

Grundy, W.E., A.L. Whitman, E.G. Rdzok, E.J. Rdzok, M.E. Hanes and J.C. Sylvester. 1952. Actithiazic acid. I. Microbiological studies. Antibiot. Chemother. *2:* 399–408.

Gupta, K.C. 1965. A new species of the genus *Streptosporangium* isolated from Indian soil. J. Antibiot. Ser. A *18:* 125–127.

Gupta, K.C. 1965. *Streptomyces tropicalensis*, a new whorl-forming species of *Streptomyces*. J. Antibiot. (Tokyo) Ser. A *18:* 53–55.

Gupta, K.C. and I.C. Chopra. 1963a. A new whorl-forming species of *Streptomyces*. Hind. Antibiot. Bull. *5:* 110–112.

Gupta, K.C. and I.C. Chopra. 1963b. *Streptomyces katrae*—a new species of *Streptomyces* isolated from soil. Indian J. Microbiol. *3:* 1–4.

Gupta, K.C., R.R. Sobti and I.C. Chopta. 1963. Actinomycin produced by a new species of *Streptomyces*. Hind. Antibiot. Bull. *6:* 12–16.

Gutmann, L., F.W. Goldstein, M.D. Kitzis, B. Hautefort, C. Darmon and J.F. Acar. 1983. Susceptibility of *Nocardia asteroides* to 46 antibiotics, including 22 β-lactams. Antimicrob. Agents Chemother. *23:* 248–251.

Guzeva, L.N., T.P. Efimova, N.S. Agre and N.A. Krasil'nikov. 1973. Fatty acids from mycelium of actinomycetes forming catenuate spores. Mikrobiologiya *42:* 26–31.

Gyllenberg, H.G. 1970. Factor analytical evaluation of patterns of correlated characteristics in streptomycetes. *In* Prauser (Editor), The Actinomycetales, Gustav Fischer Verlag, Jena, pp. 101–105.

Gyllenberg, H.G., W. Woznicka and W. Kurylowicz. 1967. Application of factor analysis in microbiology. III. A study of the 'yellow series' of streptomycetes. Ann. Acad. Sci. Fenn. Ser. A IV Biol. *114:* 1–15.

Hafeez, F., A.D.L. Akkermans and A.H. Chaudhary. 1984. Morphology, physiology and infectivity of two *Frankia* isolates An1 and An2 from root nodules of *Alnus nitida*. Plant Soil *78:* 45–59.

Hagedorn, C. 1976. Influences of soil acidity on *Streptomyces* populations inhabiting forest soils. Appl. Environ. Microbiol. *32:* 368–375.

Hagemann, G., G. Nominé and L. Pénasse. 1958. Sur un antibiotique, l'hydroxymycin, produit par un *Streptomyces*. Ann. Pharm. France *16:* 585–596.

Hägerdal, B.G.R., J.D. Ferchak and E.K. Pye. 1978. Cellulolytic enzyme system of *Thermoactinomyces* sp. grown on microcrystalline cellulose. Appl. Environ. Microbiol. *36:* 606–612.

Hall, R.B., H.S. McNabb, Jr., C.A. Maynard and T.L. Green. 1979. Toward development of optimal *Alnus glutinosa* symbiosis. Bot. Gaz. *140(Suppl):* S120–S126.

Hamada, S. 1958. A study of a new antitumor substance, cellostatin. Tohoku J. Exp. Med. *67:* 173–179.

Hanka, L.J., J.S. Evans, D.J. Mason and A. Dietz. 1967. Microbiological production of 5-azacytidine. Antimicrob. Agents Chemother. *1966:* 619–624.

Hanka, L.J., P.W. Rueckert and T. Cross. 1985. Method for isolating strains of the genus *Streptoverticillium* from soil. FEMS Microbiol. Lett. *30:* 365–368.

Hankin, L., M. Zucker and D.C. Sands. 1971. Improved solid medium for detection and enumeration of pectolytic bacteria. Appl. Microbiol. *22:* 205–209.

Hanton, W.K. 1968. *Amorphosporangium (Actinoplanaceae)*. Report of motility and additional characters. J. Gen. Microbiol. *53:* 317-320.

Hardisson, C. and J.E. Suárez. 1979. Fine structure of spore formation and germination in *Micromonospora chalcea*. J. Gen. Microbiol. *110:* 233-237.

Hardy, K. 1981. Bacterial plasmids. *In* Cole and Knowles (Editors), Aspects of Microbiology Series No. 4, Thomas Nelson and Sons, Ltd., Walton-on-Thames, United Kingdom.

Harrison, F.C. 1929. The discoloration of halibut. Can. J. Res. *1:* 214-239.

Harvey, S. and M.J. Pickett. 1980. Comparison of adansonian analysis and deoxyribonucleic acid hybridization results in the taxonomy of *Yersinia enterocolitica*. Int. J. Syst. Bacteriol. *30:* 86-102.

Harwood, C.R. 1980. Plasmids. *In* Goodfellow and Board (Editors), Microbiological Classification and Identification, Academic Press, London, pp. 27-53.

Hasegawa, T. 1985. A new actinomycete, *Actinosynnema*. Chem. Biol. (in Japanese) *23:* 143-145.

Hasegawa, T., M.P. Lechevalier and H.A. Lechevalier. 1978a. A new genus of the *Actinomycetales: Actinosynnema* gen. nov. Int. J. Syst. Bacteriol. *28:* 304-310.

Hasegawa, T., M.P. Lechevalier and H.A. Lechevalier. 1979. Phospholipid composition of motile actinomycetes. J. Gen. Appl. Microbiol. *25:* 209-213.

Hasegawa, T., S. Tanida, K. Hatano, E. Higashide and M. Yoneda. 1983. Motile actinomycetes: *Actinosynnema pretiosum* subsp. *pretiosum* sp. nov., and *Actinosynnema pretiosum* subsp. *auranticum* subsp. nov. Int. J. Syst. Bacteriol. *33:* 314-320.

Hasegawa, T., S. Tanida and H. Ono. 1986. *Thermomonospora formosensis* sp. nov. Int. J. Syst. Bacteriol. *36:* 20-23.

Hasegawa, T., T. Yamano and M. Yoneda. 1978b. *Streptomyces inusitatus* sp. nov. Int. J. Syst. Bacteriol. *28:* 407-410.

Hata, T., N. Ohki and T. Higuchi. 1952. Studies on the antibiotic substance 'luteomycin'. On the strains and the cultural conditions. J. Antibiot. Tokyo Ser. A *5:* 529-534.

Hatano, K., E. Higashide and M. Shibata. 1976. Studies on juvenimicin, a new antibiotic. I. Taxonomy, fermentation and antimicrobial properties. J. Antibiot. *29:* 1163-1170.

Hayakawa, M. and H. Nonomura. 1984. HV agar, a new selective medium for isolation of soil actinomycetes. Abstracts of papers presented at the annual meeting of the Actinomycetologists held at Osaka, Japan. p. 6 (in Japanese).

Hayward, A.C. and L.I. Sly. 1976. Dextranase activity of *Oerskovia xanthineolytica*. J. Appl. Bacteriol. *40:* 355-364.

Hecht, S.T. and W.A. Causey. 1976. Rapid method for the detection and identification of mycolic acids in aerobic actinomycetes and related bacteria. J. Clin. Microbiol. *4:* 284-287.

Hefferan, M. 1904. A comparative and experimental study of bacilli producing red pigment. Zentralbl. Bakteriol. Parasitenkd. Infektionskr. Hyg. Abt. II *11:* 397-404, 456-475.

Heim, R. 1951. Mémoire sur l'Antennopsis ectoparasite du termite de Saintonge. IV. Étude du *Streptomyces termitum* n. sp., associé a l'antennopsis. Bull. Soc. Mycol. Fr. *67:* 359-364.

Heinemann, B., M.A. Kaplan, R.D. Muir and I.R. Hooper. 1953. Amphomycin, a new antibiotic. Antibiot. Chemother. *3:* 1239-1242.

Helmke, E. and H. Weyland. 1984. *Rhodococcus marinonascens* sp. nov., an actinomycete from the sea. Int. J. Syst. Bacteriol. *34:* 127-138.

Henrici, A.T. and D.E. Johnson. 1935. Studies of freshwater bacteria. II. Stalked bacteria, a new order of Schizomycetes. J. Bacteriol. *30:* 61-93.

Henssen, A. 1957. Beiträge zur Morphologie und Systematik der thermophilen Actinomyceten. Arch. Mikrobiol. *26:* 373-414.

Henssen, A., C. Happach-Kasan, B. Renner and G. Vobis. 1983. *Pseudonocardia compacta* sp. nov. Int. J. Syst. Bacteriol. *33:* 829-836.

Henssen, A., H.W. Kothe and R.M. Kroppenstedt. 1987. Transfer of *Pseudonocardia azurea* and "*Pseudonocardia fastidiosa*" to the genus *Amycolatopsis*, with emended species description. Int. J. Syst. Bacteriol. *37:* 292-295.

Henssen, A. and D. Schäfer. 1971. Emended description of the genus *Pseudonocardia* Henssen and description of a new species *Pseudonocardia spinosa* Schäfer. Int. J. Syst. Bacteriol. *21:* 29-34.

Henssen, A. and E. Schnepf. 1967. Zur Kenntnis thermophiler Actinomyceten. Arch. Mikrobiol. *57:* 214-231.

Henssen, A., E. Weise, G. Vobis and B. Renner. 1981. Ultrastructure of sporogenesis in actinomycetes forming spores in chains. *In* Schaal and Pulverer (Editors). Actinomycetes, Proceedings of the IVth International Symposium on Actinomycete Biology, Fischer, Stuttgart, pp. 137-146.

Hesseltine, C.W., R.G. Benedict and T.G. Pridham. 1954. Useful criteria for species differentiation in the genus *Streptomyces*. Ann. N.Y. Acad. Sci. *60:* 136-151.

Hewett, M.J., A.J. Wicken, K.W. Knox and M.E. Sharpe. 1976. Isolation of lipoteichoic acids from *Butyrivibrio fibrisolvens*. J. Gen. Microbiol. *94:* 126-130.

Hickey, R.J. and H.D. Tresner. 1952. A cobalt-containing medium for sporulation of *Streptomyces* species. J. Bacteriol. *64:* 891-892.

Higashide, E., T. Hasegawa, M. Shibata, K. Mizuno and H. Akaike. 1966. Studies on the Streptomycetes. *Streptomyces cuspidosporus* nov. sp. and the antibiotics sparsomycin and tubercidin produced thereby. Ann. Rep. Takeda Res. Lab. *25:* 1-14.

Higgens, C.E. and R.E. Kastner. 1967. Nebramycin, a new broad spectrum antibiotic complex. II. Description of *Streptomyces tenebrarius*. Antimicrob. Agents Chemother. *1967:* 324-331.

Higgens, C.E. and R.E. Kastner. 1971. *Streptomyces clavuligerus* sp. nov., a β-lactam antibiotic producer. Int. J. Syst. Bacteriol. *21:* 326-331.

Higgins, M.L. 1967. Release of sporangiospores by a strain of *Actinoplanes*. J. Bacteriol. *94:* 495-498.

Higgins, M.L. and M.P. Lechevalier. 1969. Poorly lytic bacteriophage from *Dactylosporangium thailandensis* (*Actinomycetales*). J. Virol. *3:* 210-216.

Higgins, M.L., M.P. Lechevalier and H.A. Lechevalier. 1967. Flagellated actinomycetes. J. Bacteriol. *93:* 1446-1451.

Hill, L.R. and L.G. Silvestri. 1962. Quantitative methods in the systematics of Actinomycetales. III. The taxonomic significance of physiological-biochemical characters and the construction of a dignostic key. G. Microbiol. *10:* 1-28.

Hill, L.R., V.B.D. Skerman and P.H.A. Sneath. 1984. Corrigenda to the approved list of bacterial names. Int. J. Syst. Bacteriol. *34:* 508-511.

Hill, L.R., M. Turri, E. Gilardi and L.G. Silvestri. 1961. Quantitative methods in the systematics of Actinomycetales II. G. Microbiol. *9:* 56-72.

Hill, R.A. and J. Lacey. 1983. Factors determining the microflora of stored barley. Ann. Appl. Biol. *102:* 467-483.

Hinuma, Y. 1954. Zaomycin, a new antibiotic from a *Streptomyces* sp. Studies on the antibiotic substances from *Actinomyces*. III. J. Antibiot. (Tokyo) Ser. A *7:* 134-136.

Hirsch, P. 1960. Einige, weitere, von Luftverunreinigungen lebende Actinomyceten und ihre Klassifizierung. Arch. Mikrobiol. *35:* 391-414.

Hirsch, P. 1961. Wasserstoffaktivierung und Chemoautotrophie bei Actinomyceten. Arch. Mikrobiol. *39:* 360-373.

Hirsch, P. 1972. Re-evaluation of *Pasteuria ramosa* Metchnikoff 1888, a bacterium pathogenic for *Daphnia* species. Int. J. Syst. Bacteriol. *22:* 112-116.

Hobbs, G. and T. Cross. 1983. Identification of endospore-forming bacteria. *In* Hurst and Gould (Editors), The Bacterial Spore, Vol. 2, Academic Press, London, pp. 49-78.

Hodgson, D.A. and K.F. Chater. 1981. A chromosomal locus controlling extracellular agarase production by *Streptomyces coelicolor*. A3(2), and its inactivation by chromosomal integration plasmid SCPI. J. Gen. Microbiol. *124:* 339-348.

Holdeman, L.V., E.P. Cato and W.E.C. Moore (Editors). 1977. Anaerobe Laboratory Manual, 4th Ed., Virginia Polytechnic Institute and State University, Blacksburg.

Holdeman, L.V., W.E.C. Moore, P.J. Churn and J.L. Johnson. 1982. *Bacteroides oris* and *Bacteroides buccae*, new species from human periodontitis and other human infections. Int. J. Syst. Bacteriol. *32:* 125-131.

Hollick, G.E. 1982. Enzymatic profiles of selected thermophilic actinomycetes. Microbios *35:* 187-196.

Hollick, G.E., N.K. Hall and H.W. Larsh. 1979. Chemical and serological comparison of two antigen extracts of *Thermoactinomyces candidus*. Mykosen *22:* 49-59.

Hollick, G.E. and H.W. Larsh. 1979. Crossed immunoelectrophoretic analysis of two antigen extracts from *Thermoactinomyces candidus*. Infect. Immun. *26:* 1057-1064.

Hollingdale, M.R. 1974. Antibody responses in patients with farmer's lung disease to antigens from *Micropolyspora faeni*. J. Hyg. *72:* 79-89.

Holtman, D.F. 1945. *Corynebacterium equi* in chronic pneumonia of the calf. J. Bacteriol. *49:* 159-162.

Hopwood, D.A. 1957. Genetic recombination in *Streptomyces coelicolor*. J. Gen. Microbiol. *16:* ii-iii.

Hopwood, D.A. and H.M. Ferguson. 1969. A rapid method for lyophilizing *Streptomyces* cultures. J. Appl. Bacteriol. *32:* 434-436.

Hopwood, D.A. and H.M. Ferguson. 1970. Genetic recombination in a thermophilic actinomycete, *Thermoactinomyces vulgaris*. J. Gen. Microbiol. *63:* 133-136.

Hopwood, D.A. and M.J. Merrick. 1977. Genetics of antibiotic production. Bacteriol. Rev. *41:* 595-635.

Hopwood, D.A. and H.M. Wright. 1972. Transformation in *Thermoactinomyces vulgaris*. J. Gen. Microbiol. *71:* 383-398.

Hopwood, D.A., H.M. Wright, M.J. Bibb and S.N. Cohen. 1977. Genetic recombination through protoplast fusion in *Streptomyces*. Nature (Lond.) *268:* 171-173.

Horan, A.C. 1971. Mycolic acids in the classification of nocardiae, mycobacteria and corynebacteria. Ph.D. Thesis, Rutgers University, New Brunswick, New Jersey, pp. 124-125.

Horan, A.C. and B.C. Brodsky. 1982. A novel antibiotic-producing *Actinomadura, Actinomadura kijaniata* sp. nov. Int. J. Syst. Bacteriol. *32:* 195-200.

Horan, A.C. and B. Brodsky. 1986. *Actinoplanes caeruleus* sp. nov., a blue-pigmented species of the genus *Actinoplanes*. Int. J. Syst. Bacteriol. *36:* 187-191.

Horan, A.C. and B.C. Brodsky. 1986. *Micromonospora rosaria* sp. nov., nom. rev., the rosaramicin producer. Int. J. Syst. Bacteriol. *36:* 478-480.

Horrière, F. 1984. *In vitro* physiological approach to classification of *Frankia* isolates of the *Alnus* group based on urease, protease and β-glucosidase activities. Plant Soil *78:* 7-13.

Horrière, F., M.P. Lechevalier and H.A. Lechevalier. 1983. *In vitro* morphogenesis and ultrastructure of a *Frankia* sp. ArI3 (Actinomycetales) from *Alnus rubra* and a morphologically similar isolate (AirI2) from *Alnus incana* subsp. *rugosa*. Can. J. Bot. *61:* 2843-2954.

Houwers, A. and A.D.L. Akkermans. 1981. Influence of inoculation on yield of *Alnus glutinosa* in the Netherlands. Plant Soil *61:* 189-202.

Hoyer, B.H., B.J. McCarthy and E.T. Bolton. 1964. A molecular approach in the

systematics of higher organisms. Science (Washington) *144:* 959–967.
Hoysoya, S., N. Komatsu, M. Soeda and Y. Sonoda. 1952. Trichomycin, a new antibiotic produced by *Streptomyces hachijoensis* with trichomonadicidal and antifungal activity. Jpn. J. Exp. Med. *22:* 505–509.
Hsu, S.C. and J.L. Lockwood. 1975. Powdered chitin as a selective medium for enumeration of actinomycetes in water and soil. Appl. Microbiol. *29:* 422–426.
Huang, J., S. Shen, H. Liu, J. Jiang, H. Yang, Z. Zhao and B. Yang. 1984. An investigation of the resource of non-leguminous nodulating trees in China. *In* Veeger and Newton (Editors), Advances in Nitrogen Fixation Research, Martinus Nijhoff/Dr. W. Junk Pub., The Hague, p. 373.
Huang, J., Z. Zhao, G. Chen and H. Liu. 1985. Host range of *Frankia* endophytes. Plant Soil *87:* 61–65.
Huang, L.H. 1980. *Actinomadura macra* sp. nov., the producer of antibiotics CP-47,433 and CP-47,434. Int. J. Syst. Bacteriol. *30:* 565–568.
Huber, G., K.H. Wallhaüsser, L. Fries, A. Steigler and H.-L. Weidenmüller. 1962. Niddamycin ein neues Makrolid-Antibiotikum. Arzneim. Forsch. *12:* 1191–1195.
Hungate, R.E. 1946. Studies on cellulose fermentation. II. An anaerobic cellulose-decomposing actinomycete, *Micromonospora propionici* n. sp. J. Bacteriol. *51:* 51–56.
Hungate, R.E. 1966. The Rumen and Its Microbes, Academic Press, New York.
Hunter, J.C., D.E. Eveleigh and G. Casella. 1981. Actinomycetes of a salt marsh. Zentralbl. Bakteriol. Mikrobiol. Hyg. 1 Abt. Suppl. *11:* 195–200.
Huntjens, J.L.M. 1972. Amino acid composition of humic acid-like polymers produced by streptomycetes and of humic acid from pasture and arable land. Soil Biol. Biochem. *4:* 345–379.
Huss, V.A.R., H. Festl and K.H. Schleifer. 1984. Nucleic acid hybridization studies and deoxyribonucleic acid base composition of anaerobic Gram-positive cocci. Int. J. Syst. Bacteriol. *34:* 95–101.
Huss-Danell, K., W. Roelofsen, A.D.L. Akkermans and P. Meijer. 1982. Carbon metabolism of *Frankia* sp. in root nodules of *Alnus glutinosa* and *Hippophaë rhamnoides*. Physiol. Plant. *54:* 461–466.
Hussein, A.M., A.M. Allam and A.M. Ragab. 1975. Taxonomical studies on thermophilic actinomycetes of some soils in Egypt. Ann. Microbiol. *25:* 19–28.
Hussey, R.S. 1971. A technique for obtaining quantities of living *Meloidogyne* females. J. Nematol. *3:* 99–100.
Hütter, R. 1961. Zur Systematik der Actinomyceten. 5. Die Art *Streptomyces albus* (Rossi-Doria emend. Krainsky) Waksman et Henrici 1943. Arch. Mikrobiol. *38:* 367–383.
Hütter, R. 1962. Zur Systematik der Actinomyceten 8. Quirlbildende Streptomyceten. Arch. Microbiol. *43:* 365–391.
Hütter, R. 1963. Zur Systematik der Actinomyceten. 10. Streptomyceten mit *griseus*-Luftmycel. G. Microbiol. *11:* 191–246.
Hütter, R. 1964. Zur Systematik der Actinomyceten. 9. Streptomyceten mit *cinnamoneus* Luftmycel. Zentralbl. Bakteriol. Parasitenkd. Infektionskr. Hyg. Abt. II *117:* 603–661.
Hütter, R. 1967. Systematik der Streptomyceten unter besonderer Berücksiehtung der von ihnen gebildetern Antibiotica. Bibl. Microbiol. Fasc. Vol. 6, S. Karger, Basel pp. 1–382.
Hütter, R., T. Kieser, R. Crammeri and G. Hintermann. 1981. Chromosomal instability in *Streptomyces glaucescens*. Zentralbl. Bakteriol. Mikrobiol. Hyg. 1 Abt. Suppl. *11:* 551–559.
Hyman, I.S. and S.D. Chaparas. 1977. A comparative study of the 'rhodochrous' complex and related taxa by delayed type skin reactions on guinea pigs and by polyacrylamide gel electrophoresis. J. Gen. Microbiol. *100:* 363–371.
Imamura, A., M. Hori, K. Nakazawa, M. Shibata, S. Tatsuoka and A. Miyake. 1956. A new species of *Streptomyces* producing dihydrostreptomycin. Proc. Jpn. Acad. Sci. *32:* 648–653.
Imbriani, J.L. and R. Mankau. 1977. Ultrastructure of the nematode pathogen, *Bacillus penetrans*. J. Invertebr. Pathol. *30:* 337–347.
International Code of Nomenclature of Bacteria. 1975. American Society for Microbiology, Washington, D.C.
International Committee on Systematic Bacteriology Subcommittee on the Taxonomy of *Mollicutes*. 1979. Proposal of minimal standards for descriptions of new species of the class *Mollicutes*. Int. J. Syst. Bacteriol. *29:* 172–180.
Ishida, N., K. Kumagai, T. Niida, K. Hamamoto and T. Shomura. 1967. Nojirimycin, a new antibiotic. I. Taxonomy and fermentation. J. Antibiot. (Tokyo) Ser. A *20:* 62–65.
Ishiguro, E.E. and D.W. Fletcher. 1975. Characterization of *Geodermatophilus* strains isolated from high altitude Mount Everest soils. Mikrobiologija (Belgr.) *12:* 99–108.
Ishizawa, S., M. Araragi and T. Suzuki. 1969. Actinomycete flora of Japanese soils III. Actinomycete flora of paddy soils. Soil Sci. Plant Nutr. *15:* 104–112.
Isono, K., S. Yamashita, Y. Tomiyama, S. Suzuki and H. Sakai. 1957. Studies on homomycin II. J. Antibiot. (Tokyo) Ser. A *10:* 21–30.
Ivanitskaya, L.P., S.M. Singal, M.V. Bibikova and S.N. Yostrov. 1978. Direct isolation of *Micromonospora* on selective media with gentamicin. Antibiotiki *28:* 690–692.
Iwami, M., S. Kiyoto, M. Nishikawa, H. Terano, M. Kohsaka, H. Aoki and H. Imanaka. 1985. New antitumor antibiotics, FR-900405 and FR-900406. I. Taxonomy of the producing strain. J. Antibiot. *38:* 835–839.
Iwasaki, A., H. Itoh and T. Mori. 1979. A new broad spectrum aminoglycoside antibiotic complex, sporarcin. II. Taxonomic studies on the sporarcin producing strain *Saccharopolyspora hirsuta* subsp. *kobensis* nov. subsp. J. Antibiot. *32:* 180–186.
Iwasaki, A., H. Itoh and T. Mori. 1981. *Streptomyces sannanensis* sp. nov. Int. J. Syst. Bacteriol. *31:* 280–284.
Jackman, P.J.H. 1985. Bacterial taxonomy based on electrophoretic whole-cell protein patterns. *In* Goodfellow and Minnikin (Editors), Chemical Methods in Bacterial Systematics, Academic Press, London, pp. 115–129.
Jager, K., K. Marialigeti, M. Hauck and G. Barabas. 1983. *Promicromonospora enterophila* sp. nov., a new species of monospore actinomycetes. Int. J. Syst. Bacteriol. *33:* 525–531.
Jasmin, A.M., C.P. Powell and J.N. Baucom. 1972. Actinomycotic mycetoma in a Bottlenose dolphin (*Tursiops truncatus*) due to *Nocardia paraguayensis*. Vet. Med. Small Anim. Clin. *67:* 542–543.
Jeffrey, C. 1977. Biological Nomenclature, 2nd Ed., Arnold, London.
Jeffries, L., M.A. Cawthorne, M. Harris, B. Cook and A.T. Diplock. 1969. Menaquinone determination in the taxonomy of *Micrococcaceae*. J. Gen. Microbiol. *54:* 365–380.
Jensen, H.L. 1930a. Actinomycetes in Danish soils. Soil Sci. *30:* 59–77.
Jensen, H.L. 1930b. The genus *Micromonospora* Ørskov, a little known group of soil microorganisms. Proc. Linnean Soc. N.S.W. *55:* 231–248.
Jensen, H.L. 1931. Contributions to our knowledge of the Actinomycetales. II. The definition and subdivision of the genus *Actinomyces*, with a preliminary account of Australian soil Actinomycetes. Proc. Linnean Soc. N.S.W. *56:* 345–370.
Jensen, H.L. 1932. Contributions to our knowledge of the actinomycetes. III. Further observations on the genus *Micromonospora*. Proc. Linnean Soc. N.S.W. *57:* 173–180.
Jensen, H.L. 1934. Studies on saprophytic mycobacteria and corynebacteria. Proc. Linnean Soc. N.S.W. *59:* 19–61.
Jiang, C. and J. Ruan. 1982. Two new species and a new variety of *Ampullariella*. Acta Microbiol. Sin. *22:* 207–211.
Jirovec, O. 1939. *Dermocystidium vejdovskyi* n. sp., ein neuer Parasit des Hechtes, nebst einer Bemerkung über *Dermocystidium daphniae* (Rühberg). Arch. Protistenk. *92:* 137–146.
Johnson, J.L. 1973. Use of nucleic-acid homologies in the taxonomy of anaerobic bacteria. Int. J. Syst. Bacteriol. *23:* 308–315.
Johnson, J.L. 1978. Taxonomy of the *Bacteroides*. I. Deoxyribonucleic acid homologies among *Bacteroides fragilis* and other saccharolytic *Bacteroides* species. Int. J. Syst. Bacteriol. *28:* 245–256.
Johnson, J.L. 1980. Classification of anaerobic bacteria. *In* Proceedings of International Symposium on Anaerobes (Tokyo, Japan, June 22, 1980), Nippon Merck-Banyu Co., Ltd., Tokyo, p. 19.
Johnson, J.L. 1981. Genetic characterization. *In* Gerhardt et al. (Editors), Manual of Methods for General Bacteriology, American Society for Microbiology, Washington, D.C., pp. 450–472.
Johnson, J.L. and D.A. Ault. 1978. Taxonomy of the *Bacteroides*. II. Correlation of phenotypic characteristics with deoxyribonucleic acid homology groupings for *Bacteroides fragilis* and other saccharolytic *Bacteroides* species. Int. J. Syst. Bacteriol. *28:* 257–265.
Johnson, J.L. and C.S. Cummins. 1972. Cell wall composition and deoxyribonucleic acid similarities among the anaerobic coryneforms, classical propionibacteria and strains of *Arachnia propionica*. J. Bacteriol. *109:* 1047–1066.
Johnson, K.G. and P.H. Lanthier. 1986. β-Lactamases from *Actinopolyspora halophila*, an extremely halophilic actinomycete. Arch. Microbiol. *143:* 379–386.
Johnson, K.G., P.H. Lanthier and M.B. Gochnauer. 1986a. Cell walls from *Actinopolyspora halophila*, an extremely halophilic actinomycete. Arch. Microbiol. *143:* 365–369.
Johnson, K.G., P.H. Lanthier and M.B. Gochnauer. 1986b. Studies of two strains of *Actinopolyspora halophila*, an extremely halophilic actinomycete. Arch. Microbiol. *143:* 370–378.
Johnson, L.E. and A. Dietz. 1969. Lomofungin, a new antibiotic produced by *Streptomyces lomondensis* sp. n. Appl. Microbiol. *17:* 755–759.
Johnston, D.W. and T. Cross. 1976. The occurrence and distribution of actinomycetes in lakes of the English Lake District. Freshwater Biol. *6:* 457–463.
Johnstone, D.B. and S.A. Waksman. 1947. Streptomycin II, an antibiotic substance produced by a new species of *Streptomyces*. Proc. Soc. Exp. Biol. Med. *65:* 294–295.
Jones, C.W. 1980. Cytochrome patterns in classification and identification including their relevance to the oxidase test. *In* Goodfellow and Board (Editors), Microbiological Classification and Identification, Academic Press, London, New York, pp. 127–138.
Jones, D. 1975. A numerical taxonomic study of coryneform and related bacteria. J. Gen. Microbiol. *87:* 52–96.
Jones, D. 1978. Composition and differentiation of the genus *Streptococcus*. *In* Skinner and Quesnel (Editors), Streptococci, Academic Press, London, New York, pp. 1–49.
Jones, D. and M.D. Collins. 1986. Irregular, nonsporing Gram-positive rods. *In* Sneath, Mair, Sharpe and Holt (Editors), Bergey's Manual of Systematic Bacteriology, Vol. 2, The Williams and Wilkins Co., Baltimore, pp. 1383–1418.
Jones, D. and P.H.A. Sneath. 1970. Genetic transfer and bacterial taxonomy. Bacteriol. Rev. *34:* 40–81.
Jones, K.L. 1949. Fresh isolates of actinomycetes in which the presence of sporogenous aerial mycelia is a fluctuating characteristic. J. Bacteriol. *57:* 141–145.
Jones, K.L. 1952. A new *Streptomyces* that produces vitamin B_{12} actively. Papers

from the Michigan Academy of Science, Arts and Letters *37:* 47-48.

Jones, L.A. and S.G. Bradley. 1964. Phenetic classification of actinomycetes. Dev. Ind. Microbiol. *5:* 267-272.

Jones, R.T. 1976. Subcutaneous infection with *Dermatophilus congolensis* in a cat. J. Comp. Pathol. *86:* 415-421.

Juan, C. and Y. Zhang. 1974. A taxonomic study of *Actinoplanaceae*. I. Classification of *Ampullariella*. Acta Microbiol. Sin. *14:* 31-41.

Judicial Commission. 1985. Opinion 58. Confirmation of the types in the Approved Lists as nomenclatural types including recognition of *Nocardia asteroides* (Eppinger 1891) Blanchard 1896 and *Pasteuria multocida* (Lehmann and Neumann 1899) Rosenbusch and Merchant 1939 as the respective type species of the genera *Nocardia* and *Pasteurella* and rejection of the species name *Pasteurella gallicida* (Burrill 1883) Buchanan 1925. Int. J. Syst. Bacteriol. *35:* 538.

Judicial Commission. 1986. Opinion 61. Rejection of the type strain of *Pasteuria ramosa* (ATCC 27377) and conservation of the species *Pasteuria ramosa* Metchnikoff 1888 on the basis of the type descriptive material. Int. J. Syst. Bacteriol. *36:* 119.

Kalakoutskii, L.V. 1964. A new species of *Micropolyspora—Micropolyspora caesia* n. sp. Mikrobiologiya *33:* 858-862.

Kalakoutskii, L.V. and N.S. Agre. 1973. Endospores of actinomyces: dormancy and germination. *In* Sykes and Skinner (Editors), Actinomycetales: Characteristics and Practical Importance, Soc. Appl. Bacteriol. Symp. Ser. No. 2, Society for Applied Bacteriology, London, pp. 179-195.

Kalakoutskii, L.V. and N.S. Agre. 1976. Comparative aspects of development and differentiation in actinomycetes. Bacteriol. Rev. *40:* 469-524.

Kalakoutskii, L.V., N.S. Agre and N.A. Krasil'nikov. 1968. Comparative study on some oligosporic actinomycetes. Hind. Antibiot. Bull. *10:* 254-268.

Kalakoutskii, L.V., I.P. Kirillova and N.A. Krasil'nikov. 1967. A new genus of the Actinomycetales—*Intrasporangium* gen. nov. J. Gen. Microbiol. *48:* 79-85.

Kalakoutskii, L.V. and V.D. Kusnetsov. 1964. A new species of the *Actinoplanes—A. armeniacus* and some peculiarities of its mode of spore formation. Mikrobiologiya *33:* 613.

Kalakoutskii, L.V., N.I. Nikitina and O.I. Artamonova. 1969. Spore germination in Actinomycetes (in Russian). Mikrobiologiya *38:* 834-841.

Kandler, O. 1981. Archaebakterien und Phylogenie der Organismen. Naturwissenschaften *68:* 183-192.

Kandler, O. 1982. Cell wall structures and their phylogenetic implications. Zentralbl. Bakteriol. Parasitenkd. Infektionskr. Hyg. Abt. Orig. C *3:* 149-160.

Kandler, O. and K.-H. Schleifer. 1980. Taxonomy I: Systematics of bacteria. *In* Ellenberg, Esser, Kubitzki, Schnepf and Ziegler (Editors), Progress in Botany, Fortschritte der Botanik, Vol. 42, Springer-Verlag, Berlin.

Kane, W.D. 1966. A new genus of *Actinoplanaceae, Pilimelia,* with a description of two species, *Pilimelia terevasa* and *Pilimelia anulata*. J. Elisha Mitchell Sci. Soc. *82:* 220-230.

Kane Hanton, W.D. 1974. Genus *Pilimelia* Kane. *In* Buchanan and Gibbons (Editors), Bergey's Manual of Determinative Bacteriology, 8th Ed., The Williams and Wilkins Co., Baltimore, pp. 718-719.

Kaneko, T., K. Kitamura and Y. Yamamoto. 1969. *Arthrobacter luteus* nov. sp. isolated from brewery sewage. J. Gen. Appl. Microbiol. *15:* 317-326.

Käppler, W. 1965. Acetyl-Naphthylamine-Esterasen-Aktivität von Mykobakterien. Beitr. Klin. Tuberk. *130:* 1-4.

Karling, J.S. 1954. An unusual keratinophilic microorganism. Proc. Indiana Acad. Sci. *63:* 83-86.

Kasweck, K.L. and M.L. Little. 1982. Genetic recombination in *Nocardia asteroides*. J. Bacteriol. *149:* 403-406.

Kasweck, K.L., M.L. Little and S.G. Bradley. 1982. Plasmids in mating strains of *Nocardia asteroides*. Dev. Ind. Microbiol. *23:* 279-286.

Kates, M. 1978. The phytanyl ether-linked polar lipids and isoprenoid neutral lipids of extremely halophilic bacteria. Prog. Chem. Fats Other Lipids *15:* 301-342.

Kates, M., S. Porter and D.J. Kushner. 1987. *Actinopolyspora halophila* does not contain mycolic acids. Can. J. Microbiol. *33:* 822-823.

Katô, H. and T. Arai. 1957. On the production of antibiotic substances from the *Streptomyces luteoreticuli*. Annu. Rep. Inst. Food Microbiol. Chiba Univ. *10:* 52-57.

Kauffmann, F. 1966. The bacteriology of the *Enterobacteriaceae*. Munkssgaard, Copenhagen.

Kawaguchi, H., H. Tsukiura, M. Okanishi, T. Miyaki, T. Ohmori, K. Fujisawa and H. Koshiyama. 1965. Studies on coumermycin, a new antibiotic. I. Production, isolation and characterization of coumermycin A2. J. Antibiot. (Tokyo) Ser. A *18:* 1-10.

Kawamoto, I., T. Oka and T. Nara. 1981. Cell wall composition of *Micromonospora olivoasterospora, Micromonospora sagamiensis,* and related organisms. J. Bacteriol. *146:* 527-534.

Kawamoto, I., T. Oka and T. Nara. 1982. Spore resistance of *Micromonospora olivoasterospora, Micromonospora sagamiensis,* and related organisms. Agric. Biol. Chem. *46:* 221-231.

Kawamoto, I., T. Oka and T. Nara. 1983a. Carbon and nitrogen utilization by *Micromonospora* strains. Agric. Biol. Chem. *47:* 203-215.

Kawamoto, I., R. Okachi, H. Kato, S. Yamamoto, I. Takahashi, S. Takasawa and T. Nara. 1974. The antibiotic XK-41 complex. I. Production, isolation and characterization. J. Antibiot. *27:* 493-501.

Kawamoto, I., S. Takasawa, R. Okachi, M. Kohakura, I. Takahashi and T. Nara. 1975. A new antibiotic victomycin (XK 49-1-B-2). I. Taxonomy and production of the producing organism. J. Antibiotics (Tokyo) *28:* 358-365.

Kawamoto, I., M. Yamamoto and T. Nara. 1983b. *Micromonospora olivasterospora* sp. nov. Int. J. Syst. Bacteriol. *33:* 107-112.

Kawamura, T., Y. Yasuda and M. Mayama. 1981. Isolation of L-2-(1-methylcyclopropyl)glycine from *Micromonospora miyakoensis* sp. nov. I. Taxonomic studies of the producing organism. J. Antibiot. *34:* 367-369.

Kawato, N. and R. Shinobu. 1959. On *Streptomyces herbaricolor*, nov. sp. Supplement: A simple technique for the microscopic observation. Mem. Osaka Univ. Lib. Arts Educ., B. Nat. Sci. *8:* 114-119.

Keddie, R.M. and I.J. Bousfield. 1980. Cell wall composition in the classification and identification of coryneform bacteria. *In* Goodfellow and Board (Editors), Microbiological Classification and Identification, Academic Press, London, pp. 167-188.

Keddie, R.M., M.D. Collins and D. Jones. 1986. Genus *Arthrobacter* Conn and Dimmick 1947. *In* Sneath, Mair, Sharpe and Holt (Editors), Bergey's Manual of Systematic Bacteriology, Vol. 2, The Williams and Wilkins Co., Baltimore, pp. 1288-1301.

Kelly, J., A.H. Kutscher and F. Tuoti. 1959. Thiostrepton, a new antibiotic: Tube dilution sensitivity studies. Oral Surg. Oral Med. Oral Pathol. *12:* 1334-1339.

Kelly, K.L. and D.B. Judd. 1976. Color: Universal Language and Dictionary of Names, NBS special publication 440, U.S. Department of Commerce, National Bureau of Standards, Washington, D.C.

Kersters, K. 1985. Numerical methods in the classification of bacteria by protein electrophoresis. *In* Goodfellow, Jones and Priest (Editors), Computer Assisted Bacterial Systematics, Special Publications of the Society for General Microbiology 15, Academic Press, London, pp. 337-368.

Kersters, K. and J. De Ley. 1968. The occurrence of the Entner-Doudoroff pathway in bacteria. Antonie van Leeuwenhoek J. Microbiol. Serol. *34:* 393-408.

Kersters, K. and J. De Ley. 1975. Identification and grouping of bacteria by numerical analysis of their electrophoretic protein patterns. J. Gen. Microbiol. *87:* 333-342.

Kersters, K. and J. De Ley. 1980. Classification and identification of bacteria by electrophoresis of their proteins. *In* Goodfellow and Board (Editors). Microbial Classification and Identification, Society for Applied Bacteriology Symposium Series 8, Academic Press, London, pp. 273-297.

Khan, M.R. and S.T. Williams. 1975. Studies on the ecology of actinomycetes in soil. VIII. Distribution and characteristics of acidophilic actinomycetes. Soil Biol. Biochem. *7:* 345-348.

Kieser, T., G. Hintermann, R. Crameri and R. Hütter. 1981. Restriction analysis of *Streptomyces*-DNA. Zentralbl. Bakteriol. I. Abt. Suppl. *11:* 561-562.

Kikuchi, M. and D. Perlman. 1977. Bacteriophages infecting *Micromonospora purpurea*. J. Antibiot. *30:* 423-424.

Kikuchi, M. and D. Perlman. 1978. Characteristics of bacteriophages for *Micromonospora purpurea*. Appl. Environ. Microbiol. *36:* 52-55.

Kilpper-Bälz, R. and K.H. Schleifer. 1984. Nucleic acid hybridization and cell wall composition studies of pyogenic streptococci. FEMS Microbiol. Lett. *24:* 355-364.

Kilpper-Bälz, R., B.L. Williams, R. Lütticken and K.H. Schleifer. 1984. Relatedness of '*Streptococcus milleri*' with *Streptococcus anginosus* and *Streptococcus constellatus*. Syst. Appl. Microbiol. *5:* 494-500.

Kim, K.S., D.D.Y. Ryu and S.Y. Lee. 1983. Application of protoplast fusion technique to genetic recombination of *Micromonospora rosaria*. Enzyme Microb. Technol. *5:* 273-280.

Kirillova, I.P., N.S. Agre and L.V. Kalakoutskii. 1973. Conditions for initiation of *Thermoactinomyces vulgaris* spores (in Russian). Mikrobiologiya *42:* 867-872.

Kirillova, I.P., N.S. Agre and L.V. Kalakoutskii. 1974. Spore initiation and minimum temperature for growth of *Thermoactinomyces vulgaris*. Z. Allg. Mikrobiol. *14:* 69-72.

Kirsop, B. 1985. The current status of culture collections and their contribution to biotechnology. CRC Crit. Rev. Biotechnol. *2:* 287-314.

Kirsop, B.E. and J.J.S. Snell. 1984. Maintenance of Microorganisms. A Manual of Laboratory Methods. Academic Press, London.

Klein, R.A., G.P. Hazelwood, P. Kemp and R.M.C. Dawson. 1979. A new series of long chain, dicarboxylic acids with vicinal dimethyl branching found as major components of the lipids of *Butyrivibrio* spp. Biochem. J. *183:* 691-700.

Kluyver, A.J. and C.B. van Niel. 1936. Prospects for a natural system of classification of bacteria. Zentralbl. Bakteriol. Parasitenkd. Infektionskr. Hyg. Abt. II *94:* 369-403.

Knowlton, S., A. Berry and J.G. Torrey. 1980. Evidence that associated soil bacteria may influence root hair infection of actinorhizal plants by *Frankia*. Can. J. Microbiol. *26:* 971-972.

Knowlton, S. and J.O. Dawson. 1983. Effects of *Pseudomonas cepacia* and cultural factors on the nodulation of *Alnus rubra* roots by *Frankia*. Can. J. Bot. *61:* 2877-2882.

Ko, C.Y., J.L. Johnson, L.B. Barnett, H.M. McNair and J.R. Vercellotti. 1977. A sensitive estimation of the percentage of guanine plus cytosine in deoxyribonucleic acid by high performance liquid chromatography. Anal. Biochem. *80:* 183-192.

Koch, A.L. 1981. Evolution of antibiotic resistance gene function. Microbiol. Rev. *45:* 335-378.

Kokina, V.Y. and N.S. Agre. 1977a. Factors causing degeneration of *Thermoactinomyces vulgaris* cultures (in Russian). Mikrobiologiya *46:* 304-310.

Kokina, V.Y. and N.S. Agre. 1977b. Investigation of spores of *Thermoactinomyces vulgaris* degenerated cultures (in Russian). Mikrobiologiya *46:* 378-380.

Komura, I., K. Yamada, S. Otsuka and K. Komagata. 1975. Taxonomic significance of phospholipids in coryneform and nocardioform bacteria. J. Gen. Appl. Microbiol. *21:* 251-261.

Konev, I.E. and V.A. Tsyganov. 1962. A new species in the group of yellow actinomycetes. Microbiologiya *31:* 1023-1028.

Konev, I.E., V.A. Tsyganov, P. Minbaev and V.M. Morogov. 1967. A new genus of actinomycetes—*Microechinospora* gen. nov. Mikrobiologiya *36:* 308-317.

Konev, Y.E. 1981. The Actinomycetes of the Genus *Streptoverticillium* (Baldacci) Baldacci et al. (in Russian). Acad. Sci. SSSR, Pushchino.

Korn, F., B. Weingärtner and H.J. Kutzner. 1978. A study of twenty actinophages: Morphology, serological relationships and host range. *In* Freerksen, Tarnok and Thumin (Editors), Genetics of the Actinomycetales, Gustav Fischer Verlag, Stuttgart, pp. 251-270.

Koshiyama, H., M. Okanishi, T. Ohmori, T. Miyaki, H. Tsakiura, M. Matsuzaki and H. Kawaguchi. 1963. Cirramycin, a new antibiotic. J. Antibiot. (Tokyo) Ser. A *16:* 59-66.

Kosmachev, A.E. 1962. A thermophilic *Micromonospora* and its production of antibiotic T-12 under conditions of surface and submerged fermentation at 50°C-60°C (in Russian). Mikrobiologiya *31:* 66-71.

Kosmachev, A.E. 1963. Thermophilic actinomycetes in milk and dairy products (in Russian). Mikrobiologiya *32:* 136-142.

Kosmachev, A.E. 1964. A new thermophilic actinomycete *Micropolyspora thermovirida* n. sp. Mikrobiologiya *33:* 267-269.

Krainsky, A. 1914. Die Aktinomyceten und ihren Bedeutung in der Natur. Zentralbl. Bakteriol. Parasitenkd. Infektionskr. Hyg. Abt. II *41:* 649-688.

Krasil'nikov, N.A. 1938. Ray Fungi and Related Organisms—*Actinomycetales*, Akad. Nauk. SSSR Moscow.

Krasil'nikov, N.A. 1941. Guide to the Bacteria and Actinomycetes (in Russian), Akad. Nauk. SSSR Moscow.

Krasil'nikov, N.A. 1949. Guide to the Bacteria and Actinomycetes, Akad. Nauk. SSSR Moscow.

Krasil'nikov, N.A. 1958. The significance of antibiotics as specific characteristics of actinomycetes and their determination by the method of experimental transformation. Folia Biol. (Prague) *4:* 257-265.

Krasil'nikov, N.A. 1960. Taxonomic principles in the actinomycetes. J. Bacteriol. *79:* 65-71.

Krasil'nikov, N.A. 1962. A new genus of ray fungus—*Actinopycnidum* n. gen. of family *Actinomycetaceae*. Microbiologiya *31:* 250-253.

Krasil'nikov, N.A. 1964. Systematic position of ray fungi among the lower organisms. Hind. Antibiot. Bull. *7:* 1-17.

Krasil'nikov, N.A. 1970a. Pigmentation of actinomycetes and its significance in taxonomy. *In* Prauser (Editor), The Actinomycetales, Gustav Fischer Verlag, Jena, pp. 123-131.

Krasil'nikov, N.A. 1970b. Ray Fungi. Higher Forms. Nauka, Moscow.

Krasil'nikov, N.A. and N.S. Agre. 1964a. A new actinomycete genus—*Actinobifida* n. gen. yellow group—*Actinobifida dichotomica* n. sp. (in Russian). Mikrobiologiya *33:* 935-943.

Krasil'nikov, N.A. and N.S. Agre. 1964b. On two new species of *Thermopolyspora*. Hind. Antibiot. Bull. *6:* 97-107.

Krasil'nikov, N.A. and N.S. Agre. 1965. The brown group of *Actinobifida chromogena* n. sp. (in Russian). Mikrobiologiya *34:* 284-291.

Krasil'nikov, N.A., N.S. Agre and G.I. El-Registan. 1968. New thermophilic species of the genus *Micropolyspora*. Mikrobiologiya *37:* 1065-1072.

Krasil'nikov, N.A., L.V. Kalakoutskii and N.F. Kirillova. 1961a. A new genus of ray fungi—*Promicromonospora* gen. nov. (in Russian). Izv. Akad. Nauk SSSR (Ser. Biol.) *1:* 107-112.

Krasil'nikov, N.A., A.I. Korenyako, M.M. Meksina, L.V. Valedinskaya and N.M. Veselov. 1957. On the culture of actinomycete no. 111 *Actinomyces luridus* nov. sp. producing an antiviral antibiotic 'luridin'. Mikrobiologiya *26:* 558-564.

Krasil'nikov, N.A., A.I. Korenyako and N.I. Nikitina. 1965a. Actinomycetes of the yellow group. *In* Krasil'nikov (Editor), Biology of Selected Groups of Actinomycetes (in Russian), Izdatel, 'Nauka', Moscow, pp. 205-229.

Krasil'nikov, N.A., N.I. Nikitina and A.T. Korenyako. 1961b. On external features in the taxonomy of actinomycetes. Int. Bull. Bacteriol. Nomencl. Taxon. *11:* 133-159.

Krasil'nikov, N.A., E.I. Sorokina, V.A. Alferova and A.P. Bezzubenkova. 1965b. Classification of blue actinomycetes. *In* Krasil'nikov (Editor), Biology of Selected Groups of Actinomycetes (in Russian), Izdatel, 'Nauka', Moscow, pp. 74-123.

Krasil'nikov, N.A. and C.-S. Yuan. 1960. A new species in the *Actinomyces aurantiacus* group. Mikrobiologiya *29:* 482-489.

Krasil'nikov, N.A. and C.S. Yuan. 1961. *Actinosporangium*—a new genus in the family *Actinoplanaceae* (in Russian). Izv. Akad. Nauk. SSSR Ser. Biol. *8:* 113-116.

Krasil'nikov, N.A. and T. Yuan. 1965. The specific composition of the orange actinomycetes. *In* Krasil'nikov (Editor), Biology of Selected Groups of Actinomycetes, (in Russian). Izdatel, 'Nauka', Moscow, pp. 28-57.

Kretschmer, S. 1980. Transfection in *Thermoactinomyces vulgaris*. Z. Allg. Mikrobiol. *20:* 73-75.

Kretschmer, S. 1982. Alteration of interaction with virulent bacteriophage Ta 1 during differentiation of *Thermoactinomyces vulgaris*. Z. Allg. Mikrobiol. *22:* 629-637.

Kretschmer, S. 1984a. Alternative life cycles in *Thermoactinomyces vulgaris*. Z. Allg. Mikrobiol. *24:* 93-100.

Kretschmer, S. 1984b. Characterization of aerial mycelium of *Thermoactinomyces vulgaris*. Z. Allg. Mikrobiol. *24:* 101-111.

Kretschmer, S. 1984c. Intercalary growth of *Thermoactinomyces vulgaris*. Z. Allg. Mikrobiol. *24:* 211-215.

Kretschmer, S. and H.-E. Jacob. 1983. Autolysis of *Thermoactinomyces vulgaris* spores lacking carbon dioxide during germination. Z. Allg. Mikrobiol. *23:* 27-32.

Kretschmer, S., D. Körner, G. Strohbach, P. Klingenberg, H.-E. Jacob, J. Gumpet and H. Tuttloff. 1982. Physiological and cell biological characteristics of protease-forming *Thermoactinomyces vulgaris* during prolonged culture in a fermenter (in German). Z. Allg. Mikrobiol. *22:* 693-703.

Kretschmer, S. and E. Sarfert. 1980. Transfection in *Thermoactinomyces vulgaris*. Z. Allg. Mikrobiol. *20:* 73-75.

Krichevsky, M.I. and L.M. Norton. 1974. Storage and manipulation of data by computers for determinative bacteriology. Int. J. Syst. Bacteriol. *24:* 525-531.

Kriss, A.E. 1939. *Micromonospora*—an actinomycete-like organism (*Micromonospora globosa* n. sp.). Mikrobiologiya *8:* 178-185.

Kroppenstedt, R.M. 1979. Chromatographische Identifizierung von Mikroorganismen, dargestellt am Beispiel der Actinomyceten. Kontakte (Merk) *2:* 12-21.

Kroppenstedt, R.M. 1985. Fatty acid and menaquinone analysis of actinomycetes and related organisms. *In* Goodfellow and Minnikin (Editors), Chemical Methods in Bacterial Systematics, Society for Applied Bacteriology, Technical Series No. 20, Academic Press, London, pp. 173-199.

Kroppenstedt, R.M., F. Korn-Wendisch, V.J. Fowler and E. Stackebrandt. 1981. Biochemical and molecular evidence for a transfer of *Actinoplanes armeniacus* into the family *Streptomycetaceae*. Zentralbl. Bakteriol. Mikrobiol. Hyg. 1 Abt. Orig. *C2:* 254-262.

Kroppenstedt, R.M. and H.J. Kutzner. 1976. Biochemical markers in the taxonomy of the *Actinomycetales*. Experientia *32:* 318-319.

Kroppenstedt, R.M. and H.J. Kutzner. 1978. Biochemical taxonomy of some problem actinomycetes. Zentralbl. Bakteriol. Parasitenkd. Infektionskr. Hyg. Abt. 1 Suppl. *6:* 125-133.

Krüger, F. 1904. Untersuchungen über der Gürtelschorf der Zuckerruben. Arbeiten aus der Biologischen Abteilung fur Land- under Forstwirtschaft am Kaiserlichen Gesundheitsamte Band IV, Heft 3, Verlagsbuchhandlung Paul Parey, Verlagsbuchhandlung Julius Springer, Berlin, pp. 275-318.

Kruse, W. 1896. Systematik der Streptothricheen und Bakterién. *In* Flugge (Editor), Die Mikroorganismen, Vol. 2, pp. 48-66.

Krych, V.A., J.L. Johnson and A.A. Yousten. 1980. Deoxyribonucleic acid homologies among strains of *Bacillus sphaericus*. Int. J. Syst. Bacteriol. *30:* 476-484.

Kubo, H., S. Suzuki and S. Tamura. 1964. Process for obtaining a new antibiotic piericidin. Japanese Patent 9443.

Kuchaeva, A.G., N.A. Krasil'nikov, S.D. Taptykova and R.L. Gesheva. 1961. On the systematics of the *Actinomyces* of the *lavendulae* group (in Russian). Izv. Mikrobiol. Inst. Bulgar. Akad. Nauk. Biol. Sci. *13:* 103-124.

Kudo, T., K. Hatai and A. Seino. 1988. *Nocardia seriolae* sp. nov. causing nocardiosis of cultured fish. Int. J. Syst. Bacteriol. *38:* 173-178.

Kudrina, E.S. and T.S. Maksimova. 1963. Some species of thermophilic actinomycetes from the soil of China and their antibiotic properties (in Russian). Mikrobiologiya *32:* 623-631.

Kuo, M.J. and P.A. Hartman. 1966. Isolation of amylolytic strains of *Thermoactinomyces vulgaris* and production of thermophilic actinomycete amylases. J. Bacteriol. *92:* 723-726.

Kuo, M.J. and P.A. Hartman. 1967. Purification and partial characterization of *Thermoactinomyces vulgaris* amylases. Can. J. Microbiol. *13:* 1157-1163.

Kuraishi, H. 1985. Distribution of menaquinones in actinomycetes. The Actinomycetologist *46:* 7-18.

Kurtzman, C.P., M.J. Smiley, C.J. Johnson, L.B. Wickerham and G.B. Fuson. 1980. Two new and closely related heterothallic species, *Pichia amylophila* and *Pichia mississippiensis*: Characterization by hybridization and deoxyribonucleic acid reassociation. Int. J. Syst. Bacteriol. *30:* 208-216.

Kurup, V.P. 1979. Characterisation of some members of the genus *Thermomonospora*. Curr. Microbiol. *2:* 267-272.

Kurup, V.P. 1981. Taxonomic study of some members of *Micropolyspora* and *Saccharomonospora*. Microbiologica (Bologna) *4:* 249-259.

Kurup, V.P. and N.S. Agre. 1983. Transfer of *Micropolyspora rectivirgula* (Krasil'nikov and Agre, 1964) Lechevalier, Lechevalier and Becker, 1966) to *Faenia* gen. nov. Int. J. Syst. Bacteriol. *33:* 663-665.

Kurup, V.P., J.J. Barboriak, J.N. Fink and M.P. Lechevalier. 1975. *Thermoactinomyces candidus*, a new species of thermophilic actinomycetes. Int. J. Syst. Bacteriol. *25:* 150-154.

Kurup, V.P., J.J. Barboriak, J.N. Fink and G. Scribner. 1976a. Immunologic cross reactions among thermophilic actinomycetes associated with hypersensitivity pneumonitis. J. Allergy Clin. Immunol. *57:* 417-421.

Kurup, V.P. and J.N. Fink. 1977. Extracellular antigens of *Micropolyspora faeni* grown in synthetic medium. Infect. Immun. *15:* 608-613.

Kurup, V.P., J.N. Fink and D.M. Bauman. 1976b. Thermophilic actinomycetes from the environment. Mycologia *68:* 662-666.

Kurup, V.P. and R.J. Heinzen. 1978. Isolation and characterization of actinophages of *Thermoactinomyces* and *Micropolyspora*. Can. J. Microbiol. *24:* 794-797.

Kurup, V.P., G.E. Hollick and E.F. Pagan. 1980. *Thermoactinomyces intermedius*, a new species of amylase negative thermophilic actinomycetes. Science-Ciencia Bol. Cien. Sur. *7:* 104–108.

Kurup, V.P., G.E. Hollick and E.F. Pagan. 1981a. *In* Validation of the publication of new names and new combinations previously effectively published outside the IJSB. List No. 6. Int. J. Syst. Bacteriol. *31:* 215–218.

Kurup, V.P., J.E. Piechura, E.Y. Ting and J.A. Orlowski. 1983. Immunochemical characterization of *Nocardia asteroides* antigens: Support for a single species concept. Can. J. Microbiol. *29:* 425–432.

Kurup, V.P., H.S. Randhawa and N.P. Gupta. 1970. Nocardiosis: A review. Mycopathol. Mycol. Appl. *40:* 193–219.

Kurup, V.P. and J.A. Schmitt. 1973. Numerical taxonomy of *Nocardia*. Can. J. Microbiol. *19:* 1034–1048.

Kurup, V.P. and G.H. Schribner. 1981. Antigenic relationship among *Nocardia asteroides* immunotypes. Microbios *31:* 25–30.

Kurup, V.P., E.Y. Ting, J.N. Fink and N.J. Calvanico. 1981b. Characterization of *Micropolyspora faeni* antigens. Infect. Immun. *34:* 508–512.

Kurylowicz, W., A. Paszkiewicz, W. Wóznicka, W. Kurtatkowski and T. Szulga. 1975. Numerical taxonomy of streptomycetes. Classification of streptomycetes by different numerical methods. Postepy Hig. Med. Dõsw. *29:* 281–355.

Kurylowicz, W. and W. Wóznicka. 1967. *Actinomyces (Streptomyces) varsoviensis*. I. Taxonomic studies. Med. Dõsw. Mikrobiol. *19:* 1–9.

Kusakabe, Y., Y. Yamaguchi, C. Nagatsu, H. Abe, K. Akasaki and S. Shirato. 1969. Citromycin, a new antibiotic. I. Isolation and identification. J. Antibiot. *22:* 112–118.

Küster, E. 1961. Results of comparative study of criteria used in classification of the actinomycetes. Int. Bull. Bacteriol. Nomencl. Taxon. *11:* 91–98.

Küster, E. 1972. Simple working key for the classification and identification of named taxa included in the International *Streptomyces* Project. Int. J. Syst. Bacteriol. *22:* 139–148.

Küster, E. 1974a. Family *Micromonosporaceae* Krasil'nikov 1938. *In* Buchanan and Gibbons (Editors), Bergey's Manual of Determinative Bacteriology, 8th Ed., The Williams and Wilkins Co., Baltimore, pp. 846–865.

Küster, E. 1974b. Genus II. *Thermoactinomyces* Tsiklinsky 1899, 501. *In* Buchanan and Gibbons (Editors), Bergey's Manual of Determinative Bacteriology, 8th Ed., The Williams and Wilkins Co., Baltimore, pp. 855–856.

Küster, E. and R. Locci. 1963a. Studies on peat and peat microorganisms. I. Taxonomic studies on thermophilic actinomycetes isolated from peat. Arch. Mikrobiol. *45:* 188–197.

Küster, E. and R. Locci. 1963b. Transfer of *Thermoactinomyces viridis* Schuurmans et al. 1956 to the genus *Thermomonospora* as *Thermomonospora viridis* (Schuurmans, Olson and San Clemente) comb. nov. Int. Bull. Bacteriol. Nomencl. Taxon. *13:* 213–216.

Küster, E. and R. Locci. 1964. Taxonomic studies on the genus *Thermoactinomyces*. Int. Bull. Bacteriol. Nomencl. Taxon. *14:* 109–114.

Küster, E. and S.T. Williams. 1964a. Production of hydrogen sulphide by streptomycetes and methods for its detection. J. Appl. Microbiol. *12:* 46–52.

Küster, E. and S.T. Williams. 1964b. Selection of media for isolation of streptomycetes. Nature *202:* 928–929.

Kutzner, H.J. 1968. Über die Bildung von Huminstoffen durch Streptomyceten. Landwirtsch. Forsch. *21:* 48–61.

Kutzner, H.J. 1972. Storage of *Streptomyces* in soft agar and by other methods. Experientia *28:* 1395.

Kutzner, H.J. 1976. Methoden zur Untersuchung von Streptomyceten und einigen anderen Actinomyceten. Teilsammlung Darmstad am Institut für Mikrobiologie der Technischen Hochschule, Darmstadt.

Kutzner, H.J. 1981. The family *Streptomycetaceae*. *In* Starr, Stolp, Trüper, Balows and Schlegel (Editors), The Prokaryotes: A Handbook on Habitats, Isolation and Identification of Bacteria, Vol. II, Springer-Verlag, Berlin, pp. 2028–2090.

Kutzner, H.J., V. Bottinger and R.D. Heitzer. 1978. The use of physiological criteria in the taxonomy of *Streptomyces* and *Streptoverticillium*. Zentralbl. Bakteriol. Parasitenkd. Infektionskr. Hyg. I Abt. Suppl. *6:* 25–29.

Kutzner, H.J. and S.A. Waksman. 1959. *Streptomyces coelicolor* Muller and *Streptomyces violaceoruber* Waksman and Curtis, two distinctly different organisms. J. Bacteriol. *78:* 528–538.

Kuznetsov, V.D. 1962. A new species of genus *Chainia*. Mikrobiologiya *31:* 534–539.

Labeda, D.P. 1986. Transfer of *"Nocardia" aerocolonigenes* (Shinobu and Kawato 1960) Pridham 1970 into the genus *Saccharothrix* Labeda, Testa, Lechevalier and Lechevalier (1984) as *Saccharothrix aerocolonigenes* sp. nov. Int. J. Syst. Bacteriol. *36:* 109–110.

Labeda, D.P. 1987. Transfer of the type strain of *Streptomyces erythraeus* (Waksman 1923) Waksman and Henrici 1948 to the genus *Saccharopolyspora* Lacey and Goodfellow 1975 as *Saccharopolyspora eryghraea* sp. nov. and designation of a new type strain for *Streptomyces erythraeus*. Int. J. Syst. Bacteriol. *37:* 19–22.

Labeda, D.P. 1988. *Kitasatosporia mediocidica* sp. nov. Int. J. Syst. Bacteriol. *38:* 287–290.

Labeda, D.P., R.T. Testa, M.P. Lechevalier and H.A. Lechevalier. 1984. *Saccharothrix*, a new genus of the *Actinomycetales* related to *Nocardiopsis*. Int. J. Syst. Bacteriol. *34:* 426–431.

Labeda, D.P., R.T. Testa, M.P. Lechevalier and H.A. Lechevalier. 1985a. *Actinomadura yumaensis* sp. nov. Int. J. Syst. Bacteriol. *35:* 333–336.

Labeda, D.P., R.T. Testa, M.P. Lechevalier and H.A. Lechevalier. 1985b. *Glycomyces*, a new genus of the *Actinomycetales*. Int. J. Syst. Bacteriol. *35:* 417–421.

Lacey, J. 1971a. The microbiology of moist barley storage in unsealed silos. Ann. Appl. Biol. *69:* 187–212.

Lacey, J. 1971b. *Thermoactinomyces sacchari* sp. nov., a thermophilic actinomycete causing bagassosis. J. Gen. Microbiol. *66:* 327–338.

Lacey, J. 1973. Actinomycetes in soils, composts and fodders. *In* Sykes and Skinner (Editors), Actinomycetales: Characteristics and Practical Importance, Soc. Appl. Bacteriol. Symp. Ser. No. 2, Society for Applied Bacteriology, London, pp. 231–251.

Lacey, J. 1974. Moulding of sugar-cane bagasse and its prevention. Ann. Appl. Biol. *76:* 63–76.

Lacey, J. 1978. Ecology of actinomycetes in fodders and related substances. Zentralbl. Bakteriol. Parasitenkd. Infektionskr. Hyg. Abt. I Suppl. *6:* 161–170.

Lacey, J. 1981. Airborne actinomycete spores as respiratory allergens. Zentralbl. Bakteriol. Mikrobiol. Hyg. I. Abt. Suppl. *11:* 243–250.

Lacey, J. and J. Dutkiewicz. 1976a. Isolation of actinomycetes and fungi using a sedimentation chamber. J. Appl. Bacteriol. *41:* 315–319.

Lacey, J. and J. Dutkiewicz. 1976b. Methods for examining the microflora of mouldy hay. J. Appl. Bacteriol. *41:* 13–27.

Lacey, J. and M. Goodfellow. 1975. A novel actinomycete from sugar-cane bagasse, *Saccharopolyspora hirsuta* gen. et sp. nov. J. Gen. Microbiol. *88:* 75–85.

Lacey, J., M. Goodfellow and G. Alderson. 1978. The genus *Actinomadura*, Lechevalier and Lechevalier. Zentralbl. Bakteriol. Parasitenkd. Infektionskr. Hyg. Abt. 1 Suppl. *6:* 107–117.

Lacey, J., A.J. McCarthy, M. Goodfellow and T. Cross. 1984. Conservation of the name *Micropolyspora* Lechevalier, Solotorovsky, and McDurmont over *Faenia* Kurup and Agre and designation of *Micropolyspora faeni* Cross, Maciver, and Lacey as the type species of the genus. Int. J. Syst. Bacteriol. *34:* 505–507.

Lacey, J. and M.E. Lacey. 1987. Micro-organisms in the air of cotton mills. Ann. Occup. Hyg. *31:* 1–19.

Lacey, J. and D.A. Vince. 1971. Endospore formation and germination in a new *Thermoactinomyces* species. *In* Barker, Gould and Wolf (Editors), Spore Research 1971. Academic Press, London, pp. 181–187.

Lalonde, M. 1977. Infection process of the *Alnus* root nodule symbiosis. *In* Newton, Postgate and Rodriguez-Barrueco (Editors), Recent Developments in Nitrogen Fixation, Academic Press, New York, pp. 569–589.

Lalonde, M. 1978. Confirmation of the infectivity of a free-living actinomycete isolated from *Comptonia peregrina* root nodules by immunological and ultrastructural studies. Can. J. Bot. *56:* 2621–2635.

Lalonde, M. 1979. Immunological and ultrastructural demonstration of nodulation of the European *Alnus glutinosa* (L.) Gaertn. host plant by an actinomycetal isolate from the North American *Comptonia peregrina* (L.) Coult. root nodule. Bot Gaz. *140 (Suppl):* S35–S43.

Lalonde, M. 1981. Practical aspects of *Frankia* in forestry. Am. Soc. Microbiol. Seminar, Dallas, March, Session 71.

Lalonde, M. and H.E. Calvert. 1979. Production of *Frankia* hyphae and spores as an infective inoculant for *Alnus* sp. *In* Gordon, Wheeler and Perry (Editors), Symbiotic Nitrogen Fixation in the Management of Temperate Forests, Oregon State Univ., Corvallis, pp. 95–110.

Lalonde, M., H.E. Calvert and S. Pine. 1981. Isolation and use of *Frankia* strains in actinorhizae formation. *In* Gibson and Newton (Editors), Current Perspectives in Nitrogen Fixation, Australian Acad. Sci., Canberra, pp. 296–299.

Lämmler, C., G.S. Chatwal and H. Blobel. 1983. Variations in the binding of mammalian fibrinogens to streptococci of different animal origin. Med. Microbiol. Immunol. *172:* 191–196.

Lancefield, R.C. 1933. A serological differentiation of human and other groups of hemolytic streptococci. J. Exp. Med. *57:* 571–595.

Lancefield, R.C. 1934. A serological differentiation of specific types of bovine hemolytic streptococci (Group B). J. Exp. Med. *59:* 441–458.

Lancelle, S.A., J.G. Torrey, P.K. Hepler and D.A. Callaham. 1985. Ultrastructure of freeze-substituted *Frankia* strain HFPCcI3, the actinomycete isolated from root nodules of *Casuarina cunninghamiana*. Protoplasma. *127:* 64–72.

Lanéelle, G., J. Asselineau and G. Chamoiseau. 1971. Présence de mycosides C' (formes simplifées de mycoside C) dans les bacteries isolées de bovins atteints du farcin. FEBS Lett. *19:* 109–111.

Langworthy, T.A. 1982. Lipids of bacteria living in extreme environments. Curr. Top. Membr. Transp. *17:* 45–77.

Lapage, S.P. 1971. Culture collections of bacteria. Biol. J. Linnean Soc. *3:* 197–210.

Lapage, S.P. 1975. Report of the World Federation for Culture Collections. Int. J. Syst. Bacteriol. *25:* 90–94.

Lapage, S.P., S. Bascomb, W.R. Willcox and M.A. Curtis. 1973. Identification of bacteria by computer. I. General aspects and perspectives. J. Gen. Microbiol. *77:* 273–290.

Lapage, S.P., P.H.A. Sneath, E.F. Lessel, V.B.D. Skerman, H.P.R. Seeliger and W.A. Clark (Editors). 1975. International Code of Nomenclature of Bacteria, 1975 Revision, American Society for Microbiology, Washington, D.C.

Laurent, E. 1890. Sur le microbe des nodosites des legumineuses. C.R. Acad. Sci., Paris *3:* 754–756.

Laveran, M. 1906. Tumeur provoquée par un microcoque rose en zooglées. C.R. Hebd. Soc. Biol. *2:* 340–341.

Lavrova, N.V. and T.P. Preobrazhenskaya. 1975. Isolation of new species of *Actinomadura* on selective media with rubomycin. Antibiotiki *20:*

483-488.

Lavrova, N.V., T.P. Preobrazhenskaya and M.A. Sveshnikova. 1972. Isolation of soil actinomycetes on selective media with rubomycin. Antibiotiki *11:* 965-970.

Leach, B.E., K.M. Calhoun, L.E. Johnson, C.M. Teeters and W.G. Jackson. 1953. Chartreusin, a new antibiotic produced by *Streptomyces chartreusis* a new species. J. Am. Chem. Soc. *75:* 4011-4012.

Leatherwood, J.M. and M.P. Sharma. 1972. Novel anaerobic cellulolytic bacterium. J. Anim. Sci. *220:* 752-753.

Lechevalier, H.A. 1964. Principles and application in aquatic microbiology. *In* Heukelekan and Dondero (Editors), The Actinomycetes, John Wiley & Sons, New York, pp. 230-250.

Lechevalier, H.A. 1965. Priority of the generic name *Microbispora* over *Waksmania* and *Thermopolyspora.* Int. Bull. Bacteriol. Nomencl. Taxon. *15:* 139-142.

Lechevalier, H.A. 1968a. Status of the generic names *Micropolyspora* Lechevalier et al. 1961 and *Micropolispora* Shchepkina 1940 (Actinomycetales): Request for an opinion from the Judicial Commission (ICNB) conserving the generic name *Micropolyspora* Lechevalier. Int. J. Syst. Bacteriol. *18:* 203-206.

Lechevalier, H.A. 1976a. Report on the cooperative study of the generic assignment of strains labelled *Nocardia farcinica.* Biol. Actinomycetes *12:* 8-16.

Lechevalier, H.A. and P.E. Holbert. 1965. Electron microscopic observation of the sporangial structure of a strain of *Actinoplanes.* J. Bacteriol. *89:* 217-222.

Lechevalier, H.A. and M.P. Lechevalier. 1965. Classification des actinomycetes aerobes basée sur leur morphologie et leur composition chimique. Ann. Inst. Pasteur *108:* 662-673.

Lechevalier, H.A. and M.P. Lechevalier. 1967. Biology of the actinomycetes. Annu. Rev. Microbiol. *21:* 71-100.

Lechevalier, H.A. and M.P. Lechevalier. 1969. Ultramicroscopic structure of *Intrasporangium calvum* (Actinomycetales). J. Bacteriol. *100:* 522-525.

Lechevalier, H.A. and M.P. Lechevalier. 1970a. A critical evaluation of the genera of aerobic actinomycetes. *In* Prauser (Editor), The Actinomycetales, VEB Gustav Fischer Verlag, Jena, pp. 393-405.

Lechevalier, H.A. and M.P. Lechevalier. 1981a. Actinomycete genera "in search of a family". *In* Starr, Stolp, Trüper, Balows and Schlegel (Editors), The Prokaryotes: A Handbook on Habitats, Isolation and Identification of Bacteria, Springer-Verlag, New York, pp. 2118-2123.

Lechevalier, H.A. and M.P. Lechevalier. 1981b. Introduction to the order *Actinomycetales. In* Starr, Stolp, Trüper, Balows and Schlegel (Editors), The Prokaryotes: A Handbook on Habitats, Isolation and Identification of Bacteria, Springer-Verlag, New York, pp. 1915-1922.

Lechevalier, H.A., M.P. Lechevalier and B. Becker. 1966a. Comparison of the chemical composition of cell-walls of nocardiae with that of other aerobic actinomycetes. Int. J. Syst. Bacteriol. *16:* 151-160.

Lechevalier, H.A., M.P. Lechevalier and N.N. Gerber. 1971a. Chemical composition as a criterion in the classification of actinomycetes. Adv. Appl. Microbiol. *14:* 47-72.

Lechevalier, H.A., M.P. Lechevalier and P.E. Holbert. 1966b. Electron microscopic observation of the sporangial structure of strains of *Actinoplanaceae.* J. Bacteriol. *92:* 1228-1235.

Lechevalier, H.A., M. Solotorovsky and C.I. McDurmont. 1961. A new genus of *Actinomycetales*: *Micropolyspora gen. nov.* J. Gen. Microbiol. *26:* 11-18.

Lechevalier, M.P. 1968b. Identification of aerobic actinomycetes of clinical importance. J. Lab. Clin. Med. *71:* 934-944.

Lechevalier, M.P. 1972. Description of a new species, *Oerskovia xanthineolytica* and emendation of *Oerskovia* Prauser et al. Int. J. Syst. Bacteriol. *22:* 260-264.

Lechevalier, M.P. 1976b. The taxonomy of the genus *Nocardia*: Some light at the end of the tunnel? *In* Goodfellow, Brownell and Serrano (Editors), The Biology of the Nocardiae, Academic Press, New York, pp. 1-38.

Lechevalier, M.P. 1977. Lipids in bacterial taxonomy—a taxonomist's view. CRC Crit. Rev. Microbiol. *5:* 109-210.

Lechevalier, M.P. 1981. Ecological associations involving actinomycetes. Zentralbl. Bakteriol. Mikrobiol. Hyg. I Abt., Suppl. *11:* 159-166.

Lechevalier, M.P. 1982. Lipids in bacterial taxonomy. *In* Laskin and Lechevalier (Editors), CRC Handbook of Microbiology, 2nd ed., Vol. IV, Microbial Composition: Carbohydrates, Lipids and Minerals, CRC Press, Boca Raton, Florida, pp. 435-541.

Lechevalier, M.P. 1983. Cataloging *Frankia* strains. Can. J. Bot. *61:* 2964-2967.

Lechevalier, M.P. 1984. The taxonomy of the genus *Frankia.* Plant Soil *78:* 1-6.

Lechevalier, M.P., D. Baker and F. Horrière. 1983. Physiology, chemistry, serology and infectivity of two *Frankia* isolates from *Alnus incana* subsp. *rugosa.* Can. J. Bot. *61:* 2826-2833.

Lechevalier, M.P., C. De Bievre and H. Lechevalier. 1977. Chemotaxonomy of aerobic actinomycetes: phospholipid composition. Biochem. Syst. Ecol. *5:* 249-260.

Lechevalier, M.P. and N.N. Gerber. 1970. The identity of madurose with 3-*0*-methyl-D-galactose. Carbohydr. Res. *13:* 451-453.

Lechevalier, M.P., N.N. Gerber and T.A. Umbreit. 1982a. Transformation of adenine to 8-hydroxyadenine by strains of *Oerskovia xanthineolytica.* Appl. Environ. Microbiol. *43:* 367-370.

Lechevalier, M.P., A.C. Horan and H.A. Lechevalier. 1971b. Lipid composition in the classification of nocardiae and mycobacteria. J. Bacteriol. *105:* 313-318.

Lechevalier, M.P., F. Horrière and H.A. Lechevalier. 1982b. The biology of *Frankia* and related organisms. Dev. Ind. Microbiol. *23:* 51-60.

Lechevalier, M.P. and H.A. Lechevalier. 1957. A new genus of the *Actinomycetales: Waksmania* gen. nov. J. Gen. Microbiol. *17:* 104-111.

Lechevalier, M.P. and H.A. Lechevalier. 1970b. Chemical composition as a criterion in the classification of aerobic actinomycetes. Int. J. Syst. Bacteriol. *20:* 435-443.

Lechevalier, M.P. and H.A. Lechevalier. 1970c. Composition of whole-cell hydrolysates as a criterion in the classification of aerobic actinomycetes. *In* Prauser (Editor), The Actinomycetales, VEB Gustav Fischer Verlag, Jena, pp. 311-316.

Lechevalier, M.P. and H.A. Lechevalier. 1974. *Nocardia amarae* sp. nov., an actinomycete common in foaming activated sludge. Int. J. Syst. Bacteriol. *24:* 278-288.

Lechevalier, M.P. and H.A. Lechevalier. 1975. Actinoplanete with cylindrical sporangia, *Actinoplanes rectilineatus* sp. nov. Int. J. Syst. Bacteriol. *25:* 371-376.

Lechevalier, M.P. and H.A. Lechevalier. 1976. Chemical methods as criteria for the separation of nocardiae from other actinomycetes. Biol. Actinmycetes Relat. Org. *11:* 78-92.

Lechevalier, M.P. and H.A. Lechevalier. 1979. The taxonomic position of the actinomycetic endophytes. *In* Gordon, Wheeler and Perry (Editors), Symbiotic Nitrogen Fixation in the Management of Temperate Forests, Forest Research Laboratory, Corvallis, Oregon, pp. 111-122.

Lechevalier, M.P. and H.A. Lechevalier. 1980. The chemotaxonomy of actinomycetes. *In* Dietz and Thayer (Editors), Actinomycete Taxonomy, Special Publication 6, Society for Industrial Microbiology, Arlington, VA, pp. 227-291.

Lechevalier, M.P. and H.A. Lechevalier. 1984. Taxonomy of *Frankia. In* Ortiz-Ortiz, Bojalil and Yakoleff (Editors), Biological, Biochemical and Biomedical Aspects of Actinomycetes, Academic Press, New York, pp. 575-582.

Lechevalier, M.P., H.A. Lechevalier and C.E. Heintz. 1973a. Morphological and chemical nature of the sclerotia of *Chainia olivacea.* Thirumalachar and Sukapure of the order *Actinomycetales.* Int. J. Syst. Bacteriol. *23:* 57-170.

Lechevalier, M.P., H.A. Lechevalier and P.E. Holbert. 1968. *Sporichthya*, un nouveau genre de *Streptomycetaceae.* Ann. Inst. Pasteur *114:* 277-286.

Lechevalier, M.P., H. Lechevalier and A.C. Horan. 1973b. Chemical characteristics and classification of nocardiae. Can. J. Microbiol. *19:* 965-972.

Lechevalier, M.P., H. Prauser, D.P. Labeda and J.-S. Ruan. 1986. Two new genera of nocardioform actinomycetes: *Amycolata* gen. nov. and *Amycolatopsis* gen. nov. Int. J. Syst. Bacteriol. *36:* 29-37.

Lechevalier, M.P. and J.-S. Ruan. 1984. Physiology and chemical diversity of *Frankia* spp. isolated from nodules of *Comptonia peregrina* (L.) Coult. and *Ceanothus americanus* L. Plant Soil *78:* 15-22.

Lechevalier, M.P., A.E. Stern and H.A. Lechevalier. 1981. Phospholipids in the taxonomy of actinomycetes. Zentralbl. Bakteriol. Parasitenkd. Infektionskr. Hyg. I Abt. Suppl. *11:* 111-116.

Leedle, J.A.Z. and R.B. Hespell. 1980. Differential carbohydrate media and anaerobic replica plating techniques in delineating carbohydrate-utilizing subgroups in rumen bacterial populations. Appl. Environ. Microbiol. *39:* 709-719.

Lehmann, K.B. and R. Neumann. 1896. Atlas und Grundriss der Bakteriologie und Lehrbuch der speciellen bakteriologischen Diagnostik, Teil II, J. F. Lehmann, München.

Lehrer, S.B. and J.E. Salvaggio. 1978. Characterization of *Thermoactinomyces sacchari* antigens. Infect. Immun. *20:* 519-525.

Lerner, P.I. and G.L. Baum. 1973. Antimicrobial susceptibility of *Nocardia* species. Antimicrob. Agents Chemother. *4:* 85-93.

Leuchtenberger, A., P. Klingenberg and H. Ruttloff. 1979. Isolation and characterization of proteases from *Thermoactinomyces vulgaris.* III. Studies of protease formation in a small pilot plant (in German). Z. Allg. Mikrobiol. *19:* 27-35.

Leuchtenberger, A. and H. Ruttloff. 1983. Effect of oil and fatty acids on growth and enzyme formation by *Thermoactinomyces vulgaris.* III. Influence of culture vessel, strain and medium composition (in German). Z. Allg. Mikrobiol. *23:* 635-644.

Lewis, J.A. and R.L. Starkey. 1969. Decomposition of plant tannins by some soil micro-organisms. Soil Sci. *107:* 235-241.

Liegard, H. and M. Landrieu. 1911. Un cas de mycose conjonctivale. Ann. Ocul. *146:* 418-426.

Lieske, R. 1921. Morphologie und Biologie der Strahlenpilze (Actinomyceten), Gebrüder Borntraeger, Leipzig.

Lind, A., O. Ouchterlony and M. Ridell. 1980. Mycobacterial antigens. *In* Meissner and Schimiedel (Editors), Infektionskrankheiten und ihre Erreger, Bd. 4, Mycobakterien und mykobakterielle Krankheiten, Fischer Verlag, Jena, pp. 275-303.

Lind, A. and M. Ridell. 1976. Serological relationships between *Nocardia*, *Mycobacterium*, *Corynebacterium* and the 'rhodochrous' taxon. *In* Goodfellow, Brownell and Serrano (Editors), The Biology of the Nocardiae, Academic Press, London, pp. 220-235.

Lindenbein, W. 1952. Über einige chemisch interessante Aktinomyceten—stämme und ihre Klassifizierung. Arch. Mikrobiol. *17:* 361-383.

Lindenberg, A. 1909. Un nouveau mycétome. Arch. Parasitol. *13:* 265-282.

Lindner, F., R. Junk, G. Nesemann and J. Schmidt-Thomé. 1958. Gewinnung von 20 β-Hydroxysteroiden aus 17 α-21-Dihydroxy-20-Ketosteroiden durch mikrobiologische Hydrierung mit *Streptomyces hydrogerans.* Hoppe-Seyler's Z. Physiol. Chem. *313:* 117-123.

Liston, J., W. Weibe and R.R. Colwell. 1963. Quantitative approach to the study of

bacterial species. J. Bacteriol. *85:* 1061-1070.
Liu, Z., J. Ruan and X. Yan. 1984. A new species of *Nocardiopsis*. Acta Microbiol. Sin. *24:* 26-29.
Lloyd, A.B. 1969. Dispersal of streptomycetes in air. J. Gen. Microbiol. *57:* 35-40.
Locci, R. 1963. The phenomenon of autoinhibition in *Thermoactinomyces vulgaris* (in Italian). G. Microbiol. *11:* 183-189.
Locci, R. 1971. On the spore formation process in actinomycetes. IV. Examination by scanning electron microscopy of the genera *Thermoactinomyces, Actinobifida* and *Thermomonospora*. Riv. Patol. Veg. *7:* 63-80.
Locci, R. 1972. On the spore formation process in actinomycetes. IV. Examination by scanning electron microscopy of the genera *Thermoactinomyces, Actinobifida* and *Thermomonospora*. Riv. Patol. Veg. 4 Suppl. *7:* 63-80.
Locci, R. 1976. Developmental micromorphology of actinomycetes. *In* Arai (Editor), Actinomycetes: The Boundary Micro-organisms, University Park Press, Baltimore, London, pp. 249-297.
Locci, R. 1981. Micromorphology and development of actinomycetes. Zentralbl. Bakteriol. Mikrobiol. Hyg. Abt. I Orig. Suppl. *11:* 119-130.
Locci, R. 1985. New combinations and validation of some taxa of the genus *Streptoverticillium*. Ann. Microbiol. Enzimol. *35:* 231-234.
Locci, R., E. Baldacci and B. Petrolini. 1967. Contribution to the study of oligosporic actinomycetes. I. Description of a new species of *Actinobifida: Actinobifida alba* sp. nov. and revision of the genus. G. Microbiol. *15:* 79-91.
Locci, R., E. Baldacci and B. Petrolini Baldan. 1969. The genus *Streptoverticillium*. A taxonomic study. G. Microbiol. *17:* 1-60.
Locci, R., M. Goodfellow and G. Pulverer. 1982. Micro-morphological, morphogenetic and chemical characters of rhodococci. *In* Abstracts of the Fifth International Symposium of Actinomycete Biology, Oaxtepec, Mexico, pp. 118-119.
Locci, R. and B. Petrolini Baldan. 1970. Morphology and development of *Streptoverticillium* as examined by scanning electron microscopy. G. Microbiol. *18:* 69-76.
Locci, R. and B. Petrolini Baldan. 1971. On the spore formation process in actinomycetes. V. Scanning electron microscopy of some genera of *Actinoplanaceae*. Riv. Pat. Veg. Ser. IV Suppl. *7:* 81-96.
Locci, R., J. Rogers, P. Sardi and G.M. Schofield. 1981. A preliminary numerical study of named species of the genus *Streptoverticillium*. Ann. Microbiol. Enzimol. *31:* 115-121.
Locci, R. and G.P. Sharples. 1984. Morphology. *In* Goodfellow, Mordarski and Williams (Editors), The Biology of the Actinomycetes, Academic Press, London, pp. 165-199.
Lockhart, W.R. and J. Liston. 1970. Methods for Numerical Taxonomy, American Society for Microbiology, Washington, D.C.
London, J. and K. Kline. 1973. Aldolases of lactic acid bacteria: A case history in the use of an enzyme as an evolutionary marker. Bacteriol. Rev. *37:* 453-478.
Lopez, M.F. and J.G. Torrey. 1985a. Enzymes of glucose metabolism in *Frankia* sp. J. Bacteriol. *162:* 110-116.
Lopez, M.F. and J.G. Torrey. 1985b. Purification and properties of trehalase in *Frankia* ArI3. Arch. Microbiol. *143:* 209-215.
Lopez, M.F., C.S. Whaling and J.G. Torrey. 1983. The polar lipids and free sugars of *Frankia* in culture. Can. J. Bot. *61:* 2834-2842.
Louria, D.B. and R.E. Gordon. 1960. Pericarditis and pleuritis caused by a recently discovered micro-organism, *Waksmania rosea*. Am. Rev. Respir. Dis. *81:* 83-88.
Lu, Y. and X. Yan. 1983. Studies on the classification of thermophilic actinomycetes. IV. Determination of thermophilic members of *Nocardiaceae* (in Chinese). Acta Microbiol. Sin. *23:* 220-228.
Ludwig, W., K.H. Schleifer and E. Stackebrandt. 1984. 16S rRNA analysis of *Listeria monocytogenes* and *Brochothrix thermosphacta*. FEMS Microbiol. Lett. *25:* 199-204.
Ludwig, W., E. Seewaldt, R. Kilpper-Bälz, K.H. Schleifer, L. Magrum, C.R. Woese, G.F. Fox and E. Stackebrandt. 1985. The phylogenetic position of *Streptococcus* and *Enterococcus*. J. Gen. Microbiol. *131:* 543-551.
Ludwig, W., E. Seewaldt, K.H. Schleifer and E. Stackebrandt. 1981. The phylogenetic status of *Kurthia zopfii*. FEMS Microbiol. Lett. *10:* 193-197.
Luedemann, G. 1968. *Geodermatophilus*, a new genus of the *Dermatophilaceae* (Actinomycetales). J. Bacteriol. *96:* 1848-1858.
Luedemann, G.M. 1971a. Designation of neotype strains for *Micromonospora coerulea* Jensen 1932 and *Micromonospora chalcea* (Foulerton 1905) Ørskov 1923. Int. J. Syst. Bacteriol. *21:* 248-253.
Luedemann, G.M. 1971b. *Micromonospora purpureochromogenes* (Waksman and Curtis 1916) comb. nov. (subjective synonym: *Micromonospora fusca* Jensen 1932). Int. J. Syst. Bacteriol. *21:* 240-247.
Luedemann, G.M. 1971c. Species concepts and criteria in the genus *Micromonospora*. Trans. N.Y. Acad. Sci. *33:* 207-218.
Luedemann, G.M. 1974. Addendum to *Micromonosporaceae*. *In* Buchanan and Gibbons (Editors), Bergey's Manual of Determinative Bacteriology, 8th Ed., The Williams and Wilkins Co., Baltimore, pp. 848-855.
Luedemann, G.M. and B.C. Brodsky. 1964. Taxonomy of gentamicin producing *Micromonospora*. Antimicrob. Agents Chemother. *1963:* 116-124.
Luedemann, G.M. and B.C. Brodsky. 1965. *Micromonospora carbonacea* sp. n., an everninomicin producing organism. Antimicrob. Agents Chemother. *1964:* 47-52.
Luedemann, G.M. and C.J. Casmer. 1973. Electron microscope study of whole mounts and thin section of *Micromonospora chalcea* ATCC 12452. Int. J. Syst. Bacteriol. *23:* 243-255.
Lyons, A.J. and T.G. Pridham. 1965. Colorimetric determination of color of aerial mycelium of streptomycetes. J. Bacteriol. *89:* 159-169.
Lyons, A.J. and T.G. Pridham. 1966. *Streptomyces griseus* (Krainsky) Waksman and Henrici. A Taxonomic Study of Some Strains, Tech. Bull. No. 1360, Agric. Res. Service, U.S. Department of Agriculture, Washington, D.C.
Lyons, A.J. and T.G. Pridham. 1971. *Streptomyces torulopsis* sp. n. an unusual knobby-spored taxon. Appl. Microbiol. *22:* 190-193.
Macario, A.J.L. and E. Conway de Macario (Editors). 1985. Monoclonal Antibodies Against Bacteria, Vols. 1 and 2, Academic Press, New York.
Macé, E. 1897. Traité Pratique de Bactériologie, 3rd Ed., Baillière, Paris.
Mach, F. and N.S. Agre. 1970. Structure of the spores of *Actinobifida chromogena*. *In* Prauser (Editor), The Actinomycetales, VEB Gustav Fischer Verlag, Jena, pp. 221-225.
Maeda, K., Y. Okami, H. Kosaka, O. Taya and H. Umezawa. 1952a. On an antitubercular antibiotic produced by *Streptomyces cinnamonensis* n. sp. J. Antibiot. (Tokyo) *5:* 572-573.
Maeda, K., Y. Okami, O. Taya and H. Umezawa. 1952b. On new antifungal substances, moldin and phaeofacin, produced by *Streptomyces* sp. Jpn. J. Med. Sci. Biol. *5:* 237-338.
Maehr, H., C.-M. Liu, T. Herman, B.L.T. Prosser, J.M. Smallheer and N.J. Palleroni. 1980. Microbial products. IV. X-14847. A new aminoglycoside from *Micromonospora echinospora*. J. Antibiot. *33:* 1431-1437.
Maerz, A. and M.R. Paul. 1950. A Dictionary of Color, 2nd Ed., McGraw Hill Book Co., Inc., New York.
Magnusson, H. 1923. Spezifische infektiose Pneumonie beim Fohlen. Ein neuer Entreneger beim Pferde. Arch. Wiss. Prakt. Tierheilk. *50:* 22-38.
Magnusson, M. 1976. Sensitin tests in *Nocardia* taxonomy. *In* Goodfellow, Brownell and Serrano (Editors), The Biology of the Nocardiae, Academic Press, New York, pp. 236-265.
Maiorova, V.I. 1965. Composition of polysaccharides in *Chainia*. Microbiology (English Trans.) *34:* 837-840.
Makkar, N.S. and T. Cross. 1982. Actinoplanetes in soil and on plant litter from freshwater habitats. J. Appl. Bacteriol. *52:* 209-218.
Malabarba, A., P. Strazzolini, A. Depaoli, M. Landi, M. Berti and B. Cavalleri. 1984. Teicoplanin, antibiotics from *Actinoplanes teichomyceticus* nov. sp. J. Antibiotics *37:* 988-999.
Maluszyńska, G.M. and L. Janota-Bassalik. 1974. A cellulolytic rumen bacterium, *Micromonospora ruminantium* sp. nov. J. Gen. Microbiol. *82:* 57-65.
Manachini, P.L., A. Craveri and R. Craveri. 1966. *Thermoactinomyces citrina*, a new species of thermophilic actinomycete isolated from soil (in Italian). Ann. Microbiol. *16:* 83-89.
Manachini, P.L., A. Ferrari and R. Craveri. 1965. Forme termofile di *Actinoplanaceae*. Isolamento et caracteristiche di *Streptosporangium album* var. *thermophilum*. Ann. Microbiol. Enzimol. *15:* 129-144.
Mandel, M., C.L. Schildkraut and J. Marmur. 1968. Use of CsCl density gradient analysis for determining the guanine plus cytosine content of DNA. Methods Enzymol. *12B:* 184-195.
Mankau, R. 1973. Utilization of parasites and predators in nematode pest management ecology. *In* Proceedings of the Tall Timbers Conference on Ecological Animal Control by Habitat Management *4:* 129-143.
Mankau, R. 1975. *Bacillus penetrans* n. comb. causing a virulent disease of plant-parasitic nematodes. J. Invertebr. Pathol. *26:* 333-339.
Mankau, R. and J.L. Imbriani. 1975. The life cycle of an endoparasite in some Tylenchid nematodes. Nematologica *21:* 89-94.
Mankau, R. and N. Prasad. 1977. Infectivity of *Bacillus penetrans* in plant-parasitic nematodes. J. Nematol. *9:* 40-45.
Mann, J.W., T.W. Jeffries and J.D. Macmillan. 1978. Production and ecological significance of yeast cell-wall degrading enzymes from *Oerskovia*. Appl. Environ. Microbiol. *36:* 594-605.
Mara, D.D. and J.I. Oragui. 1981. Occurrence of *Rhodococcus coprophilus* and associated actinomycetes in faeces, sewage and freshwater. Appl. Environ. Microbiol. *42:* 1037-1042.
Margalith, P. and G. Beretta. 1960a. A new antibiotic producing *Streptomyces: Str. bellus* nov. sp., Mycopathol. Mycol. Appl. *12:* 189-195.
Margalith, P. and G. Beretta. 1960b. Rifomycin. XI. Taxonomic study on *Streptomyces mediterranei* nov. sp. Mycopathol. Mycol. Appl. *13:* 321-330.
Margalith, P., G. Beretta and M.T. Timbal. 1959. Matamycin, a new antibiotic. I. Biological studies. Antibiotic Chemother. *9:* 71-75.
Margherita, S.S. and R.E. Hungate. 1963. Serological analysis of *Butyrivibrio* from the bovine rumen. J. Bacteriol. *86:* 855-860.
Mariat, F. and H. Lechevalier. 1977. Actinomycètes aerobies pathogènes. *In* Dumas (Editor), Bacteriologie Médicale, Flammarion, Paris, pp. 566a-566z.
Marmur, J. and P. Doty. 1961. Thermal renaturation of deoxyribonucleic acids. J. Mol. Biol. *3:* 585-594.
Marmur, J. and P. Doty. 1962. Determination of the base composition of deoxyribonucleic acid from its thermal denaturation temperature. J. Mol. Biol. *5:* 109-118.
Martin, J.F. and A.L. Demain. 1980. Control of antibiotic synthesis. Microbiol. Rev. *44:* 230-251.
Martin, S.M. (Editor). 1963. Culture Collections: Perspectives and Problems, Proceedings of the Specialists' Conference on Culture Collections, Ottawa, 1962,

University of Toronto Press, Toronto.
Martin, S.M. and V.B.D. Skerman. 1972. World Directory of Collections of Cultures of Microorganisms, Wiley-Interscience, New York.
Mason, D.J., A. Dietz and C. De Boer. 1963a. Lincomycin, a new antibiotic. I. Discovery and biological properties. Antimicrob. Agents Chemother. *1962:* 554-559.
Mason, D.J., A. Dietz and L.J. Hanka. 1963b. U-12898, a new antibiotic. I. Discovery, biological properties and assay. Antimicrob. Agents Chemother. *1962:* 607-613.
Mason, D.J., A. Dietz and R.M. Smith. 1961. Actinospectacin, a new antibiotic. I. Discovery and biological properties. Antibiot. Chemother. *11:* 118-122.
Matsumae, A.M., M. Ohtani, H. Takeshima and T. Hata. 1968. A new genus of the *Actinomycetales: Kitasatoa* gen. nov. J. Antibiot. (Tokyo) Ser. A *21:* 616-625.
Maxam, A.M. and W. Gilbert. 1977. A new method for sequencing DNA. Proc. Nat. Acad. Sci. USA *74:* 560-564.
Mayfield, C.I., S.T. Williams, S.M. Ruddick and H.L. Hatfield. 1972. Studies on the ecology of actinomycetes in soil IV. Observations on the form and growth of Streptomycetes in soil. Soil Biol. Biochem. *4:* 79-91.
Mays, T.D., L.V. Holdeman, W.E.C. Moore, M. Rogosa and J.L. Johnson. 1982. Taxonomy of the genus *Veillonella* Prévot. Int. J. Syst. Bacteriol. *32:* 28-36.
Mazzucco, C.E. and D.R. Benson. 1984. (^{14}C) Methylammonium transport by *Frankia* sp. strain CpI1. J. Bacteriol. *160:* 636-641.
McCarthy, A.J. and P. Broda. 1984. Screening for lignin-degrading actinomycetes and characterisation of their activity against [^{14}C] lignin-labelled wheat lignocellulose. J. Gen. Microbiol. *130:* 2905-2913.
McCarthy, A.J. and T. Cross. 1981. A note on a selective isolation medium for the thermophilic actinomycete *Thermomonospora chromogena*. J. Appl. Bacteriol. *51:* 299-302.
McCarthy, A.J. and T. Cross. 1984a. A taxonomic study of *Thermomonospora* and other monosporic actinomycetes. J. Gen. Microbiol. *130:* 5-25.
McCarthy, A.J. and T. Cross. 1984b. *In* Validation of the publication of new names and new combinations previously effectively published outside the IJSB. List No. 15. Int. J. Syst. Bacteriol. *34:* 355-357.
McCarthy, A.J. and T. Cross. 1984c. Taxonomy of *Thermomonospora* and related oligosporic actinomycetes. *In* Ortiz-Ortiz, Bojalil and Yakoleff (Editors), Biological, Biochemical and Biomedical Aspects of Actinomycetes, Academic Press, San Diego, pp. 521-536.
McCarthy, A.J., T. Cross, J. Lacey and M. Goodfellow. 1983. Conservation of the name *Micropolyspora* Lechevalier, Solotorovsky and McDurmont and designation of *Micropolyspora faeni* Cross, Maciver, and Lacey as the type species of the genus. Request for an Opinion. Int. J. Syst. Bacteriol. *33:* 430-433.
McCarthy, A.J., E. Peace and P. Broda. 1985. Studies on the extracellular xylanase activity of some thermophilic actinomycetes. Appl. Microbiol. Biotechnol. *21:* 238-244.
McCarthy, B.J. and E.T. Bolton. 1963. An approach to the measurement of genetic relatedness among organisms. Proc. Nat. Acad. Sci. USA *50:* 156-164.
McClung, N.M. 1974. Genus *Nocardia* Trevisan 1889, 9. *In* Buchanan and Gibbons (Editors), Bergey's Manual of Determinative Bacteriology, 8th Ed., The Williams and Wilkins Company, Baltimore, pp. 726-746.
McGowan, V.F. and V.B.D. Skerman (Editors). 1982. World Directory of Collections of Cultures of Microorganisms, 2nd Ed., World Data Centre on Microorganisms, University of Queensland, Brisbane, Australia.
McInnis, T.M. and A. Domnas. 1970. Pigments of *Actinoplanaceae*. III. A spirillomycin-type pigment from *Spirillospora* 1309-b. Z. Allg. Mikrobiol. *10:* 129-136.
McVeigh, I. and C.R. Reyes. 1961. A new species of *Streptomyces* and its antibiotic activity. Antibiot. Chemother. *11:* 312-319.
McVittie, A., H. Wildermuth and D.A. Hopwood. 1972. Fine structure and surface topography of endospores of *Thermoactinomyces vulgaris*. J. Gen. Microbiol. *71:* 367-381.
Mertz, F.P. and C.E. Higgens. 1982. *Streptomyces capillospiralis* sp. nov. Int. J. Syst. Bacteriol. *32:* 116-124.
Mertz, F.P. and R.C. Yao. 1986. *Actinomadura oligospora* sp. nov., the producer of a new polyether antibiotic. Int. J. Syst. Bacteriol. *36:* 179-182.
Mertz, F.P. and R.C. Yao. 1988. *Kibdelosporangium philippinense* sp. nov. isolated from soil. Int. J. Syst. Bacteriol. *38:* 282-286.
Metcalf, G. and M. Brown. 1957. Nitrogen fixation by new species of *Nocardia*. J. Gen. Microbiol. *17:* 567-572.
Metchnikoff, E. 1888. *Pasteuria ramosa*, un représentant des bactéries à division longitudinale. Ann. Inst. Pasteur, Paris *2:* 165-170.
Meyer, D.J. and C.W. Jones. 1973. Distribution of cytochromes in bacteria: relationship to general physiology. Int. J. Syst. Bacteriol. *23:* 459-467.
Meyer, J. 1976. *Nocardiopsis*, a new genus of the order *Actinomycetales*. Int. J. Syst. Bacteriol. *26:* 487-493.
Meyer, J. 1979. New species of the genus *Actinomadura*. Z. Allg. Mikrobiol. *19:* 37-44.
Meyer, J. 1981. *In* Validation of the publication of new names and new combinations previously effectively published outside the IJSB. List No. 6. Int. J. Syst. Bacteriol. *31:* 215-218.
Meyer, J. and M. Sveshnikova. 1974. *Micromonospora rubra* Sveshnikova et al. *Actinomadura rubra* comb. nov. Z. Allg. Mikrobiol. *14:* 167-170.
Micoletzky, H. 1925. Die freilebenden Süsswasser und Moornematoden Dänemarks, Andr. Fred. Host & Son, København.
Migula, W. 1900. System der Bakterien, Vol. 2, Gustav Fischer, Jena.
Migula, W. 1904. Allgemeine Morphologie, Entwicklungsgeschichte, Anatomie und Systematik der Schizomyceten. *In* Lafar (Editor), Handbuch der technischen Mykologie, 2nd Ed., Verlag von Gustav Fischer, Jena, pp. 29-149.
Mikami, Y., K. Miyashita and T. Arai. 1982. Diaminopimelic acid profiles of alkalophilic and alkaline-resistant strains of actinomycetes. J. Gen. Microbiol. *128:* 1709-1712.
Millard, W.A. and S.Burr. 1926. A study of twenty-four strains of *Actinomyces* and their relation to types of common scab of potato. Ann. Appl. Biol. *13:* 580-644.
Miller, I.M. and D.D. Baker. 1985. The initiation, development and structure of root nodules in *Elaeagnus angustifolia* L. (Elaeagnaceae). Protoplasma. *128:* 107-119.
Minnikin, D.E., L. Alshamaony and M. Goodfellow. 1975. Differentiation of *Mycobacterium, Nocardia* and related taxa by thin-layer chromatographic analysis of whole-organism methanolysates. J. Gen. Microbiol. *88:* 200-204.
Minnikin, D.E., M.D. Collins and M. Goodfellow. 1978. Menaquinone patterns in the classification of nocardioform and related bacteria. Zentralbl. Bakteriol. Parasitenkd. Infektionskr. Hyg. I Abt. Suppl. *6:* 85-90.
Minnikin, D.E., M.D. Collins and M. Goodfellow. 1979. Fatty acid and polar lipid composition in the classification of *Cellulomonas, Oerskovia* and related taxa. J. Appl. Bacteriol. *47:* 87-95.
Minnikin, D.E., G. Dobson, M. Goodfellow, P. Draper and M. Magnusson. 1985. Quantitative comparison of the mycolic and fatty acid composition of *Mycobacterium leprae* and *Mycobacterium gordonae*. J. Gen. Microbiol. *131:* 2013-2021.
Minnikin, D.E. and M. Goodfellow. 1980. Lipid composition in the classification and identification of acid-fast bacteria. *In* Goodfellow and Board (Editors), Microbiological Classification and Identification, Academic Press, London, pp. 189-256.
Minnikin, D.E. and M. Goodfellow. 1981. Lipids in the classification of actinomycetes. *In* Schaal and Pulverer (Editors), Actinomycetes. Zentralbl. Bakteriol. Parasitenkd. Infektionskr. Hyg. Suppl. *11:* 99-109.
Minnikin, D.E., I.G. Hutchinson, A.B. Caldicott and M. Goodfellow. 1980. Thin layer chromatography of methanolysates of mycolic acid-containing bacteria. J. Chromatogr. *188:* 221-233.
Minnikin, D.E., S.M. Minnikin, J.M. Parlett, M. Goodfellow and M. Magnusson. 1984a. Mycolic acid patterns of some species of *Mycobacterium*. Arch. Microbiol. *139:* 225-231.
Minnikin, D.E. and A.G. O'Donnell. 1984. Actinomycete envelope lipid and peptidoglycan composition. *In* Goodfellow, Mordarski and Williams (Editors), The Biology of the Actinomycetes, Academic Press, London, pp. 337-388.
Minnikin, D.E., A.G. O'Donnell, M. Goodfellow, G. Alderson, M. Athalye, A. Schaal and J.H. Parlett. 1984b. An integrated procedure for the extraction of bacterial isoprenoid quinones and polar lipids. J. Microbiol. Meth. *2:* 233-241.
Minnikin, D.E., P.V. Patel, L. Alshamaony and M. Goodfellow. 1977a. Polar lipid composition in the classification of *Nocardia* and related bacteria. Int. J. Syst. Bacteriol. *27:* 104-117.
Minnikin, D.E., T. Pirouz and M. Goodfellow. 1977b. Polar lipid composition in the classification of some *Actinomadura* species. Int. J. Syst. Bacteriol. *27:* 118-121.
Mishra, S.K., R.E. Gordon and D.A. Barnett. 1980. Identification of nocardiae and streptomycetes of medical importance. J. Clin. Microbiol. *11:* 728-736.
Miyadoh, S., T. Shomura, T. Ito and T. Niida. 1983. *Streptomyces sulfonofaciens* sp. nov. Int. J. Syst. Bacteriol. *33:* 321-324.
Miyadoh, S., H. Tohyama, S. Amano, T. Shomura and T. Niida. 1985. *Microbispora viridis*, a new species of *Actinomycetales*. Int. J. Syst. Bacteriol. *35:* 281-284.
Miyadoh, S., S. Amano, H. Tohyama and T. Shomura. 1987. *Actinomadura atramentaria*, a new species of the *Actinomycetales*. Int. J. Syst. Bacteriol. *37:* 342-346.
Miyagawa, E. 1982. Cellular fatty acid and fatty aldehyde composition of rumen bacteria. J. Gen. Appl. Microbiol. *28:* 389-408.
Miyairi, N., M. Takashima, K. Shimizu and H. Sakai. 1966. Studies on new antibiotics, cineromycins A and B. J. Antibiot. (Tokyo) Ser. A *19:* 56-62.
Miyashita, K., Y. Mikami and T. Arai. 1984. Alkalophilic actinomycete, *Nocardiopsis dassonvillei* subsp. *prasina* subsp. nov., isolated from soil. Int. J. Syst. Bacteriol. *34:* 405-409.
Miyazawa, Y. and C.A. Thomas. 1965. Composition of short segments of DNA molecules. J. Mol. Biol. *11:* 223-237.
Mizuno, S., H. Toyama and M. Tsukamura. 1966. Susceptibility of various mycobacteria to ethambutol. Differentiation between *M. avium* and *M. terrae*. Jpn. J. Bacteriol. *21:* 672-674.
Moiroud, A., M. Fauré-Raynaud and P. Simonet. 1984. Influence de basses températures sur la croissance et la survie de souches pures de *Frankia* isolées de nodules d'Aulnes. Plant Soil *78:* 91-97.
Molina, C. 1974. Farmer's lung in France. *In* de Haller and Suter (Editors), Aspergillosis and Farmer's Lung in Man and Animal, Hans Huber Publishers, Bern, pp. 205-206.
Monson, A.M., S.G. Bradley, L.W. Enquist and G. Cruces. 1969. Genetic homologies among *Streptomyces violaceoruber* strains. J. Bacteriol. *99:* 702-706.
Moore, W.E.C., D.E. Hash, L.V. Holdeman and E.P. Cato. 1980. Polyacrylamide slab gel electrophoresis of soluble proteins for studies of bacterial floras. Appl. Environ. Microbiol. *39:* 900-907.

Moore, W.E.C. and L.V. Holdeman. 1974. Human fecal flora: The normal flora of 20 Japanese-Hawaiians. Appl. Microbiol. *27:* 961–979.

Moore, W.E.C., J.L. Johnson and L.V. Holdeman. 1976. Emendation of *Bacteriodaceae* and *Butyrivibrio* and descriptions of *Desulfomonas* gen. nov. and ten new species in the genera *Desulfomonas, Butyrivibrio, Eubacterium, Clostridium* and *Ruminococcus.* Int. J. Syst. Bacteriol. *26:* 238–252.

Moppet, C.E., D.T. Dix, F. Johnson and C. Coronelli. 1972. Structure of thermorubin A the major orange-red antibiotic of *Thermoactinomyces antibioticus.* J. Am. Chem. Soc. *94:* 3269–3272.

Mordarska, H., S. Cebrat, B. Bach and M. Goodfellow. 1978. Differentiation of nocardioform actinomycetes by lyzozyme sensitivity. J. Gen. Microbiol. *109:* 381–384.

Mordarska, H., A. Gamian and J. Carrasco. 1983. Sugar-containing lipids in the classification of representative *Actinomadura* and *Nocardiopsis* species. Arch. Immunol. Ther. Exp. *31:* 135–143.

Mordarska, H., M. Mordarski and M. Goodfellow. 1972. Chemotaxonomic characters and classification of some nocardioform bacteria. J. Gen. Microbiol. *71:* 77–86.

Mordarski, M., M. Goodfellow, K. Szyba, G. Pulverer and A. Tkacz. 1977. Classification of the 'rhodochrous' complex and allied taxa based upon deoxyribonucleic acid reassociation. Int. J. Syst. Bacteriol. *27:* 31–37.

Mordarski, M., M. Goodfellow, K. Szyba, A. Tkacz, G. Pulverer and K.P. Schaal. 1980a. Deoxyribonucleic acid reassociation in the classification of the genus *Rhodococcus.* Int. J. Syst. Bacteriol. *30:* 521–527.

Mordarski, M., M. Goodfellow, A. Tkacz, G. Pulverer and K.P. Schaal. 1980b. Ribosomal ribonucleic acid similarities in the classification of *Rhodococcus* and related taxa. J. Gen. Microbiol. *118:* 313–319.

Mordarski, M., M. Goodfellow, S.T. Williams and P.H.A. Sneath. 1986. Evaluation of species groups in the genus *Streptomyces. In* Szabó, Biró and Goodfellow (Editors), Biological, Biochemical and Biomedical Aspects of Actinomycetes, Akadémiai Kaidó, Budapest, pp. 517–525.

Mordarski, M., I. Kaszen, A. Tkacz, M. Goodfellow, G. Alderson, K.P. Schaal and G. Pulverer. 1981a. Deoxyribonucleic acid pairing in the classification of *Rhodococcus.* Zentralbl. Bakteriol. Mikrobiol. Hyg. I. Abt. Orig., Suppl. *11:* 25–31.

Mordarski, M., K. Schaal, A. Tkacz, G. Pulverer, K. Szyba and M. Goodfellow. 1978. Deoxyribonucleic acid base composition and homology studies on *Nocardia.* Zentralbl. Bakteriol. Parasitenkd. Infektionskr. Hyg. I. Abt. Suppl. *6:* 91–97.

Mordarski, M., K. Szyba, G. Pulverer and M. Goodfellow. 1976. Deoxyribonucleic acid reassociation in the classification of the 'rhodochrous' complex and allied taxa. J. Gen. Microbiol. *94:* 235–245.

Mordarski, M., A. Tkacz, M. Goodfellow, K.P. Schaal and G. Pulverer. 1981b. Ribosomal ribonucleic acid similarities in the classification of actinomycetes. Zentralbl. Bakteriol. Mikrobiol. Hyg. 1 Abt. Suppl. *11:* 79–85.

Morse, M.E. 1912. A study of the diphtheria group of organisms by the biometrical method. J. Infect. Dis. *11:* 253–285.

Mort, A., P. Normand and M. Lalonde. 1983. 2-*O*-Methyl-D-mannose, a key sugar in the taxonomy of *Frankia.* Can. J. Microbiol. *29:* 993–1002.

Mouches, C., J.C. Vignault, J.G. Tully, R.F. Whitcomb and J.M. Bove. 1979. Characterization of spiroplasmas by one- and two-dimensional protein analysis on polyacrylamide slab gels. Curr. Microbiol. *2:* 69–74.

Müller, O.F. 1773. Vermium Terrestrium et Fluviatilium, seu Animalium Infusoriorum, Helminthicorum et Testaceorum, non Marionorum, Succincta Historia. *1(1):* 1–135.

Müller, R. 1908. Eine Diphtheridee und eine *Streptothrix* mit gleichen blauen Farbstoff sowie Untersuchungen¨uber Streptothrixarten im allegemeinen. Zentralbl. Bakteriol. Parasitenkd. Infektionskr. Hyg. Abt. 1 Orig. *46:* 195–212.

Murase, M., T. Hikiji, K. Nitta, Y. Okami, T. Takeuchi and H. Umezawa. 1961. Peptimycin, a product of *Streptomyces* exhibiting apparent inhibition against Ehrlich carcinoma. J. Antibiot. (Tokyo) Ser. A *14:* 113–118.

Murray, R.G.E. 1962. Fine structure and taxonomy of bacteria. *In* Ainsworth and Sneath (Editors), Microbial Classification, Cambridge University Press, Cambridge, England.

Murray, R.G.E. 1968. Microbial structure as an aid to microbial classification and taxonomy. Spisy (Faculte des Sciences de l'Universite J.E. Purkyne, Brno) *43:* 249–252.

Murray, R.G.E. 1974. A place for bacteria in the living world. *In* Buchanan and Gibbons (Editors), Bergey's Manual of Determinative Bacteriology, 8th Ed., The Williams and Wilkins Co., Baltimore, pp. 4–9.

Murry, M.A., M.S. Fontaine and J.G. Torrey. 1984. Growth kinetics and nitrogenase induction in *Frankia* sp. HFP ArI3 grown in batch culture. Plant Soil *78:* 61–78.

Mutimer, M.D. and J.B. Woolcock. 1980. *Corynebacterium equi* in cattle and pigs. Vet. Q. *2:* 25–27.

Myhre, E.B. and P. Kuusela. 1983. Binding of human fibronectin to group A, C and G streptococci. Infect. Immun. *40:* 29–34.

Nagatsu, J., K. Anzai, K. Ohkuma and S. Suzuki. 1963. Studies on a new antibiotic, tuberin. IV. Taxonomic studies on the tuberin-producing organisms, *Streptomyces amakusaenisis.* J. Antibiot. (Tokyo) Ser. A *16:* 207–210.

Nakagawa, M., Y. Hayakawa, H. Kawai, K. Imamura, H. Inoue, A. Shimazu, H. Seto and N. Otake. 1983. A new anthracycline antibiotic *N*-formyl-13-dehydrocarminomycin. J. Antibiot. *36:* 457–458.

Nakamura, G. 1961a. Studies on antibiotic actinomycetes. I. On *Streptomyces* producing a new antibiotic tubermycin. J. Antibiot. (Tokyo) Ser. A *14:* 86–89.

Nakamura, G. 1961b. Studies on antibiotic actinomycetes. II. On *Streptomyces* producing a new antibiotic tubercidin. J. Antibiot. (Tokyo) Ser. A *14:* 90–93.

Nakamura, G. 1961c. Studies on antibiotic actinomycetes. III. On *Streptomyces* producing A-B-D-ribofuranosylpurine. J. Antibiot. (Tokyo) Ser. A *14:* 94–97.

Nakamura, G. and K. Isono. 1983. A new species of *Actinomadura* producing a polyether antibiotic, cationomycin. J. Antibiot. *36:* 1468–1472.

Nakazawa, K. 1955. *Streptomyces albireticuli* nov. sp. J. Agr. Chem. Soc. Jpn. *29:* 647–649.

Nakazawa, K. and S. Fujii. 1957. Studies on streptomycetes. On *Streptomyces sindenensis* nov. sp. Annu. Rep. Takeda Res. Lab. *16:* 109–110.

Nakazawa, K., K. Tanabe, M. Shibata, A. Miyake and T. Takewaka. 1956. Studies on streptomycetes. Cladomycin, a new antibiotic produced by *Streptomyces lilacinus* nov. sp. J. Antibiot. (Tokyo) *9(B):* 81.

Naumova, I.B., N.V. Potekhina, L.P. Terekhova, T.P. Preobrazhenskaya, K. Digimbay. 1986. Wall polyol phosphate polymers of bacteria belonging to the genus *Actinomadura. In* Szabó, Biró and Goodfellow (Editors), Biological, Biochemical and Biomedical Aspects of Actinomycetes, Akadémiai Kiadó, Budapest, pp. 561–566.

Naumova, I.B., M.S. Zaretskaya, N.F. Dmitrieva and G.M. Streshinskaya. 1978. Structural features of teichoic acids of certain *Streptomyces* species. Zentralbl. Bakteriol. Parasitenkd. Infektionskr. Hyg. Abt. 1 Suppl. *6:* 261–268.

Nesterenko, O.A., S.A. Kasumova and E.I. Kvasnikov. 1978a. Microorganisms of the *Nocardia* genus and the '*rhodochrous*' group in soils of the Ukranian SSR. Mikrobiologiya *47:* 866–870.

Nesterenko, O.A., E.I. Kvasnikov and S.A. Kasumova. 1978b. Properties and taxonomy of some spore-forming *Nocardia.* Zentralbl. Bakteriol. Parasitenkd. Infektionskr. Hyg. Abt. I Suppl. *6:* 253–260.

Nesterenko, O.A., E.I. Kvasnikov and T.M. Nogina. 1985. *Nocardioides* Prauser 1976 and "*tumescens*" taxon (Jensen 1943) Conn and Dimmick 1947. *In* Nesterenko, Kvasnikov and Nogina (Editors), Nocardioform and Coryneform Bacteria (in Russian), Naukova Dumka, Kiev, pp. 172–182.

Nesterenko, O.A., T.M. Nogina, S.A. Kasumova, E.I. Kvasnikov and S.G. Batrakov. 1982. *Rhodococcus luteus* nom. nov. and *Rhodococcus maris* nom. nov. Int. J. Syst. Bacteriol. *32:* 1–14.

Newcomb, W., D. Callaham, J.G. Torrey and R.L. Peterson. 1979. Morphogenesis and fine structure of the actinomycetous endophyte of nitrogen-fixing root nodules of *Comptonia peregrina.* Bot. Gaz. *140(Suppl):* S22–S34.

Nicolet, J. and E.N. Bannerman. 1975. Extracellular enzymes of *Micropolyspora faeni* found in mouldy hay. Infect. Immun. *12:* 7–12.

Nicolet, J., P. Paroz and M. Krawinkler. 1980. Polyacrylamide gel electrophoresis of whole-cell proteins of porcine strains of *Haemophilus.* Int. J. Syst. Bacteriol. *30:* 69–76.

Niida, T. 1965. *In* Kondo, Sezaki, Koike, Shimura, Akita, Satoh and Hara. 1965. Destomycins A and B, two new antibiotics produced by a *Streptomyces.* J. Antibiot. (Tokyo) Ser. A *18:* 38–42.

Niida, T., K. Hamamoto, T. Tsuruoka and T. Hara. 1963. Taxonomic studies on a new *Streptomyces* producing both blasticidin S and 8-azaguanine. Sci. Rep. Meiji Seika Kaisha *6:* 27–39.

Niida, T. and M. Ogasawara. 1960. Taxonomical study on a new *Streptomyces* producing taitomycin. Sci. Rep. Meiji Seika Kaisha *3:* 23–26.

Nisbet, L.J. 1980. Strain degeneration in antibiotic-producing actinomycetes. *In* Kirsop (Editor), The Stability of Industrial Organisms, Commonwealth Mycological Institute, Kew, Surrey, England, pp. 39–52.

Nishimura, H., T. Kimura and M. Kuroya. 1953. On a yellow crystalline antibiotic, identical with aureothricin, isolated from a new species of *Streptomyces*, 39a, and its taxonomic study. J. Antibiot. (Tokyo) Ser. A *6:* 57–65.

Nishimura, H., T. Kimura, K. Tawara, K. Sasaki, K. Nakajima, N. Shimaoka, S. Okamodo, M. Shimohira and J. Isono. 1957. Aburamycin, a new antibiotic. J. Antibiot. (Tokyo) Ser. A *10:* 205–211.

Nishimura, H., M. Mayama, Y. Komatsu, H. Katôo, N. Shimaoka and Y. Tanaka. 1964. Showdomycin, a new antibiotic from a *Streptomyces* sp. J. Antibiot. (Tokyo) Ser. A *17:* 148–155.

Nishimura, H., S. Okamoto, M. Mayama, H. Ohtsuka, K. Nakajima, K. Tawara, M. Shimohira and N. Shimaoka. 1961. Siomycin, a new thiostrepton-like antibiotic. J. Antibiot. (Tokyo) Ser. A *14:* 255–263.

Nitsch, B. and H.J. Kutzner. 1969. Egg-yolk agar as a diagnostic medium for streptomycetes. Experientia *25:* 220–221.

Nocard, E. 1888. Note sur la maladie de boeufs de la Guadeloupe, connue sous le nom de farcin. Ann. Inst. Pasteur, Paris *2:* 293–302.

Noel, K.D. and W.J. Brill. 1980. Diversity and dynamics of indigenous *Rhizobium japonicum* populations. Appl. Environ. Microbiol. *40:* 931–938.

Nolof, G. 1962. Beitrage zur Kenntnis des Stoffwechsels von *Nocardia hydrocarbonoxydans* n. spec. Arch. Mikrobiol. *44:* 278–297.

Nolof, G. and P. Hirsch. 1962. *Nocardia hydrocarbonoxydans* n. spec: ein oligocarbophiler Actinomycet. Arch. Mikrobiol. *44:* 266–277.

Nomi, R. 1960. Studies on the classification of streptomycetes, part XIII. The microscopic morphology and other characteristics of whorl forming strains. J. Gen. Appl. Microbiol. *5:* 191–199.

Nonomura, H. 1974a. Key for classification and identification of 458 species of the streptomycetes included in I.S.P. J. Ferment. Technol. *52:* 78–92.

Nonomura, H. 1974b. Key for classification and identification of species of rare actinomycetes isolated from soils in Japan. J. Ferment. Technol. *52:* 71–77.

Nonomura, H. 1984. Design of a new medium for isolation of soil actinomycetes.

The Actinomycetes *18:* 206-209.

Nonomura, H., S. Iino and M. Hayakawa. 1979. Classification of actinomycetes of genus *Ampullariella* from soils of Japan. J. Ferment. Technol. *57:* 79-85.

Nonomura, H. and Y. Ohara. 1957. Distribution of actinomycetes in soil. II. *Microbispora*, a new genus of *Streptomycetaceae*. J. Ferment. Technol. *35:* 307-311.

Nonomura, H. and Y. Ohara. 1960a. Distribution of actinomycetes in soil. IV. The isolation and classification of the genus *Microbispora*. J. Ferment. Technol. *38:* 401-405.

Nonomura, H. and Y. Ohara. 1960b. Distribution of the actinomycetes in soil. V. The isolation and classification of the genus *Streptosporangium*. J. Ferment. Technol. *38:* 405-409.

Nonomura, H. and Y. Ohara. 1969a. Distribution of actinomycetes in soil. VI. A culture method effective for both preferential isolation and enumeration of *Microbispora* and *Streptosporangium* strains in soil (Part 1). J. Ferment. Technol. *47:* 463-469.

Nonomura, H. and Y. Ohara. 1969b. Distribution of actinomycetes in soil. VII. A culture method effective for both preferential isolation and enumeration of *Microbispora* and *Streptosporangium* strains in soil. (Part 2). Classification of the isolates. J. Ferment. Technol. *47:* 701-709.

Nonomura, H. and Y. Ohara. 1971a. Distribution of actinomycetes in soil. VIII. Green-spore group of *Microtetraspora*, its preferential isolation and taxonomic characteristics. J. Ferment. Technol. *49:* 1-7.

Nonomura, H. and Y. Ohara. 1971b. Distribution of actinomycetes in soil. IX. New species of the genus *Microbispora* and *Microtetraspora* and their isolation methods. J. Ferment. Technol. *49:* 887-894.

Nonomura, H. and Y. Ohara. 1971c. Distribution of actinomycetes in soil. X. New genus and species of monosporic actinomycetes in soil. J. Ferment. Technol. *49:* 895-903.

Nonomura, H. and Y. Ohara. 1971d. Distribution of actinomycetes in soil. XI. Some new species of the genus *Actinomadura* Lechevalier et al. J. Ferment. Technol. *49:* 904-912.

Nonomura, H. and Y. Ohara. 1974. A new species of actinomycetes, *Thermomonospora mesouviformis* sp. nov. J. Ferment. Technol. *52:* 10-13.

Nonomura, H. and S. Takagi. 1977. Distribution of actinoplanetes in soils of Japan. J. Ferment. Technol. *55:* 423-428.

Normand, P. and M. Lalonde. 1982. Evaluation of *Frankia* strains isolated from provenances of two *Alnus* species. Can. J. Microbiol. *28:* 1133-1142.

Normand, P., P. Simonet, J.E.L. Butour, C. Rosenberg, A. Moiroud and M. Lalonde. 1983. Plasmids in *Frankia* sp. J. Bacteriol. *155:* 32-35.

Novick, R.P. 1969. Extrachromosomal inheritance in bacteria. Bacteriol. Rev. *33:* 210-235.

Nygaard, A.P. and B.D. Hall. 1963. A method for detection of RNA-DNA complexes. Biochem. Biophys. Res. Comm. *12:* 98-104.

Obi, S.K.C. and F.J.C. Odibo. 1984. Some properties of a highly thermostable α-amylase from a *Thermoactinomyces* sp. Can. J. Microbiol. *30:* 780-785.

O'Callaghan, C., A. Morris, S.M. Kirby and A.H. Shingler. 1972. Novel method for detection of β-lactamases by using a chromogenic cephalosporin substrate. Antimicrob. Agents Chemother. *1:* 283-288.

Ochi, K., M.J.M. Hitchcock and E. Katz. 1979. High-frequency fusion of *Streptomyces parvulus* or *Streptomyces antibioticus* protoplasts induced by polyethylene glycol. J. Bacteriol. *139:* 984-992.

Ochoa, A.G. 1973. Virulence of nocardiae. Can. J. Microbiol. *19:* 901-904.

O'Donnell, A.G. 1986. Chemical and numerical methods in the characterisation of novel isolates. *In* Szabó, Biró and Goodfellow (Editors), Biological, Biochemical and Biomedical Aspects of Actinomycetes, Akadémiai Kaidó, Budapest, pp. 541-549.

O'Donnell, A.G., M. Goodfellow and D.E. Minnikin. 1982a. Lipids in the classification of *Nocardioides*: Reclassification of *Arthrobacter simplex* (Jensen) Lochhead in the genus *Nocardioides* (Prauser) emend. O'Donnell et al. as *Nocardioides simplex* comb. nov. Arch. Microbiol. *133:* 323-329.

O'Donnell, A.G., D.E. Minnikin, M. Goodfellow and J.H. Parlett. 1982b. The analysis of actinomycete wall amino acids by gas chromatography. FEMS Microbiol. Lett. *15:* 75E-78E.

O'Donnell, A.G., D.E. Minnikin and M. Goodfellow. 1984. Integrated lipid and wall analysis of actinomycetes. *In* Goodfellow and Minnikin (Editors), Chemical Methods in Bacterial Systematics, Society for Applied Bacteriology, Technical Series No. 20, Academic Press, London, pp. 131-143.

O'Donnell, A.G., D.E. Minnikin, M. Goodfellow, J.H. Parlett, G.M. Schofield and K.P. Schaal. 1985. Lipid and wall amino acid composition in the classification and identification of *Arachnia propionica* Zentralbl. Bakteriol. Mikrobiol. Hyg. A *260:* 300-310.

O'Farrell, P. 1975. High resolution two-dimensional electrophoresis of proteins. J. Biol. Chem. *250:* 4007-4021.

Okami, Y. 1952a. Classification of the antagonistic ray fungi of Japan of the family *Streptomycetaceae* (in Japanese). Doctoral Dissertation, Hokaido University.

Okami, Y. 1952b. On the new *Streptomyces* isolated from soil. J. Antibiot. (Tokyo) Ser. A *5:* 477-480.

Okami, Y. and T. Okazaki. 1978. Actinomycetes in marine environments. Zentralbl. Bakteriol. Parasitenkd. Infektionskr. Hyg. Abt. 1 Suppl. *6:* 145-152.

Okami, Y., T. Okuda, T. Takeuchi, K. Nitta and H. Umezawa. 1953. Studies on antitumor substances produced by microorganisms, IV. Sarkomycin-producing *Streptomyces* and two other *Streptomyces* producing the antitumor substance No. 289 and caryomycin. J. Antibiot. (Tokyo) Ser. A *6:* 153-157.

Okami, Y., R. Utahara, S. Nakamura and H. Umezawa. 1954. Studies on antibiotic actinomycetes. IX. On *Streptomyces* producing a new antifungal substance of fungicidin-rimocidinchromin group, eurocidin group and trichomycin-ascosin-candicidin group. J. Antibiotics *7(A):* 98-103.

Okami, Y., R. Utahara, H. Oyagi, S. Nakamura, H. Umezawa, K. Yanagisawa and Y. Tsunematsu. 1955. The screening of anti-toxoplasmic substance produced by streptomycete and anti-toxoplasmic substance No. 534. J. Antibiot. (Tokyo) Ser. A *8:* 126-131.

Okanishi, Y., H. Akagawa and H. Umezawa. 1972. An evaluation of taxonomic criteria in streptomycetes on the basis of deoxyribonucleic acid homology. J. Gen. Microbiol. *72:* 49-58.

Okuda, T., T. Furumai, E. Watanabe, T. Okugawa and S. Kimura. 1966. *Actinoplanaceae* antibiotics. II. Study of sporoviridin. 2. Taxonomic study of the sporoviridin producing microorganism: *Streptosporangium viridogriseum* sp. nov. J. Antibiotics Ser. A *19:* 121-127.

Oliver, T.J., A. Goldstein, R.R. Bower, J.C. Holper and R.H. Otto. 1961. M-141, a new antibiotic. I. Antimicrobial properties, identity with actinospectacin, and production by *Streptomyces flavopersicus*, sp. n. Antimicrob. Agents Chemother. *1960-1961:* 495-502.

Oliver, T.J. and A.C. Sinclair. 1964. Antibiotic M-319. U.S. Patent 3,155,582. November 3, 1964.

Ōmura, S., Y. Iwai, Y. Takahashi, K. Kojima, K. Otoguro and R. Oiwa. 1981. Type of diaminopimelic acid different in aerial and vegetative mycelia of setamycin-producing actinomycete KM-6054. J. Antibiot. *34:* 1633-1634.

Ōmura, S., Y. Takahashi, Y. Iwai and H. Tanaka. 1982. *Kitasatosporia*, a new genus of the order *Actinomycetales*. J. Antibiot. *35:* 1013-1019.

Ōmura, S., Y. Takahashi, Y. Iwai and H. Tanaka. 1983a. *In* Validation of the publication of new names and new combinations previously effectively published outside the IJSB. List No. 11. Int. J. Syst. Bacteriol. *33:* 672-674.

Ōmura, S., Y. Takahashi, Y. Iwai and H. Tanaka. 1985. Revised nomenclature of *Kitasatosporia setalba*. Int. J. Syst. Bacteriol. *35:* 221.

Ōmura, S., H. Tanaka, Y. Tanaka, P. Spiri-Nakagawa, R. Oiwa, Y. Takahashi, K. Matsuyama and Y. Iwai. 1979. Studies on bacterial cell wall inhibitors VII. Azureomycin A and B, new antibiotics produced by *Pseudonocardia azurea* nov. sp. J. Antibiot. *32:* 985-994.

Ōmura, S., H. Tanaka, Y. Tanaka, P. Spiri-Nakagawa, R. Oiwa, Y. Takahashi, K. Matsuyama and Y. Iwai. 1983b. *In* Validation of the publication of new names and new combinations previously effectively published outside the IJSB. List No. 11. Int. J. Syst. Bacteriol. *33:* 672-674.

Orchard, V.A. 1981. The ecology of *Nocardia* and related taxa. Zentralbl. Bakteriol. Mikrobiol. Hyg. Suppl. *11:* 167-180.

Orchard, V.A. and M. Goodfellow. 1980. Numerical classification of some named strains of *Nocardia asteroides* and related isolates from soil. J. Gen. Microbiol. *118:* 295-312.

Orchard, V.A., M. Goodfellow and S.T. Williams. 1977. Selective isolation and occurrence of nocardiae in soil. Soil Biol. Biochem. *9:* 233-238.

Orla-Jensen, S. 1909. Die Hauptlinien des natürlichen Bakteriensystems. Zentralbl. Bakteriol. Parasitenkd. Infektionskr. Hyg. Abt. II *22:* 305-346.

Orlowska, B. 1969. *Thermoactinomyces albus*, a new species of thermophilic actinomycetes. *In* Annual Report of the Ludwig Hirszfeld Institute for Immunol. Exp. Ther., Wroclaw, Poland, pp. 25-26.

Ørskov, J. 1923. Investigations into the Morphology of the Ray Fungi, Levin and Munksgaard, Copenhagen, Denmark.

Ortiz-Ortiz, L., M.F. Contreras and L.F. Bojalil. 1976. *In* Goodfellow, Brownell and Serrano (Editors), The Biology of the Nocardiae, Academic Press, New York, pp. 418-428.

Ou, L.-T., J.M. Davidson and D.F. Rothwell. 1978. Responses of soil microflora to high 2,4-D applications. Soil Biol. Biochem. *10:* 443-445.

Owen, S.P., A. Dietz and G.W. Camiener. 1963. Sparsomycin, a new antitumor antibiotic. I. Discovery and biological properties. Antimicrob. Agents Chemother. *1962:* 772-779.

Pagani, H. and F. Parenti. 1978. *Kineosporia*, a new genus of the order *Actinomycetales*. Int. J. Syst. Bacteriol. *28:* 401-406.

Palleroni, N.J. 1976a. Chamber for bacterial chemotaxis experiments. Appl. Environ. Microbiol. *32:* 729-730.

Palleroni, N.J. 1976b. Chemotaxis in *Actinoplanes*. Arch. Microbiol. *110:* 13-18.

Palleroni, N.J. 1979. New species of the genus *Actinoplanes*, *Actinoplanes ferrugineus*. Int. J. Syst. Bacteriol. *29:* 51-55.

Palleroni, N.J. 1980. A chemotactic method for the isolation of *Actinoplanaceae*. Arch. Microbiol. *128:* 53-55.

Palleroni, N.J. 1983a. Biology of *Actinoplanes*. The Actinomycetes *17:* 46-65.

Palleroni, N.J. 1983b. Genetic recombination in *Actinoplanes brasiliensis* by protoplast fusion. Appl. Environ. Microbiol. *45:* 1865-1869.

Palleroni, N.J. 1984. Genus *Pseudomonas*. *In* Krieg and Holt (Editors), Bergey's Manual of Systematic Bacteriology, Vol. 1, The Williams and Wilkins Co., Baltimore, pp. 141-199.

Palleroni, N.J. and M. Doudoroff. 1972. Some properties and subdivisions of the genus *Pseudomonas*. Annu. Rev. Phytopathol. *10:* 73-100.

Palleroni, N.J., K.E. Reichett, D. Mueller, R. Epps, B. Tabenkin, D.N. Bull, W. Schüep and J. Berger. 1978. Production of a novel red pigment, rubidone by *Streptomyces echinoruber* sp. nov. I. Taxonomy, fermentation and partial purification. J. Antibiot. *31:* 1218-1225.

Palleroni, N.J., K.E. Reichett, D. Mueller, R. Epps, B. Tabenkin, D.N. Bull, W.

Schüep and J. Berger. 1981. *In* Validation of the publication of new names and new combinations previously effectively published outside the IJSB List No. 7. Int. J. Syst. Bacteriol. *31:* 382-383.

Parenti, F., G. Beretta, M.S. Berti and V. Arioli. 1978. Teichomycins, new antibiotics from *Actinoplanes teichomyceticus* nov. sp. I. Description of the producer strain, fermentation studies and biological properties. J. Antibiotics *31:* 276-283.

Parenti, F. and C. Coronelli. 1979. Members of the *Actinoplanes* and their antibiotics. Annu. Rev. Microbiol. *33:* 389-412.

Parenti, F., H. Pagani and G. Beretta. 1975. Lipiarmycin, a new antibiotic from *Actinoplanes*. I. Description of the producer strain and fermentation studies. J. Antibiotics *28:* 247-252.

Parenti, F., H. Pagani and G. Beretta. 1976. Gardimycin, a new antibiotic from *Actinoplanes*. I. Description of the producer strain and fermentation studies. J. Antibiotics *29:* 501-506.

Patel, J.J. 1969. Phages of lysogenic *Thermoactinomyces vulgaris*. Arch. Mikrobiol. *69:* 294-300.

Pearson, H.W., R. Howsley and S.T. Williams. 1982. A study of nitrogenase activity in *Mycoplana* species and free-living actinomycetes. J. Gen. Microbiol. *128:* 2073-2080.

Peattie, D.A. 1979. Direct chemical method for sequencing RNA. Proc. Nat. Acad. Sci. U.S.A. *76:* 1760-1764.

Peczynska-Czoch, W. and H. Mordarski. 1984. Transformation of xenobiotics. *In* Goodfellow, Mordarski and Williams (Editors), The Biology of the Actinomycetes, Academic Press, London, pp. 287-336.

Pepys, J., P.A. Jenkins, G.N. Festenstein, P.H. Gregory, M.E. Lacey and F.A. Skinner. 1963. Farmer's lung: Thermophilic actinomycetes as a source of "farmer's lung hay" antigen. Lancet *2:* 607-611.

Percich, J.A. and J.L. Lockwood. 1978. Interaction of atrazine with soil microorganisms: Population changes and accumulation. Can. J. Microbiol. *24:* 1145-1152.

Perez, C. 1908. Sur *Duboscqia legeri*, microsporidie nouvelle parasite die *Termes lucifugus* et sur la classification des microsporidies. Soc. Biol. (Paris) *65:* 631.

Périnet, P., J.G. Brouillette, J.A. Fortin and M. Lalonde. 1985. Large scale inoculation of actinorhizal plants with *Frankia*. Plant Soil *87:* 175-183.

Péerinet, P. and M. Lalonde. 1983. Axenic nodulation of *in vitro* propagated *Alnus glutinosa* plantlets by *Frankia* strains. Can. J. Bot. *61:* 2883-2888.

Perradin, Y., M.J. Mottet and M. Lalonde. 1983. Influence of phenolics on *in vitro* growth of *Frankia* strains. Can. J. Bot. *61:* 2807-2814.

Person, L.H. and W.J. Martin. 1940. Soil rot of sweet potatoes in Louisiana. Phytopathology *30:* 913-926.

Pether, J.V.S. and F.B. Greatorex. 1976. Farmer's lung disease in Somerset. Br. J. Ind. Med. *33:* 265-268.

Pidcock, K.A., B.S. Montenecourt and J.A. Sands. 1985. Genetic recombination and transformation in protoplasts of *Thermomonospora fusca*. Appl. Environ. Microbiol. *50:* 693-695.

Pier, A.C. and R.E. Fichtner. 1971. Serologic typing of *Nocardia asteroides* by immunodiffusion. Am. Rev. Respir. Dis. *103:* 398-707.

Pijper, A. and B.D. Pullinger. 1927. South African nocardioses. J. Trop. Med. Hyg. *30:* 153-156.

Pine, L. and L.K. Georg. 1974. Genus *Arachnia*. Pine and Georg 1969. *In* Buchanan and Gibbons (Editors), Bergey's Manual of Determinative Bacteriology, 8th Ed., The Williams and Wilkins Co., Baltimore, pp. 668-669.

Pinnert-Sindico, S. 1954. Une nouvelle espèce de *Streptomyces* productrice d'antibiotiques: *Streptomyces ambofaciens* n. sp.—caractères culturaux. Ann. Inst. Pasteur (Paris) *87:* 702-707.

Pinoy, E. 1912. Isolement et culture d'une nouvelle oospora pathogene. *In* Thiroux et Pelletier. Mycetome a grains rouges de la paroi thoracique. Bull. Soc. Path. Exot. *5:* 585-589.

Pinoy, E. 1913. Actinomycoses et mycétomes. Bull. Inst. Pasteur (Paris) *11:* 929-938.

Pittenger, R.C. and R.B. Brigham. 1956. *Streptomyces orientalis* n. sp., the source of vancomycin. Antibiot. Chemother. *6:* 642-647.

Pommer, E. 1959. Über die isolierung des endophyten aus den wurzelknollchen *Alnus glutinosa* Gaertn. und über erfolgreiche re-infektions versuche. Ber. Dtsch. Bot. Ges. *72:* 138-150.

Porter, J.N., R.I. Hewitt, C.W. Hesseltine, G. Krupka, J.A. Lowery, W.S. Wallace, N. Bohonos and J.H. Williams. 1952. Achromycin: A new antibiotic having trypanocidal properties. Antibiot. Chemother. *2:* 409-410.

Porter, J.N., J.J. Wilhelm and H.D. Tresner. 1960. Method for the preferential isolation of actinomycetes from soils. Appl. Microbiol. *8:* 174-178.

Porter, J.R. 1976. The world view of culture collections. *In* Colwell (Editor), The Role of Culture Collections in the Era of Molecular Biology, American Society for Microbiology, Washington, D.C., pp. 62-72.

Poschner, J., R.M. Kroppenstedt, A. Fischer and E. Stackebrandt. 1985. DNA:DNA reassociation and chemotaxonomic studies on *Actinomadura, Microbispora, Microtetraspora, Micropolyspora* and *Nocardiopsis*. Syst. Appl. Microbiol. *6:* 264-270.

Potekchina, L.L. 1965. *Streptosporangium rubrum* sp. nov.—a new species of the *Streptosporangium* genus. Mikrobiologiya *34:* 292-299.

Potter, L.F. and G.E. Baker. 1956. The microbiology of Flathead and Rogers Lakes. I. Preliminary survey of microbial populations. Ecology *37:* 351-355.

Prasad, N. 1971. Studies on the biology, ultrastructure, and effectiveness of a sporozoan endoparasite of nematodes. Dissertation, University of California, Riverside. [Diss. Abst. *73(10):* 388 p. 5234 B.]

Prauser, H. 1964. Aptness and application of colour codes for exact description of streptomycetes. Z. Allg. Mikrobiol. *4:* 95-98.

Prauser, H. 1966. New and rare actinomycetes and their DNA base composition. Publ. Fac. Sci. Univ. Brno *475:* 268-270.

Prauser, H. 1967. Contributions to the taxonomy of the *Actinomycetales*. Publ. Fac. Sci. Univ. Purkyne, Brno *K40:* 196-199.

Prauser, H. 1970. Character and genera arrangements in the *Actinomycetales*. *In* Prauser (Editor), The Actinomycetales, Gustav Fischer Verlag, Jena, pp. 407-418.

Prauser, H. 1974a. Host-phage relationships in nocardioform organisms. *In* Brownell (Editor), Proceedings of the International Conference on Biology of the Nocardiae, McGowen, Augusta, Georgia, pp. 84-85.

Prauser, H. 1974b. *Nocardioides* Prauser, *Nocardiopsis* J. Meyer, and *Micropolyspora fascifera* Prauser—new taxa of the *Actinomycetales*. Actinomycetologist (Tokyo) *24:* 14-15.

Prauser, H. 1976a. Host-phage relationships in nocardioform organisms. *In* Goodfellow, Brownell and Serrano (Editors), The Biology of the Nocardiae, Academic Press, New York, pp. 266-284.

Prauser, H. 1976b. *Nocardioides*, a new genus of the order *Actinomycetales*. Int. J. Syst. Bacteriol. *26:* 58-65.

Prauser, H. 1978. Considerations on taxonomic relations among Gram-positive, branching bacteria. Zentralbl. Bakteriol. Parasitenkd. Infektionskr. Hyg. Abt. 1 Suppl. *6:* 3-12.

Prauser, H. 1981a. Nocardioform organisms: General characterization and taxonomic relationships. Zentralbl. Bakteriol. Mikrobiol. Hyg. Suppl. *11:* 17-24.

Prauser, H. 1981b. Taxon specificity of lytic actinophages that do not multiply in the cells affected. Zentralbl. Bakteriol. Mikrobiol. Hyg. Suppl. *11:* 87-92.

Prauser, H. 1984a. *Nocardioides luteus* spec. nov. Z. Allg. Mikrobiol. *24:* 647-648.

Prauser, H. 1984b. One-tube method for liquid nitrogen preservation and shipping of actinomycetes. *In* Proceedings of IV International Conference on Culture Collections, Brno, Czechoslovakia, 20-24 July, 1981, World Federation for Culture Collections, pp. 109-115.

Prauser, H. 1984c. Phage host ranges in the classification and identification of Gram-positive branched and related bacteria. *In* Ortiz-Ortiz, Bojalil and Yakoleff (Editors), Biological, Biochemical, and Biomedical Aspects of Actinomycetes, Academic Press, Orlando, Florida, pp. 617-633.

Prauser, H. 1986. The *Cellulomonas, Oerskovia, Promicromonospora* complex. *In* Szabó, Biró and Goodfellow (Editors), Biological, Biochemical and Biomedical Aspects of Actinomycetes, Akademiai Kiadó, Budapest, pp. 527-539.

Prauser, H. and M. Bergholz. 1974. Taxonomy of actinomycetes and screening for antibiotic substances. Postepy. Hig. Med. Dosw. *28:* 441-457.

Prauser, H. and R. Falta. 1968. Phagensensibilität, Zellwandzusammensetzung und Taxonomie von Actinomyceten. Z. Allg. Mikrobiol. *8:* 39-46.

Prauser, H., M.P. Lechevalier and H.A. Lechevalier. 1970. Description of *Oerskovia* gen. n. to harbor Ørskov's motile *Nocardia*. Appl. Microbiol. *19:* 534.

Prauser, H. and S. Momirova. 1970. Phagensensibilitat, Zellwand-Zusammensetzung und Taxonomie einiger thermophiler Actinomyceten. Z. Allg. Mikrobiol. *10:* 219-222.

Prauser, H., L. Müller and R. Falta. 1967. On taxonomic position of the genus *Microellobosporia* Cross, Lechevalier and Lechevalier 1963. Int. J. Syst. Bacteriol. *17:* 361-366.

Preobrazhenskaya, T.P. 1957. *In* Gauze, Preobrazhenskaya, Kudrina, Blinov, Ryabova and Sveshnikova (Editors), Problems of Classification of Actinomycetes-Antagonists. Government Publishing House of Medical Literature, Medgiz, Moscow, pp. 1-165.

Preobrazhenskaya, T.P., N.V. Lavrova and N.O. Blinov. 1975a. Taxonomy of *Streptomyces luteofluorescens*. Mikrobiologiya *44:* 524-527.

Preobrazhenskaya, T.P., N.V. Lavrova, R.S. Ukholina and N.P. Nechaeva. 1975b. Isolation of new species of *Actinomadura* on selective media with streptomycin and bruneomycin. Antibiotiki *20:* 404-409.

Preobrazhenskaya, T.P. and M.A. Sveshnikova. 1974. New species of the *Actinomadura* genus. Mikrobiologiya *43:* 864-868.

Preobrazhenskaya, T.P. and M.A. Sveshnikova. 1985. *In* Validation of the publication of new names and new combinations previously effectively published outside the IJSB. List No. 17. Int. J. Syst. Bacteriol. *35:* 223-225.

Preobrazhenskaya, T.P., M.A. Sveshnikova and G.F. Gauze. 1982. On the transfer of certain species belonging to the genus *Actinomadura* Lechevalier and Lechevalier 1970 into the genus *Nocardiopsis* Meyer 1976. Mikrobiologiya *51:* 111-113.

Preobrazhenskaya, T.P., M.A. Sveshnikova and L.P. Terekhova. 1977. Key for the identification of the species of the genus *Actinomadura*. Biol. Actinomycetes Relat. Org. *12:* 30-38.

Preobrazhenskaya, T.P., M.A. Sveshnikova, L.P. Terekhova and N.T. Chormonova. 1978. Selective isolation of soil actinomycetes. *In* Mordarski, Kurylowicz and Jeljaszewicz (Editors), Nocardia and Streptomyces, Gustav Fischer Verlag Stuttgart, New York, pp. 119-123.

Preobrazhenskaya, T.P., L.P. Terekhova, A.V. Laiko, T.I. Selezneva, V.A. Zenkova and N.O. Blinov. 1976. *Actinomadura coeruleoviolaceae* sp. nov. and its antagonistic properties. Antibiotiki *21:* 779-784.

Preobrazhenskaya, T.P., L.P. Terekhova, A.V. Laiko, T.I. Selezneva, V.A. Zenkova and N.A. Blinov. 1987. *In* Validation of the publication of new names and new combinations previously effectively published outside the IJSB. List No. 23. Int. J. Syst. Bacteriol. *37:* 179-180.

Preobrazhenskaya, T.P., R.S. Ukholina, N.P. Nechaeva, V.A. Filicheva, G.V. Gavrilina, M.K. Kudinova, V.N. Borisova, N.M. Petukhova, I.N. Kovsharova, V.V. Proshlyakova and O.K. Rossolimo. 1973. A new species of the genus *Micropolyspora* and its antibiotic properties. Antibiotiki *18:* 963-968.

Pribram, E. 1919. Der gegenwärtige Bestand der vorm, Králschen Sammlung von Microorganismen, Wien, pp. 1-48.

Pribram, E. 1933. Klassification der Schizomyceten, F. Deuticke, Leipzig, pp. 1-143.

Pridham, T.G. 1965. Color and streptomycetes. Report of an international workshop on determination of color of streptomycetes. Appl. Microbiol. *13:* 43-61.

Pridham, T.G. 1970. New names and new combinations in the order *Actinomycetales* Buchanan 1917. U.S. Dept. Agric. Tech. Bull. *1424:* 1-55.

Pridham, T.G. 1974a. Genus IV *Microellobosporia* Cross et al. *In* Buchanan and Gibbons (Editors), Bergey's Manual of Determinative Bacteriology, 8th Ed., The Williams and Wilkins Co., Baltimore, pp. 843-845.

Pridham, T.G. 1974b. Micro-organism Culture Collections: Acronyms and Abbreviations, ARS-NC-17, Agricultural Research Service, U.S. Department of Agriculture, North Central Region, Peoria, Illinois.

Pridham, T.G. 1976a. Contemporary species concepts in *Actinomycetales. In* Arai (Editor), Actinomycetes: The Boundary Microorganisms, Toppan Company Ltd., Tokyo, pp. 163-174.

Pridham, T.G. 1976b. Identification of streptomycetes and streptoverticillia at the species level: Revision of 1965 system. *In* Arai (Editor), The Actinomycetes: The Boundary Microorganisms, Toppan Co. Ltd., Tokyo, pp. 175-181.

Pridham, T.G., P. Anderson, C. Foley, L.A. Lindenfelser, C.W. Hesseltine and R.G. Benedict. 1957. A selection of media for maintenance and taxonomic study of *Streptomyces.* Antibiot. Annu. *1956-57:* 947-953.

Pridham, T.G. and D. Gottlieb. 1948. The utilization of carbon compounds by some *Actinomycetales* as an aid for species determination. J. Bacteriol. *56:* 107-114.

Pridham, T.G. and C.W. Hesseltine. 1975. Culture collections and patent depositions. Adv. Appl. Microbiol. *19:* 1-23.

Pridham, T.G., C.W. Hesseltine and R.G. Benedict. 1958. A guide for the classification of streptomycetes according to selected groups. Placement of strains in morphological sections. Appl. Microbiol. *6:* 52-79.

Pridham, T.G. and A.J. Lyons. 1969. Progress in the clarification of the taxonomic and nomenclatural status of some problem actinomycetes. Dev. Ind. Microbiol. *10:* 183-221.

Pridham, T.G., A.J. Lyons and B. Phronpatima. 1973. Viability of *Actinomycetales* stored in soil. Appl. Microbiol. *26:* 441-442.

Pridham, T.G., A.J. Lyons and H.L. Seckinger. 1965. Comparison of some dried holotype and neotype specimens of streptomycetes with their living counterparts. Int. Bull. Bacteriol. Nomencl. Taxon. *15:* 191-237.

Pridham, T.G., O.L. Shotwell, F.H. Stodola, L.A. Lindenfelser, R.G. Benedict and R.V. Jackson. 1956. Antibiotics against plant disease. II. Effective agents produced by *Streptomyces cinnamoneus* forma *azacoluta* f. nov. Phytopathology *46:* 575-581.

Pridham, T.G. and H.D. Tresner. 1974a. Family *Streptomycetaceae* Waksman and Henrici. *In* Buchanan and Gibbons (Editors), Bergey's Manual of Determinative Bacteriology, 8th Ed., The Williams and Wilkins Co., Baltimore, pp. 747-748.

Pridham, T.G. and H.D. Tresner. 1974b. Genus I. *Streptomyces* Waksman and Henrici. *In* Buchanan and Gibbons (Editors), Bergey's Manual of Determinative Bacteriology, 8th Ed., The Williams and Wilkins Co., pp. 748-829.

Pringsheim, E.G. 1967. Bakterien und Cyanophyceen. Oesterr. Bot. Z. *114:* 324-340.

Prokop, J.F. 1964. Method for the preparation of a composition of matter having antitumour and antifungal activity. Offic. Gaz. US Patent Off. Patent 3,117,916.

Prosser, B. La T., and N.J. Palleroni. 1976. *Streptomyces longwoodensis* sp. nov. Int. J. Syst. Bacteriol. *26:* 319-322.

Prosser, B. La T. and N.J. Palleroni. 1981. *In* Validation of the publication of new names and new combinations previously effectively published outside the IJSB. List No. 7. Int. J. Syst. Bacteriol. *31:* 382-388.

Pulverer, G. and K.P. Schaal. 1978. Pathogenicity and medical importance of aerobic and anaerobic actinomycetes. *In* Mordarski, Kurylowicz and Jeljaszewicz (Editors), Nocardia and Streptomyces, Gustav Fischer Verlag Stuttgart, New York, pp. 417-428.

Pulverer, G., H. Schütt-Gerowitt and K.P. Schaal. 1974. Bacteriophages of *Nocardia. In* Brownell (Editor), Proceedings of the International Conference on the Biology of the Nocardiae, Merida, Venezuela, McGowan, Augusta, Georgia, p. 82.

Puppo, A., L. Dimitrijevic, H.G. Diem and Y.R. Dommergues. 1985. Homogeneity of superoxide dismutase patterns in *Frankia* strains from *Casuarinaceae.* FEMS Microbiol. Lett. *30:* 43-46.

Quispel, A. 1955. Symbiotic nitrogen fixation in non-leguminous plants. III. Experiments on the growth *in vitro* of the endophyte of *Alnus glutinosa.* Acta Bot. Neerl. *4:* 671-689.

Quispel, A. 1960. Symbiotic nitrogen fixation in non-leguminous plants V. The growth requirements of the endophyte of *Alnus glutinosa.* Acta. Bot. Neerl. *9:* 380-396.

Quispel, A. and A.J.P. Burggraaf. 1981. *Frankia,* the diazotrophic endophyte from actinorhizas. *In* Gibson and Newton (Editors), Current Perspectives in Nitrogen Fixation, Australian Acad. Sci., Canberra, pp. 229-236.

Quispel, A., A.J.P. Burggraaf, H. Borsje and T. Tak. 1983. The role of lipids in the growth of *Frankia* isolates. Can. J. Bot. *61:* 2801-2806.

Rahalkar, P.W. and M.J. Thirumalachar. 1968. Cultural characters and identity of some *Streptoverticillium* species producing polyene antibiotics. Hind. Antibiot. Bull. *11:* 90-96.

Rao, V.A., K.K. Prabhu, B.P. Sridhar, A. Venkateswarlu and P. Actor. 1987. Two new species of *Microbispora* from Indian soils: *Microbispora karnatakensis* sp. nov. and *Microbispora indica* sp. nov. Int. J. Syst. Bacteriol. *37:* 181-185.

Rao, K.V., W.S. Marsh and S.C. Brooks. 1964. Antibiotic narangomycin and method of production. US Patent 3,155,583.

Rao, K.V., W.S. Marsh and D.W. Renn. 1967. Antibiotic product and process of producing same. U.S. Patent 3,328,248, June 27, 1967.

Rast, H.G., G. Engelhardt, W. Diegler and P.R. Wallhoffer. 1980. Bacterial degradation of model compounds for lignin and chlorophenol derived lignin bound residues. FEMS Microbiol. Lett. *8:* 259-263.

Rautenshtein, Y.I. 1960. (Editor). Biology of Antibiotic-Producing Actinomycetes (in Russian). Tr. Inst. Mikrobiol. Akad. Nauk SSSR *8:* 1-344.

Ravin, A.W. 1963. Experimental approaches to the study of bacterial phylogeny. Am. Natur. *97:* 307-318.

Razin, S. and S. Rottem. 1967. Identification of *Mycoplasma* and other microorganisms by polyacrylamide gel electrophoresis of cell proteins. J. Bacteriol. *94:* 1807-1810.

Reanney, D. 1976. Extrachromosomal elements as possible agents of adaptation and development. Bacteriol. Rev. *40:* 552-590.

Reh, M. 1981. Chemolithoautotrophy as an autonomous and transferable property of *Nocardia opaca* 1b. Zentralbl. Bakteriol. Mikrobiol. Hyg. *11:* 577-583.

Reh, M. and H.G. Schlegel. 1981. Hydrogen autotrophy as a transferable genetic character of *Nocardia opaca* 1b. J. Gen. Microbiol. *126:* 327-336.

Reller, L.B., G.L. Maddoux, M.R. Eckman and G. Pappas. 1975. Bacterial endocarditis caused by *Oerskovia turbata.* Ann. Intern. Med. *83:* 664-666.

Richter, G. 1977. Routine use of thin-layer chromatography for cell wall analysis of aerobic actinomycetes, including two strains from sediments of the North Sea. Veroff. Inst. Meeresforsch. Bremerhaven. *16:* 125-138.

Ridell, M. 1974. Serological study of nocardiae and mycobacteria by using "*Mycobacterium*" *pellegrino* and *Nocardia corallina* precipitation reference systems. Int. J. Syst. Bacteriol. *24:* 64-72.

Ridell, M. 1975. Taxonomic study of *Nocardia farcinica* using serological and physiological characters. Int. J. Syst. Bacteriol. *25:* 124-132.

Ridell, M. 1977. Studies on corynebacterial precipitinogens common to mycobacteria, nocardiae and rhodochrous. Int. Arch. Allergy Appl. Immunol. *55:* 468-475.

Ridell, M. 1981a. Immunodiffusion studies of *Mycobacterium, Nocardia* and *Rhodococcus* for taxonomic purposes. Zentralbl. Bakteriol. Mikrobiol. Hyg. I Abt. Orig., Suppl. *11:* 235-241.

Ridell, M. 1981b. Immunodiffusion studies of some *Nocardia* strains. J. Gen. Microbiol. *123:* 69-74.

Ridell, M. 1983. Sensitivity to capreomycin and prothionamide in strains of *Mycobacterium, Nocardia, Rhodococcus* and related taxa for taxonomical purposes. Zentralbl. Bakteriol. Mikrobiol. Hyg. I Abt. A *255:* 309-316.

Ridell, M. 1984. Serotaxonomical analyses of strains referred to *Nocardia amarae* and *Rhodococcus equi.* Zentralbl. Bakteriol. Mikrobiol. Hyg. I. Abt. Orig. A *259:* 492-497.

Ridell, M., R. Baker, A. Lind and O. Ouchterlony. 1979. Immunodiffusion studies of ribosomes in classification of mycobacteria and related taxa. Int. Arch. Allergy Appl. Immunol. *59:* 162-172.

Ridell, M., M. Goodfellow and D.E. Minnikin. 1985. Immunodiffusion and lipid analyses in the classification of "*Mycobacterium album*" and the "*aurantiaca*" taxon. Zentralbl. Bakteriol. Mikrobiol. Hyg. A *259:* 1-10.

Ridell, M. and M. Norlin. 1973. Serological study of *Nocardia* by using mycobacterial precipitation reference systems. J. Bacteriol. *113:* 1-7.

Ridell, M., G. Wallerström and S.T. Williams. 1986. Immunodiffusion analysis of phenetically defined strains of *Streptomyces, Streptoverticillium* and *Nocardiopsis.* Syst. Appl. Microbiol. *8.* 24-27.

Ridell, M. and S.T. Williams. 1983. Serotaxonomical analysis of some *Streptomyces* and related organisms. J. Gen. Microbiol. *129:* 2857-2861.

Ridgway, R. 1912. Color Standards and Color Nomenclature, Publisher, Washington, D.C.

Rippon, J.W. 1968. Extracellular collagenase produced by *Streptomyces madurae.* Biochim. Biophys. Acta *159:* 147-152.

Ritchie, D.A., J. Swift and S.T. Williams. 1986. Screening *Streptomyces* for phenotypic changes induced by transferable plasmids. *In* Szabó, Biró and Goodfellow (Editors), Biological, Biochemical and Biomedical Aspects of Actinomycetes, Akadémiai Kiadó, Budapest, p. 157.

Roberts, D.S. 1981. The family *Dermatophiliceae. In* Starr, Stolp, Trüper, Balows and Schlegel (Editors), The Prokaryotes: A Handbook on Habitats, Isolation and Identification of Bacteria, Springer-Verlag, New York, pp. 2011-2015.

Roberts, G.P., W.T. Leps, L.E. Silver and W.J. Brill. 1980. Use of two-dimensional gel electrophoresis to identify and classify *Rhizobium* strains. Appl. Environ. Microbiol. *39:* 414-422.

Roberts, R.C., F.J. Wenzel and D.A. Emanuel. 1976a. Precipitating antibodies in a midwest dairy farming population toward the antigens associated with farmers lung disease. J. Allergy Clin. Immunol. *57:* 518-524.

Roberts, R.C., D.P. Zais and D.A. Emanuel. 1976b. The frequency of precipitins to trichloroacetic acid-extractable antigens from thermophilic actinomycetes in farmer's lung patients and asymptomatic farmers. Annu. Rev. Resp. Dis. *114:* 23-28.

Roberts, R.C., D.P. Zais, J.J. Marx and M.W. Treuhaft. 1977. Comparative electrophoresis of the proteins and proteases in thermophilic actinomycetes. J. Lab. Clin. Med. *90:* 1076-1085.

Roché, C., H. Albertyn, N.O. van Gylswyk and A. Kistner. 1973. The growth response of cellulolytic acetate-utilizing and acetate-producing butyrivibrios to volatile fatty acids and other nutrients. J. Gen. Microbiol. *78:* 253-260.

Rode, L.M., B.R.S. Genthner and M.P. Bryant. 1981. Syntrophic association by cocultures of the methanol and CO_2-H_2 utilizing species, *Eubacterium limosum*, and pectin fermenting *Lachnospira multiparus* during growth in a pectin medium. Appl. Environ. Microbiol. *42:* 20-22.

Ross, H.N.M., M.D. Collins, B.J. Tindall and W.D. Grant. 1981. A rapid procedure for the detection of archaebacterial lipids in halophilic bacteria. J. Gen. Microbiol. *123:* 75-80.

Rossi-Doria, T. 1891. Su di alcune specie di '*Streptothrix*' trovate nell'aria studate in rapporto a quelle giá note a specialmente all' '*Actinomyces*'. Ann. Ist. Igiene Sper. Univ. Roma *1:* 399-438.

Rothrock, C.S. and D. Gottlieb. 1981. Importance of antibiotic production in antagonism of selected *Streptomyces* species to two soil-borne plant pathogens. J. Antibiot. *34:* 830-835.

Rothwell, F.M. 1957. A further study of Karling's keratinophilic organism. Mycologia *49:* 68-72.

Routien, J.B. and A. Hofmann. 1951. *Streptomyces californicus* productor de viomicina. Antibiot. Chemother. *1:* 387-389.

Rowbotham, T.J. and T. Cross. 1977a. Ecology of *Rhodococcus coprophilus* and associated actinomycetes in fresh water and agricultural habitats. J. Gen. Microbiol. *100:* 231-240.

Rowbotham, T.J. and T. Cross. 1977b. Ecology of *Rhodococcus coprophilus* sp. nov.: an aerobic nocardioform actinomycete belonging to the '*rhodochrous*' complex. J. Gen. Microbiol. *100:* 123-138.

Rowbotham, T.J. and T. Cross. 1979. *In* Validation of the publication of new names and new combinations previously effectively published outside the IJSB. List No. 2. Int. J. Syst. Bacteriol. *29:* 79-80.

Ruan, J., Y. Zhang and C. Jiang. 1976. A taxonomic study of *Actinoplanaceae*. II. Four new species of *Actinoplanes*. Acta Microbiol. Sin. *16:* 291-300.

Ruan, J.-S., M.P. Lechevalier, C.-L. Jiang and H.A. Lechevalier. 1985. *Chainia kunmingensis*, a new actinomycete species found in soil. Int. J. Syst. Bacteriol. *35:* 164-168.

Ruan, J.-S., M.P. Lechevalier, C.-R. Jiang and H.A. Lechevalier. 1986. A new species of the genus *Actinoplanes*, *Actinoplanes minutisporangius* n. sp. Actinomycetes *19:* 163-175.

Ruan, J.-S. and Y.-M. Zhang. 1979. Two new species of *Nocardioides* (in Chinese). Acta Microbiol. Sin. *19:* 347-352.

Rucker, R.R. 1949. A streptomycete pathogenic to fish. J. Bacteriol. *58:* 659-664.

Ruddick, S.M. and S.T. Williams. 1972. Studies on the ecology of actinomycetes in soil V. Some factors influencing the dispersal and adsorption of spores. Soil Biol. Biochem. *4:* 93-103.

Rühberg, W. 1933. Über eine Hefeinfektion bei *Daphnia magna*. Arch. Protistenk. *80:* 72-100.

Rullmann, W. 1895. Chemisch bacteriologische Untersuchungen von Zwischendeckenfüllengen mit besondere Berücksichtung von *Cladotrix oderifera*. Inaugural Dissertation Akad. Buchdruckerei von F. Strauv., Munich, p. 1-47.

Runmao, H. 1987. *Saccharomonospora azurea* sp.nov., a new species from soil. Int. J. Syst. Bacteriol. *37:* 60-61.

Ruschmann, G. 1952. *Streptomyces mirabilis* und das Miramycin. Pharmazie *7:* 542-550.

Ruvkun, G.B. and F.M. Ausubel. 1980. Interspecies homology of nitrogenase genes. Proc. Natl. Acad. Sci. U.S.A. *77:* 191-195.

Ryu, D.D.Y., K.S. Kim, N.Y. Cho and H.S. Pai. 1983. Genetic recombination in *Micromonospora rosaria* by protoplast fusion. Appl. Environ. Microbiol. *45:* 1854-1858.

Saddler, G.S., M. Goodfellow, D.E. Minnikin and A.G. O'Donnell. 1986. Influence of the growth cycle on the fatty acid and menaquinone composition of *Streptomyces cyaneus* NCIB 9616. J. Appl. Bacteriol. *60:* 51-56.

Sakamoto, J.M., S.-I. Kondo, H. Yumoto and M. Arishima. 1962. Bundlins A and B, two antibiotics produced by *Streptomyces griseofuscus* nov. sp. J. Antibiot. (Tokyo) Ser. A *5:* 98-102.

Samsonoff, W.A., M.A. Detlefsen, A.F. Fonseca and M.R. Edwards. 1977. Deoxyribonucleic acid base composition of *Dermatophilus congolensis* and *Geodermatophilus obscurus*. Int. J. Syst. Bacteriol. *27:* 22-25.

Sandrak, N.A. 1977. Cellulose decomposition by *Micromonospora*. Mikrobiologiya *46:* 478-481.

Sanger, F., G.G. Brownless and B.G. Barrell. 1965. A two-dimensional fractionation procedure for radioactive nucleotides. J. Mol. Biol. *13:* 373-398.

Sanger, F., S. Nicklen and A.R. Coulson. 1977. DNA sequencing with chain-terminating inhibitors. Proc. Nat. Acad. Sci. U.S.A. *74:* 5463-5467.

Sarfert, E., S. Kretschmer, H. Triebel and G. Luck. 1979. Properties of *Thermoactinomyces vulgaris* phage Ta, and its extracted DNA. Z. Allg. Mikrobiol. *19:* 203-210.

Sartory, A. 1920. Champignons Parasites de l'Homme et des Animaux, V. Arsant, Saint-Nicholas-du-Port.

Sawazaki, T., S. Suzuki, G. Nakamura, M. Kawasaki, S. Yamashita, K. Isono, K. Anzai, Y. Serizawa and Y. Sekiyama. 1955. Streptomycin production by a new strain *Streptomyces mashuensis*. J. Antibiotics *8(A):* 44-47.

Sayre, R.M. 1980. Biocontrol: *Bacillus penetrans* and related parasites of nematodes. J. Nematol. *12:* 260-270.

Sayre, R.M., J.R. Adams and W.P. Wergin. 1979. Bacterial parasite of a cladoceran: Morphology, development in vivo, and taxonomic relationships with *Pasteuria ramosa* Metchnikoff 1888. Int. J. Syst. Bacteriol. *29:* 252-262.

Sayre, R.M., R.L. Gherna and W.P. Wergin. 1983. Morphological and taxonomic reevaluation of *Pasteuria ramosa* Metchnikoff 1888 and *"Bacillus penetrans"* Mankau 1975. Int. J. Syst. Bacteriol. *33:* 636-649.

Sayre, R.M. and M.P. Starr. 1985. *Pasteuria penetrans* (ex Thorne 1940) nom. rev., comb. n., sp. n., a mycelial and endospore-forming bacterium parasitic in plant-parasitic nematodes. Proc. Helminthol. Soc. Wash. *52:* 149-165.

Sayre, R.M. and M.P. Starr. 1986. *In* Validation of the publication of new names and new combinations effectively published outside the IJSB. List No. 20. Int. J. Syst. Bacteriol. *36:* 354-356.

Sayre, R.M. and M.P. Starr. 1988. Bacterial diseases and antagonisms of nematodes. *In* Poinar and Jansson (Editors), Nematode Pathology, Vol. 1, CRC Press, Boca Raton, Florida, pp. 65-101.

Sayre, R.M., M.P. Starr, A.M. Golden, W.P. Wergin and B.Y. Endo. 1988. Comparison of *Pasteuria penetrans* from *Meloidogyne incognita* with a related mycelial and endospore-forming bacterial parasite from *Pratylenchus brachyurus*. Proc. Helminthol. Soc. Wash. *55:* 28-49.

Sayre, R.M. and W.P. Wergin. 1977. Bacterial parasite of a plant nematode: Morphology and ultrastructure. J. Bacteriol. *129:* 1091-1101.

Sayre, R.M., W.P. Wergin and R.E. Davis. 1977. Occurrence in *Monia* [sic] *rectirostris* (Cladocera: Daphnidae) of a parasite morphologically similar to *Pasteuria ramosa* (Metchnikoff, 1988). Can. J. Microbiol. *23:* 1573-1579.

Schaal, K.P. 1972. Zur mikrobiologisher Diagnostik der Nocardiose. Zentralbl. Bakteriol. Parasitenkd. Infektionskr. Hyg. I Abt. Orig. *220:* 242-246.

Schaal, K.P. 1977. *Nocardia, Actinomadura and Streptomyces. In* CRC Handbook Series in Clinical Laboratory Sciences Sect. E., Clinical Microbiology Vol. 1, CRC Press, Cleveland, Ohio, pp. 131-158.

Schaal, K.P. 1984a. Identification of clinically significant actinomycetes and related bacteria using chemical techniques. *In* Goodfellow and Minnikin (Editors), Chemical Methods in Bacterial Systematics, Society for Applied Bacteriology, Technical Series No. 20, Academic Press, London, pp. 359-381.

Schaal, K.P. 1984b. Laboratory diagnosis of actinomycete diseases. *In* Goodfellow, Mordarski and Williams (Editors), The Biology of the Actinomycetes, Academic Press, London, pp. 425-456.

Schaal, K.P. 1986a. Genus *Actinomyces* Harz 1877. *In* Sneath, Mair, Sharpe and Holt (Editors), Bergey's Manual of Systematic Bacteriology, Vol. 2, The Williams and Wilkins Co., Baltimore, pp. 1383-1418.

Schaal, K.P. 1986b. Genus *Arachnia* Pine and Georg 1969. *In* Sneath, Mair, Sharpe and Holt (Editors), Bergey's Manual of Systematic Bacteriology, Vol. 2, The Williams and Wilkins Co., Baltimore, pp. 1332-1342.

Schaal, K.P. and B.L. Beaman. 1984. Clinical significance of actinomycetes. *In* Goodfellow, Mordarski and Williams (Editors), The Biology of the Actinomycetes, Academic Press, London, pp. 389-424.

Schaal, K.P. and H. Reutersberg. 1978. Numerical taxonomy of *Nocardia asteroides*. Zentralbl. Bakteriol. Parasitenkd. Infektionskr. Hyg. I Abt. Suppl. *6:* 53-62.

Schäfer, D. 1969. Eine neue *Streptosporangium*-Art aus türkischer Steppenerde. Arch. Mikrobiol. *66:* 365-373.

Schäfer, D. 1973. Beiträge zur Klassifizierung und Taxonomie der Actinoplanaceen. Dissertation, Marburg.

Schildkraut, C.L., J. Marmur and P. Doty. 1962. Determination of the base composition of deoxyribonucleic acid from its buoyant density in CsCl. J. Mol. Biol. *4:* 430-443.

Schleifer, K.-H. and O. Kandler. 1967. Zur chemischen zusammensetzung der zellwand der Streptokokken. I. Die aminosäuresequenz des mureins von *Str. thermophilus* und *Str. faecalis*. Arch. Mikrobiol. *57:* 335-364.

Schleifer, K.-H. and O. Kandler. 1972. Peptidoglycan types of bacterial cell walls and their taxonomic implications. Bacteriol. Rev. *36:* 407-477.

Schleifer, K.-H., R. Kilpper-Bälz and L.-A. Devriese. 1984a. *Staphylococcus arlettae* sp. nov., *S. equorum* sp. nov. and *S. kloosii* sp. nov.: three new coagulase-negative, novobiocin-resistant species from animals. Syst. Appl. Microbiol. *5:* 501-509.

Schleifer, K.-H., R. Kilpper-Bälz, J. Kraus and F. Gehring. 1984b. Relatedness and classification of *Streptococcus mutans* and 'mutans-like' streptococci. J. Dent. Res. *63:* 1047-1050.

Schleifer, K.-H., R. Kilpper-Bälz and L.-A. Devriese. 1985a. *In* Validation of the publication of new names and new combinations previously effectively published outside the IJSB. List No. 17. Int. J. Syst. Bacteriol. *35:* 223-225.

Schleifer, K.-H., J. Kraus, C. Dvorak, R. Kilpper-Bälz, M.D. Collins and W. Fischer. 1985b. Transfer of *Streptococcus lactis* and related streptococci to the genus *Lactococcus* gen. nov. Syst. Appl. Microbiol. *6:* 183-195.

Schleifer, K.-H. and K. Lang. 1980. Close relationship among strains of *Micrococcus conglomeratus* and *Arthrobacter* species. FEMS Microbiol. Lett. *9:* 223-226.

Schleifer, K.-H. and H.P. Seidl. 1985. Chemical composition and structure of murein. *In* Goodfellow and Minnikin (Editors), Chemical Methods in Bacteri-

al Systematics. Academic Press, London, pp. 201–219.
Schleifer, K.-H. and E. Stackebrandt. 1983. Molecular Systematics of Prokaryotes. Annu. Rev. Microbiol. *37:* 143–187.
Schmitz, H., S.B. Deak, K.E. Crook, Jr. and I.R. Hooper. 1964. Peliomycin, a new cytotoxic agent. I. Production, isolation and characterization. Antimicrob. Agents Chemother. *1963:* 89–94.
Schobert, B. and J.K. Lanyi. 1982. Halorhodopsin is a light-driven chloride pump. J. Biol. Chem. *257:* 10306–10313.
Schofield, G.M. and K.P. Schaal. 1981. A numerical taxonomic study of members of the *Actinomycetales* and related taxa. J. Gen. Microbiol. *127:* 237–259.
Schupp, T., R. Hütter and D.A. Hopwood. 1975. Genetic recombination in *Nocardia mediterranei*. J. Bacteriol. *121:* 128–135.
Schuurmans, D.M., B.H. Olson and C.L. San Clemente. 1956. Production and isolation of thermoviridin an antibiotic produced by *Thermoactinomyces viridis* n. sp. Appl. Microbiol. *4:* 61–66.
Schwartz, R.M. and M.O. Dayhoff. 1978. Origins of prokaryotes, eukaryotes, mitochondria, and chloroplasts. Science (Washington) *199:* 395–403.
Scott, J.H. and R. Schekman. 1980. Lyticase: Endogluconase and protease activities that act together in yeast cell lysis. J. Bacteriol. *142:* 414–423.
Seaward, M.R.D., T. Cross and B.A. Unsworth. 1976. Viable bacterial spores recovered from an archeological excavation. Nature (Lond.) *261:* 407–408.
Sebald, M. and A.R. Prévot. 1962. Étude d'une nouvelle espèce anáerobie stricte *Micromonospora acetoformici* n. sp. isolée de l'intestin posterieur de *Reticulitermes lucifugus* var. *saintonnensis*. Ann. Inst. Pasteur (Paris) *102:* 199–214.
Seeliger, H.P.R. and D. Jones. 1986. Genus *Listeria* Pirie 1940. *In* Sneath, Mair, Sharpe and Holt (Editors), Bergey's Manual of Systematic Bacteriology, Vol. 2, The Williams and Wilkins Co., Baltimore, pp. 1235–1245.
Sehgal, S.N. and N.E. Gibbons. 1960. Effect of some metal ions on the growth of *Halobacterium cutirubrum*. Can. J. Microbiol. *6:* 165–169.
Seidl, P.H., A.H. Faller, R. Loider and K.-H. Schleifer. 1980. Peptidoglycan types and cytochrome patterns of strains of *Oerskovia turbata* and *O. xanthineolytica*. Arch. Microbiol. *127:* 173–178.
Seino, A. 1983. Surface ornamentation of sporangia of *Actinoplanaceae*. Hakko to Kogyo (Fermentation and Industry) *41:* 3–4.
Sellstedt, A. and K. Huss-Danell. 1984. Growth, nitrogen fixation and relative efficiency of nitrogenase in *Alnus incana* grown in different cultivation systems. Plant Soil *78:* 147–158.
Sermonti, G. and S. Casciano. 1963. Sexual polarity in *Streptomyces coelicolor*. J. Gen. Microbiol. *33:* 293–301.
Serrano, J.A., R.V. Tablante, A.A. de Serrano, G.C. de San Blas and T. Imaeda. 1972. Physiological, chemical and ultrastructural characteristics of *Corynebacterium rubrum*. J. Gen. Microbiol. *70:* 339–349.
Sgorbati, B. 1979. Preliminary quantification of immunological relationships among the transaldolases of the genus *Bifidobacterium*. Antonie van Leeuwenhoek J. Microbiol. Serol. *45:* 557–564.
Sgorbati, B. and V. Scardovi. 1979. Immunological relationshps among transaldolases in the genus *Bifidobacterium*. Antonie van Leeuwenhoek J. Microbiol. Serol. *49:* 129–140.
Shah, H.N., R.A. Nash, J.M. Hardie, D.A. Weetman, D.A. Geddes and T.W. MacFarlane. 1985. Detection of acidic end-products of metabolism of anaerobic Gram-negative bacteria. *In* Goodfellow and Minnikin (Editors), Chemical Methods in Bacterial Systematics, Academic Press, London, pp. 317–340.
Shane, B.S., L. Gouws and A. Kistner. 1969. Cellulolytic bacteria occurring in the rumen of sheep conditioned to low-protein teff hay. J. Gen. Microbiol. *55:* 445–457.
Sharpell, F.H. 1980. Industrial uses of biocides in processes and products. Dev. Indust. Microbiol. *21:* 133–140.
Sharples, G.P. and S.T. Williams. 1974. Fine structure of the globose bodies of *Dactylosporangium thailandense (Actinomycetales)*. J. Gen. Microbiol. *84:* 219–222.
Sharples, G.P. and S.T. Williams. 1976. Development and fine structure of sclerotia and spores of the actinomycete *Chainia olivacea*. Microbios *15:* 37–47.
Sharples, G.P., S.T. Williams and R.M. Bradshaw. 1974. Spore formation in the *Actinoplanaceae (Actinomycetales)*. Arch. Mikrobiol. *101:* 9–20.
Shaw, N. 1975. Bacterial glycolipids and glycophospholipids. Adv. Microb. Physiol. *12:* 141–167.
Shchepkina, T.V. 1940. Description of endoparasites of cotton fibres. Bull. Acad. Sci. U.S.S.R. Classe, Sci. Biol. *5:* 643–661.
Shearer, M.C., P. Actor, B.A. Bowie, S.F. Grappel, C.H. Nash, D.J. Newman, Y.K. Oh, C.H. Pan and L.J. Nisbet. 1985. Aridicins, novel glycopeptide antibiotics. I. Taxonomy, production and biological activity. J. Antibiot. *38:* 555–560.
Shearer, M.C., P.M. Colman, R.M. Ferrin, L.J. Nisbet and C.H. Nash, III. 1986. A new genus of the *Actinomycetales: Kibdelosporangium aridum* gen. nov., sp. nov. Int. J. Syst. Bacteriol. *36:* 47–54.
Shearer, M.C., P.M. Colman and C.H. Nash III. 1983a. *Nocardiopsis mutabilis*, a new species of nocardioform bacteria. Int. J. Syst. Bacteriol. *33:* 369–374.
Shearer, M.C., P.M. Colman and C.H. Nash. 1983b. *Streptosporangium fragile* sp. nov. Int. J. Syst. Bacteriol. *33:* 364–368.
Shearer, M.C., A.J. Giovenella, S.F. Grappel, R.D. Hedde, R.J. Mehta, Y.K. Oh, C.H. Pan, D.H. Pitkin and L.J. Nisbet. 1986. Kibdelins, novel glycopeptide antibiotics. I. Discovery, production, and biological evaluation. J. Antibiot. *39:* 1386–1394.
Sherris, J.C., J.G. Shoesmith, M.T. Parker and D. Breckon. 1959. Tests for the rapid breakdown of arginine by bacteria: Their use in the identification of pseudomonads. J. Gen. Microbiol. *21:* 389–396.
Shibata, M. 1959. On a new streptomycin-producing species. *Streptomyces rameus*, n. sp. J. Antibiot. (Tokyo) Ser. B *12:* 389–400.
Shibata, M., E. Higashide, T. Kanzaki, H. Yamamoto and K. Nakazawa. 1961. Studies on Streptomycetes Part I: *Streptomyces pulveraceus* nov. sp., producing new antibiotics zygomycin A and B. J. Agric. Chem. Soc. Jpn. *25:* 171–176.
Shibata, M., E. Higashide, H. Yamamoto and K. Nakazawa. 1962. Studies on Streptomycetes. Part I. *Streptomyces atratus* nov. sp., producing new antituberculous antibiotics rugomycin A and B. Agric. Biol. Chem. *26:* 228–233.
Shibata, M., M. Honso, Y. Tokui and N. Nakazawa. 1954. On a new antifungal and antiyeast substance, candicidin produced by a streptomyces. J. Antibiot. (Tokyo) Ser. B *7:* 168.
Shibata, M., K. Nakazawa, A. Miyake, M. Inoue and A. Okabori. 1957. Studies on streptomycetes. Croceomycin, a new antituberculous substance. Annu. Rep. Takeda Res. Lab. *16:* 32–37.
Shimazu, A., S. Higara, T. Endō, K. Furuhata, N. Takizawa and N. Ōtake. 1984. The Society for Actinomycetes, Japan, Abstract Paper 9.
Shimizu, M., M. Kanno, M. Tamura and M. Suckane. 1978. Purification and some properties of a novel α-amylase produced by a strain of *Thermoactinomyces vulgaris*. Agric. Biol. Chem. *42:* 1681–1688.
Shimo, M., T. Shiga, T. Tomosugi and I. Kamoi. 1959. Studies on taitomycin, a new antibiotic produced by *Streptomyces*, sp. No. 772 (*S. afghaniensis*) I. Studies on the strain and production of taitomycin. J. Antibiot. (Tokyo) Ser. A *12:* 1–6.
Shinobu, R. 1955. On *Streptomyces hiroshimensis* nov. sp. Seibutsugakkaishi *6:* 43–46.
Shinobu, R. 1956. Three new species of *Streptomyces* forming whirls. Mem. Osaka Univ. Lib. Arts Educ. *5B:* 84–93.
Shinobu, R. 1957. Two new species of *Streptomyces*. Mem. Osaka Univ. Lib. Arts Educ. Ser. B Nat. Sci. *6:* 63–73.
Shinobu, R. 1958. On *Streptomyces spiroverticillatus* nov. sp. Bot. Mag. (Tokyo) *71:* 87–93.
Shinobu, R. 1962. A new *Streptomyces* species producing fluorescent-yellow soluble pigment. Mem. Osaka Univ. Lib. Arts Educ. Ser. B Nat. Sci. *11:* 115–122.
Shinobu, R. 1965. Taxonomy of the whirl forming *Streptomycetaceae*. Mem. Osaka Univ. Lib. Arts Educ. B Nat. Sci. *14:* 72–210.
Shinobu, R. and M. Kawato. 1959. On *Streptomyces massasporeus* nov. sp. Bot. Mag. (Tokyo) *72:* 283–288.
Shinobu, R. and M. Kawato. 1960a. On *Streptomyces aerocolonigenes*, nov. sp., forming the secondary colonies on the aerial mycelia. Bot. Mag. *73:* 212–216.
Shinobu, R. and M. Kawato. 1960b. On *Streptomyces indigoferus* nov. sp., producing blue to green soluble pigment on some synthetic media. Mem. Osaka Univ. Lib. Arts Educ. Ser. B Nat. Sci. *9:* 49–53.
Shinobu, R. and Y. Kayamura. 1964. On a new whirl-forming species of *Streptomyces*. Bot. Mag. (Tokyo) *77:* 176–180.
Shinobu, R. and Y. Shimada. 1962. On a new whirl-forming species of *Streptomyces*. Bot. Mag. (Tokyo) *75:* 170–175.
Shipton, W.A. and A.J.P. Burggraaf. 1982a. A comparison of the requirements for various carbon and nitrogen sources and vitamins in some *Frankia* isolates. Plant Soil *69:* 149–161.
Shipton, W.A. and A.J.P. Burggraaf. 1982b. *Frankia* growth and activity as influenced by water potential. Plant Soil *69:* 293–297.
Shipton, W.A. and A.J.P. Burggraaf. 1983. Aspects of the behaviour of *Frankia* and possible ecological implications. Can. J. Bot. *61:* 2783–2792.
Shirling, E.B. and D. Gottlieb. 1966. Methods for characterization of *Streptomyces* species. Int. J. Syst. Bacteriol. *16:* 313–340.
Shirling, E.B. and D. Gottlieb. 1968a. Cooperative description of type cultures of *Streptomyces*. II. Species description from the first study. Int. J. Syst. Bacteriol. *18:* 69–189.
Shirling, E.B. and D. Gottlieb. 1968b. Cooperative description of type cultures of *Streptomyces*. III. Additional species descriptions from first and second studies. Int. J. Syst. Bacteriol. *18:* 279–392.
Shirling, E.B. and D. Gottlieb. 1969. Cooperative description of type cultures of *Streptomyces*. IV. Species descriptions from the second, third and fourth studies. Int. J. Syst. Bacteriol. *19:* 391–512.
Shirling, E.B. and D. Gottlieb. 1970. Report of the International *Streptomyces* Project. Five years collaborative research. *In* Prauser (Editor), The Actinomycetales, Gustav Fischer Verlag, Jena, pp. 79–90.
Shirling, E.B. and D. Gottlieb. 1972. Cooperative description of type strains of *Streptomyces*. V. Additional descriptions. Int. J. Syst. Bacteriol. *22:* 265–394.
Shirling, E.B. and D. Gottlieb. 1977. Retrospective evaluation of International *Streptomyces* Project taxonomic criteria. *In* Arai (Editor), Actinomycetes: The Boundary Microorganisms, University Park Press, Baltimore, pp. 9–41.
Shomura, T., S. Amano, H. Tohyama, J. Yoshida, T. Ito, and T. Niida. 1985. *Dactylosporangium roseum* sp. nov. Int. J. Syst. Bacteriol. *35:* 1–4.
Shomura, T., S. Amano, J. Yoshida, N. Ezaki, T. Ito and T. Niida. 1983a. *Actinosporangium vitaminophilum* sp. nov. Int. J. Syst. Bacteriol. *33:* 557–564.
Shomura, T., S. Amano, J. Yoshida and M. Kojima. 1986. *Dactylosporangium fulvum* sp. nov. Int. J. Syst. Bacteriol. *36:* 166–169.
Shomura, T., S. Gomi, M. Ito, J. Yoshida, E. Tanaka, S. Amano, H. Watabe, S. Ohuchi, J. Itoh, M. Sezaki, H. Takebe and K. Uotani. 1987. Studies of new antibiotics SF 2415. I. Taxonomy, fermentation, isolation, physico-chemical

properties and biological activities. J. Antibiot. *40:* 732–739.

Shomura, T., M. Kojima, J. Yoshida, M. Ito, S. Amano, K. Totsugawa, T. Niwa, S. Inouye, T. Ito and T. Niida. 1980. Studies on a new aminoglycoside antibiotic, dactimicin. I. Producing organism and fermentation. J. Antibiot. *33:* 924–930.

Shomura, T., M. Kojima, J. Yoshida, M. Ito, S. Amaro, K. Totsugawa, T. Niwa, S. Inouye, T. Ito and T. Niida. 1983b. *In* Validation of the Publication of New Names and New Combinations Previously Effectively Published Outside the IJSB. List No. 11. Int. J. Syst. Bacteriol. *33:* 672–674.

Shomura, T., J. Yoshida, S. Miyadoh, T. Ito and T. Niida. 1983c. *Dactylosporangium vinaceum* sp. nov. Int. J. Syst. Bacteriol. *33:* 309–313.

Shomura, T. et al. 1986. Int. J. Syst. Bacteriol. *36:* 166–169.

Shomura, T. et al. 1987. J. Antibiot. *40:* 732–739.

Sierra, G. 1957. A simple method for the detection of lipolytic activity of microorganisms and some observations on the influence of the contact between cells and fatty substrates. Antonie van Leeuwenhoek J. Microbiol. Serol. *23:* 15–22.

Silvestri, L., M. Turri, L.R. Hill and E. Gilardi. 1962. A quantitative approach to the systematics of *Actinomycetales* based on overall similarity. *In* Ainsworth and Sneath (Editors), Microbial Classification, Symp. Soc. Gen. Microbiol. *12:* 333–360.

Simonet, P., A. Capellano, E. Novarro, R. Bardin and A. Moiroud. 1984. An improved method for lysis of *Frankia* with achromopeptidase allows detection of new plasmids. Can. J. Microbiol. *30:* 1292–1295.

Simonet, P., P. Normand, A. Moiroud and M. Lalonde. 1985. Restriction enzyme digestion patterns of *Frankia* plasmids. Plant Soil *87:* 49–60.

Sing, P.J. and R.S. Mehrotra. 1980. Biological control of *Rhizoctonia bataticola* on grain by coating seed with *Bacillus* and *Streptomyces* spp. and their influence on plant growth. Plant Soil *56:* 475–483.

Sitrin, R.D., G.W. Chan, J.J. Dingerdissen, W. Holl, J.R.E. Hoover, J.R. Valenta, L. Webb and K.M. Snader. 1985. Aridicins, novel glycopeptide antibiotics. II. Isolation and characterization. J. Antibiot. *38:* 561–571.

Skarbek, J.D. and L.R. Brady. 1978. *Streptomyces cavourensis* sp. nov. (nom. rev.) and *Streptomyces cavourensis* subsp. *washingtonensis* subsp. nov., a chromomycin-producing sub-species. Int. J. Syst. Bacteriol. *28:* 43–53.

Skerman, V.B.D. 1967. A Guide to the Identification of the Genera of Bacteria, 2nd Ed., The Williams and Wilkins Co., Baltimore.

Skerman, V.B.D., V. McGowan and P.H.A. Sneath. 1980. Approved Lists of Bacterial Names, American Society for Microbiology, Washington, D.C. (Reprinted from Int. J. Syst. Bacteriol. *30:* 225–420, 1980.)

Slepecky, R.E. and E.R. Leadbetter. 1983. On the prevalence and roles of sporeforming bacteria and their spores in nature. *In:* Hurst and Gould (Editors), The Bacterial Spore, Vol. 2, Academic Press, London, pp. 79–99.

Smibert, R.M. and N.R. Krieg. 1981. General characterization. *In* Gerhardt, Murray, Costilow, Nester, Wood, Krieg and Phillips (Editors), Manual of Methods for General Bacteriology, American Society for Microbiology, Washington, D.C., pp. 409–443.

Smith, C.G., A. Dietz, W.T. Sokolski and G.M. Savage. 1956. Streptonivicin, a new antibiotic. I. Discovery and biologic studies. Antibiot. Chemother. *6:* 135–142.

Smith, E.F. 1905. Bacteria in Relation to Plant Diseases, Vol. 1, Carnegie Institution of Washington, Washington, D.C.

Sneath, P.H.A. 1957. The application of computers to taxonomy. J. Gen. Microbiol. *17:* 201–226.

Sneath, P.H.A. 1962. Longevity of micro-organisms. Nature (Lond) *195:* 643–646.

Sneath, P.H.A. 1972. Computer taxonomy. *In* Norris and Ribbons (Editors), Methods in Microbiology, Vol. 7A, Academic Press, London, pp. 29–98.

Sneath, P.H.A. 1974. Phylogeny of microorganisms. Symp. Soc. Gen. Microbiol. *24:* 1–39.

Sneath, P.H.A. 1977a. A method for testing the distinctness of clusters: A test for the disjunction of two clusters in euclidean space as measured by their overlap. J. Int. Assoc. Math. Geol. *9:* 123–143.

Sneath, P.H.A. 1977b. The maintenance of large numbers of strains of microorganisms, and the implications for culture collections. FEMS Microbiol. Lett. *1:* 333–334.

Sneath, P.H.A. 1978. Classification of microorganisms. *In* Norris and Richmond (Editors), Essays in Microbiology, John Wiley, Chichester, United Kingdom, pp. 9/1–9/31.

Sneath, P.H.A. 1979a. BASIC program for a significance test for two clusters in euclidean space as measured by their overlap. Comput. Geosci. *5:* 143–155.

Sneath, P.H.A. 1979b. BASIC program for a significance test for clusters in UPGMA dendrograms obtained from square euclidean distances. Comput. Geosci. *5:* 127–137.

Sneath, P.H.A. 1979c. BASIC program for character separation indices from an identification matrix of percent positive characters. Comput. Geosci. *5:* 349–357.

Sneath, P.H.A. 1979d. BASIC Program for identification of an unknown with presence-absence data against an identification matrix of percent positive characters. Comput. Geosci. *5:* 195–213.

Sneath, P.H.A. 1980a. BASIC program for determining the best identification scores possible for the most typical example when compared with an identification matrix of percent positive characters. Comput. Geosci. *6:* 27–34.

Sneath, P.H.A. 1980b. BASIC program for the most diagnostic properties of groups from an identification matrix of percent positive characters. Comput. Geosci. *6:* 21-26.

Sneath, P.H.A. 1982. Status of nomenclatural types in the approved lists of bacterial names. Request for an opinion. Int. J. Syst. Bacteriol. *32:* 459–460.

Sneath, P.H.A. and A.O. Chater. 1978. Information content of keys for identification. *In* Street (Editor), Essays in Plant Taxonomy, Academic Press, London, pp. 79–95.

Sneath, P.H.A. and R.R. Sokal. 1973. Numerical Taxonomy: The Principles and Practice of Numerical Classification, W.H. Freeman, San Francisco, pp. 1–573.

Sneath, P.H.A. and M. Stevens. 1967. A divided Petri dish for use with multipoint inoculators. J. Appl. Bacteriol. *30:* 495–497.

Snijders, E.P. 1924. Cavia-scheefkopperij, een nocardiose. Geneesk. Tijdschr. Ned. Ind. *64:* 85–87.

Söhngen, N.L. 1913. Benzin, Petroleum, Paraffinöl und Paraffin als Kohlenstoff- und Energie-quelle für Mikroben. Zentralbl. Bakteriol. Parasitenkd. Infektionskr. Hyg. Abt. 2. *37:* 595–609.

Soina, V.S., A.A. Sokolov and N.S. Agre. 1975. Ultrastructure of mycelium and spores of *Actinomadura fastidiosa* sp. nov. Mikrobiologiya *44:* 883–887.

Sokal, R.R. and C.D. Michener. 1958. A statistical method for evaluating systematic relationships. Univ. Kans. Sci. Bull. *38:* 1409–1438.

Solovieva, N.K. 1972. Actinomycetes of littoral and sub-littoral zones of the White Sea. Antibiotiki *17:* 387–392.

Solovieva, N.K. and E.M. Singal. 1972. Some data on ecology of *Micromonospora*. Antibiotiki *17:* 778–781.

Sonea, S. 1971. A tentative unifying view of bacteria. Rev. Can. Biol. *30:* 239–244.

Sonea, S. and M. Panisset. 1976. Pour une nouvelle bacteriologie. Rev. Can. Biol. *35:* 103–167.

Sonea, S. and M. Panisset. 1980. Introduction a la Nouvelle Bacteriologie, Les Presses de l'Université de Montréal et Masson, Montréal.

Sottnek, F.O., J.M. Brown, R.E. Weaver and G.F. Carroll. 1977. Recognition of *Oerskovia* species in the clinical laboratory: Characterization of 35 isolates. Int. J. Syst. Bacteriol. *27:* 263–270.

Southey, J.F. (Editor). 1986. Laboratory Methods for Work with Plant and Soil Nematodes. 6th Ed. [Reference Book 402, Ministry of Agriculture, Fisheries and Food, United Kingdom], Her Majesty's Stationery Office, London.

Stackebrandt, E. 1986. The significance of "wall types" in phylogenetically based taxonomic studies on actinomycetes. *In* Szabó, Biró and Goodfellow (Editors), Biological, Biochemical and Biomedical Aspects of Actinomycetes, Academiai Kiadó, Budapest, pp. 497–506.

Stackebrandt, E., F. Fiedler and O. Kandler. 1978. Peptidoglycantyp und Zusammensetzung der Zellwandpolysaccharide von *Cellulomonas cartalyticum* und einigen coryneformen Organismen. Arch. Mikrobiol. *117:* 115–118.

Stackebrandt, E., V.J. Fowler and C.R. Woese. 1983a. A phylogenetic analysis of lactobacilli, *Pediococcus pentosaceus* and *Leuconostoc mesenteroides*. Syst. Appl. Microbiol. *4:* 326–337.

Stackebrandt, E., M. Haringer and K.H. Schleifer. 1980a. Molecular genetic evidence for the transfer of *Oerskovia* species into the genus *Cellulomonas*. Arch. Microbiol. *127:* 179–185.

Stackebrandt, E. and O. Kandler. 1979. Taxonomy of the genus *Cellulomonas*, based on phylogenetic characters and deoxyribonucleic acid-deoxyribonucleic acid homology, and proposal of seven neotype strains. Int. J. Syst. Bacteriol. *29:* 273–282.

Stackebrandt, E. and R.M. Kroppenstedt. 1987. Union of the genera *Actinoplanes* Couch, *Ampullariella* Couch, and *Amorphosporangium* Couch in a redefined genus *Actinoplanes*. Syst. Appl. Microbiol. *9:* 110–114.

Stackebrandt, E., R.M. Kroppenstedt and V.J. Fowler. 1983b. A phylogenetic analysis of the family *Dermatophilaceae*. J. Gen. Microbiol. *129:* 1831–1838.

Stackebrandt, E., B.J. Lewis and C.R. Woese. 1980b. The phylogenetic structure of the coryneform group of bacteria. Zentralbl. Bakteriol. Mikrobiol. Hyg. Abt. II Orig. C *1:* 137–149.

Stackebrandt, E., W. Ludwig, E. Seewaldt and K.-H. Schleifer. 1983c. Phylogeny of sporeforming members of the order *Actinomycetales*. Int. J. Syst. Bacteriol. *33:* 173–180.

Stackebrandt, E., C. Scheurlein and K.H. Schleifer. 1983d. Phylogenetic and biochemical studies on *Stomatococcus mucilaginosus*. Syst. Appl. Microbiol. *4:* 207–217.

Stackebrandt, E. and K.-H. Schleifer. 1984. Molecular systematics of actinomycetes and related organisms. *In* Ortiz-Ortiz, Bojalil and Yakoleff (Editors), Biological, Biochemical and Biomedical Aspects of Actinomycetes, Academic Press, Orlando, Florida, pp. 485–504.

Stackebrandt, E., H. Seiler and K.-H. Schleifer. 1982a. Union of the genera *Cellulomonas* Bergey et al. and *Oerskovia* Prauser et al. in a redefined genus *Cellulomonas*. Zentralbl. Bakteriol. Mikrobiol. Hyg. I Abt. Orig. C *3:* 401–409.

Stackebrandt, E., B. Wittek, E. Seewaldt and K.H. Schleifer. 1982b. Physiological, biochemical and phylogenetic studies on *Gemella haemolysans*. FEMS Microbiol. Lett. *13:* 361–365.

Stackebrandt, E. and C.R. Woese. 1979. A phylogenetic dissection of the family *Micrococcaceae*. Curr. Microbiol. *2:* 317–322.

Stackebrandt, E. and C.R. Woese. 1981a. The evolution of prokaryotes. *In* Carlile, Collins and Moseley (Editors), Molecular and Cellular Aspects of Microbial Evolution. Symp. Soc. Gen. Microbiol. *32:* 1–31.

Stackebrandt, E. and C.R. Woese. 1981b. Towards a phylogeny of the actinomycetes and related organisms. Curr. Microbiol. *5:* 197–202.

Stackebrandt, E., B. Wunner-Füssl, V.J. Fowler and K.-H. Schleifer. 1981. Deoxyribonucleic acid homologies and ribosomal ribonucleic acid similarities among sporeforming members of the order *Actinomycetales*. Int. J. Syst. Bacteriol. *31:* 420–431.

Staley, J.T. 1973. Budding bacteria of the *Pasteuria-Blastobacter* group. Can. J. Microbiol. *19:* 609-614.

Staneck, J.L. and G.D. Roberts. 1974. Simplified approach to identification of aerobic actinomycetes by thin-layer chromatography. Appl. Microbiol. *28:* 226-231.

Stanier, R.Y. 1961. La place des bactéries dans le monde vivant. Ann. Inst. Pasteur (Paris) *101:* 297-303.

Stanier, R.Y. 1970. Some aspects of the biology of cells and their possible evolutionary significance. *In* Charles and Knight (Editors), Organization and Control in Procaryotic and Eucaryotic Cells, Cambridge University Press, Cambridge, England.

Stanier, R.Y., N.J. Palleroni and M. Doudoroff. 1966. The aerobic pseudomonads: A taxonomic study. J. Gen. Microbiol. *43:* 159-271.

Stanier, R.Y. and C.B. van Niel. 1962. The concept of a bacterium. Arch. Mikrobiol. *42:* 17-35.

Stanier, R.Y., D. Wachter, D. Gasser and A.C. Wilson. 1970. Comparative immunological studies of two *Pseudomonas* enzymes. J. Bacteriol. *102:* 351-362.

Stapley, E.O., J.M. Mata, I.M. Miller, T.C. Demny and H.B. Woodruff. 1964. Antibiotic MSD-235. I. Production by *Streptomyces avidinii* and *Streptomyces lavendulae*. Antimicrob. Agents Chemother. *1963-64:* 20-27.

Starr, M.P. 1949. The nutrition of phytopathogenic bacteria. III. The Gram-positive phytopathogenic *Corynebacterium* species. J. Bacteriol. *57:* 253-258.

Starr, M.P. and R.M. Sayre. 1988a. *Pasteuria thornei* sp. nov. and *Pasteuria penetrans sensu stricto* emend., mycelial and endospore-forming bacteria parasitic, respectively, on plant-parasitic nematodes of the genera *Pratylenchus* and *Meloidogyne*. Ann. Inst. Pasteur/Microbiol. *139:* 11-31.

Starr, M.P. and R.M. Sayre. 1988b. *In* Validation of the publication of new names and new combinations previously effectively published outside the IJSB. List No. 26. Int. J. Syst. Bacteriol. *38:* 328-329.

Starr, M.P., R.M. Sayre and J.M. Schmidt. 1983. Assignment of ATCC 27377 to *Planctomyces staleyi* sp. nov. and conservation of *Pasteuria ramosa* Metchnikoff 1888 on the basis of type descriptive material. Request for an Opinion. Int. J. Syst. Bacteriol. *33:* 666-671.

Starr, M.P. and J.M. Schmidt. 1981. Prokaryote diversity. *In* Starr, Stolp, Trüper, Balows and Schlegel (Editors), The Prokaryotes, A Handbook on Habitats, Isolation and Identification of Bacteria, Springer-Verlag, Berlin, pp. 3-42.

Starr, M.P., H. Stolp, H.G. Trüper, A. Balows and H.G. Schlegel (Editors). 1981. The Prokaryotes. A Handbook on Habitats, Isolation and Identification of Bacteria. Springer-Verlag, Berlin.

Staub, F., S.K. Mishra and A. Blisse. 1980. Interaction between aspergilli and streptomycetes in the soil of potted indoor plants: A preliminary report (contributing to the epidemiology of aspergillosis). Mycopathologia *70:* 9-12.

Steiner, G. 1938. Opuscula miscellanea nematologica VII. Proc. Helminthol. Soc. Wash. *5:* 35-40.

Sterba, G. and W. Naumann. 1970. Untersuchungen über *Dermocystidium granulosum* n. sp. bei *Tetraodon palembangensis* (Bleeker 1852). Arch. Protistenk. *112:* 106-118.

Sternberg, G.M. 1892. Manual of Bacteriology, Wood and Co., New York.

Stevens, R.T. 1975. Fine structure of sporogenesis and septum formation in *Micromonospora globosa* Kriss and *M. fusca* Jensen. Can. J. Microbiol. *21:* 1081-1088.

Stirling, G.R. 1984. Biological control of *Meloidogyne javanica* with *Bacillus penetrans*. Phytopathology *74:* 55-60.

Stirling, G.R. and M.F. Wachtel. 1980. Mass production of *Bacillus penetrans* for the biological control of root-knot nematodes. Nematologica *26:* 308-312.

Stowers, M.D., R.K. Kulkarni and D.B. Steele. 1986. Intermediary carbon metabolism in *Frankia*. Arch. Microbiol. *143:* 319-324.

Stuttard, C.S. 1979. Transduction of auxotrophic markers in a chloramphinicol-producing strain of *Streptomyces*. J. Gen. Microbiol. *110:* 479-482.

Stutzenberger, F.J. 1971. Cellulase production by *Thermomonospora curvata* isolated from municipal solid waste compost. Appl. Microbiol. *22:* 147-152.

Stutzenberger, F.J. 1972. Cellulolytic activity of *Thermomonospora curvata:* Optimal conditions, partial purification and product of the cellulase. Appl. Microbiol. *24:* 83-90.

Stutzenberger, F.J., A.J. Kaufman and R.D. Lossin. 1970. Cellulolytic activity in municipal solid waste composting. Can. J. Microbiol. *16:* 553-560.

Suarez, J., C. Barbes and C. Hardisson. 1980. Germination of spores of *Micromonospora chalcea*: Physiological and biochemical changes. J. Gen. Microbiol. *121:* 159-167.

Sugai, T. 1956. New antibiotics 229 and 229B of colorless, water-soluble and basic nature. J. Antibiot. (Tokyo) Ser. B *9:* 170-179.

Sugawara, R. and M. Onuma. 1957. Melanomycin, a new antitumour substance from *Streptomyces*. II. Description of the strain. J. Antibiot. (Tokyo) Ser. A *10:* 138-142.

Sukapure, R.S., M.P. Lechevalier, H. Reber, M.L. Higgins, H.A. Lechevalier and H. Prauser. 1970. Motile nocardoid *Actinomycetales*. Appl. Microbiol. *19:* 527-533.

Sundmann, V. and H.V. Gyllenberg. 1967. Application of factor analysis. I. General aspects on the use of factor analysis in microbiology. Ann. Acad. Sci. Fenn. Ser. A IV Biol. *112:* 1-32.

Sutherland, J.B., R.A. Blanchette, D.L. Crawford and A.L. Pometto. 1979. Breakdown of Douglas fir phloem by a lignocellulose-degrading *Streptomyces*. Curr. Microbiol. *2:* 123-126.

Suzuki, S., K. Asahi, J. Nagatsu, Y. Kawashima and I. Suzuki. 1967. Triculamin, a new antituberculosis substance. J. Antibiot. (Tokyo) Ser. A *20:* 126.

Suzuki, K. and K. Komagata. 1983. *Pimelobacter* gen. nov.—a new genus of coryneform bacteria with LL-diaminopimelic acid in the cell wall. J. Gen. Appl. Microbiol. *29:* 59-71.

Suzuki, M. 1957. Studies on an antitumor substance, gancidin. Mycological study on the strain AAK-84 and production, purification (sic) of active fractions (in Japanese). J. Chiba Med. Soc. *33:* 535-542.

Suzuki, S., G. Nakamura, K. Okuma and Y. Tomiyama. 1958. Cellocidin, a new antibiotic. J. Antibiot. (Tokyo) Ser. A *11:* 81-83.

Sveshnikova, M.A., N.T. Chormonova, N.V. Lavrova, L.P. Terekhova and T.P. Preobrazhenskaya. 1976. Isolation of soil actinomycetes on selective media with novobiocin. Antibiotiki *21:* 784-787.

Sveshnikova, M.A., T.S. Maksimova and E.S. Kudrina. 1969. The species belonging to the genus *Micromonospora* Ørskov, 1923 and their taxonomy. Mikrobiologiya *38:* 883-893.

Sweet, L.C., J.A. Anderson, Q.C. Callies and E.O. Coates. 1971. Hypersensitivity pneumonitis related to home furnace humidifier. J. Allergy *48:* 171-178.

Switalski, L.M., A. Ljungh, C. Rydén, K. Rubin, M. Höök and T. Wadström. 1982. Binding of fibronectin to the surface of groups A, C and G streptococci isolated from human infections. Eur. J. Clin. Microbiol. *1:* 381-387.

Sykes, I.K. and S.T. Williams. 1978. Interactions of actinophage and clays. J. Gen. Microbiol. *108:* 97-102.

Szabó, I.M., K. Marialigeti, C.T. Loc, K. Jager, J. Szabó, E. Contreras, K. Ravasz, M. Heydrich and E. Palic. 1986. On the ecology of nocardioform intestinal actinomycetes of millipedes *(Diplopoda)*. *In* Szabó, Biro and Goodfellow (Editors), Biological, Biochemical and Biomedical Aspects of Actinomycetes, Academiai Kiado, Budapest, pp. 701-714.

Szabó, I. and M. Marton. 1958. A *Streptomyces vastus* és *Streptomyces viridoniger* új sugárgomba fajokról (Adatok a szikestalajok mikrobiologiájahoz). Agrokem. Talajtan. *7:* 243-262.

Szabó, I.M. and M. Marton. 1976. Evaluation of criteria used in the ISP cooperative description of type strains of *Streptomyces* and *Streptoverticillium* species. Int. J. Syst. Bacteriol. *26:* 105-110.

Szabó, I., M. Marton, I. Buti and G. Pártai. 1963. *Actinomyces finlayi* n. sp. Acta Microbiol. Acad. Sci. Hung. *10:* 207-214.

Szabó, I., M. Marton, I. Buti and C. Fernandez. 1975. A diagnostic key for the identification of species of *Streptomyces* and *Streptoverticillium* included in the International *Streptomyces* Project. Acta Bot. Acad. Sci. Hung. *21:* 387-418.

Szabó, I., M. Marton, L. Ferenczy and I. Buti. 1967. Intestinal microflora of the larvae of St. Mark's fly. II. Computer analysis of intestinal actinomycetes from the larvae of a bibio population. Acta Microbiol. Acad. Sci. Hung. *14:* 239-249.

Szabó, I.M., M. Marton, G. Kulcsar and I. Buti. 1976. Taxonomy of primycin producing actinomycetes. I. Description of the type strain of *Thermomonospora galeriensis*. Acta Microbiol. Acad. Sci. Hung. *23:* 371-376.

Szabo, Z. and C. Fernandez. 1984. *Micromonospora brunnea* Sveshnikova, Maksimova, and Kudrina 1969 is a junior subjective synonym of *Micromonospora purpureochromogenes* (Waksman and Curtis 1916) Luedemann 1971. Int. J. Syst. Bacteriol. *34:* 463-464.

Szaniszlo, P.J. 1968. The nature of the intramycelial pigmentation of some *Actinoplanaceae*. J. Elisha Mitchell Sci. Soc. *84:* 24-26.

Szaniszlo, P.J. and H. Gooder. 1967. Cell wall composition in relation to the taxonomy of some *Actinoplanaceae*. J. Bacteriol. *94:* 2037-2047.

Szegi, J. and F. Gulyas. 1968. Data on the humus-decomposing activity of some streptomycetes and microscopic fungi. Agrokem. Talajtan *17:* 109-119.

Szulga, T. 1978. A critical evaluation of taxonomic procedures applied in *Streptomyces*. Zentralbl. Bakteriol. Parasitenkd. Infektionskr. Hyg. Abt. 1 Orig. *6:* 31-42.

Szvoboda, G., T. Lang, I. Gado, G. Ambrus, C. Kari, K. Fodor and L. Alfoldi. 1980. Fusion of *Micromonospora* protoplasts. *In* Ferenczy and Farkas (Editors), Advances in Protoplast Research, Pergamon Press, Oxford, England, pp. 235-240.

Taber, W.A. 1960. Evidence for the existence of acid-sensitive actinomycetes in soil. Can. J. Microbiol. *6:* 503-514.

Tacquet, A., M.T. Plancot, J. Debruyne, B. Devulder, M. Joseph and J. Losfeld. 1971. Études préliminaires sur la classification numérique des mycobactéries et des nocardias. 1) Relations taxonomiques entre *Mycobacterium rhodochrous*, *Mycobacterium pellegrino* et les genres *Mycobacterium* et *Nocardia*. Ann. Inst. Pasteur (Lille) *XXII:* 121-135.

Takahashi, A., K. Hotta, N. Saito, M. Morioka, Y. Okami and H. Umezawa. 1986. Production of novel antibiotic, dopsisamine, by a new subspecies of *Nocardiopsis mutabilis* with multiple antibiotic resistance. J. Antibiotics *39:* 175-183.

Takahashi, Y., Y. Iwai and S. Ōmura. 1983. Relationship between cell morphology and the types of diaminopimelic acid in *Kitasatosporia setalba*. J. Gen. Appl. Microbiol. *29:* 459-465.

Takahashi, Y., Y. Iwai and S. Ōmura. 1984a. Two new species of the genus *Kitasatosporia*, *Kitasatosporia phosalacinea* sp. nov. and *Kitasatosporia griseola* sp. nov. J. Gen. Appl. Microbiol. *30:* 377-387.

Takahashi, Y., Y. Iwai and S. Ōmura. 1985. *In* Validation of the publication of new names and new combinations previously effectively published outside the IJSB. List No. 19. Int. J. Syst. Bacteriol. *35:* 535.

Takahashi, Y., T. Kuwana, Y. Iwai and S. Ōmura. 1984b. Some characteristics of aerial and submerged spores of *Kitasatosporia setalba*. J. Gen. Appl. Microbiol. *30:* 223-229.

Takamiya, A. and K. Tubaki. 1956. A new form of streptomyces capable of growing autotrophically. Arch. Mikrobiol. *25:* 58-64.

Takita, T. 1959. Studies on purification and properties of phleomycin. J. Antibiot. Jpn. Ser. A. *12:* 285-289.

Tamura, A., R. Furuta, H. Kotani and S. Naruto. 1973a. Antibiotic AB-64, a new indicator-pigment antibiotic from *Actinomadura roseoviolaceae* var. *rubescens*. J. Antibiot. *26:* 492-500.

Tamura, A., R. Furuta, S. Naruto and H. Ishii. 1973b. Actinotiocin, a new sulfur-containing peptide antibiotic from *Actinomadura pusilla*. J. Antibiot. *26:* 343-350.

Tamura, A., I. Takeda, S. Naruto and Y. Yoshimura. 1971. Chloramphenicol from *Streptosporangium viridogriseum* var. *kofuense*. J. Antibiotics *24:* 270.

Tanner, R.S., E. Stackebrandt, G.E. Fox and C.R. Woese. 1981. A phylogenetic analysis of *Actinobacterium woodii, Clostridium barkeri, Clostridium butyricum, Clostridium lutuseburense, Eubacterium limosum* and *Eubacterium tenus*. Curr. Microbiol. *5:* 35-38.

Taptykova, S.D., L.V. Kalakoutskii and N.S. Agre. 1969. Cytochromes in spores of actinomycetes. J. Gen. Appl. Microbiol. *15:* 383-386.

Tárnok, I. 1976. Metabolism in nocardiae and related bacteria. *In* Goodfellow, Brownell and Serrano (Editors), Biology of Nocardiae, Academic Press, New York, pp. 451-500.

Taufel, A., U. Behnke and H. Ruttloff. 1979. Isolation and characterization of proteases from *Thermoactinomyces vulgaris*. IV. Extracellular protease spectrum during the course of culture (in German). Z. Allg. Mikrobiol. *19:* 129-138.

Terao, M., K. Furuya and R. Emokita. 1965. Studies on antibiotics from thermophilic actinomycetes. I. An antibiotic produced by strain no. BT3-3. Annu. Rep. Sankyo Res. Lab. *17:* 110-117.

Terekhova, L.P., O.A. Galatenko and T.P. Preobrazhenskaya. 1982. *Actinomadura fulvescens* sp. nov. and *A. turkmeniaca* sp. nov. and their antagonistic properties. Antibiotiki *27:* 87-92.

Terekhova, L.P., O.A. Galatenko and T.P. Preobrazhenskaya. 1987. *In* Validation of the publication of new names and new combinations previously effectively published outside the IJSB. List No. 23. Int. J. Syst. Bacteriol. *37:* 179-180.

Terekhova, L.P., T.P. Preobrazhenskaya and O.A. Galtenko. 1986. Occurrence of galactose in the whole cell hydrolysates of *Actinomadura* strains. The Actinomycetes *19:* 73-81.

Terekhova, L.P., O.A. Sadikova and T.P. Preobrazhenskaya. 1977. *Actinoplanes cyaneus* new species and its antagonistic properties. Antibiotiki (Moscow) *22:* 1059-1063.

Terho, E.O. and J. Lacey. 1979. Microbiological and serological studies on farmer's lung in Finland. Clin. Allergy *9:* 43-52.

Thiemann, J.E. 1967. A new species of the genus *Amorphosporangium* isolated from Italian soil. Mycopathol. Mycol. Appl. *33:* 233-240.

Thiemann, J.E. 1970a. *Dactylosporangium thailandensis* should be *D. thailandense*. Int. J. Syst. Bacteriol. *20:* 59.

Thiemann, J.E. 1970b. Study of some new genera and species of the *Actinoplanaceae*. *In* Prauser (Editor), The Actinomycetales, VEB Gustav Fischer Verlag, Jena, pp. 245-257.

Thiemann, J.E. 1974a. Genus *Dactylosporangium* Thiemann, Pagani and Beretta. *In* Buchanan and Gibbons (Editors), Bergey's Manual of Determinative Bacteriology, 8th Ed., The Williams and Wilkins Co., Baltimore, pp. 721-722.

Thiemann, J.E. 1974b. Genus *Planobispora* Thiemann and Beretta. *In* Buchanan and Gibbons (Editors), Bergey's Manual of Determinative Bacteriology, 8th Ed., The Williams and Wilkins Co., Baltimore, pp. 720-721.

Thiemann, J.E. 1974c. Genus *Planomonospora* Thiemann, Pagani and Beretta. *In* Buchanan and Gibbons (Editors), Bergey's Manual of Determinative Bacteriology, 8th Ed., The Williams and Wilkins Co., Baltimore, pp. 719-720.

Thiemann, J.E. and G. Beretta. 1966. Alanosine, a new antiviral and antitumor antibiotic from *Streptomyces*. Description of the strain and antibiotic production. J. Antibiot. (Tokyo) Ser. A *19:* 155-160.

Thiemann, J.E. and G. Beretta. 1968. A new genus of the *Actinoplanaceae: Planobispora*, gen. nov. Arch. Mikrobiol. *62:* 157-166.

Thiemann, J.E., G. Beretta, C. Coronelli and H. Pagani. 1969a. Antibiotic production by new form-genera of the *Actinomycetales*, II. Antibiotic A/672 isolated from a new species of *Actinoplanes: Actinoplanes brasiliensis* nov. spec. J. Antibiotics *22:* 119-125.

Thiemann, J.E., C. Coronelli, H. Pagani, G. Beretta, G. Tamoni and V. Arioli. 1968a. Antibiotic production by new form-genera of the *Actinomycetales* I. Sporangiomycin, an antibacterial agent isolated from *Planomonospora parontospora* var. *antibiotica* var. nov. J. Antibiotics *21:* 525-531.

Thiemann, J.E., C. Hengeller, A. Virgilio, O. Buelli and G. Licciardello. 1964. Rifamycin 33. Isolation of actinophages active on *Streptomyces mediterranei* and characteristics of phage-resistant strains. Appl. Microbiol. *12:* 261-268.

Thiemann, J.E., H. Pagani and G. Beretta. 1967a. A new genus of the *Actinoplanaceae: Dactylosporangium*, gen. nov. Arch. Mikrobiol. *58:* 42-52.

Thiemann, J.E., H. Pagani and G. Beretta. 1967b. A new genus of the *Actinoplanaceae: Planomonospora*, gen. nov. G. Microbiol. *15:* 27-38.

Thiemann, J.E., H. Pagani and G. Beretta. 1968b. A new genus of *Actinomycetales: Microtetraspora* gen. nov. J. Gen. Microbiol. *50:* 295-303.

Thiemann, J.E., G. Zucco and G. Pelizza. 1969b. A proposal for the transfer of *Streptomyces mediterranei* Margalith and Beratta 1960 to the genus *Nocardia* as *Nocardia mediterranea* (Margalith and Beretta) comb. nov. Arch. Mikrobiol. *67:* 147-155.

Thirumalachar, M.J. 1955. *Chainia*, a new genus of the *Actinomycetales*. Nature (London) *176:* 934-935.

Thirumalachar, M.J. 1968. *In* Rahalkar, and Thirumalachar. 1968. Cultural characteristics and identity of some *Streptoverticillium* species producing polyene antibiotics. Hind. Antibiot. Bull. *11:* 90-96.

Thirumalachar, M.J. and V.V. Bhatt. 1960. Some *Streptomyces* species producing oxytetracycline. Hind. Antibiot. Bull. *3:* 61-63.

Thirumalachar, M.J., P.W. Rahalkar, P.V. Desmukh and R.S. Sukapure. 1965. Production of aburamycin by *Chainia minutisclerotica* a new species of actinomycetes. Hind. Antibiot. Bull. *8:* 6-9.

Thirumalachar, M.J. and R.S. Sukapure. 1964. Studies on species of the genus *Chainia* from India. Hind. Antibiot. Bull. *6:* 157-166.

Thirumalachar, M.J., R.S. Sukapure, P.W. Rahalkar and K.S. Gopalkrishnan. 1966. Studies on species of the genus *Chainia* from India, II. Hind. Antibiot. Bull. *9:* 10-14.

Thorne, G. 1940. *Duboscqia penetrans* n. sp. (Sporozoa: Microsporidia, Nosematidae), a parasite of the nematode *Pratylenchus pratensis* (de Man) Filipjev. Proc. Helminthol. Soc. Wash. *7:* 51-53.

Thorne, G. 1961. Principles of Nematology, McGraw-Hill Book Co., New York.

Tilford, P.E. 1936. Fasciation of sweet peas caused by *Phytomonas fascians* n. sp. J. Agr. Res. *53:* 383-394.

Tille, D., H. Prauser, K. Szyba and M. Mordarski. 1978. On the taxonomic position of *Nocardioides albus* Prauser by DNA/DNA-hybridization. Z. Allg. Mikrobiol. *18:* 459-462.

Timm, R.W. 1969. The genus *Isolaimium* Cobb, 1920 (Order Isolaimida: Isolaimidae New Family). J. Nematol. *1:* 97-106.

Tisa, L., M. McBride and J.C. Ensign. 1983. Studies of growth and morphology of *Frankia* strains EAN_{pec}, EUI1, CpI1, and $ACN1^{AG}$. Can. J. Bot. *61:* 2768-2773.

Tisdall, P.A., G.D. Roberts and J.P. Anhalt. 1979. Identification of clinical isolates of mycobacteria with gas-liquid chromatography alone. J. Clin. Microbiol. *10:* 506-514.

Tjepkema, J.D., W. Ormerod and J.G. Torrey. 1980. Vesicle formation and acetylene reduction activity in *Frankia* sp. CpI1 cultured in defined nutrient media. Nature. *287:* 633-635.

Tohyama, H., S. Miyadoh, M. Ito, T. Shomura, T. Ito and T. Ishikawa. 1984. A new indole *N*-glycoside antibiotic SF-2140 from *Actinomadura*. I. Taxonomy and fermentation of producing microorganism. J. Antibiot. *37:* 1144-1148.

Tomita, K., Y. Hoshino, T. Sasahira, K. Hasegawa and M. Akiyama. 1980a. Taxonomy of the antibiotic BU 2313 producing organism, *Microtetraspora caesia* sp. nov. J. Antibiotics *33:* 1491-1501.

Tomita, K., Y. Hoshino, T. Sasahira and H. Kawaguchi. 1980b. BBM-928, a new antitumor antibiotic complex. II. Taxonomic studies on the producing organism. J. Antibiot. *33:* 1098-1102.

Tomita, K., S. Kobaru, M. Hanada, and H. Tsukiara, Bristol-Myers Company. 1977. Fermentation process. U.S. Patent 4,026,766. May 31.

Tomita, K., Y. Nakakita, Y. Hoshino, K. Numata and H. Kawaguchi. 1987. New genus of the *Actinomycetales: Streptoalloteichus hindustanus* gen. nov., nom. rev.; sp. nov., nom. rev. Int. J. Syst. Bacteriol. *37:* 211-213.

Tomita, K., Y. Uenoyama, K. Numata, T. Sasahira, Y. Hoshino, K. Fugisawa, H. Tsukiura and H. Kawaguchi. 1978. *Streptoalloteichus*, a new genus of the family *Actinoplanaceae*. J. Antibiotics *31:* 497-510.

Tomiyasu, I., S. Tcriyama, I. Yano and M. Masui. 1981. Changes in molecular species composition of nocardomycolic acids in *Nocardia rubru* by the growth temperature. Chem. Phys. Lipids *28:* 41-54.

Torrey, J.G. and D. Callaham. 1982. Structural features of the vesicle of *Frankia* sp. CpI1 in culture. Can. J. Microbiol. *28:* 749-757.

Torrey, J.G. and J.D. Tjepkema. 1983. International conference on the biology of *Frankia*. Introduction. Can. J. Bot. *61:* 2765-2767.

Torrey, J.G., J.D. Tjepkema, G.L. Turner, F.J. Bergerson and A.H. Gibson. 1981. Dinitrogen fixation by cultures of *Frankia* sp. CpI1 demonstrated by $^{15}N_2$ incorporation. Plant Physiol. *68:* 983-984.

Toyama, H., M. Okanishi and H. Umezawa. 1974. Heterogeneity among whorl-forming streptomycetes determined by DNA reassociation. J. Gen. Microbiol. *80:* 507-514.

Trejo, W. and R.E. Bennett. 1963. *Streptomyces* species comprising the blue-spore series. J. Bacteriol. *85:* 676-690.

Trejo, W.H. 1970. An evaluation of some concepts and criteria used in the speciation of streptomycetes. Trans. N.Y. Acad. Sci. Ser. II *32:* 989-997.

Trejo, W.H., L.D. Dean, J. Pluscec, E. Meyers and W.E. Brown. 1977. *Streptomyces laurentii*, a new species producing thiostrepton. J. Antibiot. (Tokyo) Ser. A *30:* 639-643.

Trejo, W.H., L.D. Dean, J. Pluscec, E. Meyers and W.E. Brown. 1979. *In* Validation of the publication of new names and new combinations previously effectively published outside the IJSB. List No. 2. Int. J. Syst. Bacteriol. *29:* 79-80.

Tresner, H.D. and E.J. Backus. 1956. A broadened concept of the characteristics of *Streptomyces hygroscopicus*. Appl. Microbiol. *4:* 243-250.

Tresner, H.D. and E.J. Backus. 1963. System of color wheels for streptomycete taxonomy. Appl. Microbiol. *11:* 335-338.

Tresner, H.D., F. Danga and J.N. Porter. 1960. Long-term maintenance of *Streptomyces* in deep-freeze. J. Appl. Microbiol. *8:* 339-341.

Tresner, H.D., M.C. Davies and E.J. Backus. 1961. Electron microscopy of *Streptomyces* spore morphology and its role in species differentiation. J. Bacteriol. *81:* 70-80.

Tresner, H.D., J.A. Hayes and E.J. Backus. 1966. *Streptomyces prasinosporus* sp. nov. a new green-spored species. Int. J. Syst. Bacteriol. *16:* 161-169.

Tresner, H.D., J.A. Hayes and E.J. Backus. 1967. Morphology of submerged growth of streptomycetes as a taxonomic aid. 1. Morphological development in *Streptomyces aureofaciens* in agitated liquid media. Appl. Microbiol. *15:* 1185-1191.

Tresner, H.D., J.A. Hayes and E.J. Backus. 1968. Differential tolerance of streptomycetes to sodium chloride as a taxonomic aid. Appl. Microbiol. *16:* 1134-1136.

Treuhaft, M.W. 1977. Isolation of bacteriophage from *Thermoactinomyces*. J. Clin. Microbiol. *6:* 420-424.

Treuhaft, M.W., J.G. Green, R. Arusel and A. Borge. 1980. Role of *Saccharomonospora viridis* in hypersensitivity pneumonitis. Am. Rev. Respir. Dis. *121:* 100.

Trevisan, V. 1889. I Generi e le Specie delle Batteriacee, Zanaboni and Gabuzzi, Milano.

Tribe, H.T. and S.M. Abu El-Souod. 1979. Colonization of hair in soil-water cultures, with especial reference to the genera *Pilimelia* and *Spirillospora (Actinomycetales)*. Nova Hedwigia *31:* 789-805.

Trolldenier, G. 1967. Isolierung und Zählung von Bodenactinomyceten auf Erdplatten mit Membranfiltern. Plant Soil *27:* 285-288.

Tsao, P.H., C. Leben and G.W. Keitt. 1960. An enrichment method for isolating actinomycetes that produce diffusible antifungal antibiotics. Phytopathology *50:* 88-89.

Tsiklinsky, P. 1899. On the thermophilic moulds (in French). Ann. Inst. Pasteur *13:* 500-505.

Tsukamura, M. 1962. Differentiation of *Mycobacterium tuberculosis* from other mycobacteria by sodium salicylate susceptibility. Am. Rev. Resp. Dis. *86:* 81-83.

Tsukamura, M. 1965. Differentiation of mycobacteria by picric acid tolerance. Am. Rev. Resp. Dis. *92:* 491-492.

Tsukamura, M. 1966. Adansonian classification of mycobacteria. J. Gen. Microbiol. *45:* 253-273.

Tsukamura, M. 1971. Proposal of a new genus, *Gordona*, for slightly acid-fast organisms occurring in sputa of patients with pulmonary disease and in soil. J. Gen. Microbiol. *68:* 15-26.

Tsukamura, M. 1972. Susceptibility of *Mycobacterium intracellulare* to rifampicin: A trial of ecological observation. Jpn. J. Microbiol. *16:* 444-446.

Tsukamura, M. 1973. A taxonomic study of some strains received as *"Mycobacterium" rhodochrous*. Description of *Gordona rhodochroa* (Zopf; Overbeck; Gordon et Mihn) Tsukamura comb. nov. Jpn. J. Microbiol. *17:* 189-197.

Tsukamura, M. 1974a. A further numerical taxonomic study of the rhodochrous group. Jpn. J. Microbiol. *18:* 37-44.

Tsukamura, M. 1974b. Differentiation of the *Mycobacterium rhodochrous* group from nocardiae by β-galactosidase activity. J. Gen. Microbiol. *80:* 553-555.

Tsukamura, M. 1975a. Identification of Mycobacteria, The National Chubu Hospital, Obu, Aichi, Japan.

Tsukamura, M. 1975b. Numerical analysis of the relationship between *Mycobacterium*, rhodochrous group and *Nocardia* by use of hypothetical median organisms. Int. J. Syst. Bacteriol. *25:* 329-335.

Tsukamura, M. 1977. Extended numerical taxonomy study of *Nocardia*. Int. J. Syst. Bacteriol. *27:* 311-323.

Tsukamura, M. 1978. Numerical classification of *Rhodococcus* (formerly *Gordona*) organisms recently isolated from sputa of patients: Description of *Rhodococcus sputi* Tsukamura sp. nov. Int. J. Syst. Bacteriol. *28:* 169-181.

Tsukamura, M. 1981a. Differentiation between the genera *Mycobacterium*, *Rhodococcus* and *Nocardia* by susceptibility to 5-fluorouracil. J. Gen. Microbiol. *125:* 205-208.

Tsukamura, M. 1981b. Tests from susceptibility to mitomycin C as aids in differentiating the genus *Rhodococcus* from the genus *Nocardia* and for differentiating *Mycobacterium fortuitum* and *Mycobacterium chelonei* from other rapidly growing mycobacteria. Microbiol. Immunol. *25.* 1197-1199.

Tsukamura, M. 1982a. Differentiation between the genera *Rhodococcus* and *Nocardia* and between species of the genus *Mycobacterium* by susceptibility to bleomycin. J. Gen. Microbiol. *128:* 2385-2388.

Tsukamura, M. 1983. *In* Validation of the publication of new names and new combinations previously effectively published outside the IJSB. List No. 12. Int. J. Syst. Bacteriol. *33:* 896-897.

Tsukamura, M. 1982b. Numerical analysis of the taxonomy of nocardiae and rhodococci. Division of *Nocardia asteroides* sensu stricto into two species and descriptions of *Nocardia paratuberculosis* sp. nov. Tsukamura (formerly the Kyoto-I group of Tsukamura). *Nocardia nova* sp. nov. Tsukamura, *Rhodococcus aichiensis* sp. nov. Tsukamura, *Rhodococcus chubuensis* sp. nov. Tsukamura, and *Rhodococcus obuensis* sp. nov. Tsukamura. Microbiol. Immunol. *26:* 1101-1119.

Tsukamura, M. 1982d. Rejection of the name *Nocardia farcinica* Trevisan 1889 (Approved Lists 1980). Request for an opinion. Int. J. Syst. Bacteriol. *32:* 235-236.

Tsukamura, M., S. Mizuno and H. Murata. 1975. Numerical taxonomy study of the taxonomic position of *Nocardia rubra* reclassified as *Gordona lentifragmenta* Tsukamura nom. nov. Int. J. Syst. Bacteriol. *25:* 377-382.

Tsukamura, M. and S. Mizuno. 1971. A new species *Gordona aurantiaca* occurring in sputa of patients with pulmonary disease. Kekkaku *46:* 93-98.

Tsukamura, M., S. Mizuno, S. Tsukamura and J. Tsukamura. 1979. Comprehensive numerical classification of 369 strains of *Mycobacterium, Rhodococcus* and *Nocardia*. Int. J. Syst. Bacteriol. *29:* 110-129.

Tsukamura, M. and S. Tsukamura. 1968. Differentiation of mycobacteria by susceptibility to nitrite and propylene glycol. Am. Rev. Resp. Dis. *98:* 505-506.

Tsukamura, M. and I. Yano. 1985. *Rhodococcus sputi* sp. nov., nom. rev., and *Rhodococcus aurantiacus* sp. nov., nom. rev. Int. J. Syst. Bacteriol. *35:* 364-368.

Tsukiura, H., M. Okanishi, H. Koshiyama, T. Ohmori, T. Miyaki and H. Kawaguchi. 1964a. Proceomycin, a new antibiotic. J. Antibiot. (Tokyo) Ser. A *17:* 224-229.

Tsukiura, H., M. Okanishi, T. Ohmori, H. Koshiyama, T. Miyaki, H. Kitazima and H. Kawaguchi. 1964b. Danomycin, a new antibiotic. J. Antibiot. (Tokyo) Ser. A *17:* 39-47.

Tsyganov, V.A., V.P. Namestnikova and V.A. Krassykova. 1966. DNA composition in various genera of the *Actinomycetales*. Mikrobiologiya *35:* 92-95.

Tsyganov, V.A., R.A. Zhukova and K.A. Timofeeva. 1964. Morphological and biochemical peculiarities of a new species, actinomycetes 2732/3. Mikrobiologiya *33:* 863-869.

Turfitt, G.E. 1944. Microbiological agencies in the degradation of steriods. I. The Cholesterol-decomposing organisms of soil. J. Bacteriol. *47:* 487-493.

Uchida, K. and K. Aida. 1977. Acyl type of bacterial cell wall: its simple identification by colorimetric method. J. Gen. Appl. Microbiol. *23:* 249-260.

Uchida, K. and K. Aida. 1979. Taxonomic significance of cell-wall acyl type in *Corynebacterium-Mycobacterium-Nocardia* group by a glycolate test. J. Gen. Appl. Microbiol. *25:* 169-183.

Uchida, T., L. Bonen, H.W. Schaup, B.J. Lewis, L. Zablen and C.R. Woese. 1974. The use of ribonuclease U_2 in RNA sequence determination: some corrections in the catalog of oligomers produced by ribonuclease T1 digestion of *Escherichia coli* 16s ribosomal RNA. J. Mol. Evol. *3:* 63-77.

Ullman, J.S. and B.J. McCarthy. 1973. The relationship between mismatched base pairs and the thermal stability of DNA duplexes. II. Effects of deamination of cytosine. Biochim. Biophys. Acta *294:* 416-424.

Uma, B.N. and P.L. Narasimha Rao. 1959. Actinomycetes. I. Distribution of streptomycetes in Indian soils. Formation of antifungal antibiotics by *Streptomyces champavati* n. sp. *In* Indian Inst. Sci. Golden Jubilee Res., Vol. 1909-1959, Indian Institute of Science, Bangalore, India, pp. 130-141.

Umbreit, W.W. and E. McCoy. 1940. The occurrence of actinomycetes of the genus *Micromonospora* in inland lakes. Symposium on Hydrobiology, University of Wisconsin, pp. 106-114.

Umezawa, H., S. Hayano, K. Maeda, Y. Ogata and Y. Okami. 1950a. On a new antibiotic, griseolutein, produced by *Streptomyces*. Jpn. Med. J. *3:* 111-117.

Umezawa, H., T. Takeuchi, Y. Okami and T. Tazaki. 1953. On screening of antiviral substances produced by *Streptomyces* and on an antiviral substance achromoviromycin. Jpn. J. Med. Sci. Biol. *6:* 261-268.

Umezawa, H., T. Tazaki and S. Fukuyama. 1951. On anti-viral substances, abikoviromycin, produced by *Streptomyces* species. Jpn. Med. J. *4:* 331-346.

Umezawa, H., T. Tazaki, Y. Okami and S. Fukuyama. 1950b. On the new source of chloromycetin, *Streptomyces omiyaensis*. J. Antibiot. (Tokyo) *3:* 292-296.

Umezawa, H., M. Ueda, K. Maeda, K. Yagashita, S. Kondo, Y. Okami, R. Utahara, Y. Osato, N. Nitta and T. Takeuchi. 1957. Production and isolation of a new antibiotic, kanamycin. J. Antibiot. (Tokyo) Ser. A *10:* 181-188.

Unsworth, B.A. 1978. The genus *Thermoactinomyces* Tsiklinsky, Ph.D. Thesis, University of Bradford, U.K.

Unsworth, B.A. and T. Cross. 1980. Thermophilic actinomycetes implicated in Farmer's Lung: numerical taxonomy of *Thermoactinomyces* species. *In* Goodfellow and Broad (Editors), Microbiological Classification and Identification, Academic Press, London, pp. 389-390.

Unsworth, B.A., T. Cross, M.R.D. Seaward and R.E. Sims. 1977. The longevity of thermoactinomycete endospores in natural substrates. J. Appl. Bacteriol. *42:* 45-52.

van Brummelen, J. and J.C. Bent. 1957. *Streptosporangium* isolated from forest litter in the Netherlands. Antonie van Leeuwenhoek J. Microbiol. Serol. *23:* 385-392.

VandenBosch, K.A. and J.G. Torrey. 1984. Production of sporangia by the actinomycetous endophyte in root nodules of *Comptonia peregrina*: Development and consequences for nodule function. *In* Veeger and Newton (Editors), Advances in Nitrogen Fixation Research, Martinus Nijhoff/Dr. W. Junk Publ., The Hague, p. 376.

Van Dijk, C. 1978. Spore formation and endophyte diversity in root nodules of *Alnus glutinosa* (L.) VIII. New Phytol. *81:* 601-615.

Van Dijk, C. 1979. Endophyte distribution in the soil. *In* Gordon, Wheeler and Perry (Editors), Symbiotic Nitrogen Fixation in the Management of Temperate Forests, Oregon State Univ., Corvallis, pp. 84-94.

Van Gylswyk, N.O. 1980. *Fusobacterium polysaccharolyticum* sp. nov., a Gram-negative rod from the rumen that produces butyrate and ferments cellulose and starch. J. Gen. Microbiol. *116:* 157-163.

Van Saceghem, R. 1915. Dermatose contagieuse (Impétigo contagieux). Bull. Soc. Pathol. Exot. *8:* 354-359.

Vavra, J.J. and A. Dietz. 1965. U-13, 714, a new antiviral agent. I. Discovery and biological properties. Antimicrob. Agents Chemother. *1964:* 75-79.

Vickers, J.C., S.T. Williams and G.W. Ross. 1984. A taxonomic approach to selective isolation of streptomycetes from soil. *In* Ortiz-Ortiz, Bojalil and Yakoleff

(Editors), Biological, Biochemical and Biomedical Aspects of Actinomycetes, Academic Press, Orlando, Florida, pp. 553-561.

Villax, I. 1963. *Streptomyces lusitanus* and the problem of classification of the various tetracycline-producing streptomyces. Antimicrob. Agents Chemother. *1962:* 661-668.

Vincent, H. 1894. Étude sur le parasite du pied le madura. Ann. Inst. Pasteur *8:* 129-151.

Virgilio, A. and C. Hengeller. 1960. Produzione di Tetraciclina con *Streptomyces psammoticus*. Farm. Ed. Sci. *15:* 164-174.

Vobis, G. 1984. Sporogenesis in the *Pilimelia* species. *In* Ortiz-Ortiz, Bojalil, and Yakoleff (Editors), Biological, Biochemical, and Biomedical Aspects of Actinomycetes, Academic Press, Inc., Orlando, Florida, pp. 423-439.

Vobis, G. and H.-W. Kothe. 1985. Sporogenesis in sporangiate actinomycetes. *In* Mukerji, Pathak and Singh (Editors), Frontiers in Applied Microbiology, Vol. I, Print House (India), Lucknow, pp. 25-47.

Vobis, G., D. Schäfer, H.-W. Kothe and B. Renner. 1986a. Description of *Pilimelia columellifera* (ex Schäfer 1973) nom. rev. and *Pilimelia columellifera* subsp. *pallida* (ex Schäfer 1973) nom. rev. Syst. Appl. Microbiol. *8:* 67-74.

Vobis, G., D. Schäfer, H.-W. Kothe and B. Renner. 1986b. *In* Validation of the publication of new names and new combinations previously published outside the IJSB. List No. 22. Int. J. Syst. Bacteriol. *36:* 573-576.

Von Tubeuf, K. 1895. Pflanzenkrankheiten durch Kryptograme Parasiten verursacht, Verlag von Julius Springer, Berlin, pp. 1-599.

Vuillemin, P. 1913. Genera Schizomycetum. Ann. Mycol. Berlin *11:* 512-527.

Vuillemin, P. 1931. Les Champignons Parasites et les Mycoses de l'Homme, Paul Lechevalier et Fils, Paris.

Wagman, G.H., R.T. Testa, J.A. Marquez and M.J. Weinstein. 1974. Antibiotic G418, a new *Micromonospora*-produced aminoglycoside with activity against protozoa and helminths: Fermentation, isolation and preliminary characterization. Antimicrob. Agents Chemother. *6:* 144-149.

Waitz, J.A., A.C. Horan, M. Kalyanpur, B.K. Lee, D. Loebenberg, J.A. Marquez, G. Miller and M.G. Patel. 1981. Kijanimicin (Sch 25663), a novel antibiotic produced by *Actinomadura kijaniata* SCC 1256. J. Antibiot. *34:* 1101-1106.

Wakisaka, Y., Y. Kawamura, Y. Yasuda, K. Koizumi and Y. Nishimoto. 1982. A selective isolation procedure for *Micromonospora*. J. Antibiot. *35:* 822-836.

Waksman, S.A. 1919. Cultural studies of species of *Actinomyces*. Soil Sci. *8:* 71-215.

Waksman, S.A. 1923. Genus III. *Actinomyces* Harz, *In* Bergey, Harrison, Breed, Hammer and Huntoon (Editors), Bergey's Manual of Determinative Bacteriology, 1st Ed., The Williams and Wilkins Co., Baltimore, pp. 339-371.

Waksman, S.A. 1950. The Actinomycetes—Their Nature, Occurrence, Activities and Importance. Chronica Botanica Co., Waltham, Massachusetts.

Waksman, S.A. 1957a. Family III. *Streptomycetaceae* Waksman and Henrici. *In* Breed, Murray and Smith (Editors), Bergey's Manual of Determinative Bacteriology, 7th Ed., The Williams and Wilkins Co., Baltimore, pp. 744-825.

Waksman, S.A. 1957b. Species concept among the actinomycetes with special reference to the genus *Streptomyces*. Bacteriol. Rev. *21:* 1-29.

Waksman, S.A. 1959. Strain specificity and production of antibiotic substances. X. Characterization and classification of species within the *Streptomyces griseus* Group. Proc. Nat. Acad. Sci. U.S.A. *45:* 1043-1047.

Waksman, S.A. 1961. The Actinomycetes. Classification, Identification and Descriptions of Genera and Species, Vol. 2, The Williams and Wilkins Co., Baltimore, pp. 1-363.

Waksman, S.A. 1967. The Actinomycetes. A Summary of Current Knowledge, The Ronald Press Co., New York.

Waksman, S.A. and C.T. Corke. 1953. *Thermoactinomyces* Tsiklinsky, a genus of thermophilic actinomycetes. J. Bacteriol. *66:* 377-378.

Waksman, S.A. and R.E. Curtis. 1916. The actinomyces of the soil. Soil Sci. *1:* 99-134.

Waksman, S.A. and F.J. Gregory. 1954. Actinomycin-II. Classification of organisms producing different forms of actinomycin. Antibiot. Chemother. *4:* 1050-1056.

Waksman, S.A. and A.T. Henrici. 1943. The nomenclature and classification of the actinomycetes. J. Bacteriol. *46:* 337-341.

Waksman, S.A. and A.T. Henrici. 1948a. Family II. *Actinomycetaceae* Buchanan. *In* Breed, Murray and Hitchens (Editors), Bergey's Manual of Determinative Bacteriology, 6th Ed., The Williams and Wilkins Co., Baltimore, pp. 892-928.

Waksman, S.A. and A.T. Henrici. 1948b. Family III. *Streptomycetaceae* Waksman and Henrici. *In* Breed, Murray and Hitchens (Editors), Bergey's Manual of Determinative Bacteriology, 6th Ed., The Williams and Wilkins Co., Baltimore, pp. 929-980.

Waksman, S.A., E.S. Horning, M. Welsch and H.B. Woodruff. 1942. Distribution of antagonistic actinomycetes in nature. Soil Sci. *54:* 281-296.

Waksman, S.A. and H.A. Lechevalier. 1953. Guide to the Classification and Identification of Actinomycetes and Their Antibiotics, The Williams and Wilkins Co., Baltimore.

Waksman, S.A. and W.A. Taber. 1953. *In* Waksman and Lechevalier. 1953, Guide to the Classification and Identification of Actinomycetes and Their Antibiotics. The Williams and Wilkins Co., Baltimore, pp. 1-162.

Waksman, S.A., W.W. Umbreit and T.C. Cordon. 1939. Thermophilic actinomycetes and fungi in soils and in composts. Soil Sci. *47:* 37-61.

Waksman, S.A. and H.B. Woodruff. 1940. The soil as a source of microorganisms antagonistic to disease-producing bacteria. J. Bacteriol. *40:* 581-600.

Waksman, S.A. and H.B. Woodruff. 1941. *Actinomyces antibioticus*, a new soil organism antagonistic to pathogenic and non-pathogenic bacteria. J. Bacteriol. *42:* 231-249.

Walbaum, S., J. Biquet and P. Tran Van Ky. 1969. Structure antigenique de *Thermopolyspora polyspora*: repercussions pratiques sur le diagnostic du poumon du fermier. Ann. Inst. Pasteur *117:* 673-693.

Walbaum, S., T. Vaucelle and J. Biquet. 1973. Analyse de l'extrait de *Micropolyspora faeni* par immunoelectrophorese en double dimension. Localisation des activities chymotrypsique. Path. Biol. Paris *21:* 555-558.

Walker, J.T., C.H. Specht and J.F. Bekker. 1966. Nematocidal activity to *Pratylenchus penetrans* by culture fluids from actinomycetes and bacteria. Can. J. Microbiol. *12:* 347-353.

Wallhäusser, K.-H., G. Huber, G. Nesemann, P. Präve and K. Zepf. 1964. Die antibiotica FF 3582A und B und ihre Identität mit nonactin und seinen Honologen. Arzneim. Forsch. *14:* 356-360.

Wallhäusser, K.-H., G. Nesemann, P. Präve and A. Steigler. 1966. Moenomycin, a new antibiotic. I. Fermentation and isolation. Antimicrob. Agents Chemother. *1965:* 734-736.

Wallick, H., D.A. Harris, M.A. Reagan, M. Ruger and H.B. Woodruff. 1956. Discovery and antimicrobial properties of cathomycin, a new antibiotic produced by *Streptomyces spheroides*, n. sp. Antibiot. Annu. *1955-56:* 909-917.

Walter, M.R. (Editor). 1977. Life in the Precambrian. Precambrian Res. *5(2):* 105-219.

Wang, E.L., M. Hamada, Y. Okami and H. Umezawa. 1966. A new antibiotic, spinamycin. J. Antibiot. (Tokyo) Ser. A *19:* 216-221.

Watanabe, K., T. Okuda, K. Yokose, T. Furumai and H.B. Maruyama. 1983. *Actinosynnema mirum*, a new producer of nocardicin antibiotics. J. Antibiot. *36:* 321-324.

Watanabe, K., T. Tanaka, K. Fukuhara, N. Miyairi, H. Yonehara and H. Umezawa. 1957. Blastomycin, a new antibiotic from *Streptomyces* sp. J. Antibiot. (Tokyo) *10(A):* 39-45.

Watson, E.T. and S.T. Williams. 1974. Studies on the ecology of actinomycetes in soil. VII. Actinomycetes in a coastal sand belt. Soil Biol. Biochem. *6:* 43-52.

Wayne, L.G. 1982. Actions of the Judicial Commission of the International Committee of Systematic Bacteriology on requests for opinions published between July 1979 and April 1981. Int. J. Syst. Bacteriol. *32:* 464-465.

Wayne, L.G. 1986. Actions of the Judicial Commission of the International Committee on Systematic Bacteriology on requests for opinions published in 1983 and 1984. Int. J. Syst. Bacteriol. *36:* 357-358.

Webley, D.M. 1954. The effect of oxygen on the growth and metabolism of the aerobic thermophilic actinomycete *Micromonospora vulgaris*. J. Gen. Microbiol. *11:* 114-122.

Webley, D.M. 1958. A defined medium for the growth of the thermophilic actinomycete *Micromonospora vulgaris*. J. Gen. Microbiol. *19:* 402-406.

Weinstein, M.J., G.M. Luedemann, E.M. Oden and G.H. Wagman. 1968. Halomicin, a new *Micromonospora*-produced antibiotic. Antimicrob. Agents Chemother. *1967:* 435-441.

Weiser, J. 1943. Beiträge zur Entwicklungsgeschichte von *Dermocystidium daphniae* Jirovec. Zool. Anz. *142:* 200-205.

Weitzman, P.D.J. 1980. Citrate synthase and succinate thiokinase in classification and identification. *In* Goodfellow and Board (Editors), Microbiological Classification and Identification, Academic Press, London, pp. 107-125.

Wellington, E.M.H. and S.T. Williams. 1978. Preservation of actinomycete inoculum in frozen glycerol. Microbios Lett. *6:* 151-157.

Wellington, E.M.H. and S.T. Williams. 1981a. Host ranges of phages isolated to *Streptomyces* and other genera. Zentralbl. Bakteriol. Mikrobiol. Hyg. I Abt. Suppl. *11:* 93-98.

Wellington, E.M.H. and S.T. Williams. 1981b. Transfer of *Actinoplanes armeniacus* Kalakoutskii and Kuznetsov to *Streptomyces: Streptomyces armeniacus* (Kalakoutskii and Kuznetsov) comb. nov. Int. J. Syst. Bacteriol. *31:* 77-81.

Wenzel, F.J., D.A. Emanuel and B.R. Lawton. 1967. Pneumonitis due to *Micromonospora vulgaris* (farmer's lung). Am. Rev. Respir. Dis. *95:* 652-655.

Wenzel, F.J., R.L. Gray, R.C. Roberts and D.A. Emanuel. 1974. Serologic studies in farmers' lung. Precipitins to the thermophilic actinomycetes. Am. Rev. Respir. Dis. *109:* 464-468.

Wetmur, J.G. 1976. Hybridization and renaturation kinetics of nucleic acids. Annu. Rev. Biophys. Bioeng. *5:* 337-361.

Wetmur, J.G. and N. Davidson. 1968. Kinetics of renaturation of DNA. J. Mol. Biol. *31:* 349-370.

Weyland, H. 1969. Actinomycetes in North Sea and Atlantic Ocean sediments. Nature *223:* 858.

Weyland, H. 1981. Distribution of actinomycetes on the sea floor. Zentralbl. Bakteriol. Mikrobiol. Hyg. I. Abt. Orig. Suppl. *11:* 185-193.

Wheeler, C.T., A. Crozier and G. Sandberg. 1984. The biosynthesis of indole-3-acetic acid by *Frankia*. Plant Soil *78:* 99-107.

Whittaker, R.H. and L. Margulis. 1978. Protist classification and the kingdoms of organisms. BioSystems *10:* 3-18.

Wildermuth, H. 1972. The surface structure of spores and aerial hyphae in *Streptomyces viridochromogenes*. Arch. Microbiol. *81:* 309-320.

Willcox, W.B., S.P. Lapage, S. Bascomb and M.A. Curtis. 1973. Identification of bacteria by computer: theory and programming. J. Gen. Microbiol. *77:* 317-330.

Willcox, W.R., S.P. Lapage and B. Holmes. 1980. A review of numerical methods in bacterial identification. Antonie van Leeuwenhoek J. Microbiol. Serol. *46:* 233-299.

Williams, J.R. 1960. Studies on the nematode soil fauna of sugarcane fields of Mauritius. 5. Notes upon a parasite of root-knot nematodes. Nematologica *5:* 37–42.

Williams, S.T. 1967. Sensitivity of streptomycetes to antibiotics as a taxonomic character. J. Gen. Microbiol. *46:* 115–160.

Williams, S.T. 1970. Further investigations of actinomycetes by scanning electron microscopy. J. Gen. Microbiol. *62:* 67–73.

Williams, S.T. 1978. Streptomycetes in the soil ecosystem. Zentralbl. Bakteriol. Parasitenkd. Infektionskr. Hyg. Abt. 1 Suppl. *6:* 137–144.

Williams, S.T. 1982. Are antibiotics produced in soil? Pedobiologia *23:* 427–435.

Williams, S.T., R.M. Bradshaw, J.W. Costerton and A. Forge. 1972a. Fine structure of the spore sheath of some *Streptomyces* species. J. Gen. Microbiol. *72:* 249–258.

Williams, S.T. and T. Cross. 1971. Isolation, purification, cultivation and preservation of actinomycetes. Methods Microbiol. *4:* 295–334.

Williams, S.T. and F.L. Davies. 1965. Use of antibiotics for selective isolation and enumeration of actinomycetes in soil. J. Gen. Microbiol. *38:* 251–261.

Williams, S.T. and F.L. Davies. 1967. Use of a scanning electron microscope for the examination of actinomycetes. J. Gen. Microbiol. *48:* 171–177.

Williams, S.T., F.L. Davies, C.I. Mayfield and M.R. Khan. 1971. Studies on the ecology of actinomycetes II. The pH requirements of streptomycetes from two acid soils. Soil Biol. Biochem. *3:* 187–195.

Williams, S.T. and T.H. Flowers. 1978. The influence of pH on starch hydrolysis by neutrophilic and acidophilic streptomycetes. Microbios *20:* 99–106.

Williams, S.T., M. Goodfellow, G. Alderson, E.M.H. Wellington, P.H.A. Sneath and M.J. Sackin. 1983a. Numerical classification of *Streptomyces* and related genera. J. Gen. Microbiol. *129:* 1743–1813.

Williams, S.T., M. Goodfellow, E.M.H. Wellington, J.C. Vickers, G. Alderson, P.H.A. Sneath, M.J. Sackin and A.M. Mortimer. 1983b. A probability matrix for identification of streptomycetes. J. Gen. Microbiol. *129:* 1815–1830.

Williams, S.T., S. Lanning and E.M.H. Wellington. 1984a. Ecology of actinomycetes. *In* Goodfellow, Mordarski and Williams (Editors), The Biology of the Actinomycetes, Academic Press, London, pp. 481–528.

Williams, S.T., R. Locci, J. Vickers, G.M. Schofield, P.H.A. Sneath and A.M. Mortimer. 1985a. Probabilistic identification of *Streptoverticillium* species. J. Gen. Microbiol. *131:* 1681–1689.

Williams, S.T. and C.I. Mayfield. 1971. Studies on the ecology of actinomycetes in soil. III. The behaviour of neutrophilic streptomycetes in acid soil. Soil Biol. Biochem. *3:* 197–208.

Williams, S.T. and C.S. Robinson. 1981. The role of streptomycetes in decomposition of chitin in acidic soils. J. Gen. Microbiol. *127:* 55–63.

Williams, S.T., M. Shameemullah, E.T. Watson and C.I. Mayfield. 1972b. Studies on the ecology of actinomycetes in soil VI. The influence of moisture tension on growth and survival. Soil Biol. Biochem. *4:* 215–225.

Williams, S.T. and G.P. Sharples. 1976. *Streptosporangium corrugatum* sp. nov., an actinomycete with some unusual morphological features. Int. J. Syst. Bacteriol. *26:* 45–52.

Williams, S.T., G.P. Sharples and R.M. Bradshaw. 1973. The fine structure of the *Actinomycetales. In* Sykes and Skinner (Editors), Actinomycetales: Characteristics and Practical Importance, Academic Press, London, pp. 113–130.

Williams, S.T., G.P. Sharples and R.M. Bradshaw. 1974. Spore formation in *Actinomadura dassonvillei* (Brocq-Rousseu) Lechevalier and Lechevalier. J. Gen. Microbiol. *84:* 415–419.

Williams, S.T., G.P. Sharples, J.A. Serrano, A.A. Serrano and J. Lacey. 1976. The micromorphology and fine structure of nocarioform organisms. *In* Goodfellow, Brownell and Serrano (Editors) The Biology of the Nocardiae, Academic Press, London, pp. 102–140.

Williams, S.T., J.C. Vickers, M. Goodfellow, G. Alderson, E.M.H. Wellington, P.H.A. Sneath, M.J. Sackin and A.M. Mortimer. 1984b. Numerical classification and identification of streptomycetes. *In* Ortiz-Ortiz, Bojalil and Yakoleff (Editors), Biological, Biochemical and Biomedical Aspects of Actinomycetes, Academic Press, Orlando, Florida, pp. 537–551.

Williams, S.T., J.C. Vickers and M. Goodfellow. 1985b. Application of new theoretical concepts to the identification of streptomycetes. *In* Goodfellow, Jones and Priest (Editors), Computer-assisted Bacterial Systematics, Spec. Publ. Soc. Gen. Microbiol. 15, Academic Press, London, pp. 289–306.

Williams, S.T. and E.M.H. Wellington. 1980. Micromorphology and fine structure of actinomycetes. *In* Goodfellow and Board (Editors), Microbiological Classification and Identification, Academic Press, London, pp. 139–165.

Williams, S.T. and E.M.H. Wellington. 1981. The genera *Actinomadura, Actinopolyspora, Excellospora, Microbispora, Micropolyspora, Microtetraspora, Nocardiopsis, Saccharopolyspora*, and *Pseudonocardia. In* Starr, Stolp, Trüper, Balows, and Schlegel (Editors), The Prokaryotes. A Handbook of Habitats, Isolation and Identification of Bacteria, Vol. II. Springer-Verlag, Heidelburg, pp. 2103–2117.

Williams, S.T. and E.M.H. Wellington. 1982a. Actinomycetes. *In* Page, Miller and Keeney (Editors), Methods of Soil Analysis, Part 2, Chemical and Microbiological Properties, American Society of Agronomy and Soil Science, Madison, Wisconsin, pp. 969–987.

Williams, S.T. and E.M.H. Wellington. 1982b. Principles and problems of selective isolation of microbes. *In* Bu'lock, Nisbet and Winstanley (Editors), Bioactive Microbial Products: Search and Discovery, Academic Press, London, pp. 9–26.

Williams, S.T., E.M.H. Wellington, M. Goodfellow, G. Alderson, M. Sackin and P.H.A. Sneath. 1981. The genus *Streptomyces*—a taxonomic enigma. Zentralbl. Bakteriol. Parasitenkd. Infektionskr. Hyg. I Abt. Suppl. *11:* 47–57.

Williams, S.T., E.M.H. Wellington and L.S. Tipler. 1980. The taxonomic implications of the reactions of representative *Nocardia* strains to actinophage. J. Gen. Microbiol. *119:* 173–178.

Willoughby, L.G. 1966. A conidial *Actinoplanes* isolate from Blelham Tarn. J. Gen. Microbiol. *44:* 69–72.

Willoughby, L.G. 1968. Aquatic *Actinomycetales* with particular reference to the *Actinoplanaceae*. Veroeff. Inst. Meersforsch. Bremerhaven *3:* 19–26.

Willoughby, L.G. 1969a. A study of aquatic actinomycetes of Blelham Tarn. Hydrobiologia *34:* 465–483.

Willoughby, L.G. 1969b. A study of aquatic actinomycetes, the allochthonous leaf component. Nova Hedwigia *18:* 45–113.

Willoughby, L.G. and C.D. Baker. 1969. Humic and fulvic acids and their derivatives as growth and sporulation media for aquatic actinomycetes. Verh. Int. Verein. Limnol. *17:* 795–801.

Willoughby, L.G., C.D. Baker and S.E. Foster. 1968. Sporangium formation in the *Actinoplanaceae* induced by humic acids. Experientia *24:* 730–731.

Willoughby, L.G., S.M. Smith and R.M. Bradshaw. 1972. Actinomycete virus in fresh water. Freshwater Biol. *2:* 19–26.

Wilson, A.C., S.S. Carlson and T.J. White. 1977. Biochemical evolution. Annu. Rev. Biochem. *46:* 573–639.

Wingender, W., H. von Hugo, W. Frommer and D. Schäfer. 1975. A protease inhibitor isolated from *Planomonospora parontospora*. J. Antibiotics *28:* 611–612.

Winogradsky, S. 1949. Microbiologie du Sol, Masson et Cie, Paris.

Winslow, C.-E.A. and A. Winslow. 1908. The systematic relationships of the *Coccaceae*, John Wiley and Sons, New York.

Woese, C.R. 1981. Archaebacteria. Sci. Am. *244:* 98–122.

Woese, C.R. 1987. Bacterial evolution. Microbiol. Rev. *51:* 221–271.

Woese, C.R. and G.E. Fox. 1977a. Phylogenetic structure of the prokaryotic domain: The primary kingdoms. Proc. Nat. Acad. Sci. U.S.A. *74:* 5088–5090.

Woese, C.R. and G.E. Fox. 1977b. The concept of cellular evolution. J. Mol. Evol. *10:* 1–6.

Woese, C.R., G.E. Fox, L. Zablen, T. Uchida, L. Bonen, K. Pechman, B.J. Lewis and D. Stahl. 1975. Conservation of primary structure in 16s ribosomal RNA. Nature (Lond) *254:* 83–86.

Woese, C.R., L.J. Magrum and G.E. Fox. 1978. Archaebacteria. J. Mol. Evol. *11:* 245–252.

Woese, C.R., J. Maniloff and L.B. Zablen. 1980. Phylogenetic analysis of the Mycoplasmas. Proc. Nat. Acad. Sci. U.S.A. *77:* 494–498.

Woese, C.R., E. Stackebrandt, W.G. Weisburg, B.J. Paster, M.D. Madigan, V.J. Fowler, C.M. Hain, P. Blanz, R. Gupta, K.H. Nealson and G.E. Fox. 1984. The phylogeny of purple bacteria: The alpha subdivision. Syst. Appl. Microbiol. *5:* 315–326.

Wollenweber, H.W. 1920. Der Kartoffelschorf, Arbeiten des Forschunginstitutes für Kartoffelbau Verlagsbuchandlung Paul Parey, Berlin, No. 2.

Wong, P.T.W. and D.M. Griffin. 1974. Effect of osmotic potential on streptomycete growth, antibiotic production and antagonism to fungi. Soil Biol. Biochem. *6:* 319–325.

Wood, S., S.T. Williams and W.R. White. 1983. Microbes as a source of earthy flavours in potable water—a review. Int. Biodeterior. Bull. *19:* 83–97.

Woolcock, J.B., A.M.T. Farmer and M.D. Mutimer. 1979. Selective medium for *Corynebacterium equi* isolation. J. Clin. Microbiol. *9:* 640–642.

Woronin, M. 1866. Ueber die bei der Schwarzerle (*Alnus glutinosa*) und der gewöhnlichen Garten-Lupine (*Lupinus mutabilis*) auftretenden wurzelanschwellungen. Mém. Acad. Sci. St. Pétersberg, Série 7, *10:* 1–10.

Wóznicka, W. 1965. Taxonomic and Taxonometric Studies on 'Yellow Series' of Actinomycetes (in Polish), Panstowowz Zaklad Higieny, Warsaw.

Wóznicka, W. 1967. Trials of classification of the 'Yellow Series' of Actinomycetes. II. Taxonometric studies. Exp. Med. Microbiol. *21:* 143–151.

Yamada, Y., K. Aoki and Y. Tahara. 1982a. The structure of hexahydrogenated isoprenoid side-chain menaquinone with nine isoprene units isolated from *Actinomadura madurae*. J. Gen. Appl. Microbiol. *28:* 321–329.

Yamada, Y., C.F. Hou, J. Sasaki, Y. Tahara and H. Yoshioka. 1982b. The structure of the octahydrogenated isoprenoid side-chain menaquinone with nine isoprene units isolated from *Streptomyces albus*. J. Gen. Appl. Microbiol. *28:* 519–529.

Yamada, Y., G. Inouye, Y. Tahara and K. Kondo. 1976. The menaquinone system in the classification of coryneform and nocardioform bacteria and related organisms. J. Gen. Appl. Microbiol. *22:* 203–214.

Yamada, Y., T. Ishikawa, Y. Tahara and K. Kondo. 1977a. The menaquinone system in the classification of the genus *Nocardia*. J. Gen. Appl. Microbiol. *23:* 207–216.

Yamada, Y., M. Yamashita, Y. Tahara and K. Kondo. 1977b. The menaquinone system in the classification of the genus *Actinomadura*. J. Gen. Appl. Microbiol. *23:* 331–335.

Yamaguchi, H., Y. Nakayama, K. Takeda, K. Tawara, K. Maeda, T. Takeuchi and H. Umezawa. 1957. A new antibiotic, althiomycin. J. Antibiot. (Tokyo) Ser. A *10:* 195–200.

Yamaguchi, T. 1965. Comparison of the cell wall composition of morphologically distinct actinomycetes. J. Bacteriol. *89:* 444–453.

Yamaguchi, T. 1967. Similarity in DNA of various morphologically distinct actinomycetes. J. Gen. Appl. Microbiol. *13:* 63–71.

Yamaguchi, T. and Y. Saburi. 1955. Studies on the anti-trichomonal actinomycetes and their classification. J. Gen. Appl. Microbiol. *1:* 201–235.

Yamamoto, H., K. Nakazawa, S. Horii and A. Miyake. 1960. Studies on agricultural antibiotic folimycin, a new antifungal antibiotic produced by *Streptomyces neyagawaensis* nov. sp. J. Agric. Chem. Soc. Jpn. *34:* 268–272.

Yano, I., K. Kageyama, Y. Ohno, M. Masui, E. Kusunose, M. Kusunose and N. Akimori. 1978. Separation and analysis of molecular species of mycolic acids in *Nocardia* and related taxa by gas chromatography mass spectrometry. Biomed. Mass. Spectrom. *5:* 14–24.

Young, J.M., D.W. Dye, J.F. Bradbury, C.G. Panagopoulos and C.F. Robbs. 1978. A proposed nomenclature and classification for plant pathogenic bacteria. N.Z. J. Agric. Res. *21:* 153–177.

Yuan, C.-S. 1962. Biology of the group of orange-coloured actinomycetes. Diss. Inst. Microbiol. Akad. Nauk SSSR (Abst. by Konova, I.V. 1962 Mikrobiologiya *31:* 188–189).

Zhang, G., G. Xing and X. Yan. 1984a. Studies on classification of *Streptomycetaceae*, III. A new genus *Streptomycoides* in the *Streptomycetaceae*. Acta Microbiol. Sinica *24:* 189–194.

Zhang, Z., M.F. Lopez and J.G. Torrey. 1984b. A comparison of cultural characteristics and infectivity of *Frankia* isolates from root nodules of *Casuarina* species. Plant Soil *78:* 79–90.

Zhang, Z. and J.G. Torrey. 1985. Studies of an effective strain of *Frankia* from *Allocasuarina lehmanniana* of the *Casuarinaceae*. Plant Soil *87:* 1–16.

Zhukova, R.A., V.A. Tsyganov and V.M. Morozov. 1968. A new species of *Micropolyspora—Micropolyspora angiospora* (sp. nov.). Mikrobiologiya *97:* 724–728.

Ziegler, P. and H.J. Kutzner. 1973. Hippurate hydrolysis as a taxonomic criterion in the genus *Streptomyces* (Order *Actinomycetales*). Z. Allg. Mikrobiol. *13:* 265–272.

Zopf, W. 1889. Über das Mikrochemische Verhalten von Fettfarbstoff-haltigen. Organen. Z. Wiss. Mikrosk. *6:* 172–177.

Zuckerkandl, E. and L. Pauling. 1965. Molecules as documents of evolutionary history. J. Theoret. Biol. *8:* 357–366.

Zuckerman, B.M., W.F. Mai and M.B. Harrison (Editors). 1985. Laboratory Manual for Plant Nematology, Massachusetts Agricultural Experiment Station, Amherst.

Zviagintzev, D.G., I.V. Aseyeva, I.P. Babieva and T.G. Mirchink. 1980. The Methods of Soil Microbiology and Biochemistry (in Russian). Lomonosov State University, Moscow.

Cumulative Index (Volumes 1–4) of Scientific Names of Bacteria

Key to the fonts and symbols used in this index:

Nomenclature

Lower case, Roman:	Genera, species, and subspecies of bacteria. Every bacterial name mentioned in the *Manual* is listed in the "Index." Specific epithets are listed individually and also under the genus.*
CAPITALS, ROMAN:	Names of taxa higher than genus (tribes, families, orders, classes, divisions, kingdoms).

Pagination

Roman:	Pages on which taxa are mentioned.
Boldface:	Indicates page on which the description of a taxon is given.†

*Infrasubspecific names, such as serovars, biovars, and pathovars, are not listed in the Cumulative "Index."

†A description may not necessarily be given in the *Manual* for a taxon that is considered as *incertae sedis* or that is listed in an addendum or note added in proof; however, the page on which the complete citation of such a taxon is given is indicated in boldface type.

Cumulative Index (Volumes 1–4) of Scientific Names of Bacteria